THE IMMUNE SYSTEM

FOURTH EDITION

THE IMMUNE SYSTEM

PETER PARHAM

FOURTH EDITION

The Immune System is adapted from *Janeway's Immunobiology*, also published by Garland Science.

Garland Science
Vice President: Denise Schanck
Assistant Editor: Allie Bochicchio
Editorial Assistants: Alina Yurova and Allison Grinberg-Funes
Text Editor: Eleanor Lawrence
Production Editor and Layout: Emma Jeffcock of EJ Publishing Services
Illustration and Design: Nigel Orme
Copyeditor: Bruce Goatly
Senior Production Editor: Georgina Lucas
Cover Photographer: © Getty Images/Bartosz Hadyniak
Indexer: Medical Indexing Ltd.

ISBNs: 978-0-8153-4466-7 (paperback) 978-0-8153-4526-8 (looseleaf) 978-0-8153-4527-5 (pb ise).

Library of Congress Cataloging-in-Publication Data
Parham, Peter, 1950- author.
The immune system / Peter Parham. -- Fourth edition.
pages cm
Includes bibliographical references and index.
ISBN 978-0-8153-4466-7 (paperback) -- ISBN 978-0-8153-4526-8 (looseleaf) -- ISBN 978-0-8153-4527-5 (pb ise) 1. Immune system. 2. Immunopathology. I. Janeway, Charles. Immunobiology. Based on: II. Title.
QR181.P335 2014
616.07'9--dc23
2014024879

Published by Garland Science, Taylor & Francis Group, LLC, an informa business, 711 Third Avenue, New York, NY 10017, USA, and 3 Park Square, Milton Park, Abingdon, OX14 4RN, UK.

Printed in the United States of America
15 14 13 12 11 10 9 8 7 6 5 4 3 2 1

Visit our website at http://www.garlandscience.com

Peter Parham is on the faculty at Stanford University where he is a Professor in the Department of Structural Biology and the Department of Microbiology and Immunology.

Preface

This book is aimed at students of all types who are coming to immunology for the first time. The guiding principle of the book is a focus on human immune systems—how they work and how their successes, compromises, and failures affect the daily life of every one of us. In providing the beginning student with a coherent, concise, and contemporary narrative of the mechanisms used by the immune system to control invading microbes, the emphasis has had to be on what we know, rather than how we know it. In other words, our emphasis here is more on the work of nature than on the work of immunologists.

Nevertheless, since the third edition of *The Immune System* was published in 2009 the work of immunologists has dramatically advanced the boundaries of knowledge. Following close behind the discovery of immunological mechanisms has been the rational design of new drugs and therapies based on this knowledge. Other important developments have been an increasing understanding of the numerous idiosyncrasies of human immune systems and the importance of studying immune-system cells in the tissues where they function. While working on this fourth revision of *The Immune System* I was not infrequently struck and excited by the extent to which phenomena that were loose ends in 2009 are now connected and making sense in ways that were unpredictable. As a result, substantial changes have been made in this fourth edition. For readers and instructors familiar with the third edition, what follows is a guide to the major changes. For those who are new to the book it will provide an overview of its contents.

Chapter 1 provides a focused introduction to the cells and tissues of the immune system, and to their place and purpose within the human body. The two following chapters describe the innate immune response to infection. These replace the single chapter in the previous edition, reflecting how innate immunity continues to be a rich area for discovery. Particularly relevant is the now widespread appreciation that the vast majority of microorganisms inhabiting human bodies are essential for human health, for the development of the immune system, and for preventing the growth and invasion of pathogenic microorganisms. These concepts are introduced in Chapter 2, along with the immediate, front-line defenses of complement, defensins, and other secreted proteins. The induced cellular defenses of innate immunity—macrophages, neutrophils, and natural killer cells—are the topic of Chapter 3. In the previous edition of the book, there was an introductory chapter on adaptive immunity at this point. This has been dropped in the fourth edition, partly because of overlap with Chapter 1 and partly on the advice of the book's users.

The next six chapters cover the fundamental biology of the adaptive immune response. Chapters 4 and 5 describe how B lymphocytes and T lymphocytes detect the presence of infection. These chapters introduce antibodies, the variable antigen-binding receptors of B cells and T cells, and the polymorphic major histocompatibility complex (MHC) class I and II molecules that present peptide antigens to T-cell receptors.

Chapters 6 and 7 describe and compare the development of B cells and T cells, including the gene rearrangements that generate the antigen receptors and the selective processes that eliminate cells with potential for causing autoimmunity. At the end of these two chapters, mature but naive B cells and T cells enter the circulation of the blood and the lymph in the quest for their specific antigens. Chapters 8 and 9 describe how these naive lymphocytes respond to infections and use diverse effector mechanisms to get rid of them. Here we look in detail at the dendritic cells that activate naive T cells, how immune responses are generated in secondary lymphoid organs, the differentiation of activated T cells into various effector subsets, and the generation of antibodies by B cells. The order and scope of these six chapters are the same as in previous editions of the book, but they have undergone significant revision, particularly to account for the increased knowledge and understanding of the functional diversity of both CD4 T cells and the classes and subclasses of human antibodies.

In the previous edition, Chapter 10 was divided into three parts that dealt with mucosal immunity, immunological memory, and the connection between innate and adaptive immunity. These three important areas have been given a chapter each in this edition. Chapter 10 now describes the nature of the immune response in mucosal tissue, where most immune activity takes place, and the ways in which it differs from the systemic immune response, with emphasis on the gut and the mucosal immune system's interactions with commensal microorganisms.

Chapter 11 is a new chapter that combines two related topics—immunological memory and vaccination—that were in different chapters in the previous edition. Users of the book have for some years suggested bringing these two topics together. Now is an opportune time to do so, because vaccine research and development is undergoing a renaissance after a period of considerable decline.

The more we learn about the immune system, the more blurred the distinction between innate and adaptive immunity becomes. On reflection this should not be surprising, because the two systems have been coevolving in vertebrate bodies for the past 400 million years. The largely new content of Chapter 12, entitled 'Coevolution of Innate and Adaptive Immunity,' concentrates on several populations of lymphocyte that combine characteristics of innate and adaptive immunity. These include natural killer cells, γ:δ cells, natural killer T cells, and mucosa-associated invariant T cells. After years of being a cipher, the ligands that bind to the variable antigen receptors of γ:δ are now being discovered and defined.

The first part of Chapter 13, 'Failures of the Body's Defenses,' describes the ways in which some pathogens change and avoid the immunological memories gained by their human hosts during previous infections. The second part of the chapter describes the inherited genetic defects that segregate in human populations and cause a wide range of immunodeficiency diseases. An invaluable by-product of identifying such patients and treating their diseases has been the ability to define the physiological functions of the component of the human immune system that is missing or nonfunctional in each different immunodeficiency disease. The third part of the chapter is devoted to the human immunodeficiency virus (HIV). At this time there is renewal of hope for HIV vaccines and immunotherapies based upon the results of studying the successful immune responses in exceptional individuals who maintain health despite having been infected with HIV.

Chapter 14 in this edition, 'IgE-mediated Immunity and Allergy,' has evolved from Chapter 12 in the previous edition, 'Over-reactions of the Immune System.' After introducing the four types of hypersensitivity reaction, the chapter focuses on the immunology of IgE and how it provides protection against parasitic worms in the people of developing countries and causes type I hypersensitivity reactions (allergies) in the people of industrialized countries. Much of this chapter is new and explains how IgE and its powerful receptor on mast cells, eosinophils, and basophils constitute an entire arm of the immune system that evolved specifically to control multicellular parasites, notably helminth worms. In-depth consideration of the type II, III, and IV hypersensitivity reactions is now given in Chapter 15, 'Transplantation of Tissues and Organs,' and Chapter 16, 'Disruption of Healthy Tissue by the Adaptive Immune Response,' which cover transplantation and autoimmunity, respectively. As users of the book have pointed out, different forms of transplant rejection and different types of autoimmune disease provide good examples of the type II, III, and IV hypersensitivity reactions. In these two chapters and also Chapter 17, on 'Cancer and its Interactions with the Immune System,' the amount of clinical description has been reduced so as to accommodate examples of promising new immunotherapies that are being used to treat transplant rejection, graft-versus-host disease, autoimmune disease, and various types of cancer. Although the order of the chapters on transplantation and autoimmunity has been changed in the fourth edition, the scope of these chapters has not changed.

In addition to these major changes, all chapters have been subject to revision aimed at bringing the content up to date and improving its clarity. Exemplifying the extent of these changes, about 20% of the figures are new and they include new images generously donated by colleagues.

I thank and acknowledge the authors of *Janeway's Immunobiology* and of *Case Studies in Immunology* for giving me license with the text and figures of their books. I have been fortunate to work with a collegial team of experts on this fourth edition. Sheryl L. Fuller-Espie (Cabrini College, Radnor, Pennsylvania) superbly composed the questions and answers for the end-of-chapter questions. Eleanor Lawrence expertly edited the text and the figures as well as the end-of-chapter questions. Nigel Orme created all the new illustrations for this edition, Bruce Goatly was a critical, creative copyeditor, and Yasodha Natkunam provided some superb new micrographs. Emma Jeffcock did wonders with the layout. I am indebted to Janet Foltin for her valuable contributions to this revision and to Denise Schanck, who has led the team and orchestrated the entire operation. Frances Brodsky has not only been a loyal user of the book but has generously given of her advice, suggestions, and much else to this Fourth Edition of *The Immune System*.

Acknowledgments

The author and publisher would like to thank the following reviewers for their thoughtful comments and guidance:

Carla Aldrich, Indiana University School of Medicine-Evansville; Igor C. Almeida, University of Texas at El Paso; Ivica Arsov, C.U.N.Y. York College; Roberta Attanasio, University of Georgia; Susanne Brix Pedersen, Technical University of Denmark; Eunice Carlson, Michigan Technological University; Peter Chimkupete, De Montfort University; Michael Chumley, Texas Christian University; My Lien Dao, University of South Florida; Karen Duus, Albany Medical Center; Michael Edidin, The Johns Hopkins University; Randle Gallucci, The University of Oklahoma; Michael Gleeson, Loughborough University; Gail Goodman Snitkoff, Albany College-Pharmacy & Health Sciences; Elaine Green, Coventry University; Neil Greenspan, Case Western Reserve University; Robin Herlands, Nevada State College; Cheryl Hertz, Loyola Marymount University; Allen L. Honeyman, Baylor College of Dentistry; Susan H. Jackman, Marshall University School of Medicine; Deborah Lebman, Virginia Commonwealth University; Lisa Lee-Jones, Manchester Metropolitan University; Lindsay Marshall, Aston University; Mehrdad Matloubian, University of California, San Francisco; Mark Miller, University of Tennessee; Debashis Mitra, Pune University India; Ashley Moffett, University of Cambridge; Carolyn Mold, University of New Mexico School of Medicine; Marc Monestier, Temple University; Kimberly J. Payne, Loma Linda University; Edward Roy, University of Illinois Urbana-Champaign; Ulrich Sack, Universitat Leipzig; Paul K. Small, Eureka College; Brian Sutton, King's College London; Richard Tapping, University of Illinois; John Taylor, Newcastle University; Ruurd Torensma, The Radboud University Nijmegen Medical Centre; Alan Trudgett, Queen's University Belfast; Alexander Tsygankov, Temple University; Bart Vandekerckhove, Universiteit Gent; Paul Whitley, University of Bath; Laurence Wood, Texas Tech University Health Center.

Resources for Instructors and Students

Case Studies in Immunology

by Raif Geha and Luigi Notarangelo

The companion book, *Case Studies in Immunology*, provides an additional, integrated discussion of clinical topics to reinforce and extend the basic science. In *The Immune System* diseases covered in *Case Studies* are indicated by a clipboard symbol in the margin. *Case Studies in Immunology* is sold separately.

INSTRUCTOR RESOURCES

Instructor resources are available on the Garland Science Instructor's Resource Site, located at http://www.garlandscience.com/instructors. The password-protected website provides access to the teaching resources for both this book and all other Garland Science textbooks. Qualified instructors can obtain access to the site from their sales representative or by emailing science@garland.com.

Art of *The Immune System*, Fourth Edition

The images from the book are available in two convenient formats: PowerPoint® and JPEG. They have been optimized for display on a computer. Figures are searchable by figure number, by figure name, or by keywords used in the figure legend from the book.

Figure-integrated Lecture Outlines

The section headings, concept headings, and figures from the text have been integrated into PowerPoint presentations. These will be useful for instructors who would like a head start in creating lectures for their course. Like all of our PowerPoint presentations, the lecture outlines can be customized. For example, the content of these presentations can be combined with videos and questions from the book or 'Question Bank,' to create unique lectures that facilitate interactive learning.

Question Bank

Written by Sheryl L. Fuller-Espie, PhD, DIC, Cabrini College, the revised and expanded question bank includes a variety of question formats: multiple-choice, true-false, matching, essay, and challenging 'thought' questions.

USMLE-style questions help prepare students for medical licensing examinations. There are more than 900 questions, and a large number of the multiple-choice questions are suitable for use with personal response systems (that is, clickers). The questions are organized by book chapter and provide a comprehensive sampling of concepts that can be used either directly or as inspiration for instructors to write their own test questions.

Diploma® Test Generator Software

The questions from the question bank have been loaded into the Diploma test generator software. The software is easy to use and can scramble questions to create multiple tests. Questions are organized by chapter and type, and can be additionally categorized by the instructor according to difficulty or subject. Existing questions can be edited and new ones added. It is compatible with several course management systems, including Blackboard® .

STUDENT RESOURCES

The resources for students are available on *The Immune System* Student Website, located at http://www.garlandscience.com/IS4-students.

Flashcards

Each chapter contains a set of flashcards, built into the website, that allow students to review key terms from the text.

Glossary

The complete glossary from the book is available on the website and can be searched or browsed.

Contents

Chapter 1	Elements of the Immune System and their Roles in Defense	1
Chapter 2	Innate Immunity: the Immediate Response to Infection	29
Chapter 3	Innate Immunity: the Induced Response to Infection	47
Chapter 4	Antibody Structure and the Generation of B-Cell Diversity	81
Chapter 5	Antigen Recognition by T Lymphocytes	113
Chapter 6	The Development of B Lymphocytes	149
Chapter 7	The Development of T Lymphocytes	177
Chapter 8	T Cell-Mediated Immunity	199
Chapter 9	Immunity Mediated by B Cells and Antibodies	231
Chapter 10	Preventing Infection at Mucosal Surfaces	267
Chapter 11	Immunological Memory and Vaccination	295
Chapter 12	Coevolution of Innate and Adaptive Immunity	329
Chapter 13	Failures of the Body's Defenses	365
Chapter 14	IgE-Mediated Immunity and Allergy	401
Chapter 15	Transplantation of Tissues and Organs	433
Chapter 16	Disruption of Healthy Tissue by the Adaptive Immune Response	473
Chapter 17	Cancer and Its Interactions with the Immune System	509
Answers to Questions		A:1
Glossary		G:1
Figure Acknowledgments		F:1
Index		I:1

Detailed Contents

Chapter 1

Elements of the Immune System and their Roles in Defense 1

1-1 Numerous commensal microorganisms inhabit healthy human bodies 2
1-2 Pathogens are infectious organisms that cause disease 3
1-3 The skin and mucosal surfaces form barriers against infection 4
1-4 The innate immune response causes inflammation at sites of infection 8
1-5 The adaptive immune response adds to an ongoing innate immune response 10
1-6 Adaptive immunity is better understood than innate immunity 12
1-7 Immune system cells with different functions all derive from hematopoietic stem cells 12
1-8 Immunoglobulins and T-cell receptors are the diverse lymphocyte receptors of adaptive immunity 16
1-9 On encountering their specific antigen, B cells and T cells differentiate into effector cells 17
1-10 Antibodies bind to pathogens and cause their inactivation or destruction 18
1-11 Most lymphocytes are present in specialized lymphoid tissues 19
1-12 Adaptive immunity is initiated in secondary lymphoid tissues 20
1-13 The spleen provides adaptive immunity to blood infections 23
1-14 Most secondary lymphoid tissue is associated with the gut 25
Summary to Chapter 1 26
Questions 27

Chapter 2

Innate Immunity: the Immediate Response to Infection 29

2-1 Physical barriers colonized by commensal microorganisms protect against infection by pathogens 29
2-2 Intracellular and extracellular pathogens require different types of immune response 30
2-3 Complement is a system of plasma proteins that mark pathogens for destruction 31
2-4 At the start of an infection, complement activation proceeds by the alternative pathway 32
2-5 Regulatory proteins determine the extent and site of C3b deposition 34
2-6 Phagocytosis by macrophages provides a first line of cellular defense against invading microorganisms 36
2-7 The terminal complement proteins lyse pathogens by forming membrane pores 37
2-8 Small peptides released during complement activation induce local inflammation 39
2-9 Several classes of plasma protein limit the spread of infection 39
2-10 Antimicrobial peptides kill pathogens by perturbing their membranes 41
2-11 Pentraxins are plasma proteins of innate immunity that bind microorganisms and target them to phagocytes 43
Summary to Chapter 2 43
Questions 44

Chapter 3

Innate Immunity: the Induced Response to Infection 47

3-1 Cellular receptors of innate immunity distinguish 'non-self' from 'self' 47
3-2 Tissue macrophages carry a battery of phagocytic and signaling receptors 49
3-3 Recognition of LPS by TLR4 induces changes in macrophage gene expression 51
3-4 Activation of resident macrophages induces a state of inflammation at sites of infection 53
3-5 NOD-like receptors recognize bacterial degradation products in the cytoplasm 54
3-6 Inflammasomes amplify the innate immune response by increasing the production of IL-1β 55
3-7 Neutrophils are dedicated phagocytes and the first effector cells recruited to sites of infection 56
3-8 Inflammatory cytokines recruit neutrophils from the blood to the infected tissue 57
3-9 Neutrophils are potent killers of pathogens and are themselves programmed to die 59
3-10 Inflammatory cytokines raise body temperature and activate the liver to make the acute-phase response 62
3-11 The lectin pathway of complement activation is initiated by the mannose-binding lectin 63
3-12 C-reactive protein triggers the classical pathway of complement activation 66
3-13 Toll-like receptors sense the presence of the four main groups of pathogenic microorganisms 66
3-14 Genetic variation in Toll-like receptors is associated with resistance and susceptibility to disease 67
3-15 Internal detection of viral infection induces cells to make an interferon response 68
3-16 Plasmacytoid dendritic cells are factories for making large quantities of type I interferons 71
3-17 Natural killer cells are the main circulating lymphocytes that contribute to the innate immune response 71
3-18 Two subpopulations of NK cells are differentially distributed in blood and tissues 72
3-19 NK-cell cytotoxicity is activated at sites of virus infection 73
3-20 NK cells and macrophages activate each other at sites of infection 75
3-21 Interactions between dendritic cells and NK cells influence the immune response 76
Summary to Chapter 3 78
Questions 78

Chapter 4

Antibody Structure and the Generation of B-Cell Diversity 81

The structural basis of antibody diversity 82
4-1 Antibodies are composed of polypeptides with variable and constant regions 82
4-2 Immunoglobulin chains are folded into compact and stable protein domains 83
4-3 An antigen-binding site is formed from the hypervariable regions of a heavy-chain V domain and a light-chain V domain 85
4-4 Antigen-binding sites vary in shape and physical properties 86
4-5 Monoclonal antibodies are produced from a clone of antibody-producing cells 88
4-6 Monoclonal antibodies are used as treatments for a variety of diseases 90
Summary 91

Generation of immunoglobulin diversity in B cells before encounter with antigen 91
4-7 The DNA sequence encoding a V region is assembled from two or three gene segments 91
4-8 Random recombination of gene segments produces diversity in the antigen-binding sites of immunoglobulins 92
4-9 Recombination enzymes produce additional diversity in the antigen-binding site 95
4-10 Developing and naive B cells use alternative mRNA splicing to make both IgM and IgD 96
4-11 Each B cell produces immunoglobulin of a single antigen specificity 96
4-12 Immunoglobulin is first made in a membrane-bound form that is present on the B-cell surface 97
Summary 98

Diversification of antibodies after B cells encounter antigen 98
4-13 Secreted antibodies are produced by an alternative pattern of heavy-chain RNA processing 98
4-14 Rearranged V-region sequences are further diversified by somatic hypermutation 100
4-15 Isotype switching produces immunoglobulins with different C regions but identical antigen specificities 101

4-16 Antibodies with different C regions have different effector functions 103
4-17 The four subclasses of IgG have different and complementary functions 105
Summary 107
Summary to Chapter 4 107
Questions 110

Chapter 5
Antigen Recognition by T Lymphocytes 113
T-cell receptor diversity 114
5-1 The T-cell receptor resembles a membrane-associated Fab fragment of immunoglobulin 114
5-2 T-cell receptor diversity is generated by gene rearrangement 115
5-3 The RAG genes were key elements in the origin of adaptive immunity 117
5-4 Expression of the T-cell receptor on the cell surface requires association with additional proteins 117
5-5 A distinct population of T cells expresses a second class of T-cell receptor with γ and δ chains 118
Summary 119
Antigen processing and presentation 120
5-6 T-cell receptors recognize peptide antigens bound to MHC molecules 121
5-7 Two classes of MHC molecule present peptide antigens to two types of T cell 122
5-8 The two classes of MHC molecule have similar three-dimensional structures 123
5-9 MHC molecules bind a variety of peptides 124
5-10 MHC class I and MHC class II molecules function in different intracellular compartments 125
5-11 Peptides generated in the cytosol are transported to the endoplasmic reticulum for binding to MHC class I molecules 126
5-12 MHC class I molecules bind peptides as part of a peptide-loading complex 127
5-13 Peptides presented by MHC class II molecules are generated in acidified intracellular vesicles 129
5-14 Invariant chain prevents MHC class II molecules from binding peptides in the endoplasmic reticulum 130
5-15 Cross-presentation enables extracellular antigens to be presented by MHC class I 131
5-16 MHC class I molecules are expressed by most cell types, MHC class II molecules are expressed by few cell types 132
5-17 The T-cell receptor specifically recognizes both peptide and MHC molecule 132
Summary 133
The major histocompatibility complex 135
5-18 The diversity of MHC molecules in the human population is due to multigene families and genetic polymorphism 135
5-19 The HLA class I and class II genes occupy different regions of the HLA complex 137
5-20 Other proteins involved in antigen processing and presentation are encoded in the HLA class II region 138
5-21 MHC polymorphism affects the binding of peptide antigens and their presentation to T cells 138
5-22 MHC diversity results from selection by infectious disease 140
5-23 MHC polymorphism triggers T-cell reactions that can reject transplanted organs 143
Summary 144
Summary to Chapter 5 144
Questions 145

Chapter 6
The Development of B Lymphocytes 149
The development of B cells in the bone marrow 150
6-1 B-cell development in the bone marrow proceeds through several stages 150
6-2 B-cell development is stimulated by bone marrow stromal cells 151
6-3 Pro-B-cell rearrangement of the heavy-chain locus is an inefficient process 152
6-4 The pre-B-cell receptor monitors the quality of immunoglobulin heavy chains 153
6-5 The pre-B-cell receptor causes allelic exclusion at the immunoglobulin heavy-chain locus 154
6-6 Rearrangement of the light-chain loci by pre-B cells is relatively efficient 155
6-7 Developing B cells pass two checkpoints in the bone marrow 157
6-8 A program of protein expression underlies the stages of B-cell development 157
6-9 Many B-cell tumors carry chromosomal translocations that join immunoglobulin genes to genes that regulate cell growth 160
6-10 B cells expressing the glycoprotein CD5 express a distinctive repertoire of receptors 161
Summary 162
Selection and further development of the B-cell repertoire 163
6-11 The population of immature B cells is purged of cells bearing self-reactive B-cell receptors 164

6-12 The antigen receptors of autoreactive immature B cells can be modified by receptor editing 165
6-13 Immature B cells specific for monovalent self antigens are made nonresponsive to antigen 166
6-14 Maturation and survival of B cells requires access to lymphoid follicles 167
6-15 Encounter with antigen leads to the differentiation of activated B cells into plasma cells and memory B cells 168
6-16 Different types of B-cell tumor reflect B cells at different stages of development 170
Summary 170
Summary to Chapter 6 172
Questions 173

Chapter 7

The Development of T Lymphocytes 177

7-1 T cells develop in the thymus 178
7-2 Thymocytes commit to the T-cell lineage before rearranging their T-cell receptor genes 180
7-3 The two lineages of T cells arise from a common thymocyte progenitor 181
7-4 Gene rearrangement in double-negative thymocytes leads to assembly of either a γ:δ receptor or a pre-T-cell receptor 183
7-5 Thymocytes can make four attempts to rearrange a β-chain gene 184
7-6 Rearrangement of the α-chain gene occurs only in pre-T cells 185
7-7 Stages in T-cell development are marked by changes in gene expression 186
Summary 188

Positive and negative selection of the T-cell repertoire 188
7-8 T cells that recognize self-MHC molecules are positively selected in the thymus 189
7-9 Continuing α-chain gene rearrangement increases the chance for positive selection 190
7-10 Positive selection determines expression of either the CD4 or the CD8 co-receptor 191
7-11 T cells specific for self antigens are removed in the thymus by negative selection 192
7-12 Tissue-specific proteins are expressed in the thymus and participate in negative selection 192
7-13 Regulatory CD4 T cells comprise a distinct lineage of CD4 T cells 193
7-14 T cells undergo further differentiation in secondary lymphoid tissues after encounter with antigen 193
Summary 194
Summary to Chapter 7 194
Questions 196

Chapter 8

T Cell-Mediated Immunity 199

Activation of naive T cells by antigen 199
8-1 Dendritic cells carry antigens from sites of infection to secondary lymphoid tissues 200
8-2 Dendritic cells are adept and versatile at processing pathogen antigens 202
8-3 Naive T cells first encounter antigen presented by dendritic cells in secondary lymphoid tissues 203
8-4 Homing of naive T cells to secondary lymphoid tissues is determined by chemokines and cell-adhesion molecules 204
8-5 Activation of naive T cells requires signals from the antigen receptor and a co-stimulatory receptor 206
8-6 Signals from T-cell receptors, co-receptors, and co-stimulatory receptors activate naive T cells 207
8-7 Proliferation and differentiation of activated naive T cells are driven by the cytokine interleukin-2 209
8-8 Antigen recognition in the absence of co-stimulation leads to a state of T-cell anergy 210
8-9 Activation of naive CD4 T cells gives rise to effector CD4 T cells with distinctive helper functions 211
8-10 The cytokine environment determines which differentiation pathway a naive T cell takes 213
8-11 Positive feedback in the cytokine environment can polarize the effector CD4 T-cell response 214
8-12 Naive CD8 T cells require stronger activation than naive CD4 T cells 215
Summary 217

The properties and functions of effector T cells 218
8-13 Cytotoxic CD8 T cells and effector CD4 T_H1, T_H2, and T_H17 work at sites of infection 218
8-14 Effector T-cell functions are mediated by cytokines and cytotoxins 220
8-15 Cytokines change the patterns of gene expression in the cells targeted by effector T cells 221

8-16 Cytotoxic CD8 T cells are selective and serial killers of target cells at sites of infection 222
8-17 Cytotoxic T cells kill their target cells by inducing apoptosis 223
8-18 Effector T_H1 CD4 cells induce macrophage activation 224
8-19 T_{FH} cells, and the naive B cells that they help, recognize different epitopes of the same antigen 225
8-20 Regulatory CD4 T cells limit the activities of effector CD4 and CD8 T cells 226
Summary 227
Summary to Chapter 8 227
Questions 228

Chapter 9
Immunity Mediated by B Cells and Antibodies **231**

Antibody production by B lymphocytes 231
9-1 B-cell activation requires cross-linking of surface immunoglobulin 232
9-2 B-cell activation requires signals from the B-cell co-receptor 232
9-3 Effective B cell-mediated immunity depends on help from CD4 T cells 234
9-4 Follicular dendritic cells in the B-cell area store and display intact antigens to B cells 235
9-5 Antigen-activated B cells move close to the T-cell area to find a helper T_{FH} cell 236
9-6 The primary focus of clonal expansion in the medullary cords produces plasma cells secreting IgM 238
9-7 Activated B cells undergo somatic hypermutation and isotype switching in the specialized microenvironment of the primary follicle 239
9-8 Antigen-mediated selection of centrocytes drives affinity maturation of the B-cell response in the germinal center 241
9-9 The cytokines made by helper T cells determine how B cells switch their immunoglobulin isotype 243
9-10 Cytokines made by helper T cells determine the differentiation of antigen-activated B cells into plasma cells or memory cells 244
Summary 245

Antibody effector functions 245
9-11 IgM, IgG, and monomeric IgA protect the internal tissues of the body 246
9-12 Dimeric IgA protects the mucosal surfaces of the body 246
9-13 IgE provides a mechanism for the rapid ejection of parasites and other pathogens from the body 247
9-14 Mothers provide protective antibodies to their young, both before and after birth 250
9-15 High-affinity neutralizing antibodies prevent viruses and bacteria from infecting cells 251
9-16 High-affinity IgG and IgA antibodies are used to neutralize microbial toxins and animal venoms 253
9-17 Binding of IgM to antigen on a pathogen's surface activates complement by the classical pathway 255
9-18 Two forms of C4 tend to be fixed at different sites on pathogen surfaces 256
9-19 Complement activation by IgG requires the participation of two or more IgG molecules 257
9-20 Erythrocytes facilitate the removal of immune complexes from the circulation 258
9-21 Fcγ receptors enable effector cells to bind and be activated by IgG bound to pathogens 258
9-22 A variety of low-affinity Fc receptors are IgG-specific 260
9-23 An Fc receptor acts as an antigen receptor for NK cells 261
9-24 The Fc receptor for monomeric IgA belongs to a different family than the Fc receptors for IgG and IgE 262
Summary 263
Summary to Chapter 9 263
Questions 264

Chapter 10
Preventing Infection at Mucosal Surfaces **267**

10-1 The communication functions of mucosal surfaces render them vulnerable to infection 267
10-2 Mucins are gigantic glycoproteins that endow the mucus with the properties to protect epithelial surfaces 269
10-3 Commensal microorganisms assist the gut in digesting food and maintaining health 269
10-4 The gastrointestinal tract is invested with distinctive secondary lymphoid tissues 272
10-5 Inflammation of mucosal tissues is associated with causation not cure of disease 273

10-6 Intestinal epithelial cells contribute to innate immune responses in the gut 275
10-7 Intestinal macrophages eliminate pathogens without creating a state of inflammation 276
10-8 M cells constantly transport microbes and antigens from the gut lumen to gut-associated lymphoid tissue 277
10-9 Gut dendritic cells respond differently to food, commensal microorganisms, and pathogens 278
10-10 Activation of B cells and T cells in one mucosal tissue commits them to defending all mucosal tissues 279
10-11 A variety of effector lymphocytes guard healthy mucosal tissue in the absence of infection 281
10-12 B cells activated in mucosal tissues give rise to plasma cells secreting IgM and IgA at mucosal surfaces 282
10-13 Secretory IgM and IgA protect mucosal surfaces from microbial invasion 283
10-14 Two subclasses of IgA have complementary properties for controlling microbial populations 285
10-15 People lacking IgA are able to survive, reproduce, and generally remain healthy 286
10-16 T_H2-mediated immunity protects against helminth infections 288

Summary to Chapter 10 290

Questions 292

Chapter 11

Immunological Memory and Vaccination 295

Immunological memory and the secondary immune response 296

11-1 Antibodies made in a primary immune response persist for several months and provide protection 296
11-2 Low levels of pathogen-specific antibodies are maintained by long-lived plasma cells 297
11-3 Long-lived clones of memory B cells and T cells are produced in the primary immune response 297
11-4 Memory B cells and T cells provide protection against pathogens for decades and even for life 299
11-5 Maintaining populations of memory cells does not depend upon the persistence of antigen 299
11-6 Changes to the antigen receptor distinguish naive, effector, and memory B cells 300
11-7 In the secondary immune response, memory B cells are activated whereas naive B cells are inhibited 300
11-8 Activation of the primary and secondary immune responses have common features 301
11-9 Combinations of cell-surface markers distinguish memory T cells from naive and effector T cells 302
11-10 Central and effector memory T cells recognize pathogens in different tissues of the body 304
11-11 In viral infections, numerous effector CD8 T cells give rise to relatively few memory T cells 305
11-12 Immune-complex-mediated inhibition of naive B cells is used to prevent hemolytic anemia of the newborn 305
11-13 In the response to influenza virus, immunological memory is gradually eroded 306

Summary 307

Vaccination to prevent infectious disease 308

11-14 Protection against smallpox is achieved by immunization with the less dangerous cowpox virus 308
11-15 Smallpox is the only infectious disease of humans that has been eradicated worldwide by vaccination 309
11-16 Most viral vaccines are made from killed or inactivated viruses 310
11-17 Both inactivated and live-attenuated vaccines protect against poliovirus 311
11-18 Vaccination can inadvertently cause disease 312
11-19 Subunit vaccines are made from the most antigenic components of a pathogen 313
11-20 Invention of rotavirus vaccines took at least 30 years of research and development 313
11-21 Bacterial vaccines are made from whole bacteria, secreted toxins, or capsular polysaccharides 314
11-22 Conjugate vaccines enable high-affinity antibodies to be made against carbohydrate antigens 315
11-23 Adjuvants are added to vaccines to activate and enhance the response to antigen 316
11-24 Genome sequences of human pathogens have opened up new avenues for making vaccines 316
11-25 The ever-changing influenza virus requires a new vaccine every year 318

11-26 The need for a vaccine and the demands placed upon it change with the prevalence of disease 319
11-27 Vaccines have yet to be made against pathogens that establish chronic infections 322
11-28 Vaccine development faces greater public scrutiny than drug development 323
Summary 324
Summary to Chapter 11 325
Questions 326

Chapter 12

Coevolution of Innate and Adaptive Immunity **329**

Regulation of NK-cell function by MHC class I and related molecules 330
12-1 NK cells express a range of activating and inhibitory receptors 330
12-2 The strongest receptor that activates NK cells is an Fc receptor 332
12-3 Many NK-cell receptors recognize MHC class I and related molecules 333
12-4 Immunoglobulin-like NK-cell receptors recognize polymorphic epitopes of HLA-A, HLA-B, and HLA-C 335
12-5 NK cells are educated to detect pathological change in MHC class I expression 336
12-6 Different genomic complexes encode lectin-like and immunoglobulin-like NK-cell receptors 339
12-7 Human KIR haplotypes uniquely come in two distinctive forms 340
12-8 Cytomegalovirus infection induces proliferation of NK cells expressing the activating HLA-E receptor 341
12-9 Interactions of uterine NK cells with fetal MHC class I molecules affect reproductive success 342
Summary 345

Maintenance of tissue integrity by γ:δ T cells 347
12-10 γ:δ T cells are not governed by the same rules as α:β T cells 347
12-11 γ:δ T cells in blood and tissues express different γ:δ receptors 348
12-12 $V_\gamma9$:$V_\gamma2$ T cells recognize phosphoantigens presented on cell surfaces 350
12-13 $V_\gamma4$:$V_\gamma5$ T cells detect both virus-infected cells and tumor cells 351
12-14 V_γ:$V_\gamma1$ T-cell receptors recognize lipid antigens presented by CD1d 352
Summary 354

Restriction of α:β T cells by non-polymorphic MHC class I-like molecules 354
12-15 CD1-restricted α:β T cells recognize lipid antigens of mycobacterial pathogens 354
12-16 NKT cells are innate lymphocytes that detect lipid antigens by using α:β T-cell receptors 356
12-17 Mucosa-associated invariant T cells detect bacteria and fungi that make riboflavin 357
Summary 359
Summary to Chapter 12 360
Questions 361

Chapter 13

Failures of the Body's Defenses **365**

Evasion and subversion of the immune system by pathogens 365
13-1 Genetic variation within some species of pathogens prevents effective long-term immunity 366
13-2 Mutation and recombination allow influenza virus to escape from immunity 366
13-3 Trypanosomes use gene conversion to change their surface antigens 368
13-4 Herpesviruses persist in human hosts by hiding from the immune response 369
13-5 Some pathogens sabotage or subvert immune defense mechanisms 371
13-6 Bacterial superantigens stimulate a massive but ineffective CD4 T-cell response 373
13-7 Subversion of IgA action by bacterial IgA-binding proteins 374
Summary 375

Inherited immunodeficiency diseases 375
13-8 Rare primary immunodeficiency diseases reveal how the human immune system works 375
13-9 Inherited immunodeficiency diseases are caused by dominant, recessive, or X-linked gene defects 377
13-10 Recessive and dominant mutations in the IFN-γ receptor cause diseases of differing severity 378
13-11 Antibody deficiency leads to poor clearing of extracellular bacteria 379

13-12 Diminished production of antibodies also results from inherited defects in T-cell help 380

13-13 Complement defects impair antibody-mediated immunity and cause immune-complex disease 381

13-14 Defects in phagocytes result in enhanced susceptibility to bacterial infection 382

13-15 Defects in T-cell function result in severe combined immune deficiencies 383

13-16 Some inherited immunodeficiencies lead to specific disease susceptibilities 385

Summary 386

Acquired immune deficiency syndrome 386

13-17 HIV is a retrovirus that causes a slowly progressing chronic disease 388

13-18 HIV infects CD4 T cells, macrophages, and dendritic cells 388

13-19 In the twentieth century, most HIV-infected people progressed in time to get AIDS 389

13-20 Genetic deficiency of the CCR5 co-receptor for HIV confers resistance to infection 391

13-21 HLA and KIR polymorphisms influence the progression to AIDS 392

13-22 HIV escapes the immune response and develops resistance to antiviral drugs by rapid mutation 393

13-23 Clinical latency is a period of active infection and renewal of CD4 T cells 394

13-24 HIV infection leads to immunodeficiency and death from opportunistic infections 395

13-25 A minority of HIV-infected individuals make antibodies that neutralize many strains of HIV 396

Summary 397

Summary to Chapter 13 398

Questions 398

Chapter 14

IgE-Mediated Immunity and Allergy 401

14-1 Different effector mechanisms cause four distinctive types of hypersensitivity reaction 401

Shared mechanisms of immunity and allergy 403

14-2 IgE-mediated immune responses defend the body against multicellular parasites 404

14-3 IgE antibodies emerge at early and late times in the primary immune response 404

14-4 Allergy is prevalent in countries where parasite infections have been eliminated 406

14-5 IgE has distinctive properties that contrast with those of IgG 406

14-6 IgE and FcεRI supply each mast cell with a diversity of antigen-specific receptors 407

14-7 FcεRII is a low-affinity receptor for IgE Fc regions that regulates the production of IgE by B cells 407

14-8 Treatment of allergic disease with an IgE-specific monoclonal antibody 409

14-9 Mast cells defend and maintain the tissues in which they reside 410

14-10 Tissue mast cells orchestrate IgE-mediated reactions through the release of inflammatory mediators 411

14-11 Eosinophils are specialized granulocytes that release toxic mediators in IgE-mediated responses 413

14-12 Basophils are rare granulocytes that initiate T_H2 responses and the production of IgE 415

Summary 415

IgE-mediated allergic disease 416

14-13 Allergens are protein antigens, some of which resemble parasite antigens 416

14-14 Predisposition to allergic disease is influenced by genetic and environmental factors 418

14-15 IgE-mediated allergic reactions consist of an immediate response followed by a late-phase response 419

14-16 The effects of IgE-mediated allergic reactions vary with the site of mast-cell activation 420

14-17 Systemic anaphylaxis is caused by allergens in the blood 421

14-18 Rhinitis and asthma are caused by inhaled allergens 423

14-19 Urticaria, angioedema, and eczema are allergic reactions in the skin 424

14-20 Food allergies cause systemic effects as well as gut reactions 426

14-21 Allergic reactions are prevented and treated by three complementary approaches 427

Summary 428

Summary to Chapter 14 428

Questions 429

Chapter 15

Transplantation of Tissues and Organs 433

Allogeneic transplantation can trigger hypersensitivity reactions 433

15-1 Blood is the most common transplanted tissue 434
15-2 Before blood transfusion, donors and recipients are matched for ABO and the Rhesus D antigens 434
15-3 Incompatibility of blood group antigens causes type II hypersensitivity reactions 435
15-4 Hyperacute rejection of transplanted organs is a type II hypersensitivity reaction 436
15-5 Anti-HLA antibodies can arise from pregnancy, blood transfusion, or previous transplants 437
15-6 Transplant rejection and graft-versus-host disease are type IV hypersensitivity reactions 438
Summary 439

Transplantation of solid organs 440
15-7 Organ transplantation involves procedures that inflame the donated organ and the transplant recipient 440
15-8 Acute rejection is a type IV hypersensitivity caused by effector T cells responding to HLA differences between donor and recipient 441
15-9 HLA differences between transplant donor and recipient activate numerous alloreactive T cells 442
15-10 Chronic rejection of organ transplants is caused by a type III hypersensitivity reaction 443
15-11 Matching donor and recipient HLA class I and II allotypes improves the success of transplantation 445
15-12 Immunosuppressive drugs make allogeneic transplantation possible as routine therapy 445
15-13 Some treatments induce immuno-suppression before transplantation 447
15-14 T-cell activation can be targeted by immunosuppressive drugs 448
15-15 Alloreactive T-cell co-stimulation can be blocked with a soluble form of CTLA4 451
15-16 Blocking cytokine signaling can prevent alloreactive T-cell activation 452
15-17 Cytotoxic drugs target the replication and proliferation of alloantigen-activated T cells 453
15-18 Patients needing a transplant outnumber the available organs 455
15-19 The need for HLA matching and immunosuppressive therapy varies with the organ transplanted 456
Summary 457

Hematopoietic cell transplantation 458
15-20 Hematopoietic cell transplantation is a treatment for genetic diseases of blood cells 459
15-21 Allogeneic hematopoietic cell transplantation is the preferred treatment for many cancers 461
15-22 After hematopoietic cell transplantation, the patient is attacked by alloreactive T cells in the graft 461
15-23 HLA matching of donor and recipient is most important for hematopoietic cell transplantation 462
15-24 Minor histocompatibility antigens trigger alloreactive T cells in recipients of HLA-identical transplants 464
15-25 Some GVHD helps engraftment and prevents relapse of malignant disease 465
15-26 NK cells also mediate graft-versus-leukemia effects 466
15-27 Hematopoietic cell transplantation can induce tolerance of a solid organ transplant 467
Summary 467

Summary to Chapter 15 468

Questions 469

Chapter 16

Disruption of Healthy Tissue by the Adaptive Immune Response 473

16-1 Every autoimmune disease resembles a type II, III, or IV hypersensitivity reaction 474
16-2 Autoimmune diseases arise when tolerance to self antigens is lost 477
16-3 HLA is the dominant genetic factor affecting susceptibility to autoimmune disease 478
16-4 HLA associations reflect the importance of T-cell tolerance in preventing autoimmunity 480
16-5 Binding of antibodies to cell-surface receptors causes several autoimmune diseases 481
16-6 Organized lymphoid tissue sometimes forms at sites inflamed by autoimmune disease 484
16-7 The antibody response to an autoantigen can broaden and strengthen by epitope spreading 485
16-8 Intermolecular epitope spreading occurs in systemic autoimmune disease 487
16-9 Intravenous immunoglobulin is a therapy for autoimmune diseases 489

16-10 Monoclonal antibodies that target TNF-α and B cells are used to treat rheumatoid arthritis 490
16-11 Rheumatoid arthritis is influenced by genetic and environmental factors 491
16-12 Autoimmune disease can be an adverse side-effect of an immune response to infection 492
16-13 Noninfectious environmental factors affect the development of autoimmune disease 494
16-14 Type 1 diabetes is caused by the selective destruction of insulin-producing cells in the pancreas 495
16-15 Combinations of HLA class II allotypes confer susceptibility and resistance to type 1 diabetes 496
16-16 Celiac disease is a hypersensitivity to food that has much in common with autoimmune disease 498
16-17 Celiac disease is caused by the selective destruction of intestinal epithelial cells 498
16-18 Senescence of the thymus and the T-cell population contributes to autoimmunity 501
16-19 Autoinflammatory diseases of innate immunity 502
Summary to Chapter 16 503
Questions 506

Chapter 17
Cancer and Its Interactions With the Immune System 509

17-1 Cancer results from mutations that cause uncontrolled cell growth 510
17-2 A cancer arises from a single cell that has accumulated multiple mutations 510
17-3 Exposure to chemicals, radiation, and viruses facilitates progression to cancer 512
17-4 Certain common features distinguish cancer cells from normal cells 513
17-5 Immune responses to cancer have similarities with those to virus-infected cells 514
17-6 Allogeneic differences in MHC class I molecules enable cytotoxic T cells to eliminate tumor cells 515
17-7 Mutations acquired by somatic cells during oncogenesis can give rise to tumor-specific antigens 516
17-8 Cancer/testis antigens are a prominent type of tumor-associated antigen 517
17-9 Successful tumors evade and manipulate the immune response 518
17-10 Vaccination against human papillomaviruses can prevent cervical and other genital cancers 519
17-11 Vaccination with tumor antigens can cause cancer to regress but it is unpredictable 520
17-12 Monoclonal antibodies that interfere with negative regulators of the immune response can be used to treat cancer 521
17-13 T-cell responses to tumor cells can be improved with chimeric antigen receptors 522
17-14 The antitumor response of $\gamma{:}\delta$ T cells and NK cells can be augmented 524
17-15 T-cell responses to tumors can be improved by adoptive transfer of antigen-activated dendritic cells 525
17-16 Monoclonal antibodies are valuable tools for the diagnosis of cancer 526
17-17 Monoclonal antibodies against cell-surface antigens are increasingly used in cancer therapy 528
Summary to Chapter 17 529
Questions 530

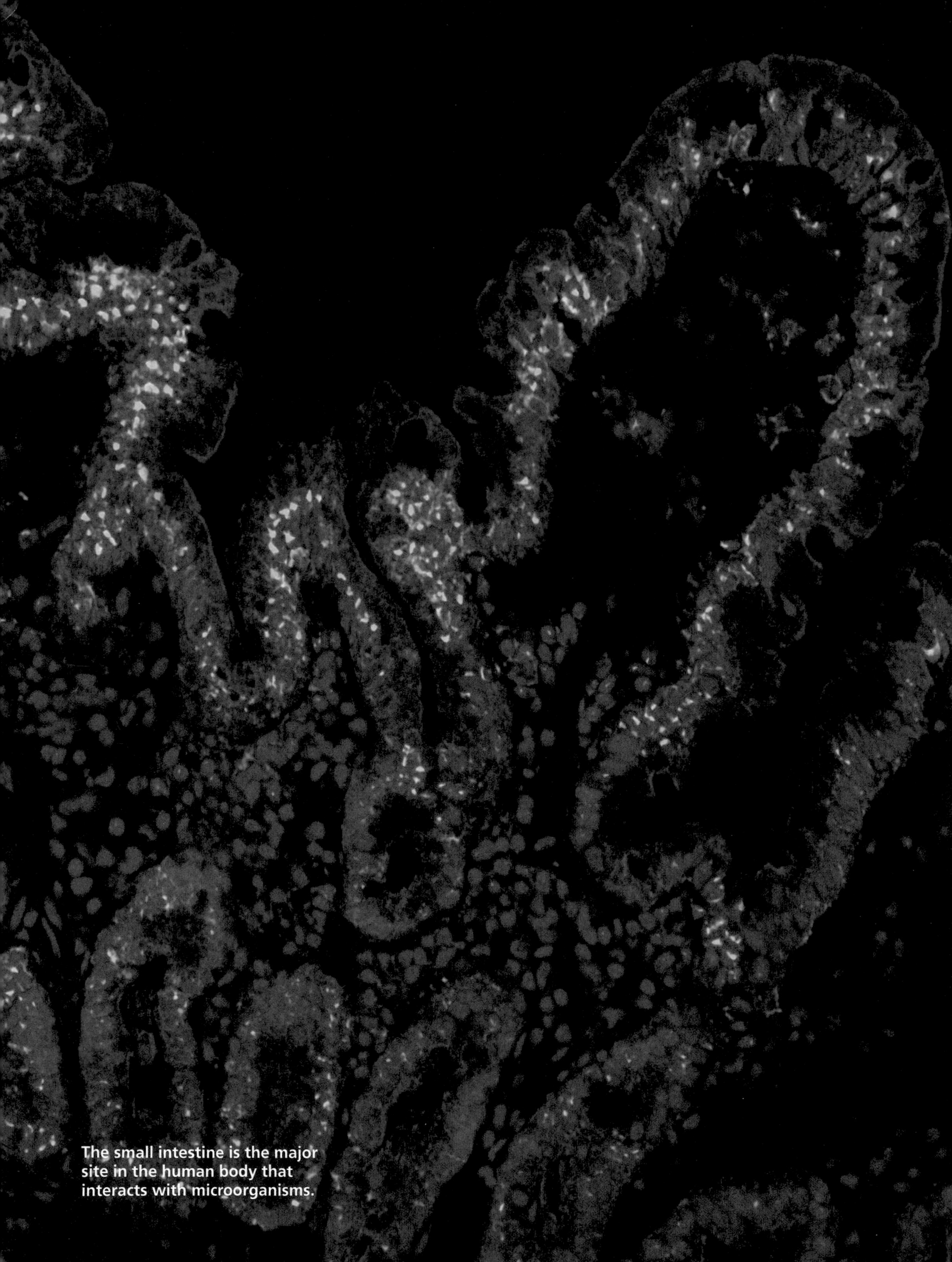

The small intestine is the major site in the human body that interacts with microorganisms.

Chapter 1

Elements of the Immune System and their Roles in Defense

Immunology is the study of the physiological mechanisms that humans and other animals use to defend their bodies from invasion by all sorts of other organisms. The origins of the subject lie in the practice of medicine and in historical observations that people who survived the ravages of epidemic disease were untouched when faced with that same disease again—they had become **immune** to infection. Infectious diseases are caused by microorganisms, which have the advantage of reproducing and evolving much more rapidly than their human hosts. During the course of an infection, the microorganism can pit a vast population of its species against an individual *Homo sapiens*. In response, the human body invests heavily in cells dedicated to defense, which collectively form the **immune system**.

The immune system is crucial to human survival. In the absence of a working immune system, even minor infections can take hold and prove fatal. Without intensive treatment, children born without a functional immune system die in early childhood from the effects of common infections. However, in spite of their immune systems, all humans suffer from infectious diseases, especially when young. This is because the immune system takes time to build up its strongest response to an invading microorganism, time during which the invader can multiply and cause disease. To provide **immunity** that will give future protection from the disease, the immune system must first do battle with the microorganism. This places people at highest risk during their first infection with a microorganism and, in the absence of modern medicine, leads to substantial child mortality, as witnessed in the developing world. When entire populations face a completely new infection, the outcome can be catastrophic, as experienced by indigenous Americans who were killed in large numbers by European diseases to which they were suddenly exposed after 1492. Today, infection with human immunodeficiency virus (HIV) and the acquired immune deficiency syndrome (AIDS) it causes are having a similarly tragic impact on the populations of several African countries.

In medicine the greatest triumph of immunology has been **vaccination**, or **immunization**, a procedure whereby severe disease is prevented by prior exposure to the infectious agent in a form that cannot cause disease. Vaccination provides the opportunity for the immune system to gain the experience needed to make a protective response with little risk to health or life. Vaccination was first used against smallpox, a viral scourge that once ravaged populations and disfigured the survivors. In Asia, small amounts of smallpox virus had been used to induce protective immunity for hundreds of years before 1721, when Lady Mary Wortley Montagu introduced the method into Western Europe. Subsequently, in 1796, Edward Jenner, a doctor in rural

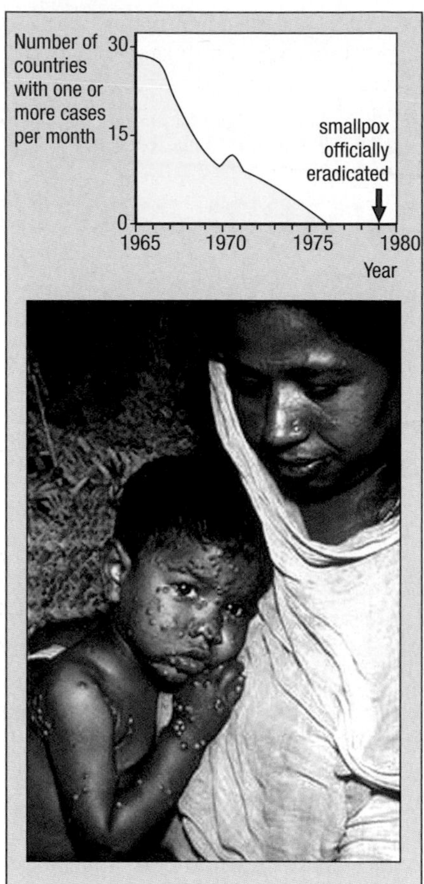

Figure 1.1 The eradication of smallpox by vaccination. Upper panel: smallpox vaccination was started in 1796. In 1979, after 3 years in which no case of smallpox was recorded, the World Health Organization announced that the virus had been eradicated. Since then the proportion of the human population that has been vaccinated against smallpox, or has acquired immunity from an infection, has steadily decreased. The result is that the human population has become increasingly vulnerable should the virus emerge again, either naturally or as a deliberate act of human malevolence. Lower panel: photograph of a child with smallpox and his immune mother. The distinctive rash of smallpox appears about 2 weeks after exposure to the virus. Photograph courtesy of the World Health Organization.

England, showed how inoculation with cowpox virus offered protection against the related smallpox virus with less risk than the earlier methods. Jenner called his procedure vaccination, after vaccinia, the name given to the mild disease produced by cowpox, and he is generally credited with its invention. Since his time, vaccination dramatically reduced the incidence of smallpox worldwide until it was eventually eliminated. The last cases of smallpox were seen by physicians in the 1970s (Figure 1.1).

Effective vaccines have been made from only a fraction of the agents that cause disease, and some are of limited availability because of their cost. Most of the widely used vaccines were first developed many years ago by a process of trial and error, before very much was known about the workings of the immune system. That approach is no longer so successful for developing new vaccines, perhaps because all the easily won vaccines have been made. But deeper understanding of the mechanisms of immunity is spawning new ideas for vaccines against infectious diseases and even against other types of disease such as cancer. Much is now known about the molecular and cellular components of the immune system and what they can do in the laboratory. Current research seeks to understand the contributions of these immune components to fighting infections in the world at large. The new knowledge is also being used to find better ways of manipulating the immune system to prevent the unwanted immune responses that cause allergies, autoimmune diseases, and rejection of organ transplants.

In this chapter we first consider the microorganisms that infect human beings and then the defenses they must overcome to start and propagate an infection. The individual cells and tissues of the immune system are described, and how they integrate their functions with the rest of the human body. The first line of defense is innate immunity, which includes physical and chemical barriers to infection, and responses that are ready and waiting to halt infections before they can barely start. Most infections are stopped by these mechanisms, but when they fail, the more flexible and forceful defenses of the adaptive immune response are brought into play. The adaptive immune response is always targeted to the specific problem at hand and is made and refined during the course of the infection. When successful, it clears the infection and provides long-lasting immunity that prevents its recurrence.

1-1 Numerous commensal microorganisms inhabit healthy human bodies

The main purpose of the immune system is to protect the human body from infectious disease. Almost all infectious diseases of humans are caused by microorganisms smaller than a single human cell. For both benign and dangerous microorganisms alike, the human body constitutes a vast resource-rich environment in which to live, feed, and reproduce. More than 1000 different microbial species live in the healthy adult human gut and contribute about 10 pounds (4.5 kilograms) to the body's weight; they are called **commensal** species, meaning they 'eat at the same table.' The community of microbial species that inhabits a particular niche in the human body—skin, mouth, gut, or vagina—is called the **microbiota**, for example the 'gut microbiota.' Many of these species have not yet been studied properly because they cannot be propagated in the laboratory, growing only under the special conditions furnished by their human hosts.

Animals have evolved along with their commensal species and in so doing have become both tolerant of them and dependent upon them. Commensal organisms enhance human nutrition by processing digested food and making several vitamins. They also protect against disease, because their presence helps to prevent colonization by dangerous, disease-causing microorganisms. In addition to competing for their space, *Escherichia coli*, a major bacterial component of the normal mammalian gut flora, secretes antibacterial proteins called colicins that incapacitate other bacteria and prevent them from colonizing the gut. When a patient with a bacterial infection takes a course of antibiotic drugs, much of the normal gut microbiota is killed along with the disease-causing bacteria. After such treatment the body is recolonized by a new population of microorganisms; in this situation, opportunistic disease-causing bacteria, such as *Clostridium difficile*, can sometimes establish themselves, causing further disease and sometimes death (Figure 1.2). *C. difficile* produces a toxin that causes diarrhea and, in some cases, an even more serious gastrointestinal condition called pseudomembranous colitis.

1-2 Pathogens are infectious organisms that cause disease

Any organism with the potential to cause disease is known as a **pathogen**. This definition includes not only microorganisms such as the influenza virus or the typhoid bacillus that habitually cause disease if they enter the body, but also ones that can colonize the human body to no ill effect for much of the time but cause illness if the body's defenses are weakened or if the microbe gets into the 'wrong' place. The latter microorganisms are known as **opportunistic pathogens**.

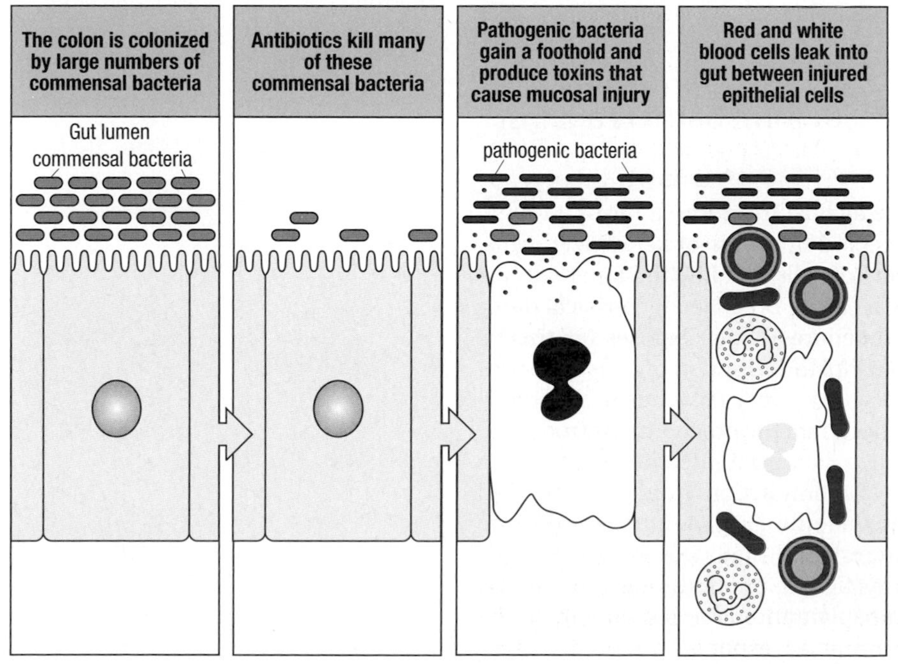

Figure 1.2 Antibiotic treatments disrupt the natural ecology of the colon. When antibiotics are taken orally to counter a bacterial infection, beneficial populations of commensal bacteria in the colon are also decimated. This provides an opportunity for pathogenic strains of bacteria to populate the colon and cause further disease. *Clostridium difficile* is an example of such a bacterium; it produces a toxin that can cause severe diarrhea in patients treated with antibiotics. In hospitals, acquired *C. difficile* infections are an increasing cause of death for elderly patients.

Pathogens can be divided into four kinds: **bacteria**, **viruses**, and **fungi**, each of which is a group of related microorganisms, and internal **parasites**, a less precise term used to embrace a heterogeneous collection of unicellular protozoa and multicellular invertebrates, mainly worms. In this book we consider the functions of the human immune system principally in the context of controlling infections. For some pathogens this necessitates their complete elimination, but for others it is sufficient to limit the size and location of the pathogen population within the human host. Figure 1.3 illustrates the variety in shape and form of the four kinds of pathogen. Figure 1.4 lists common or well-known infectious diseases and the pathogens that cause them. Reference to many of these diseases and the problems they pose for the immune system will be made in the rest of this book.

Over evolutionary time, the relationship between a pathogen and its human hosts inevitably changes, affecting the severity of the disease produced. Most pathogenic organisms have evolved special adaptations that enable them to invade their hosts, replicate in them, and be transmitted. However, the rapid death of its host is rarely in a microbe's interest, because this destroys its home and source of food. Consequently, those organisms with the potential to cause severe and rapidly fatal disease often evolve toward an accommodation with their hosts. In complementary fashion, human populations evolve a degree of built-in genetic resistance to common disease-causing organisms, as well as acquiring lifetime immunity to endemic diseases. Endemic diseases are those, such as measles, chickenpox, and malaria, that are ubiquitous in a given population and to which most people are exposed in childhood. Because of the interplay between host and pathogen, the nature and severity of infectious diseases in the human population are always changing.

Influenza is a good example of a common viral disease that, although severe in its symptoms, is usually overcome successfully by the immune system. The fever, aches, and lassitude that accompany infection can be overwhelming, and it is difficult to imagine overcoming foes or predators at the peak of a bout of influenza. Nevertheless, despite the severity of the symptoms, most strains of influenza pose no great danger to healthy people in populations in which influenza is endemic. Warm, well-nourished, and otherwise healthy people usually recover in a couple of weeks and take it for granted that their immune system will accomplish this task. Pathogens new to the human population, in contrast, often cause high mortality in those infected—between 60% and 75% in the case of the Ebola virus.

Figure 1.3 Many different microorganisms can be human pathogens. (a) Human immunodeficiency virus (HIV), the cause of AIDS. (b) Influenza virus. (c) *Staphylococcus aureus*, a bacterium that colonizes human skin, is the common cause of pimples and boils, and can also cause food poisoning. (d) *Streptococcus pneumoniae*, is the major cause of pneumonia and is also a common cause of meningitis in children and the elderly. (e) *Salmonella enteritidis*, the bacterium that commonly causes food poisoning. (f) *Mycobacterium tuberculosis*, the bacterium that causes tuberculosis. (g) A human cell (colored green) containing *Listeria monocytogenes* (colored yellow), a bacterium that can contaminate processed food, causing disease (listeriosis) in pregnant women and immunosuppressed individuals. (h) *Pneumocystis carinii*, an opportunistic fungus that infects patients with acquired immunodeficiency syndrome (AIDS) and other immunosuppressed individuals. The fungal cells (colored green) are in lung tissue. (i) *Epidermophyton floccosum*, the fungus that causes ringworm. (j) The fungus *Candida albicans*, a normal inhabitant of the human body that occasionally causes thrush and more severe systemic infections. (k) Red blood cells and *Trypanosoma brucei* (colored orange), a protozoan that causes African sleeping sickness. (l) *Schistosoma mansoni*, the helminth worm that causes schistosomiasis. The adult intestinal blood fluke forms are shown: the male is thick and bluish, the female thin and white. All the photos are false-colored electron micrographs, with the exception of (l), which is a light micrograph.

1-3 The skin and mucosal surfaces form barriers against infection

The skin is the human body's first defense against infection. It forms a tough, impenetrable barrier of **epithelium** protected by layers of keratinized cells. Epithelium is a general name for the layers of cells that line the outer surface and the inner cavities of the body. The skin can be breached by physical damage, such as wounds, burns, or surgical procedures, which exposes soft tissues and renders them vulnerable to infection. Until the adoption of antiseptic procedures in the nineteenth century, surgery was a very risky business, principally because of the life-threatening infections that the procedures introduced. For the same reason, far more soldiers have died from infection acquired on the battlefield than from the direct effects of enemy action. Ironically, the need to conduct increasingly sophisticated and wide-ranging warfare has been the major force driving improvements in surgery and medicine. As an example from immunology, the burns suffered by fighter pilots during the Second World War stimulated studies on skin transplantation that led directly to the understanding of the cellular basis of the immune response.

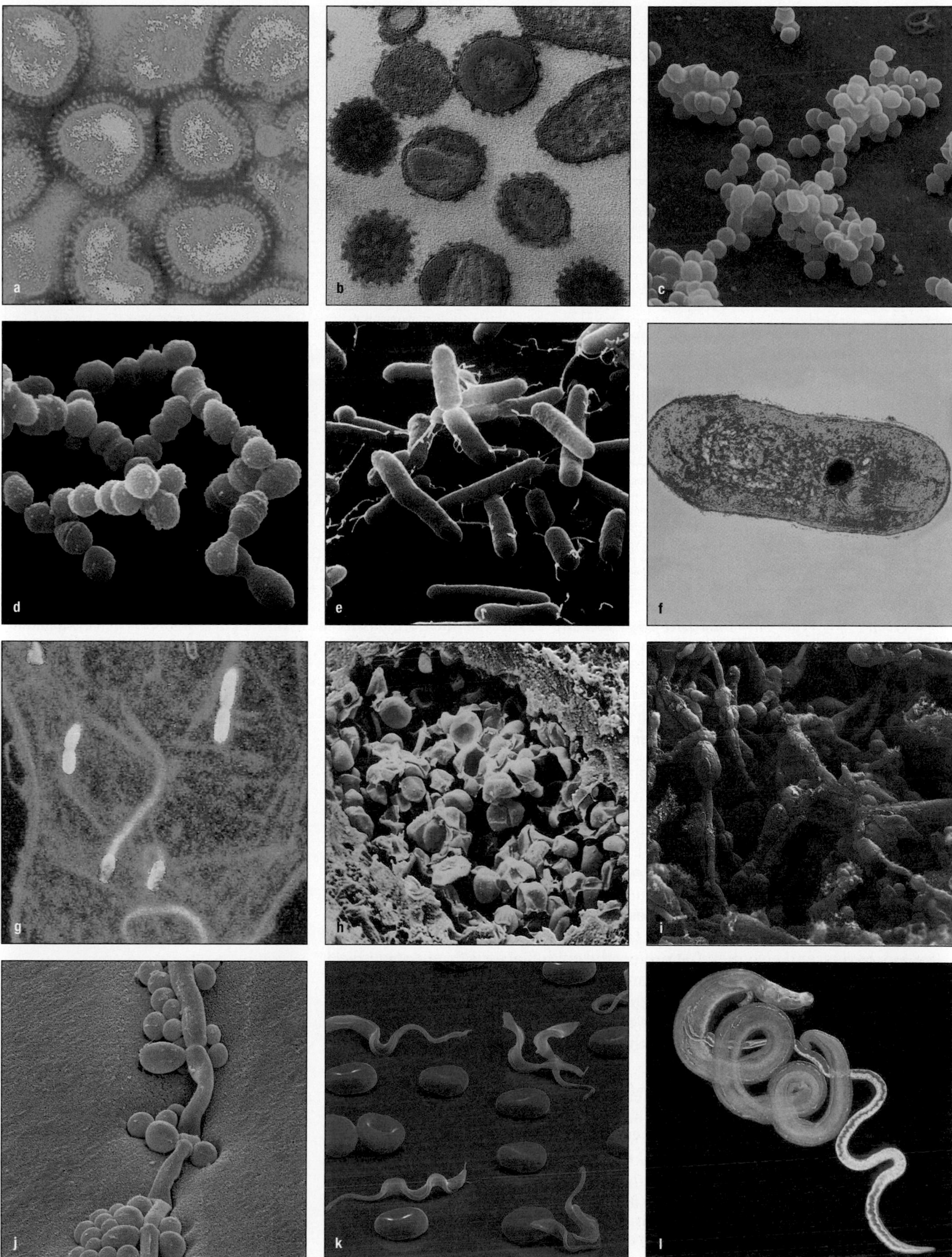
a
b
c
d
e
f
g
h
i
j
k
l

Type	Disease	Pathogen	General classification*	Route of infection
Viruses	Severe acute respiratory syndrome	SARS virus	Coronaviruses	Oral/respiratory/ocular mucosa
	West Nile encephalitis	West Nile virus	Flaviviruses	Bite of an infected mosquito
	Yellow fever	Yellow fever virus	Flaviviruses	Bite of infected mosquito (*Aedes aegypti*)
	Hepatitis B	Hepatitis B virus	Hepadnaviruses	Sexual transmission; infected blood
	Chickenpox	Varicella-zoster	Herpes viruses	Oral/respiratory
	Mononucleosis	Epstein–Barr virus	Herpes viruses	Oral/respiratory
	Influenza	Influenza virus	Orthomyxoviruses	Oral/respiratory
	Measles	Measles virus	Paramyxoviruses	Oral/respiratory
	Mumps	Mumps virus	Paramyxoviruses	Oral/respiratory
	Poliomyelitis	Polio virus	Picornaviruses	Oral
	Jaundice	Hepatitis A virus	Picornaviruses	Oral
	Smallpox	Variola	Pox viruses	Oral/respiratory
	AIDS	Human immunodeficiency virus	Retroviruses	Sexual transmission, infected blood
	Rabies	Rabies virus	Rhabdoviruses	Bite of an infected animal
	Common cold	Rhinoviruses	Rhinoviruses	Nasal
	Diarrhea	Rotavirus	Rotaviruses	Oral
	Rubella	Rubella	Togaviruses	Oral/respiratory

Figure 1.4 ***(opposite page and above)*** **Diverse microorganisms cause human disease.** Pathogenic organisms are of four main types—viruses, bacteria, fungi, and parasites, which are mostly protozoans or worms. Some important pathogens in each category are listed along with the diseases they cause. *The classifications given are intended as a guide only and are not taxonomically consistent; families are given for the viruses; general groupings often used in medical bacteriology for the bacteria; and higher taxonomic divisions for the fungi and parasites. The terms Gram-negative and Gram-positive refer to the staining properties of the bacteria; Gram-positive bacteria stain purple with the Gram stain, Gram-negative bacteria do not.

Continuous with the skin are the epithelia lining the respiratory, gastrointestinal, and urogenital tracts (Figure 1.5). On these internal surfaces, the impermeable skin gives way to tissues that are specialized for communication with their environment and are more vulnerable to microbial invasion. Such surfaces are known as **mucosal surfaces**, or **mucosae**, as they are continually bathed in the **mucus** they secrete. This thick fluid layer contains glycoproteins, proteoglycans, and enzymes that protect the epithelial cells from damage and help to limit infection. In the respiratory tract, mucus is continuously removed through the action of epithelial cells that bear beating cilia and is replenished by mucus-secreting goblet cells. The respiratory mucosa is thus continually cleansed of unwanted material, including infectious microorganisms that have been inhaled.

All epithelial surfaces secrete antimicrobial substances. The sebum secreted by sebaceous glands associated with hair follicles contains fatty acids and lactic acids, both of which inhibit bacterial growth on the surface of the skin. All epithelia produce **antimicrobial peptides** that kill bacteria, fungi, and enveloped viruses by perturbing their membranes. Tears and saliva contain lysozyme, an enzyme that kills bacteria by degrading their cell walls. Microorganisms are also deterred by the acidic environments within the stomach, the vagina, and the skin.

Type	Disease	Pathogen	General classification*	Route of infection
Bacteria	Trachoma	*Chlamydia trachomatis*	Chlamydias	Oral/respiratory/ocular mucosa
	Bacillary dysentery	*Shigella flexneri*	Gram-negative bacilli	Oral
	Food poisoning	*Salmonella enteritidis, S. typhimurium*	Gram-negative bacilli	Oral
	Plague	*Yersinia pestis*	Gram-negative bacilli	Infected flea bite, respiratory
	Tularemia	*Pasteurella tularensis*	Gram-negative bacilli	Handling infected animals
	Typhoid fever	*Salmonella typhi*	Gram-negative bacilli	Oral
	Gonorrhea	*Neisseria gonorrhoeae*	Gram-negative cocci	Sexually transmitted
	Meningococcal meningitis	*Neisseria meningitidis*	Gram-negative cocci	Oral/respiratory
	Meningitis, pneumonia	*Haemophilus influenzae*	Gram-negative coccobacilli	Oral/respiratory
	Legionnaire's disease	*Legionella pneumophila*	Gram-negative coccobacilli	Inhalation of contaminated aerosol
	Whooping cough	*Bordetella pertussis*	Gram-negative coccobacilli	Oral/respiratory
	Cholera	*Vibrio cholerae*	Gram-negative vibrios	Oral
	Anthrax	*Bacillus anthracis*	Gram-positive bacilli	Oral/respiratory by contact with spores
	Diphtheria	*Corynebacterium diphtheriae*	Gram-positive bacilli	Oral/respiratory
	Tetanus	*Clostridium tetani*	Gram-positive bacilli (anaerobic)	Infected wound
	Boils, wound infections	*Staphylococcus aureus*	Gram-positive cocci	Wounds; oral/respiratory
	Pneumonia, scarlet fever	*Streptococcus pneumoniae*	Gram-positive cocci	Oral/respiratory
	Tonsillitis	*Streptococcus pyogenes*	Gram-positive cocci	Oral/respiratory
	Leprosy	*Mycobacterium leprae*	Mycobacteria	Infected respiratory droplets
	Tuberculosis	*Mycobacterium tuberculosis*	Mycobacteria	Oral/respiratory
	Respiratory disease	*Mycoplasma pneumoniae*	Mycoplasmas	Oral/respiratory
	Typhus	*Rickettsia prowazekii*	Rickettsias	Bite of infected tick
	Lyme disease	*Borrelia burgdorferi*	Spirochetes	Bite of infected deer tick
	Syphilis	*Treponema pallidum*	Spirochetes	Sexual transmission
Fungi	Aspergillosis	*Aspergillus* species	Ascomycetes	Opportunistic pathogen, inhalation of spores
	Athlete's foot	*Tinea pedis*	Ascomycetes	Physical contact
	Candidiasis, thrush	*Candida albicans*	Ascomycetes (yeasts)	Opportunistic pathogen, resident flora
	Pneumonia	*Pneumocystis carinii*	Ascomycetes	Opportunistic pathogen, resident lung flora
Protozoan parasites	Leishmaniasis	*Leishmania major*	Protozoa	Bite of an infected sand fly
	Malaria	*Plasmodium falciparum*	Protozoa	Bite of an infected mosquito
	Toxoplasmosis	*Toxoplasma gondii*	Protozoa	Oral, from infected material
	Trypanosomiasis	*Trypanosoma brucei*	Protozoa	Bite of an infected tsetse fly
Helminth parasites (worms)	Common roundworm	*Ascaris lumbricoides*	Nematodes (roundworms)	Oral, from infected material
	Schistosomiasis	*Schistosoma mansoni*	Trematodes	Through skin by bathing in infected water

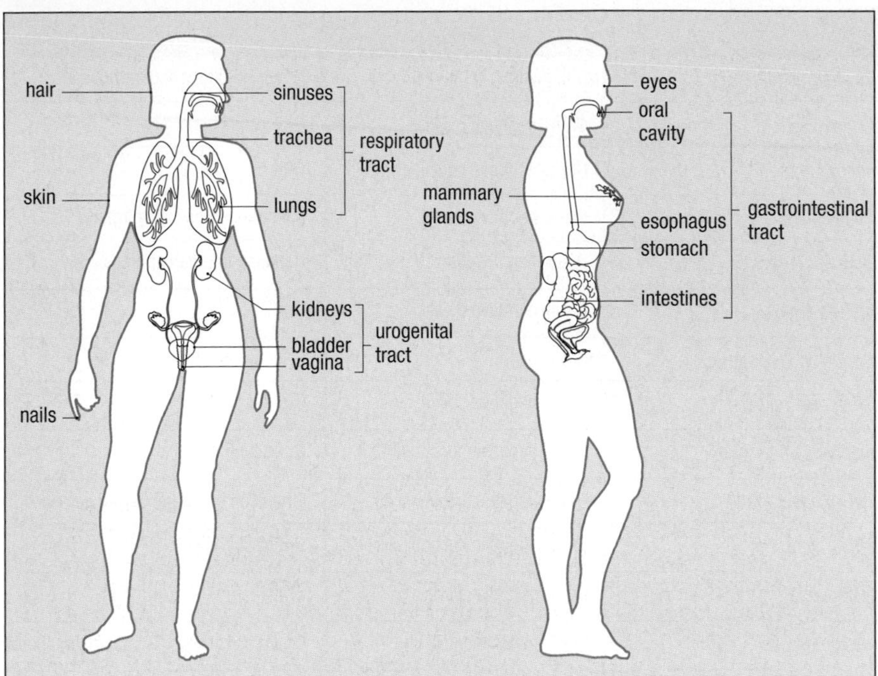

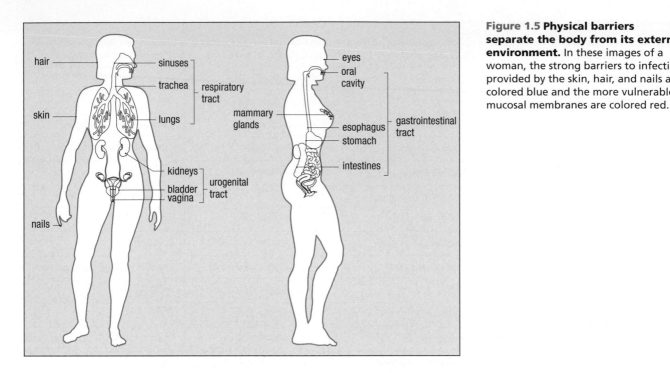

Figure 1.5 Physical barriers separate the body from its external environment. In these images of a woman, the strong barriers to infection provided by the skin, hair, and nails are colored blue and the more vulnerable mucosal membranes are colored red.

The fixed defenses of skin and mucosa provide well-maintained mechanical, chemical, and microbiological barriers that prevent most pathogens from gaining access to the cells and tissues of the body. When those barriers are breached and pathogens gain entry to the body's soft tissues, the defenses of the innate immune system are brought into play.

1-4 The innate immune response causes inflammation at sites of infection

Cuts, abrasions, bites, and wounds provide routes for pathogens to get through the skin. Touching, rubbing, picking and poking the eyes, nose, and mouth help pathogens to breach mucosal surfaces, as does breathing polluted air, eating contaminated food, and being around infected people. With very few exceptions, infections remain highly localized and are extinguished within a few days without illness or incapacitation. Such infections are controlled and terminated by the **innate immune response**, which is ready to react quickly. This response consists of two parts (Figure 1.6). The first is recognition that a pathogen is present. This involves soluble proteins and cell-surface receptors that bind either to the pathogen and its products or to human cells and serum proteins that become altered in the presence of the pathogen. Once the pathogen has been recognized, the second part of the response involves the recruitment of destructive **effector mechanisms** that kill and eliminate the pathogen. The effector mechanisms are provided by **effector cells** of various types that engulf bacteria, kill virus-infected cells, or attack protozoan parasites, and a battery of serum proteins called **complement** that help the effector cells by marking pathogens with molecular flags but also attack pathogens in their own right. Collectively, these defenses are called **innate immunity**. The word 'innate' refers to qualities a person is born with, and innate immunity comprises a genetically programmed set of responses that can be mobilized immediately an infection occurs. Many families of receptor proteins contribute to the recognition of pathogens in the innate immune response. They are of several different structural types and bind to chemically diverse ligands: peptides, proteins, glycoproteins, proteoglycans, peptidoglycan, carbohydrates, glycolipids, phospholipids, and nucleic acids.

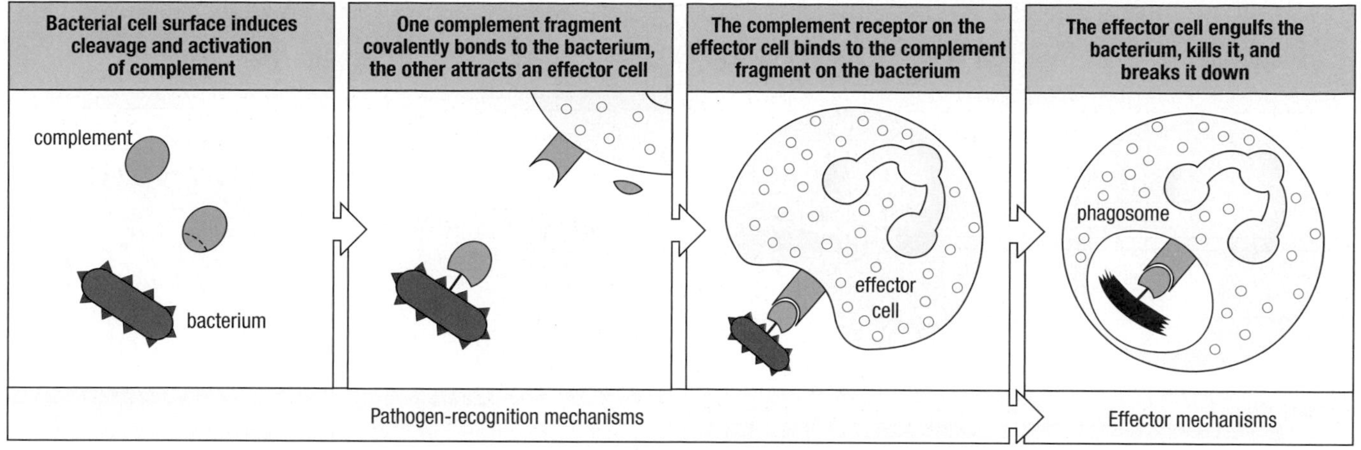

Bacterial cell surface induces cleavage and activation of complement
One complement fragment covalently bonds to the bacterium, the other attracts an effector cell
The complement receptor on the effector cell binds to the complement fragment on the bacterium
The effector cell engulfs the bacterium, kills it, and breaks it down
complement
bacterium
effector cell
phagosome
Pathogen-recognition mechanisms
Effector mechanisms

Figure 1.6 Immune defense involves recognition of pathogens followed by their destruction. Almost all components of the immune system contribute to mechanisms for either recognizing pathogens or destroying pathogens, or to mechanisms for communicating between these two activities. This is illustrated here by a fundamental process used to get rid of pathogens. Serum proteins of the complement system (turquoise) are activated in the presence of a pathogen (red) to form a covalent bond between a fragment of complement protein and the pathogen. The attached piece of complement marks the pathogen as dangerous. The soluble complement fragment summons a phagocytic white blood cell to the site of complement activation. This effector cell has a surface receptor that binds to the complement fragment attached to the pathogen. The receptor and its bound ligand are taken up into the cell by phagocytosis, which delivers the pathogen to an intracellular vesicle called a phagosome, where it is destroyed. A phagocyte is a cell that eats, 'phago' being derived from the Greek word for eat.

An infection that would typically be cleared by innate immunity is that experienced by skateboarders when they tumble onto a San Francisco sidewalk. On returning home the graze is washed, which removes most of the dirt and the associated pathogens of human, soil, pigeon, dog, cat, raccoon, skunk, and possum origin. Of the bacteria that remain, some begin to divide and set up an infection. Cells and proteins in the damaged tissue sense the presence of bacteria, and the cells send out soluble proteins called **cytokines** that interact with other cells to trigger the innate immune response. The overall effect of the innate immune response is to induce a state of **inflammation** in the infected tissue. Inflammation is an ancient concept in medicine that has traditionally been defined by the Latin words *calor, dolor, rubor,* and *tumor*: for heat, pain, redness, and swelling, respectively. These symptoms, which are part of everyday human experience, are not due to the infection itself but to the immune system's response to the pathogen.

Cytokines induce the local dilation of blood capillaries, which by increasing the blood flow causes the skin to warm and redden. Vascular dilation (vasodilation) introduces gaps between the cells of the **endothelium**, the thin layer of specialized epithelium that lines the interior of blood vessels. This makes the endothelium permeable and increases the leakage of blood plasma into the connective tissue. Expansion of the local fluid volume causes **edema** or swelling, putting pressure on nerve endings and causing pain. Cytokines also change the adhesive properties of the vascular endothelium, inviting white blood cells to attach to it and move from the blood into the inflamed tissue (Figure 1.7). White blood cells that are usually present in inflamed tissues and release substances that contribute to the inflammation are called **inflammatory cells.** Infiltration of cells into the inflamed tissue increases the swelling, and some of the molecules they release contribute to the pain. The benefit of the discomfort and disfigurement caused by inflammation is that it enables cells and molecules of the immune system to be brought rapidly and in large numbers into the infected tissue. The mechanisms of innate immunity are considered in Chapters 2 and 3.

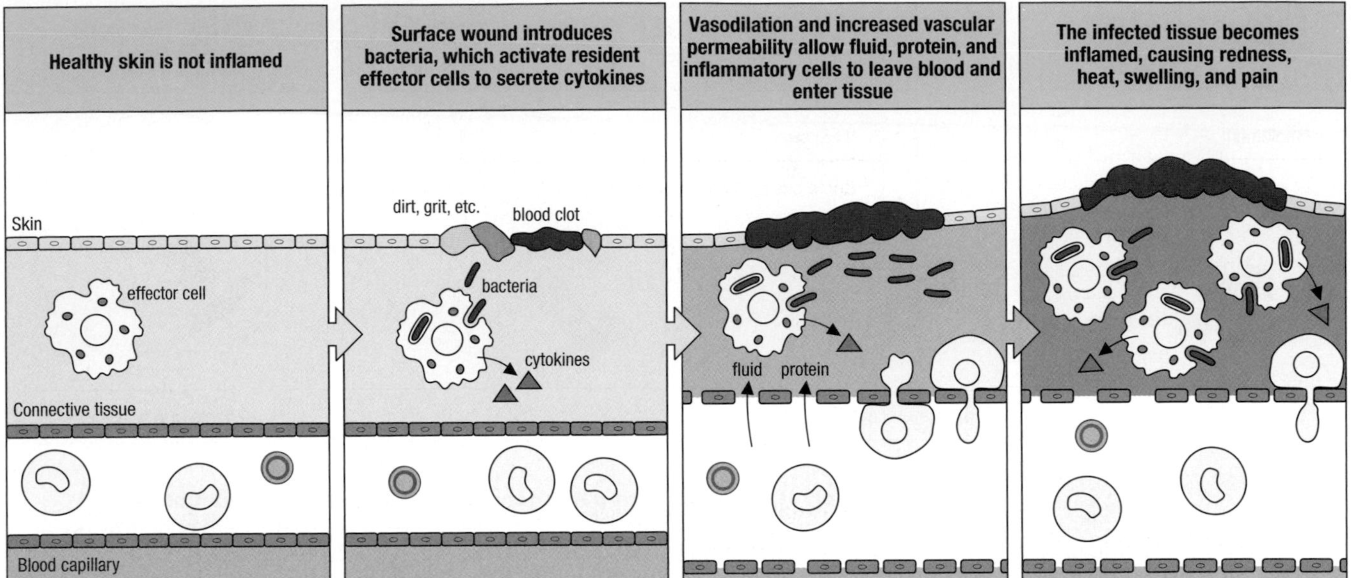
Healthy skin is not inflamed
Surface wound introduces bacteria, which activate resident effector cells to secrete cytokines
Vasodilation and increased vascular permeability allow fluid, protein, and inflammatory cells to leave blood and enter tissue
The infected tissue becomes inflamed, causing redness, heat, swelling, and pain
Skin
effector cell
Connective tissue
Blood capillary
dirt, grit, etc.
blood clot
bacteria
cytokines
fluid
protein

Figure 1.7 Innate immune mechanisms establish a state of inflammation at sites of infection. Illustrated here are the events following an abrasion of the skin. Bacteria invade the underlying connective tissue and stimulate the innate immune response.

1-5 The adaptive immune response adds to an ongoing innate immune response

Human beings are exposed to pathogens every day. The intensity of exposure and the diversity of the pathogens encountered increase with crowded city living and the daily exchange of people and pathogens from international airports. Despite this exposure, innate immunity keeps most people healthy for most of the time. Nevertheless, some infections outrun the innate immune response, an event more likely in people who are malnourished, poorly housed, deprived of sleep, or stressed in other ways. When this occurs, the innate immune response works to slow the spread of infection while it calls upon white blood cells called **lymphocytes** that increase the power and focus of the immune response. Their contribution to defense is the **adaptive immune response**. It is so called because it is organized around an ongoing infection and adapts to the nuances of the infecting pathogen. Consequently, the long-lasting **adaptive immunity** that develops against one pathogen provides a highly specialized defense that is of little use against infection by a different pathogen. Adaptive immunity evolved only in vertebrates, and in these species it complements the mechanisms of innate immunity that vertebrates share with other, invertebrate, animals.

The effector mechanisms used in the adaptive immune response are similar to those used in the innate immune response; the important difference is in the way in which lymphocytes recognize pathogens (Figure 1.8). The receptors of innate immunity comprise many different types. They each recognize features shared by groups of pathogens and are not specific for a particular pathogen. In contrast, lymphocytes recognize pathogens by using cell-surface receptors of just one molecular type. These proteins can, however, be made in billions of different versions, each capable of binding a different ligand. This means that the adaptive immune response can be made specific for a particular pathogen by using only those lymphocyte receptors that bind to the infecting pathogen. The lymphocyte receptors are not encoded by conventional genes but by genes that are cut, spliced, and modified during lymphocyte development. In this way, each lymphocyte is programmed to make one variant of the basic receptor type, but among the population of lymphocytes are represented billions of different receptor variants.

During infection, only those lymphocytes bearing receptors that recognize the infecting pathogen are selected to participate in the adaptive response. These

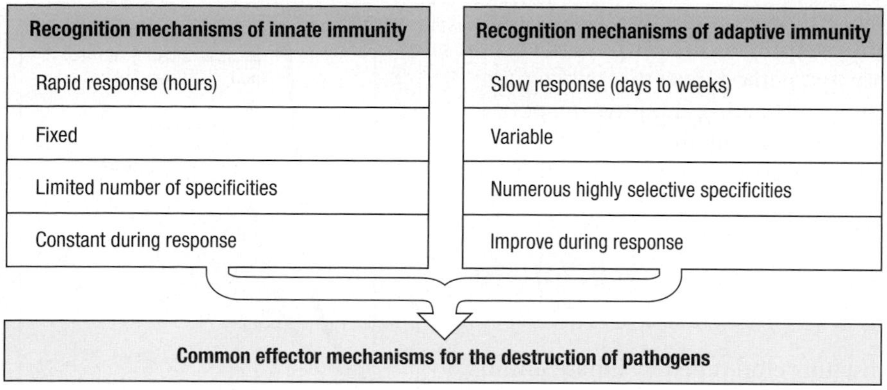

Recognition mechanisms of innate immunity	Recognition mechanisms of adaptive immunity
Rapid response (hours)	Slow response (days to weeks)
Fixed	Variable
Limited number of specificities	Numerous highly selective specificities
Constant during response	Improve during response

Common effector mechanisms for the destruction of pathogens

Figure 1.8 The principal characteristics of innate and adaptive immunity.

then proliferate and differentiate to produce large numbers of effector cells specific for that pathogen (Figure 1.9). These processes, which select the small subset of pathogen-specific lymphocytes for proliferation and differentiation into effector lymphocytes, are called **clonal selection** and **clonal expansion**, respectively. Because these processes take time, the benefit of an adaptive immune response only begins to be felt about a week after the infection has started.

The value of the adaptive immune response is well illustrated by influenza, the disease caused by infection of epithelial cells in the lower respiratory tract with influenza virus. The debilitating symptoms start 3 or 4 days after the start of infection, when the virus has begun to outrun the innate immune response. The disease persists for 5–7 days while the adaptive immune response is being organized and put to work. As the adaptive immune response gains the upper hand, fever subsides and a gradual convalescence begins in the second week after infection.

Some of the lymphocytes selected during an adaptive immune response persist in the body and provide long-term **immunological memory** of the pathogen. These memory cells allow subsequent encounters with the same pathogen to elicit a stronger and faster adaptive immune response, which terminates infection with minimal illness. The adaptive immunity provided by immunological memory is also called **acquired immunity** or **protective immunity**. For some pathogens such as measles virus, one full-blown infection can provide immunity for decades, whereas for influenza virus the protection is less effective. This is not because the immunological memory is faulty but because the influenza virus changes on a yearly basis to escape the immunity acquired by its human hosts.

The first time that an adaptive immune response is made to a given pathogen it is called the **primary immune response**. The second and subsequent times

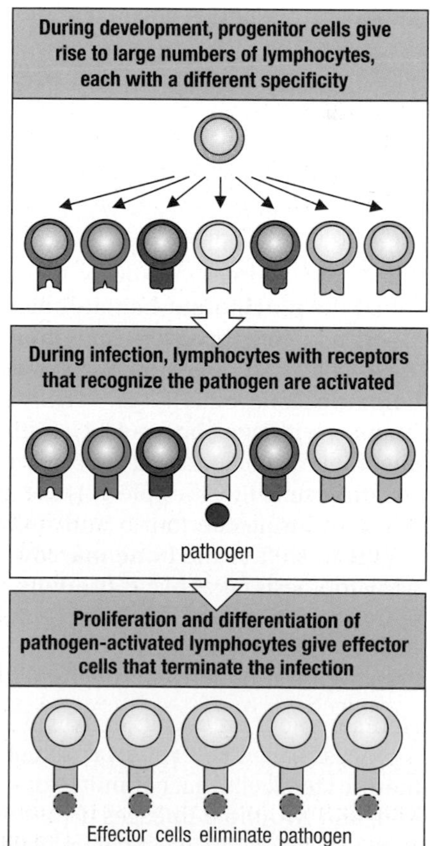

Figure 1.9 Selection of lymphocytes by a pathogen. Top panel: during its development from a progenitor cell (gray), a lymphocyte is programmed to make a single species of cell-surface receptor that recognizes a particular molecular structure. Each lymphocyte makes a receptor of different specificity, so that the population of circulating lymphocytes includes many millions of such receptors, all recognizing different structures, which enables all possible pathogens to be recognized. Lymphocytes with different receptor specificities are represented by different colors. Center panel: upon infection by a particular pathogen, only a small subset of lymphocytes (represented by the yellow cell) will have receptors that bind to the pathogen or its components. Bottom panel: these lymphocytes are stimulated to divide and differentiate, thereby producing an expanded population of effector cells from each pathogen-binding lymphocyte.

that an adaptive immune response is made, and when immunological memory applies, it is called a **secondary immune response**. The purpose of vaccination is to induce immunological memory to a pathogen so that subsequent infection with the pathogen elicits a strong, fast-acting adaptive response. Because all adaptive immune responses are contingent upon an innate immune response, vaccines must induce both innate and adaptive immune responses.

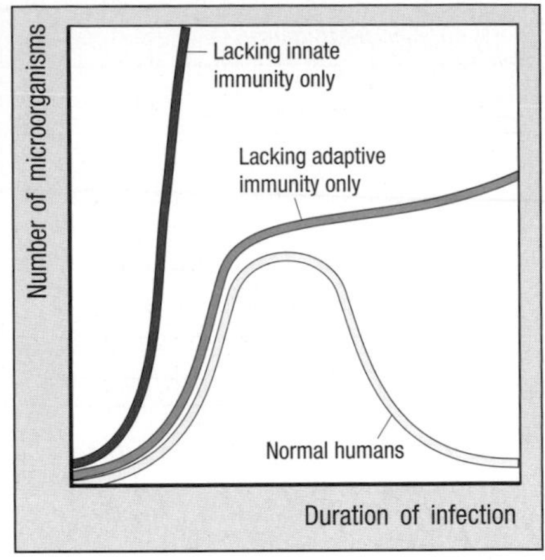

Figure 1.10 The benefits of having both innate and adaptive immunity. In normal individuals, a primary infection is cleared from the body by the combined effects of innate and adaptive immunity (yellow line). In a person who lacks innate immunity, uncontrolled infection occurs because the adaptive immune response cannot be deployed without the preceding innate response (red line). In a person who lacks adaptive immune responses, the infection is initially contained by innate immunity but cannot be cleared from the body (green line).

1-6 Adaptive immunity is better understood than innate immunity

The proportion of infections that are successfully eliminated by innate immunity is difficult to assess, mainly because such infections are overcome before they have caused symptoms severe enough to command the attention of those infected or of their physicians. Intuitively, it seems likely to be a high proportion, given the human body's capacity to sustain vast populations of resident microorganisms without these causing symptoms of disease. The importance of innate immunity is also implied by the rarity of inherited deficiencies in innate immune mechanisms and the considerable impairment of protection when these deficiencies do occur (Figure 1.10).

Much of medical practice is concerned with the small proportion of infections that innate immunity fails to terminate, and in which the spread of the infection results in overt disease such as pneumonia, measles, or influenza and stimulates an adaptive immune response. In such situations the attending physicians and the adaptive immune response work together to effect a cure, a partnership that has historically favored the scientific investigation of adaptive immunity over innate immunity. Consequently, less has been learned about innate immunity than about adaptive immunity. Now that immunologists realize that innate immunity mechanisms are fundamental to every immune response, this glaring gap in our knowledge is being impressively filled.

1-7 Immune system cells with different functions all derive from hematopoietic stem cells

The cells of the immune system are principally the white blood cells or **leukocytes**, and the tissue cells related to them. Along with the other blood cells, they are continually being generated by the body in the developmental process known as **hematopoiesis**. Leukocytes derive from a common progenitor called the **pluripotent hematopoietic stem cell**, which also gives rise to red blood cells (**erythrocytes**) and **megakaryocytes**, the source of **platelets**. All these cell types, together with their precursor cells, are collectively called **hematopoietic cells** (Figure 1.11). The anatomical site for hematopoiesis changes with age (Figure 1.12). In the early embryo, blood cells are first produced in the yolk sac and later in the fetal liver. From the third to the seventh month of fetal life, the spleen is the major site of hematopoiesis. As the bones develop during the fourth and fifth months of fetal growth, hematopoiesis begins to shift to the **bone marrow** and by birth this is where practically all hematopoiesis takes place. In adults, hematopoiesis occurs mainly in the bone marrow of the skull, ribs, sternum, vertebral column, pelvis, and femurs. Because blood cells are short-lived, they have to be continually renewed, and hematopoiesis is active throughout life.

Figure 1.11 *(opposite page)* Types of hematopoietic cell. The different types of hematopoietic cell are depicted in schematic diagrams, which indicate their characteristic morphological features, and in accompanying light micrographs. Their main functions are indicated. We use these schematic representations for these cells throughout the book. Megakaryocytes (k) reside in bone marrow and release tiny non-nucleated, membrane-bound packets of cytoplasm, which circulate in the blood and are known as platelets. Red blood cells (erythrocytes) (l) are smaller than the white blood cells and have no nucleus. Original magnification ×15,000. Photographs courtesy of Yasodha Natkunam.

Hematopoietic stem cells can divide to give further hematopoietic stem cells, a process called **self renewal**; daughter cells can alternatively become more mature stem cells that commit to one of three cell lineages: the erythroid, myeloid, and lymphoid lineages (Figure 1.13). The erythroid progenitor gives rise to the erythroid lineage of blood cells—the oxygen-carrying erythrocytes and

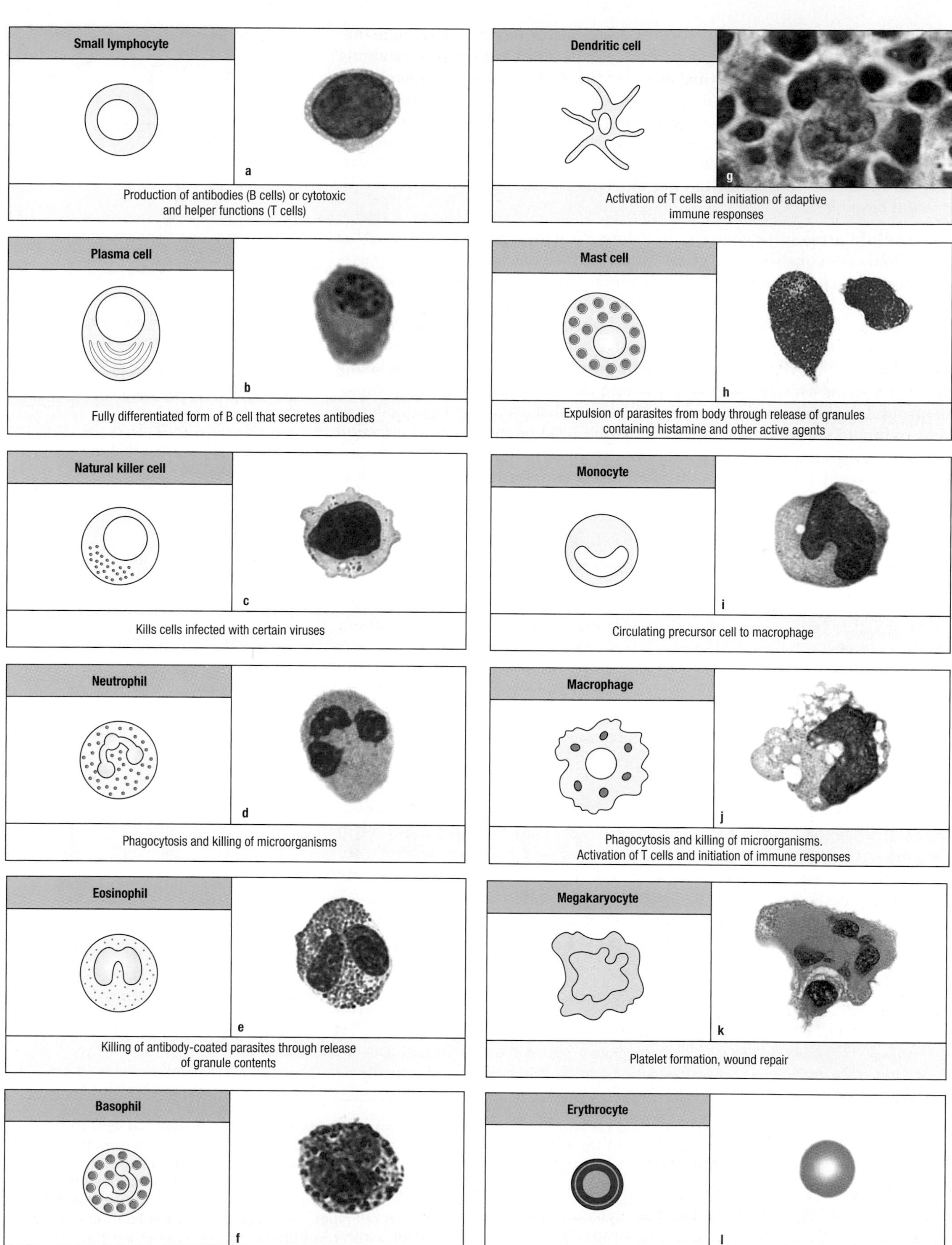
Small lymphocyte
a
Production of antibodies (B cells) or cytotoxic and helper functions (T cells)
Plasma cell
b
Fully differentiated form of B cell that secretes antibodies
Natural killer cell
c
Kills cells infected with certain viruses
Neutrophil
d
Phagocytosis and killing of microorganisms
Eosinophil
e
Killing of antibody-coated parasites through release of granule contents
Basophil
f
Controlling immune responses to parasites
Dendritic cell
g
Activation of T cells and initiation of adaptive immune responses
Mast cell
h
Expulsion of parasites from body through release of granules containing histamine and other active agents
Monocyte
i
Circulating precursor cell to macrophage
Macrophage
j
Phagocytosis and killing of microorganisms. Activation of T cells and initiation of immune responses
Megakaryocyte
k
Platelet formation, wound repair
Erythrocyte
l
Oxygen transport

the platelet-producing megakaryocytes. Megakaryocytes are giant cells that arise from the fusion of multiple precursor cells and have nuclei containing multiple sets of chromosomes (megakaryocyte means 'cell with giant nucleus'). Megakaryocytes are permanent residents of the bone marrow. Platelets are small packets of membrane-enclosed cytoplasm that break off from these cells. They are small non-nucleated cell fragments of plate-like shape and their function is to maintain the integrity of blood vessels. Platelets initiate and participate in the clotting reactions that block badly damaged blood vessels to prevent blood loss.

The **myeloid progenitor** gives rise to the **myeloid lineage** of cells. One group of myeloid cells consists of the **granulocytes**, which have prominent cytoplasmic granules containing reactive substances that kill microorganisms and enhance inflammation. Because granulocytes have irregularly shaped nuclei with two to five lobes, they are also called **polymorphonuclear leukocytes**. Most abundant of the granulocytes, and of all white blood cells, is the **neutrophil** (Figure 1.14), which is specialized in the capture, engulfment and killing of microorganisms. Cells with this function are called **phagocytes**, of which neutrophils are the most numerous and most lethal. Neutrophils are effector cells of innate immunity that are rapidly mobilized to enter sites of infection and can work in the anaerobic conditions that often prevail in damaged tissue. They are short-lived and die at the site of infection, forming **pus**, the stuff of

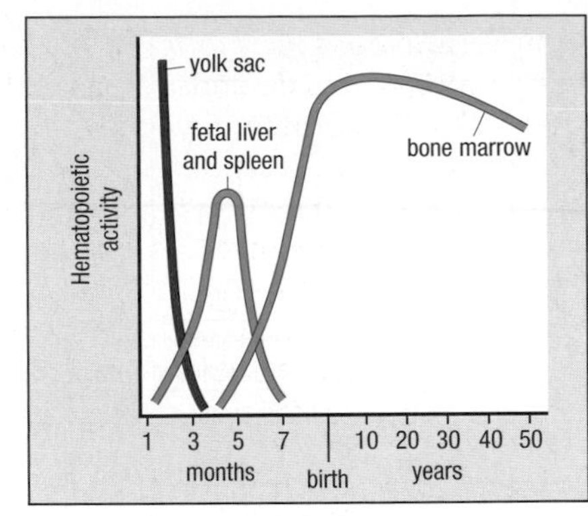

Figure 1.12 The site of human hematopoiesis changes during development. Blood cells are first made in the yolk sac of the embryo and subsequently in the embryonic liver and spleen. They start to be made in the bone marrow before birth, and by the time of birth this is the only tissue in which hematopoiesis occurs.

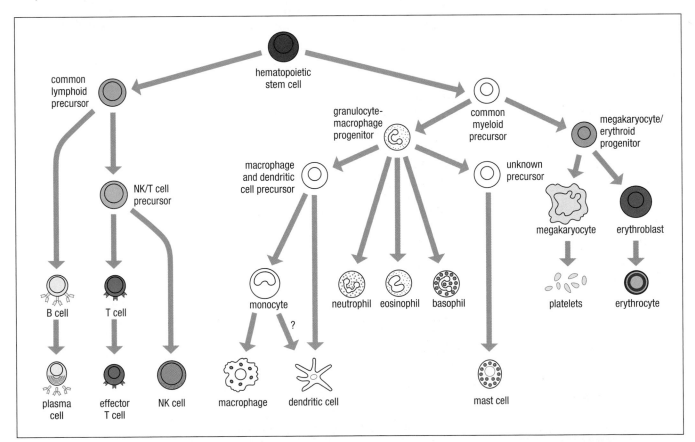

Figure 1.13 Blood cells and certain tissue cells derive from a common hematopoietic stem cell. The pluripotent stem cell (brown) divides and its progeny differentiate into more specialized progenitor cells that give rise to the lymphoid, myeloid, and erythroid lineages of blood cells. The common lymphoid progenitor divides and differentiates to give B cells (yellow), T cells (blue), and NK cells (purple). On activation by infection, B cells divide and differentiate into plasma cells, whereas T cells differentiate into various types of effector T cell. The myeloid progenitor cell divides and differentiates to produce at least six cell types. These are: the three types of granulocyte—the neutrophil, the eosinophil, and the basophil; the mast cell, which takes up residence in connective and mucosal tissues; the circulating monocyte, which gives rise to the macrophages resident in tissues; and the dendritic cell. The word myeloid means 'of the bone marrow.' MDP, macrophage and dendritic cell precursor.

pimples and boils (Figure 1.15). The second most abundant granulocyte is the **eosinophil**, which defends against helminth worms and other intestinal parasites. The least abundant granulocyte, the **basophil**, is also implicated in regulating the immune response to parasites, but is so rare that relatively little is known of its contribution to immune defense. The names of the granulocytes refer to the staining of their cytoplasmic granules with commonly used histological stains: the eosinophil's granules contain basic substances that bind the acidic stain eosin, the basophil's granules contain acidic substances that bind basic stains such as hematoxylin, and the contents of the neutrophil's granules bind to neither acidic nor basic stains.

Cell type	Proportion of leukocytes (%)
Neutrophil	40–75
Eosinophil	1–6
Basophil	<1
Monocyte	2–10
Lymphocyte	20–50

Figure 1.14 The relative abundance of the leukocyte cell types in human peripheral blood. The values for each cell type give the normal range for venous blood donated by healthy individuals.

The second group of myeloid cells consists of monocytes, macrophages, and dendritic cells. **Monocytes** are leukocytes that circulate in the blood. They are distinguished from the granulocytes by being bigger, by having a distinctive indented nucleus, and by all looking the same: hence the name monocyte. Monocytes are the mobile progenitors of sedentary tissue cells called **macrophages**. They travel in the blood to tissues, where they mature into macrophages and take up residence. The name macrophage means 'large phagocyte,' and like the neutrophil, which was historically called the microphage, the macrophage is well equipped for phagocytosis. Tissue macrophages are large, irregularly shaped cells characterized by an extensive cytoplasm with numerous vacuoles, often containing engulfed material (Figure 1.16). They are the general scavenger cells of the body, phagocytosing and disposing of dead cells and cell debris as well as invading microorganisms.

If neutrophils are the short-lived infantry of innate immunity, then macrophages are the long-lived commanders who provide warning to other cells and orchestrate the local response to infection. Macrophages resident in the infected tissues are generally the first cell to sense an invading microorganism. As part of their response to the pathogen, macrophages secrete the cytokines that recruit neutrophils and other leukocytes into the infected area.

Dendritic cells are resident in the body's tissues and have a distinctive star-shaped morphology. Although they have many properties in common with macrophages, their unique function is to act as cellular messengers that are sent to call up an adaptive immune response when it is needed. At such times, dendritic cells that reside in the infected tissue will leave the tissue with a cargo of intact and degraded pathogens and take it to one of several lymphoid organs that specialize in making adaptive immune responses.

The last type of myeloid cell that will concern us is the **mast cell**, which is resident in all connective tissues. It has granules like those of the basophil, but it is

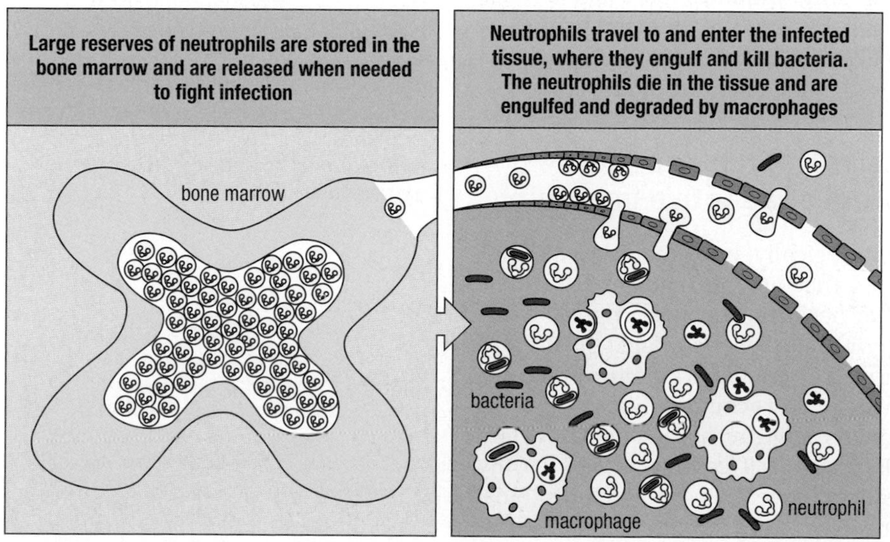

Figure 1.15 Neutrophils are stored in the bone marrow and move in large numbers to sites of infection, where they act and then die. After one round of ingestion and killing of bacteria, a neutrophil dies. The dead neutrophils are eventually mopped up by long-lived tissue macrophages, which break them down. The creamy material known as pus is composed of dead neutrophils.

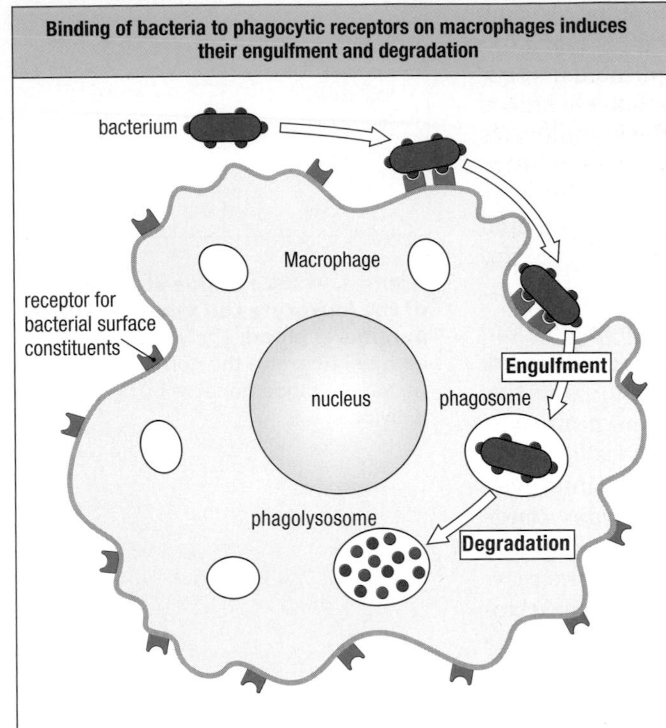
Binding of bacteria to phagocytic receptors on macrophages induces their engulfment and degradation
bacterium
Macrophage
receptor for bacterial surface constituents
nucleus
Engulfment
phagosome
phagolysosome
Degradation

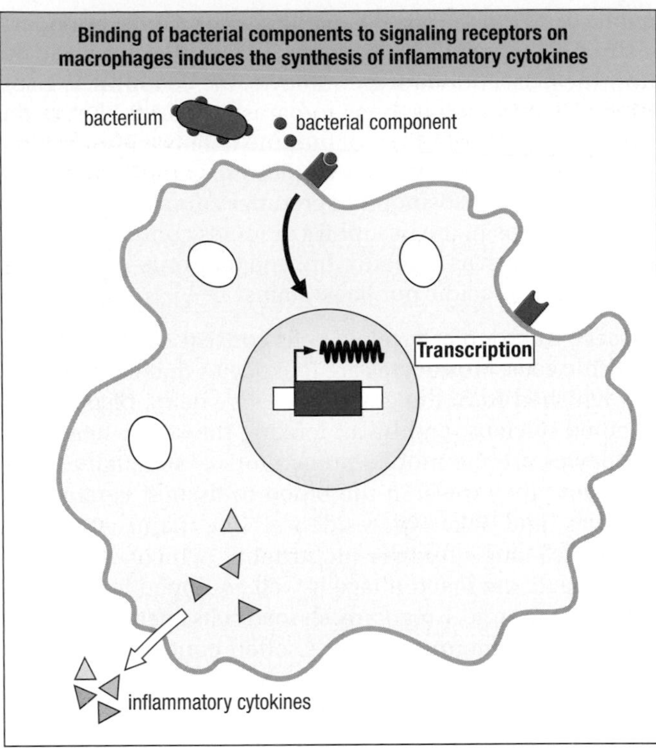
Binding of bacterial components to signaling receptors on macrophages induces the synthesis of inflammatory cytokines
bacterium
bacterial component
Transcription
inflammatory cytokines

Figure 1.16 Macrophages respond to pathogens by using different receptors to stimulate phagocytosis and cytokine secretion. The left panel shows receptor-mediated phagocytosis of bacteria by a macrophage. The bacterium (red) binds to cell-surface receptors (blue) on the macrophage, inducing engulfment of the bacterium into an internal vesicle called a phagosome within the macrophage cytoplasm. Fusion of the phagosome with lysosomes forms an acidic vesicle called a phagolysosome, which contains toxic small molecules and hydrolytic enzymes that kill and degrade the bacterium. The right panel shows how a bacterial component binding to a different type of cell-surface receptor sends a signal to the macrophage's nucleus that initiates the transcription of genes for inflammatory cytokines. The cytokines are synthesized in the cytoplasm and secreted into the extracellular fluid.

not closely related in development to the basophil and the identity of its blood-borne progenitor has yet to be discovered (see Figure 1.13). The activation and degranulation of mast cells at sites of infection make major contributions to inflammation.

The **lymphoid progenitor** gives rise to the **lymphoid lineage** of white blood cells. Two populations of blood lymphocytes are distinguished morphologically: large lymphocytes with a granular cytoplasm, and small lymphocytes with almost no cytoplasm. The **large granular lymphocytes** are effector cells of innate immunity called **natural killer cells** or **NK cells**. NK cells are important in the defense against viral infections. They enter infected tissues, where they prevent the spread of infection by killing virus-infected cells and secreting cytokines that impede viral replication in infected cells. The **small lymphocytes** are the cells responsible for the adaptive immune response. They are small because they circulate in a quiescent and immature form that is functionally inactive. Recognition of a pathogen by small lymphocytes drives a process of lymphocyte selection, growth, and differentiation that after 1–2 weeks produces a powerful response tailored to the invading organism.

1-8 Immunoglobulins and T-cell receptors are the diverse lymphocyte receptors of adaptive immunity

The small lymphocytes, although morphologically indistinguishable from each other, comprise several sub-lineages that are distinguished by their cell-surface receptors and the functions they perform. The most important difference is between **B lymphocytes** and **T lymphocytes**, also called **B cells** and **T cells**. For B cells, the cell-surface receptors for pathogens are **immunoglobulins**, whereas those of T cells are known as **T-cell receptors**. Effector B cells, called **plasma cells**, secrete soluble forms of these immunoglobulins, which are known as **antibodies** (Figure 1.17). In contrast, T-cell receptors are only ever expressed as cell-surface receptors, never as soluble proteins. Immunoglobulins and T-cell receptors are structurally similar molecules, the products of genes that are cut, spliced, and modified during lymphocyte development. The structures of these receptors and their generation during lymphocyte development are discussed in Chapters 4–7. As a consequence of these processes, each B cell expresses a single type of immunoglobulin and each T cell expresses a single type of T-cell receptor. Many millions of different immunoglobulins and T-cell receptors are represented within the population of small lymphocytes in one human being.

Any molecule, macromolecule, virus particle, or cell that contains a structure recognized and bound by an immunoglobulin or T-cell receptor is called its corresponding **antigen**. Surface immunoglobulins and T-cell receptors are thus also referred to as the **antigen receptors** of lymphocytes. Differences in the amino acid sequences of the variable regions of immunoglobulins and T-cell receptors create a vast variety of binding sites that are specific for different antigens and thus for different pathogens. A consequence of this specificity is that the adaptive immune response made against one pathogen provides no immunity to another. For example, antibodies made in response to a measles infection bind to measles virus but not to influenza virus; conversely, antibodies specific for influenza virus do not bind to measles virus.

1-9 On encountering their specific antigen, B cells and T cells differentiate into effector cells

On encountering the antigen recognized by their antigen receptors, B cells differentiate into antibody-producing plasma cells, and this is their only effector function (discussed in Chapter 9). Antigen-activated effector T cells, however, undertake a variety of functions within the immune response (discussed in Chapter 8). Effector T cells are subdivided into two main kinds, called **cytotoxic T cells** and **helper T cells.** Cytotoxic T cells kill cells that are infected with viruses or with certain bacteria that live inside human cells. NK cells and cytotoxic T cells have similar effector functions, the former providing these functions during the innate immune response, the latter during the adaptive immune response. Helper T cells secrete cytokines that help other cells of the immune system become fully activated effector cells. For example, one subset of helper T cells helps macrophages become more functionally active in phagocytosis, whereas another subset helps activate B cells to become antibody-secreting plasma cells. A third subset of helper T cells comprises the **regulatory T cells** that control the activities of the cytotoxic and other types of T cell, thereby preventing unnecessary tissue damage and stopping the immune response once the pathogen has been defeated.

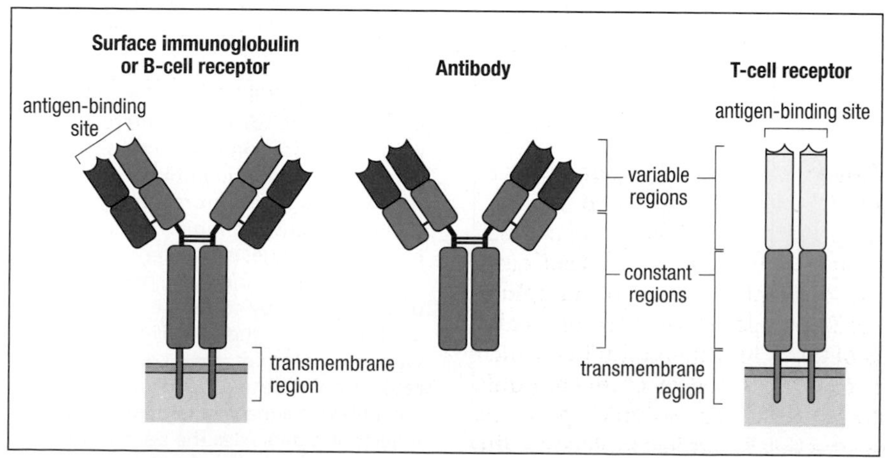

Figure 1.17 Comparison of the B-cell antigen receptor, antibody, and the T-cell receptor. The B-cell receptor for antigen is a Y-shaped immunoglobulin molecule with a transmembrane tail that anchors it in the plasma membrane. It has two identical antigen-binding sites. When a B cell differentiates into a plasma cell, it secretes a soluble form of this receptor, called an antibody, which lacks the transmembrane portion but is otherwise identical. The T-cell receptor for antigen is a membrane protein with one antigen-binding site. There is no secreted form of the T-cell receptor.

1-10 Antibodies bind to pathogens and cause their inactivation or destruction

The antibodies secreted by plasma cells circulate in the blood and can enter infected tissues. Immunity due to antibodies and their actions is often known as **humoral immunity**. This is because antibodies were first discovered circulating in body fluids such as blood and lymph, and **humors** is an old-fashioned term for the body fluids. One way in which antibodies reduce infection is by binding tightly to a site on a pathogen so as to inhibit pathogen growth, replication, or interaction with human cells. This mechanism is called **neutralization**. For example, certain antibodies when bound to influenza virus prevent the virus from infecting human cells. Similarly, the lethal effect that a bacterial toxin has on human cells can be abrogated by an antibody that binds tightly to the toxin and covers up its active site (Figure 1.18).

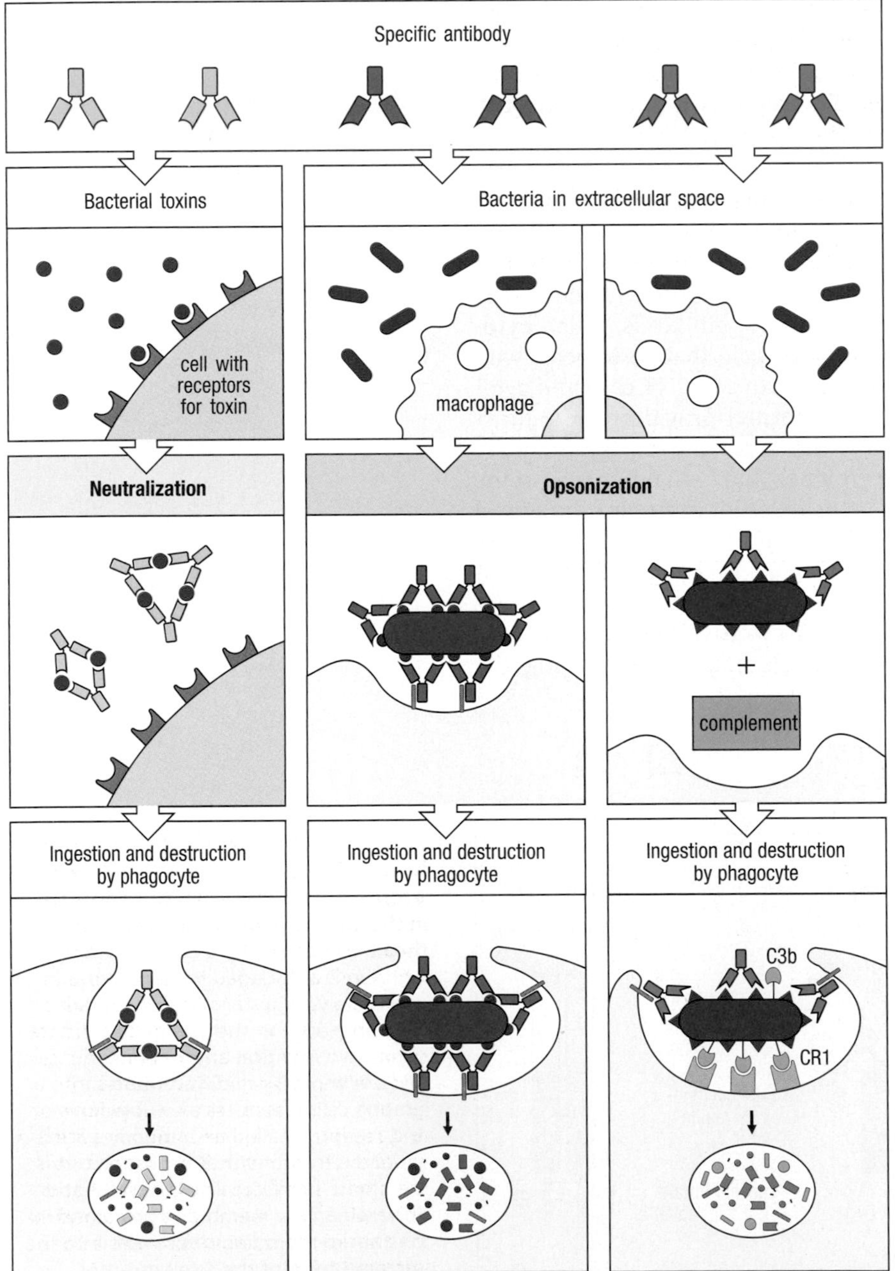

Figure 1.18 Mechanisms by which antibodies combat infection. Left panels: antibodies bind to a bacterial toxin and neutralize its toxic activity by preventing the toxin from interacting with its receptor on human cells. The complex of toxin and antibodies binds to macrophage receptors through the antibody's constant region. Finally, the macrophage ingests and degrades the complex. Right panels: the opsonization of a bacterium by coating it with antibody. When the bacterium is coated with IgG molecules, their constant regions point outward and can bind to receptors on a macrophage, which then ingests and degrades the bacterium.

The most important function of antibodies is to facilitate the engulfment and destruction of extracellular microorganisms and toxins by phagocytes. Because pathogen surfaces consist of relatively few different molecules that are at high density, an antibody specific for one of these antigens can coat the entire surface of the pathogen. The phagocytic neutrophils and macrophages have cell-surface receptors that bind to the antibody molecule at a site apart from the antigen-binding site that is attached to the pathogen. Thousands of receptors on the phagocyte will then bind to thousands of antibodies on the pathogen, which cannot escape a rapid engulfment and death. A bacterium coated with antibody thus becomes more efficiently phagocytosed than an uncoated bacterium. This phenomenon by which a coating of antibody facilitates phagocytes is called **opsonization** (see Figure 1.18).

1-11 Most lymphocytes are present in specialized lymphoid tissues

Although doctors and immunologists usually sample and study human lymphocytes from blood samples taken from their patients and voluntary donors, the vast majority of lymphocytes are to be found in specialized tissues known as **lymphoid tissues** or **lymphoid organs**. The major lymphoid organs are bone marrow, thymus, spleen, adenoids, tonsils, appendix, lymph nodes, and Peyer's patches (Figure 1.19). Less organized lymphoid tissue is also found lining the extensive mucosal surfaces of the respiratory, gastrointestinal, and urogenital tracts. The lymphoid tissues are functionally divided into two types. **Primary** or **central lymphoid tissues** are where lymphocytes develop and

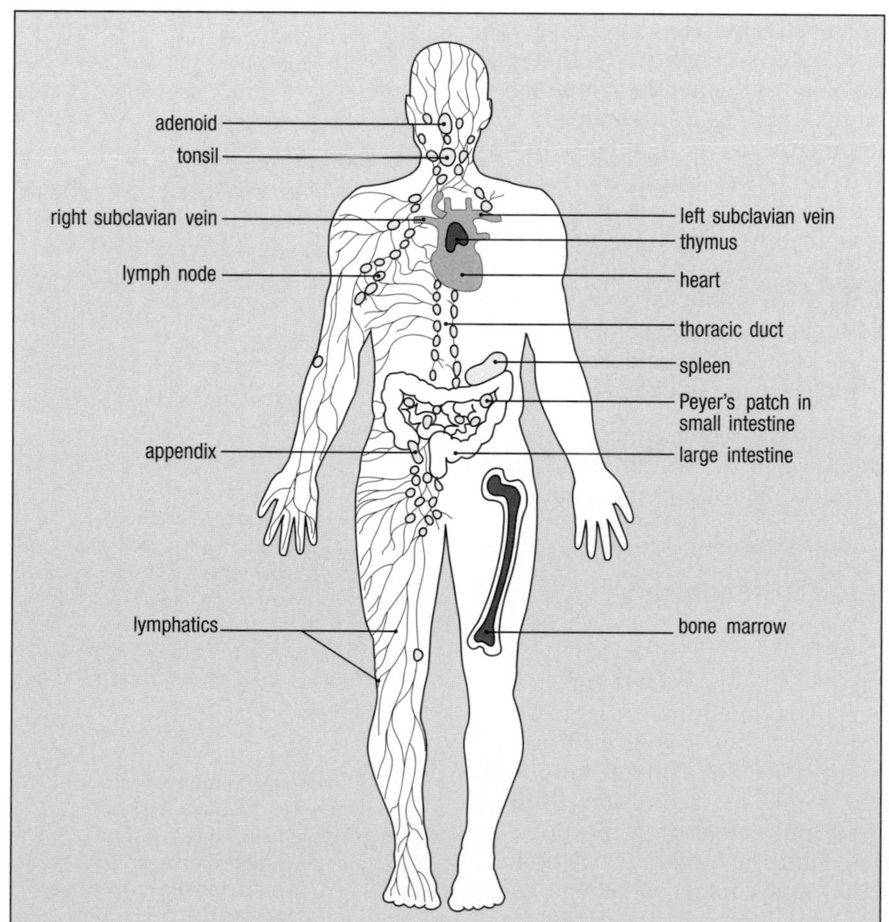

Figure 1.19 The sites of the principal lymphoid tissues within the human body. Lymphocytes arise from stem cells in the bone marrow. B cells complete their maturation in the bone marrow, whereas T cells leave at an immature stage and complete their development in the thymus. The bone marrow and the thymus are the primary lymphoid tissues and are shown in red. The secondary lymphoid tissues are shown in yellow, and the thin black branching lines are the lymphatics. Plasma that has leaked from the blood is collected by the lymphatics as lymph and is returned to the blood via the thoracic duct, which empties into the left subclavian vein.

mature to the stage at which they are able to respond to a pathogen. The bone marrow and the **thymus** are the primary lymphoid tissues; B and T lymphocytes both originate from lymphoid precursors in the bone marrow (see Figure 1.13), but B cells complete their maturation in the bone marrow before entering the circulation, whereas T cells leave the bone marrow at an immature stage and migrate in the blood to the thymus where they mature. Apart from the bone marrow and the thymus, all other lymphoid tissues are known as **secondary** or **peripheral lymphoid tissues**; they are the sites where mature lymphocytes become stimulated to respond to invading pathogens.

Lymph nodes lie at the junctions of an anastomosing network of **lymphatic vessels** called the **lymphatics**, which originate in the connective tissues throughout the body and collect the plasma that continually leaks out of blood vessels and forms the extracellular fluid. The lymphatics eventually return this fluid, called **lymph**, to the blood, chiefly via the thoracic duct, which empties into the left subclavian vein in the neck. Unlike the blood, the lymph is not driven by a dedicated pump and its flow is comparatively sluggish. One-way valves within lymphatic vessels, and the lymph nodes placed at their junctions, ensure that the net movement of the lymph is always in a direction away from the peripheral tissues and toward the ducts in the upper body where the lymph empties into the blood. The flow of lymph is driven by the continual movements of one part of the body with respect to another. In the absence of such movement, as when a patient is confined to bed for a long period of time, lymph flow slows and fluid accumulates in tissues, causing the swelling known as edema.

A unique property of mature B and T cells, which distinguishes them from other blood cells, is that they move through the body in both blood and lymph. Lymphocytes are the only cell type present in lymph in any numbers; hence their name. When small lymphocytes leave the primary lymphoid tissues in which they have developed, they enter the bloodstream. When they reach the blood capillaries that invest a lymph node, or other secondary lymphoid tissue, small lymphocytes are able to leave the blood and enter the lymph node proper. If a lymphocyte becomes activated by a pathogen it remains in the lymph node; otherwise it will spend some time there and then leave in the efferent lymph and eventually be returned to the blood. This means that the population of lymphocytes within a node is in a continual state of flux, with new lymphocytes entering from the blood while others leave in the efferent lymph. This pattern of movement between blood and lymph is termed **lymphocyte recirculation** (Figure 1.20). It allows the lymphocyte population to continually survey the secondary lymphoid organs for evidence of infection. An exception to this pattern is the spleen, which has no connections to the lymphatic system. Lymphocytes both enter and leave the spleen in the blood.

The secondary lymphoid organs are dynamic tissues in which lymphocytes are constantly arriving from the blood and departing in the lymph. At any one time, only a very small fraction of lymphocytes are in the blood and lymph; the majority are in lymphoid organs and tissues.

1-12 Adaptive immunity is initiated in secondary lymphoid tissues

Experiments involving the deliberate infection of volunteers have shown that a large initial dose of pathogenic microorganisms is usually necessary to cause disease. To establish an infection, a microorganism must colonize a tissue in sufficient numbers to overwhelm the cells and molecules of innate immunity that are promptly recruited from the blood to the site of invasion. Even in these circumstances, the effects will usually be minor unless the infection can spread within the body. Frequent sites of infection are the connective tissues, which pathogens penetrate as a result of skin wounds. Intact pathogens, components of pathogens, and pathogen-infected dendritic cells are carried from

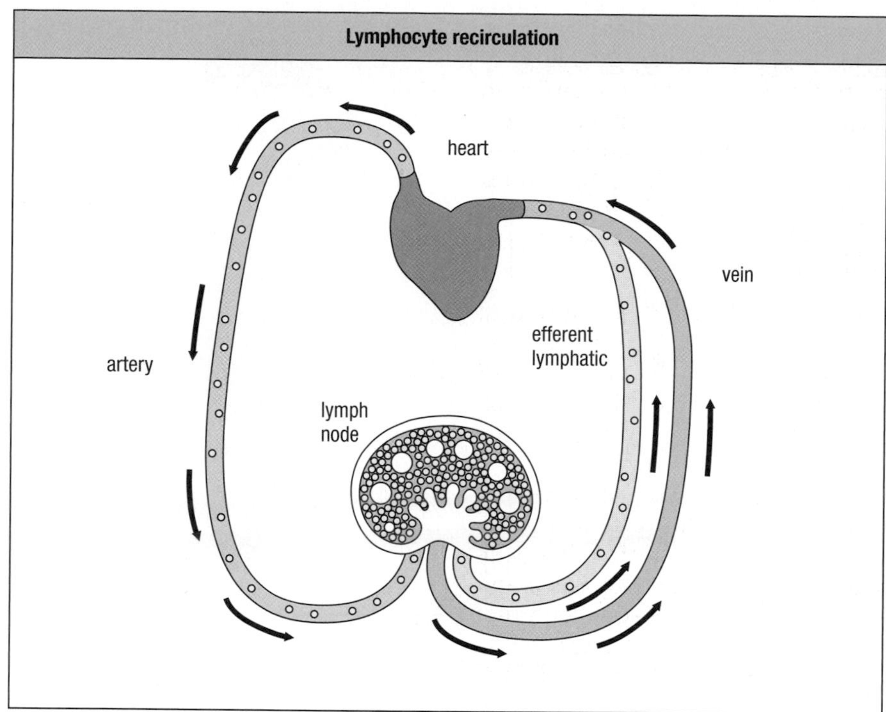

Figure 1.20 Lymphocyte recirculation. Small lymphocytes are unique among blood cells in traveling through the body in the lymph as well as the blood. That is why they were named lymphocytes. Lymphocytes leave the blood through the walls of fine capillaries in secondary lymphoid organs. A lymph node is illustrated here. After spending some time in the lymph node, lymphocytes leave in the efferent lymph and return to the blood at the left subclavian vein. If a lymphocyte in a lymph node encounters a pathogen to which its cell-surface receptor binds, it stops recirculating.

such sites to the nearest **lymph node** by the lymphatics. The lymph node receiving the fluid collected at an infected site is called the **draining lymph node** (Figure 1.21).

The anatomy of the lymph node provides meeting places where lymphocytes coming from the blood encounter pathogens and their products brought from

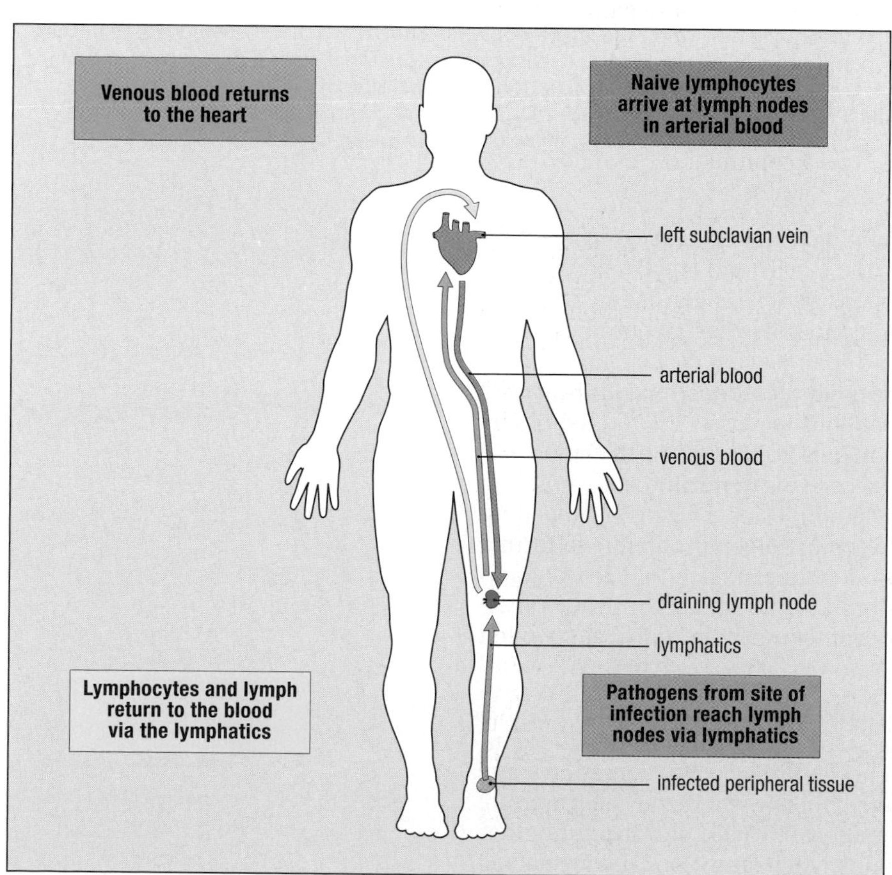

Figure 1.21 Circulating lymphocytes meet lymph-borne pathogens in draining lymph nodes. Lymphocytes leave the blood and enter lymph nodes, where they can be activated by pathogens in the afferent lymph draining from a site of infection. The circulation pertaining to a site of infection in the left foot is shown here. When activated by pathogens, lymphocytes stay in the node to divide and differentiate into effector cells. If lymphocytes are not activated, they leave the node in the efferent lymph and are carried by the lymphatics to the thoracic duct (see Figure 1.19), which empties into the blood at the left subclavian vein. Lymphocytes recirculate all the time, irrespective of infection. Every minute, 5×10^6 lymphocytes leave the blood and enter secondary lymphoid tissues.

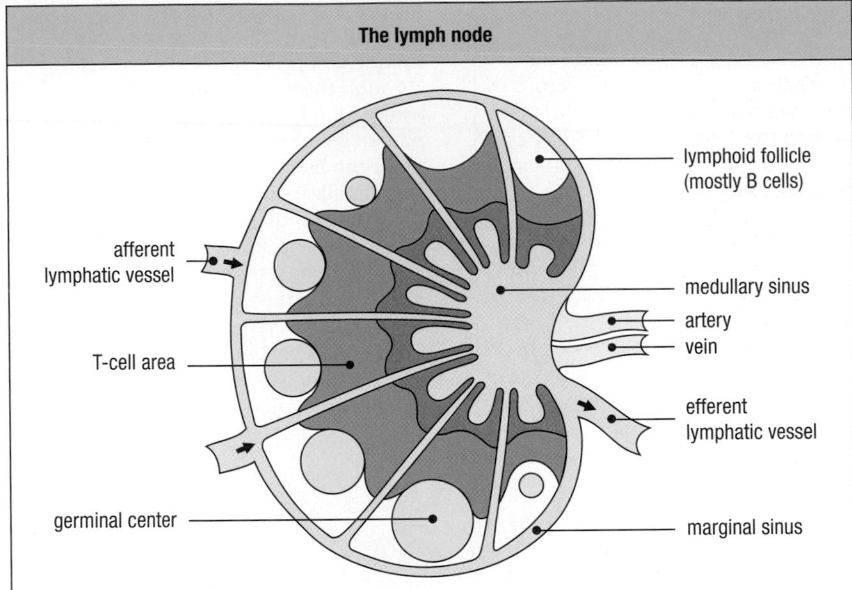

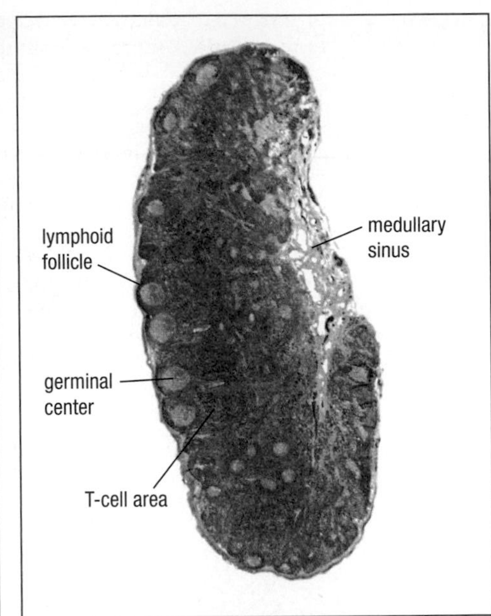

Figure 1.22 Architecture of the lymph node, the site where blood-borne lymphocytes respond to lymph-borne pathogens. Human lymph nodes are small kidney-shaped organs about 1.2 centimeters long and weighing 1 gram or less. They are located at junctions in the lymphatic system where a number of afferent lymph vessels bringing lymph from the tissues unite to form a single, larger efferent lymph vessel that takes the lymph away from the lymph node. A micrograph of a lymph node in longitudinal section is shown on the right and a lymph node is shown schematically in longitudinal section in the panel on the left. A lymph node is composed of a cortex (the yellow and blue areas on this panel) and a medulla (shown in dark pink). Lymph arrives via the afferent lymphatic vessels (pale pink); during an infection, pathogens and pathogen-laden dendritic cells arrive from infected tissue in the draining afferent lymph. The lymph node is packed with lymphocytes, macrophages, and other cells of the immune system, between which the lymph percolates until it reaches the marginal sinus and leaves by the efferent lymphatic vessel. Lymphocytes arrive at lymph nodes in the arterial blood. They enter the node by passing between the endothelial cells that line the fine capillaries within the lymph node (not shown). Within the lymph node, B cells and T cells tend to congregate in anatomically discrete areas. T cells populate the inner cortex (paracortex) and B cells form lymphoid follicles in the outer cortex. During an infection, the expansion of pathogen-specific B cells forms a spherical structure called a germinal center within each follicle. Germinal centers are clearly visible in the photograph. Original magnification ×40. Photograph courtesy of Yasodha Natkunam.

infected connective tissue (Figure 1.22). Arriving lymphocytes segregate to different regions of the lymph node: T cells to the T-cell areas and B cells to B-cell areas known as **lymphoid follicles**. Pathogens and pathogen-laden dendritic cells from the infected tissue arrive at a lymph node in **afferent lymphatic vessels**. Several of these unite at the node and then leave it as a single **efferent lymphatic vessel**. As the lymph passes through the node, the dendritic cells settle there and pathogens and other extraneous materials are filtered out by macrophages. This prevents infectious organisms from reaching the blood and provides a depot of pathogen and pathogen-carrying dendritic cells within the lymph node that can be used to activate lymphocytes. During an infection, pathogen-specific B cells that have bound the pathogen proliferate to form a dense spherical structure called a **germinal center** in each follicle. A lymph node draining a site of infection increases in size as a result of the proliferation of activated lymphocytes, a phenomenon sometimes referred to as 'swollen glands.'

In the lymph node, the small fraction of B and T cells bearing receptors that bind to the pathogen or its products will be stimulated to divide and differentiate into effector cells. T cells are activated by dendritic cells, whereupon some of the T cells move to the associated lymphoid follicle where they help activate the B cells to become plasma cells. Other effector T cells and the antibodies secreted by plasma cells are carried by efferent lymph and blood to the infected

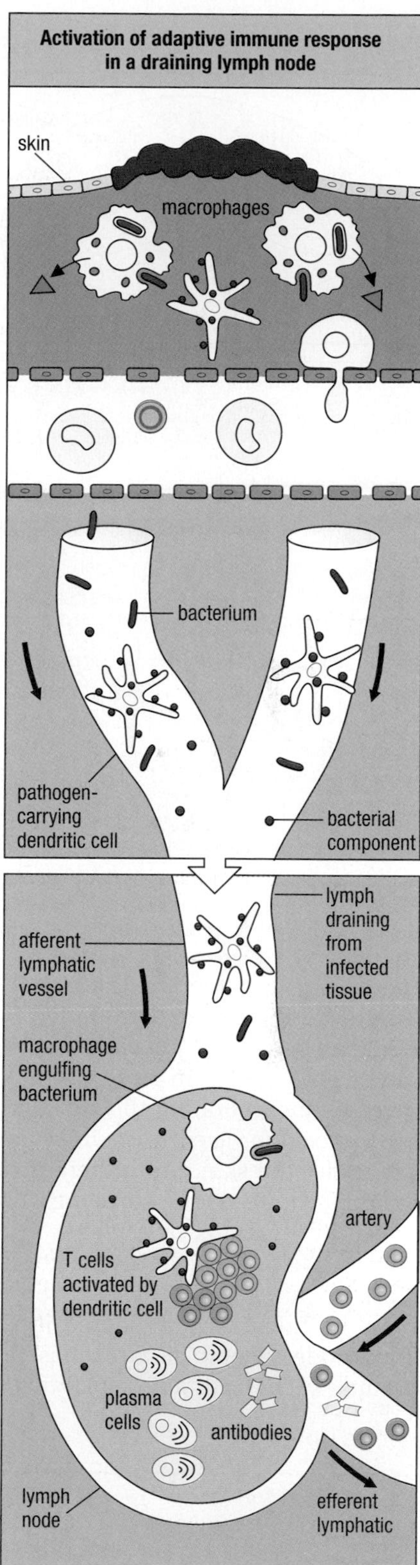

Figure 1.23 Activation of adaptive immunity in the draining lymph node. Pathogens, pathogen components, and dendritic cells carrying pathogens and molecules derived from them arrive in the afferent lymph draining the site of infection. Free pathogens and debris are removed by macrophages; the dendritic cells become resident in the lymph node and move to the T-cell areas, where they meet small lymphocytes that have entered the node from the blood (green). The dendritic cells specifically stimulate the division and differentiation of pathogen-specific small lymphocytes into effector lymphocytes (blue). Some helper T cells and cytotoxic T cells leave in the efferent lymph and travel to the infected tissue via the lymph and blood. Other helper T cells remain in the lymph node and stimulate the division and differentiation of pathogen-specific B cells into plasma cells (yellow). Plasma cells move to the medulla of the lymph node, where they secrete pathogen-specific antibodies, which are taken to the site of infection by the efferent lymph and subsequently the blood. Some plasma cells leave the lymph node and travel via the efferent lymph and the blood to the bone marrow, where they continue to secrete antibodies.

tissues (Figure 1.23). There the effector cells and molecules of adaptive immunity work with their counterparts of innate immunity to subdue the infection. Recovery from an infection involves the clearance of infectious organisms from the body and repair of the damage caused by both infection and the immune response. A cure is not always possible. Infection can overwhelm the immune system, with death as the consequence. In the United States, some 36,000 deaths each year are associated with influenza. For other pathogens, the infection can often persist, but its pathological effects are controlled by the adaptive immune response, as usually occurs with herpes virus infections.

1-13 The spleen provides adaptive immunity to blood infections

Pathogens can enter the blood directly, as occurs when blood-feeding insects transmit disease or when lymph nodes draining infected tissue have failed to remove all microorganisms from the lymph returned to the blood. The **spleen** is the lymphoid organ that serves as a filter for the blood. One purpose of the filtration is to remove damaged or senescent red cells; the second function of the spleen is that of a secondary lymphoid organ that defends the body against blood-borne pathogens. Any microorganism in the blood is a potential pathogen and source of dangerous systemic infection. Microorganisms and microbial products in the blood are taken up by splenic macrophages and dendritic cells, which then stimulate the B and T cells that arrive in the spleen from the blood. The spleen is made up of two different types of tissue: the red pulp, where red blood cells are monitored and removed, and the white pulp, where white blood cells gather to provide adaptive immunity. The organization and functions of splenic white pulp are similar to those of the lymph node, the main difference being that both pathogens and lymphocytes enter and leave the spleen in the blood (Figure 1.24).

A small number of children are born without a spleen, a condition called asplenia (Figure 1.25). This condition runs in families and is caused by deleterious mutations in the gene for ribosomal protein SA, a component of the small subunit of the ribosome. Inheriting a single defective copy of the SA gene is sufficient to cause asplenia, showing that two functional copies of the SA gene are required for development of the spleen, although that seems not to be the case for other organs, which appear normal in these patients. Children with asplenia are unusually susceptible to infections, many of which prove fatal, with bacteria such as *Streptococcus pneumoniae* (the pneumococcus) or

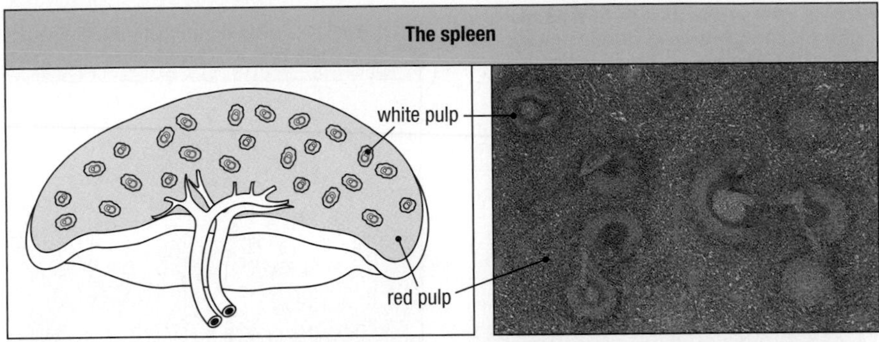

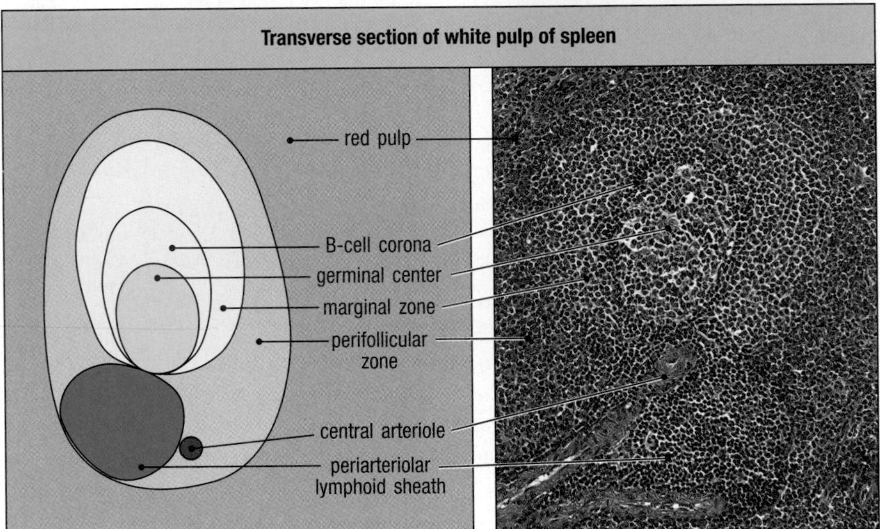

Figure 1.24 The spleen has aggregations of lymphocytes similar to those in lymph nodes. The human spleen is a large lymphoid organ in the upper left part of the abdomen, weighing about 150 grams. The upper diagram depicts a section of spleen in which nodules of white pulp are scattered within the more extensive red pulp. The red pulp is where old or damaged red cells are removed from the circulation; the white pulp is secondary lymphoid tissue, where lymphocyte responses to blood-borne pathogens are made. The bottom diagram shows a nodule of white pulp in transverse section. It comprises a sheath of lymphocytes surrounding a central arteriole. The sheath is called the periarteriolar lymphoid sheath (PALS). The lymphocytes close to the arteriole are mostly T cells (blue region); B cells (yellow regions) are placed more peripherally. Lymphoid follicles comprise a germinal center, a B-cell corona, and a marginal zone containing macrophages and differentiating B cells. Both follicle and PALS are surrounded by a perifollicular zone abutting the red pulp and containing a variety of cells, including erythrocytes, macrophages, T cells, and B cells. Photographs courtesy Yasodha Natkunam.

Haemophilus influenzae. These are examples of so-called **encapsulated bacteria**, whose cells are surrounded by a thick polysaccharide capsule. Figure 1.3d is an image of *S. pneumoniae*. If diagnosed in time, children with asplenia can be protected from these life-threatening infections by immunization with vaccines incorporating the capsular polysaccharides of these bacteria. For good effect the vaccines are injected subcutaneously, into the connective tissue under the skin. From there they stimulate an immune response in the draining lymph nodes, secondary lymphoid organs that have normal immunological functions in asplenic individuals. Asplenia is an example of an **immunodeficiency disease**, a condition in which the inheritance of one or more mutant genes leads to defects in one or more aspects of the immune system's functions (considered in Chapter 13). The discovery and study of immunodeficiency diseases has largely been the work of pediatricians, because such conditions usually show up early in childhood.

Congenital asplenia

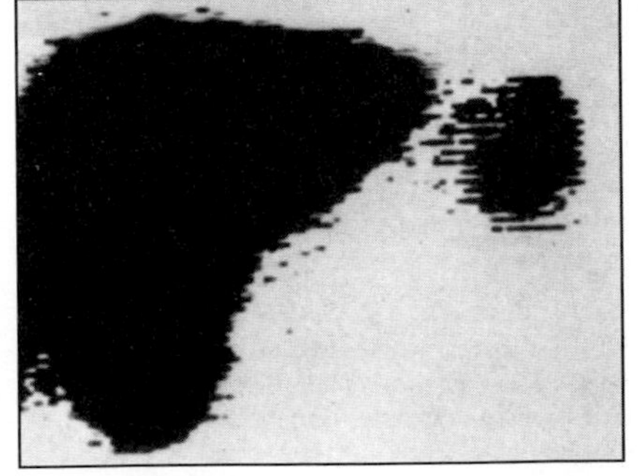

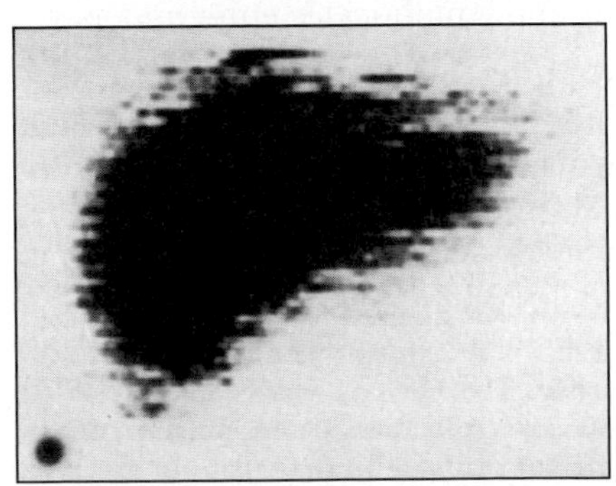

Figure 1.25 Diagnosis of a child with asplenia. The photos show scintillation scans of the abdomen of a mother (left panel) and child (right panel) after intravenous injection with radioactive colloidal gold. The large irregularly shaped organ to the left is the liver, which is seen in both panels. The smaller more rounded organ to the right in the mother is the spleen, which is not present in the child. Photographs courtesy of F. Rosen and R. Geha.

When a person's spleen becomes damaged as a consequence of a traumatic accident or wound, it will often be removed surgically to prevent life-threatening loss of blood into the abdominal cavity. Children who have this procedure, called splenectomy, can be as vulnerable to bacterial infections as asplenic children. For adults, who have already been infected by these pathogens and have developed protective immunity, the consequences of splenectomy are usually slight, but protective vaccination against some pathogens, especially *S. pneumoniae*, is advised.

1-14 Most secondary lymphoid tissue is associated with the gut

The parts of the body that harbor the largest and most diverse populations of microorganisms are the respiratory and gastrointestinal tracts. The extensive mucosal surfaces of these tissues make them particularly vulnerable to infection and they are therefore heavily invested with secondary lymphoid tissue. The **gut-associated lymphoid tissues** (**GALT**) include the **tonsils**, **adenoids**, and **appendix**, and the **Peyer's patches** which line the small intestine (see Figure 1.5). Similar but less organized aggregates of secondary lymphoid tissue line the respiratory epithelium, where they are called **bronchial-associated lymphoid tissue** (**BALT**), and other mucosal surfaces, including the gastrointestinal tract. The more diffuse mucosal lymphoid tissues are known generally as **mucosa-associated lymphoid tissue** (**MALT**).

Although different from the lymph nodes or spleen in outward appearance, the mucosal lymphoid tissues are similar to them in their microanatomy (Figure 1.26) and in their function of trapping pathogens to activate lymphocytes. The differences are chiefly in the route of pathogen entry and the migration patterns of their lymphocytes (mucosal immunity is discussed in Chapter 10). Pathogens arrive at mucosa-associated lymphoid tissues by direct delivery across the mucosa, mediated by specialized cells of the mucosal epithelium called **M cells**. Lymphocytes first enter mucosal lymphoid tissue from the blood and, if not activated, leave via lymphatics that connect the mucosal tissues to draining lymph nodes. Lymphocytes activated in mucosal tissues tend to stay within the mucosal system, either moving directly out from the lymphoid tissue into the lamina propria and the mucosal epithelium, where they perform their effector actions, or reentering the mucosal tissues from the blood as effector cells after being recirculated.

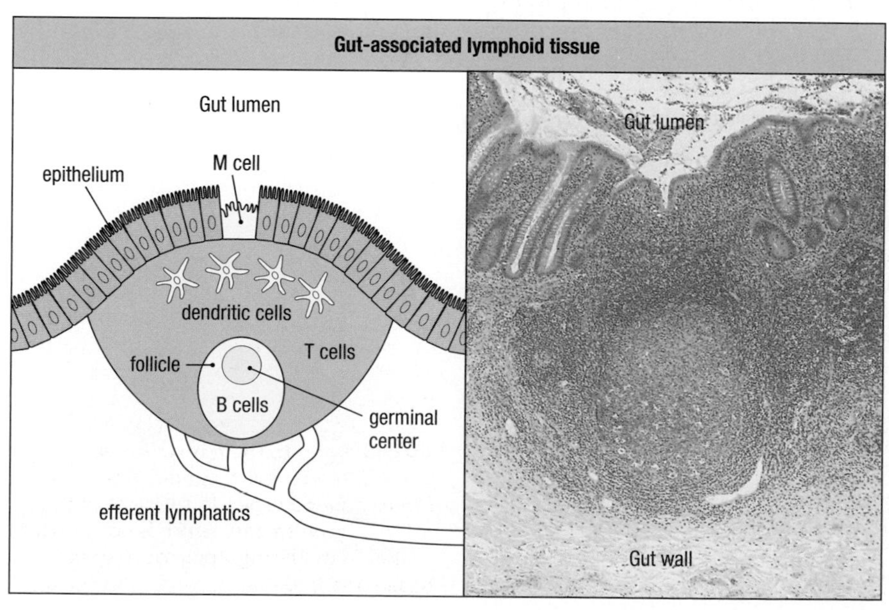

Figure 1.26 A typical region of gut-associated lymphoid tissue (GALT). Schematic diagram (left panel) and a light micrograph (right panel) of a typical region of organized GALT such as a Peyer's patch. M cells in the gut epithelium deliver pathogens from the gut lumen to the lymphoid tissue within the gut wall. The GALT is organized similarly to the lymph node and the white pulp of the spleen, with distinctive B- and T-cell zones, lymphoid follicles, and germinal centers. Lymphocytes, are delivered from the blood through the walls of small blood capillaries, as in the lymph node. Lymphocytes activated in the GALT leave in the efferent lymphatics and are delivered via the mesenteric lymph nodes (not shown) into the thoracic duct and back into the blood, from which they specifically return to the gut as effector lymphocytes. Photograph courtesy of N. Rooney.

Summary to Chapter 1

Throughout their evolutionary history, multicellular animals have been colonized and infected by microorganisms. To restrict the nature, size, and location of microbial infestation, animals have evolved a series of defenses, which humans still use today. The skin and contiguous mucosal surfaces provide physical and chemical barriers that confine microorganisms to the external surfaces of the body. When pathogens manage to breach the barriers and gain entry to the soft tissues, they are sought out and destroyed by the immune system. The cells of the immune system are principally the various types of lymphocyte or lymphoid cells and the myeloid cells, which all derive from a common stem cell in the bone marrow.

In responding to infection, the immune system starts with innate immune mechanisms that are fast, fixed in their mode of action, and effective in stopping most infections at an early stage. The cells and molecules of innate immunity identify common classes of pathogen and destroy them. Four key elements of innate immunity are: pathogen receptors that bind noncovalently to the surfaces of pathogens; proteins such as complement that bond covalently to pathogen surfaces, forming ligands for receptors on phagocytes; phagocytic cells that engulf and kill pathogens; and cytotoxic cells that kill virus-infected cells.

The defenses of adaptive immunity are brought into play when innate immunity fails to stop an infection. Although slow to start, the adaptive immune response eventually becomes powerful enough to terminate almost all of the infections that outrun innate immunity. The mechanisms of adaptive immunity are ones that improve pathogen recognition rather than pathogen destruction. They involve the T lymphocytes and B lymphocytes, which collectively have the ability to recognize the vast array of potential pathogens. Adaptive immune responses are initiated in specialized lymphoid tissues such as lymph nodes and spleen, to which infections that elude innate immunity spread. In these secondary lymphoid organs, small recirculating B lymphocytes and T lymphocytes with receptors that bind to pathogens or their macromolecular components are selected and activated. Because each individual B or T lymphocyte expresses receptors of a single and unique binding type, a pathogen stimulates only the small subset of lymphocytes that express receptors for the pathogen, focusing the adaptive immune response on that pathogen. When successful, an adaptive immune response terminates infection and provides long-lasting protective immunity against the pathogen that provoked the response. Failures to develop a successful response can arise from inherited deficiencies in the immune system or from the pathogen's ability to escape, avoid, or subvert the immune response. Such failures can lead to debilitating chronic infections or death.

Adaptive immunity builds on the mechanisms of innate immunity to provide a powerful response that is tailored to the pathogen at hand and can be rapidly reactivated on future challenge with that same pathogen, providing lifelong immunity to many common diseases. Adaptive immunity is an evolving process within a person's lifetime, in which each infection changes the make-up of that individual's lymphocyte population. These changes are neither inherited nor passed on but, during the course of a lifetime, they determine a person's fitness and their susceptibility to disease.

Questions

1–1
A. Name the primary lymphoid tissues in mammals and the main types of secondary lymphoid tissue.
B. What is the difference in function between primary and secondary lymphoid tissues, and what are the principal events that take place in each?

1–2 One reason that pathogenic microorganisms have an advantage in the host they infect is because they _______.
a. have previously been encountered through natural exposure
b. have previously been encountered through vaccination
c. strengthen the host's immune response
d. reproduce and evolve more rapidly than the host can eliminate them
e. reproduce and evolve more slowly than the host can eliminate them.

1–3 Match the term in column A with its description in column B.

Column A	Column B
a. defensins	1. bathed in antimicrobial fluids that protect underlying epithelial surfaces
b. lymph	2. membrane-disrupting cytotoxic peptides
c. mucosae	3. accumulation of fluid across permeable endothelium
d. edema	4. a combination of cells and fluid that is transported to the lymphatics
e. complement	5. proteins in the serum that tag pathogens for destruction by effector cells

1–4 Which of the following is not a characteristic of inflammation?
a. inactivation of macrophages
b. increased vascular permeability and edema
c. vasodilation
d. pain
e. influx of leukocytes.

1–5 The processes of clonal selection and clonal expansion occur during _____.
a. adaptive immune responses
b. innate immune responses
c. hematopoiesis
d. self renewal
e. immunodeficiency diseases.

1–6 The _____ is (are) the lymphoid organ(s) that filter(s) the blood.
a. spleen
b. tonsils
c. Peyer's patches
d. appendix
e. adenoids.

1–7 Which of the following best describes the movement of a T cell through a lymph node?
a. It enters via efferent lymphatics and exits via the bloodstream.
b. It enters via afferent lymphatics and exits via the bloodstream.
c. It enters via the bloodstream and exits via afferent lymphatics.
d. It enters via the bloodstream and exits via the bloodstream.
e. It enters via the bloodstream and exits via efferent lymphatics.

1–8 Which of the following pairs is mismatched?
a. T-cell activation: cell division and differentiation
b. effector B cell: plasma cell
c. plasma cell: antibody secretion
d. helper T cell: kills pathogen-infected cells
e. helper T cell: facilitates the differentiation of B cells.

1–9 Which of the following pairs is mismatched?
a. plasma cell: mediation of phagocytosis and killing of microorganisms in the plasma
b. megakaryocyte: formation of platelets
c. dendritic cell: activation of adaptive immune responses
d. natural killer cell: develops from a common lymphoid progenitor
e. neutrophil: formation of pus
f. regulatory T cell: inhibition of T-cell activity.

1–10 The specialized cells that provide the interface between the lumen of the gut and the underlying lymphoid tissue _____.
a. make up the germinal center
b. are called the M cells
c. are located in the periarteriolar lymphoid sheath
d. are specialized in killing encapsulated bacteria
e. are progenitors of monocytes.

1–11 Explain how the adaptive and innate immune systems work together to generate an effective immune response.

1–12 Two-year-old Janice Tumminello survived an episode of *Haemophilus influenzae* septicemia at 4 months by intravenous antibiotic therapy. Her immunizations are up to date. Three days ago her adoptive parents became concerned when she became lethargic, had several bouts of vomiting, and developed a fever, and they took her to the emergency room. She had a rapid pulse and low blood pressure, her peripheral areas were cold, and purpura began to develop on her fingers and toes. Blood cultures tested positive for *Streptococcus pneumoniae.* Aggressive antibiotic treatment and fluid-replacement therapy were successful in preventing further dissemination of the bacteremia. However, amputation of three digits on the left hand and debridement of her toes on both feet was required. A diagnosis of overwhelming severe pneumococcal sepsis was made. DNA analysis showed that Janice had a deleterious mutation in a gene encoding _____, which is associated with congenital asplenia.
a. an immunoglobulin
b. a defensin
c. a ribosomal protein
d. a complement protein
e. a T-cell receptor.

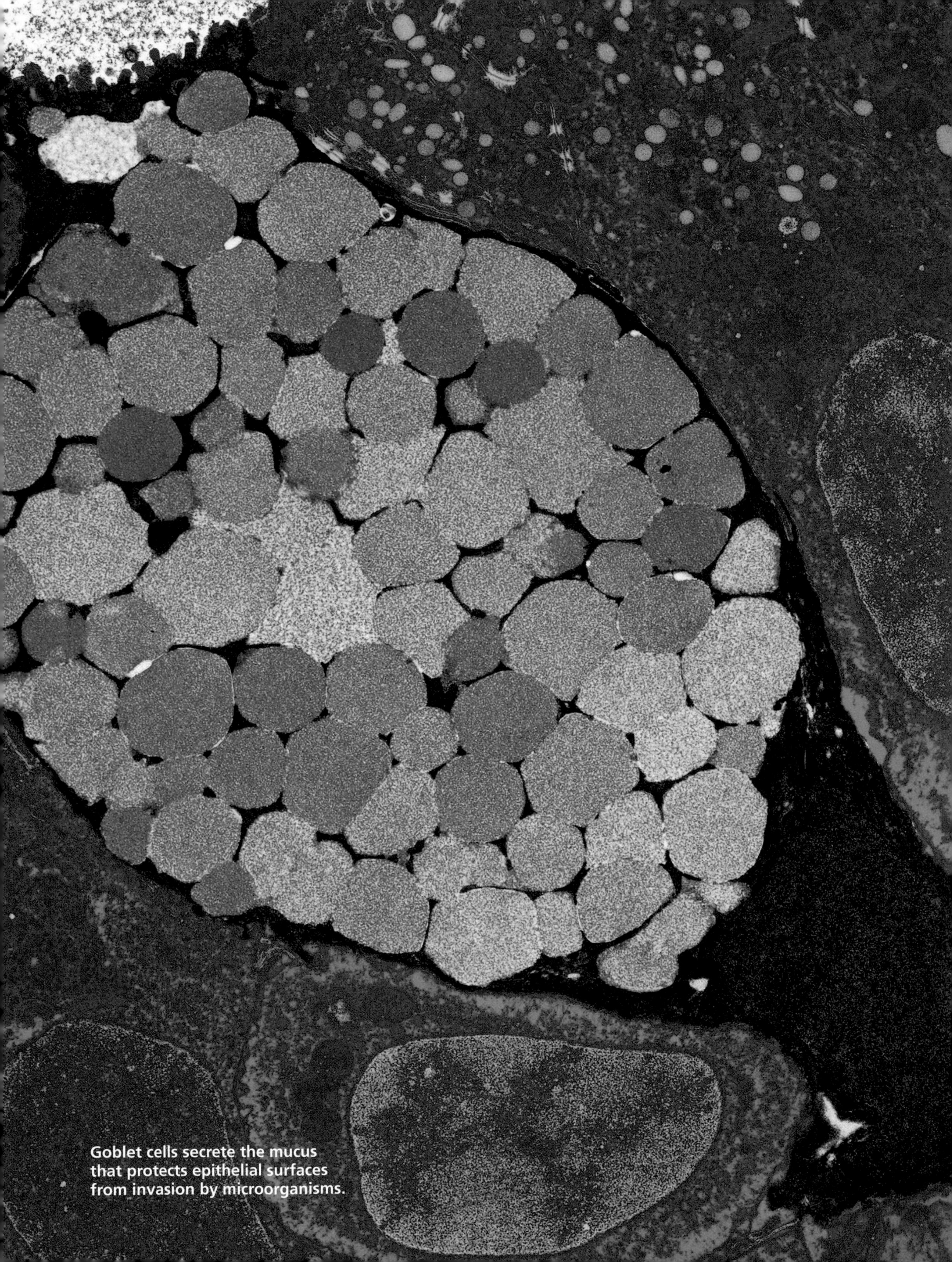

Goblet cells secrete the mucus that protects epithelial surfaces from invasion by microorganisms.

Chapter 2

Innate Immunity: the Immediate Response to Infection

Although we become conscious of infectious agents only when we are suffering from the diseases they cause, microorganisms are always with us. Fortunately, almost all of the microorganisms we come into contact with are prevented from ever causing an infection or disease. To prevent and control infectious disease, the immune system has three lines of defense that operate at different times as an infection progresses. This chapter examines the first line of defense: the mechanisms of **innate immunity**, which are ready for action at all times and function from the very beginning of an infection. The second line of defense comprises innate immune mechanisms that are mobilized once cells of the immune system have detected the presence of infection and have turned on the gene expression and protein synthesis necessary to make the response. These induced mechanisms of innate immunity, which require from a few hours to four days of development to become fully functional, are the subject of Chapter 3. The third line of defense, the adaptive immune response, is reserved for the small minority of infections that innate immunity cannot subdue. Although slow to mobilize, the adaptive immune response is more powerful than the innate response and has the added advantage of providing long-lasting immunity to further infection by the pathogen. The mechanisms of adaptive immunity are covered in Chapters 4–11 of this book.

2-1 Physical barriers colonized by commensal microorganisms protect against infection by pathogens

The outside surface of the human body comprises the skin and the mucosal epithelia that line the digestive, respiratory, and urogenital tracts. These epithelial tissues provide effective physical and chemical barriers that prevent pathogens in the external environment from gaining access to the internal tissues and organs (Figure 2.1). Further deterring pathogen invasion are the large communities of **commensal microorganisms** that colonize the skin and mucosal surfaces of healthy individuals (see Section 1-1). To expand its numbers and become an infection, any aspiring pathogen must compete successfully with the resident commensals for nutrients and space. Pathogenic strains are not always competitive, as exemplified by *Clostridium difficile* infections, which can only take hold in unhealthy individuals whose populations of commensal bacteria have been decimated by one or more courses of antibiotics (see Figure 1.2).

Before birth, mammalian babies have no commensal microorganisms. Starting at birth and contact with the mother's vagina, the infant's skin and mucosal

	Skin	Gut	Lungs	Eyes/nose/oral cavity
Mechanical	Epithelial cells joined by tight junctions			
	Longitudinal flow of air or fluid		Movement of mucus by cilia	Tears Nasal cilia
Chemical	Fatty acids	Low pH Antimicrobial enzymes	Pulmonary surfactant	Antimicrobial enzymes in tears and saliva
	Antimicrobial peptides	Antimicrobial peptides	Antimicrobial peptides	Antimicrobial peptides
Microbiological	Normal microbiota			

Figure 2.1 Many barriers prevent pathogens from crossing epithelia and colonizing tissues. Surface epithelia provide mechanical, chemical, and microbiological barriers to infection.

surface begins to be populated by commensals coming from family members, friends, and pets. The mucosal surfaces provide a much larger and richer habitat for commensal microorganisms than the skin; the gut is a particularly good habitat because it provides an abundant and reliable source of food. More than 1000 species of bacteria inhabit the human gut, and the combined population of the bacterial cells outnumbers the cells in the human body by ten to one. Having been co-evolving with placental mammals for more than 100 million years, many commensal bacteria have developed symbiotic relationships with their hosts, which mutually improve their nutrition, metabolism, and general health. Direct evidence for the benefit of having and maintaining commensal relationships is the poor health of laboratory mice when raised from birth in a completely germ-free environment. The population of commensal microorganisms, known as the microbiota, is increasingly considered to be an important and integral part of a healthy human body that needs looking after. Because the immune system is relatively underdeveloped at birth, its continuing development after birth is influenced and shaped by the acquisition of the microbiota.

2-2 Intracellular and extracellular pathogens require different types of immune response

Pathogens are a diverse collection of organisms that exploit the human body in numerous ways. A key difference, which determines what type of an immune response can be effective, is between pathogens that live and replicate in the spaces between human cells to produce **extracellular infections** and pathogens that replicate inside human cells to produce **intracellular infections** (Figure 2.2). Extracellular pathogens are accessible to soluble, secreted molecules of the immune system, whereas intracellular pathogens are not. The strategy used to attack intracellular pathogens is to kill the human cells in which the pathogens are living. This sacrifice interferes with the pathogen's life cycle and exposes any pathogens released from the dead cells to the soluble molecules of the immune system.

Many pathogens have life cycles that involve both extracellular and intracellular forms. For example, when particles of influenza virus first enter the respiratory tract they are extracellular and susceptible to attack by soluble molecules of the immune system. Once the viral particles have infected the epithelial cells that line the respiratory tract, that is no longer the case. At this stage in the viral life cycle, the infected cells can be recognized and killed by effector cells of the immune system, which detect virus-induced changes on the surface of

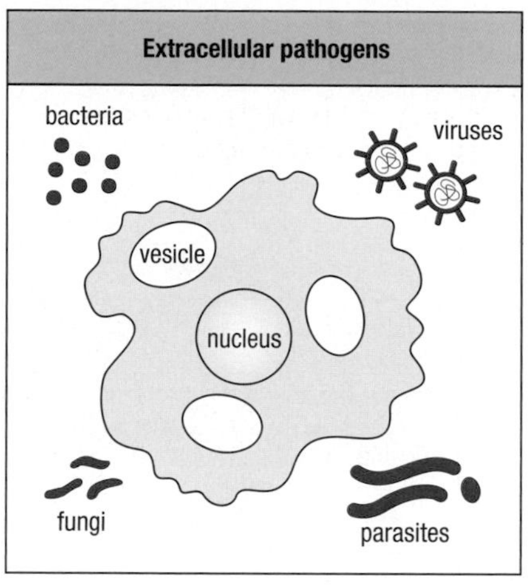

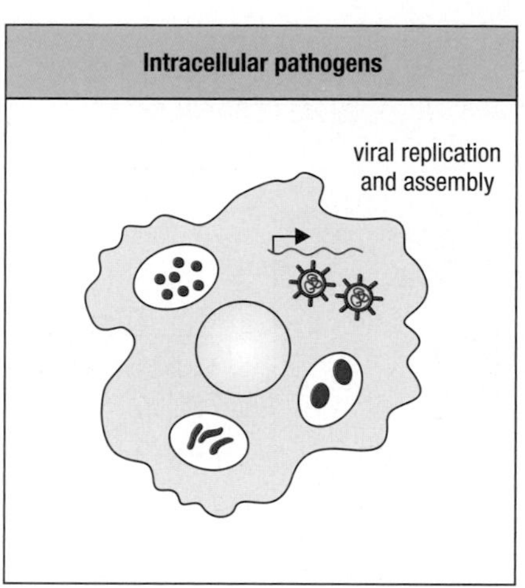

Figure 2.2 Pathogens can be present outside and inside human cells. On invading the human body all four classes of pathogen—bacteria, viruses, fungi, and parasites—can be present in the extracellular spaces of the infected tissue (left panel). Viruses and some bacteria need to get inside human cells in order to replicate (right panel).

the infected cells. With replication of the virus in the epithelial cell and the release of viral particles into the extracellular space, the virus again becomes a target for soluble effector molecules.

2-3 Complement is a system of plasma proteins that mark pathogens for destruction

As soon as a pathogen penetrates an epithelial barrier and starts to live in a human tissue, the defense mechanisms of innate immunity are brought into play. One of the first weapons to fire is a system of soluble proteins made constitutively by the liver and present in the blood, lymph, and extracellular fluids. These plasma proteins are collectively known as the **complement system** or just **complement**. Complement coats the surface of bacteria and extracellular virus particles and makes them more easily phagocytosed (see Section 1-10). Without such a coating, many bacteria resist phagocytosis, especially those that are enclosed in thick polysaccharide capsules.

Many complement components are proteolytic enzymes, or proteases, that circulate in functionally inactive forms known as **zymogens**. Infection triggers **complement activation**, which proceeds by a series, or cascade, of enzymatic reactions, in which each protease cleaves and activates the next protease in the pathway. Each enzyme is highly specific for the complement component it cleaves, and cleavage is usually at a single site. Most of these enzymes belong to the large family of serine proteases, which also includes the digestive enzymes chymotrypsin and trypsin.

Although more than 30 proteins make up the complement system, **complement component 3** (**C3**) is by far the most important. People lacking other complement components have relatively minor immunodeficiencies, but people lacking C3 are prone to successive severe infections. When the complement system is activated by infection it leads to the cleavage of C3 into a small C3a fragment and a large C3b fragment. In the process, some of the C3b fragments become covalently bound to the pathogen's surface (Figure 2.3). This attachment of C3b to pathogen surfaces is the essential function of the complement system; it is called **complement fixation** because C3b becomes firmly fixed to the pathogen. The bound C3b tags the pathogen for destruction by phagocytes and can also organize the formation of protein complexes that damage the pathogen's membrane. The soluble C3a fragment also contributes to the body's defenses by acting as a chemoattractant to recruit effector cells, including phagocytes, from the blood to the site of infection.

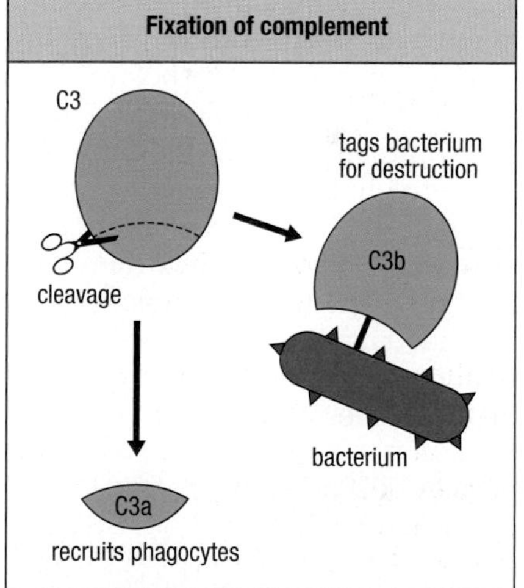

Figure 2.3 Complement activation results in covalent attachment of C3b to a pathogen's surface. The key event in complement activation in response to a pathogen is the proteolytic cleavage of complement fragment C3. This cleavage produces a large C3b fragment and a small C3a fragment. C3b is chemically reactive and becomes covalently attached, or fixed, to the pathogen's surface, thereby marking the pathogen as dangerous. C3a recruits phagocytic cells to the site of infection.

The unusual feature of C3 that underlies its unique and potent function is a high-energy thioester bond within the glycoprotein. C3 is made and enters the

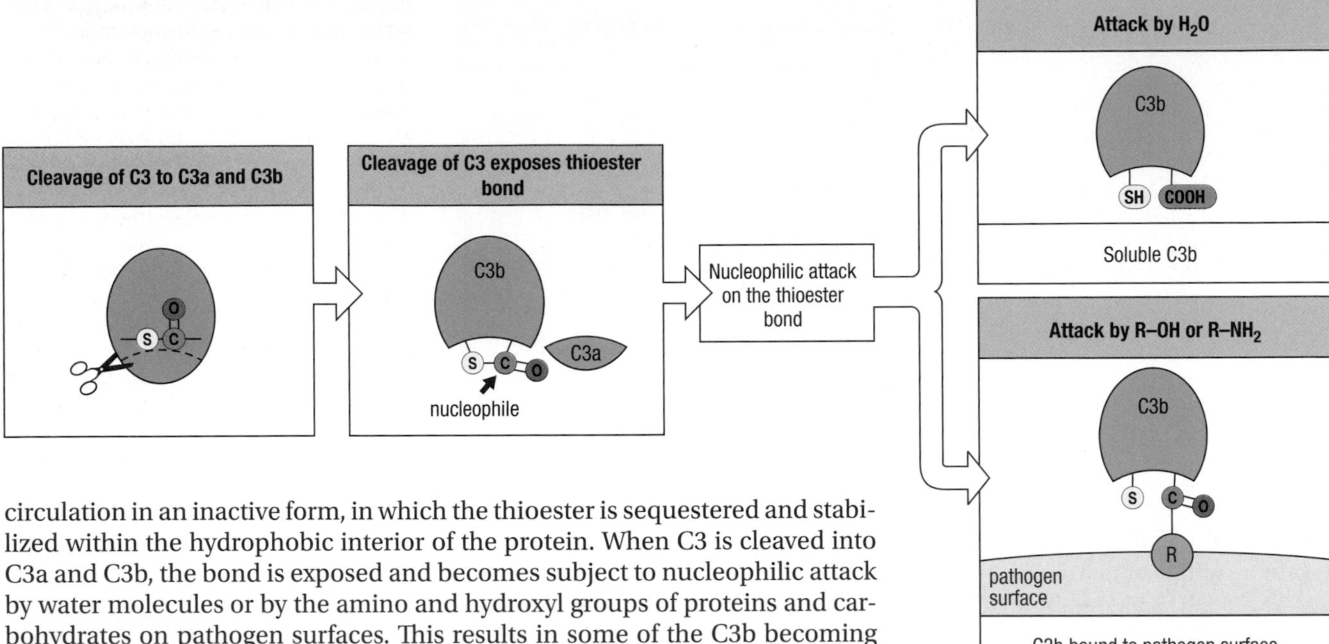

circulation in an inactive form, in which the thioester is sequestered and stabilized within the hydrophobic interior of the protein. When C3 is cleaved into C3a and C3b, the bond is exposed and becomes subject to nucleophilic attack by water molecules or by the amino and hydroxyl groups of proteins and carbohydrates on pathogen surfaces. This results in some of the C3b becoming covalently bonded to the pathogen (Figure 2.4). The thioester bonds of the vast majority of C3b molecules are attacked by water, and so most C3b remains in solution in an inactive hydrolyzed form.

Figure 2.4 Cleavage of C3 exposes a reactive thioester bond that covalently attaches the C3b fragment to the pathogen surface. Circulating C3 is an inactive serine protease consisting of α and β polypeptide chains in which a thioester bond in the α chain is protected from hydrolysis within the hydrophobic interior of the protein. The thioester bond is denoted in the left-hand two panels by the circled letters S, C, and O. The C3 molecule is activated by cleavage of the α chain to give fragments C3a and C3b. This exposes the thioester bond of C3b to the hydrophilic environment. The thioester bonds of most of the C3b fragments will be spontaneously hydrolyzed by water as shown in the upper right panel, but a minority will react with hydroxyl and amino groups on molecules on the pathogen's surface, bonding C3b to the pathogen surface, as shown in the lower right panel.

Three pathways of complement activation are defined. Although differing in how they are triggered and in the first few reactions in the cascade, they all lead to C3 activation, the deposition of C3b on the pathogen's surface, and the recruitment of similar effector mechanisms for pathogen destruction (Figure 2.5). The pathway that works at the start of infection is the **alternative pathway of complement activation**. A second pathway, the **lectin pathway of complement activation**, is also a part of innate immunity but is induced by infection and requires some time before it gains strength. The third pathway, the **classical pathway of complement activation**, is a part of both innate and adaptive immunity and requires the binding of either antibody or an innate immune-system protein called C-reactive protein to the pathogen's surface. The names of the pathways reflect the order of their scientific discovery: the classical pathway was discovered first, then the alternative pathway and lastly the lectin pathway. The name complement was coined because the effector functions provided by these proteins were seen to 'complement' the pathogen-binding function of antibodies in the classical pathway of complement activation and pathogen destruction.

2-4 At the start of an infection, complement activation proceeds by the alternative pathway

We shall start by describing the alternative pathway of complement activation, which is one of the first responses of the innate immune system, especially to bacterial infection. C3 is made in the liver and secreted into the blood in a conformation that sequesters the thioester bond in an inactive form within the hydrophobic interior of the protein. At a slow rate, and without being cleaved, C3 spontaneously changes its conformation and exposes the thioester bond. In the aqueous environment of the blood, the thioester bond becomes active and quickly makes a covalent bond that attaches C3 to another molecule that has either an amino or a hydroxyl group. This usually involves a water molecule, because they are so plentiful, and gives a form of C3 called **iC3** or **C3(H_2O)**. This hydrolysis is the first step in the alternative pathway of complement activation.

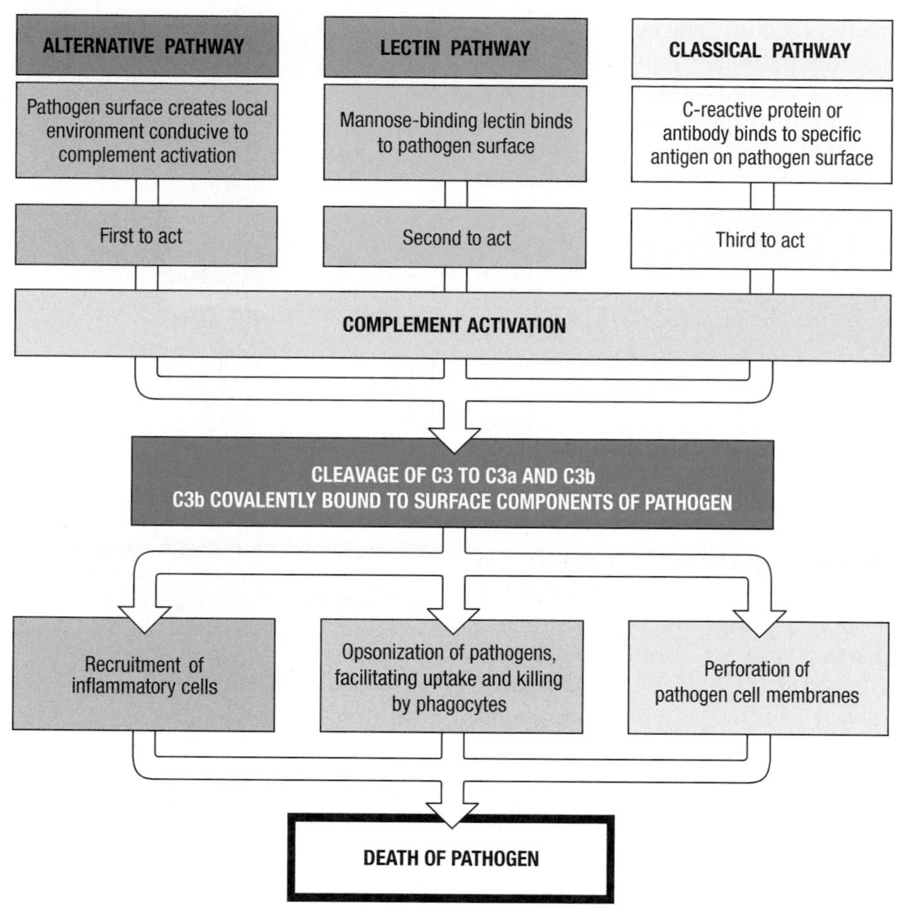
ALTERNATIVE PATHWAY
LECTIN PATHWAY
CLASSICAL PATHWAY
Pathogen surface creates local environment conducive to complement activation
Mannose-binding lectin binds to pathogen surface
C-reactive protein or antibody binds to specific antigen on pathogen surface
First to act
Second to act
Third to act
COMPLEMENT ACTIVATION
CLEAVAGE OF C3 TO C3a AND C3b
C3b COVALENTLY BOUND TO SURFACE COMPONENTS OF PATHOGEN
Recruitment of inflammatory cells
Opsonization of pathogens, facilitating uptake and killing by phagocytes
Perforation of pathogen cell membranes
DEATH OF PATHOGEN

Figure 2.5 The three pathways of complement activation. The alternative pathway of complement activation is triggered by changes in the local physicochemical environment that are caused by the constituents of some bacterial surfaces. The alternative pathway acts at the earliest times during infection. The lectin-mediated pathway is initiated by the mannose-binding lectin of plasma, which binds to carbohydrates found on bacterial cells and other pathogens. The lectin-mediated pathway is induced by infection and contributes to innate immunity. The classical pathway is initiated in the innate immune response by the binding of C-reactive protein to bacterial surfaces, and in the adaptive immune response by the binding of antibodies to pathogen surfaces.

The environment near the surface of certain pathogens, particularly bacteria, increases the rate at which C3 is hydrolyzed to give iC3. Also facilitating production of iC3 is the high concentration of C3 in blood (about 1.2 mg/ml). iC3 binds to the inactive complement **factor B**, making factor B susceptible to cleavage by the protease **factor D**. This reaction produces a small fragment, Ba, which is released, and a large fragment, Bb, that has protease activity and remains bound to iC3. The iC3bBb complex is a protease that specifically and efficiently cleaves C3 into the C3a and C3b fragments, with exposure of the thioester bond that is in C3b. With large numbers of C3 molecules being cleaved and activated, some C3b fragments become covalently attached to amino and hydroxyl groups of the pathogen's outer surface (Figure 2.6).

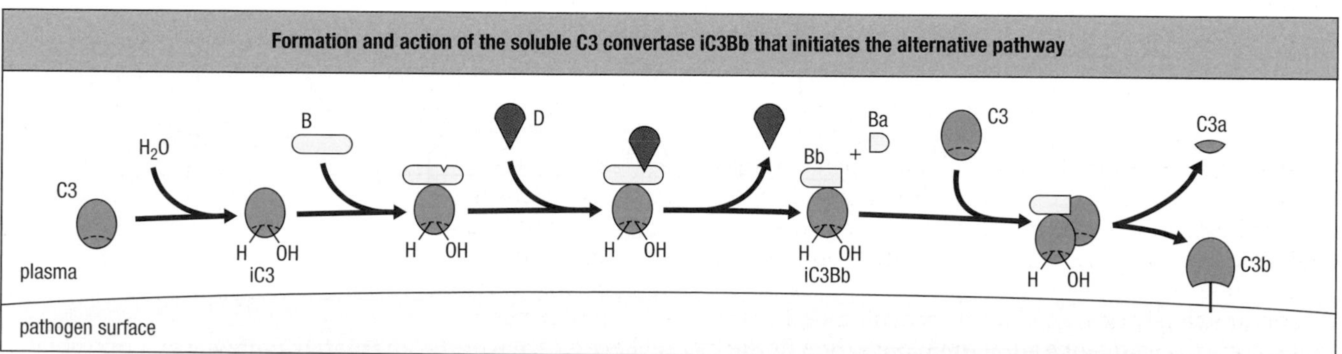
Formation and action of the soluble C3 convertase iC3Bb that initiates the alternative pathway
C3
H_2O
B
D
Ba
Bb
+
C3
C3a
H OH
iC3
iC3Bb
C3b
plasma
pathogen surface

Figure 2.6 Formation and action of the soluble C3 convertase that initiates the alternative pathway of complement activation. In the plasma close to a microbial surface the thioester bond of C3 spontaneously hydrolyzes at low frequency. This activates the C3, which then binds factor B. Cleavage of B by the serine protease factor D produces a soluble C3 convertase, called iC3Bb, which then activates C3 molecules by cleavage into C3b and C3a. Some of the C3b fragments become covalently attached to the microbial surface.

Proteases that cleave and activate C3 are called **C3 convertases**, iC3Bb being an example of a soluble C3 convertase. Like iC3, pathogen-bound C3b binds factor B and facilitates the cleavage of factor B by factor D. This reaction leads to the release of Ba and the formation of a C3bBb complex on the microbial surface. C3bBb is a potent C3 convertase, called the **alternative C3 convertase**, which works right at the surface of the pathogen (Figure 2.7). C3bBb binds C3 and cleaves it into C3a and C3b with activation of the thioester bond. Because this convertase is situated at the pathogen's surface and is unable to diffuse away like iC3Bb, a larger proportion of the C3b fragments it produces become fixed to the pathogen. Once some C3 convertase molecules have been assembled, they cleave more C3 and fix more C3b at the microbial surface, leading to the assembly of yet more convertase. This positive-feedback process, in which the C3b product of the enzymatic reaction can assemble more enzyme, is one of progressive amplification of C3 cleavage. From the initial deposition of a few molecules of C3b, the pathogen rapidly becomes coated with C3b (Figure 2.8).

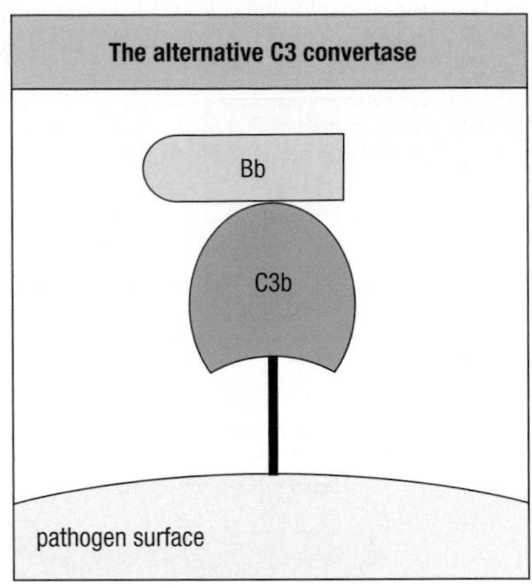

Figure 2.7 The C3 convertase of the alternative pathway is a complex of C3b and Bb. In this complex the Bb fragment of factor B provides the protease activity to cleave C3, and the C3b fragment of C3 locates the enzyme to the pathogen's surface.

2-5 Regulatory proteins determine the extent and site of C3b deposition

As we saw in Section 2-4, the alternative C3 convertase, C3bBb, is capable of rapid and runaway reactions because one molecule of C3bBb can catalyze the formation of numerous additional molecules of C3bBb. Two broad categories of **complement control proteins** have evolved to regulate these reactions, which they do mainly by stabilizing or degrading C3b at cell surfaces. One class comprises plasma proteins that interact with C3b attached to human and microbial cell surfaces; the other includes membrane proteins on human cells that prevent complement fixation at the cell surface.

The plasma protein **properdin** (**factor P**) increases the speed and power of complement activation by binding to the C3 convertase C3bBb on microbial surfaces and preventing its degradation by proteases (Figure 2.9, upper panel). Countering the effect of properdin is plasma protein **factor H**, which binds to C3b and facilitates its further cleavage to a form called iC3b by the plasma serine protease **factor I** (Figure 2.9, center panel). Fragment iC3b cannot assemble a C3 convertase, so the combined action of factors H and I is to decrease the number of C3 convertase molecules on the pathogen surface.

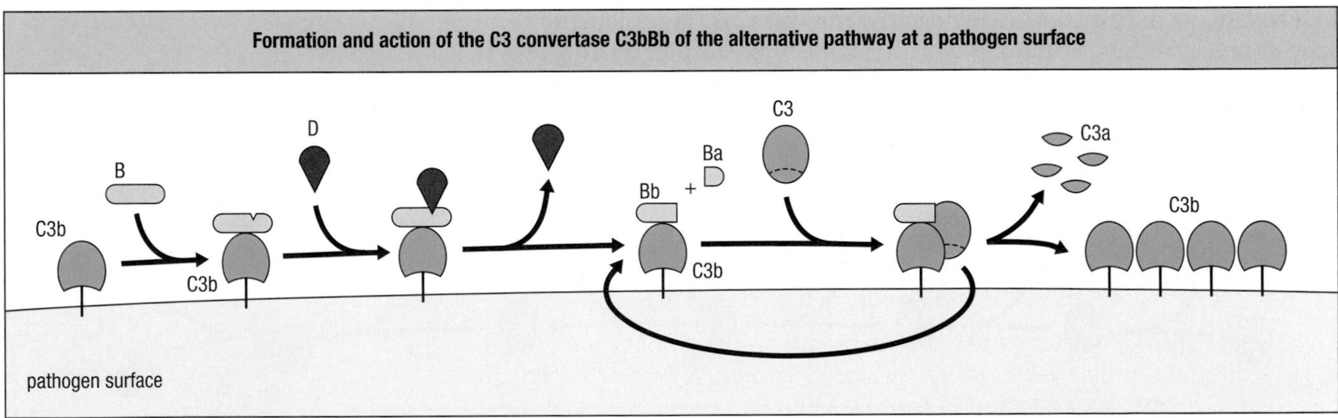

Figure 2.8 Formation and action of the C3 convertase C3bBb of the alternative pathway at a microbial surface. Through the action of iC3Bb, the soluble C3 convertase, C3b fragments are bound to the microbial surface (see Figure 2.6). These bind factor B, which is then cleaved by factor D to produce C3bBb, the surface-bound convertase of the alternative pathway. This enzyme cleaves C3 to produce further C3b fragments bound to the microbe and small soluble C3a fragments. The C3b fragments can be used either to make more C3 convertase, which amplifies the activation of C3, or to provide ligands for the receptors of phagocytic cells. The small, soluble C3a fragments attract phagocytes to sites of complement fixation.

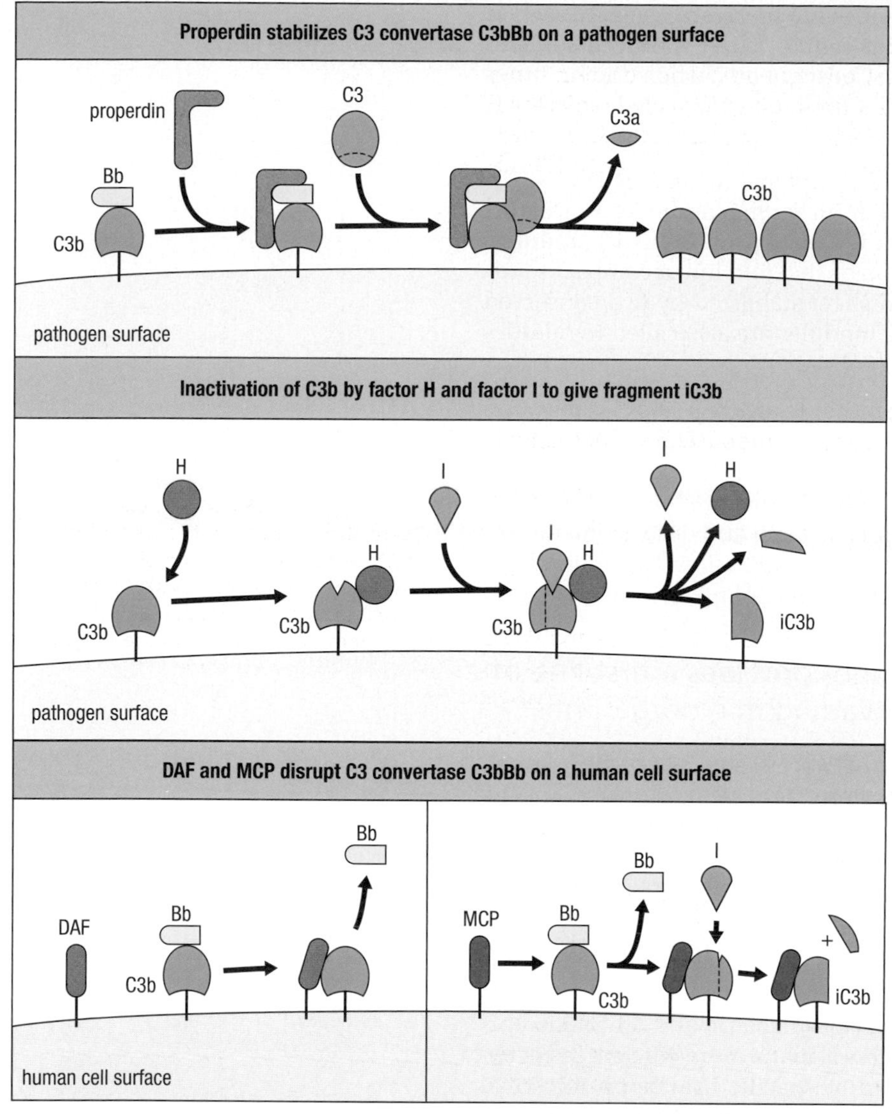

Figure 2.9 Formation and stability of the alternative C3 convertase on cell surfaces is determined by complement control proteins. Top panel: the soluble protein properdin (factor P) binds to C3bBb, extending its lifetime on the microbial surface. Middle panel: factor H binds C3b and changes its conformation to one susceptible to cleavage by factor I. The product of this cleavage is the iC3b fragment of C3, which remains attached to the pathogen surface but cannot form a C3 convertase. Bottom panel: when C3bBb is formed on a human cell surface, it is rapidly disrupted by the action of one of two membrane proteins: decay-accelerating factor (DAF) or membrane cofactor protein (MCP). In combination, these regulatory proteins ensure that much complement is fixed to pathogen surfaces and little is fixed to human cell surfaces.

The importance of the negative regulation by factors H and I is illustrated by the immunodeficiency suffered by patients who, for genetic reasons, lack factor I. In these people, formation of the C3 convertase C3bBb runs away unchecked until it has depleted the reservoir of C3 in blood, extracellular fluid, and lymph. When faced with bacterial infections, people with factor I deficiency fix abnormally small amounts of C3b on bacterial surfaces, resulting in less efficient bacterial clearance by phagocytes. Consequently, these people are more susceptible than usual to ear infections and abscesses caused by encapsulated bacteria. These bacteria have a thick polysaccharide capsule and are phagocytosed more efficiently when coated with complement.

The second category of complement-control proteins comprises membrane proteins of human cells that interfere with complement activation at human cell surfaces. **Decay-accelerating factor** (**DAF**) binds to the C3b component of the alternative C3 convertase, causing its dissociation and inactivation. **Membrane cofactor protein** (**MCP**) also has this function, but the binding of MCP to C3b makes it also susceptible to cleavage and inactivation by factor I (Figure 2.9, bottom panel). The functions of MCP are similar to those of the soluble complement regulator, factor H, which can also become membrane-associated; factor H has a binding site for sialic acid, a component of human cell-surface carbohydrates that is absent from most bacteria. As a strategy to evade the actions of complement, some species of bacteria, such as *Streptococcus pyogenes* and *Staphylococcus aureus*, cover their cell surfaces with sialic acid and so mimic human cells. Consequently, when C3b becomes deposited on the surface of these bacteria it is readily inactivated by factor H bound to the bacterial sialic acid.

Many of the diverse proteins that regulate complement, such as DAF, MCP, and factor H, are elongated structures built from varying numbers of structurally similar modules known as **complement control protein (CCP) modules**. Each module consists of about 60 amino acids that fold into a compact sandwich formed from two slices of β-pleated sheet stabilized by two conserved disulfide bonds. Proteins made up of CCP modules are also called **regulators of complement activation (RCA)**.

The combined effect of the reactions that promote and regulate C3 activation is to ensure that in practice C3b is deposited only on the surfaces of pathogenic microorganisms and not on human cells. In this manner the complement system provides a simple and effective way of distinguishing human cells from microbial cells, and for guiding mechanisms of death and destruction toward invading pathogens and away from healthy cells and tissues. In immunology, this type of distinction is called the discrimination of non-self from self.

2-6 Phagocytosis by macrophages provides a first line of cellular defense against invading microorganisms

When a pathogen invades a human tissue, the first effector cells of the immune system it encounters are the resident macrophages. **Macrophages** are the mature forms of circulating monocytes (see Figure 1.11, p. 13) that have left the blood and taken up residence in the tissues. They are prevalent in the connective tissues, the linings of the gastrointestinal and respiratory tracts, the alveoli of the lungs, and in the liver, where they are known as **Kupffer cells**. Macrophages are long-lived phagocytic cells that participate in both innate and adaptive immunity.

Although macrophages are able to phagocytose bacteria and other microorganisms in a nonspecific fashion, the process is made more efficient by receptors on the macrophage surface that bind to specific ligands on microbial surfaces. One such receptor binds to C3b fragments that have been deposited at high density on the surface of a pathogen through activation of the alternative pathway of complement. This receptor is called **complement receptor 1** or **CR1**. The interaction of an array of C3b fragments on a pathogen with an array of CR1 molecules on the macrophage facilitates the engulfment and destruction of the pathogen. Bacteria coated with C3b are more efficiently phagocytosed than uncoated bacteria: the coating of a pathogen with a protein that facilitates phagocytosis is called **opsonization** (Figure 2.10).

CR1 also serves to protect the surface of cells on which it is expressed. Like MCP and factor H, CR1 disrupts the C3 convertase by making C3b susceptible to cleavage by factor I. During phagocytosis, some of a macrophage's CR1 molecules have this protective role, whereas others engage the C3b fragments deposited on the pathogen's surface. Like MCP and factor H, CR1 is made up of CCP modules.

Two other macrophage receptors, **complement receptor 3 (CR3)** and **complement receptor 4 (CR4)**, bind to iC3b fragments on microbial surfaces. Although the iC3b fragment has no C3 convertase activity, it facilitates phagocytosis and pathogen destruction by serving as the ligand for CR3 and CR4. These receptors are structurally unrelated to CR1, being members of a family of surface glycoproteins, the integrins, that contribute to adhesive interactions between cells. The CR1, CR3, and CR4 receptors work together more

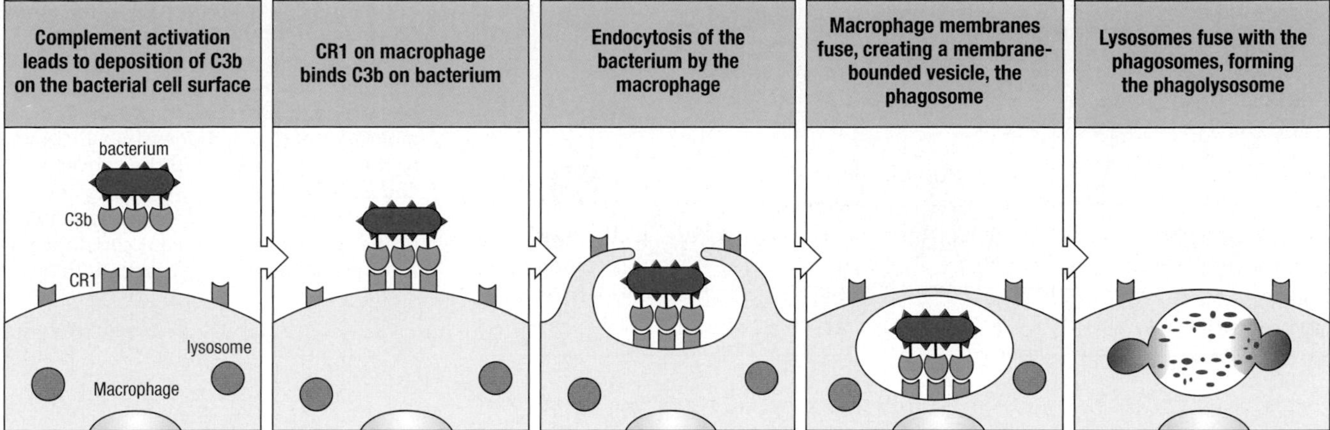

Complement activation leads to deposition of C3b on the bacterial cell surface
CR1 on macrophage binds C3b on bacterium
Endocytosis of the bacterium by the macrophage
Macrophage membranes fuse, creating a membrane-bounded vesicle, the phagosome
Lysosomes fuse with the phagosomes, forming the phagolysosome
bacterium
C3b
CR1
lysosome
Macrophage

Figure 2.10 Complement receptors on phagocytes trigger the uptake and breakdown of C3b-coated pathogens. Covalently attached C3b fragments coat the pathogen surface, here a bacterium, and bind to complement receptor 1 (CR1) molecules on the phagocyte surface. This tethers the bacterium to the phagocyte. Intracellular signals generated by CR1 enhance the phagocytosis of the bacterium and fusion of the phagosome with lysosomes that contain toxic molecules and degradative enzymes. Ultimately, the bacterium is killed.

effectively in the phagocytosis of complement-coated pathogens than does each receptor on its own. The combination of opsonization by complement activated through the alternative pathway and subsequent phagocytosis by macrophages allows pathogens to be recognized and destroyed from the very beginning of an infection.

2-7 The terminal complement proteins lyse pathogens by forming membrane pores

As we have seen, the most important product of complement activation is C3b bonded to pathogen surfaces. However, the cascade of complement reactions does extends beyond this stage, involving five additional complement components (Figure 2.11). C3b binds to the alternative C3 convertase to produce an enzyme that acts on the C5 component of complement and is called the **alternative C5 convertase**; it consists of Bb plus two C3b fragments and is designated $C3b_2Bb$ (Figure 2.12).

Complement component C5 is structurally similar to C3 but lacks the thioester bond and thus has a very different function. It is cleaved by the C5 convertase into a smaller C5a fragment and a larger C5b fragment (see Figure 2.12). The function of C5b is to initiate the formation of a **membrane-attack complex**, which can make holes in the membranes of bacterial pathogens and eukaryotic cells. In succession, C6 and C7 bind to C5b—interactions that expose a

The terminal complement components that form the membrane-attack complex		
Protein	**Concentration in serum (μg/ml)**	**Function**
C5	85	On activation the soluble C5b fragment initiates assembly of the membrane-attack complex in solution
C6	60	Binds to and stabilizes C5b. Forms a binding site for C7
C7	55	Binds to C5b6 and exposes a hydrophobic region that permits attachment to the cell membrane
C8	55	Binds to C5b67 and exposes a hydrophobic region that inserts into the cell membrane
C9	60	Polymerization on the C5b678 complex to form a membrane-spanning channel that disrupts the cell's integrity and can result in cell death

Figure 2.11 The terminal components of the complement pathway.

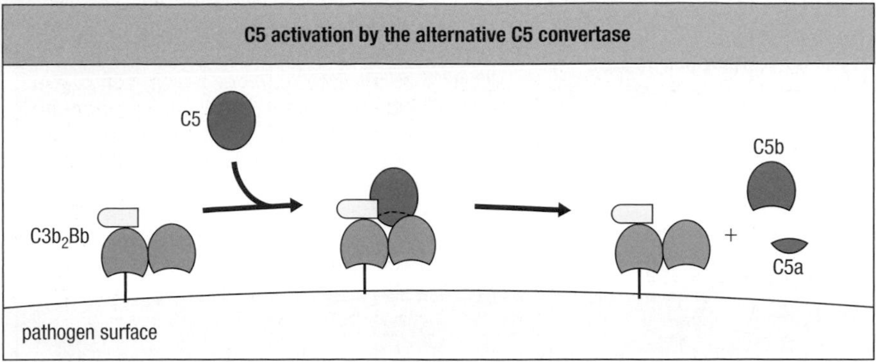

Figure 2.12 Complement component C5 is cleaved by C5 convertase to give a soluble active C5b fragment. The C5 convertase of the alternative pathway ($C3b_2Bb$) consists of two molecules of C3b and one of Bb. C5 binds to the C3b component of the convertase and is cleaved into fragments C5a and C5b, of which C5b initiates the assembly of the terminal complement components to form the membrane-attack complex.

hydrophobic site in C7, which inserts into the lipid bilayer. When C8 binds to C5b a hydrophobic site in C8 is exposed and, on insertion into the membrane, this part of C8 initiates the polymerization of C9, the component that forms the transmembrane pores (Figure 2.13). Figure 2.11 lists the components of the membrane-attack complex and summarizes their activities.

In the laboratory the perforation of membranes by the membrane-attack complex is dramatic, but clinical evidence demonstrating the importance of the components C5–C9 remains limited. The clearest effect of deficiency in any of these components is to increase susceptibility to infection by bacteria of the genus *Neisseria*, different species of which cause the sexually transmitted disease gonorrhea and a common form of bacterial meningitis. Inherited deficiency for some complement components is not uncommon. For example, 1 in 40 Japanese people are heterozygous for C9 deficiency, predicting that 1 in 1600 of them will be completely deficient in C9.

The activity of the terminal complement components on human cells is regulated by soluble and surface-associated proteins. The soluble proteins **S protein**, **clusterin**, and **factor J** prevent the soluble complex of C5b with C6 and C7 from associating with cell membranes. At the human cell surface, the proteins **homologous restriction factor** (**HRF**) and **CD59** (also called **protectin**) prevent the recruitment of C9 by the complex of C5b, C6, C7, and C8

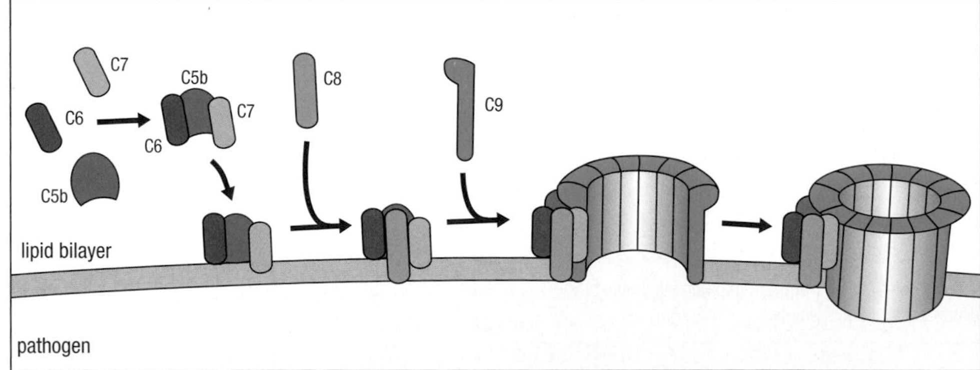

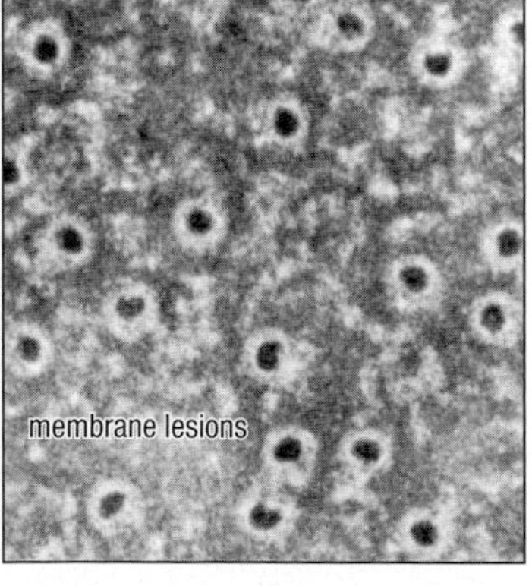

Figure 2.13 The membrane-attack complex assembles to generate a pore in the lipid bilayer membrane. The sequence of steps and their approximate appearance are shown here in schematic form. C5b is generated by the cleavage of C5 by the alternative C5 convertase $C3b_2Bb$. C5b then forms a complex by the successive binding of one molecule each of C6, C7, and C8. In forming the complex, C7 and C8 undergo conformational changes that expose hydrophobic sites, which insert into the membrane. This complex causes some membrane damage and also induces the polymerization of C9. As each molecule of C9 is added to the polymer, it exposes a hydrophobic site and inserts into the membrane. Up to 16 molecules of C9 can be added to generate a transmembrane channel 100 Å in diameter. The channel disrupts the bacterial outer membrane, killing the bacterium. In the laboratory, the erythrocyte is a convenient cell with which to measure complement-mediated lysis. The electron micrograph shows erythrocyte membranes with membrane-attack complexes seen end-on. Photograph courtesy of S. Bhakdi and J. Tranum-Jensen.

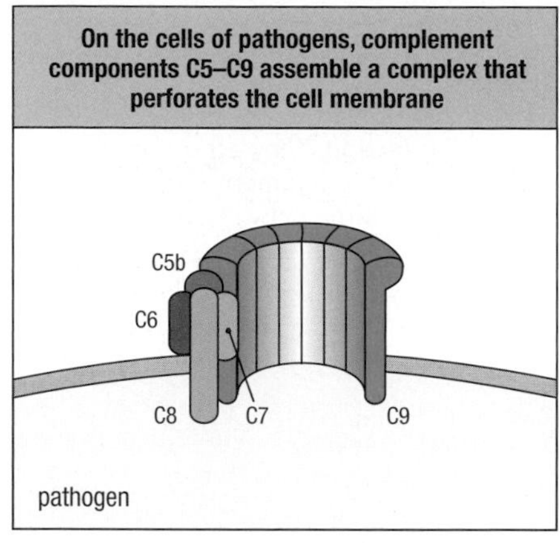

On the cells of pathogens, complement components C5–C9 assemble a complex that perforates the cell membrane
C5b
C6
C8
C7
C9
pathogen

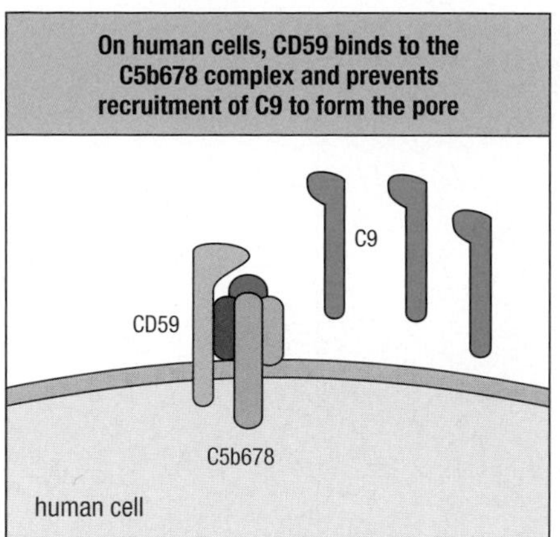

On human cells, CD59 binds to the C5b678 complex and prevents recruitment of C9 to form the pore
C9
CD59
C5b678
human cell

Figure 2.14 CD59 prevents assembly of the membrane attack complex on human cells. Left panel: the formation of a pore by the membrane attack complex (MAC) on a pathogenic microorganism. Right panel: how the human cell-surface protein CD59 prevents pore formation on human cells. By binding to the C5b678 complex, CD59 prevents the polymerization of C9 in the membrane to form a pore. Homologous restriction factor (HRF, not shown) works in the same way.

(Figure 2.14). DAF, HRF, and CD59 are all linked to the plasma membrane by glycosylphosphatidylinositol lipid tails. Impaired synthesis of this tail is the common cause of **paroxysmal nocturnal hemoglobinuria**, a disease characterized by episodes of complement-mediated lysis of red blood cells that lack cell-surface DAF, HRF, or CD59.

2-8 Small peptides released during complement activation induce local inflammation

During complement activation, C3 and C5 are each cleaved into two fragments, of which the larger (C3b and C5b) continue the pathway of complement activation. The smaller soluble C3a and C5a fragments are also physiologically active, increasing inflammation at the site of complement activation through binding to receptors on several cell types. Inflammation (see Section 1-4, p. 8) is a major consequence of the innate immune response to infection, which is also called the inflammatory response. In some circumstances, the C3a and C5a fragments induce anaphylactic shock, an acute inflammatory reaction that occurs simultaneously in tissues throughout the body; they are therefore referred to as **anaphylatoxins**. Of the anaphylatoxins, C5a is more stable and more potent than C3a. Phagocytes, endothelial cells, and mast cells have specific receptors for C5a and C3a. The two receptors are related and of a type that is embedded in the cell membrane and signals through the activation of a guanine-nucleotide-binding protein.

The anaphylatoxins induce the contraction of smooth muscle and the degranulation of mast cells and basophils, with the consequent release of histamine and other vasoactive substances that increase capillary permeability. They also have direct vasoactive effects on local blood vessels, increasing blood flow and vascular permeability. These changes make it easier for plasma proteins and cells to pass out of the blood into the site of an infection (Figure 2.15).

C5a acts directly on neutrophils and monocytes to increase their adherence to blood vessel walls, and serves as a chemoattractant to direct their migration toward sites of complement fixation. It also increases the phagocytic capacity of these cells, as well as raising the expression of CR1 and CR3 on their surfaces. In these ways, the anaphylatoxins act in concert with other complement components to speed the destruction of pathogens by phagocytes.

2-9 Several classes of plasma protein limit the spread of infection

In addition to complement, several other types of plasma protein impede the invasion and colonization of human tissues by microorganisms. Damage to

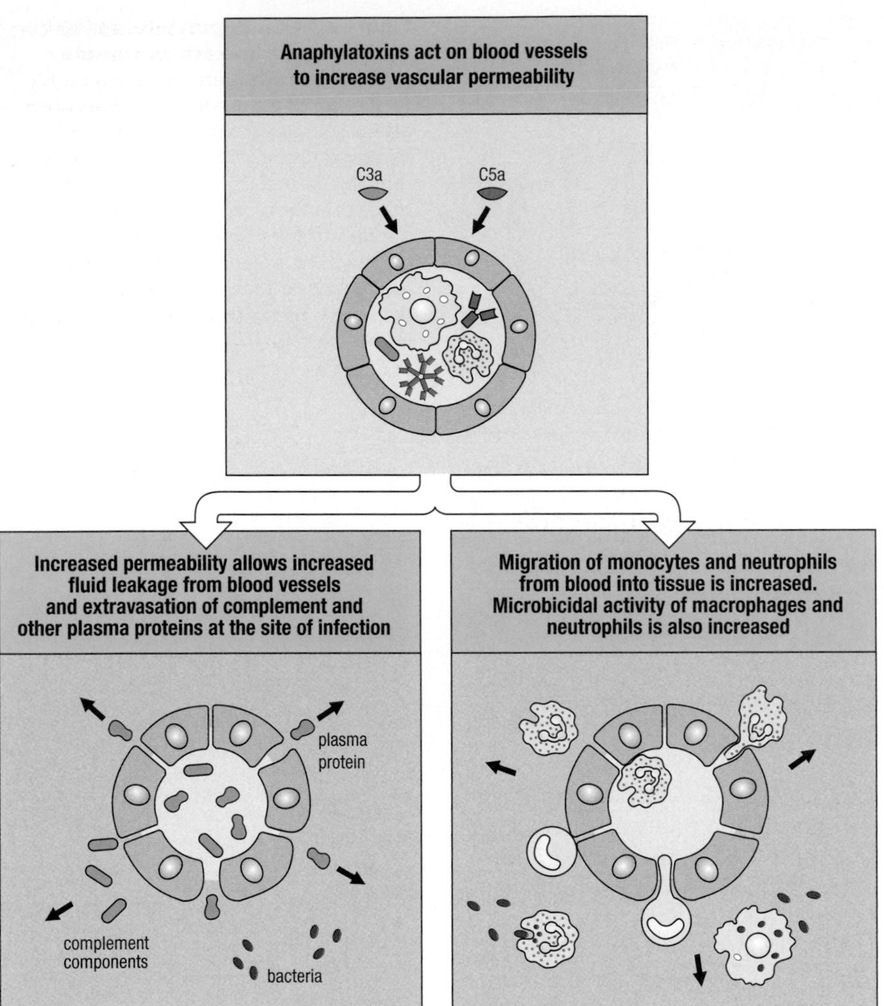

Figure 2.15 C3a and C5a contribute to local inflammatory responses. These small anaphylatoxic peptides are byproducts of complement fixation at sites of infection. They act on nearby blood vessels to augment the local inflammatory response.

blood vessels activates the **coagulation system**, a cascade of plasma enzymes that forms blood clots. Microorganisms are immobilized in the clots, which prevent them from entering the blood and lymph, as well as decreasing the loss of blood and fluid. Platelets are a major component of blood clots, and during clot formation they release a variety of highly active substances from their storage granules. These include prostaglandins, hydrolytic enzymes, growth factors, and other mediators that stimulate various cell types to contribute to antimicrobial defense, wound healing, and inflammation. Further mediators, including the vasoactive peptide bradykinin, are produced by the **kinin system**, a second enzymatic cascade of plasma proteins that is triggered by tissue damage. By causing vasodilation, bradykinin increases the supply of the soluble and cellular materials of innate immunity to the infected site.

As part of their invasive mechanism, many pathogens make cell-surface or secreted proteases. These enzymes break down human tissues and aid the pathogen's dissemination, and can inactivate antimicrobial proteins. In some instances the proteases are made by the pathogen; in others the pathogen hijacks a human protease for its own purposes. One example is the bacterium *Streptococcus pyogenes*, which acquires the human protease plasmin on its surface. In response, human secretions and plasma contain **protease inhibitors**, which constitute about 10% of serum proteins. Among these are the **α_2-macroglobulins**, glycoproteins with a molecular mass of 180 kDa that circulate as monomers, dimers, and trimers and inhibit a broad range of proteases. α_2-Macroglobulins have structural similarities to complement component C3, including the presence of an internal thioester bond. The

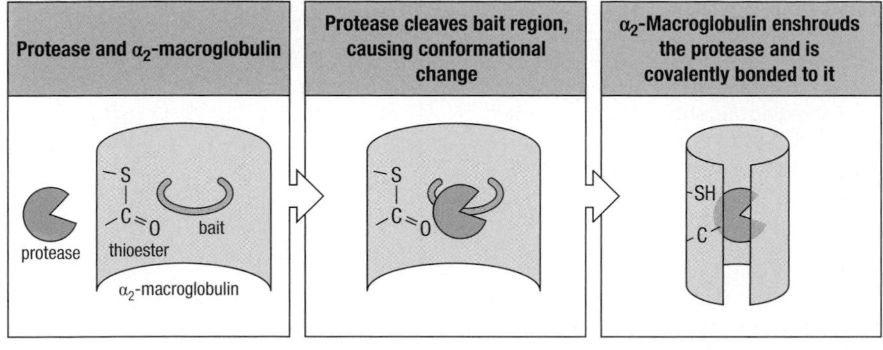

Figure 2.16 α_2-Macroglobulin inhibits potentially damaging proteases. The α_2-macroglobulins contain a highly reactive thioester bond (left panel) and inhibit microbial proteases. An α_2-macroglobulin first traps the microbial protease with a 'bait' region. When the protease cleaves the bait, the α_2-macroglobulin binds the protease covalently through activation of the thioester group (middle panel). It enshrouds the protease so that it cannot access other protein substrates, even though the protease is still catalytically active (right panel).

α_2-macroglobulin molecule lures a protease with a bait region that it is allowed to cleave. This activates the α_2-macroglobulin, producing two effects: first, the thioester is used to attach the protease covalently to the α_2-macroglobulin; second, the α_2-macroglobulin undergoes a conformational change by which it envelops the protease and prevents it attacking other substrates (Figure 2.16). The resulting complexes of protease and α_2-macroglobulin are rapidly cleared from the circulation by a receptor present on hepatocytes, fibroblasts, and macrophages.

2-10 Antimicrobial peptides kill pathogens by perturbing their membranes

Antimicrobial peptides are soluble effector molecules of innate immunity that kill a wide range of pathogens. The major family of human antimicrobial peptides is the **defensins**, peptides of 35–40 amino acids that are rich in arginine residues and which characteristically have three intra-chain disulfide bonds. They divide into two classes—the **α-defensins** and the **β-defensins**. The defensin molecule is amphipathic in character, meaning that its surface has both hydrophobic and hydrophilic regions. This property allows defensin molecules to penetrate microbial membranes and disrupt their integrity—the mechanism by which they destroy bacteria, fungi, and enveloped viruses (Figure 2.17).

Defensins are constitutively secreted at mucosal surfaces, where they protect the epithelium from infection by enteric pathogens and maintain the normal gut flora and microbiota. The α-defensins HD5 and HD6 (also called **cryptdins**) are secreted by Paneth cells, specialized epithelial cells of the small intestine that are situated at the base of the crypts between the intestinal villi (Figure 2.18). In addition, Paneth cells secrete other antimicrobial agents, including lysozyme, that contribute to innate immunity. The β-defensins are produced by a broad range of epithelial cells, in particular those of the skin, the respiratory tract, and the urogenital tract.

Defensins are also produced by neutrophils, the predominant phagocytes of innate immunity. Neutrophils are recruited to sites of infection throughout the body as a component of the induced innate immune response. The defensins are packaged in the neutrophil's granules and used to kill pathogens that have

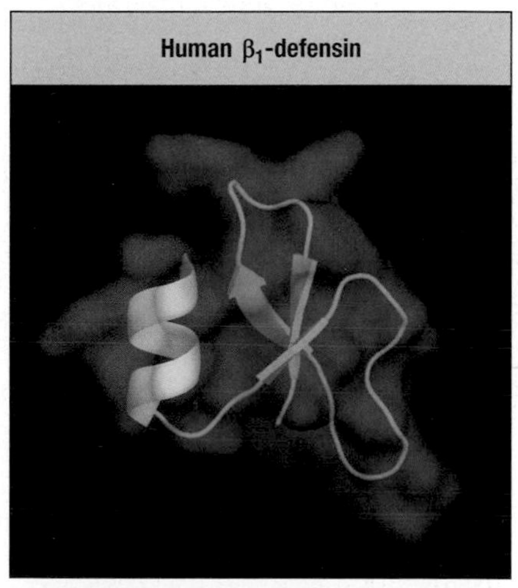

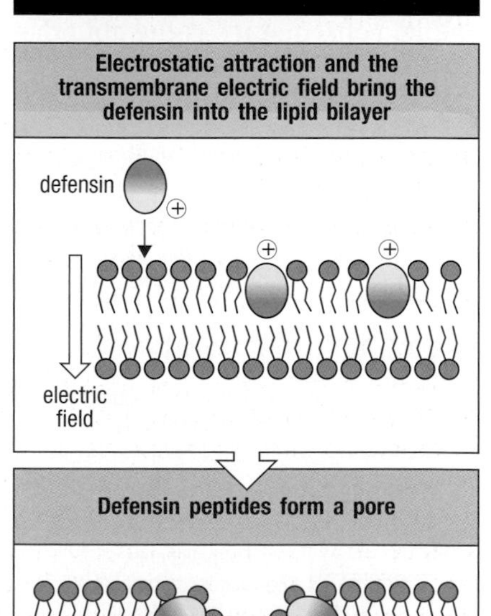

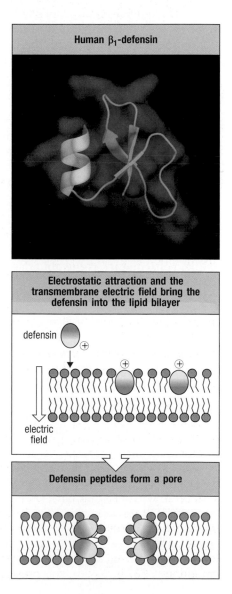

Figure 2.17 Defensins disrupt microbial membranes. The structure of human β_1-defensin is shown in the top panel. It is composed of a short segment of α helix (yellow) resting against three strands of antiparallel β sheet (green), generating an amphipathic peptide with separate regions having charged or hydrophobic residues. This feature allows defensins to interact with the charged surface of the cell membrane and then insert into the lipid bilayer (middle panel). This leads to the formation of pores and a loss of membrane integrity (bottom panel).

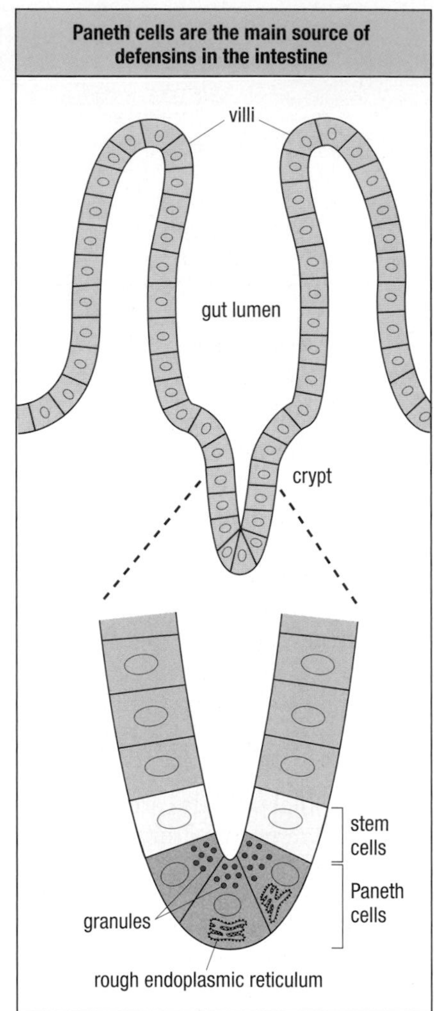

Figure 2.18 Paneth cells are located in the crypts of the small intestine. The α-defensins HD5 and HD6, also known as cryptdins, are made only by Paneth cells. The upper part of the diagram shows the location of a crypt between two villi in the distal part of the small intestine (ileum). The lower part of the diagram shows the Paneth cells at the base of the crypt and the epithelial stem cells that give rise to them. Paneth cells also secrete other antimicrobial factors, including lysozyme and phospholipase A2. Although they are of epithelial, not hematopoietic, origin, Paneth cells can be considered cells of the immune system.

been phagocytosed by the neutrophil. To prevent defensins from disrupting human cells, they are synthesized as part of longer, inactive polypeptides that are then cleaved to release the active fragment. Even then, they function poorly in the physiological conditions under which they were produced, needing the lower ionic strength of sweat, tears, the gut lumen, or the phagosome to become fully active.

The set of defensins that can be recruited to an innate immune response varies from one individual to another. There are at least six α-defensins and four β-defensins (Figure 2.19). The regions of the human genome that encode the defensins are even more variable because individuals differ in their number of copies of a defensin gene: from 2 to 14 copies for α-defensin genes and from 2 to 12 copies for β-defensin genes. The gene copy number determines the amount of protein made, with the result that the arsenal of defensins that a neutrophil carries varies from one person to another. Variation in the amino acid sequences of defensins correlates with their different skills in killing microorganisms. For example, the β-defensin HBD2 specializes in killing Gram-negative bacteria, whereas its relative HBD3 kills Gram-positive and Gram-negative bacteria. The purple Gram stain distinguishes between two large classes of medically important bacteria (see Figure 1.4). Defensins can also differ in the epithelial surfaces they protect: the α-defensin HD5 is secreted in the female urogenital tract, whereas β-defensin HBD1 is secreted in the respiratory tract as well as the urogenital tract.

Under pressure from human immunity, pathogens evolve ways to escape the defensins. In return, the pressures those pathogens impose on the human immune system select for new variants of human defensins that kill the pathogens more efficiently. This evolutionary cat-and-mouse game never ends, and its consequence is the abundance and variability of the defensin genes

Defensin		Site of synthesis	Tissues defended	Regulation of synthesis
Class	Name			
α	HNP1	Neutrophils > monocytes, macrophages, NK cells, B cells, and some T cells	Intestinal epithelium, placenta, and cervical mucus plug	Constitutive
α	HNP2			
α	HNP3			
α	HNP4	Neutrophils	Not determined	Constitutive
α	HD5	Paneth cells > vaginal epithelial cells	Salivary glands, gastrointestinal tract, eyes, female genital tract, and breast milk	Constitutive and induced by sexually transmitted infection
α	HD6	Paneth cells	Salivary glands, gastrointestinal tract, eyes, and breast milk	
β	HBD1	Epithelial cells > monocytes, macrophages, dendritic cells, and keratinocytes	Gastrointestinal tract, respiratory tract, urogenital tract, skin, eyes, salivary glands, kidneys, and blood plasma	Constitutive and induced by infection
β	HBD2			
β	HBD3			
β	HBD4	Epithelial cells	Stomach (gastric antrum) and testes	

Figure 2.19 Human defensins are variable antimicrobial peptides. Defensins are small antimicrobial peptides present at epithelial surfaces and in the granules of neutrophils. They form two families: the α-defensins and the β-defensins. HNP, human neutrophil protein; HD, human defensin; HBD, human β-defensin. The gastric antrum is that part of the stomach nearer the outlet and does not secrete acid.

Type	Name	Source	Ligands
Short pentraxin	Serum amyloid P component	Liver hepatocytes	Bacteria, viruses, fungi, parasites
Long pentraxin	PTX3	Monocytes Macrophages Dendritic cells Endothelial cells Epithelial cells	Bacteria, viruses, fungi

Figure 2.20 The two types of pentraxins. Pentraxins are effector molecules of innate immunity that play a similar role to the antibodies of adaptive immunity.

accumulated by the global human population. In comparison with most other genes in the genome, the defensin genes evolve rapidly. Such instability, or plasticity, is not unique to the defensin genes but also occurs in other gene families that encode pathogen-binding proteins of innate immunity.

2-11 Pentraxins are plasma proteins of innate immunity that bind microorganisms and target them to phagocytes

The **pentraxins** are a family of cyclic multimeric proteins that circulate in the blood and lymph and bind to the surfaces of various pathogens and target them for destruction. Defining the pentraxins is a characteristic 200-residue pentraxin domain at the carboxy-terminal end of the polypeptide. Two subfamilies of pentraxins are distinguished. The short pentraxins are made by hepatocytes in the liver and are represented by serum amyloid P component (SAP). The long pentraxins, represented by PTX3, are made by a variety of cells, including myeloid, endothelial, and epithelial cells, but not by liver hepatocytes (Figure 2.20). The pentraxins function as bridging molecules that bind pathogens with one binding site on their molecular surface and use a second site to bind to human cell-surface receptors, for example CD89 on phagocytes. When the cell-surface receptors are cross-linked by the pentraxin-coated pathogen, the phagocyte is signaled to engulf and destroy the pathogen. The pentraxins have a similar role in the innate immune response to that of antibodies in the adaptive immune response, and both pentraxins and antibodies bind to the same surface receptors on phagocytes.

Summary to Chapter 2

In confronting the microbial universe, the human body has several types of defense, which must all be overcome if a pathogen is to establish an infection and then exploit its human host for the rest of that unfortunate person's life. First and foremost are the protective epithelial tissues of the body—the skin and mucosal surfaces—and their commensal microorganisms, which successfully prevent most pathogens from ever gaining entry to the rich resources of the body's interior. Any pathogen that successfully penetrates an epithelial surface is immediately faced by the standing army of the innate immune response. These troops are largely soluble peptides and proteins, which are made constitutively and are present at mucosal surfaces or in the blood and extracellular spaces. Families of antimicrobial peptides called defensins provide a simple but effective mechanism for destroying some types of pathogen, but not all. Much more elaborate is the complement system, which provides a general means for tagging any pathogen with a molecular marker that ensures its destruction by resident macrophages in the infected tissue. Pentraxins can also consign pathogens to phagocyte-mediated destruction by simultaneously binding to pathogens and to cell-surface receptors on phagocytes. Inhibiting the pathogen's progress in colonizing tissues and spreading infection are the protease inhibitors, the blood-clotting cascade, and the kinin reactions. In total, an army of plasma proteins and cell-surface molecules provide systems for identifying microbiological invaders and distinguishing them from human cells. These immediate defenses of innate immunity are always available and do not improve with repeated exposure to the same pathogen.

Questions

2–1 Which of the following statements is incorrect?
a. Mucosal surfaces are better than skin surfaces at supporting colonization by commensal microorganisms.
b. The skin provides a larger surface area than mucosal surfaces for commensal microorganisms.
c. There are ten times as many bacteria residing in the intestinal tract than the number of cells in the human body.
d. It is common for commensal bacteria to live in symbiosis with their human hosts.
e. During gestation in mammals, a fetus does not have any commensal microorganisms on their skin or mucosal surfaces.

2–2 All of the following are associated with the skin except _____.
a. blood vessels
b. a low surface density of putative pathogens
c. specialized lymphoid tissues
d. lymphatics
e. an impenetrable physical barrier to microorganisms.

2–3 Which of these pairs are mismatched?
a. cytosol: intracellular pathogen
b. surface of epithelium: extracellular pathogen
c. nucleus: intracellular pathogen
d. lymph: intracellular pathogen

2–4 Explain why the type of innate immune response used by the host during an infection is dependent upon the particular location in which the pathogen resides.

2–5 All of the following are true of some or all complement proteins except _____.
a. they are soluble and bind to pathogen surfaces
b. they participate only in innate immunity
c. they are present in extracellular fluids, blood, and lymph
d. they facilitate the phagocytosis of pathogens
e. they are made as zymogens.

2–6 A(n) _____ is an inactive form of an enzyme that frequently participates in a cascade of enzymatic reactions during complement activation.
a. regulator of complement activation
b. convertase
c. complement control protein module
d. zymogen
e. opsonin.

2–7 Which of the following does not describe the outcome of complement activation?
a. chemotaxis
b. opsonization
c. vasoconstriction
d. proteolysis
e. membrane permeabilization.

2–8 All of the following are involved in the alternative pathway of complement activation except _____.
a. factor B
b. factor D
c. factor P (properdin)
d. C4
e. C5.

2–9 Match the term in column A with its description in column B.

Column A	Column B
a. classical pathway of complement activation	1. activated by C-reactive protein or antibody
b. $C3b_2Bb$	2. produces C5a and C5b
c. C3	3. contains a thioester bond that is susceptible to nucleophilic attack
d. properdin	4. protects C3bBb from protease degradation
e. anaphylatoxins	5. induce local and systemic inflammatory responses

2–10 The primary role of complement control proteins that operate in the early stages of complement activation is to _____.
a. regulate the expression of complement proteins
b. ensure that the components of the membrane-attack complex assemble in the correct sequence
c. facilitate the secretion of complement proteins to extracellular locations
d. stabilize complement proteins, thus extending their half-lives in serum
e. ensure that C3b is deposited on appropriate surfaces.

2–11 Which of the following is not an example of a complement control protein?
a. decay-accelerating factor (DAF)
b. factor H
c. factor B
d. membrane cofactor protein (MCP)
e. factor P (properdin)
f. factor I.

2–12 Match the term in column A with its description in column B.

Column A	Column B
a. α_2-macroglobulin	1. amphipathic peptides making up two classes (α and β) that penetrate and disrupt microbial membranes
b. defensins	2. vasoactive peptide facilitating the recruitment of innate immunity mediators
c. coagulation system	3. a soluble protease inhibitor found in human plasma
d. bradykinin	4. inhibits dissemination of pathogens by forming blood clots
e. cryptdins	5. α-defensins made by Paneth cells

2–13 At which anatomical location do Paneth cells reside?
a. crypts of the intestinal tract
b. the lining of blood vessels
c. in the urogenital tract
d. in the liver adjacent to Kupffer cells
e. alveolar spaces of the respiratory tract.

2–14 John Binstead, 27 years old, has noticed that his urine was cola-colored in the morning for the past 5 days, and today he experienced a sudden onset of severe abdominal pain. Physical exam, blood tests, and urine analysis show that John is anemic as the result of intravascular hemolysis. A Coombs test is negative, ruling out autoimmune hemolytic anemia. Flow cytometric analysis of peripheral blood cells for CD235a (glycophorin A, present on erythroid precursors and mature red blood cells) together with CD55 (decay-accelerating factor) and CD59 (protectin) reveals the absence of CD55 and CD59 on 54% of his red blood cells. Molecular analysis confirms a defective *PIGA* gene, which encodes the protein phosphatidylinositol glycan class A, which is involved in producing glycophosphatidylinositol anchors for many different proteins, including CD55 and CD59. John is diagnosed with paroxysmal nocturnal hemoglobinuria. A long-term treatment regimen including the administration of intravenous eculizumab, an anti-C5 humanized monoclonal antibody, is recommended. Which of the following provides the rationale for selecting eculizumab for the long-term management of John's condition?
a. The inhibition of anaphylatoxins such as C5 decreases the inflammation that is responsible for hemolytic anemia.
b. By blocking the activity of C5, the early events of complement will not be able to produce a functional C5 convertase.
c. GPI anchors will form correctly if C5 is inhibited.
d. A defect in CD59 function renders red blood cells especially susceptible to lysis if the membrane-attack complex is able to form.
e. None of the above.

2–15 Jonathan Miller, aged 6 years, was brought to emergency room by his parents; he was presenting with fever, severe headache, a petechial rash, stiff neck and vomiting. Jonathan had a history of recurrent sinusitis and otitis media, all caused by pyogenic bacteria and treated successfully with antibiotics. Suspecting bacterial meningitis, the attending physician began an immediate course of intravenous antibiotics and requested a lumbar puncture. *Neisseria meningitidis* was grown from the cerebrospinal fluid. The physician was concerned about the recurrence of infections caused by pyogenic bacteria, and he suspected an immunodeficiency. He ordered blood tests and found the serum complement profiles to have low C3, factor B, and factor H, and undetectable factor I. Which of the following explains why a factor I deficiency is associated with infections caused by pyogenic bacteria?
a. Elevated levels of C3 convertase C3bBb interfere with the activation of the classical pathway of complement activation.
b. Rapid turnover and consumption of C3 in the serum causes inefficient fixation of C3b on the surface of pathogens, compromising opsonization and phagocytosis.
c. Factor I is an opsonin that facilitates phagocytosis.
d. Factor I is a chemokine and is important for the recruitment of phagocytes.
e. Factor I is required for assembly of the terminal components of the complement pathway.

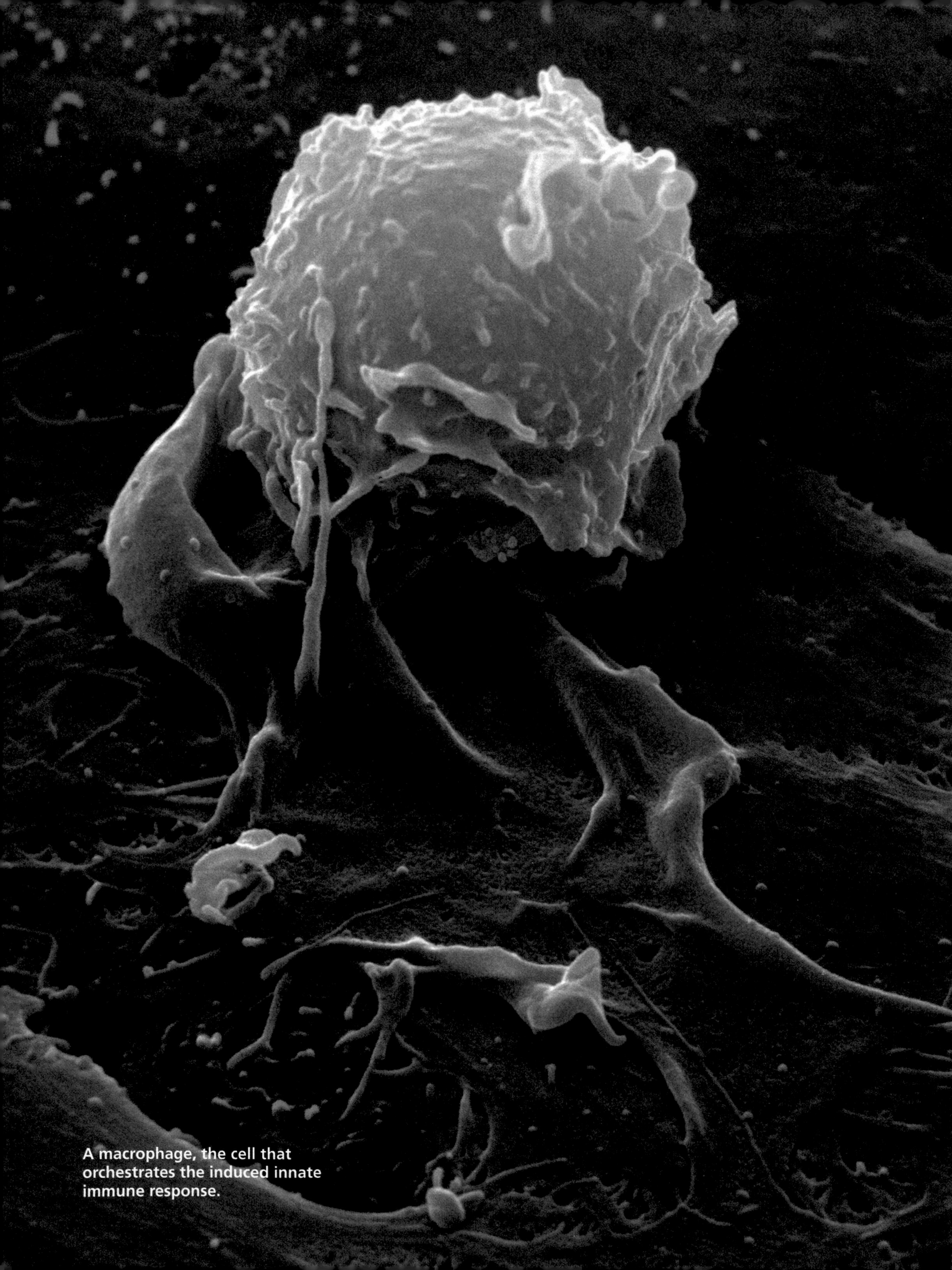

A macrophage, the cell that orchestrates the induced innate immune response.

Chapter 3

Innate Immunity: the Induced Response to Infection

The human body has several successive lines of defense, all of which must be overcome if a pathogen is to establish a long-term infection in a human host. As described in Chapter 2, when a pathogen breaches one of the protective epithelial surfaces of the body and gains access to the soft internal tissues, it encounters a standing army in the form of the immediate defenses of innate immunity. If these cellular and molecular defences are successful in repelling the pathogen within a few hours, the effects of the infection, and of the immune response against it, are small, and may not even be perceived by the human host. Most people have these minor skirmishes with microorganisms on a daily basis, particularly in childhood. This chapter considers what happens if the pathogen is able to outrun the immediate defenses of innate immunity, simply by expanding its population at a greater rate than the defenses can kill the pathogens or eliminate them from the human body.

The next lines of defense, the subject of this chapter and placed in context in Figure 3.1, have to be induced by the infection. They take about 4 days to be ordered up, marshaled and brought into service, processes that require considerable investment of resources and time. This later phase of innate immunity involves soluble and cellular receptors that detect the presence of infecting organisms and then recruit leukocytes to do something about it. The integrity of blood vessels is deliberately damaged by inflammation so that a vast army of destructive cells can freely enter the infected tissue to confront the infecting pathogen. Now a battleground, the infected tissue becomes flooded with plasma and the temperature rises. Even the most stoical person becomes aware that something is wrong.

The internal conflict begins to interfere with daily life, as material and energy resources normally used for work, thought, and play are diverted to fighting the infection. The symptoms of disease are felt and they are almost all indirect consequences of the innate immune response, rather than effects wrought by the pathogen itself. Although there is discomfort, and damage to the infected tissue, which will take time to be repaired, most infections reaching this stage are resolved successfully by the induced innate immune response. For this reason, there are built-in mechanisms for preventing the stimulation of the even more expensive and time-consuming adaptive immune response until it is absolutely necessary.

3-1 Cellular receptors of innate immunity distinguish 'non-self' from 'self'

The innate immune system has evolved an extensive array of receptors to cope with the numerous microbial species that inhabit and exploit the human body.

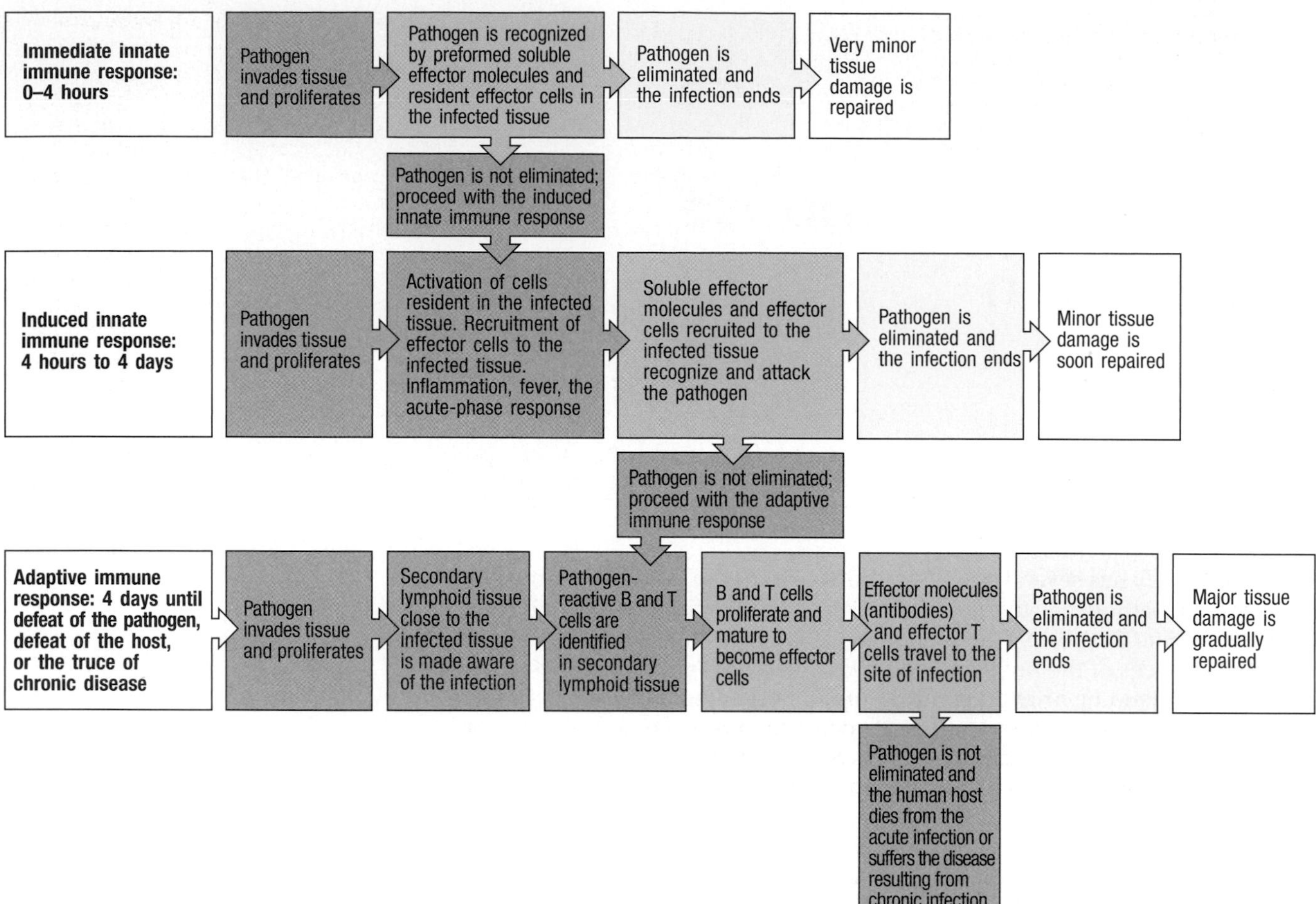

Figure 3.1 The human immune response to an invading pathogen can consist of one, two, or three phases, depending upon the severity of the infection. The immediate innate immune response, covered in Chapter 2, is the initial phase and begins as soon as a pathogen invades. This is sufficient to terminate most infections, but when that does not happen, the induced innate immune response, the topic for this chapter of the book, is brought into play as the second phase. If after 4 days of infection the combined actions of the immediate and induced innate immune responses have not been sufficient to subdue the infection, the immune system commits to make an adaptive immune response. This response, which is made to order for the infection at hand, is more powerful than the forces of innate immunity alone, and efficiently terminates almost all infections that get to this third phase. It is also more damaging to the tissues in which it is working. If adaptive immunity fails to suppress the pathogen, the infected human host will either die of the acute infection, as frequently occurs for Ebola virus infections, or develop chronic infection, as occurs with human immunodeficiency virus (HIV), associated with debilitating disease and premature death.

These receptors are expressed by macrophages, NK cells, and other cells of innate immunity. A general property of these innate immune receptors is they recognize structural features that distinguish microbial carbohydrates, lipids, proteins, and nucleic acids from their mammalian counterparts. Consequently, the receptors and the cells that carry them are able to distinguish between the healthy cells of the human body—or 'self'—and 'non-self,' which includes all bacteria, fungi, viruses, and parasites, as well as unhealthy infected human cells (Figure 3.2). This discrimination enables the induced innate immune response to be directed with precision at the invading organism and those human cells that become corrupted by its presence.

Each type of receptor can recognize multiple pathogenic species that share, for example, a particular type of cell-surface carbohydrate. This enables a wide range of pathogens to be recognized by using a relatively limited number of receptors. Receptor cross-reactions usually occur between microbial species that are closely related evolutionarily, but they can also involve more distantly

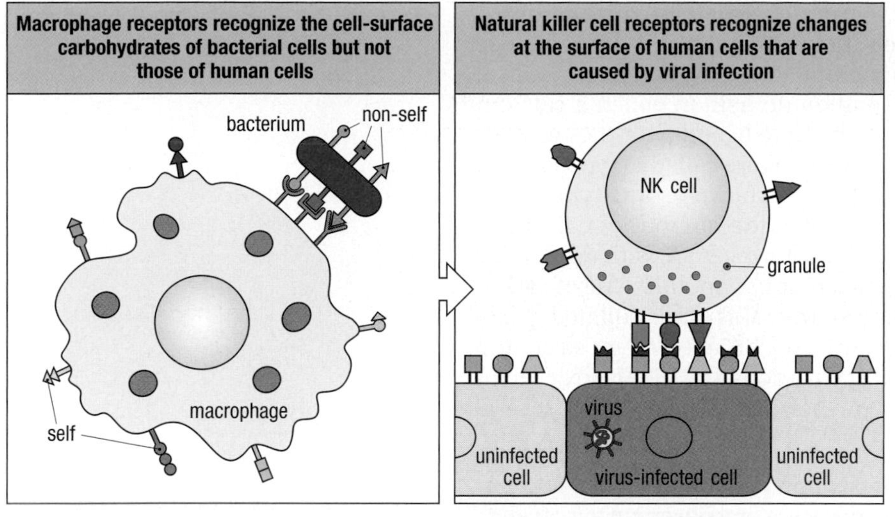

Figure 3.2 Cell-surface receptors of innate immune cells that detect pathogens distinguish between 'self' and 'non-self.' In the left-hand panel, the surface receptors of a macrophage are shown binding to cell-surface carbohydrates that abound on bacterial surfaces but are not on human cell surfaces. In the right-hand panel, the surface receptors of a natural killer (NK) cell, a type of lymphocyte that mediates innate immunity, are shown interacting with molecules that are present on the surface of virus-infected cells but not on uninfected cells. NK-cell receptors are able to detect the features that distinguish virus-infected cells from healthy uninfected cells.

related species that have some structural feature in common. There are more than 100 different innate immune receptors, and each type of innate immune cell expresses only a subset of them. Even within the effector cells of a particular functional type, for example macrophages or NK cells, individual cells can express different combinations of receptors, giving rise to functional heterogeneity in the population of cells. Such diversity increases the probability that some fraction of the cells will be able to respond effectively to any given pathogen, even one to which the human population has had no previous exposure.

3-2 Tissue macrophages carry a battery of phagocytic and signaling receptors

All tissues of the body contain resident macrophages, which are ready to attack any encroaching microorganism, whether it is a commensal or a pathogen. Any commensal gaining access to a tissue becomes a potential pathogen. An array of receptors on the macrophage surface, including the complement receptors CR1 and CR2, work together to bind the invaders and trigger their phagocytosis by the macrophage. Many of the ligands recognized by these so-called **phagocytic receptors** are bacterial carbohydrates and lipids (Figure 3.3).

Macrophage receptor	Phago-cytosis	Signaling	Target pathogens	Ligand
Mannose receptor (CD206)	+	–	Bacteria	LPS, CPs, ManLam
Complement receptors 3 and 4 (Mac-1, CD11b/CD18)	+	–	Bacteria	Oligosaccharides, proteins
			Fungi	β-Glucans
Dectin-1	+	–	Bacteria	Mycobacterial ligand
			Fungi	β-Glucans
Macrophage receptor with collagenous structure (MARCO)	+	–	Bacteria	LPS, proteins
Scavenger receptor A (SR-A)	+	–	Bacteria	LPS, LTA, proteins, CpG DNA
Scavenger receptor B (SR-B) (CD36)	+	–	Bacteria	Diacylated lipopeptide
Lipopolysaccharide receptor (CD14)	+	–	Bacteria	Peptidoglycan, LPS, LTA, mannuronic acid
Toll-like receptors (TLR)	–	+	A family of 10 receptors with variable specificity for a range of pathogens	

Figure 3.3 Some phagocytic and signaling receptors of the macrophage. Lipopolysaccharide (LPS), bacterial capsular polysaccharides (CPs), mannosylated lipoarabinomannan (ManLam), lipoteichoic acid (LTA).

As a group, cell-surface receptors and plasma proteins that recognize carbohydrates are called **lectins**, of which the **mannose receptor** and **dectin-1** (see Figure 3.3) are two examples present on macrophages. They belong to a subset of lectins that share a carbohydrate-recognition domain in which a calcium ion coordinates the interaction of the carbohydrate ligand with the receptor protein. Because of the role of calcium, these lectins are called **C-type lectins**, and their characteristic ligand-binding domain is known as a **C-type lectin domain** (**CTLD**). Different C-type lectins differ in the number of CTLDs, as exemplified by dectin-1, with one, and the mannose receptor, with eight. Over and above the eight CTLDs, the mannose receptor has another type of carbohydrate-recognition domain at its amino terminus. This binds sulfated galactosamine residues and is called an R-type lectin because of its structural similarity to the binding domain of ricin, the infamous and highly toxic lectin of the castor bean that has been sent in 'poison pen' letters to several politicians.

Another group of cell-surface receptors present on macrophages, the **scavenger receptors** (**SR**), are so named because they were first observed to scavenge damaged molecules of low-density lipoprotein from the blood. Only subsequently were they shown to recognize an assortment of negatively charged microbial ligands. Scavenger receptors **SR-A** and **SR-B** are structurally unrelated, as are their microbial ligands. SR-B recognizes lipopeptides, whereas SR-A recognizes the lipopolysaccharide (LPS) of Gram-negative bacteria and the teichoic acids of Gram-positive bacteria, as well as CpG-rich bacterial DNA. SR-A illustrates how cell-surface receptors of innate immunity can bind physically similar but chemically different ligands.

Segments of collagen-like triple helix are a structural feature of several innate immunity receptors, including SR-A and the **macrophage receptor with collagenous structure** (**MARCO**), another scavenger receptor that recognizes Gram-negative and Gram-positive bacteria. The triple helices form rigid rods that serve as binding sites for some microbial ligands, but more generally they allow the carbohydrate-recognition domains and other ligand-binding sites to be held at a distance from the cell surface, away from the clutter of macromolecules at the macrophage's plasma membrane. This gives the receptor greater access to pathogen surfaces.

In addition to binding to the complement fragment iC3b (see Section 2-5), the complement receptors **CR3** and **CR4** recognize a range of ligands on microbial surfaces. These include LPS, the lipophosphoglycan of the protozoan parasite *Leishmania*, the filamentous hemagglutinin protein of the bacterium *Bordetella pertussis*, and polymers of glucose called glucans on pathogenic yeasts such as *Candida* and *Histoplasma*. CR3 and CR4 are members of a family of structurally similar proteins called **integrins** that contribute to adhesive interactions between cells. Such interactions enable cells of the immune system to communicate with each other and with other types of cell.

Many of the ligands recognized by macrophage receptors are arrayed at high density on the microbial surface. This feature enables numerous receptor molecules to bind their ligands in a cooperative fashion. The consequence is an irreversible attachment between the pathogen and the macrophage. Having securely captured the pathogen, the macrophage then initiates a process of engulfment called **receptor-mediated endocytosis**, in which the receptor-bound pathogen is surrounded by the macrophage membrane and internalized into a membrane-bound vesicle called an **endosome** or a **phagosome**. Phagosomes then fuse with the cellular organelles called **lysosomes** to form **phagolysosomes**, acidic vesicles that are loaded with degradative enzymes and toxic substances for destroying the pathogens (Figure 3.4).

Complementing the phagocytic receptors are signaling receptors that, by recognizing the pathogen, instruct the macrophage to recruit additional cells of

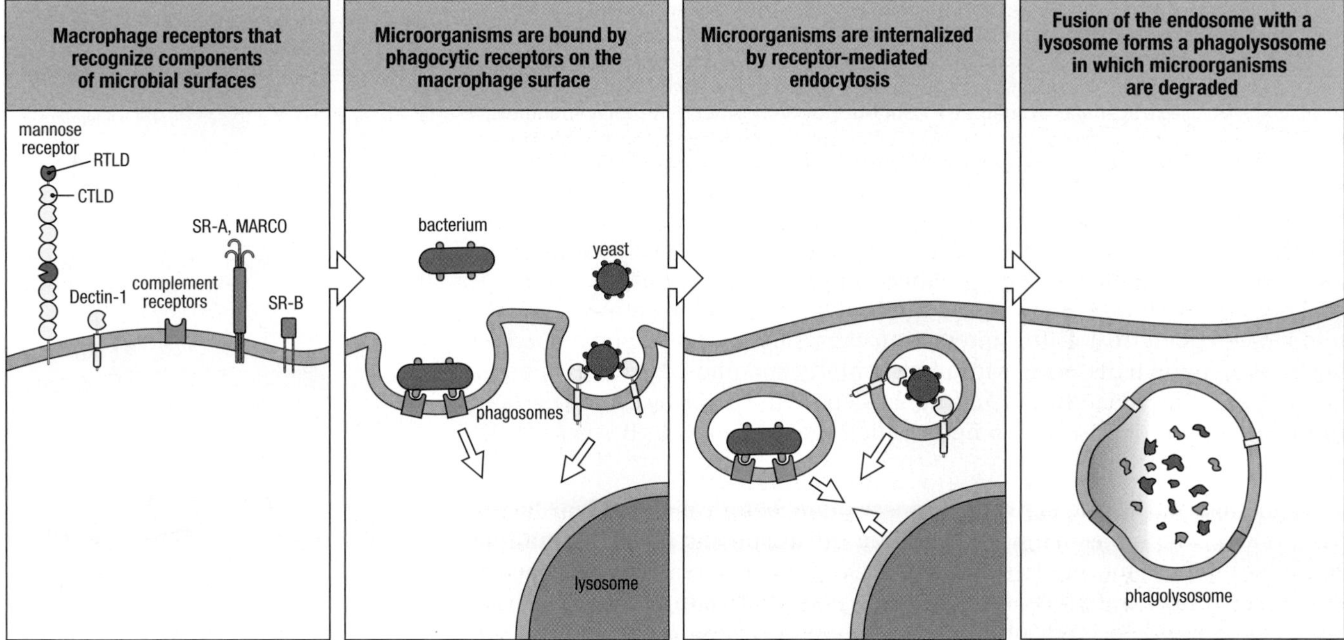
Macrophage receptors that recognize components of microbial surfaces
Microorganisms are bound by phagocytic receptors on the macrophage surface
Microorganisms are internalized by receptor-mediated endocytosis
Fusion of the endosome with a lysosome forms a phagolysosome in which microorganisms are degraded
mannose receptor
RTLD
CTLD
SR-A, MARCO
Dectin-1
complement receptors
SR-B
bacterium
yeast
phagosomes
lysosome
phagolysosome

Figure 3.4 **Phagocytosis and degradation of invading microorganisms by a tissue macrophage.** The ricin-like lectin domain (RTLD) and one of the C-type-lectin domains (CTLD) of the mannose receptor are indicated.

innate immunity to the infected tissue. Well-studied signaling receptors are the family of **Toll-like receptors (TLRs)**, which recognize a variety of microbial ligands and are expressed by different types of innate immune cell. Of the Toll-like receptors, TLR4 is expressed by macrophages and associates with CD14 to recognize LPS. In the next section, TLR4 is used to exemplify the structure and function of Toll-like receptors.

3-3 Recognition of LPS by TLR4 induces changes in macrophage gene expression

A family of 10 genes encodes the Toll-like receptor proteins TLR1–TLR10. These transmembrane proteins have a variable extracellular domain for recognizing pathogens, and a conserved, cytoplasmic signaling domain for conveying this information to the inside of the cell. The signaling domain is of a type called a **TIR** (**T**oll **I**nterleukin-1 **R**eceptor) domain because it was first observed in Toll-like receptors and the receptor for the inflammatory cytokine interleukin-1, and subsequently in other proteins. The pathogen-recognition domain of the Toll-like receptor protein consists of a repeated sequence motif of 20–29 amino acid residues, which is rich in the hydrophobic amino acid leucine and is termed a **leucine-rich repeat region (LRR)**. Variation in the number of LLRs, as well as their amino acid sequences, gives the Toll-like receptors their different specificities for microbial ligands. The functional Toll-like receptors all consist of two Toll-like receptor proteins, which can be either homodimers or heterodimers. The TLR4 protein only associates with itself, to give the homodimeric TLR4 receptor. Each of the two pathogen-recognition domains of TLR4 has a horseshoe-shaped structure and can bind one molecule of LPS (Figure 3.5).

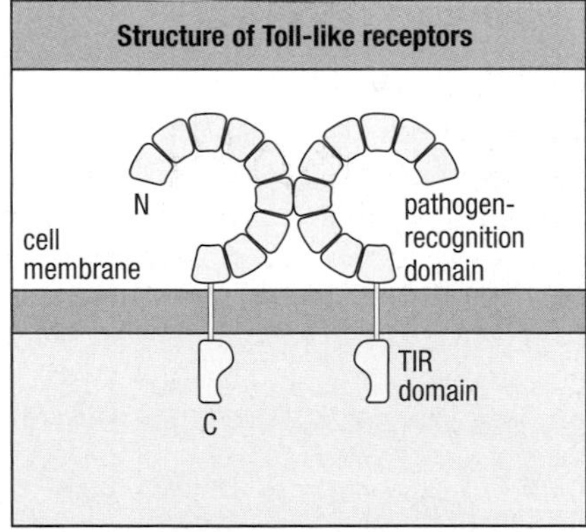
Structure of Toll-like receptors
N
cell membrane
pathogen-recognition domain
TIR domain
C

Figure 3.5 **Toll-like receptors sense infection by means of a horseshoe-shaped structure.** A Toll-like receptor (TLR) protein is a transmembrane polypeptide with a Toll–interleukin receptor (TIR) signaling domain on the cytoplasmic side of the membrane and a horseshoe-shaped sensor domain on the other side. Functional receptors can be homodimers (as shown here) or heterodimers of TLR polypeptides. The ten Toll-like receptor proteins, TCR 1–10, are also called CD281–CD290.

In a tissue infected with Gram-negative bacteria, macrophages bind the bacteria with their phagocytic receptors. When LPS is released from the bacterial surfaces, it is bound by a protein on the macrophage surface called CD14, which acts as a co-receptor to TLR4. Alternatively, LPS can be picked up by a soluble LPS-binding protein in the plasma and delivered to CD14 on the macrophage surface. The TLR4 dimer associates with a protein called MD2, and together they form a complex with CD14 and LPS (Figure 3.6). Such extracellular recognition of the pathogen causes the cytoplasmic TIR domain of TLR4 to

Figure 3.6 TLR4 recognizes bacterial lipopolysaccharide with help from other proteins. Bacterial lipopolysaccharide (LPS) is recognized by a complex of the TLR4, MD2, and CD14 proteins at the cell surface. MD2 is a soluble protein that associates with the extracellular domains of TLR4 but not with other TLR family members, and confers sensitivity to LPS. The soluble lipopolysaccharide-binding protein (LBP) can also deliver LPS to this cell-surface complex.

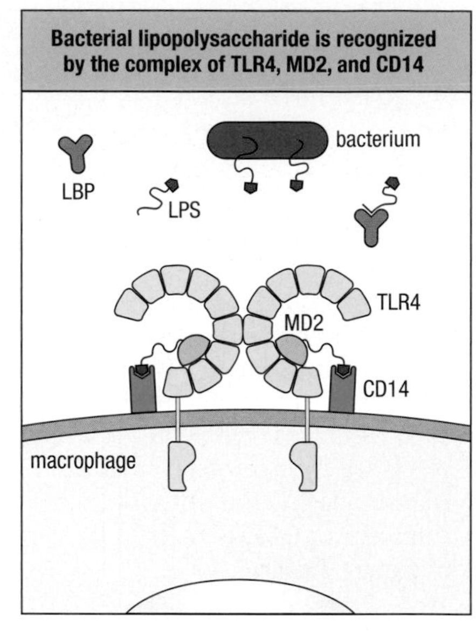

bind to a similar TIR domain of the MyD88 protein. MyD88 is an example of an **adaptor protein**, which brings together two signaling components. The second domain of MyD88 now binds to the next member of the pathway—a protein kinase called IRAK4. This interaction causes the kinase to phosphorylate itself, whereupon it dissociates from the complex and phosphorylates another adaptor protein called TRAF6. Additional steps in the pathway eventually lead to the activation of a kinase complex called the inhibitor of κB kinase (IKK) (Figure 3.7).

The function of IKK is to activate a transcription factor called **nuclear factor κB (NFκB)**, which has major roles in both the innate and adaptive immune responses. When not required, NFκB is held in the cytoplasm in an inactive complex with the inhibitor of κB (IκB). By phosphorylating IκB, IKK causes NFκB to be released from inhibition, whereupon it is translocated from the

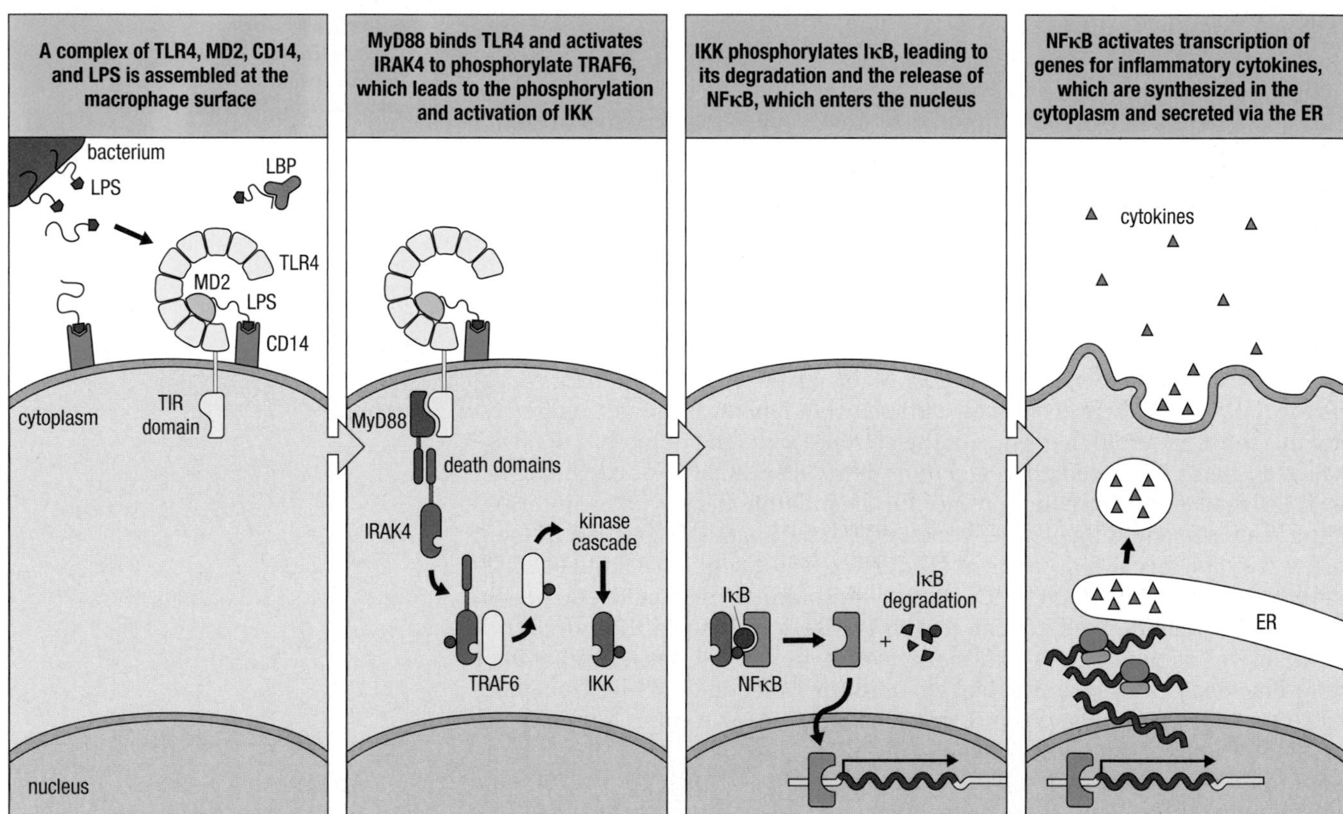

Figure 3.7 Sensing of LPS by TLR4 on macrophages leads to activation of the transcription factor NFκB and the synthesis of inflammatory cytokines. First panel: LPS is detected by the complex of TLR4, CD14, and MD2 on the macrophage surface. Second panel: the activated receptor binds the adaptor protein MyD88, which binds the protein kinase IRAK4. IRAK4 binds and phosphorylates the adaptor TRAF6, which leads via a kinase cascade to the activation of IKK. Third panel: in the absence of a signal, the transcription factor NFκB is bound by its inhibitor, IκB, which prevents it from entering the nucleus. In the presence of a signal, activated IKK phosphorylates IκB, which induces the release of NFκB from the complex; IκB is degraded. NFκB then enters the nucleus, where it activates genes encoding inflammatory cytokines. Fourth panel: cytokines are synthesized from cytokine mRNA in the cytoplasm and secreted via the endoplasmic reticulum (ER). This MyD88–NFκB pathway is also stimulated by the receptors for cytokines IL-1 and IL-18.

cytoplasm to the nucleus. Once there, it initiates the transcription of genes for cytokines, adhesion molecules, and other proteins that are necessary to establish a state of inflammation in the infected tissue.

Children with a rare genetic disease called **X-linked hypohidrotic ectodermal dysplasia and immunodeficiency**, or **NEMO deficiency**, lack one of the subunits of IKK and thus have impaired activation of NFκB. This makes them susceptible to bacterial infections because their capacity to activate macrophages through TLR4 signaling is abnormally low. The gene for the kinase subunit, called IKKγ or NFκB essential modulator (NEMO), is on the X chromosome, and so this syndrome is more frequent in boys than girls. Boys inherit only one copy of the X chromosome, whereas girls inherit two and thus both would have to be defective for disease to be manifest. NFκB has functions in development as well as in immunity, and so other consequences of IKKγ deficiency are abnormalities in the development of tissues derived from embryonic ectoderm: the skin, teeth, and hair (Figure 3.8). These more visible symptoms of disease help in the identification and diagnosis of this immunodeficiency disease.

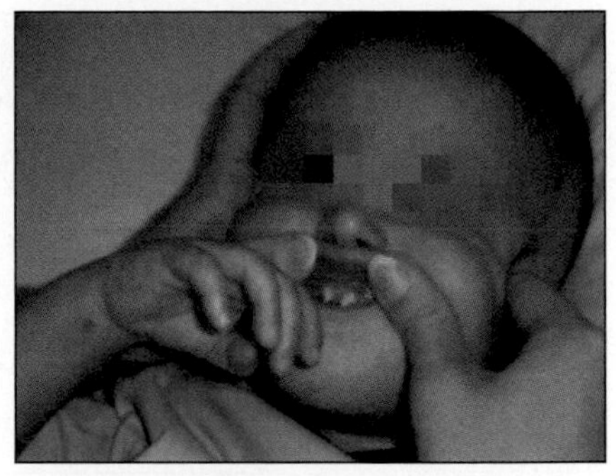

Figure 3.8 Infant with X-linked ectodermal dysplasia and immunodeficiency. This genetic disease is caused by impairment of NFκB activation due to lack of IKKγ. This causes immunological and developmental defects. Among the unusual physical features of these patients are conical or missing teeth. Photograph courtesy of F. Rosen and R. Geha.

3-4 Activation of resident macrophages induces a state of inflammation at sites of infection

Cytokines are small soluble proteins used as a means of communication between cells. In response to an external stimulus such as infection, one type of cell secretes a cytokine that binds to a specific cytokine receptor on the surface of another type of cell. Binding of cytokine to the receptor induces intracellular signals that change the behavior of the second cell. In general, cytokines are short-lived molecules that exert their influence within a short distance from the cell that made them, and in some instances the cytokine-secreting cell makes direct contact with the cell it will influence. These properties of cytokines ensure that the immune response, which inevitably causes collateral damage to tissues, is restricted to where it is needed.

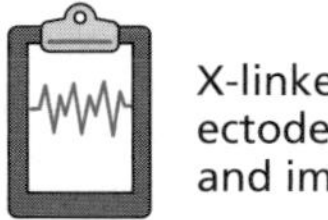

X-linked hypohidrotic ectodermal dysplasia and immunodeficiency

When infection has been detected in a tissue, the resident macrophages become activated to recruit other cells to the infected tissue by secreting several cytokines. Prominent among these are **interleukin-1β (IL-1β)**, **interleukin-6 (IL-6)**, **interleukin-12 (IL-12)**, **CXCL8**, and **tumor necrosis factor-α (TNF-α)**. Collectively, these cytokines are called **inflammatory cytokines**, or **pro-inflammatory cytokines**, because their combined effect is to create a state of **inflammation** in the infected tissue (Figure 3.9). Inflammation causes the infected tissue to swell and become red, painful, and hot. Although these are the symptoms we associate with infections and wounds, they are actually caused by the immune response to those insults. Inflammation is a familiar and universal experience for us all, but especially for children with their frequent coughs and colds as well as daily bangs, bumps, and scrapes. The contributions of the five inflammatory cytokines will be summarized here and expanded upon in subsequent sections.

In general, the cells of the immune system have some limited access to healthy tissue, but for neutrophils—the destructive phagocytes that do the hardest work in clearing infection—all access is denied. A major task of the inflammatory cytokines is to reverse this situation. When that happens, not only are neutrophils able to migrate from blood to the infected tissue, but they are actively encouraged to do so and guided to their destination. Cytokines IL-1β and TNF-α induce changes to the endothelial cells of blood vessels in the infected tissue that allow fluid and cells to leave the blood. This involves an increase in the diameter of the blood vessels (dilation), with a commensurate decrease in the blood flow. Cytokine-induced changes to the surface molecules of the vascular endothelium instruct neutrophils and other leukocytes to

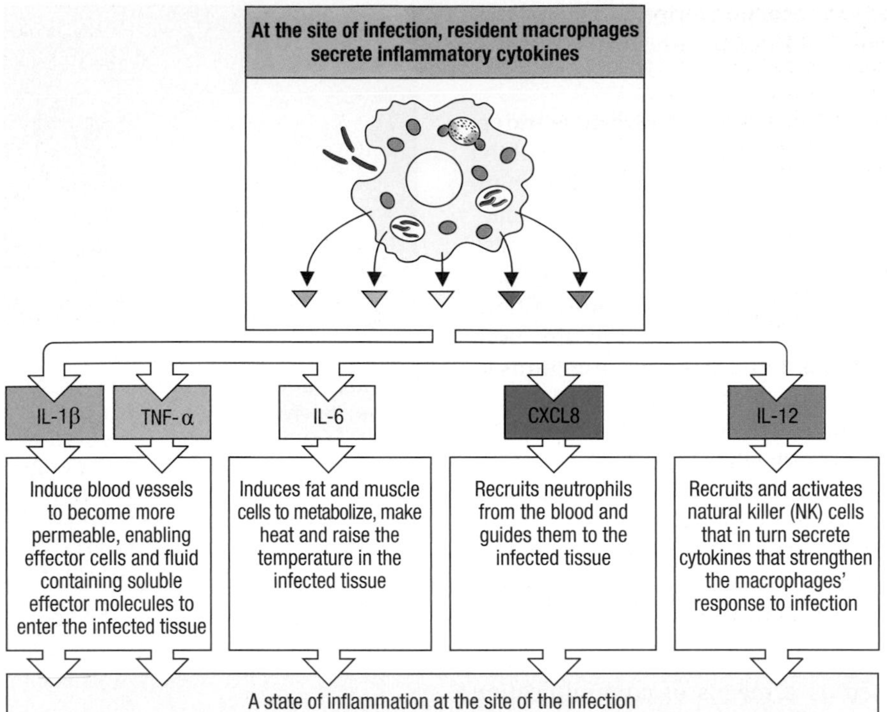

Figure 3.9 Macrophages respond to infection by secreting inflammatory cytokines. IL-1β, TNF-α, IL-6, CXCL8, and IL-12 are the five key cytokines that macrophages secrete. The cytokines recruit effector cells and plasma proteins to the infected tissue, where they work together to create a state of inflammation.

stop and exit from the blood. CXCL8 is a chemoattractant cytokine, or **chemokine**, which attracts neutrophils away from blood and toward the infected area where macrophages are secreting inflammatory cytokines. This is achieved by a chemokine receptor on neutrophils that binds CXCL8, which then signals the cells to move up the concentration gradient of the chemokine.

Monocytes (macrophage precursors) and NK cells are also recruited to the infected tissue, although in much smaller numbers than neutrophils. Cytokine IL-12 induces NK cells to proliferate and secrete cytokines that sustain macrophage activation. Monocytes recruited from the blood mature into macrophages in the infected tissue, adding to the resident population. As the inflammation develops, a major task of these macrophages is the phagocytosis and disposal of the immense number of neutrophils that die in the cause of containing infection. The increased passage of fluid and cells from the blood into the surrounding connective tissues causes the characteristic swelling, reddening, and pain that is associated with inflammation. The increase in temperature is a consequence of the cytokine IL-6, which acts on local muscle and fat cells, causing them to adjust their metabolism and generate more heat.

3-5 NOD-like receptors recognize bacterial degradation products in the cytoplasm

Complementing the transmembrane Toll-like receptors is a family of receptors in the cytoplasm that detect products derived from the intracellular degradation of phagocytosed pathogens. Well-studied examples of these **NOD-like receptors** (**NLRs**) are **NOD1** and **NOD2**, which recognize components of bacterial cell walls. NOD-like receptors are so called because of their central domain, called a nucleotide-binding oligomerization domain (NOD), by means of which the receptors form oligomers. On the carboxy-terminal side of the NOD domain is a pathogen-recognition domain made up of LRRs. The pathogen-recognition domain of **NOD1** recognizes γ-glutamyl diaminopimelic acid, a degradation product of the peptidoglycan of Gram-negative

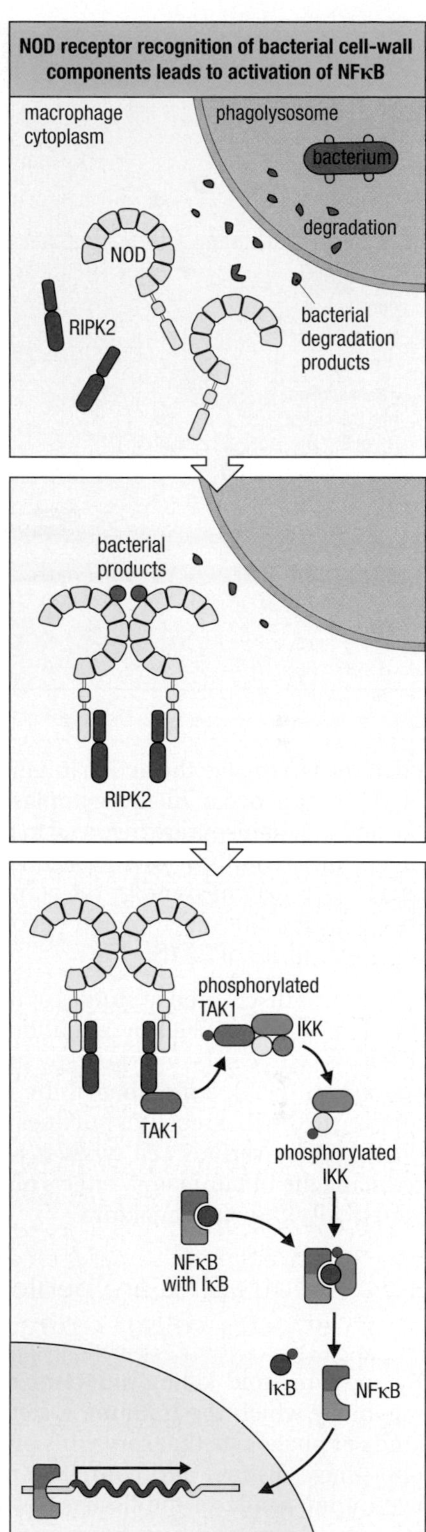

Figure 3.10 NOD receptors are intracellular sensors of bacterial infections. Top panel: degradation of phagocytosed bacteria in the lysosomes of the macrophage leads to delivery of bacterial cell-wall components to the cytoplasm. Middle panel: on recognizing bacterial products, cytoplasmic NOD receptors dimerize and bind the kinase RIPK2. Bottom panel: RIPK2 phosphorylates the kinase TAK1, initiating reactions that lead to the activation of NFκB. This contributes to macrophage activation.

bacteria, whereas that of **NOD2** recognizes muramyl dipeptide, a degradation product derived from the peptidoglycan of most bacteria.

On the amino-terminal side of the NOD domain is a **caspase-recruitment domain** (**CARD**), which is structurally similar to a domain on some signaling proteins that is used to recruit proteases called **caspases**. In NOD receptors the CARD domains do not recruit proteases; instead, they recruit signaling proteins that have the same type of CARD domain as the NOD receptors. When a NOD receptor recognizes its ligand, the CARD domain of the receptor dimerizes with the CARD domain of a serine–threonine kinase called RIPK2. This interaction causes RIPK2 to phosphorylate and activate the kinase TAKI, which in turn phosphorylates and activates IKK. IKK activates NFκB (see Section 3-3), leading to cellular activation (Figure 3.10).

3-6 Inflammasomes amplify the innate immune response by increasing the production of IL-1β

It is no coincidence that Toll-like receptors and the IL-1 receptor share the same TIR signaling domain. Like the Toll-like receptors, signaling through the IL-1 receptor is central to innate immunity and inflammation, and IL-1 is the cytokine most involved in these processes. IL-1 is, in fact, a family of 11 related cytokines, of which IL-1β has been the most studied because of its major contribution to chronic inflammatory diseases such as rheumatoid arthritis. Because of its destructive potential, the production and secretion of IL-1β are strictly regulated. It is synthesized on ribosomes in the cytoplasm and kept there as an inactive precursor protein of 35 kDa called proIL-1β. Active IL-1β is generated after cleavage of this precursor by the protease caspase 1. The active cytokine of 17 kDa, one of two cleavage products, can then be released from the cell. The supply of caspase 1 is also controlled; it is synthesized as the precursor procaspase 1, which must undergo proteolytic cleavage to become active. The enzymatic activity of caspases depends on a cysteine residue in the active site, and their site of cleavage is next to an aspartic acid residue, hence the name caspase.

Once a macrophage has been activated by infection, it can accelerate the production of IL-1β by a feedback mechanism in which secreted IL-1β binds to the IL-1 receptors on the macrophage. Signaling through the IL-1 receptor then increases the transcription and translation of proIL-1β (Figure 3.11). Other signals cause the supply of caspase 1 to increase. As a consequence of the release of ATP by the macrophage, ion channels are activated to lower the intracellular concentration of potassium ions. This ionic change facilitates the assembly of a protein complex called the **inflammasome**, through the oligomerization of a NOD-like receptor called NLRP3, or cryopyrin. NLRP3 has no CARD domain, this function being provided by an adaptor protein that links procaspase 1 to NLRP3. Procaspase 1 is bound by the CARD of the adaptor protein and becomes oligomerized as the inflammasome assembles (see Figure 3.11). This creates a high local concentration of procaspase 1 in the inflammasome, a condition leading to autoproteolysis of procaspase 1 at three aspartic acid residues and the generation of active caspase 1. In turn, caspase 1 cleaves proIL-1β on the carboxy-terminal side of the aspartic acid residue at

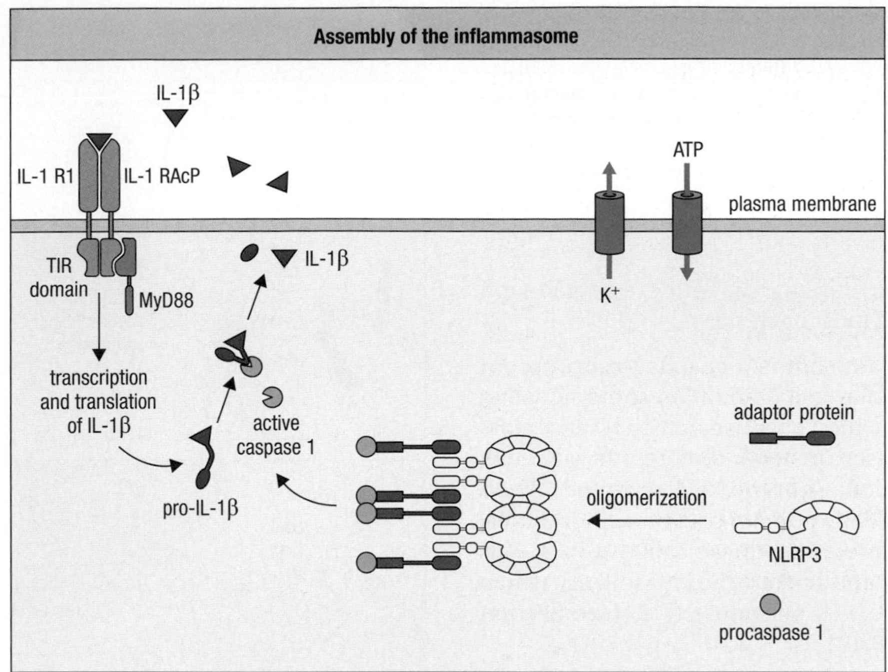

Figure 3.11 The supply and secretion of IL-β depend on the activation of caspase 1 by the inflammasome. IL-1β is an inflammatory cytokine made by activated macrophages. Shown here is the feedback mechanism by which an activated macrophage amplifies its own production of IL-1β. The binding of secreted IL-1β to its receptor on the macrophage signals the macrophage to make large quantities of pro-IL-1β, which accumulates in the cytoplasm. The generation of IL-1β and its secretion require pro-IL-β to be cleaved by the protease caspase 1. This, in turn, is produced by the self-cleavage of procaspase 1. This reaction is catalyzed within the inflammasome, an oligomeric assembly of procaspase 1 with NLRP3. The uptake of ATP and release of potassium ions by the macrophage induce assembly of the inflammasome. IL-1 RAcP, IL-1 receptor accessory protein.

position 116 to give the active inflammatory cytokine, IL-1β. This maturation of IL-1β can occur in the cytoplasm or in specialized secretory granules. Several mechanisms are involved in the release of IL-1β from the macrophage, including shedding of microvesicles from the plasma membrane, exocytosis of secretory lysosomes, and by the activity of specific membrane transporters. In this manner, the macrophage becomes supercharged to make and release large quantities of IL-1β.

NLRP3 was discovered in studies of patients with a range of chronic inflammatory conditions caused by mutations in the *NLRP3* gene. The symptoms of these diseases can be alleviated by subcutaneous injection of the IL-1 receptor agonist (IL-1RA), a member of the IL-1 family that competes with IL-1β for binding to the IL-1 receptor but does not induce intracellular signals. Secretion of IL-1RA by various cell types is a natural mechanism for controlling and reducing the inflammatory effects of IL-1. The clinically used form of IL-1RA is generically known as anakinra.

3-7 Neutrophils are dedicated phagocytes and the first effector cells recruited to sites of infection

By engulfing and killing microorganisms, phagocytic cells are the principal means by which the immune system destroys invading pathogens. The two kinds of phagocyte that serve this purpose—the macrophage and the neutrophil—have distinct and complementary properties. Macrophages are long-lived: they reside in the tissues, work from the very beginning of infection, raise the alarm, and have functions other than phagocytosis. **Neutrophils**, in contrast, are short-lived dedicated killers that circulate in the blood, awaiting a call from a macrophage to enter infected tissue. Thus neutrophils are the first population of effector cells to be recruited to an infected tissue.

Neutrophils are a type of granulocyte, having numerous granules in the cytoplasm, and are also known as **polymorphonuclear leukocytes** because of the variable and irregular shapes of their nuclei (see Figure 1.11, p. 13). Neutrophils were historically called 'microphages' because they are smaller than macrophages. What they lack in size they more than make up for in number. They are the most abundant white blood cells, a healthy adult having some 50 billion (5×10^{10}) in circulation at any time. Mature neutrophils are kept in the bone marrow for about 5 days before being released into the circulation; this constitutes a large reserve of neutrophils that can be called on at times of infection.

Neutrophils are excluded from healthy tissue, but at infected sites the release of inflammatory mediators attracts neutrophils to leave the blood and enter the infected area in large numbers, where they soon become the dominant phagocytic cell. Every day, some 3 billion neutrophils enter the tissues of the mouth and throat, the most contaminated sites in the body. The arrival of neutrophils is the first of a series of reactions, called the **inflammatory response**, by which cells and molecules of innate immunity are recruited into sites of wounding or infection. Although neutrophils are specialized for working under the anaerobic conditions that prevail in damaged tissues, they still die within a few hours after entry. In doing so, they form the creamy **pus** that characteristically develops at infected wounds and other sites of infection. This is why extracellular bacteria such as *Staphylococcus aureus*, which are responsible for the superficial infections and abscesses that neutrophils tackle in large numbers, are known as pus-forming or **pyogenic** bacteria.

3-8 Inflammatory cytokines recruit neutrophils from the blood to the infected tissue

CXCL8 is an inflammatory cytokine made by pathogen-activated macrophages and is one of the family of about 40 chemoattractant cytokines called chemokines. Chemokines serve to direct the flow of leukocyte traffic throughout the body. The function of CXCL8 is to recruit neutrophils that are circulating in the blood and set them to work at the site of infection. Such movements of leukocytes between blood and tissue are crucial to all aspects of the immune response and are determined by interactions between complementary pairs of **adhesion molecules**, one partner expressed on the leukocyte surface and the other on a tissue cell surface. Adhesion molecules are grouped according to four structural types (Figure 3.12).

In the absence of infection, neutrophils travel rapidly through the narrow blood capillaries and do not interact with the vascular endothelium. Blood vessels dilate in the presence of infection and inflammatory cytokines, and the endothelial cells are activated to express adhesion molecules called selectins. The blood flow decreases, and neutrophils make contact with the vascular endothelium in the post-capillary venules. Contact involves transient interactions between selectin adhesion molecules on the endothelium and sialyl-Lewis^{x} carbohydrate groups of glycoproteins on the surface of the neutrophils. These weak interactions slow the neutrophils down relative to the blood flow, causing them to roll along the endothelial surface (Figure 3.13).

The presence of TNF-α also induces the vascular endothelium to express **ICAM-1** and **ICAM-2**, adhesion molecules that function as ligands for complement receptor CR3 and LFA-1 on neutrophils. CR3 and LFA-1 are both members of the integrin family of adhesion molecules. Their interactions with ICAM-1 and ICAM-2 are initially weak, but they strengthen when chemokine CXCL8 binds to CXCR1 and CXCR2, the chemokine receptors on the neutrophil. Chemokine receptors contain seven membrane-spanning helices that

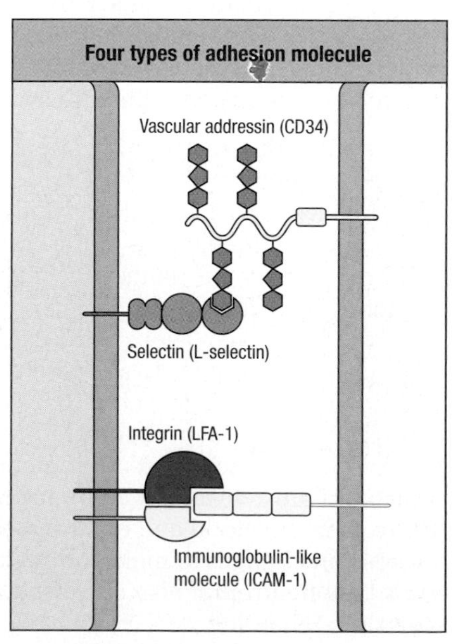

Figure 3.12 Adhesion of leukocytes to vascular endothelium involves interactions between adhesion molecules of four structurally different types. These are the vascular addressins, the selectins, the integrins, and proteins containing immunoglobulin-like domains.

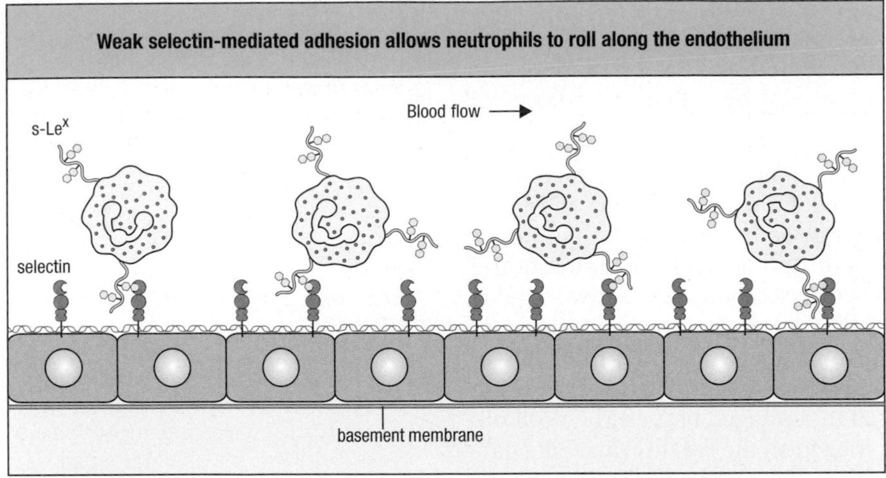

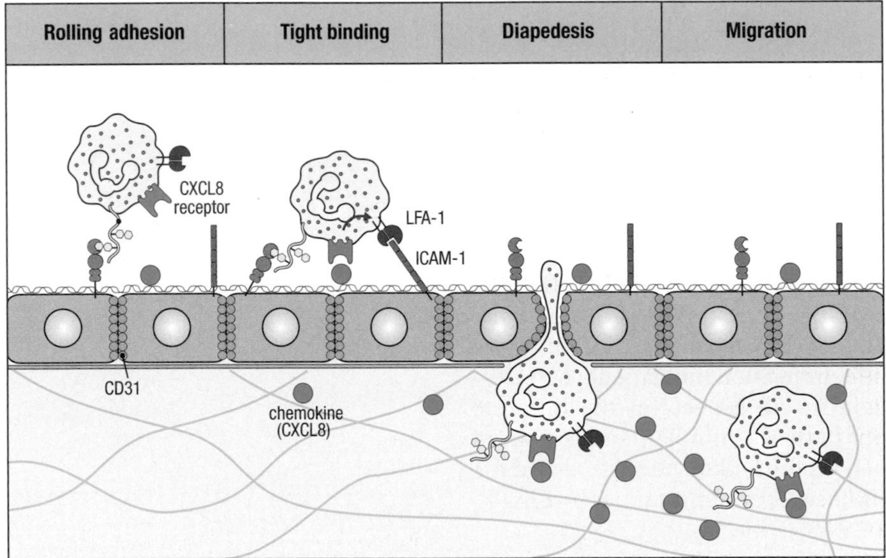

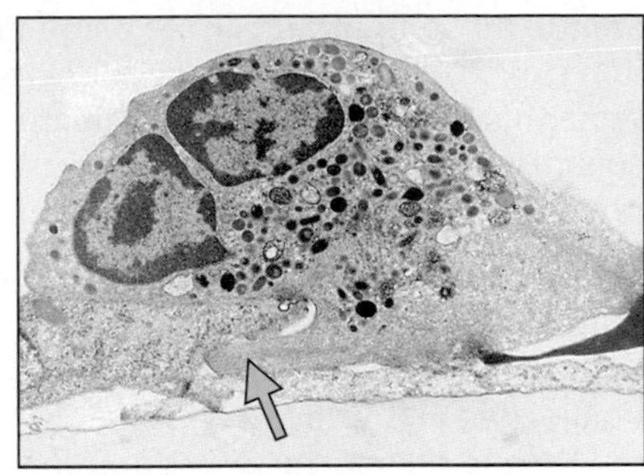

Figure 3.13 Neutrophils are directed to sites of infection through interactions between adhesion molecules. Cytokines induce the expression of selectin on vascular endothelium, enabling it to bind leukocytes. The upper panel shows the rolling interaction of a neutrophil with vascular endothelium as a result of transient interactions between selectin on the endothelium and sialyl-Lewis^{x} (s-Le^{x}) on the leukocyte. The lower panel shows the conversion of rolling adhesion into tight binding and subsequent migration of the leukocyte into the infected tissue. The four stages of extravasation are shown. Rolling adhesion is converted into tight binding by interactions between integrins on the leukocyte (LFA-1 is shown here) and adhesion moleules on the endothelium (ICAM-1). Expression of these adhesion molecules is also induced by cytokines. A strong interaction is induced by the presence of chemoattractant cytokines (the chemokine CXCL8 is shown here) that have their source at the site of infection. They are held on proteoglycans of the extracellular matrix and cell surface to form a gradient along which the leukocyte can travel. Under the guidance of these chemokines, the neutrophil squeezes between the endothelial cells and penetrates the connective tissue (diapedesis). It then migrates to the center of infection along the CXCL8 gradient. The electron micrograph at the right shows a neutrophil that has just started to migrate between adjacent endothelial cells but has yet to break through the basement membrane, which is at the bottom of the photograph. The blue arrow points to the pseudopod that the neutrophil is inserting between the endothelial cells. The dark mass in the bottom right-hand corner is an erythrocyte that has become trapped under the neutrophil. Photograph (× 5500) courtesy of I. Bird and J. Spragg.

develop intracellular signals through an associated GTP-binding protein (Figure 3.14). These signals lead to conformational changes in CR3 and LFA-1, which strengthen their hold on ICAM-1 and ICAM-2 and immobilize the neutrophil on vascular endothelium within the infected tissue.

At this stage the neutrophil leaves the blood by squeezing through the gaps between neighboring endothelial cells. This movement, termed **extravasation**, involves LFA-1 and CR3 as well as the CD31 protein expressed by neutrophils and endothelial cells at the surface where they interact with each other. By these means the neutrophil reaches the basement membrane, which it breaches by secreting the protease elastase and other enzymes that degrade the laminins and collagens of the basement membrane. Giving direction to the neutrophil's migration is a gradient of CXCL8 that has been formed by the chemokine sticking to cell surfaces and proteoglycans in the extracellular matrix. By using their chemokine receptors to bind CXCL8, the neutrophils move up the concentration gradient toward the sources of the chemokine, which are the activated macrophages in the infected tissue (see Section 3-4). On leaving the blood and entering the tissue, the neutrophil's pattern of gene expression changes to make it more active in phagocytosis and killing pathogens. The neutrophil also secretes CXCL8, which recruits additional neutrophils into the tissue.

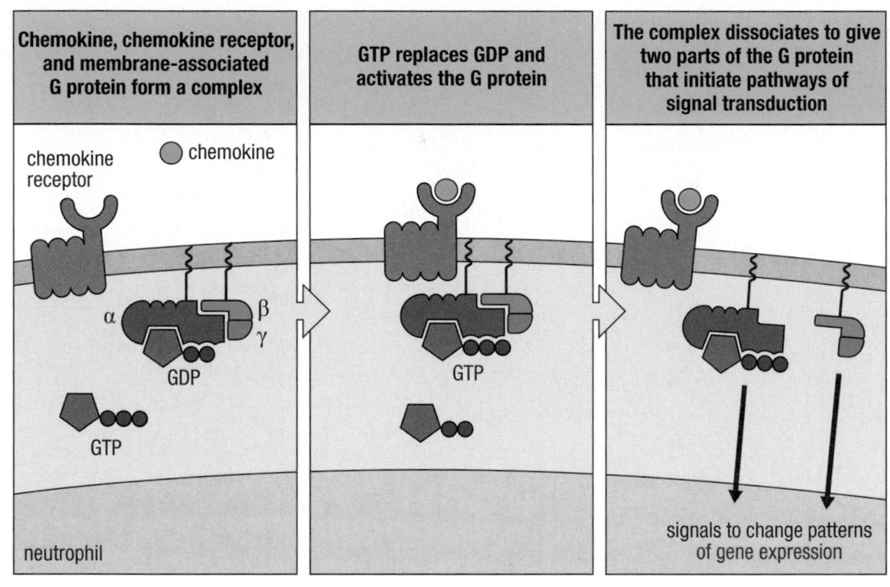

Figure 3.14 Chemokine receptors are G-protein-coupled receptors. The receptors for chemokines belong to a family of transmembrane receptors that have seven transmembrane helices. When a chemokine such as CXCL8 binds to its receptor, the receptor associates with an intracellular GTP-binding (G) protein, which in its inactive state consists of three polypeptides (α, β, and γ) and has GDP bound to it. On association with the chemokine receptor, GDP is replaced by GTP, which leads to dissociation of the α chain of the G protein from the β and γ chains. The α chain, and to a lesser extent the β and γ chains, bind to other cellular proteins that generate signals to change the cell's pattern of gene expression.

3-9 Neutrophils are potent killers of pathogens and are themselves programmed to die

Neutrophils phagocytose microorganisms by mechanisms similar to those used by macrophages. Neutrophils have a range of phagocytic receptors that recognize microbial products; they also have complement receptors that facilitate the phagocytosis of pathogens opsonized by complement fixation (Figure 3.15). The range of particulate material that neutrophils engulf is greater than that tackled by macrophages, as is the diversity of the microbicidal substances stored in their granules. Because mature neutrophils are programmed to die young, they devote more of their resources to the storage and delivery of antimicrobial weaponry than do the longer-lived macrophages.

Almost immediately after a pathogen has been engulfed by a neutrophil and taken into a phagosome, a battery of degradative enzymes and other toxic substances is brought to bear upon the pathogen, which is rapidly killed. Phagosomes containing recently captured microorganisms are fused with three types of preformed neutrophil granule: **primary** (or **azurophilic**) **granules** marked by the presence of the enzyme myeloperoxidase; **secondary** (or **specific**) **granules** marked by the protein lactoferrin; and **tertiary** (or **gelatinase**) **granules** marked by the enzyme gelatinase (Figure 3.16). Primary, secondary, and tertiary refers to the order in which they are made by the neutrophil. The primary granules are packed with proteins and peptides that disrupt and digest microbes. These include lysozyme, defensins, myeloperoxidase, neutral proteases such as cathepsin G, elastase and proteinase 3, and a bactericidal permeability-increasing protein that binds LPS and kills Gram-negative bacteria. Binding these proteins and peptides together in the granule is a negatively charged matrix of sulfated proteoglycans. The combination of this matrix and the acidity of the granule's interior sequesters the weaponry in a safe, inactive form until it is needed.

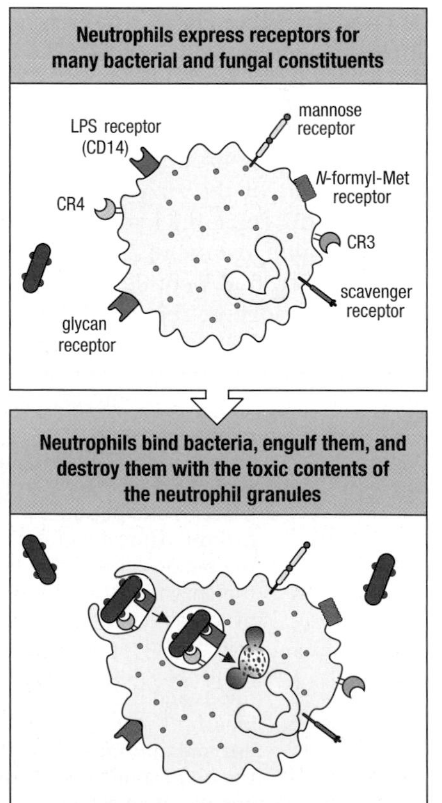

Figure 3.15 The binding of bacteria to neutrophil receptors induces phagocytosis and microbial killing. Upper panel: the neutrophil has several different receptors for microbial products. Lower panel: the mechanism of phagocytosis for two such receptors, CD14 and CR4, which are specific for bacterial lipopolysaccharide (LPS). A bacterium binding to these receptors stimulates its own phagocytosis and degradation.

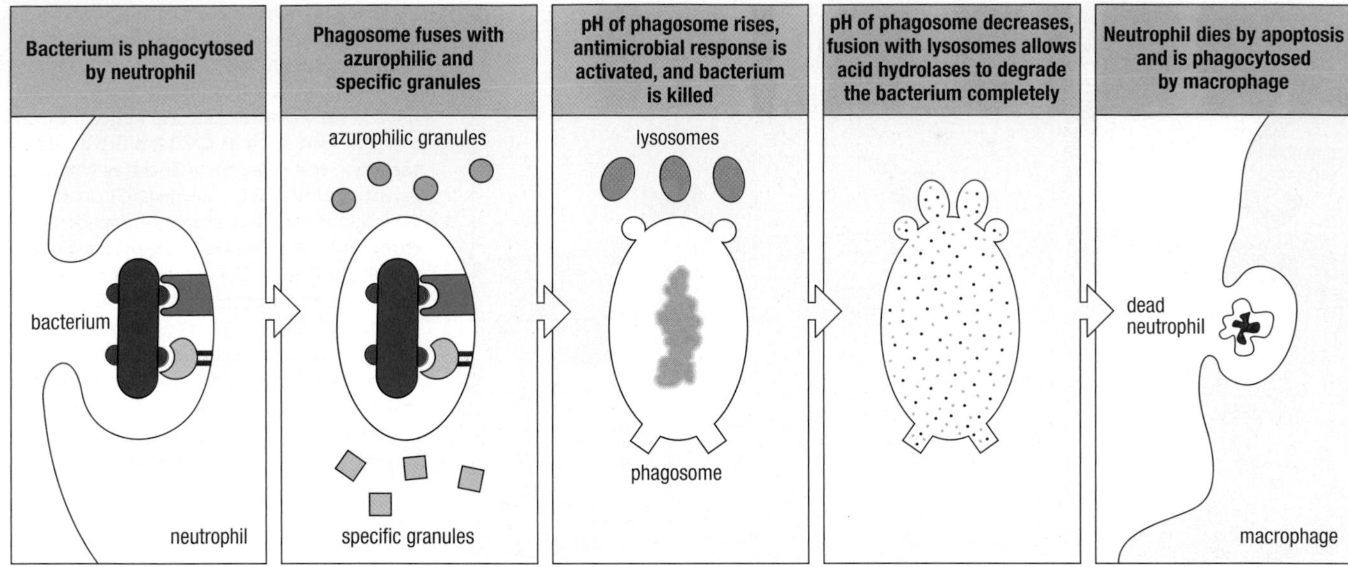

Figure 3.16 Killing of bacteria by neutrophils involves the fusion of two types of granule and lysosomes with the phagosome. After phagocytosis (first panel), the bacterium is held in a phagosome inside the neutrophil. The neutrophil's azurophilic granules and specific granules fuse with the phagosome, releasing their contents of antimicrobial proteins and peptides (second panel). NAPDH oxidase components contributed by the specific granules enable the respiratory burst to occur, which raises the pH of the phagosome (third panel). Antimicrobial proteins and peptides are activated, and the bacterium is damaged and killed. A subsequent decrease in pH and the fusion of the phagosome with lysosomes containing acid hydrolases results in complete degradation of the bacterium (fourth panel). The neutrophil dies and is phagocytosed by a macrophage (fifth panel).

Gelatinase in the tertiary granules is a metal-containing protease that restricts the growth of bacteria by sequestering iron. Unsaturated lactoferrin in the secondary granules similarly competes with pathogens for metal ions. The secondary granules also contain lysozyme and several membrane proteins, including components of **NADPH oxidase**, an essential enzyme for neutrophil function that is assembled in the phagosome after fusion with the three types of granule.

NADPH oxidase produces superoxide radicals that are converted into hydrogen peroxide by the enzyme superoxide dismutase (Figure 3.17). These reactions rapidly consume hydrogen ions and have the direct effect of raising the pH of the phagosome to 7.8–8.0 within 3 minutes of phagocytosis. At this pH, the antimicrobial peptides and proteins become activated and attack the trapped pathogens. The pH of the phagosome then slowly goes down, reaching neutrality (pH 7.0) after 10–15 minutes. At this point some of the neutrophil's lysosomes fuse with the phagosome to form the phagolysosome. The

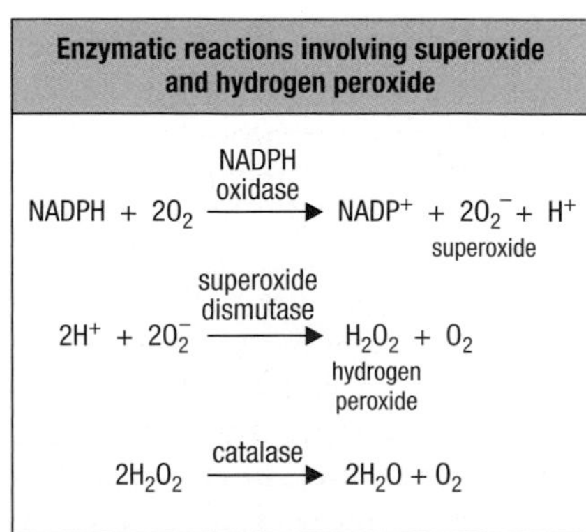

Figure 3.17 Killing of bacteria by neutrophils is dependent on a respiratory burst. In the absence of infection, the antimicrobial proteins and peptides in neutrophil granules are kept inactive at low pH. After the granules fuse with the phagosome, the pH within the phagosome is raised through the first two reactions, involving the enzymes NADPH oxidase and superoxide dismutase. Each round of these reactions eliminates a hydrogen ion, thereby reducing the acidity of the phagosome. A product of the two reactions is hydrogen peroxide, which has the potential to damage human cells. (In hair salons and in the manufacture of paper, it is used as a powerful bleach.) The third reaction, involving catalase, the most efficient of all enzymes, promptly gets rid of the hydrogen peroxide produced during the neutrophil's respiratory burst, raising the pH of the phagosome and enabling activation of the antimicrobial peptides and proteins.

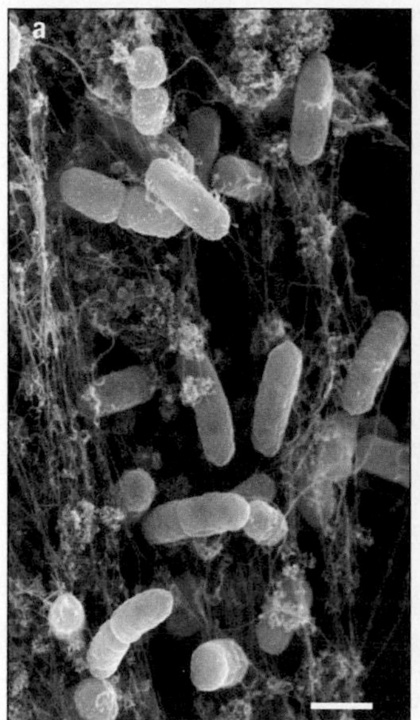

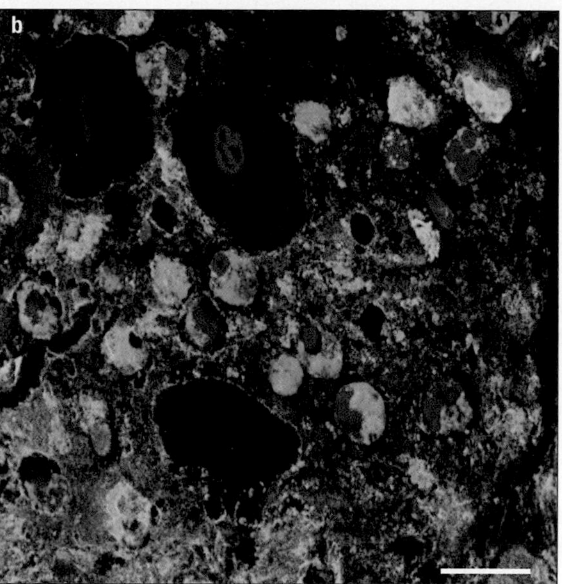

Figure 3.18 Dying neutrophils form neutrophil extracellular traps that capture infecting bacteria and viruses. The left panel shows *Salmonella* bacteria that have been caught in neutrophil extracellular traps (NETs). The scale bar corresponds to 1 μm. *Salmonella* causes typhoid fever and gastroenteritis. The right panel is an image of pus arising from molluscum contagiosum, a disease of the skin caused by a poxvirus called molluscum contagiosum virus. Many neutrophils at various stages of NETosis are shown. Chromatin is stained red, and the protein fibers of the NETs are stained green. The scale bar corresponds to 20 μm.

lysosomes contribute a variety of degradative enzymes, collectively called acid hydrolases, that are active at the lower pH of the phagolysosome and ensure the continued and complete breakdown of the pathogen's macromolecules.

Powering the neutrophil's ferocious intracellular attack, which kills both Gram-positive and Gram-negative bacteria and also fungi, is a transient increase in oxygen consumption called the **respiratory burst**. The products of the respiratory burst are several toxic oxygen species that diffuse out of the cell and damage other host cells. To limit the damage, the respiratory burst is accompanied by the synthesis of enzymes that inactivate these potent small molecules: one such enzyme is catalase, which degrades hydrogen peroxide to water and oxygen (see Figure 3.17).

The mature neutrophil cannot replenish its granule contents; once they are used up, the neutrophil dies. Some neutrophils die by apoptosis and are then phagocytosed by a macrophage. A second way in which neutrophils die is by a process called **netosis**, which produces **neutrophil extracellular traps** (**NETs**) that trap and kill pathogens (Figure 3.18). In netosis the nucleus swells and bursts, and the chromatin dissolves and becomes extruded from the cell in a network of decondensed DNA decorated with histones and other cationic proteins derived from the neutrophil granules. Among the proteins present in these NETs are bactericidal defensins, several proteases, and calprotectin, which impairs the growth of fungi. Thus, even after their own demise, neutrophils continue to trap microorganisms and kill them. If the neutrophils succeed in eliminating the bacteria and thus terminate the infection, the resident macrophages halt further neutrophil recruitment and initiate the process of repairing the damage that the neutrophils have done to the tissue.

The dependence of the body's defenses on neutrophils is well illustrated by **chronic granulomatous disease**, a genetic syndrome caused by defective forms of the genes encoding NADPH oxidase subunits. In the absence of functional NADPH oxidase, there is no respiratory burst after phagocytosis and the pH of the neutrophil's phagosome cannot be raised to the level needed to activate a successful attack by antimicrobial peptides and proteins. Bacteria and fungi are not cleared, and persist as chronic intracellular infections of neutrophils and macrophages. Because of the actions of other mechanisms of innate and adaptive immunity, the infections become contained in localized nodules, called **granulomas**, which imprison the infected macrophages that have eaten more than their fill of infected neutrophils. The bacteria and fungi that

Chronic granulomatous disease

most commonly cause infections in chronic granulomatous disease include organisms such as *Escherichia coli* that form part of the normal microbiota of healthy people (Figure 3.19).

3-10 Inflammatory cytokines raise body temperature and activate the liver to make the acute-phase response

Inflammation at a localized infection in one tissue can produce wide-ranging changes throughout the body. One systemic effect of the inflammatory cytokines IL-1β, IL-6, and TNF-α is to raise the body temperature, causing **fever**. The cytokines act on temperature-control sites in the hypothalamus, and on muscle and fat cells, altering energy mobilization to generate heat (Figure 3.20). The cytokines and other molecules that induce fever are called **pyrogens**. Some pathogen products also raise the body's temperature and generally do so through inducing the production of these cytokines. In this context, the bacterial products are called 'exogenous' pyrogens because they originate outside the body, and the cytokines are called 'endogenous' pyrogens because they originate inside the body.

On balance, a raised body temperature helps the immune system fight infection, because most bacterial and viral pathogens grow and replicate faster at temperatures lower than that of the human body, and adaptive immunity becomes more potent at higher temperatures. In addition, human cells become more resistant to the deleterious effects of TNF-α when experiencing fever. Cytokines also induce the lethargy, somnolence, and anorexia that accompany fever, which may represent tactics for fighting infection by not wasting energy on other activity.

Another systemic effect of IL-6 and, to lesser extent, IL-1β and TNF-α, is to change the spectrum of soluble plasma proteins that are made and secreted by the hepatocytes of the liver. Concentrations increase for some 30 plasma proteins involved in the response to infection. The concentrations of some other plasma proteins go down, including albumin, the most abundant plasma protein. The sum of these changes is referred to as the **acute-phase response**, and those proteins that either increase or decrease their concentration by 25% or more are described as **acute-phase proteins**. Among the acute-phase proteins, C-reactive protein (CRP) and serum amyloid A protein clearly stand out, because their concentrations rise several hundredfold. The effect is so strong and predictable that C-reactive protein concentration is used clinically as a diagnostic test for infection, inflammation, and tissue damage (Figure 3.21).

C-reactive protein (**CRP**), a member of the pentraxin family of proteins, contains five identical subunits that form a slab-like pentameric molecule with a

Fungus
Aspergillus fumigatus
Bacteria
Staphylococcus aureus
Chromobacterium violaceum
Burkholderia cepacia
Nocardia asteroides
Salmonella typhimurium
Serratia marcescens
Mycobacterium fortuitum
Several species of *Klebsiella*
Escherichia coli
Several species of *Actinomyces*
Legionella bosmanii
Clostridium difficile

Figure 3.19 The species of fungi and bacteria most commonly responsible for infections in chronic granulomatous disease.

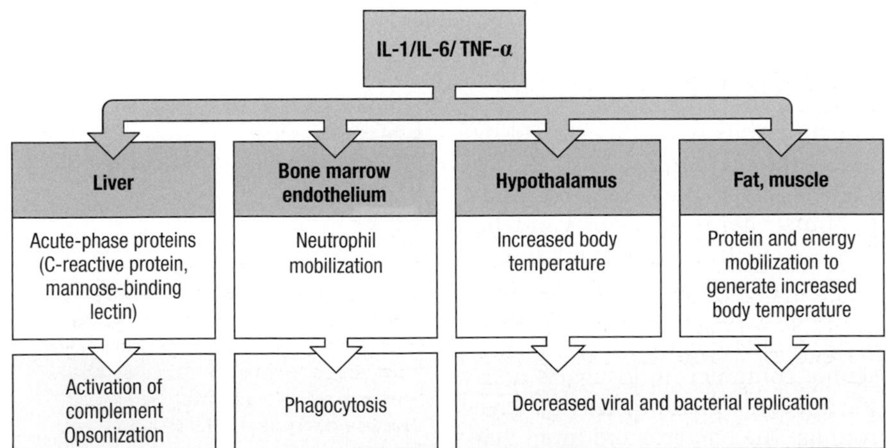

Figure 3.20 The macrophage-produced cytokines TNF-α, IL-1, and IL-6 have a spectrum of biological activity.

Acute-phase protein	Function
C-reactive protein	pathogen recognition
Mannose-binding lectin	
Lipopolysaccharide-binding protein	
Complement components C3, C4, C9, factor B	pathogen elimination
Granulocyte colony-stimulating factor, serum amyloid A, secreted phospholipase A2	inflammatory response
Fibrinogen, plasminogen, tissue plasminogen factor	coagulation

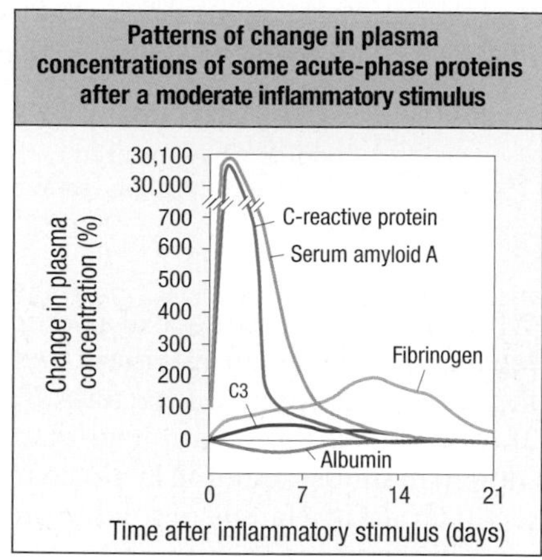

Figure 3.21 The kinetics of acute-phase proteins in the blood. The left panel shows representative examples, and their functions, of the different types of plasma protein that increase during the acute-phase response. The graph on the right shows the change in concentration of five proteins in blood plasma after the initiation of an inflammatory response. C-reactive protein and serum amyloid A are massively increased acute-phase proteins, whereas C3 and fibrinogen are moderately increased. Serum albumin, the most abundant plasma protein, is reduced in concentration during the acute phase.

hole in the middle (Figure 3.22). C-reactive protein binds to bacteria and was originally named for its property to bind the C-polysaccharide of *Streptococcus pneumoniae*. The target was phosphorylcholine, a component of lipopolysaccharides in the bacterial cell wall. As well as bacteria, C-reactive protein binds to fungi, yeasts, and the parasites that cause malaria and leishmaniasis. On binding to pathogens, C-reactive protein acts as an opsonin, triggering the classical pathway of complement fixation in the absence of specific antibody. C-reactive protein also binds to the surface of phagocytes, suggesting that it can deliver pathogens to these cells for elimination. Thus the greatly raised level of C-reactive protein in the acute-phase response can be viewed as a strategy of overwhelming force to overcome the population of pathogens.

Serum amyloid A protein is a small protein of around 100 amino acids that associates with high-density lipoprotein protein particles. It can interact with various cell-surface receptors, including Toll-like receptors and the CD36 scavenger receptor (see Figure 3.21), to activate cells to produce inflammatory cytokines. Thus the greatly increased concentration of serum amyloid A protein in the acute-phase response can be viewed as a way of amplifying the state of inflammation first triggered by the pathogen.

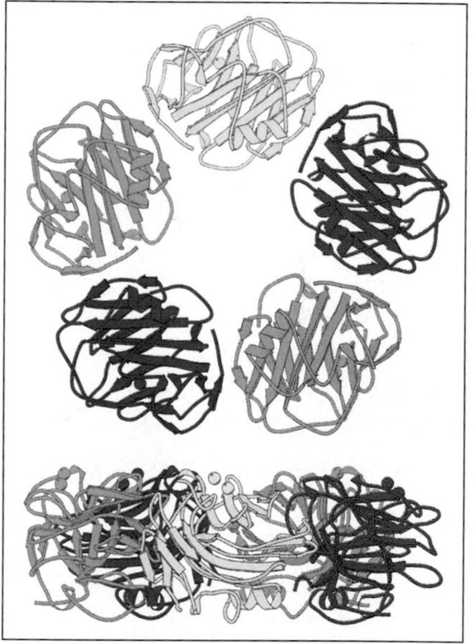

Figure 3.22 The structure of C-reactive protein. C-reactive protein belongs to the pentraxin family, so called because these proteins are composed of five identical subunits. The polypeptide backbones of the five subunits are traced by ribbons of different color. Overall, C-reactive protein resembles a pentagonal slab with a hole in the middle, as is seen by comparing a view from above (upper image) with one from the side (lower image). Images courtesy of Annette Shrive and Trevor Greenhough.

3-11 The lectin pathway of complement activation is initiated by the mannose-binding lectin

Mannose-binding lectin (MBL) is a C-type lectin that binds to mannose-containing carbohydrates of bacteria, fungi, protozoa, and viruses; it is also an acute-phase protein. The structure of MBL resembles a bouquet of flowers in which each stalk is a triple helix made from three identical polypeptides. These helices are just like those found in collagen molecules and fibers. Each polypeptide contributes a carbohydrate-recognition domain, the three together forming a 'flower' (Figure 3.23). Each molecule of mannose-binding lectin has five or six flowers, giving it either 15 or 18 potential sites for attachment to a pathogen's surface. Even relatively weak individual interactions with a carbohydrate structure can be developed into an overall strong binding through the use of such multipoint attachments. Although some carbohydrates on human cells contain mannose, they do not bind mannose-binding lectin because their geometry does not permit multipoint attachment.

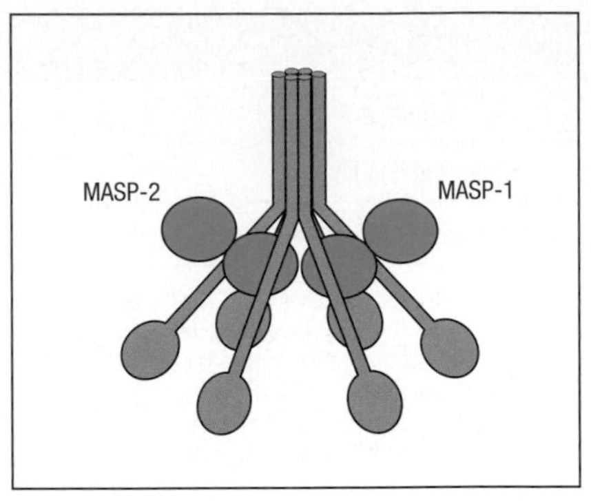

Figure 3.23 Structure of mannose-binding lectin. It resembles a bunch of flowers, with each flower composed of three identical polypeptides. The stalks are rigid triple helices like collagen, with a single bend; each flower comprises three carbohydrate-binding domains. Associated with the mannose-binding lectin (blue) are the mannose-binding lectin associated serine proteases (MASP) 1 and 2.

When bound to the surface of a pathogen, mannose-binding lectin triggers the **lectin pathway of complement activation**; it also serves as an opsonin that facilitates the uptake of bacteria by monocytes in the blood (Figure 3.24). These cells lack the macrophage mannose receptor but have receptors that can bind to mannose-binding lectin coating a bacterial surface. Mannose-binding lectin is part of a family of soluble proteins called the **collectins** because these proteins combine the properties of collagen and lectins. The pulmonary surfactant proteins A and D (SP-A and SP-D) are also collectins; they defend the lungs by opsonizing pathogens such as *Pneumocystis jirovecii*. The MARCO and SR-A receptors on the macrophage surface also combine the properties of collagen and lectins (see Section 3-2).

Mannose-binding lectin circulates in plasma as a complex with two serine protease zymogens: MBL-associated serine protease (MASP) 1 and 2. Two molecules each of MASP-1 and MASP-2 associate with the main stalk of MBL (see Figure 3.23). When the MBL complex binds to mannose-containing macromolecules at a pathogen surface, one molecule of MASP-2 is induced to become enzymatically active and cut itself. It then cuts the second MASP-2 molecule. It is not known whether MASP-1 has an enzymatic role in lectin-mediated complement activation. Substrates for the activated MASP-2 proteases are the **C4** and **C2** complement components. C4 is similar to C3 in its structure, function, and thioester bond, whereas C2 is a serine protease zymogen similar to factor B (see Figure 2.8).

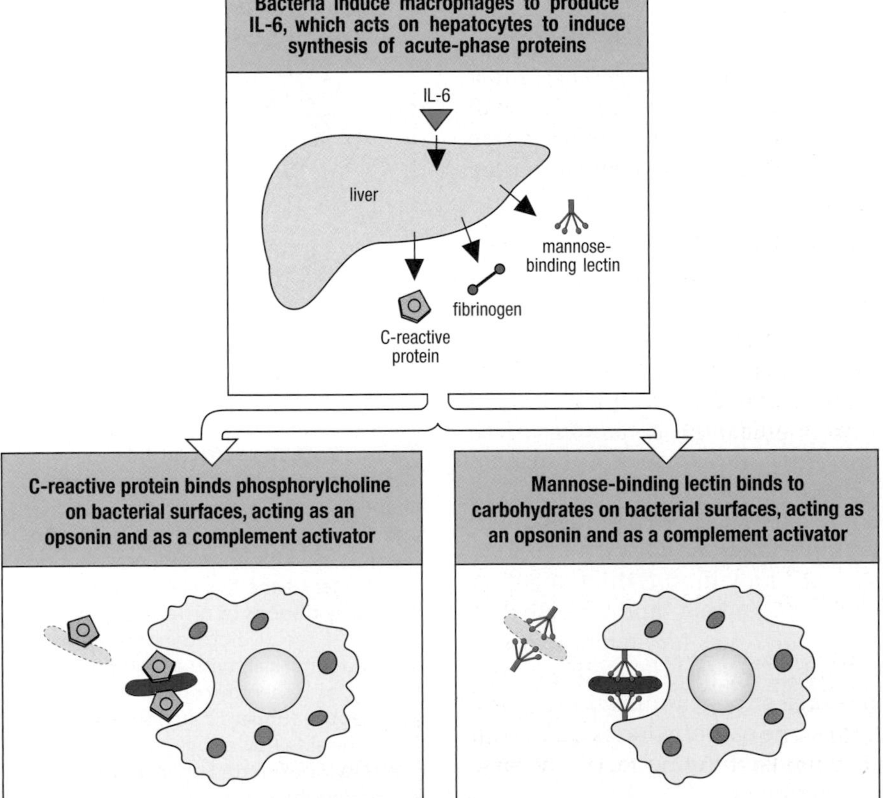

Figure 3.24 The acute-phase response increases the supply of the recognition molecules of innate immunity. Acute-phase proteins are produced by liver cells in response to the cytokines released by phagocytes in the presence of bacteria. They include C-reactive protein, fibrinogen, and mannose-binding lectin. Both C-reactive protein and mannose-binding lectin bind to structural features of bacterial cell surfaces that are not on human cells. On binding to bacteria, they act as opsonins. They also activate complement, facilitating phagocytosis and direct lysis (dashed lines) of the bacteria by the terminal complement components (not shown).

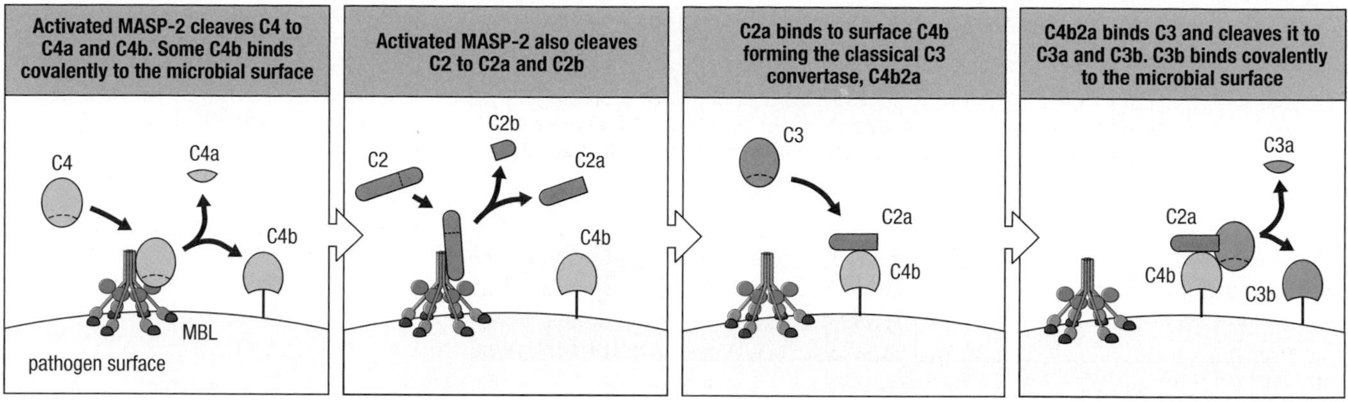

Figure 3.25 The activated MBL complex cleaves C4 and C2 to produce C4b and C2a, which associate to form the classical C3 convertase. First panel: a complex of MBL and MASP-1 and MASP-2 binds to the pathogen surface. This activates MASP-2, which binds and cleaves C4 to reveal the thioester bond of the C4b fragment. C4b becomes covalently bound to the microbial surface. Second panel: C2 binds to the MBL complex and is cleaved by activated MASP-2. Third panel: the C2a fragment binds to C4b to form the classical C3 convertase, C4bC2a. Fourth panel: C3 is bound and cleaved by C4bC2a. The thioester bond of the C3b fragment is exposed, and C3b becomes covalently bound to the microbial surface.

When a C4 molecule interacts with activated MASP-2, it is cleaved into a large C4b fragment and a small C4a fragment. This cleavage exposes the thioester bond of C4b, which leads to the fixation of some C4b fragments to the pathogen surface (Figure 3.25). The soluble C4a fragment is an anaphylatoxin that can recruit leukocytes to the site of C4b fixation, but its activity is weaker than that of either C3a or C5a (see Section 2-7). When a C2 molecule interacts with activated MASP-2, it is cleaved into a larger enzymatically active fragment called C2a, which binds to pathogen-bonded C4b, and a small inactive fragment called C2b. (For historical reasons, the small cleavage product of C2 is called C2b and the larger product is called C2a, whereas for other complement components the larger fragment is called 'b' and the smaller 'a.') The complex of C4b and C2a, designated C4bC2a, is a C3 convertase. Although called the **classical C3 convertase**, it is actually a component of both the lectin and classical pathways of complement activation, for it is at this stage that the lectin and classical pathways converge. The unique aspects of the lectin pathway are the contribution of mannose-binding lectin to binding pathogens, and the activation of C4 and C2 by the MASP proteins.

The classical C3 convertase, C4bC2a, binds and cleaves C3 to yield C3b fragments attached to the pathogen surface. These in turn bind and activate factor B to assemble molecules of the alternative C3 convertase, C3bBb (Figure 3.26). It is at this stage that the lectin and classical pathways converge on the alternative pathway of complement activation described in Chapter 2. Because C3 is present at much higher concentrations in plasma than C4, the contribution of the alternative convertase to the fixation of complement far exceeds that of the classical convertase.

Alleles encoding nonfunctional variants of MBL are present at frequencies greater than 10% in human populations. Consequently, deficiency of MBL is common and causes increased susceptibility to infection. Individuals who carry two nonfunctional alleles are more likely to develop severe meningitis

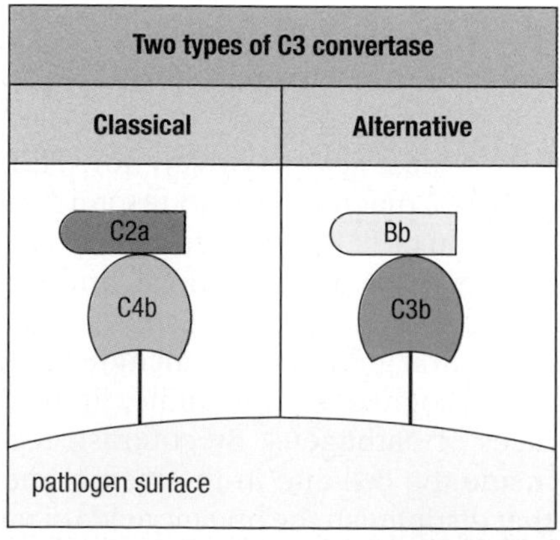

Figure 3.26 The two types of C3 convertase have similar structures and functions. In the C3 convertase produced by the classical pathway, C4bC2a, the activated protease C2a cleaves C3 to C3b and C3a (not shown). In the analogous C3 convertase of the alternative pathway, C3bBb, the activated protease Bb carries out exactly the same reaction.

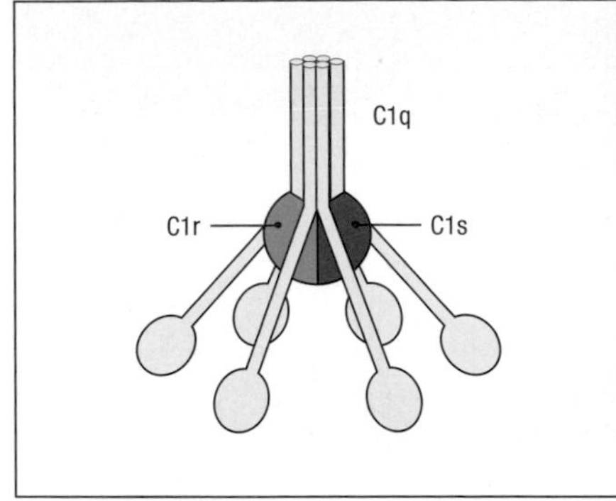

Figure 3.27 The complement component C1. The C1 molecule consists of a complex of C1q, C1r, and C1s. The C1q component consists of six identical subunits, each with one binding site for the Fc region of IgM or IgG and extended amino-terminal stalk regions that interact with each other and with two molecules each of the proteases C1r and C1s. The electron micrograph on the right contains images of three C1q molecules. Photograph courtesy of K.B.M. Reid.

caused by *Neisseria meningitidis*, a bacterium that is carried as a harmless commensal by about 1% of the population. Similar susceptibility is observed in people who are deficient for a terminal complement component, showing that complement-mediated killing of the bacteria is the mechanism by which healthy carriers keep their *N. meningitidis* in order.

3-12 C-reactive protein triggers the classical pathway of complement activation

Once C-reactive protein has bound to a bacterium it can also interact with **C1**, the first component of the classical pathway of complement activation. C1 has an organization and structure like that of the complex of MBL with MASP-1 and MASP-2 (Figure 3.27). In the C1 molecule, a bouquet of six flowers is formed from 18 C1q polypeptides and 2 molecules each of C1r and C1s, which are inactive serine proteases similar to MASP-1 and MASP-2. Each stalk is formed by a collagen-like triple helix of three C1q molecules. C-reactive protein binds to the C1q stalks and causes one molecule of C1r to cut itself, the other molecule of C1r, and both molecules of C1s. In this manner C1s becomes an active protease. It cleaves C4, leading to the covalent attachment of C4b to the pathogen surface (Figure 3.28). It also cleaves C2, leading to the formation of the classical C3 convertase C4bC2a. At this stage the classical and lectin pathways converge. Unique features in the classical pathway of complement fixation are the use of C1q in binding pathogens and the use of C1r and C1s to activate C4 and C2.

At the start of infection, complement activation is mainly by the alternative pathway. As the inflammatory response develops and acute-phase proteins are produced, MBL and C-reactive protein provide increased activation of complement via the lectin and classical pathways, respectively. All three pathways contribute to innate immunity, and they work together to produce quantities of C3b fragments and C3 convertases at the pathogen surface.

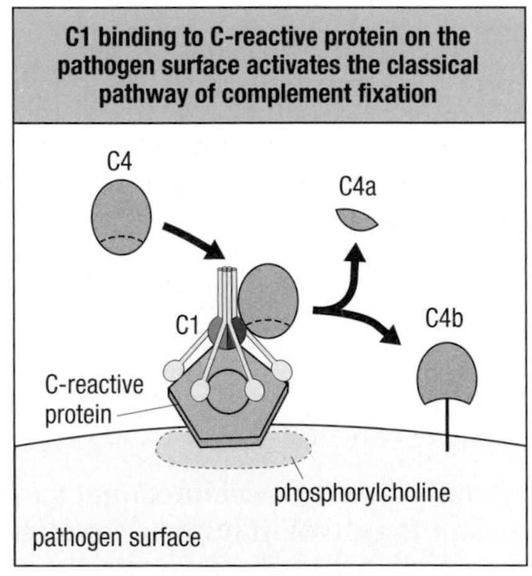

Figure 3.28 C-reactive protein can initiate the classical pathway of complement activation. C-reactive protein bound to phosphorylcholine on bacterial cell surfaces binds complement component C1, resulting in the cleavage of C4 and opsonization of the bacterial surface with C4b.

3-13 Toll-like receptors sense the presence of the four main groups of pathogenic microorganisms

In previous sections we saw how TLR4 recognizes bacterial products, and how it signals macrophages to respond to bacterial infection. Complementing this function of TLR4, other members of the family of Toll-like receptors recognize the presence of viral, fungal, and parasitic infections (Figure 3.29). The Toll-like receptors form two distinctive groups in terms of their function. Receptors in the first group, which includes TLR4, are located on the plasma membrane and recognize carbohydrate, lipid, and protein structures on the outer surfaces of pathogens. By contrast, receptors in the second group are located inside the cell and in the membranes of endosomes, and recognize features that distinguish the nucleic acids of pathogens from the nucleic acids of human

Recognition of microbial products through Toll-like receptors						
	Receptor	Chromosome	Ligands	Microorganisms recognized	Cells carrying receptor	Cellular location of receptor
I	TLR1:TLR2 heterodimer	4	Lipopeptides Glycosylphosphatidylinositol	Bacteria Parasites	Monocytes, dendritic cells, eosinophils, basophils, mast cells	Plasma membrane
I	TLR2:TLR6 heterodimer	4	Lipoteichoic acid Zymosan	Gram-positive bacteria Fungi	Monocytes, dendritic cells, eosinophils, basophils, mast cells	Plasma membrane
I	TLR10 homodimer and heterodimers with TLR1 and 2	4	Unknown		Plasmacytoid dendritic cells, basophils, eosinophils, B cells	Unknown
I	TLR4 homodimer	9	Lipopolysaccharide	Gram-negative bacteria	Macrophages, dendritic cells, mast cells, eosinophils	Plasma membrane
II	TLR7 homodimer	X	Single-stranded viral RNAs	RNA viruses	Plasmacytoid dendritic cells, NK cells, eosinophils, B cells	Endosomes
II	TLR8 homodimer	X	Single-stranded viral RNAs	RNA viruses	NK cells	Endosomes
II	TLR9 homodimer	3	Unmethylated CpG-rich DNA	Bacteria DNA viruses	Plasmacytoid dendritic cells, B cells, eosinophils, basophils	Endosomes
III	TLR3 homodimer	4	Double-stranded viral RNA	RNA viruses	NK cells	Endosomes
IV	TLR5 homodimer	1	Flagellin, a protein	Bacteria	Intestinal epithelium	Plasma membrane

Figure 3.29 The human Toll-like receptors allow the detection of different types of infection. Each of the known Toll-like receptors (TLRs) seems to recognize one or more characteristic features of microbial macromolecules, but TLR5 is the only one for which direct interaction with a microbial product, the bacterial protein flagellin, has been demonstrated. The 10 human TLR genes encode distinctive TLR polypeptides. Some TLRs are heterodimers of two polypeptides; others, such as TLR4, are homodimers. The Toll-like receptors take their name from their structural similarities to a receptor called Toll in the fruitfly *Drosophila melanogaster*, which is involved in the adult fly's defense against infection.

cells. For example, TLR3 recognizes double-stranded RNA, which characterizes many viral infections, and TLR9 detects the unmethylated CpG nucleotide motifs that are abundant in bacterial and viral genomes but not in human genomic DNA (Figure 3.30). Uptake of pathogen-derived nucleic acids from the extracellular environment delivers them to endosomes for recognition by these Toll-like receptors.

Unlike some other immune-system gene families, whose genes are all clustered together on one chromosome, the 10 human *TLR* genes are distributed between five chromosomes. This reflects the ancient, invertebrate origin of the Toll-like receptors, which were present before the two genome-wide duplications that occurred during the early evolution of the vertebrates around 500 million years ago. On the basis of sequence similarities, the Toll-like receptors form four evolutionary lineages (I, II, II, and IV) that are descendants of the four Toll-like receptor loci formed by these two ancient genome duplications. Only the lineage I proteins can form heterodimeric Toll-like receptors, whereas proteins in all four groups can form homodimers. Although the number of human Toll-like receptors is limited, they recognize characteristic features of all four groups of pathogenic microorganisms. Thus they can detect many, if not all, species of pathogen that cause infectious disease in humans.

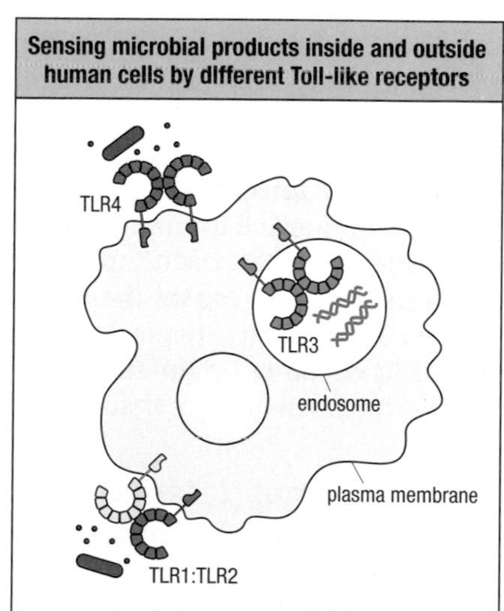

Figure 3.30 Different Toll-like receptors sense bacterial infection outside the cell and viral infection inside the cell. The TLR4 homodimer and the TLR1:TLR2 heterodimer at the cell surface sense bacterial infection. The homodimer of TLR3 in an endosomal vesicle detects viral infection.

3-14 Genetic variation in Toll-like receptors is associated with resistance and susceptibility to disease

The variation achieved by having 10 different Toll-like receptor genes is increased by the presence in the human population of alleles of these genes,

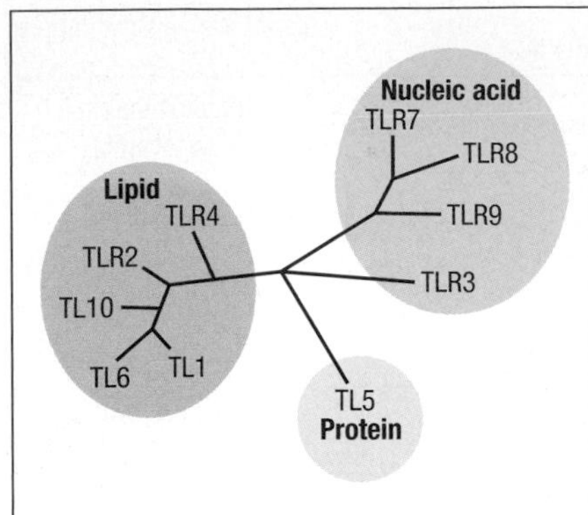

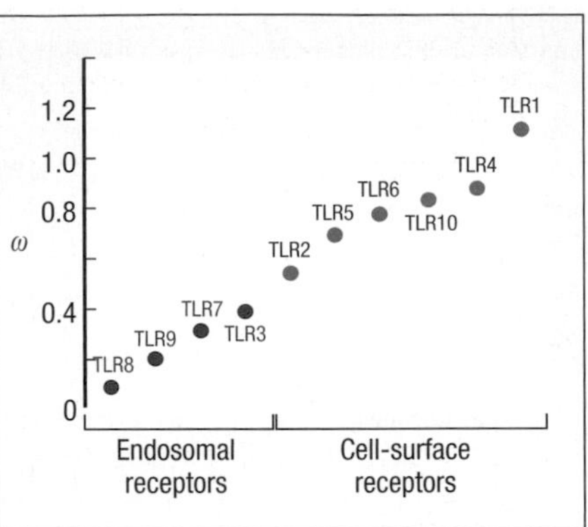

Figure 3.31 The structural similarities between Toll-like receptors and the relative evolutionary pressures upon them. The left panel shows a phylogenetic tree constructed from the nucleotide sequences of the human Toll-like receptors. The colors distinguish the receptors that recognize lipids, nucleic acids, and proteins. On the *y*-axis of the graph on the right is plotted ω, a measure of the variability of the Toll-like receptors in human populations. On the *x*-axis the Toll-like receptors are ordered according to increasing ω. Red and blue data points denote the endosomal and cell-surface receptors, respectively.

which encode variant receptor proteins. The proteins encoded by different alleles of the same gene are called **allotypes**. There is more of this variation, called **genetic polymorphism**, in the Toll-like receptors that recognize surface determinants of pathogens than in the Toll-like receptors that recognize nucleic acids (Figure 3.31). This difference is consistent with the greater structural diversity of carbohydrates, lipids, and proteins compared to nucleic acids.

Genetic polymorphism can influence the function of a Toll-like receptor, as seen from epidemiological studies of **septic shock**, a life-threatening condition that can occur when a bacterial infection spreads to the blood and becomes systemic. In this circumstance, macrophages in the liver, spleen, and other organs are activated to release TNF-α, causing widespread dilation of blood vessels and massive leakage of fluid into tissues throughout the body. This produces the state of septic shock, in which the blood supply is severely perturbed and vital organs such as the kidneys, heart, lungs, and liver can fail. Septic shock is most commonly caused by Gram-negative bacteria, which prompted the study of the distribution of two alleles of TLR-4 in patients with septic shock. The frequency of patients having the rare TLR-4 allotype with glycine at position 299 in the sequence was greater than the frequency of this variant in the population at large. This shows that the rare form of TLR-4 confers susceptibility to septic shock, whereas the common allotype, with asparagine at position 299, confers resistance. Because of its rarity, individuals who are homozygous for the rare variant, having inherited it from both their parents, are extremely uncommon. The one individual known to have this genotype died in adolescence from septic shock after a kidney infection with the Gram-negative commensal bacterium *E. coli*. The substitution of glycine for asparagine at position 299 causes the rare TLR-4 variant to be less responsive to the LPS of Gram-negative bacteria, making an infection less likely to be contained within its tissue of origin and more likely to become systemic. Septic shock causes the death of more than 100,000 people in the United States each year.

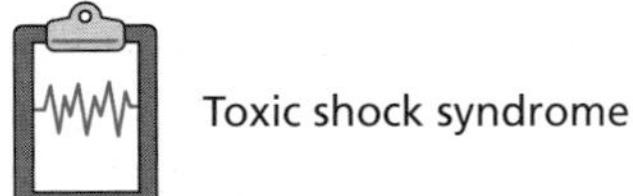

3-15 Internal detection of viral infection induces cells to make an interferon response

So far in this chapter we have focused on innate immunity against extracellular pathogens, principally bacteria. We shall now examine the mechanisms of innate immunity that cells use against viral infection. When a virus infects a human cell, it uses the machinery of the cell to transcribe, translate, and replicate its genome, resulting in the presence in the cytoplasm of viral nucleic acids. Human cells have sensor proteins in the cytoplasm that can detect viral nucleic acids and initiate a defensive response. This response results in the production of cytokines called **type I interferons**, or simply **interferon**, that interfere with the propagation of infection.

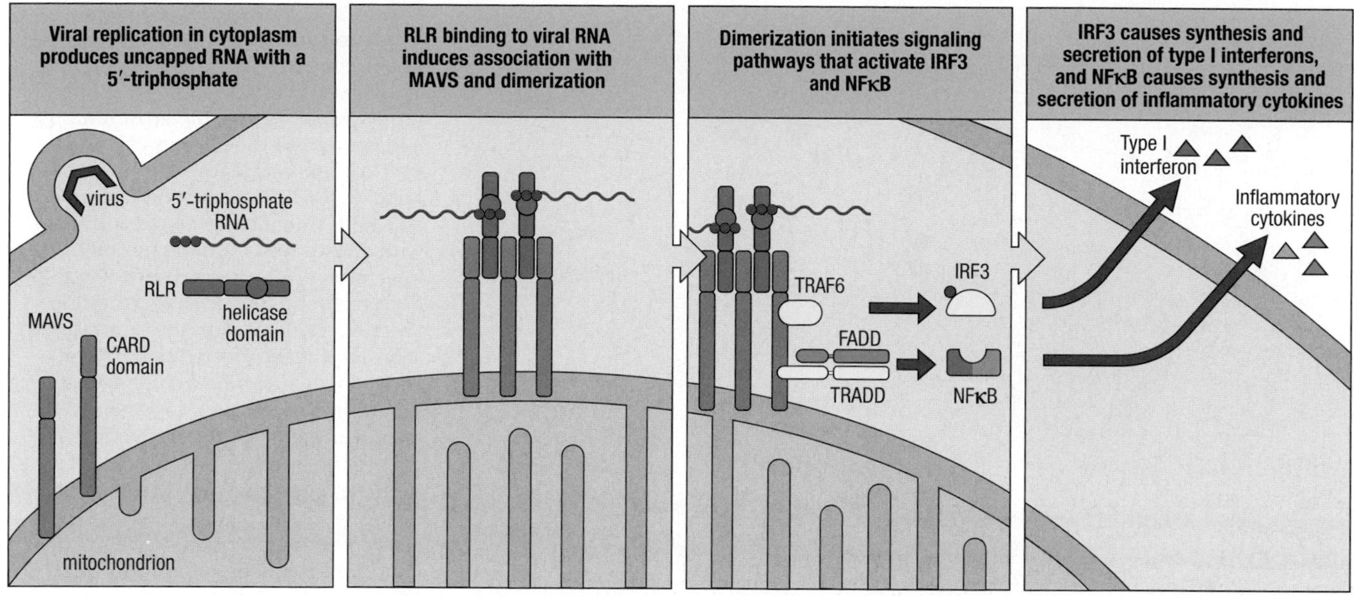
Viral replication in cytoplasm produces uncapped RNA with a 5′-triphosphate
virus
5′-triphosphate RNA
RLR
helicase domain
MAVS
CARD domain
mitochondrion
RLR binding to viral RNA induces association with MAVS and dimerization
Dimerization initiates signaling pathways that activate IRF3 and NFκB
TRAF6
IRF3
FADD
TRADD
NFκB
IRF3 causes synthesis and secretion of type I interferons, and NFκB causes synthesis and secretion of inflammatory cytokines
Type I interferon
Inflammatory cytokines

Figure 3.32 Recognition of viral nucleic acid by RIG-1-like receptors (RLRs) initiate an inflammatory response. First panel: a cell that is infected with virus is beginning to produce uncapped viral RNA with a 5′-triphosphate. At this stage the RLRs are in the cytoplasm and the MAVS adaptor molecules are associated with the mitochondrial membrane. Second panel: after the helicase domain of the RLR has recognized the 5′ triphosphate of the viral RNA and its CARD domain has recognized the CARD domain of MAVS molecules, a dimeric structure is formed on the mitochondrial membrane. One site on MAVS engages TRAF6, which leads to the activation of IRF3 (third panel) and the secretion of type I interferon (fourth panel). A second site on MAVS interacts with a complex of TRADD (TNF-receptor-associated death domain protein) and FADD (Fas-associated death domain protein) that leads to the activation of NFκB and the secretion of inflammatory cytokines.

The immediate effects of type I interferon are to interfere with viral replication by the infected cell, and to signal to neighboring uninfected cells that they, too, should prepare to resist a viral infection. Further effects of type I interferon are to alert cells of the immune system that an infection is about, and to make virus-infected cells more vulnerable to attack by killer lymphocytes. As all types of human cell are susceptible to viral infections, they are all equipped to make type I interferons and their receptors. The receptor is always present on cell surfaces, ready to bind interferon newly made in response to infection. Although type I interferon is barely detectable in the blood of healthy people, upon infection it becomes abundant.

Viral RNAs in the cytoplasm are detected by proteins called **RIG-I-like receptors (RLRs)**—**RIG-I** and **MDA-5** being examples (Figure 3.32). Sensors for viral DNA are less well characterized. RIG-I (retinoic-acid-inducible gene 1) and MDA-5 (melanoma differentiation-associated protein 5) consist of an RNA helicase-like domain that recognizes RNA, and two CARD domains that interact with the mitochondrial antiviral signaling protein (MAVS) in the outside of the mitochondrial membrane. This interaction causes dimerization of the RIG-like helicases and initiation of a signaling pathway that leads to phosphorylation of the transcription factor interferon-response factor 3 (IRF3) in the cytoplasm. IRF3 then dimerizes and enters the nucleus to help initiate transcription of the IFN-β genes, which also requires the transcription factors NFκB and AP-1. A family of genes on chromosome 9 encodes the type I interferons, which comprise IFN-β, multiple forms of IFN-α, and several additional types: IFN-δ, IFN-κ, IFN-λ, IFN-θ, and IFN-ω. Once IFN-β is secreted, it acts both in an **autocrine** fashion, by binding to receptors on the cell that made it, and a **paracrine** fashion, by binding to receptors on uninfected cells nearby (Figure 3.33).

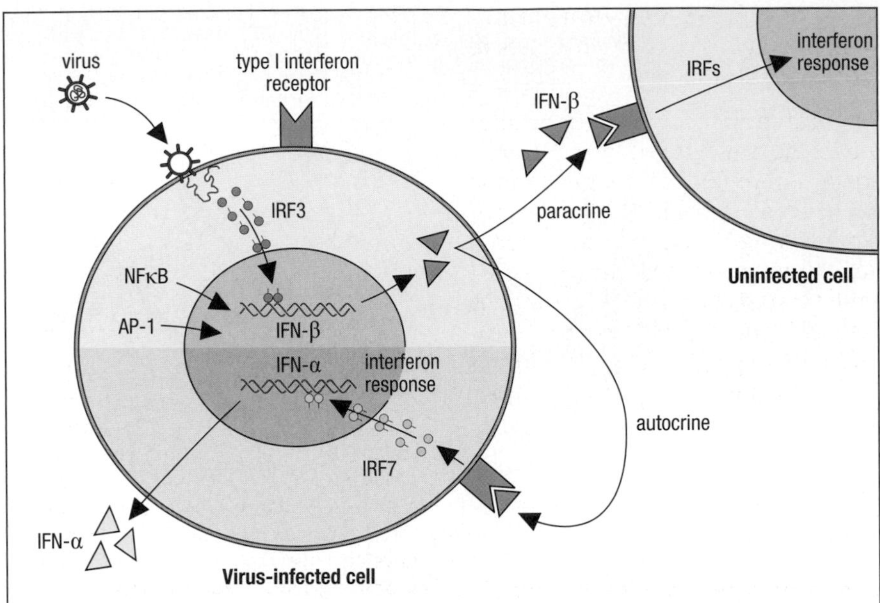

Figure 3.33 Virus-infected cells produce type I interferons. The cell on the left is infected with a virus that triggers signals that lead to the phosphorylation, dimerization, and passage to the nucleus of the transcription factor interferon-response factor 3 (IRF3). Transcription factors NFκB and AP-1 are mobilized and coordinate with IRF3 to turn on transcription of the interferon (IFN)-β gene. These events are depicted in the upper half of the cell. Secreted IFN-β binds to the interferon receptor on the infected cell surface, acting in an autocrine fashion to mobilize other interferon-response factors and change patterns of gene expression to give the interferon response. These events are depicted in the lower half of the cell, being exemplified by IRF7 turning on transcription of the IFN-α gene, which it does without the need for AP-1 or NFκB. Secreted IFN-β will also bind to the interferon receptor expressed by nearby cells (top right) that are not infected by the virus, acting in a paracrine fashion to induce the interferon response that helps these cells to resist infection.

When interferon binds to its receptor, the intracellular Jak1 and Tyk2 kinases associated with the receptor initiate reactions that change the expression of a variety of human genes, a process called the **interferon response** (Figure 3.34). Among the cellular proteins induced by interferon are some that interfere directly with viral genome replication. An example is the enzyme oligoadenylate synthetase, which polymerizes ATP by 2′–5′ linkages rather than the 3′–5′ linkages normally present in human nucleic acids. These unusual oligomers activate an endoribonuclease that degrades viral RNA. Also activated by IFN-α and IFN-β is a serine–threonine protein kinase called protein kinase R (PKR) that phosphorylates and inhibits the protein synthesis initiation factor eIF-2, thereby preventing viral protein synthesis and the production of new infectious virions.

Several of the interferon-induced proteins are transcription factors similar to IRF3, the only one of the group that is made constitutively. These other interferon-response factors are instrumental in turning on the transcription of many different genes, including those for type I interferons other than IFN-β. Interferon-response factor 7 (IRF7) initiates the transcription of IFN-α, which

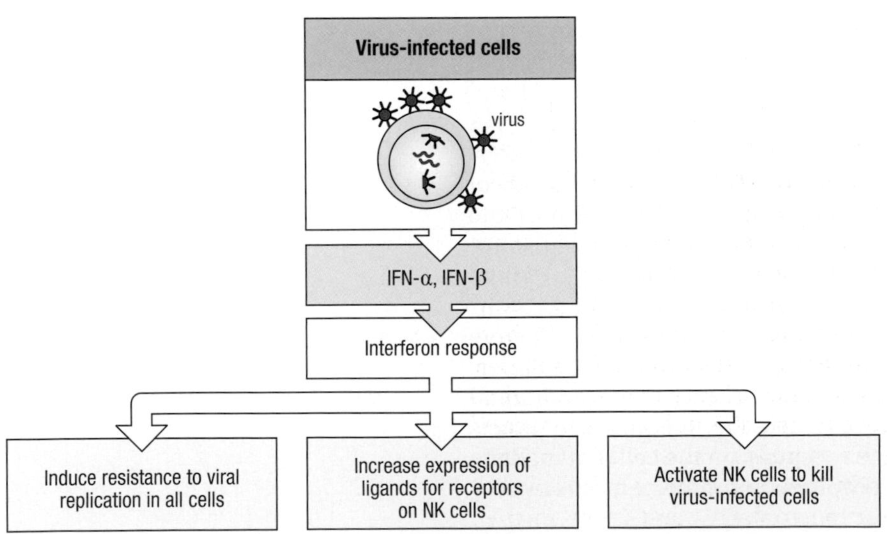

Figure 3.34 Major functions of the type I interferons. Interferon-α and interferon-β (IFN-α and IFN-β) have three major functions. First, they prevent viral replication by activating host genes that destroy viral mRNA and inhibit translation of viral proteins. Second, they increase the expression of ligands for NK-cell receptors. Third, they activate NK cells to kill virus-infected cells.

does not require the participation of NFκB and AP-1. In this manner, a positive feedback loop develops, in which a small initial amount of interferon acts to increase both the size and range of its future production.

As well as interfering with viral replication, interferon also induces cellular changes that make the infected cell more likely to be attacked by killer lymphocytes. NK cells are lymphocytes of innate immunity that defend against viral infections by secreting cytokines and killing infected cells. When IFN-α or IFN-β binds to the interferon receptors on circulating NK cells, these become activated and are drawn into infected tissues, where they attack virus-infected cells. Because of its power to boost the immune response, type I interferon has been explored as treatment for human disease. It ameliorates several conditions: infections with hepatitis B or C viruses; the degenerative autoimmune disease multiple sclerosis, which affects the central nervous system; and certain leukemias and lymphomas.

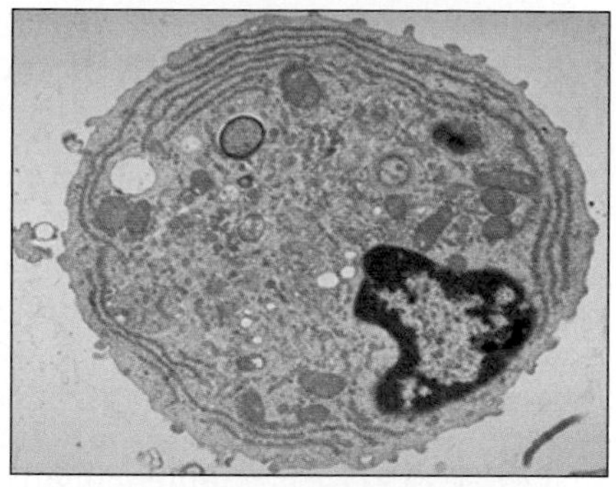

Figure 3.35 Plasmacytoid dendritic cell from human peripheral blood. Note the extensive rough endoplasmic reticulum that is similar in appearance to that of a plasma cell and is due to the massive synthesis and secretion of interferon by these cells. These cells have also been called type I interferon-producing cells. Image courtesy of Yong-Jun Liu.

3-16 Plasmacytoid dendritic cells are factories for making large quantities of type I interferons

Although almost all human cells can secrete some type I interferon, specialized cells called **plasmacytoid dendritic cells** (**PDCs**) secrete up to 1000-fold more interferon than other cells. They are therefore considered to be professional interferon-producing cells. Plasmacytoid dendritic cells combine phenotypic features of lymphocytes and myeloid dendritic cells in a way that does not fit easily into conventional schemes of branching hematopoietic cell development (see Figure 1.13). They are present in the blood, where they comprise fewer than 1% of the total leukocytes, and are also found in the lymphoid tissues but not in other tissues. Plasmacytoid dendritic cells are distinguished by a cytoplasm resembling that of an antibody-producing plasma cell, another cell committed to the mass production of a single secreted protein (Figure 3.35). Plasmacytoid dendritic cells detect the presence of viral infection by using TLR7 and TLR9, which then signal through MyD88 as seen for TLR4 (see Figure 3.7). These signals lead to the activation of IRF7 and its translocation to the nucleus, where it orchestrates the mass production of type I interferon. Within 6 hours of activation, some 60% of the transcription is devoted to producing interferon, the entire family of interferon genes being expressed. The copious quantities of type I interferons made by plasmacytoid dendritic cells spread systemically in the lymph and the blood. By stimulating cells throughout the body, the type I interferon serves to prevent the systemic spread of infection.

3-17 Natural killer cells are the main circulating lymphocytes that contribute to the innate immune response

Three main populations of lymphocytes circulate in the blood: the B cells and T cells of late-acting adaptive immunity, and the **natural killer cells** (**NK cells**) of early-acting innate immunity. Reflecting this temporal division in labor, B cells and T cells are small quiescent lymphocytes that require several days of proliferation and differentiation to become useful in resisting infection, whereas NK cells are large and active lymphocytes that are more speedily induced to respond to infection, cancer, or other form of stress. NK cells have two distinctive functions in the innate immune response to viral infection. The first is to kill cells infected with virus. This sacrifice of cells hosting the virus impedes virus replication and thus its spread to neighboring cells. The second function of NK cells is to maintain, and even increase, the state of inflammation in the infected tissue. This is achieved by their secretion of inflammatory cytokines that act mainly on the resident macrophages and increase their capacity to secrete inflammatory cytokines and to phagocytose viral particles and microorganisms in the extracellular environment. NK cells thus contribute to defense against both intracellular and extracellular pathogens.

To a rough approximation, the functions performed by NK cells in innate immunity are like those of T cells in adaptive immunity. Because of these functional similarities, many of the effector and cell-surface molecules of NK cells are also found in T cells. Reflecting this overlap, no single marker has been found to define NK cells. In other words, no intracellular protein or cell-surface receptor is exclusively expressed by NK cells and expressed by all of them. For immunologists, the working definition of a human NK cell is a lymphocyte that expresses CD56, a protein of unknown function expressed by all NK cells, and lacks CD3, a cell-surface protein present on all T cells.

People who are immunodeficient because they lack NK cells suffer from persistent viral infections, particularly of herpes viruses, which the patients are unable to terminate by themselves despite making a normal adaptive immune response to the virus. For these patients, courses of antiviral drugs are used to terminate the infections. The case histories of these rare individuals reveal the importance of NK cells in managing and containing viral infections during the period before an adaptive immune response becomes effective. Such patients also show how, in immunocompetent individuals, the innate NK-cell response works in partnership with the adaptive cytotoxic T-cell response to control and vanquish viral infections (Figure 3.36).

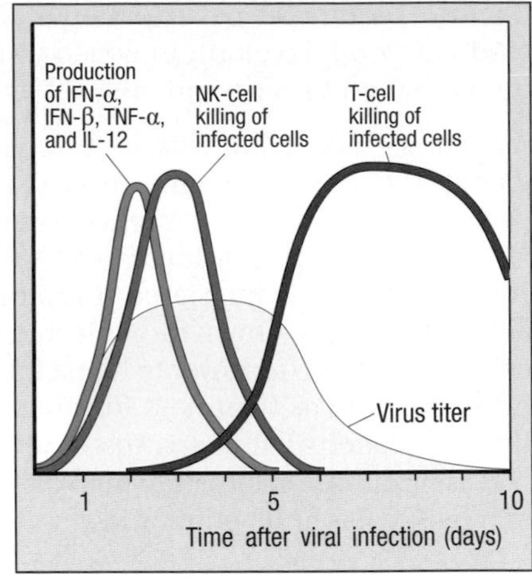

Figure 3.36 NK cells provide an early response to virus infection. The kinetics of the immune response to an experimental virus infection of mice are shown. As a result of infection, a burst of cytokines is secreted, including IFN-α, IFN-β, TNF-α, and IL-12 (green curve). These induce the proliferation and activation of NK cells (blue curve), which are seen as a wave emerging after cytokine production. NK cells control virus replication and the spread of infection while effector killer T cells (red curve) are developing. The level of virus (the virus titer) is given by the curve described by the yellow shading.

3-18 Two subpopulations of NK cells are differentially distributed in blood and tissues

In healthy people, NK cells are abundant in the blood, where they account for 5–25% of the lymphocytes, depending on the individual. They are also present in most, if not all, other tissues. NK cells have been divided into two subpopulations on the basis of the abundance of the CD56 glycoprotein on the cell surface and their cytotoxic potential. The $CD56^{dim}$ NK cells, which comprise more than 90% of the blood NK cells, have fewer CD56 molecules than the $CD56^{bright}$ cells, but have a greater capacity for killing cells. In tissues other than blood, the relative sizes of the two NK-cell subpopulations are reversed. Here, $CD56^{bright}$ cells predominate, as exemplified by the lungs, where more than 80% of NK cells are of the $CD56^{bright}$ type. It has been much debated whether the $CD56^{dim}$ and $CD56^{bright}$ NK cells represent different types of effector NK cell or different stages in NK-cell development. Currently, the evidence favors the latter. The early stages of NK-cell development take place in the bone marrow, but the later stages appear to take place in secondary lymphoid organs and other tissues.

With regard to development, a specialized population of NK cells that contributes to human reproduction is informative. In a woman's uterus, NK cells called **uterine NK cells (uNK)** are the most abundant leukocyte. Their numbers fluctuate with the menstrual cycle and they are essential to forming the placenta, which occurs at an early stage of pregnancy. The role of the uterine NK cells is not to kill cells and develop inflammation but to cooperate with fetal trophoblast cells in enlarging maternal blood vessels so that they have the capacity to supply the placenta, and thus the growing fetus, with sufficient oxygen and nourishment throughout the 9 months of pregnancy. Almost all uterine NK cells are of the $CD56^{bright}$ type, poor at killing cells but good at secreting growth factors and non-inflammatory cytokines. These properties depend on the unique environment of the uterus, and change if uterine NK cells are removed from that environment.

These observations raise the possibility that NK cells in general adapt to the tissues in which they take up residence. Like B cells and T cells, NK cells

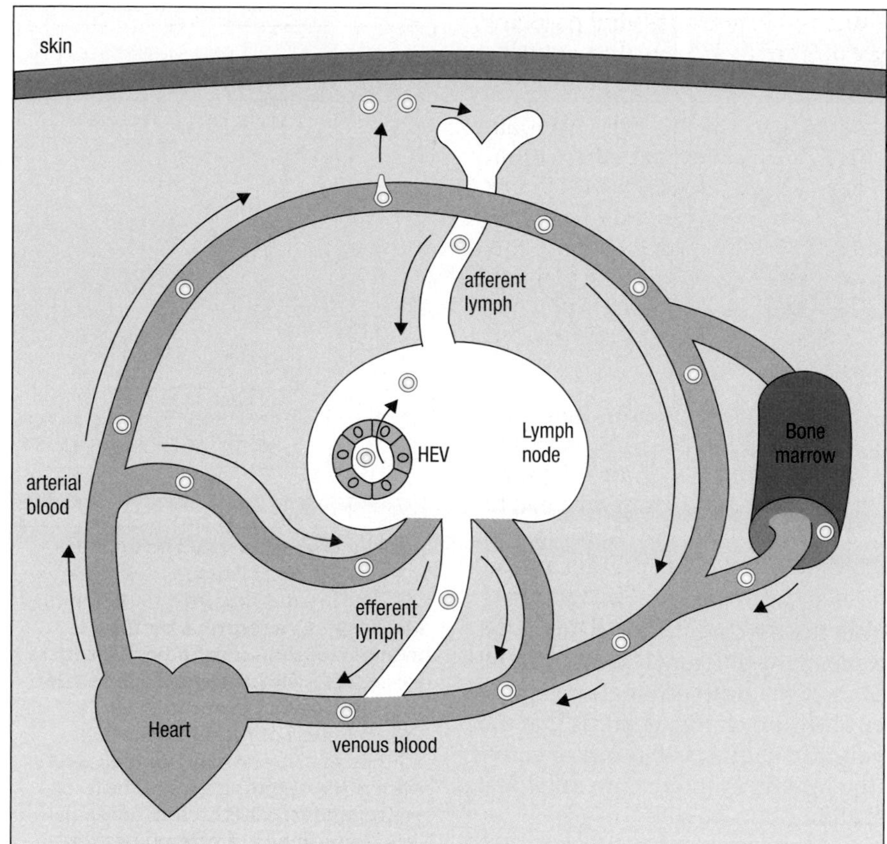

Figure 3.37 NK cells circulate between blood and tissues. NK cells originate and develop in the bone marrow, which they leave to circulate in the blood. They can enter secondary lymphoid tissues at high endothelial venules and return to the blood in the lymph. NK cells can also leave the blood to enter other tissues, such as the skin, and return to the circulation via the draining lymph. On entering lymphoid and other tissues, NK cells can further differentiate into various types of effector cells. For example, some specialize in secreting cytokines, whereas others focus on killing infected or damaged cells.

circulate in blood, enter tissues, and are present in the lymph vessels that allow lymphocytes to travel from secondary lymphoid organs and tissues to the blood. Although the recirculation of NK cells has not been as thoroughly studied as that of B cells and T cells, the growing body of evidence is consistent with the pathways being similar (Figure 3.37).

An irony of immunology is that the order in which the three main types of lymphocyte have been studied is the reverse of that in which they contribute to the immune response, and the reverse of their likely evolution. The most recently evolved lymphocytes are thought to be the B cells, and they and their antibodies were studied for more than half a century before T cells came under scrutiny in the 1960s. Although studied morphologically much earlier by pathologists, NK cells were 'rediscovered' by cancer immunologists in the 1970s, and awareness of their role in innate immunity to infection began only in the 1990s. For this reason, knowledge of NK-cell biology is less advanced than that of B and T cells, even though NK cells were probably the first to evolve.

3-19 NK-cell cytotoxicity is activated at sites of virus infection

In the absence of infection, more than 100 million NK cells are circulating at speed through the human body, fully loaded with poisons whose sole function is to kill human cells. This dangerous state of affairs necessitates strict regulation of the activation and implementation of NK-cell cytotoxicity. Various safety features have evolved to prevent NK cells from attacking healthy cells while preserving their efficient killing of infected cells. The first feature is that NK cells are unable to discharge their weapons from a distance but need to be in the most intimate contact with a target cell. This also limits killing to one

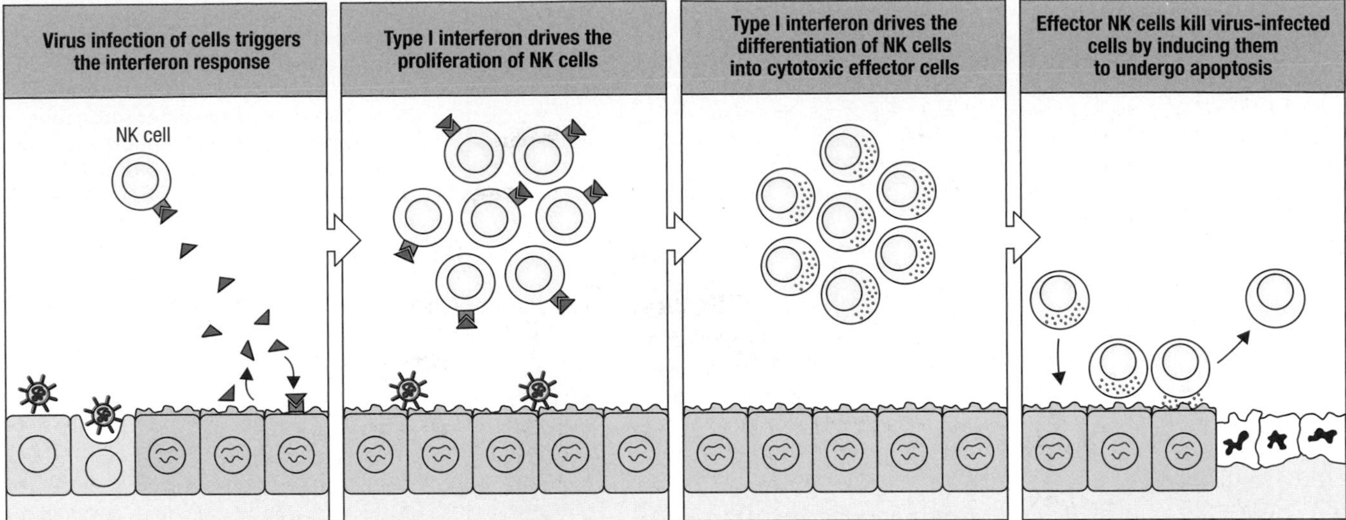
Virus infection of cells triggers the interferon response
Type I interferon drives the proliferation of NK cells
Type I interferon drives the differentiation of NK cells into cytotoxic effector cells
Effector NK cells kill virus-infected cells by inducing them to undergo apoptosis
NK cell

Figure 3.38 Activation by type I interferon induces the proliferation of NK cells and their differentiation into cytotoxic cells. The example shown here is of a virus-infected epithelium. The infected epithelial cells respond by secreting type I interferon, which binds to receptors on epithelial cells, inducing an interferon response, and also to receptors on NK cells. This interaction activates the NK cells, causing them to proliferate and differentiate into potent cytotoxic effector cells. The effector NK cells kill the virus-infected epithelial cells by inducing them to undergo apoptosis. By killing the infected cells, the NK cells interrupt the replication of the virus, thus inhibiting the spread of infection.

target cell at a time. The second feature is that the decision to kill or not to kill is made by the sum of the interactions between many different types of NK-cell receptors and their cognate ligands on the target cell, with no single receptor–ligand interaction dominating the decision. The third characteristic is that the ground state for the interaction of an NK cell with a target cell is one of active inhibition based on inhibitory receptors. The inhibitory interaction must first be overcome by activating receptors before killing can occur.

Within a virus-infected tissue, IFN-α and IFN-β are the first cytokines to which the resident NK cells are exposed. Their effect is, first, to induce NK-cell mitosis and proliferation, and second, to induce NK-cell differentiation into cytotoxic effector cells that kill virus-infected cells (Figure 3.38). In the infected tissue, an NK cell makes initial transient contacts with any tissue cell it encounters. If the cell is healthy, the NK cell moves on; however, if the cell is infected, then firmer adhesion involving the CR3 and LFA-1 adhesion molecules occurs in a manner similar to that seen for neutrophil extravasation (see Section 3-8). The interacting sets of receptors and ligands become concentrated at localized regions at the surfaces of the two cells, where together they form an immunological synapse that helps hold the cells together and through which information and material can be exchanged. This synapse is called the **NK-cell synapse**. Many of the interactions of lymphocytes with each other and other cell types involve similar synapses. Diverse activating and inhibitory receptors are recruited into the NK-cell synapse, where they can engage their ligands if they are present on the prospective target cell. Activating and inhibitory signals from the various ligand–receptor interactions are integrated by the NK cell. If inhibitory signals predominate, the target cell is released unharmed; if activating signals predominate, execution proceeds.

Once committed to killing the target cell, the NK cell strengthens the adhesive interactions at the synapse and undergoes an intracellular reorganization that facilitates a quick, precise, and clean delivery of its toxic cargo. This involves a reorganization of the intracellular network of actin microfilaments that aligns the microtubule-organizing center and the Golgi apparatus toward the synapse. The membrane-bound lytic granules are transported along microtubule tracks to the synapse where, on fusion with the NK cell's plasma membrane, the granule contents are shot out onto the outer surface of the target cell.

The various enzymes, proteins, and proteoglycans present in the granules disrupt the plasma membrane, and by selective proteolytic attack on the target cell they induce it to commit suicide by a process called **programmed cell**

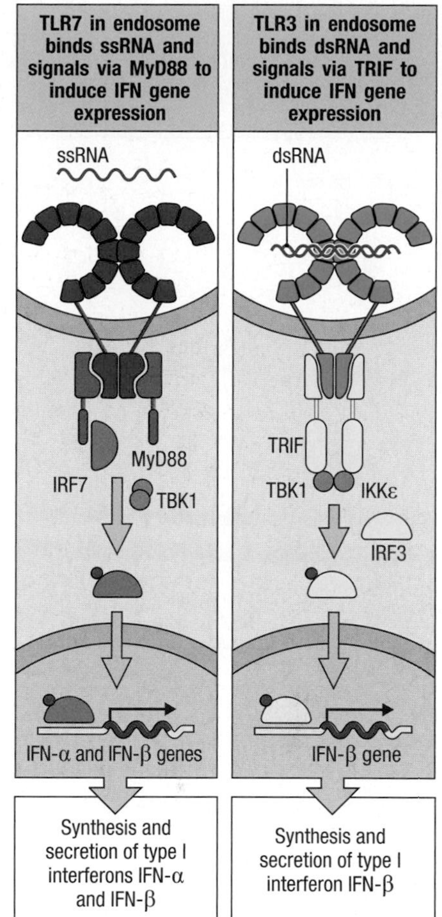

Figure 3.39 Two pathways of pathogen recognition and intracellular signaling that give an interferon response. Left panel: TLR7 detects single-stranded RNA (ssRNA) and signals through MyD88 to phosphorylate and activate IRF7. This leads to the production of interferon-α (IFN-α) and IFN-β. Right panel: TLR3 senses double-stranded viral RNA (dsRNA) and signals through TRIF, which is structurally related to MyD88. TRIF activates the serine–threonine kinases IκKε (IKKε) and TBK1, leading to IRF3 activation. This pathway leads to the production of IFN-β but not IFN-α.

death or **apoptosis**. After the NK cell has detached from the target, the target cell's DNA starts to be fragmented by the cell's own nucleases. Eventually the nucleus becomes disrupted, and there is loss of membrane integrity and normal cell morphology (see Figure 3.38). The cell shrinks by the shedding of membrane-enclosed vesicles and continued degradation of the cell contents until little is left. The remains of the apoptotic cell are disposed of by a macrophage.

NK cells express 3 of the 10 Toll-like receptors. These comprise TLR3, which recognizes double-stranded viral RNA, and TLR7 and TLR8, which recognize single-stranded RNA. TLR3 and TLR8 are expressed only by NK cells and are dedicated to the NK-cell response to infection. On recruitment to a site of infection, NK cells expressing these Toll-like receptors can recognize their ligands and be further activated by signaling coming through TLR3, TLR7, and TLR8. The nucleic-acid-sensing receptors TLR7, TLR8, and TLR9 signal through a unique MyD88-dependent pathway that leads to activation of the IRF7 transcription factor and to the production of IFN-α and IFN-β (Figure 3.39, left panels). TLR3 is the only Toll-like receptor that does not signal using the MyD88 adaptor molecule but instead uses another signaling pathway that leads to activation of the transcription factor IRF3 and to the production of IFN-β only (Figure 3.39, right panels).

This pathway uses an adaptor molecule called Toll-receptor-associated activator of interferon (TRIF). The Toll-like receptors divide into three groups according to the adaptor molecules that their signaling pathways use. TLR3 uses only TRIF, TLR4 can use MyD88 or TRIF, and the other eight receptors use only MyD88. In summary, Toll-like receptor engagement and signaling have the effect of turning on the synthesis and secretion of type I interferon at the site of infection. In turn, this increases the activation of NK-cell cytotoxicity (see Figure 3.38).

3-20 NK cells and macrophages activate each other at sites of infection

Of the inflammatory cytokines secreted by activated macrophages in infected tissue, IL-12 directs the activation and recruitment of NK cells; it was previously named NK-cell stimulatory factor. NK cells resident in the infected tissue that were not activated by type I interferon are potentially the first to be activated by IL-12, followed by NK cells entering from the blood under the influence of inflammatory cytokines. Activation by IL-12 is most effective when an NK cell and an activated macrophage contact each other directly through an NK-cell synapse similar to that described earlier (see Section 3-19). The intimate contact between the two cells allows the macrophage to deliver the IL-12 directly onto the NK cell's surface (Figure 3.40).

NK cells have a surface receptor for IL-12 that consists of two polypeptides called IL-12Rβ1 and IL-12Rβ2. In the absence of IL-12 the two polypeptides cannot interact with each other, but in the presence of IL-12 they are brought together because IL-12 has a separate binding site for each receptor polypeptide. The resulting juxtaposition of the receptor polypeptides generates

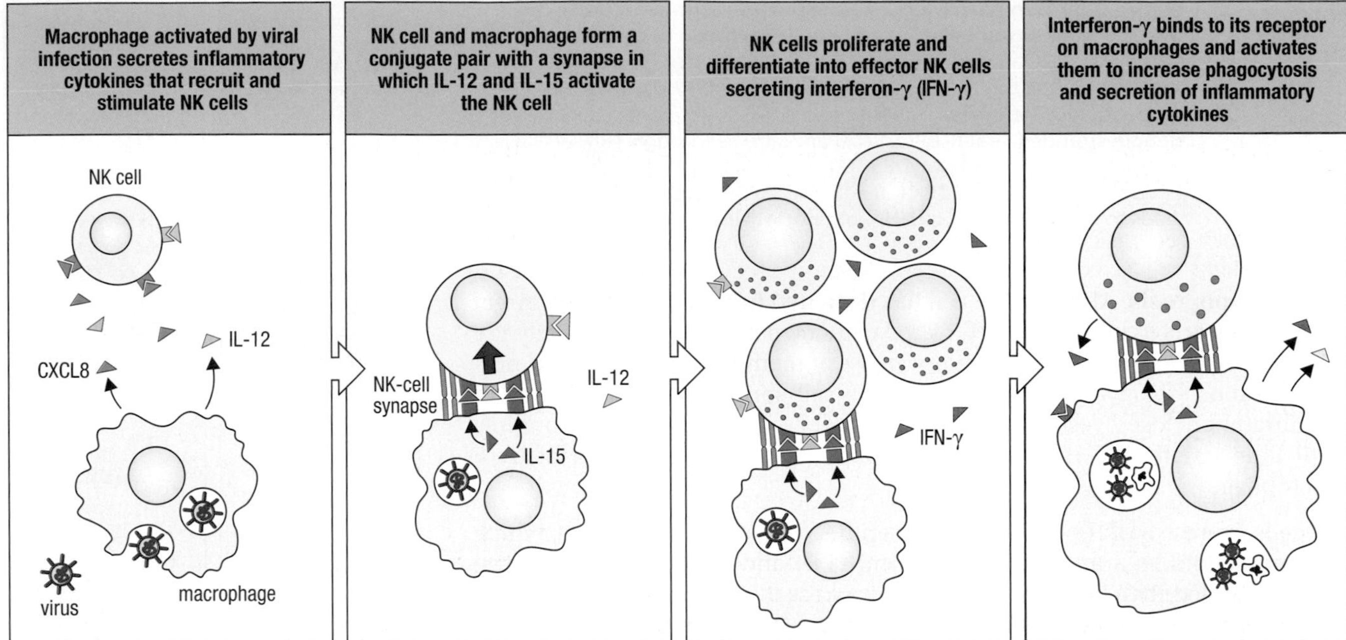

Macrophage activated by viral infection secretes inflammatory cytokines that recruit and stimulate NK cells
NK cell
CXCL8
IL-12
virus
macrophage
NK cell and macrophage form a conjugate pair with a synapse in which IL-12 and IL-15 activate the NK cell
NK-cell synapse
IL-12
IL-15
NK cells proliferate and differentiate into effector NK cells secreting interferon-γ (IFN-γ)
IFN-γ
Interferon-γ binds to its receptor on macrophages and activates them to increase phagocytosis and secretion of inflammatory cytokines

Figure 3.40 Interaction between a macrophage and an NK cell induces NK-cell proliferation and differentiation into effector cells secreting interferon-γ (IFN-γ), which further activates the macrophage. Activated macrophages at a site of viral infection secrete inflammatory cytokines. Of importance for NK cells are the chemokine CXCL8 and IL-12. By binding to its receptor on NK cells, CXCL8 recruits NK cells from the blood into the infected tissue (first panel). An NK cell and a macrophage form a conjugate pair in which IL-12 secreted by the macrophage binds to its receptor on the NK cell (second panel). This activates the NK cell, inducing its proliferation and differentiation into effector cells secreting IFN-γ (third panel). The binding of IFN-γ to its receptor on the macrophage induces increased activation of the macrophage. This is manifested in improved phagocytosis of viral particles and virus-infected cells that have been killed by NK cells, and increased secretion of inflammatory cytokines (fourth panel).

intracellular signals that lead to NK-cell activation. IL-15 is another cytokine made by macrophages that is critical for the activation, proliferation, and survival of NK cells. Its effect is particularly dependent upon contact between the NK cell and the macrophage, because IL-15 is not secreted into the extracellular space but is retained at the macrophage surface by an IL-15 receptor on the macrophage and then bound by a different IL-15 receptor on the NK cell. Stimulation by IL-12 disfavors the differentiation of NK cells into cytotoxic effector cells and favors their differentiation into effector NK cells, which produce cytokines that act on macrophages and other cells in the infected tissue to produce a heightened state of inflammation (see Figure 3.40)

The key cytokine secreted by NK cells is a potent inflammatory cytokine called **interferon-γ (IFN-γ)**. Also known as **type II interferon**, IFN-γ is structurally and functionally different from the type I interferons. Unlike the diversity of the type I interferons, there is only one form of IFN-γ. In the innate immune response, NK cells are the principal source of IFN-γ. In the infected tissue, the principal targets for IFN-γ are macrophages, both those resident at the start of the infection and those that derive from monocytes recruited from the blood during the course of infection. The interaction of IFN-γ with the IFN-γ receptor on a macrophage generates signals that further activate the macrophage so that it becomes more efficient in the phagocytosis and destruction of pathogens, as well as in secreting inflammatory cytokines (see Figure 3.40). The secretion of IL-12 by macrophages to activate NK cells, and the secretion of IFN-γ by NK cells to activate macrophages, creates a positive feedback loop that increases the strength of the innate immune response and the likelihood that the viral infection will be terminated without necessitating an adaptive immune response.

3-21 Interactions between dendritic cells and NK cells influence the immune response

Myeloid dendritic cells, which we shall refer to throughout the book simply as **dendritic cells**, are developmentally related to macrophages, and similarly found as resident cells in all tissues of the body (see Section 1-7). Macrophages, however, are effector cells that kill pathogens and cause inflammation, whereas dendritic cells are sentinels that sample the environment to detect infection and then use that information to stimulate the responses of lymphocytes against the pathogen. In innate immunity, dendritic cells interact with NK cells, whereas in adaptive immunity they turn their attention to T cells.

Dendritic cells take up pathogens and their products from the environment and can be infected with viruses and other intracellular pathogens. These events induce changes in the proteins on dendritic-cell surfaces that are monitored by the array of cell-surface receptors on NK-cell surfaces. When one or more NK-cell receptors detects pathogen-induced changes on the dendritic-cell surface, the two cells form a strong interaction with a synapse between them. Signals induced in the dendritic cell lead to the expression of IL-15, which induces NK cells to proliferate, differentiate, and survive.

In laboratory experiments, the effect of the NK cells on the dendritic cells varies with the conditions. When NK cells outnumber dendritic cells, the NK cells become cytotoxic cells and kill the dendritic cells. In contrast, when dendritic cells outnumber NK cells, the NK cells secrete cytokines that induce the dendritic cells to differentiate into a form that leaves the infected tissue and migrates in the lymph to the draining secondary lymphoid tissue. The migration of dendritic cells from infected tissue to lymphoid tissue is the event that initiates the adaptive immune response. This occurs when innate immunity is insufficient to terminate infection, a state for which too few NK cells could be a symptom. In contrast, an overpowering force of NK cells that prevents dendritic cells from differentiation and migration could represent the situation when innate immunity alone is able to stop the infection, without the need for an adaptive immune response (Figure 3.41).

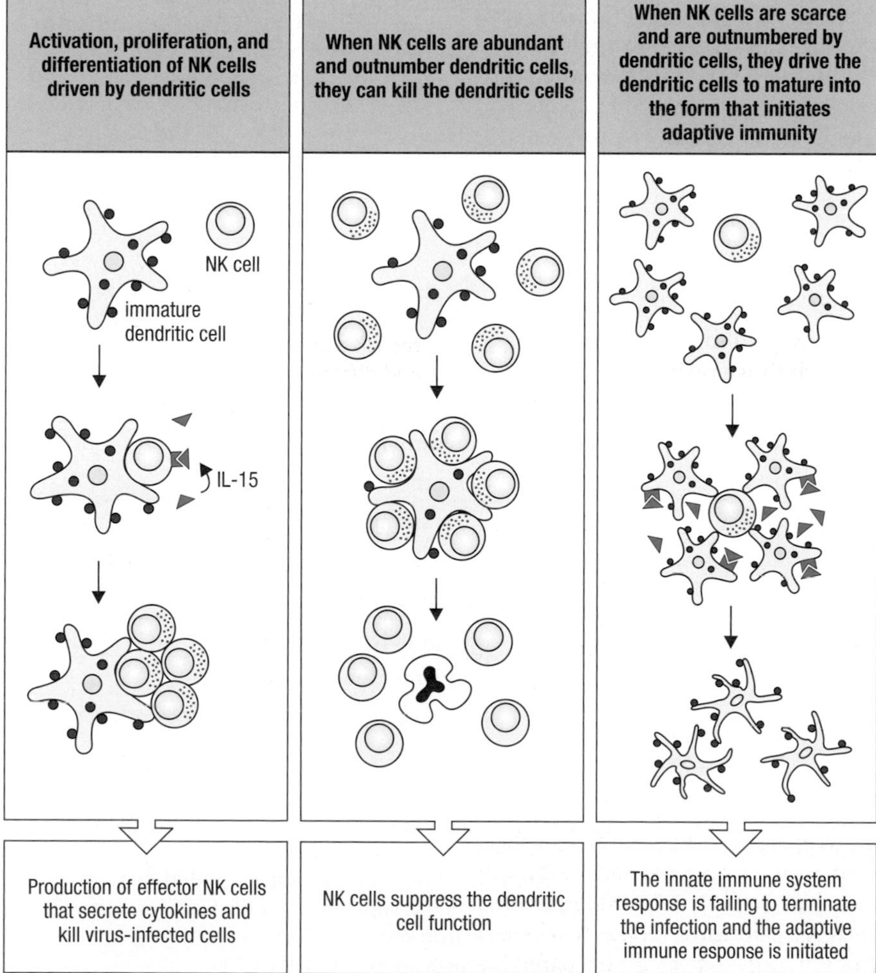

Activation, proliferation, and differentiation of NK cells driven by dendritic cells
When NK cells are abundant and outnumber dendritic cells, they can kill the dendritic cells
When NK cells are scarce and are outnumbered by dendritic cells, they drive the dendritic cells to mature into the form that initiates adaptive immunity
NK cell
immature dendritic cell
IL-15
Production of effector NK cells that secrete cytokines and kill virus-infected cells
NK cells suppress the dendritic cell function
The innate immune system response is failing to terminate the infection and the adaptive immune response is initiated

Figure 3.41 Interactions of NK cells with dendritic cells. The interaction of NK cells with dendritic cells can have several outcomes. In a virus-infected tissue, an immature dendritic cell expressing viral antigens can activate NK cells and drive their differentiation into effector cells (left panels). When activated cytotoxic NK cells are abundant and innate immunity is overcoming infection, NK cells kill dendritic cells and prevent their activation of adaptive immunity (center panels). When NK cells are scarce and innate immunity cannot control the infection, NK cells induce dendritic cells to differentiate into the form that travels to secondary lymphoid tissue in order to initiate the adaptive immune response (right panels).

Summary to Chapter 3

The macrophages resident in an infected tissue are responsible for orchestrating the induced innate response to infection. They express a variety of cell-surface, endosomal, and cytoplasmic receptors that recognize the distinctive chemical properties of the sugars, fats, proteins and nucleic acids of pathogens. After an infection has been detected, the macrophages secrete inflammatory cytokines and chemokines that alter the permeability of local blood vessels and actively recruit neutrophils, monocytes, and NK cells into the infected tissue from the blood circulation. This activity creates a localized state of inflammation with associated heat, pain, redness, and swelling. The inflammatory cytokines also induce the acute-phase response, which changes the patterns of protein synthesis in the liver, the source of many plasma proteins. The overall effect is to increase the production of proteins contributing to innate immunity at the expense of other proteins. One such protein whose concentration rises markedly is C-reactive protein, which recognizes pathogens and initiates complement fixation by the classical pathway. In extracellular infections, the powerful but short-lived neutrophil is the major effector cell that kills the pathogens. Disposing of the numerous neutrophils that die at sites of infection is another challenge faced by the resident macrophages.

Almost all cell types respond to viral infections by secreting type I interferons. This general cellular response is complemented by the plasmacytoid dendritic cell, which is dedicated to making huge quantities of these cytokines. Type I interferons activate NK cells to kill virus-infected cells, thereby preventing viral replication and the spread of infection. Killing involves the formation of a synapse between NK cells and the virus-infected target cells; this synapse is the conduit through which lethal poisons are accurately delivered. NK cells form similar synapses with resident macrophages that result in their mutual activation and the secretion of interferon-γ by the NK cells. Most infections that recruit an induced innate immune response are terminated by its powerful defense mechanisms. The minority of infections that are not resolved at this stage then call up the adaptive immune response. The interactions between the NK cell and the dendritic cell are implicated in making this decision.

Questions

3–1 Which of the following participates in an immediate rather than an induced innate immune response to infection in a human host? (Select all that apply.)

a. scavenger receptors
b. commensal microorganisms
c. dectin-1
d. alternative pathway of complement activation
e. MyD88
f. CXCL8
g. defensins
h. NOD1 and NOD2
i. lectin pathway of complement activation.

3–2 Inflammatory cytokines _____.

a. are stimulated through scavenger receptors and not Toll-like receptors
b. facilitate the migration of neutrophils from blood to infected tissues
c. are produced only by phagocytic cells
d. mediate direct dissociation of IκB from NFκB
e. act as opsonins.

3–3 Match the receptor in column A with its ligand in column B.

Column A	Column B
a. NOD2	1. carbohydrate
b. TLR4	2. muramyl dipeptide
c. dectin-1	3. γ-glutamyl diaminopimelic acid
d. CR3/CR4	4. teichoic acid
e. NOD1	5. lipopolysaccharide (LPS)
f. SR-A	6. filamentous hemagglutinin
g. TLR9	7. unmethylated CpG-rich nucleotide motif

3–4 Which of the following pairs is mismatched?

a. catalase: hydrogen peroxide production
b. bactericidal permeability-increasing protein: binds LPS and kills Gram-negative bacteria
c. elastase: basement membrane degradation
d. NADPH oxidase: superoxide radical production
e. gelatinase: iron sequestration
f. paracrine: acts on neighboring cells.

3–5 The positive-feedback loop established between macrophages and dendritic cells in infected tissues involves the secretion of _____ by the macrophage and _____ by the NK cell.
a. type I interferons; activating receptors
b. IL-15; IL-12
c. IL-12; IFN-γ
d. IFN-γ; cytotoxic granules
e. ICAM-1; LFA-1.

3–6 Describe in chronological order the steps involved in the recruitment of neutrophils to infected tissue sites during an innate immune response. Use the following terms in your description: rolling adhesion, tight binding, extravasation, migration, inflammatory mediators, integrins, adhesion molecules, chemokines, selectins, sialyl-Lewis^{x}, and basement membrane proteases.

3–7
A. Using the table below, match the local and systemic effects in column A with the appropriate cytokine in column B. Note that more than one answer in column B may be used.
B. (i) Which of these cytokines are produced by macrophages? (ii) Which cells produce the other(s)?

Column A	Column B
a. activation of blood-vessel endothelium	1. IL-1
b. fever	2. IL-6
c. induction of IL-6 synthesis	3. CXCL8
d. increase in vascular permeability	4. IL-12
e. localized tissue destruction	5. TNF-α
f. production of acute-phase proteins by hepatocytes	6. type I interferons
g. induction of resistance to viral replication	
h. activation of NK cells	
i. leukocyte chemotaxis	
j. activation of binding by β2 integrins (LFA-1, CR3)	
k. septic shock	
l. mobilization of metabolites	

3–8 The early events of inflammation are characterized by all of the following except _____.
a. vasodilation
b. extravasation
c. pain
d. apoptosis
e. increased vascular permeability
f. chemotaxis.

3–9 Identify the mismatched pair.
a. MyD88: adaptor protein
b. ICAM-1: LFA-1
c. chemokine receptor: G protein
d. CD14: LPS
e. NK cells: IFN-γ
f. selectins: protein ligands.

3–10 Which of the following statements regarding TLR4 is false? (Select all that apply.)
a. TLR4 is a Toll-like receptor expressed by macrophages, but not by NK cells.
b. TLR4 is an example of a phagocytic receptor.
c. TLR4 associates with CD14, which acts as a co-receptor to TLR4.
d. The ligand for TLR4 is lipopolysaccharide (LPS).
e. TLR4 is expressed as a homodimer on the cell surface.
f. The extracellular domain of TLR4 associates with soluble MD2.
g. TLR4 does not contain a cytoplasmic TIR domain.

3–11 Describe the properties of cytokines that help to protect the host from tissue damage during immune responses to infection.

3–12 Identify two ways in which IL-1β is regulated so as to minimize its destructive potential in the human body.

3–13 Systemic distribution of TNF-α in the blood circulation may cause _____.
a. respiratory burst
b. the interferon response
c. activation of the lectin pathway of complement
d. hypertension
e. septic shock.

3–14 Two-year-old Jaxon Markowski presented with a hyperpigmented lacy rash on his torso. He had a history of recurrent bacterial infections, had been underweight for age since birth, and had frequent bouts of diarrhea and vomiting. He had a healthy 7-year-old sister and his mother had a stillbirth male baby 4 years ago. Jaxon's temperature runs higher than normal and he did not sweat. Closer examination revealed diffuse alopecia (thinning and absence of hair) and frontal bossing (protruding forehead). Jaxon's recently erupted front teeth were cone-shaped, and he exhibited periorbital wrinkling and darkening. Blood tests revealed an abonormally high white blood cell count, low IgG and IgA, and normal IgM concentrations. Flow cytometry showed low NK-cell numbers. Skin biopsy revealed diffuse granulomatous inflammation and tested positive for *Mycobacterium avium*. Antibiotic combination therapy was begun and intravenous immunoglobulin (IVIG) was administered, to which Jaxon responded favorably. Molecular testing confirmed an insertion of cytosine at nucleotide 1167 in exon 10 of the X-chromosome-encoded IKKγ gene (also known as NFκB essential modifier, or NEMO), causing a frameshift mutation affecting protein–protein interactions of the zinc-finger domain. Jaxon's mother was confirmed as a heterozygous carrier. Which of the following best describes the consequence of this mutation?
a. The respiratory burst is impaired in phagocytes owing to insufficient hydrogen peroxide production.
b. NFκB is retained in the cytosol of macrophages in an inactive state in a complex with its inhibitor IκB.
c. Increased susceptibility to lysis of red blood cells through unregulated assembly of the membrane attack complex.
d. Loss of ability of macrophages to secrete cryptdins.
e. Inability of host cells to synthesize IFN-α.

Individual molecules of human immunoglobulin G, the predominant antibody in blood and connective tissue

Chapter 4

Antibody Structure and the Generation of B-Cell Diversity

Pathogens and their products can be found either within infected cells or outside cells in the intercellular fluid and blood. In an adaptive immune response, the body is cleared of extracellular pathogens and their toxins by means of **antibodies**, the secreted form of the **B-cell receptor** for antigen. Antibodies are produced by the effector B lymphocytes, or plasma cells, of the immune system in response to infection. They circulate as a major component of the plasma in blood and lymph and are always present at mucosal surfaces. Antibodies can recognize all types of biological macromolecule, but in practice, proteins and carbohydrates are the antigens most commonly encountered. Binding of antibody to a bacterium or virus particle can disable the pathogen and also render it susceptible to destruction by other components of the immune system. Antibodies are the best source of protective immunity, and the most successful vaccines protect through stimulating the production of high-quality antibodies.

As we learned in Chapter 1, antibodies are very variable proteins. An individual antibody molecule is said to be **specific**; that is, it can bind to only one antigen or a very small number of different antigens. The immune system has the potential to make antibodies against a vast number of different substances, paralleling the many different foreign antigens that a person is likely to be exposed to during their lifetime. Collectively, therefore, antibodies are diverse in their antigen-binding specificities; the total number of different specific antibodies that can be made by an individual is known as the **antibody repertoire** and it might be as high as 10^{16}. In practice, the number of B cells limits the actual repertoire to closer to 10^{9}.

The cell-surface B-cell antigen receptors and the secreted anibodies are known more generally as **immunoglobulins**. During its development, each individual B cell becomes committed to producing immunoglobulin of a unique antigen specificity; thus, mature B cells collectively produce immunoglobulins of many different specificities. Before it encounters antigen, a mature B cell only expresses immunoglobulin in the membrane-bound form that serves as the cell's receptor for antigen. When antigen binds to this receptor, the B cell is stimulated to proliferate and to differentiate into **plasma cells**, which then secrete large amounts of antibodies with the same specificity as that of the membrane-bound immunoglobulin (Figure 4.1). This is an example of the principle of **clonal selection**, on which adaptive immune responses are based (see Section 1-5). Antibody production is the single effector function of the B lymphocytes of the immune system. An inability to make immunoglobulins, called **agammaglobulinemia**, renders people highly susceptible to certain types of infection.

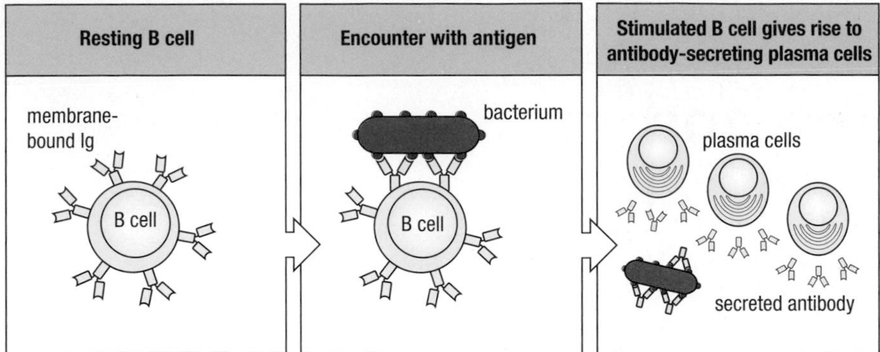

Figure 4.1 Plasma cells secrete antibody of the same antigen specificity as the antigen receptor of their B-cell precursor. A mature B cell expresses membrane-bound immunoglobulin (Ig) of a single antigen specificity. When a foreign antigen first binds to this immunoglobulin, the B cell is stimulated to proliferate. Its progeny differentiate into plasma cells that secrete antibody of the same specificity as the membrane-bound immunoglobulin.

The first part of this chapter describes the general structure of immunoglobulins, and the second part considers how immunoglobulin diversity is generated in developing B cells. After a mature B cell has encountered its specific antigen, changes are made in the specificity and effector functions of the antibodies that it produces; these are discussed in the final part of the chapter.

The structural basis of antibody diversity

The function of an antibody molecule in host defense is to recognize and bind its corresponding antigen, and then to deliver the bound antigen to other components of the immune system. These then destroy the antigen or clear it from the body. Antigen binding and interaction with other immune-system cells and molecules are carried out by different parts of the antibody molecule. One part is highly variable in that its amino acid sequence differs considerably from antibody to antibody. This variable part contains the site of antigen binding and confers specificity on the antibody. The rest of the molecule is far less variable in its amino acid sequence. This constant part of the antibody interacts with other immune-system components.

Immunoglobulins group into five isotypes, or classes—IgG, IgM, IgD, IgA, and IgE. These are distinguished on the basis of structural differences in the constant part of the molecule. The structural differences confer distinctive effector functions on the isotypes, which means that they each interact with a different subset of immune-system proteins. We first consider the general structure of immunoglobulins and then look more closely at the structural features that confer antigen specificity. IgG, the most abundant antibody in blood and lymph, is used to illustrate the common structural features of immunoglobulins.

4-1 Antibodies are composed of polypeptides with variable and constant regions

Antibodies are glycoproteins that are built from a basic unit of four polypeptide chains. This unit consists of two identical **heavy chains** (**H chains**) and two identical, smaller, **light chains** (**L chains**), which assemble into a structure that looks like the letter Y (Figure 4.2, top panel). An IgG molecule has a

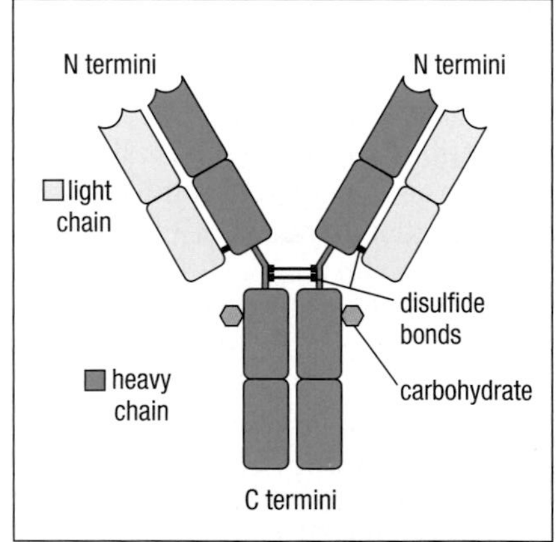

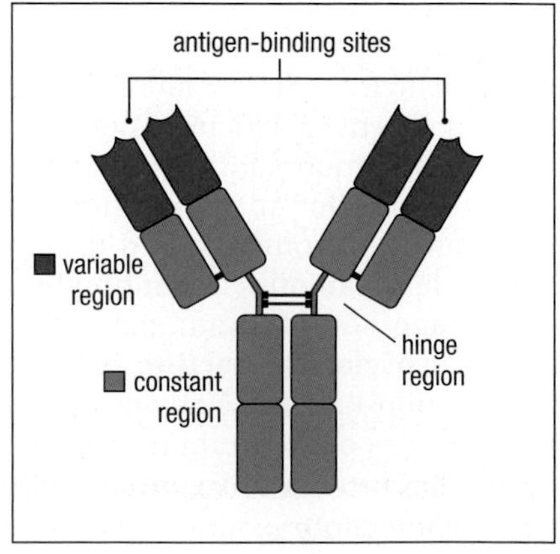

Figure 4.2 The immunoglobulin G (IgG) molecule. As shown in the top panel, each IgG molecule consists of two identical heavy chains (green) and two identical light chains (yellow). Carbohydrate (turquoise) is attached to the heavy chains. The lower panel shows the location of the variable (V) and constant (C) regions in the IgG molecule. The amino-terminal regions (red) of the heavy and light chains are variable in sequence from one IgG molecule to another; the remaining regions are constant in sequence (blue). The carbohydrate is omitted from this panel and from most subsequent figures for simplicity. In IgG a flexible hinge region is located between the two arms and the stem of the Y.

Figure 4.3 **The Y-shaped immunoglobulin molecule can be dissected with a protease.** By using a protease to cleave the hinge of each heavy chain, as shown by the scissors, and a reducing agent to break the disulfide bonds that connect the two hinges, the IgG molecule is dissected into three pieces: two Fab fragments and one Fc fragment.

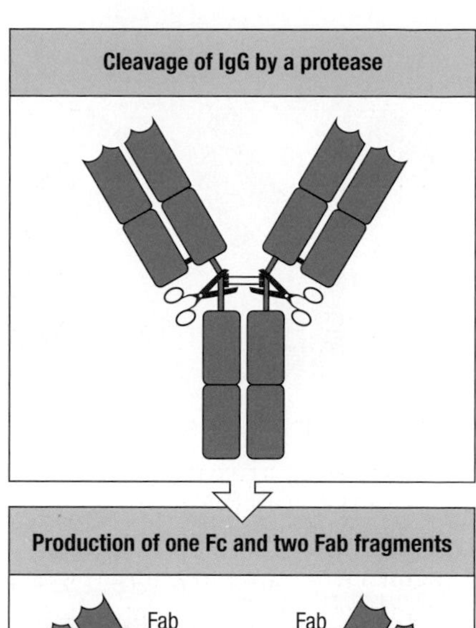

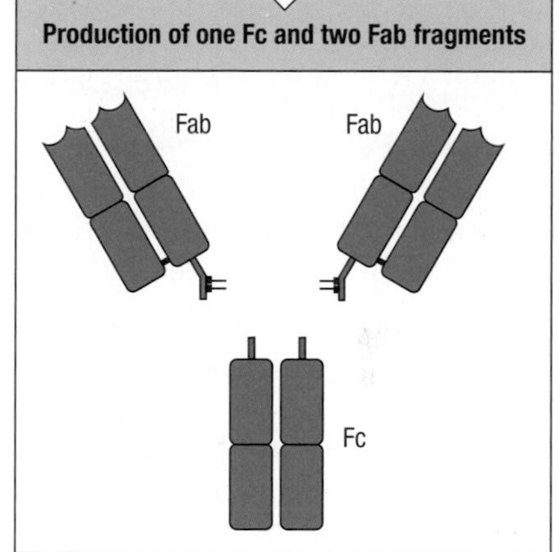

molecular weight of about 150 kDa, to which each heavy chain contributes approximately 50 kDa and each light chain approximately 25 kDa. Each arm of the Y is made up of a complete light chain paired with the amino-terminal part of a heavy chain, covalently linked by a disulfide bond. The stem of the Y consists of the paired carboxy-terminal portions of the two heavy chains. The two heavy chains are also linked to each other by disulfide bonds.

The polypeptide chains of different antibodies vary greatly in amino acid sequence, and the sequence differences are concentrated in the amino-terminal region of each type of chain; this is known as the **variable region** or **V region**. This variability is the basis for the great diversity of antigen-binding specificities that antibodies have. Forming the **antigen-binding site** is a pair of variable regions, one from a heavy and one from a light chain (Figure 4.2, bottom panel). Every Y-shaped antibody molecule has two identical antigen-binding sites, one at the end of each arm. The remaining parts of the light chain and the heavy chain have limited variation in amino acid sequence between different antibodies and are therefore known as the **constant regions** or **C regions**.

In IgG, a relatively unstructured portion in the middle of the heavy chain forms a flexible hinge region at which the molecule can be cleaved with proteases to produce two types of antibody fragment having complementary functions (Figure 4.3). The fragments corresponding to the arms are called **Fab** (**F**ragment **a**ntigen **b**inding) because they bind antigen. The fragment corresponding to the stem, called **Fc** (**F**ragment **c**rystallizable) because it readily crystallizes, mediates the effector functions of the antibody molecule by binding to serum proteins and cell-surface receptors. The stem of a complete antibody molecule is therefore often known as the **Fc region** or **Fc piece**, and the arms as Fab. The hinge of the IgG molecule allows the two Fabs to adopt many different spatial orientations with respect to each other. This flexibility enables antigens spaced at different distances apart on the surfaces of pathogens to be bound tightly by both Fab arms of an IgG molecule (Figure 4.4).

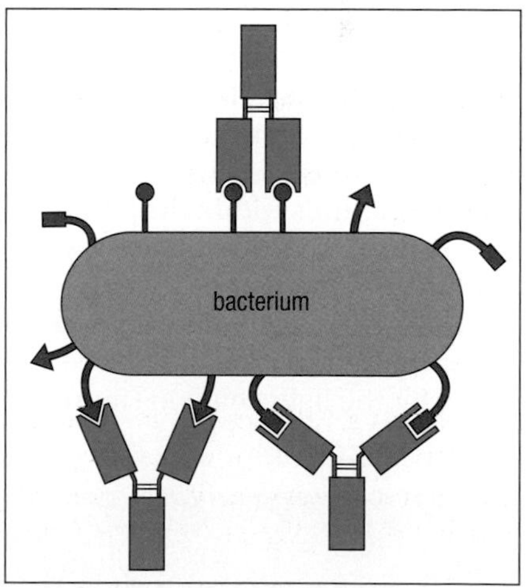

Figure 4.4 **The flexible hinge of the IgG molecule allows it to bind with both arms to many different arrangements of antigens on the surfaces of pathogens.** Three IgG molecules are shown binding with both arms to antigens that are located at different distances apart on the surface of a bacterium.

Differences in the heavy-chain C regions define five main **isotypes** or **classes** of immunoglobulin, which have different functions in the immune response. They are **immunoglobulin G (IgG)**, **immunoglobulin M (IgM)**, **immunoglobulin D (IgD)**, **immunoglobulin A (IgA)**, and **immunoglobulin E (IgE)** (Figure 4.5). Their heavy chains are denoted by the corresponding lower-case Greek letter (gamma (γ), mu (μ), delta (δ), alpha (α), and epsilon (ε), respectively).

The light chain has only two isotypes or classes—**kappa (κ)** and **lambda (λ)**. No functional difference between antibodies carrying κ light chains and those carrying λ light chains has been found; light chains of both isotypes pair with all the heavy-chain isotypes. Each antibody, however, contains either κ or λ light chains, not both. The relative abundances of κ and λ light chains vary with the species of animal. In humans, two-thirds of the antibody molecules contain κ chains and one-third have λ chains.

4-2 Immunoglobulin chains are folded into compact and stable protein domains

Antibodies function in extracellular environments in the presence of infection, where they can encounter variations in pH, salt concentration, proteolytic

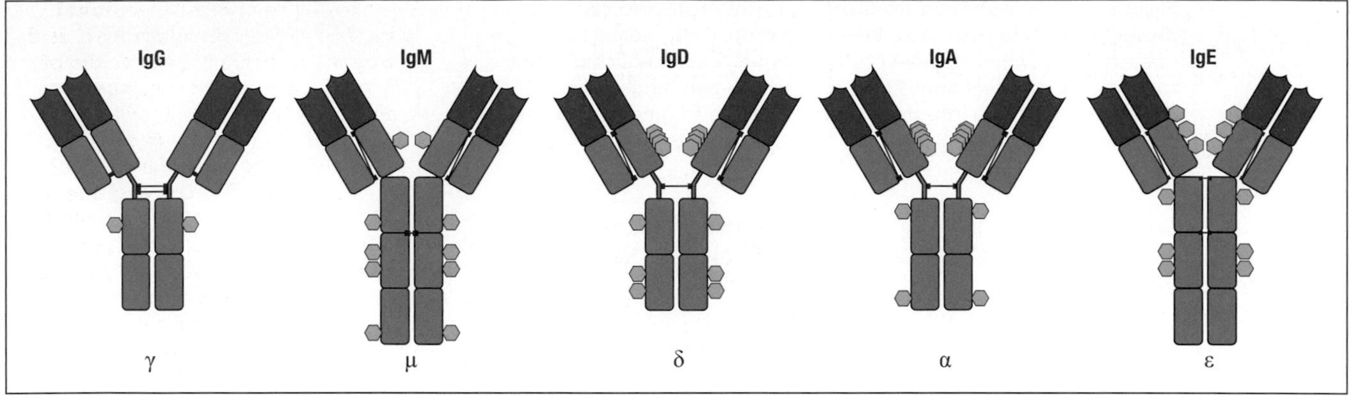

Figure 4.5 The structures of the human immunoglobulin classes. In particular, note the differences in length of the heavy-chain C regions, the locations of the disulfide bonds linking the chains, and the presence of a hinge region in IgG, IgA, and IgD, but not in IgM and IgE. The heavy-chain isotype in each antibody is indicated by the Greek letter. The isotypes also differ in the distribution of *N*-linked carbohydrate groups (turquoise). All these immunoglobulins occur as monomers in their membrane-bound form. In their soluble, secreted form, IgD, IgE, and IgG are always monomers, IgA forms monomers and dimers, and IgM forms only pentamers.

enzymes, and other potentially destabilizing factors. Their structure, however, helps them to withstand such harsh conditions. Heavy and light chains each consist of a series of similar sequence motifs; a single motif is about 100–110 amino acids long and folds up into a compact and exceptionally stable protein domain called an **immunoglobulin domain**. Each immunoglobulin chain is composed of a linear series of these domains.

The V region at the amino-terminal end of each heavy or light chain is composed of a single **variable domain** (**V domain**): V_H in the heavy chain and V_L in the light chain. A V_H and a V_L domain together form an antigen-binding site (Figure 4.6). The other domains have little or no sequence diversity within a particular class of antibody and are termed the **constant domains** (**C domains**), which make up the C regions. The constant region of a light chain is composed of a single C_L domain, whereas the constant region of a heavy chain is composed of three or four C domains, depending on the isotype. The γ heavy chains of IgG, IgD, and IgA have three domains—C_H1, C_H2, and C_H3. IgM and IgE also have a fourth C domain—C_H4 (see Figure 4.5). In the complete IgG molecule the pairing of the four polypeptide chains produces three compact globular regions, corresponding to the two Fab arms and the Fc stem, each composed of four immunoglobulin domains (see Figure 4.6).

The structure of a single immunoglobulin domain resembles a bulging sandwich. Two β sheets (the slices of bread) are held together by strong hydrophobic interactions between their constituent amino acid side chains (the filling). The structure is stabilized by a disulfide bond between the two β sheets. The adjacent strands within the β sheets are connected by loops of the polypeptide chain. This arrangement of β sheets provides the stable structural framework of all immunoglobulin domains, whereas the amino acid sequence of the loops can be varied to confer different binding properties on the domain. The structural similarities and differences between the C and V domains are illustrated for the light chain in Figure 4.7.

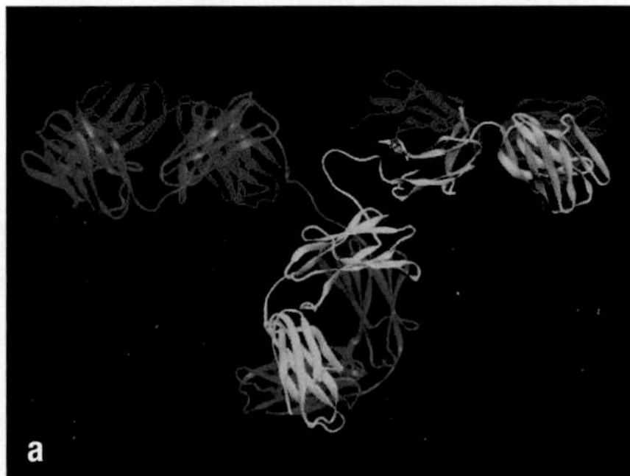

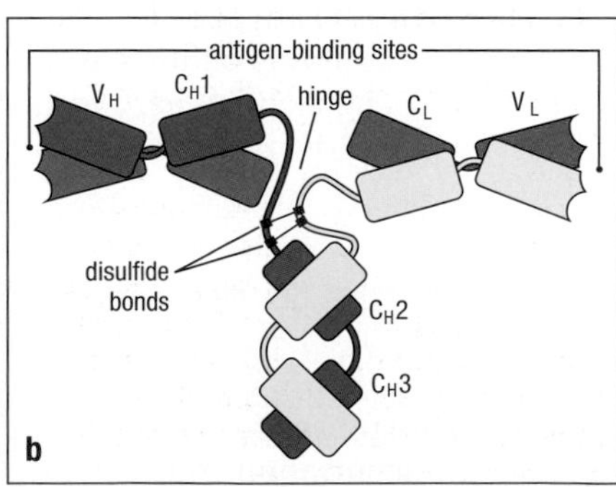

Figure 4.6 IgG is built from twelve similar shaped immunoglobulin domains. The diagram in panel a is based upon three-dimensional crystallographic structures. The polypeptide backbones of the heavy chains (one chain is yellow and the other purple) and the light chains (both red) are shown as ribbons. Each Fab and Fc fragment is made up of four immunoglobulin domains that produce a similar shaped globular structure. Panel b shows how the different types of domain are organized and connected within the three-dimensional structure. Panel a courtesy of L. Harris.

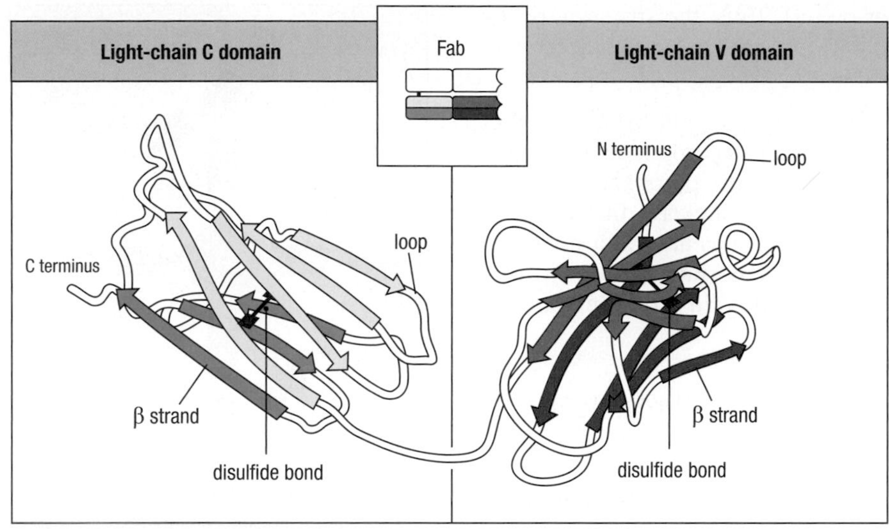

Figure 4.7 The three-dimensional structure of immunoglobulin C and V domains. Shown is the structure of one light chain with its constituent C domain (left panel) and V domain (right panel). The inset shows the location of the light chain in a Fab of IgG. The folding of the polypeptide is depicted as a ribbon diagram in which the thick colored arrows correspond to the parts of the chain that form the β strands of the β sheet. The arrows point from the amino terminus to the carboxy terminus. Adjacent strands in each β sheet run in opposite directions (antiparallel), as shown by the arrows, and are connected by loops. The C domain has four β strands in the upper β sheet (yellow) and three in the lower (green). The V domain has five strands in the upper sheet (blue) and four in the lower (red).

Although immunoglobulin domains were first discovered in antibodies, very similar domains, known generally as **immunoglobulin-like domains**, were subsequently found in many other proteins. They are particularly common in cell-surface and secreted proteins of the immune system. Collectively these proteins form the **immunoglobulin superfamily**.

4-3 An antigen-binding site is formed from the hypervariable regions of a heavy-chain V domain and a light-chain V domain

Comparison of the V domains of heavy and light chains from different antibody molecules shows that the differences in amino acid sequence are concentrated within particular regions called **hypervariable regions (HV)**, which are flanked by much less variable **framework regions**, as illustrated for the light chain in Figure 4.8 (top panel). Three hypervariable regions are found in each V domain. When mapped onto the structure of a V domain, the hypervariable regions locate to three of the loops that are exposed at the end of the domain farthest from the constant region. The framework regions correspond to the β strands and the remaining loops. This relationship is illustrated for the

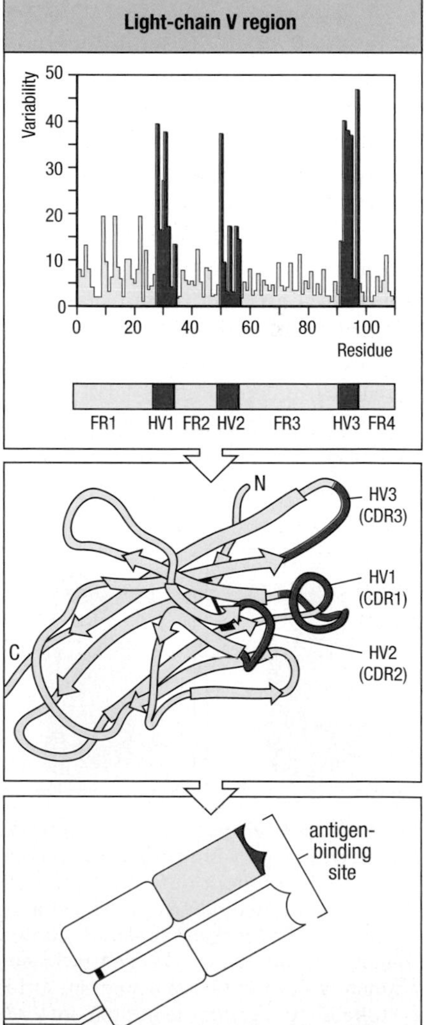

Figure 4.8 The hypervariable regions of antibody V domains lie in discrete loops at one end of the domain structure. The top panel shows the variability plot for the 110 positions within the amino acid sequence of a light-chain V domain. It is obtained from comparison of many light-chain sequences. Variability is the ratio of the number of different amino acids found at a position to the frequency of the most common amino acid at that position. The maximum value possible for the variability is 400, the square of 20, the number of different amino acids found in antibodies. The minimum value is 1. Three hypervariable regions (HV1, HV2, and HV3) can be discerned (red) flanked by four framework regions (FR1, FR2, FR3, and FR4) (yellow). The center panel shows the correspondence of the hypervariable regions to three loops at the end of the V domain farthest from the C region. The location of hypervariable regions in the heavy-chain V domain is similar (not shown). The hypervariable loops contribute much of the antigen specificity of the antigen-binding site located at the tip of each arm of the antibody molecule. Hypervariable regions are also known as complementarity-determining regions: CDR1, CDR2, and CDR3. The bottom panel shows the location of the light-chain V region in the Fab part of the IgG molecule.

light-chain V domain in Figure 4.8 (center panel). Thus, the structure of the immunoglobulin domain provides the capacity for localized variability within a structurally constant framework.

The pairing of a heavy and a light chain in an antibody molecule brings together the hypervariable loops from each V domain to create a composite hypervariable surface, which forms the antigen-binding site at the tip of each Fab arm (Figure 4.8, bottom panel). The differences in the loops between different antibodies create both the specificity of antigen-binding sites and their diversity. The three hypervariable loops are also called **complementarity-determining regions** (**CDR1**, **CDR2**, and **CDR3**) because they provide a binding surface that is complementary to that of the antigen.

4-4 Antigen-binding sites vary in shape and physical properties

The function of antibodies is to bind to microorganisms and facilitate their destruction or ejection from the body (see Section 1-10). The antibodies that are most effective against infection are generally those that bind to the exposed and accessible molecules making up the surface of a pathogen. The part of an antigen to which an antibody binds is called an **antigenic determinant** or **epitope**. In nature, these structures are usually either carbohydrate or protein, because the surface molecules of pathogens are commonly glycoproteins, polysaccharides, glycolipids, and peptidoglycans. Complex macromolecules such as these will usually contain several different epitopes, each of which can be bound by a different antibody. Individual epitopes are typically composed of a cluster of amino acids or a small part of a polysaccharide chain. An antigen that contains more than one epitope, or more than one copy of the same epitope, is known as a **multivalent** antigen (Figure 4.9).

The antigen-binding site of immunoglobulins is a versatile structure that can accommodate epitopes with a range of different structures as illustrated in Figure 4.10. An antibody that binds a small compact epitope, such as the terminal sugar at the end of a polysaccharide chain has a deep pocket between the heavy- and light-chain variable regions with which it grasps the sugars (Figure 4.10, first panel). By contrast an antibody that binds to several adjacent sugars in a polysaccharide chain uses a long, shallow groove as its antigen-binding site (Figure 4.10, second panel). Antibodies that bind to the surface of a globular protein make contact over an extended area (around 700–900Å). This is not accomplished with either a pocket or a groove but with

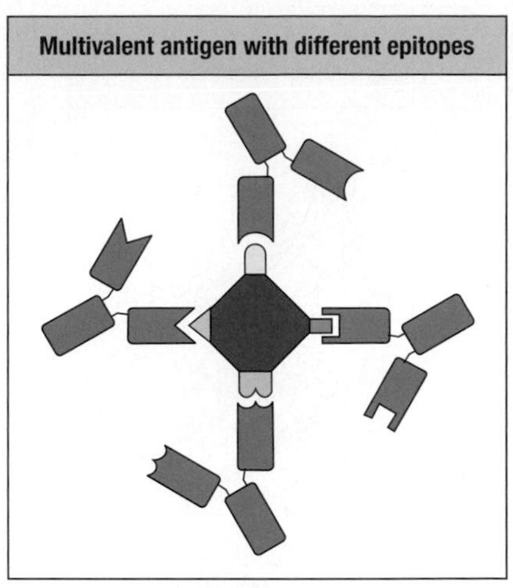

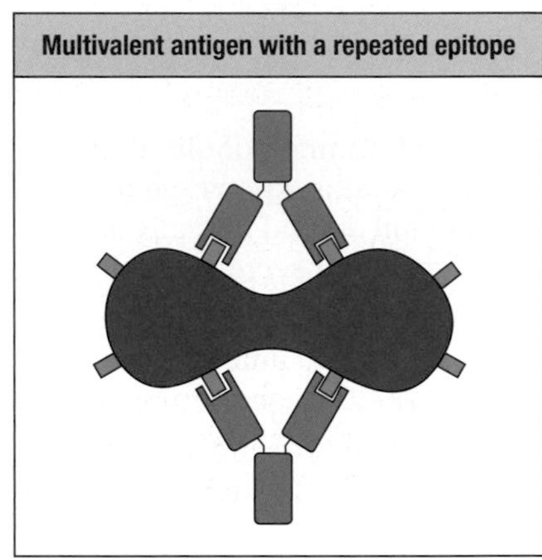

Figure 4.9 Two kinds of multivalent antigen. Many soluble protein antigens have several different epitopes, but each is represented only once on the surface of the protein. This situation is depicted in the left panel, where four IgG molecules with different specificities all bind to the protein antigen using a single Fab arm. On pathogen surfaces there are numerous copies of the same epitope. This situation, depicted in the right panel, allows many IgG molecules with identical antigenic specificity to bind to the multivalent antigen with both Fab arms.

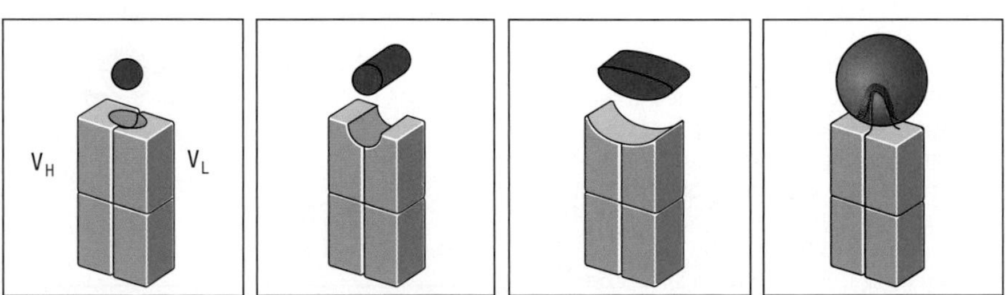

Figure 4.10 Epitopes can bind to pockets, grooves, extended surfaces, or knobs in antigen-binding sites. Shown are schematic representations of antibodies (blue) binding to four different types of epitope (red). The first panel shows a small compact epitope binding to a pocket in the antigen-binding site; in the second panel an epitope consisting of a relatively unfolded part of a polypeptide chain binds to a shallow groove; the third panel shows an epitope with an extended surface binding to a similarly sized but complementary surface in the antibody; and in the fourth panel the epitope is a pocket in the antigen into which the antigen-binding site of the antibodies intrudes. Only one Fab arm of each antibody is shown.

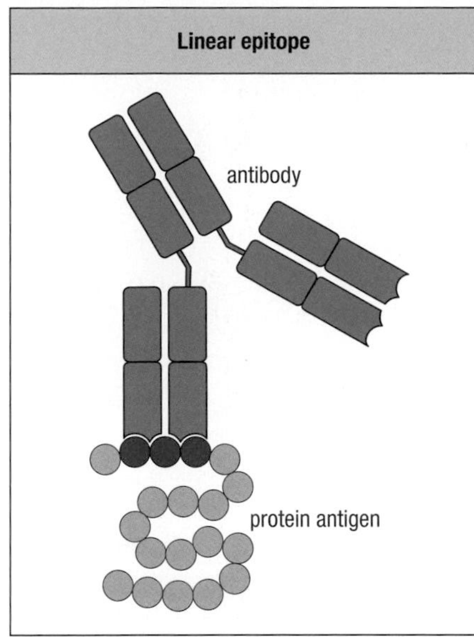

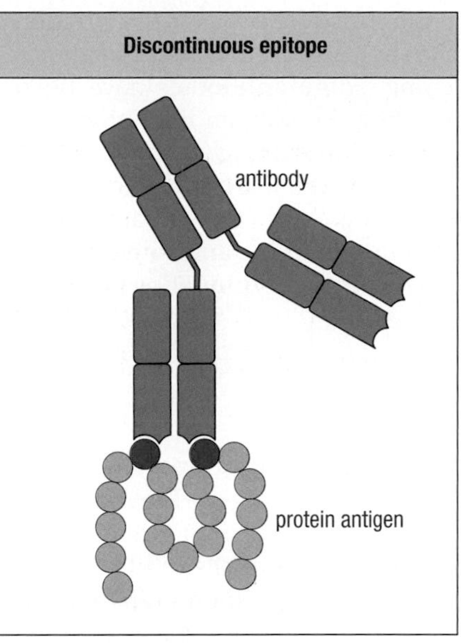

Figure 4.11 Linear and discontinuous epitopes. A linear epitope of a protein antigen is formed from contiguous amino acids. A discontinuous epitope is formed from amino acids from different parts of the polypeptide that are brought together when the chain folds.

an interacting surface that has the same area as the epitope (Figure 4.10, third panel). In some instances the epitope is actually a pocket or groove in the antigenic protein and the antigen-binding site of the antibody is a knob or projecting spike that fits into the pocket or groove (Figure 4.10, fourth panel). In this last situation the antigen appears to be binding the antibody. The antibody binding site has six complementarity determining regions which vary in length, sequence, and chemical properties. The number of these that are used to contact a given antigen varies upon the size, shape, and chemical properties of the antigen.

The epitopes of protein antigens fall into two broad groups (Figure 4.11). Epitopes composed of several successive amino acids in a protein sequence are called **linear epitopes** (Figure 4.11, left panel). Such epitopes can be formed by accessible loops of a protein antigen. Antibodies that recognize linear epitopes of a native protein often bind to synthetic peptides that contain the amino acid sequence of the epitope. Contrasting with linear epitopes are the **discontinuous epitopes** (Figure 4.11, right panel).These are formed by two or more parts of the protein antigen that are separated in the amino acid sequence but are brought together in the folded protein. Because discontinuous epitopes are dependent upon three-dimensional structure, antibodies that recognize discontinous epitopes rarely interact with synthetic peptides corresponding to the parts of the protein that form the epitope.

The binding of antigens to antibodies is based solely on noncovalent forces—electrostatic forces, hydrogen bonds, van der Waals forces, and hydrophobic interactions. The antigen-binding sites of antibodies are unusually rich in aromatic amino acids, which can participate in many van der Waals and hydrophobic interactions. The better the general fit between the interacting surfaces of antigen and antibody, the stronger are the bonds formed by these short-range forces. Thus, several different antibodies may recognize the same epitope, but the small differences in the shapes and chemical properties of the binding sites give these antibodies different binding strengths, or **affinities**, for the epitope. Binding caused by van der Waals forces and hydrophobic interactions is complemented by the formation of electrostatic interactions (salt bridges) and hydrogen bonds between particular chemical groups on the antigen and particular amino acid residues of the antibody.

In the immune response, the effective antibodies are those that bind tightly to an antigen and do not let go. This behavior contrasts with that of enzymes, which bind a substrate, chemically change the substrate's structure, and then release it. However, the same set of 20 amino acids is used to make enzymes and antibodies and, after intensive searching, some antibodies have been found that catalyze chemical reactions involving the antigen they bind. The application of such **catalytic antibodies** in medicine is being explored. A possible use for a catalytic antibody would be to convert a toxic chemical within the body into an innocuous product. Toward this end, catalytic antibodies have been made that bind to cocaine or methamphetamine and convert these highly addictive drugs to derivatives that have no psychostimulating activity.

4-5 Monoclonal antibodies are produced from a clone of antibody-producing cells

The specificity and strength of an antibody for its antigen make antibodies useful reagents for detecting and quantifying a particular antigen. They are used in this way in a range of clinical and laboratory tests and in biological research generally. The traditional method for making antibodies of desired specificity is to immunize animals with the appropriate antigen and then prepare **antisera** from their blood. The specificity and quality of such antisera are highly dependent upon the purity of the immunizing antigen preparation because antibodies will be made against all the foreign components it contains. Consequently, antisera specific for a single protein or carbohydrate can only be made when the antigen is already available in a purified form.

A more modern method for making antibodies does not require a purified antigen. In this method, B cells are isolated from the immunized animals, usually mice, and immortalized by fusion with a tumor cell to form **hybridoma** cell lines that grow and produce antibodies indefinitely. The individual hybridoma cells are then separated from each other, and the cells making antibodies of the desired specificity are identified and selected for further propagation. The antibodies made by a hybridoma cell line are all identical and are therefore called **monoclonal antibodies** (Figure 4.12).

Since the invention of hybridomas in the 1970s, monoclonal antibodies have replaced antisera as reagents in many clinical assays. They have also been used to identify numerous previously unknown surface proteins of cells of the immune system. All of the more than 300 cell-surface proteins in the CD series (for example, CD4 and CD8) were discovered by using monoclonal antibodies. The reactions of monoclonal antibodies with different types of cell are measured by the technique of **flow cytometry** (Figure 4.13).

In medicine, flow cytometry is used to analyze the cell populations in peripheral blood and detect perturbations caused by disease. For example, children can inherit a gene defect that results in them having no B cells or antibodies. At birth and during the first year of life the defect is not apparent, because the children are protected from infection by the antibodies they obtain from their mothers in breast milk. Soon after they stop being breastfed, these children start to lose the protection that was provided by their mothers' antibodies. Infants then become particularly susceptible to infections with encapsulated

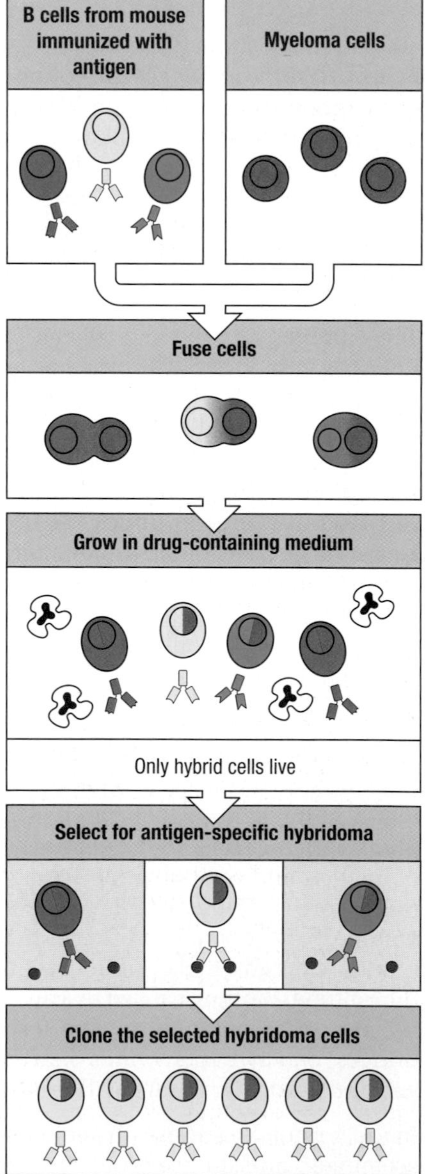

Figure 4.12 Production of a mouse monoclonal antibody. Lymphocytes from a mouse immunized with the antigen are fused with myeloma cells by using polyethyleneglycol. The cells are then grown in the presence of drugs that kill myeloma cells but permit the growth of hybridoma cells. Unfused lymphocytes also die. Individual cultures of hybridomas are tested to determine whether they make the desired antibody. The cells are then cloned to produce a homogeneous culture of cells making a monoclonal antibody. Myelomas are tumors of plasma cells; those used to make hybridomas were selected not to express heavy and light chains. Thus, hybridomas only express the antibody made by the B-cell partner.

bacteria such as *Streptococcus pneumoniae* or *Haemophilus influenzae*, which if not treated with antibiotics would eventually lead to death. Knowledge that a child is B-cell deficient determines how they can be successfully protected from future infections. This is achieved by regular injections of IgG antibodies purified from the pooled sera of healthy blood donors. Because such antibodies are used to treat many types of immunodeficiency, as well as other

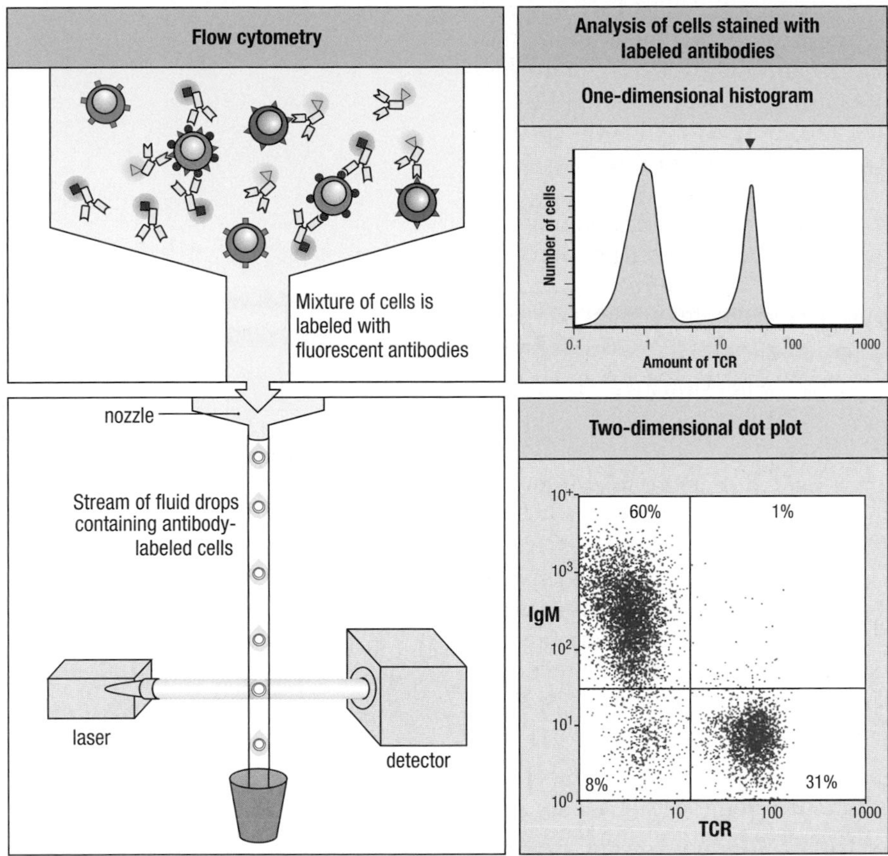

Figure 4.13 The flow cytometer allows cells to be distinguished by their cell-surface molecules. The left-hand panels illustrate the principles of flow cytometry. Human cells are labeled with mouse monoclonal antibodies specific for human cell-surface proteins, antibodies of different specificity being tagged with fluorescent dyes of different colors. The labeled cells pass through a nozzle, forming a stream of droplets each containing one cell. The stream of cells passes through a laser beam, which causes the fluorescent dyes to emit light of different wavelengths. The emitted signals are analyzed by a detector: cells with particular characteristics are counted and the abundance of each type of labeled cell-surface molecule is measured. The right-hand panels show how the data obtained can be represented, as exemplified by the presence of IgM and T-cell receptor (TCR) on lymphocytes from the human spleen. The expression of IgM and TCR distinguishes between B cells and T cells, respectively. When the presence of one type of molecule is analyzed, the data are usually displayed as a histogram, as shown in the upper right panel for TCR. Two populations of cells are distinguished in the histogram. The larger peak to the left consists of lymphocytes (mostly B cells) that do not bind the anti-TCR monoclonal antibody; it comprises 58% of the total cell number. The smaller peak to the right corresponds to the T cells that bind the anti-TCR antibody and comprises 32% of the total. Two-dimensional plots (lower right panel) are used to compare the expression of two cell-surface molecules. The horizontal axis represents the amount of fluorescent anti-TCR antibody bound by a cell; the vertical axis represents the amount of fluorescent anti-IgM antibody bound. Each dot represents the values obtained for a single cell. Many thousands of cells are usually analyzed, and the dots can blend into each other in those parts of the plot that are heavily populated. The dot plot is divided into four quadrants that roughly correspond to the four cell populations distinguished by analysis with two antibodies. In the top left quadrant are B cells; these bind anti-IgM antibody but not anti-TCR antibody and comprise 60% of the cells. In the bottom right quadrant are the T cells; these bind anti-TCR antibody but not anti-IgM antibody and comprise 31% of the cells. In the bottom left quadrant are cells that bind neither anti-IgM nor anti-TCR antibody; these comprise 8% of the cells and probably include NK cells and some contaminating non-lymphoid leukocytes. Theoretically, cells in the upper right quadrant would correspond to lymphocytes that bind both anti-IgM and anti-TCR. As no lymphocytes express both IgM and TCR at their surface, these double reactions (1% of the total) arise from imprecision in using quadrants to separate the cell population and from experimental artifact, for example the nonspecific sticking of antibody molecules to cells that do not express their specific antigen. Flow cytometry is used both to analyze and purify cell populations. Although this example shows the simultaneous analysis of two cell-surface molecules, the technology has advanced to where up to 18 cell-surface markers can be analyzed simultaneously.

diseases, it is prepared in large batches using blood from 10,000–60,000 donors. This therapeutic antibody is called **intravenous immunoglobulin**.

4-6 Monoclonal antibodies are used as treatments for a variety of diseases

Monoclonal antibodies are used in therapy as well as in diagnosis. The first successful therapeutic application was of a mouse monoclonal antibody specific for the CD3 antigen on human T cells. The antibody was used to prevent rejection of transplanted kidneys in patients showing early symptoms of T-cell-mediated graft rejection. Although effective, the antibody could only be used once for each patient, because after the first treatment patients made antibodies against the C regions of the mouse antibody. Interaction of the human antibodies with a subsequent dose of mouse monoclonal antibodies diminishes the latter's therapeutic effect and increases the likelihood of complications. To reduce this problem, genetic engineering has been used to make **chimeric monoclonal antibodies** that combine mouse V regions with human C regions (Figure 4.14).

One chimeric monoclonal antibody of this sort, called rituximab, is used to treat several diseases including some types of non-Hodgkin B-cell lymphoma. Rituximab is specific for the cell-surface protein CD20, which is present on normal and malignant human B cells and influences B-cell proliferation and survival. Rituximab binds strongly to both normal and malignant B cells and targets their destruction by NK cells. The Fc of the rituximab coating the B cells is bound by a receptor on NK cells that activates its cell-killing machinery. Despite their temporary loss of normal B cells, treated patients continue to make antibodies because plasma cells lack CD20 and are therefore unaffected by rituximab. Normal populations of circulating B cells are usually reconstituted within a year after treatment. Since its approval for medical use in 1997, more than 1 million patients have been treated with rituximab.

Another approach to reducing the tendency for useful mouse monoclonal antibodies to provoke human antibodies against them is to **humanize** the mouse antibodies by genetic engineering. Sequences encoding the CDR loops of the mouse antibody's heavy and light chains are used to replace the corresponding CDR sequences in a human immunoglobulin gene (see Figure 4.14). Cells expressing the humanized heavy-chain and light-chain genes make an antibody in which only the CDR loops are of mouse origin and the rest is human. Omalizumab is a humanized antibody specific for human IgE that is used to treat cases of severe allergic asthma. By binding to the constant region of IgE, the therapeutic antibody prevents the recruitment of mast cells, eosinophils, and basophils to the allergic response.

Now it is possible to avoid the problems associated with mouse immunoglobulins and make fully human monoclonal antibodies for therapeutic use. In one approach, the antibodies are made by mice whose mouse immunoglobulin genes have been replaced with the human counterparts. Adalimumab is a

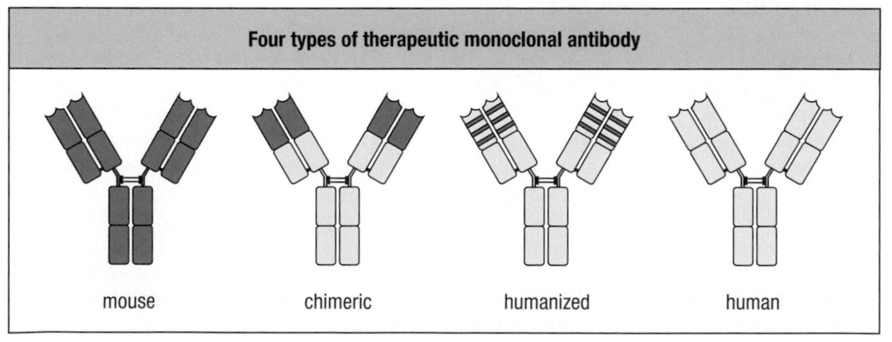

Figure 4.14 Monoclonal antibodies as treatments for disease. Mouse antibodies were the first monoclonal antibodies used therapeutically. Their drawback was that patients responded to the foreign mouse protein by making antibodies against them, which reduced their clinical effect. This problem was reduced by making chimeric antibodies, in which the constant regions of the mouse antibody (blue) are replaced with human constant regions (yellow). Extending this approach are humanized antibodies, in which only the CDR loops are of mouse origin. Fully human monoclonal antibodies are made by immunizing mice in which the mouse immunoglobulin genes were replaced by human immunoglobulin genes. All four types of antibody are being used therapeutically.

widely used all-human monoclonal antibody specific for the inflammatory cytokine TNF-α (see Section 3-4). It is used in the treatment of rheumatoid arthritis, a condition in which chronic inflammation of the joints is perpetuated by TNF-α.

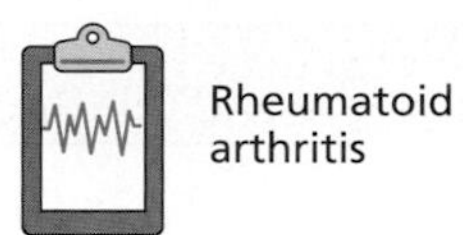

Summary

IgG antibodies are made of four polypeptide chains—two identical heavy chains and two identical light chains. Each chain has a V region that contributes to the antigen-binding site and a C region that, in the heavy chain, determines the antibody isotype and its specialized effector functions. The IgG molecule is Y-shaped, in which the stem and the arms are of comparable sizes and can flex with respect to each other. Each arm contains an antigen-binding site, whereas the stem contains binding sites for cells and effector molecules that enable bound antigen to be cleared from the body. Immunoglobulin chains are built from a series of structurally related immunoglobulin domains. Within the V domain, sequence variability is localized to three complementarity-determining regions (CDRs) corresponding to the three loops that are clustered at one end of the domain. In the antibody molecule the CDRs of the heavy and light chains form a variable surface that binds antigen. The type of antigen bound by an antibody depends on the shape of the antigen-binding site: antigens that are small molecules can be bound within deep pockets; linear epitopes from proteins or carbohydrates can be bound within clefts or grooves; and the binding of conformational epitopes of folded proteins takes place over an extended surface area. Monoclonal antibodies are antibodies of a single specificity that are derived from a clone of identical antibody-producing cells, which are used in diagnostic tests and as therapeutic agents.

Generation of immunoglobulin diversity in B cells before encounter with antigen

The number of different antibodies that can be produced by the human body is virtually limitless. To achieve this, the immunoglobulin genes are organized differently from other genes. In all cells, except B cells, the immunoglobulin genes are in a fragmented form that cannot be expressed; instead of containing a single complete gene, immunoglobulin heavy-chain and light-chain loci consist of families of **gene segments**, sequentially arrayed along the chromosome, with each set of segments containing alternative versions of parts of the immunoglobulin V region. The immunoglobulin genes are inherited in this form through the germline (egg and sperm); this arrangement is therefore called the **germline form** or **germline configuration**.

For an immunoglobulin gene to be expressed, individual gene segments must first be rearranged to assemble a functional gene, a process that occurs only in developing B cells. Immunoglobulin-gene rearrangements occur during the development of B cells from B-cell precursors in the bone marrow. When gene rearrangements are complete, heavy and light chains are produced, and membrane-bound immunoglobulin appears at the B-cell surface. The B cell can now recognize and respond to an antigen through this receptor. Much of the diversity within the mature antibody repertoire is generated during this process of gene rearrangement, as we shall see in this part of the chapter.

4-7 The DNA sequence encoding a V region is assembled from two or three gene segments

In humans, the immunoglobulin genes are found at three chromosomal locations: the heavy-chain locus on chromosome 14, the κ light-chain locus on

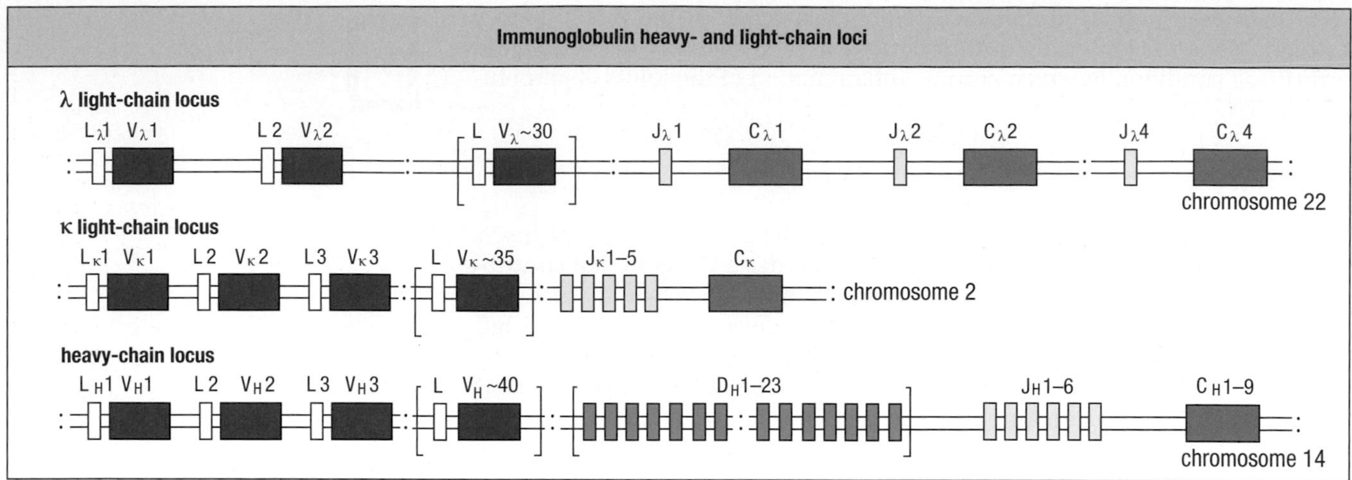

Figure 4.15 The germline organization of the human immunoglobulin heavy-chain and light-chain loci. The upper row shows the λ light-chain locus, which has about 30 functional V_λ gene segments and four pairs of functional J_λ gene segments and C_λ gene segments. The κ locus (center row) is organized in a similar way, with about 35 functional V_κ gene segments accompanied by a cluster of five J_κ gene segments but with a single C_κ gene segment. In approximately half of the human population, the entire cluster of V_κ gene segments is duplicated (not shown, for simplicity). The heavy-chain locus (bottom row) has about 40 functional V_H gene segments, a cluster of about 23 D segments and 6 J_H gene segments. For simplicity, a single C_H gene (C_H1–9) is shown in this diagram to represent the nine C genes. The diagram is not to scale: the total length of the heavy-chain locus is more than 2 megabases (2 million bases), whereas some of the D segments are only six bases long. L, leader sequence.

chromosome 2, and the λ light-chain locus on chromosome 22. Different gene segments encode the leader peptide (L), the V region (V), and the constant region (C) of the heavy and light chains. Gene segments encoding C regions are commonly called C genes (Figure 4.15). Within the heavy-chain locus are C genes for all the different heavy-chain isotypes.

The gene segments encoding the leader peptides and the C regions consist of exons and introns, like those found in other human genes, and they are ready to be transcribed. In contrast, the V regions are encoded by two (V_L) or three (V_H) gene segments that must be selected from arrays of similar gene segments and rearranged to produce an exon that can be transcribed. The two types of gene segment that encode the light-chain V region are called **variable (V)** and **joining (J) gene segments.** The heavy-chain locus includes an additional set of **diversity (D) gene segments** that lies between the arrays of V and J gene segments (see Figure 4.15).

The V region of a light chain is encoded by the combination of one V and one J segment, whereas the C region is encoded by a single C gene segment. The main differences between V gene segments are in the sequences that encode CDR1 and CDR2 of the V domain (see Section 4-3); CDR3 is determined by the sequence diversity introduced at the junction between the V and J segments. In an individual B cell, only one of the light-chain loci (κ or λ) gives rise to a functional light-chain gene. The V region of a heavy chain is encoded by the combination of one V, one D, and one J segment. Diversity in CDR1 and CDR2 is determined by differences in the sequences of the heavy-chain V gene segments; CDR3 is determined by differences in the D gene segments and the junctions they make with the V and J gene segments. The C region of a heavy chain is encoded by one of the C genes.

4-8 Random recombination of gene segments produces diversity in the antigen-binding sites of immunoglobulins

During the development of B cells, the arrays of V, D, and J segments are cut and re-spliced by DNA recombination. This process is called **somatic recombination** and it brings together a single gene segment of each type to form a DNA sequence encoding the V region of an immunoglobulin chain. It is called somatic recombination because B cells are cells of the soma, a term that embraces all of the cells of the body except the germ cells. For light chains, a single recombination occurs, between a V_L and a J_L segment (Figure 4.16, left panels) whereas for heavy chains, two recombinations are needed, the first to join a D and a J_H segment, and the second to join the combined DJ segment to

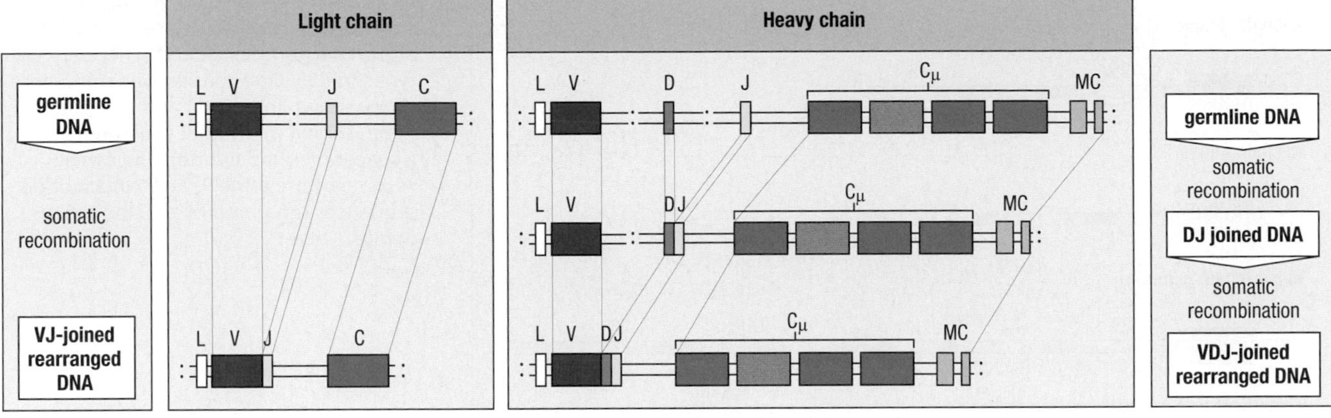

Figure 4.16 V-region sequences are constructed from gene segments. Light-chain V-region genes are constructed from two segments (left panel). A variable (V) and a joining (J) gene segment in the genomic DNA are joined to form a complete light-chain V-region (V_L) exon. After rearrangement, the light-chain gene consists of three exons, encoding the leader (L) peptide, the V region, and the C region, which are separated by introns. Heavy-chain V regions are constructed from three gene segments (right panels). First the diversity (D) and J gene segments join, then the V gene segment joins to the combined DJ sequence, forming a complete heavy-chain V-region (V_H) exon. For simplicity, only the first of the heavy-chain genes, Cμ, is shown here. Each immunoglobulin domain is encoded by a separate exon, and two additional membrane-coding exons (MC, colored light blue) specify the hydrophobic sequence that will anchor the heavy chain to the B-cell membrane.

a V_H segment (Figure 4.16, right panels). In each case, the particular V, D, and J gene segments that are joined together are selected at random. Because of the multiple gene segments of each type, numerous different combinations of V, D, and J gene segments are possible. Thus, the gene rearrangement process generates many different V-region sequences in the B-cell population. This is one of several factors contributing to the diversity of immunoglobulin V regions.

For the human κ light chain, approximately 35 V_κ gene segments and 5 J_κ gene segments (Figure 4.17) can be recombined in 175 (35 × 5) different ways. Similarly, some 30 V_λ gene segments and 4 J_λ gene segments can be recombined in 120 (30 × 4) different ways. At the heavy-chain locus, 40 V_H segments, some 23 D segments and 6 J_H segments can produce 5520 (40 × 23 × 6) combinations. If recombination between different gene segments were the only mechanism producing V-region diversity, then a maximum of 295 different light chains and 5520 different heavy chains could be made. With random combination of heavy and light chains, this diversity could make 1.6 million different antibodies.

Somatic recombination is carried out by enzymes that 'cut and paste' the DNA, and it exploits some of the mechanisms more universally used by cells for DNA recombination and repair. The recombination of V, J, and D gene segments is directed by sequences called **recombination signal sequences (RSSs)**, which flank the 3′ side of the V segment, both sides of the D segment, and the 5′ side of the J segment (Figure 4.18). There are two types of RSS, and recombination can occur only between different types. One type of RSS comprises a defined heptamer [7 base pairs (bp); CACAGTG] sequence and a defined nonamer (9 bp; ACAAAAACC) separated by a 12-bp spacer; the other comprises the heptamer and the nonamer sequences separated by a 23-bp spacer. As well as providing recognition sites for the enzymes that cut and rejoin the DNA, RSSs ensure that the gene segments are joined in the correct order, as shown for light-chain V_J recombination in Figure 4.19. Because of the requirements of RSS recombination, known as the **12/23 rule**, recombination in the heavy-chain DNA cannot join a V_H directly to a J_H without the involvement of D_H, because the V_H and J_H segments are flanked by the same type of RSS (see Figure 4.18).

Number of gene segments in the three human immunoglobulin loci

Segment	Light chains		Heavy chain
	κ	λ	H
Variable (V)	34–38	29–33	38–46
Diversity (D)	0	0	23
Joining (J)	5	4–5	6
Constant (C)	1	4–5	9

Figure 4.17 The numbers of functional gene segments available to construct the variable and constant regions of human immunoglobulin heavy chains and light chains.

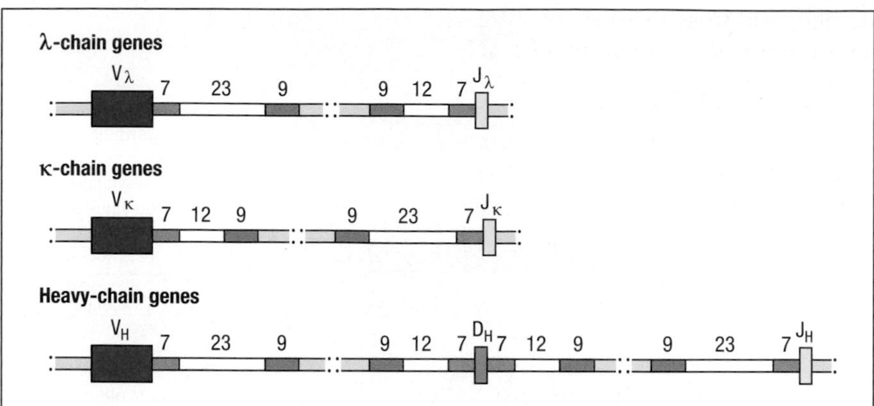

Figure 4.18 Each V, D, or J gene segment is flanked by recombination signal sequences (RSSs). There are two types of RSS. One consists of a nonamer (9 bp, shown in purple) and a heptamer (7 bp, shown in orange) separated by a spacer of 12 bp (white). The other consists of the same 9- and 7-nucleotide sequences separated by a 23-bp spacer (white).

The set of enzymes needed to recombine V, D, and J segments is called the **V(D)J recombinase**. Two of the component proteins are made only in lymphocytes, and are specified by the **recombination-activating genes (*RAG1* and *RAG2*)**. The other components are present in all nucleated cells and have activities that repair double-stranded DNA, bend DNA, or modify the ends of broken DNA strands. The RAG-1 and RAG-2 proteins interact with each other and with other proteins, known as the high-mobility group of proteins, to form a RAG complex. In the first step in recombination, a RAG complex binds to one

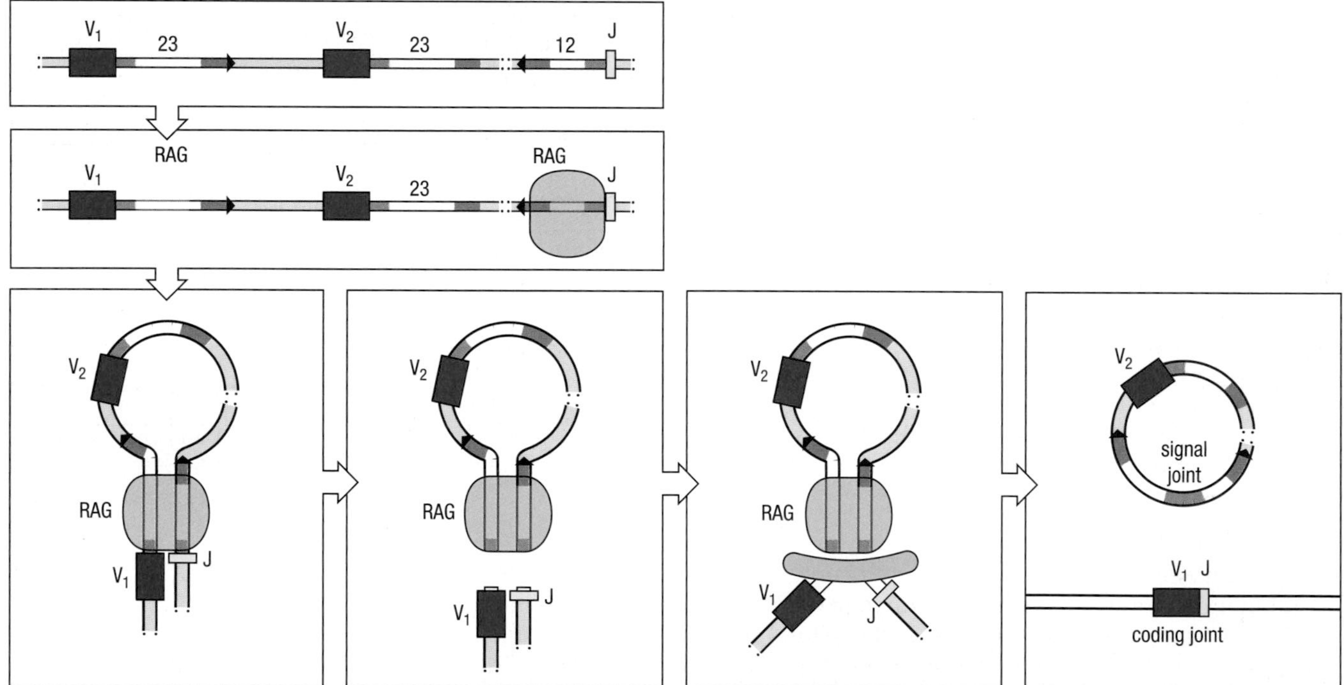

Figure 4.19 Gene segments encoding the variable region are joined by recombination at recombination signal sequences. The recombination between a V (red) and a J (yellow) segment of a light-chain gene is shown here. A RAG complex (green) binds to one of the recombination signal sequences (RSSs) flanking the coding sequences to be joined. The RAG complex then recruits the other RSS. This process brings an RSS containing a 12-bp spacer together with one containing a 23-bp spacer. This is known as the 12/23 rule and ensures that gene segments are joined in the correct order. The DNA molecules are broken at the ends of the heptamer sequences (orange) and are then joined together with different topologies. The region of DNA that was originally between the V and J segments to be joined is excised as a small circle of DNA that has no function. The joint made in forming this circle is called the signal joint. Within the chromosomal DNA, the V and J segments are joined to form the coding joint. Formation of this joint involves the introduction of additional variability into the nucleotide sequence around the joint. Enzymes other than RAG involved in these processes are represented in blue. Nonamers are shown in purple, heptamers in orange, spacers in white.

of the RSSs flanking the sequences to be joined, and then recruits the other RSS to the complex (see Figure 4.19). The endonuclease functions of RAG cleave the DNA to generate a clean break at the ends of the two heptamer sequences. The RSSs are aligned and held in place by the RAG complex while the broken ends are rejoined by DNA repair enzymes in a process called non-homologous end-joining. This brings together the ends of two gene segments to form a so-called **coding joint** in the chromosome while joining the ends of the removed DNA in a **signal joint**.

4-9 Recombination enzymes produce additional diversity in the antigen-binding site

The recombination process initiated by the RAG complex in a developing B cell introduces changes in nucleotide sequence and adds nucleotides at random at the coding joints, thus introducing additional diversity into the variable regions. This diversity is not encoded in the sequence of the germline DNA and is generated by the sequential action of the enzymes that repair the DNA breaks, as shown in Figure 4.20 for a D to J rearrangement. The initial cleavage of the DNA within the RAG complex generates single-stranded structures known as DNA hairpins at the ends of the cleaved gene segments (Figure 4.20, first and second panels). DNA repair enzymes then open the hairpins. The nick that opens the hairpin can occur at any of several positions, which is one source of diversity. It generates a single-stranded end in which bases that were complementary in the two DNA strands are now on the same strand. This creates a new nucleotide sequence that would be a palindrome in double-stranded DNA—in DNA terms a palindrome is a sequence of base pairs that is identical when read from either end (Figure 4.20, third panel). The additional nucleotides are thus called **P nucleotides** (for **p**alindromic nucleotides). The ends of the opened hairpins can then be variably modified by exonucleases that remove germline-encoded nucleotides and by the enzyme **terminal deoxynucleotidyl transferase** (**TdT**), which randomly adds nucleotides, generating further diversity. These added nucleotides need not correspond to the germline sequence in either order or number, and are called **N nucleotides** because they are **n**ot encoded in germline DNA (Figure 4.20, fourth panel). Once the single-stranded tails of the two gene segments are able to pair, the single-strand gaps are filled in with complementary nucleotides to complete the coding joint (Figure 4.20, last three panels).

The contribution of P and N nucleotides to the coding joints is called **junctional diversity** and increases diversity in the resulting amino-acid sequence of the CDR3s of immunoglobulin chains. In the light-chain V domain, CDR3 is

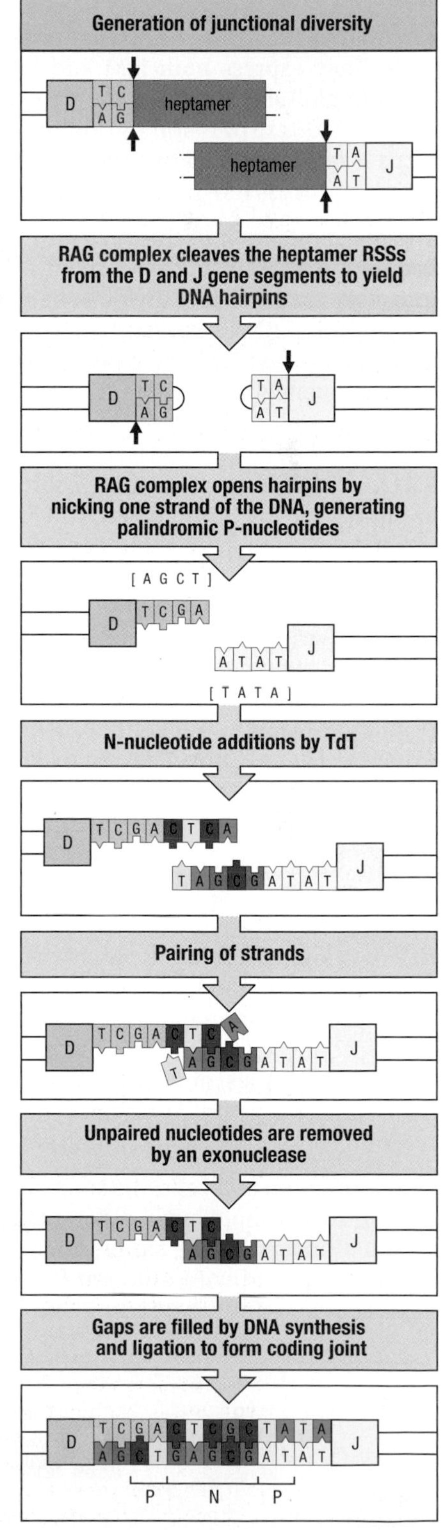

Figure 4.20 The generation of junctional diversity during gene rearrangement. The process is illustrated for a D to J rearrangement. The RSSs are brought together and the RAG complex cleaves (arrows) between the heptamer sequences and the gene segments (top panel). This leads to excision of the DNA that separates the D and J segments. The ends of the two strands of the DNA double helix in the D and J segments are joined to form structures known as 'hairpins'. Further cleavage (arrows) on one DNA strand of the D and the J segments opens the hairpins and generates short single-stranded sequences at the ends of the D and J segments. The extra nucleotides are known as P nucleotides because they make a palindromic sequence in the final double-stranded DNA (as indicated on the diagram). Terminal deoxynucleotidyl transferase (TdT) adds nucleotides randomly to the ends of the single strands. These nucleotides, which are not encoded in the germline, are known as N nucleotides. The single strands pair, and through the action of exonuclease, DNA polymerase, and DNA ligase the double-stranded DNA molecule is repaired to give the coding joint.

determined by the junction formed between the V and J segments (see Figure 4.8); in the heavy-chain V domain it is formed by the D segment and its two junctions with the rearranged V and J segments. Junctional diversity is an important source of immunoglobulin variability, having the potential to increase overall diversity by a factor of up to 3×10^7.

4-10 Developing and naive B cells use alternative mRNA splicing to make both IgM and IgD

Circulating B cells that have yet to encounter antigen are known as **naive B cells**. They express both IgM and IgD on their surfaces, which are the only immunoglobulin isotypes that are ever produced simultaneously by a B cell. Simultaneous expression of both μ and δ chains from the same heavy-chain locus is accomplished by differential splicing of the same primary RNA transcript, a process that involves no rearrangement of genomic DNA.

The rearrangement of the V, D, and J segments of the heavy-chain locus that occurs during B-cell development brings a gene promoter and an enhancer into closer juxtaposition, which enables the rearranged gene to be transcribed. The resulting RNA transcript is then spliced and processed and the mRNA is translated to give a heavy-chain protein. In the rearranged heavy-chain locus, the exons encoding the leader peptide and the V region are on the 5′ side (upstream) of the DNA encoding the nine different C regions. Closest to the rearranged V-region gene is the μ gene, which is followed by the δ gene (Figure 4.21). In each C gene, separate exons encode each immunoglobulin domain, as shown for the μ and δ genes in Figure 4.22. In mature naive B cells, transcription of the heavy-chain gene starts upstream of the exons encoding the leader peptide and the V region, continues through the μ and δ C genes and terminates downstream of the δ gene, before the γ3 C gene. This long primary RNA transcript is then spliced and processed in two different ways: one yields mRNA for the μ heavy chain (Figure 4.22, left panel) and one yields mRNA for the δ heavy chain (Figure 4.22, right panel). In making μ-chain mRNA from the primary transcript, the entire δ-gene RNA is removed along with the introns from the μ gene. Conversely, in making δ-chain mRNA the entire μ-gene RNA is removed as well as the δ-gene introns.

4-11 Each B cell produces immunoglobulin of a single antigen specificity

In a developing B cell, the process of immunoglobulin-gene rearrangement is tightly controlled so that only one heavy chain and one light chain are finally expressed, a phenomenon known as **allelic exclusion**. This ensures that each B cell produces IgM and IgD of a single antigen specificity. Although every B cell has two copies, or alleles, of the heavy-chain locus and two copies of each light-chain locus, only one heavy-chain locus and one light-chain locus are rearranged to produce functional genes.

In a population of B cells, the same functional light-chain gene rearrangement is found associated with different functional heavy-chain gene rearrangements. Conversely, the same functional heavy-chain gene rearrangement is found associated with different functional light-chain gene rearrangements. Because each antigen-binding site is formed by a heavy chain and a light

Figure 4.21 Rearrangement of V, D, and J segments produces a functional heavy-chain gene. The assembled VDJ sequence lies some distance from the cluster of C genes. Only functional C genes are shown here. The four different γ genes specify four different subtypes of the γ heavy chain, whereas the two α genes specify two subtypes of the α heavy chain. For simplicity, individual exons in the C genes are not shown. The diagram is not to scale.

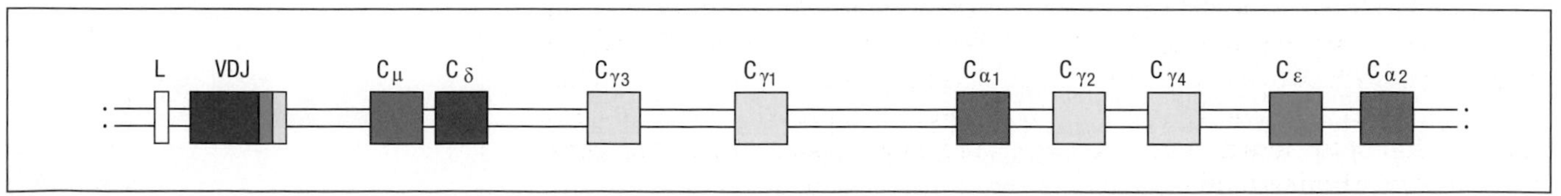

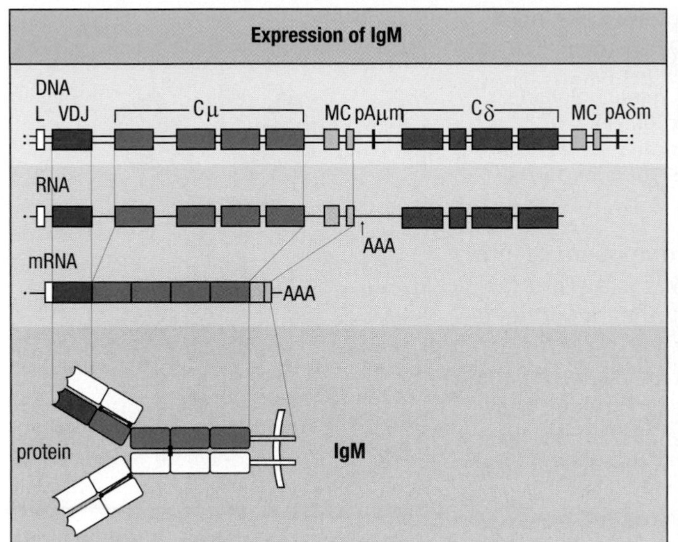

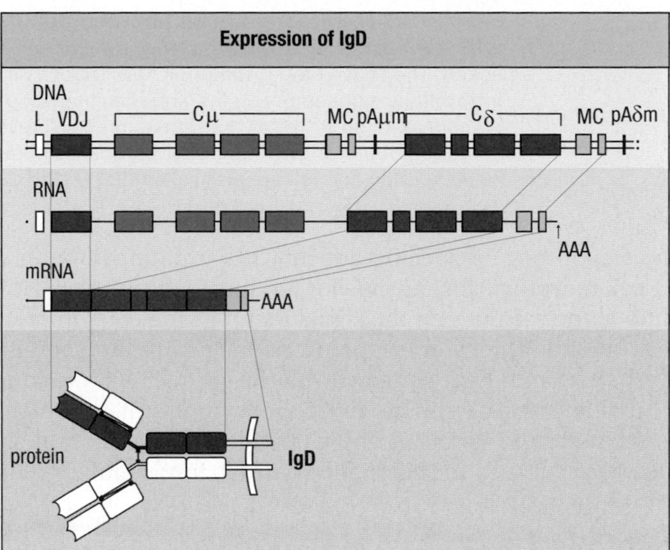

Figure 4.22 Coexpression of IgD and IgM is regulated by RNA processing. In mature B cells, transcription initiated at the V_H promoter extends through both the C_μ and C_δ genes. For simplicity we have not shown all the individual C-gene exons but only those of relevance to the production of IgM and IgD. The long primary transcript is then processed by cleavage, polyadenylation, and splicing. Cleavage and polyadenylation at the μ site (pAμm; the 'm' denotes that this site produces membrane-bound IgM) and splicing between C_μ exons yields an mRNA encoding the μ heavy chain (left panel). Cleavage and polyadenylation at the δ site (pAδm) and a different pattern of splicing that removes the C_μ exons yields mRNA encoding the δ heavy chain (right panel). AAA designates the poly(A) tail. MC, exons that encode the transmembrane region of the heavy chain.

chain, the random combinatorial association of heavy and light chains makes an important contribution to the overall diversity of immunoglobulins (see Section 4-8).

The fact that B cells are monospecific means that an encounter with a given pathogen engages a subset of B cells that will make antibodies that bind only to the pathogen. This clonal selection is one of the central principles of adaptive immunity (see Chapter 1). It ensures that the specificity of the antibody made in response to an infection is specific for the invading pathogen and explains why vaccination against diphtheria, for example, gives protection against diphtheria but not influenza.

4-12 Immunoglobulin is first made in a membrane-bound form that is present on the B-cell surface

When a B cell first makes IgM and IgD, the heavy chains (but not the light chains) have a hydrophobic sequence near the carboxy terminus by which the immunoglobulins associate with cell membranes. Like all proteins destined for the cell surface, immunoglobulin chains enter the endoplasmic reticulum as soon as they are synthesized. There they associate with each other to form immunoglobulin molecules attached to the endoplasmic reticulum membrane. By themselves, these immunoglobulin molecules cannot be transported to the cell surface. For that to happen they must associate with two additional transmembrane proteins called **Igα** and **Igβ** (Figure 4.23). Unlike immunoglobulins, these proteins are encoded by genes that do not rearrange and so do not vary in sequence between different clones of B cells. They travel to the B-cell surface in a complex with the immunoglobulin molecule, and this complex of IgM, Igα, and Igβ forms the receptor for antigen on naive B cells. The function of the less abundant complex of IgD with Igα and Igβ is by comparison poorly understood.

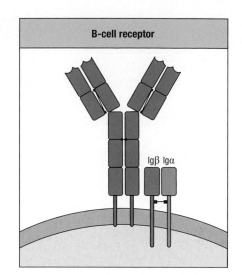

Figure 4.23 Membrane-bound immunoglobulins are associated with two other proteins, Igα and Igβ. Igα and Igβ are disulfide-linked. They have long cytoplasmic tails that can interact with intracellular signaling proteins, and the complex of immunoglobulin with Igα and Igβ serves as the functional B-cell receptor. The immunoglobulin shown here is IgM, but all isotypes can serve as B-cell receptors. Igα and Igβ are also called CD79a and CD79b, respectively.

In responding to specific antigen, the immunoglobulin component of the B-cell receptor is responsible for binding to the antigen. However, this interaction alone cannot give the signal that tells the cell's interior when an antigen has bound. The cytoplasmic portions of immunoglobulin heavy chains are very short and do not interact with the intracellular proteins that signal B cells to divide and differentiate. The longer cytoplasmic tails of the Igα and Igβ components of the B-cell receptor provide that function of signal transduction. They contain amino acid motifs to which intracellular signaling proteins bind.

Summary

In the human genome, the immunoglobulin heavy-chain and light-chain genes are in a form that is incapable of being expressed. In developing B cells, however, the immunoglobulin genes undergo structural rearrangements that permit their expression. The V domains of immunoglobulin light and heavy chains are encoded in two (V and J) or three (V, D, and J) different kinds of gene segment, respectively, that are brought into juxtaposition by recombination reactions. One mechanism contributing to the diversity in V-region sequences is the random combination of different V and J segments in light-chain genes and of different V, D, and J segments in rearranged heavy-chain genes. A second mechanism is the introduction of additional nucleotides (P and N nucleotides) at the junctions between gene segments during the process of recombination. A third mechanism that creates diversity in the antigen-binding sites of antibodies is the association of heavy and light chains in different combinations. Gene rearrangement in an individual B cell is strictly controlled so that only one type of heavy chain and one type of light chain are expressed, resulting in an individual B cell and its progeny expressing only immunoglobulin of a single antigen specificity. A naive B cell that has never encountered antigen expresses only membrane-bound immunoglobulin of the IgM and IgD classes. The μ and δ heavy chains are produced from a single transcriptional unit that undergoes alternative RNA processing and splicing.

Diversification of antibodies after B cells encounter antigen

A mature B cell's first encounter with antigen marks a watershed in its development. Binding of antigen to the surface immunoglobulin of a mature naive B cell triggers the cell's proliferation and differentiation, and ultimately the secretion of antibodies. As the immune response progresses, antibodies with different properties are made. In this part of the chapter we look at the structural and functional changes that occur in immunoglobulins after antigen has been encountered, and how antibodies with different functions are made.

4-13 Secreted antibodies are produced by an alternative pattern of heavy-chain RNA processing

Gene rearrangement in an immature B cell leads to the expression of functional heavy and light chains and to the production of membrane-bound IgM and IgD on the mature B cell. After an encounter with antigen, these isotypes

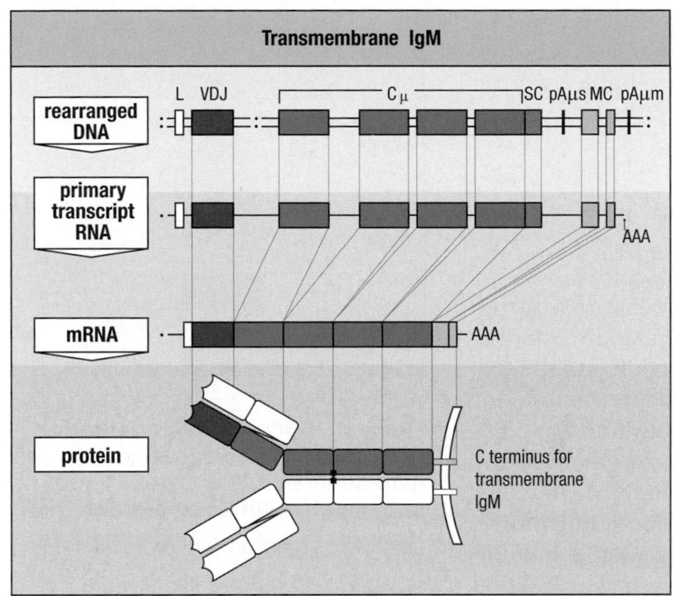

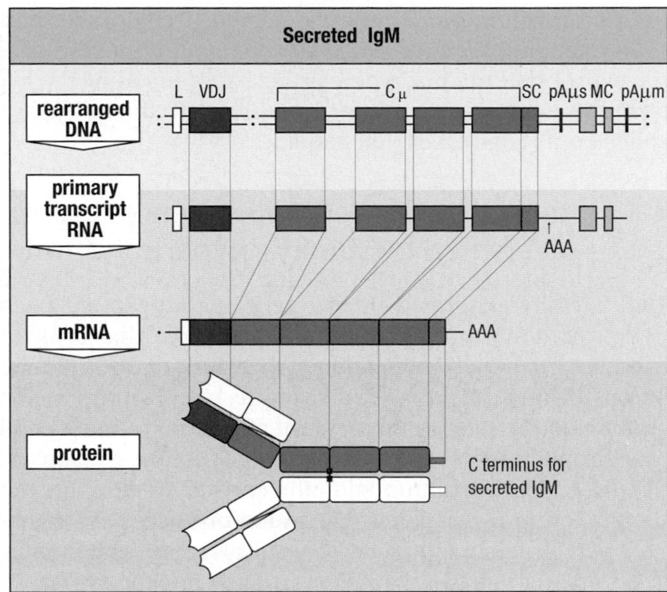

Figure 4.24 The surface and secreted forms of an immunoglobulin are derived from the same heavy-chain gene by alternative RNA processing. Each heavy-chain C gene has two exons (membrane-coding (MC), light blue) encoding the transmembrane region and cytoplasmic tail of the surface form of that isotype, and a secretion-coding (SC) sequence (orange) encoding the carboxy terminus of the secreted form. The events that dictate whether a heavy-chain RNA will result in a secreted or transmembrane immunoglobulin occur during processing of the initial transcript and are shown here for IgM. Each heavy-chain C gene has two potential polyadenylation sites (shown as pAμs and pAμm). In the left panel, the transcript is cleaved and polyadenylated at the second site (pAμm). Splicing between a site located between the fourth C_{μ} exon and the SC sequence, and a second site at the 5′ end of the MC exons, removes the SC sequence and joins the MC exons to the fourth C_{μ} exon. This generates the transmembrane form of the heavy chain. In the right panel, the primary transcript is cleaved and polyadenylated at the first site (pAμs), eliminating the MC exons and giving rise to the secreted form of the heavy chain. AAA designates the poly(A) tail.

are produced as secreted antibodies. IgM antibodies are produced in large amounts and are important in protective immunity against a range of infections, whereas IgD antibodies are produced mainly within the respiratory tract against the bacteria that commonly cause infections there.

All classes of immunoglobulin are made in two forms: one bound to the cell membrane that serves as the B-cell receptor for antigen, and the other, the antibody, that is a secreted and soluble effector molecule. During their differentiation from naïve B cells to antibody-producing plasma cells, B cells go from making only the membrane-bound form to making only the soluble, secreted form. The difference between membrane-bound and secreted immunoglobulin lies at the carboxy terminus of the heavy chain: here, membrane-associated immunoglobulin has a hydrophobic anchor sequence that is inserted into the membrane, whereas antibody has a hydrophilic sequence. This difference is determined by differential RNA splicing and processing of the same primary RNA transcript, and involves no rearrangement of the underlying genomic DNA. The alternative patterns of splicing for membrane-bound and secreted IgM are compared in Figure 4.24.

The hydrophilic carboxy terminus of the secreted μ chain is encoded at the 3′ end of the exon encoding the fourth C-region domain, whereas the hydrophobic membrane anchor of the membrane-associated μ chain is encoded by two small, separate exons downstream. The splicing to give secreted μ chain is the simpler pattern: the sequence encoding the hydrophilic carboxy terminus is retained and the sequences 3′ of that, including the exons encoding the hydrophobic membrane anchor, are discarded (Figure 4.24, right panel). To produce mRNA encoding membrane-associated μ chain, alternative splicing in the exon encoding the fourth C-region domain removes the sequence encoding the hydrophilic anchor, whereas the exons encoding the hydrophobic carboxy terminus are retained and incorporated into the mRNA when the introns are spliced out (Figure 4.24, left panel).

4-14 Rearranged V-region sequences are further diversified by somatic hypermutation

The diversity generated during gene rearrangement is concentrated in the CDR3s of the V_H and V_L domains. Once a B cell has been activated by antigen, however, further diversification of the whole of the V-domain coding sequence occurs through a process of **somatic hypermutation**. This introduces single-nucleotide substitutions (point mutations) almost randomly and at a high rate throughout the rearranged V regions of heavy-chain and light-chain genes (Figure 4.25). The immunoglobulin constant regions are not affected, and neither are other B-cell genes. The mutations occur at a rate of about one mutation per V-region sequence per cell division, which is more than a million times greater than the ordinary mutation rate for a gene.

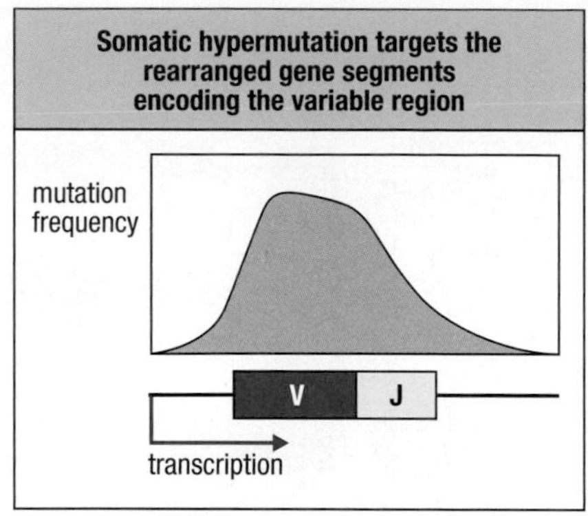

Figure 4.25 Somatic hypermutation is targeted to the rearranged gene segments that encode immunoglobulin V regions. The frequency of mutations at positions in and around the rearranged VJ sequence of an expressed light-chain gene is shown here.

Somatic hypermutation is dependent on the enzyme **activation-induced cytidine deaminase** (**AID**), which is made only by proliferating B cells. AID converts cytosine in single-stranded DNA to uracil, a normal component of RNA but not DNA. AID acts during transcription, when the two DNA strands of the immunoglobulin gene become temporarily separated. Other enzymes, which are not specific to B cells but are components of general pathways of DNA repair and modification, then convert the uracil to any one of the four bases of normal DNA, thus potentially causing a mutation.

Somatic hypermutation gives rise to B cells bearing mutant immunoglobulin molecules on their surface. Some of these mutant immunoglobulins have substitutions in the antigen-binding site that increase its affinity for the antigen. B cells bearing these mutant high-affinity immunoglobulin receptors compete most effectively for binding to antigen and are preferentially selected to mature into antibody-secreting plasma cells. The mutant antibodies that emerge from this selection do not have a random distribution of amino-acid substitutions. The changes are concentrated at positions in the heavy-chain and light-chain CDR loops that form the antigen-binding site and directly contact antigen (Figure 4.26). As the adaptive immune response to infection proceeds, antibodies of progressively higher affinity for the infecting pathogen are produced—a phenomenon called **affinity maturation**.

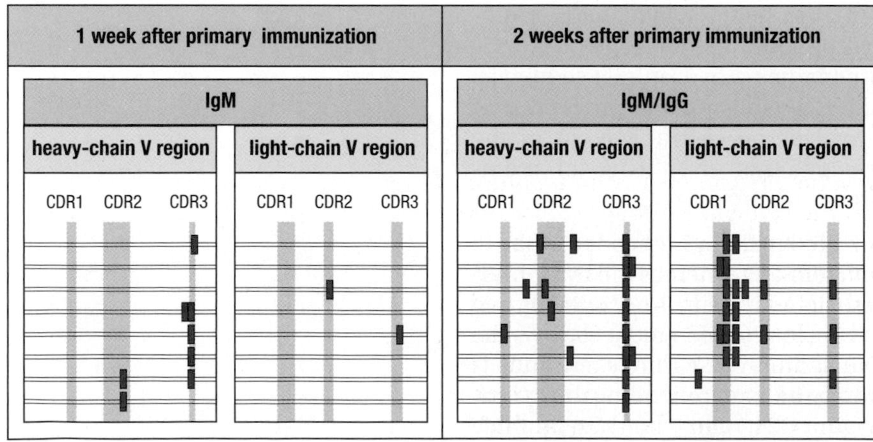

Figure 4.26 Somatic hypermutation permits selection of B cells making higher affinity antibodies. B cells were collected 1 and 2 weeks after immunization with the same epitope and used to make hybridomas secreting monoclonal antibodies. The amino acid sequences of the heavy and light chains of the epitope-specific monoclonal antibodies were determined. Each line represents one variant and the red bars represent amino acid positions that differ from the prototypic sequence. One week after primary immunization, most of the B cells make IgM, which shows some new sequence variation in the V region. This variation is confined to the CDRs, which form the antigen-binding site. Two weeks after immunization, both IgG- and IgM-producing B cells are present and their antibodies show increased variation that now involves all six CDRs of the antigen-binding site.

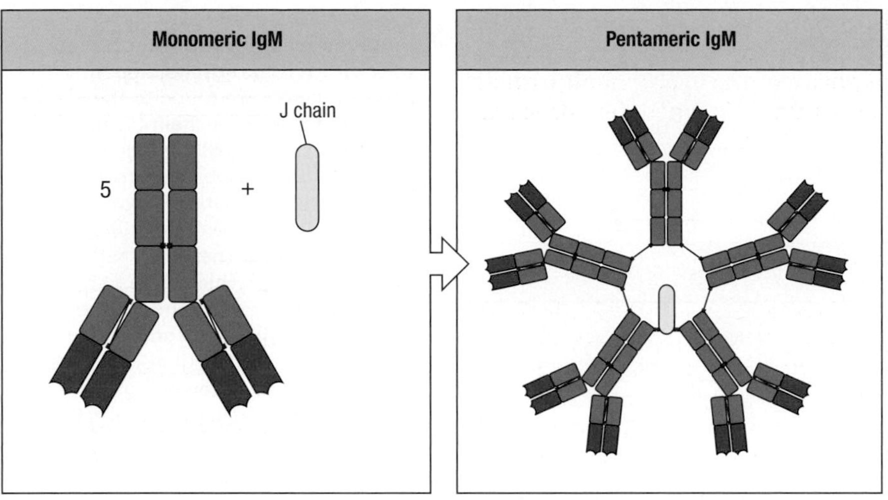

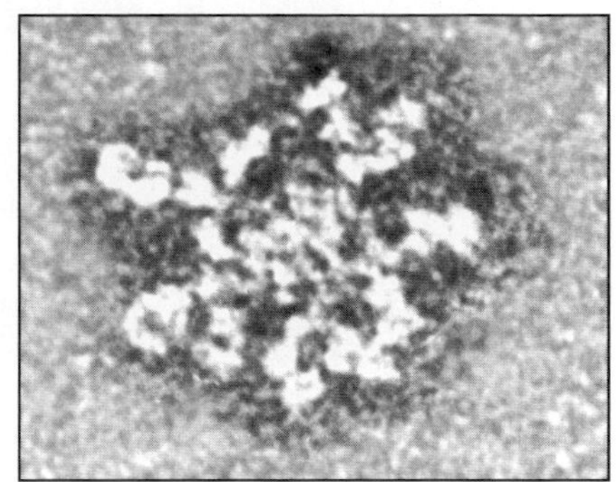

Figure 4.27 IgM is secreted as a pentamer of immunoglobulin monomers. The left two panels show schematic diagrams of the IgM monomer and pentamer. The IgM pentamer is held together by a polypeptide called the J chain, for joining chain (not to be confused with a J segment). The monomers are cross-linked by disulfide bonds to each other and to the J chain. The right panel shows an electron micrograph of an IgM pentamer, showing the arrangement of the monomers in a flat disc. The lack of a hinge region in the IgM monomer makes the molecule less flexible than, say, IgG, but this is compensated for by the pentamer having five times as many antigen-binding sites as IgG. Faced with a pathogen having multiple identical epitopes on its surface, IgM can usually attach to it with several binding sites simultaneously. Photograph (× 900,000) courtesy of K.H. Roux and J.M. Schiff.

Affinity maturation is a process of evolution in which variant immunoglobulins generated in a random manner are subjected to selection for improved binding to a pathogen. It achieves in a few days what would require thousands, if not millions, of years of classical Darwinian evolution in a conventional gene. This capacity for extraordinarily rapid evolution in pathogen-binding immunoglobulins is a major factor in allowing the human immune system to keep up with the generally faster-evolving pathogens.

4-15 Isotype switching produces immunoglobulins with different C regions but identical antigen specificities

IgM is the first class of antibody made in a primary immune response. Whereas the surface IgM of the B-cell receptor is monomeric, secreted effector IgM consists of a circular pentamer of the Y-shaped immunoglobulin monomers (Figure 4.27). Having 10 antigen-binding sites, IgM binds strongly to the surface of pathogens with multiple repetitive epitopes, but has limitations due to its bulk, low-affinity binding site and restricted recruitment of effector mechanisms. Antibodies with other effector functions are produced by the process of **isotype switching** or **class switching**, in which further DNA recombination events enable the rearranged V-region coding sequence to be used with other heavy-chain C genes. Isotype switching, like somatic hypermutation, is dependent on AID and similarly occurs only in B cells proliferating in response to antigen.

Isotype switching is accomplished by a recombination within the cluster of C genes that excises the previously expressed C gene and brings a different one into juxtaposition with the assembled V-region sequence. Thus, the antigen specificity of the antibody remains unchanged, even though its isotype changes. An example of isotype switching from IgM (plus IgD) to IgG1 is shown in Figure 4.28. Flanking the 5′ side of each C gene, with the exception of the δ gene, are highly repetitive sequences that mediate recombination. They are called **switch sequences** or **switch regions.** The first step is the initiation of transcription of the C-region gene to which the cell will be switched, in this case $C_{\gamma1}$ and its flanking switch region, $S_{\gamma1}$. AID then targets the cytosines in the $S_{\mu1}$ and the $S_{\gamma1}$ switch regions for deamination to uracil (see Figure 4.28). Next, uracil is removed by the enzyme uracil-DNA-glycosylase (UNG), leaving a nucleotide lacking a base. A specific endonuclease, APE1, then excises this abasic nucleotide, leaving a nick in the DNA strand. Nicks on both strands of the DNA in both the switch sequences facilitate their recombination, by which the μ, δ, and $C_{\gamma3}$ genes are excised as a circular DNA molecule, bringing the V

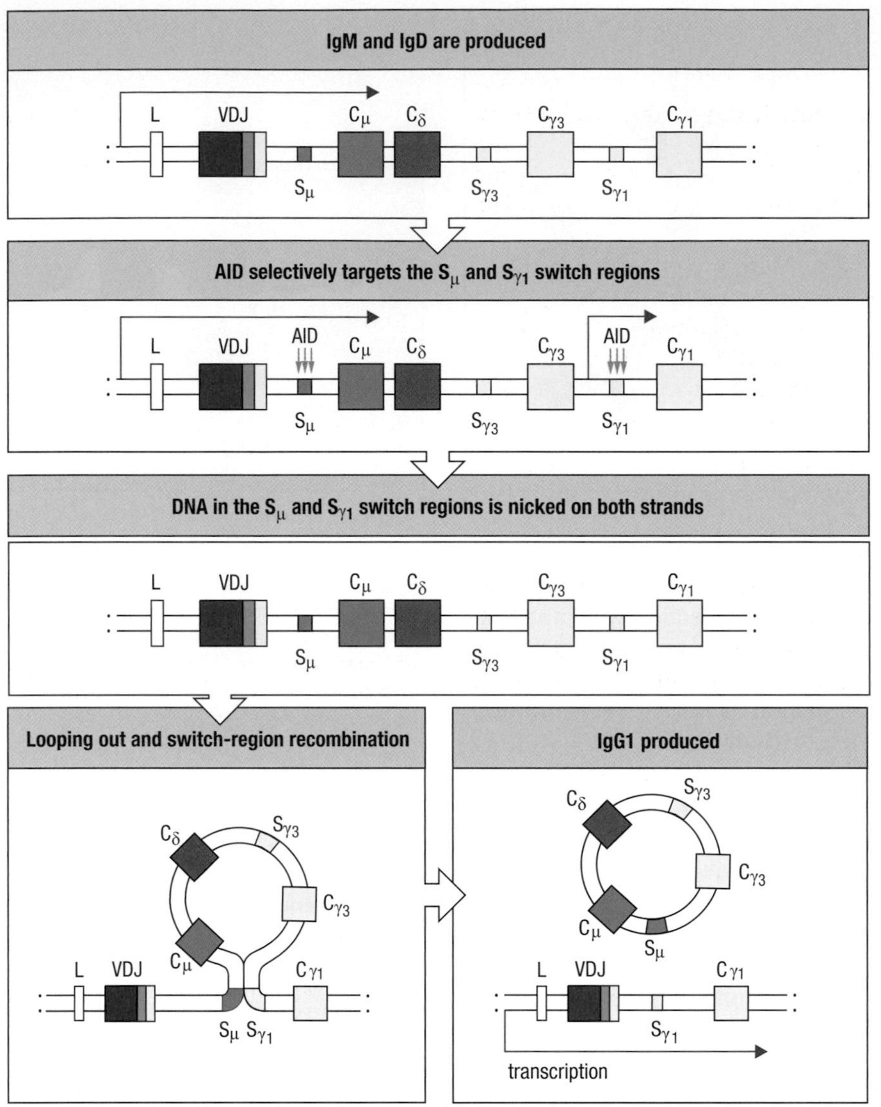

Figure 4.28 Isotype switching involves recombination between specific switch regions. Repetitive DNA sequences are found to the 5′ side of each of the heavy-chain C genes, with the exception of the δ gene. Immunoglobulin isotype switching occurs by recombination between these switch regions (S), with deletion of the intervening DNA. The switch regions are targeted by AID, which leads to nicks being made in both strands of the DNA. These nicks facilitate recombination between the switch regions, which leads to excision of the intervening DNA as a non-functional circle of DNA and brings the rearranged VDJ segments into juxtaposition with a different C gene. The first switch a clone of B cells makes is from the μ isotype to another isotype. A switch from μ to the γ1 isotype is shown here. Further switching to other isotypes can take place subsequently.

region into juxtaposition with the $C_{\gamma1}$ gene. The mRNA for the new immunoglobulin is then produced as described in Section 4-9. Switching can take place between the μ switch region and that of any other isotype. Sequential switching can also occur, for example from μ to γ_1 to α_1. Isotype switching only occurs during an active immune response, and the patterns of isotype switching are regulated by cytokines secreted by antigen-activated T cells.

Knowledge of the molecular mechanisms underlying isotype switching and somatic hypermutation has been aided by the study of patients deficient in enzymes involved in the mutation and modification of DNA in the V-region gene and in the switch sequences of the C-region genes. In particular, the immunoglobulin genes in the B cells of patients lacking a functional AID gene do not undergo somatic hypermutation or isotype switching. The only antibodies that these patients make are low-affinity IgM, which are produced in greater amounts than in normal individuals. For this reason their condition is called **hyper-IgM syndrome**. The main consequence of this immunodeficiency is susceptibility to infection with pyogenic bacteria, particularly in the sinuses, ears, and lungs. These infections are usually treatable with antibiotics and can be prevented by regular injections of intravenous immunoglobulin.

4-16 Antibodies with different C regions have different effector functions

The five classes of human immunoglobulin are IgG, IgM, IgD, IgA, and IgE (see Figure 4.5). IgG has four subclasses (IgG1, IgG2, IgG3, and IgG4), which are numbered according to their relative abundance in plasma, IgG1 being the most abundant (Figure 4.29). IgA has two subclasses (IgA1 and IgA2). The heavy chains of the human IgA subclasses are designated α1 and α2 and the heavy chains of the human IgG subclasses by γ1, γ2, γ3, and γ4. The α, δ, and γ heavy-chain C regions are made up of three C domains, whereas the μ and ε heavy chains have four (see Figure 4.5). Each C domain is encoded by a separate exon in the relevant C gene. Additional exons are used to encode the hinge region and the carboxy terminus, depending on the isotype. The physical properties of the human immunoglobulin isotypes are given in Figure 4.29.

Antibodies can clear pathogens from the body in various ways. **Neutralizing antibodies** directly inactivate a pathogen or a toxin and prevent it from interacting with human cells. Neutralizing antibodies against viruses, for example, bind to a site on the virus that is normally used to gain entry to cells. Another function of antibodies is opsonization, a term used to describe the coating of pathogens with a protein that facilitates its elimination (see Section 2-6). The common **opsonins** are antibodies and complement proteins. Opsonized pathogens are more efficiently ingested by phagocytes, which have receptors for the Fc regions of some antibodies and for certain complement proteins. Activation of complement by antibodies bound to a bacterial surface can also lead to the direct lysis of the bacterium. IgM is the first antibody produced in an immune response against a pathogen. It is made principally by plasma cells resident in lymph nodes, spleen, and bone marrow, and it circulates in blood and lymph. At the beginning of the immune response, the antibodies bind antigen with low **affinity**, the term immunologists use to describe the strength with which a single antigen-binding site of an antibody binds to antigen. In this situation the 10 antigen-binding sites of IgM are needed to coat a pathogen with IgM. Although each of the binding sites has a low-affinity antigen, IgM makes a cooperative multipoint attachment to the pathogen in which the overall binding is strong and effectively irreversible. The overall strength of binding at multiple sites is called the **avidity** of an antibody. When bound to antigen, sites exposed in the constant region of IgM initiate reactions with complement that can kill microorganisms directly or facilitate their phagocytosis. Because somatic hypermutation leads to antibodies of increased affinity for the antigen, two antigen-binding sites are then sufficient to produce strong binding. By switching isotype, different effector functions can be brought into play while preserving antigen specificity (Figure 4.30).

Monomeric IgA is made by plasma cells in lymph nodes, spleen, and bone marrow and is secreted into the bloodstream. IgA can also be made as a dimer,

	Immunoglobulin class or subclass								
	IgM	IgD	IgG1	IgG2	IgG3	IgG4	IgA1	IgA2	IgE
Heavy chain	μ	δ	γ_1	γ_2	γ_3	γ_4	α_1	α_2	ε
Molecular mass (kDa)	970	184	146	146	165	146	160	160	188
Serum level (mean adult mg/ml)	1.5	0.03	9	3	1	0.5	2.0	0.5	5×10^{-5}
Half-life in serum (days)	5	3	21	20	7	21	6	6	3

Figure 4.29 The physical properties of the human immunoglobulin isotypes. The molecular mass given for IgM is that of the pentamer (see Figure 4.27), the predominant form in serum. The molecular mass given for IgA is that of the monomer. Large amounts of IgA are also produced in the form of dimers, which is the form found in secretions at mucosal surfaces.

Function/property	IgM	IgD	IgG1	IgG2	IgG3	IgG4	IgA	IgE
Neutralization	+	–	+++	+++	+++	+++	+++	–
Opsonization	–	–	+++	*	++	+	+	–
Sensitization for killing by NK cells	–	–	++	–	++	–	–	–
Sensitization of mast cells	–	–	+	–	+	–	–	+++
Sensitization of basophils	–	+++	–	–	–	–	–	++
Activation of complement system	+++	–	++	+	+++	–	+	–
Transport across epithelium	+	–	–	–	–	–	+++ (dimer)	–
Transport across placenta	–	–	+++	+	++	++	–	–
Diffusion into extravascular sites	+/–	–	+++	+++	+++	+++	++ (monomer)	+

Figure 4.30 Each human immunoglobulin isotype has specialized functions correlated with distinctive properties. The major effector functions of each isotype (+++) are shaded in dark red; lesser functions (++) are shown in dark pink, and very minor functions (+) in pale pink. Other properties are similarly marked. Opsonization refers to the ability of the antibody itself to facilitate phagocytosis. Antibodies that activate the complement system indirectly cause opsonization via complement. The properties of IgA1 and IgA2 are similar and are given here under IgA. *IgG2 acts as an opsonin in the presence of one genetic variant of its phagocyte Fc receptor, which is found in about 50% of Caucasians. NK cells, natural killer cells.

two monomers being joined by a J chain identical to that in pentameric IgM (Figure 4.31). Dimeric IgA is made principally in the lymphoid tissues underlying mucosal surfaces and is the antibody secreted into the lumen of the gut; it is also the main antibody in other secretions, including milk, saliva, sweat, and tears. The mucosal surface of the gastrointestinal tract provides an extensive surface of contact between the human body and the environment, and the transport processes involved in the uptake of food make it vulnerable to infection. In total, more IgA is made than any other isotype. Some of it is directed against the numerous species of resident microorganism that colonize mucosal surfaces, keeping their population in check.

The IgE class of antibodies is highly specialized toward recruiting the effector functions of mast cells in epithelium, activated eosinophils present at mucosal surfaces, and basophils in blood. These cell types carry a high-affinity receptor that binds IgE in the absence of antigen. The presence of antigen, which binds to the IgE, triggers strong physical and inflammatory reactions that can expel and kill infecting parasites. In countries where parasite infections are rare, the major impact of the IgE response is from the allergies and asthma caused when IgE is made against otherwise harmless antigens.

Although the δ gene lacks a conventional switch region, a cryptic switch region between the μ and δ genes is used to produce B cells that make IgD as their only secreted immunoglobulin. These B cells are concentrated in the upper airways of the bronchial tract, particularly in the tonsils, and they make antibodies that interact with commensal and pathogenic respiratory bacteria. Basophils have a high-affinity receptor that binds to IgD in the absence of antigen, which is different from the IgE receptor. When antigen is present and binds to the IgD it triggers basophils to orchestrate a local immune response that eliminates the bacteria.

IgG is the most abundant antibody in the internal body fluids, including blood and lymph. Like IgM, it is made principally in the lymph nodes, spleen, and bone marrow and circulates in lymph and blood. IgG is smaller and more flexible than IgM, properties that give it easier access to antigens in the extracellular spaces of damaged and infected tissues. A key feature of the IgG molecule that contributes to its potency and versatility is conformational flexibility. This

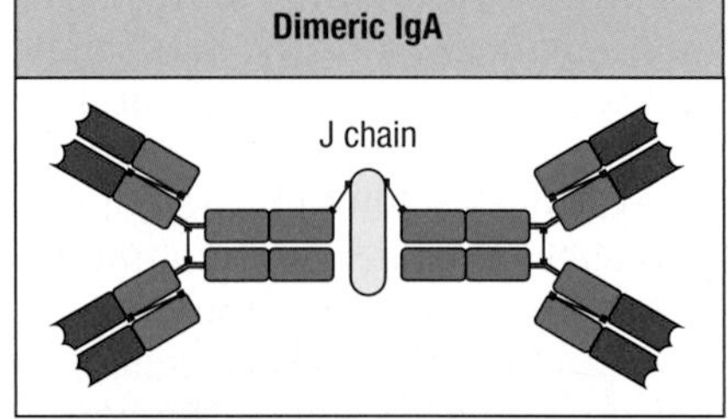

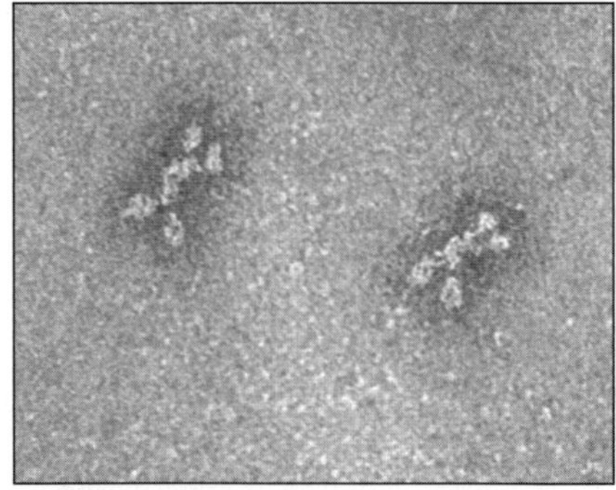

Figure 4.31 IgA molecules can form dimers. In mucosal lymphoid tissue, IgA is synthesized as a dimer in association with the same J chain as that found in pentameric IgM. In dimeric IgA, the monomers have disulfide bonds to the J chain but not to each other. The bottom panel shows an electron micrograph of dimeric IgA. Photograph (× 900,000) courtesy of K.H. Roux and J.M. Schiff.

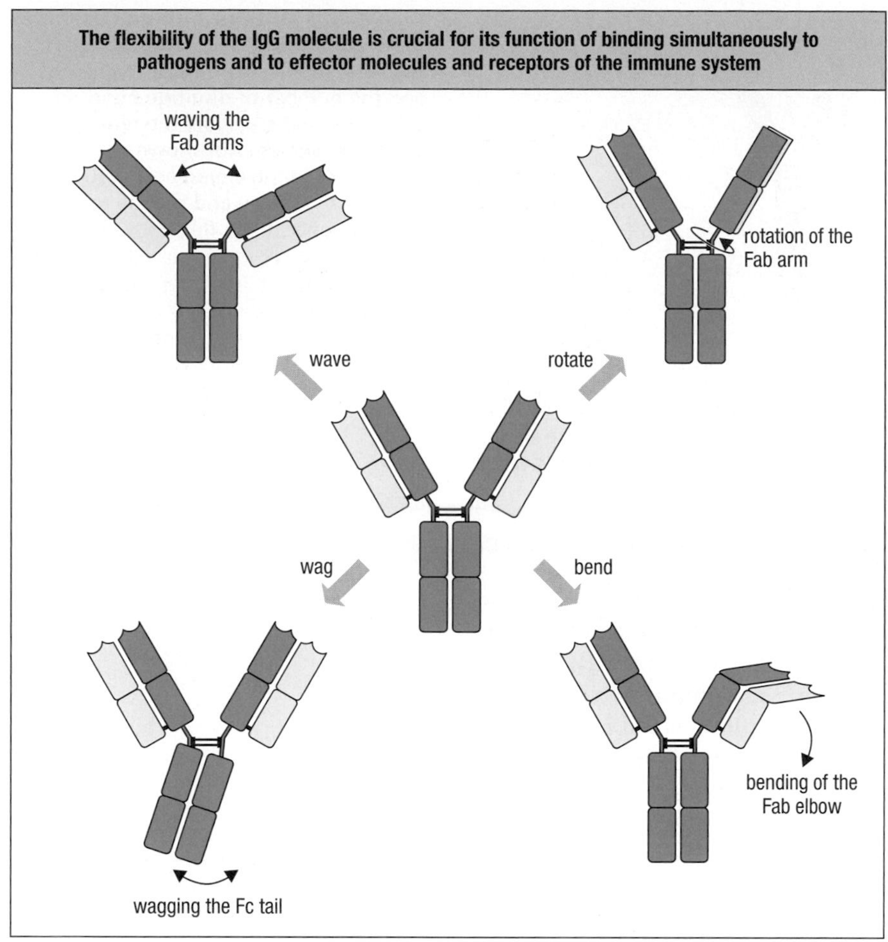

Figure 4.32 IgG is a highly flexible molecule. The most flexible part of the IgG molecule is the hinge, which allows the Fab arms to wave and rotate and thus accommodate the antibody to the orientation of epitopes on pathogen surfaces. Adding to further flexibility in binding antigen is the 'elbow' within the Fab that allows the variable domains to bend with respect to the constant domains. Similarly, the wagging of the Fc tail allows IgG molecules that have bound to antigen to accommodate to the binding of C1q and other effector molecules. Shown is a molecule of the IgG_1 subclass, which has represented IgG in figures throughout this book.

allows the antigen-binding sites in the two Fabs and the effector-binding sites in the Fc to move in a partly independent manner and assume a wide range of different positions with respect to each other (Figure 4.32). This flexibility, due mainly to the hinge region, greatly improves the likelihood that an IgG molecule can simultaneously bind to two antigens on the surface of a pathogen and to effector molecules such as complement component C1 (see Section 3-12) and the **Fc receptors** of phagocytes, the cell-surface receptors that bind the Fc portions of antibodies.

4-17 The four subclasses of IgG have different and complementary functions

For IgG the advantages of a flexible hinge are offset by the susceptibility of this loosely folded part of the heavy chain to proteolytic cleavage, which compromises IgG function by separating the Fc from the Fabs. In response to these conflicting pressures, the four different IgG subclasses—IgG1, IgG2, IgG3, and IgG4—have evolved. They differ in the constant region of the heavy chain and many of the differences are within the hinge (Figure 4.33).

Some of the functional differences between the IgG subclasses are listed in Figure 4.34. IgG1 is the most abundant and versatile of the four subclasses and is intermediate in its flexibility, susceptibility to proteolysis, and capacity to activate complement. It is a good all-rounder that constitutes most of the antibody made against protein antigens. In IgG2, the second most abundant subclass, the hinge is of similar length to that of IgG1 but contains additional disulfide bonds that reduce the flexibility, the susceptibility to proteolysis, and the capacity to activate complement. IgG2 is preferentially made against the

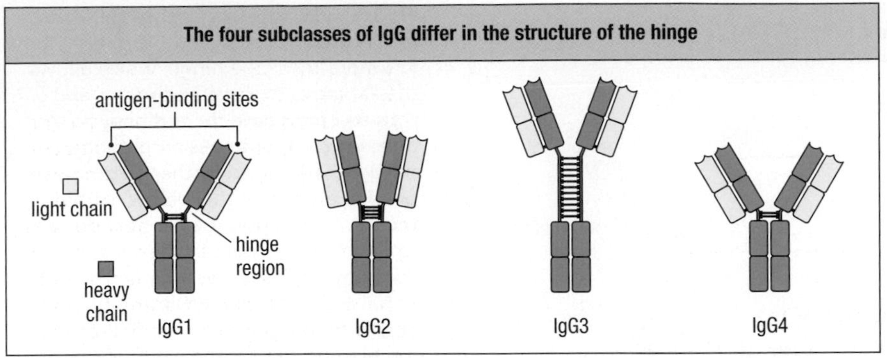

Figure 4.33 Different hinge structures distinguish the four subclasses of IgG. The relative lengths of the hinge and the number of disulfide bonds in the hinge that cross-link the two heavy chains are shown. Not shown are other differences in the amino acid sequence, particularly glycine and proline residues, that influence hinge flexibility.

highly repetitive carbohydrate antigens of microbial surfaces, which put fewer demands on an antibody's flexibility. Consistent with this function for IgG2, infections with encapsulated bacteria are poorly controlled in individuals who have IgG2 deficiency (see Figure 4.34).

Of the four IgG subclasses, IgG3 is the best at activating complement. It differs from the other subclasses by having a much longer hinge region: four times the length of the IgG1 hinge. The long hinge gives IgG3 greater flexibility in binding to antigens and also makes the Fc more accessible for binding to C1. Both these qualities contribute to its superior activation of complement. The disadvantage of the long hinge is that IgG3 is particularly susceptible to cleavage by proteases, as is reflected by its half-life in the circulation, which is one-third that of the other three subclasses (see Figure 4.29). IgG3 deficiency is associated with recurrent infection leading to chronic lung disease.

By binding to pathogens and activating complement, the IgG1, IgG2, and IgG3 subclasses of antibodies facilitate the activity of phagocytic cells and increase inflammation at a site of infection. In contrast, IgG4, the least abundant subclass of IgG, has an Fc region that does not activate complement because it binds C1q poorly. Further distinguishing the IgG4 molecule is a unique capacity to exchange a module composed of one heavy chain and one light chain with that from another IgG4 molecule. Because of the frequency with which this occurs, most IgG4 molecules have two different heavy chains, two different light chains and two antigen-binding sites of different specificity (Figure 4.35). This means that IgG4 is functionally monovalent and can only impede pathogens by the mechanism of neutralization. All these properties make IgG4 anti-inflammatory in its effects. In allergic individuals, the levels of

	IgG subclass			
	IgG1	IgG2	IgG3	IgG4
Proportion of total IgG (%)	45–75	16–48	2–8	1–12
Length of heavy-chain hinge (amino acids)	15	12	62	12
Number of disulfide bonds in the hinge	2	4	11	2
Susceptibility of hinge to proteolytic cleavage	++	+	+++	+
Half-life in serum (days)	21	21	7	21
Capacity to activate complement system	++	+	+++	+
Response to protein antigens	++	+	++	+
Response to carbohydrate antigens	+	++	-	-
Response to allergens	+	-	-	++

Figure 4.34 The four subclasses of IgG have different and complementary functions.

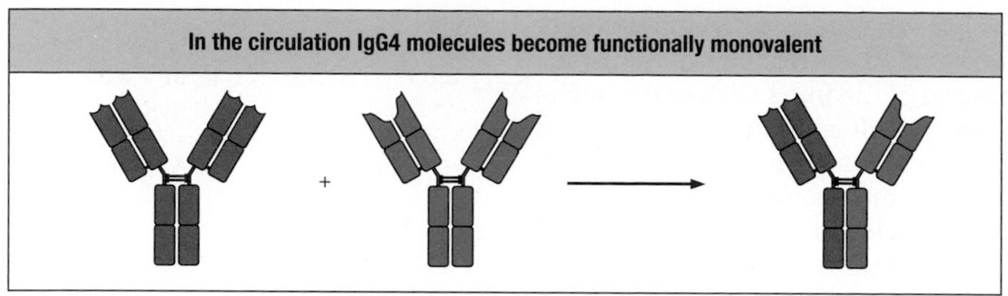

In the circulation IgG4 molecules become functionally monovalent
+

Figure 4.35 IgG4 is present in the circulation in a functionally monovalent form. Like other IgG subclasses, IgG4 is synthesized in a form that has two heavy chains, two light chains, and two identical antigen-binding sites. Unlike other IgGs, however, molecules of IgG4 can interact in the circulation and exchange one heavy chain and its associated light chain. Because of this property, most IgG4 molecules in the circulation have two different binding sites for antigen. Thus they only interact with a pathogen or a protein antigen through one binding site.

IgG4 are frequently elevated, as well as those of IgE. When IgG4 binds to an allergen it can block the binding of IgE and so reduce the severity of the allergic reaction. Thus, the function of IgG4 seems to be to dampen down the immune response.

Before the unique exchange reaction of IgG4 was fully understood, IgG4 was considered the best isotype to use for a therapeutic monoclonal antibody because it would minimize the damaging effects of inflammation. The monovalency of IgG4 is, however, a disadvantage in a therapeutic antibody, because its effective strength of binding is less than that of a bivalent IgG. The way round this problem is to mutate the C_H3 region of IgG4 so that it can no longer exchange binding sites but preserves its poor activation of complement.

Summary

Membrane-bound immunoglobulins of the IgM and IgD classes serve as antigen receptors, or B-cell receptors, for mature, naive B cells. When a pathogen binds to a B-cell receptor, the cell receives signals that cause cell proliferation and the differentiation of some of the resulting B cells into plasma cells. Although they have no immunoglobulin on their surface, plasma cells secrete copious quantities of soluble IgM antibody and some IgD. This change in production, from membrane-associated immunoglobulin to secreted soluble antibody, is caused by altered splicing of heavy-chain mRNA that eliminates the sequence encoding the hydrophobic transmembrane anchor. Other members of the clone of proliferating B cells are subject to somatic hypermutation, which introduces mutations throughout the heavy-chain and light-chain V regions. After hypermutation, the B cells carrying cell-surface immunoglobulins with the highest affinity for antigen are selected to become plasma cells. As antibody affinity increases, the dependence on pentameric IgM for strong binding to antigen relaxes. It then becomes advantageous to cease production of IgM and change the immunoglobulin to another isotype, either IgG, IgE, IgA, or IgD, that has effector functions more suited to the type of infecting pathogen and the anatomical site of infection. This process of isotype switching is accomplished by further somatic recombination of the expressed heavy-chain gene, in which the rearranged VDJ segment is moved next to a 'new' C region, with the excision of the μ and the intervening C genes. Both somatic hypermutation and isotype switching depend upon activation-induced cytidine deaminase (AID), a highly specific enzyme of adaptive immunity that is made only by B cells proliferating in response to antigen.

Summary to Chapter 4

The principal function of B lymphocytes is to produce antibodies, which are secreted immunoglobulins that bind tightly to infectious agents and tag them for destruction or elimination. Each antibody is highly specific for its corresponding antigen; the antibody repertoire of each person is enormous because it is composed of many millions of different antibodies that can bind a wide variety of different antigens. Antibodies can also be divided into five different structural classes—IgM, IgG, IgD, IgA, and IgE—that have different functions in the immune response. This chapter has provided an overview of the structure and function of the antibody molecule and of the unusual genetic mechanisms that create this diversity in specificity and effector function. Within an antibody molecule, the V regions that bind antigen are physically separated from the C region that interacts with effector molecules and cells of the immune system, such as complement, phagocytes, and other leukocytes. Antigen binding is the property of the paired V domains of the heavy and light chains, which can form an almost unlimited number of different binding sites with structural complementarity to a vast range of molecules. The immunoglobulin genes (heavy chain, κ light chain, and λ light chain) are expressed only in B cells, and their expression involves an unusual process of DNA rearrangement in which somatic DNA recombination assembles a V-region coding sequence from sets of gene segments that are present in the unrearranged gene. The random selection of gene segments for assembly creates much of the collective diversity of antigen-binding sites. The lymphocyte-specific proteins and ubiquitous enzymes of DNA repair and recombination are involved in the recombination machinery. Imprecision is inherent in some of their reactions, which creates additional diversity at the junctions between gene segments. In an individual B cell, only one rearranged heavy-chain gene and one rearranged light-chain gene become functional, ensuring that each B cell expresses immunoglobulin of a single specificity. The series of gene rearrangements that result in the production of membrane-bound IgM, the first immunoglobulin produced, is summarized in Figure 4.36.

Figure 4.36 Gene rearrangement and the synthesis of cell-surface IgM in B cells. Before immunoglobulin light-chain (center panel) and heavy-chain (right panel) genes can be expressed, rearrangements of gene segments are needed to produce exons encoding the V regions. Once this has been achieved, the genes are transcribed to give primary transcripts containing both exons and introns. The latter are spliced out to produce mRNAs that are translated to give κ or λ light chains and μ heavy chains that assemble inside the cell and are expressed as membrane-bound IgM at the cell surface. The main stages in the biosynthesis of the heavy and light chains are shown in the panel on the left.

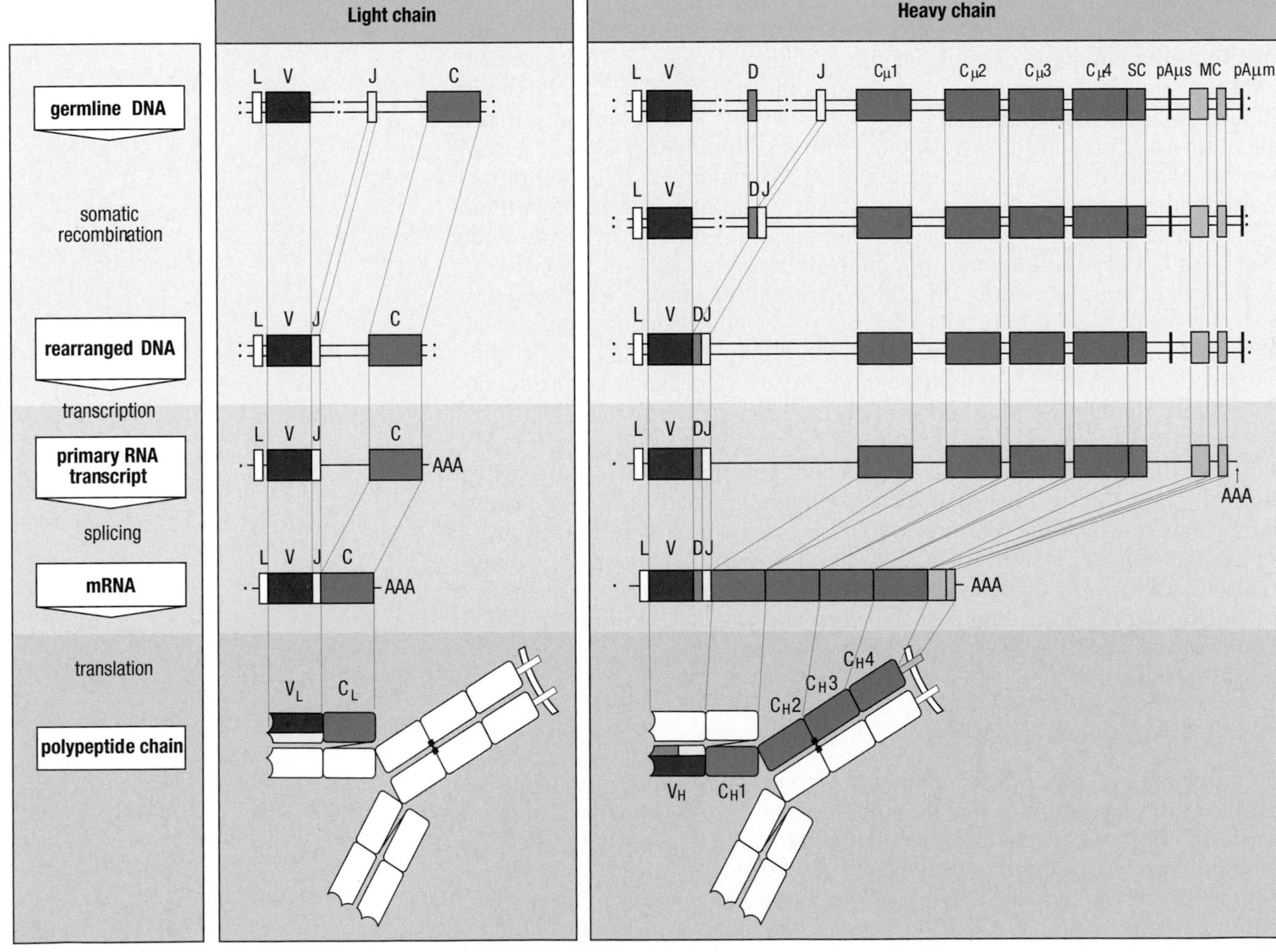

Changes in immunoglobulin genes during a B cell's life			
	Event	Mechanism	Nature of change to the B cell's genome
1	V-region assembly from gene fragments	Somatic recombination of genomic DNA	Irreversible
2	Generation of junctional diversity	Imprecision in joining rearranged DNA segments adds nongermline nucleotides (P and N) and deletes germline nucleotides	Irreversible
3	Assembly of transcriptional controlling elements	Promoter and enhancer are brought closer together by V-region assembly	Irreversible
4	Transcription activated with coexpression of surface IgM and IgD	Two patterns of splicing and processing RNA are used	Reversible and regulated
5	Synthesis changes from membrane Ig to secreted antibody	Two patterns of splicing and processing RNA are used	Reversible and regulated
6	Somatic hypermutation	Point mutation of genomic DNA	Irreversible
7	Isotype switch	Somatic recombination of genomic DNA	Irreversible

Figure 4.37 Changes in the immunoglobulin genes that occur during a B cell's lifetime.

On mature B cells, the membrane-bound immunoglobulin functions as the specific receptor for antigen; on encountering antigen, the B cell is stimulated to proliferate and differentiate into plasma cells that secrete antibody of the same specificity as the membrane-bound immunoglobulin. This ensures that an immune response is directed only against the invading pathogen or immunizing antigen. On the stimulation of a B cell by its specific antigen, a mechanism of somatic hypermutation introduces point mutations into the rearranged V-region DNA, diversifying the clone of proliferating B cells. Further selection of B cells by antigen increases the overall binding strength of the antibodies for antigen. Thus, the diversity of antibodies is due in part to inherited variation that is encoded in the genome and in part to non-inherited diversity that develops in B cells during an individual's lifetime. The first antibody produced after an encounter with antigen is always IgM. As the immune response proceeds, the process of isotype switching further rearranges the expressed heavy-chain gene so that the protein made has the same V region but a different C region. The functional effect of isotype switching is to produce antibody molecules with the same antigen specificity but different effector functions—IgG, IgA, and IgE. The changes in the immunoglobulin genes that occur throughout the life of a B cell are summarized in Figure 4.37.

Questions

4–1 The antigen-binding site of an immunoglobulin is formed from
a. the V regions of light chains only
b. the C regions of heavy chains only
c. paired V regions of a single heavy chain and a single light chain
d. paired V regions of two light chains
e. paired C regions of two heavy chains.

4–2 When IgG is cleaved with a protease that targets the hinge region, which of the following is generated?
a. two Fab fragments and one Fc fragment
b. two Fc fragments and two Fab fragments
c. Fc fragments that bind antigen
d. Fab fragments that facilitate antibody effector function
e. a membrane-bound form of antibody.

4–3 Match the antibody term in Column A with its correct characteristics in Column B. Use each answer only once.

Column A	Column B
a. light chain	1. the part of the antibody that binds antigen
b. hypervariable region	2. contributes around 50 kDa to the overall molecular weight of IgG
c. constant region	3. comprises the β strands and loops not involved in antigen binding
d. heavy chain	4. the most conserved region of the molecule with limited variation between antibodies
e. framework region	5. located within a V domain and varies greatly between different antibodies
f. V domain	6. pairs with the amino-terminal part of a heavy chain to form one of the two arms of an antibody

4–4 A protein epitope formed as a result of three-dimensional folding of the protein, and which is destroyed if the protein denatures, is called a _______ epitope.
a. linear
b. multivalent
c. conformational
d. complementarity
e. framework.

4–5 The numbered events listed below participate in the generation of junctional diversity. Put them in chronological order.
a. DNA strands pair, and unpaired nucleotides are removed by exonuclease activity.
b. P-nucleotides are generated after nicking of one DNA strand.
c. DNA polymerase fills in gaps, and DNA ligation forms a coding joint.
d. The RAG complex cleaves heptamer RSSs, and DNA hairpins are formed.
e. Terminal deoxynucleotidyl transferase adds N-nucleotides to the 3′ end of the stretch of P-nucleotides.

4–6 Which of the following statements regarding the process of differential splicing of primary RNA transcripts encoding immunoglobulins is incorrect?
a. It does not require rearrangement of genomic DNA sequences.
b. It underlies the production of both membrane-bound IgM and IgD in naive B cells.
c. It is the mechanism involved in isotype switching.
d. It occurs during B-cell differentiation into plasma cells when antibodies are produced in their secreted form.
e. It permits the production of different types of protein originating from the same RNA transcript.

4–7 Which of the following recombinations is not permitted during somatic recombination in the heavy-chain and light-chain immunoglobulin loci? (Select all correct answers.)
a. D_H:J_H
b. V_λ:J_λ
c. D_κ:V_H
d. V_H:J_H
e. V_H:D_H.

4–8 Which of the following does not contribute to generating the diversity of antigen-binding specificities among immunoglobulins?
a. somatic hypermutation
b. random combination of heavy and light chains
c. somatic recombination
d. activation-induced cytidine deaminase (AID)
e. alternative splicing of heavy-chain RNA transcripts.

4–9 The phenomenon of allelic exclusion ensures that B cells
a. use only one V, D, and J segment during somatic recombination
b. express only one type of heavy chain and one type of light chain
c. do not undergo alternative splicing until cell proliferation commences
d. do not secrete antibody until antigen is encountered
e. carry out affinity maturation directed at heavy chains and not light chains
f. derived from B-cell lymphomas are heterogeneous.

4–10
A. Which of the IgG subclasses is most efficient at activating complement?
B. Explain why.
C. Which is least efficient and why?

4–11 Match the antibody in Column A with its description or function in Column B. There may be more than one answer for each antibody.

Column A	Column B
a. IgA	1. opsonin
b. IgD	2. complement activation
c. IgE	3. transport across placenta
d. IgG	4. most abundant in serum
e. IgM	5. most abundant in mucosal secretions (for example, colostrum)
	6. sensitization of mast cells
	7. sensitization of basophils
	8. sensitization for killing by NK cells
	9. more made in body than any other antibody
	10. least abundant in serum

4–12 Humanized monoclonal antibodies are best described as
a. antibodies made in mice in which the mouse antibody genes have been replaced with human equivalents
b. human antibodies in which the CDR loops have been replaced by mouse-derived CDRs of the desired specificity
c. antibodies made in culture by human hybridomas
d. antibodies containing mouse Fab regions of both the heavy and light chain and human Fc regions
e. antibodies containing human Fab regions of both the heavy and light chain and mouse Fc regions.

4–13 Which of the IgG subclasses would you think was in principle most desirable for use as a therapeutic monoclonal antibody, and why? Are there any disadvantages to using this subclass and how might they be overcome?

4–14 Which of the following statements best explains why the immune system can continue to make antibodies after treatment with the anti-CD20 antibody rituximab?
a. New CD20-positive B cells are reconstituted so quickly that antibody concentration during and after treatment is unaffected.
b. Rituximab stimulates B-cell proliferation, so for a short while after its administration there is actually an increase in antibody concentration.
c. Rituximab is a mouse monoclonal antibody and therefore its Fc region is unable to bind to the surface receptors on human NK cells that bind to Fc.
d. Plasma cells do not express CD20 on their cell surface, and antibody production by these cells continues unhampered.
e. Rituximab stimulates anti-antibody production, leading to its rapid clearance by the body.

4–15 Three-week-old Xavier Capelleto was brought to the emergency room with a bright scaly rash that first developed on his legs and then spread to his trunk and face. He also had blisters on his palms and the soles of his feet. Xavier's parents said that he had been experiencing looser bowel movements than expected, a large amount of yellow pus had been accumulating around his swollen eyelids, and he showed signs of oral thrush. Tests revealed that Xavier's lymphocyte count was only 8% of total white blood cells (normal = 50%), all immunoglobulins were markedly decreased except for IgE, and no thymic shadow was detected on a chest X-ray. Eosinophilia was also detected. His parents were told that Xavier had an autosomal recessive form of severe combined immunodeficiency (SCID) known as Omenn syndrome, which affects the development of both B cells and T cells. A bone marrow transplant was recommended; however, Xavier died from respiratory failure due to an opportunistic bacterial infection. This history is consistent with a genetic deficiency in
a. α- or β-defensins
b. activation-induced cytidine deaminase (AID)
c. MHC class I
d. RAG-1 or RAG-2
e. Toll-like receptors.

4–16 Aliya Agassi, a 3-year-old girl with pneumonia, a temperature of 40.8°C, respirations 42 per minute (normal 20), and blood oxygen saturation of 90% (normal is more than 98%), was admitted to the hospital. Her neck and armpit lymph nodes were enlarged, and X-rays confirmed inflammation in the lower lobe of her right lung. Her medical history revealed two previous cases of pneumonia and six middle-ear infections that were treated successfully with antibiotics. A blood culture grew *Haemophilus influenzae.* Blood tests showed elevated levels of IgM above normal, whereas IgA and IgG were not detected. Her father had normal levels of serum IgA, IgG, and IgM. Which of the following is the most likely cause of her symptoms?
a. acute lymphoblastic leukemia
b. IgA deficiency
c. X-linked agammaglobulinemia
d. severe combined immunodeficiency
e. X-linked hyper IgM syndrome
f. AID deficiency
g. myeloma.

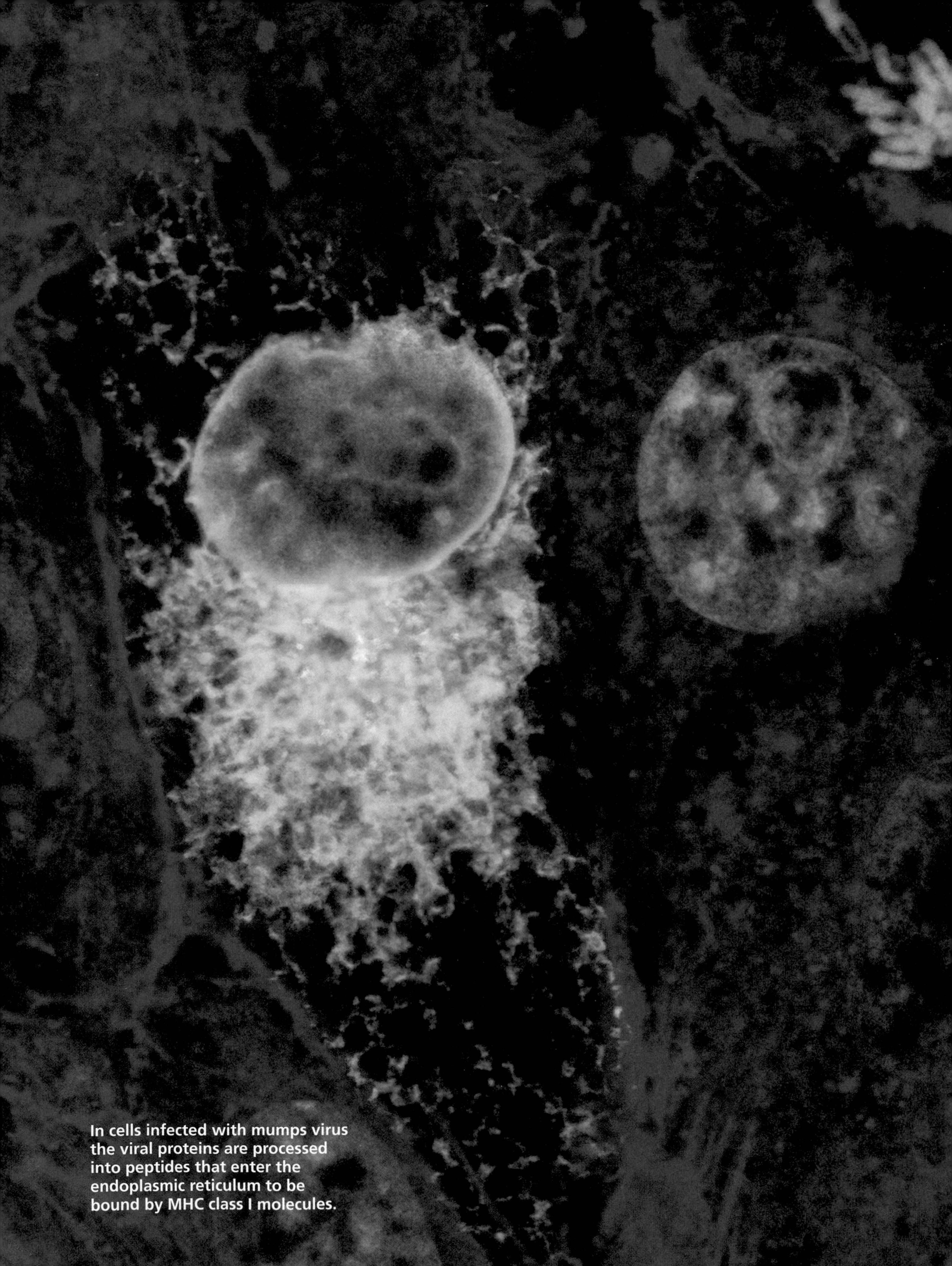

In cells infected with mumps virus the viral proteins are processed into peptides that enter the endoplasmic reticulum to be bound by MHC class I molecules.

Chapter 5

Antigen Recognition by T Lymphocytes

Like B cells, T lymphocytes—or T cells—recognize and bind antigen through highly variable antigen-specific receptors. Whereas the sole function of B cells is to make antibodies (see Chapter 4), T cells have more diverse roles, all of which involve interactions with other cells. The same basic principles underlie antigen recognition by B cells and T cells but because of the distinct functions of B and T cells there are important differences between them, particularly in the type of antigen they recognize.

In the first part of the chapter we describe the structure of the T-cell antigen receptor and the mechanisms that generate its antigen specificity and diversity. The antigen receptor on T cells is commonly referred to simply as the **T-cell receptor** or **TCR**. T-cell receptors have much in common with the immunoglobulins. They have a similar structure, are produced as a result of gene rearrangement, and are highly variable and diverse in their antigen specificity. Like B cells, each clone of T cells expresses a single species of antigen receptor, and thus different clones possess different and unique antigen specificities. This clonal distribution of diverse T-cell receptors is produced by genetic mechanisms similar to those that generate immunoglobulins during B-cell development.

The antigens recognized by T cells are quite distinct from those recognized by immunoglobulins. Immunoglobulins bind epitopes on a wide range of intact molecules, such as proteins, carbohydrates, and lipids. These kinds of epitope are present on the surfaces of bacteria, viruses, and parasites, and also on soluble protein toxins. T-cell receptors, in contrast, recognize and bind mainly to peptide antigens, which are derived from the pathogen's proteins.

The effector functions of T cells, in contrast to that of B cells, are carried out through antigen-specific interactions with other cells of the body. This is reflected in the type of antigen that T-cell receptors recognize. They recognize a composite structure on a human cell surface that consists of a peptide antigen derived from a pathogen that is bound by a human glycoprotein called an MHC molecule. The ligand for a T-cell receptor is therefore not simply a peptide antigen but the combination of peptide and MHC molecule on a cell surface. In the second part of the chapter we trace the pathway of antigen-processing by which peptides derived from the proteins of infecting microorganisms become bound to MHC molecules and are displayed on cell surfaces.

The genes that encode the MHC molecules are clustered in a chromosomal region called the **major histocompatibility complex** (**MHC**). Some of the genes for MHC proteins are highly **polymorphic**—that is, there are numerous different variants of these genes, and therefore of the MHC molecules they encode, in the human population. In clinical transplantation of organs and tissues, the differences between the MHC molecules of the donor and those of the recipient are the major cause of tissue incompatibility and transplant rejection. The third part of the chapter deals with the MHC and with the immunological implications of the exceptional genetic polymorphism of the MHC genes.

T-cell receptor diversity

The T-cell receptor is a membrane-bound glycoprotein that closely resembles a single antigen-binding arm of an immunoglobulin molecule. It is composed of two different polypeptide chains and has one antigen-binding site. T-cell receptors are always membrane bound and there is no secreted form as there is for immunoglobulins. Like immunoglobulins, each chain has a variable region, which binds antigen, and a constant region. During T-cell development, gene rearrangement produces sequence variability in the variable regions of the T-cell receptor by the same mechanisms that are used in B cells to produce the variable regions of immunoglobulins. However, after the T cell is stimulated with antigen, there is no further change in the T-cell receptor and therefore no equivalent of somatic hypermutation of the antigen-binding site or switching of the constant-region isotype as occurs for immunoglobulins (see Chapter 4). These differences reflect the use of T-cell receptors only as antigen receptors, whereas immunoglobulins serve as both antigen receptors and effector molecules.

5-1 The T-cell receptor resembles a membrane-associated Fab fragment of immunoglobulin

A T-cell receptor consists of two different polypeptide chains, termed the **T-cell receptor α chain (TCRα)** and the **T-cell receptor β chain (TCRβ)**. The genes encoding the α and β chains have a germline organization similar to that of the immunoglobulin heavy-chain and light-chain genes, comprising sets of gene segments that must be rearranged to form a functional gene. As a consequence of the gene rearrangements that occur during T-cell development, each mature T cell expresses one functional α chain and one functional β chain, which together define a unique T-cell receptor molecule. Within the population of T cells in a healthy human there are many millions of different T-cell receptors, each of which defines a clone of T cells and a single antigen-binding specificity.

Comparison of the amino acid sequences of the α and β chains from different T-cell clones shows that each chain has a variable region (V region) and a constant region (C region) like those found in immunoglobulin chains. The α and β chains are folded into discrete protein domains resembling those in immunoglobulin chains. Each chain consists of an amino-terminal V domain, followed by a C domain, and then a membrane-anchoring domain. The antigen-recognition site of T-cell receptors is formed from the V_α and V_β domains and is the most variable part of the molecule, as in immunoglobulins. The three-dimensional structure of the four extracellular domains of the T-cell receptor is very similar to that of the antigen-binding Fab fragment of IgG (Figure 5.1).

Comparison of the amino acid sequences of V domains from different clones of T cells shows that sequence variation in the α and β chains is clustered into regions of hypervariability, which correspond to loops of the polypeptide chain at the end of the domain farthest from the T-cell membrane. These loops form the binding site for antigen and are termed complementarity-determining regions (CDRs), as in the immunoglobulins. The V domains of the T-cell

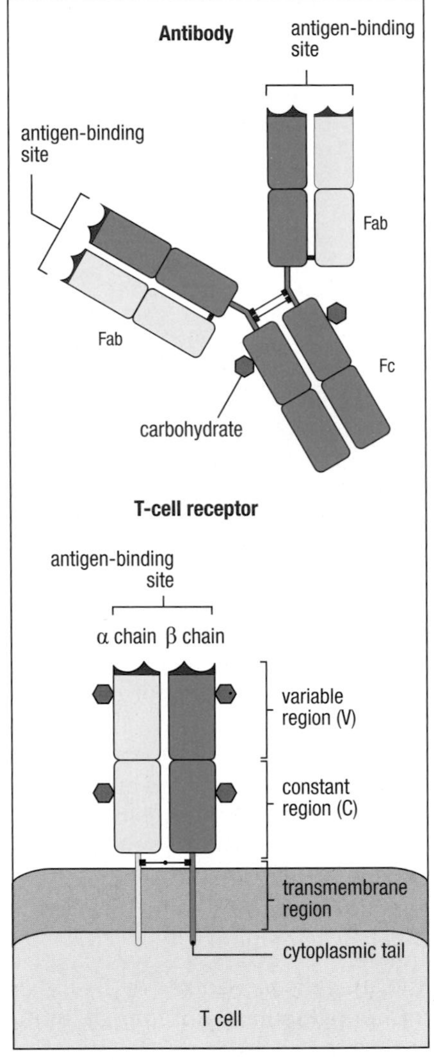

Figure 5.1 The T-cell receptor resembles a membrane-bound Fab fragment. Comparison of the T-cell receptor with an IgG antibody. The T-cell receptor is a membrane-bound heterodimer composed of an α chain of 40–50 kDa and a β chain of 35–46 kDa. The extracellular portion of each chain consists of two immunoglobulin-like domains: the one nearest to the membrane is a C domain and the domain farthest from the membrane is a V domain. The α and β chains both span the cell membrane and have very short cytoplasmic tails. The three-dimensional structure formed by the four immunoglobulin-like domains of the T-cell receptor resembles that of the antigen-binding Fab fragment of antibody.

receptor α chain and the β-chain each have three CDR loops, called CDR1, CDR2, and CDR3 (Figure 5.2).

Immunoglobulins possess two or more binding sites for antigen; this strengthens the interactions of soluble antibody with the repetitive antigens found on the surfaces of microorganisms. In contrast, T-cell receptors possess a single binding site for antigen and are used only as cell-surface receptors for antigen, never as soluble antigen-binding molecules. Antigen binding to T-cell receptors occurs always in the context of two opposing cell surfaces, where multiple copies of the T-cell receptor bind to multiple copies of the antigen:MHC complex on the opposing cell, thus achieving multipoint attachment.

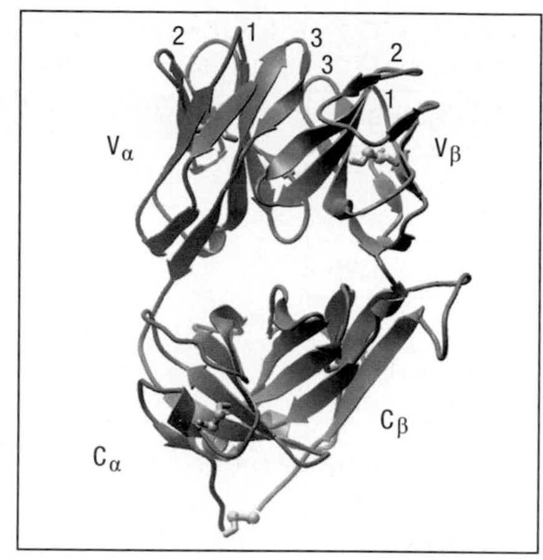

Figure 5.2 Three-dimensional structure of the T-cell receptor showing the antigen-binding CDR loops. The ribbon diagram shows the α chain (in magenta) and the β chain (in blue). The receptor is viewed from the side as it would sit on a cell surface, with the highly variable CDR loops, which bind the peptide:MHC molecule ligand, arrayed across its relatively flat top surface. The CDR loops are numbered 1–3 for each chain. Courtesy of I.A. Wilson.

5-2 T-cell receptor diversity is generated by gene rearrangement

In Chapter 4 we divided the mechanisms that generate immunoglobulin diversity into two categories: those operating before the B cell is stimulated with specific antigen, and those operating afterward. In the first category are the gene rearrangements that generate the V-region sequence, whereas in the second category are changes in mRNA splicing that produce a secreted immunoglobulin, C-region DNA rearrangements that switch the heavy-chain isotype, and somatic hypermutation of the V-region gene to produce antibodies of higher affinity. In T cells the mechanisms that generate diversity before antigen stimulation are essentially the same as those in B cells, but after antigen stimulation the picture is quite different: whereas immunoglobulin genes continue to diversify, the genes encoding T-cell receptors remain unchanged. This fundamental difference reflects the use of the T-cell receptor only for the recognition of antigen and not for mediating effector functions, which are handled by other proteins produced by T cells.

The human T-cell α-chain locus is on chromosome 14 and the β-chain locus is on chromosome 7. The organization of the gene segments encoding T-cell receptor α and β chains is essentially like that of the immunoglobulin gene segments (Figure 5.3). The main difference is the simplicity of the T-cell receptor C region: there is only one C_α gene, and, although there are two C_β genes, no functional distinction between them is known. The T-cell receptor α-chain locus is otherwise similar to an immunoglobulin light-chain locus, containing only sets of V and J gene segments; the β-chain locus is similar to an immunoglobulin heavy-chain locus, containing D gene segments in addition to V and J gene segments. The V domain of the T-cell receptor α chain is thus encoded by a V gene segment and a J gene segment; that of the β chain is encoded by a D gene segment in addition to a V and a J gene segment.

T-cell receptor gene rearrangement occurs during T-cell development in the thymus, and the mechanism is similar to that outlined in Chapter 4 for the immunoglobulin genes. In the α-chain gene, a V gene segment is joined to a J gene segment by somatic DNA recombination to make the V-region sequence; in the β-chain gene, recombination first joins a D and a J gene segment, which are then joined to a V gene segment (see Figure 5.3). The T-cell receptor gene segments are flanked by recombination signal sequences similar to those found in the immunoglobulin genes, and the RAG complex and the same DNA-modifying enzymes are involved in the recombination process (see Sections 4-8 and 4-9). During recombination, additional, nontemplated P and N nucleotides are inserted in the junctions between the V, D, and J gene segments of the T-cell receptor β-chain coding sequence and between the V and J gene segments of the α-chain sequence. These mechanisms contribute junctional diversity in the CDR3 to T-cell receptor α and β chains.

Genetic defects that result in an absence of RAG proteins are one cause of a rare syndrome called **severe combined immunodeficiency disease (SCID)**.

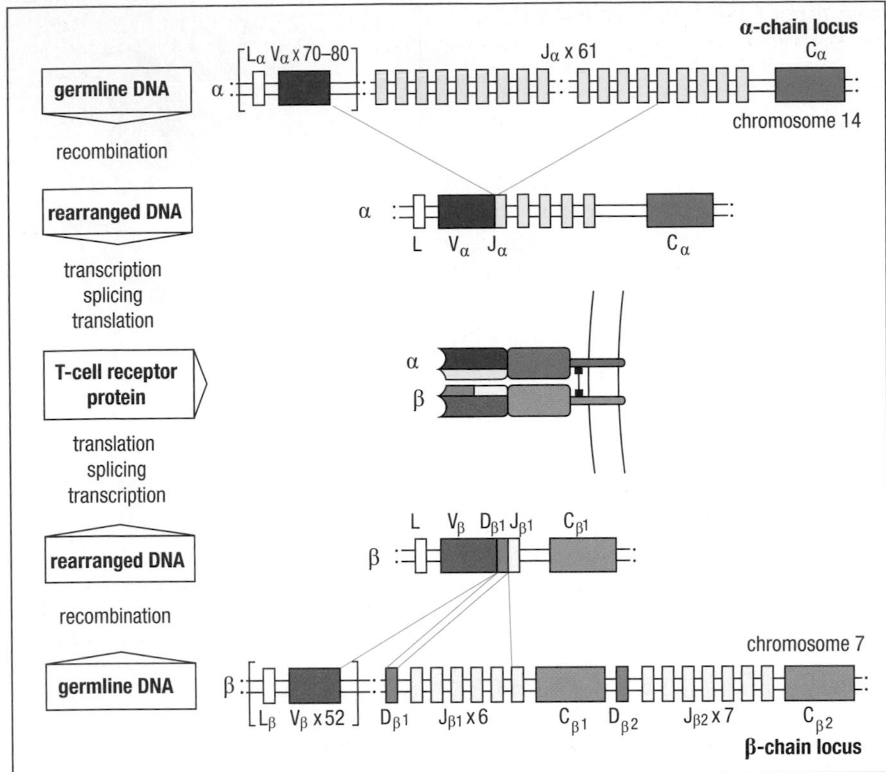

Figure 5.3 Organization and rearrangement of the T-cell receptor genes. The top and bottom rows of the figure show the germline arrangement of the variable (V), diversity (D), joining (J), and constant (C) gene segments at the T-cell receptor α-chain and β-chain loci. During T-cell development, a V-region sequence for each chain is assembled by DNA recombination. For the α chain (top), a V_α gene segment rearranges to a J_α gene segment to create a functional exon encoding the V domain. For the β chain (bottom), rearrangement of a V_β, a D_β, and a J_β gene segment creates the functional V-domain exon. The assembled genes are transcribed and spliced to produce mRNA (not shown) encoding the α and β chains. Exons encoding the membrane-spanning regions are not shown. L, leader sequence.

The disease is called 'combined' because functional B and T lymphocytes are both absent, and it is 'severe' compared with immunodeficiencies in which only B cells are lacking. Without a bone marrow transplant or other medical intervention, children with SCID die in infancy from common infections (Figure 5.4, upper panel). Rare cases of missense mutations that produce RAG proteins with partial enzymatic activity have also been found in humans; these also cause a rapidly fatal immunodeficiency that differs from SCID in some of its symptoms and is known as **Omenn syndrome** (Figure 5.4, lower panel).

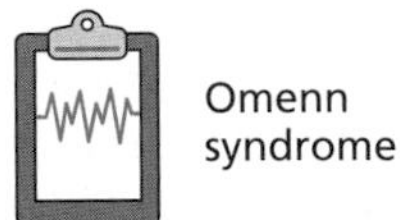

Omenn syndrome

After gene rearrangement, functional α-chain and β-chain genes consist of exons encoding the leader peptide, the V region, the C region, and the membrane-spanning region. The exons are separated by introns, which in the case of the intron between the V-domain and the C-domain exons may contain unrearranged gene segments (see the α-chain gene in Figure 5.3). When the gene is transcribed, the primary RNA transcript is spliced to remove the introns and is processed to give mRNA. Translation of the α-chain and β-chain mRNA produces α and β chains, respectively. Like all proteins destined for the cell

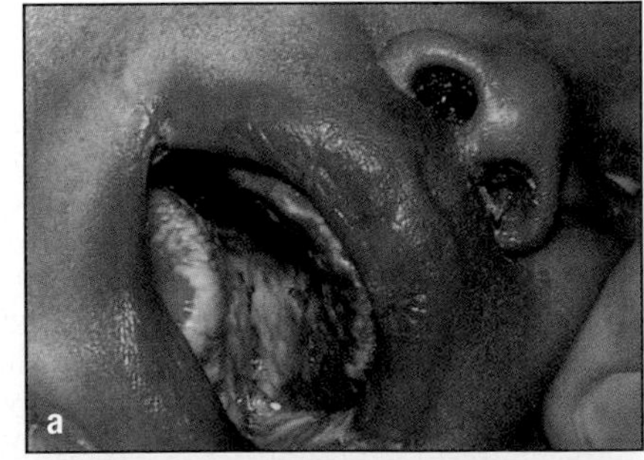

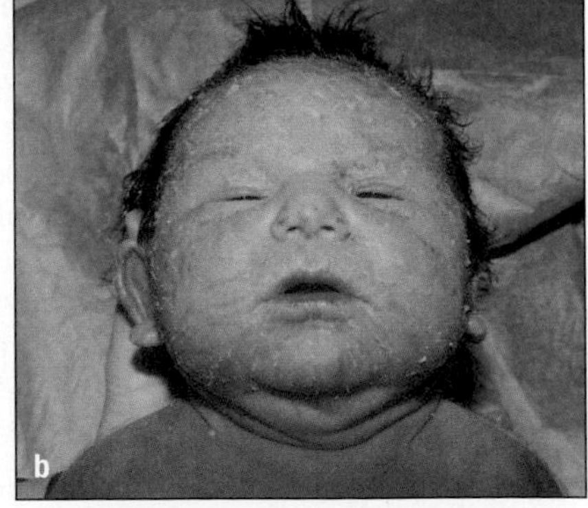

Figure 5.4 Severe combined immunodeficiency syndrome (SCID). Infants with SCID lack functional T cells and B cells and cannot make an adaptive immune response. They typically show infections with opportunistic pathogens, like the *Candida albicans* infection in the mouth of this infant with SCID (panel a). SCID can be caused by various genetic defects, one of which is complete loss of RAG function. A related immunodeficiency is Omenn syndrome, in which a mutation in a *RAG* gene reduces RAG activity by 80%. The bright red rash seen on the face and shoulders of infants with Omenn syndrome (panel b) is caused by chronic inflammation, a characteristic of this condition. Babies diagnosed with SCID or Omenn syndrome will die in infancy unless they receive a bone marrow transplant from a healthy donor. This procedure replaces the impaired immune system of the child with a competent immune system generated from the hematopoietic stem cells in the transplant. Panel a courtesy of Fred Rosen; panel b courtesy of Luigi Notarangelo.

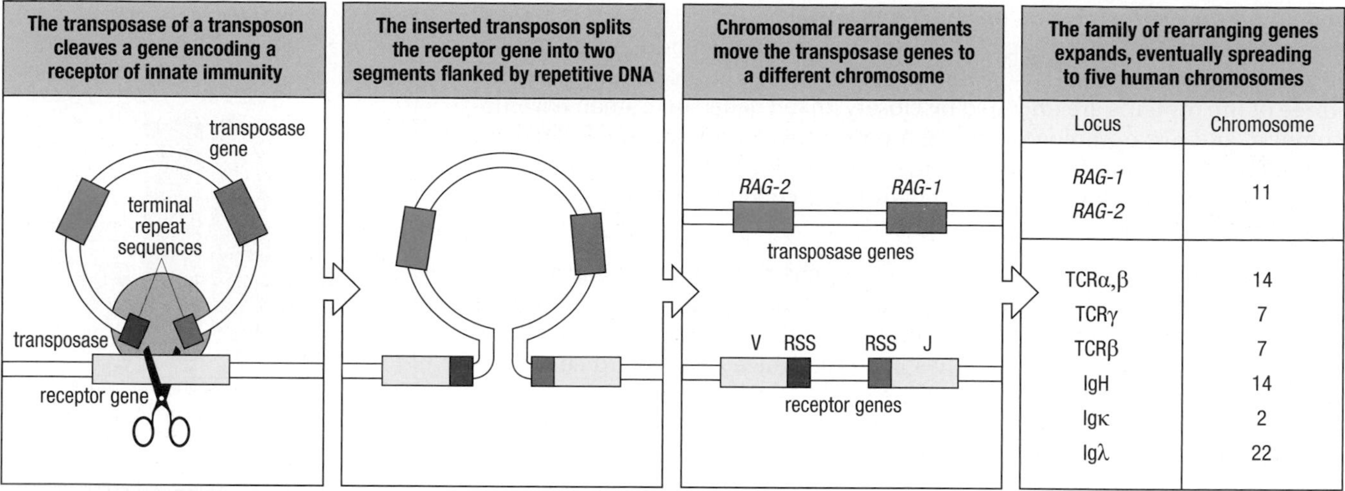
The transposase of a transposon cleaves a gene encoding a receptor of innate immunity
transposase gene
terminal repeat sequences
transposase
receptor gene
The inserted transposon splits the receptor gene into two segments flanked by repetitive DNA
Chromosomal rearrangements move the transposase genes to a different chromosome
RAG-2
RAG-1
transposase genes
V RSS
RSS J
receptor genes
The family of rearranging genes expands, eventually spreading to five human chromosomes

Locus	Chromosome
RAG-1 *RAG-2*	11
TCRα,β	14
TCRγ	7
TCRβ	7
IgH	14
Igκ	2
Igλ	22

Figure 5.5 Components of a transposon could have evolved to become the *RAG* genes and the recombination signal sequences of immunoglobulin and T-cell receptor genes. The first event in the evolution of rearranging antigen-receptor genes is thought to have occurred more than 400 million years ago, when a transposon integrated into a gene encoding an innate immune receptor protein (first panel). The transposon separated the gene into two segments, each flanked by a piece of repetitive transposon DNA (second panel). Subsequently, chromosomal rearrangements placed the transposase gene (or genes) onto a different chromosome from the primordial rearranging gene. The repetitive DNAs became the RSSs of a primordial rearranging gene and the transposase genes became the ancestral *RAG-1* and *RAG-2* genes (third panel). Over 400 million years of evolution, the family of rearranging genes has expanded and is now spread over five different human chromosomes (fourth panel).

membrane, newly synthesized α and β chains enter the endoplasmic reticulum. There they pair to form the **α:β T-cell receptor** (see Figure 5.3).

5-3 The *RAG* genes were key elements in the origin of adaptive immunity

T cells and B cells use identical mechanisms of gene rearrangement, often called **V(D)J recombination**, to generate the clonal diversity in their antigen receptors. Key to this recombination are the two subunits of the RAG recombinase, which are made only by lymphocytes (see Section 4-9). The RAG proteins are thus specific to adaptive immunity and are essential for its function, as shown by the detrimental effects of their absence or malfunction (see Figure 5.4). These properties are consistent with the emergence of *RAG* genes in a common vertebrate ancestor being a formative event in the evolution of the adaptive immune system.

The *RAG* genes lack the introns that characterize eukaryotic genes. In this unusual feature they resemble the transposase gene of a transposon, a type of genetic element that can make and move copies of itself to different chromosomal locations. The essential components of a transposon are a transposase—an enzyme that cuts double-stranded DNA—and regions of repetitive DNA, called the terminal repeat sequences, that are recognized by the transposase (Figure 5.5). These two features allow the transposon to be excised from one location and inserted into another. The similarity of the RAG recombinase to a transposase has led to the hypothesis that the mechanism now used to rearrange immunoglobulin and T-cell receptor gene segments originated in a vertebrate ancestor with the insertion of a transposon into a gene encoding a receptor of innate immunity. The inserted transposase genes evolved to encode the RAG proteins, and the terminal repeat sequences evolved to become the recombination signal sequences for the first rearranging gene segments. During this evolution, the transposase gene and the long terminal repeats of the transposon were separated and become components of different genes, both expressed specifically in lymphocytes. Today, the human *RAG* genes are on chromosome 11 and on four other chromosomes are the much-expanded sets of rearranging antigen-receptor genes.

5-4 Expression of the T-cell receptor on the cell surface requires association with additional proteins

T-cell receptors are diverse and specific receptors for antigen. By themselves, however, heterodimers of α and β chains are unable to leave the endoplasmic reticulum and be expressed on the T-cell surface. In this respect the T-cell receptor resembles immunoglobulin. Before leaving the endoplasmic reticulum, an α:β heterodimer associates with four invariant membrane proteins. Three of the proteins are encoded by closely linked genes on human chromosome 11 and are homologous to each other: these proteins are collectively termed the **CD3 complex** and are individually called CD3γ, CD3δ, and CD3ε. The fourth protein is known as the ζ chain and is encoded by a gene on human chromosome 1.

At the cell surface the CD3 proteins and the ζ chain remain in stable association with the T-cell receptor and form the functional **T-cell receptor complex** (Figure 5.6). In this complex the CD3 proteins and the ζ chain transmit signals to the cell's interior after antigen has been recognized by the α:β chain heterodimer. Like the Igα and Igβ proteins of the B-cell receptor complex, the cytoplasmic domains of the CD3 proteins and the ζ chain contain sequences that associate with intracellular signaling molecules. In contrast, the T-cell receptor α and β chains have very short cytoplasmic tails that lack signaling function. In people lacking functional CD3δ or CD3ε chains, transport of T-cell receptors to the cell surface is inefficient and their T cells have abnormally low numbers of receptors that do not signal effectively. As a consequence of both the low receptor numbers and impaired signal transduction, these people are immunodeficient.

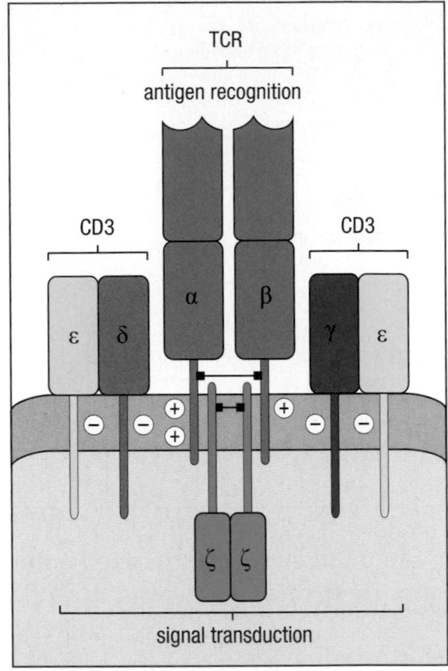

Figure 5.6 Polypeptide composition of the T-cell receptor complex. The functional antigen receptor on the surface of T cells is composed of eight polypeptides and is called the T-cell receptor complex. The α and β chains bind antigen and form the core T-cell receptor (TCR). They associate with one copy each of CD3γ and CD3δ and two copies each of CD3ε and the ζ chain. These associated invariant polypeptides are necessary for the transport of newly synthesized TCR to the cell surface and for the transduction of signals to the cell's interior after the TCR has bound antigen. The transmembrane domains of the α and β chains contain positively charged amino acids (+), which form strong electrostatic interactions with negatively charged amino acids (–) in the transmembrane regions of the CD3γ, δ, and ε chains.

5-5 A distinct population of T cells expresses a second class of T-cell receptor with γ and δ chains

A second type of T-cell receptor is similar in overall structure to the α:β receptor but is formed of two different protein chains called the **T-cell receptor γ chain (TCRγ**, not to be confused with CD3γ) and the **T-cell receptor δ chain (TCRδ**). TCRγ is most structurally similar to the α chain, and TCRδ to the β chain (Figure 5.7). All T cells express either an α:β receptor or a γ:δ receptor but never both. These two types of T-cell receptor define two fundamental, distinctive, and ancient T-cell lineages that are present in all jawed vertebrates. T cells expressing α:β T-cell receptors are called **α:β T cells**, whereas those expressing γ:δ receptors are called **γ:δ T cells**.

The organization of the γ and δ loci resembles that of the α and β loci, but there are some important differences (Figure 5.8). The δ gene segments are situated within the α-chain locus on chromosome 14, between the V_α and J_α gene

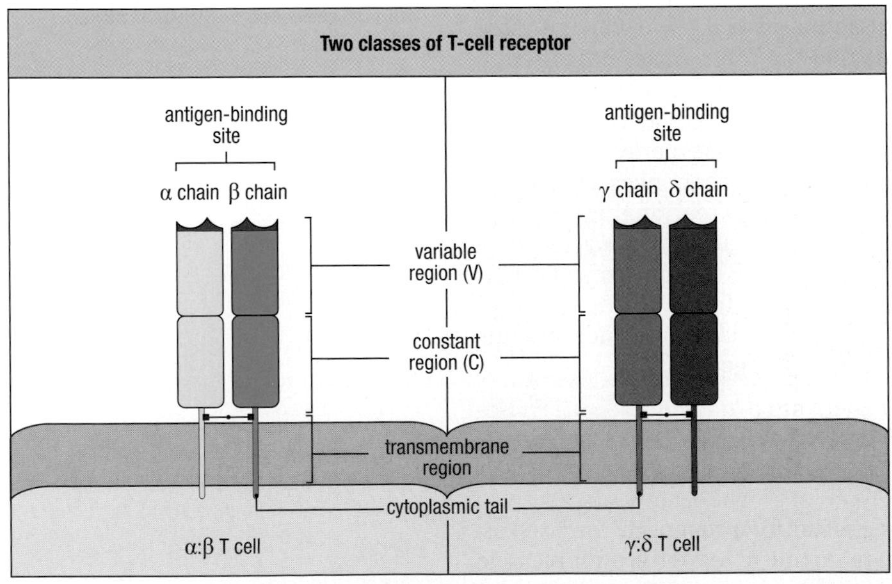

Figure 5.7 There are two classes of T-cell receptor. The α:β T-cell receptor (left panel) and the γ:δ T-cell receptor (right panel) have similar structures, but they are encoded by different sets of rearranging gene segments and have different functions.

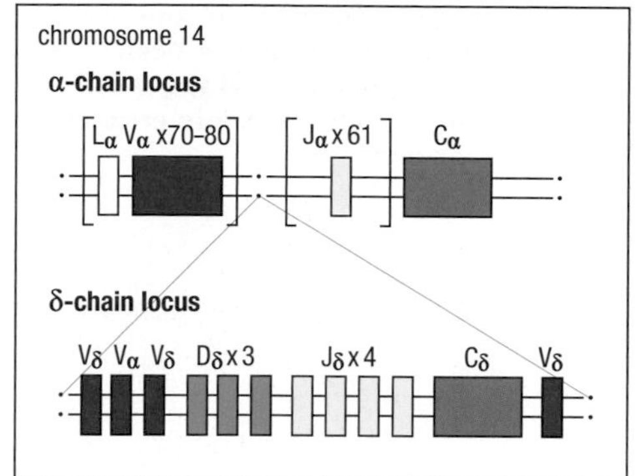

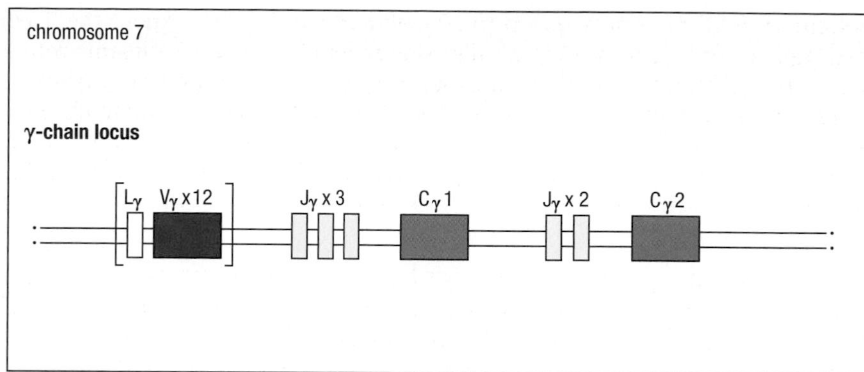

Figure 5.8 The organization of the human T-cell receptor γ-chain and δ-chain loci. The TCR γ and δ loci, like the α and β loci, contain sets of variable (V), diversity (D), joining (J), and constant (C) gene segments. The δ locus is located within the α-chain locus on chromosome 14, lying between the clusters of V_α and J_α gene segments. There are at least three V_δ gene segments, three D_δ gene segments, three J_δ gene segments, and a single C_δ gene segment. V_δ segments are interspersed among V_α and other gene segments. The γ locus, on chromosome 7, resembles the β locus, with a set of V segments and two C gene segments each with its own set of J segments.

segments. This location means that DNA rearrangement within the α-chain locus inevitably results in the deletion and inactivation of the δ-chain locus. The human γ-chain locus is on chromosome 7. The γ-chain and δ-chain loci contain fewer V gene segments than the α-chain or β-chain loci. Consequently γ:δ T-cell receptors are less diverse than α:β T-cell receptors. For the δ chain this is partly compensated for by increased junctional diversity. Rearrangement at the γ and δ loci proceeds as for the α and β loci, with the exception that during δ-gene rearrangement two D segments can be incorporated into the final gene sequence. This increases the variability of the δ chain in two ways. First, the potential number of combinations of gene segments is increased. Second, extra N nucleotides can be added at the junction between the two D segments, as well as at the VD and DJ junctions.

Whether in fish, birds, mammals or any other jawed vertebrate, α:β T cells recognize the same type of antigen—that is, a peptide antigen bound to an MHC molecule. In contrast, antigen recognition and other aspects of γ:δ T-cell function are not conserved, and there are major differences between species, notably between humans and the mice that immunologists use as their principal animal model. As a consequence, the study of mouse α:β T cells has valuably predicted much, though not all, about human α:β T-cell biology. That is not the case for γ:δ T-cell biology, where species differences seem to outweigh similarities. Important common features are that γ:δ T cells are more abundant in tissues than in the circulation, that γ:δ T cells share properties with NK cells and other innate lymphocytes, and that γ:δ T-cell receptor recognition is not dependent on MHC molecules and their bound peptides, although they can be involved. Because of the complexities and the species differences, much more is known about the functions of α:β T cells than about γ:δ T cells. For simplicity, in the rest of this book, T cells will refer to α:β T cells and T-cell receptor will refer to the α:β T-cell receptor, unless specified otherwise.

Summary

T cells recognize antigen through a cell-surface receptor known as the T-cell receptor, which has structural and functional similarities to the membrane-bound immunoglobulin that serves as the B-cell receptor for antigen. T-cell receptors are heterodimeric glycoproteins in which each polypeptide chain consists of a V domain and a C domain, similar to those found in immunoglobulin chains, and a membrane-spanning region. The three-dimensional structure of the extracellular domains of the T-cell receptor resembles that of the antigen-binding Fab fragment of IgG. There are two types of T-cell receptor: one made up of α and β chains and expressed by T cells whose function is understood, and another made up of γ and δ chains and carried by T cells whose function remains elusive. The four types of T-cell receptor chain are all encoded by genes that resemble the immunoglobulin genes, and require similar mechanisms of DNA rearrangement in order to be expressed. The

Element	Immunoglobulin		α:β T-cell receptors	
	H	κ+λ	β	α
Variable segments (V)	40	70	52	~70
Diversity segments (D)	23	0	2	0
D segments read in three frames	rarely	–	often	–
Joining segments (J)	6	5(κ) 4(λ)	13	61
Joints with N- and P-nucleotides	2	50% of joints	2	1
Number of V gene pairs	1.9×10^6		5.8×10^6	
Junctional diversity	$\sim 3 \times 10^7$		$\sim 2 \times 10^{11}$	
Total diversity	$\sim 5 \times 10^{13}$		$\sim 10^{18}$	

Figure 5.9 Comparison of the potential diversity in the T-cell receptor repertoire and the B-cell receptor repertoire before encounter with antigen.

presence of the T-cell receptor at the T-cell surface requires its association with proteins of the CD3 complex. These proteins transmit signals to the interior of the cell when the T-cell receptor binds antigen.

The critical difference between T-cell receptors and immunoglobulins is that T-cell receptors serve only as cell-surface receptors and are not secreted as soluble proteins with effector function; T cells use other molecules for effector function. This explains why the T-cell receptor has only a single binding site for antigen that does not change its affinity on encountering antigen, and a simple constant region that does not switch isotype. Despite the fact that T-cell receptors do not undergo further diversification after encounter with antigen, the T-cell receptor repertoire is more diverse than that of immunoglobulins (Figure 5.9) because there is greater capacity for diversification during V(D)J recombination.

Antigen processing and presentation

The antigen receptors of naive B cells are cell-surface IgM molecules that have the capacity to bind with low affinity to any natural or synthetic chemical compound or macromolecule. The antigen receptors of naive T cells have antigen-binding sites that are structurally similar to those of immunoglobulin, but unlike antibodies they do not recognize a broad range of structures. The ligands recognized by T-cell receptors are highly constrained, being variations on a single structural theme. The foreign antigen is a short peptide produced by intracellular degradation of a pathogen's proteins. Within the endoplasmic reticulum or an endosome, the peptide antigen is bound by a specialized type of human protein that needs to bind a peptide in order to fold properly, attain a stable structure, and travel to the cell surface. In this way, the human protein brings the peptide to the cell surface where it can be recognized and engaged by a T-cell receptor. This interaction involves contact between the T-cell receptor and the human protein as well as with the peptide antigen. So the interactions of T-cell receptors with the universe of pathogen-derived antigens are also variations on a single structural theme. Antigen processing is the production of peptide antigens inside human cells. Antigen presentation is the display of peptide antigens on the surface of human cells in the precise manner that can be bound by a T-cell receptor. This part of the chapter will describe the intracellular mechanisms of antigen processing and presentation.

5-6 T-cell receptors recognize peptide antigens bound to MHC molecules

The antigens recognized by T-cell receptors are short peptides, from 8–25 amino acids in length, generated by the degradation of a pathogen and its proteins. This strategy simplifies antigen recognition by T cells in two ways: one because it concentrates on only one type of macromolecule—proteins—and the other because it ignores the complexity of the three-dimensional structures of proteins and concentrates instead on targeting small, linear elements of primary structure. A variety of antigenic peptides can be generated from a typical protein, and the diversity of T-cell receptors in the T-cell population is such that a specific, adaptive T-cell response can be made to any pathogen.

The peptides that are recognized by T-cell receptors are produced in human cells by the degradation of pathogens and their products. The uptake and killing of pathogens by macrophages and dendritic cells is the major source of peptides from bacteria and other extracellular pathogens. Peptides arising from the degradation of viral proteins are produced in all human cells that become infected with a virus. The process by which peptide antigens are produced from pathogens and their products is called **antigen processing**.

After an antigenic peptide has been generated inside a human cell it must be brought to the cell surface in order to be bound by the receptor of an antigen-specific T cell. This is accomplished by having the peptide bind to a membrane-associated protein inside the cell that then carries the antigen to the cell surface. The dedicated human glycoproteins that perform this function of binding and transporting peptide antigens are called **major histocompatibility complex molecules** or **MHC molecules**. Each MHC molecule can bind a single peptide, and only when a peptide has been bound can an MHC molecule travel from inside the cell to the surface. At the cell surface the complex of the peptide antigen and the MHC molecule is recognized by the antigen-specific T-cell receptor (Figure 5.10). The MHC molecule is said to **present** the peptide antigen to the T-cell receptor, a process that is called **antigen presentation**. A cell that can present antigens is an **antigen-presenting cell**.

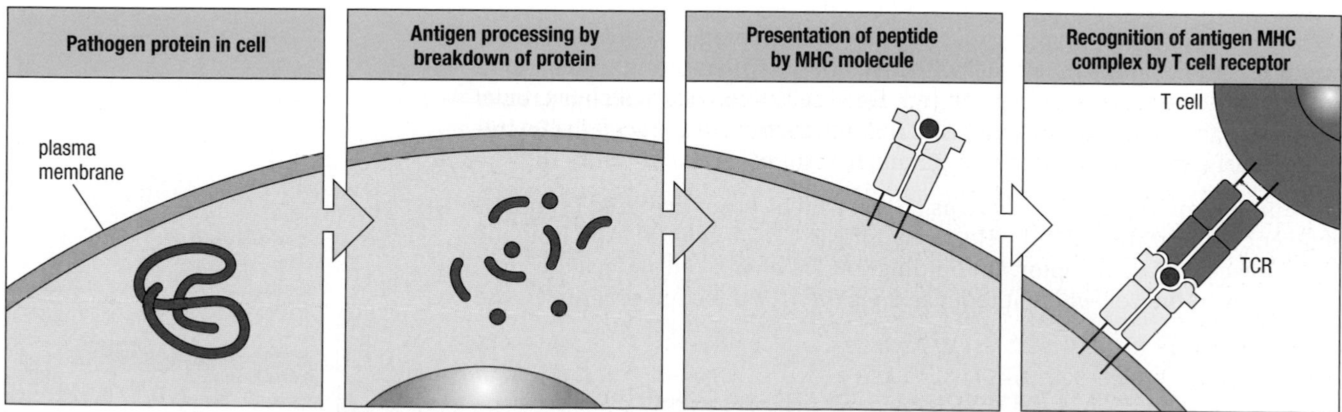

Figure 5.10 Antigen processing and presentation. The antigens recognized by T cells are peptides that arise from the breakdown of macromolecular structures, the unfolding of individual proteins, and their cleavage into short fragments. These events constitute antigen processing. For a T-cell receptor to recognize a peptide antigen, the peptide must be bound by an MHC molecule and displayed at the cell surface, a process called antigen presentation.

5-7 Two classes of MHC molecule present peptide antigens to two types of T cell

Microorganisms that infect human tissues are of two broad kinds: extracellular pathogens, such as many bacteria, that live and replicate in the spaces between human cells, and intracellular pathogens, such as viruses, that live and replicate inside human cells. Because of this topological difference, proteins from extracellular pathogens are degraded within endocytic vesicles and lysosomes, whereas proteins from intracellular pathogens are degraded in the cytosol, also called the cytoplasm. To aim T-cell responses at these different infections, two types of MHC molecule are used: **MHC class I** presents antigens from intracellular pathogens and **MHC class II** presents antigens from extracellular pathogens (Figure 5.11). Peptides produced by the cytosolic degradation of intracellular pathogens are delivered to the endoplasmic reticulum, where MHC class I molecules bind them and take them to the cell surface. In contrast, MHC class II molecules first travel to endosomal vesicles, where they bind peptides produced by lysosomal degradation of extracellular pathogens, before moving to the cell surface.

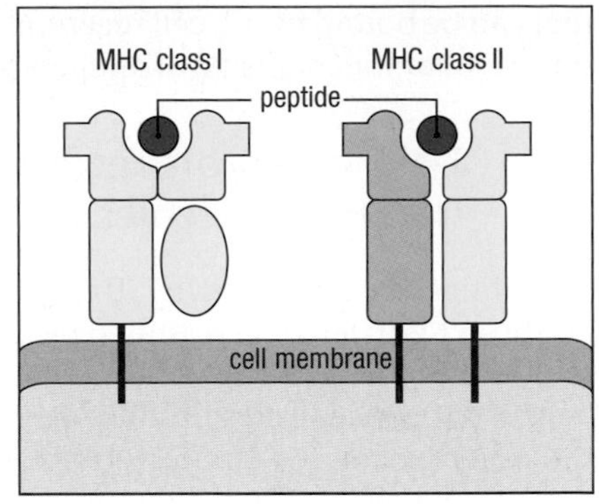

Figure 5.11 The two types of MHC molecule. MHC class I and MHC class II molecules have similar overall three-dimensional structures. Where they differ is in their constituent polypeptide chains. An MHC class II molecule is made of two similarly sized polypeptides that contain two extracellular domains and are anchored in the plasma membrane. In the MHC class I molecule, a larger polypeptide contains three extracellular domains and is anchored in the plasma membrane; a smaller polypeptide comprises the fourth extracellular domain and is not attached to the membrane.

Corresponding to the two classes of MHC class I molecule are two specialized types of effector T cell: the **cytotoxic T cell** that recognizes antigens presented by MHC class I molecules and defends against intracellular infection and the **helper T cell** that recognizes antigens presented by MHC class II molecules and defends against extracellular infection. The cytotoxic and helper T cells are easily distinguished because cytotoxic T cells all express the cell-surface protein **CD8**, whereas helper T cells all express the cell-surface protein **CD4**. Conversely, no cytotoxic T cells express CD4 and no helper T cells express CD8.

CD4 and CD8 are called **T-cell co-receptors** because they cooperate with the T-cell receptor in the recognition of complexes of peptide antigens and MHC molecules. CD4 binds to a site on the MHC class II molecules that is topographically separate from the site bound by the T-cell receptor. Likewise, CD8 binds to a site on the MHC class I molecule that does not overlap with the site recognized by the T-cell receptor. Although CD4 and CD8 have similar functions, their structures are different. CD4 is a single polypeptide with four extracellular immunoglobulin-like domains and CD8 is heterodimer of two chains, each having a single extracellular immunoglobulin-like domain (Figure 5.12).

The main function of cytotoxic CD8 T cells is to kill cells that are infected with a virus, a bacterium, or some other intracellular pathogen. The help that CD4 T cells give to tissue macrophages facilitates macrophage activation, improving their capacity to phagocytose extracellular pathogens and to secrete cytokines and chemokines that will drive the adaptive immune response. Equally important is the help given by CD4 T cells to B cells, which is crucial for making the high-affinity antibodies that bind tightly to extracellular bacteria and viruses and ensure their elimination (Figure 5.13).

The human immunodeficiency virus (HIV), which causes acquired immunodeficiency syndrome (AIDS), selectively infects CD4 T cells by exploiting the CD4 molecule as its receptor. On binding to CD4 on a T-cell surface the virus gains entry to the cell where it replicates. As the HIV infection progresses, the

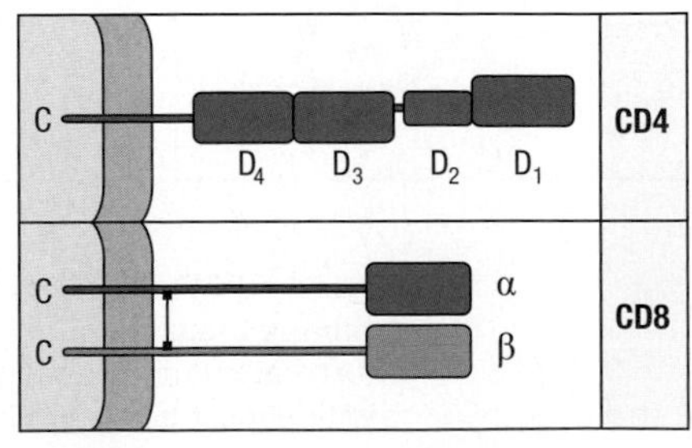

Figure 5.12 The structures of the CD4 and CD8 glycoproteins. The co-receptors CD4 and CD8 are members of the immunoglobulin superfamily of proteins. CD4 has four extracellular immunoglobulin-like domains (D_1–D_4) with a hinge between the D_2 and D_3 domains. CD8 consists of an α and a β chain, which both have an immunoglobulin-like domain that is connected to the membrane-spanning region by an extended stalk. C denotes the carboxy terminus.

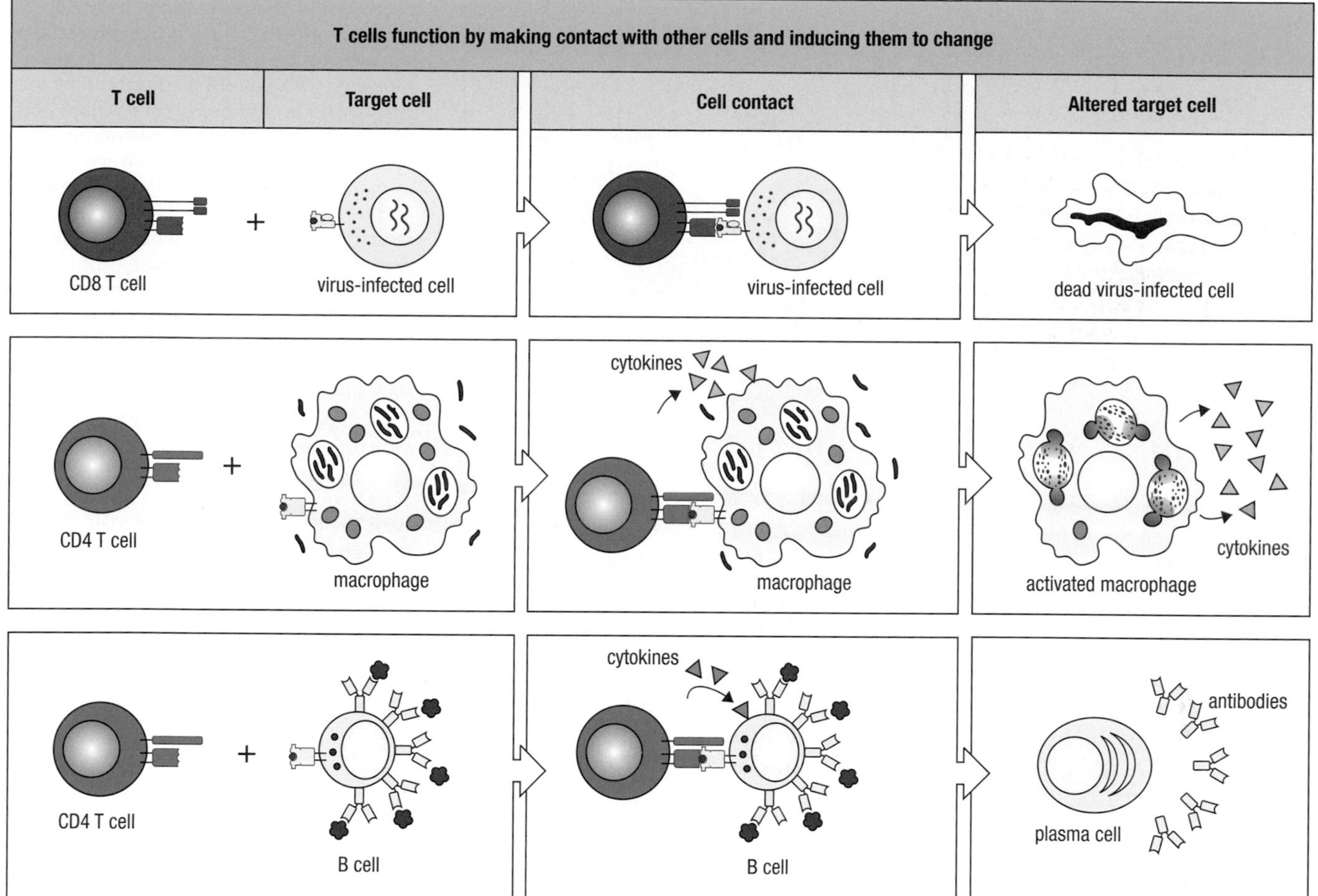

Figure 5.13 T cells function by making contact with other cells. Top panels: a cytotoxic CD8 T cell makes contact with a virus-infected cell, recognizes that it is infected, and kills it. Middle panels: a CD4 helper T cell contacts a macrophage that is engaged in the phagocytosis of bacteria and secretes cytokines that increase the microbicidal power of the macrophage and its secretion of inflammatory cytokines. Bottom panels: a CD4 helper T cell contacts a B cell that is binding its specific antigen and secretes cytokines that cause the B cell to differentiate into an antibody-secreting plasma cell.

number of circulating CD4 T cells gradually declines and, in the absence of effective treatment, will eventually reach a level at which the adaptive immune response to other types of infection becomes fatally compromised.

Before we consider how the appropriate MHC molecules become associated with peptides from different sources, we shall look at their structure and general peptide-binding properties, and how they associate with CD4 and CD8.

5-8 The two classes of MHC molecule have similar three-dimensional structures

MHC class I and class II molecules are membrane glycoproteins that function by binding peptide antigens and presenting them to T cells. Underlying this common function is a similar three-dimensional structure, which is formed in different ways in the two MHC classes.

An MHC class I molecule is made up of a transmembrane heavy chain, or α chain, which is noncovalently complexed with the small protein **β_2-microglobulin** (Figure 5.14). The heavy chain has three extracellular domains (α_1, α_2, and α_3). The peptide-binding site is formed by the folding of α_1 and α_2, the domains farthest from the membrane. Supporting the peptide-binding site are two immunoglobulin-like domains, the α_3 domain of the heavy chain and β_2-microglobulin. The MHC class I heavy chain is encoded by a gene in the MHC, whereas β_2-microglobulin is not.

In contrast, an MHC class II molecule consists of two transmembrane chains (α and β), each contributing one domain to the peptide-binding site and one

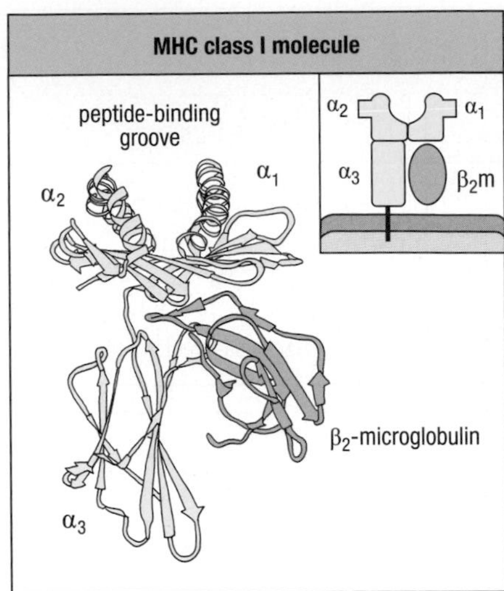

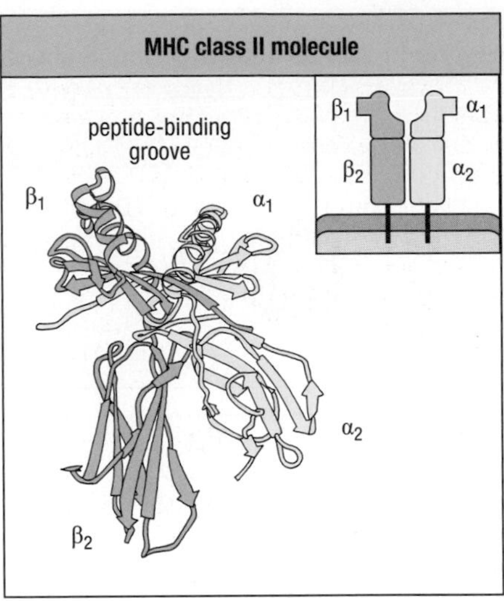

Figure 5.14 The structures of MHC class I and MHC class II molecules are variations on a theme. An MHC class I molecule (left panels) is composed of one membrane-bound heavy (or α) chain and noncovalently bonded β_2-microglobulin. The heavy chain has three extracellular domains, of which the amino-terminal α_1 and α_2 domains resemble each other in structure and form the peptide-binding site. An MHC class II molecule (right panels) is composed of two membrane-bound chains, an α chain (which is a different protein from MHC class I α) and a β chain. These have two extracellular domains each, the amino-terminal two (α_1 and β_1) resembling each other in structure and forming the peptide-binding site. The β_2 domain of MHC class II molecules should not be confused with the β_2-microglobulin of MHC class I molecules. The ribbon diagrams in the lower panels trace the paths of the polypeptide backbone chains.

immunoglobulin-like supporting domain (see Figure 5.14). Both these chains are encoded by genes in the MHC. Thus, the two classes of MHC molecule are seen to have similar three-dimensional structures.

The immunoglobulin-like domains of MHC class I and II molecules are not just a support for the peptide-binding site; they also provide specific binding sites for the CD4 and CD8 co-receptors. The sites on an MHC molecule that interact with the T-cell receptor and the co-receptor are separated, allowing the MHC molecule to engage both a T-cell receptor and a co-receptor simultaneously (Figure 5.15).

5-9 MHC molecules bind a variety of peptides

The peptide-binding site is a deep groove on the surface of the MHC molecule (Figure 5.16), within which a single peptide is held tightly by noncovalent bonds. There are constraints on the length and amino acid sequence of peptides that are bound by MHC molecules. These are dictated by the structure of the peptide-binding groove, which differs between MHC class I and class II. Within these constraints MHC class I molecules can bind thousands of peptides having different amino acid sequences. MHC class I and II molecules are said to have **promiscuous binding specificity** or **promiscuous specificity**.

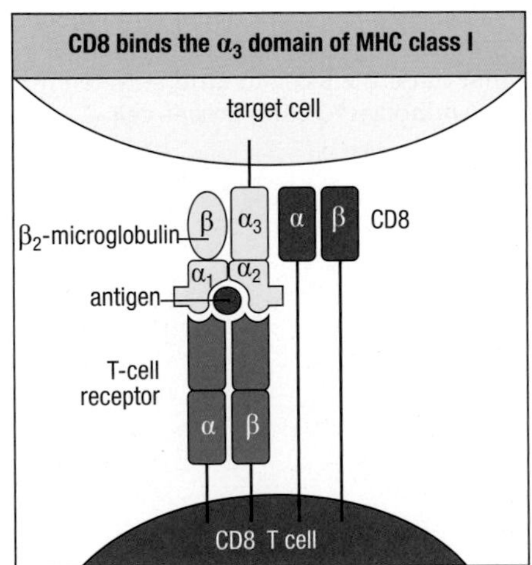

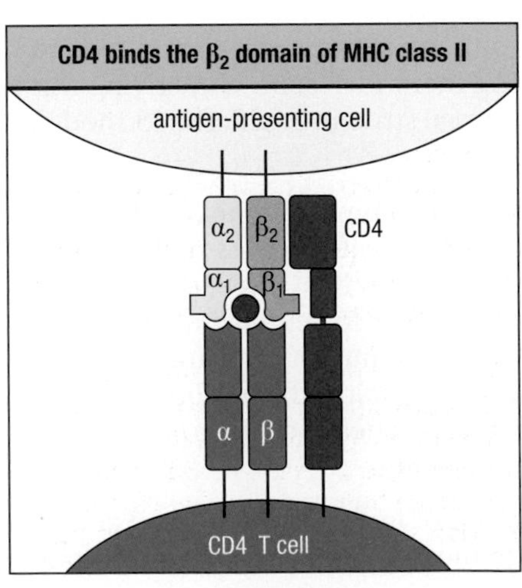

Figure 5.15 MHC class I molecules bind to CD8, and MHC class II molecules bind to CD4. The CD8 co-receptor binds to the α_3 domain of the MHC class I heavy chain, ensuring that MHC class I molecules present peptides only to CD8 T cells (left panel). In a complementary fashion, the CD4 co-receptor binds to the β_2 domain of MHC class II molecules, ensuring that peptides bound by MHC class II stimulate only CD4 T cells (right panel).

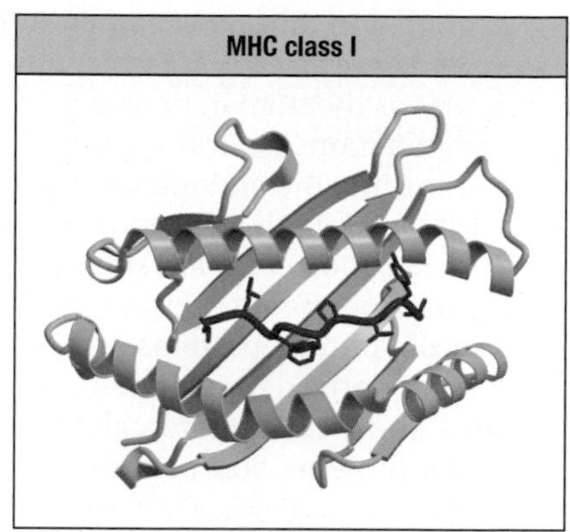

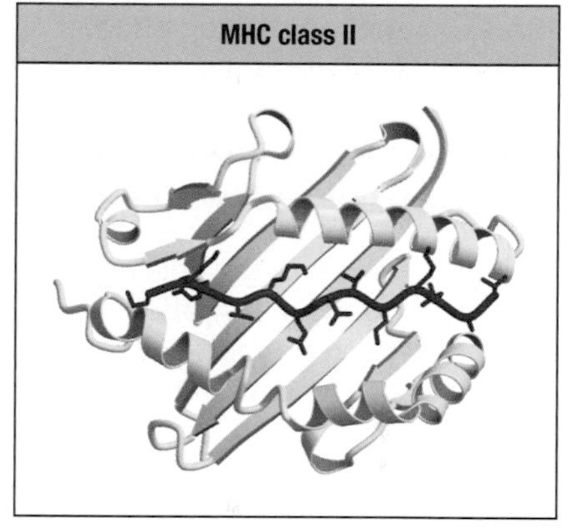

Figure 5.16 The peptide-binding groove of MHC class I and MHC class II molecules. Shown is a T-cell receptor's view of the peptide-binding groove, with a peptide bound. In the MHC class I molecule (top panel) the groove is formed by the α_1 and α_2 domains of the MHC class I heavy chain; in the MHC class II molecule (bottom panel) it is formed by the α_1 domain of the class II α chain and the β_1 domain of the class II β chain. Amino acid residues of the MHC molecule that are important for interacting with bound peptide are shown. Peptides bind to MHC class I molecules by their ends (top panel), whereas in MHC class II molecules the peptide extends beyond the peptide-binding groove and is held by interactions along its length (bottom panel). Structures courtesy of R.I. Stanfield and I.A. Wilson.

The length of the peptides bound by MHC class I molecules is limited because the two ends of the peptide are grasped by pockets situated at the ends of the peptide-binding groove (see Figure 5.16, left panel). The vast majority of peptides that bind MHC class I molecules are eight, nine, or ten amino acids in length; most are nine amino acids. The differences in length are accommodated by making a kink in the extended conformation of the bound peptide. Features common to all peptides—the amino terminus, the carboxy terminus, and the peptide backbone—interact with binding pockets present in all MHC class I molecules, and these form the basis for all peptide–MHC class I interactions. Most peptides bound by MHC class I molecules have either a hydrophobic or a basic residue at the carboxy terminus. The two groups of peptides bind to different forms of MHC class I that have a binding pocket complementary to either hydrophobic or basic amino acid side chains.

In MHC class II molecules, the two ends of the peptide are not pinned down into pockets at each end of the peptide-binding groove (see Figure 5.16, right panel). As they can extend out from each end of the groove, the peptides bound by MHC class II molecules are both longer and more variable in length than peptides bound by MHC class I. Peptides that bind to MHC class II molecules are usually 13–25 amino acids long, and some are much longer.

5-10 MHC class I and MHC class II molecules function in different intracellular compartments

Human cells consist of two topological compartments defined by the cellular membranes. One compartment is sealed off from the outside of the cell and comprises the nucleus and the cytosol (also called cytoplasm), which are connected by nuclear pores. The other compartment, called the vesicular system, is contiguous with the outside of the cell and comprises the endoplasmic reticulum, the Golgi apparatus, the lysosomes and a variety of smaller endocytic and exocytic vesicles (Figure 5.17).

Translation occurs in the cytosol and is an inefficient process sometimes producing incomplete and improperly folded human proteins that are targeted

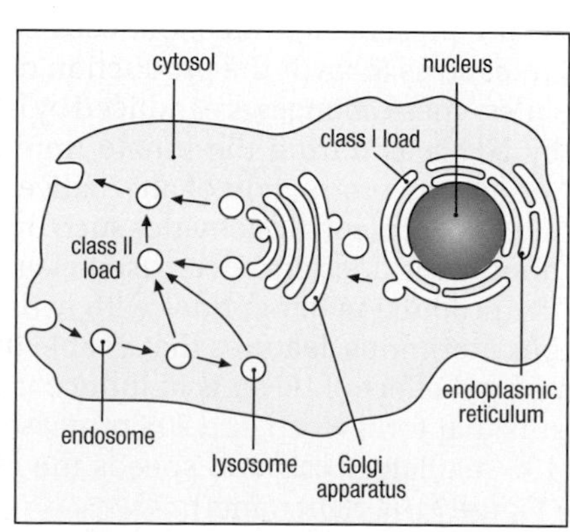

Figure 5.17 MHC class I and MHC class II molecules load peptides that are produced in two different compartments of the cell. One compartment consists of the cytosol and the nucleus. They are connected by the pores in the nuclear membrane. The other compartment is the vesicular system, consisting of the endoplasmic reticulum, the Golgi apparatus, endocytic vesicles, and lysosomes. Different groups of pathogens exploit the two cellular compartments. Peptides produced by protein degradation in the cytosol are pumped into the lumen of the endoplasmic reticulum where they are loaded onto MHC class I molecules and then taken by secretory vesicles through the Golgi to the plasma membrane. Peptides produced by protein degradation in the lysosomes are transported to endosomes where they are loaded onto MHC class II molecules and then taken to the plasma membrane.

for degradation and recycling by intracellular proteases. On virus infection, the synthesis of viral proteins becomes subject to the same scrutiny and quality control as the human proteins. Some of the peptides produced by this protein degradation are transported from the cytosol across the intracellular membrane into the endoplasmic reticulum, which is where they bind to MHC class I, if their length and sequence are right. Complexes of peptides bound to MHC class I molecules are subsequently exported via the Golgi apparatus to the plasma membrane. In healthy cells all the peptides bound by MHC class I derive from human proteins, whereas for infected cells they will be a mixture of peptides derived from human and viral proteins. In this context, the human proteins and peptides are called **self proteins** and **self peptides**, whereas pathogen proteins and peptides are called **non-self proteins** and **non-self peptides**.

A major function of the vesicular system is the uptake by phagocytosis and endocytosis of nutrients, molecular signals, and damaged molecular and cellular components of the circulation. As these materials move inward to the lysosomes they are degraded, producing large numbers of peptides. At a junction connecting the endocytic and exocytic pathways, peptides of compatible sequence can bind to MHC class II molecules that are held there waiting for a suitable peptide. In the absence of extracellular infection, MHC class II molecules bind only self peptides, but in the presence of infection, commonly caused by bacteria, non-self peptides derived from the uptake and degradation of bacteria and their products will also be bound by MHC class II molecules.

Because the two cellular compartments are subject to infection by different types of microorganism, the immune system has evolved two parallel pathways for monitoring their health. These involve two types of MHC molecule, binding peptides in two different cellular compartments and presenting them to two types of T cell.

5-11 Peptides generated in the cytosol are transported to the endoplasmic reticulum for binding to MHC class I molecules

The degradation in the cytosol of proteins that are damaged, poorly folded, or no longer needed is accomplished by a large barrel-shaped protein complex called the **proteasome**, which makes up 1% of cellular protein. Forming the barrel, or core, of the proteasome are four rings of seven polypeptide subunits, each of 20–30 kDa molecular mass. The β_1, β_2, and β_5 subunits of the inner two rings have a variety of protease activities that in combination efficiently degrade proteins held inside the catalytic chamber of the core. At either end of the barrel are two identical protein complexes, called 19S regulatory caps, which recognize proteins destined for degradation and steer them into the catalytic chamber (Figure 5.18, left panel).

In the presence of infection, cells can modify the structure of their proteasomes so as to favor the production of peptides that bind to MHC class I molecules. These changes are induced by interferon-γ (IFN-γ), the cytokine secreted by NK cells during the innate immune response. One effect of IFN-γ is to induce the production of alternative subunits that replace the β subunits and alter the proteolytic activities such that there is increased cleavage after hydrophobic residues and decreased cleavage after acidic residues. This increases the production of peptides with hydrophobic or basic amino acids at the carboxy terminus, features that enable them to bind to MHC class I molecules. A second effect of IFN-γ is to induce the production of PA28α and PA28β subunits that form a cap, the 20S proteasome-activator complex, that replaces the 19S regulatory cap and speeds the release of peptides from the proteasome (Figure 5.18, right panel).

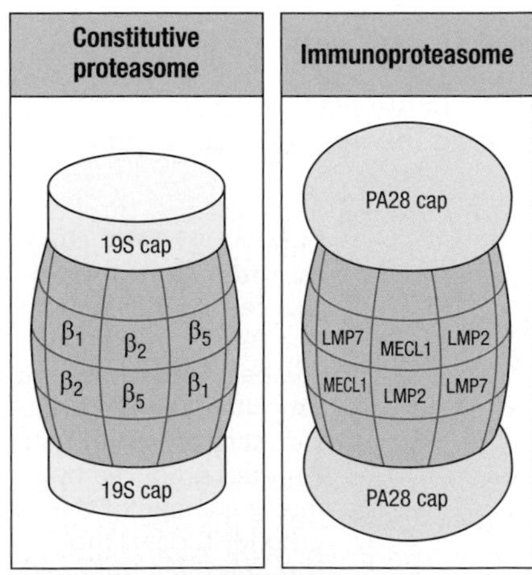

Figure 5.18 Structures of the constitutive proteasome and the immunoproteasome.

The modified form of the proteasome in cells exposed to IFN-γ is called the **immunoproteasome** whereas the form present in the absence of infection and IFN-γ is called the **constitutive proteasome.**

Once formed, the antigenic peptides are transported out of the cytosol and into the endoplasmic reticulum (Figure 5.19). Transport of peptides across the endoplasmic reticulum membrane is accomplished by a protein that is embedded in the membrane and is called the **transporter associated with antigen processing** (**TAP**). TAP is a heterodimer of two structurally related polypeptide chains, TAP-1 and TAP-2. TAP is a member of a large family of transporters that depend on the binding and hydrolysis of ATP to carry out their transport functions. The types of peptide that are preferentially transported by TAP are similar to those that bind MHC class I molecules: they are eight or more amino acids long and have either hydrophobic or basic residues at the carboxy terminus. Most of the peptides transported by TAP are not successful in binding an MHC class I molecule; they are transported out of the endoplasmic reticulum and back to the cytosol.

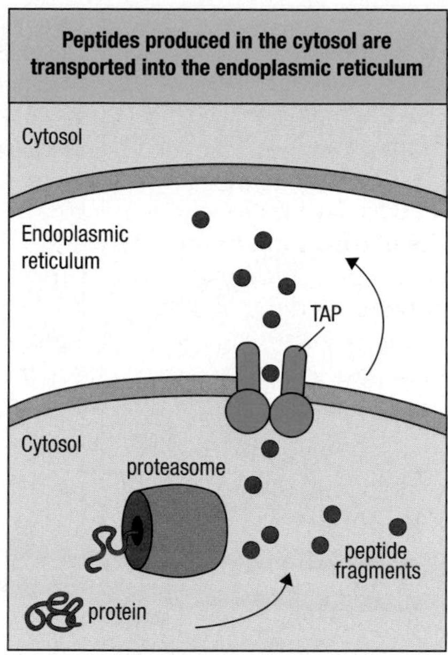

Figure 5.19 Formation and transport of peptides that bind to MHC class I molecules. In all cells, proteasomes degrade cellular proteins that are poorly folded, damaged, or unwanted. When a cell becomes infected, pathogen-derived proteins in the cytosol are also degraded by the proteasome. Peptides are transported from the cytosol into the lumen of the endoplasmic reticulum by the protein called transporter associated with antigen processing (TAP), which is embedded in the endoplasmic reticulum membrane.

In one form of a rare genetic disease called **bare lymphocyte syndrome type 1** or **MHC class I deficiency**, the TAP protein is nonfunctional and so there is no supply of peptides to the endoplasmic reticulum. Although not completely bare of MHC class I molecules, cells from patients with this defect have less than 1% of the normal level on their surfaces, demonstrating how peptides delivered by TAP are essential components of MHC class I molecules. They are also critical for immune function: these patients mount feeble CD8 T-cell responses to viruses and suffer chronic respiratory infections from a young age.

5-12 MHC class I molecules bind peptides as part of a peptide-loading complex

Newly synthesized MHC class I heavy chains and β_2-microglobulin are also translocated into the endoplasmic reticulum, where they partly complete their folding, associate with each other, and finally bind peptide, at which point folding is completed. Correct folding and peptide loading of MHC class I molecules is aided by proteins known as chaperones. These are proteins that assist in the correct folding and subunit assembly of other proteins while keeping them out of harm's way until they are ready to enter cellular pathways and carry out their functions.

When MHC class I heavy chains first enter the endoplasmic reticulum they bind the membrane-bound chaperone known as **calnexin**, which retains the partly folded heavy chain in the endoplasmic reticulum (Figure 5.20). Calnexin is a calcium-dependent lectin that binds to the asparagine-linked oligosaccharides of multisubunit glycoproteins, including T-cell receptors and immunoglobulins, and retains them in the endoplasmic reticulum until their subunits have folded and associated correctly.

When the MHC class I heavy chain has folded, formed its disulphide bonds, and bound β_2-microglobulin, calnexin is released and the heterodimer is incorporated into the **peptide-loading complex**. The central component of this complex is **tapasin**, a bridging protein that brings the heterodimer of β_2-microglobulin and the class I heavy chain into close proximity to TAP. The two extracellular immunoglobulin-like domains of tapasin bind to the α_2 and α_3 domains of the MHC class I heavy chain, whereas TAP and tapasin interact via their membrane-spanning regions. The effect of this bridge is that the empty peptide-binding groove of the MHC class I molecule is positioned to receive peptides from TAP. Two other proteins contribute to the peptide-loading complex: **calreticulin**, a soluble chaperone related to calnexin that binds to the MHC class I heavy chain; and **ERp57**, a thiol reductase that is linked to tapasin through a disulfide bond. ERp57 stabilizes the peptide-loading

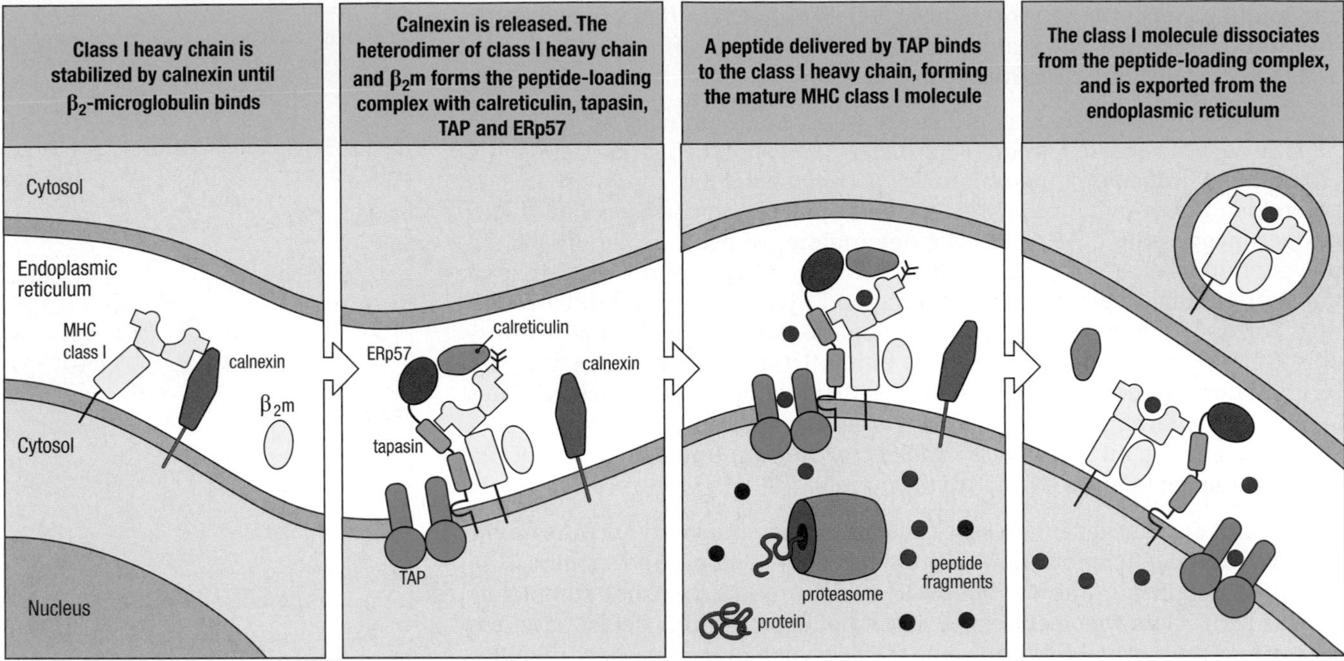

complex physically by interacting with calreticulin and enzymatically by protecting the disulfide bond of the α_2 domain of the class I heavy chain from being broken by chemical reduction.

Interaction of tapasin with the MHC class I heavy chain causes the peptide-binding groove to be in a more open conformation than exists in peptide-bound MHC class I molecules. This reduces the overall strength of the peptide-binding site, making it more selective for peptides that bind tightly. Many of the peptides that TAP delivers to the binding site do not form a stable complex and soon dissociate. The arrival of a peptide that binds tightly to the MHC class I molecule causes conformational changes that break the hold of tapasin, allowing the MHC class I molecule to leave the peptide-loading complex and to exit from the endoplasmic reticulum in a membrane-enclosed vesicle. It travels through the Golgi stacks, where MHC class I glycosylation is completed, and then to the plasma membrane.

The process by which tapasin helps the heterodimer of class I heavy chain and β_2-microglobulin to try on a variety of peptides to find one that fits really well is one aspect of what is called **peptide editing**. Sometimes a peptide binds strongly to the peptide-binding site through contacts made by its carboxy terminus but it is too long to fit snugly in the groove (Figure 5.21). In that case an enzyme called **endoplasmic reticulum aminopeptidase** (**ERAP**) will sequentially remove amino acids from the amino-terminal end until the fit is good. This mechanism also contributes to peptide editing.

Despite peptide editing in the peptide-loading complex, some MHC class I molecules emerge from the endoplasmic reticulum carrying a weakly bound peptide that falls out of the peptide-binding groove during transport. When this happens, calreticulin in the vesicle membrane can stabilize the heterodimer of the MHC class I heavy chain and β_2-microglobulin, and there is a mechanism that can recycle this 'empty' MHC class I molecule back to the endoplasmic reticulum, where it can try again to acquire a tightly binding peptide. This is an additional mode of peptide editing. If an MHC class I molecule is to be effective at presenting peptide antigens to T cells it needs to reach the cell surface and hang on to its peptide for an extended period. The various mechanisms of peptide editing help to satisfy this requirement.

Figure 5.20 Proteins of the peptide-loading complex aid the assembly and peptide loading of MHC class I molecules in the endoplasmic reticulum. MHC class I heavy chains assemble in the endoplasmic reticulum with the membrane-bound protein calnexin. When this complex binds β_2-microglobulin (β_2m) the partly folded MHC class I molecule is released from calnexin and then associates with the TAP via a complex of tapasin, ERp57, and calreticulin to form the peptide-loading complex. Degradation of cytosolic proteins by the proteasome produces some peptides (red circles) that are delivered into the endoplasmic reticulum through TAP, and others (black circles) that are not transported by TAP. The MHC class I molecule is retained in the endoplasmic reticulum until it binds a peptide, which completes the folding of the molecule. The peptide:MHC class I molecule complex is then released from the other proteins and leaves the endoplasmic reticulum for transport to the cell surface.

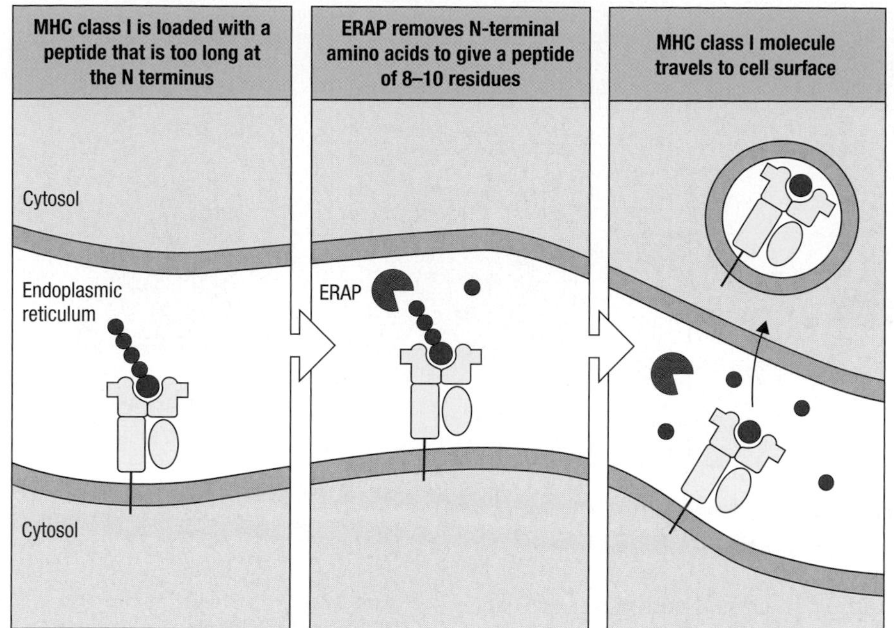
MHC class I is loaded with a peptide that is too long at the N terminus
ERAP removes N-terminal amino acids to give a peptide of 8–10 residues
MHC class I molecule travels to cell surface
Cytosol
Endoplasmic reticulum
Cytosol
ERAP

Figure 5.21 An aminopeptidase in the endoplasmic reticulum can trim peptides bound to MHC class I to improve their binding affinity. The endoplasmic reticulum aminopeptidase (ERAP) binds to MHC class I molecules in which the amino terminus of an overlong peptide hangs out of the binding site. It removes the accessible amino acid residues to leave a peptide of 8–10 amino acids with an improved fit to the peptide-binding groove. It is not known whether ERAP acts on the peptide when the class I molecule is part of the peptide-loading complex, or subsequently, as illustrated here, or both.

5-13 Peptides presented by MHC class II molecules are generated in acidified intracellular vesicles

Proteins of extracellular bacteria, extracellular virus particles, and soluble protein antigens are processed by a different intracellular pathway from that followed by cytosolic proteins, and their peptide fragments end up bound to MHC class II molecules. This pathway exploits the mechanisms that cells use to take up nutrients, hormones, and other molecular signals from their immediate environment and which is highly elaborated in phagocytes for the purposes of defense. Most cells are continually internalizing extracellular fluid and material bound at their surface by the process of endocytosis. Phagocytic cells—dendritic cells, macrophages, and neutrophils—also have surface receptors that bind to components of pathogen surfaces and promote their phagocytosis (see Section 2-6). All these uptake mechanisms produce intracellular vesicles in which the vesicular membrane is derived from the plasma membrane and the lumen contains extracellular material. The uptake vesicles that are produced in all cells are called endosomes, and the much larger vesicles formed by phagocytes, which can accommodate whole microrganisms and apoptotic human cells, are called phagosomes.

These membrane-enclosed vesicles become part of an interconnected vesicular system that carries materials to and from the cell surface. As vesicles travel inward from the plasma membrane, their interiors become acidified by the action of proton pumps in the vesicle membrane and they fuse with other vesicles, such as lysosomes, that contain proteases and hydrolases that are active in acid conditions. Within the phagolysosomes formed by this fusion, enzymes degrade the vesicle contents and produce peptides derived from the proteins and glycoproteins of the internalized pathogens.

Microorganisms in the extracellular environment are taken up by macrophages and dendritic cells by phagocytosis, and are then degraded within phagolysosomes. B cells also bind specific antigens via their surface immunoglobulin, and then internalize these antigens by receptor-mediated endocytosis. These antigens are similarly degraded within the vesicular system. Peptides produced within phagolysosomes become bound to MHC class II molecules

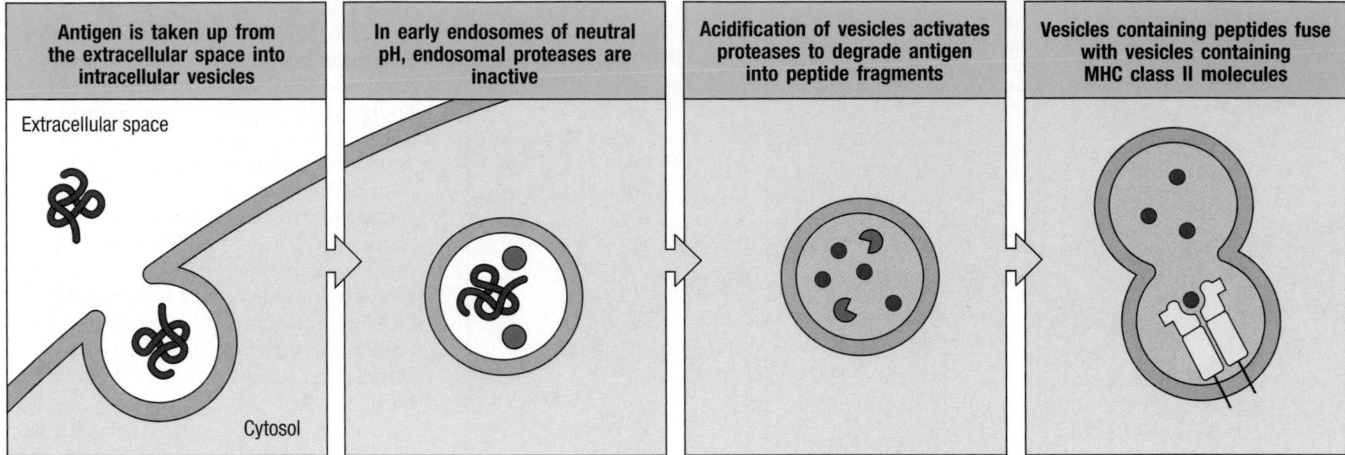

Figure 5.22 Peptides that bind to MHC class II molecules are generated in acidified endocytic vesicles. In the case illustrated here, extracellular foreign antigens, such as bacteria or bacterial antigens, have been taken up by an antigen-presenting cell such as a macrophage or immature dendritic cell. In other cases, the source of the peptide antigen can be bacteria or parasites that have invaded the cell to replicate in intracellular vesicles. In both situations the antigen-processing pathway is the same. The pH of the endosomes containing the ingested pathogens decreases progressively as they move inward, activating proteases within the vesicles that degrade the engulfed material. At some point on their pathway to the cell surface, newly synthesized MHC class I molecules pass through such acidified vesicles and bind peptide fragments of the antigen, transporting them to the cell surface.

within the vesicular system, and the peptide: MHC class II complexes are carried to the cell surface by outward-going vesicles. Thus, the MHC class II pathway samples the extracellular environment, complementing the MHC class I pathway, which samples the intracellular environment (Figure 5.22).

The mycobacteria that cause leprosy and tuberculosis exploit the vesicular system as a protected site for intracellular growth and replication. Because their proteins do not enter the cytosol, they are not processed and presented to cytotoxic CD8 T cells by MHC class I molecules. They protect themselves from degradation by lysosomal enzymes by preventing the fusion of the phagosome with a lysosome. This also prevents the presentation of mycobacterial antigens by MHC class II molecules.

5-14 Invariant chain prevents MHC class II molecules from binding peptides in the endoplasmic reticulum

Newly synthesized MHC class II α and β chains are translocated from the ribosomes into the membranes of the endoplasmic reticulum. There, an α chain and a β chain associate with a third chain, the **invariant chain** (Figure 5.23). So called because it is identical in all individuals, whereas the α and β chains are variable. One function of the invariant chain is to prevent MHC class II molecules from binding peptides in the endoplasmic reticulum. The invariant chain also stabilizes the conformation of the MHC class II molecule until it binds peptide. A third function of the invariant chain is to deliver MHC class II molecules to specialized endocytic vesicles where they bind peptides.

These vesicles, called **MIIC**, for **MHC class II compartment**, contain cathepsin S and other proteases that selectively attack the invariant chain. A series of cleavages leaves a small fragment called **class II-associated invariant-chain peptide** (**CLIP**) that covers up the MHC class II peptide-binding site. The removal of CLIP and the binding of another peptide are aided by the interaction of the MHC class II molecule with a glycoprotein in the vesicle membrane called **HLA-DM** (see Figure 5.22). HLA-DM is a specialized form of MHC class II molecule that cannot bind peptides, is absent from the cell surface, and is concentrated in the MIIC. The part it plays in loading MHC class II molecules with peptides is analogous to that of tapasin in loading MHC class I molecules. On binding to an MHC class II molecule, HLA-DM induces conformational changes that open up the peptide-binding groove, facilitating the dissociation of CLIP and the binding of other peptides. While remaining bound to HLA-DM, the MHC class II molecule can sample a succession of peptides until one binds

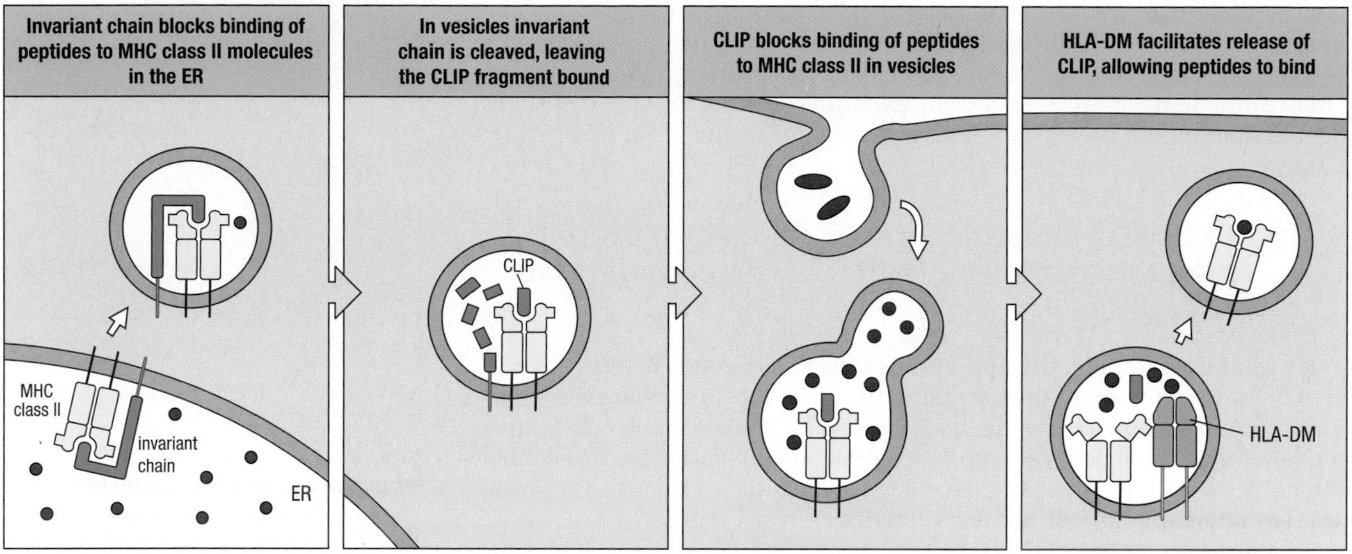

Invariant chain blocks binding of peptides to MHC class II molecules in the ER
MHC class II
invariant chain
ER
In vesicles invariant chain is cleaved, leaving the CLIP fragment bound
CLIP
CLIP blocks binding of peptides to MHC class II in vesicles
HLA-DM facilitates release of CLIP, allowing peptides to bind
HLA-DM

Figure 5.23 The invariant chain prevents peptides from binding to an MHC class II molecule until it reaches the site of extracellular protein breakdown. In the endoplasmic reticulum (ER), MHC class II α and β chains are assembled with an invariant chain that fills the peptide-binding groove; this complex is transported to the acidified vesicles of the endocytic system. The invariant chain is broken down, leaving a small fragment called class II-associated invariant-chain peptide (CLIP) attached in the peptide-binding site. The vesicle membrane protein HLA-DM catalyzes the release of the CLIP fragment and its replacement by a peptide derived from endocytosed antigen that has been degraded within the acidic interior of the vesicles.

sufficiently tightly to induce the conformational change that releases HLA-DM. Once loaded with peptide, MHC class II molecules leave the MIIC compartment in vesicles that deliver them to the plasma membrane.

HLA-DO is a second specialized MHC class II molecule that cannot bind peptides, is absent from the cell surface, and confined to intracellular vesicles. The function of HLA-DO is to antagonize HLA-DM. When HLA-DO binds to HLA-DM, the latter cannot release CLIP from MHC class II molecules. The balance between HLA-DM and HLA-DO is controlled by IFN-γ, which increases the expression of HLA-DM but not HLA-DO. In the presence of infection, when IFN-γ is made, first by NK cells and later by effector T cells, HLA-DM is made in excess of HLA-DO and therefore increases the presentation of antigenic peptides by MHC class II molecules.

5-15 Cross-presentation enables extracellular antigens to be presented by MHC class I

Throughout this part of the chapter the emphasis has been on the segregation of the cell into two discrete compartments that are associated with distinctive infections, MHC molecules, and T cells. This situation is not absolute, as became apparent from the study of viral infections. People infected with hepatitis C virus make a cytotoxic CD8 T-cell response that kills hepatocytes infected with the virus. Stimulating naive T cells to make such a response requires that dendritic cells or macrophages present virus-derived peptides on MHC class I molecules to naive CD8 T cells. Because hepatitis C virus infects only hepatocytes and not Kupffer cells, the macrophage-like cells of the liver, the only way that the latter can acquire viral proteins is by phagocytosis or endocytosis of dead, dying, or disintegrating infected hepatocytes. This is the conventional pathway for presentation by MHC class II, and the eventual presentation of these antigens to naive CD8 T cells implies that antigens located on the class II pathway of antigen presentation can be transferred to the class I pathway of antigen presentation. This phenomenon is called **cross presentation** (Figure 5.24). When an immune response is initiated by cross-presentation this is known as **cross-priming** of the immune response.

Although the phenomena of cross-presentation and cross-priming are well established, the underlying mechanisms remain poorly defined. One proposal is that phagocytosis of cells allows complexes of MHC class I molecules and viral peptides to be transferred from the dead hepatocyte to the membranes of intracellular vesicles inside the Kupffer cell, with subsequent transport to the cell surface where they can engage CD8 T-cell receptors. Another idea is that viral components are delivered from endocytic vesicles into the cytosol for degradation by the proteasome and presentation by the usual MHC class I pathway.

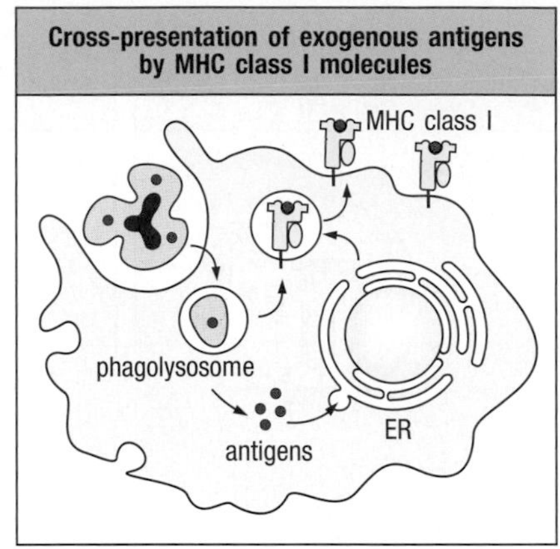

Figure 5.24 Cross-presentation of extracellular antigens by MHC class I molecules. The molecular pathway of cross-presentation is still unclear. One route could involve the translocation of ingested proteins from the phagolysosome into the cytosol, where they would undergo degradation by the proteasome, enter the endoplasmic reticulum via TAP, and be loaded on to MHC class I molecules in the usual way. Another route could involve the transport of antigens directly from the phagolysosome into a vesicular compartment (without passage through the cytosol) where peptides are allowed to bind to mature MHC class I molecules.

5-16 MHC class I molecules are expressed by most cell types, MHC class II molecules are expressed by few cell types

T-cell responses are guided in appropriate directions by the differential expression of the two classes of polymorphic MHC molecules in human cells (Figure 5.25). Virtually all the cells of the body express MHC class I molecules constitutively. Because all cell types are susceptible to virus infection, this enables the destruction of any type of infected cell by cytotoxic effector CD8 T cells. The erythrocyte is one of the few human cell types that lack MHC class I molecules.

MHC class II molecules are, in contrast, constitutively expressed by only a few cell types, notably B cells, macrophages, and dendritic cells. These cells of the immune system specialize in the uptake, processing, and presentation of antigens from the extracellular environment. This distribution of MHC class II molecules is consistent with their function of alerting CD4 T cells to the presence of extracellular infections. To be effective, this function need not be performed by every cell within a tissue or organ, but just by a sufficient number of specialized cells equipped to guard the extracellular territory. In recognition of this function B cells, macrophages, and dendritic cells are called **professional antigen-presenting cells**.

5-17 The T-cell receptor specifically recognizes both peptide and MHC molecule

Once a peptide:MHC molecule complex appears on the cell surface, it can be recognized by its corresponding T-cell receptor. When the T-cell receptor binds to a peptide:MHC complex it makes contacts with both the peptide and the surrounding surface of the MHC molecule. Thus, each peptide:MHC complex forms a unique ligand for a T-cell receptor.

In both classes of MHC molecule, the floor of the peptide-binding groove is formed by eight strands of antiparallel β-pleated sheet, on which lie two antiparallel α helices (see Figure 5.16). The peptide lies between the helices and parallel to them, such that the top surfaces of the helices and peptide form a roughly planar surface to which the T-cell receptor binds. The amino acid residues of the peptide that contact the MHC molecule lie deep in the peptide-binding groove and are inaccessible to the T-cell receptor; side chains of other residues of the peptide stick out of the binding site and contact the T-cell receptor.

In its overall organization the T-cell receptor antigen-binding site resembles that of an antibody (see Figure 5.2). The interaction between T-cell receptors and ligands comprising peptides bound to MHC molecules has been visualized by X-ray crystallography. Analysis of numerous complexes has revealed broadly similar interactions for peptides bound to either MHC class I or class

Tissue/cell	MHC	
	class I	class II
Hematopoietic		
T cells	+++	+*
B cells	+++	+++
Macrophages	+++	++
Dendritic cells	+++	+++
Neutrophils	+++	–
Erythrocytes	–	–
Non-hematopoietic		
Liver hepatocytes	+	–
Kidney epithelium	+	–
Brain	+	–†

Figure 5.25 Tissue distribution of MHC molecules. *Activated T cells express MHC class II molecules, whereas resting T cells do not. †Most cell types in the brain are MHC class II-negative, but microglial cells, which are related to macrophages, are MHC class II-positive.

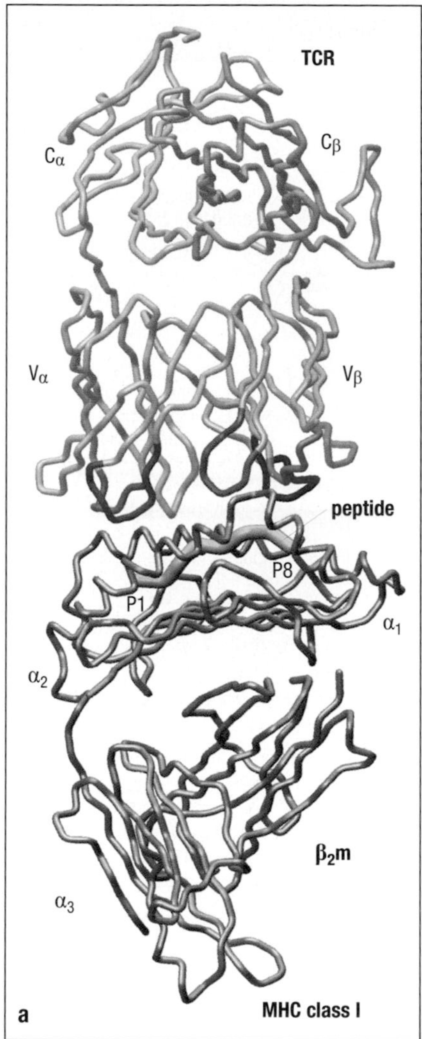

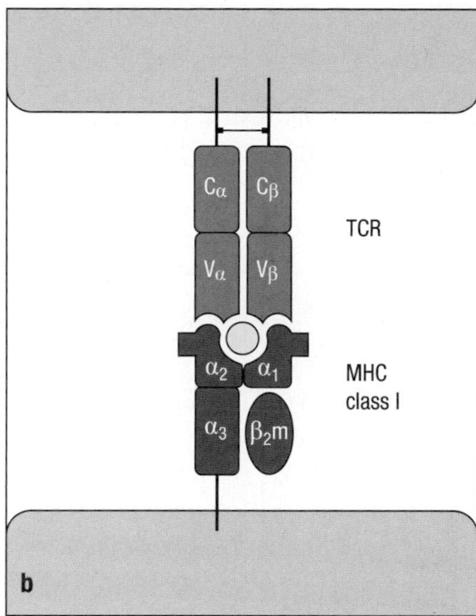

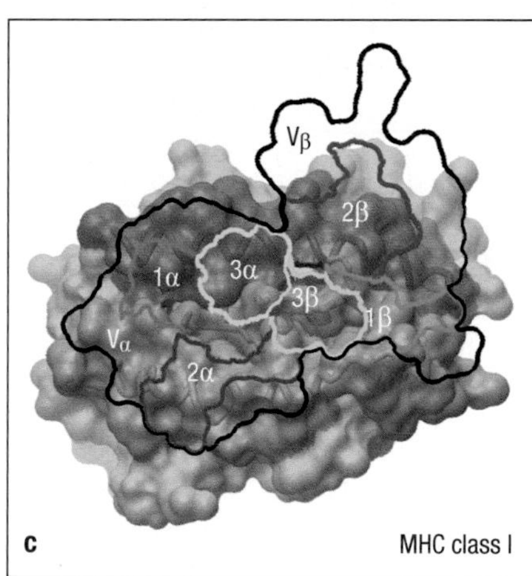

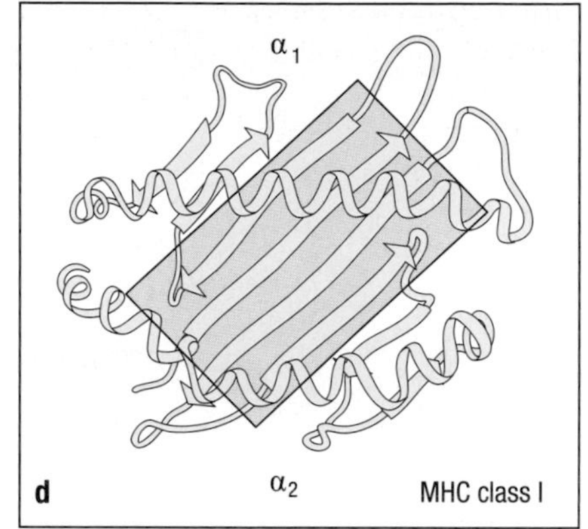

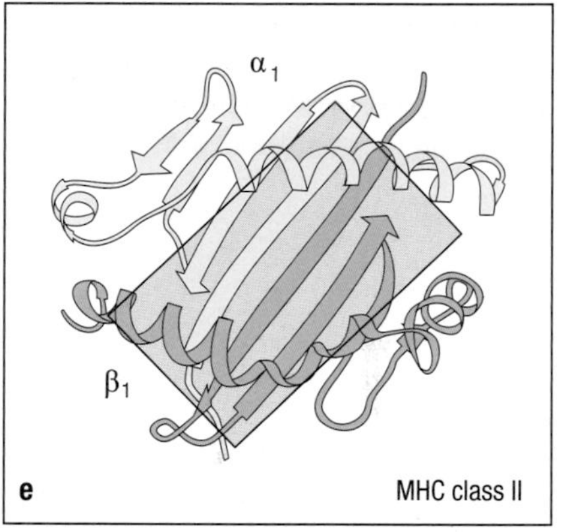

Figure 5.26 The peptide: MHC:T-cell receptor complex. Panel a shows a diagram of the polypeptide backbone of the complex of a T-cell receptor (TCR) bound to its peptide:MHC class I ligand. Panel b shows a schematic representation of this view of the receptor:ligand complex. In panel a the T-cell receptor's CDRs are colored: the α-chain CDR1 and CDR2 are light and dark blue respectively, while the β-chain CDR1 and CDR2 are light and dark purple respectively. The α-chain CDR3 is yellow and the β-chain CDR3 is dark yellow. The eight-amino-acid peptide is colored yellow and the positions of the first (P1) and last (P8) amino acids are indicated. Panel c is a view rotated 90° from that of panel a and shows the surface of the peptide:MHC class I ligand and the footprint made on it by the T-cell receptor (outlined in black). Within this footprint the contributions of the CDRs are outlined in different colors and labeled. In panels d and e the diagonal orientations of the T-cell receptor with respect to the peptide-binding grooves of MHC class I and class II molecules, respectively, are shown in schematic diagrams. The T-cell receptor is represented by the black rectangle superimposed on the ribbon diagram of the peptide-binding domains of the MHC molecules. Panels a and c courtesy of I.A. Wilson.

II molecules, as is illustrated here for the MHC class I interaction with peptide (Figure 5.26). A T-cell receptor binds to the peptide:MHC class I complex with the long axis of its binding site oriented diagonally across the peptide-binding groove of the MHC class I molecule (see Figure 5.26, panel d). A T-cell receptor binds to a peptide:MHC class II complex in a similar orientation (see Figure 5.26, panel e). The CDR3 loops of the T-cell receptor α and β chains form the central part of the binding site and they grasp the side chain of one of the amino acids in the middle of the peptide. The CDR1 and CDR2 loops form the periphery of the binding site and contact the α helices of the MHC molecule. The CDR3 loops that directly contact the peptide are the most variable part of the T-cell receptor antigen-recognition site; the α-chain CDR3 includes the joint between the V and J sequences, and the β chain CDR3 includes the joints between V and D, the whole of the D segment, and the joint between D and J.

Summary

T cells expressing α:β T-cell receptors recognize peptides presented at cell surfaces by MHC molecules, the third type of antigen-binding molecule in the adaptive immune system. Unlike immunoglobulins and T-cell receptors, however, MHC molecules have promiscuous binding sites and each MHC molecule can bind peptides of many different amino acid sequences. The peptides are produced by the intracellular degradation of proteins of infectious agents and of self proteins. An α:β T cell either expresses the CD8 co-receptor

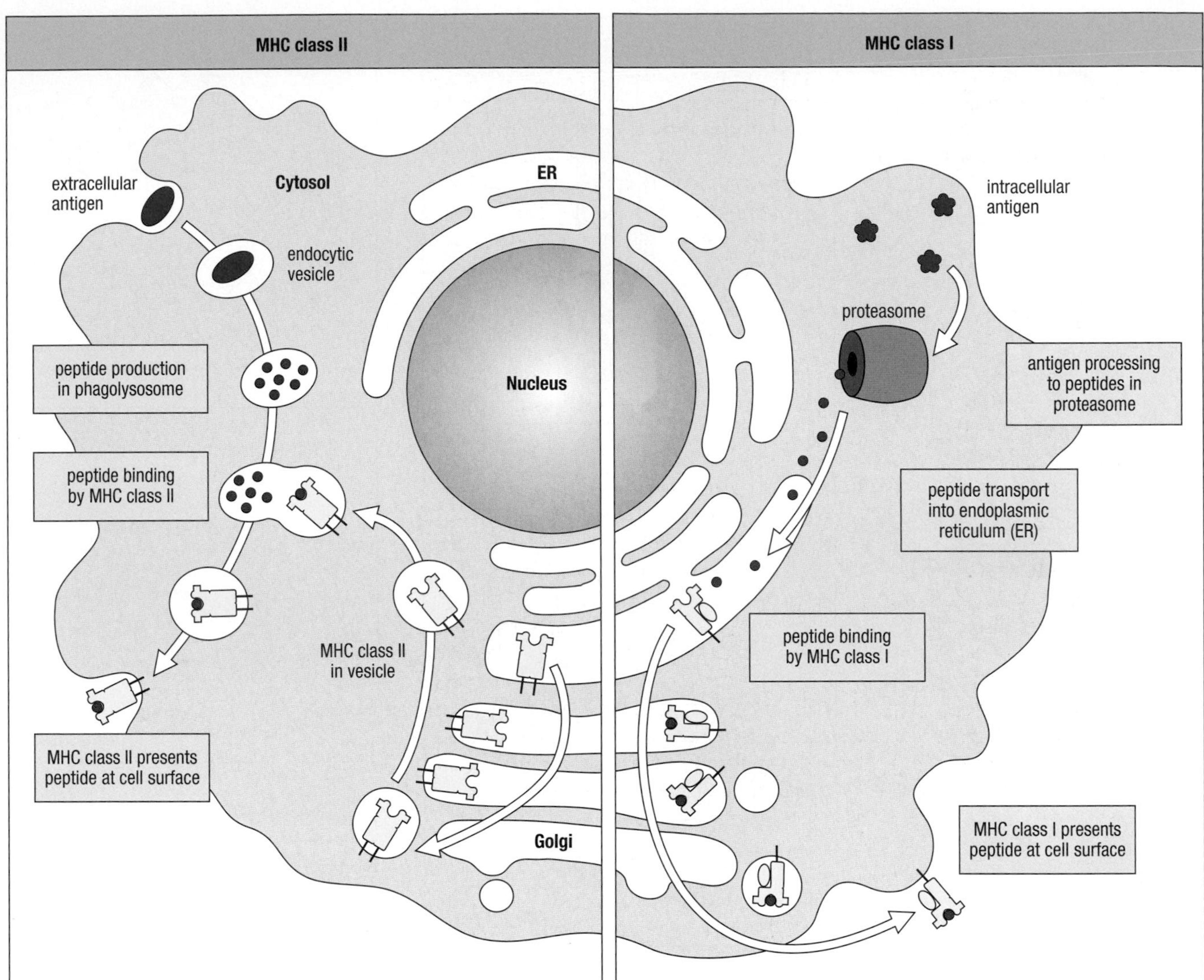

Figure 5.27 Processing of antigens for presentation by MHC class II or MHC class I molecules occurs in different cellular compartments. The left half of the figure shows the fate of peptides derived from extracellular antigens and pathogens. Extracellular material is taken up by endocytosis and phagocytosis into the vesicular system of the cell, in this case a macrophage. Proteases in these vesicles break down proteins to produce peptides that are bound by MHC class II molecules, which have been transported to the vesicles via the endoplasmic reticulum (ER) and the Golgi apparatus. The peptide:MHC class II complex is transported to the cell surface in outgoing vesicles. The right half of the figure shows the fate of peptides generated in the cytosol as a result of infection with viruses or intracytosolic bacteria. Proteins from such pathogens are broken down in the cytosol by the proteasome to peptides, which enter the ER. There the peptides are bound by MHC class I molecules. The peptide:MHC class I complex is transported to the cell surface via the Golgi apparatus.

and recognizes peptides presented by MHC class I molecules, or expresses the CD4 co-receptor and recognizes peptides presented by MHC class II molecules. The CD8 co-receptor interacts specifically with MHC class I molecules, and the CD4 co-receptor interacts specifically with MHC class II molecules. Protein antigens from intracellular and extracellular sources are processed into peptides by two different pathways. Peptides generated in the cytosol from viruses and other intracytosolic pathogens enter the endoplasmic reticulum, where they are bound by MHC class I molecules (Figure 5.27). Thus, these peptides are recognized by CD8 T cells, which are specialized to fight intracellular infections. All human cells are susceptible to infection, and so MHC class I molecules are expressed by most cell types. Extracellular material that has been taken up by endocytosis is degraded into peptides in endocytic vesicles and these peptides are bound by MHC class II molecules within the vesicular system. The resulting complex is recognized by CD4 T cells that are specialized to fight extracellular sources of infection by mobilizing other cells of the immune system, such as B cells and macrophages. MHC class II molecules are expressed by a few cell types of the immune system, the professional antigen-presenting cells, that are specialized to take up extracellular antigens efficiently and activate CD4 T cells. Cross-presentation of extracellular antigens by MHC class I molecules enables professional antigen-presenting cells to stimulate cytotoxic CD8 T-cell responses against viruses that do not infect them directly.

The major histocompatibility complex

MHC molecules and other proteins involved in antigen processing and presentation are encoded in a cluster of closely linked genes, which in humans is located on chromosome 6. This region is called the major histocompatibility complex (MHC) because it was first recognized as the site of genes that cause T cells to reject tissues transplanted from unrelated donors to recipients. We now know that these genes encode the MHC class I and class II molecules that present antigens to T cells. For some MHC class I and class II molecules, numerous genetic variants are present in the human population, each of which is different in the peptides it binds and the T-cell receptors that will recognize it. These differences, individually and collectively, have helped the human species to survive predation by diverse and numerous pathogens. Although the magnitude of MHC diversity is much smaller than that of immunoglobulins or T-cell receptors, this diversity has a major impact on the immune response, disease susceptibility, and the practice of medicine. In fact, the MHC is the region of the genome most strongly correlated with human disease and with both the broadest range and highest number of diseases. Because of its impact on medical practice, particularly in tissue transplantation, MHC diversity has been extensively studied in the human population, far beyond that of any other species. In this part of the chapter we consider the nature and function of MHC diversity and the immunological consequences for humans.

5-18 The diversity of MHC molecules in the human population is due to multigene families and genetic polymorphism

The human MHC is called the **human leukocyte antigen complex** (**HLA complex**) because the antibodies used to identify human MHC molecules react with the white cells of the blood—the leukocytes—but not with the red cells, which lack MHC molecules. This observation distinguished the MHC from the other known systems of cell-surface antigens, for example the ABO system that has to be matched in blood transfusion, which all involve antigens on the surface of red blood cells. Human MHC class I and II molecules are called **HLA class I molecules** and **HLA class II molecules**, respectively.

In contrast to immunoglobulins and T-cell receptors, the MHC class I and II molecules are encoded by conventionally stable genes that neither rearrange nor undergo any other developmental or somatic process of structural change. The inherited diversity of MHC molecules has two sources. The first is the presence of **gene families**, consisting of multiple similar genes encoding the MHC class I heavy chains, MHC class II α chains, and MHC class II β chains. The second source is **genetic polymorphism**, which is the presence within the population of multiple alternative forms of a gene, or alleles.

The protein products of the different genes in a MHC class I or class II family are called **isotypes**. The different forms of any given gene are called alleles,

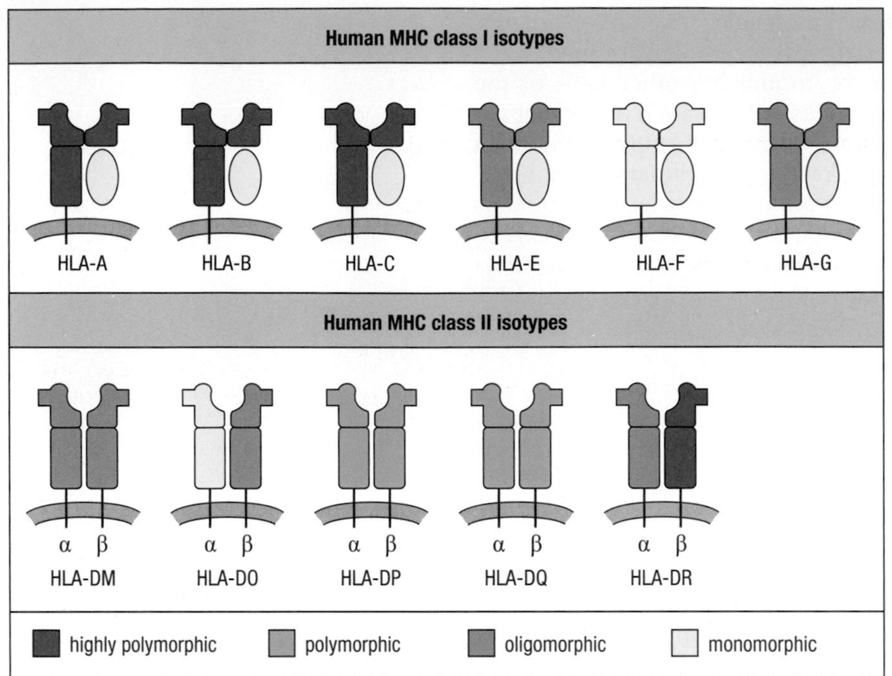

Figure 5.28 Human MHC class I and II isotypes differ in function and in the extent of their polymorphism. Of the human MHC class I isotypes, HLA-A, HLA-B, and HLA-C are highly polymorphic. They present peptide antigens to CD8 T cells and also interact with NK-cell receptors. HLA-E and HLA-G are oligomorphic and interact with NK-cell receptors. HLA-F is intracellular and of unknown function, and occurs as a single isotype. Of the human MHC class II isotypes, HLA-DP, HLA-DQ, and HLA-DR are polymorphic and present peptide antigens to CD4 T cells, whereas HLA-DM and HLA-DO occur in only a few isotypes, are intracellular, and regulate the loading of peptides onto HLA-DP, HLA-DQ, and HLA-DR.

and their encoded proteins are called **allotypes**. When considering the diversity of MHC class I or II molecules that arises from the combination of multiple genes and multiple alleles, the term **isoform** can be useful to denote any particular MHC protein. The numerous alleles of certain MHC class I and II genes, and the many differences that distinguish them, make these MHC genes stand out from other polymorphic genes, and they are said to be **highly polymorphic**. MHC class I and II genes that have no polymorphism are described as **monomorphic**, and genes having a few alleles are described as **oligomorphic**. An important consequence of high genetic polymorphism is that most children inherit a different allele of the gene from each parent, in which case the child is said to be **heterozygous**; children who inherit the same allele from both parents are said to be **homozygous**.

In humans there are six MHC class I isotypes, namely **HLA-A**, **HLA-B**, **HLA-C**, **HLA-E**, **HLA-F**, and **HLA-G**, and five MHC class II isotypes, namely **HLA-DM**, **HLA-DO**, **HLA-DP**, **HLA-DQ**, and **HLA-DR** (Figure 5.28). Of the class I isotypes, HLA-A, HLA-B, and HLA-C are highly polymorphic and their function is to present antigens to CD8 T cells and to form ligands for receptors on natural killer (NK) cells. HLA-E and HLA-G are oligomorphic and form ligands for NK-cell receptors. HLA-F seems to act as a chaperone that retrieves other HLA class I molecules that lose their peptides at the cell surface and brings them back inside the cell. The human MHC class II isotypes also display a range of properties. The three highly polymorphic molecules, **HLA-DP**, **HLA-DQ**, and **HLA-DR**, present peptide antigens directly to CD4 T cells, whereas the oligomorphic **HLA-DM** and **HLA-DO** molecules supervise peptide loading of HLA-DP, HLA-DQ, and HLA-DR. The numbers of alleles currently known for each HLA locus are listed in Figure 5.29.

The polymorphism of the HLA-A, B, and C molecules is the property of the heavy chain, as the β_2-microglobulin is monomorphic. By contrast, the polymorphic HLA class II isotypes differ in the diversity contributed by their α and β chains. In HLA-DR the α chain contributes almost no diversity and the β chain is highly polymorphic, whereas the α and β chains of HLA-DP and HLA-DQ are both polymorphic. Overall, there is greater diversity in HLA class I than in HLA class II.

HLA polymorphism

MHC class	HLA locus	Number of allotypes
MHC class I	A	1939
	B	2577
	C	1595
	E	6
	F	4
	G	16
MHC class II	DMA	4
	DMB	7
	DOA	3
	DOB	5
	DPA1	17
	DPB1	286
	DQA1	32
	DQB1	399
	DRA	2
	DRB1	1158
	DRB3	46
	DRB4	8
	DRB5	17

Figure 5.29 Some HLA class I and class II genes are highly polymorphic. The number of known functional alleles in the human population for each gene is shown. Data are from the website http://www.ebi.ac.uk/imgt/hla/

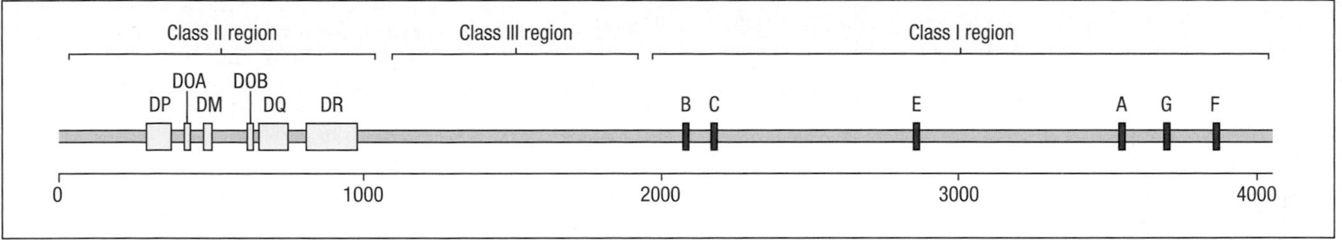

5-19 The HLA class I and class II genes occupy different regions of the HLA complex

The HLA complex (the human MHC) consists of about 4 million base pairs of DNA on the short arm of chromosome 6 and is divided into three regions (Figure 5.30). The **class I region**, at the end of the complex farthest from the chromosome's centromere, contains the six expressed HLA class I genes as well as several nonfunctional class I genes and gene fragments. At the other end of the complex is the **class II region**, which contains all the expressed class II genes and several nonfunctional class II genes. Separating the class I and class II regions is a stretch of around 1 million base pairs called the **class III region** or the **central MHC**; although dense with other types of gene, it contains no class I or II genes. Notably absent from the HLA complex is the gene encoding β_2-microglobulin, the invariant light chain of HLA class I molecules, which is located on human chromosome 15.

For each HLA class II isotype, the genes encoding the α and β chains are called *A* and *B* respectively, for example *HLA-DMA* and *HLA-DMB*. When there is more than one gene, including nonfunctional genes, a number in series is added, for example *HLA-DQA1* and *HLA-DQA2*. The genes for the α and β chains of the HLA-DM, HLA-DP, HLA-DQ, and HLA-DR class II isotypes cluster together in different subregions within the class II region of the MHC (see Figure 5.30). An exception is HLA-DO, for which the *HLA-DOA* and *HLA-DOB* genes are separated by *HLA-DM* and other genes. There are two pairs of genes for both HLA-DP and HLA-DQ, one pair being functional (*HLA-DPA1, DPB1* and *HLA-DQA1, DQB1*) and the other nonfunctional (*HLA-DPA2, DPB2* and *HLA-DQA2, DQB2*). For HLA-DR there is a single *HLA-DRA* gene, but four different genes encoding HLA-DR β chains (*DRB1, DRB3, DRB4*, and *DRB5*) and several nonfunctional genes (*DRB2, DRB6, DRB7, DRB8*, and *DRB9*). Only *DRB1* is present on all chromosomes 6, and for some people this is the only DRB gene expressed. Three other types of chromosome 6 carry either *DRB3, DRB4,* or *DRB5* in addition to *DRB1* (Figure 5.31).

The particular combination of HLA alleles found on a given chromosome 6 is known as the **haplotype**. Within the HLA complex, meiotic recombination occurs at a frequency of about 2%. This low frequency means that in most families, the parental HLA haplotypes are inherited intact from one generation to the next. However, over the course of human history, comprising some 10,000 generations, HLA alleles have been recombined into many thousands of different haplotypes. The presence of two HLA haplotypes in each individual thus means that millions of different combinations of HLA class I and II molecules are represented in the present human population. Individuals homozygous for the HLA complex are rare, but they are usually healthy. They express three class I (HLA-A, B, and C) and three class II (HLA-DP, DQ, and DR) isoforms that present antigens to their T cells. HLA heterozygous individuals can express up to six class I and sixteen class II isoforms; the maximum number is when each HLA haplotype contains two functional DRB genes and contributes a different allele for all the polymorphic HLA class I and II genes.

Figure 5.30 The MHC is divided into three regions containing different types of gene. The positions within the HLA complex (the human MHC) of the HLA class I and II genes are shown here. The class I genes (red) are all contained in the class I region, and the class II genes (yellow) are all contained in the class II region. Separating the class I and class II regions is the class III region, which contains a variety of genes (not shown), none of which contributes to antigen processing and presentation. For HLA-DM, HLA-DP, HLA-DQ, and HLA-DR, the α-chain and β-chain genes are close together and are shown as a single yellow block; for HLA-DO the α and β genes (DOA and DOB, respectively) are separated by the DM genes and are therefore shown separately. Approximate distances are given in thousands of base pairs (kb).

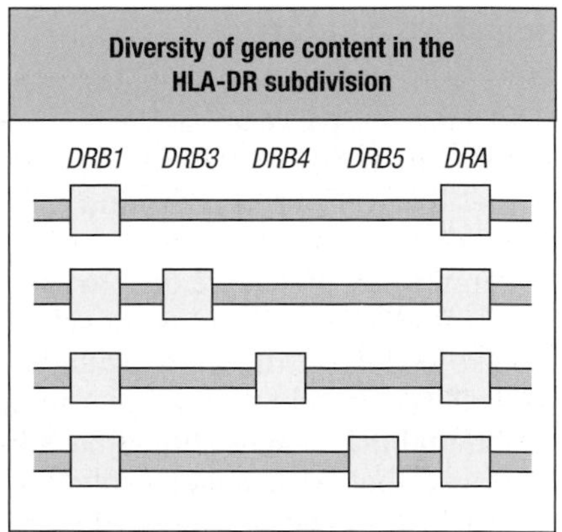

Figure 5.31 Human MHC regions differ in their number of DR genes. The MHC on every human chromosome 6 contains one gene (*DRA*) for the HLA class II DRα chain and one gene (*DRB1*) for the DRβ chain. In addition, some MHCs have either *DRB3* or *DRB4* or *DRB5*. Any DRβ chain can pair with the DRα chain to form a class II molecule.

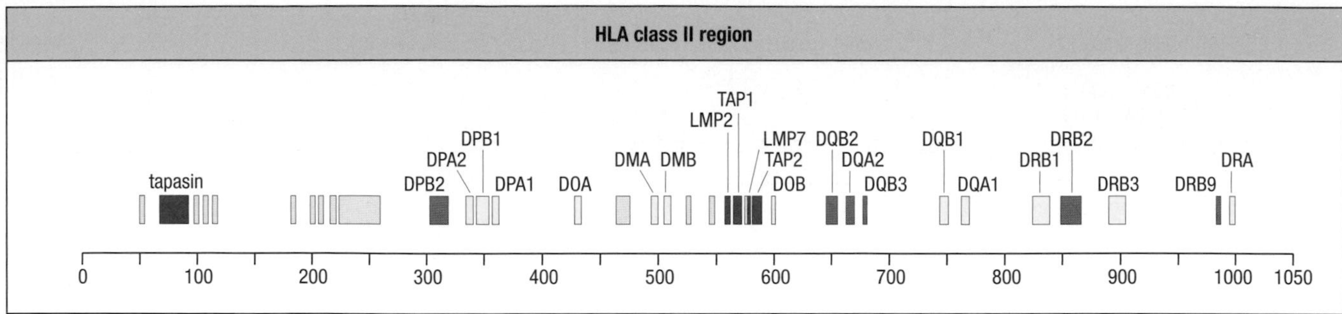

5-20 Other proteins involved in antigen processing and presentation are encoded in the HLA class II region

Figure 5.32 **Almost all of the genes in the HLA class II region are involved in the processing and presentation of antigens to T cells.** A detailed map of the HLA class II region is shown. Genes shown in dark gray are pseudogenes that are related to functional genes but are not expressed. Unnamed genes in light gray are not involved in immune system function. In addition to genes encoding the MHC class II isoforms, the class II region includes genes for the peptide transporter (TAP), proteasome components (LMP) and tapasin. Approximate distances are given in thousands of base pairs (kb).

The HLA complex as a whole contains more than 200 genes, of which the HLA class I and II genes are a minority. The other genes embrace a variety of functions including several that are important for the immune system. Particularly striking is the fact that the class II region of the HLA is almost entirely dedicated to genes involved in the processing of antigens and their presentation to T cells (Figure 5.32). In addition to genes encoding the α and β chains of the five HLA class II isotypes, the class II region contains genes encoding the two polypeptides of the TAP peptide transporter, the gene for tapasin, and genes encoding two of the three proteolytic subunits, LMP2 and LMP7, that are specific to the immunoproteasome (see Figure 5.18). The gene encoding the invariant chain is an exception; it is on chromosome 5.

Genes encoding proteins that work together in antigen processing and presentation are coordinately regulated by the cytokines IFN-α, -β, and -γ, which are produced at sites of infection at an early stage in the immune response. These cytokines stimulate cells in the vicinity to increase their expression of HLA class I heavy chains, β_2-microglobulin, TAP, and the LMP2 and LMP7 proteasome subunits. Expression of the HLA-DM, HLA-DP, HLA-DQ, HLA-DR, and invariant-chain genes is coordinated by the cytokine IFN-γ. These genes are turned on by a transcriptional activator known as **MHC class II transactivator** (**CIITA**), which is itself induced by IFN-γ. Inherited impaired CIITA function leads to MHC class II deficiency in which HLA class II molecules are not made and CD4 T cells cannot function.

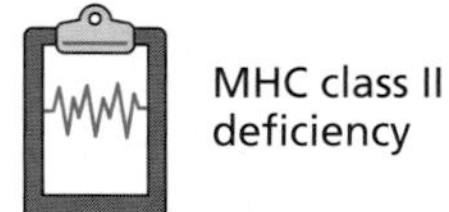
MHC class II deficiency

The majority of genes in the class I region are not involved in the immune system; neither do the class I genes form as compact a cluster as the class II genes (see Figure 5.30). Whereas genes encoding class II molecules are only present in the class II region of the HLA, genes encoding class I molecules and related class I-like molecules are found on several different chromosomes. A further difference is that HLA class II molecules are dedicated components of adaptive immunity that serve only to present antigen to T cells, whereas HLA class I and class I-like molecules encompass a broader range of functions, including uptake of IgG in the gut, regulation of iron metabolism, and regulation of NK-cell function in the innate immune response. Together, these genetic and functional differences show that MHC class I is the older form of MHC molecule and that MHC class II evolved more recently from MHC class I. Consistent with this proposition, MHC class II is not always essential for a vertebrate immune system, unlike MHC class I. The Atlantic cod manages very well with only MHC class I genes.

5-21 MHC polymorphism affects the binding of peptide antigens and their presentation to T cells

The alleles of highly polymorphic MHC genes encode proteins that differ by 1–50 amino acid substitutions. Such differences are unmatched by any other

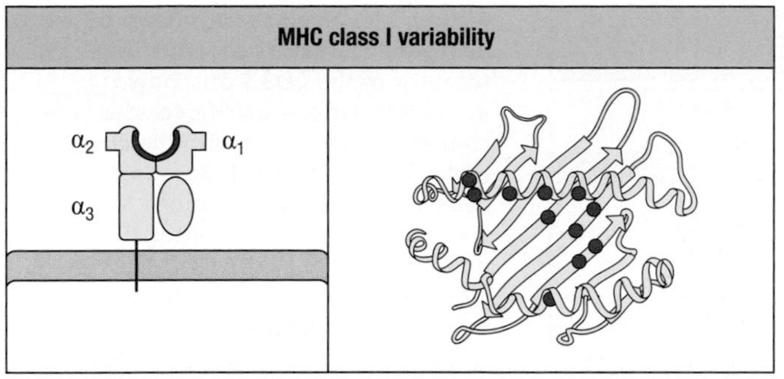

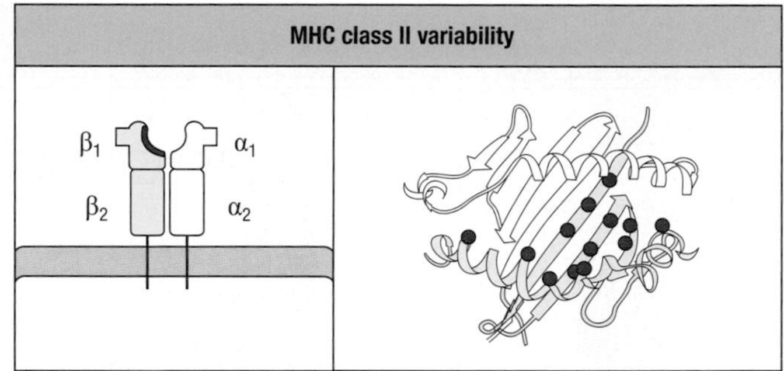

Figure 5.33 Variation between MHC allotypes is concentrated in the sites that bind peptide and T-cell receptor. In the HLA class I molecule (left), allotype variability is clustered in specific sites (shown in red) within the α_1 and α_2 domains. These sites line the peptide-binding groove, lying either in the floor of the groove, where they influence peptide binding, or in the α helices that form the walls, which are also involved in binding the T-cell receptors. In the HLA class II molecule illustrated (right), which is a DR molecule, variability is found only in the β_1 domain because the α chain is monomorphic.

human genes. Neither are the substitutions randomly distributed within the sequence, for they are mainly located in the domains that bind peptide and interact with the T-cell receptor: the α_1 and α_2 domains of MHC class I, and the α_1 and β_1 domains of MHC class II. More specifically, the substitutions are focused at positions within those domains that contact either bound peptide or the T-cell receptor (Figure 5.33). Not all the contact residues vary, as is apparent from the HLA-DR molecule, in which the α_1 domain is invariant. In contrast, variability occurs in both the α_1 and β_1 domains of HLA-DP and HLA-DQ molecules.

Variation in the peptide-contact residues on the floor and sides of the peptide-binding groove determines the types of peptide that each isoform binds. At certain positions within the peptide sequence, most of the peptides that bind an MHC isoform have the same amino acid, or one of a few chemically similar amino acids. The preferences arise because the side chains of amino acids at these positions are bound by complementary pockets within the binding groove. These amino acids are described as **anchor residues** because they anchor the peptide to the MHC molecule. The combination of anchor residues that binds to a particular MHC isoform is called its **peptide-binding motif**. For MHC class I molecules, which bind mostly nonamer peptides, positions 2 and 9 are the usual anchor residues. For MHC class II the anchor residues are less clearly defined, in part because of the heterogeneity in length of the bound peptides (Figure 5.34).

Figure 5.34 Peptide-binding motifs of some MHC isoforms and the sequences of peptides bound. For the HLA-A and HLA-B isoforms, both the peptide-binding motif of the MHC molecule and the complete amino acid sequence of one peptide presented by that isoform are given. Blank boxes in the peptide-binding motifs are positions at which the identity of the amino acid can vary. For the HLA-DR and HLA-DQ isoforms, only the sequence of a self peptide that is bound by the isoform is shown. Anchor residues are in green circles. Peptide-binding motifs for MHC class II molecules are not readily defined. The one-letter code for amino acids is used here. In the nomenclature used to distinguish the allotypes of an HLA class I or II gene the two numbers following the asterisk distinguish the major groups of allotypes, whereas the numbers following the colon distinguish variants within each major group, many of which differ by single amino-acid substitutions.

	MHC molecule	Amino acid sequence of peptide-binding motifs and bound peptides		Source of bound peptide
		Position in peptide sequence	N–1 2 3 4 5 6 7 8 9–C	
Class I	HLA-A*02:01	Peptide-binding motif	□ L/M □ □ □ V □ □ V/L	
		Bound peptide	I L K E P V H G V	HIV reverse transcriptase
	HLA-B*27:05	Peptide-binding motif	□ R □ □ □ □ □ □ R/K	
		Bound peptide	S R Y W A I R T R	Influenza A nucleoprotein
Class II	HLA-DRB1*04:01	Self peptide	G V Y F Y L Q W G R S T L V S V S	Igκ light chain
	HLA-DQA1*05:01 HLA-DQB1*03:01	Self peptide	I P E L N K V A R A A A	Transferrin receptor

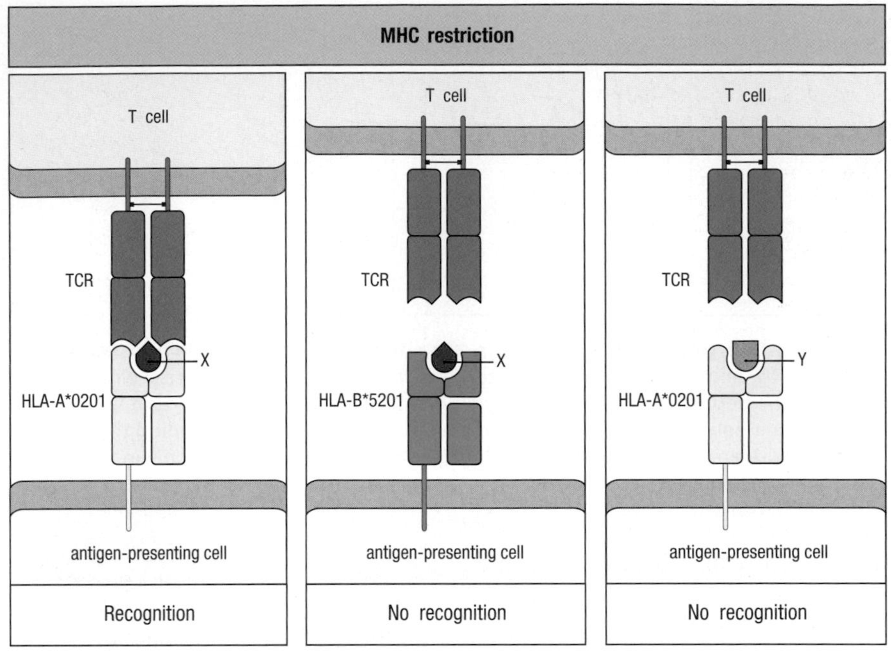

Figure 5.35 T-cell recognition of antigens is MHC restricted. The receptor of the CD8 T cell shown in the left panel is specific for the complex of peptide X with the class I molecule HLA-A*02:01. Because of this co-recognition, which is called MHC restriction, the T-cell receptor (TCR) does not recognize the same peptide when it is bound to a different class I molecule, HLA-B*52:01 (middle panel). Nor does the T-cell receptor recognize the complex of HLA-A*02:01 with a different peptide, Y (right panel). X is HIV-1 Nef residues 190–198, AFHHVAR. Y is influenza A matrix protein residues 58–68, GILGFVFTL.

The number of peptide-binding motifs is limited and so MHC allotypes that differ by only a few amino acids often bind overlapping populations of peptides. To a rough approximation, the greater the difference in sequence between two MHC allotypes, the more different will be the sets of peptides they bind.

In the complex of peptide bound to an MHC molecule, the anchor residues are buried and inaccessible to the T-cell receptor. In contrast, the peptide's other positions, which are occupied by a much greater diversity of amino acids, are available for contact with T-cell receptors. These form part of the planar surface that interacts with the T-cell receptor and which also includes variable residues on the upper surfaces of the α helices of the MHC molecule. Any given T-cell receptor is therefore specific for the complex of a particular peptide bound to a particular MHC molecule. This basic principle of T-cell biology is known as **MHC restriction**, because the antigen-specific T-cell response is restricted to a particular MHC type. As a consequence of MHC restriction, a T cell that responds to a peptide presented by one MHC allotype will not respond to another peptide bound by that same MHC allotype or to the same peptide bound to another MHC allotype (Figure 5.35).

5-22 MHC diversity results from selection by infectious disease

We have just seen that the amino acid substitutions that distinguish different MHC isoforms are concentrated at sites that affect peptide binding and presentation. Further nonrandomness is present in the gene sequences, where the frequency of nucleotide substitutions that result in a change of amino acid is much greater than would be generated by chance. The inescapable conclusion is that MHC diversity is due to natural selection, and because of the immunological functions of MHC molecules the likely sources of the selection are the infections caused by pathogens.

For an individual, the advantage to having multiple MHC class I and class II genes is that they contribute different peptide-binding specificities, allowing a greater number of pathogen-derived peptides to be presented during any infection. This improves the strength of the immune response against the

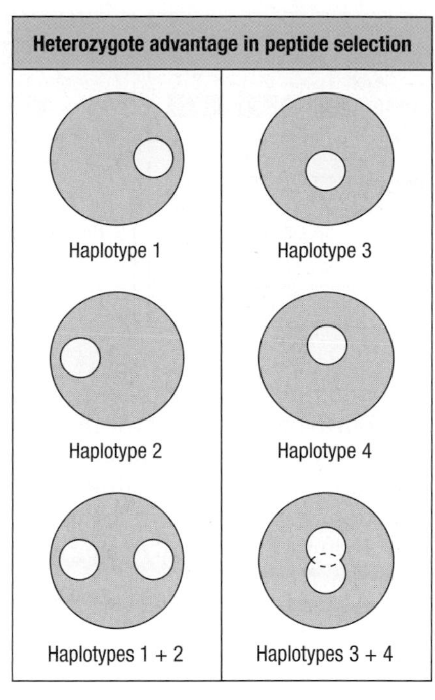

Figure 5.36 The advantage of being heterozygous at the MHC. The large gray circles represent the total number of antigenic peptides derived from a pathogen that can be presented by human MHC class I and II molecules. The small yellow circles represent the subpopulation of peptides that can be presented by the MHC class I and II molecules encoded by the genes of particular MHC haplotypes. These subpopulations differ between the haplotypes. In general, heterozygous individuals (that is, those with haplotypes 1 + 2 and haplotypes 3 + 4 in our example), will have a set of MHC class I and II molecules able to present a wider range of pathogen-derived peptides than a homozygote. The extent of this benefit varies, however. A person who has the divergent haplotypes 1 and 2 will, on average, present a larger number of different peptides from any pathogen than a person who has the more closely related haplotypes 3 and 4.

pathogen by increasing the number of activated pathogen-specific T cells. The same argument applies to polymorphism of an MHC gene, the advantage to the heterozygote being that two different peptide-binding specificities can be brought into play compared with one for the homozygote (Figure 5.36). Moreover, the high degree of polymorphism of the antigen-presenting HLA isotypes ensures that most members of the population are heterozygotes. The advantage conferred by heterozygosity will of course vary depending on the peptide-binding specificities of the two allotypes (see Figure 5.36). All these selective processes act to maintain a variety of MHC isoforms in the population, which is described as **balancing selection** (Figure 5.37, first two panels).

A different mode of selection favors certain MHC alleles, or combinations of alleles, at the expense of others and is imposed by specific, epidemic disease. Here, presentation of particular pathogen-derived peptides by particular MHC allotypes is advantageous and in the extreme case makes the difference between life and death. As a consequence, the selected alleles are driven to

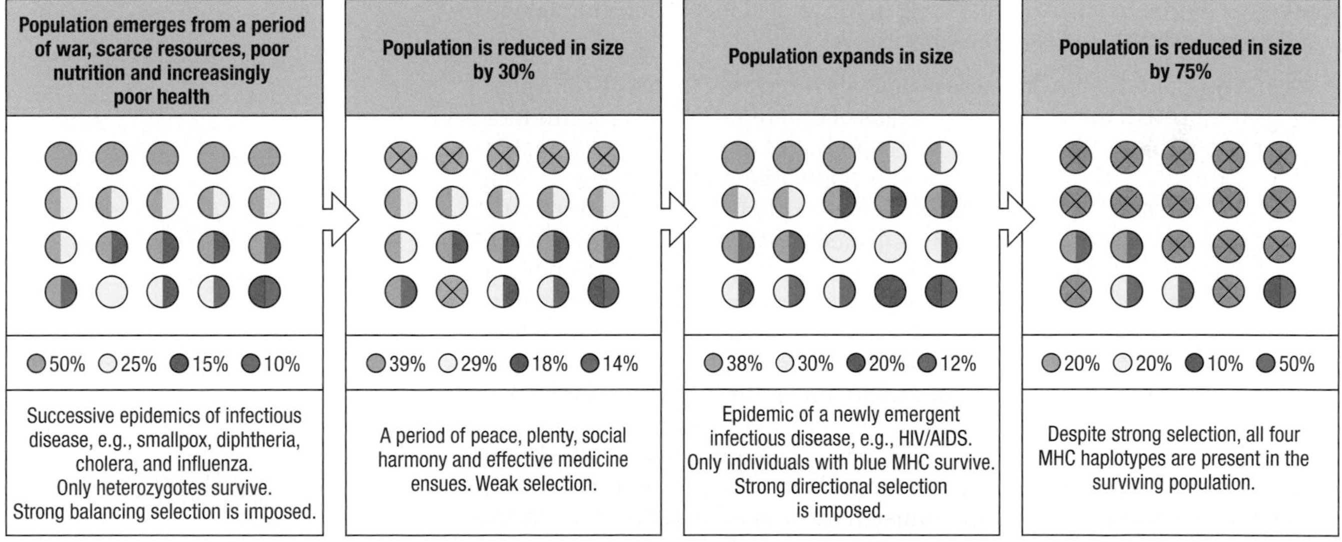

Figure 5.37 Exposure to pathogens selects for MHC polymorphism. In our example population there are four different MHC haplotypes, each represented by a different color. The frequencies of different genotypes are represented by the 20 circles in each panel, the frequencies of the four haplotypes being given underneath. First panel: the population experiences a period characterized by balancing selection arising from successive epidemic infections, after which only heterozygotes survive (second panel) and in which 30% of the population die, as indicated by circles containing an X. After recovery of the population during a period of relative calm and health (third panel), it becomes subject to directional selection by a new and particularly nasty infection. Only individuals with the blue MHC haplotype survive and 75% of the population dies (fourth panel). As a result of these selections the frequencies of the MHC haplotypes change considerably, but all four MHC haplotypes are retained within the population.

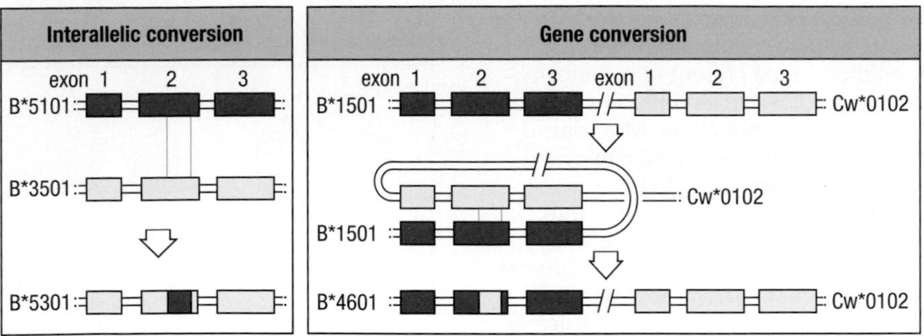

Figure 5.38 New MHC alleles are generated by interallelic conversion or gene conversion. Recombination between alleles of the same gene (HLA-B*51:01 and HLA-B*35:01), as shown on the left, and between alleles of different genes (HLA-B*15:01 and HLA-Cw*01:02), as shown on the right, can both result in the formation of a new allele in which a small block of DNA sequence has been replaced. B*53:01 is characteristic of African populations and has been associated with resistance to severe malaria. B*46:01 is specific to southeast Asian populations and has been associated with susceptibility to nasopharyngeal carcinoma.

higher frequency while other alleles decrease in frequency (Figure 5.37, third and fourth panels). This type of selection disrupts the balance and is therefore called **directional selection.** The numerous HLA differences between human populations of different ethnicity and geographical origins are evidence for directional selection. Only a minority of HLA alleles are common to all human populations, the majority being of recent origin and specific to particular ethnic groups.

Because pathogens adapt to the MHC of their host populations, it has been argued that rare, recently formed MHC alleles to which the pathogen has not adapted will more probably confer advantage on the host and be selected during disease epidemics. New variants of HLA class I and II alleles arise through point mutation and several types of recombination, which can involve alleles of the same gene or alleles of two different genes in the same family (Figure 5.38). Particularly favored seem to be new HLA alleles in which a small segment of one allele has been replaced by the homologous section of another, with the introduction of several amino acid substitutions that change contact residues in the peptide-binding groove (see Figure 5.33). The recombination mechanism that produces such variants has been termed **interallelic conversion** or **segmental exchange** (Figure 5.38, left panel). Selection for new alleles of this type has been particularly strong in the HLA-B locus in the indigenous populations of South and Central America.

In large city populations, where most clinical HLA typing for tissue transplantation is performed, there can be hundreds of different HLA alleles. This large number is not the result of natural selection but of the way that city populations are formed, by the mixture of various populations who migrated in from somewhere else and brought their different sets of HLA alleles with them. A better assessment of the amount of HLA diversity that can be sustained by natural selection and that provides for the long-term survival of human populations is obtained from the study of indigenous peoples. Because indigenous American populations are descended from small migrant populations from Siberia, they have less genetic diversity throughout the genome than other human populations. Despite this, they maintain around six alleles for HLA-A, -B, -C, and DRB1 (Figure 5.39), sufficient for a good majority of the population to be heterozygotes and for there to be a variety of heterozygote combinations, which reduces the probability that the population will succumb to any particular infection. Conversely, when human populations lose HLA diversity to a level significantly below that shown in Figure 5.39, their chance of survival severely diminishes unless they merge with another population.

In industrialized countries, the current epidemic of HIV infection provides a unique opportunity to study the effects of HLA polymorphism on an infectious disease and vice versa. In addition to a general advantage of HLA heterozygosity (Figure 5.40), certain families of alleles (HLA-B14, B27, B57, HLA-C8, and C14) are associated with slow progression of the disease, whereas others are associated with rapid progression (HLA-A29, HLA-B22, B35, HLA-C16, and HLA-DR11). Almost all the correlations are with HLA class I; this is consistent

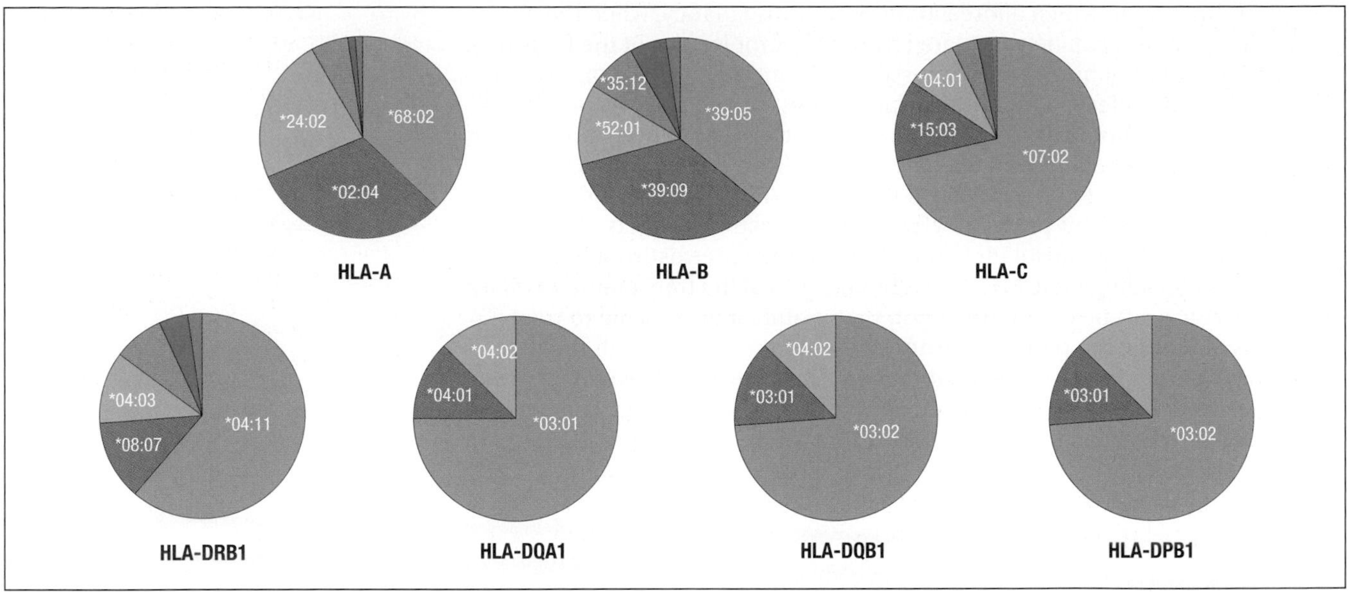

Figure 5.39 Human populations must maintain a diversity of HLA class I and II allotypes if they are to survive. Each pie chart corresponds to one of the polymorphic HLA class I or II genes and shows the number of alleles present in the Yucpa population of South Amerindians and their relative frequencies. For each gene, the most common allele is colored blue, the second most common allele brown, and the third most common allele green. Similarly, different colors denote the fourth, fifth, and sixth most common alleles. In surviving many epidemics of infectious disease and population bottleneck the Yucpa have maintained polymorphism at all of these seven HLA genes.

with the principal mechanism for controlling the infection being killing of virus-infected cells by cytotoxic CD8 T cells.

5-23 MHC polymorphism triggers T-cell reactions that can reject transplanted organs

During T-cell development, any cells having T-cell receptors that respond to complexes of self peptide and MHC class I and class II molecules at healthy cell surfaces are eliminated. This quality-control mechanism, which prevents a person's T cells from attacking their own healthy tissue and causing disease, encompasses only the MHC isoforms expressed by that person and not other MHC isoforms. In this context, the **self-MHC** isoforms are described as **autologous**, and all other MHC isoforms are described as **allogeneic**. Because of the diversity of receptors generated during T-cell development, in every person's circulation there are T cells that can respond to complexes of peptide and allogeneic MHC class I and II molecules, and so will react against the cells of another individual. T cells with this property are called **alloreactive T cells**, and those reactive against any given allogeneic cell constitute between 1% and 10% of circulating T cells.

When allogeneic kidneys are transplanted to patients with renal failure, the major danger is that the kidney graft will be rejected by the patient's immune system. One cause of rejection is that alloreactive T cells in the patient's circulation are activated by allogeneic HLA molecules in the graft, leading to an **alloreaction**, a powerful T-cell response that attacks the graft. To reduce the probability of graft rejection, donors may be selected who have combinations of HLA alleles identical or similar to that of the patient. The combination of HLA alleles that a person has is called their **HLA type**. Immunosuppressive drugs are also used to preempt the alloreactive T-cell response and to treat rejection when it occurs.

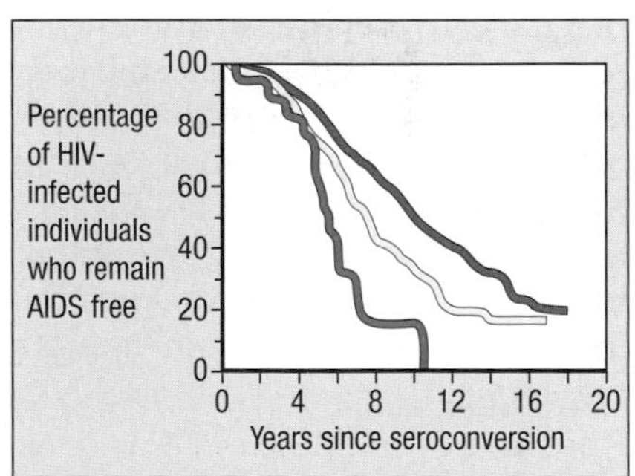

Figure 5.40 MHC heterozygosity delays the progression to AIDS in people infected with HIV-1. When people who have been infected with HIV-1 start to make detectable antibodies against the virus they are said to have undergone seroconversion. The onset of overt symptoms of AIDS occurs years after seroconversion. The rate of progress to AIDS decreases with the extent of HLA heterozygosity, as shown here by a comparison of individuals who are heterozygous for all the highly polymorphic HLA class I and II loci (red) with those who are homozygous for one locus (yellow) or for two or three loci (blue).

A natural situation in which alloreactions occur is pregnancy, when the mother's immune system can be stimulated by the HLA molecules of the fetus that derive from the father but are not expressed on the mother's cells. This response leads to alloantibodies in the mother's circulation with specificity for paternal MHC molecules. **Alloantibody** is the name given to any antibody raised in one member of a species against an allotypic protein from another member of the same species. Although of no harm to the fetus, which is protected, the alloantibodies produced by pregnancy can have disastrous effects should the mother need a kidney transplant in the future. If these preexisting alloantibodies react with the allogeneic MHC class I molecules of the transplanted kidney cells, they cause a type of graft rejection that is almost impossible to treat. To avoid this outcome, the patient's serum is tested for reactivity with the potential donor's leukocytes, and transplantation is undertaken only when the reaction is acceptably low. Before molecular genetic methods were available, the HLA types of transplant recipients and donors were determined using the anti-HLA class I and class II alloantibodies present in sera obtained from women who had had several children.

Summary

In humans, the highly polymorphic MHC class I and II genes are closely linked in the HLA region on chromosome 6, which comprises the human MHC. In contrast to immunoglobulin and T-cell receptor genes, MHC class I and class II genes have a conventional organization and do not rearrange. In humans, the MHC class I genes encode the heavy (α) chains of three different class I molecules—HLA-A, HLA-B, and HLA-C—whereas the MHC class II genes encode the α and β chains of three different MHC class II molecules—HLA-DP, HLA-DQ, and HLA-DR. β_2-Microglobulin, the light chain of MHC class I molecules, is encoded outside the MHC, on chromosome 15. Certain MHC class I and class II genes are highly polymorphic; some have several thousand alleles worldwide. Genes encoding other proteins involved in antigen presentation are located in the MHC and, like the MHC class I and II genes, their expression is regulated by the interferons produced during an immune response. The strategy used by MHC molecules to bind diverse antigens contrasts with that of the T-cell receptors. MHC molecules have highly promiscuous binding sites for peptides; an MHC molecule is therefore usually able to present a diversity of peptide antigens to a large number of T-cell receptors with highly specific binding sites. Polymorphism in families of MHC class I heavy-chain genes and MHC class II α-chain and β-chain genes is a secondary strategy that serves to increase the breadth and strength of T-cell immunity. It also diversifies T-cell immunity within human populations; this helps them survive epidemic disease, but is also the main immunological barrier to clinical organ transplantation.

Summary to Chapter 5

The general structure of T-cell antigen receptors resembles that of the membrane-bound immunoglobulins of B cells and they are encoded by similarly organized genes that undergo gene rearrangement before they are expressed. As with B cells, this gives rise to a population of T cells each expressing a unique receptor. Differences between immunoglobulins and T-cell receptors reflect the fact that the T-cell receptor is used only as a membrane-bound receptor, whereas immunoglobulins are also used as secreted effector molecules.

T-cell receptors are more limited than immunoglobulins in the antigens they bind, recognizing only short peptides bound to MHC molecules on a cell surface. The two main classes of T cell—CD8 and CD4—are specialized to respond to intracellular and extracellular pathogens, respectively. They respond to an antigen from the appropriate source by specific interactions between the MHC molecules displaying the peptide antigen and the CD4 or CD8 glycoprotein on the T-cell surface. Processing of extracellular and intracellular pathogen antigens into peptides, and their binding to MHC molecules, occur inside the cells of the infected host. Cytotoxic CD8 T cells recognize peptides presented by MHC class I molecules, and their function is to destroy the antigen-presenting cell. Most cells express MHC class I molecules and so can present pathogen-derived peptides to CD8 T cells if infected with a virus or other pathogen that penetrates the cytosol. In contrast, the function of CD4 T cells is to interact with and help activate a limited set of immune effector cells, principally macrophages and B cells. CD4 T cells recognize peptides presented by MHC class II molecules, which are usually expressed only on specialized cells that take up and process material from the extracellular environment, and which include macrophages and B cells.

MHC molecules have promiscuous peptide-binding sites, a feature that enables the relatively small number of different MHC molecules present in each individual to bind peptides of many different sequences. The diversity of peptides that can be presented by the human population as a whole is further increased by the highly polymorphic nature of MHC class I and II genes. Each person differs from almost all other individuals in some or all of the MHC alleles that they possess. Thus, the population is able to respond to the pathogens it encounters with a wide diversity of individual immune responses. Because of the polymorphism of the MHC, organs or tissues transplanted between individuals of different MHC type provoke strong T-cell responses directed against the 'foreign' MHC molecules.

Questions

5–1 In which of the following ways are T-cell receptors distinct from immunoglobulins? (Select all that apply.)

a. T-cell receptors are generated through somatic recombination.
b. The T-cell repertoire encompasses a very large degree of diversity.
c. T-cell receptors are never secreted subsequent to antigen encounter.
d. The variable region of T-cell receptors contains complementarity-determining regions that interact with antigen.
e. T-cell receptors are used solely for the purpose of antigen recognition and not for effector function.

5–2 Match the T-cell component in Column A with its description in Column B. Use each description only once.

Column A	Column B
a. four extracellular domains	1. regions exhibiting the highest amount of diversity between T-cell receptors
b. T-cell receptor complex	2. resembles a Fab fragment of IgG
c. complementarity-determining regions (CDRs)	3. T-cell receptor, CD3 complex plus ζ chains
d. V_α and V_β domains	4. found on small subset of T-cell population
e. γ:δ T-cell receptor	5. membrane-distal loops forming antigen-binding site

5–3 All of the following have been observed in individuals deficient in either CD3δ or CD3ε except

a. their T-cell receptors fail to signal effectively
b. there is an increase in the number of γ:δ T cells in the circulation
c. their T-cell receptors are not transported effectively to the cell surface
d. their T cells express a lower number of T-cell receptors on the cell surface
e. they are immunodeficient.

5–4 A primary difference between how B cells recognize antigen and how T cells recognize antigen is that

a. T-cell receptors can bind antigen only after secretion of the T-cell receptor from the surface of the T cell
b. antibodies can bind only to denatured proteins
c. T-cell receptors can bind to carbohydrate groups or clusters of amino acids
d. B cells recognize degraded proteins bound to major histocompatibility molecules
e. T cells recognize degraded proteins bound to major histocompatibility molecules.

5–5 Which of the following statements regarding CD4 is incorrect?

a. MHC class II molecules present antigens to CD4 T cells.
b. CD4 is the receptor used for HIV entry into CD4 T cells.
c. CD4 is made up of two separate membrane-bound chains.
d. Late in the progression of an HIV infection, the number of CD4 T cells in the circulation diminishes.
e. CD4 is referred to as a T-cell co-receptor.

5–6 Match the term in Column A with its description in Column B.

Column A	Column B
a. invariant chain	1. presentation of extracellular sources of antigen by class I MHC molecules
b. cross-presentation	2. present peptides to CD8 T cells
c. MHC class I	3. one of the soluble MHC-associated chains that is highly conserved
d. MHC class II	4. transports MHC molecules to endocytic vesicles
e. β_2-microglobulin	5. present peptides to CD4 T cells

5–7 In regard to antigen presentation, MHC class I molecules usually present peptides derived from _____, whereas MHC class II molecules usually present peptides derived from _____.

a. intracellular cytosolic sources; vesicular system
b. phagolysosome; proteasomes
c. MIIC; self proteins
d. CLIP; HLA-DM
e. endocytic vesicles; endoplasmic reticulum.

5–8 Match the molecule or domains of molecules in Column A to what it binds in Column B.

Column A	Column B
a. α_1 and β_1 domains	1. MHC class I-associated peptide
b. β_2-microglobulin	2. CD4
c. CLIP	3. MHC class II molecules in MIIC
d. β_2 domain of MHC class II	4. MHC class I α chain
e. α_1 and α_2 domains of MHC class I molecules	5. MHC class II-associated peptide
f. α_3 domain of MHC class I molecules	6. CD8

5–9 Identify which of the following statements regarding peptide–MHC molecule interactions are correct. (Select all that apply.)

a. Peptides with different amino acid sequences may be able to bind to the same type of MHC molecule.
b. Covalent bonds hold the peptide in the groove of the MHC molecule.
c. The length of peptides bound by MHC class I molecules is shorter than that of those bound by MHC class II molecules.
d. The groove of the MHC molecule is deep enough to accommodate two or more peptides.
e. Binding pockets of MHC molecules anchor the side chains of only certain amino acids of the peptide.
f. The amino- and carboxy-terminal amino acids of peptides are used for binding to MHC class I molecules, whereas amino acids along the length of the peptide are used for binding to MHC class II molecules.
g. Only non-self peptides form stable interactions with MHC molecules.
h. The type of MHC molecule presenting a non-self protein is informative regarding whether the pathogen originated from an intracellular or extracellular source.

5–10 MHC class I molecules present peptide antigens derived from a(n) ______ compartment, whereas MHC class II molecules present peptide antigens derived from a(n) ______ compartment:

a. extracellular; intracellular
b. intracellular; extracellular
c. opsonization; neutralization
d. neutralization; opsonization
e. none of the above.

5–11 Amino acid variation among MHC class II allotypes that present antigens to CD4 T cells is concentrated

a. where MHC contacts the co-receptors CD4 or CD8
b. in the β chain, because the α chain is monomorphic
c. where the MHC molecule contacts peptide and the T-cell receptor
d. in the α chain, because the β chain is monomorphic
e. throughout both the α and β chains in all domains.

5–12 The T-cell co-receptor CD4 interacts with ______ bound to the surface of ______:

a. MHC class I; antigen-presenting cells
b. MHC class I; T cells
c. MHC class II; antigen-presenting cells
d. MHC class II; T cells
e. none of the above.

5–13 The high degree of polymorphism in MHC class I molecules that present antigens to CD8 T cells is found in ______ because ______ is/are monomorphic:

a. β_2-microglobulin; the heavy chain
b. both the α and β chains; none
c. HLA-DOβ; HLA-DOα
d. the heavy chain; β_2-microglobulin
e. HLA-E and HLA-G; HLA-F.

5–14 At 5 months of age, Christina Kitchenman was admitted to the hospital with a fever and a severe non-productive cough that was subsequently determined by immunofluorescence staining to be pneumocystosis, a form of pneumonia caused by an opportunistic yeast-like fungus. Her CD4 T-cell count (220 μl^{-1}) was much lower than expected, at only one-third of her CD8 T-cell count (650 μl^{-1}). Her B-cell count was slightly higher than normal. An immunodeficiency was suspected, so T-lymphocyte functionality tests were carried out, proving that Christina's T cells did not respond to a specific antigen stimulus involving tetanus toxoid as a recall antigen, even though she had received routine vaccination for tetanus (DPT vaccine) several months previously. Normal T-cell proliferation responses, however, were detected upon exposure of peripheral blood mononuclear cells to either the plant lectin mitogen phytohemagglutinin or to allogeneic B lymphocytes. Further tests revealed hypogammaglobulinemia, a deficiency of all MHC class II isotypes on her white blood cells, but no deficiency of MHC class I isotypes. Christina was successfully treated with pentamidine and received a bone marrow transplant without complication after being diagnosed with bare lymphocyte syndrome, type II (BLSII). A genetic defect in which of the following would support this diagnosis?

a. HLA-DQ
b. CIITA (MHC class II transactivator)
c. RAG-1
d. TAP-1
e. CD3ε.

5–15 Brittany Hudson, 16 years of age, was seen by her physician after the development of a small pustule around her nostrils that had expanded and was now showing signs of ulceration typical of chronic granulomatous inflammation. In the previous year, Brittany had experienced a similar lesion on her left thigh that healed slowly, leaving a hyperpigmented scar. She also had a history of chronic bacterial infections of the upper and lower respiratory tract. Flow-cytometric analysis of peripheral blood revealed abnormally low numbers of MHC class I molecules on cell surfaces and abnormally low numbers of CD8 T cells. A diagnosis of type I bare lymphocyte syndrome was made. A deficiency in which of the following would explain this etiology?

a. HLA-DM
b. invariant chain
c. class II-associated invariant chain peptide (CLIP)
d. TAP-1 or TAP-2
e. MHC class II transactivator (CIITA).

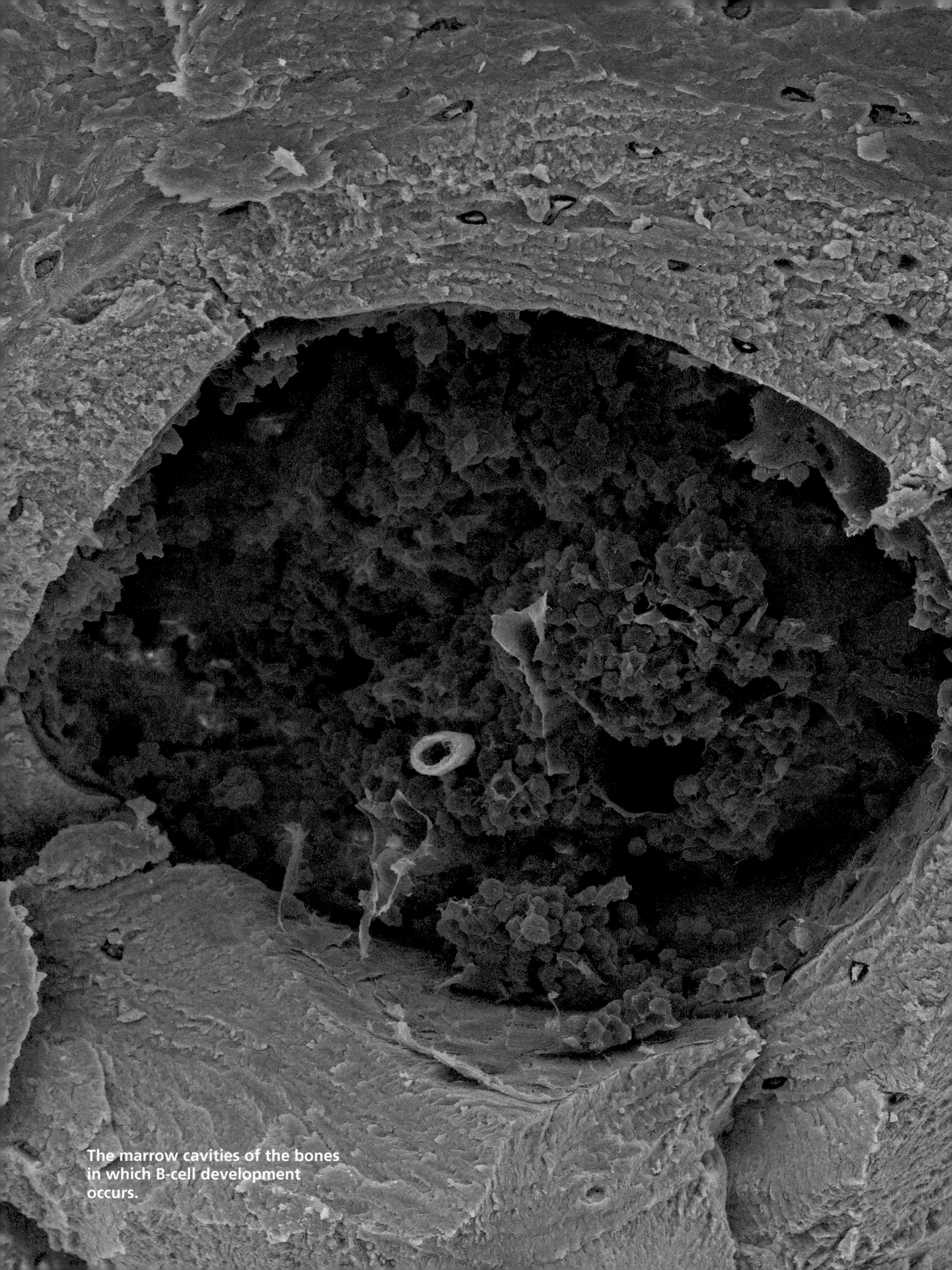

The marrow cavities of the bones in which B-cell development occurs.

Chapter 6

The Development of B Lymphocytes

The B cells of the human immune system have the capacity to make immunoglobulins specific for almost every nuance of chemical structure, which gives each person the potential to make antibodies against all the infectious microorganisms that could possibly be encountered in a lifetime. The body does not, however, stockpile all the B cells needed to do this. That would probably mean devoting the vast majority of the body's resources to the immune system, leaving little for the immune system to protect. Instead, the body carries a less complete inventory of B cells, but one that expands and contracts its individual clones according to need and circumstance. Fueling this system are stem cells in the bone marrow, which generate more than 60 billion new B cells every day of your life.

The development of B cells from bone marrow stem cells through to antibody-producing plasma cells is the subject of this chapter. B-cell development can be divided into six functionally distinct phases (Figure 6.1). The first part of the chapter covers phase 1 of development, in which B-cell precursors in the bone marrow acquire functional antigen receptors through the immunoglobulin gene rearrangements described in Chapter 4. Although each mature B cell expresses immunoglobulin of just one antigen specificity as its B-cell receptor, the B-cell population as a whole represents a vast repertoire of immunoglobulins with different binding specificities. The second part of this chapter describes how this repertoire is modified in various ways in phases 2–6 as B cells mature and travel from the bone marrow to the secondary lymphoid organs, where they can be activated by antigen to function in the body's defense.

Phase 2 of B-cell development is one of negative selection, which prevents the emergence of mature B cells bearing receptors that bind to normal constituents of the human body. Such cells are potentially dangerous because of their potential to attack healthy tissue and cause autoimmune disease. Negative selection begins in the bone marrow and continues as the immature B cells

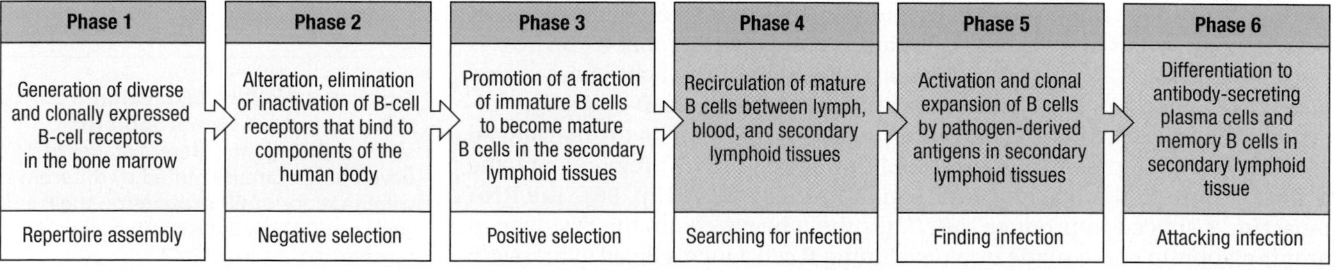

Phase 1
Generation of diverse and clonally expressed B-cell receptors in the bone marrow
Repertoire assembly
Phase 2
Alteration, elimination or inactivation of B-cell receptors that bind to components of the human body
Negative selection
Phase 3
Promotion of a fraction of immature B cells to become mature B cells in the secondary lymphoid tissues
Positive selection
Phase 4
Recirculation of mature B cells between lymph, blood, and secondary lymphoid tissues
Searching for infection
Phase 5
Activation and clonal expansion of B cells by pathogen-derived antigens in secondary lymphoid tissues
Finding infection
Phase 6
Differentiation to antibody-secreting plasma cells and memory B cells in secondary lymphoid tissue
Attacking infection

Figure 6.1 The development of B cells can be divided into six functionally distinct phases. The first three phases (yellow boxes) correspond to development in the bone marrow; the last three phases (pink boxes) correspond to development in the secondary lymphoid tissues.

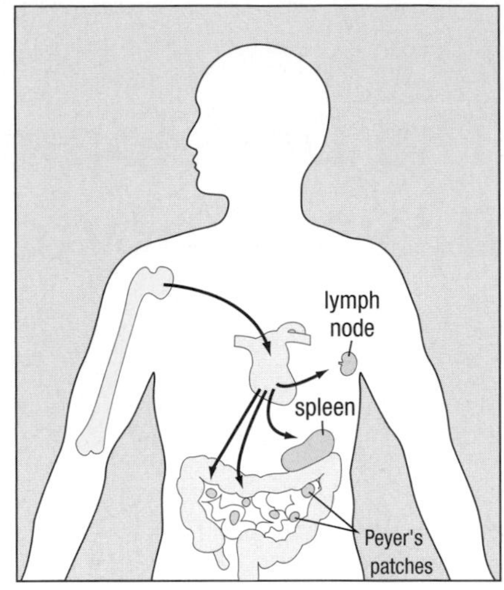

Figure 6.2 B cells develop in bone marrow and then migrate to secondary lymphoid tissues. B cells leaving the bone marrow (yellow) are carried in the blood to lymph nodes, the spleen, Peyer's patches (all shown in red), and other secondary lymphoid tissues such as those lining the respiratory tract (not shown).

leave the bone marrow and travel to the secondary lymphoid organs. Phase 3 is one of positive selection, in which the immature B cells compete for the limited number of sites in the follicles of the secondary lymphoid tissues where they must go to complete their maturation (Figure 6.2). In phase 4, the mature B cells travel between secondary lymphoid tissues in the lymph and blood, patrolling for infections and the pathogen-derived antigens to which the B-cell receptors bind. Activation of B cells by antigen in phase 5 leads to the proliferation and clonal expansion of the antigen-specific B cells. In phase 6, differentiation and diversification of B cells within each expanded clone gives rise to plasma cells that provide antibodies to attack ongoing infection and memory B cells for the speedy elimination of any future infections by the same pathogen. At every phase in B-cell development, significant numbers of cells are, for various reasons, prevented from advancing to the next stage. Overall, the efficiency of B-cell development is very low, but that is the price paid for having an almost infinite capacity to make antibodies against any possible pathogen.

The development of B cells in the bone marrow

The sole purpose of a B cell is to make an immunoglobulin. Consequently, B-cell development in the bone marrow can be divided into stages that correspond to successive steps in the rearrangement and expression of the immunoglobulin genes. The quality of the gene rearrangements and the proteins they make are assessed at two key checkpoints, and further development is halted if they are found wanting. In this part of the chapter we shall see how gene rearrangements are ordered and controlled to produce an immature B cell that makes only one type of heavy chain and one type of light chain, and thus expresses immunoglobulin of a single antigen specificity on its surface.

6-1 B-cell development in the bone marrow proceeds through several stages

In the bone marrow, pluripotent hematopoietic stem cells give rise to the common lymphoid progenitor cells, which have the potential to produce both B cells and T cells (see Figure 1.13, p. 14). Some of these cells develop into precursor cells that are committed to becoming B cells (Figure 6.3). All these undifferentiated precursor cells can be distinguished by cell-surface markers. One of these is the protein CD34, which is present on all human hematopoietic stem cells and is exploited medically, through the use of anti-CD34 monoclonal antibodies, to separate hematopoietic stem cells from the other bone marrow cells for use in therapeutic transplantation.

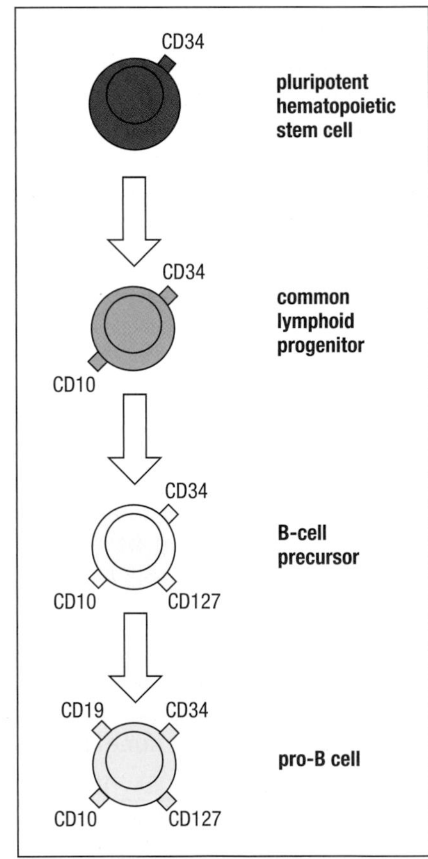

Figure 6.3 Pro-B cells develop from the pluripotent hematopoietic stem cell. Cells at different stages of development are identified by different combinations of CD proteins on their surface. CD127 is the α chain of the receptor for interleukin-7.

The earliest identifiable cells of the B-cell lineage are called **pro-B cells** (see Figure 6.3). These progenitor cells retain a limited capacity for self-renewal, dividing to produce more pro-B cells and also cells that go on to develop further. The main event in the pro-B-cell stage is the rearrangement of the heavy-chain genes, which always precedes rearrangement of the light-chain genes. Joining of D_H and J_H gene segments occurs at the **early pro-B-cell** stage, followed at the **late pro-B-cell** stage by joining of a V_H segment to the rearranged DJ_H. The rearranged gene is transcribed through to the μ C-region gene, the nearest C gene to the rearranged V region (see Figure 4.21, p. 96). The RNA transcript is spliced to produce mRNA for the μ heavy chain, the first type of immunoglobulin chain made by a developing B cell. Once a B cell expresses a

	Stem cell	Early pro-B cell	Late pro-B cell	Large pre-B cell	Small pre-B cell	Immature B cell
H-chain genes	Germline	D–J rearrangement	V–DJ rearrangement	VDJ rearranged	VDJ rearranged	VDJ rearranged
L-chain genes	Germline	Germline	Germline	Germline	V–J rearranging	VJ rearranged
Ig status	None	None	None	μ heavy chain is made	μ chain in endoplasmic reticulum	μ heavy chain. λ or κ light chain. IgM on surface.

Figure 6.4 The development of B cells in the bone marrow proceeds through stages defined by the rearrangement and expression of the immunoglobulin genes. In the stem cell, the immunoglobulin (Ig) genes are in the germline configuration. The first rearrangements are of the heavy-chain (H-chain) genes. Joining D_H to J_H defines the early pro-B cell, which becomes a late pro-B cell on joining V_H to DJ_H. Expression of a functional μ chain defines the large pre-B cell. Large pre-B cells proliferate, producing small pre-B cells in which rearrangement of the light-chain (L-chain) gene occurs. Successful light-chain gene rearrangement and expression of IgM on the cell surface define the immature B cell.

μ chain it is known as a **pre-B cell**. Pre-B cells represent two stages in B-cell development: the less mature **large pre-B cells** and the more mature **small pre-B cells** (Figure 6.4). Large pre-B cells have successfully rearranged a heavy-chain gene and make a μ heavy chain; they have stopped rearrangements of the heavy-chain genes but have yet to commence rearrangement of the light-chain genes.

Rearrangement of the light-chain genes occurs in the small pre-B cells. The κ light-chain genes are the first to rearrange, and only if those rearrangements fail to make a viable κ chain are the λ light-chain genes rearranged. When successful joining of light-chain V and J segments is achieved, a light-chain protein is synthesized and assembled in the endoplasmic reticulum with the μ chains to form membrane-bound IgM. The IgM further associates with Igα and Igβ to form a functional B-cell receptor complex, which is then transported to the cell surface (see Section 4-12). Rearrangement of the light-chain genes stops and the small pre-B cell becomes an **immature B cell**.

6-2 B-cell development is stimulated by bone marrow stromal cells

B-cell development in the bone marrow is dependent on a network of non-lymphoid **stromal cells**, which provide specialized microenvironments for B cells at various stages of maturation (Figure 6.5). Stroma is the name given to the supportive cells and connective tissue in any organ. The stromal cells perform

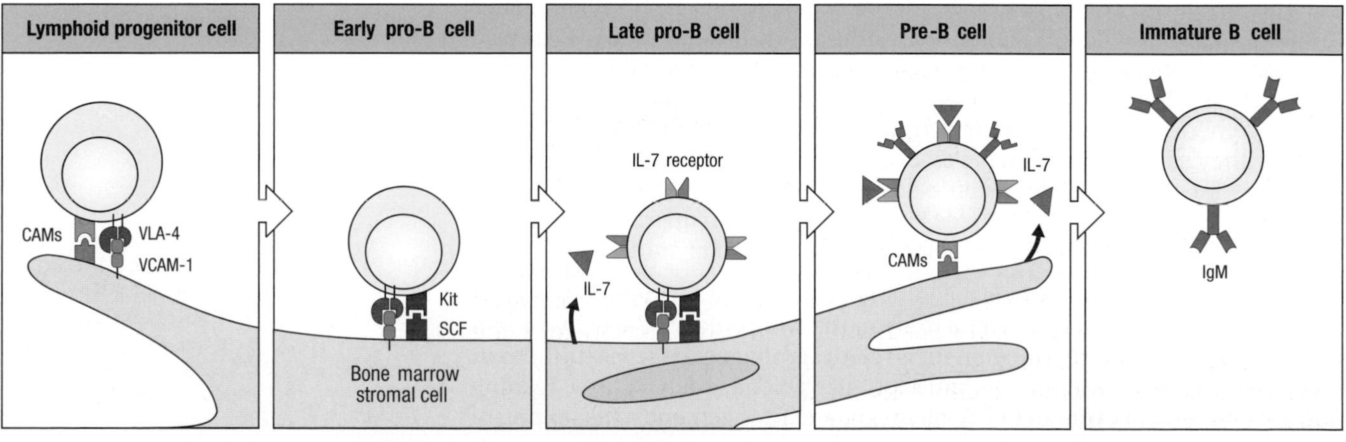

Figure 6.5 The early stages of B-cell development are dependent on bone marrow stromal cells. The panels show the interactions of developing B cells with bone marrow stromal cells. Stem cells and early pro-B cells use the integrin VLA-4 to bind to the adhesion molecule VCAM-1 on stromal cells. This and interactions between other cell-adhesion molecules (CAMs) promote the binding of the receptor Kit on the B cell to stem-cell factor (SCF) on the stromal cell. Activation of Kit causes the B cell to proliferate. B cells at a later stage of maturation require interleukin-7 (IL-7) to stimulate their growth and proliferation.

two distinct functions. First, they make specific contacts with the developing B cells through the interaction of adhesion molecules and their ligands. Second, they produce growth factors that act on the attached B cells, for example the membrane-bound stem-cell factor (SCF), which is recognized by a receptor called Kit on maturing B cells. Another important growth factor for B-cell development is interleukin-7 (IL-7), a cytokine secreted by stromal cells that acts on late pro-B and pre-B cells.

The most immature stem cells lie in a region of the bone marrow called the subendosteum, which is adjacent to the interior surface of the bone. As the B cells mature, they move, while maintaining contact with stromal cells, to the central axis of the marrow cavity. Later stages of maturation are less dependent on contact with stromal cells, allowing the B cells eventually to leave the bone marrow. The next stages of development occur after an immature B cell leaves the bone marrow, and they take place in a secondary lymphoid organ such as a lymph node, the spleen, or a Peyer's patch. It is here that the immature B cell becomes a mature B cell that can respond to its specific antigen.

6-3 Pro-B-cell rearrangement of the heavy-chain locus is an inefficient process

Quality control is absolutely necessary in B-cell development because the process of gene rearrangement is inherently imprecise and inefficient. These complications come mainly from the random addition of N and P nucleotides, which occurs on forming the joints between V, D, and J gene segments (see Section 4-9). Such additions can change the reading frame of the DNA sequence so that it no longer encodes a functional heavy chain. Gene rearrangements that do not translate into a useful protein are called **nonproductive rearrangements**. Those rearrangements that preserve a correct reading frame and give rise to a complete and functional immunoglobulin chain are known as **productive rearrangements**. At each rearrangement event, there is only a one in three chance of maintaining the correct reading frame.

Every B cell has two copies of the immunoglobulin heavy-chain locus, which increases the likelihood that a pro-B cell will make a productive heavy-chain gene rearrangement. The two copies of each locus are on homologous chromosomes; one is inherited from the mother and the other from the father. In the developing B cell, gene rearrangements can be made on both homologous chromosomes. Therefore, a B cell that makes a nonproductive rearrangement on one chromosome still has a chance of producing a heavy chain if it makes a productive rearrangement at the locus on the other homologous chromosome. If developing B cells make nonproductive rearrangements at both copies of the heavy-chain gene they lose their potential to make immunoglobulin; they do not develop further and they die in the bone marrow.

For an early pro-B cell to rearrange the immunoglobulin heavy-chain genes it must express the recombination-activating genes *RAG1* and *RAG2*, as well as other DNA-modifying enzymes needed to cut, paste, and add to the DNA (see Section 4-9). Turning on the genes that encode this machinery for gene rearrangement is a network of transcription factors, in which E2A and EBF are central. They cause expression of Pax-5, a transcription factor that switches on the genes for many proteins expressed only in B cells, including Igα and a cell-surface protein called CD19 (see Figure 6.3), which in mature B cells forms part of the B-cell co-receptor that helps the cell respond to antigen.

The first rearrangement event is the joining of a D_H gene segment to a J_H gene segment, which occurs at the same time on the two copies of the heavy-chain locus (Figure 6.6). This rearrangement is relatively efficient because the human D segments give a functional protein sequence when read in all three reading frames. The second event in heavy-chain gene rearrangement is the joining of

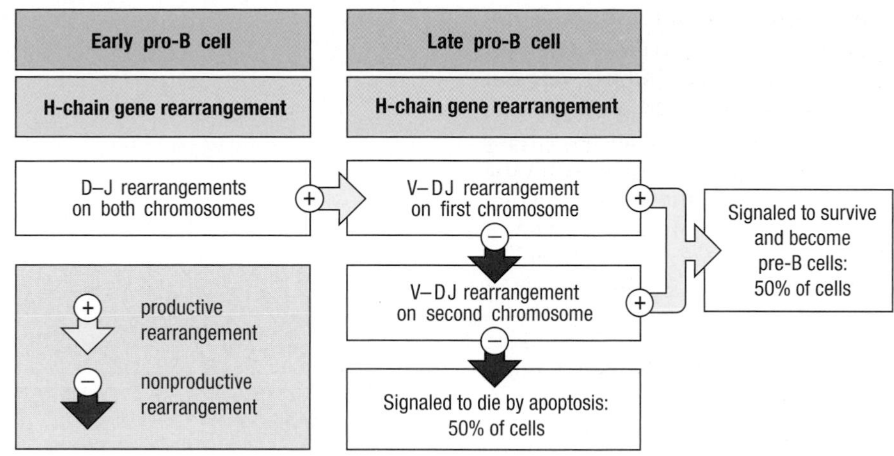

Figure 6.6 Immunoglobulin heavy-chain gene rearrangement in pro-B cells gives rise to productive and nonproductive rearrangements. A productive rearrangement enables the B cell to proceed to the next stage of development. Rearrangements occur at the H-chain genes on both chromosomes, and if neither is successful the cell dies.

a V_H gene segment to the rearranged DJ_H sequence. This rearrangement is first targeted to just one of the heavy-chain loci. Only when V_H to DJ_H rearrangement on the first chromosome is nonproductive does rearrangement proceed to the second chromosome. The two-thirds failure rate in maintaining the reading frame, and the independent chances for success offered by two chromosomes, mean that just over half the total number of pro-B cells make a functional heavy-chain gene. These cells advance along the developmental pathway to become large, dividing pre-B cells producing μ chains. Pro-B cells that fail to make a μ chain die by apoptosis in the bone marrow. The cells consigned to die include pro-B cells that make two nonproductive V_H to DJ_H rearrangements and also ones that make two nonproductive D_H to J_H rearrangements. Despite their failure, the latter cells are allowed to proceed to V_H to DJ_H rearrangement even though it might have no useful effect. A general feature of lymphocyte development is for apoptosis to be the 'default' pathway unless a positive signal for survival and further differentiation is received. Such signals are called **survival signals**.

6-4 The pre-B-cell receptor monitors the quality of immunoglobulin heavy chains

One criterion that a pro-B cell must satisfy in order to survive is that it makes a μ heavy chain. The second criterion is that the μ chain must demonstrate the ability to combine with an immunoglobulin light chain. At this stage of B-cell development there are no genuine light chains with which to make such a test, but pro-B cells synthesize two proteins, called **VpreB** and **λ5**, which bind to μ heavy chains in a manner that mimics the immunoglobulin light chain. VpreB is structurally similar to the variable region, and λ5 is like a constant region; together they form what is known as the **surrogate light chain**. The VpreB and λ5 proteins are encoded by conventional genes that do not rearrange and are separate from the immunoglobulin loci. Transcription of the VpreB and λ5 genes is controlled by the E2A and EBF transcription factors.

In the endoplasmic reticulum of the pro-B cell the μ chain forms disulfide-bonded homodimers and has the opportunity to assemble with the components of the surrogate light chain and Igβ to form a complex resembling the B-cell receptor and called the **pre-B-cell receptor** (Figure 6.7).

If the μ chain assembles a functional pre-B-cell receptor, the signals it sends through Igβ shut down gene rearrangements at the immunoglobulin heavy-chain locus and drive the pro-B cell into several rounds of cell division. This gives rise to a small clonal population of what are now large pre-B cells. On the other hand, if the μ chain in a pro-B cell assembles poorly with the surrogate

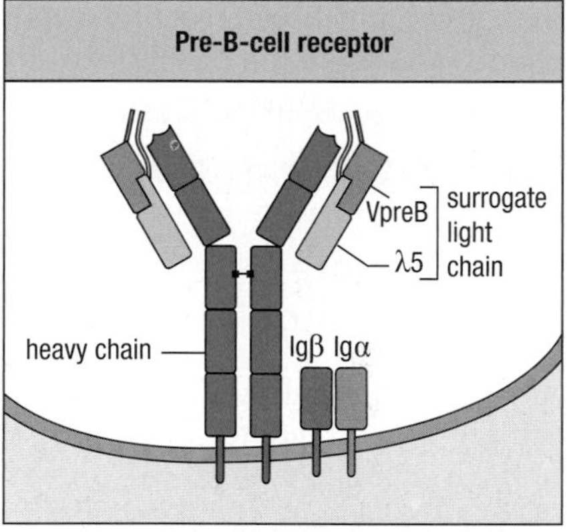

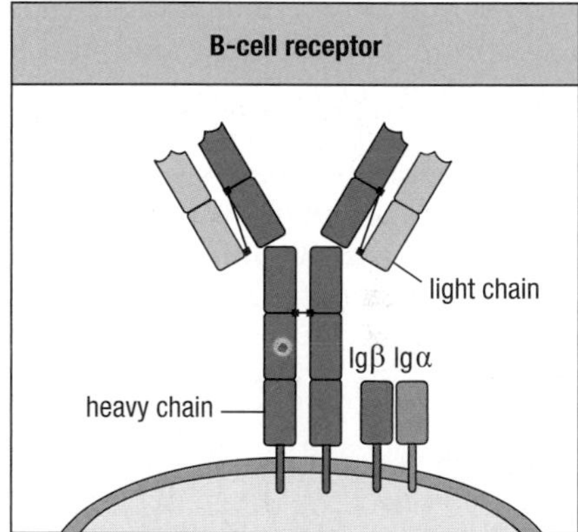

Figure 6.7 The pre-B-cell receptor resembles the B-cell receptor. Distinguishing the pre-B-cell receptor from the B-cell receptor is the absence of a κ or λ immunoglobulin light chain, and the presence instead of the surrogate light chain composed of the VpreB and λ5 polypeptides. In low abundance at the cell surface, the pre-B-cell receptor is largely retained inside the cell in membrane-enclosed vesicles, from where it generates signals that lead to the cessation of heavy-chain gene rearrangements. In addition to forming the two Ig-like domains of the surrogate light chain, VpreB and λ5 have extensions that cause oligomerization of pre-B-cell receptors and the transduction of signals necessary for pre-B-cell survival.

light chain and fails to make a functional pre-B-cell receptor, the cell is not given the signal to survive, and dies by apoptosis. Unlike the B-cell receptor (IgM), the pre-B-cell receptor does not have an antigen-binding site, nor is it well represented at the cell surface. At the C-terminal end of VpreB and the N-terminal end of λ5 are extensions beyond the Ig-like domain that have unique sequences. These extensions enable pre-B-cell receptors to form the dimers and higher oligomers that are needed to transmit the signals needed for pre-B-cell survival. They also regulate the abundance of pre-B-cell receptor at the cell surface and mediate binding of the pre-B-cell receptor to galectin-1 and heparan sulphate, potential ligands on the surface of bone marrow stromal cells.

The importance of the pre-B-cell receptor for B-cell development is highlighted by the case of a young patient who inherited defective alleles of the λ5 gene from both parents. None of the developing B cells in this child could assemble a pre-B-cell receptor, even those that were making fully functional μ chains, and so all the B cells were consigned to die by apoptosis at the pro-B-cell stage. The consequence for the patient was a profound B-cell immunodeficiency. This resulted in persistent bacterial infections, which were treated with antibiotics and intravenous immunoglobulin.

6-5 The pre-B-cell receptor causes allelic exclusion at the immunoglobulin heavy-chain locus

As well as eliminating B cells that fail to make functional μ chains, assembly of the pre-B-cell receptor prevents a B cell from making more than one functional μ chain. In a pro-B cell where rearrangement of the first immunoglobulin locus succeeds, the synthesis of μ chain and assembly of the pre-B-cell receptor quickly signals for *RAG* gene transcription to stop. It also signals for RAG proteins to be degraded and for the structure of the chromatin of the heavy-chain locus to be reorganized into a state that resists gene rearrangement. These three effects synergize to prevent rearrangement of the second immunoglobulin heavy-chain locus and production of a second μ chain. This phenomenon, whereby a cell expresses only one of its two copies of a gene, is called **allelic exclusion**. Although individual B cells express only one allele, the two alleles are expressed equally within the B-cell population.

The advantage of allelic exclusion can best be appreciated by considering the alternative. A mature B cell making two μ chains with different antigen-binding specificities in equal quantities would express three types of B-cell receptor, of which the most abundant would have two different μ chains and thus two binding sites of different antigenic specificity. These receptors would be functionally substandard because they could not make strong bivalent attachments to multivalent antigens (Figure 6.8). The problems caused by this heterogeneity would be exponentially compounded when such a B cell produced plasma cells secreting pentameric IgM (see Figure 4.27, p. 101). The antibodies would be very heterogeneous, with less than 0.1% of them containing 10 identical μ chains. By this simple account, the antibodies made by B cells under allelic exclusion are estimated to be more than 1000-fold more effective than those made by hypothetical B cells without allelic exclusion.

These considerations were of practical importance in the early development of monoclonal antibodies for diagnostic and therapeutic applications (see Sections 4-5 and 4-6). The very first monoclonal antibodies were made by hybridomas that expressed two different heavy chains and two different light chains. One heavy chain and one light chain pair derived from a B cell of the immunized mouse, and together they form an IgG specific for the immunizing antigen. The second heavy and light chains came from the fusion partner, a myeloma cell line, and together form an IgG of unknown specificity. In the first generation of hybridoma cells there was random assembly of the two heavy

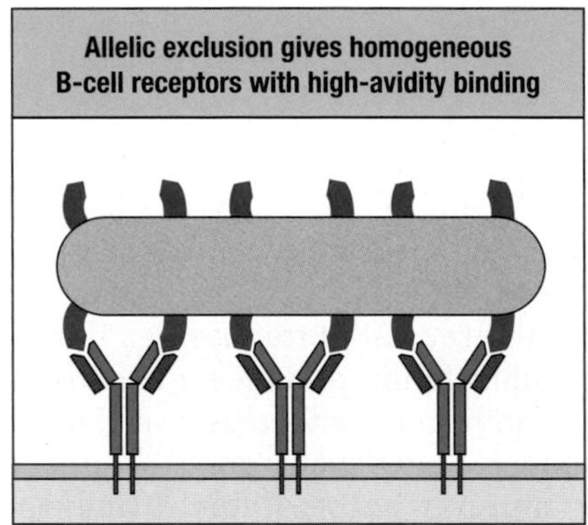

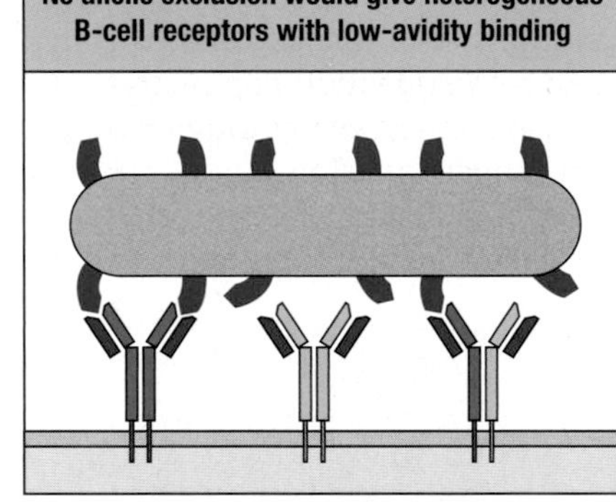

Figure 6.8 Allelic exclusion at the immunoglobulin loci gives rise to B cells having antigen receptors of a single specificity. The top panel shows the binding to antigen of B-cell receptors produced in a B cell expressing immunoglobulin from one immunoglobulin heavy-chain locus and one immunoglobulin light-chain locus only. All the receptors have identical antigen-binding sites and bind their antigen with high avidity. The bottom panel shows the B-cell receptors formed in a hypothetical B cell expressing immunoglobulin from both the immunoglobulin heavy-chain loci and one light-chain locus. Hybrid immunoglobulins are formed with disparate antigen-binding sites and bind the antigen poorly and with low avidity. The disparities would be even greater in a B cell expressing two heavy-chain and two light-chain genes.

and light chains to give 10 different forms of IgG. Of these only one form had two identical binding sites specific for the desired antigen, and three forms had one specific binding site. The forms could not easily be separated, and so much of the antibody produced (7 of the 10 forms) was inactive. A major advance in the production of monoclonal antibodies came with the use of myeloma fusion partners in which the immunoglobulin genes have been made nonfunctional. Hybridomas made with these fusion partners secrete homogeneous immunoglobulin that is specific for the immunizing antigen.

6-6 Rearrangement of the light-chain loci by pre-B cells is relatively efficient

Every large pre-B cell goes through several rounds of cell division, yielding a clone of around 100 small, resting pre-B cells that express identical μ chains and no longer make the surrogate light chain and the pre-B-cell receptor. At this stage the B cells turn their attention to making a proper light chain. The *RAG* genes, which were turned off in the dividing pre-B cells, are now reactivated; RAG proteins are produced and the immunoglobulin light-chain genes begin to rearrange.

Rearrangements take place one light-chain locus at a time, with the κ locus usually being rearranged before the λ locus. Whereas two recombination events are required to bring together the V, D, and J segments of the heavy-chain locus, only a single event joining V with J is needed to rearrange a light-chain gene. Although this reduces the diversity achieved by combining different gene segments it has the advantage that several attempts to rearrange the same light-chain gene can be made by using V and J gene segments not involved in previous rearrangements (Figure 6.9). This is not possible at the heavy-chain locus, because the first two rearrangements that join a V to D and J excise all the other D segments. Several attempts to rearrange the κ light-chain locus on one or other chromosome can therefore be made before commencing rearrangement of a λ-chain locus on a different chromosome. The existence of four light-chain loci, and the opportunity for making several

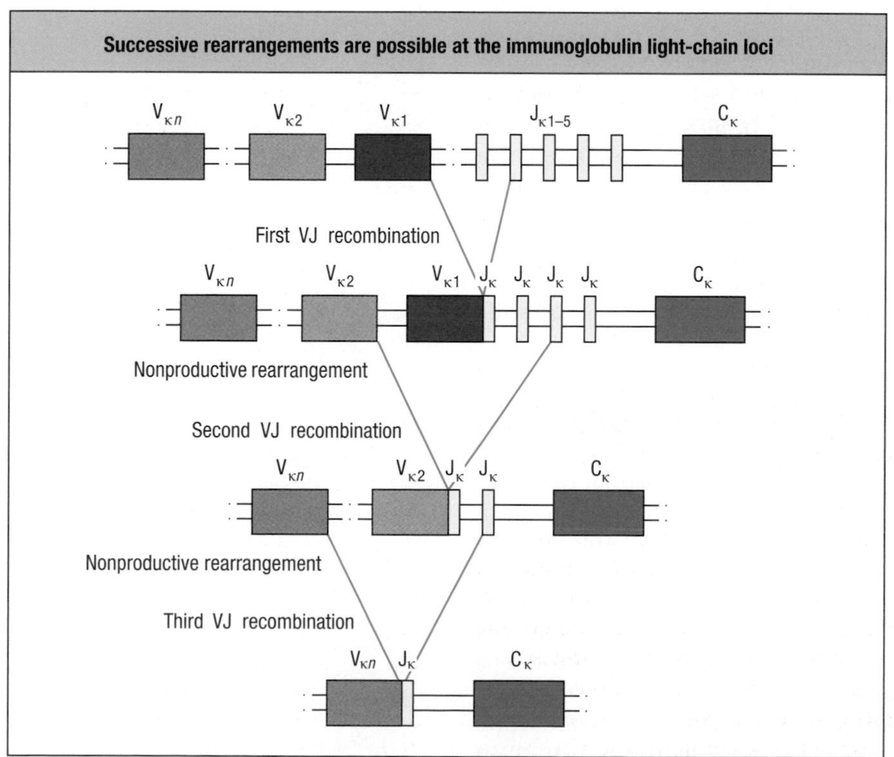

Figure 6.9 Nonproductive light-chain gene rearrangements can be superseded by further gene rearrangement. The organization of the light-chain loci allows initial nonproductive rearrangements at one locus to be followed by further rearrangements of that same locus that can lead to the production of a functional light chain. This type of rescue is shown for the κ light-chain gene. After a nonproductive rearrangement of V_κ to a J_κ, in which the translational reading frame has been lost, a second rearrangement can be made by $V_{\kappa 2}$, or any other V_κ that is on the 5′ side of the first joint, with a J_κ that is on the 3′ side of the first joint. When the second joint is made, the intervening DNA containing the first joint is excised. There are five J_κ gene segments and many more V_κ gene segments, so that as many as five successive attempts at productive rearrangement can be made at each of the two κ light-chain loci.

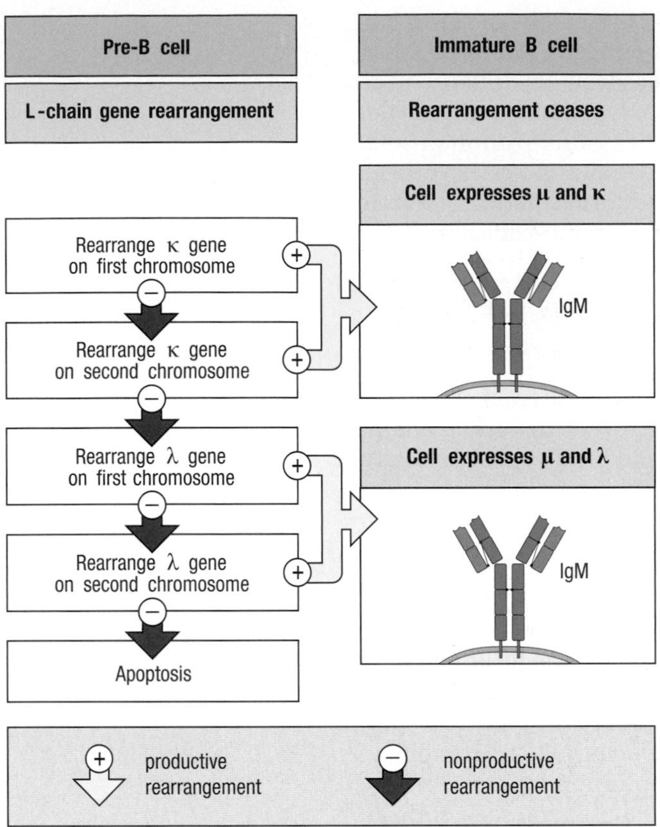

Figure 6.10 Rearrangement of the immunoglobulin light-chain genes in pre-B cells leads to the expression of cell-surface IgM.

attempts to rearrange each locus, means that about 85% of the pre-B cell population makes a successful rearrangement of a light-chain gene (Figure 6.10).

Overall, less than half of the B-lineage cells produced in the bone marrow end up producing functional immunoglobulin heavy and light chains. Although there are no functional differences between κ and λ light-chain isotypes, the benefit of having the two isotypes is that it increases the chance that small pre-B cells will succeed in making immunoglobulin. That two-thirds of human immunoglobulins have κ chains and one-third have λ chains indicates that having the second isotype improves the success rate by 50%. Although the use of κ light chains outnumbers λ light chains in the total B-cell population, the IgD-secreting B cells in the upper respiratory tract use λ chains almost exclusively. These IgD antibodies recognize bacteria such as *Haemophilus influenzae* and recruit basophils to secrete antibacterial peptides. The cause of the bias in light-chain use is unknown. One possible explanation is that the combination of λ light chains with δ heavy chains makes for particularly good antibodies against these bacteria.

When a functional light-chain gene has been formed, light chains are made and assemble with μ chains in the endoplasmic reticulum to form IgM. Association with the Igα and Igβ signaling components forms the mature B-cell receptor, which then moves to the cell surface (see Figure 6.7). The presence of a functional B-cell receptor at the surface sends signals to the cell's interior that quickly shut down light-chain gene rearrangement. This control, which is analogous to that exerted by the pre-B-cell receptor on the heavy-chain locus, ensures that a B cell expresses only one form of immunoglobulin light chain. Because there are two light-chain isotypes, the regulation consists of both allelic exclusion and isotype exclusion. With a functional B-cell receptor on its surface and recombination switched off in the nucleus, the pre-B cell becomes an immature B cell. If a cell makes a light chain that fails to associate with the heavy chain, it will usually be due to a fault in the light chain, because the heavy chain's competence has already been ensured by the surrogate light chain. In these circumstances, recombination continues until either a functional light chain is made or no further rearrangements are possible because all V or J gene segments have been used up.

Light-chain rearrangement occurs independently in each of the 100 or so small pre-B cells that are the descendants of a single large pre-B cell and so express the same μ chain. Different light-chain rearrangements will be made in each of these cells. A success rate of 85% in light-chain manufacture means that around 85 of these small pre-B cells will become immature B cells making IgM. Two benefits arise from the clonal expansion of the large pre-B cells. First, it guarantees that the investment made to obtain each functional heavy chain is never lost through failure to make a light chain. Second, it creates a diverse population of immature B cells that have the same μ chain but a different κ or λ light chain. Clonal expansion has thereby allowed one successful heavy-chain gene rearrangement to give rise to up to 85 clones of immature B cells with different antigenic specificities.

6-7 Developing B cells pass two checkpoints in the bone marrow

As the principal function of B cells is to make immunoglobulins, the early development of B cells in the bone marrow is organized around the gene rearrangements that are necessary to make heavy and light chains. Because of the inherent inefficiency of the recombination process, B cells evolved a conservative strategy for their development in the bone marrow. First, they concentrate on making a heavy chain, and only when that is successful does a B cell turn to making a light chain. After each step the cells go through a **checkpoint**, when the quality of the immunoglobulin chains is assessed. As we saw in Section 6-4, the first checkpoint occurs at the late pro-B-cell stage, with the formation, or not, of a functional pre-B-cell receptor. During the pro-B-cell stage, a mixture of productive and nonproductive gene rearrangements is randomly introduced into the immunoglobulin heavy-chain genes. The genetic variation this imparts to the B-cell population is then subjected to stringent selection for the capacity of cells to make μ chains that assemble a functional pre-B-cell receptor. Cells that can, survive and multiply their kind; cells that cannot, die. Selection by the pre-B-cell receptor is the first checkpoint in B-cell development (Figure 6.11).

The second checkpoint is at the stage of the small pre-B cell, after both productive and nonproductive rearrangements have been randomly made in the light-chain genes. Pre-B cells that survive this checkpoint are those with the capacity to make light chains that will bind to the existing μ chains and assemble a functional B-cell receptor. If they do not, they die in the bone marrow; if they do, they survive to become immature B cells (see Figure 6.11).

Commensurate with their principal function of making immunoglobulins, the only B cells that pass both checkpoints are those that successfully make an immunoglobulin. This is only the beginning, because further checkpoints will be met as B cells develop further when they leave the bone marrow, enter the circulation, and participate in an immune response.

6-8 A program of protein expression underlies the stages of B-cell development

The successive steps in B-cell development are defined by changes in the expression of proteins directly involved in rearranging the immunoglobulin genes, testing the quality of the immunoglobulin chains, and driving cellular proliferation. Driving these changes is an extensive network of transcription

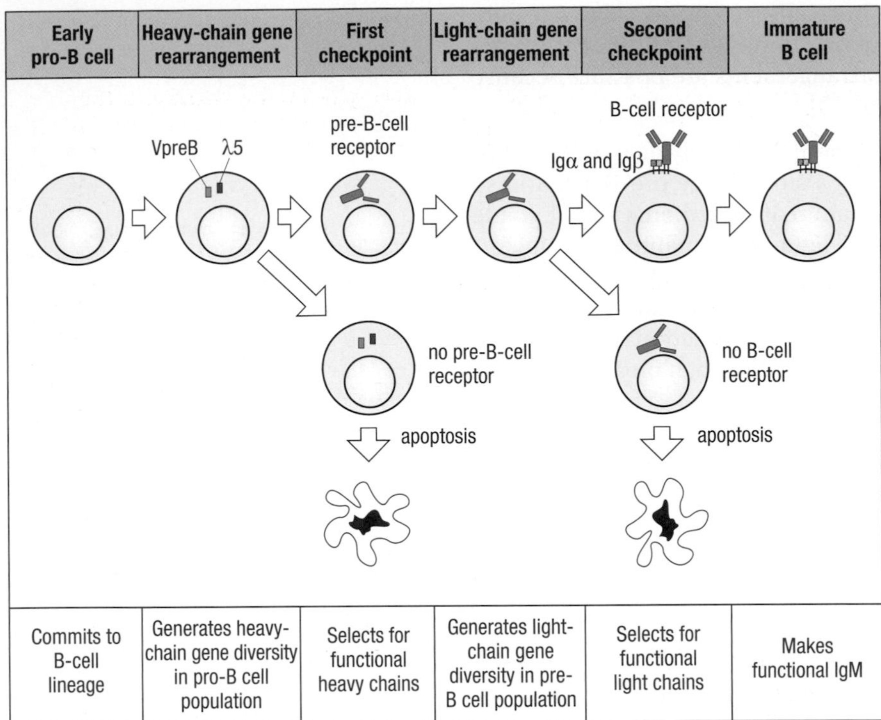

Early pro-B cell
Heavy-chain gene rearrangement
First checkpoint
Light-chain gene rearrangement
Second checkpoint
Immature B cell
VpreB λ5
pre-B-cell receptor
B-cell receptor
Igα and Igβ
no pre-B-cell receptor
apoptosis
no B-cell receptor
apoptosis
Commits to B-cell lineage
Generates heavy-chain gene diversity in pro-B cell population
Selects for functional heavy chains
Generates light-chain gene diversity in pre-B cell population
Selects for functional light chains
Makes functional IgM

Figure 6.11 There are two fate-determining checkpoints during B-cell development in the bone marrow. Both checkpoints are marked by whether a functional receptor is made, which is a test of whether a functional heavy chain (at the first checkpoint) or a functional light chain (at the second checkpoint) has been produced. Cells that fail either of these checkpoints die by apoptosis.

factors that ensures that the genes encoding these proteins are turned on and off at the appropriate times (Figure 6.12). The RAG-1 and RAG-2 proteins are essential components of the recombination machinery (see Section 4-9), and their genes are specifically turned on at the stages in B-cell development when heavy-chain or light-chain gene rearrangement occurs. Conversely, RAG proteins are absent at the large pre-B-cell stage, when the quality of heavy-chain gene rearrangements is examined and the successful cells divide before rearranging the light-chain genes. Terminal deoxynucleotidyl transferase (TdT), the enzyme that adds N nucleotides at the junctions between rearranging gene segments, is expressed in pro-B cells when the heavy-chain genes begin rearrangement and is turned off in small pre-B cells when the light-chain genes begin to rearrange; this explains why N nucleotides are found in all VD and DJ joints of rearranged human heavy-chain genes but in only about half of the VJ joints of rearranged human light-chain genes.

Several proteins contribute to B-cell development by transducing signals from cell-surface receptors. Synthesis of the Igα and Igβ polypeptides is turned on at the pro-B-cell stage and continues throughout the life of a B cell, which means they are always ready to associate with immunoglobulin chains and form pre-B-cell and B-cell receptors that can send signals to the cell. This is important for allelic exclusion, because it ensures that gene rearrangement can be halted once a functional heavy chain, and then a light chain, has been made. Only when antigen-stimulated B cells differentiate into antibody-secreting plasma cells, and their antigen receptors are no longer required, are the Igα and Igβ genes turned off and immunoglobulin ceases to be present on the cell surface. For the same reason, the λ5 and VpreB components of the surrogate light chain are ready and waiting to form pre-B-cell receptors throughout the pro-B-cell stage (see Figure 6.12).

Only a few of the transcription factors contributing to immunoglobulin gene rearrangement and B-cell development are shown in Figure 6.12. Their coordinated activity ensures that the enzymatic machinery for gene rearrangement, which is common to B and T cells, is used in B cells to rearrange and express the immunoglobulin genes and not the T-cell receptor genes. Of particular

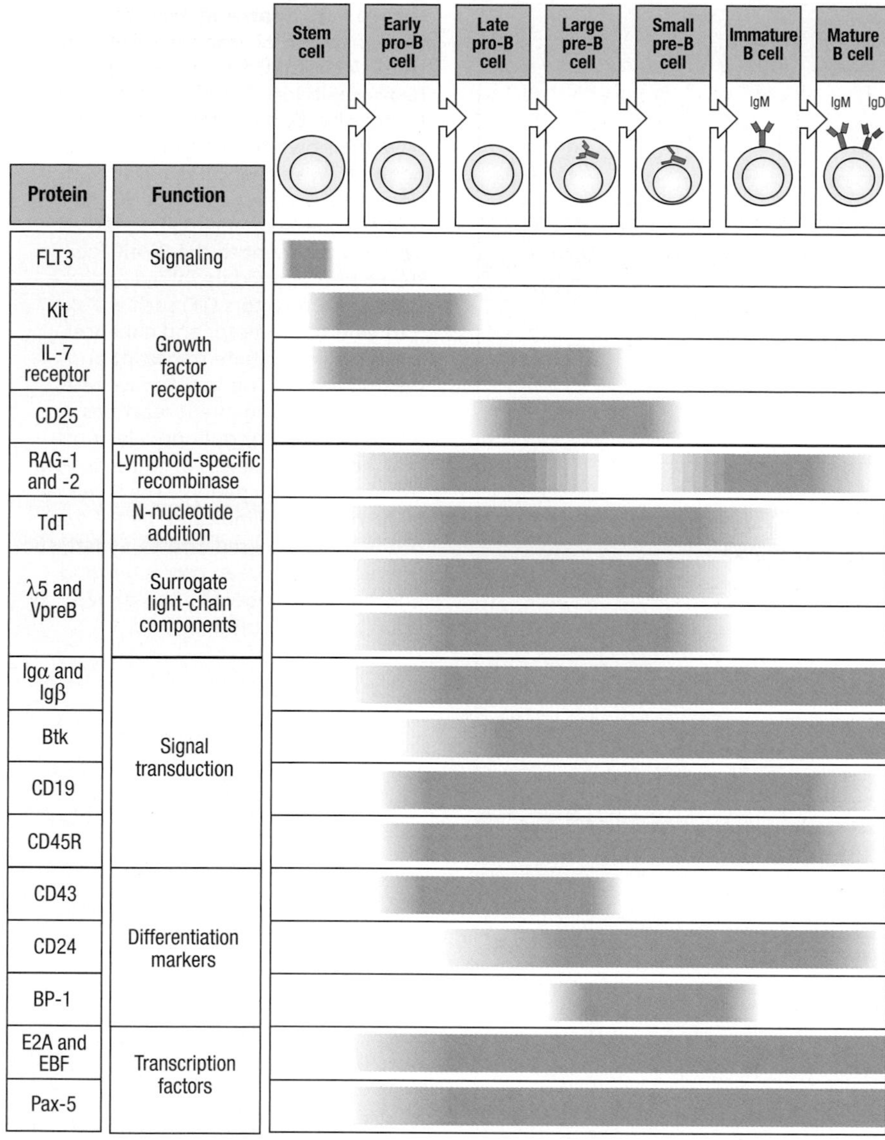

Stem cell
Early pro-B cell
Late pro-B cell
Large pre-B cell
Small pre-B cell
Immature B cell
Mature B cell
IgM
IgM IgD
Protein
Function
FLT3
Signaling
Kit
IL-7 receptor
Growth factor receptor
CD25
RAG-1 and -2
Lymphoid-specific recombinase
TdT
N-nucleotide addition
λ5 and VpreB
Surrogate light-chain components
Igα and Igβ
Btk
Signal transduction
CD19
CD45R
CD43
CD24
Differentiation markers
BP-1
E2A and EBF
Transcription factors
Pax-5

Figure 6.12 The timing of proteins involved in immunoglobulin gene rearrangement and expression during B-cell development. The rearrangement of immunoglobulin genes and the expression of the pre-B-cell receptor and cell-surface IgM requires several categories of specialized proteins at different times during B-cell development. Examples of such proteins are listed here, with their expression during B-cell development shown by red shading. FLT3 is a cell-surface receptor protein kinase that receives signals from stromal cells for the differentiation of common lymphoid progenitors. CD19 is a subunit of the B-cell co-receptor, which cooperates with the antigen receptor to produce activating signals when antigen is bound. CD45 is a cell-surface protein phosphatase that modulates signals from an antigen-recognizing B-cell receptor. CD43, CD24, and BP-1 are cell-surface markers that are useful for distinguishing different stages in B-cell development. The contributions of the other proteins listed here are explained in the text.

importance in defining the B-cell lineage is the B-cell-specific transcription factor Pax-5, which is first synthesized in the early pro-B cell and continues to be synthesized throughout B-cell development and the lifetime of a mature B cell. Pax-5 binds to the regulatory elements of the immunoglobulin genes and also to regulatory elements in the genes for many B-cell-specific proteins, such as λ5, VpreB, and CD19.

In all human cells except B cells the chromatin containing the immunoglobulin loci is kept in a 'closed' form that is never transcribed. In the early pro-B cell, Pax-5 binds to enhancer sequences located 3′ of the heavy-chain C-region genes. This opens up the chromatin and allows the access of other transcription factors that are not B-cell specific (Figure 6.13). A low level of transcription now occurs from the promoters upstream of the D and J segments. Transcription facilitates access of the RAG complex, which then brings the D and J segments together and allows the rearrangement and joint to be made. In turn, this initiates transcription upstream of a V segment, which favors rearrangement of the V segment to DJ. As a consequence of the two rearrangements, the promoter that is 5′ of the V-region segment is brought nearer to the enhancer that is 3′ of the C-region gene, resulting in a further increase in transcription. We can therefore see that gene rearrangement controls the expression of the immunoglobulin genes as well as generating their diversity.

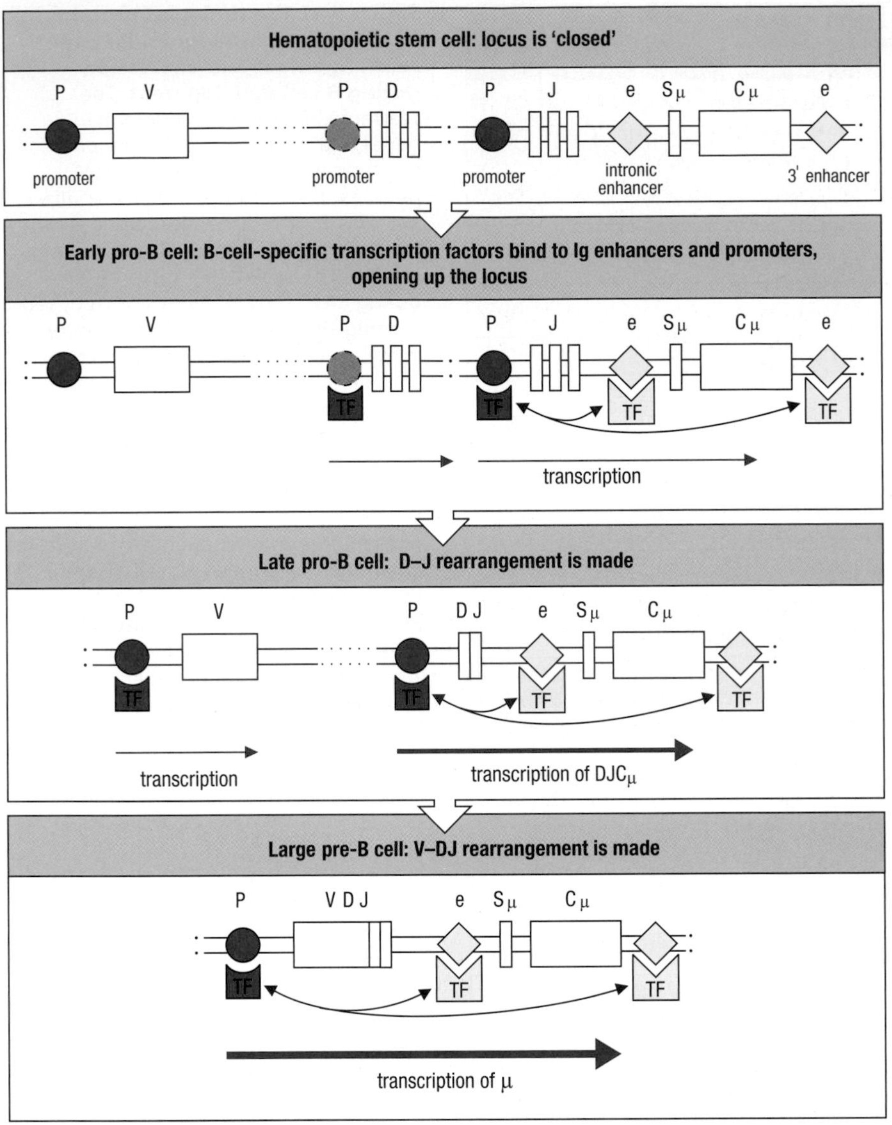

Figure 6.13 Rearrangement and transcription of immunoglobulin genes are linked processes. The heavy-chain locus is shown here. The blocks of V, D, and J segments each have their own promoter (P), and in addition two enhancers (e) are located flanking the C_μ gene and its isotype switch signal sequence S_μ (top panel). With commitment to the B-cell lineage the locus is opened up, allowing transcription factors (TF) such as Pax-5 to bind to promoter and enhancer elements and initiate transcription (second panel). This in turn facilitates the gene rearrangement reactions, which juxtapose regulatory elements and increase the levels of transcription (third and fourth panels). The levels of transcription are indicated by the width of the horizontal red arrows. Synergistic interactions between promoters and enhancers are shown by the curved, double-headed, black arrows.

One of the signaling proteins shown in Figure 6.12, Bruton's tyrosine kinase (Btk), is essential for B-cell development beyond the pre-B-cell stage. Patients who lack a functional *Btk* gene have almost no circulating antibodies because their B cells are blocked at the pre-B-cell stage. The immune deficiency suffered by these patients is called **X-linked agammaglobulinemia** (**XLA**) and leads to recurrent infections from common extracellular bacteria such as *Haemophilus influenzae*, *Streptococcus pneumoniae*, *Streptococcus pyogenes*, and *Staphylococcus aureus*. The infections respond to treatment with antibiotics and are prevented by infusion of intravenous immunoglobulin. The *Btk* gene is located on the X chromosome and the defective gene is recessive, so X-linked agammaglobulinemia is seen mostly in boys.

6-9 Many B-cell tumors carry chromosomal translocations that join immunoglobulin genes to genes that regulate cell growth

As B cells cut, splice, and mutate their immunoglobulin genes in the normal course of events, it is hardly surprising that this process sometimes goes awry to produce a mutation that helps convert a B cell into a tumor cell. The process of malignant transformation involves an accumulation of mutations that ultimately release the cell from the normal restraints on its growth. For B-cell tumors, the disruption of regulated growth is often associated with an aberrant immunoglobulin gene rearrangement that joined an immunoglobulin gene to a gene on a different chromosome. Events that fuse part of one chromosome with another are called **translocations** and, in B-cell tumors, the immunoglobulin gene is often joined to a gene involved in the control of cellular growth. Genes that cause cancer when their function or expression is perturbed are collectively called **proto-oncogenes**. Many of them were discovered in the study of RNA tumor viruses that can transform cells directly. The viral genes responsible for transformation were named **oncogenes**, and it was only later realized that they had evolved from cellular genes of the virus's vertebrate host that control cell growth, division, and differentiation.

The translocations between immunoglobulin genes and proto-oncogenes in B-cell tumors can be seen in metaphase chromosomes examined under the light microscope. Certain translocations define particular types of tumor and are valuable in diagnosis. In Burkitt's lymphoma, for example, the *MYC* proto-oncogene on chromosome 8 is joined by translocation to either an immunoglobulin heavy-chain gene on chromosome 14, a κ light-chain gene on chromosome 2, or a λ light-chain gene on chromosome 22 (Figure 6.14). The Myc protein is normally involved in regulating the cell cycle, but in B cells carrying these translocations its expression is abnormal, which removes some of the restraints on cell division. Because single genetic changes are not sufficient to produce malignant transformation, the onset of Burkitt's lymphoma requires further mutations elsewhere in the genome.

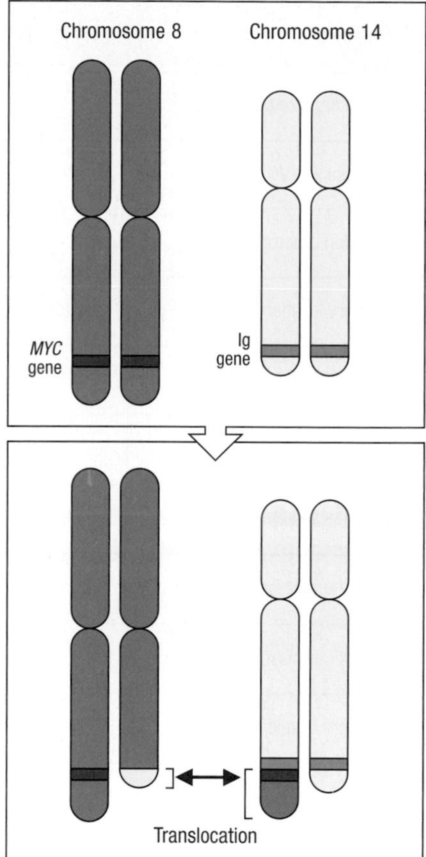

Figure 6.14 Chromosomal rearrangements in Burkitt's lymphoma. In this example of Burkitt's lymphoma, parts of chromosome 8 and chromosome 14 have been exchanged. The sites of breakage and rejoining are in the proto-oncogene *MYC* on chromosome 8 and the immunoglobulin heavy-chain gene on chromosome 14. In these tumors it is usual for the second immunoglobulin gene to be productively rearranged and for the tumor to express cell-surface immunoglobulin. The translocation probably occurred during the first attempt to rearrange a heavy-chain gene. This counted as a nonproductive rearrangement and so the other gene was then rearranged. In cases where the second rearrangement is also nonproductive, the cell dies and cannot give rise to a tumor.

Another translocation found in B-cell tumors is the fusion of an immunoglobulin gene to the proto-oncogene *BCL2*. Normally, the Bcl-2 protein prevents premature apoptosis in B-lineage cells. Overproduction of Bcl-2 in the mutant B cells having this translocation enables them to live longer than normal, during which time they accumulate additional mutations that can lead to malignant transformation. In addition to their value in immunological research, B-cell tumors have provided fundamental advances in knowledge of the proteins involved in the mechanism and regulation of cell division.

6-10 B cells expressing the glycoprotein CD5 express a distinctive repertoire of receptors

Not all B cells conform exactly to the developmental pathway described above. A subset of human B cells that arises early in embryonic development is distinguished from other B cells by the expression of CD5, a cell-surface glycoprotein that is otherwise considered a marker for the human T-cell lineage. B cells in this minority subset are termed **B-1 cells** because their development precedes that of the majority subset, for which we have traced the development in previous sections and in this context are called **B-2 cells** (Figure 6.15). B-1 cells are also characterized by having little or no surface IgD and a distinctive repertoire of antigen receptors. The B-1 cells are also called **CD5 B cells**. Although comprising only around 5% of the B cells in humans and mice, B-1 cells are the main type of B cell in rabbits.

B-1 cells arise from a stem cell that is most active in the prenatal period. Their immunoglobulin heavy-chain gene rearrangements are dominated by the use of the V_H gene segments that lie closest to the D gene segments in the germline. As TdT is not expressed early in the prenatal period, the rearranged heavy-chain genes of B-1 cells are characterized by a lack of N nucleotides, and their VDJ junctions are less diverse than those in rearranged heavy-chain genes of B-2 cells. Consequently, the antibodies secreted by B-1 cells tend to be of low affinity and each binds to many different antigens, a property known as **polyspecificity**. B-1 cells contribute to the antibodies made against common

Property	B-1 cells	Conventional B-2 cells
When first produced	Fetus	After birth
N-regions in VDJ junctions	Few	Extensive
V-region repertoire	Restricted	Diverse
Primary location	Peritoneal and pleural cavities	Secondary lymphoid organs
Mode of renewal	Self-renewing	Replaced from bone marrow
Spontaneous production of immunoglobulin	High	Low
Isotypes secreted	IgM >> IgG	IgG > IgM
Requirement for T-cell help	No	Yes
Somatic hypermutation	Low–none	High
Memory development	Little or none	Yes

Figure 6.15 Comparison of the properties of B-1 cells and B-2 cells. B-1 cells develop in the omentum, a part of the peritoneum, as well as in the liver in the fetus, and are produced by the bone marrow for only a short period around the time of birth. A pool of self-renewing B-1 cells is then established, which does not require the microenvironment of the bone marrow for its survival. The limited diversity of the antibodies made by B-1 cells and their tendency to be polyspecific and of low affinity suggests that B-1 cells produce a simpler, less adaptive immune response than that involving B-2 cells.

bacterial polysaccharides and other carbohydrate antigens but are of little importance in making antibodies against protein antigens.

The B-1 cells that develop postnatally use a more diverse repertoire of V gene segments and their rearranged immunoglobulin genes have abundant N nucleotides. With time, B-1 cells are no longer produced by the bone marrow and, in adults, the population of B-1 cells is maintained by the division of existing B-1 cells at sites in the peripheral circulation. This self-renewal is dependent on the cytokine IL-10. Most cases of chronic lymphocytic leukemia (CLL) arise from B-1 cells, a propensity that could stem directly from their capacity for self-renewal. In the remainder of this chapter, and in the rest of this book, B cell will refer to the population of B-2 cells, unless otherwise specified.

Summary

B cells are generated throughout life from stem cells in the bone marrow. Different stages in B-cell maturation correlate with molecular changes that accompany the gene rearrangements required to make functional immunoglobulin heavy and light chains. These changes are summarized in Figure 6.16. Many rearrangements are nonproductive, an inefficiency compensated for, in part, by the presence of two heavy-chain loci and four light-chain loci in each B cell, all of which can be rearranged in the quest to obtain a functional heavy chain and light chain. The heavy-chain genes are rearranged first, and only when this gives a functional protein is a B cell allowed to rearrange its light-chain genes. After a successful heavy-chain gene rearrangement, the μ chain and surrogate light chain form the pre-B-cell receptor, which signals the cell to halt heavy-chain gene rearrangement and to proceed to light-chain gene rearrangement. Similarly, after successful light-chain gene rearrangement, the assembly of IgM signals the cell to halt light-chain gene rearrangement. These feedback mechanisms, whereby protein product regulates the progression of gene rearrangement, ensure that each B cell expresses only one heavy chain and one light chain and thus produces immunoglobulin of a single defined

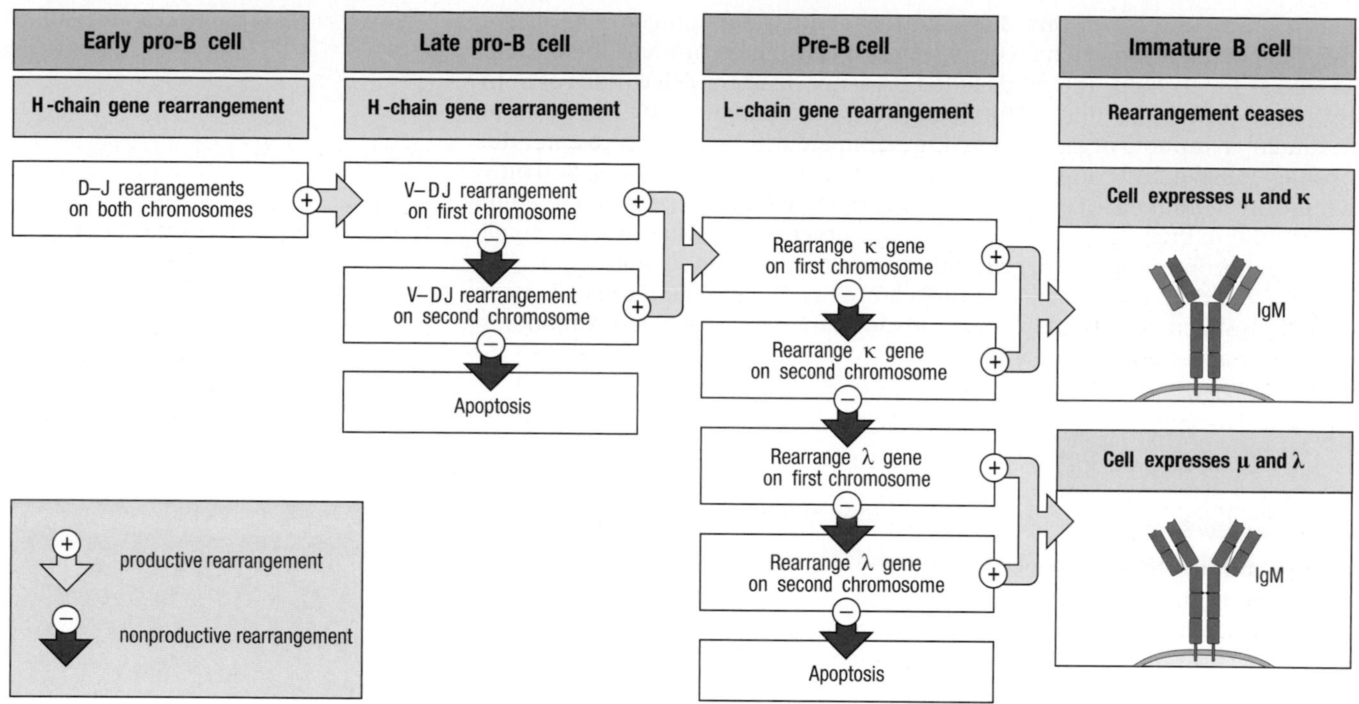

Figure 6.16 Summary of the order of gene rearrangements that lead to the expression of cell-surface immunoglobulin. The heavy-chain genes are rearranged before the light-chain genes. Developing B cells are allowed to proceed to the next stage only when a productive rearrangement has been made. If a nonproductive rearrangement is made on one chromosome of a homologous pair, then rearrangement is attempted on the second chromosome.

antigen specificity. In this manner, the program of gene rearrangement produces the monospecificity essential for the efficient operation of clonal selection. The success rate in obtaining a functional light-chain gene rearrangement is much higher than that for the heavy-chain gene because only two gene segments are involved (compared with three for the heavy-chain gene) and the B cell has four light-chain genes (two κ and two λ) with which to work. Errors in immunoglobulin gene rearrangement give rise to chromosomal translocations that predispose B cells to malignant transformation. A minority class of B cells, termed B-1 cells, develops early in embryonic life. They produce antibodies that bind bacterial polysaccharides, but are polyspecific, of low affinity, and mainly IgM. These properties of B-1 cells suggest they represent a simpler and evolutionarily older lineage of B cell than the majority B-cell population.

Selection and further development of the B-cell repertoire

In the previous part of this chapter we saw how the population of immature B cells acquires a vast repertoire of receptors with different antigen specificities. This repertoire includes some immunoglobulins that bind to normal constituents of the human body and have the potential to initiate a damaging immune response. Receptors and cells that react against a person's own tissues are described as being **self-reactive** or **autoreactive**, and the components to which they bind are called **self antigens**. To prevent such responses, self-reactive immature B cells that encounter their self antigens, either in the bone marrow or in the peripheral circulation, are prevented from advancing from the immature B-cell stage to the mature B-cell stage.

For an immature B cell that does not react with a self antigen, alternative mRNA splicing of heavy-chain gene transcripts will then produce the δ chain as well as the μ chain. This leads to the presence on the B-cell surface of both IgD and IgM (see Section 4-10). At this stage the immature B cells travel to secondary lymphoid organs, where they compete with each other to enter the primary follicles and complete their maturation. Competition is stiff, and only a minority of immature B cells become the **mature B cells** that are competent to respond to their specific antigens. In this process of maturation, the B-cell receptor becomes capable of generating positive signals upon binding specific antigen. But only a tiny fraction of the mature B-cell population will ever meet specific antigen and deploy their effector function of making antibodies against a pathogenic microorganism.

6-11 The population of immature B cells is purged of cells bearing self-reactive B-cell receptors

The mechanisms of gene rearrangement and somatic mutation described in the first part of this chapter produce a population of immature B cells that express surface IgM embracing a wide range of antigenic specificities. This repertoire of immature B-cell receptors includes many with affinity for self antigens that are components of healthy human tissue. Activation of mature B cells carrying such receptors by these antigens would produce self-reactive and potentially harmful antibodies that could disrupt the body's normal functions and cause disease. To prevent this from happening, the B-cell receptors of immature B cells are wired to generate negative intracellular signals when they bind to antigen. These signals cause the B cell either to die by apoptosis or to become inactivated. In this way, the immature B-cell population is subject to a process of negative selection that prevents the maturation of self-reactive B cells. As a consequence, the resulting population of mature B cells in a healthy individual does not respond to self antigens and is said to be **self-tolerant**. As many as 75% of immature B cells have some affinity for self antigens, and so once they have been eliminated, the receptor repertoire of mature B cells is substantially different from that of immature B cells.

The lives of B cells are spent largely in the blood, the lymph, the secondary lymphoid organs, and the bone marrow. In these tissues, B-cell receptors can interact with molecules exposed on cell surfaces and in connective tissue and with soluble components of blood plasma. Because of their accessibility, these self antigens are the ones most likely to provoke an autoimmune B-cell response. Multivalent antigens (see Figure 4.9, p. 86) are most effective at linking B-cell receptors together and activating B cells, and many of the abundant glycoproteins, proteoglycans, and glycolipids present on human cells can act as multivalent self antigens.

To prevent the emergence of mature B cells specific for these common multivalent self antigens, the immature B-cell population is subject to a further process of selection that identifies such B cells and prevents their maturation. This selection starts in the bone marrow, where immature B cells are exposed to the wide variety of self antigens expressed by stromal cells, hematopoietic cells and macromolecules circulating in the blood plasma. Only immature B cells bearing receptors that do not interact with any of these self antigens are allowed to leave the bone marrow and continue their maturation in the peripheral circulation (Figure 6.17). Maturation first involves the alternative splicing of the heavy-chain mRNA so that the cell makes IgD as well as IgM (see Section 4-10). This is followed by a shift from an excess of IgM over IgD on the cell surface to an excess of IgD over IgM in the mature B cell.

In contrast, immature B cells with receptors that bind a multivalent self antigen are signaled to arrest their developmental progression. They are retained in the bone marrow and given the opportunity to lose their self-reactivity for self antigen by altering their B-cell receptor.

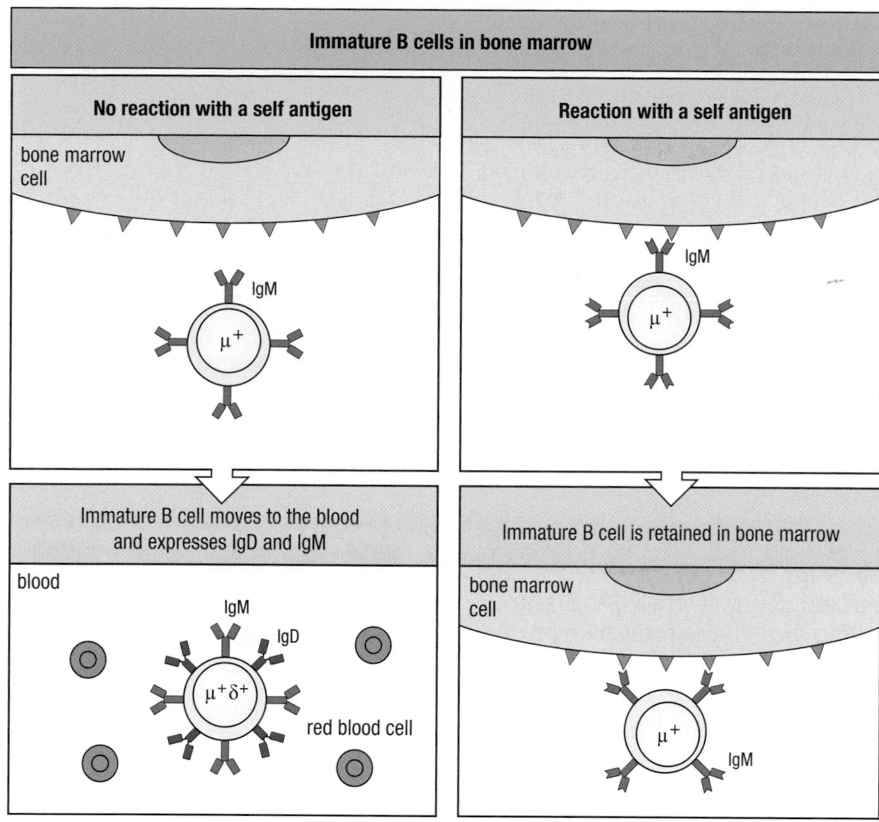

Figure 6.17 Immature B cells with specificity for multivalent self antigens are retained in the bone marrow. Immature B cells that are not specific for a self antigen in bone marrow mature further to express IgM and IgD and leave the bone marrow (left panels). Immature B cells specific for a self antigen on bone marrow cells are retained in the bone marrow (right panels).

6-12 The antigen receptors of autoreactive immature B cells can be modified by receptor editing

If the cell-surface IgM of an immature B cell binds a multivalent self antigen in the bone marrow, this signals the B cell to reduce the amount of IgM on its surface and to maintain expression of the RAG proteins, which permits the B cell to continue rearranging its light-chain loci. This further rearrangement leads to excision of the existing rearrangement, and so the original light chain is no longer made. Continuing rearrangement may also produce a new functional light chain that assembles with the old heavy chain to make a new B-cell receptor. If this new receptor does not react with a self antigen, then the B cell can continue its developmental progression and become a mature B cell.

If self-reactivity is retained, then further light-chain gene rearrangements are tried. This process of assessing the compatibility of receptors produced from successive gene rearrangements is called **receptor editing** (Figure 6.18). The multiplicity of V and J segments in the κ and λ light-chain genes provides considerable capacity for receptor editing. Inevitably, some B cells exhaust all the possibilities for light-chain gene rearrangement without ever gaining a receptor that does not react with self. These cells are signaled to die by apoptosis in the bone marrow where they are phagocytosed by macrophages. The selective death of developing lymphocytes, with consequent elimination of their self-reactive receptor specificities from the B-cell repertoire, is called **clonal deletion**. Every day some 55 billion B cells die in the bone marrow, either because they fail to make a functional immunoglobulin or because they are guilty of reacting with self.

By enabling immature B cells to exchange a self-reactive light chain for one that is not self-reactive, receptor editing increases the overall efficiency of B-cell development and the probability that successful heavy-chain gene rearrangements will not be lost from the mature B-cell receptor repertoire.

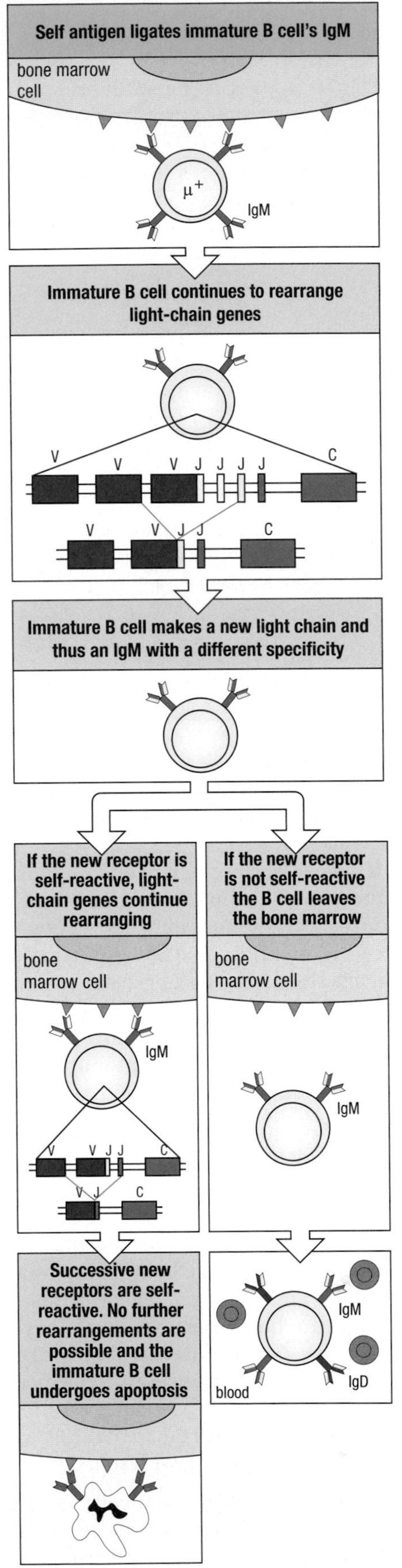

Figure 6.18 By changing their antigen specificities receptor editing rescues many self-reactive B cells. Binding to a self antigen in the bone marrow causes the immature B cell to continue to rearrange its light-chain loci. This deletes the existing self-reactive rearrangement and may also produce a new light-chain rearrangement giving immunoglobulin that is not specific for self antigen. Successive rearrangements occur until either a new non-self-reactive rearrangement is made, and the B-cell continues its development (right panels), or no more rearrangements are possible and the B cell dies (left panels).

Numerous B cells are rescued by receptor editing, which is the dominant mechanism for establishing self-tolerance in the developing B-cell population.

6-13 Immature B cells specific for monovalent self antigens are made nonresponsive to antigen

In contrast with the multivalent self antigens, many other common self antigens are soluble proteins that carry only one copy of an epitope. Self-reactive B cells that bind to monovalent self antigens are neither signaled to continue light-chain gene rearrangement nor signaled to die by apoptosis. Instead they become inactivated and unresponsive to their specific antigen. This state of developmental arrest is called **anergy** (Figure 6.19). Anergic cells make both IgD and IgM, but unlike the situation in mature B cells, the IgM is prevented from assembling a functional B-cell receptor and is largely retained within the cell. Although normal amounts of IgD are present on the cell surface, they cannot activate the B cell on binding to antigen. Anergic B cells are allowed to enter the peripheral circulation but their life-span of 1–5 days is short compared with the approximately 40-day half-life of mature B cells (cellular life-span is often expressed as the **half-life**, which is the period of time over which a population of cells reduces to half its original size).

B cells reactive to self antigens in the bone marrow experience one of three fates. They either survive through receptor editing, which gets rid of their self-reactivity, die by apoptosis, or become anergic. These three mechanisms ensure that B cells leaving the bone marrow are tolerant to all self antigens present in the bone marrow, many of which are also present in other tissues. This type of immunological tolerance is called **central tolerance** because it is induced in a primary lymphoid organ—the bone marrow in the case of B cells.

Not all self antigens are encountered in the bone marrow. Most organs and tissues of the body express tissue-specific cell-surface proteins, secreted proteins and other antigens that are accessible to circulating B cells. An immature B cell tolerant toward bone-marrow self antigens but reactive to a self antigen in the blood will encounter these specific self antigens soon after leaving the bone marrow. Receptor editing is not an option for these cells, because by this stage the machinery for rearranging their immunoglobulin genes has been shut down permanently. Instead, self-reactive B cells outside the bone marrow either die by apoptosis or are rendered anergic when they engage self antigens. Tolerance induced to antigens outside the bone marrow is called **peripheral tolerance**, and it removes circulating B cells reactive against the self antigens of all tissues other than bone marrow.

Central and peripheral tolerance do not remove B cells that are reactive to self antigens that are present in places inaccessible to B cells, such as the insides of cells, or are present in amounts insufficient to trigger the B cell to undergo apoptosis or become anergic. In situations of stress, disease, or trauma, self antigens to which the B-cell population has not become tolerant can become accessible and provoke an autoimmune response. One example is DNA, against which antibodies are made by patients with the autoimmune disease systemic lupus erythematosis.

6-14 Maturation and survival of B cells requires access to lymphoid follicles

After developing B cells leave the bone marrow, they begin to recirculate between the blood, the secondary lymphoid tissues and the lymph (see Figure 1.20, p. 21). At this stage of development the B cells are still not fully mature; they express high levels of surface IgM and low levels of surface IgD, whereas mature B cells have low surface IgM and high IgD. The final stage in B-cell maturation occurs when the immature B cells enter a secondary lymphoid tissue. Because the lymph node, spleen, Peyer's patches, and other secondary lymphoid organs have a similar microanatomy and function, we focus here on the maturation of B cells as they circulate through a lymph node (Figure 6.20). B cells enter lymph nodes from the blood through the walls of high endothelial venules; stromal cells in the lymph node cortex secrete a chemokine called **CCL21**, for which B cells express a receptor, called **CCR7**. The mechanism underlying this homing of B cells to secondary lymphoid tissues is essentially the same as that used by neutrophils to enter infected tissue (see Section 3-8) but uses different cell adhesion molecules and chemokines. Dendritic cells within the lymph node also secrete CCL21 and another chemokine, **CCL19**, to which B cells are responsive. The gradient of chemokines attracts blood-borne B cells to the high endothelial venules, where they enter the lymph node by squeezing between the high endothelial cells (Figure 6.21).

Once inside the lymph node, B cells are further guided by chemokines to congregate in organized structures called **primary lymphoid follicles**. These areas consist mainly of B cells enmeshed in a network of specialized stromal cells called **follicular dendritic cells** (**FDCs**). Despite their name, which comes from the long cell processes with which they touch B cells and each other, FDCs are not of hematopoietic origin and therefore are unrelated to either the myeloid dendritic cells that present antigens to T cells or the plasmacytoid dendritic cells that make type I interferons. FDCs attract B cells into the follicle by secreting the chemokine **CXCL13** (see Figure 6.21). The interaction between B cells and FDCs is mutually beneficial, providing signals that allow B cells to mature and survive while also preserving the integrity of the FDC network. The latter is mediated by a surface protein on B cells called lymphotoxin (LT), which is structurally related to the inflammatory cytokine

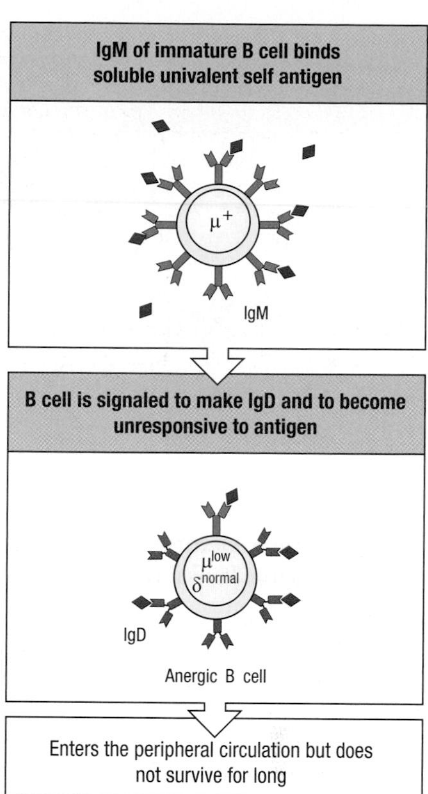

Figure 6.19 Immature B cells specific for monovalent self antigens develop a state of anergy.

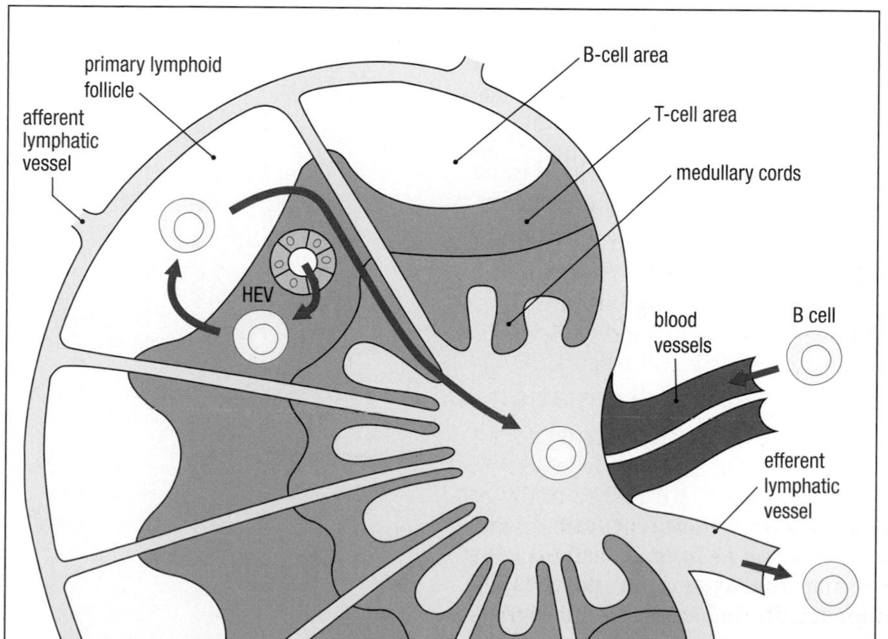

Figure 6.20 The general route of B-cell circulation through a lymphoid tissue. B cells circulating in the blood enter the lymph node cortex via a high endothelial venule (HEV). From there they pass into a primary lymphoid follicle. If they do not encounter their specific antigen, they leave the follicle and exit from the lymph node in the efferent lymph. The circulation route is the same for immature and mature B cells, which all compete with each other to enter primary follicles.

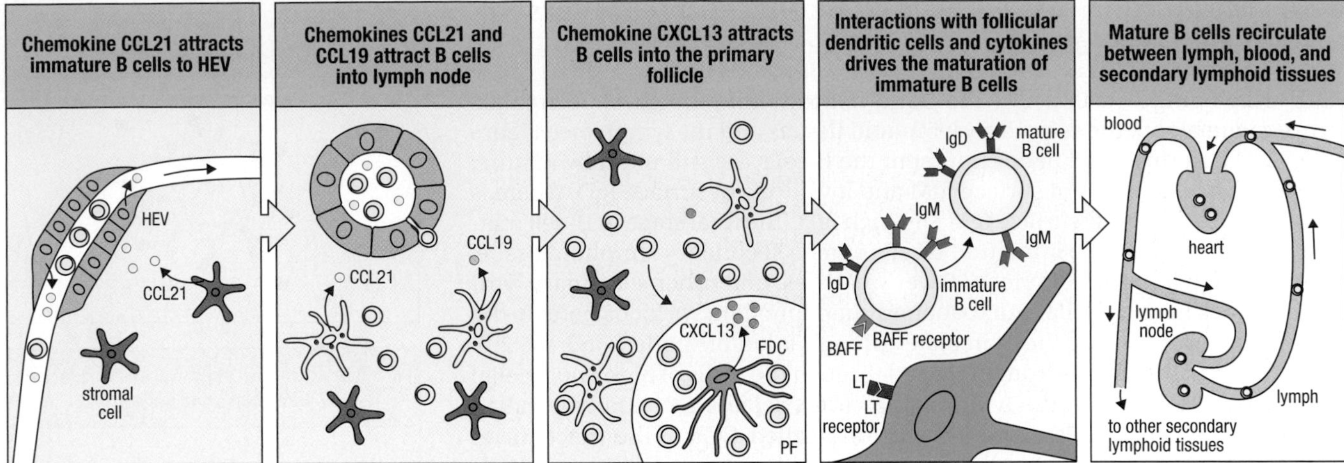

Figure 6.21 Immature B cells must pass through a primary follicle in a secondary lymphoid tissue to become mature B cells. Immature B cells enter the secondary lymphoid tissue through the walls of HEVs, attracted by the chemokines CCL21 and CCL19, and compete with other immature and mature B cells to enter a primary follicle (PF). Follicular dendritic cells (FDCs, turquoise) secrete the chemokine CXCL13, which attracts B cells into the follicle. Immature B cells that enter a follicle interact with proteins on FDCs that signal their final maturation into mature B cells. If they do not encounter their specific antigen in the follicle, mature B cells leave the lymph node and continue recirculating through the secondary lymphoid tissues via lymph and blood. Immature B cells that fail to enter a follicle also continue recirculation but soon die.

tumor necrosis factor-α (TNF-α) and binds to a receptor on FDCs. Another member of the TNF family called **BAFF** (B-cell activating factor in the TNF family) also promotes B-cell survival. BAFF is made by several cell types in secondary lymphoid organs and is bound by the BAFF receptor on B cells.

In the absence of its specific antigen, the B cell detaches from the FDC and leaves the lymph node in the efferent lymph to continue its circulation (see Figure 6.21). It is now a mature B cell, and if it meets its specific antigen it will respond by proliferating and undergoing differentiation into antibody-producing plasma cells, rather than becoming anergic or undergoing apoptosis. Mature B cells that have not yet encountered their antigen are called naive B cells. The survival of circulating mature naive B cells is dependent upon regular passage through the primary follicles of secondary lymphoid tissues.

Throughout life the production of B cells is enormous. In the bone marrow of a healthy young adult, about 2.5 billion (2.5×10^9) cells per day embark on the program of B-cell development. From the progeny of these progenitors, some 30 billion B cells leave the bone marrow to join the circulation. These immature B cells have to compete with the existing population of mature, circulating B cells for access to a limited number of follicular sites; B cells that fail to gain access to a follicle die. This competition is weighted in favor of mature B cells and is so intense that most immature B cells fail to enter a follicle and die by apoptosis after only a few days in the peripheral circulation. Immature B cells that gain access to primary follicles and become mature B cells live longer, but they too disappear from the system with a half-life of around 100 days, unless they are activated by encounter with their specific antigen.

To reach the primary lymphoid follicles, B cells entering from the blood must pass through areas where T cells congregate (see Figure 6.20). Although anergic B cells can enter lymph nodes, they fail to reach a primary follicle and instead concentrate at the boundary between the follicle and the T-cell zone. This exclusion from the follicle means they do not receive the necessary survival signals and they eventually die by apoptosis.

6-15 Encounter with antigen leads to the differentiation of activated B cells into plasma cells and memory B cells

Secondary lymphoid tissues provide the sites where mature, naive B cells encounter specific antigen. When that happens, the antigen-specific B cells are detained in the T-cell areas and become activated by antigen-specific CD4 helper T cells, as briefly touched on in Chapter 1. These T cells provide signals that activate the B cells to proliferate and differentiate further. In lymph nodes and spleen, some of the activated B cells proliferate and differentiate immediately into plasma cells, which secrete IgM antibody (see Figure 6.21). Antibody secretion is effected by a change in the processing of the heavy-chain mRNA, which leads to the synthesis of the secreted form of immunoglobulin rather than the membrane-bound form (see Section 4-13).

Plasma cells are dedicated to the continuous synthesis and secretion of antibody; the cellular organelles involved in protein synthesis and secretion become highly developed, and 10–20% of the total cell protein made is antibody. As part of this program of terminal differentiation, plasma cells cease to divide and no longer express cell-surface immunoglobulin and MHC class II molecules. In this way plasma cells become unresponsive to antigen and to interaction with T cells.

Other activated B cells migrate to a nearby primary follicle, which changes in morphology to become a **secondary lymphoid follicle** containing a **germinal center** (Figure 6.22). Here, the activated B cells become large proliferating lymphoblasts called **centroblasts**; these mature into more slowly dividing B cells called **centrocytes**, which have undergone isotype switching and somatic hypermutation. Those B cells making surface immunoglobulins with the highest affinity for antigen are selected in the process of affinity maturation (see Section 4-14), which occurs in germinal centers. A few weeks after a germinal center is formed, the intense cellular activity of the germinal center reaction dies down and the germinal center shrinks in size.

Cells that survive the selection process of affinity maturation undergo further proliferation as lymphoblasts and then migrate from the germinal center to other sites in the secondary lymphoid tissue or to the bone marrow, where they complete their differentiation into plasma cells secreting high-affinity, isotype-switched antibodies. As the primary immune response subsides, germinal center B cells also develop into quiescent resting **memory B cells** possessing high-affinity, isotype-switched antigen receptors. The production of memory cells after a successful encounter with antigen establishes antigen specificities of proven usefulness permanently in the B-cell repertoire.

Memory B cells can persist throughout a person's lifetime, and in their recirculation through the body they require only intermittent stimulation in the follicular environment. They are much more easily activated on recognizing antigen than naive B cells. Their rapid activation and differentiation into

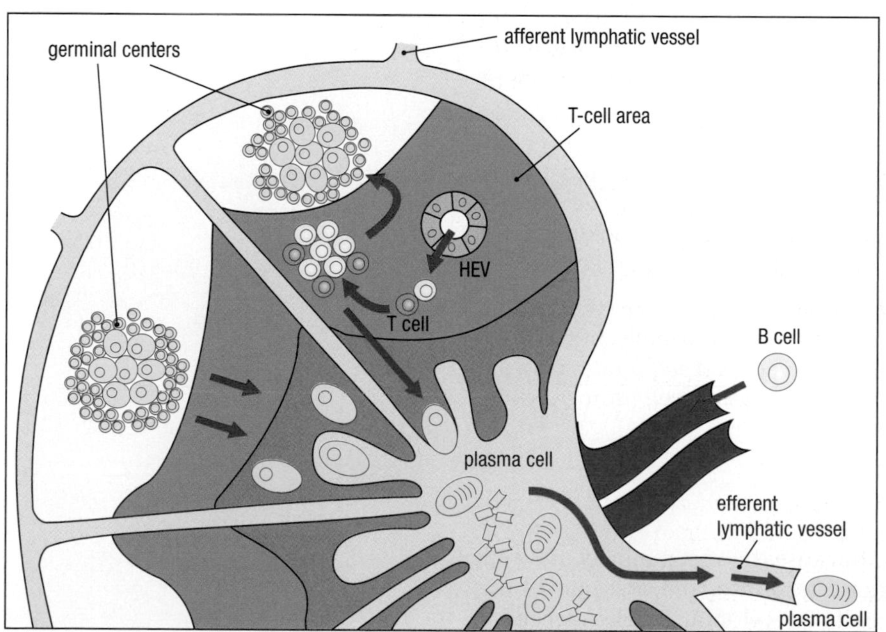

Figure 6.22 Antigen-mediated B cell activation and differentiation in secondary lymphoid tissue. A lymph node is illustrated here. A mature naive B cell enters the lymph node through an HEV. In the cortex of the lymph node the B cell encounters antigen delivered in the afferent lymph draining from infected tissue. At the border between a lymphoid follicle and the T-cell area, the B cell is activated by a CD4 helper T cell (blue) to form a primary focus of dividing cells. Some of these B cells migrate to the medullary cords and differentiate into antibody-secreting plasma cells; others migrate to a primary follicle and form a germinal center where they further divide and differentiate. B cells from the germinal center migrate either to the medullary cords or the bone marrow to complete their maturation into plasma cells.

plasma cells on a subsequent encounter with antigen enable a secondary antibody response to an antigen to develop more quickly and become stronger than the primary immune response. It also explains why IgG and antibodies of isotype other than IgM predominate in secondary responses.

When B cells become committed to differentiation into plasma cells, they migrate to particular sites in the lymphoid tissues. In the lymph nodes these are the medullary cords and in the spleen the red pulp. In the gut-associated lymphoid tissues, prospective plasma cells migrate to the lamina propria, which lies immediately under the gut epithelium. Prospective plasma cells also migrate from lymph nodes and spleen to the bone marrow, which becomes a major site of antibody production. So, in one sense, the life of a B cell both starts and ends in the bone marrow.

6-16 Different types of B-cell tumor reflect B cells at different stages of development

The study of B-cell tumors has provided fundamental insights into both B-cell development and the control of cell growth generally. B-cell tumors arise from both the B-1 and B-2 lineages and from B cells at different stages of maturation and differentiation. The general principle that a tumor represents the uncontrolled growth of a single transformed cell is illustrated vividly by tumors derived from B-lineage cells. In a B-cell tumor, every cell has an identical immunoglobulin-gene rearrangement, which is proof of their derivation from the same ancestral cell. Although the cells of an individual B-cell tumor are homogeneous, the B-cell tumors from different patients have different rearrangements, which reflects the multiplicity of rearrangements found in the normal B cells of a healthy person.

Tumors retain characteristics of the cell type from which they arose, especially when the tumor is relatively differentiated and slow growing. This principle is exceptionally well illustrated by the B-cell tumors. Human tumors corresponding to all the stages of B-cell development have been described, from the most immature progenitor to the highly differentiated plasma cell (Figure 6.23). Among the characteristics retained by the tumors is their location at defined sites in the lymphoid tissues. Tumors derived from mature, naive B cells grow in the follicles of lymph nodes and form **follicular center cell lymphoma**, whereas plasma-cell tumors, called **myelomas**, propagate in the bone marrow.

Although **Hodgkin's disease** was one of the first tumors to be successfully treated by radiotherapy, the tumor's origin as a germinal center B cell was discovered only recently. As a result of a somatic mutation, the tumor cells no longer have an antigen receptor. They often have a strange dendritic morphology and have been known as Reed–Sternberg cells. The disease presents clinically in several forms. In some patients the disease is dominated by the proliferation of nonmalignant T cells that respond to the tumor cells, whereas in others there are both Reed–Sternberg cells and more normal-looking malignant B cells that have identical immunoglobulin-gene rearrangements.

B-cell tumors have proved invaluable in the study of the immune system. They uniquely provide cells that well represent what happens normally, but they are obtained in large quantity, which is unusual. The very first sequences of heavy and light chains were obtained from patients with plasma-cell tumors, whose plasma is dominated by a single species of antibody.

Summary

B-cell development is inherently wasteful because the vast majority of B cells die without ever contributing to an immune response. The first part of this chapter described how only around one-half of developing B cells succeed in

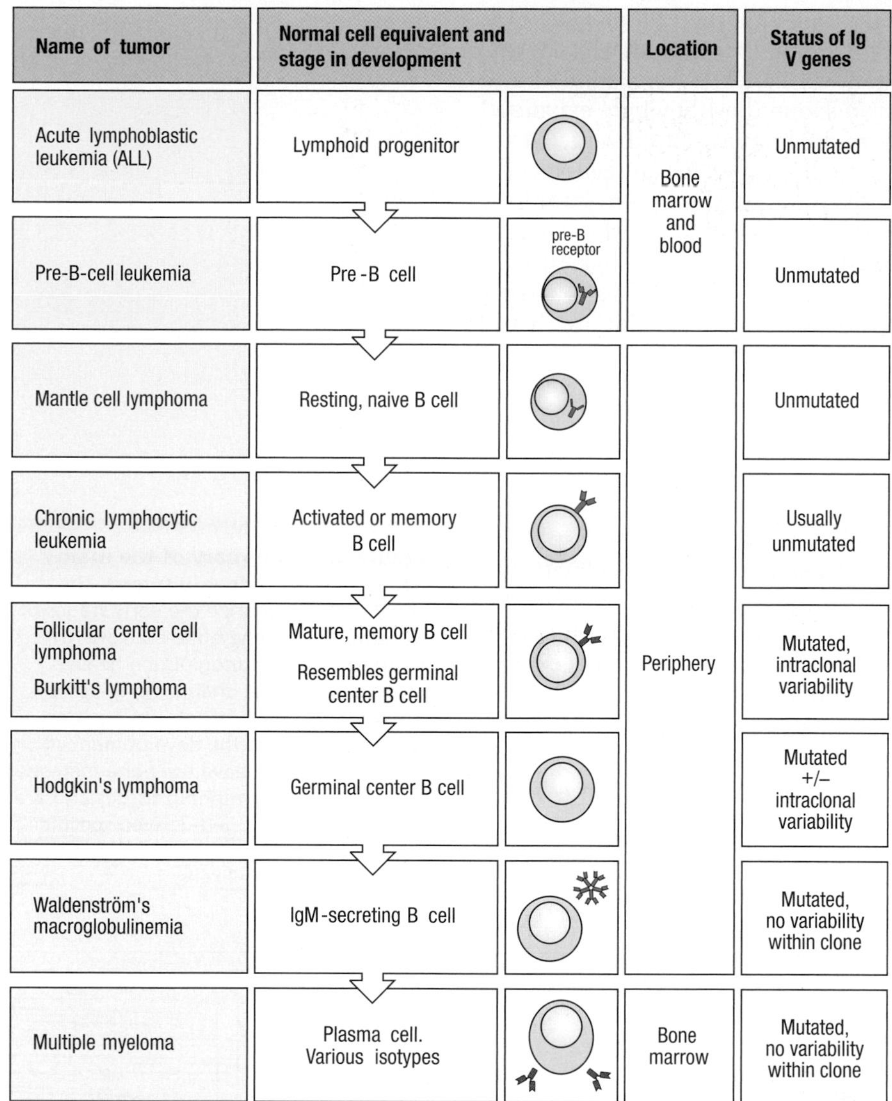

Name of tumor	Normal cell equivalent and stage in development	Location	Status of Ig V genes
Acute lymphoblastic leukemia (ALL)	Lymphoid progenitor	Bone marrow and blood	Unmutated
Pre-B-cell leukemia	Pre-B cell (pre-B receptor)	Bone marrow and blood	Unmutated
Mantle cell lymphoma	Resting, naive B cell	Periphery	Unmutated
Chronic lymphocytic leukemia	Activated or memory B cell	Periphery	Usually unmutated
Follicular center cell lymphoma Burkitt's lymphoma	Mature, memory B cell Resembles germinal center B cell	Periphery	Mutated, intraclonal variability
Hodgkin's lymphoma	Germinal center B cell	Periphery	Mutated +/– intraclonal variability
Waldenström's macroglobulinemia	IgM-secreting B cell	Periphery	Mutated, no variability within clone
Multiple myeloma	Plasma cell. Various isotypes	Bone marrow	Mutated, no variability within clone

Figure 6.23 The different B-cell tumors reflect the heterogeneity of developmental and differentiation states in the normal B-cell population. Each type of tumor corresponds to a normal state of B-cell development or differentiation. Tumor cells have similar properties to their normal cell equivalent, they migrate to the same sites in the lymphoid tissues, and they have similar patterns of expression of cell-surface glycoproteins.

making immunoglobulin and reaching the stage of an immature B cell bearing IgM at its surface. In this second part we have seen that the immunoglobulins made by a majority of immature B cells are autoreactive and rather than being a benefit for human health and survival are a potential cause of autoimmune disease. In the bone marrow, receptor editing enables many of the autoreactive B cells specific for multivalent self antigens to change their light chains, thereby modifying their immunoglobulins so that they no longer react with self antigens; those that fail to find a suitable light chain are eliminated by apoptosis. Autoreactive B cells specific for univalent self antigens are rendered anergic, a state in which their signaling is impaired and they die by neglect. As the immature B cells move out of the bone marrow and into other tissues they encounter additional self antigens not present in the bone marrow. B cells specific for these self antigens die quickly by apoptosis or slowly by anergy, but cannot be saved by receptor editing. The surviving B cells, which now express a little surface IgD as well as much IgM, remain immature and to complete their maturation they must enter a primary lymphoid follicle and interact with follicular dendritic cells. The competition between B cells for access to the follicles is such that only around 20% of the immature B cells mature and survive beyond a few days.

The mature B cells, which now have a high level of IgD and a low level of IgM on their surfaces, travel between the secondary lymphoid tissues in search of infection and pathogen-derived antigens that will bind their B-cell receptors. If in a period of 50–100 days they fail to find specific antigen, the B cells die. Only if a B cell encounters specific antigen will its clone be expanded and maintained for the rest of a person's life. No sooner has the clone expanded than individual B cells in the clone begin to differentiate and diverge. Some become plasma cells, which service the immediate need for antigen-specific antibody, while others become memory cells providing long-term immunity for the future. Over and above these changes, B cells within the clone hypermutate their immunoglobulins (see Section 4-14). Most of the B cells expressing mutant immunoglobulins are abandoned, and only the few making antibodies of highest affinity are selected to survive and provide the B-cell response to the pathogen.

Summary to Chapter 6

B lymphocytes are highly specialized cells whose sole function is to recognize foreign antigens by means of cell-surface immunoglobulins and then to differentiate into plasma cells that secrete antibodies of the same antigen specificity. Each B cell expresses immunoglobulin of a single antigen specificity, but as a population B cells express a diverse repertoire of immunoglobulins. This enables the B-cell response to any antigen to be highly specific. Immunoglobulin diversity is a result of the unusual arrangement and mode of expression of the immunoglobulin genes. In B-cell progenitors the immunoglobulin genes are in the form of arrays of different gene segments that can be rearranged in many different combinations. Gene rearrangement occurs in bone marrow and does not depend on B-cell encounter with specific antigen. The gene rearrangements needed for the expression of a functional surface immunoglobulin follow a program; the successive steps define the antigen-independent stages of B-cell development as shown in Figure 6.24. The

Figure 6.24 Summary of the main stages in B-cell development. The top panels summarize the early stages of development in the bone marrow. The status of the immunoglobulin heavy-chain (μ) and light-chain (κ/λ) genes is shown below each panel. The bottom panels summarize the development of B cells after they leave the bone marrow, enter secondary lymphoid tissues and are activated by pathogen-derived specific antigen. The diagram refers only to the development of B-2 cells.

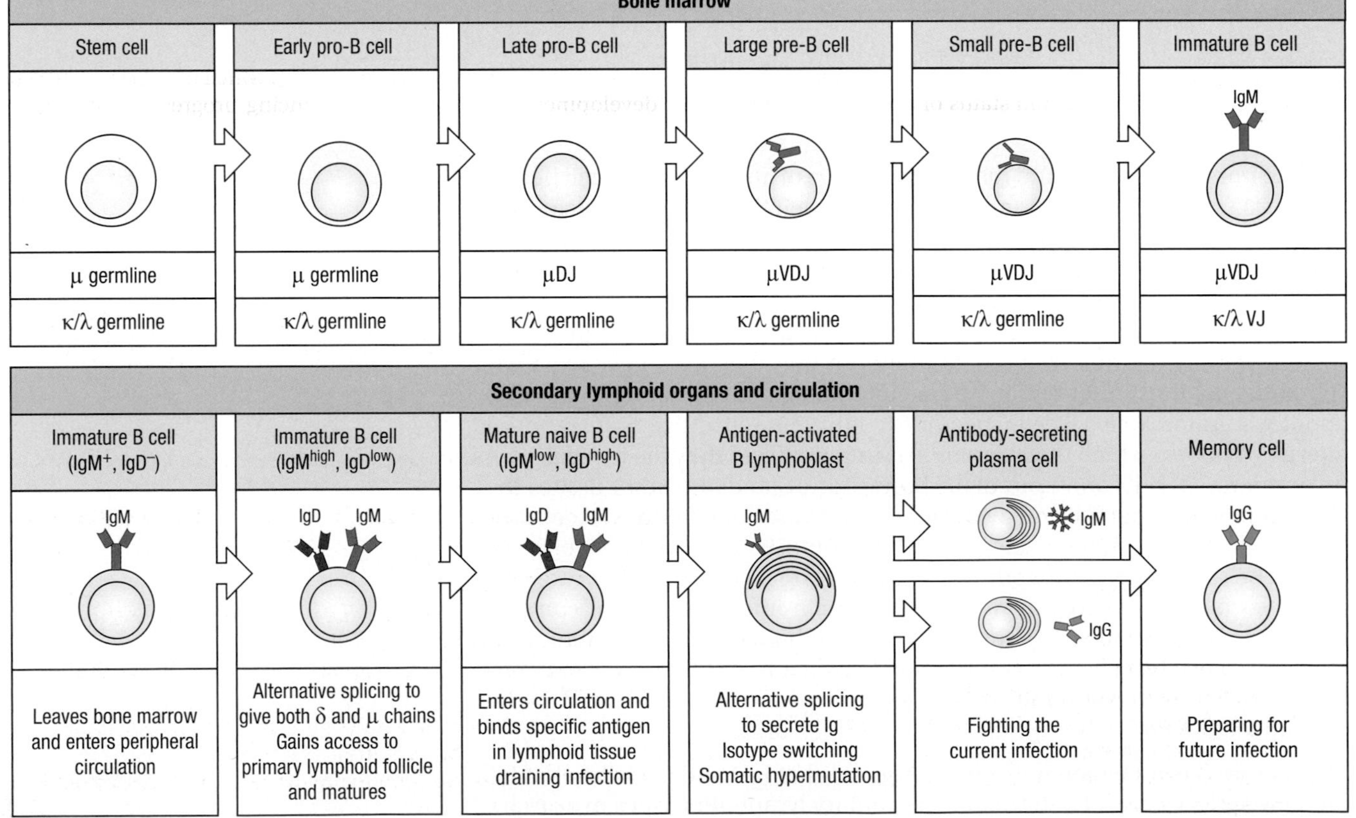

success rate for individual rearrangements is far from optimal, but the use of a stepwise series of reactions allows the quality of the products to be tested at critical checkpoints after heavy-chain and light-chain gene rearrangement. Failure at any step leads to the withdrawal of positive signals and death of the B cell by apoptosis. Gene rearrangement is controlled so that only one functional heavy-chain gene and one functional light-chain gene are produced in each cell. Thus, individual B cells produce immunoglobulins of a single antigen specificity. For a short period after successful immunoglobulin-gene rearrangement, any interaction with specific antigen leads to the elimination or inactivation of the immature B cell, thus rendering the mature B-cell population tolerant of the normal constituents of the body.

Every day the bone marrow pushes out billions of new mature, naive B cells into the peripheral circulation. There is, however, limited space for B cells in the secondary lymphoid tissues and unless naive B cells encounter specific antigen they are likely to be dead within weeks. Thus, the B-cell repertoire is never static, and newly generated specificities are continually being tested against the antigens of microorganisms causing infection. Binding of antigen to the cell-surface immunoglobulin of a B cell in secondary lymphoid tissue initiates an antigen-dependent program of cellular proliferation and development and subsequent terminal differentiation, as shown in Figure 6.24. After encounter with antigen in secondary lymphoid tissues, B cells either differentiate into IgM-secreting plasma cells or undergo somatic hypermutation, isotype switching, and affinity maturation in germinal centers before differentiation into plasma cells or long-lived memory B cells. The end product of B-cell development is the plasma cell, which no longer expresses surface immunoglobulin and devotes its resources to the synthesis and secretion of antibody.

Questions

6–1 Match the stage of B-cell development in Column A with the rearrangement or expression status of the immunoglobulin genes in Column B.

Column A	Column B
a. immature B cell	1. V–J rearranging
b. small pre-B cell	2. germline configuration of heavy- and light-chain genes
c. large pre-B cell	3. V–DJ rearranging
d. stem cell	4. μ heavy chain plus λ or κ light chains on cell surface
e. early pro-B cell	5. VDJ rearranged but light-chain gene in germline configuration
f. late pro-B cell	6. D–J rearranging

6–2 All of the following occur after B cells leave the bone marrow except

a. peripheral tolerance
b. receptor editing
c. differentiation into plasma cells
d. formation of follicular center cell lymphomas
e. production of memory B cells
f. recirculation between lymph, blood, and secondary lymphoid tissues.

6–3 Stromal cells produce ________, which is a secreted B-cell development growth factor influencing progression of B cells from the late pro-B-cell to the pre-B-cell stage.

a. stem-cell factor
b. Kit
c. IL-7
d. VLA-4
e. VCAM-1.

6–4 Which of the following events is not associated with the establishment of allelic exclusion of immunoglobulin heavy-chain loci?

a. active transcription of RAG1 and RAG2
b. degradation of RAG proteins
c. expression of only one productively rearranged heavy-chain locus
d. remodeling of chromatin structure of heavy-chain locus so that it resists gene rearrangement
e. formation of a functional pre-B-cell receptor.

6–5 Multiple successive gene rearrangements at the immunoglobulin heavy-chain locus

a. increase the likelihood that a functional heavy chain is formed
b. always follow successive gene rearrangement of the immunoglobulin light-chain locus
c. occur after the completion of the second checkpoint in small pre-B cells

d. are not possible because the D segments are removed following the joining between D to J and V to DJ
e. may result in the production of B cells expressing two different types of heavy chain.

6–6 Developing B cells that fail to make productive D to J heavy-chain rearrangements on both homologous chromosomes
a. die by apoptosis in the bone marrow
b. will rearrange heavy-chain loci multiple times until a productive rearrangement is made
c. undergo clonal proliferation
d. upregulate expression of transcription factors E2A and EBF
e. fail to rearrange V to DJ.

6–7 All of the following statements regarding immunoglobulin gene rearrangement are true except
a. Non-B cells retain their chromatin containing the immunoglobulin loci in a 'closed' configuration.
b. T-cells never transcribe immunoglobulin genes.
c. There is only one promoter on the heavy-chain locus that is 5′ of the V segments.
d. Pax-5, a transcription factor, interacts with enhancer sequences positioned 3′ of the heavy-chain C-region genes.
e. Transcription of immunoglobulin loci precedes gene rearrangement.

6–8 All of the following are properties of B-1 cells except
a. They exhibit polyspecificity.
b. They arise during fetal development.
c. They express CD5 on their cell surface.
d. They are commonly associated with the cause of chronic lymphocytic leukemia (CLL).
e. They have extensive N nucleotide diversity in VDJ junctions.

6–9 Which of the following statements regarding negative selection of B cells is correct?
a. Negative selection is a process that occurs in secondary lymphoid organs.
b. Negative selection is a process that occurs in the bone marrow but not in secondary lymphoid organs.
c. Negative selection ensures that B cells bearing receptors for pathogens that will not be encountered in a person's lifetime are eliminated to make room for B cells bearing useful receptors.
d. Negative selection eliminates B cells at the end of an infection as a means of terminating an immune response once the pathogen has been removed from the body.
e. Negative selection ensures that autoreactive B cells are prohibited from emerging in the body.

6–10 _______ involves successive rearrangement at the light-chain loci when reactivity to self antigen occurs during B-cell development.
a. receptor editing
b. somatic hypermutation
c. chromosomal translocation
d. clonal deletion
e. anergy.

6–11 Match the term/description in Column A with the process it induces/deploys in Column B.

Column A	Column B
a. nonlymphoid stromal cells of the bone marrow	1. expression of only one heavy-chain locus and one light-chain locus in a B cell
b. allelic exclusion	2. alteration of light-chain gene
c. multivalent self antigens in the bone marrow	3. chemokine-mediated recruitment of blood-borne B cells to high endothelial venules
d. non-lymphoid stromal cells of lymph node cortex	4. establishment of an anergic state
e. monovalent self antigens in the bone marrow	5. provision of adhesion molecules and growth factors for B-cell development

6–12 Immature B cells specific for monovalent self antigens encountered in the bone marrow
a. enter a state of developmental arrest known as anergy
b. express abnormally high levels of IgM on the cell surface
c. can become activated but only if IgD is no longer expressed on the cell surface
d. never leave the bone marrow and die by apoptosis
e. engage in the process of receptor editing.

6–13 B-cell tumors originate during different developmental stages of B cells during their maturation in the bone marrow or after maturation and export to the periphery.
A. Explain why B cells isolated from a particular B-cell tumor all express the same immunoglobulin.
B. How might the immunoglobulin expressed on pre-B-cell leukemia cells be different from that expressed on immature B cells?

6–14 Yasuo Yamagata, a 63-year-old, experienced severe back pain for several weeks before visiting his family physician. He also complained of fatigue and looked pale. Blood analysis revealed a red blood cell count of $3.2 \times 10^6/\mu l$ (normal $4.2–5.0 \times 10^6/\mu l$), a white blood cell count of 2800/μl (normal 5000/μl), a sedimentation rate of 30 mm/h (normal <20 mm/h) and a serum IgG of 4500 mg/dl (normal 600–1500 mg/dl). IgA and IgM levels were well below normal. Skeletal survey showed lytic lesions in vertebrae, ribs and skull. A bone marrow sample revealed 75% infiltration with plasma cells. Elevated protein in urine was confirmed to be Bence-Jones protein (immunoglobulin light chains). The patient was diagnosed with IgG λ multiple myeloma and began an immediate chemotherapy regime. Which of the following would be consistent with this type of malignant tumor of plasma cells?
a. Serum IgG is polyclonal.
b. Anemia and neutropenia are present as the result of plasma-cell infiltration in the bone marrow and consequent limitation of space.
c. Susceptibility to pyogenic infections is unaffected.
d. Serum IgG consists of IgG1, IgG2, IgG3, and IgG4 in approximately equal proportions.
e. κ and λ light chains are found in excessive quantities in the urine.

6–15 Thomas Harrison, born at 39 weeks after an uneventful pregnancy, developed normally and thrived for the first 8 months of infancy. During the next 12 months he received antibiotics for a number of infections including two episodes of acute otitis media (middle ear infection), sinusitis, and superficial cellulitis (streptococcal skin infection) on his left thigh. Two days ago, after an upper respiratory infection, Thomas developed a fever, became drowsy, and experienced a seizure. Lumbar puncture confirmed *Haemophilus influenzae* type B (HiB) in the cerebrospinal fluid. Intravenous ceftriaxone was administered and his condition stabilized after 24 hours. Laboratory results showed his total lymphocyte count to be normal; however, flow cytometry using antibodies for the B-cell marker CD19 revealed an absence of B cells. Antibodies against CD3, CD8, and CD4 revealed the presence of cytotoxic and helper T cells. Serum IgG and IgM levels were remarkably low at 75 mg/dl and 10 mg/dl, respectively (IgG normally 600–1500 mg/dl; IgM 75–150 mg/dl), and IgA was undetectable. Family history showed that neither of Thomas's two older sisters (ages 4 and 7) had experienced recurrent infections like this in infancy. Thomas's maternal aunt had a boy of about the same age as Thomas who had also been experiencing repeated episodes of otitis media, sinusitis, and, most recently, a severe case of pneumonia, all caused by pyogenic bacteria and treated successfully with antibiotics. Thomas began intravenous IgG replacement therapy administered at 3–4-week intervals, and his repeated episodes of infection abated. These findings are most consistent with a diagnosis of

a. hyper-IgM syndrome
b. X-linked agammaglobulinemia (XLA)
c. chronic granulomatous disease (CGD)
d. DiGeorge syndrome
e. MHC class I deficiency.

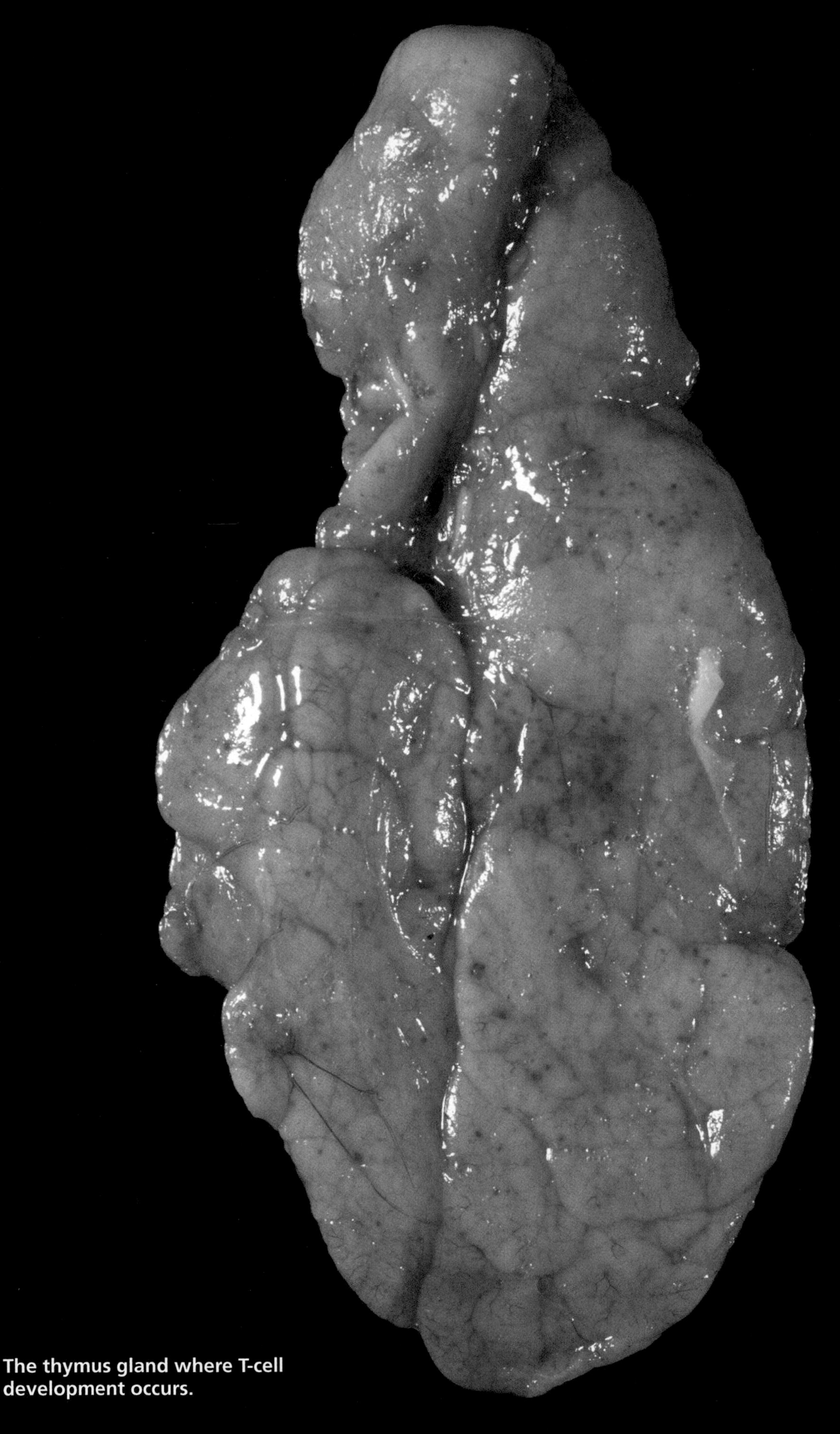

The thymus gland where T-cell development occurs.

Chapter 7

The Development of T Lymphocytes

The developmental pathways of T and B lymphocytes have much in common: both cell types derive from bone marrow stem cells and, during development, they must undergo gene rearrangement to produce their antigen receptors. However, whereas B cells rearrange their immunoglobulin genes while remaining in the bone marrow, the precursors of T cells have to leave the bone marrow and enter another primary lymphoid organ—the thymus—before they rearrange their T-cell receptor genes. Gene rearrangements in developing T cells proceed in a broadly similar fashion to those in B cells, but with key differences. The main one being the formation of two distinct T-cell lineages: one expressing α:β receptors and the other γ:δ receptors (see Section 5-5).

As we saw in Chapter 5, T-cell receptors do not recognize peptide antigens in isolation but in complexes with MHC molecules (see Section 5-7). A major function of the thymus is to ensure that a person's mature T cells bear T-cell receptors that recognize peptides in the context of the particular MHC class I and class II isoforms expressed by that person, their self MHC. This is achieved by a process of positive selection in the thymus, which gives a survival signal to immature T cells bearing receptors that interact with a self-MHC molecule and consigns immature T cells that lack such receptors to die by neglect. The immature T cells chosen by positive selection then undergo an additional selective process, called negative selection, which induces the death of T cells bearing receptors that bind too strongly to a self-MHC molecule and thus are autoreactive. Because of positive and negative selection, the mature T cells that leave the thymus to circulate through the secondary lymphoid organs are tolerant of self antigens, responsive to foreign antigens presented by self-MHC molecules, and ready to fight infection.

The first part of this chapter traces the stages in gene rearrangement that produce the primary repertoire of T-cell receptors. The second part of the chapter describes the processes of positive and negative selection that act on this repertoire in the thymus to produce the peripheral population of mature naive T cells.

The development of T cells in the thymus

T cells are lymphocytes that originate from bone marrow stem cells but emigrate to mature in the thymus (Figure 7.1). With the discovery of this developmental pathway, these lymphocytes were called **thymus-dependent lymphocytes**, which soon became shortened to **T lymphocytes** or **T cells**. Two lineages of T cells develop in the thymus: the majority are α:β T cells and

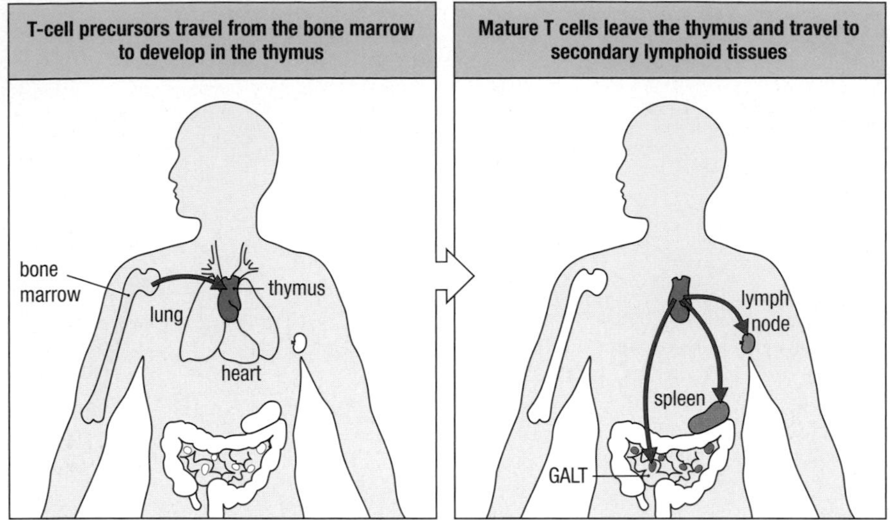
T-cell precursors travel from the bone marrow to develop in the thymus
Mature T cells leave the thymus and travel to secondary lymphoid tissues
bone marrow
thymus
lung
heart
lymph node
spleen
GALT

Figure 7.1 T-cell precursors migrate from the bone marrow to the thymus to mature. T cells derive from bone marrow stem cells whose progeny migrate in the blood from the bone marrow to the thymus (left panel), where the development of T cells occurs. Mature T cells leave the thymus in the blood, from where they enter secondary lymphoid tissues (right panel) and then return to the blood in the lymph. In the absence of activation by specific antigen, mature T cells continue to recirculate between the blood, secondary lymphoid tissues, and lymph. GALT, gut-associated lymphoid tissue.

the minority are γ:δ T cells. These lineages develop in parallel from a common **T-cell precursor**. During their development in the thymus, immature T cells also start to express other cell-surface glycoproteins related to their eventual functions in the immune response. Prominent among these are CD4 and CD8, the co-receptors that distinguish the two sublineages of α:β T cells that recognize peptide antigens presented by MHC class II and class I, respectively.

DiGeorge syndrome

7-1 T cells develop in the thymus

The **thymus** is a lymphoid organ in the upper anterior thorax just above the heart. It is dedicated to T-cell development and contains immature T cells, called **thymocytes**, which are embedded in a network of epithelial cells known as the **thymic stroma** (Figure 7.2). Together with other cell types, these elements form an outer close-packed cortex and an inner, less dense, medulla (Figure 7.3). The thymus is designated a primary lymphoid organ because it is concerned with the production of useful lymphocytes, not with their application to the problems of infection. Unlike the secondary lymphoid organs, which perform the latter function, the thymus is not involved in lymphocyte recirculation; neither does it receive lymph from other tissues. The blood is the only route by which progenitor cells enter the thymus and by which mature T cells leave.

In the embryonic development of the thymus, the epithelial cells of the cortex arise from ectodermal cells, whereas those of the medulla derive from endodermal cells. Together, these two types of epithelial cell form a rudimentary thymus, called the **thymic anlage**, which subsequently becomes colonized by progenitor cells from the bone marrow. The progenitor cells give rise to thymocytes and to dendritic cells, the latter populating the medulla of the thymus (see Figure 7.3). Independently of these progenitors, the thymus is also colonized by bone marrow derived macrophages, which, although concentrated in the medulla, are also scattered throughout the cortex (see Figure 7.3). The T-cell progenitors enter the thymus at the junction between the cortex and the medulla. As they differentiate, the thymocytes first move out through the cortex to the subcapsular region and then move progressively back from the outer cortex to the inner cortex and the medulla.

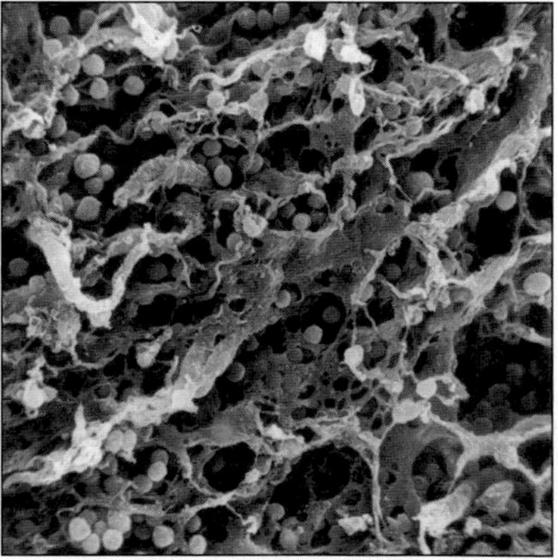

Figure 7.2 The epithelial cells of the thymus form a network surrounding developing thymocytes. In this scanning electron micrograph of the thymus, the developing thymocytes (the spherical cells) occupy the interstices of an extensive network of epithelial cells. Micrograph courtesy of W. van Ewijk.

The importance of the thymus in establishing a functional T-cell repertoire is clearly demonstrated by patients who have complete **DiGeorge syndrome**. In this genetic disease, the thymus fails to develop, T cells are absent although

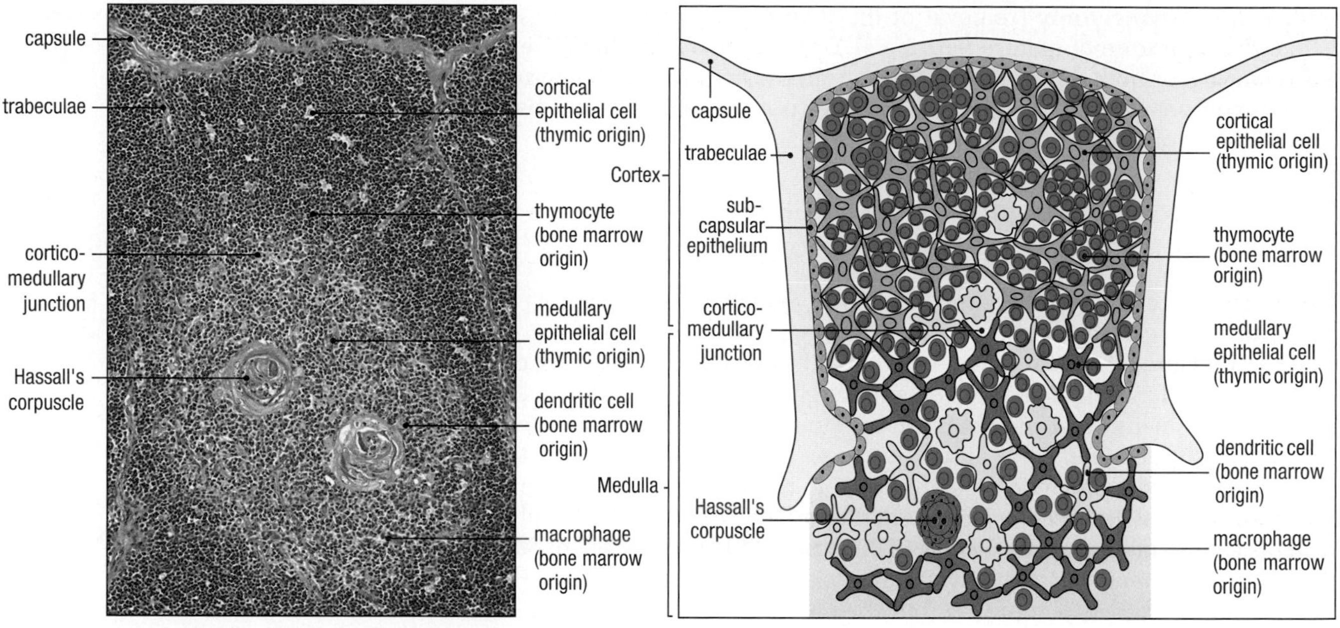

Figure 7.3 The cellular organization of the thymus. The thymus is made up of several lobules. A section through a lobule stained with hematoxylin and eosin and viewed with the light microscope is shown in the left panel. The cells in this view are shown in diagrammatic form in the right panel. In the left panel, the darker staining of cortex compared with the medulla can be discerned. As shown in the right panel, the cortex consists of immature thymocytes (blue), branched cortical epithelial cells (light orange), and a few macrophages (yellow). The medulla consists of mature thymocytes (blue), medullary epithelial cells (orange), dendritic cells (yellow), and macrophages (yellow). One of the functions of the macrophages in both cortex and medulla is to remove the many thymocytes that fail to mature properly. A characteristic feature of the medulla is Hassall's corpuscles, which are believed to be sites of cell destruction. Photograph courtesy of Yasodha Natkunam.

B cells are made, and the resulting patients suffer from a wide range of opportunistic infections. In effect these patients have no adaptive immune system.

The human thymus is fully developed before birth, and by one year after birth it begins to degenerate, with fat gradually claiming the areas once packed with thymocytes. This process, which continues steadily throughout life, is called the involution of the thymus (Figure 7.4). The reduced production of new T cells by the thymus with age does not grossly impair T-cell immunity, and

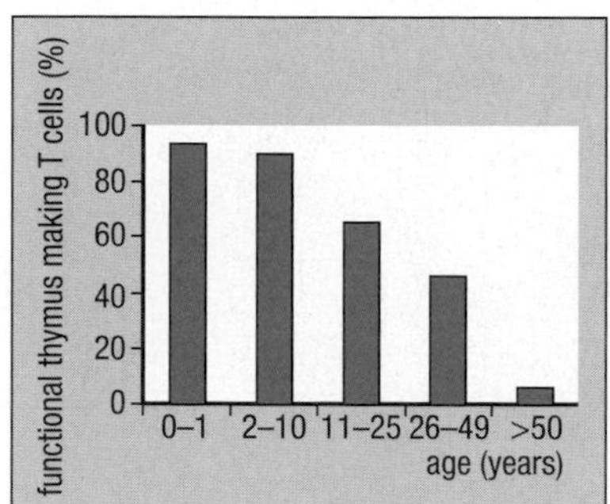

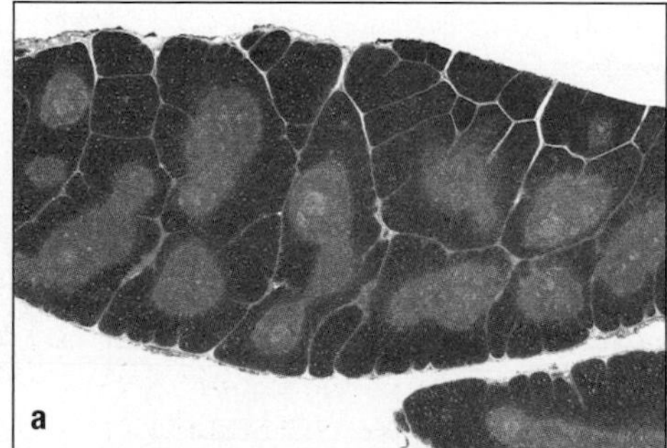

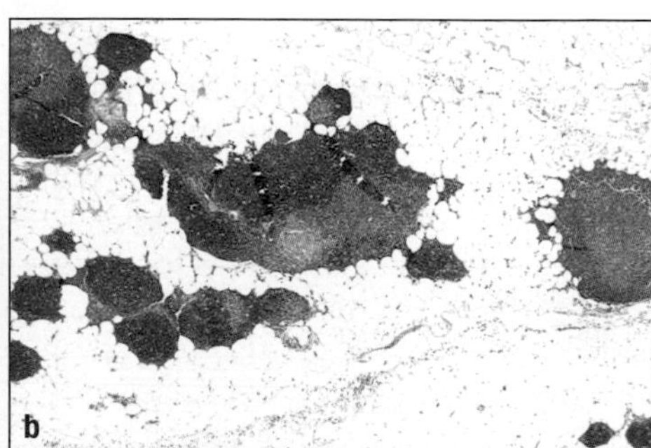

Figure 7.4 The proportion of the thymus that produces T cells decreases with age. Starting at birth, the T-cell-producing tissue of the thymus is gradually replaced by fatty tissue. This process is called the involution of the thymus. The graph shows the percentage of thymic tissue that is still producing T cells at different ages. The micrograph in panel a shows a section through the thymus of a 3-day-old infant; the micrograph in panel b shows a section through the thymus from a 70-year-old person for comparison. Tissue is stained with hematoxylin and eosin (red and blue). Magnification × 20. Micrographs courtesy of Yasodha Natkunam.

neither does **thymectomy** (removal of the thymus) of adults. Once established, the repertoire of mature peripheral T cells seems to be long lived or self-renewing, or both. In this it is different from the mature B-cell repertoire, which is composed of short-lived cells that are continually being replenished from the bone marrow.

7-2 Thymocytes commit to the T-cell lineage before rearranging their T-cell receptor genes

The progenitor cells that will eventually give rise to mature T cells are not committed to the T-cell lineage when they enter the thymus. At that time they express CD34 and other cell-surface glycoproteins that characterize stem cells, and they lack all the characteristic cell-surface glycoproteins of mature T cells. On interaction with thymic stromal cells, the progenitor cells are signaled to divide and differentiate. After around a week, the cells have lost stem-cell markers and become thymocytes committed to the T-cell lineage, as seen by their expression of the T-cell-specific adhesion molecule CD2 and other glycoproteins expressed by T cells, such as CD5 (Figure 7.5). At this stage of development, these precursor thymocytes still lack any component of the T-cell receptor complex (see Section 5-4) but they are beginning to rearrange the T-cell receptor genes. These cells express neither CD4 nor CD8 and so are called **double-negative thymocytes** or **DN thymocytes**.

A critical cytokine in T-cell development is interleukin-7 (IL-7), which is secreted by thymic stromal cells and binds to the interleukin-7 receptor on the CD34-expressing progenitor cells. The importance of this interaction is seen from the absence of T cells in immunodeficient patients who have inherited two defective alleles of the interleukin-7 receptor. Another major regulator of T-cell development is Notch1, a cell-surface receptor on thymocytes that interacts with transmembrane ligands on thymic epithelial cells. At all stages in the

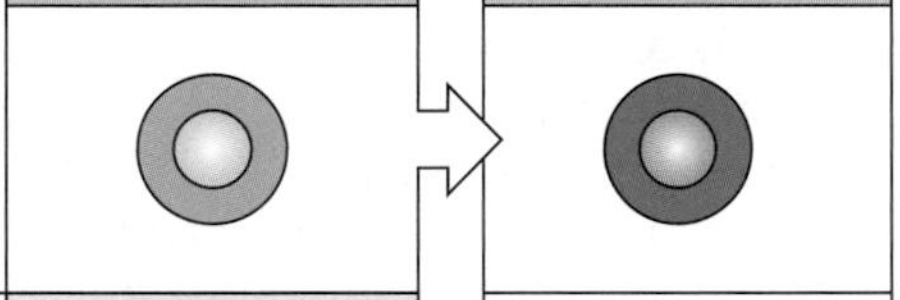

		Uncommitted progenitor cell	Double-negative thymocyte committed to the T-cell lineage
CD34	stem-cell surface marker	+	-
CD44	adhesion	+	-
CD2	adhesion and signaling	-	+
CD5	adhesion and signaling	-	+
IL-7 receptor (CD127)	cytokine receptor	-	+
CD1A	MHC class-I-like molecule	-	+
CD4	co-receptor	-	-
CD8	co-receptor	-	-
TCR genes	antigen receptor	germline	beginning to rearrange

Figure 7.5 Commitment to the T-cell lineage involves changes in gene expression and in cell-surface markers.

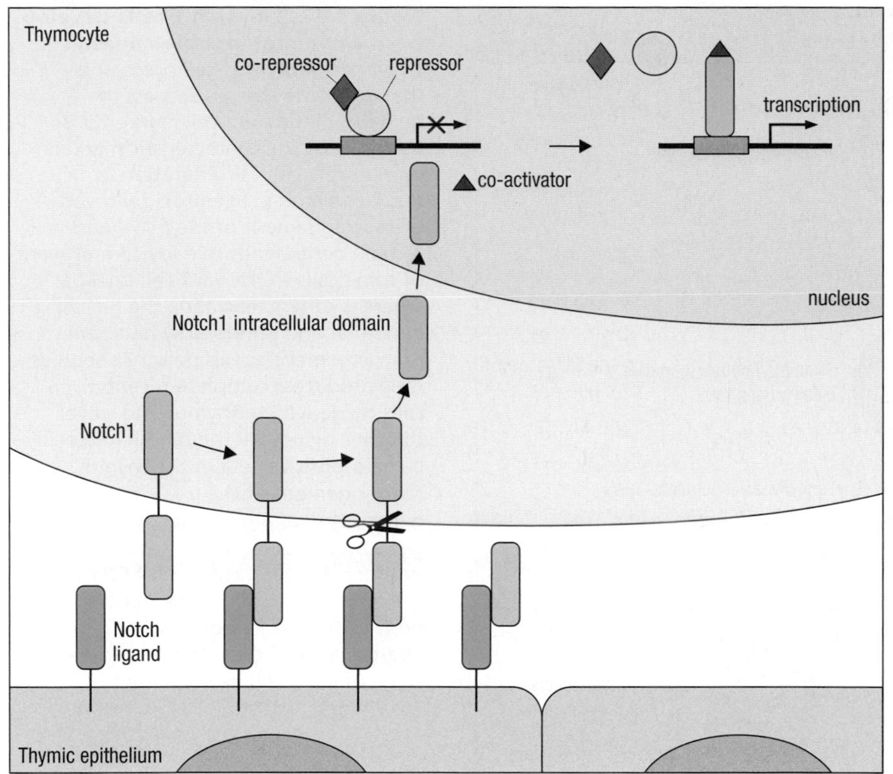

Figure 7.6 T-cell development is driven by the receptor Notch1. The membrane-associated receptor called Notch1 on the suface of the thymocyte binds to its ligand on thymic epithelium. This interaction induces a protease to cleave the intracellular domain, releasing it from the plasma membrane. The soluble intracellular domain is translocated to the nucleus, where it turns on the expression of genes essential for T-cell development by removing repressive transcription factors and recruiting co-activating transcription factors.

early development of T cells, signals generated through Notch1 are necessary to drive the cells along the pathway of T-cell differentiation. Notch1 is one of a family of four Notch receptors in humans that control the development of many different cell types by deciding between one of two fates. In the unique environment of the thymus, Notch1 keeps thymocytes on the pathway of T-cell differentiation and away from the pathway of B-cell differentiation. In this regard, Notch1 has a role in T-cell development which parallels that of Pax-5 in B-cell development (see Section 6-8).

Notch proteins are transmembrane receptors in which the extracellular and intracellular domains have distinct and complementary functions. When the extracellular domain of Notch1 binds to its ligand on thymic epithelium, it initiates a proteolytic cleavage that releases the intracellular domain of Notch1 from the membrane. The intracellular domain translocates to the thymocyte nucleus, where it becomes part of a transcription factor complex that initiates the transcription of genes needed for T-cell development. To do this it displaces repressive transcription factors from gene promoters and recruits activating factors (Figure 7.6).

7-3 The two lineages of T cells arise from a common thymocyte progenitor

T-cell development gives rise to two functionally different lineages of T cells. These are distinguished by the expression of an α:β or a γ:δ T-cell receptor. Although their later stages of development in the thymus are very different, α:β and γ:δ T cells both derive from the common double-negative thymocyte precursor in which rearrangement of the genes encoding both of the T-cell receptors is initiated (Figure 7.7).

How T cells commit to either the α:β or γ:δ lineage is a complicated business, because individual cells are not restricted to rearranging either the α and β genes or the γ and δ genes. Instead, the double-negative thymocytes are

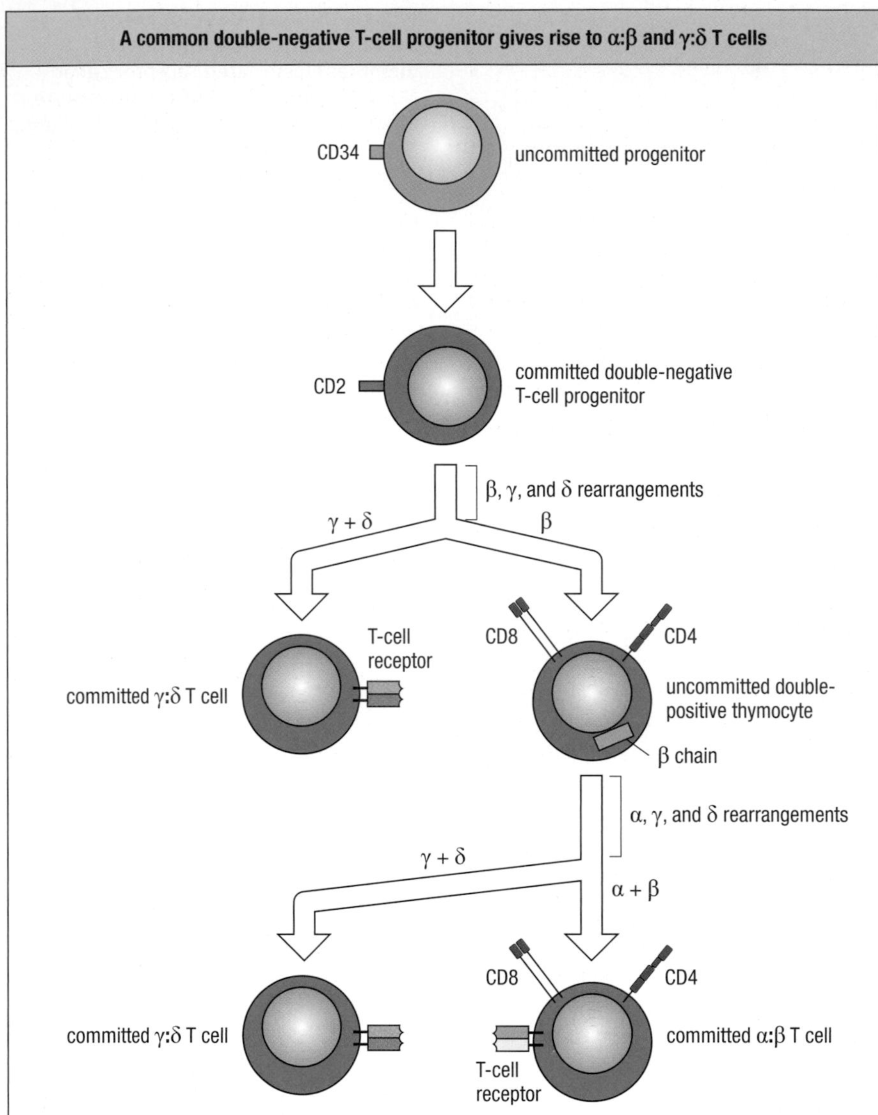

Figure 7.7 α:β and γ:δ T cells develop from a common double-negative T-cell progenitor. T-cell precursors that enter the thymus express the hematopoietic stem-cell marker CD34 but none of the characteristic markers of mature T cells. Proliferation of these common progenitors followed by rearrangement of the δ-, γ-, and β-chain genes leads to early commitment of some cells to the γ:δ T-cell lineage, whereas others rearrange the β-chain gene first and temporarily halt gene rearrangement at this point. As soon as they produce a complete receptor, γ:δ cells can leave the thymus and travel to other tissues via the blood. In the β-chain-positive cells in the thymus, rearrangement of the α-, γ-, and δ-chain genes resumes and productive α-chain gene rearrangements in these cells produce double-positive CD4 CD8 α:β cells. A minority of the double-positive thymocytes give rise to additional γ:δ T cells. This ends the early stage of α:β T-cell development.

programmed to begin rearrangements at the γ, δ, and β loci at roughly the same time. In effect, the γ and δ loci compete with the β locus in a race to make productive gene rearrangements and functional T-cell receptor chains (see Figure 7.7). If the thymocyte makes a functional γ:δ receptor before a functional β chain, then it commits to becoming a γ:δ T-cell. If a functional β chain is made before a γ:δ receptor it is incorporated into a protein called the pre-T-cell receptor. Although this outcome favors the α:β lineage it does not commit the cell to that lineage. Gene rearrangement stops at this point, while the cell proliferates and expresses the CD4 and CD8 co-receptors. Because they express both co-receptors, cells at this stage of development are called **double-positive thymocytes** or **DP thymocytes**. Rearrangement of the α-chain genes is now allowed and rearrangement at the γ- and δ-chain genes can also continue. Further competition ensues. If a double-positive thymocyte makes an α:β receptor before a γ:δ receptor it commits to the α:β lineage. Conversely, if a γ:δ receptor is made first, then the cell commits to the γ:δ lineage (see Figure 7.7).

Cells that fail to make productive T-cell receptor gene rearrangements die by apoptosis and are phagocytosed by macrophages in the thymic cortex (Figure 7.8). Apoptosis is the fate of all except a very few thymocytes (around

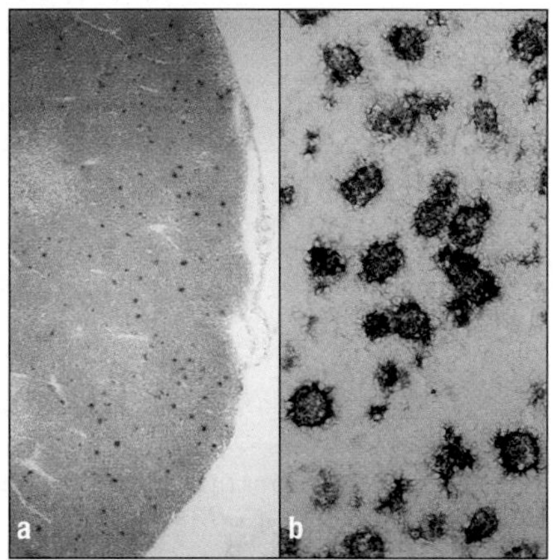

Figure 7.8 Immature T cells that undergo apoptosis are ingested by macrophages in the thymic cortex. Panel a shows a section through the thymic cortex (to the right side of the panel) and part of the medulla, in which cells have been stained for apoptosis with a red dye. Apoptotic cells are scattered throughout the cortex but are rare in the medulla. Panel b shows a section of thymic cortex at higher magnification that has been stained red for apoptotic cells and blue for macrophages. Apoptotic cells can be seen within the macrophages. Magnifications: panel a, ×45; panel b, ×164. Photographs courtesy of J. Sprent and C. Surh.

2%), and the macrophages of the thymus are continually removing dead and dying cells while not interfering with ongoing thymocyte development.

7-4 Gene rearrangement in double-negative thymocytes leads to assembly of either a γ:δ receptor or a pre-T-cell receptor

Like the immunoglobulin genes, T-cell receptor genes can make rearrangements that are either productive or nonproductive, and rearrangements can be attempted on both copies of each locus. The type of rearrangement is also analogous to those made by the immunoglobulin genes. For the β- and δ-chain loci, which contain V, D, and J segments, the first rearrangement joins D to J and the second joins V to DJ; for the α- and γ-chain loci, which contain only V and J segments, a single rearrangement joins V to J (see Figure 5.3, p. 116, and Figure 5.8, p. 119).

The developmental pathway taken by a T cell is determined by the sequence in which the T-cell receptor genes make productive rearrangements. If a thymocyte makes productive γ- and δ-chain gene rearrangements before a productive β-chain gene rearrangement, then a γ:δ heterodimer is made. This assembles with the CD3 signaling complex (see Section 5-4), moves to the cell surface and signals the cell to stop β-chain rearrangement. The cell is then committed to being a γ:δ T cell. As γ:δ T-cell receptors are not subject to the stringent positive and negative selection imposed on the α:β receptor repertoire, γ:δ T cells soon leave the thymus in the blood and enter the circulation (Figure 7.9).

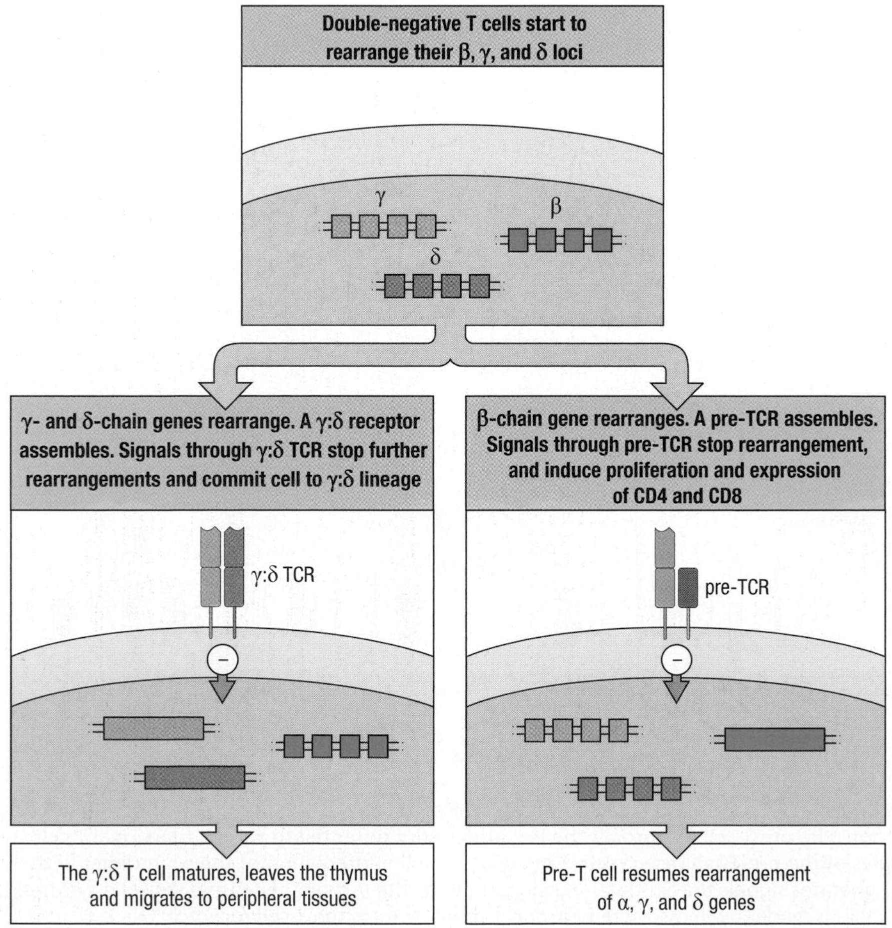

Figure 7.9 T-cell receptor gene rearrangements in double-negative thymocytes can lead to expression of either a γ:δ receptor or a pre-T-cell receptor. In double-negative thymocytes, the β, γ, and δ genes rearrange (top panel). If successful γ- and δ-chain gene rearrangements occur first, then a γ:δ receptor is expressed and the cell is signaled to differentiate into a mature γ:δ cell (bottom left panel). If a successful β-chain gene rearrangement is made before the γ and δ genes have both made successful rearrangements, a pre-T-cell receptor (pre-TCR) assembles and signals the cell to proliferate, express CD4 and CD8, and become a pre-T cell. At this stage the pre-T cell turns on the recombination machinery to rearrange the α, γ, and δ genes.

The more frequent outcome of the competition between the β-, γ- and δ-chain genes is for a productive β-chain gene rearrangement to be made before both the γ- and δ-chain genes achieve a productive rearrangement. In this situation the β chain is made, translocated to the endoplasmic reticulum and tested there for its capacity to bind to an invariant polypeptide called **pTα**, which acts as a surrogate α chain analogous to the surrogate light chain of the pre-B-cell receptor (see Section 6-4). If the β chain binds to pTα, this heterodimer assembles with the CD3 complex and ζ chain to form a protein complex called the **pre-T-cell receptor** (Figure 7.10), which performs a similar function to the pre-B-cell receptor. For γ:δ T cells, in which there is no fixed order for rearranging the γ and δ chain genes, there is no equivalent surrogate chain and no equivalent to the pre-T-cell receptor.

The functional form of the pre-T-cell receptor is a superdimer comprising two heterodimers of a β chain and pTα. In the superdimer, pTα interacts extensively with both the C and V regions of the β chain. Forming the superdimer is thus a stringent test of the β chain's conformation and its potential for associating with α chains and making functional T cell receptors. In forming the superdimer, the heterodimer of pTα and the β chain can be seen to serve as both a receptor and a ligand. As no exogenous ligand is required, the pre-T-cell receptor constutively generates intracellular signals through the CD3 complex and the cytoplasmic protein Lck, a tyrosine kinase. Assembly of a pre-T-cell receptor signals the cell to halt rearrangement of the β-, γ-, and δ-chain genes (see Figure 7.9). Although the pre-B-cell receptor and the pre-T-cell receptor have the same function of quality control, they have differences as well as similarities in how they achieve this end. Assembly of the pre-T-cell receptor is a checkpoint in T-cell development that determines whether the β chain made by a developing T cell has the potential to bind to α chains, and if not, further development of the cell is stopped. Thymocytes that pass this test are called **pre-T cells** and proceed to the next stage of development.

7-5 Thymocytes can make four attempts to rearrange a β-chain gene

The competition to recruit thymocytes to the two T-cell lineages is biased to favor the α:β lineage, because commitment to that lineage requires only one productive gene rearrangement, whereas commitment to the γ:δ lineage requires a minimum of two productive rearrangements. Further increasing the probability of commitment to the α:β lineage is the potential for nonproductive β-chain gene rearrangements to be rescued by the further rearrangement

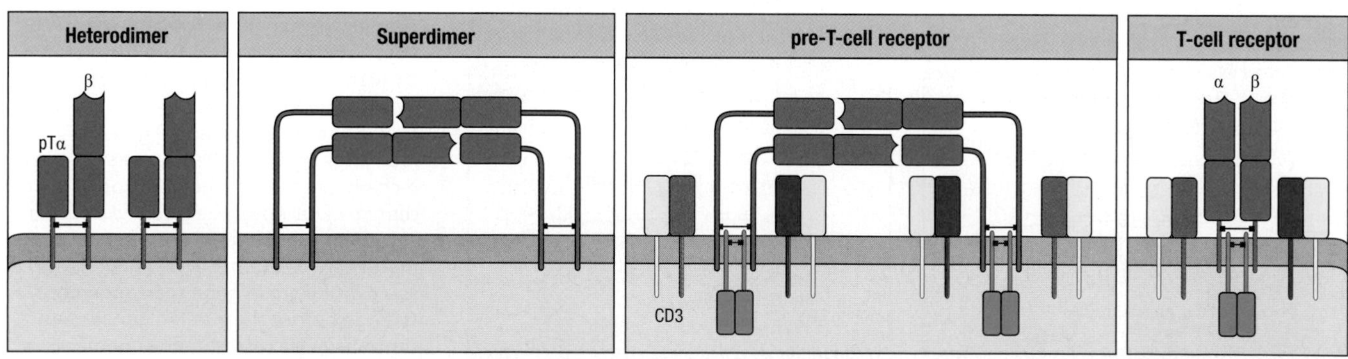

Figure 7.10 Comparison of the structures of the pre-T-cell receptor and the T-cell receptor. A heterodimer of a β chain and the pTα chain forms the pre-T-cell receptor. If the β chain has the potential to form a functional T-cell receptor it allows two molecules of the pre-T-cell receptor to form a superdimer. Interaction of the superdimer with the CD3 signaling molecules generates signals that initiate rearrangement at the α-chain genes and stop the synthesis of pTα. When a functional α chain is made it will associate with the β chain to form the T-cell receptor.

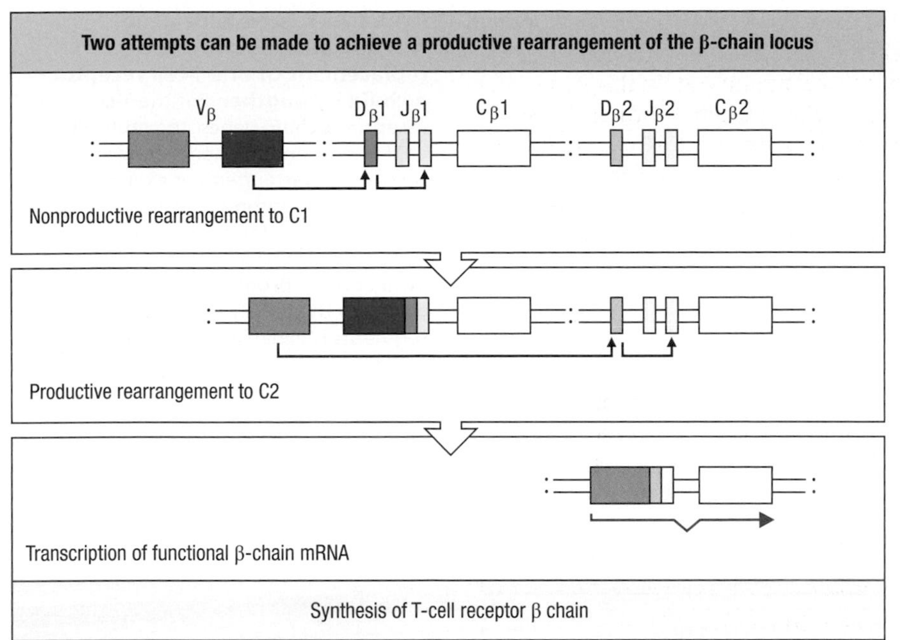

Figure 7.11 Rescue of nonproductive rearrangements of the β-chain locus. Successive rearrangements can rescue an initial nonproductive β-chain gene rearrangement, but only if that rearrangement involved D and J gene segments associated with the $C_\beta 1$ gene segment. A second rearrangement is then possible, in which a second V_β gene segment rearranges to a DJ segment associated with the $C_\beta 2$ gene segment, deleting $C_\beta 1$ and the nonproductively rearranged gene segments.

made possible by the two C_β genes and their associated D_β and J_β segments (Figure 7.11). If a rearrangement at one β-chain locus is nonproductive, a thymocyte can attempt a rearrangement at the β-chain locus on the other, homologous, chromosome. A nonproductively rearranged β-chain gene can also be rescued by a second rearrangement at the same locus. This possibility is not available to the immunoglobulin heavy-chain genes, because a nonproductive rearrangement deletes all the nonrearranged D segments. The potential for trying out up to four β-chain gene rearrangements means that 80% of T cells make a productive rearrangement of the β-chain gene, compared with a 55% success rate for heavy-chain gene rearrangement by B cells.

7-6 Rearrangement of the α-chain gene occurs only in pre-T cells

Successful rearrangement of a β-chain gene followed by signaling through the pre-T-cell receptor induces the pre-T cell to stop gene rearrangements by suppressing the expression of the *RAG1* and *RAG2* recombination-activating genes; the same phenomenon occurs during B-cell development (see Section 6-5). This ensures that only one β-chain gene has a productive rearrangement and is expressed, so that there is allelic exclusion at the β-chain locus. The pre-T cell is also induced to proliferate, which creates a clone of cells all expressing the same β chain. Proliferation is accompanied by the expression first of CD4 and then CD8 to give so-called double-positive thymocytes. These cells, which constitute the majority of thymocytes, are found predominantly in the inner cortex of the thymus, where they interact intimately with the branching network of epithelial cells. On ceasing to proliferate, the large double-positive thymocytes become small double-positive thymocytes in which the recombination machinery is reactivated and targeted to the α-chain locus, as well as to the γ and δ loci, but not to the β-chain locus.

The T-cell receptor α-chain genes are comparable to the immunoglobulin κ and λ light-chain genes in that they do not have D segments and are rearranged only after their partner receptor chain has been expressed. As with the immunoglobulin light-chain gene, repeated attempts at α-chain gene rearrangement are possible. The presence of multiple V_α gene segments and about 60 J_α gene segments spread over some 80 kb of DNA allows many

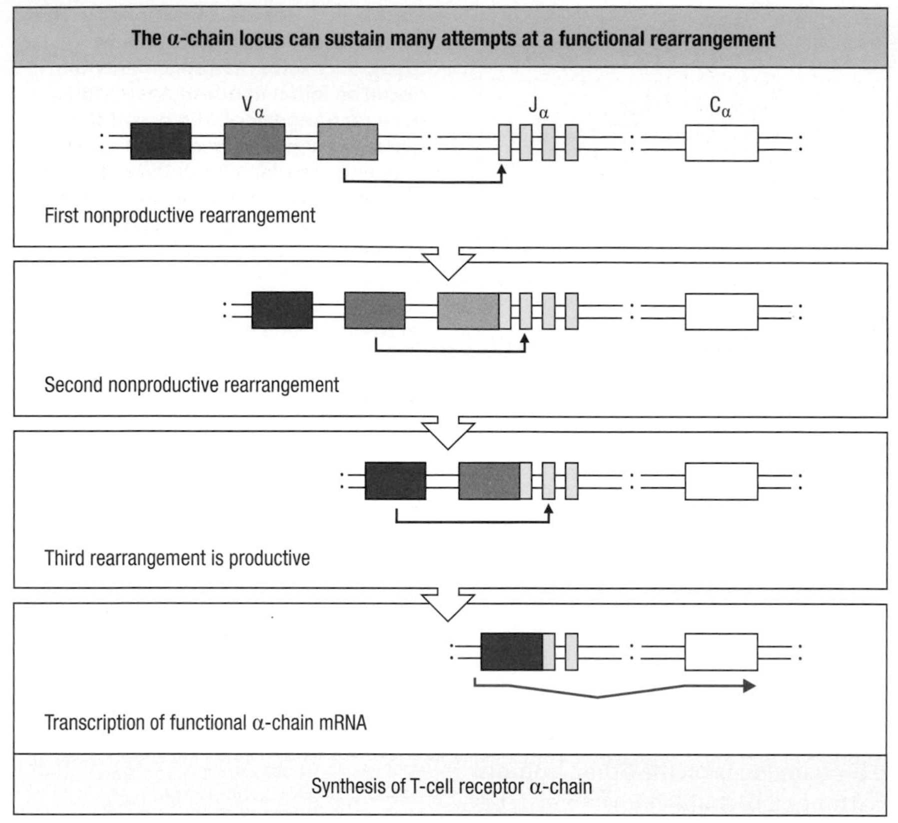
The α-chain locus can sustain many attempts at a functional rearrangement
Vα
Jα
Cα
First nonproductive rearrangement
Second nonproductive rearrangement
Third rearrangement is productive
Transcription of functional α-chain mRNA
Synthesis of T-cell receptor α-chain

Figure 7.12 Successive gene rearrangements allow the replacement of one T-cell receptor α chain by another. For the T-cell receptor α-chain genes, the multiplicity of V and J gene segments allows successive rearrangement events to jump over nonproductively rearranged VJ segments, deleting the intervening gene segments. This process continues until either a productive rearrangement occurs or the supply of V and J gene segments is exhausted, whereupon the cell dies.

successive V_α to J_α rearrangements to take place at both α-chain alleles. This means that T cells with an initial nonproductive α-gene rearrangement are highly likely to be rescued by a subsequent rearrangement (Figure 7.12).

When an α-chain gene is rearranged, the δ locus situated within the α locus will be automatically deleted, irrespective of whether the rearrangement is productive or not (Figure 7.13); see also Figure 5.8 (p. 119) for a detailed diagram of the arrangement of the α and δ loci. This arrangement greatly reduces the probability that a T cell committed to the α:β lineage will end up expressing a γ:δ receptor as well as an α:β receptor.

When a double-positive T cell makes a productive α-chain gene rearrangement, the gene is transcribed and an α chain is made. After translocation to the endoplasmic reticulum, the α chain is tested for its capacity to bind the β chain and assemble a T-cell receptor. This represents a second checkpoint during T-cell development. If successful, the double-positive cell is signaled to survive and to proceed to positive selection, the next step in the developmental pathway. If the α chain does not properly assemble with the β chain, further α-chain gene rearrangements are made until either a functional α chain is produced or the possibilities for gene rearrangement are exhausted. In the latter case, the pre-T cell will die by apoptosis.

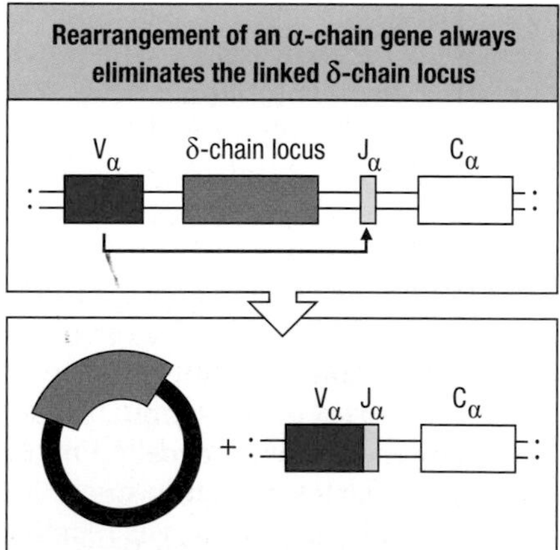
Rearrangement of an α-chain gene always eliminates the linked δ-chain locus
Vα
δ-chain locus
Jα
Cα
Vα Jα
Cα
+

Figure 7.13 The δ-chain gene is sequestered within the α-chain gene and is deleted upon α-chain gene rearrangement. Upon rearrangement of an α-chain gene, the δ-chain gene it contains is eliminated as part of an extrachromosomal circle.

7-7 Stages in T-cell development are marked by changes in gene expression

The outcome of this first part of α:β T-cell development is a set of rearranged T-cell receptor genes and a diverse population of immature cells that each carry CD4, CD8, and a correctly folded and potentially useful α:β T-cell receptor at the cell surface. The stages in this development, which are usually defined by their readily detectable cell-surface phenotypes, are also marked by changes in gene expression. Relevant genes are those encoding intracellular

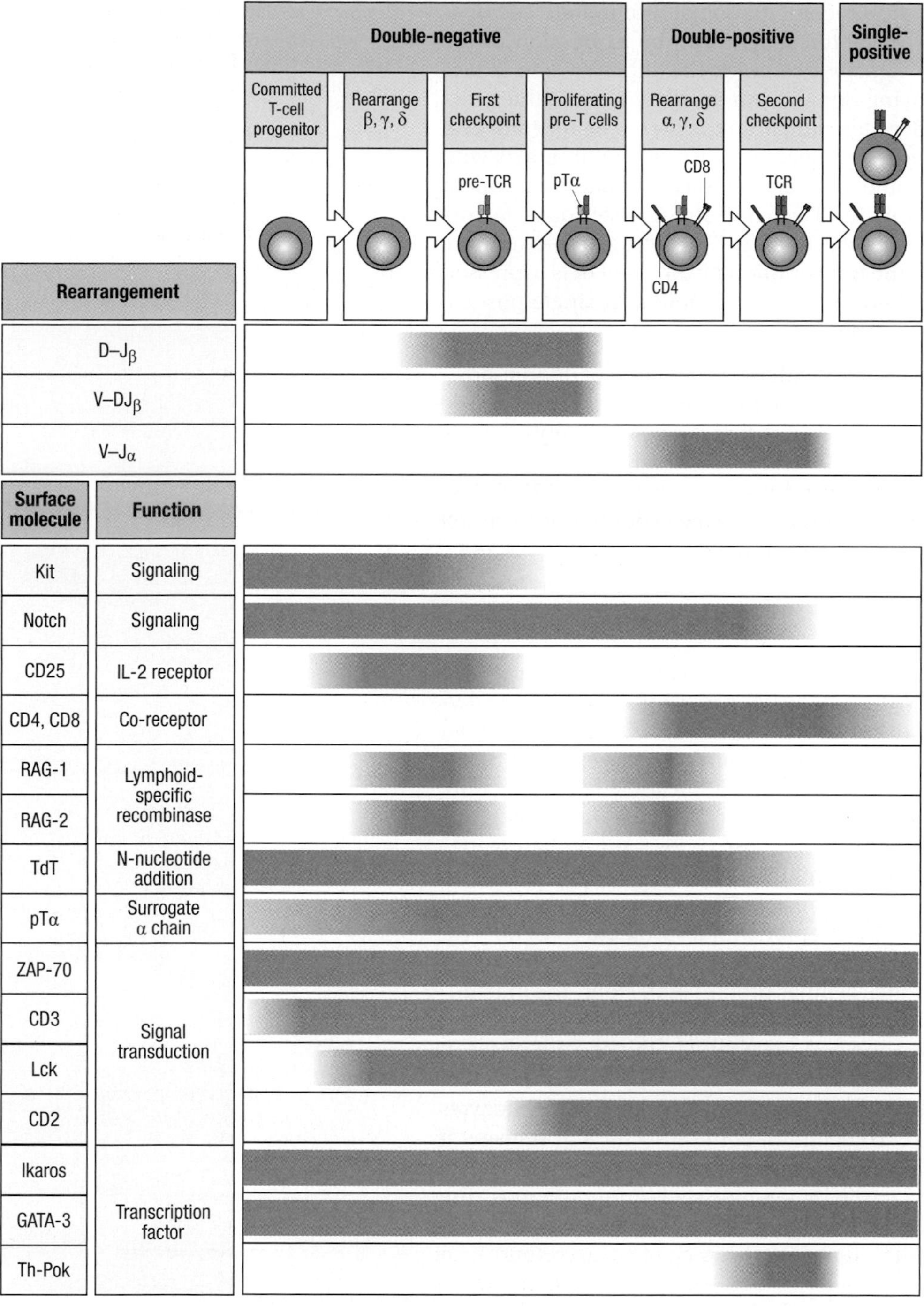

Figure 7.14 Stages of α:β T-cell development in the thymus correlate with T-cell receptor gene rearrangement and the expression of particular proteins by the developing T cell. All these proteins are described in the text.

molecules that contribute to the rearrangement and transcription of T-cell receptor genes, to formation of the pre-T-cell receptor, to signal transduction from the pre-T-cell and T-cell receptors, and to the regulated display of the CD4 and CD8 co-receptors (Figure 7.14).

The RAG-1 and RAG-2 proteins are essential for gene rearrangement and are selectively expressed at the two stages where the β and the α gene rearrangements, respectively, are made. With RAG-1 and RAG-2 controlling when gene rearrangement occurs, other enzymes involved in somatic recombination, such as the TdT that inserts the N nucleotides, are expressed throughout this phase of development. The unique component of the pre-T-cell receptor, pTα, is also expressed throughout the period when rearrangements occur, so that a newly made β chain can be quickly assembled into a pre-T-cell receptor that signals the interior of the cell to stop recombination and to initiate cell division. Signals from the pre-T-cell receptor depend on the expression of the co-receptors CD4 and CD8, the signaling complex CD3, the tyrosine kinase ZAP-70, which is involved in relaying signals from the receptor, and the tyrosine kinase Lck, which is involved in signaling from the pre-T-cell receptor and the co-receptors. CD2 is an adhesion molecule on T cells that interacts with the cell-surface protein CD58 on other cells and generates signals that work in conjunction with those coming from the T-cell receptor. Transcription factors called Ikaros and GATA-3 are expressed in early T-cell progenitors and are essential for T-cell development. The transcription factor Th-POK is expressed late in development and is required for the development of single-positive CD4 T cells from double-positive thymocytes.

The early development of α:β T cells ends with the production of double-positive thymocytes that express CD4, CD8, and a functional T-cell receptor. The advantage of expressing both CD4 and CD8 is that it gives each immature T cell the option to use either co-receptor depending on whether its T-cell receptor is more fitted to recognizing peptide antigens presented by self-MHC class I or self-MHC class II. This increases the probability that a double-positive cell will complete its maturation by a factor of around two.

Summary

The thymus provides a sequestered and organized environment dedicated to T-cell development. Progenitor cells from the bone marrow migrate to the thymus, where they go through phases of division and differentiation associated with the rearrangement of T-cell receptor genes and the expression of other cell-surface glycoproteins involved in T-cell recognition and effector function. Commitment of thymocytes to either the γ:δ or the α:β lineage occurs as a consequence of the gene rearrangements made. The β-, γ-, and δ-chain genes rearrange simultaneously, and if a γ:δ receptor is made first, the cell is committed to the γ:δ lineage. If a β-chain gene rearranges productively before this occurs, the β chain forms a pre-T-cell receptor with a surrogate α chain, pTα. This stops recombination, initiates cell division, and induces the expression of the CD4 and CD8 co-receptors. Once cellular proliferation stops, the machinery for T-cell receptor gene rearrangement is reactivated and now works on the α-chain gene in addition to the γ- and δ-chain genes. In this second phase of gene rearrangements, lineage commitment is determined by the expression of either γ:δ or α:β receptors on the cell surface. More than 90% of thymocytes that complete successful gene rearrangement commit to the α:β lineage. If cells fail to make successful gene rearrangements, and thus do not express a T-cell receptor, they are signaled to die by apoptosis within the thymus. The development of double-positive α:β T cells in the thymus is summarized in Figure 7.15.

Positive and negative selection of the T-cell repertoire

In the first phase of T-cell development just described, the role of the thymus is to produce T-cell receptors, irrespective of their antigenic specificity. The second phase of T-cell development involves a critical examination of the receptors produced and the selection of those that can work effectively with the individual's own MHC molecules in the recognition of pathogen-derived peptides. These selection processes involve only the α:β T cells and not the γ:δ T cells, which do not depend on MHC molecules for detecting infection, damage, and other forms of stress. Once the γ- and δ-chain genes are rearranged productively, the thymic development of γ:δ T cells seems to be complete.

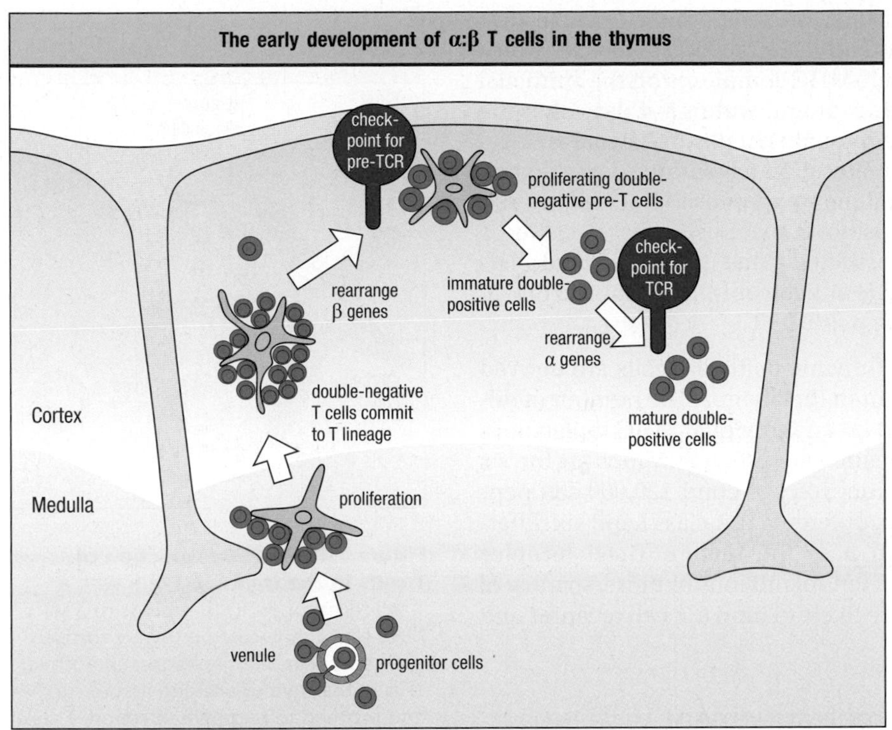

Figure 7.15 The development in the thymus of double-positive α:β T cells. This figure summarizes the early development of α:β T cells (round dark blue cells) in the thymus, from uncommitted progenitor cell to a double-positive cell bearing an α:β T-cell receptor. The two checkpoints of T-cell development are indicated. The paler blue cells are resident thymic cells.

In this part of the chapter we see how the population of α:β double-positive thymocytes undergoes two types of screening. In the first screen, positive selection favors T cells that can recognize peptides presented by a self-MHC molecule; in the second, negative selection eliminates potentially autoreactive cells that could be activated by the peptides normally presented by MHC molecules on the surface of healthy cells.

7-8 T cells that recognize self-MHC molecules are positively selected in the thymus

T-cell receptors and MHC molecules have been co-evolving under natural selection for around 500 million years. Consequently, the primary T-cell receptor repertoire has a bias toward interaction with MHC molecules that is due to specificities built into the V gene segments. Thus, gene rearrangement provides an extensive repertoire of T-cell receptors that can be used with the thousands of MHC class I and II isoforms present in the human population (see Chapter 5). However, the T-cell receptor genes of a given individual are not tailored specifically toward making receptors that interact with the particular forms of MHC molecule expressed by the same individual. Only a small subpopulation of the double-positive thymocytes, at most 2% of the total, have receptors that can interact with one of the MHC class I or II isoforms expressed by the individual and will therefore be able to respond to antigens presented by these MHC molecules. **Positive selection** is the name given to the process whereby that small subpopulation is selected and signaled to mature further, leaving the vast majority of double-positive cells to die by apoptosis in the thymic cortex.

Positive selection takes place in the cortex of the thymus (see Figure 7.3). It is mediated by the complexes of self peptides and self-MHC molecules present on the surface of the cortical epithelial cells. As noted in Chapter 5, in the absence of infection, MHC molecules assemble with self peptides derived from the normal breakdown of the body's own proteins. Thymic cortical epithelium expresses both MHC class I and class II molecules and thus presents self peptides in both MHC contexts. The cortical epithelial cells form a web of cell processes that envelop and make contact with the double-positive CD4 CD8 thymocytes. At regions of contact, potential interactions of the α:β receptor of a thymocyte with the self-peptide:self-MHC complexes on the epithelial cell are tested. If a peptide:MHC complex is bound within 3–4 days after the thymocyte first expresses a functional receptor, then a positive signal is delivered to the thymocyte, which permits its maturation to continue. Cells that do not receive such a signal within this period die by apoptosis (Figure 7.16) and are removed by macrophages. The MHC isoform that positively selects the T cell is the one to which it becomes MHC restricted—that is, the T cell and all its progeny will recognize their specific peptide antigen only in the context of that particular MHC molecule (see Figure 5.35, p. 140).

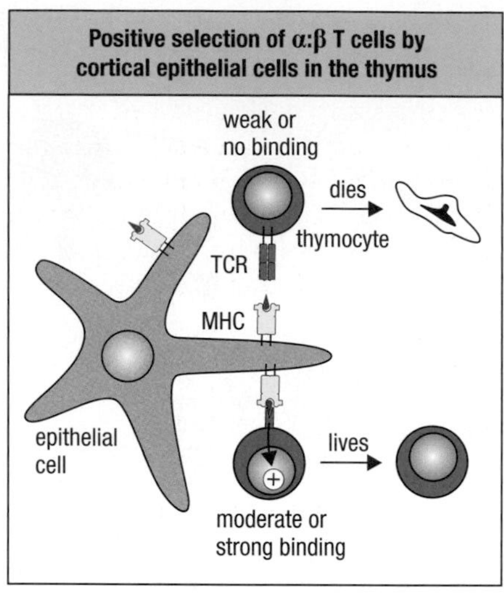

Figure 7.16 Positive selection of T cells in the thymus. T cells with a T-cell receptor (TCR) that binds to a self-MHC class I molecule on thymic cortical epithelial cells, macrophages, and other cells in the thymus are signaled to survive and proceed to negative selection. T cells with a T-cell receptor that binds to no self-MHC class I molecules are signaled to die.

The peptides presented on the surface of thymic epithelial cells are derived from the set of self proteins that are present in the thymus. The number of different self peptides that can be presented by an individual MHC molecule is estimated to be about 10,000. So for someone who is heterozygous for the six major polymorphic HLA genes (see Section 5-21), around 120,000 self peptides in total could be presented by the six different HLA class I and six different HLA class II molecules that person has. The mature T-cell receptor repertoire is estimated to be of the order of tens of millions or more, so most of these self-peptide:self-MHC complexes are likely to bind a T-cell receptor and contribute to positive selection.

7-9 Continuing α-chain gene rearrangement increases the chance for positive selection

If the first α-chain gene rearrangement that a pre-T cell makes is productive and leads to the assembly of an α:β T-cell receptor that interacts with a self-MHC molecule, positive selection will occur within a few hours. The cell is then signaled to turn off the recombination machinery, by degrading the RAG proteins and stopping transcription of the *RAG* genes, and to enter a phase of proliferation. In contrast, if the T-cell receptor does not bind a self-MHC molecule within this timeframe, rearrangement of the α-chain genes continues. This allows the T cell to try out a second α chain in its search for a functional T-cell receptor. A second productive rearrangement produces a different α chain that assembles with the β chain to give a T-cell receptor with a different binding-site specificity. The second T-cell receptor can then be assessed for its capacity to bind self MHC. Further rearrangements at the α-chain locus can continue throughout the 3–4 days of positive selection, during which time a double-positive thymocyte can explore the usefulness of several receptors in succession. This process improves the chance that the T cell will be positively selected.

After successfully rearranging a β-chain gene, a developing T cell immediately switches off the recombination machinery, with the result that there is allelic exclusion at the β-chain locus (see Section 7-6). Because the α-chain gene is not subject to allelic exclusion in the same way as the β-chain gene, rearrangements can occur at both copies of the α-chain locus, and double-positive thymocytes can express two α chains and, thus, two types of T-cell receptor. Such T cells can be positively selected through the engagement of either of these receptors, and around one-third of all mature α:β T cells carry two T-cell receptors. But the proportion of T-cell receptors that succeed in positive selection is so small that the proportion of T cells with two T-cell receptors that can both be activated by peptides presented by self-MHC molecules will be at most 1–2%. Thus, in almost all mature T cells with two receptors, one receptor will be nonfunctional, and for most practical purposes T cells can be said to have a single working antigen receptor. For immunoglobulins, which contain two or more heavy and light chains, allelic exclusion of both the heavy- and light-chain genes is essential to preserve their functional integrity (see Section 6-5).

Lack of allelic exclusion at the T-cell receptor α-chain locus is not so disruptive because each T-cell receptor only contains one α chain.

The process by which a T-cell receptor tries out different α chains in order to become reactive with a self-MHC molecule is one of receptor editing, analogous to that used by B cells to change the light chain of their B-cell receptors (see Section 6-12). The difference is that the B cell uses receptor editing to get rid of reactivity against self antigens, including self MHC, whereas the T cell uses receptor editing to acquire reactivity to self MHC.

7-10 Positive selection determines expression of either the CD4 or the CD8 co-receptor

Positive selection not only selects a repertoire of cells that can interact with an individual's own MHC allotypes, but it is also instrumental in determining whether a double-positive T cell will become a CD4 T cell or a CD8 T cell. As a result of positive selection, double-positive thymocytes mature into cells that express just one or other of the two co-receptors. They are then known as **single-positive thymocytes**.

As we saw in Chapter 5, CD4 interacts only with MHC class II molecules, and CD8 interacts only with MHC class I molecules. During positive selection, the double-positive CD4 CD8 T cell interacts through its α:β receptor with a particular peptide:MHC complex. When the interacting MHC molecule is class I, CD8 molecules are recruited into the interaction, whereas CD4 molecules are excluded. Conversely, when the selecting MHC molecule is class II, CD4 is recruited and CD8 excluded (Figure 7.17). The specific interactions between

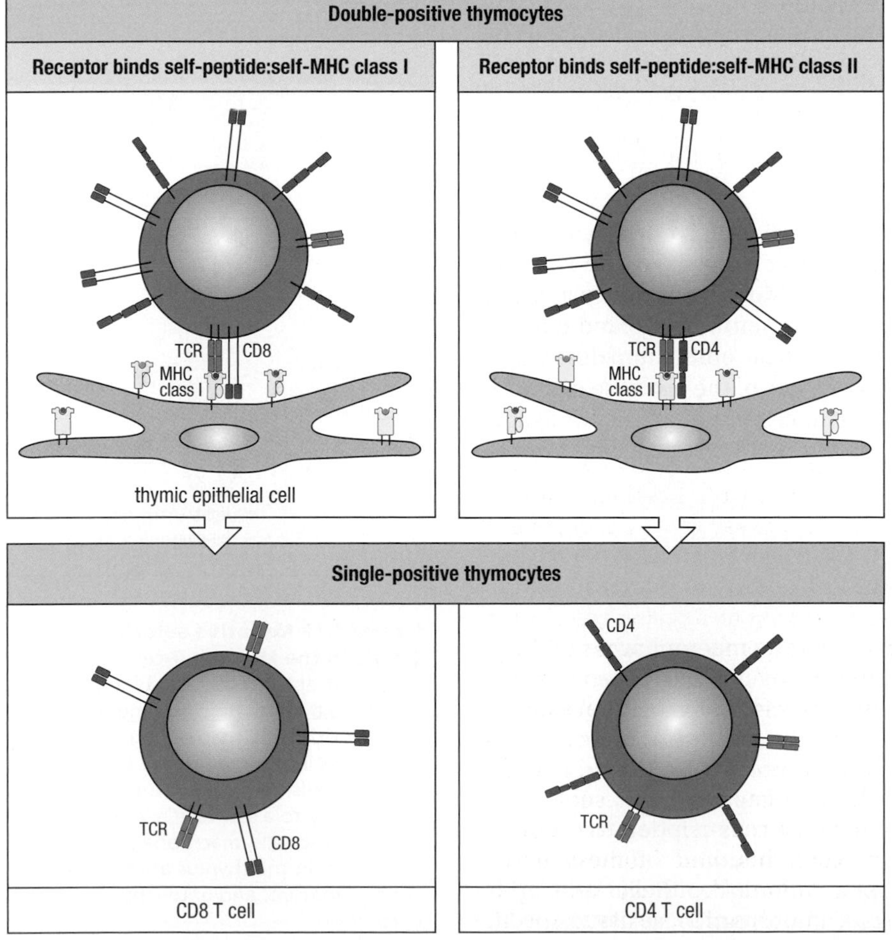

Figure 7.17 Interaction of a double-positive T cell with a self-peptide:self-MHC complex during positive selection determines whether the T cell will become a CD4 or a CD8 T cell. The left panels show the selection of a T cell whose T-cell receptor (TCR) interacts with peptide:MHC class I complexes on a thymic epithelial cell. The right panels show the outcome for a cell bearing a receptor that interacts with peptide:MHC class II complexes.

T-cell receptor, co-receptor, and MHC molecule and the differential signaling that is generated by the two co-receptors, particularly through the protein kinase Lck (see Section 7-7), commit the cell to either the CD4 or CD8 lineage and halt the synthesis of the other co-receptor. The single-positive cells initiate a program of gene expression that provides CD4 T cells with the capacity for helper function and CD8 T cells with the capacity for cytotoxic function. For example, the transcription factor Th-POK is essential for the development of CD4 T cells from double-positive thymocytes (see Figure 7.14).

The importance of MHC molecules in co-receptor selection and subsequent T-cell development is demonstrated by human immunodeficiency diseases called bare lymphocyte syndromes, which are characterized by a lack of expression of either MHC class I or MHC class II molecules by lymphocytes and thymic epithelial cells. Patients who lack MHC class I expression have CD4 T cells but no CD8 T cells, whereas patients lacking MHC class II have plenty of CD8 T cells but only a few abnormal CD4 T cells.

7-11 T cells specific for self antigens are removed in the thymus by negative selection

In Chapter 4 we saw how immature B cells whose immunoglobulin receptors bind to self antigens on the surface of bone marrow cells are eliminated from the repertoire by clonal deletion. A similar mechanism, called **negative selection**, deletes T cells whose antigen receptors bind too strongly to the complexes of self peptides and self-MHC molecules presented by cells in the thymus. Such T cells are potentially autoreactive, and if allowed to enter the peripheral circulation they could cause tissue damage and autoimmune disease. Whereas epithelial cells in the cortex of the thymus are the exclusive mediators of positive selection, negative selection is mediated by several cell types, including thymocytes themselves. The most important cells responsible for negative selection are, however, the bone marrow derived dendritic cells and macrophages (Figure 7.18). When the receptors of an autoreactive T cell engage the MHC molecules of one of these specialized thymic antigen-presenting cells, the T cell is induced to undergo apoptosis. The dead thymocyte is then phagocytosed by macrophages.

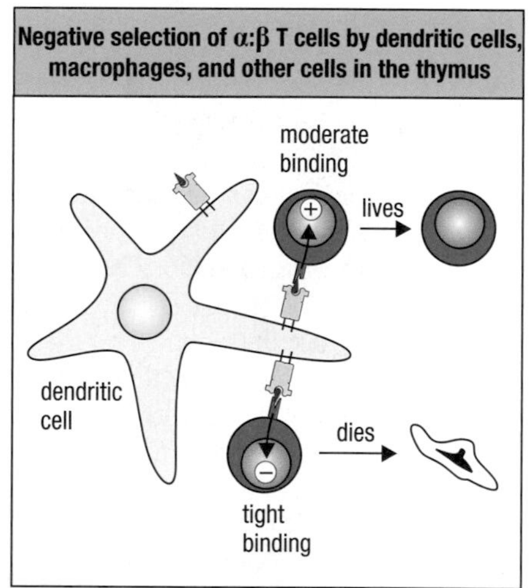

Figure 7.18 Negative selection of T cells in the thymus. T cells with a T-cell receptor (TCR) that binds too tightly to a self-MHC class I molecule on dendritic cells, macrophages, and other cells in the thymus are signaled to die. T cells with a receptor that binds moderately to a self-MHC class I molecule on dendritic cells, macrophages, and other cells in the thymus are signaled to survive, mature, and enter the peripheral circulation.

The mechanisms of positive and negative selection used by the thymus both involve the screening of interactions between T-cell receptors and self peptides bound to MHC molecules. How these interactions lead to the widely differing endpoints of cell death or cell growth has yet to be worked out. The processing and presentation of self antigens by thymic epithelium differs from that in other cells in several ways, including the use of the protease cathepsin L, whereas other cells use cathepsin S. Differences in the affinity of ligand–receptor interactions and the type of signaling pathways activated are also possible contributors.

7-12 Tissue-specific proteins are expressed in the thymus and participate in negative selection

The MHC molecules expressed by dendritic cells and macrophages in the thymus present peptides derived from all the proteins made by these cells and other cells they phagocytose, and also soluble proteins taken up from extracellular fluids. Negative selection will eliminate T cells specific for these self antigens, which include the ubiquitously expressed proteins. To extend negative selection to proteins that are specific to one or a few cell types, such as the insulin made only by the β cells of the pancreas, a transcription factor called **autoimmune regulator** (**AIRE**) causes several hundred of these tissue-specific genes to be transcribed by a subpopulation of epithelial cells in the medulla of the thymus. The presence of small amounts of these tissue-specific proteins in the thymus means that peptides derived from these proteins can be bound by MHC class I molecules to form complexes that participate in negative selection of the T-cell repertoire. AIRE was discovered as a result of the symptoms displayed by children who lack a functional *AIRE* gene. In these patients, T cells specific for tissue-specific antigens are not eliminated by negative selection and are permitted to mature and enter the peripheral circulation. Here they attack cells in a variety of tissues, causing a broad-spectrum autoimmune disease known as **autoimmune polyendocrinopathy–candidiasis–ectodermal dystrophy** (**APECED**) or autoimmune polyglandular syndrome type 1.

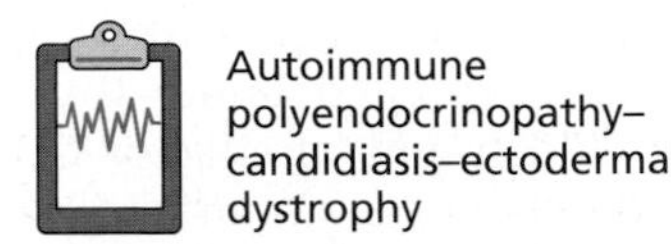

Negative selection in the thymus produces so-called central tolerance in the T-cell repertoire (see Section 7-11). There are also mechanisms of peripheral tolerance that prevent the activation of self-reactive T cells that elude negative selection in the thymus and enter the peripheral circulation. As with B cells, one such mechanism is to render the self-reactive T cell anergic (see Section 6-13). In general, a T cell outside the thymus that encounters a self antigen in the absence of infection will receive signals that cause it either to be inactivated or to be briefly activated and then die; this latter mechanism is called activation-induced cell death. In patients lacking AIRE, the mechanisms of peripheral tolerance become overwhelmed by the unusually large number of self-reactive T cells that are present in the peripheral circulation.

7-13 Regulatory CD4 T cells comprise a distinct lineage of CD4 T cells

So far in this book we have considered only the helper functions of CD4 T cells, which drive the immune response to infection by activating macrophages and B cells. Another function carried out by a distinct lineage of CD4 T cells is to suppress the response of self-reactive CD4 T cells to their specific self antigens: despite negative selection and the activity of AIRE, self-reactive CD4 T cells are present in the circulation of all people, even the most healthy. This subset of CD4 T cells is called **regulatory CD4 T cells** or just **regulatory T cells** (**T_{reg}**). Regulatory T cells have T-cell receptors specific for self antigens and are distinguished from other CD4 T cells by the expression of CD25 on the cell surface and the unique use of a transcriptional repressor protein called FoxP3. On contacting self antigens presented by MHC class II molecules, the regulatory T cells do not proliferate but respond by suppressing the proliferation of naive T cells responding to self antigens presented on the same antigen-presenting cell (Figure 7.19). These suppressive effects require contact between the two T cells and the secretion of cytokines that inhibit the activation and differentiation of effector T cells. Although once controversial, this active form of tolerance mediated by regulatory T cells is now considered one of the main mechanisms for protecting the integrity of the body's tissues and organs. The basis for the selection and development of this lineage of CD4 T cells in the thymus has yet to be defined.

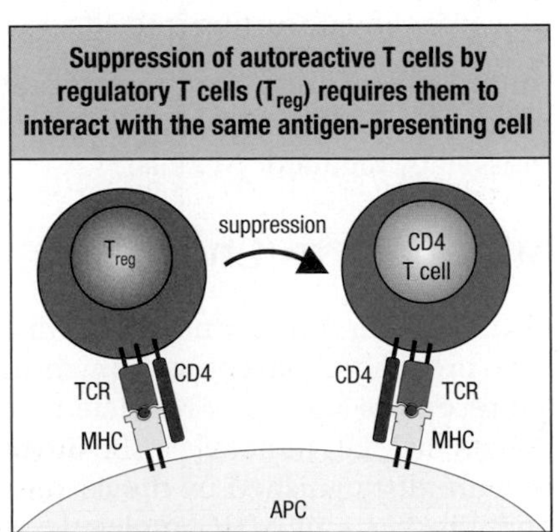

Figure 7.19 Autoreactive regulatory CD4 T cells prevent the proliferation of autoreactive helper CD4 T cells. Suppression of an autoreactive CD4 T cell by a regulatory T cell (T_{reg}) is dependent on the interaction of both T cells with the same antigen-presenting cell (APC).

7-14 T cells undergo further differentiation in secondary lymphoid tissues after encounter with antigen

Only a small fraction of α:β T cells survive the obstacle course of positive and negative selection and leave the thymus as mature naive T cells. Like mature B cells, these mature T cells recirculate through the tissues of the body, passing from blood to secondary lymphoid tissues to lymph and then back to the blood. Mature T cells are longer lived than mature B cells and, in the absence of their specific antigen, continue to circulate for many years.

The T-cell rich areas of secondary lymphoid tissues provide specialized sites where naive T cells are activated by their specific antigens. Encounter with antigen provokes the final phases of T-cell development and differentiation: the mature T cells divide and differentiate into effector T cells, some of which stay in the lymphoid tissues whereas others migrate to sites of infection.

Unlike B cells, which have just one terminally differentiated state—the antibody-secreting plasma cell—there are several different types of effector T cell. On activation by antigen, CD8 T cells become activated cytotoxic T cells, whereas CD4 T cells can be regulatory T cells and various types of helper cell that have differentiated under the influence of different combinations of cytokines. Which type of effector CD4 T cell predominates depends on the nature of the pathogen and the type of immune response required to clear it (see Chapter 5).

Summary

Once a developing thymocyte expresses an $\alpha:\beta$ receptor and CD4 and CD8 on its surface, it undergoes two types of selection—positive and negative—both of which involve testing the receptor's interactions with the complexes of self peptides bound by self-MHC molecules on the surface of thymic cells. Positive selection is the responsibility of epithelial cells in the cortex of the thymus. Double-positive thymocytes whose receptors engage self-peptide:self-MHC complexes on these cells continue their maturation. During positive selection the α-chain genes continue to rearrange, allowing the T-cell to try different receptors in order to find one that engages self MHC. Despite this receptor editing, the vast majority of double-positive thymocytes fail positive selection and die by apoptosis. The class of MHC molecule that drives positive selection determines which co-receptor—CD4 or CD8—is maintained on the single-positive T cell.

Negative selection is effected by other cells of the thymus, most importantly the dendritic cells and macrophages, which derive from bone marrow progenitors. Negative selection eliminates autoreactive cells having receptors that bind too tightly to a self-peptide:self-MHC complex, thereby helping to create a mature T-cell repertoire that does not react to the peptide:MHC complexes of normal healthy cells. Many tissue-specific proteins are also expressed by a subpopulation of thymic epithelial cells, which improves the efficiency of negative selection. Having survived both positive and negative selection, the now mature T cells leave the thymus in the blood and circulate through the secondary lymphoid organs, where they encounter foreign antigen. On activation by antigen they differentiate further into effector T cells of various types.

Autoreactive T cells that avoid negative selection and enter the peripheral circulation are either anergized on contact with self antigen or actively suppressed by regulatory T cells.

Summary to Chapter 7

In the thymus, three functionally distinct types of T cell develop from a common progenitor that comes from the bone marrow. One type of T cell expresses $\gamma:\delta$ receptors and is not restricted to the recognition of peptide antigens presented by MHC molecules. The other two types of T cell express $\alpha:\beta$ receptors and are distinguished by the co-receptors that they express—CD4 or CD8—and the class of MHC molecule to which their receptors are restricted. Commitment to either the $\gamma:\delta$ or the $\alpha:\beta$ T-cell lineage is determined by which pair of T-cell receptor loci successfully rearranges first. Subsequent phases of thymic development concern only the $\alpha:\beta$ T cells. The primary repertoire of T-cell receptors arising from gene rearrangement is acted upon by both positive and negative selection to produce a functional repertoire of mature naive T cells whose receptors can be activated by pathogen-derived peptides presented by self-MHC molecules but cannot be activated by the self peptides derived from the body's normal constituents. Positive selection tests the ability of α:β T-cell receptors to interact with self-MHC molecules expressed on the surface of cells in the thymus. Thymocytes with receptors that interact with self-peptide:self-MHC complexes are signaled to continue their maturation. Negative selection then eliminates those cells whose receptors interact too strongly with self-peptide:self-MHC complexes. The small fraction of thymocytes that survive both positive and negative selection leave the thymus to become mature circulating α:β T cells. These phases in T-cell development are summarized in Figure 7.20.

T-cell development produces several types of T cell and is therefore more complicated than B-cell development. It is also more wasteful of cells. The vast majority of developing thymocytes die without doing anything useful, and

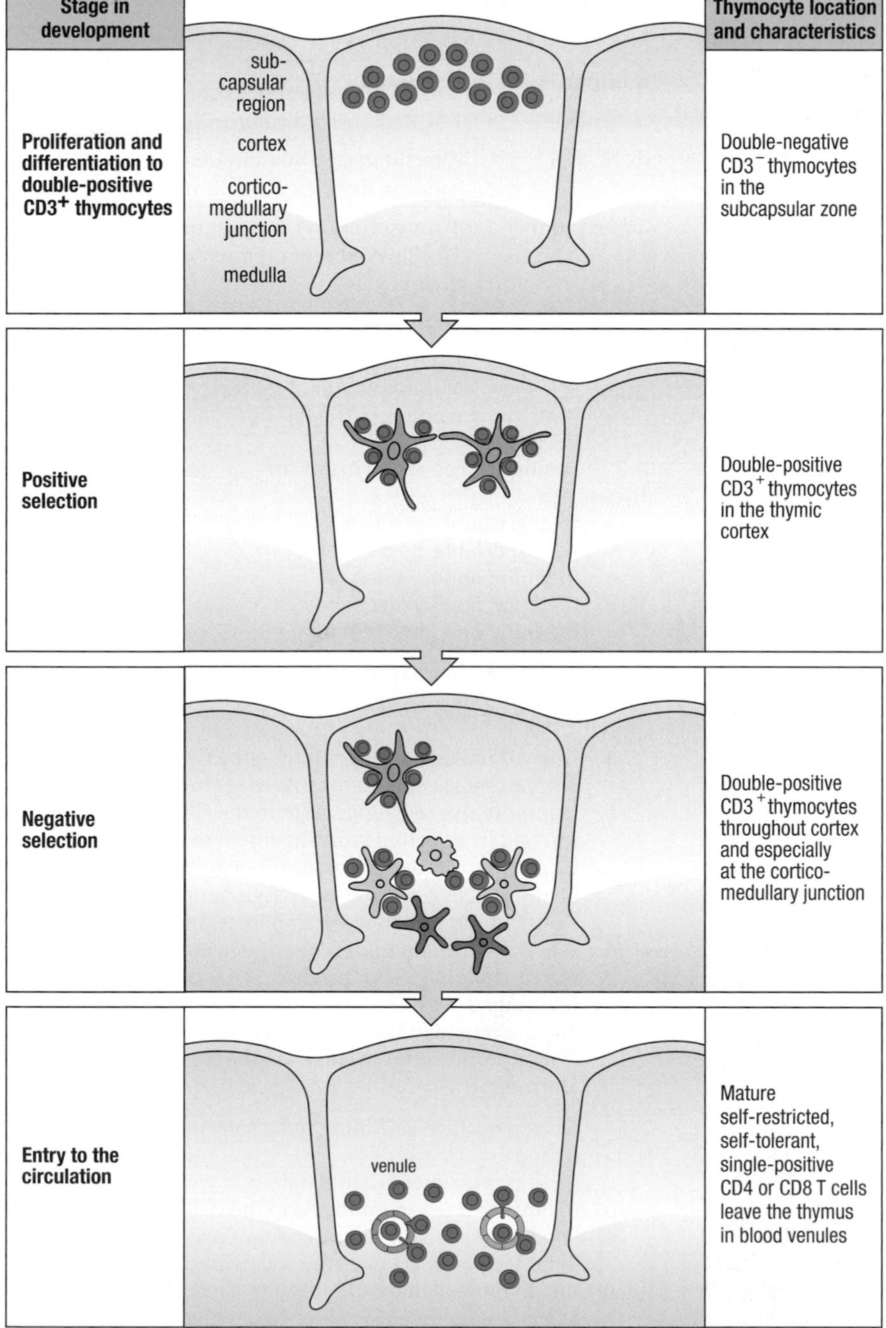

Figure 7.20 T cells develop in the thymus in a series of stages. In the first phase of development, thymocyte progenitors enter the thymus from the blood and migrate to the subcapsular region. At this stage, they do not express the antigen receptor, the CD3 complex, or the CD4 and CD8 co-receptors, and are known as double-negative thymocytes (top panel). These cells proliferate and begin to rearrange their β, γ, and δ T-cell receptor genes, which leads to the production of γ:δ cells and to cells expressing the pre-T-cell receptor. Stimulation through the pre-T-cell receptor leads to cell proliferation and the expression of both CD4 and CD8 co-receptors, producing double-positive cells. As the cells mature, they move deeper into the thymus. In the second phase of development, the double-positive thymocytes rearrange their α-chain genes, express an α:β T-cell receptor and the CD3 complex, and become sensitive to interaction with self-peptide:self-MHC complexes. Double-positive cells undergo positive selection in the thymic cortex through intimate contact with cortical epithelial cells (second panel). During positive selection, matching between the receptor specificity for MHC and the co-receptor molecules starts to take place, which eventually leads to single-positive CD4 or CD8 T cells. In the third phase of development, double-positive cells undergo negative selection for self-reactivity. Negative selection is believed to be most stringent at the cortico-medullary junction, where the nearly mature thymocytes encounter a high density of dendritic cells (third panel). Thymocytes that survive both positive and negative selection leave the thymus in the blood as mature single-positive CD4 or CD8 T cells and enter the circulation (bottom panel).

only a few percent of thymocytes fulfill the stringent requirements of selection and leave the thymus to enter the circulation. Whereas the bone marrow is continually turning over the B-cell repertoire during the whole of a person's lifetime, the thymus works principally during youth, when it serves to accumulate a repertoire of T cells that can then be used throughout life. This difference might reflect the magnitude of the body's investment in the development of each useful T cell, and the savings to be made by gradually shutting down the thymus with age.

Questions

7–1 Which of the following describes the developmental pathway for T cells?

a. thymus → bone marrow → spleen
b. thymus → bone marrow → thymus
c. bone marrow → thymus → lymph nodes
d. lymph nodes → thymus → spleen
e. bone marrow → lymph nodes → thymus.

7–2 Which of the following statements is false concerning T-cell development?

a. γ and δ rearrangements may occur at two different stages, once at the committed double-negative stage and again at the uncommitted double-positive stage.
b. If α-chain rearrangement is successful, it is still possible to produce a committed γ:δ T cell.
c. Neither γ nor δ chains associate with pTα.
d. T-cell development is biased in favor of α:β lineage commitment.
e. A maximum of four attempts can be made to produce a functional β chain because there are two Cβ genes and associated Dβ and Jβ segments per β-chain locus.
f. γ:δ T cells are not subject to positive or negative selection processes in the thymus.

7–3 Which of the following statements regarding γ:δ T cells is incorrect? (Select all that apply.)

a. They do not require CD3 proteins for cell-surface expression of their antigen receptors.
b. They comprise the minority subpopulation of T lymphocytes.
c. They rearrange their antigen receptor genes in the periphery.
d. The do not pair the γ- or δ-chains with pTα at any stage in their developmental pathway.
e. They originate from the same bone marrow-derived precursor that gives rise to α:β T cells.

7–4 An adult who has been thymectomized will have a T-cell repertoire that _____.

a. resembles a patient with DiGeorge syndrome
b. is MHC-independent
c. is dominated by γ:δ T cells
d. fails to provide protective T-cell immunity
e. is self-renewing and long lived and does not require replenishment from the thymus.

7–5 During the early developmental stages of α:β T cells in the thymus, there are two key checkpoints that must be satisfied to permit the progression of T-cell development. Explain what occurs at each checkpoint.

7–6 Match the term in Column A with its correct description in Column B.

Column A	Column B
a. Th-POK	1. β chain/pTα:β chain/pTα
b. Notch1	2. facilitates expression of tissue-specific genes in thymic epithelium
c. superdimer	3. transcription factor required for CD4 T-cell development from double-positive thymocytes
d. FoxP3	4. facilitates the expression of genes required for T-cell development
e. AIRE	5. transcriptional repressor protein of regulatory T cells

7–7 Positive selection occurs in the _____ and involves _____ thymocytes

a. cortex; double-positive
b. cortex; double-negative
c. medulla; double-positive
d. medulla; double-negative
e. medulla; uncommitted progenitors of.

7–8 Which of the following statements about the successive rearrangement of α:β T-cell receptors is correct? (Select all that apply.)

a. Unproductive rearrangement to the Cβ1 locus can be rescued by a second rearrangement to the Cβ2 locus.
b. Unproductive rearrangement to the Cβ2 locus can be rescued by a second rearrangement to the Cβ1 locus.
c. Unproductive rearrangements between Vα and Jα can be rescued by a second rearrangement between upstream Vα and downstream Jα segments.
d. When the α chain undergoes a successive gene rearrangement, the unproductive intervening rearranged gene segments are deleted.
e. The δ-chain locus is permanently deleted as an extrachromosomal circle when the α-chain gene rearranges.

7–9 During positive selection, if the selecting MHC molecule is class I, then

a. gene expression of CD8 is halted
b. the thymocyte is committed to the CD8 lineage
c. continued α-chain rearrangement occurs on the other locus to produce a mature T cell with two receptors
d. the thymocyte differentiates into a CD4 T cell
e. gene expression of CD4 is enhanced.

7–10 MHC restriction is best described as
a. elimination of thymocytes bearing T-cell receptors that are unable to interact with self-MHC molecules
b. preferential survival and proliferation of thymocytes that survive negative selection
c. T-cell recognition of a peptide antigen only when it is bound to a particular form of MHC molecule
d. a state of non-responsiveness to a peptide antigen
e. a condition in which either MHC class I or class II molecules are not expressed on cells.

7–11 Match the term in Column A with its correct description in Column B.

Column A	Column B
a. T-cell receptor editing	1. absence of either CD4 T cells or CD8 T cells in the circulation
b. positive selection	2. elimination of autoreactive thymocytes
c. bare lymphocyte syndrome	3. continued α-chain gene rearrangement until a functional T-cell receptor is generated
d. negative selection	4. T-cell repertoire selection based on interaction with the individual's MHC allotypes
e. MHC restriction	5. activation of a T cell only if an appropriate MHC molecule presents antigenic peptide

7–12 If a T-cell receptor on a double-positive thymocyte binds to a self-peptide:self-MHC class I complex with low affinity the result is
a. negative selection and apoptosis
b. cell proliferation
c. rearrangement of the second β-chain locus
d. positive selection of a CD4 T cell
e. positive selection of a CD8 T cell.

7–13 The expression of MHC class II molecules is restricted to a small number of cell types.
A. What are these cell types?
B. Which of these cell types populate the thymus or circulate through it, and what role do they play in mediating positive and/or negative selection?
C. Can you explain why it would be detrimental for non-circulating cells that populate tissues and glands to express MHC class II molecules?

7–14 Generally, in healthy individuals and in the absence of infection, a mature naive T cell that encounters a self antigen outside the thymus may _____. (Select all that apply.)
a. induce an autoimmune attack on the cell presenting that antigen
b. be suppressed by a regulatory T cell
c. undergo activation-induced cell death
d. be rendered anergic
e. continue to rearrange its α-chain genes.

7–15 All of the following are characteristic of regulatory T cells except
a. they express CD25
b. they facilitate the proliferation of naive T cells interacting with the same antigen-presenting cell
c. they express CD4
d. they express the transcriptional repressor FoxP3
e. they express MHC class II-restricted T-cell receptors specific for self antigen.

7–16 Raija Berglund and her parents emigrated from Finland to the United States when she was 6 months old. From the age of 2 years, Raija experienced chronic mucocutaneous candidiasis involving her mouth, skin, and nails, which was treated successfully with superficial antifungal therapy on flare-ups. One severe episode of esophageal candidiasis at age 5 required systemic therapy with ketoconazole. At that time blood tests revealed normal numbers of B and T lymphocytes and normal concentrations of IgM, IgG, and IgA, but low calcium and parathyroid hormone levels. A diagnosis of hypoparathyroidism was made, which is managed with calcium and vitamin D supplements. After her 13th birthday, Raija began to experience dizzy spells when standing (orthostatic hypotension), complained of loss of appetite and not having much energy, and lost weight. [Q1] Laboratory tests showed decreased levels of adrenal hormones, and a diagnosis of Addison's disease (chronic adrenal insufficiency) was made. A course of prednisone therapy was initiated. The endocrinologist suspected an inherited disease known as autoimmune polyendocrinopathy–candidiasis–ectodermal dystrophy (APECED), which would explain her medical history. He told Raija's parents that she should be closely monitored for future changes in her medical condition because of the high risk of developing additional autoimmune manifestations. The underlying cause of this disease is
a. overexpression of the autoimmune regulator AIRE
b. a breakdown in the development of normal immune tolerance
c. a deficiency in B-cell function leading to susceptibility to numerous opportunistic infections
d. CD40 ligand deficiency
e. DiGeorge syndrome.

The lymph node. An example of a secondary lymphoid tissue where adaptive immune responses are produced.

Chapter 8

T Cell-Mediated Immunity

In Chapter 7 we saw how T cells develop in the thymus to give a diverse and highly selected population of mature naive T cells. These long-lived cells circulate between the blood and the lymph, passing through secondary lymphoid tissues where they can meet and be activated by antigens. Dendritic cells transport pathogens and their antigens from sites of infection to the T-cell areas of the draining lymph node. When a T-cell receptor recognizes the complex of a peptide antigen and an MHC molecule on a dendritic cell, it activates the T cell, which then undergoes proliferation and differentiation to become a clone of effector T cells. Effector T cells are diverse in function and phenotype. Some remain in the secondary lymphoid tissue to help B cells become antibody-producing cells; others travel to sites of infection where they work in various ways to terminate infection.

The first part of this chapter considers what happens when a naive T cell first encounters its specific antigen and is instructed to divide, proliferate, and differentiate into effector T cells. This process of **T-cell activation**—also referred to as **T-cell priming**—is the first stage of a primary adaptive immune response. Effector T-cell functions are examined more closely in the second part of the chapter. Activated CD8 T cells inevitably become cytotoxic T cells that kill pathogen-infected target cells. Activated CD4 T cells become effector cells that comprise five different functional subtypes. What the subtypes have in common is the secretion of cytokines that activate or suppress other immune-system cells. Because the function of effector CD4 T cells is principally to help other cell types acquire and exert their effector functions, CD4 T cells are often called **helper T cells**.

Activation of naive T cells by antigen

If an infecting pathogen successfully resists the innate immune response, the adaptive immune response is brought into play with the activation of T cells. This requires bringing the small number of pathogen-specific naive T cells in the circulation into contact with the pathogen's antigens. Such encounters are made in the unique microanatomy of the secondary lymphoid tissue. In this part of the chapter we examine the activation of naive T cells by dendritic cells within secondary lymphoid tissues and their differentiation into various types of effector T cell. This involves the selection of T cells that will attack the pathogen, and the suppression of T cells that might turn against the human body. Once activated, some types of effector T cell travel to infected tissue to work on the front line of defense, and others stay in the lymphoid tissues to continue developing more weaponry. Overall, this first phase of the primary immune

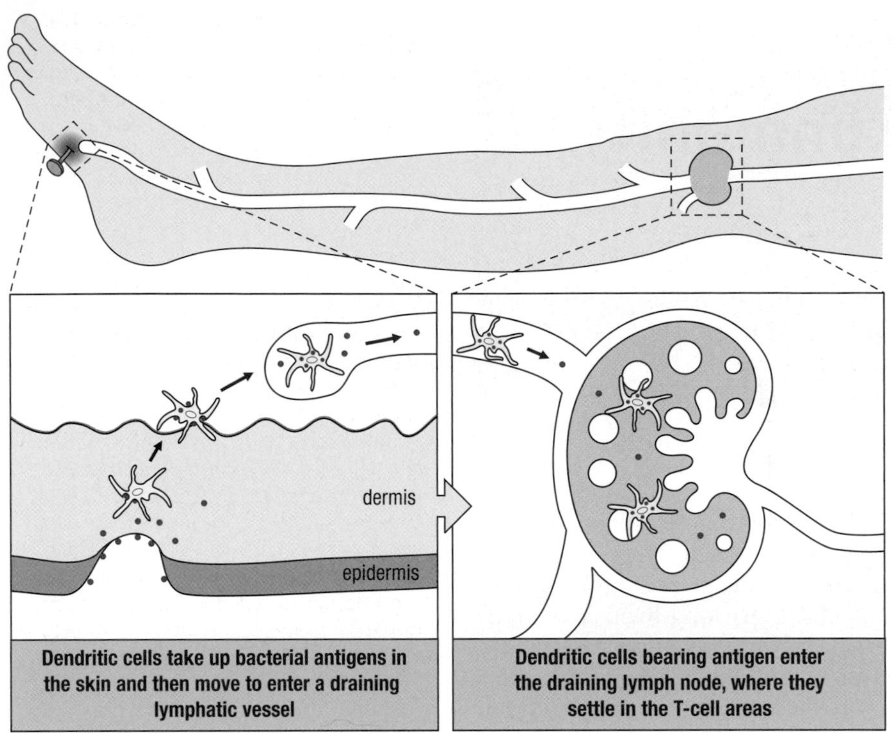

dermis
epidermis
Dendritic cells take up bacterial antigens in the skin and then move to enter a draining lymphatic vessel
Dendritic cells bearing antigen enter the draining lymph node, where they settle in the T-cell areas

Figure 8.1 Dendritic cells take up antigens at a site of wounding and infection in the skin and carry them to the draining lymph node for presentation to naive T cells. Dendritic cells in the skin are immature and specialized in the uptake of pathogens and their antigens (red dots). On migration to the lymph node, they settle in the T-cell areas and differentiate into mature dendritic cells that are specialized in activating naive T cells. The immature dendritic cells in the skin, also known as Langerhans cells, are distinguished morphologically by their Birbeck granules (not shown), which are part of the system of endosomal vesicles.

response produces large numbers of effector T cells that are ready to fight the invading pathogen but are tolerant of self antigens.

8-1 Dendritic cells carry antigens from sites of infection to secondary lymphoid tissues

The immune system does not initiate adaptive immune responses at the innumerable sites in the body where infections can occur. Instead, the strategy is to capture some of the pathogen and sequester it in secondary lymphoid tissue, an environment dedicated to the generation of adaptive immune responses. Central to this strategy are the myeloid dendritic cells, which capture antigens at the site of infection, transport them to the secondary lymphoid tissue, and present them to naive T cells. This course of events is essentially the same wherever an infection occurs. The T-cell response to infections of the skin, and other peripheral tissues, is made in the draining lymph nodes (Figure 8.1). Blood infections stimulate T-cell responses in the spleen, whereas infections in the mucosal tissues, such as the respiratory, gastrointestinal, and reproductive tracts, are made in the closely associated mucosal secondary lymphoid tissues. In each of the secondary lymphoid tissues a similar sequence of events occurs, which will be illustrated throughout this chapter by the response in a lymph node to an infection caused by a skin wound.

The movement of a dendritic cell from infected skin tissue to the lymph node that drains the infection is accompanied by changes in the dendritic cell's surface molecules, functions, and morphology (Figure 8.2). Dendritic cells in the skin are active in the capture, uptake, and processing of antigens, but they lose these properties on moving to secondary lymphoid tissue while gaining the capacity to activate naive T cells. The dendritic cells in the skin and other peripheral tissues are called **immature dendritic cells**, whereas those in lymph nodes are called **mature dendritic cells** or **activated dendritic cells**. On maturation, the finger-like processes called dendrites, for which the dendritic cell is named, become highly elaborated, which facilitates extensive

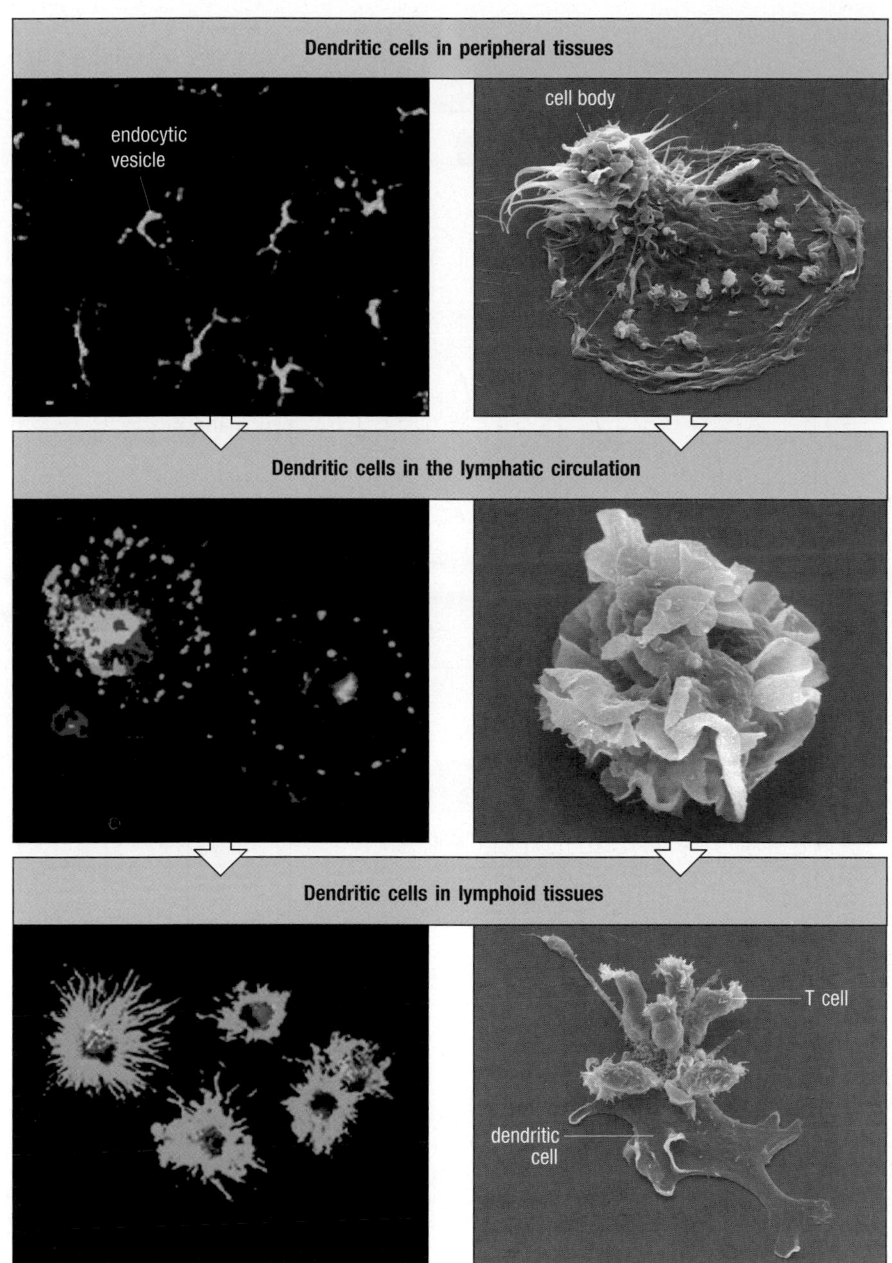

Figure 8.2 Maturation of dendritic cells is associated with changes in form and function. In these images, the dendritic cells are shown in peripheral tissue (top panels), lymph (middle panels), and lymph node (bottom panels). The left-hand panels are fluorescent micrographs of several cells in which their MHC class II molecules have been stained green and a lysosomal protein has been stained red. The right-hand panels are scanning electron micrographs of single cells. The dendritic cell body is clearly seen in the top right panel, but in the top left panel the cell bodies are difficult to discern. The observed yellow staining shows endocytic vesicles within the dendrites that contain both MHC class II molecules and the lysosomal protein; the combination of the red and green stains gives the yellow color. On activation and passage to the lymph, the morphology of the dendritic cell changes as shown in the center row. Phagocytosis has now stopped, a change indicated in the center left panel by a partial separation of the green MHC class II staining from the red lysosomal protein staining. This shows that peptide-loaded MHC class II molecules are moving out of the endocytic vesicles and onto the cell surface. On reaching the T-cell area of a lymph node (bottom panels), the now mature dendritic cells concentrate on antigen presentation and T-cell stimulation instead of the uptake and processing of antigens. Here, the MHC class II molecules, which are at high density on the surface of the numerous dendrites, are completely separated from the lysosomal protein in the intracellular vesicles. Photographs courtesy of I. Mellman, P. Pierre, and S. Turley.

interaction with T cells in the cortex of the lymph node. Consistent with their dedication to T-cell responses, dendritic cells are confined to the outermost part of the cortex of the lymph node, which is where naive T cells congregate.

Macrophages are present throughout the cortex and the medulla of the lymph node and perform complementary functions to those of the dendritic cells. While dendritic cells focus on activating naive T cells, the macrophages are responsible for removing pathogens and their breakdown products from the afferent lymph that arrives from the site of infection. This macrophage-mediated lymph filtration is highly efficient and absolutely critical, because it prevents infectious microorganisms from passing through the lymph node and gaining access to the blood via the efferent lymph (see Figure 1.20, p. 21). In this way, lymph-node macrophages prevent infections from spreading to the blood and becoming both systemic and life threatening. Another unique function that macrophages serve in the lymph node is to eliminate the many

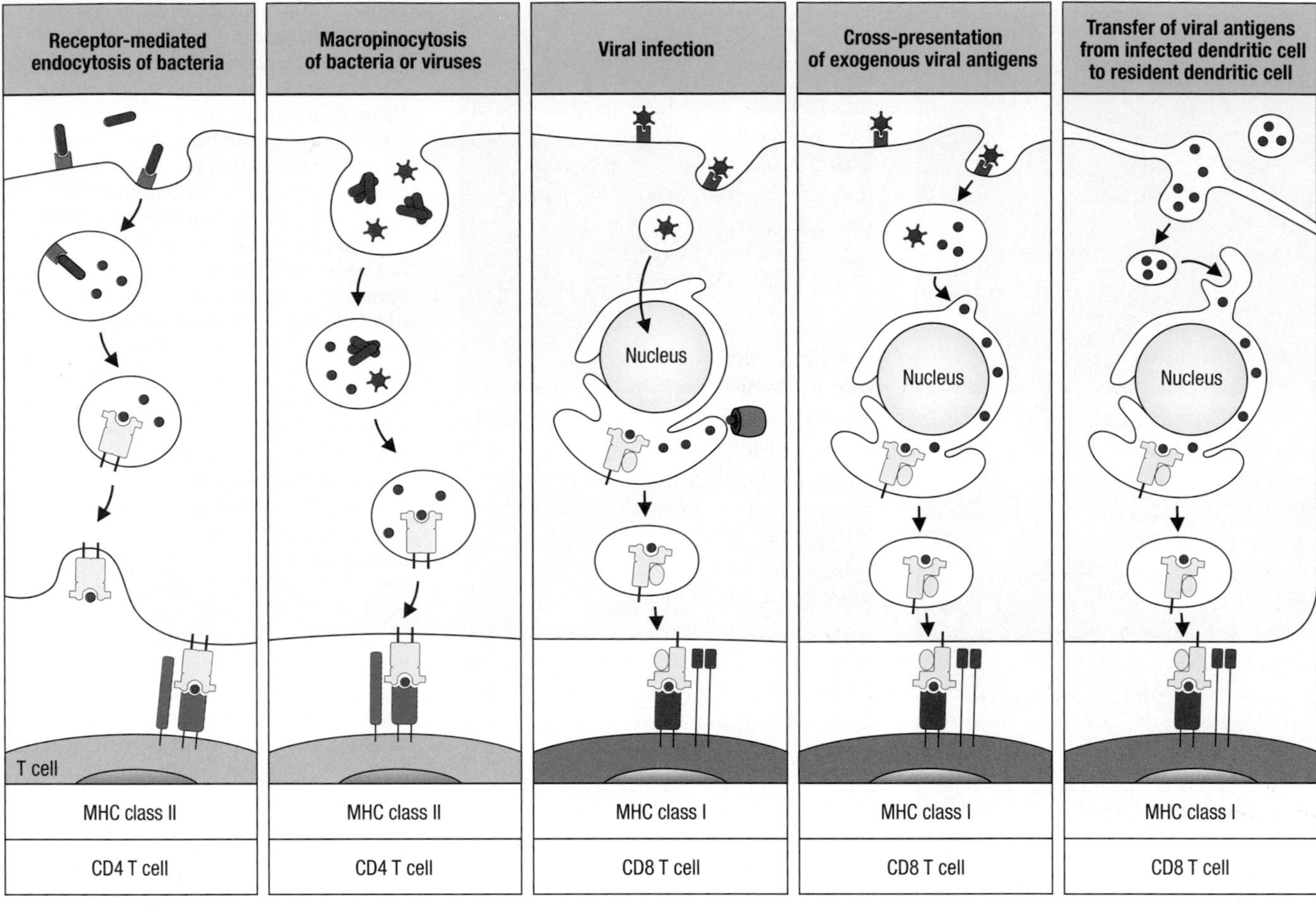

Figure 8.3 Dendritic cells use several pathways to process and present protein antigens. Uptake of antigens by phagocytosis or macropinocytosis delivers antigens to endocytic vesicles for presentation by MHC class II molecules to CD4 T cells (first two panels). Viral infection of the dendritic cell delivers peptides processed in the cytosol to the endoplasmic reticulum for presentation by MHC class I molecules to CD8 T cells (third panel). Viral particles taken up by the 'class II' pathways of phagocytosis and macropinocytosis can be delivered to the cytosol for processing and presentation to CD8 T cells by MHC class I (fourth panel). The mechanism of this cross-presentation is poorly understood. Lastly, antigens taken up by one dendritic cell can be delivered to a second dendritic cell for presentation by MHC class I molecules to CD8 T cells (fifth panel).

lymphocytes that are signaled to die by apoptosis in the course of the production of large numbers of antigen-specific effector T and B cells.

8-2 Dendritic cells are adept and versatile at processing pathogen antigens

To activate naive T cells, dendritic cells must present pathogen-derived peptides on MHC class II molecules. For this purpose, dendritic cells have a variety of endocytic and signaling receptors that recognize the components of pathogens and contribute to innate immunity. Receptor-mediated endocytosis is used to capture bacteria and virus particles from the extracellular fluid and target them for processing in the lysosomes. This mechanism of internalization is also called **micropinocytosis** because it involves the internalization of small volumes of extracellular fluid. **Macropinocytosis**, which involves the nonspecific ingestion of larger volumes of extracellular fluid, is used to capture pathogens not recognized by any endocytic receptor. Antigens captured by both these routes enter the endosomal system, where they are processed into small peptides, loaded onto the dendritic cell's MHC class II molecules, and presented to naive CD4 T cells (see Figure 5.27, p. 134) (Figure 8.3).

During some viral infections, the dendritic cells themselves become infected and make viral proteins that are processed into peptides that enter the endosomal reticulum and are presented on MHC class I molecules (see Figure 5.27, p. 134). If the infection is not lethal to dendritic cells, they can carry the virus to the draining lymph node and activate naive virus-specific CD8 T cells. Even if the virus does kill cells, the infected dendritic cells may still reach the secondary lymphoid tissue but be too sick to activate a T cell. In such instances, virus released by the dying dendritic cells can infect healthy dendritic cells resident in the lymph node. These newly infected dendritic cells are then able to present viral antigens on their MHC class I molecules and activate naive CD8 T cells. For those viruses that do not infect dendritic cells, cross-presentation is used to stimulate a CD8 T-cell response. The dendritic cells take up and degrade viral particles by the endocytic pathway and then transfer the virus-derived peptides to the exocytic pathway for cross-presentation by MHC class I molecules (see Section 5-15).

Dendritic cells carry all the Toll-like receptors except TLR9, making them highly sensitive to the presence of all manner of pathogens (see Chapter 3). Signals from the Toll-like receptors change the pattern of gene expression in a dendritic cell, leading to the cell's activation. One effect of activation is to increase the efficiency with which antigens are taken up, processed, and presented by MHC class II molecules. Activation also involves the appearance on the dendritic cell surface of CCR7, the receptor for the CCL21 chemokine made in secondary lymphoid tissue. Signals induced by the interaction of CCL21 with CCR7 cause the pathogen-loaded and migratory dendritic cells to leave the lymph and enter the tissue of the draining lymph node. These signals also induce the maturation of dendritic cells, so that by the time they reach the lymphoid tissue they no longer take up and process antigens. Instead, they now concentrate on the presentation of antigens to naive T cells. During maturation, the expression of MHC class I and II molecules increases, leading to an abundance of stable, long-lived peptide:MHC complexes on the surface of the fully mature dendritic cell.

8-3 Naive T cells first encounter antigen presented by dendritic cells in secondary lymphoid tissues

The specialized microanatomy of the lymph node allows antigen-bearing dendritic cells to interrogate millions of naive T cells, to identify the few that will be recruited to the adaptive immune response. Naive T cells, the most numerous of blood lymphocytes, arrive at a lymph node in the arterial blood that provides oxygen and nutrients to the lymph node's cells and tissues. Once in the capillaries around the lymph node, the T cells bind to the endothelial cells of the thin-walled high endothelial venules (HEV). They then squeeze through the vessel wall, entering the cortex of the node in the outermost region called the **T-cell area** or the **T-cell zone**. Passing through this crowded tissue, the naive T cells encounter mature dendritic cells, and their T-cell receptors examine the peptide:MHC complexes on the dendritic-cell surface. When a T-cell receptor binds to a peptide:MHC complex, the T cell is selected for activation and is retained in the lymph node by the dendritic cell (Figure 8.4).

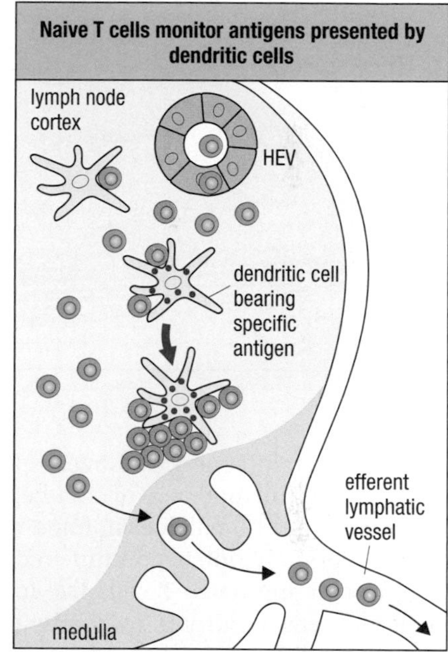

Figure 8.4 Naive T cells encounter antigen during their recirculation through secondary lymphoid organs. Naive T cells (blue and green) recirculate through secondary lymphoid organs, such as the lymph node shown here. They leave the blood at high endothelial venules and enter the lymph node cortex, where they mingle with professional antigen-presenting cells (mainly dendritic cells and macrophages). T cells that do not encounter their specific antigen (green) leave the lymph node in the efferent lymph and eventually rejoin the bloodstream. T cells that encounter antigen (blue) on antigen-presenting cells are activated to proliferate and to differentiate into effector cells. These effector T cells can also leave the lymph node in the efferent lymph and enter the circulation.

An alternative route for naive T cells to arrive at the T-cell area of a lymph node is in the afferent lymph. One function of lymph nodes is as junctions where several afferent lymphatic vessels unite and form a single efferent lymphatic vessel. The lymph draining from peripheral tissues often passes through several nodes on its way to the blood. Naive T cells that entered an 'upstream' lymph node from the blood, but did not encounter specific antigen, leave that node in the efferent lymph and are carried to the 'downstream' lymph node, which they enter through one of the afferent lymphatic vessels. Here they enter the T-cell area directly, without having to cross a vessel wall, and may find their specific antigen in this downstream node. With this route of entry to the lymph node, the T cells arrive at the node in a different afferent lymphatic vessel from that carrying pathogens, antigens, and dendritic cells from the infected tissue (Figure 8.5).

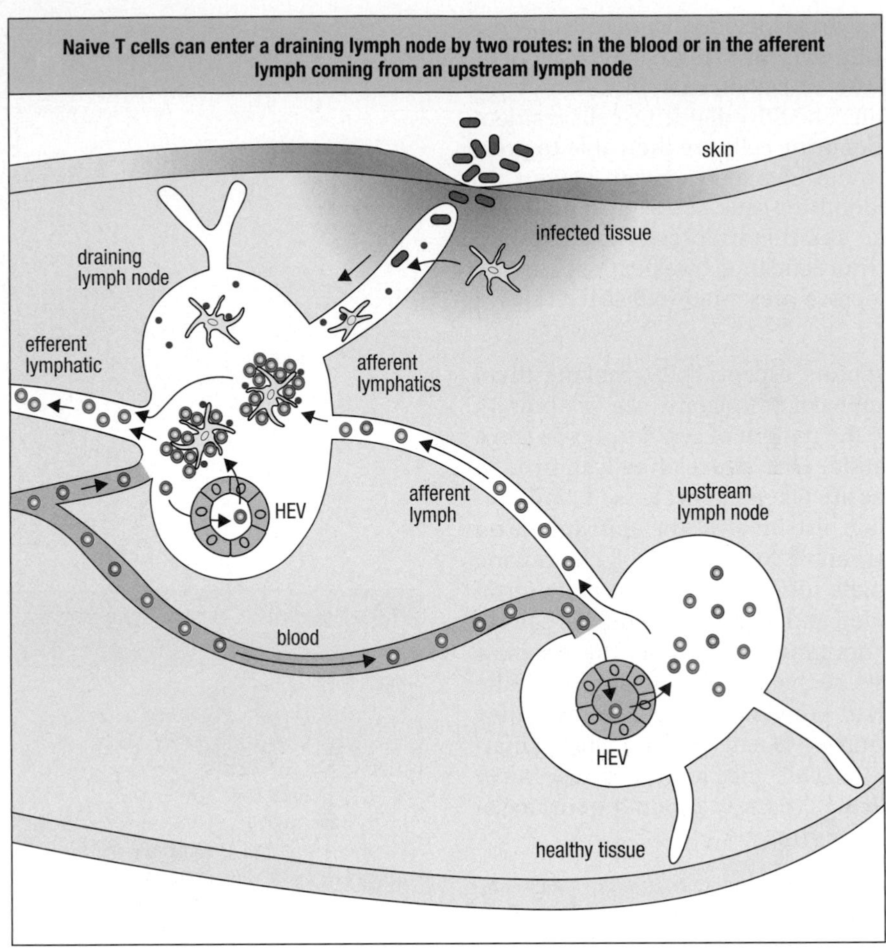

Figure 8.5 Naive T cells can enter lymph nodes from the blood or from the lymph. Recirculating naive T cells can enter a lymph node either directly from the blood or by moving from one lymph node to another via the lymphatics that connect them. In the case illustrated here, pathogen-specific T cells (blue) in the blood enter a lymph node that is draining an infected tissue. They encounter pathogen antigens, are activated, and leave as effector cells in the efferent lymphatic. At the same time, other pathogen-specific T cells (green) in the blood enter an 'upstream' lymph node that is draining healthy tissue. They do not encounter their antigen there but can be carried to the infected lymph node via a connecting lymphatic vessel. There, they too will become activated by dendritic cells presenting pathogen antigens.

For any given infection, pathogen-specific naive T cells represent only 1 in 10^4 to 1 in 10^6 of the total pool of circulating T cells. During most passages through a lymph node, a T cell does not find its specific antigen and leaves the medulla in the efferent lymph to continue recirculation. In the absence of specific antigen, circulating naive T cells live for many years as small nondividing cells with condensed chromatin, scanty cytoplasm, and little synthesis of RNA or proteins. With time, the entire population of circulating T cells will pass through any given lymph node. The trapping of pathogens and their antigens in the draining lymph node creates a concentrated depot of processed and presented antigens, and enables the small subpopulation of T cells specific for those antigens to be efficiently pulled out of the circulating T-cell pool, activated, and their number multiplied. Once an antigen-specific T cell has been trapped in a lymph node and activated, it takes several days for the cell to proliferate and for its progeny to differentiate into effector T cells. This accounts for much of the delay between the onset of an infection and the appearance of a primary adaptive immune response.

8-4 Homing of naive T cells to secondary lymphoid tissues is determined by chemokines and cell-adhesion molecules

The process by which naive T cells leave the bloodstream and enter the T-cell zone of a lymph node is called **homing**, and its mechanisms are similar to those of neutrophil homing to sites of infection (see Figure 3.13, p. 58). T-cell homing is guided by CCL21 and CCL19. These chemokines are secreted by

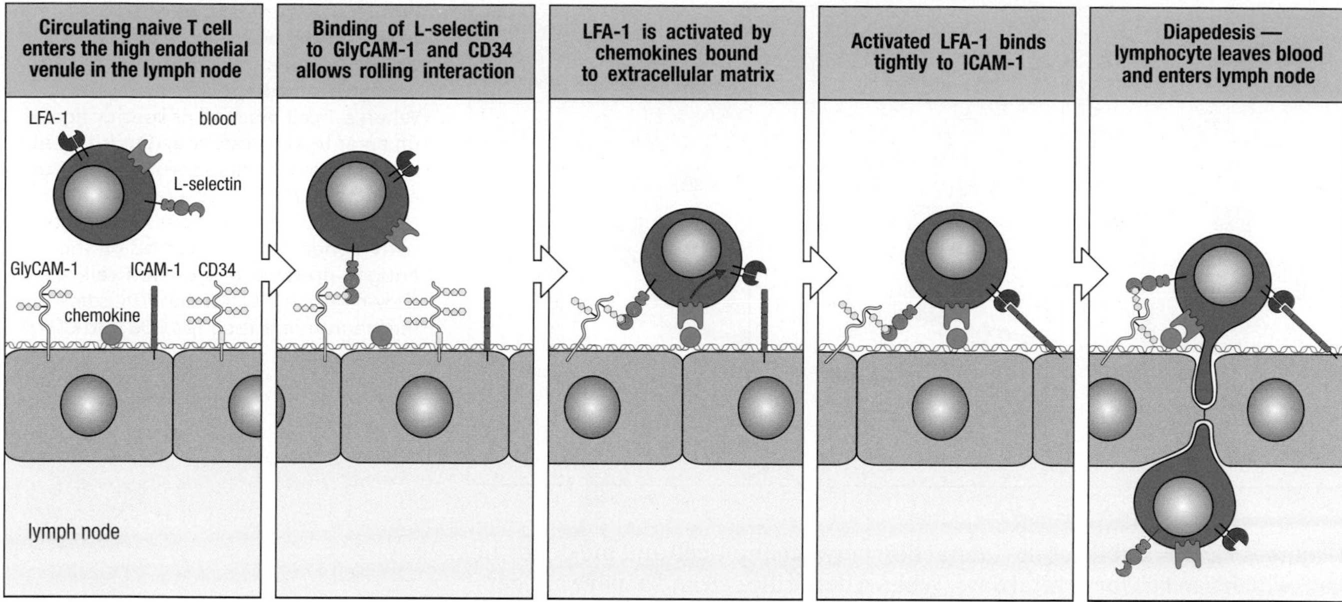

Circulating naive T cell enters the high endothelial venule in the lymph node
Binding of L-selectin to GlyCAM-1 and CD34 allows rolling interaction
LFA-1 is activated by chemokines bound to extracellular matrix
Activated LFA-1 binds tightly to ICAM-1
Diapedesis — lymphocyte leaves blood and enters lymph node
LFA-1
blood
L-selectin
GlyCAM-1
ICAM-1
CD34
chemokine
lymph node

Figure 8.6 Homing of circulating naive T cells to the lymph node draining the infection. Lymphocytes bind to high endothelium in the lymph node through the interaction of L-selectin with the vascular addressins GlyCAM-1 and CD34. Chemokines, which are also bound to the endothelium, activate the integrin LFA-1 on the lymphocyte surface, enabling it to bind tightly to ICAM-1 on the endothelial cell. Establishment of tight binding allows the lymphocyte to squeeze between two endothelial cells, leaving the lumen of the blood vessel and entering the lymph node proper.

stromal cells and dendritic cells in the T-cell area and are bound to the surface of the high endothelial cells of the venules, where they establish a concentration gradient along the endothelial surface. Naive T cells express the chemokine receptor CCR7, which binds to CCL21 and CCL19 and causes the cells to be guided up the chemokine gradient and toward its source in the lymph node.

The initial contact between the T cell and the vascular endothelium allows interactions to be established between complementary adhesion molecules on the surfaces of the two cell types. These interactions involve the four types of cell-adhesion molecule (see Section 3-8). **L-selectin** on the T-cell surface binds to the sulfated sialyl-Lewis^{x} carbohydrate of the vascular addressins **CD34** and **GlyCAM-1** on the surface of the high endothelial venules. The cooperative effect of many such interactions causes the naive T cell to slow down and attach to the high endothelial surface (Figure 8.6). After the initial attachment, contact between the naive T cell and the endothelium is strengthened by additional interactions between the integrin LFA-1 on the T-cell surface and the adhesion molecules ICAM-1 and ICAM-2 on the vascular endothelium. The interaction between chemokines secreted by the lymph node and chemokine receptors on the T cell sends a signal that changes the conformation of LFA-1 so that it strengthens its hold on the ICAM. Through this array of cell-surface interactions the naive T cell is held at the surface of the vascular endothelium, where it gradually moves toward the source of the chemokines. Eventually the T cell squeezes between the endothelial cells to enter the lymph-node cortex (see Figure 8.6).

As naive T cells negotiate their way through the packed cells of the cortex, they bind transiently to the dendritic cells they meet. These interactions involve the binding of T-cell LFA-1 to ICAM-1 and ICAM-2 on the dendritic cell, and the binding of dendritic-cell LFA-1 to a third ICAM—**ICAM-3**—on the T-cell surface. ICAM-3 also binds to the lectin **DC-SIGN**, an adhesion molecule unique to activated dendritic cells. Adhesion is strengthened by interaction between T-cell **CD2** and dendritic-cell **LFA-3**. These transitory cell–cell interactions enable the T cell to assess the peptide:MHC complexes on the surface of dendritic cells in the search for one that engages the T-cell receptor. When a naive T cell encounters a specific peptide:MHC complex, the T-cell receptor starts signaling the cell to prepare for activation. A conformational change is induced in the T cell's LFA-1 molecules that increases their affinity for ICAMs

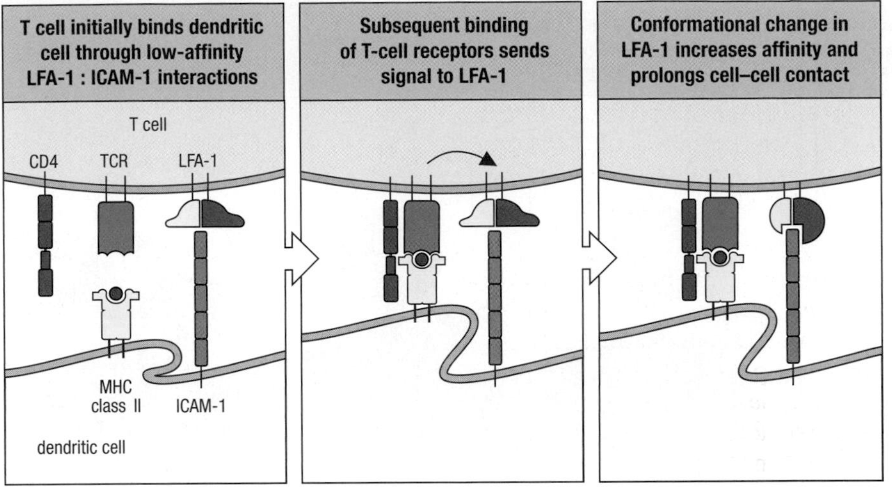
T cell initially binds dendritic cell through low-affinity LFA-1 : ICAM-1 interactions
Subsequent binding of T-cell receptors sends signal to LFA-1
Conformational change in LFA-1 increases affinity and prolongs cell–cell contact
T cell
CD4
TCR
LFA-1
MHC class II
ICAM-1
dendritic cell

Figure 8.7 Transient adhesive interactions between T cells and dendritic cells are stabilized by the recognition of specific antigen. When a T cell binds to its specific ligand on an antigen-presenting dendritic cell, intracellular signaling through the T-cell receptor (TCR) induces a conformational change in LFA-1 that causes it to bind with higher affinity to ICAMs on the antigen-presenting cell. The T cell shown here is a CD4 T cell. The same mechanisms are used for CD8 and CD4 T-cell homing.

(Figure 8.7). This stabilizes the interaction between the T cell and the dendritic cell, which can last for several days. When a dendritic cell and a naive antigen-specific T cell couple together, they are said to form a **conjugate** or a **cognate pair**. Under the influence of the dendritic cell, the T cell proliferates and the progeny differentiate to form a clone of effector cells. During this process, the dendritic cell maintains contact with all the T cells in the clone. This feat is made possible by the stellate morphology of the mature dendritic cell, which gives it an extensive cell-surface area. The mature dendritic cell thus provides a nursery for the propagation and maturation of effector T cells.

The vast majority of naive T cells will not find their antigens in the T-cell area and so pass from the cortex to the medulla and leave the lymph node in the efferent lymph. This is also the route for effector T cells to leave the lymph node once their differentiation is complete. These departures are controlled by a receptor on the T-cell surface that recognizes **sphingosine 1-phosphate (S1P)**, a lipid with chemotactic activity like that of a chemokine that is made by all cells. Within the lymph node a gradient of S1P is established. The S1P concentration is lowest in the T-cell areas and increases in the direction toward the medulla and the efferent lymphatic vessel. Under the influence of the S1P gradient, T cells expressing a S1P receptor move out of the lymph node into the efferent lymph and rejoin the circulation. The S1P gradient does not affect the naive T cells being nurtured by a dendritic cell, because upon recognizing antigen they are induced to express CD69, a protein that sequesters S1P receptors inside the cell and thus prevents them from sensing external S1P. Once the activated T cells mature into effector T cells, they no longer express CD69. Consequently, S1P receptors return to the cell surface and the effector T cells leave the lymph node as directed by the S1P gradient.

8-5 Activation of naive T cells requires signals from the antigen receptor and a co-stimulatory receptor

Intracellular signals generated by ligation of the T-cell receptor and co-receptor (either CD4 or CD8) with a peptide:MHC complex are necessary for activating a naive T cell, but they are not sufficient. A further signal, known as the **co-stimulatory signal**, is also required. Without this signal the T cell can neither divide nor survive. The receptor that delivers the co-stimulatory signal to the T cell is called **CD28**, and its ligand on the dendritic cell is the **B7 molecule**. B7 is described as a **co-stimulator molecule** or a **co-stimulatory molecule** and CD28 is a **co-stimulatory receptor**. At the region of contact between the naive T cell and its conjugated dendritic cell, B7 engages CD28 while peptide:MHC complexes engage the T-cell receptor and the co-receptor

(Figure 8.8). The combination of intracellular signals being simultaneously generated by the antigen receptor, the co-receptor, and the co-stimulatory receptor are necessary for the naive T cell to proliferate and differentiate into effector cells. In some discussions of T-cell signaling, the signals from the antigen receptor and co-receptor are called 'signal 1' and the co-stimulatory signal is called 'signal 2'. A decade before research on the pathways of lymphocyte signaling began, these two signals were predicted to exist on purely theoretical grounds.

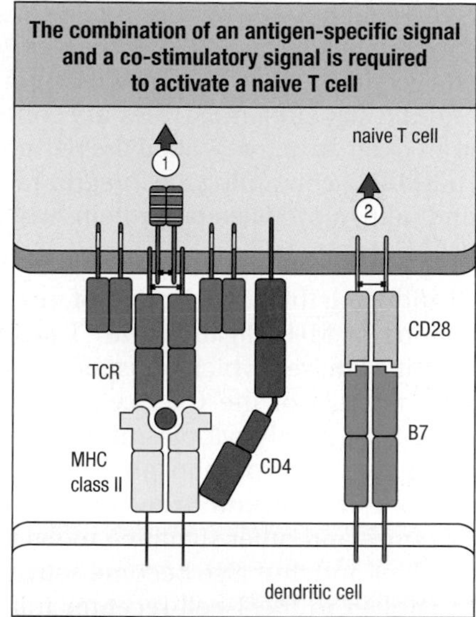

Figure 8.8 Activation of a naive T cell requires an antigen-specific signal and a co-stimulatory signal. The antigen-specific signal (arrow 1) is delivered when the T-cell receptor and its co-receptor CD4 recognize the peptide:MHC class II complex on the dendritic cell surface. The co-stimulatory signal (arrow 2) is delivered when the T-cell co-stimulatory receptor CD28 binds to the B7 co-stimulatory molecule expressed on the dendritic cell. Activation and clonal expansion of the antigen-specific naive T cell occurs only if these two signals are delivered together. This applies to both naive CD8 and CD4 T cells. Both CD28 and B7 are members of the immunoglobulin superfamily. There are two forms of B7, called B7.1 (CD80) and B7.2 (CD86), but no functional difference between them is known.

Only dendritic cells, macrophages, and B cells have co-stimulatory molecules on their surface. A further restriction is that these 'professional' antigen-presenting cells only express B7 in the presence of infection and not constitutively. When dendritic cells at infected sites take up pathogens and antigens using Toll-like receptors and other receptors of innate immunity, signals are generated that induce B7 expression. So when a dendritic cell carries antigens from the site of infection it arrives at the draining lymph node already expressing co-stimulatory molecules. Macrophages and B cells do not contribute to the activation of naive T cells. On the contrary, the effector T cells produced in the primary immune response are responsible for activating both macrophages and naive pathogen-specific B cells. Only then do B cells and macrophages express co-stimulatory molecules.

Although CD28 is the only B7 receptor on naive T cells, an additional and complementary B7 receptor is expressed once T cells are activated. This receptor, called **CTLA4**, is structurally similar to CD28 but binds B7 twentyfold more strongly and functions as a brake on CD28. Whereas B7 binding to CD28 activates a T cell to divide, the engagement of CTLA4 inhibits both the activation and proliferation of T cells.

The importance of CD28 in T-cell activation was tragically illustrated in 2006 by a phase 1 clinical trial in which six healthy volunteers were given a humanized anti-CD28 monoclonal antibody. This type of antibody had been shown to ameliorate autoimmune disease in a rat model: on binding to CD28, the antibody mimicked the natural B7 ligand, inducing signals from CD28 that activated regulatory T cells, which then acted to suppress inflammatory effector T cells in a nonspecific fashion. On the basis of these encouraging results in rats, the humanized anti-CD28 antibody was being tested as a possible therapy in humans to reduce the T cell-mediated inflammation that characterizes arthritis and other autoimmune diseases. The phase 1 trial was a necessary preliminary test to determine the effects of the antibody on healthy humans, before giving it to patients. The antibody was never given to patients, because of its catastrophic effects on the healthy volunteers—the opposite of what was expected. Instead of inducing a suppressive response, the antibody activated an army of effector T cells that produced a storm of cytokines and chemokines, causing systemic, life-threatening inflammation and autoimmunity. All six volunteers ended up in hospital as patients themselves; one remained in a coma for three weeks after experiencing failure of the heart, liver, and kidney, as well as suffering from septicemia, pneumonia, and gangrene. Such effects had not been observed when the antibody was similarly given to healthy monkeys. One of many consequences emanating from this catastrophe was investigation to explore the biological differences that distinguish human CD28 from its rat and monkey counterparts.

8-6 Signals from T-cell receptors, co-receptors, and co-stimulatory receptors activate naive T cells

When a T cell binds the peptide:MHC complexes on a dendritic cell, the interactions between receptors and ligand occur at localized, opposed areas of the two cell membranes. This region of contact and communication between the two cells is called the **T-cell synapse** and is analogous to the NK-cell synapse (see Section 3-19). Within the synapse an inner structure called the central supramolecular activation complex (c-SMAC) is where the T-cell receptors, co-receptors, co-stimulatory receptors (CD28), the CD2 adhesion molecule, and the signaling molecules are concentrated. Maintaining this organization is an outer structure, called the peripheral supramolecular activation complex (p-SMAC), containing the integrin LFA-1, the cell-adhesion molecule ICAM-1, and talin, a cytoskeletal protein, which together form a tight seal around the SMAC (Figure 8.9).

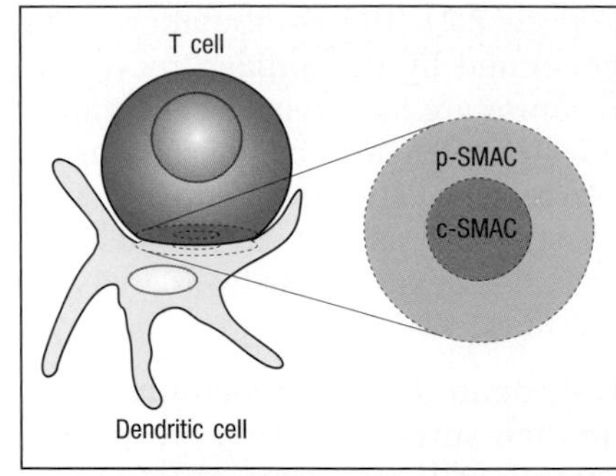

c-SMAC	p-SMAC
TCR CD2 CD4 CD8 CD28 PKC-θ	LFA-1 ICAM-1 talin

Figure 8.9 Protein–protein interactions in the area of contact between a T cell and an antigen-presenting cell form an ordered structure called the immunological synapse. The area of contact is divided into the central supramolecular activation complex (c-SMAC) and the peripheral supramolecular activation complex (p-SMAC), in which different sets of T-cell proteins are segregated.

By inducing this organization of molecules in the T-cell membrane, the interactions of MHC ligands with T-cell receptors activate cytoplasmic protein tyrosine kinases, which phosphorylate particular tyrosine residues in the cytoplasmic tails of the CD3 cell-surface proteins and the associated ζ chain (CD247), a purely cytoplasmic protein (see Figure 5.6, p. 118). The tyrosine residues that become phosphorylated are part of short amino-acid sequence motifs called **immunoreceptor tyrosine-based activation motifs (ITAMs)**. Enzymes and other signaling molecules bind to the phosphorylated tyrosine residues and thus also become activated. In this way, the extracellular binding of antigen to the T-cell receptor initiates pathways of intracellular signaling that lead to alterations in gene expression and end with T-cell differentiation.

Signals from the T-cell receptor and the CD4 or CD8 co-receptor synergize to activate the T cell (Figure 8.10). The cytoplasmic tails of CD4 and CD8 associate with **Lck**, one of several protein kinases that phosphorylate the CD3 ITAMS. On formation of the T-cell synapse, Lck phosphorylates and activates another protein kinase, **ZAP-70** (ζ chain-associated protein of 70 kDa molecular mass), which then binds to the phosphorylated tyrosines of the ζ chain. Produced only in T cells, ZAP-70 is necessary for the initiation of the major pathways of T-cell signaling. Emphasizing the crucial role of ZAP-70 is the severity of the immunodeficiency caused by the absence of functional ZAP-70. Although these infants have normal numbers of CD4 and CD8 T cells, engagement of their antigen receptors fails to induce intracellular signals. Without functional T cells these children begin to suffer recurrent bacterial, viral, and fungal infections in their first year and die in their second year unless rescued by a hematopoietic stem cell transplant.

Once activated, ZAP-70 triggers three signaling pathways that are common to many types of cell (Figure 8.11). In naive T cells, they lead to changes in gene

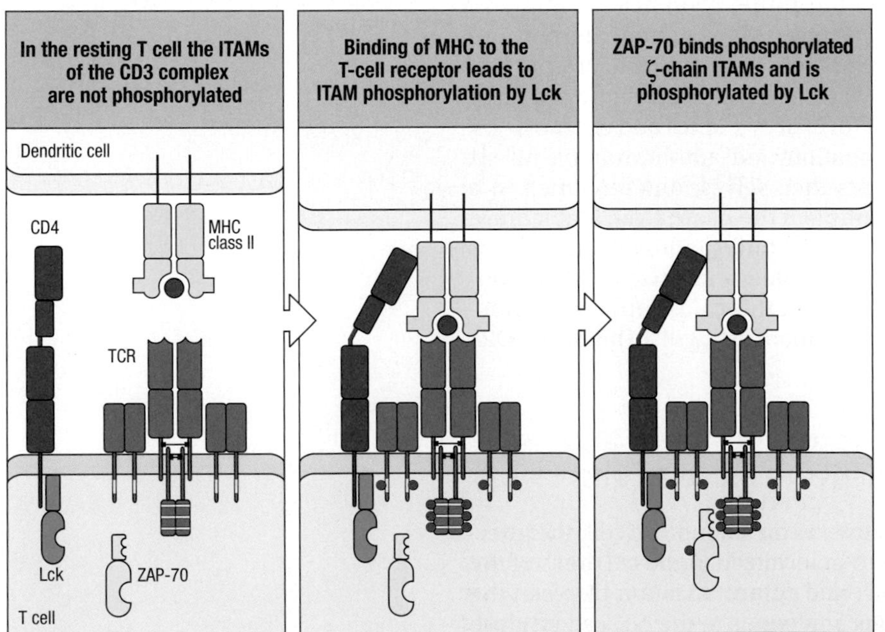

Figure 8.10 Clustering of the T-cell receptor and a co-receptor initiates signaling within the T cell. First panel: in a resting T cell, the ITAMs (yellow boxes) on CD3γ, δ, and ε, and the ζ chains of the T-cell receptor are not phosphorylated. Second panel: when T-cell receptors and co-receptors become clustered on binding to peptide:MHC complexes on the surface of an antigen-presenting cell, the ITAMs become phosphorylated (phosphorylated tyrosines are shown as small red circles) by kinases such as Lck that are associated with the complex. Third panel: the tyrosine kinase ZAP-70 recognizes and binds to the phosphorylated ITAMs on the ζ chain and is activated by phosphorylation by Lck. Activated ZAP-70 transmits the signal onward along the T-cell signaling pathway.

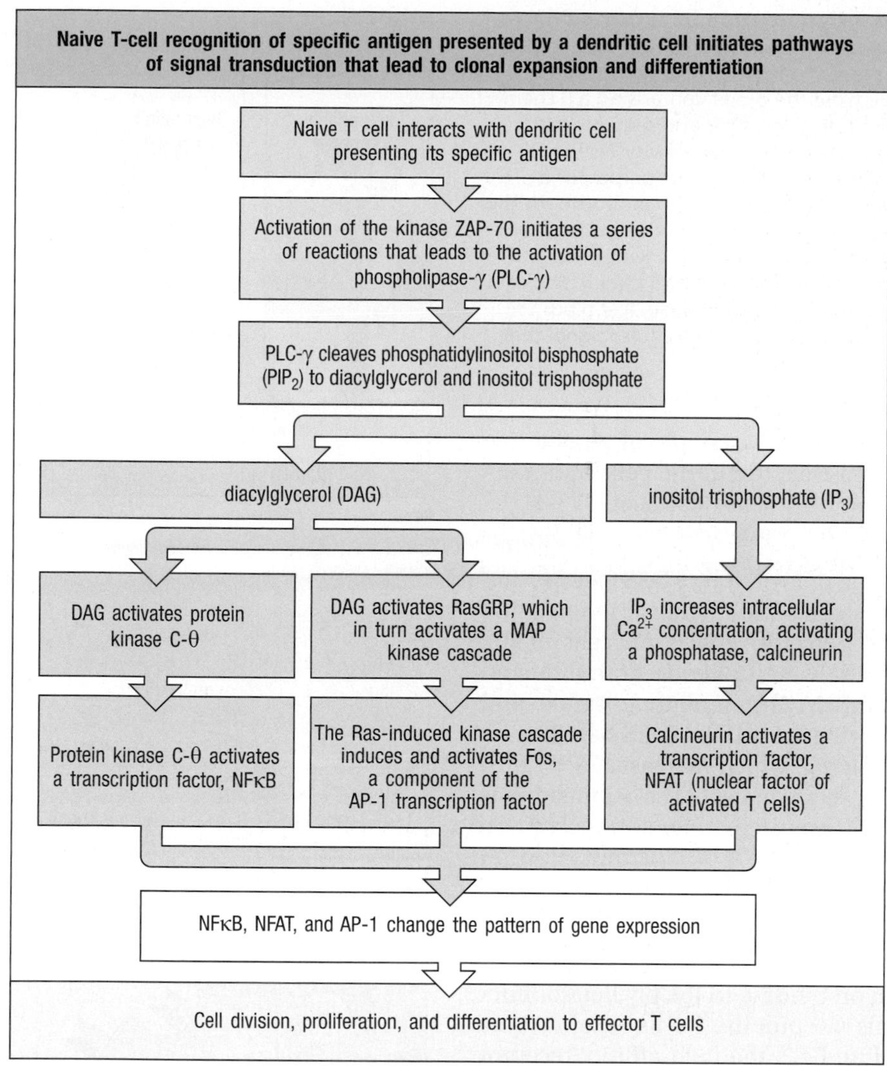

Figure 8.11 Summary of the intracellular signaling pathways initiated by the T-cell receptor complex, the CD4 co-receptor, and the CD28 co-stimulatory receptor. Similar pathways operate in CD8 T cells, because CD8, like CD4, interacts with Lck.

expression produced by the transcriptional activator **NFAT** (**nuclear factor of activated T cells**) in combination with other transcription factors. One pathway initiated by ZAP-70 leads via the second messenger inositol trisphosphate to the activation of NFAT. A second pathway leads to the activation of protein kinase C-θ, which induces the transcription factor **NFκB**. The third signaling pathway initiated by ZAP-70 involves Ras, a GTP-binding protein, and leads to the activation of a nuclear protein called Fos, one component of the transcription factor **AP-1**. The other component of AP-1 is called Jun, and its activation is one consequence of co-stimulatory signals delivered through CD28. The combined effect of NFAT, AP-1, and NFκB is to initiate transcription of genes that direct T-cell proliferation, differentiation, and acquisition of effector functions.

8-7 Proliferation and differentiation of activated naive T cells are driven by the cytokine interleukin-2

Among the genes switched on in activated T cells, the one encoding the cytokine **interleukin-2 (IL-2)** is particularly important. IL-2 is essential for the proliferation and differentiation of T cells and was first named T-cell growth factor. It is synthesized and secreted by the activated T cells themselves and acts back on the cells that make it. This type of action is called **autocrine** action. The alternative is **paracrine** action, in which the cytokine acts on a

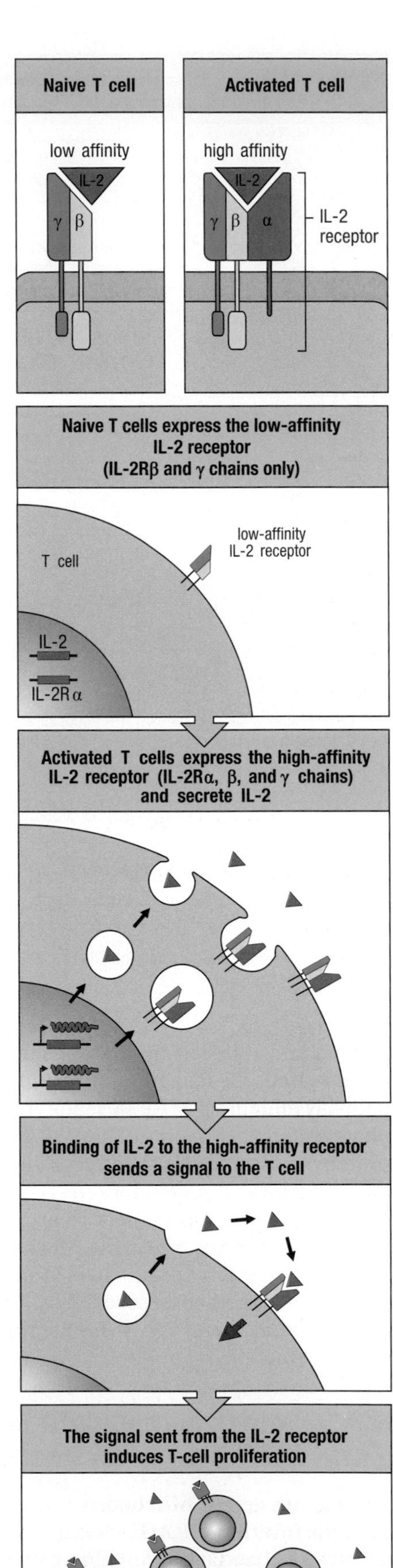

Figure 8.12 Proliferation and differentiation of activated naive T cells is driven by the cytokine interleukin-2 (IL-2). The receptor for IL-2 is made from three different chains: α, β, and γ. A receptor with low affinity for IL-2 is made from the dimer comprising a β chain and a γ chain, whereas the high-affinity receptor also requires the α chain (top panels). Naive T cells express the low-affinity receptor (second panel). Activation of a naive T cell by the recognition of a peptide:MHC complex accompanied by co-stimulation induces both the synthesis and secretion of IL-2 (orange triangles) and the synthesis of the IL-2 receptor α chain (blue). The α chain combines with the β and γ chains to make a high-affinity receptor for IL-2 (third panel). The cell also enters the first phase (G1) of the cell-division cycle. IL-2 binds to the IL-2 receptor (fourth panel), producing an intracellular signal that promotes T-cell proliferation (fifth panel).

different type of cell from the one that made it. An example of a paracrine cytokine is IL-12, which is made by myeloid cells (dendritic cells and macrophages) but acts on lymphocytes (NK cells and T cells) (see Section 3-4).

The production of IL-2 requires signals delivered by the T-cell receptor:co-receptor complex and the co-stimulatory signal delivered by CD28 (see Section 8-5). Through the activation of NFAT, these signals lead to transcription of the *IL-2* gene. Like many cytokines, IL-2 has powerful effects on cells of the immune system, and its production is precisely controlled in both time and space. To achieve such control, cytokine mRNA is inherently unstable, and sustained production of IL-2 requires stabilizing the mRNA. Such stabilization is one consequence of the co-stimulatory signal, which increases T-cell production of IL-2 by twentyfold to thirtyfold. A second effect of co-stimulation is the production of transcription factors that increase the rate of transcription from the *IL-2* gene by threefold. The principal effect of co-stimulation on T cells is to increase their synthesis of IL-2 by about one hundredfold.

The IL-2 receptor on naive T cells is a heterodimer of a β and a γ chain that binds IL-2 with low affinity. On activation, naive T cells synthesize a third component of the receptor, the α chain, which on binding to the β:γ heterodimer forms a high-affinity IL-2 receptor. With this receptor the T cells become more responsive to IL-2 (Figure 8.12). On binding IL-2, the high-affinity receptor triggers the T cells to progress through cell division. For about a week the cells divide two to three times daily, so that one activated T cell can give rise to several thousand identical daughter cells. The T-cell area of the lymph node is thus seen to function as a T-cell farm in which rare, naive antigen-specific T cells are propagated to produce an army of effector T cells. Consistent with this picture, during the peak response to certain viral infections nearly 50% of the CD8 T cells are specific for the same viral peptide:MHC complex.

The importance of IL-2 for activating the adaptive immune response is reflected in the mode of action of the immunosuppressive drugs cyclosporin A (also called cyclosporine), tacrolimus (also called FK506), and rapamycin (also called sirolimus), which are used to prevent the T cell-mediated rejection of organ transplants. Cyclosporin A and tacrolimus inhibit IL-2 production by disrupting signals from the T-cell receptor, whereas rapamycin inhibits signaling from the IL-2 receptor. By suppressing the activation and differentiation of naive T cells, these drugs prevent transplant recipients from making an adaptive immune response against the non-self antigens of the grafted organ, which came from another person.

8-8 Antigen recognition in the absence of co-stimulation leads to a state of T-cell anergy

In the naive T-cell population there are always some cells with specificity for self antigens. These T cells escape negative selection in the thymus because

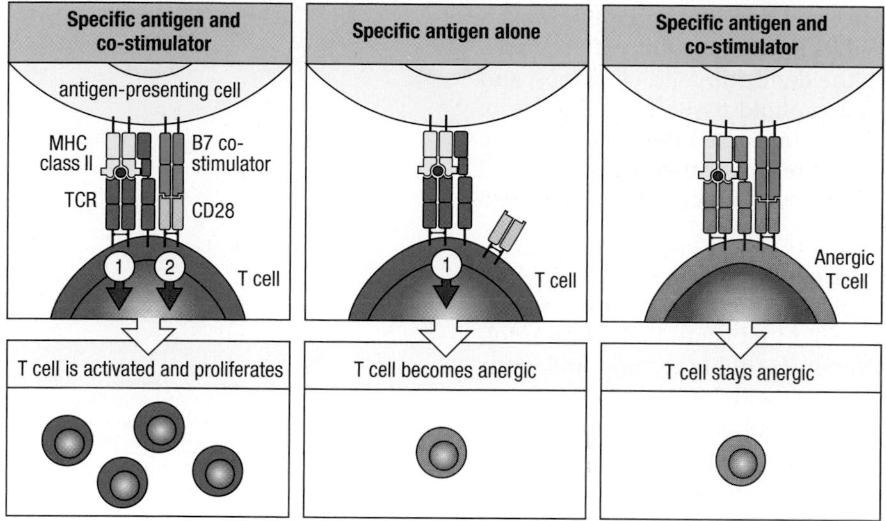

Figure 8.13 Naive T cells that recognize self antigens are prevented from responding to those antigens. During infection, naive pathogen-specific T cells are activated by mature dendritic cells expressing B7 co-stimulators (left panel). In the absence of infection, dendritic cells do not express B7. Naive self-reactive T cell antigens that escape negative thymic selection and enter the circulation recognize their antigens presented by cells in healthy tissue that have no B7 (center panel). By giving an antigen-specific signal without a co-stimulatory signal, this interaction induces anergy in the self-reactive T cells. If, during an infection, the anergic T cells encounter their self antigen on a mature dendritic cell, they cannot signal to activate the cell (right panel).

the self antigens they recognize are not expressed in the thymus. If such T cells encounter their self antigen it will almost always be on a cell that does not express B7, because this co-stimulator is expressed on few cell types and its expression occurs only during periods of infection and inflammation. In the absence of B7 and co-stimulation, engagement of the peptide:MHC complex by the T-cell receptor and co-receptor on the self-reactive T cell gives a signal that causes the cell to enter a state in which it cannot respond to any external signal. Once a T cell enters this state, called **anergy** or **T-cell anergy**, it cannot be revived, even if its specific antigen is subsequently presented in the context of co-stimulation by a mature dendritic cell (Figure 8.13). The defining characteristic of anergic T cells is their failure to make IL-2, the essential cytokine for sustaining T-cell growth and differentiation. The induction of anergy is thus seen as an irreversible mechanism of self-tolerance that works in the peripheral circulation to render harmless those self-reactive T cells that manage to leave the thymus.

Immunologists confronted anergy long before they gave it a name. A puzzling observation was that immunizing animals with pure proteins rarely led to an adaptive immune response. Trial and error showed that good responses could be obtained if the protein antigens were deliberately contaminated with bacteria or their breakdown products. Decades later, it was discovered that the microbial components, known in this context as **adjuvants**, were inducing co-stimulatory activity in the dendritic cells, a phenomenon that occurs naturally during the innate immune response to infection (see Section 8-5). Conversely, immunization with pure protein antigens induces anergy in antigen-specific T cells and thus actively prevents the generation of an adaptive immune response. There is a logic to this phenomenon: unlike infections, pure proteins rarely pose a threat to the human body.

8-9 Activation of naive CD4 T cells gives rise to effector CD4 T cells with distinctive helper functions

Effector CD4 T cells do not directly attack the pathogens that cause infection, but help other cells of the immune system to achieve that goal. In providing this help, CD4 T cells form a cognate pair with the other cell, which in this context is called the **target cell**. The type of immune response that is effective at terminating infection varies with the nature of the pathogen and the site in the human body that is infected. Thus bacterial infection in the skin, viral infection of the airways, and an intestinal infection with helminth worms all require different forms of adaptive immune response. Such shaping of adaptive immunity to the specifications of the infection and its location begins with the activation of naive T cells and their exposure to the dendritic cells, cytokines, and other materials coming into the secondary lymphoid tissue from the site of infection. In this way, the tissue of origin of the activated dendritic cell, the nature of the pathogen, and the innate immune response made against it will promote the differentiation of helper CD4 T cells with functions best suited to the infection at hand.

The population of effector CD4 T cells is extremely heterogeneous, as seen from the diverse combination of cell-surface markers they express and the cytokines they secrete. Since the discovery of CD4 in the late 1970s, immunologists have sought to bring order to this diversity by defining subpopulations based on phenotypic characteristics that correlate with different helper functions. This approach met with limited success until correlations were also made with the transcription factors driving CD4 differentiation and heterogeneity. Currently, five types of helper CD4 T cell are distinguished according to the cytokines that induce their differentiation, the transcription factors that determine their differentiation, the cytokines they make, and the cells that they help. The different helper cells are called **T_H1 cells, T_H17 cells, T_H2 cells, T follicular helper cells (T_{FH})**, and **regulatory T cells (T_{reg})** (Figure 8.14).

T_H1 cells help macrophages respond to intracellular bacterial infections and viral infections, and their differentiation is controlled by the T-bet transcription factor. T_H17 cells help neutrophils respond to extracellular bacterial and fungal infections, and their differentiation is controlled by the transcription factor RORγT. Differentiation of T_H2 cells, which help eosinophils, basophils, mast cells, and B cells respond to parasite infections, is controlled by the GATA3 transcription factor. T_{FH} cells, which are differentiated by the Bcl6 transcription factor, are responsible for the activation of naive B cells and their differentiation into antibody-producing cells. The differentiation of regulatory T cells, which control and limit the activities of other types of effector CD4 and CD8 T cell, is controlled by the FoxP3 transcription factor. Each of the transcription factors is described as the **master regulator** of the differentiation pathway it controls.

	T_H1 cells	T_H17 cells	T_H2 cells	T_{FH} cells	T regulatory cells (T_{reg})
Effector CD4 T cell	T_H1	T_H17	T_H2	T_{FH}	T_{reg}
Cytokines that induce differentiation	IL-12 IFN-γ	IL-6 IL-21	IL-4	IL-16 TGF-β IL-23	TGF-β
Defining transcription factor	T-bet	RORγT	GATA3	Bcl6	FoxP3
Characteristic cytokines	IL-12 IFN-γ	IL-17 IL-6	IL-4 IL-5	IL-21	TGF-β IL-10
Function	Activate macrophages	Enhance neutrophil response	Activate cellular and antibody response to parasites	Activate B cells Maturation of antibody response	Suppress other effector T cells

Figure 8.14 Five functional classes of effector CD4 T cell are produced by activation and differentiation in different cytokine environments. Summarized here are the cytokines that induce the different pathways of differentiation, the transcription factors uniquely associated with each pathway, the cytokines made by each type of effector CD4 T cell, and the roles of these cells in the immune response.

Whereas mature CD4 and CD8 T cells represent two distinctive, stable, and mutually exclusive lineages of T cells, that is not the case for the five subsets of CD4 T cells. The properties of these effector CD4 T cells can change with the local environment and the cytokines it contains. A major challenge to studying effector CD4 T cells is the nagging possibility that observations made on isolated and cultured cells in the laboratory do not reflect their properties when present in their natural environment, the human body.

8-10 The cytokine environment determines which differentiation pathway a naive T cell takes

The pathway of differentiation that an activated naive T cell takes is determined by the cytokines in the immediate environment of the T cell and the attached dendritic cell that guides its development. This environment is created by the innate immune response at the site of infection and the delivery of activated dendritic cells, pathogens, antigens, and cytokines to the responding secondary lymphoid tissue.

T_H1 CD4 T cells increase inflammation within infected tissues and are a major defense against intracellular viral and bacterial infections. Stimulating the production of T_H1 cells are the cytokines IL-12 and IFN-γ, which are first made during the innate immune response: IL-12 by dendritic cells and macrophages, and IFN-γ by NK cells (see Section 3-20). By binding to their receptors on the CD4 T cell, these cytokines induce changes in gene expression and the production of T-bet (see Figure 8.14), the master regulator that commits naive T cells to becoming T_H1 cells. T-bet causes the T-cell's own IFN-γ gene to be turned on, leading to the local production and secretion of IFN-γ in the secondary lymphoid tissue. The increased IFN-γ concentration caused by the first CD4 T_H1 cells creates a cytokine environment that favors further differentiation of activated CD4 T cells along the T_H1 pathway. By this positive-feedback mechanism, the entire clone of T cells that a dendritic cell is nurturing can become T_H1 cells. It is also possible that the IFN-γ secreted by one clone of T_H1 cells influences naive CD4 cells attached to another dendritic cell by recruiting them to the T_H1 pathway.

T_H2 CD4 T cells orchestrate the defense against parasites that colonize the tissues and surfaces of the human body. These biologically diverse organisms include many that are multicellular and much larger than the cells of the human immune system. The defense mechanisms that work against parasites are therefore different from those that control infections by microorganisms. In general, T_H2 cells do not generate inflammation, but instead promote the repair and recovery of tissues damaged by infection. T_H2 cells favor the production of pathogen-specific IgE antibodies, that work with basophils, mast cells, and eosinophils to eliminate parasites from the respiratory, gastrointestinal, and urogenital tracts, as well as from within tissues. The crucial cytokine for a T_H2 response is IL-4, which induces naive T cells to differentiate into T_H2 effector cells (see Figure 8.14). Binding of IL-4 to the IL-4 receptor of an activated naive CD4 T cell induces expression of the transcription factor GATA3, which commits the cell to the T_H2 pathway. GATA3 turns on the genes for IL-4 and IL-5, the characteristic cytokines secreted by T_H2 cells. Once CD4 T_H2 cells start secreting IL-4, the local environment becomes one that promotes the further differentiation of activated CD4 T cells along the T_H2 pathway. The cells of innate immunity that provide the IL-4 needed to start the T_H2 pathway of differentiation have yet to be identified. The basophil is a good candidate: basophils make IL-4, interact with T_H2 cells, are few in number and thus suited to a 'catalytic' function, and are recruited into secondary lymphoid tissues responding to infection.

T_H17 cells secrete IL-17 that binds to the IL-17 receptor on epithelial and stromal cells, inducing them to secrete CXCL8 and other cytokines that recruit neutrophils to infected tissue. Patients lacking either IL-17 or its receptor suffer chronic infections with the fungus *Candida albicans*, a common commensal organism of the human body. Although the signature cytokines made by T_H17 effector cells are those of the IL-17 family, the cytokines that nurture the differentiation of T_H17 cells are IL-16 and transforming growth factor-β (TGF-β) (see Figure 8.14). They induce the activated T cell to make IL-21 (an IL-17 family member), which acts in an autocrine fashion to turn on expression of STAT3, a transcription factor essential for T_H17-cell differentiation. Transcription factor RORγT, the master regulator of T_H17 differentiation, is necessary for turning on the synthesis and secretion of IL-17 (see Figure 8.14). T_H17 cells are numerous in the skin and the gut mucosa.

T_{FH} cells are the helper T cells that cooperate with naive B cells to initiate the antibody response to infection. The differentiation of T_{FH} cells from naive CD4 T cells is induced by IL-6 and is distinguished by the transcription factor Bcl6. As well as CD28 interaction with B7 molecules, co-stimulation of T_{FH} cells by dendritic cells involves the **inducible T-cell co-stimulator** (**ICOS**), which binds to the ICOS ligand on dendritic cells and is structurally related to CD28 and CTLA4. Bcl6 is required for T_{FH} cells to express CXCR5, the receptor for the CXCL13 chemokine produced by the stromal cells of the B-cell follicle. The binding of CXCL13 to CXCR5 causes T_{FH} cells to leave the T-cell areas of the secondary lymphoid tissue and take up residence in the follicles of the B-cell areas. Cytokines secreted by T_{FH} cells guide the switching of immunoglobulin isotype in the B cells, so as to make the class of antibody most suited for clearing the particular infection. Thus a parasite infection will stimulate IgE production, whereas an extracellular bacterial infection will stimulate opsonizing IgG (see Section 4-16).

T regulatory cells (T_{reg}) do not contribute to generation of the primary immune response to a pathogen, but help to keep it under control and wind it down once the pathogen poses no further threat. By limiting the tissue damage caused by the immune response, T_{reg}s facilitate healing and reduce the chance of secondary infections. T_{reg}s work by interacting with the other types of effector T cell and instructing them to cease effector functions. **Natural regulatory cells** commit to regulatory function during thymic development (see Section 7-13). **Induced regulatory T cells** emerge in the course of the immune response to infection, when naive T cells are activated in the presence of TGF-β but in the absence of IL-6 and other pro-inflammatory cytokines. Both types of regulatory T cell are distinguished from other effector T cells by their expression of the transcription factor FoxP3, as well as the CD4 and CD25 cell-surface proteins. Induced regulatory T cells produce TGF-β and IL-10, which act to inhibit inflammation and the immune response.

8-11 Positive feedback in the cytokine environment can polarize the effector CD4 T-cell response

For both T_H1 and T_H2 cells, the cytokine that drives their differentiation—IFN-γ for T_H1, and IL4 for T_H2—is central to their effector function and secreted in quantity. This situation creates the opportunity for positive feedback in which the functioning effector T cells drive further differentiation of the same type of effector T cell. This can lead to the rapid expansion of a population of pathogen-specific CD4 T cells with either T_H1 or T_H2 cells becoming dominant. When this happens the T-cell response is described as being **polarized.** A polarized T_H1 response corresponds to what immunologists traditionally called **cell-mediated immunity**, a response dominated by the effector cells of the immune system. In contrast, a polarized T_H2 response, in which antibodies dominate, corresponds to the traditional description of **humoral immunity**—'humors' being an alternative term for the body fluids in which antibodies circulate.

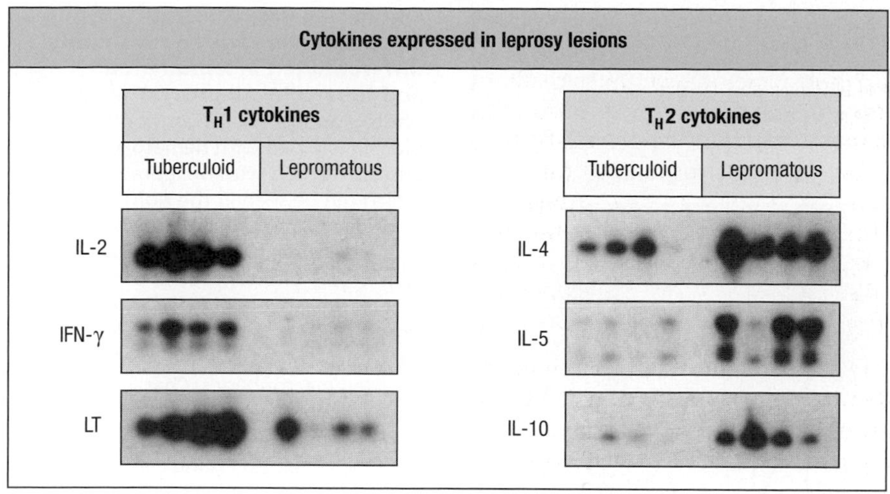

Figure 8.15 T_H1 cytokines characterize tuberculoid leprosy, whereas T_H2 cytokines characterize lepromatous leprosy. Shown here is a Northern blot analysis of mRNA for six cytokines in biopsied tissue from lesions of four patients with lepromatous leprosy and four patients with tuberculoid leprosy. Typical T_H1 cytokines predominate in the tuberculoid form, whereas typical T_H2 cells predominate in the lepromatous form. Lymphotoxin (LT) is a characteristic T_H1 cytokine that is structurally and functionally similar to TNF-α. Cytokine blots courtesy of R.L. Modlin.

Leprosy is a disease caused by the persistent infection of the vesicular system of macrophages with *Mycobacterium leprae*. About 95% of humans are able to resist infection by *M. leprae* without significant symptoms of disease. In patients with leprosy, the effector CD4 T-cell response becomes polarized toward either T_H1 or T_H2 cells, a choice that profoundly influences the progression of disease (Figure 8.15). In patients with a T_H1-biased response, the cytokines made by the T_H1 cells help the infected macrophages to suppress the growth and dissemination of the bacteria. Although the chronic inflammation damages the skin and peripheral nerves, the disease progresses slowly and the patients usually do not die from it. For patients making a T_H2-biased response, the outcome is quite different: although large amounts of pathogen-specific antibodies are made, they are ineffective against the bacteria hiding inside the macrophages. By growing unchecked, the bacteria become disseminated to other sites in the body, causing gross tissue destruction, which is eventually fatal. The visible symptoms of disease in patients with leprosy who are making a T_H1 or T_H2 response are so dissimilar that the conditions are given different names: tuberculoid leprosy and lepromatous leprosy, respectively (Figure 8.16). Both these conditions are chronic infections in which the immune system fails to eliminate the mycobacterial pathogen.

8-12 Naive CD8 T cells require stronger activation than naive CD4 T cells

CD4 helper T cells are functionally diverse but interact with relatively few target cell types: lymphocytes, phagocytes, and granulocytes. In contrast, cytotoxic CD8 T cells are functionally homogeneous but have to interact with a wide range of target cells. These comprise all the cells that are susceptible to viral or microbial infection, which to first approximation is all the cells in the human body. CD8 T cells are crucial elements of defense against intracellular infections, most of which are viral infections. Because of the inherently destructive nature of CD8 T cells, the activation of naive CD8 T cells in the secondary lymphoid tissues is not done lightly and requires stronger co-stimulatory activity than the activation of naive CD4 T cells.

Naive CD8 T cells recognize antigens presented on the MHC class I molecules of mature dendritic cells. For some viral infections, the interaction of a virus-specific naive CD8 T cell with a dendritic cell that presents the peptide:MHC class I complex is sufficient for activation and differentiation to proceed. When activated by antigen and the dendritic cell's co-stimulatory

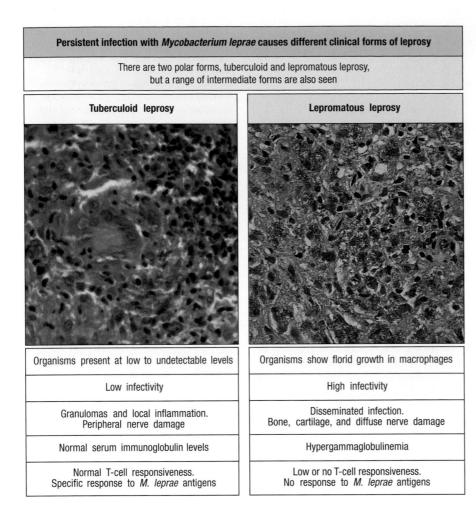

Persistent infection with *Mycobacterium leprae* causes different clinical forms of leprosy

There are two polar forms, tuberculoid and lepromatous leprosy, but a range of intermediate forms are also seen

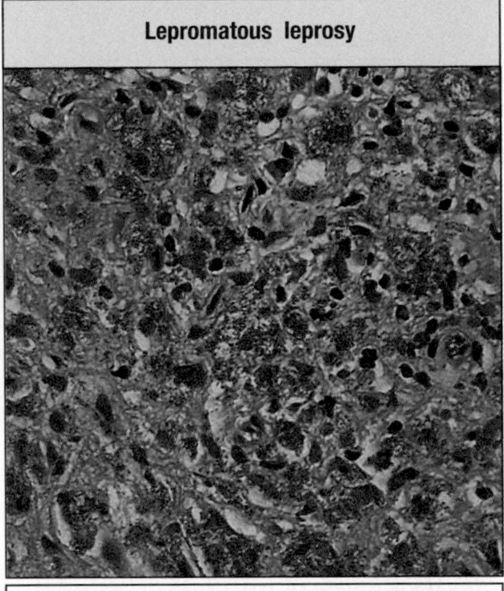

Tuberculoid leprosy	Lepromatous leprosy
Organisms present at low to undetectable levels	Organisms show florid growth in macrophages
Low infectivity	High infectivity
Granulomas and local inflammation. Peripheral nerve damage	Disseminated infection. Bone, cartilage, and diffuse nerve damage
Normal serum immunoglobulin levels	Hypergammaglobulinemia
Normal T-cell responsiveness. Specific response to *M. leprae* antigens	Low or no T-cell responsiveness. No response to *M. leprae* antigens

Figure 8.16 Responses to *Mycobacterium leprae* are sharply differentiated in lepromatous and tuberculoid leprosy. The photographs show sections of lesion biopsies stained with hematoxylin and eosin. Infection with *M. leprae* bacilli, which can be seen in the right-hand photograph as numerous small dark red dots inside macrophages, can lead to two very different forms of the disease. In tuberculoid leprosy (left), growth of the microorganism is well controlled by T_H1-like cells that activate infected macrophages. Characteristic of the tuberculoid lesions are small nodules called granulomas in which the macrophages organize in such a way as to contain both infection and inflammation to a localized area. Consequently, in this disease there is only local damage to peripheral nerves. In lepromatous leprosy (right), infection is widely disseminated and the bacilli grow uncontrolled in macrophages. In the late stages of this disease, there is severe damage to connective tissues and to the peripheral nervous system. Between the polar forms of clinical leprosy there is a spectrum of intermediate forms. Photographs courtesy of G. Kaplan.

molecules, the naive CD8 T cells synthesize both the cytokine IL-2 and its high-affinity receptor, which together induce the CD8 T cells to proliferate and differentiate (Figure 8.17, left panels).

For other viral infections, dendritic cells alone are insufficient to activate naive CD8 T cells, and they solicit the help of virus-specific effector CD4 T cells, which give the necessary IL-2 to 'jump-start' the activation. The dendritic cell must interact simultaneously with both the naive CD8 T cell and the effector CD4 T cell, the former recognizing a viral-peptide:MHC class I complex on the dendritic cell surface and the latter a viral-peptide:MHC cell class II complex (Figure 8.17, right panels). The effector CD4 T cell is activated to make and secrete IL-2, which then binds to the IL-2 receptors induced on the CD8 T cell by its antigen-specific interaction with the dendritic cell. The combination of intracellular signals generated from the IL-2 receptor, the T-cell receptor, the CD8 co-receptor, and the CD28 co-stimulatory receptor then drives the naive CD8 T cell to proliferate and differentiate.

The more stringent requirement for the activation of naive CD8 T cells means that they are activated only when the evidence of infection is unambiguous. Cytotoxic T cells inevitably inflict damage on any target tissue to which they are directed, and their actions are only of benefit to the host if the pathogen is eliminated in the process. Even then, the actions of cytotoxic T cells can have damaging effects. For example, in fighting viral infections of the airways, cytotoxic T cells prevent viral replication by destroying the epithelial layer, which then makes the underlying tissue vulnerable to secondary bacterial infection.

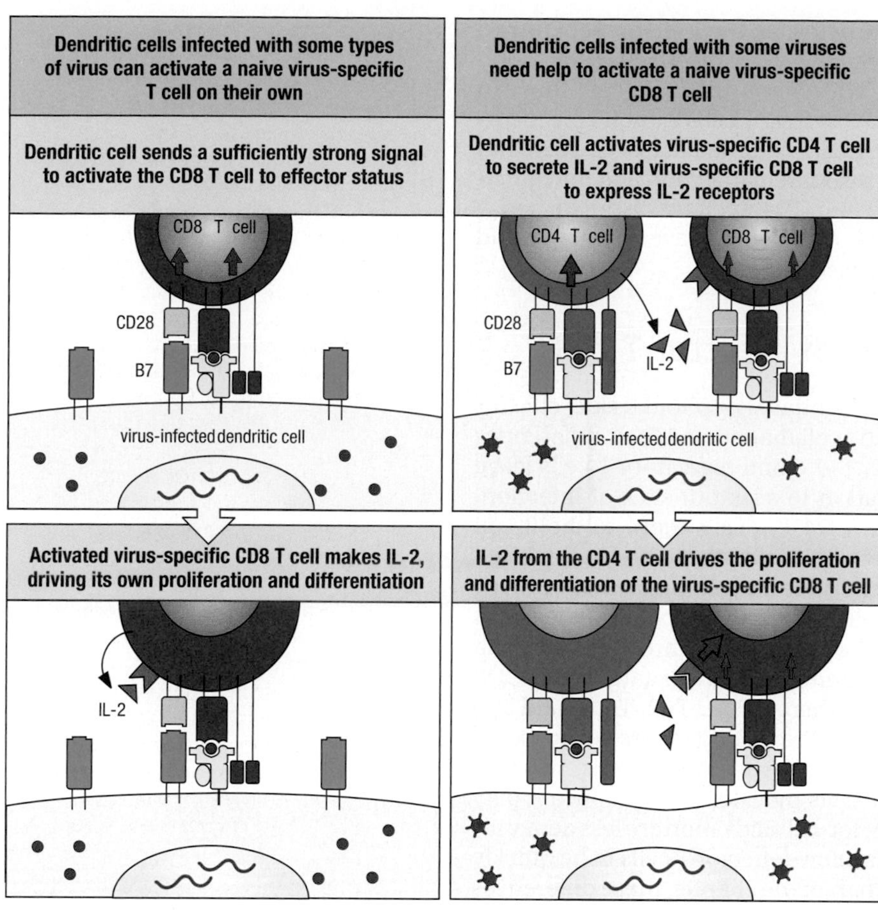

Dendritic cells infected with some types of virus can activate a naive virus-specific T cell on their own
Dendritic cell sends a sufficiently strong signal to activate the CD8 T cell to effector status
CD8 T cell
CD28
B7
virus-infected dendritic cell
Activated virus-specific CD8 T cell makes IL-2, driving its own proliferation and differentiation
IL-2
Dendritic cells infected with some viruses need help to activate a naive virus-specific CD8 T cell
Dendritic cell activates virus-specific CD4 T cell to secrete IL-2 and virus-specific CD8 T cell to express IL-2 receptors
CD4 T cell
CD8 T cell
CD28
B7
IL-2
virus-infected dendritic cell
IL-2 from the CD4 T cell drives the proliferation and differentiation of the virus-specific CD8 T cell

Figure 8.17 Two ways to activate a naive CD8 T cell. The left panels show how a naive CD8 T cell can be activated directly by a virus-infected dendritic cell. The right panels show how a dendritic cell (or other virus-infected cell expressing MHC class II molecules) that induces insufficient co-stimulation can be helped by CD4 effector T cells to activate naive virus-specific CD8 T cells. In this situation, the IL-2 secreted by the CD4 T cell acts directly on the naive CD8 T cell interacting with the same dendritic cell, and provides the necessary boost for activating the CD8 T cell (right panels).

Summary

All forms of the adaptive immune response are started by the activation of pathogen-specific naive T cells. Their proliferation and differentiation to form large clones of pathogen-specific effector T cells make up the first stage of a primary immune response. T-cell activation takes place in the secondary lymphoid tissues, where dendritic cells carrying antigen from the site of infection meet antigen-specific naive T cells coming from the blood or lymph. T-cell activation and differentiation requires signals from the T-cell receptor and CD3 complex, the co-receptor, and the CD28 co-stimulatory receptor. These signals are delivered on engagement of a naive T cell with a dendritic cell presenting the appropriate antigen and expressing B7 co-stimulatory molecules. The combination of these signals induces many changes in the T cells' pattern of gene expression and the proteins they make, including the production of IL-2, a cytokine that is essential for their clonal expansion and differentiation into effector T cells. The whole process of T-cell activation and differentiation occurs in the immediate environment of the dendritic cell, during which time the naive T cell and its numerous progeny maintain contact with the dendrites of the dendritic cell.

The co-stimulatory B7 molecules, which interact with CD28 on the T cell, are not expressed ubiquitously by dendritic cells but are induced by the presence of infection and the signals generated when innate immune receptors on the dendritic cell sense microbial products. This ensures that naive T cells do not respond to their specific antigens in the absence of infection, and provides a mechanism of peripheral tolerance that prevents the activation of naive T cells expressing receptors that bind self antigens in the absence of infection. When such a T cell engages its specific antigen, the lack of co-stimulation induces anergy, a permanent state of tolerance.

Naive CD4 T cells and naive CD8 T cells are activated by similar mechanisms. Activation of naive CD8 T cells produces a relatively homogeneous population of cytotoxic effector T cells that can kill the wide variety of target cells that can become infected by a virus. In contrast, effector CD4 T cells exhibit considerable functional diversity, but their interactions are confined to helping myeloid and lymphoid cells of the immune system respond to infection and to regulating that response. Five main subtypes of effector CD4 T cell are distinguished—T_H1, T_H2, T_H17, T_{FH}, and regulatory T cells—all of which have distinctive and complementary roles in the immune response.

The properties and functions of effector T cells

After completing their differentiation in a secondary lymphoid tissue, effector T cells detach themselves from the dendritic cell that nursed their differentiation. CD8 effector cells and CD4 T_H1, T_H2, T_H17, and regulatory T cells leave the lymphoid tissue and enter the circulation to seek out sites of infection. Staying in the secondary lymphoid tissue, CD4 T_{FH} cells move to the B-cell area, where they activate the B-cell response to infection. All effector T-cell functions are activated by recognition of a peptide antigen presented by MHC class I or class II molecules on the target cell. The effector T cell and the target cell then form a conjugate pair connected by a synapse, through which the T cell delivers effector molecules that profoundly affect the target cell. CD8 effector cells induce their target cells to die, whereas T_H1, T_H2, T_H17, and T_{FH} cells help their target cells attack the infection. Preventing excessive attack are the regulatory T cells, whose targets are the other effector T cells. Because effector T cells are the progeny of naive T cells that were chosen by a highly selective process, the requirements for effector-cell activation are less demanding than those applied to naive T cells. This allows effector T cells to be quickly mobilized to confront an infection. In this part of the chapter, these differences between effector T cells and naive T cells will be considered first, followed by a comparison of the specialized effector functions of the various effector T cells.

8-13 Cytotoxic CD8 T cells and effector CD4 T_H1, T_H2, and T_H17 work at sites of infection

After their differentiation in secondary lymphoid tissue, CD8 cytotoxic T cells and CD4 T_H1, T_H2, and T_H17 cells travel to infected tissue to do their work. To equip them for this purpose, the differentiation of effector T cells from naive T cells involves changes in cell-surface molecules that facilitate travel to the site of infection and performance of their functions there (Figure 8.18).

Some of the cell-surface differences that distinguish effector T cells from naive T cells affect adhesion molecules. These changes allow effector T cells to leave the secondary lymphoid tissue in the lymph, travel to the blood stream, and

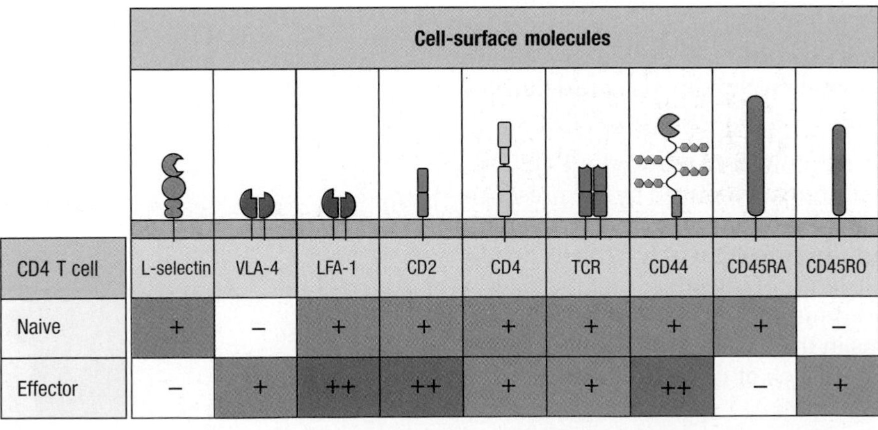

	Cell-surface molecules								
CD4 T cell	L-selectin	VLA-4	LFA-1	CD2	CD4	TCR	CD44	CD45RA	CD45RO
Naive	+	–	+	+	+	+	+	+	–
Effector	–	+	++	++	+	+	++	–	+

Figure 8.18 Activation of T cells changes the expression of several cell-surface molecules. Resting naive T cells express L-selectin allowing them to home to lymph nodes. Effector T cells express VLA-4, enabling them to enter infected tissue. Increased amounts of the LFA-1, CD2, and CD44 adhesion molecules allows effector T cells to make stronger interactions with target cells. Changes in CD45 increase the capacity of effector T cells to interact with target cells. The cell surface of a CD4 T cell is shown, CD8 T cells exhibit similar properties.

then leave the blood at the inflamed tissue. L-selectin, which permits naive T cells to gain entry to lymph nodes from the blood, is replaced in effector T cells by the integrin **VLA-4**, which binds to the adhesion molecule VCAM-1 (Figure 8.19). The activated endothelial cells of blood vessels at inflammatory sites express VCAM-1 and so can halt passing effector T cells and direct them to enter the infected tissue. Through these alterations of the T-cell surface, effector T cells are excluded from returning to secondary lymphoid tissue and ordered to enter the infected tissues where their services are needed.

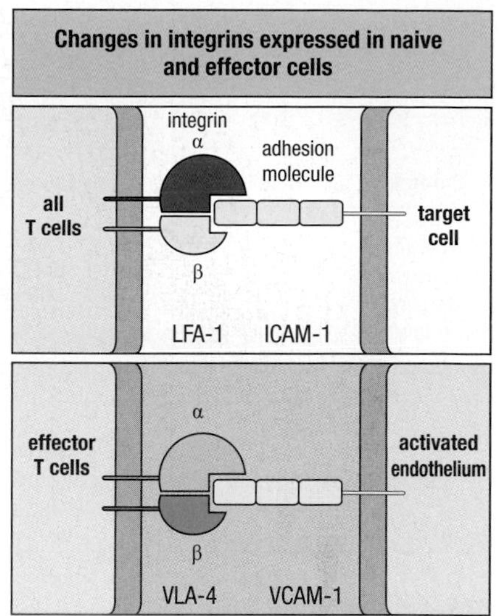

Figure 8.19 Integrin VLA-4 enables effector T cells to home to inflamed tissue. The integrin LFA-1 is an adhesion molecule that binds to ICAMs and is expressed by all leukocytes. LFA-1 contributes to T-cell interactions with a variety of types of target cell (upper panel). In addition to LFA-1, effector T cells express a second integrin, VLA-4, which binds to the VCAM-1 adhesion molecule (lower panel). VCAM-1, which is selectively expressed on the endothelium of blood vessels in inflamed tissue, recruits effector T cells from the blood into the infected tissue.

On entering an infected site, an effector T cell forms transient contacts with the tissue cells, searching for a target cell that is presenting its specific antigen. To facilitate the formation of these transient contacts, effector T cells express two to four times more CD2 and LFA-1 cell-adhesion molecules than naive T cells do, which makes them more sensitive than professional antigen-presenting cells to the smaller amounts of ICAM-1 and LFA-3 on most tissue cells (see Figure 8.18). Without specific engagement of the T-cell receptor, an effector T cell's interaction with a target cell is short-lived. But if the T-cell receptor succeeds in binding a peptide:MHC complex, a conformational change is induced in LFA-1 that strengthens its binding to ICAM-1 and ensures a long-lived interaction between the T cell and the now-attached target cell.

Naive and effector T cells differ markedly in their recognition of antigen. The signals coming from the T-cell receptor and co-receptor are sufficient to activate effector T cells, whereas such signals alone induce anergy in naive T cells (see Section 8-8). Thus, co-stimulation through CD28 and B7 molecules is not needed to activate the effector T-cell response. For CD8 T cells, this relaxation in activation requirements is of particular importance, because it means that they are able to kill any type of cell that becomes infected with a virus (Figure 8.20). For CD4 effector T cells, the range of potential target cells is also expanded. Naive T cells are activated only by recognition of antigen on dendritic cells, whereas effector CD4 T cells are able to recognize antigens on all cells that express MHC class II molecules. As well as the professional antigen-presenting cells, these include many other cell types, including T cells and vascular endothelial cells, that are induced to express MHC class II by the IFN-γ secreted by NK cells and effector T cells at sites of infection. Co-stimu-lation is thus seen as the mechanism that restricts the initiation of T-cell responses to the ordered microenvironment of the secondary lymphoid tissues. Relaxing the need for co-stimulation allows effector T cells to work quickly and efficiently in the chaotic environment of infected and inflamed tissues.

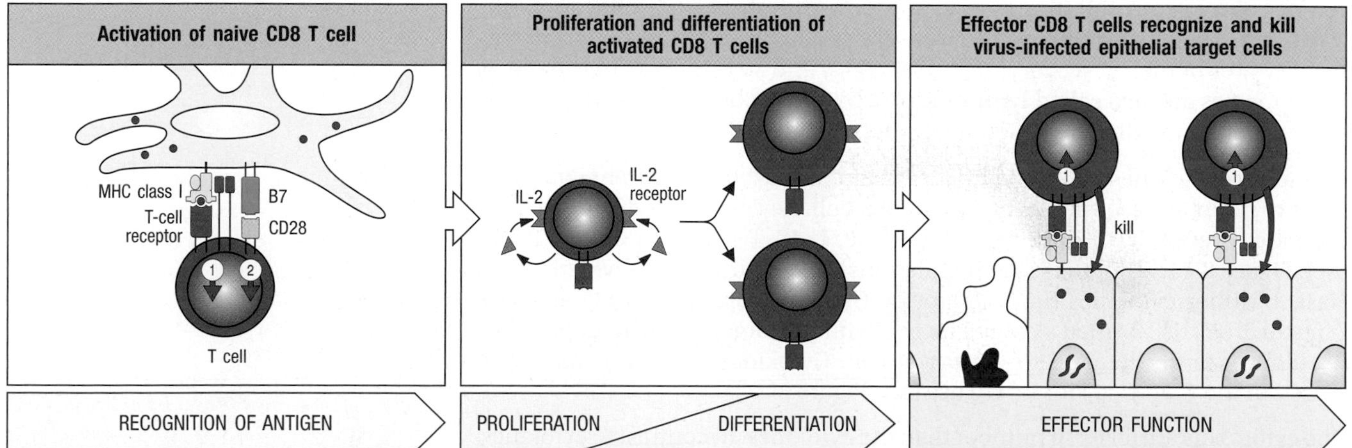

Figure 8.20 Co-stimulatory signals are required for the activation of naive T cells but not for the activation of effector T cells. Activation of naive T cells requires intracellular signals from the CD28 co-stimulatory receptor as well as from the T-cell receptor and co-receptor (left panel). After the proliferation and differentiation of the activated naive T cell (center panel), the resulting effector cells can recognize and respond to targets that present the specific antigen but lack the B7 co-stimulatory molecules that engage CD28 (right panel). Neither CD4 nor CD8 effector T cells require co-stimulatory signals for activation. Shown here is a CD8 T cell.

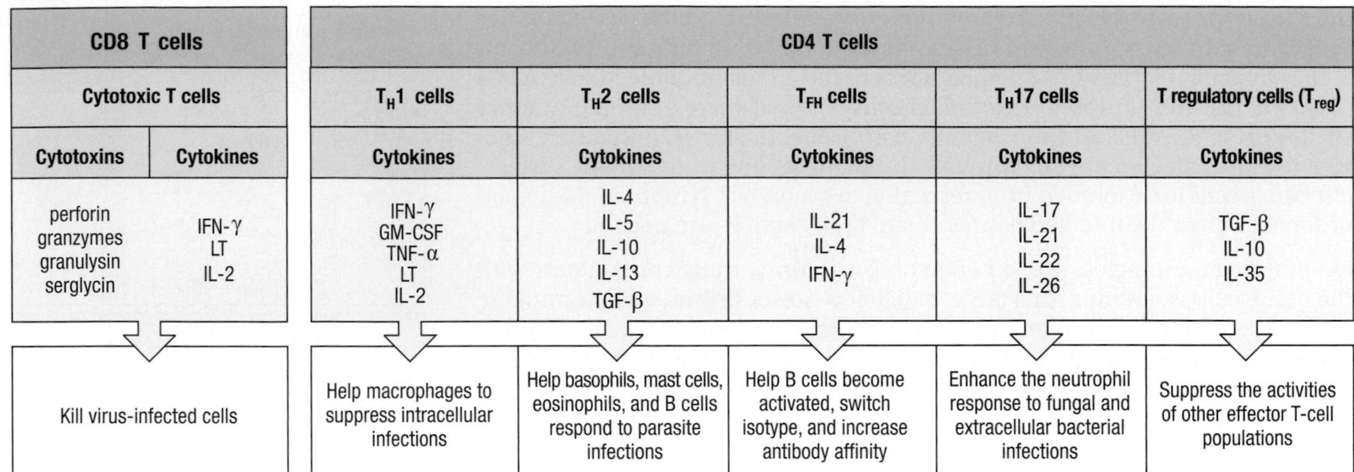

CD8 T cells
Cytotoxic T cells
Cytotoxins
Cytokines
perforin
granzymes
granulysin
serglycin
IFN-γ
LT
IL-2
Kill virus-infected cells
CD4 T cells
T_H1 cells
Cytokines
IFN-γ
GM-CSF
TNF-α
LT
IL-2
Help macrophages to suppress intracellular infections
T_H2 cells
Cytokines
IL-4
IL-5
IL-10
IL-13
TGF-β
Help basophils, mast cells, eosinophils, and B cells respond to parasite infections
T_{FH} cells
Cytokines
IL-21
IL-4
IFN-γ
Help B cells become activated, switch isotype, and increase antibody affinity
T_H17 cells
Cytokines
IL-17
IL-21
IL-22
IL-26
Enhance the neutrophil response to fungal and extracellular bacterial infections
T regulatory cells (T_{reg})
Cytokines
TGF-β
IL-10
IL-35
Suppress the activities of other effector T-cell populations

Figure 8.21 The effector molecules of cytotoxic CD8 T cells and helper CD4 T cells.

8-14 Effector T-cell functions are mediated by cytokines and cytotoxins

Effector T cells always exert their functions in the context of a conjugate pair with a target cell. At the localized area of contact between the two cells a T-cell synapse is formed in which the T-cell receptor and co-receptor bind the complex of peptide antigen and MHC molecule. At the synapse, the effector molecules made by the T cell are delivered to the target cell. The molecules that mediate T-cell effector functions are of two types: cytokines that alter the behavior of target cells; and cytotoxic proteins, or **cytotoxins**, that kill the target cells. All effector T cells make cytokines, whereas cytotoxins are made only by CD8 effector T cells (Figure 8.21).

Cytokines are small secreted and membrane-bound proteins that are essential components of innate (see Section 3-4) and adaptive immunity. They control all aspects of the immune response by binding to cell-surface receptors called cytokine receptors. Intracellular signals generated by these interactions induce changes of gene expression within the target cell. By autocrine action, cytokines can amplify an immune response, as seen in extreme form in the polarized T-cell responses in the two forms of leprosy (see Section 8-11). By paracrine action, cytokines provide communication and cooperation between different cell types in the immune response. Many of the cytokines made by T cells are called **interleukins** and were assigned numbers in series according to the order of their discovery: IL-1, IL-2, IL-3, and so on. Cytokines made by lymphocytes are also called **lymphokines**, but in this book we use the general term cytokine for all these functionally related molecules.

Numerous cytokines are made by T cells and are used in different combinations to influence the various types of target cell. Key cytokines distinguish the six main types of CD4 helper T cell—T_H1, T_H2, T_H17, T_{FH}, and regulatory T cells (see Figure 8.14)—but these subpopulations can be further subdivided on the basis of other cytokines that they make. Cytokines are made only after the effector T cell has formed a conjugate pair with the target cell, whereupon they are delivered on demand to the target cell. Cytokines are never made and stored in the CD4 T cell for future use (see Figure 8.21).

The cytotoxins are fewer in number than the cytokines. In contrast to cytokines, effector CD8 T cells manufacture cytotoxins in quantity and store them in **lytic granules** before an encounter with a target cell. The cytotoxins comprise the **granzymes** (a family of five serine proteases), a membrane-disrupting protein called **perforin**, the proteoglycan **serglycin**, and **granulysin**, a detergent-like protein that associates with membranes. The killing mechanisms used by the

CD8 T cells of adaptive immunity and the NK cells of innate immunity are very similar, as are the morphology and contents of the lytic granules (see Section 3-19). Where CD8 T cells and NK cells differ is in the way in which they distinguish infected cells from healthy cells. For CD8 T cells, this function is served by a single dedicated receptor, the T-cell receptor, whereas NK cells use a variety of many different receptors.

8-15 Cytokines change the patterns of gene expression in the cells targeted by effector T cells

Cytokines work in the immediate vicinity of the effector T cell and for short periods of time. Membrane-bound cytokines on the T cell affect the target cell only in the immediate area of the immunological synapse. The secretion of soluble cytokines is also focused at the immunological synapse. Membrane-associated cytokines can achieve their effect with smaller quantity and lower affinity for the cytokine receptor than secreted cytokines can; they can also contribute to the strength of adhesion between T cell and target. Being soluble proteins, secreted cytokines have the possibility of diffusing away from the T cell that made them and exerting their effect on cells other than the T cell's target. Some cytokines, TNF-α for example, are made in both membrane-associated and soluble forms.

Many cytokine receptors comprise two membrane proteins that together form the cytokine-binding site but do not interact with each other in the absence of cytokine. Presence of the cytokine drives the formation of the complex of cytokine and cytokine receptor (Figure 8.22). Associated with the cytoplasmic tails of the two cytokine receptor polypeptides are inactive forms of protein kinases known as **Janus kinases (JAKs)**. The dimerization of receptor polypeptides induced by cytokine binding also dimerizes the Janus kinases, making them into active enzymes. The kinases now phosphorylate proteins called **STATs** (**s**ignal **t**ransducers and **a**ctivators of **t**ranscription). The phosphorylated STATs then dimerize, which allows them to move from the cytoplasm to the nucleus, where they change the pattern of gene expression in the target cell. Both the JAKs and the STATs are families of proteins, the members being used in distinctive combinations with the various cytokine receptors to effect different changes in gene expression. The JAK–STAT signaling pathways are short, direct, and efficient, which enables target cells to respond rapidly to stimulation by cytokines.

The effects of cytokines can also be quickly turned off. One mechanism is for intracellular phosphatases to remove the phosphate groups from JAKs, STATs, and cytokine receptors that are critical for signaling. A second mechanism involves a family of inhibitory proteins called suppressors of cytokine

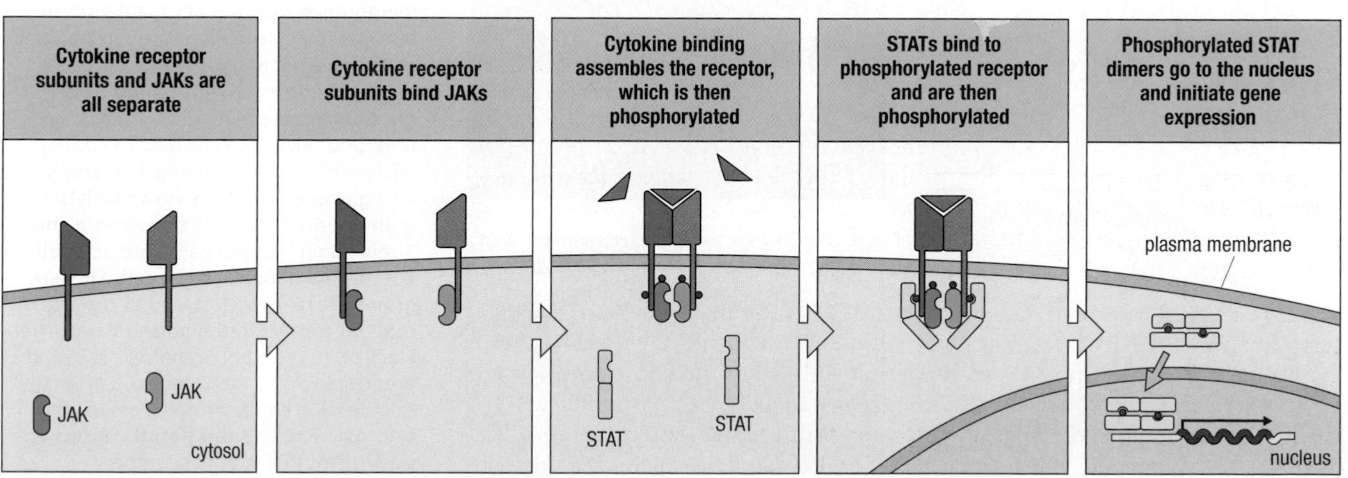

Figure 8.22 Cytokine-induced assembly of a cytokine receptor and associated JAK and STAT signaling molecules quickly changes gene expression. Cytokine receptor subunits do not associate in the absence of cytokine (first panel). The cytoplasmic tail of each cytokine receptor subunit binds a JAK kinase (second panel). In the presence of cytokine, the receptor assembles. The JAKs are brought together, become activated, and phosphorylate the receptor subunits (third panel). Two STATs bind to the phosphorylated receptor and they, too, are phosphorylated by the JAKs (fourth panel). The phosphorylated STATs dimerize, dissociate from the cytokine receptor, and go to the nucleus, where they activate transcription from various genes that contribute to the adaptive immune response (fifth panel). This pathway is followed by many cytokines and using different combinations of JAK and STAT family members.

signaling (SOCS) proteins. By binding directly to the phosphorylated tyrosine residues of cytokine receptors and JAKs, SOCS proteins inhibit the signaling pathway and also target its components for degradation.

8-16 Cytotoxic CD8 T cells are selective and serial killers of target cells at sites of infection

Once inside a cell, a pathogen becomes inaccessible to antibodies and other soluble proteins of the immune system. It can be eliminated either through the efforts of the infected cell itself or by direct attack on the infected cell by the immune system. The function of cytotoxic CD8 T cells is to kill cells that are overwhelmed by intracellular infection. Overall, the sacrifice of the infected cells serves to prevent the spread of infection to healthy cells. People lacking functional cytotoxic T cells suffer from persistent viral infections, showing the importance of cytotototoxic T cells for defense against viruses.

After activation by specific antigen and as part of their differentiation in the secondary lymphoid tissue, CD8 T cells synthesize cytotoxins in inactive forms and package them into membrane-bound lytic granules. Effector CD8 T cells then migrate to sites of infection to search for infected cells presenting their specific antigens. At sites of infection, cytotoxic CD8 T cells and infected target cells are surrounded by healthy tissue cells and cells of the immune system that have infiltrated the infected tissue. Because of their antigen specificity, cytotoxic T cells pick out only infected cells for attack and leave healthy cells alone. The T cell focuses granule secretion at the small, localized area of the synapse where it is attached to the target cell (Figure 8.23). Here the granule membrane fuses with the plasma membrane of the T cell and discharges its

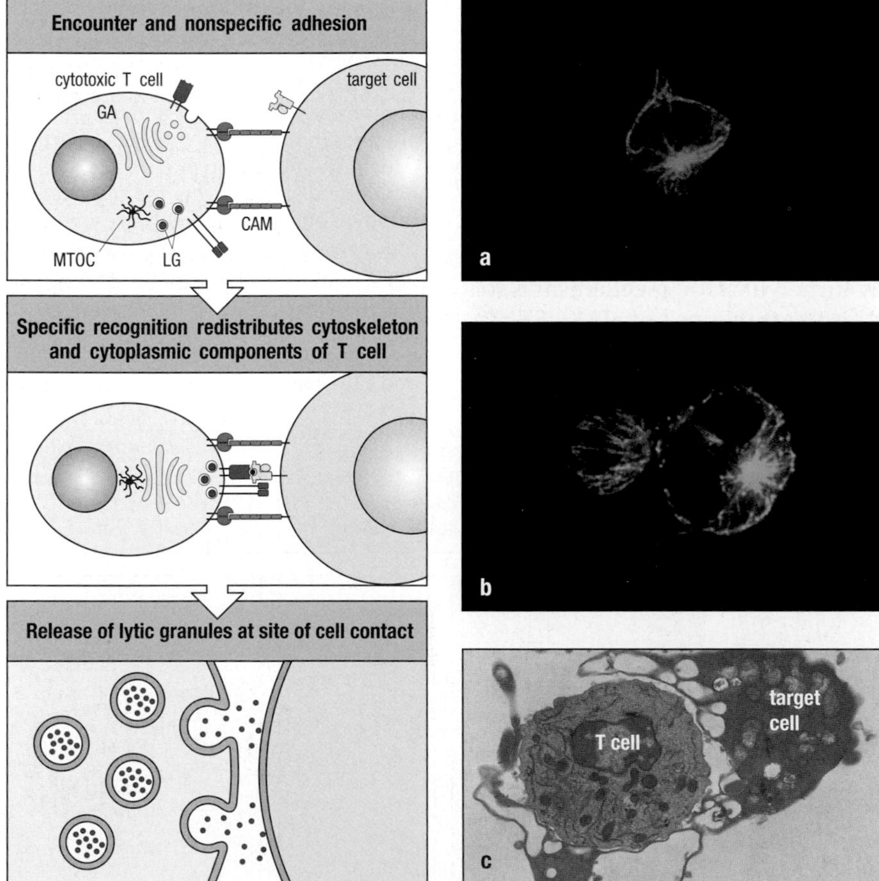

Figure 8.23 CD8 T cells kill infected cells by secretion of lytic granules and focused delivery of cytotoxins onto the target cell surface. Initial nonspecific adhesion of the CD8 T cell to a target cell does not perturb the lytic granules (LG) (top left panel). Engagement of the antigen receptor causes the T cell to become polarized: the cortical actin cytoskeleton at the site of contact reorganizes, enabling the microtubule-organizing center (MTOC), the Golgi apparatus (GA), and the lytic granules to align toward the target cell (center left panel). Fusion of the granule membranes with the plasma membrane releases the cytotoxins stored in the lytic granules directly onto the targetcell membrane (bottom left panel). The photomicrograph in panel a shows an unbound, isolated cytotoxic T cell. The microtubules are stained green and the lytic granules red. Note how the lytic granules are dispersed throughout the T cell. Panel b depicts a cytotoxic T cell bound to a (larger) target cell. The lytic granules are now clustered at the site of cell–cell contact in the bound T cell. The electron micrograph in panel c shows the deterioration of the target cell after the release of granules from its associated cytotoxic T cell. Photographs courtesy of G. Griffiths and J. Stinchcombe.

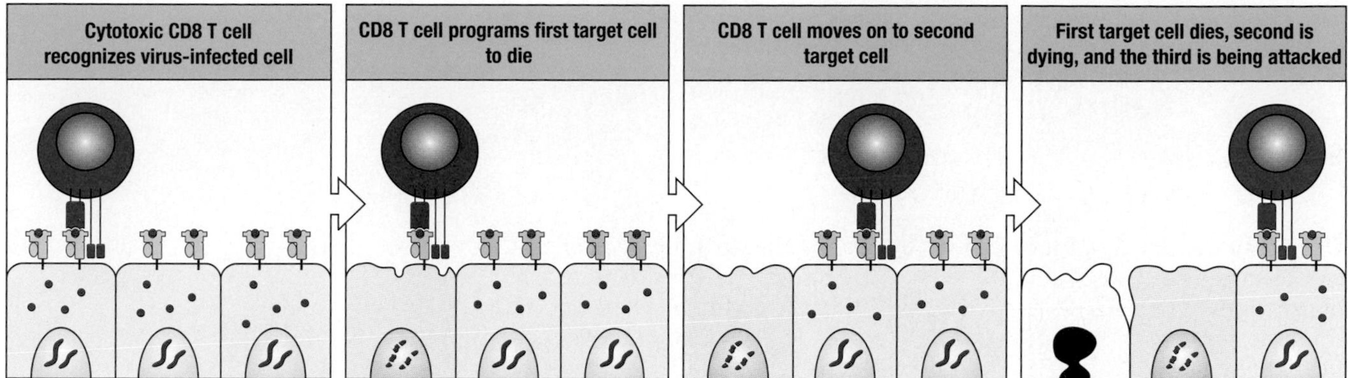

Figure 8.24 Cytotoxic CD8 T cells can kill several infected target cells in succession. When a CD8 T cell recognizes a peptide:MHC class I complex on an infected cell (first panel), it programs the infected cell to die by apoptosis (second panel). The T cell then detaches from the first target cell, synthesizes a new set of lytic granules, and then seeks out a second target cell to attack (third panel). The cycle is repeated with an attack on a third target cell (fourth panel).

contents onto the target cell surface. In this way, cytotoxic granules neither attack healthy neighbors of an infected cell nor kill the T cell itself. As the target cell starts to die, the cytotoxic T cell is released from the target cell and starts to make new granules. Once new granules have been made, the cytotoxic T cell is able to kill another target cell. In this manner, one cytotoxic T cell can kill many infected cells in succession (Figure 8.24).

Beside their cytotoxic action, CD8 T cells also contribute to the immune response by secreting cytokines. One of these is IFN-γ, which inhibits the replication of viruses in infected cells and increases the processing of viral antigens and their presentation by MHC class I molecules. IFN-γ also activates macrophages in the vicinity of the cytotoxic T cells. The macrophages get rid of the dying infected cells, allowing the T cells more room for maneuver and also helping the damaged tissue to heal and regenerate.

8-17 Cytotoxic T cells kill their target cells by inducing apoptosis

Cells killed by cytotoxic CD8 T cells do not lyse or disintegrate, unlike cells undergoing necrosis due to physical or chemical injury, but die by **apoptosis** (Figure 8.25). This form of cell death, which is widely used by the immune system to eliminate unwanted or dangerous cells (see, for example, Section 6-3), induces a cell to commit suicide from within. The cell shrivels and shrinks while retaining its contents, leaving a neat and tidy corpse. Apoptosis, also known as **programmed cell death**, prevents not only pathogen replication but also the release of infectious bacteria or virus particles from the infected cell.

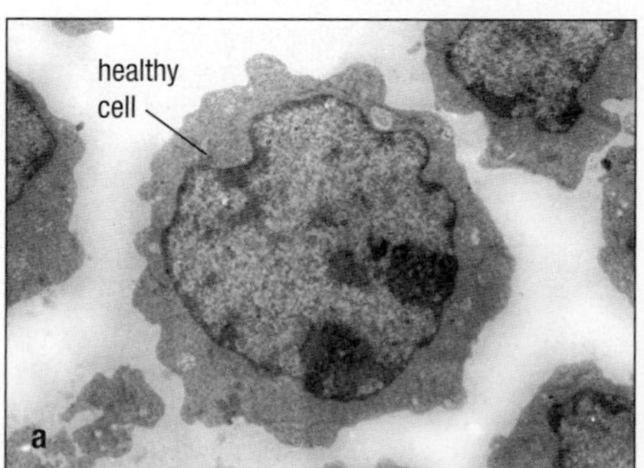

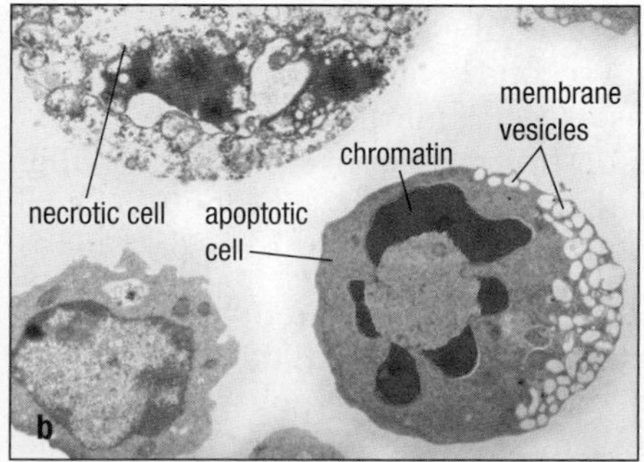

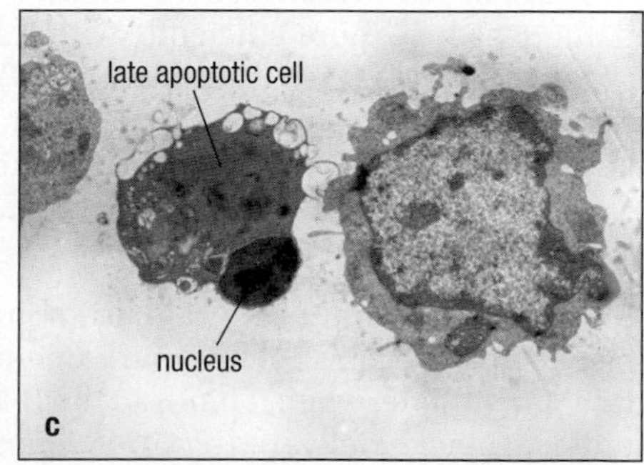

Figure 8.25 Apoptosis, or programmed cell death. Panel a shows an electron micrograph of a healthy cell with a normal nucleus. In the bottom right of panel b is a cell at an early stage of apoptosis. The chromatin in the nucleus has condensed (shown in red); the plasma membrane is well defined and is shedding membrane vesicles. In contrast, the plasma membrane of the cell dying by necrosis shown in the upper left part of panel b is poorly defined. The middle cell shown in panel c is at a late stage in apoptosis. It has a very condensed nucleus and no mitochondria, and the cytoplasm and cell membranes have largely been lost through vesicle shedding. Photographs courtesy of R. Windsor and E. Hirst.

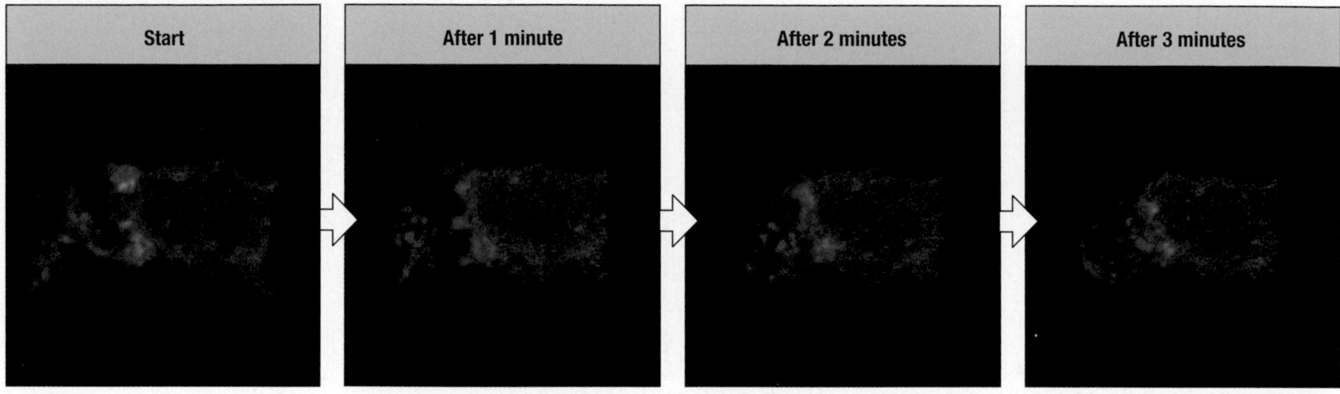

After deposition onto the surface of the target cell (see Section 8-16), the combination of perforin, granulysin, and serglycin makes pores in the target cell membrane and delivers the granzymes to the cell's interior. Once inside the target cell, the five granzymes, which are serine proteases, initiate a cascade of proteolytic cleavage reactions that leads to the activation of nucleases within the target cell. The nucleases degrade the target cell's DNA by cleaving between the nucleosomes to give DNA fragments that are multiples of 200 base pairs in length and characteristic of apoptosis. Eventually the nucleus becomes disrupted, with accompanying loss of membrane integrity and normal cell morphology. The cell destroys itself from within; it shrinks by the shedding of membrane-enclosed vesicles (Figure 8.25, center panel) and the degradation of cell contents until little is left. Changes in the plasma membrane caused by apoptosis are recognized by phagocytes, which take up and eliminate the dying cells. The apoptotic mechanisms that degrade the infected cell also act on the infecting pathogen. In particular, the breakdown of viral nucleic acids prevents the assembly of infectious virus particles that might cause further infection if they were to escape from the dying cell. A 5-minute contact between a cytotoxic T cell and its target cell is all it takes for the target cell to be programmed to die (Figure 8.26), even though visible evidence of death takes longer to become obvious.

Figure 8.26 Time course of cytotoxin release after a cytotoxic T cell encounters its target. The four panels show time-lapse photographs of a cytotoxic T cell (on the left) attacking a target cell (stained blue, on the right). The lytic granules of the T cell are labeled with a red fluorescent dye; the actin tubules at the area of contact of T cell and target are stained green. In the first panel, the T cell has just made contact with the target cell and this event is designated as Start. At this time, the T-cell granules are distant from the point of contact with the target cell. After 1 minute (second panel), the granules have begun to move toward the point of attachment of the target cell, a move that is essentially completed after 3 minutes (last panel). Note the increasing focusing with time of the cytotoxic granules toward a small area on the target cell surface. Photographs courtesy of A.T. Ritter and G. Griffiths.

8-18 Effector T_H1 CD4 cells induce macrophage activation

The principal function of T_H1 CD4 cells is to help macrophages at sites of infection to become more proficient in the uptake and killing of pathogens. When T_H1 cells first arrive at an infected tissue, the resident macrophages have been phagocytosing and degrading pathogens as part of the innate immune response and are presenting pathogen-derived peptides on their MHC class II molecules. If the antigen receptor of a T_H1 cell recognizes its antigen on a macrophage surface, the T_H1 cell and the macrophage form a conjugate pair with a synapse at which information and material are exchanged (see Section 8-14). Cytokines secreted by the T_H1 cell induce changes in the macrophage that improve its performance. One benefit is that phagosomes containing captured pathogens fuse more efficiently with lysosomes, the source of hydrolytic degradative enzymes (see Figure 3.16, p. 60). A second benefit is increased synthesis of potent microbicidal agents such as oxygen radicals, nitric oxide (NO), and proteases, which work together to destroy the captive pathogens. The overall enhancement of macrophage function induced by effector T cells is called **macrophage activation**.

Macrophages require two signals for activation, which can both be delivered by effector T_H1 cells. The primary signal is provided by IFN-γ, the characteristic T_H1 cytokine, when it binds to the IFN-γ receptor on macrophages. The secondary signal is delivered by **CD40 ligand**, a membrane-bound cytokine of the

T_H1 cell that binds to its receptor, **CD40**, on the macrophage. The combination of intracellular signals coming from CD40 and IFN-γ receptors induces the changes in gene expression that activate the macrophage (Figure 8.27). The IFN-γ secreted by CD8 T cells can also contribute to the macrophage activation.

After forming a cognate pair with a macrophage, an effector T_H1 cell takes several hours to initiate the transcription of cytokine genes and to synthesize the soluble and cell-surface cytokines. Throughout this time, the cells sustain strong contact. The newly made cytokines are translocated to the endoplasmic reticulum of the T_H1 cell and then delivered by secretory vesicles to the synapse with the macrophage. By this delivery mechanism, only the cytokine receptors on the macrophage attached to the T_H1 cell become loaded with cytokines and subject to activation. This ensures that macrophage activation is an antigen-specific process, because only macrophages that present pathogen-derived antigens are selected for activation. The value of this strategy is that the immune response is focused on where it is needed—that is, sites of active infection—which prevents unnecessary damage and disruption to healthy tissue by activated macrophages and inflammatory T_H1 cells.

The toxic microbicidal substances produced by activated macrophages are also harmful to human cells and tissues, which inevitably suffer collateral damage from the macrophage response to the pathogen. Reducing such damage are regulatory mechanisms that suppress macrophage activation. In general, the cytokines secreted by T_H2 cells, including TGF-β, IL-4, IL-10, and IL-13, inhibit macrophage activation, and thus counter the effects of T_H1 cells. Such suppression is evident in patients with lepromatous leprosy, whose macrophages become overwhelmed by proliferating bacteria. By contrast, in patients with tuberculoid leprosy the macrophages suppress bacterial growth effectively, and the symptoms of disease are the consequences of having chronically activated macrophages (see Section 8-11).

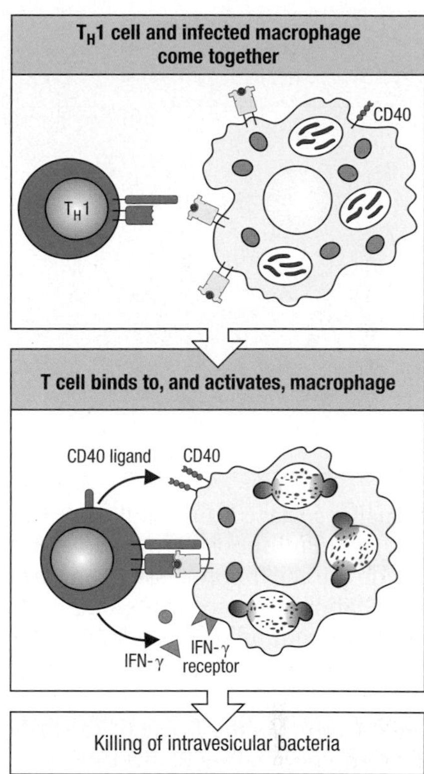
T_H1 cell and infected macrophage come together
CD40
T_H1
T cell binds to, and activates, macrophage
CD40 ligand
CD40
IFN-γ
IFN-γ receptor
Killing of intravesicular bacteria

Figure 8.27 T_H1 CD4 cells activate macrophages to become highly microbicidal. When a T_H1 cell specific for a bacterial peptide contacts a macrophage that presents the peptide, the T_H1 cell is induced to secrete the macrophage-activating cytokine interferon-γ (IFN-γ) and also to express CD40 ligand at its surface. Together, these newly synthesized proteins activate the macrophage to kill the bacteria living inside its vesicles.

8-19 T_{FH} cells, and the naive B cells that they help, recognize different epitopes of the same antigen

Some of the pathogen-specific T cells activated in the draining lymph node differentiate into T_{FH} cells. Their function is to help B cells make antibodies against the pathogen. Such cooperation between T cells and B cells is at the core of the adaptive immune response and was the first helper activity of T cells to be discovered by immunologists. To perform this helper function, the T_{FH} cells move from the T-cell area of the secondary lymphoid tissue to near the B-cell area. Here, they can interact with circulating naive B cells. Like their T-cell counterparts, these naive B cells enter the lymph node from the blood at high endothelial venules, attracted by the chemokines CCL21 and CCL19 (see Section 8-3).

The naive B cells are exposed to pathogens and their antigens arriving in the afferent lymph from the site of infection. For those few naive B cells that recognize one of the pathogen's antigens, the B-cell receptor binds the antigen and actively internalizes it by receptor-mediated endocytosis. Inside the B cell's vesicular system the antigenic proteins are degraded into peptides, some of which bind to MHC class II molecules and are presented on the B-cell surface (see Chapter 4). A pathogen-specific T_{FH} cell surveys these antigen-presenting naive B cells, searching for the peptide:MHC class II complex that is the ligand for its T-cell receptor. If the T_{FH}-cell receptor binds its specific antigen on a naive B cell, the two cells form a cognate pair and a synapse (Figure 8.28). The T_{FH} cell now synthesizes CD40 ligand, which binds to CD40 on the B cell. This interaction starts the developmental pathway by which the naive pathogen-specific B cell proliferates, differentiates, and ultimately matures into an antibody-secreting plasma cell.

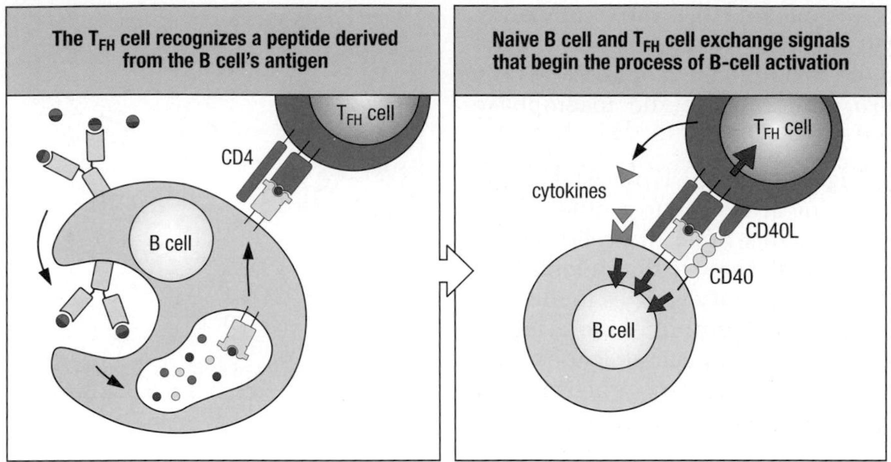
The T_{FH} cell recognizes a peptide derived from the B cell's antigen
T_{FH} cell
CD4
B cell
Naive B cell and T_{FH} cell exchange signals that begin the process of B-cell activation
T_{FH} cell
cytokines
CD40L
CD40
B cell

Figure 8-28 Activation of a naive B cell by T_{FH} cell. First panel: naive B cells bind specific antigen with their surface immunoglobulin, the B-cell receptor (BCR). This antigen is internalized by receptor-mediated endocytosis and processed. Peptides from the antigen are presented on the B cell's MHC class II molecules. A T_{FH} cell specific for a peptide:MHC complex presented by the B cell forms a cognate pair with the B cell. Second panel: cognate interaction leads to the expression of CD40 ligand (CD40L) on the T cell, which interacts with CD40 on the B cell, and to the secretion of the cytokines IL-4, IL-5, and IL-6 by the T_{FH} cell.

By using surface immunoglobulin as an endocytic receptor, the B cell extracts antigen continuously from the extracellular environment and concentrates it inside the cell for processing and presentation by MHC class II molecules. This application of receptor-mediated endocytosis increases the efficiency of antigen presentation by more than 10,000-fold over what could be achieved without it. Because of this huge difference, B cells with receptors that bind strongly to an antigen are able to present sufficient pathogen-derived peptide to engage an antigen-specific T_{FH} cell. The important consequence of this mechanism is that the cooperating B and T cells are specific for epitopes of the same antigen that was first bound and internalized by the B-cell receptor. This phenomenon is described as **linked recognition**. The B-cell and T-cell epitopes can derive from the same polypeptide, from different chains in a multimeric protein or from the various types of macromolecule present in viruses and subcellular particles. The epitope recognized by the T_{FH} cell is constrained to be a peptide, but the epitope recognized by the B cell can be protein, carbohydrate, lipid, nucleic acid, or some combination of these.

8-20 Regulatory CD4 T cells limit the activities of effector CD4 and CD8 T cells

Although not uniquely defined by one effector molecule or cell-surface marker, regulatory CD4 T cells express high levels of CD25 (the α chain of the IL-2 receptor), and usually make cytokines that are immunosuppressive and anti-inflammatory, such as IL-4, IL-10, and TGF-β.

Suppression depends on physical contact between the regulatory CD4 T cell and its target cells. One way in which regulatory T cells are thought to dampen the effector T-cell response is to enter the secondary lymphoid tissue that is generating the response. There they interact with the dendritic cells, preventing them from interacting with and activating additional naive T cells. A second way in which regulatory T cells function is through direct contact with effector T cells.

On infection with hepatitis B virus, only a minority of people eliminate the virus, whereas the rest develop a chronic liver disease associated with a chronic and ineffective adaptive immune response. The liver and blood of these people contain an unusually high level of regulatory T cells that are specific for hepatitis B virus antigens. Their suppressive activity involves the secretion of TGF-β, an immunosuppressive cytokine, and direct contact with the effector T cells. In these patients the regulatory T cells seem to have prematurely suppressed the effector T cells before the infection was terminated.

The property that uniquely defines human regulatory T cells is the transcription factor FoxP3. Infants who lack functional FoxP3 have a normal number of

T cells with the characteristic cell-surface phenotype of regulatory T cells, but these cells cannot function as regulatory cells. That deficiency of FoxP3 is a fatal disorder that causes death in infancy shows how important regulatory T cells are for human life and maintaining an orderly immune system. The absence of regulatory cells permits immune responses that are directed at self antigens and which attack a variety of the body's tissues including the gut, the skin, and the endocrine glands. The only treatment for this condition is a hemtopoietic stem cell transplant.

Summary

The six types of effector T cell—cytotoxic CD8 cells and helper CD4 T_H1 cells, T_H2, T_H17, T_{FH}, and T regulatory cells—have complementary roles in the immune response to infection. The common principle by which they function is formation of a cognate pair with a target cell that presents a pathogen-derived peptide that is recognized by the T cell's antigen receptor. At the synapse connecting the two cells, the T cell delivers effector molecules that induce changes in the target cell. The cytotoxins delivered by a cytotoxic CD8 T cells kills the target cell, whereas helper CD4 T cells deliver cytokines that inform the target cell to either increase or decrease its immune response against the pathogen. Increasing the response are interactions of T_H1 cells with macrophages; T_H2 cells with basophils, mast cells, and eosinophils; T_H17 cells with neutrophils; and T_{FH} cells with B cells. Decreasing the immune response are T regulatory cells. They interact with the other populations of effector T cells to prevent too excessive an immune response and to shut it down when the pathogen has been eliminated.

During their differentiation in secondary lymphoid tissue, cytotoxic CD8 cells and helper CD4 T_H1, T_H2, and T_H17 cells acquire changes in the cell-surface components and activation requirement that allow them to travel to the infected tissue and begin working with their respective target cells. CD8 T cells deliver lytic granules that are loaded with cytotoxins to the surface of the target cell, thereby inducing cells infected with bacteria or viruses to die by apoptosis. This well-controlled mechanism of cell death prevents the release of infective bacteria or viral particles from the dying cell that could then propagate the infection to healthy cells. After killing one cell, a CD8 T cell rapidly synthesizes a new arsenal of lytic granules that allows it to kill several target cells in succession. The cytokines that CD4 T cells deliver to the surface of their target cells are made to order once the cognate pair has been formed, and are delivered immediately their synthesis is complete. Secretion of IFN-γ by T_H1 cells increases the phagocytic function of macrophages infected with a virus or bacterium that exploits the phagocytic pathway of macrophages for their own survival. IFN-γ is the characteristic cytokine of T_H1 cells and is instrumental in activating macrophages. Secretion of IL-4 by T_H2 cells acts on basophils, mast cells, and eosinophils to respond to parasite infections. Secretion of IL-17 by T_H17 cells serves to activate and recruit neutrophils to the sites of infection with extracellular bacteria and fungi.

In contrast to other types of effector T cell, T_{FH} cells remain in the secondary lymphoid tissue, where they activate naive pathogen-specific B cells to differentiate into antibody-producing cells. Using their B-cell receptor for receptor-mediated endocytosis, naive B cells efficiently take up their specific antigen, which is then processed, presented by MHC class II molecules on the B-cell surface. and recognized by the T_{FH} cells. As a consequence of this mechanism, the naive B cell and the T_{FH} cell that form a cognate pair always recognize different epitopes of the same antigen.

Summary to Chapter 8

The adaptive immune response is initiated by dendritic cells, which take pathogens and their antigens from the site of infection to the T-cell area of a secondary lymphoid tissue. There the dendritic cells screen the circulating population of naive T cells, selecting those that are pathogen-specific for recruitment to the immune response. Each naive T cell conjugates with a dendritic cell that guides its activation and differentiation into a clone of several hundred effector T cells. Activation requires a combination of signals from the T-cell receptor, the co-receptor (CD4 or CD8), and the co-stimulatory receptor CD28, which engages B7 molecules on the dendritic cell. The proliferation of activated T cells is driven by the cytokine IL-2. The differentiation of the proliferating T cells is influenced by the nature of the pathogen, the anatomical site of infection, and the innate immune response stimulated by the pathogen. These factors influence the cytokine environment in the secondary lymphoid tissue and drive the production of effector T cells that are best suited for controlling and eliminating the pathogen. Effector T cells are more readily activated than naive T cells, which enables them to perform their functions immediately they encounter an antigen-presenting target cell.

Two fundamental lineages of effector T cells are distinguished by the CD4 and CD8 co-receptors. Cytotoxic CD8 T cells are a relatively homogeneous population of effector cells that counter intracellular infections by killing the infected cells. Because most human cells are susceptible to infection by viruses, CD8 T cells have to interact with and be able to kill, a wide range of different human cells. Helper CD4 T cells exhibit great diversity in their functions and phenotype, but they interact only with myeloid and lymphoid cells of the immune system. Pathways of differentiation associated with different transcription factors define five different subpopulations of helper CD4 T cells. T_H1 cells secrete IFN-γ and activate macrophages to eliminate intracellular infections. T_H2 cells secrete IL-4 and help basophils, mast cells, eosinophils, and B cells control the wide range of unicellular and multicellular parasites that can inhabit the human body. T_H17 cells provide defense against extracellular bacterial and fungal infections by increasing the numbers and power of neutrophils through the secretion of IL-17. By providing help to B cells within the secondary lymphoid tissue, T_{FH} cells are responsible for initiating the antibody response against the infecting pathogen. Regulatory T cells, the fifth type of CD4 effector T cell, help suppress unwanted immune responses and keep the adaptive immune response under control.

Questions

8–1 All of the following are true in reference to T-cell priming except _____.

a. it occurs in primary lymphoid organs
b. it transforms naive T cells into differentiated effector T cells
c. it is the first stage of a primary adaptive immune response
d. it requires interaction between naive T cells and antigen-presenting cells
e. it takes place in many locations including, but not limited to, lymph nodes, Peyer's patches, and the tonsils.

8–2 Which of the following cell types is unable to interact with naive T cells and induce their activation? (Select all that apply.)

a. B cells
b. macrophages resident in infected tissues
c. macrophages resident in secondary lymphoid tissue
d. interdigitating reticular cells
e. immature dendritic cells.

8–3 When an immature dendritic cell becomes an activated dendritic cell, all of the following changes occur except _____.

a. confinement to T-cell regions of lymph node cortex
b. development of elaborate finger-like processes
c. upregulated expression of MHC class II at the cell surface
d. loss of antigen-processing functions
e. expression of Toll-like receptor TLR9.

8–4 Identify which of the following statements regarding naive T cells is incorrect.

a. Naive T cells enter lymph nodes in two different ways, from the blood or from the lymph.
b. Naive T cells can only be activated in secondary lymphoid tissues.
c. Naive T cells differentiate into effector T cells after T-cell priming has occurred.
d. Naive T cells occupy both the cortex and the medulla of lymph nodes.
e. Naive T cells are only activated by dendritic cells, not by macrophages or B cells.

8–5 Homing of effector T cells to inflamed tissue is facilitated by the upregulation of _____ on the surface of the effector T cell.
a. VLA-4
b. L-selectin
c. CD28
d. VCAM-1
e. B7.

8–6 Identify which of the following statements regarding naive T cells is incorrect.
a. Naive T cells have the capacity to survive for many years as nondividing circulating cells in the absence of specific antigen.
b. Naive T cells express LFA-1 molecules that change their conformation after encountering a specific peptide:MHC complex.
c. Naive T cells exit from lymph nodes using the same route as effector T cells.
d. Naive T cells express ICAM-3, which binds to DC-SIGN on dendritic cells with high affinity.
e. Naive T cells express high levels of S1P receptors on their surface.

8–7 Which of the following statements regarding T cells activated by specific antigen is incorrect?
a. They receive co-stimulatory signals through CD28.
b. They suppress expression of sphingosine 1-phosphate (SIP).
c. They take several days before differentiating into effector T cells.
d. They cease to secrete and respond to interleukin-2 (IL-2).
e. They begin to express CTLA4, which serves to limit T-cell proliferation.

8–8 Which of the following explains why dendritic cells, but not macrophages or B cells, contribute to the activation of naive T cells?
a. Macrophages and B cells do not express MHC class II molecules until they are activated.
b. Dendritic cells upregulate B7 after engaging innate immunity receptors at sites of infection.
c. Dendritic cells express higher levels of CTLA4.
d. Macrophages and B cells do not process antigen.
e. Dendritic cells use Toll-like receptors to hold antigen in place for extended periods of time.

8–9 If CD4 or CD8 co-receptors are absent, then approximately _____ specific peptide:MHC complexes are needed to activate a T cell.
a. twice the number of
b. 10 times the number of
c. 20 times the number of
d. 100 times the number of
e. 1000 times the number of.

8–10 Cyclosporin A is an immunosuppressive drug commonly used in transplant patients to prevent graft rejection by alloreactive T cells. It acts by interfering with the signaling pathway that leads from the T-cell receptor to transcription in the nucleus of the genes for the cytokine IL-2 and the α chain of the IL-2 receptor. Why does preventing the transcription of these genes lead to immunosuppression?

8–11 Antigen recognition by T cells in the absence of co-stimulation results in
a. upregulation of B7
b. expression of the high-affinity IL-2 receptor
c. T-cell anergy
d. T-cell apoptosis
e. phosphorylation of immunoreceptor tyrosine-based activation motifs.

8–12 _____ cells remain in, rather than leave, the secondary lymphoid organs in which they differentiated.
a. CD4 T_H1 cells
b. CD4 T_{FH} cells
c. CD8 cytotoxic T cells
d. Regulatory CD4 T cells
e. All of the above.

8–13 Match the T-cell type in column A with its correct description in column B.

Column A	Column B
a. regulatory T cells (T_{reg})	1. help basophils, mast cells, eosinophils and B cells respond to parasitic infections
b. CD8 T cells	2. facilitate neutrophil response to fungal and extracellular bacterial infections
c. CD4 T_H1 cells	3. suppress effector CD4 and CD8 T-cell function
d. CD4 T_H2 cells	4. differentiate under the influence of IL-12 and IFN-γ
e. CD4 T_H17 cells	5. induce apoptosis of target cells after targeted delivery of cytotoxins

8–14 Linked recognition is best described as _____.
a. the delivery of the co-stimulatory signal by antigen-presenting cells to T cells
b. cooperation between T follicular helper (T_{FH}) and naive B cells bearing specificity for different epitopes of the same antigen and where the B cell serves as the antigen-presenting cell
c. the linking of Toll-like receptors and the innate immune response
d. the interaction between processed antigenic peptides and MHC molecules during antigen presentation
e. the region of contact between T cells and antigen-presenting cells involving cell-adhesion molecules and other cell-surface receptor-ligand pairs.

8–15 Vijay Kumar, a 19-year-old male who had emigrated to the United States from South India two years before, developed lesions in his nasal mucosa, and nodular skin lesions on his cheeks and buttocks. Examination of a stained biopsy of a skin lesion revealed numerous clumps of mycobacteria. T cells in the skin lesions were secreting IL-4, IL-5, and IL-10. What would be the most likely diagnosis?
a. tuberculosis
b. lepromatous leprosy
c. leishmaniasis
d. tuberculoid leprosy
e. allergic dermatitis.

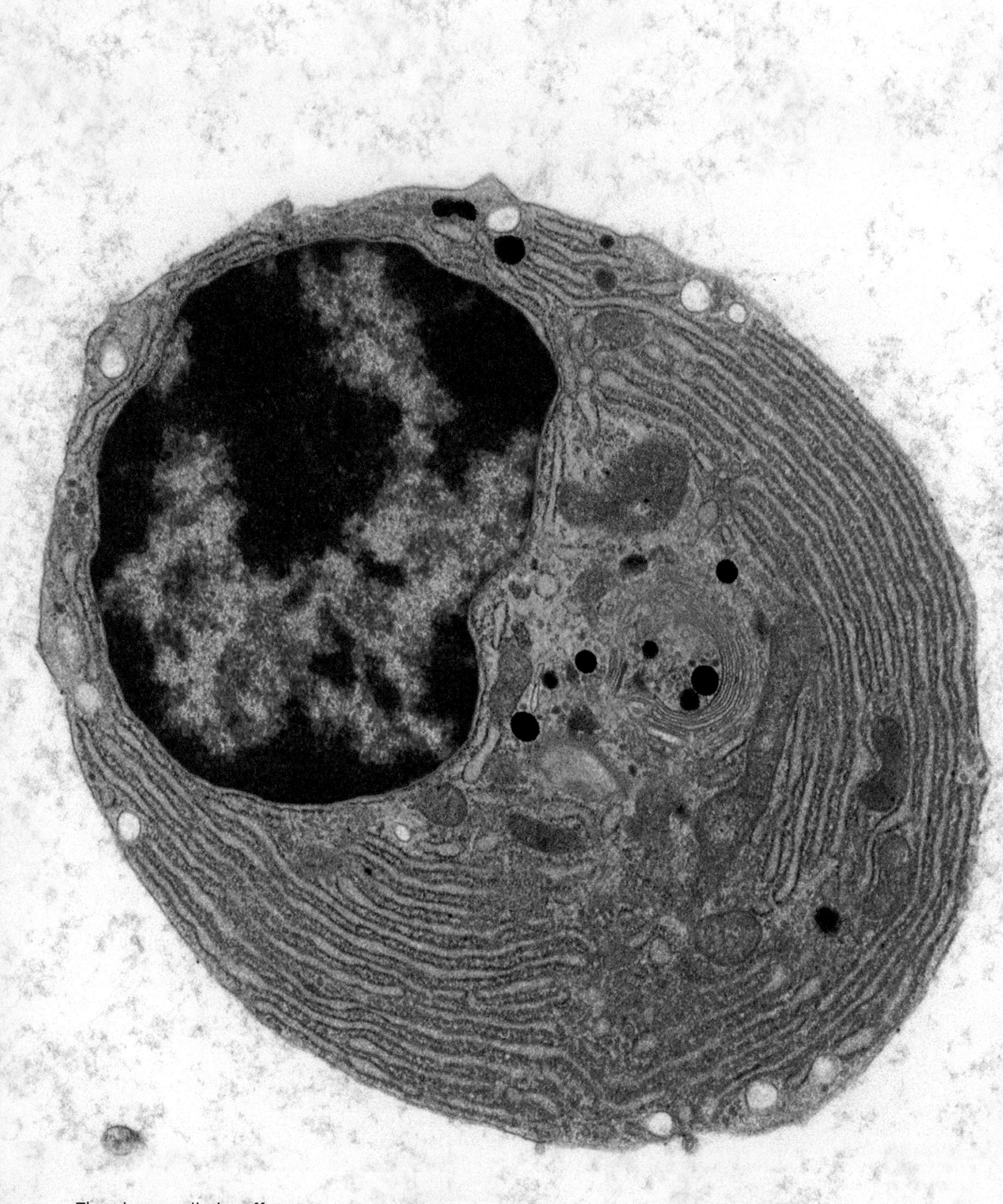

The plasma cell, the effector B cell that makes antibodies, the most potent weapon of adaptive immunity.

Chapter 9

Immunity Mediated by B Cells and Antibodies

The production of antibodies is the sole function of the B-cell arm of the immune system. Antibodies are present in the blood, lymph, and extracellular fluids, where they bind to extracellular bacteria and virus particles. Antibodies are also present at mucosal surfaces, where they control the populations of commensal microorganisms, as well as pathogenic microorganisms and the larger multicellular parasites. Antibodies are not inherently toxic or destructive to pathogens; their principal role is to bind to the pathogen, and the best antibodies are those that bind tightly and do not let go. Antibodies also act as molecular adaptors that bind to pathogens using their variable regions (V regions) and to complement components and receptors on effector cells using their constant regions (C regions). The differences between the classes and subclasses of antibody enable antibody-coated pathogens to be delivered to different types of effector cell. Coating a bacterium with IgG and complement promotes its uptake and destruction by a phagocyte, whereas coating a parasite with IgE activates mast cells to induce violent reactions that expel the parasites from the body. Some antibodies have a direct effect on the course of infection by covering up sites on a pathogen's surface that are necessary for growth or replication. For example, antibodies against the hemagglutinin and neuraminidase glycoproteins of influenza virus can prevent infection because the virus uses these glycoproteins to bind to human cells and infect them. Such antibodies are said to **neutralize** the pathogen. In the development of a vaccine against an infectious agent or its toxic products, the gold standard that a company aims for is the induction of a **neutralizing antibody**.

The structure, specificity, and diversity of antibodies were discussed in Chapter 4, and the development of B cells from bone marrow precursors into antibody-secreting plasma cells was the subject of Chapter 6. This chapter focuses on the participation of B cells and their antibodies in the primary immune response to infection. The first part of the chapter examines cellular interactions in the secondary lymphoid tissues that lead to the mass production of antibody and to improvements in the quality and performance of the antibody. The second part of the chapter describes how antibodies are used by the immune system to control or terminate infection.

Antibody production by B lymphocytes

In Chapter 8, we saw how the primary adaptive immune response begins with dendritic cells that carry antigens from infected tissue to the draining lymph node, where they activate pathogen-specific naive T cells. This results in the proliferation of different types of effector T cell. Of importance for this chapter are the T_{FH} cells, which stay in the lymph node to activate the B-cell arm of adaptive immunity. This part of the chapter examines the cellular and molecular interactions that occur in the secondary lymphoid tissue and give rise to plasma cells secreting large quantities of pathogen-specific antibody. In this process the first priority is speed of production, which results in low-affinity IgM antibody that is useful but not optimal. The second priority is to improve the quality of the antibody, which is achieved in two complementary ways. The first is to increase the affinity of the antibody through somatic hypermutation (see Section 4-14) and selection of those B cells subsequently making antibody that binds the antigen most strongly. The second is to change the isotype of the antibody to ones that will recruit the effector cells and mechanisms that are most capable of bringing the infection to an end (see Section 4-15). This choice is determined by the cytokines secreted by the T_{FH} cells.

9-1 B-cell activation requires cross-linking of surface immunoglobulin

On binding to the protein or carbohydrate epitopes arrayed on the surface of a microorganism, the surface IgM molecules of a naive, mature B cell become physically cross-linked to each other by the multimeric antigen and are drawn together in a localized area of B-cell contact with the microbe. This clustering and aggregation of B-cell receptors sends signals from the receptor complex to the inside of the cell. Signal transduction from the B-cell receptor complex resembles, in many ways, the signaling from the T-cell receptor complex discussed in Section 8-6. Both types of receptor are associated with cytoplasmic protein tyrosine kinases that are activated by receptor clustering, and both receptors activate similar intracellular signaling pathways (Figure 9.1).

Interaction of antigen with surface immunoglobulin is communicated to the interior of the B cell by the proteins Igα and Igβ, which are associated with IgM in the B-cell membrane to form the functional B-cell receptor. Like the CD3 polypeptides of the T-cell receptor complex, the cytoplasmic tails of Igα and Igβ each contain two ITAMs with which the tyrosine kinases Blk, Fyn, and Lyn associate (Figure 9.1, center panel). The ITAMs become phosphorylated on tyrosine residues, which allows the tyrosine kinase Syk to bind to Igβ tails that are doubly phosphorylated. Interaction between bound Syk molecules initiates intracellular signaling pathways that lead to changes in gene expression in the nucleus (Figure 9.1, bottom panel).

9-2 B-cell activation requires signals from the B-cell co-receptor

Cross-linking of the B-cell receptor by antigen generates a signal that is necessary, but not sufficient, to activate a naive B cell. Additional signals are required and delivered in various ways. One set of signals is delivered when the B-cell

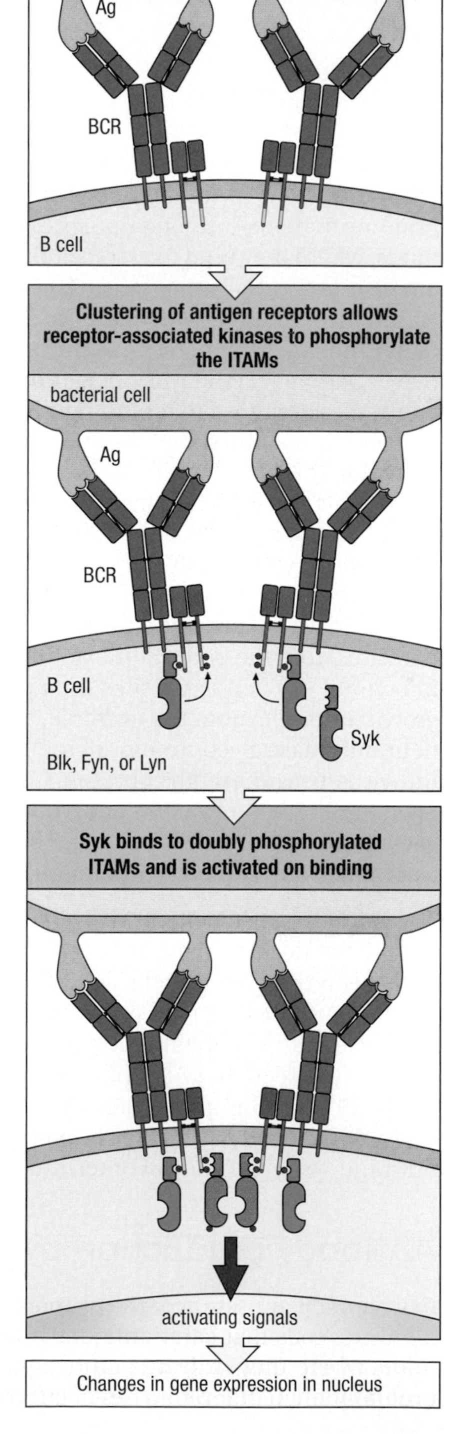

Figure 9.1 Cross-linking of B-cell receptors by antigens initiates a cascade of intracellular signals. Top panel: the B-cell receptor on a mature, naive B cell is composed of monomeric IgM that binds antigen and associated Igα and Igβ chains, which transduce intracellular signals. The IgM is shown binding repetitive antigens (Ag) on the surface of a bacterium. Center panel: on cross-linking and clustering of the receptors, the receptor-associated tyrosine kinases Blk, Fyn, and Lyn phosphorylate tyrosine residues in the ITAMs of the cytoplasmic tails of Igα (blue) and Igβ (orange). Bottom panel: subsequently, Syk binds to the phosphorylated ITAMs of the B-cell receptor Igβ chains, which are in close proximity within the cluster and activate each other by transphosphorylation, thus initiating further signaling. Ultimately, the signals are relayed to the nucleus of the B cell, where they induce the changes in gene expression that initiate B-cell activation.

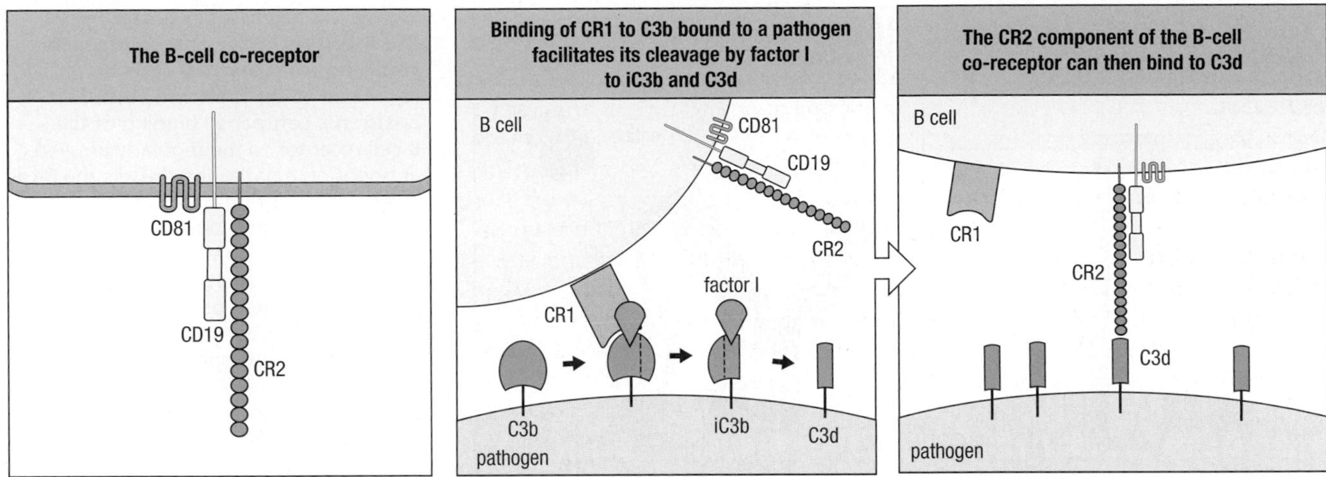

Figure 9.2 Structure and function of the B-cell co-receptor. First panel: the B-cell co-receptor is composed of the CR2, CD19, and CD81 protein subunits. CR2 is the receptor for complement fragments iC3b and C3d fixed on pathogen surfaces, CD19 is the signaling component, and CD81 brings CD19 to the surface and organizes the B-cell co-receptor within the membrane. Second panel: B cells carry both CR1 and CR2. On binding to C3b fragments on a pathogen's surface, CR1 facilitates their cleavage by factor I, first to form the iC3b intermediate and then to form C3d. In this sequence of reactions, CR1 acts as a cofactor for CR2 because it facilitates the production of C3d, the ligand for CR2. Third panel: the CR2 component of the B-cell co-receptor binds to C3d on the pathogen surface. Intracellular signals generated by the B-cell co-receptor binding C3d and by the B-cell receptor binding antigen combine in synergy to activate the B cell.

receptor becomes closely associated with the **B-cell co-receptor**, another protein complex on the B-cell surface. The B-cell co-receptor comprises three proteins: the first is complement receptor 2 (CR2 or **CD21**), which recognizes the iC3b and C3d derivatives of the C3b fragments deposited on a pathogen; the second is the protein **CD19**, which is the signaling chain of the co-receptor; and the third is the **CD81 protein**, which binds to CD19 and is essential for bringing it to the B-cell surface. CD81, a member of the tetraspanin family of proteins, organizes the interactions of the B-cell receptor and co-receptor within functional microdomains in the plasma membrane (Figure 9.2, left panel).

Generation of the iC3b and C3d ligands for the B-cell co-receptor involves complement receptor CR1, which, like CR2, is present on B cells. C3b, the ligand for CR1, will have been deposited on the pathogen's surface as a consequence of complement activation during the innate immune response to the infection (see Chapter 2). When CR1 on a B cell binds to C3b it becomes susceptible to cleavage by factor I, first to give the iC3b fragment and then the more stable C3d fragment (Figure 9.2, center panel). By this cooperative process, CR1 increases the abundance of ligands for the B-cell co-receptor on the pathogen's surface (Figure 9.2, right panel). When the B-cell receptor binds to its specific antigen on the pathogen, the CR2 component of the B-cell co-receptor binds to a nearby C3d fragment, which brings the B-cell receptor and co-receptor into juxtaposition (Figure 9.3). This sets the stage for Lyn, a tyrosine kinase bound to Igα, to phosphorylate the cytoplasmic tail of CD19. Interaction of phosphorylated CD19 with intracellular signaling molecules generates signals that synergize with those coming from the B-cell receptor. Simultaneous ligation of the B-cell receptor and co-receptor increases the overall signal 1000–10,000-fold, greatly increasing the B cell's sensitivity to antigen. The importance of this synergy, which can be achieved with multivalent antigens such as a bacterium (Figure 9.3, left panel) or with a soluble monovalent antigen (Figure 9.3, right panel), is pointedly illustrated by the immunodeficiency of patients lacking the B-cell co-receptor, because of

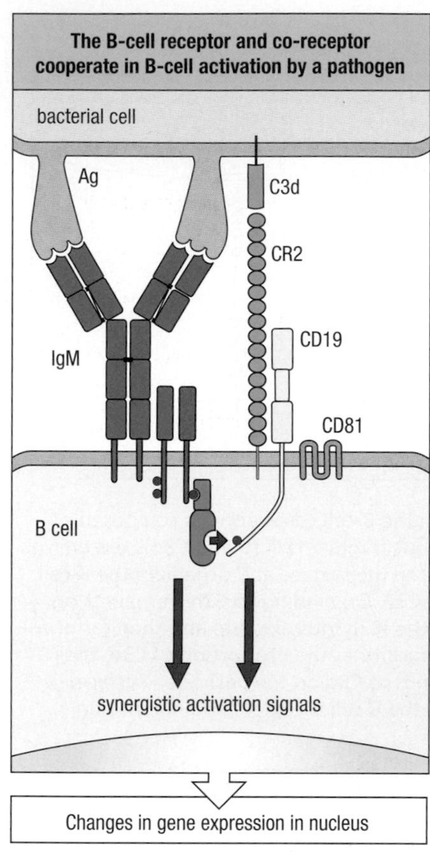

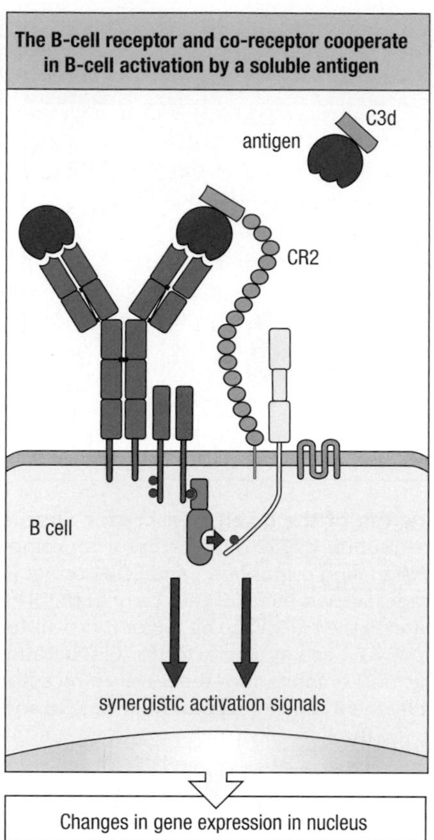

Figure 9.3 Signals generated from the B-cell receptor and co-receptor combine to activate B cells in response to surface and soluble antigens. Left panel: binding of the B-cell receptor to the regularly arrayed antigens on a pathogen clusters the B-cell receptors and brings them into close jutaposition with B-cell co-receptors, in which the CR2 component binds to C3d on the pathogen. The cytoplasmic tail of CD19 is then phosphorylated by tyrosine kinases associated with the B-cell receptor. Phosphorylated CD19 generates intracellular signals that synergize with those generated by the B-cell receptor. Right panel: shown here is the interaction of the B-cell receptor and co-receptor with a soluble antigen, which could either have been secreted by the pathogen or be a product of pathogen degradation. If the soluble antigen is tagged with C3d it cross-links the B-cell receptor to the co-receptor, thereby increasing the strength and efficiency of intracellular signaling. The more elongated and flexible structure of CR2, compared with IgM, facilitates the formation of such cross-links with a variety of antigens of different shape and size.

defective CD19 or CD81 genes. Such patients have low levels of antibody, almost no isotype switching, and generally poor B-cell responses to infections and vaccines.

9-3 Effective B cell-mediated immunity depends on help from CD4 T cells

In primary immune responses, the activation of almost all naive B cells depends on conjugation with a helper CD4 T_{FH} cell that recognizes pathogen-derived peptides presented by the B cell's MHC class II molecules (see Figure 8.28, p. 226). During this interaction the helper T cell delivers cytokines and signals to the B cell that induce the B cell to divide and differentiate. The degree to which B cells depend on T cells is demonstrated by the study of infants with complete DiGeorge syndrome, who lack a thymus and have almost no T cells in their circulation. Although these infants have normal numbers of B cells they cannot make an effective antibody response against most antigens. Consequently, they suffer from opportunistic infections and usually die from infection within the first two years of life unless given a thymus transplant.

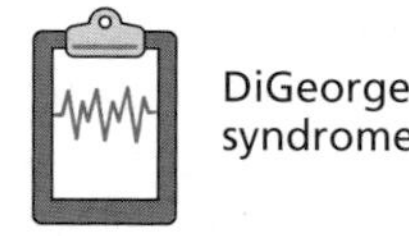

Although patients with DiGeorge syndrome cannot make effective antibody responses against infection, some of their B cells are activated to produce low-affinity IgM antibodies. These are predominantly of the minority CD5-expressing B-1 population of B cells, which do not require T-cell help and do not undergo isotype switching and affinity maturation (see Section 6-10, p. 161). This feature of DiGeorge syndrome argues that the majority population of B cells (the B-2 cells) is non-functional in these patients and thus inherently dependent on T-cell help.

The antigens recognized by the IgM antibodies from patients with DiGeorge syndrome are typically composed of repetitive carbohydrate or protein epitopes present at high density on the surface of a microorganism, such as the cell-wall polysaccharides of the bacterium *Streptococcus pneumoniae*. Because this category of antigens has the capacity to cross-link B-cell receptors and co-receptors extensively, it has been proposed that the signals they generate are sufficient to activate the B cell in the absence of additional signals (Figure 9.4). Because antibodies against these antigens are present in athymic individuals, the antigens have been called **thymus-independent antigens** (**TI antigens**). This is a confusing term, because it is not the antigens that are thymus-independent but the population of B cells that responds to them.

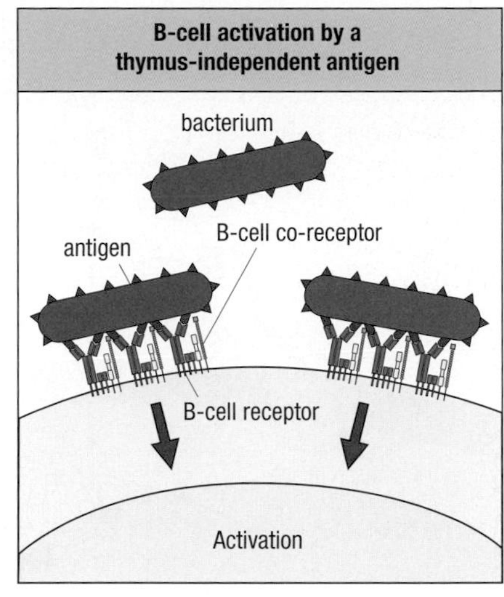

Figure 9.4 The signals generated by B-cell receptors and co-receptors are sufficient to activate a minority of B cells. The antigens to which these B cells respond are usually ones that occur in dense and regular arrays on pathogen surfaces. This situation produces dense clustering of B-cell receptors and co-receptors at the B-cell surface, which produces sufficient signaling to stimulate B-cell proliferation and differentiation. Because these antigens do not require help from T cells they are called thymus-independent antigens, or TI antigens. The B cells that respond to TI antigens are predominantly of the B-1 lineage.

9-4 Follicular dendritic cells in the B-cell area store and display intact antigens to B cells

In Chapter 6 we saw how interactions between B cells and follicular dendritic cells (FDCs) in the primary follicles of secondary lymphoid tissues are critical for the maturation and survival of B cells. This reflects the general dependence of B cells on FDCs, accessory cells that are dedicated to B-cell development and function. As well as being functionally different from the myeloid dendritic cells that present antigens to naive T cells, and from the plasmacytoid dendritic cells that make type I interferons, FDCs are also developmentally distinct: they seem not to be hematopoietic cells but stromal cells that develop from fibroblast-like cells in the bone marrow. The differentiation of FDCs depends on cytokines of the TNF family made by lymphocytes, such as TNF-α and the lymphotoxins LT-α and LT-β, that interact with their corresponding receptors on the FDC precursors.

The dense network formed by FDCs and their extensively interdigitating dendrites organizes the B-cell area of a lymph node into primary follicles. A major function of the FDCs is to serve as a vast depository of intact antigens that have not been subject to degradation and are available for interaction with the antigen receptors of circulating B cells. Two features of FDC make them well suited to this function: the first is the extensive surface area of the dendrites that allows large quantities of antigens and even intact viral particles to be accumulated; the second is that FDCs lack phagocytic activity, which preserves the antigens intact on the cell surface over long periods from months to years (Figure 9.5).

FDCs have several receptors with which they take up antigens from the lymph. For the activation of naive B cells during the primary immune response, the complement receptors CR2 and, to smaller extent, CR1 are the most important. Complement activation during the innate immune response leads to the attachment of C3b (the CR1 ligand) and its breakdown product C3d (the CR2 ligand) to pathogens and their antigens. These C3d- and C3b-tagged antigens are taken up by the respective complement receptors and held by them at the surface of the FDC. The extracellular part of CR2 consists of 15 CCP modules (see Figure 9.2), of which the outermost two modules bind to C3d. The other 14 modules form a long and flexible stalk that enables CR2 to fish for and catch C3d-tagged proteins and particles in the stream of the lymph (Figure 9.6).

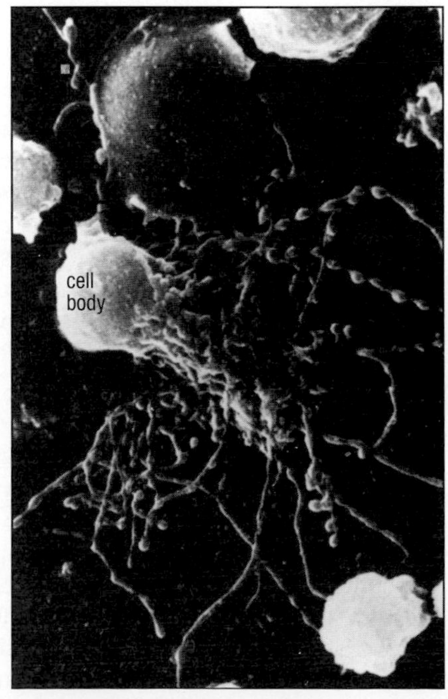

Figure 9.5 The dendrites of follicular dendritic cells take up intact pathogens and antigens and preserve them for long periods. The scanning electron micrograph shows an FDC with a well-defined cell body and numerous dendritic processes. The dendrites initially have a thread-like or filiform appearance. When pathogens and antigens are bound to complement receptors on the FDC surface they become clustered, forming prominent beads at intervals along the dendrites as seen here. Photograph courtesy of A.K. Szakal.

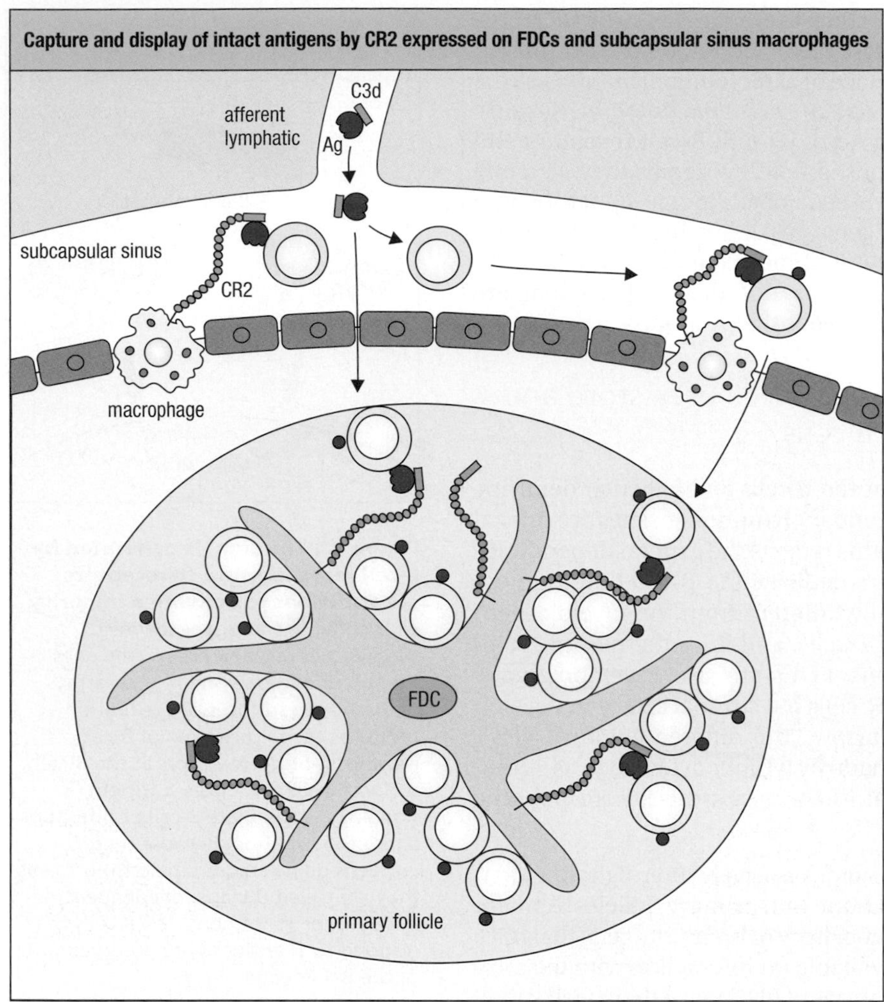

Figure 9.6 Naive B cells recognize antigens captured by subcapsular sinus macrophages and follicular dendritic cells (FDCs). Antigens tagged with C3d are brought to the lymph node from the infected tissue in the afferent lymph. By binding to C3d, CR2 on subcapsular sinus macrophages and FDCs tethers the antigen to the cell surface, where it can be screened by naive B cells arriving either from the blood via a high endothelial venule or in the afferent lymph.

Augmenting the FDCs of the B-cell follicles is a special type of macrophage, called the **subcapsular sinus macrophage**, which populates the subcapsular sinus of the lymph node. In certain respects these macrophages resemble FDCs; they have little phagocytic activity and have CR1 and CR2 for taking up antigens tagged with C3d or C3b and then holding them at the cell surface.

Lymph carrying antigens from the infected tissue will pass from the afferent lymphatic vessel into the subcapsular sinus of the lymph node, where complement-tagged antigens can be taken up by subcapsular macrophages. On subsequently passing through the cortex of the lymph node, complement-tagged antigens will be taken up by the FDCs in the B-cell area (see Figure 9.6). In the sinuses of the medulla there is a second type of macrophage, **the medullary sinus macrophage**, which is highly phagocytic and filters the lymph before it leaves the node by removing and destroying the remaining pathogens and their antigens.

9-5 Antigen-activated B cells move close to the T-cell area to find a helper T_{FH} cell

Circulating naive B cells home to lymph nodes from the blood or lymph, using the same mechanisms as those used by naive T cells (see Sections 8-3 and 8-4). Naive B cells arriving from the blood via a high endothelial venule will be attracted into the T-cell area by the chemokines CCL21 and CCL19, and then into a B-cell follicle by the chemokine CXCL13. Naive B cells arriving in the

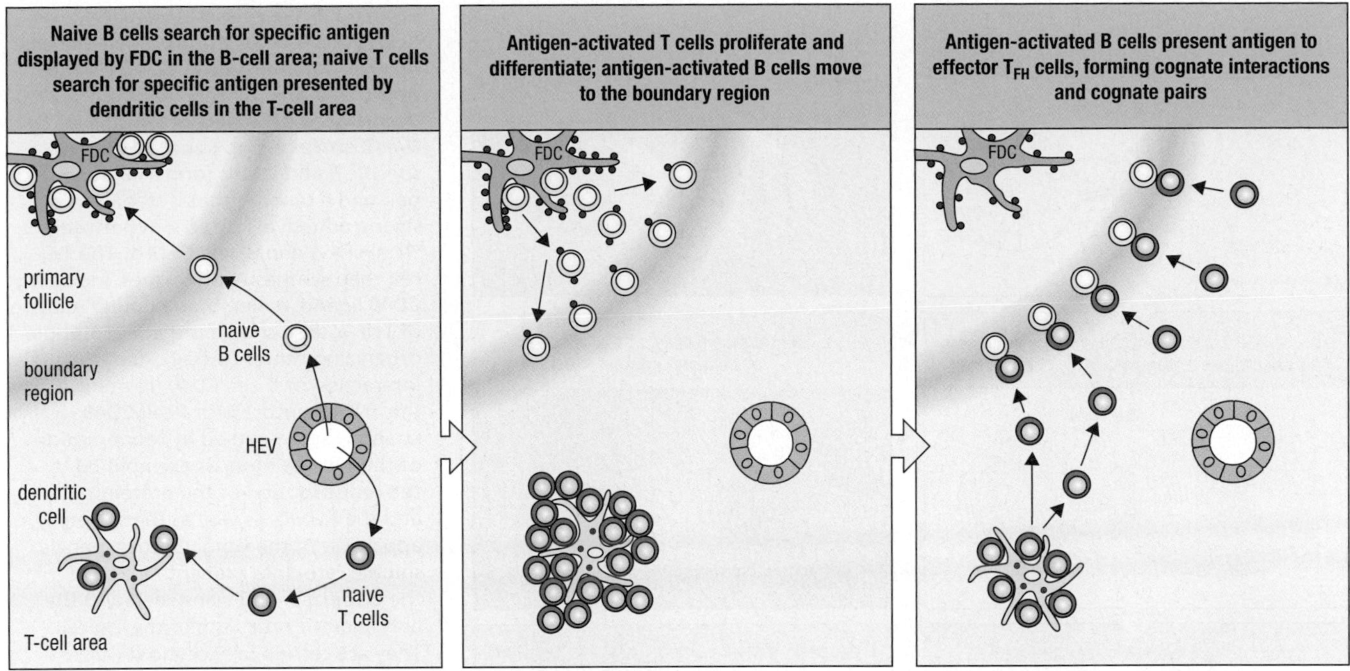
Naive B cells search for specific antigen displayed by FDC in the B-cell area; naive T cells search for specific antigen presented by dendritic cells in the T-cell area
FDC
primary follicle
naive B cells
boundary region
HEV
dendritic cell
naive T cells
T-cell area
Antigen-activated T cells proliferate and differentiate; antigen-activated B cells move to the boundary region
FDC
Antigen-activated B cells present antigen to effector T_{FH} cells, forming cognate interactions and cognate pairs
FDC

Figure 9.7 B cells activated by antigen in the B-cell area move to the area boundary, where they are helped by antigen-specific effector T_{FH} cells coming from the T-cell area. First panel: circulating naive B and T cells leave the blood at a high endothelial venule (HEV) to enter the lymph node draining an infection. Myeloid dendritic cells in the T-cell area activate antigen-specific naive CD4 T cells, while in the B-cell area (the primary follicle), intact antigen (red dots) deposited on follicular dendritic cells (FDCs) is taken up by naive antigen-specific B cells and initiates their activation. Second panel: in the T-cell area, activated naive T cells proliferate to become clones of effector T_{FH} cells. In the B-cell area, antigen-specific B cells process endocytosed antigen and present antigen-derived peptides on MHC class II molecules. These antigen-activated B cells move to the boundary region. Third panel: effector T_{FH} cells also move to the boundary region, where they screen the peptide:MHC class II complexes presented by the antigen-activated B cells. Conjugates are formed between a T_{FH} cell and a B cell that presents its specific antigen.

lymph will enter the node at the subcapsular sinus, where they can screen the antigens held at the surface of subcapsular sinus macrophages for ones that engage their antigen receptors. If specific antigen is found, the B cell enters the B-cell area of the follicle to interact with T_{FH} cells and complete its activation.

In the follicle, naive B cells that have yet to encounter antigen survey the antigens deposited on the FDCs (Figure 9.7). If antigen is recognized by the B-cell receptor, signaling is initiated that activates the B cell and induces expression of CD69. This protein prevents expression of the SIP receptor (see Secion 8-4), which keeps the activated B cells in the lymphoid tissue to continue its differentiation. In contrast, naive B cells that do not find their antigen will express the SIP receptor and be drawn out of the B-cell area into the medulla and then on to the efferent lymph under the influence of the SIP gradient. The antigen-activated B cells begin to endocytose and process the complexes of B-cell receptor with antigen and then present peptides from the degraded antigen on MHC class II molecules (see Figure 8.28, p. 226). B-cell expression of the chemokine receptor CCR7 is induced, which binds to CCL21 and CCL19 and draws the activated B cell to the boundary between the B- and T-cell areas. In this location, antigen-stimulated B cells are well placed to interact with newly differentiated helper T_{FH} cells (see Figure 9.7).

Effector T_{FH} cells that were activated by antigens presented by dendritic cells in the T-cell area reduce their secretion of CCR7. This facilitates movement of the T_{FH} cells to the boundary of the primary follicle where the antigen-activated B cells are located. The receptors of the T_{FH} cells sample the antigens presented by MHC class II molecules of the activated B cells; if they find their

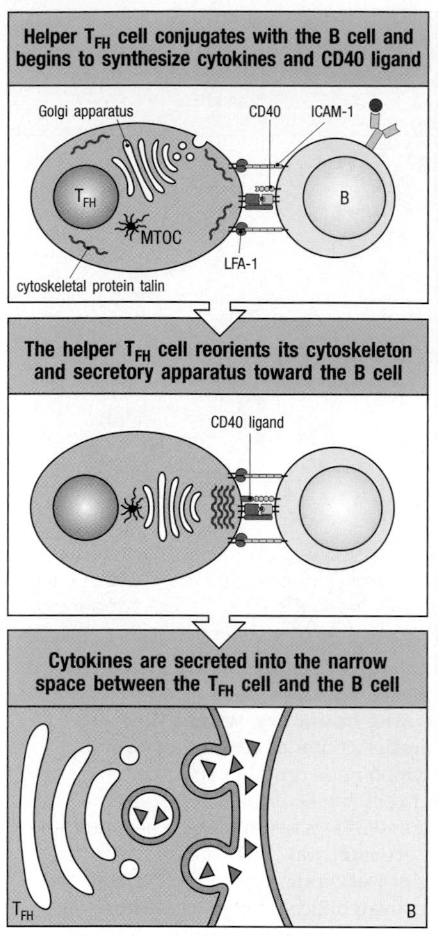

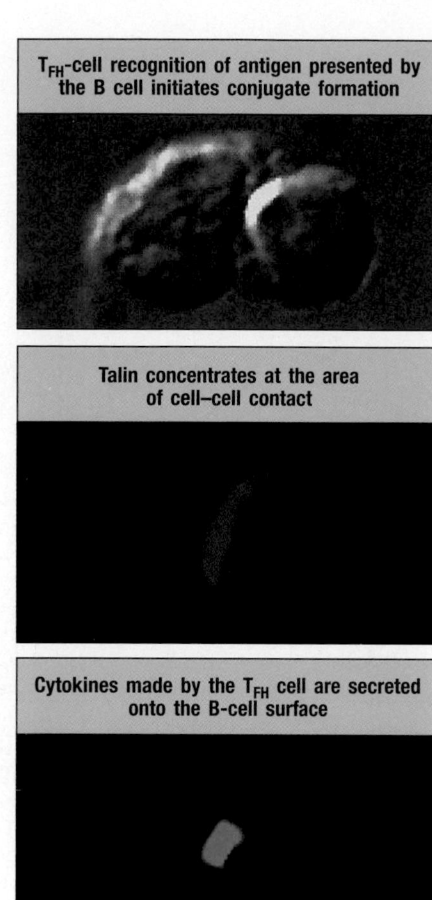

Figure 9.8 T_{FH} cells help antigen-activated B cells through cell-surface interactions between CD40 ligand and CD40 and by the targeted delivery of secreted cytokines to the B-cell surface. Top panels: Antigen-specific B and T cells form a cognate pair and a synapse that is stabilized by strong adhesive interactions between T-cell LFA-1 and B-cell ICAM-1. The T_{FH} cell then synthesizes cytokines, including CD40 ligand, a membrane-bound relative of TNF-α. Also shown is the microtubule-organizing center (MTOC). Center panels: interaction of B-cell CD40 (related to the TNF receptor) with T-cell CD40 ligand is accompanied by reorientation of the cytoskeleton, as exemplified by the redistribution of the protein talin and the MTOC, as well as the secretory apparatus of the Golgi. Bottom panels: soluble cytokines are synthesized in the cytoplasm and translocated to the endoplasmic reticulum of the T_{FH} cell. They are carried in exocytic vesicles through the Golgi and secreted, by vesicle fusion with the plasma membrane, onto a localized area of the B-cell surface. Photographs courtesy of A. Kupfer.

specific antigen, the T cell and the B cell form a conjugate pair (see Figure 9.7). This interaction induces the T cell to express CD40 ligand, which then binds the B-cell's CD40 (see Figure 8.28, p. 226). The B cell activates the transcription factor NFκB and increases the surface expression of the adhesion molecule ICAM-1, which engages integrin LFA-1 on the T cell (see Section 8-4). These developments strengthen the cognate interaction between the B cell and the helper T cell. A synapse is elaborated at the area of contact, and the T cell's cytoskeleton and Golgi apparatus are reorganized, facilitating the efficient and focused delivery of cytokines onto the B cell (Figure 9.8).

9-6 The primary focus of clonal expansion in the medullary cords produces plasma cells secreting IgM

Once antigen-specific B and T cells form a conjugate pair, they move together out of the T-cell area in the cortex and into the medullary cords. Here both cells begin to divide, forming what is called the **primary focus** of clonal expansion (Figure 9.9, left panel). This period of cellular proliferation lasts for several days and gives rise to dividing B lymphoblasts secreting IgM. The antibody leaves the node in the efferent lymph and is delivered to the blood, which rapidly carries it to the site of infection.

Some B lymphoblasts stay in the medullary cords, where they differentiate into plasma cells under the influence of the cytokines IL-5 and IL-6 secreted by T_{FH} cells. The terminal differentiation of lymphoblasts to plasma cells is determined by a transcription factor called B-lymphocyte-induced maturation protein 1 (BLIMP-1), which stops transcription of the genes necessary for

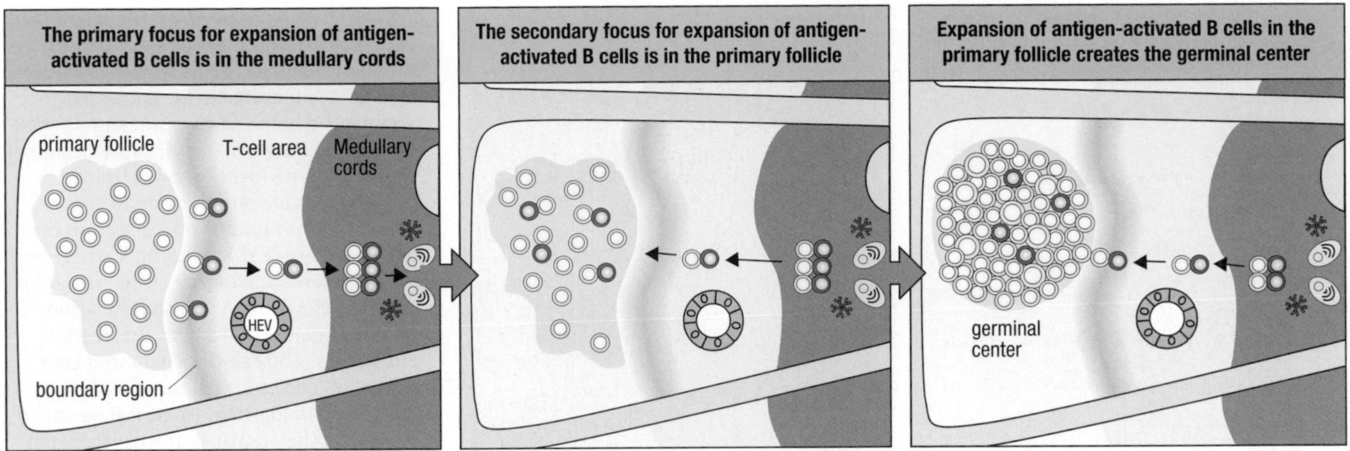

Figure 9.9 B cells activated by antigen and T-cell cytokines differentiate into plasma cells in two waves, occurring at different sites in the lymph node. First panel: cognate pairs of mutually activated antigen-specific B cells and T_{FH} cells move from the boundary region to the medullary cords, where they proliferate to form large clones of identical cognate pairs. This is the primary focus of B-cell expansion. Second panel: some members of each clone of cognate pairs differentiate to give plasma cells secreting antigen-specific IgM, while the other cognate pairs return to the primary follicle in the cortex. Third panel: the returning cognate pairs proliferate in the follicle, which becomes the second focus for B-cell expansion. A germinal center is formed by the expanding population of antigen-specific B cells, and this is where the B cells undergo affinity maturation and isotype switching of their antibodies.

proliferation. The cells then increase expression of the immunoglobulin chains and of factors involved in their synthesis and secretion. The morphological effects of activation are striking: the small resting B cell, which in appearance is all nucleus and no cytoplasm, gives rise to plasma cells whose large active cytoplasm packed with rough endoplasmic reticulum is testimony to their function as antibody factories, in which 20% of the protein made is immunoglobulin (Figure 9.10).

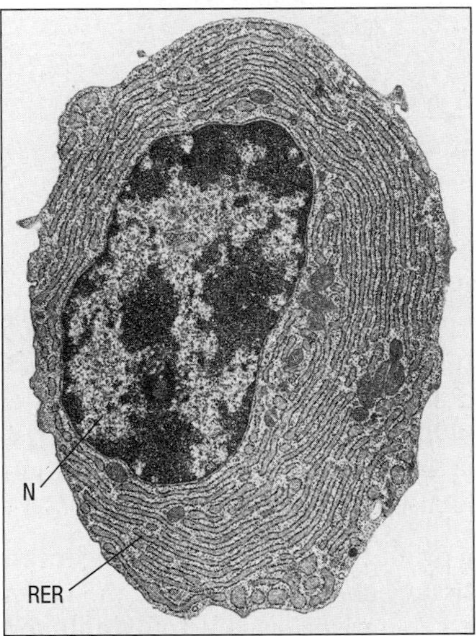

Figure 9.10 Plasma cells have a distinctive appearance that distinguishes them from all other forms of B cell. The nucleus (N) of a plasma cell has a characteristic 'clockface' pattern, resembling the hands and face of a clock. Further distinguishing plasma cells from other forms of B cells is their extensive rough endoplasmic reticulum (RER), a typical feature of cells that synthesize and secrete large quantities of proteins. For each plasma cell, the secreted protein is a homogeneous monoclonal antibody. Photograph courtesy of C. Grossi.

9-7 Activated B cells undergo somatic hypermutation and isotype switching in the specialized microenvironment of the primary follicle

Whereas some B lymphoblasts go directly from the primary focus to become IgM-secreting plasma cells in the medullary cords, other B lymphoblasts leave the primary focus and move into the primary follicles of the B-cell area while still attached to their cognate helper T_{FH} cells. There they proliferate to form a germinal center (see Figure 9.9, center and right panels). Cytokines IL-6, IL-15, 8D6, and BAFF made by the FDCs force the B cells to divide rapidly—about once every 6 hours—and to become large, metabolically active **centroblasts**. The helper T cells also divide, making cytokines and interacting via their CD40 ligand with the CD40 on B cells. This induces the B cells to produce activation-induced cytidine deaminase (AID), a DNA-modifying enzyme that is essential for both somatic hypermutation and isotype switching (see Sections 4-14 and 4-15), processes now at work in the proliferating centroblasts. During this proliferative stage the centroblast no longer expresses cell-surface immunoglobulin because the purpose of this stage is not antigen-mediated selection but expansion of a large population of B cells with switched isotype and V-region mutations. The massive proliferation of antigen-specific B cells in a primary follicle changes its morphology into what is described as a **secondary follicle**. The dominant feature is now the **germinal center** containing all the rapidly dividing B cells and T cells (Figure 9.11, upper panels). Surrounding the germinal center and pushed to the periphery of the follicle are the naive B cells that are passing through the lymph node in search of specific antigen and survival signals. These B cells now form a zone called the mantle zone. In

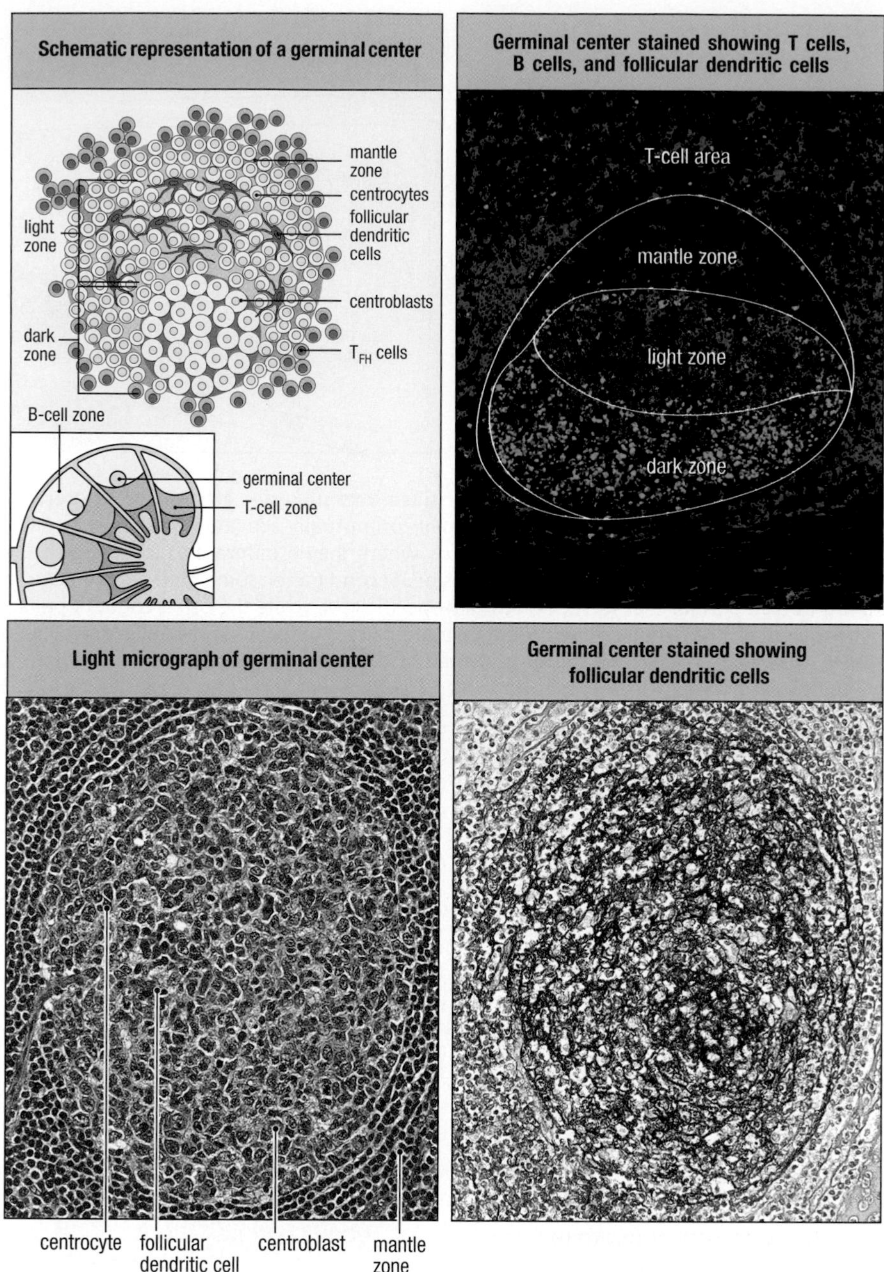

Figure 9.11 Anatomy of the germinal center: the place where activated B cells undergo affinity maturation and isotype switching. The top left panel is a schematic diagram showing the distribution of cells within a germinal center and the location of germinal centers within a lymph node (inset). The top right panel shows a section of a germinal center that roughly corresponds to the schematic diagram. The germinal center is viewed at low magnification and has been stained with fluorescent antibodies. The rapidly dividing B cells are called centroblasts and they form the so-called 'dark zone' of the germinal center. As they mature, the centroblasts stop dividing and become small centrocytes that interact with antigen-bearing FDCs to form the 'light zone.' In the top right panel, the centroblasts are stained green and the follicular dendritic cells (FDCs) are stained red. Stained blue are the numerous T cells that populate the T-cell area. The smaller number of T cells in the light zone are T_{FH} cells activating B cells. The bottom left panel shows a germinal center from a human tonsil at high magnification. The bottom right panel shows a section of the same germinal center that has been stained to reveal the network of FDCs. Photographs courtesy of Yasodha Natkunam.

a primary immune response, germinal centers appear in secondary lymphoid tissues about a week after the infection starts, and they are the cause of the characteristic swelling of lymph nodes draining an infected site. The cellular and morphological events that form the germinal center and take place there are called the **germinal center reaction.** With time, the vast majority of lymphocytes present in a germinal center are clones derived from one or a few founder pairs of antigen-specific B cells and T_{FH} cells.

As they divide, the centroblasts become increasingly closely packed and form a region that is darkly staining in histological sections and is called the **dark zone** of the germinal center (Figure 9.11, lower panels). The centroblasts give rise to **centrocytes**; these divide more slowly and begin to re-express cell-surface immunoglobulin, which is now mutated and switched in isotype. The centrocytes leave the dark zone and move into the **light zone**, where there is a lower density of B cells and a higher density of FDCs and T_{FH} cells. In germinal centers, about 80% of the cells are centroblasts, which have a $CD38^+$ $CD44^-$ $CD77^+$ phenotype (CD38, CD44, and CD77 are cell-surface glycoproteins used as markers to distinguish centroblasts and centrocytes), and 10–20% are centrocytes, which have a C38$^+$ CD44$^+$ CD77$^+$ phenotype.

Centrocytes are programmed to die by apoptosis within a short time unless their surface immunoglobulin is bound by antigen and their CD40 is bound by the CD40 ligand of a helper T cell. To survive, the centrocytes must compete with each other, first for access to antigen on FDCs and then for antigen-specific helper T cells.

9-8 Antigen-mediated selection of centrocytes drives affinity maturation of the B-cell response in the germinal center

As we saw in Chapters 6–8, a common theme in lymphocyte development is for a phase of activation and proliferation to be followed by one of selection. This is precisely what happens to the B cells maturing in a germinal center. The surface immunoglobulin expressed by an individual centrocyte after hypermutation can have an affinity for its specific antigen that is higher, lower, or the same as that of the unmutated immunoglobulin. Thus, somatic hypermutation in a clone of expanding B cells produces centrocytes with a diversity of B-cell receptors with a range of affinities for the antigen to which the founder B cell was specific. The antigen available to the centrocytes is that deposited on the FDCs. Newly formed centrocytes move from the dark zone of the germinal center and enter the light zone, where they compete for access to the limited amount of antigen displayed on the dendrites of the FDCs. In this competition the centrocytes having the antigen receptors of highest affinity are more likely to be activated by the limiting amount of antigen.

If a centrocyte receptor binds antigen with sufficient strength, a synapse is formed with the FDC through which antigen and survival signals are delivered to the centrocyte. While moving to the outer regions of the light zone, where helper T cells concentrate, the centrocyte processes the antigen and presents the peptides on its MHC class II molecules. Engagement of peptide:MHC class II complexes by the T-cell receptor, and of centrocyte CD40 by CD40 ligand on the T_{FH} cell, induces the centrocyte to express the protein Bcl-x_L, which prevents the cell dying by apoptosis. These centrocytes continue their differentiation to become plasma cells (Figure 9.12, right panel). Some of the plasma cells remain in the lymph node to produce antibody that will combine with the antigens arriving in the lymph from the site of infection. These plasma cells are relatively short lived. The remaining plasma cells migrate to the bone marrow; they provide a systemic source of antibody and are longer lived.

If a centrocyte fails to obtain and present antigen to a T_{FH} cell, it dies by apoptosis (Figure 9.12, left panel). Apoptotic centrocytes are phagocytosed by macrophages in the germinal center. Macrophages that have recently engulfed apoptotic centrocytes are a characteristic feature of germinal centers and, because of their densely staining contents, are called **tingible body macrophages**—tingible simply meaning 'capable of being stained.' Somatic hypermutation sometimes produces centrocytes bearing immunoglobulin that reacts with a self antigen on cell surfaces in the germinal center. When this happens, contact with helper T cells or other cells in the germinal center renders such centrocytes inactive or anergic, by mechanisms similar to those used during B cell development in the bone marrow (see Section 6-13).

Only the centrocytes with the highest-affinity antigen receptors are selected for survival and differentiation into either antibody-producing plasma cells or long-lived memory cells. In this way, the average affinity of antibodies for antigen increases during the course of an immune response and in subsequent exposures to the same antigen. This process of affinity maturation ensures that the B cells concentrate their resources on making the best possible antibodies.

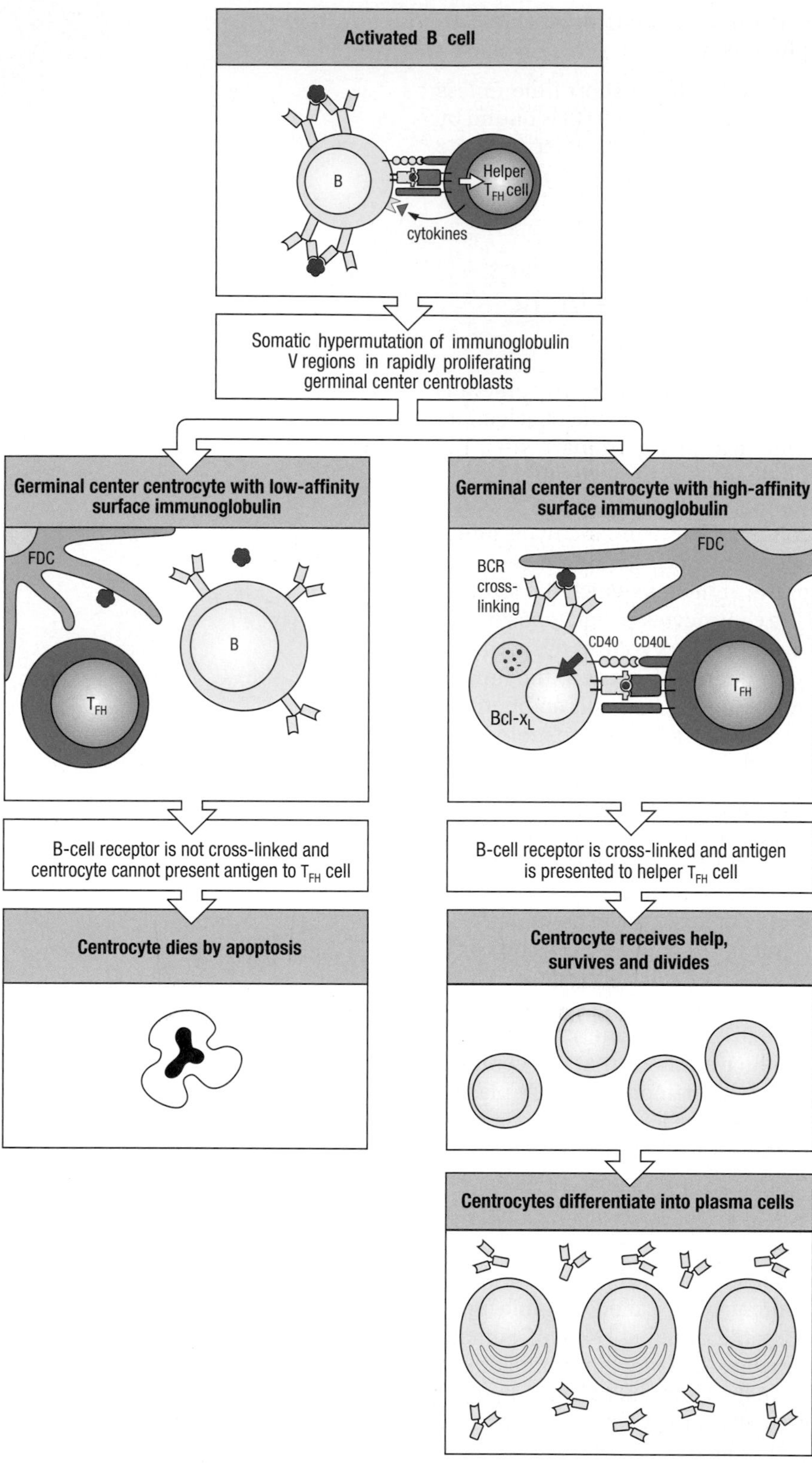

Figure 9.12 After undergoing somatic hypermutation, centrocytes with high-affinity receptors for antigen are rescued from apoptosis. Top panel: in the germinal center, T_{FH} cells induce dividing centroblasts to undergo somatic hypermutation. The centrocytes expressing mutant IgM then test their B-cell receptors (BCR) against the intact antigens displayed by the FDCs. Left panels: Centrocytes whose B-cell receptors have acquired mutations that reduce the affinity are induced to die by apoptosis. Right panels: centrocytes with receptors having superior affinity for antigen are induced to express Bcl-x_L, an intracellular protein that prevents apoptosis and ensures the cell's survival. These centrocytes go on to become plasma cells.

In tonsillitis a secondary lymphoid organ, the tonsil, becomes overwhelmed by a bacterial pathogen. Such abundance of antigen results in less stringent selection for better antibodies, and those that are made cannot terminate the infection, which now follows a chronic course. Tonsillitis can usually be overcome by a course of antibiotics, but may need to be combined with surgical removal of the infected lymphoid tissue.

The proliferating B cells of the germinal center are subject to mutational processes that change the sequence of the immunoglobulin genes (somatic hypermutation) and recombine their sequences with deletion of intervening DNA (during isotype switching). Although these events are largely confined to the immunoglobulin genes, they do occur at a low frequency in other genes and in ways that favor malignant transformation and cancer. Increasing the probability for such progression is the nurturing environment provided by the FDCs. As a consequence, most B-cell lymphomas originate with a germinal center cell.

9-9 The cytokines made by helper T cells determine how B cells switch their immunoglobulin isotype

In Chapter 4 we saw how the first immunoglobulins made by B cells are of the IgM and IgD classes, but that after activation by antigen, B cells can switch their heavy-chain isotype to produce IgG, IgA, or IgE. Isotype switching takes place in activated B cells mainly within the germinal center, and the isotype to which an individual B cell switches is determined by its cognate interactions with T_{FH} cells. The isotype to which a switch is made depends on the cytokines secreted by the T_{FH} cell, which in turn is determined by the nature of the infection and the dendritic cell that nurtured the T_{FH} cell (see Section 8-19). The roles of individual cytokines in switching the isotype of mouse immunoglobulin heavy chains are summarized in Figure 9.13. Knowledge of the human counterparts is incomplete. IFN-γ switches human B cells to making the strongly opsonizing IgG1 antibody, whereas IL-4 switches human B cells to make IgE.

T-cell cytokines induce isotype switching by stimulating transcription from the switch regions that lie 5′ to each heavy-chain C gene (see Figure 4.28, p. 102). For example, when activated B cells are exposed to IL-4, transcription from a site upstream of the switch regions of $C_{\gamma}1$ and C_{ε} can be detected a day or two before switching occurs. As with the low-level transcription that occurs in immunoglobulin loci before rearrangement (see Section 6-8), this transcription could be opening up the chromatin and making the switch regions accessible to the somatic recombination machinery that will place a new C gene in juxtaposition to the V-region sequence.

Isotype switching requires CD40 on the B cell to be bound by CD40 ligand on the T_{FH} cell, as is evident from the immunodeficiencies suffered by people who lack CD40 ligand and have the condition called hyper-IgM syndrome. Because their B cells cannot switch immunoglobulin isotype, these patients have abnormally high amounts of IgM in their circulation, but almost no IgG and IgA. In general these patients make poor antibody responses to most antigens, and their secondary lymphoid tissues contain no germinal centers

	Influence of cytokines on antibody isotype switching						
Cytokine	IgM	IgG3	IgG1	IgG2b	IgG2a	IgA	IgE
IL-4	Inhibits	Inhibits	Induces		Inhibits		Induces
IL-5						Augments production	
IFN-γ	Inhibits	Induces	Inhibits		Induces		Inhibits
TGF-β	Inhibits	Inhibits		Induces		Induces	

Figure 9.13 Different cytokines induce B cells to switch to different immunoglobulin isotypes. Cytokines can either induce (green), augment (yellow), or inhibit (red) the switching of immunoglobulin isotype. The inhibitory effects are due to the positive effect of a cytokine on switching to another isotype. This figure shows isotype switching by mouse B cells because the human system is not well worked out. There are known differences. For example, switching to IgA in humans involves TGF-β and IL-10, not IL-5 as in mouse.

(Figure 9.14). Aspects of cell-mediated immunity are also impaired in these patients, who are mostly male because the gene for CD40 ligand is on the X chromosome.

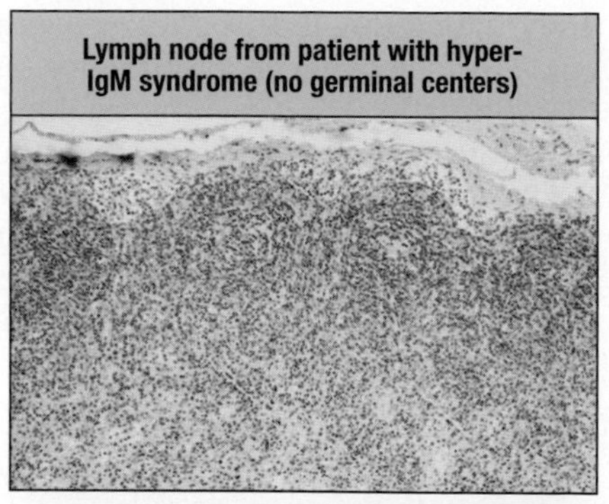

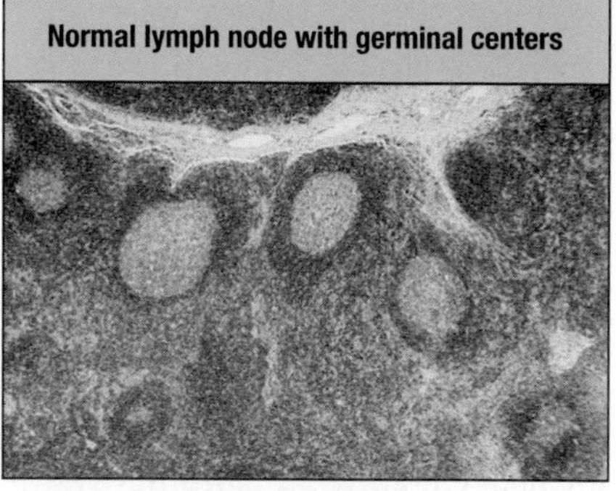

Figure 9.14 Lymph node in a patient with hyper-IgM syndrome compared with a normal lymph node. The lymph node from the patient is shown in the upper panel; the normal lymph node is shown in the bottom panel. Bottom panel photograph courtesy of Antonio Perez-Atayde.

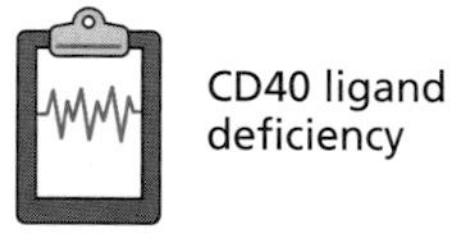

9-10 Cytokines made by helper T cells determine the differentiation of antigen-activated B cells into plasma cells or memory cells

For mutated centrocytes that survive selection in the germinal center, the interaction with an antigen-specific helper T cell serves several purposes. The mutual engagement of ligands and receptors on the two cells generates an exchange of signals that induces the further proliferation of both B and T cells. This serves to expand the population of selected high-affinity, isotype-switched B cells. Individual B cells are also directed along pathways of differentiation leading to either plasma cells or **memory B cells**.

At the height of the adaptive immune response, when the main need is for large quantities of antibodies to fight infection, the centrocytes that win in this selection leave the germinal center and differentiate under the influence of IL-10 into antibody-producing plasma cells (Figure 9.15, left). In the later stages of a successful immune response, as the infection subsides, centrocytes differentiate under the influence of IL-4 into long-lived, memory B cells, which now possess isotype-switched, high-affinity antigen receptors (Figure 9.15, right). These cells will then remain largely quiescent until reactivated by a subsequent encounter with the same antigen. Plasma cells are the effector B cells that provide antibody for dealing with today's infection, whereas the memory B cells represent an investment in preventing future infection by the same pathogen, should the current infection be successfully resolved.

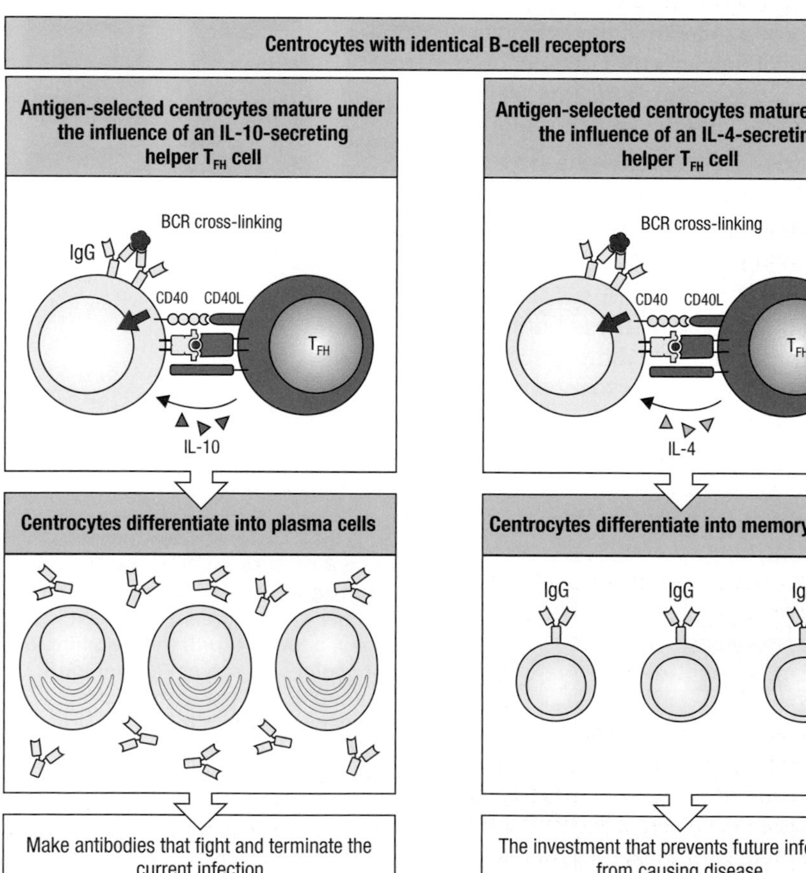

Figure 9.15 Cytokines made by helper T_{FH} cells determine whether centrocytes differentiate into plasma cells or memory B cells. Centrocytes with identical high-affinity immunoglobulins differentiate into either plasma cells (left panels) or memory B cells (right panels) depending on the cytokines secreted by their cognate T_{FH} cells.

	Property					
	Intrinsic			Inducible		
B-lineage cell	Surface Ig	Surface MHC class II	High-rate Ig secretion	Growth	Somatic hyper-mutation	Isotype switch
Naive B cell	Yes	Yes	No	Yes	Yes	Yes
Plasma cell	No	No	Yes	No	No	No

Figure 9.16 Comparison of key properties of naive B cells and plasma cells. In the course of the adaptive immune response, quiescent naive B cells evolve into highly active plasma cells. Shown here are key properties and capacities that distinguish naive B cells from plasma cells.

Summary

In responding to specific antigen, naive B cells undergo activation, proliferation, and differentiation to become plasma cells that synthesize and secrete antibody in massive amounts. The main differences between naive B cells and plasma cells are summarized in Figure 9.16. Activation of a mature but naive B cell requires signals delivered through simultaneous ligation of both its antigen receptor (the B-cell receptor) and the B-cell co-receptor. For a minority of B cells, which are usually B-1 cells that recognize high-density epitopes (thymus-independent antigens) on microbial surfaces, these signals are sufficient for activation and do not require T-cell help. The activation of most, if not all, the remaining B cells (B-2 cells) requires an extensive cognate interaction with a helper T_{FH} cell. In the B-cell area of a lymph node draining infection, naive B cells will recognize their specific antigens on follicular dendritic cells and subcapsular sinus macrophages and internalize them for processing and presentation on MHC class II molecules. The antigen-activated B cell moves to the boundary between the B- and T-cell areas where it forms a conjugate with a T_{FH} cell that recognizes a peptide:MHC complex on the B cell. This cognate interaction drives the differentiation of the B cell to become an antibody-producing plasma cell. The first focus for B-cell expansion is in the medulla of the lymph node and produces plasma cells that secrete IgM antibodies; the second focus for B-cell expansion is in the germinal center in the B-cell area of the cortex, where somatic hypermutation and isotype switching produce antibodies that bind more tightly to the pathogen and are more able to recruit effector cells and molecules.

Antibody effector functions

As the B-cell response to infection gets under way, isotype switching diversifies antibody function by changing the heavy-chain C region. The differences between the isotypes serve several purposes. By varying the number of antigen-binding sites and the flexibility of their movement, isotypic differences alter the strength with which antibodies bind to pathogens and antigens. They also vary the ability of antibodies to activate complement and deliver pathogens to phagocytes for destruction (see Section 4-16). Other cell types carry surface receptors that bind to the C regions—the Fc portion—of antibodies of a particular class or subclass, irrespective of the antibody's antigen specificity. Some of these receptors are expressed by epithelium and deliver antibody to anatomical sites that would otherwise be inaccessible. Others are carried by effector cells and enable them to capture and subjugate antibody-bound pathogens. In this part of the chapter we consider how the complementary functions of antibodies of different isotypes provide effective immunity against the pathogens found in the many different environments in the human body.

9-11 IgM, IgG, and monomeric IgA protect the internal tissues of the body

In any antibody response, IgM is the first antibody to be produced. It is secreted as a pentamer by plasma cells in the bone marrow, the spleen, and the medullary cords of lymph nodes. IgM enters the blood and is carried to sites of infection, inflammation, and tissue damage throughout the body. With 10 antigen-binding sites, IgM binds strongly to microorganisms and particulate antigens and quickly activates the complement cascade by the classical pathway. This coats pathogens with C3b, facilitating their uptake and destruction by phagocytes. The main disadvantage to IgM is that its bulk limits the extent to which this antibody isotype can passively leave the blood and penetrate infected tissues.

As the immune response proceeds and B cells undergo somatic hypermutation and isotype switching, the need for the bulky IgM dwindles. Isotype switching and affinity maturation produce IgG and monomeric IgA molecules with two high-affinity binding sites for antigen that are just as effective as the 10 antigen-binding sites of IgM. These antibodies also have the advantage of being smaller and better able to get into infected tissue. IgG is the dominant blood-borne antibody, but monomeric IgA made by B cells activated in the lymph nodes or spleen also makes a contribution.

To improve the delivery of IgG to tissues, it is actively transported from the blood into the extracellular spaces within tissues. The endothelial cells of blood vessels perform pinocytosis—the ingestion of small amounts of extracellular fluid—by which they take up plasma proteins for degradation in lysosomes. IgG is uniquely spared this fate, because in the acidic conditions of the endocytic vesicles the IgG associates with a membrane receptor that binds to the Fc portion of IgG. The receptor diverts the IgG away from the lysosomes and takes it to the basolateral surface of the cell, where the basic pH of the extracellular fluid induces the receptor to release the IgG (Figure 9.17). This transport receptor is called **FcRn**, or the Brambell receptor (FcRB) after the scientist who first described its function. FcRn is similar in structure to an MHC class I molecule, with the α_1 and α_2 domains forming a site that binds to the Fc region of the antibody. In the antibody:receptor complex, two molecules of FcRn bind to the Fc region of one IgG molecule. As well as maintaining a high level of IgG in the extracellular fluids of connective tissues, FcRn selectively protects IgG from the processes of degradation to which other plasma proteins are subject. As a consequence, IgG molecules have a longer half-life than most other plasma proteins. FcRn is just one of many different cellular receptors that bind to the Fc part of immunoglobulins and are collectively called Fc receptors.

Besides providing a defense to all the tissues reached by the blood, an important function of circulating IgM, IgG, and IgA is to prevent blood-borne infection—septicemia—and the spread of microorganisms by neutralizing those that enter the blood. Because the blood circulation is so effective at distributing cells and molecules to all parts of the body, infections of the blood itself can have grave consequences.

9-12 Dimeric IgA protects the mucosal surfaces of the body

Whereas IgM, IgG, and monomeric IgA provide antigen-binding functions within the fluids and tissues of the body, dimeric IgA protects the surfaces of the mucosal epithelia that communicate with the external environment and

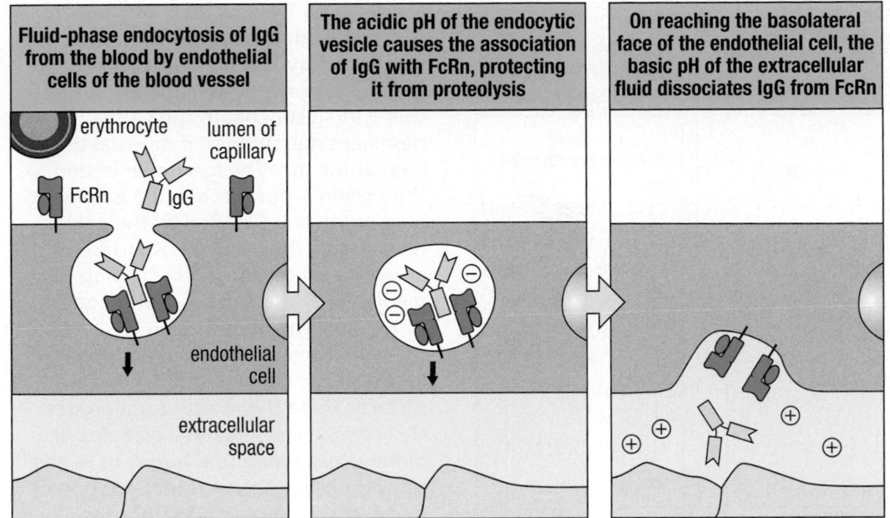

Fluid-phase endocytosis of IgG from the blood by endothelial cells of the blood vessel
The acidic pH of the endocytic vesicle causes the association of IgG with FcRn, protecting it from proteolysis
On reaching the basolateral face of the endothelial cell, the basic pH of the extracellular fluid dissociates IgG from FcRn
erythrocyte
lumen of capillary
FcRn
IgG
endothelial cell
extracellular space

Figure 9.17 The receptor FcRn transports IgG from the bloodstream into the extracellular spaces of tissues. At the apical (luminal) side of the endothelial cell, IgG and other serum proteins are actively taken up by fluid-phase pinocytosis. In the endocytic vesicle, the pH becomes acidic and each IgG molecule associates with two molecules of FcRn. The IgG is carried by FcRn to the basolateral face of the cell and away from the degradative activity of the lysosomes. At the basal side of the cell, the more basic pH dissociates the complex of IgG and FcRn, and the IgG is released into the extracellular space.

are particularly vulnerable to infection. These epithelia include the linings of the gastrointestinal tract, the eyes, nose, and throat, the respiratory, urinary, and genital tracts, and the mammary glands. Dimeric IgA is made in patches of mucosal-associated lymphoid tissue present in the lamina propria, the connective tissue underlying the mucosal epithelium. In these mucosal lymphoid tissues, antigen-specific B-cell and T-cell responses to local infections are developed. However, in this situation, the IgA-secreting plasma cells are on one side of the mucosal epithelium but their target pathogens are on the other. To reach their targets, the dimeric IgA molecules are transported individually across the epithelium by means of a receptor on the epithelial cells.

The dimeric form of IgA, but not the monomer, binds to a cell-surface receptor found on the basolateral surface of epithelial cells, which is called the **polymeric immunoglobulin receptor** or the **poly-Ig receptor** because of its specificity for IgA dimers and IgM pentamers (Figure 9.18). The poly-Ig receptor itself is made up of a series of immunoglobulin-like domains and covalently binds to IgM and dimeric IgA via their J chains (see Figure 4.27, p. 102 and Figure 4.31, p. 104), with which it makes a disulfide bond. After being bound, the IgA dimer is taken into the cell by receptor-mediated endocytosis, and the antibody:receptor complex is carried across the cell to the apical surface in endocytic vesicles. Receptor-mediated transport of a macromolecule from one side of a cell to the other is known as **transcytosis**. Once receptor-bound IgA appears on the apical surface, a protease cleaves the poly-Ig receptor at sites between the membrane-anchoring region and the IgA-binding site. Dimeric IgA is released from the membrane still bound to a small fragment of the receptor, which is called the **secretory component**, or **secretory piece**, of IgA. The IgA is then held at the mucosal surface, being bound to mucins—the glycoproteins in mucus—by the carbohydrate of the secretory piece. When IgA molecules bind to microorganisms at a mucosal surface, they prevent microbial attachment to and colonization of the mucosal epithelium, and thus facilitate the expulsion of pathogens in feces, sputum, tears, and other secretions.

9-13 IgE provides a mechanism for the rapid ejection of parasites and other pathogens from the body

Distinguishing IgE from the other antibodies (IgM, IgG, and monomeric IgA) made by B cells activated in lymph nodes and spleen is the minute quantity made and the transience of IgE antibodies in the circulation. The reason for these differences is that IgE does not function by binding to antigens as a

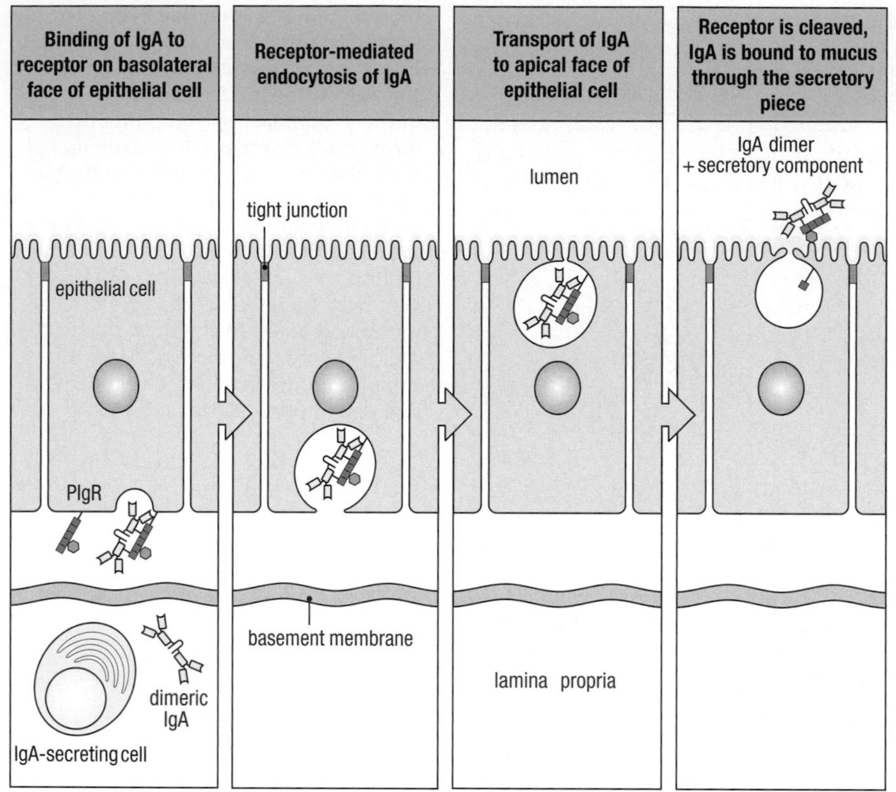
Binding of IgA to receptor on basolateral face of epithelial cell
Receptor-mediated endocytosis of IgA
Transport of IgA to apical face of epithelial cell
Receptor is cleaved, IgA is bound to mucus through the secretory piece
tight junction
lumen
IgA dimer + secretory component
epithelial cell
PIgR
basement membrane
lamina propria
dimeric IgA
IgA-secreting cell

Figure 9.18 Transcytosis of dimeric IgA antibody across epithelia is mediated by the poly-Ig receptor (pIgR). Dimeric IgA is made by plasma cells lying just beneath the epithelial basement membranes of mucosal tissues, such as the gut. The IgA dimer bound to the J chain diffuses across the basement membrane and is bound by the poly-Ig receptor on the basolateral surface of an epithelial cell. Receptor binding is mediated by the C_H3 constant domains of the IgA. The bound complex crosses the cell (transcytosis) in a membrane vesicle and is delivered to the apical surface. There the receptor undergoes cleavage, which releases a complex of dimeric IgA bound to a fragment of the receptor called the secretory component or secretory piece. Importantly, the carbohydrate (blue hexagon) of the secretory piece tethers the IgA to the mucus that coats the apical surface, thus preventing the antibody from being washed away into the gut lumen and beyond. The residual membrane-bound fragment of the poly-Ig receptor is nonfunctional and is degraded.

soluble antibody but is used as a cell-surface receptor for antigen. Because IgE thus works in the constrained two-dimensional environment of the cell surface, and not the three-dimensional environment of the blood, lymph, and extracellular fluids, far less antibody is required to be functionally effective. The Fc region of IgE is bound by an Fc receptor, called **FcεRI**, which is carried by mast cells, basophils, and activated eosinophils. FcεRI binds only to IgE antibodies, and the binding is so strong that IgE cannot dissociate once it has bound. Thus the small amounts of IgE that are secreted by plasma cells become quickly attached to the surfaces of mast cells resident in connective tissue, activated eosinophils at mucosal surfaces, and circulating basophils, where these cells are ready and waiting to bind pathogens and their antigens.

When a pathogen binds to the IgE on a mast cell and cross-links two or more FcεRI molecules in the mast-cell surface, it activates the cell to secrete active mediators that, among other effects, act on smooth muscle to cause violent reactions such as sneezing, coughing, vomiting, and diarrhea that forcibly eject pathogens from the respiratory and gastrointestinal tracts. IgE is the antibody isotype that works in the connective tissue, particularly that underlying the mucosal surfaces, and is specialized in causing the physical ejection of pathogens and toxic substances.

The major function of IgE is to provide defense against parasite infections. Parasites are a heterogeneous set of organisms that include unicellular protozoa and multicellular invertebrates, notably the helminths—intestinal worms and the blood, liver, and lung flukes—and ectoparasitic arthropods such as ticks and mites. As a group, parasites establish long-lasting, persistent infections in human hosts and are well practiced in avoiding and subverting human immune systems. Most parasites are much larger than any microbial pathogen. The largest human parasite is the tapeworm *Diphyllobothrium latum*, which can reach 9 m in length and lives in the small intestine, causing vitamin B_{12} deficiency and, in some patients, megaloblastic anemia. The larger parasites cannot be controlled by the cellular and molecular mechanisms of destruction that work for microorganisms, so a different strategy based on IgE has evolved.

IgE antibodies against a wide variety of different antigens are normally present in small amounts in all humans. Consequently, each cell that expresses FcεRI carries a variety of IgE molecules that are specific for different antigens. Mast cells are sentinels posted throughout the body's tissues, particularly in the connective tissues underlying the mucosa of the gastrointestinal and respiratory tracts and in the connective tissues along blood vessels—especially those in the dermis of the skin. The cytoplasm of the resting mast cell is filled with large granules containing **histamine** and other molecules that contribute to inflammation, known generally as **inflammatory mediators**. Mast cells become activated to release their granules when antigen binds to the IgE molecules bound to FcεRI on the mast-cell surface (Figure 9.19). To activate the cell, the antigen must cross-link at least two IgE molecules and their associated receptors, which means that the antigen must have at least two topographically separate epitopes recognized by the cell-bound IgE. Cross-linking of FcεRI generates the signal that initiates the release of the mast-cell granules. After degranulation, the mast cell synthesizes and packages a new set of granules.

Inflammatory mediators secreted into the tissues by activated mast cells, basophils, and eosinophils increase the permeability of the local blood vessels, enabling other cells and molecules of the immune system to move out of the bloodstream and into tissues. This causes a local accumulation of fluid, and the swelling, reddening, and pain that characterize inflammation. Inflammation in response to an infection is beneficial because it recruits cells and proteins required for host defense into the sites of infection.

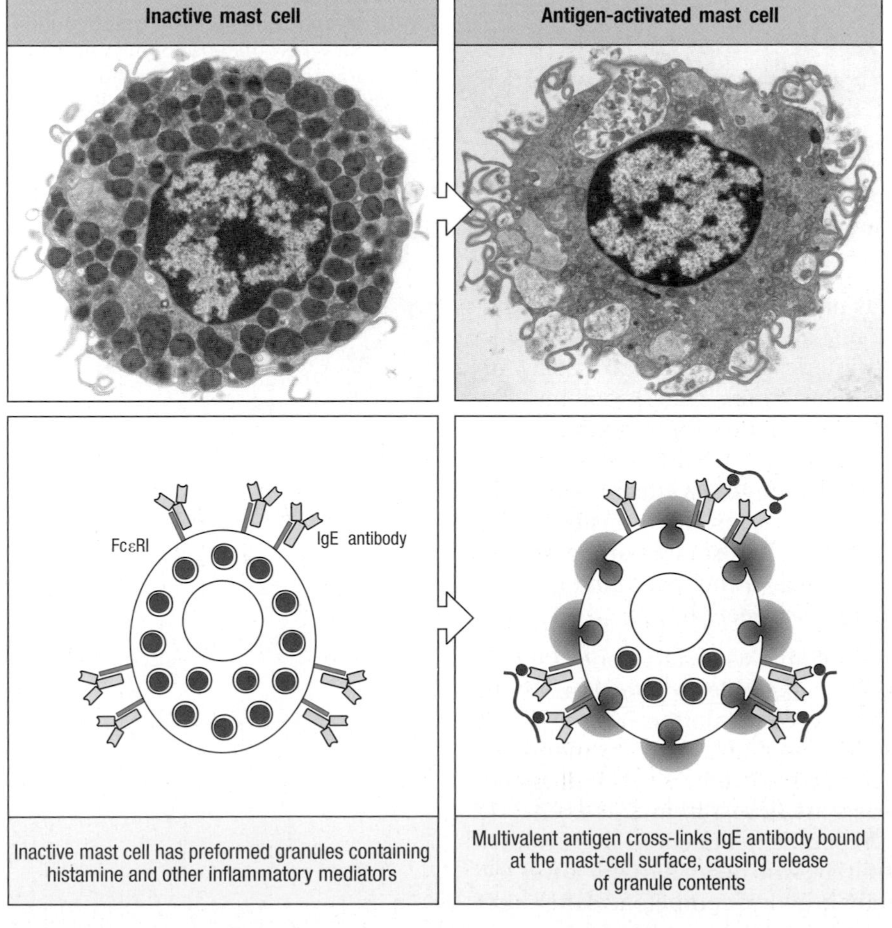

Inactive mast cell
Antigen-activated mast cell
FcεRI
IgE antibody
Inactive mast cell has preformed granules containing histamine and other inflammatory mediators
Multivalent antigen cross-links IgE antibody bound at the mast-cell surface, causing release of granule contents

Figure 9.19 Cross-linking IgE on mast-cell surfaces leads to the rapid release of mast-cell granules containing inflammatory mediators. Mast cells have numerous granules containing inflammatory mediators such as histamine and serotonin. The cells have high-affinity Fc receptors (FcεRI) on their surface that are occupied by IgE molecules in the absence of antigen (left panels). When antigen cross-links the complexes of IgE and FcεRI, the mast cell is activated, leading to degranulation and the release of inflammatory mediators into the surrounding tissue, as shown in the right panels. Photographs courtesy of A.M. Dvorak.

The prepackaged granules and the high-affinity FcεRI receptor already armed with IgE make the mast cell's response to antigen impressively fast. The infections that are the 'natural' targets of IgE-activated mast cells and eosinophils are those caused by parasites.

Inflammatory mediators released by mast cells, basophils, and eosinophils in response to parasite antigens cause smooth muscle contraction and increased permeability of blood vessels. Violent muscular contractions in airways and the gut expel parasites from these sites, while the increased permeability of local blood vessels supplies an outflow of fluid across the epithelium, which can help to flush out parasites. The combined actions of IgE, mast cells, basophils, and eosinophils are directed to the physical removal of parasites.

Eosinophils use their Fcε receptors to act directly against multicellular parasites. Such organisms, even small ones like the blood fluke *Schistosoma mansoni,* which causes schistosomiasis, cannot be ingested by phagocytes. However, if the parasite induces an antibody response and becomes coated with IgE, activated eosinophils will bind to the parasite through FcεRI and then pour the toxic contents of their granules directly onto its surface (Figure 9.20).

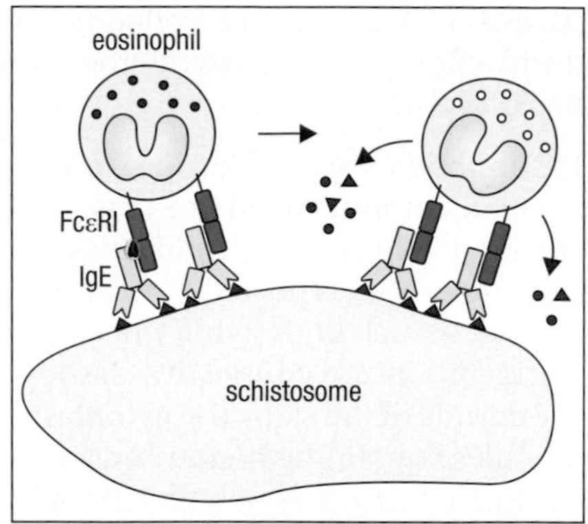

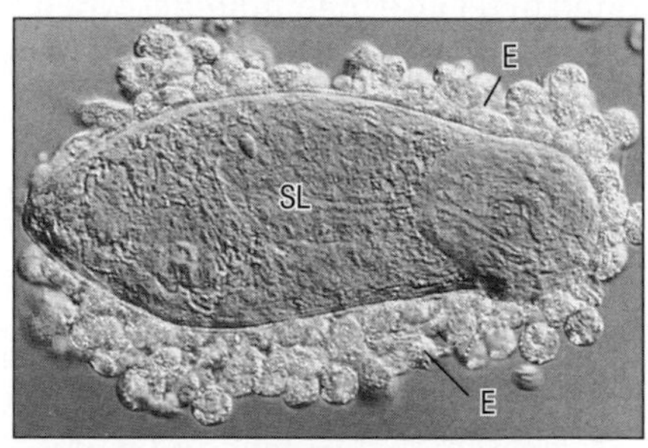

Figure 9.20 Activated eosinophils attack parasites coated with IgE. Large parasites, such as worms or the schistosome larva (SL) shown here, cannot be ingested by phagocytes. However, they can be attacked by activated eosinophils (E) that are coated with anti-parasite IgE antibodies bound to FcεRI. When the eosinophils encounter the parasite, the parasite's antigens will cross-link the IgE bound to FcεRI and activate the eosinophils to secrete the toxic and crippling contents of its granules directly on to the surface of the parasite. Photograph courtesy of A. Butterworth.

For human populations in developed countries, where parasite infections are rare, the mast cell's response is most frequently seen as a detriment, because its actions are the cause of allergy and asthma. People with these conditions make IgE in response to relatively innocuous substances, for example grass pollens or shellfish, that are often either airborne or eaten. Such substances are known as allergens. Having made specific IgE, any subsequent encounter with the allergen leads to massive mast-cell degranulation and a damaging response that is quite inappropriate to the threat posed by the antigen or its source. In extreme cases, the ingestion of an allergen can lead to a systemic life-threatening inflammatory response called anaphylaxis.

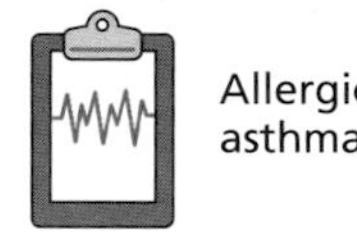

Allergic asthma

9-14 Mothers provide protective antibodies to their young, both before and after birth

During pregnancy, IgG from the maternal circulation is transported across the placenta and is delivered directly into the fetal bloodstream. This mechanism is so efficient that, at birth, human babies have as high a level of IgG in their plasma as their mothers, and as wide a range of antigen specificities. IgG is transported across the placenta by FcRn (Figure 9.21).

As a result of this mechanism, the baby has protection at birth against pathogens that provoke an IgG response but not against pathogens that infect mucosal surfaces and provoke a secreted, dimeric IgA response. To satisfy this need, infants obtain dimeric IgA in their mother's milk, which contains antibodies against the microorganisms to which the mother has mounted an IgA response. On breast-feeding, the IgA is transferred to the baby's gut (see Figure 9.21), where it binds to microorganisms, preventing their attachment to the gut epithelium and facilitating their expulsion in feces. The transfer of preformed IgA from mother to child in breast milk is an example of the **passive transfer of immunity**; another is the intravenous immunoglobulin given to patients with genetic defects in B-cell function (see Section 6-8).

During the first year of life there is a window of time when all infants are relatively deficient in antibodies and especially vulnerable to infection. As the maternally derived IgG is catabolized and the consumption of breast milk diminishes, the antibody level gradually decreases until about 6 months of age, when the infant's own immune system starts to produce substantial antibody (Figure 9.22). Consequently, IgG levels are lowest in infants aged 3–12 months, and this is when they are most susceptible to infection. This problem is particularly acute in babies born prematurely, who begin life with lower levels of maternal IgG and take longer to attain immune competence after birth

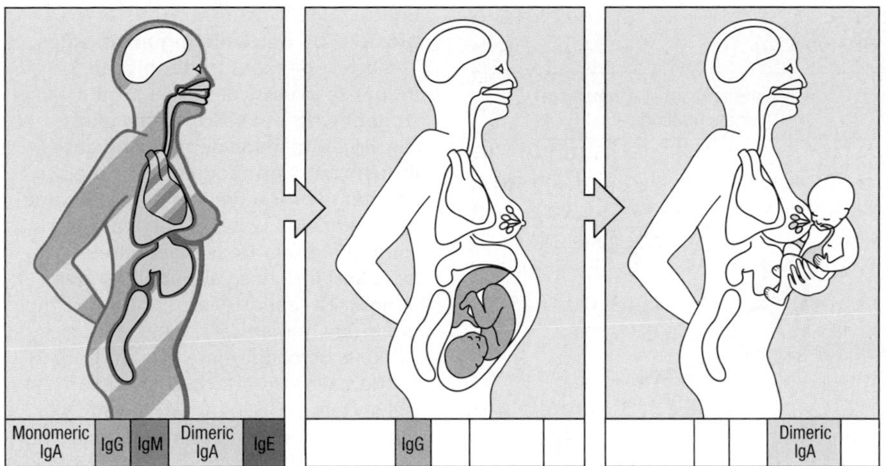

Figure 9.21 Immunoglobulin isotypes are distributed selectively in the human body and some are passed by mothers to their young. Left panel: in healthy women (and men), IgM, IgG, and monomeric IgA predominate in the blood (indicated schematically here by the coloring in the heart), whereas IgG and monomeric IgA are the major isotypes in the extracellular fluid. Dimeric IgA predominates in the secretions at mucosal epithelia. IgE is associated mainly with the mast cells in the connective tissue beneath epithelial surfaces, particularly the skin and the respiratory and gastrointestinal tracts. The brain is devoid of immunoglobulin. Center panel: during prenatal development in the mother's womb, the fetus cannot make its own immunoglobulin. To provide the fetus with protective antibody, IgG is selectively transported from the maternal to the fetal circulation by FcRn in the placenta. This IgG provides protection against the infections that the mother has experienced, and will be enriched for IgG against the most recent and current infections. Right panel: after birth and with breast-feeding, the gastrointestinal tract of the infant is supplied with protective maternal dimeric IgA, a major constituent of breast milk.

than babies born at term. This period of vulnerability is also the time when the symptoms of genetic deficiencies in immune system genes begin to appear.

9-15 High-affinity neutralizing antibodies prevent viruses and bacteria from infecting cells

The first step in a microbial infection involves attachment of the organism to the outside surface of the human body, either some part of the skin or the mucosal surfaces. These attachments are established by specific interactions between a component on the microbial surface, which acts as a ligand, and a complementary component on the surface of human epithelial cells that the pathogen exploits as its receptor. High-affinity antibodies that bind to the microbial ligand and prevent the microbe's attachment to human epithelium stop the infection before it starts. For this reason, such antibodies are called neutralizing antibodies. Because many infections start at mucosal surfaces, neutralizing antibodies are often dimeric IgA.

The influenza virus, which usually enters the body by inhalation, infects epithelial cells of the respiratory tract by binding to the oligosaccharides on their surface glycoproteins. The virus binds to oligosaccharides on cell surfaces though a major protein component of its outer envelope. This is called the **influenza hemagglutinin**, because the protein can **agglutinate**, or clump together, red blood cells by binding to the oligosaccharides on the red cell surface. Neutralizing antibodies that have been developed during primary immune responses to influenza and other viruses are the most important

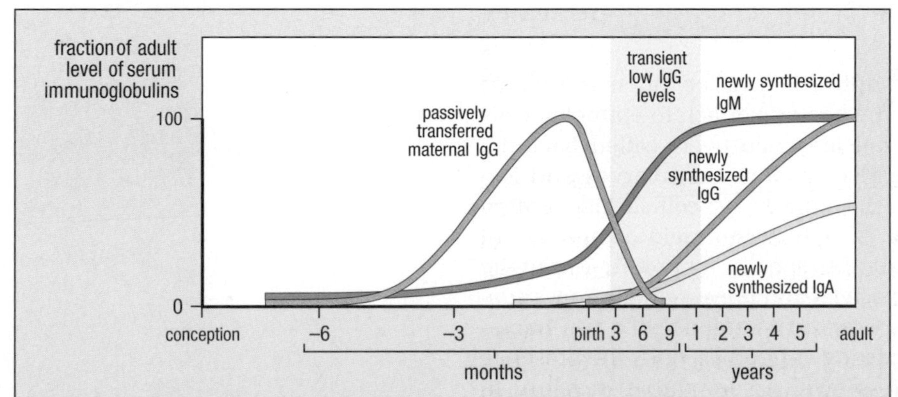

Figure 9.22 The first year of life is a period when infants have a limited supply of IgG and are particularly vulnerable to infections. Before birth, high levels of IgG are provided by the mother; after birth, maternally derived IgG declines. Although infants produce IgM soon after birth, the production of IgG antibodies does not begin for about 6 months. The concentration of IgG in the blood reaches a minimum within the first year and then gradually increases until adulthood.

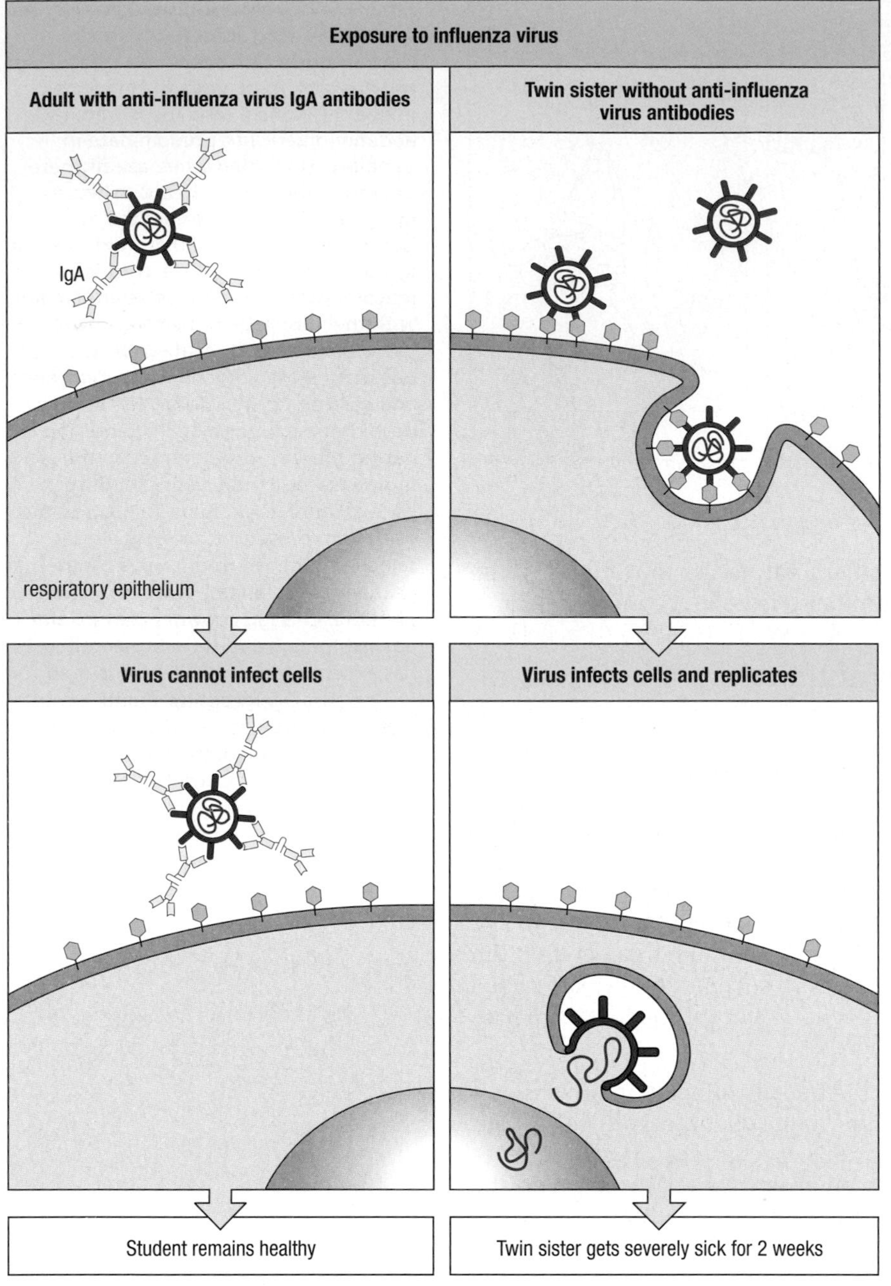
Exposure to influenza virus
Adult with anti-influenza virus IgA antibodies
Twin sister without anti-influenza virus antibodies
IgA
respiratory epithelium
Virus cannot infect cells
Virus infects cells and replicates
Student remains healthy
Twin sister gets severely sick for 2 weeks

Figure 9.23 Viral infections are blocked by neutralizing antibodies. The effect of exposure to this year's influenza virus is compared for a student who was vaccinated against the virus and made neutralizing anti-influenza IgA antibodies (left panels), and her identical twin sister at the same university who was too busy studying for a final exam to get vaccinated and lacks anti-influenza antibodies (right panels). To replicate itself, influenza must get inside human cells, which it does by using its hemagglutinin protein to bind to the sialic acid attached to human cell-surface proteins. Internalization of the virus, with subsequent fusion of the viral and endosomal membranes, releases the viral RNA into the cytoplasm, where replication occurs. This entire process can be halted at the very first step by the presence of neutralizing antibodies against the viral hemagglutinin that cover up its binding site for sialic acid. Because influenza infects epithelial cells of the respiratory tract, the effective antibodies are IgA dimers, as shown here.

aspect of subsequent immunity to these viruses. Such antibodies coat the virus, inhibit its attachment to human cells, and prevent infection (Figure 9.23).

The mucosal surfaces of the human body provide a variety of environments that are colonized by different bacteria. The bacteria attach to epithelial cells by using diverse surface components that are collectively called bacterial **adhesins**. For *Streptococcus pyogenes*, which inhabits the pharynx and is a common cause of a sore throat, the bacterial adhesin is a cell-surface protein called protein F that binds to fibronectin, a large glycoprotein component of the extracellular matrix. Secreted IgA antibodies specific for protein F limit the growth of the resident population of *S. pyogenes* and prevent them from causing disease. Only when this control is lost—by the emergence of a new bacterial strain that is not neutralized by the existing antibodies, or by the presence of an additional infection or other stress—will a sore throat develop. In

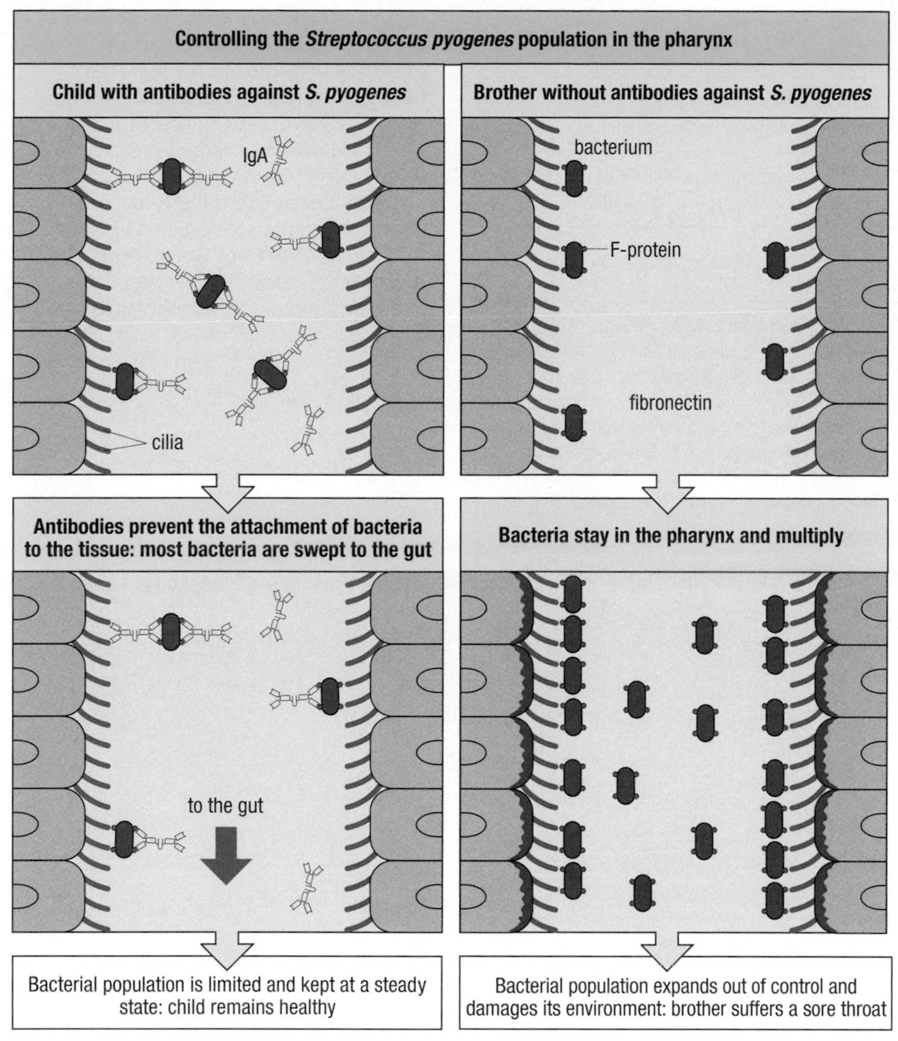

Figure 9.24 Disease-causing bacterial infections at mucosal surfaces are prevented by neutralizing antibodies. The left panels depict the pharynx of a child who has previously suffered from a 'strep throat' and is making neutralizing IgA against the causal bacterium, *Streptococcus pyogenes*. The antibodies coat the bacteria and impair their ability to attach to the fibronectin in the extracellular matrix and remain in the pharynx. This mechanism keeps the bacterial population down to a size that does not cause disease. The right panels depict the pharynx of the child's younger brother, who has never previously experienced a strep throat and has no neutralizing antibodies against *S. pyogenes*. In their absence, the size of the bacterial population is not strictly controlled; under favorable circumstances it can expand, causing damage to the mucosal surface and inducing inflammation. An unpleasantly sore throat ensues, as well as an adaptive immune response that produces neutralizing antibodies against *S. pyogenes*.

general, IgA antibodies against adhesins limit bacterial populations within the gastrointestinal, respiratory, urinary, and reproductive tracts and prevent disease-causing infections in these tissues (Figure 9.24).

9-16 High-affinity IgG and IgA antibodies are used to neutralize microbial toxins and animal venoms

Many bacteria secrete protein toxins that cause disease by disrupting the normal function of human cells (Figure 9.25). To have this effect, a bacterial toxin must first bind to a specific receptor on the surface of the human cell. In some toxins, for example the diphtheria and tetanus toxins, the receptor-binding activity is carried by one polypeptide chain and the toxic function by another. Antibodies that bind to the receptor-binding polypeptide can be sufficient to neutralize a toxin (Figure 9.26), and the vaccines for diphtheria and tetanus work on this principle. They are modified toxin molecules, called **toxoids**, in which the toxic chain has been denatured to remove its toxicity. On immunization, protective neutralizing antibodies are made against the receptor-binding chain.

Bacterial toxins are potent at low concentrations: a single molecule of diphtheria toxin is sufficient to kill a cell. To neutralize a bacterial toxin, an antibody must be of high affinity and essentially irreversible in its binding to the toxin. It

Disease	Organism	Toxin	Effects *in vivo*
Tetanus	*Clostridium tetani*	Tetanus toxin	Blocks inhibitory neuron action, leading to chronic muscle contraction
Diphtheria	*Corynebacterium diphtheriae*	Diphtheria toxin	Inhibits protein synthesis, leading to epithelial cell damage and myocarditis
Gas gangrene	*Clostridium perfringens*	Clostridial α-toxin	Phospholipase activation, leading to cell death
Cholera	*Vibrio cholerae*	Cholera toxin	Activates adenylate cyclase, elevates cAMP in cells, leading to changes in intestinal epithelial cells that cause loss of water and electrolytes
Anthrax	*Bacillus anthracis*	Anthrax toxic complex	Increases vascular permeability, leading to edema, hemorrhage, and circulatory collapse
Botulism	*Clostridium botulinum*	Botulinum toxin	Blocks release of acetylcholine, leading to paralysis
Whooping cough	*Bordetella pertussis*	Pertussis toxin	ADP-ribosylation of G proteins, leading to lymphocytosis
		Tracheal cytotoxin	Inhibits ciliar movement and causes epithelial cell loss
Scarlet fever	*Streptococcus pyogenes*	Erythrogenic toxin	Causes vasodilation, leading to scarlet fever rash
		Leukocidin Streptolysins	Kill phagocytes, enabling bacteria to survive
Food poisoning	*Staphylococcus aureus*	Staphylococcal enterotoxin	Acts on intestinal neurons to induce vomiting. Also a potent T-cell mitogen (SE superantigen)
Toxic shock syndrome	*Staphylococcus aureus*	Toxic-shock syndrome toxin	Causes hypotension and skin loss. Also a potent T-cell mitogen (TSST-1 superantigen)

Figure 9.25 Many common diseases are caused by bacterial toxins. Several examples of exotoxins, or secreted toxins, are shown here. Bacteria also make endotoxins, or nonsecreted toxins, which are usually only released when the bacterium dies. Endotoxins, such as bacterial lipopolysaccharide (LPS), are important in the pathogenesis of disease, but their interactions with the host are more complicated than those of the exotoxins and are less clearly understood.

must also be able to penetrate tissues and reach the sites where toxins are being released. High-affinity IgG is the main source of neutralizing antibodies for the tissues of the human body, whereas high-affinity IgA dimers serve a similar purpose at mucosal surfaces.

Poisonous snakes, scorpions, and other animals introduce venoms containing toxic polypeptides into humans through a bite or sting. For some venoms, a single exposure is sufficient to cause severe tissue damage or even death, and in such situations the primary response of the immune system is too slow to

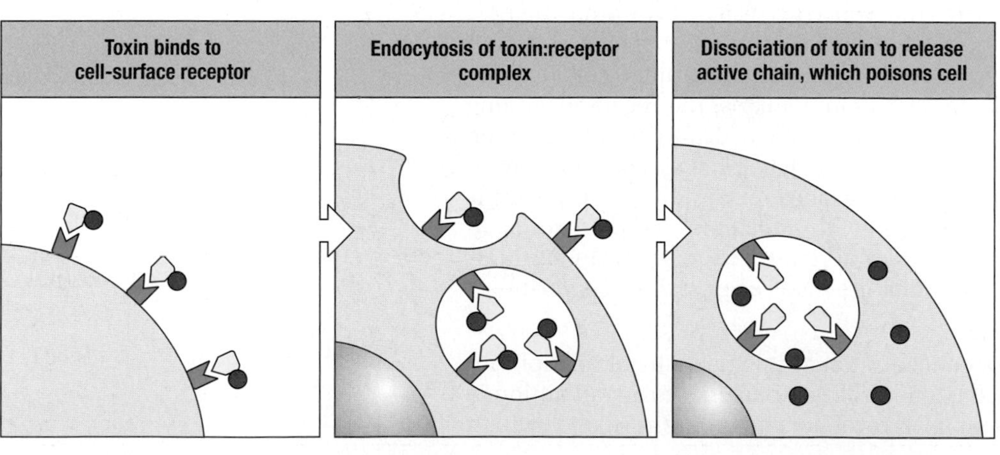

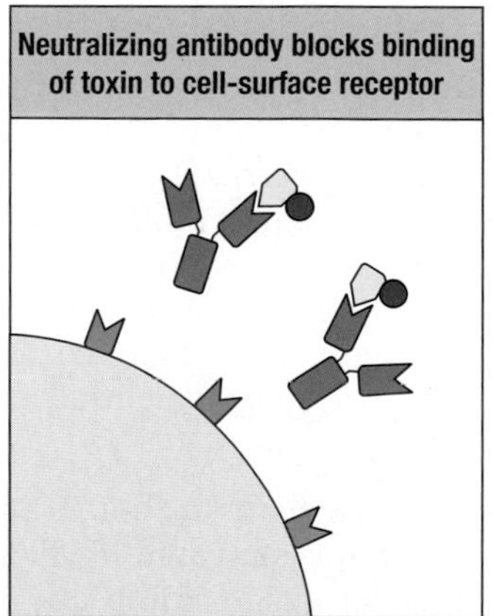

Figure 9.26 Neutralization of toxins by IgG antibodies protects cells from toxin action. The protein toxins produced by many bacteria comprise two functional modules. The first module binds to a component at the surface of a human cell, which allows the toxin to be internalized. The second module is a poison that interferes with a vital function of the cell. High-affinity neutralizing IgG antibodies cover up the binding site in the toxin's first module, thus preventing its attachment to human cells.

help. Because exposure to such venoms is rare, protective vaccines against them have not been developed. For patients who have been bitten by poisonous snakes or other venomous creatures, the preferred therapy is to infuse them with antibodies specific for the venom. These antibodies are produced by immunizing large domestic animals—such as horses—with the venom. Transfer of protective antibodies in this manner is known as **passive immunization** and is analogous to the way in which newborn babies acquire passive immunity from their mothers (see Section 9-14).

9-17 Binding of IgM to antigen on a pathogen's surface activates complement by the classical pathway

Only a fraction of antibodies have a direct inhibitory effect on a pathogen's capacity to live and replicate in the human body. The more common outcome is that antibodies bound to the pathogen recruit other molecules and cells of the immune system, which then kill the pathogen or eject it from the body. One route by which antibodies target pathogens for destruction is by activating complement through the classical pathway. In Chapter 3 we saw how the classical pathway is initiated when C-reactive protein binds to a bacterial surface. Activation of the classical pathway also occurs when antibodies of some isotypes, but not all, bind to pathogen surfaces. The most effective antibodies at activating complement are IgM and IgG3 (Figure 9.27).

For IgM, the first antibody made in a primary immune response, activating complement is the major mechanism by which it recruits effector cells to the sites of infection. By itself, pentameric IgM does not activate complement, because it is in a planar conformation that cannot bind the C1q component of C1, the necessary first step in the classical pathway. On binding to the surface of a pathogen, the conformation of IgM changes to what is called the 'staple' form. In this form, the binding site for C1q on the Fc part of each IgM monomer becomes accessible to C1q. Multipoint attachment of C1q to IgM is necessary to obtain a stable interaction, but this is readily accomplished because IgM has five binding sites for C1q, and C1q has six binding sites for IgM (Figure 9.28).

Binding of IgM to C1q activates the C1r and C1s serine proteases that are the enzymatically active components of C1. First activated is C1r, which then cleaves and activates C1s. Activated C1s is the protease that binds, cleaves and activates both the C4 and C2 components of the classical pathway. C4b fragments that become covalently bonded to the pathogen surface bind C2a to assemble the classical C3 convertase, C4bC2a. The classical C3 convertase cleaves C3 to C3b and C3a, and once C3b has become covalently bonded to

Antibody isotype	Relative capacity to fix complement
IgM	+++
IgD	–
IgG1	++
IgG2	+
IgG3	+++
IgG4	–
IgA1	+
IgA2	+
IgE	–

Figure 9.27 Classes and subclasses of antibodies differ in their capacity to activate and fix complement. The IgM and IgG3 isotypes are the most effective at activating the complement cascade; IgG1 is the runner-up.

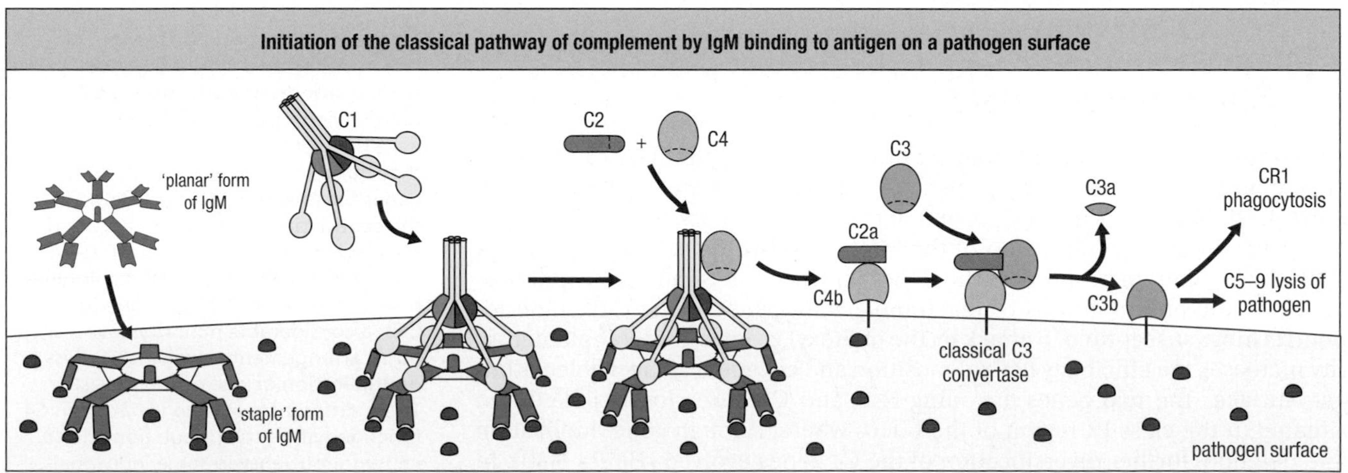

Figure 9.28 Binding of IgM to antigen on a pathogen's surface initiates the classical pathway of complement activation. When soluble pentameric IgM in the 'planar' conformation has established multipoint binding to antigens on a pathogen surface, it adopts the 'staple' conformation and exposes its binding sites for the C1q component of C1. Activated C1 then cleaves C2 and C4, and the C2a and C4b fragments form the classical C3 convertase on the pathogen surface. Conversion of C3 to C3b leads to the attachment of C3b to the pathogen surface and the recruitment of effector functions.

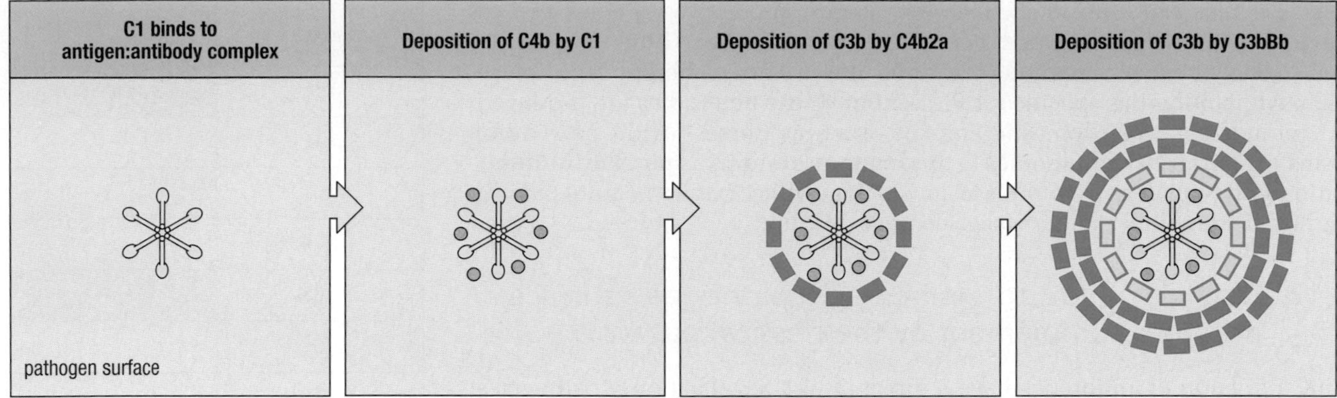

Figure 9.29 A bird's-eye view of the fixation of C4b and C3b fragments on a pathogen surface around an antigen:antibody complex. Antibody bound to an antigen on a microbial surface binds C1 (first panel), which leads to the deposition of C4b (pink circles) around the antigen:antibody complex (second panel). When C4b binds C2a to form the classical C3 convertase, a limited number of C3b molecules (green rectangles) are produced (third panel). These can bind Bb to form the alternative convertase, C3bBb (yellow rectangles), which leads to the deposition of many more C3b fragments on the microbial surface (green rectangles, fourth panel).

the pathogen it exponentially amplifies the response by assembling the C3 convertase of the alternative pathway (C3bBb). By the end of the complement-fixing reaction, most of the C3b fragments that cover the pathogen surface around the initiating antigen:antibody complex have been produced by the alternative convertase (Figure 9.29). This is another example of a situation in which adaptive immunity provides specificity to the reaction, and innate immunity provides the strength. Once the pathogen is coated with C3b it can be efficiently phagocytosed by a neutrophil or macrophage using its CR1 complement receptors. Alternatively, for some pathogens, such as the bacterium *Neisseria*, the classical pathway of complement activation continues, leading to assembly of the membrane-attack complex and the death of the pathogen through perforation of its outer membrane (see Section 2-7).

IgM fixes complement efficiently because its five binding sites for C1q allow each IgM molecule bound to a pathogen to fix complement independently. The drawbacks of IgM are its size, which restricts the extent to which it can penetrate infected tissues, and its limited capacity to recruit effector functions that are not dependent on complement. Once a primary immune response progresses to the point at which B cells have undergone isotype switching, affinity maturation, and have become plasma cells, the smaller and more versatile IgG molecules make up for the deficiencies of IgM.

9-18 Two forms of C4 tend to be fixed at different sites on pathogen surfaces

The complement components that are uniquely used by the classical pathway are the binding protein C1, the protease C2, and the thioester-containing protein C4. Of these, C4 is usually present in two forms that are encoded by separate genes in the central region of the MHC. The two types of C4—C4A and C4B—have different properties. The C4A thioester bond is more susceptible to attack by the amino groups of macromolecules, whereas the C4B thioester bond is more susceptible to attack by the hydroxyl groups. This complementarity increases the efficiency of C4 deposition and coverage of the whole pathogen surface. The two genes encoding C4A and C4B are closely linked and situated in the class III region of the MHC, where, through gene duplication and deletion, further diversification of the C4 genes evolved (Figure 9.30). In

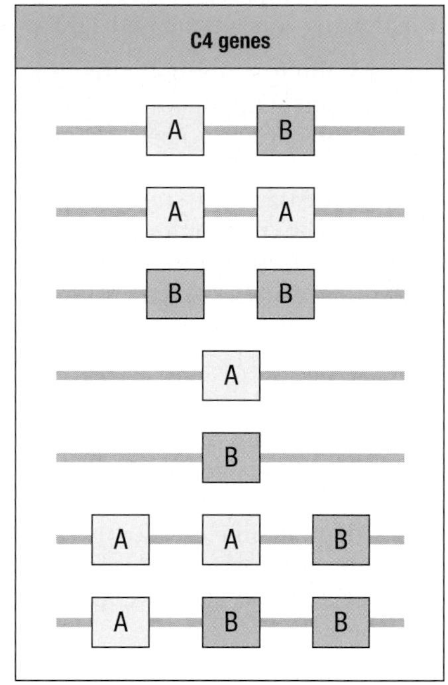

Figure 9.30 Humans differ in the number and type of genes for complement component C4. Complement proteins C4A and C4B have differences in the way in which they bond to pathogen surfaces. The genes for C4A and C4B are located in the central part of the MHC, between the class I region and the class II region. Although a majority of MHC haplotypes have one gene for C4A and one for C4B, a considerable minority have other arrangements involving the loss or duplication of one of the genes. These differences lead to variation in C4 function within the population and to immunodeficiency in some individuals.

humans, 13% of chromosomes lack a functional C4A gene and 18% of chromosomes lack a functional C4B gene; thus, more than 30% of the human population is deficient for one or other form of C4, and a partial lack of C4 is the most common human immunodeficiency. Reflecting the complementary functions of the two forms of C4, deficiency in C4A is associated with susceptibility to the autoimmune disease systemic lupus erythematosus (SLE), whereas deficiency in C4B is associated with a lowered resistance to infection. As well as the simple presence or absence of the C4A and C4B genes, there are more than 40 different alleles of the C4 genes, some of which are likely to be associated with differences in C4 function.

9-19 Complement activation by IgG requires the participation of two or more IgG molecules

The binding of IgG to antigen on a pathogen's surface can also trigger the classical pathway of complement activation. Each IgG molecule has a single binding site for C1q in its Fc region. Unlike IgM, IgG does not need to sequester its C1q-binding site in the absence of antigen. Interaction of one IgG molecule with C1q is insufficient to activate C1, making it necessary for C1q to cross-link two or more IgG molecules bound to antigen on the pathogen's surface, and the antigen-bound IgG molecules must be sufficiently close for C1q to span them (Figure 9.31, left panels). As a consequence, complement activation by IgG depends more on the amount and density of antibodies bound to the pathogen surface than complement activation by IgM. After the complexes of antigen and IgG have activated C1, the classical pathway proceeds in exactly the same way as when it is activated by IgM.

Because the IgG antigen-binding sites have higher affinity than those of IgM, they form stable immune complexes with soluble multivalent antigens—for example, toxins secreted by pathogens or breakdown products from their death and degradation. These soluble immune complexes also activate the

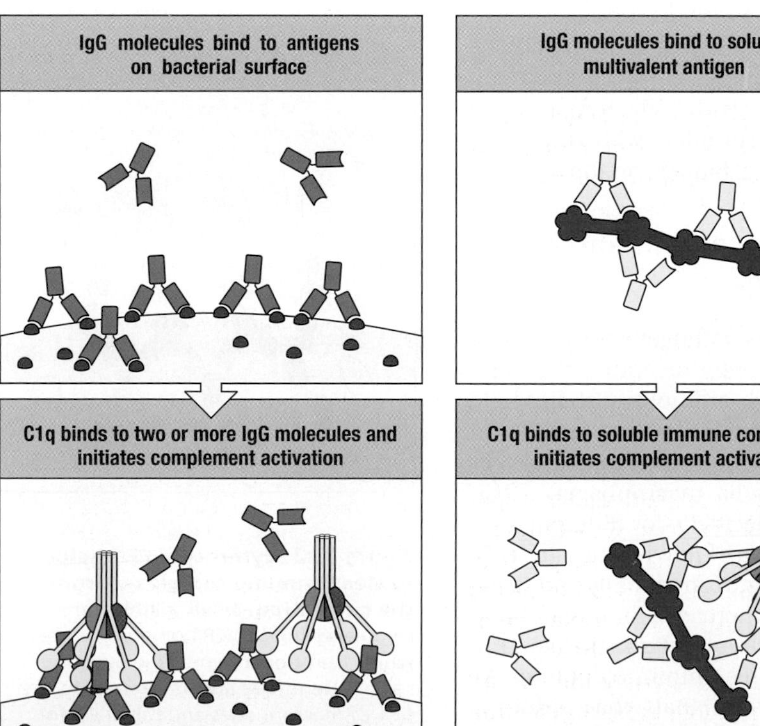

Figure 9.31 At least two molecules of IgG bound to pathogens or soluble antigens are required to activate the complement cascade. The left panels show the activation of complement by IgG bound to antigens on a pathogen surface. The C1q molecule needs to find pathogen-bound IgG molecules that are close enough to each other for the C1q molecule to span them. The right panels show the activation of complement by C1q binding to two IgG molecules in a soluble immune complex.

classical pathway (Figure 9.31, right panels), leading to the deposition of C3b on the antigen and antibody molecules in the complex. Using their complement receptors and Fc receptors, phagocytic cells readily take up these complexes of antigen, antibody, and complement from blood, lymph, and tissue fluids and clear them from the system.

9-20 Erythrocytes facilitate the removal of immune complexes from the circulation

Immune complexes that are covered with C3b fragments can be bound by circulating cells that express the complement receptor CR1. The most numerous of these is the erythrocyte, so that the vast majority of immune complexes become bound to the surface of red blood cells. During their circulation in the blood, erythrocytes pass through areas of the liver and the spleen where tissue macrophages remove and degrade the complexes of complement, antibody, and antigen from the erythrocyte surface while leaving the erythrocyte unscathed (Figure 9.32). Although the vital function of erythrocytes is the transport of gases between lungs and other tissues, they have also acquired functions in the defense and protection of tissues, of which the disposal of immune complexes is a crucial one.

If immune complexes are not removed, they tend to enlarge by aggregation and then to precipitate at the basement membrane of small blood vessels, most notably those of the kidney glomeruli, where blood is filtered to form urine and is under particularly high pressure. Immune complexes that pass through the basement membrane bind to CR1 receptors expressed by podocytes, specialized epithelial cells that cover the capillaries. Deposition of immune complexes within the kidney occurs at some level all the time. Mesangial cells within the glomerulus are specialized in eliminating immune complexes and stimulating repair of the tissue damage they cause.

A feature of the autoimmune disease SLE is a level of immune complexes in the blood sufficient to cause massive deposition of antigen, antibody, and complement on the renal podocytes. These deposits damage the glomeruli, and kidney failure is the principal danger for patients with this disease. A similar deposition of immune complexes can also be a major problem for patients who have inherited deficiencies in the early components of the complement pathway and cannot tag their immune complexes with C4b or C3b. Such patients cannot clear immune complexes; these accumulate with successive antibody responses to infection and inflict increasing damage on the kidneys.

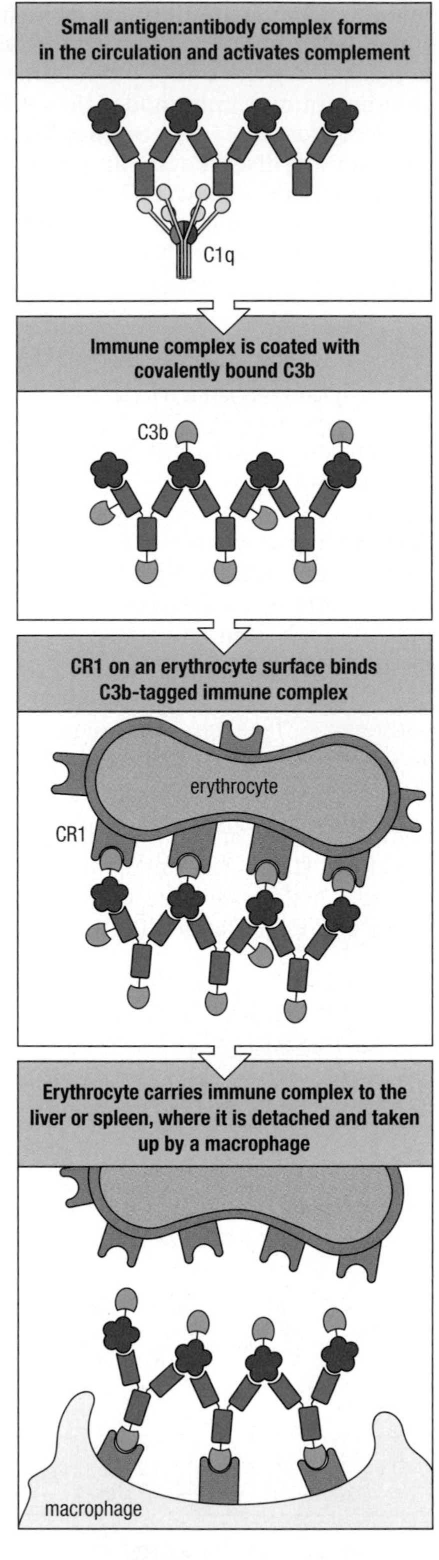

Figure 9.32 Erythrocyte CR1 helps to clear immune complexes from the circulation. Small soluble immune complexes bind to CR1 on erythrocytes, which transport them to the liver and spleen. Here they are transferred to the CR1 of macrophages and taken up for degradation.

9-21 Fcγ receptors enable effector cells to bind and be activated by IgG bound to pathogens

The power of the complement system is that it coats pathogens permanently with C3b fragments, delivering them to the complement receptors of phagocytes for opsonization, uptake, and elimination. High-affinity antibodies hold onto pathogens almost as tenaciously as C3b, and they too are ligands for receptors that are present on a range of cells of the immune system. These cells include neutrophils, eosinophils, basophils, mast cells, macrophages, FDCs, and NK cells. The receptors are diverse and have specificity for different antibody isotypes, but they all bind to the Fc region and are known generally as Fc receptors. These Fc receptors of hematopoietic cells are functionally and structurally distinct from the FcRn of endothelial cells. Whereas FcRn has an MHC class I-like structure and transports antibodies across epithelium, the hematopoietic Fc receptors are made up of two or three immunoglobulin-like domains, and they deliver activating or inhibitory signals that induce a response in the cells that express them.

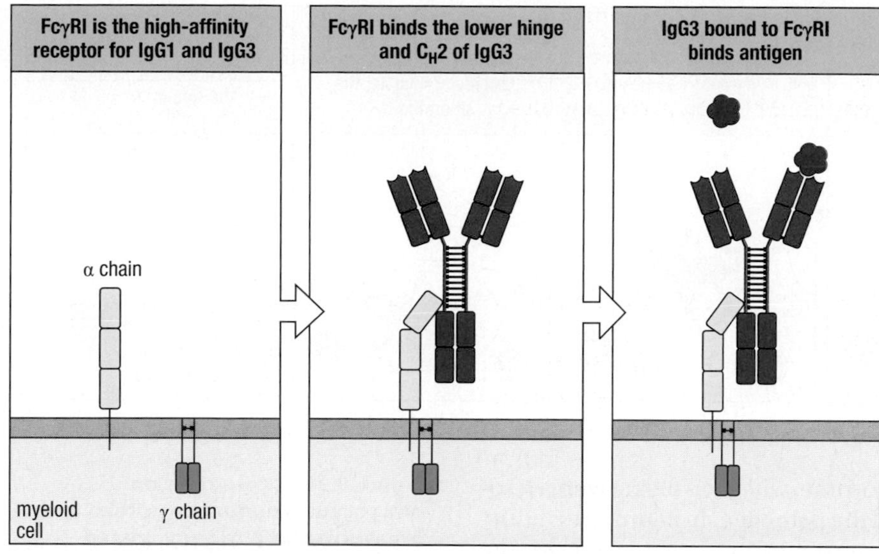
FcγRI is the high-affinity receptor for IgG1 and IgG3
FcγRI binds the lower hinge and C_H2 of IgG3
IgG3 bound to FcγRI binds antigen
α chain
myeloid cell
γ chain

Figure 9.33 FcγRI on myeloid cells is an Fc receptor that binds with high affinity to IgG1 and IgG3. The IgG-binding function of FcγRI is a property of the three extracellular, immunoglobulin-like domains of the receptor α chain. The signaling function of the receptor is the property of the γ chain, which forms a homodimer (left panel). Of the Fc receptors for IgG, only FcγRI can bind to IgG in the absence of antigen, as shown here for IgG3 (center panel). This enables the IgG bound to FcγRI at the surfaces of macrophages, dendritic cells, and neutrophils to trap pathogens and target them for uptake and degradation.

FcγRI is an Fc receptor specific for IgG and is constitutively expressed by monocytes, macrophages, and dendritic cells. At sites of infection and inflammation, FcγRI expression is also induced in neutrophils and eosinophils. Expression of FcγRI is thus specific to myeloid cells. The α chain is anchored to the membrane and has three extracellular immunoglobulin-like domains that are all necessary for binding to the C_H2 domain and the lower part of the hinge of IgG (Figure 9.33). Associated with the α chain is a dimer composed of a second type of polypeptide, the γ chain, which transduces activating signals and is closely related to the ζ chain of the T-cell receptor complex; like that chain it contains ITAM motifs. It is quite different from the γ chain associated with some cytokine receptors, such as the IL-2 receptor (see Section 8-7). FcγRI binds with different affinities to the four IgG subclasses. The hierarchy of binding—IgG3 > IgG1 > IgG4—reflects structural differences in the hinge and the C_H2 domain of the various IgG subclasses (see Figure 4.33, p.106).

The major function of FcγRI is to facilitate the uptake and degradation of pathogens by phagocytes and professional antigen-presenting cells. In Chapter 2 we saw how a variety of phagocytic and signaling receptors, including complement receptors, can enhance these processes during the innate immune response. In the adaptive immune response, antibodies made against surface antigens will coat the pathogen with their Fc regions pointing outward and free to bind to the FcγRI expressed on myeloid cell surfaces. On contact with a phagocyte, multiple ligand–receptor interactions are made, which produce a stable interaction and the clustering of receptors that is required to initiate intracellular signaling and phagocytosis (Figure 9.34). The necessity for antigen cross-linking of the bound IgG before a signal can be produced is appreciated from the behavior of IgG3 and IgG1. These subclasses have such high affinities for FcγRI that individual IgG molecules bind to the receptor in the absence of antigen and are held transiently at the cell surface. Such interactions do not send a signal to the cell, however, because that requires the cross-linking of complexes of FcγRI and IgG by antigen.

After the pathogen has been tethered to the phagocyte, interactions between antibody Fc regions and the FcγRI send signals that induce the phagocyte to engulf the antibody-coated pathogen (see Figure 9.34). The surface of the phagocyte gradually extends around the surface of the opsonized pathogen through cycles of binding and release between the Fc receptors of the phagocyte and the Fc regions projecting from the pathogen surface. Eventually the

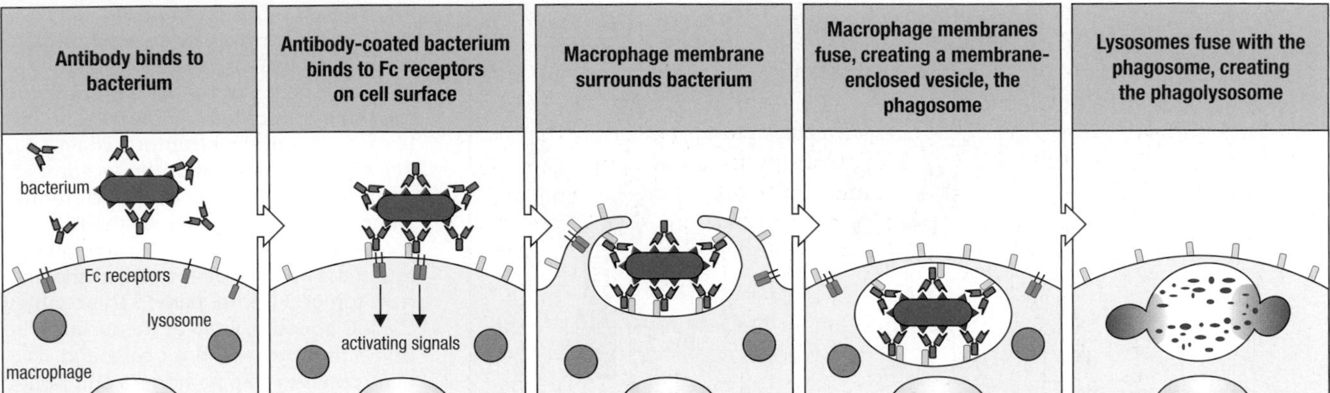

Antibody binds to bacterium
Antibody-coated bacterium binds to Fc receptors on cell surface
Macrophage membrane surrounds bacterium
Macrophage membranes fuse, creating a membrane-enclosed vesicle, the phagosome
Lysosomes fuse with the phagosome, creating the phagolysosome
bacterium
Fc receptors
lysosome
macrophage
activating signals

Figure 9.34 Fc receptors on phagocytes trigger the uptake and breakdown of antibody-coated pathogens. Specific IgG molecules bind their antigens on the surface of the pathogen, here a bacterium, and expose their Fc regions. The Fc receptors of phagocytic cells bind to the Fc regions, tethering the bacterium to the phagocyte's surface. Signals from the Fc receptors enhance both the phagocytosis and subsequent destruction of the bacterium in the lysozomes. The Fc receptors are seen to provide mechanisms for eliminating pathogens that have been opsonized with a coating of specific antibody. These Fc receptor-mediated mechanisms often work together with the complement receptors (CR1 and CR2 receptors) that recognize the C3 fragments that become attached to pathogens as a consequence of the classical pathway of complement activation (see Figure 9.28).

pathogen is completely engulfed in a phagosome, which on fusing with a lysosome leads to the death and degradation of the pathogen. In addition to reducing pathogen load, antibody-mediated opsonization also increases the efficiency with which antigens are processed and presented to pathogen-specific T cells.

As an effector mechanism of adaptive immunity, antibody-mediated opsonization enhances the phagocytic mechanisms of innate immunity (see Section 2-6) by increasing the speed at which pathogens are detected and devoured. A coating of IgG gives different types of microorganism a more uniform appearance, which enables Fc-receptor-bearing phagocytes to deal with different pathogens by using the same ligand—the Fc region of IgG—and the same receptor, FcγRI.

9-22 A variety of low-affinity Fc receptors are IgG-specific

Complementing FcγRI, the high-affinity Fc receptor, are two other classes of Fcγ receptor, namely FcγRII and FcγRIII, which bind IgG with much lower affinity. The structural basis for the difference in binding strengths is the presence of three Ig-like domains in FcγRI and only two in FcγRII and FcγRIII (Figure 9.35). As a consequence of their weaker binding to IgG, FcγRII or FcγRIII will only form stable interactions with two or more IgG molecules that have been cross-linked with antigen. This makes effector cells that carry these receptors more sensitive to the presence of antigen and also means that Fc receptors are not occupied by IgG that does not recognize the infecting pathogen and prevents the access of IgG that does.

Receptor	FcγRI	FcγRIIA	FcγRIIB2	FcγRIIB1	FcγRIIIA	FcγRIIIB
	CD64	CD32			CD16	
Structure	α 72 kDa; γ	α 40 kDa; γ-like domain	ITIM	ITIM	α 70 kDa; γ or ζ	α 50 kDa; γ or ζ
IgG subclass specificity	3>1>4>>2	R131: 3>1>>2,4 H131: 3>1,2>>4	3>1>4>>2	3>1>4>>2	1,3>>2,4	1,3>>2,4
Relative binding strength to IgG1	200	4	4	4	1	1
Effect of ligation	Activation	Activation	Inhibition	Inhibition	Activation	Activation

Figure 9.35 The family of human Fc receptors that bind to IgG. The subunit structure, the relative binding strengths for the different IgG isotypes, and the activating or inhibitory function of the Fcγ receptors are shown here. Each column corresponds to the products of a different gene. Thus the three receptors in the CD32 subfamily are encoded by different genes, and for FcγRIIA there are two common allotypes that differ in their specificity for IgG2 and their associations with disease. FcγRIII, the only Fc receptor expressed by NK cells, is encoded by two genes, each having two major alleles. These four different forms of FcγRIII differ in their specificity and avidity for the different IgG isotypes.

The FcγRII class comprises one activating and two inhibitory receptors. The activating receptor, **FcγRIIA**, promotes the uptake and destruction of pathogens by a wide range of myeloid cells. The two inhibitory receptors have similar properties but are expressed by different cell types. **FcγRIIB2** is expressed on macrophages, neutrophils, and eosinophils. By antagonizing the actions of activating Fc receptors in these cells, FcγRIIB2 controls their inflammatory responses. **FcγRIIB1** has a similar role in mast cells and B cells. These inhibitory receptors bear **immunoreceptor tyrosine-based inhibitory motifs (ITIMs)** in their cytoplasmic tails, which associate with intracellular proteins that develop the inhibitory signals.

IgG2 is the second most abundant IgG isotype and the one that is enriched in the antibodies made against bacterial polysaccharides. IgG2 cannot activate complement and binds only to FcγRIIA. Two FcγRIIA variants are common in the population, but only one of them (called H131) binds effectively to pathogens coated with IgG2 (see Figure 9.35). Individuals homozygous for the other variant (R131) comprise around 25% of African and European populations, and their neutrophils are less effective at phagocytosing and killing IgG2-coated bacteria. Homozygosity for R131 is also associated with an increased risk of fulminant meningococcal disease or septic shock on infection with *Neisseria meningitidis*, indicating a role for IgG2 in protection against this bacterium. In Japan and other parts of East Asia, less than 4% of the population is homozygous for the R131 variant, indicating that in those populations there has been selection against this variant. Such differences between human populations in disease associations are common and reflect the differences in demographic history as well as the history of exposure to infectious disease.

9-23 An Fc receptor acts as an antigen receptor for NK cells

FcγRIII is an activating receptor for IgG and is the only Fc receptor expressed by NK cells. Although also expressed on macrophages, neutrophils, and eosinophils, the functions of FcγRIII have mainly been studied in the context of killing by NK cells. NK cells can recognize and kill human cells that have been coated with IgG1 or IgG3 antibodies specific for cell-surface components. This mechanism, called **antibody-dependent cell-mediated cytotoxicity (ADCC)**, is the way in which the therapeutic anti-CD20 monoclonal antibody rituximab eliminates some types of B-cell tumor. Rituximab binds to CD20 on the tumor cells with its Fab arms, and to FcγRIII on the NK cell with its Fc region. Although FcγRIII has a low affinity for Fc, the combined effect of many antibody molecules binding to both CD20 and FcγRIII is to bind the tumor cell tightly to the NK cell. Signals from FcγRIII then activate the NK cell to kill the tumor cell (Figure 9.36). When used therapeutically, rituximab-mediated ADCC also kills

Figure 9.36 The Fc receptor expressed by NK cells recognizes IgG-coated target cells and signals them to die. This mechanism is illustrated here by the clinical use of anti-CD20 monoclonal antibodies as a common and effective therapy for certain types of B-cell tumor. When the therapeutic antibodies bind to the CD20 antigen on tumor cells, the FcγRIII receptors on the NK cell bind to the Fc regions of the cell-bound IgG. Multiple interactions between IgG and FcγRIII molecules establish stable binding between the NK cell and its target. Signals from FcγRIII activate the NK cell to form a conjugate pair and a synapse with the tumor cell. By secreting the contents of its lytic granules onto the surface of the tumor cell, the NK cell condemns the tumor cell to die by apoptosis.

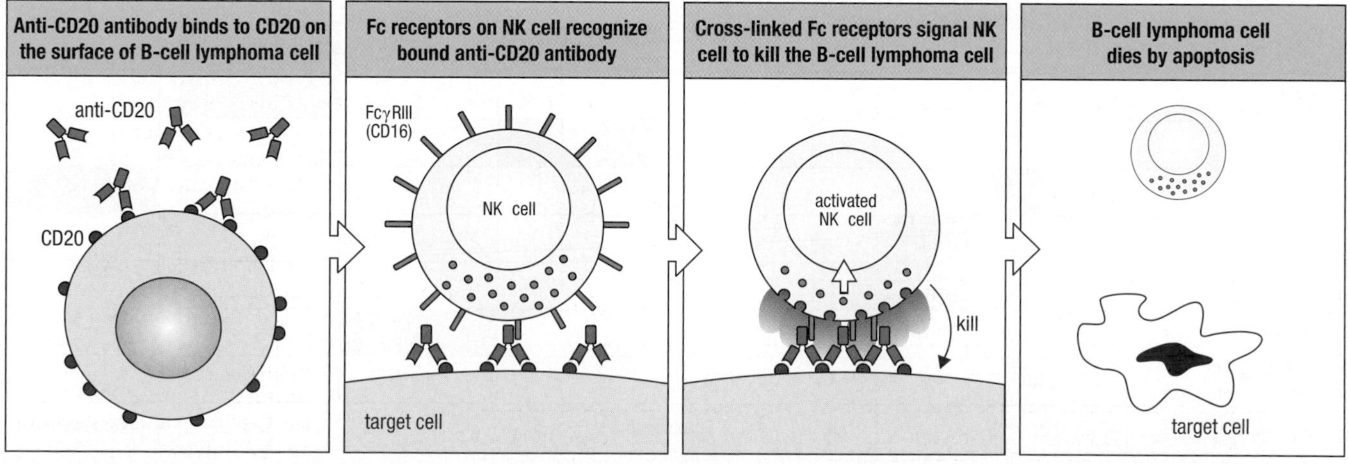
Anti-CD20 antibody binds to CD20 on the surface of B-cell lymphoma cell
anti-CD20
CD20
Fc receptors on NK cell recognize bound anti-CD20 antibody
FcγRIII (CD16)
NK cell
target cell
Cross-linked Fc receptors signal NK cell to kill the B-cell lymphoma cell
activated NK cell
kill
B-cell lymphoma cell dies by apoptosis
target cell

healthy B cells that express CD20, a disadvantage that is outweighed by the benefit of killing the tumor cells (see Section 4-6).

NK cells are major contributors to the innate immune response to infection, in which they use mechanisms other than ADCC to kill virus-infected cells (see Section 3-19). In the primary adaptive immune response, ADCC will only come into play as a mechanism of NK-cell killing once pathogen-specific IgG antibodies are available. In an influenza infection, for example, IgG specific for the viral membrane glycoproteins can bind to influenza-infected cells and recruit NK cells to kill them.

9-24 The Fc receptor for monomeric IgA belongs to a different family than the Fc receptors for IgG and IgE

Cells of the myeloid lineage also express an Fc receptor that binds to the monomeric form of IgA present in blood and lymph. This medium-affinity receptor, called FcαRI, has an α chain containing two immunoglobulin-like domains and it relies on the common γ chain for signal transduction. The main function of FcαRI is the same as that of FcγRI, with the difference that it facilitates the phagocytosis of pathogens coated with IgA, not IgG. Despite the functional similarities of these two receptors, they have different evolutionary histories. The Fc receptors for IgG and IgE are encoded by a family of genes on human chromosome 1, in which all the members derive from a common ancestral gene (Figure 9.37). In contrast, the gene for FcαRI is on chromosome 19, where

Figure 9.37 Comparison of structure and cellular distribution of the Fc receptors for IgG, IgE, and IgA. The Fc receptors for IgG and IgE are related in structure and encoded by closely linked genes. Although it has a similar structure comprising immunoglobulin-like domains, the IgA receptor FcαRI is only distantly related to the IgE and IgG receptors and is encoded by a gene on a different chromosome. Constitutive expression of an Fc receptor by a cell type is denoted by dark pink shading of the box and +; inducible expression is denoted by lighter pink shading and (+). White boxes denote no expression. The binding specificities of the Fcγ receptors are shown in Figure 9.35. As well as the Fc-binding chain and the γ signaling chain, FcεRI has an additional transmembrane β chain, which is involved in signaling.

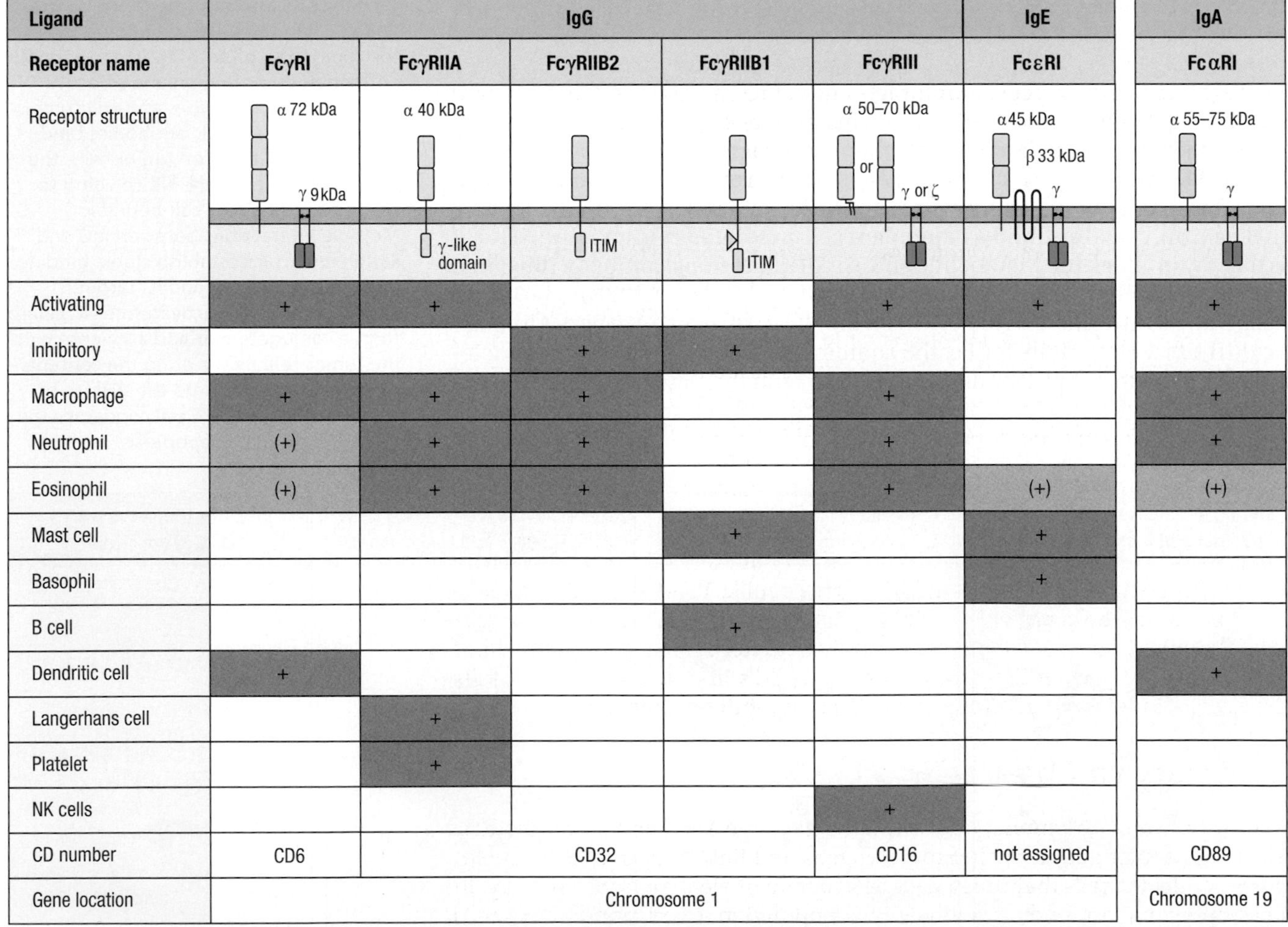

Ligand	IgG					IgE	IgA
Receptor name	FcγRI	FcγRIIA	FcγRIIB2	FcγRIIB1	FcγRIII	FcεRI	FcαRI
Receptor structure	α 72 kDa, γ 9kDa	α 40 kDa, γ-like domain	ITIM	ITIM	α 50–70 kDa, or, γ or ζ	α45 kDa, β 33 kDa, γ	α 55–75 kDa, γ
Activating	+	+			+	+	+
Inhibitory			+	+			
Macrophage	+	+	+		+		+
Neutrophil	(+)	+	+		+		+
Eosinophil	(+)	+	+		+	(+)	(+)
Mast cell				+		+	
Basophil						+	
B cell				+			
Dendritic cell	+						+
Langerhans cell		+					
Platelet		+					
NK cells					+		
CD number	CD6	CD32			CD16	not assigned	CD89
Gene location	Chromosome 1						Chromosome 19

it is part of a large, densely packed complex of gene families that encode receptors made up of immunoglobulin-like domains and expressed on myeloid cells or NK cells. This gene complex is called the leukocyte-receptor complex or LRC.

Fc receptors are not restricted to the adaptive immune response and the binding of immunoglobulin. We saw in Chapter 3 how C-reactive protein, an acute-phase protein of the innate immune response, binds to phosphorylcholine on bacterial surfaces and activates complement by the classical pathway (see Figure 3.28, p.66). C-reactive protein also binds to FcγRI and FcγRII, and by this mechanism it delivers *S. pneumoniae* for uptake and degradation by phagocytes. That these modern Fc receptors have the dual function of binding immunoglobulin and C-reactive protein indicates that the common ancestor of the IgG-specific and IgE-specific Fc receptors was a receptor for the C-reactive protein of innate immunity.

Summary

Secreted antibodies are the only effector molecules produced by B cells; their principal function is to serve as adaptor molecules that neutralize pathogens and deliver them to effector cells for destruction. Antibodies bind antigens with their variable Fab arms, whereas the Fc regions have conserved binding sites that initiate complement activation and interaction with Fc receptors on various types of effector cell. The isotype of an antibody and its synthesis as monomers or oligomers determine where in the human body an antibody will seek out antigens and the type of effector functions it can engage. During the course of an infection, the first pathogen-specific antibody to be made and distributed through the blood and tissues is pentameric IgM, which fixes complement on the surface of a pathogen and delivers it to the complement receptors of phagocytes. With maturation of the adaptive immune response, the pathogen can also be coated with high-affinity IgG and monomeric IgA and delivered to the Fc receptors of phagocytes that are specific for these antibody isotypes. The four subclasses of IgG, which are all transported from blood to the extracellular fluid in tissues by FcRn, differ in their capacities to activate complement, engage Fc receptors, and respond to different types of antigen.

Dimeric IgA is made by lymphoid tissue lining mucosal surfaces and is transported across the epithelium by the polymeric immunoglobulin receptor. In this way, the mucosal surfaces of the respiratory, gastrointestinal, and reproductive tracts have a continual supply of IgA, which controls the populations of microorganisms that inhabit and infect those tissues. The Fc receptor for IgE on mast cells has such high affinity that IgE is bound in the absence of its antigen. In the presence of antigen, the mast cells, which underlie the skin and mucosal surfaces, stimulate violent muscular contractions that expel pathogens from the body through coughing, sneezing, vomiting, or diarrhea.

During pregnancy, FcRn transports IgG from the mother's blood to the fetal circulation. After birth, the gastrointestinal tract of the infant is supplied with dimeric IgA, an important constituent of breast milk. These supplies of maternal IgG and IgA help protect the child against infections to which the mother has become immune. As the child's supply of maternal IgG and IgA declines during the first year of life, and before the child's immune system has fully developed, there is a period of enhanced susceptibility to infectious disease.

Summary to Chapter 9

The response of B lymphocytes to infection is the secretion of antibodies. These molecular adaptors bind to pathogens and link them to effector molecules or effector cells that cause their destruction or ejection from the body. In developing an antibody response, the population of responding B cells combines a quick but suboptimal response in the short term with a more powerful response that takes time to develop. Representing the short-term strategies are B-1 cells, which do not require T-cell help, and IgM antibodies, which have low-affinity antigen-binding sites. In contrast, B-cell activation driven by cognate T-cell help, with resultant isotype switching and somatic hypermutation, is the long-term strategy that provides effective protection from subsequent infection by the pathogen. One function of the isotypic diversification of immunoglobulins is to provide antibody responses in different compartments of the human body. IgM, IgG, and monomeric IgA work in the blood, lymph, and connective tissues, providing antibody responses to infections within the body's tissues. IgE also works in connective tissue, where it is bound tightly by the IgE receptor on the surface of mast cells. In contrast, dimeric IgA is transported outside the body's tissues to mucosal surfaces, such as the luminal side of the gut wall, where it controls the microorganisms that colonize these surfaces. A second function of antibody isotypes is to recruit different effector functions into the immune response: IgG and IgA fix complement and target pathogens to destruction by phagocytes, whereas IgE leads to mast-cell degranulation and the violent physical reactions that expel parasites and microorganisms from the respiratory and intestinal tracts.

Questions

9–1 Cross-linking of immunoglobulin by antigen is essential but not always sufficient to initiate the signal cascade for B-cell activation. Involvement of the B-cell co-receptor is also necessary for the full activation and differentiation of naive B cells. Describe this receptor and its ligand and explain how it facilitates B-cell activation.

9–2 All of the following are associated with B-cell activation except _____.

a. close association with complement receptor 1 (CR1), CD19, and CD81
b. clustering of surface immunoglobulin
c. activation of cytoplasmic protein tyrosine kinases
d. participation of cytoplasmic tails of Igα and Igβ
e. ITAM phosphorylation.

9–3

A. Explain how a B cell that has recently recognized its specific antigen becomes trapped in the lymph node at the boundary of the T-cell and B-cell zones.
B. Why is this necessary with respect to B-cell activation?

9–4 Follicular dendritic cells produce _____, which are needed for rapid B-cell proliferation and differentiation into centroblasts.

a. CD40 ligand and IL-4
b. TNF-α, LT-α, and LT-β
c. CCL21 and CCL19
d. CD44, CD38, and CD77
e. BAFF, IL-15, IL-6, and 8D6.

9–5 Individuals born without a thymus _____.

a. are unable to make antibodies
b. have similar B-cell function to someone who possesses a thymus
c. have elevated levels of IgG compared with IgM
d. have abnormally reduced numbers of B cells
e. do not mediate effective isotype switching in their B cells.

9–6 Which of the following are properties of plasma cells that are not shared with naive B cells? (Select all that apply.)

a. no surface immunoglobulin expression
b. elevated levels of MHC class II expression on the cell surface
c. high rate of immunoglobulin secretion
d. can be induced to switch isotype
e. can be induced to undergo cellular proliferation.

9–7 Which of the following is incorrect regarding the Fc receptor FcαRI?

a. It is a medium-affinity receptor.
b. It requires the common γ chain to mediate signaling.
c. It binds to dimeric IgA.
d. It is encoded on a different chromosome from the genes encoding the Fc receptors for IgG and IgE.
e. It mediates the phagocytosis of pathogens coated with IgA.

9–8 Match the receptor in column A with its description in Column B.

Column A	Column B
a. poly-Ig receptor	1. transports IgG out of the bloodstream to extracellular spaces in tissues
b. FcγRIII (CD16)	2. facilitates antibody-dependent cell-mediated cytotoxicity in NK cells
c. FcεRI	3. cross-links antibody:antigen complexes on mast cells followed by degranulation
d. FcRn	4. binds to dimeric IgA and facilitates its transcytosis
e. B-cell co-receptor	5. CD21/CD19/CD81
f. FcγRII-B1 (CD32)	6. inhibition of B-cell activation

9–9 Mucosal epithelia of the gastrointestinal tract, eyes, nose, throat, the respiratory, urinary, and genital tracts, and the mammary glands are protected by

a. IgG
b. IgM
c. IgE
d. monomeric IgA
e. dimeric IgA.

9–10 Which of the following accounts for the relatively low levels of IgE in the circulation?

a. Isotype switching to IgE is a relatively rare event in germinal centers.
b. IgE is rapidly degraded by serum proteases.
c. FcRn binds with high affinity to IgE and ushers it into extracellular fluids of connective tissues.
d. IgE is rapidly phagocytosed by neutrophils whether or not it is bound to antigen.
e. IgE binds with high affinity to mast cells, basophils, and activated eosinophils in the absence of antigen.

9–11 The poly-Ig receptor on the basolateral surface of epithelial cells binds to _____ via the _____. (Select all that apply.)

a. dimeric IgA; J chain
b. monomeric IgA; J chain
c. monomeric IgA; CH2 domain
d. IgM; J chain
e. IgG; CH3 domain.

9–12 Which of the following are examples of passive transfer of immunity? (Select all that apply.)

a. antibody production after vaccination
b. transplacental acquisition of IgG during fetal development
c. provision of IgA through breast milk
d. antibody production after an influenza infection
e. intravenous immunoglobulin therapy for immunocompromised individuals
f. receiving antivenom after being bitten by a poisonous snake.

9–13 Erythrocytes are equipped to facilitate the removal of small immune complexes from the circulation using receptors that bind to _____.

a. IgG
b. C1q
c. C3b
d. CR2
e. F-protein.

9–14

A. Explain how the receptor FcRn transports IgG antibodies across cellular barriers, and specify the type of cell barrier involved.
B. What is the final location of transported material?

9–15 Anthony Hoffnagle was well until 6 months of age, when he was hospitalized for pneumonia. In the following year he was hospitalized five times for additional episodes of pneumonia, septic arthritis, and febrile convulsion. Yesterday, Anthony was diagnosed with pneumonia caused by Pneumocystis jirovecii, and consultation with an immunologist was requested. Anthony's tests revealed neutropenia, normal numbers of B cells and T cells, slightly elevated IgM, but a marked decrease in IgG and IgA compared with normal. Autoantibodies against neutrophils were not detected. Liver function tests were normal. Bone marrow aspiration indicated severe maturational arrest of the myeloid lineage at the promyelocyte–myelocyte stage. A diagnosis of X-linked hyper-IgM syndrome (XHIGM) was made. Anthony's parents were informed that their son would require long-term treatment with intravenous immunoglobulin (IVIG), prophylactic antibiotics, and periodic injection of granulocyte colony-stimulating factor (G-CSF) for episodes of neutropenia. Genetic analysis of the gene encoding _____ revealed a frameshift and stop codon mutation causing aberrant transcription of this gene.

a. CD40 ligand
b. CD3
c. CD19
d. RAG1
e. CD81.

Commensal bacteria in the small intestine of the human gut.

Chapter 10

Preventing Infection at Mucosal Surfaces

Most infectious diseases suffered by humans are caused by pathogens much smaller than a human cell. For these microbes, the human body constitutes a vast resource-rich environment in which to live and reproduce. In facing such threats, the body deploys a variety of defense mechanisms that have accumulated over hundreds of millions of years of invertebrate and vertebrate evolution. In considering mechanisms of innate immunity in Chapters 2 and 3 and of adaptive immunity in Chapters 4–11, we principally used the example of a bacterial pathogen that enters the body through a skin wound, causing an innate immune response in the infected tissue that then leads to an adaptive immune response in the draining lymph node. The merits of this example are that it is simple and involves a tissue for which we have all observed the effects of wounds, infection, and inflammation. Until recently, these were the only responses studied by most immunologists, who usually administered their experimental antigens by subcutaneous injection. But in the real world, only a fraction of human infections are caused by pathogens that enter the body's tissues by passage through the skin. Many more infections, including all of those caused by viruses, make their entry by passage through one of the mucosal surfaces. Although the immune response to infection of mucosal tissue has strategies and principles in common with those directed at infections of skin and connective tissue, there are important differences, both in the cells and molecules involved, as well as the ways in which they are used. Appreciation of the extent of these differences has led to the concept that the human immune system actually consists of two semi-autonomous parts: the systemic immune system, which defends against pathogens penetrating the skin, and the mucosal immune system, which defends against pathogens breaching mucosal surfaces. This chapter focuses on mucosal immunity and how it differs from systemic immunity.

10-1 The communication functions of mucosal surfaces render them vulnerable to infection

Mucosal surfaces or the **mucosae** (singular **mucosa**) are found throughout much of the body, except the limbs, but they are predominantly out of sight. Continually bathing the mucosae is a layer of the thick, viscous fluid called **mucus**, which is secreted by the mucosae and gives them their name. Mucus contains glycoproteins, proteoglycans, peptides, and enzymes that protect the epithelial cells from damage and help to limit infection. Mucosal epithelia line the gastrointestinal, respiratory, and urogenital tracts, and are also present in the exocrine glands associated with these organs: the pancreas, the conjunctivae and lachrymal glands of the eye, the salivary glands, and the mammary

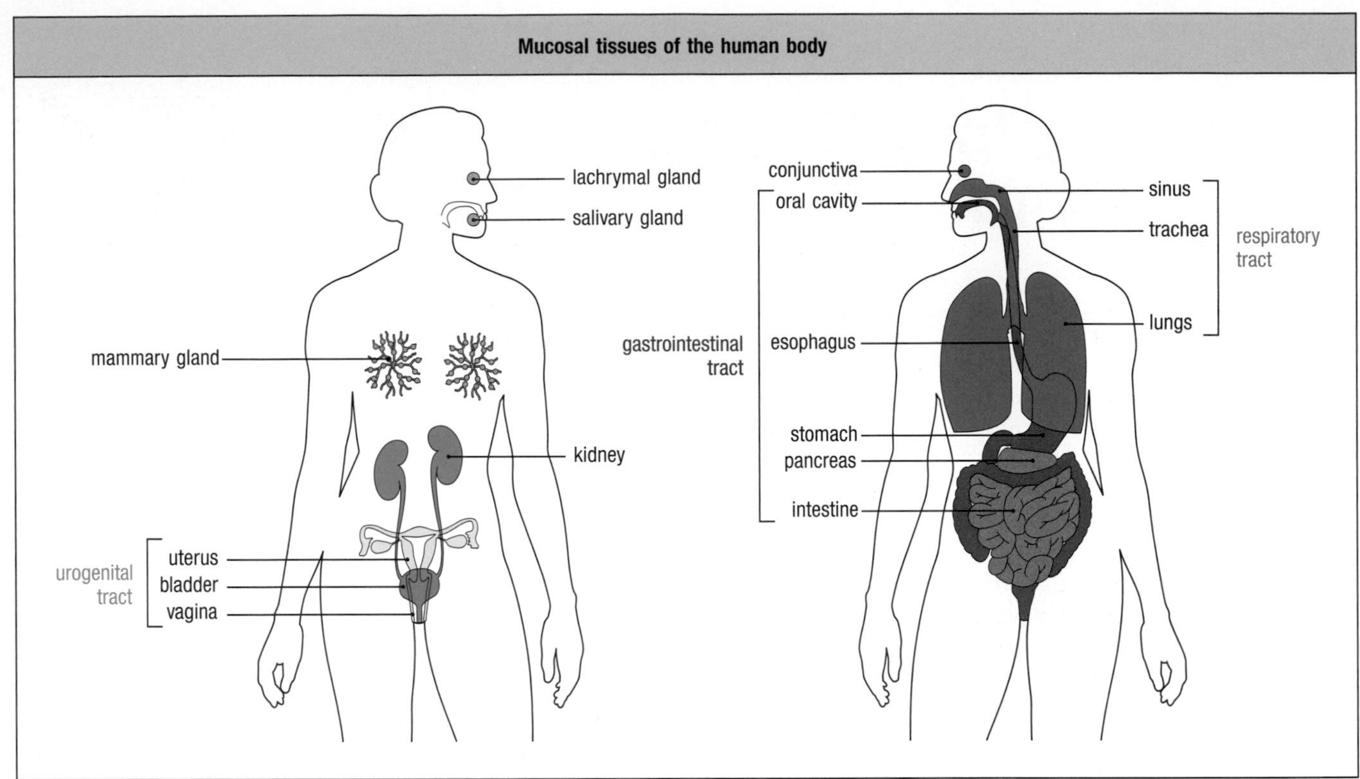

Figure 10.1 Distribution of mucosal tissues. This diagram of a woman shows the mucosal tissues. The mammary glands are a mucosal tissue only after pregnancy, when the breast is lactating. Red, gastrointestinal tract; blue, respiratory tract; green, urinary tract; yellow, genital tract; orange, secretory glands.

glands of the lactating breast (Figure 10.1). These tissues are all sites of communication, where material and information are passed between the body and its environment. Because of their physiological functions of gas exchange (lungs), food absorption (gut), sensory activity (eyes, nose, mouth, and throat), and reproduction (uterus, vagina, and breast), the mucosal surfaces are by necessity dynamic, thin, permeable barriers to the interior of the body. These properties make the mucosal tissues particularly vulnerable to subversion and breach by pathogens. This fragility, combined with the vital functions of mucosae, has driven the evolution of specialized mechanisms for their defense.

The combined area of a person's mucosal surfaces is vastly greater than that of the skin: the small intestine alone has a surface area 200 times that of the skin. Reflecting this difference, three-quarters of the body's lymphocytes and plasma cells are to be found in secondary lymphoid tissues serving mucosal surfaces. A similar proportion of all the antibodies made by the body is secreted at mucosal services as the dimeric form of IgA, also known as **secretory IgA** or **SIgA** (see Chapter 9). A distinctive feature of the gut mucosa is its constant contact with the large populations of **commensal** microorganisms that inhabit the lumen of the gut and constitute the gut microbiota. Other major contents of the gut are the proteins, carbohydrates, lipids, and nucleic acids derived from the plants and animals that contribute to our diet. In this situation, the major challenge is to make immune responses that eliminate pathogenic microorganisms and restrict the growth and location of commensal microorganisms, but do not interfere with our food and nutrition. As most research on mucosal immunity has been on the gut, this will provide our principal example of a mucosal tissue, but first we will examine the constituents and properties of the mucus.

10-2 Mucins are gigantic glycoproteins that endow the mucus with the properties to protect epithelial surfaces

In every mucosal tissue, a layer of epithelial cells joined by tight junctions separates the outside environment from the inside of the body. The epithelial layer provides a formidable barrier that prevents commensal and pathogenic organisms from gaining access to the internal issues. Adding to this defense is the mucus, which prevents microorganisms and other environmental material, such as smoke and smog particles, from gaining access to the epithelium. The molecular basis for the viscosity and protective properties of mucus is a family of glycoproteins called **mucins** that are secreted by the epithelium. These proteins are huge, their polypeptide chains reaching lengths of more than 10,000 amino acids, but they are constructed from simple sequence motifs repeated many times over. The motifs are rich in serine and threonine residues that are glycosylated with short, negatively charged glycans. This carbohydrate comprises more than 70% of the weight of the mucin glycoprotein. The extensive glycosylation forces the mucin polypeptides into extended conformations. Globular domains at the ends of the polypeptides contain cysteine residues that make disulfide bonds between the stretched-out polypeptides, forming polymers and molecular networks that reach sizes greater than 1 million daltons (1 MDa) (Figure 10.2). The intertwining of these gigantic proteins is what makes mucus viscous, so that it physically impedes the movement of microorganisms and particles. The extensive glycosylation of mucins causes mucus to be heavily hydrated and thus able to protect epithelial surfaces by retaining water and preventing dehydration. A major constituent of the mucin glycans is sialic acid, which gives mucins a polyanionic surface. Through this they can bind the positively charged soluble effector molecules of innate immunity, such as defensins and other antimicrobial peptides, and of adaptive immunity, notably secretory IgA. Bacteria negotiating their way through mucus can thus be trapped by IgA and killed by defensins. Mucosal epithelia are dynamic tissues in which the epithelial cell layer turns over every 2 days or so, and mucus with its content of microorganisms is continuously being expelled from the body.

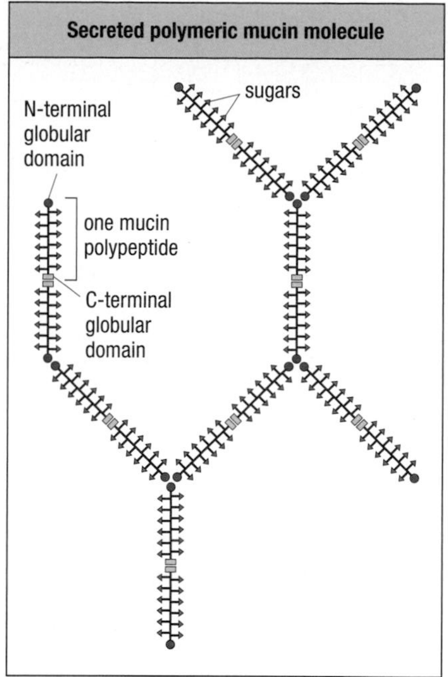

Figure 10.2 The structure of mucins gives mucus its characteristic protective properties. Mucins secreted by goblet cells are long polypeptides densely arrayed with short carbohydrates attached to serine and threonine residues. Through cysteine residues in the globular domains at the N- and C-termini, the mucin polypeptides become cross-linked into the gigantic extended polymeric networks that form the mucus. This unusual structure gives mucus its viscosity, which lubricates mucosal surfaces and prevents the approach of commensal and pathogenic microorganisms. The free cysteine residues of the mucin polypeptides are used to form covalent bonds with molecules of secreted IgA and defensins. The former are used to bind microorganisms approaching a mucosal surface; the latter are used to kill them.

The viscoelastic properties of mucus vary with the mucosal tissue and its state of health or disease. This is achieved by varying the mucin polypeptides that are incorporated into the mucus and the extent of their cross-linking. In the human genome, seven genes encode secreted mucin polypeptides and are expressed in different mucosal tissues; an additional 13 genes encode mucin molecules that are membrane glycoproteins (Figure 10.3). These are expressed on the surface of epithelial cells and are not cross-linked like the secreted mucins. Although not so well characterized as the secretory mucins, these membrane mucins are believed to form a mucus-like environment at the epithelial cell surface that has similar protective properties. Because they are so much bigger than other components of the plasma membrane, the membrane mucins stand out from the cell surface, giving them the potential to trap and kill approaching microorganisms before they can interact with other components at the surface.

10-3 Commensal microorganisms assist the gut in digesting food and maintaining health

The gastrointestinal tract extends from the mouth to the anus and is about 9 meters in length in an adult human being (Figure 10.4). Its physiological purpose is to take in food and process it into nutrients that are absorbed by the body and into waste that is eliminated from the body. Alimentation means giving nourishment; hence the older alternative name of alimentary canal for the

Mucin polypeptide	Gene location (chromosome)	Mode of action	Tissues where expressed
MUC2	11	Secreted	Small intestine, colon
MUC5A	11	Secreted	Airways, stomach
MUC5B	11	Secreted	Airways, salivary glands
MUC6	11	Secreted	Stomach, small intestine, gall bladder
MUC8	12	Secreted	Airways
MUC19	12	Secreted	Salivary glands, trachea
MUC7	4	Secreted	Salivary glands
MUC9	1	Secreted and membrane-bound	Fallopian tubes
MUC1	1	Membrane-bound	Breast, pancreas
MUC16	19	Membrane-bound	Ovarian epithelium
MUC20	3	Membrane-bound	Placenta, colon, lung, prostate
MUC4	3	Membrane-bound	Airways, colon
MUC3A	7	Membrane-bound	Small intestine, gall bladder, colon
MUC3B	7	Membrane-bound	Small intestine, gall bladder, colon
MUC17	7	Membrane-bound	Stomach, small intestine, colon
MUC11	7	Membrane-bound	Colon, airwaves, reproductive tract
MUC12	7	Membrane-bound	Colon, pancreas, prostate, uterus
MUC13	3	Membrane-bound	Trachea, small intestine, colon
MUC15	11	Membrane-bound	Airways, small intestine, prostate, colon
MUC18	4	Membrane-bound	Lung, breast

Figure 10.3 Mucosal tissues differ in the mucins they produce. In the human genome are genes encoding 20 mucin polypeptides. Six of these encode secreted mucins, 12 encode membrane-bound mucins and 1 encodes both secreted and membrane-bound mucins. Shown are the mucosal tissues in which the mucins are expressed and the chromosomal location of their genes.

gastrointestinal tract. Segments of the gastrointestinal tract serve different specialized functions and are populated to different extents by commensal bacteria. In the mouth, food is physically broken down by chewing in an environment populated by more than 750 species of bacteria. In the stomach, acid and enzymes are used to chemically degrade the masticated food in an environment that is relatively unfriendly for microbes. Here, the main function of the mucus is to protect and buffer the epithelium from the corrosive effects of hydrochloric acid secreted by the stomach. Enzymatic degradation continues the digestive process in the small intestine (the duodenum, jejunum, and ileum), which is the major site for the absorption of nutrients. In the large intestine (the colon), waste is stored, compacted, and periodically eliminated. The cecum is a pouch-like structure that connects the small and large intestines.

As food travels along the gastrointestinal tract and becomes increasingly degraded, it passes through environments with increasing numbers of resident bacteria. Starting in the stomach at 1000 bacteria per milliliter of gut contents, numbers increase to 10^5 to 10^8 per milliliter in the small intestine and

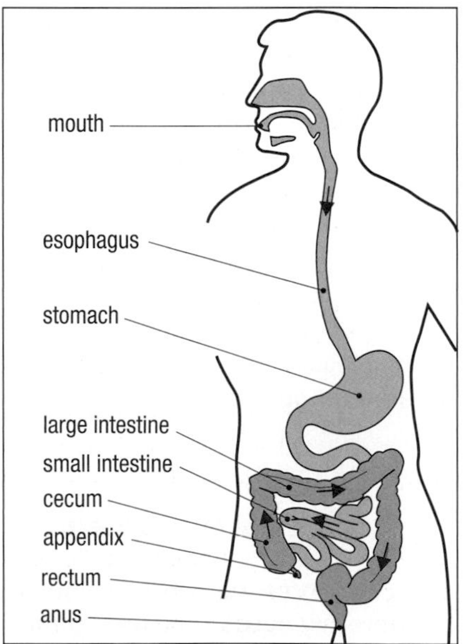

Figure 10.4 The human gastrointestinal tract.

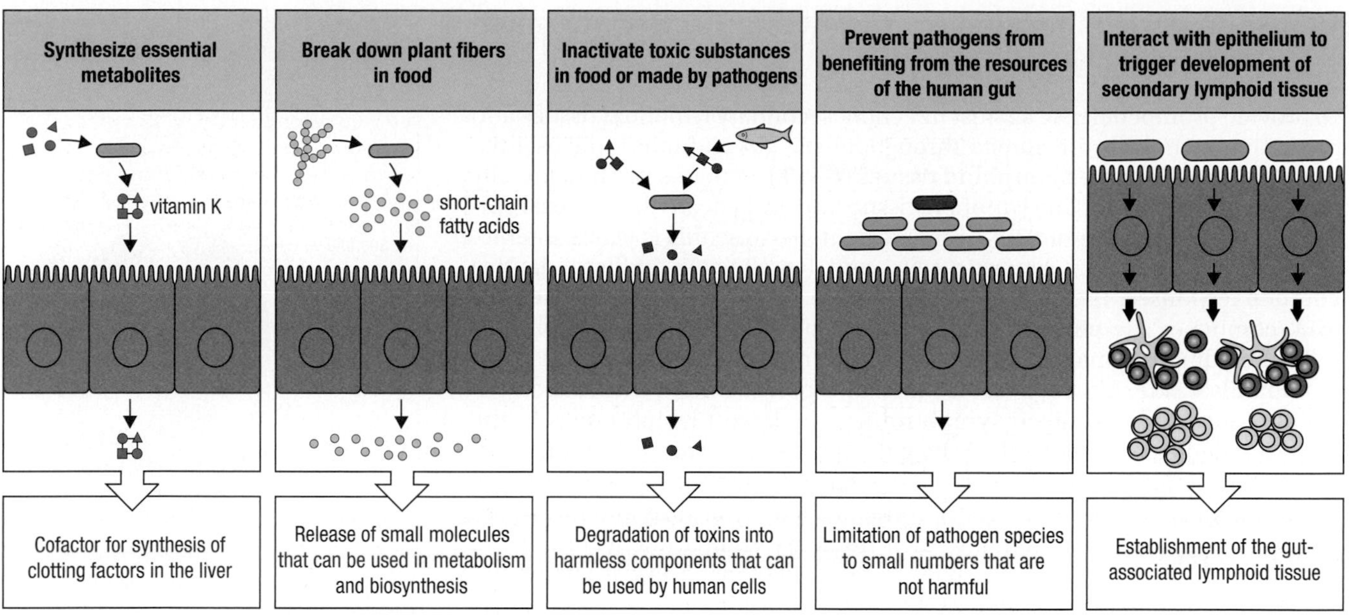
Synthesize essential metabolites
vitamin K
Cofactor for synthesis of clotting factors in the liver
Break down plant fibers in food
short-chain fatty acids
Release of small molecules that can be used in metabolism and biosynthesis
Inactivate toxic substances in food or made by pathogens
Degradation of toxins into harmless components that can be used by human cells
Prevent pathogens from benefiting from the resources of the human gut
Limitation of pathogen species to small numbers that are not harmful
Interact with epithelium to trigger development of secondary lymphoid tissue
Establishment of the gut-associated lymphoid tissue

Figure 10.5 Five ways in which the commensal gut microbiota benefit their human hosts.

reach 10^{12} per milliliter in the colon. Digestion is a highly dynamic process in which the flow from stomach to anus is driven by peristalsis in the intestines. The growth of the populations of resident commensal organisms is equally dynamic, and to contain this population at a manageable size, vast numbers of commensals are forced out of the human body each day.

Commensal microorganisms have co-evolved with their human hosts in a symbiotic relationship, which benefits the host in various ways (Figure 10.5). Bacteria provide metabolic building blocks that are essential for human health but cannot be made by human cells. One example is the menaquinone precursors used to make vitamin K, a cofactor in the synthesis of blood-clotting factors. Bacteria also increase the efficiency with which humans digest certain foods, by providing enzymes that convert plant fibers, which are indigestible by human enzymes, into energy-rich metabolites. Other microbial enzymes render toxic substances present in food or secreted by pathogens into innocuous derivatives. The presence of large, healthy populations of commensal microorganisms also prevents the emergence and proliferation of pathogenic variants by depriving them of food and space. In fact, the normal development of the gut lymphoid tissues depends on the presence of a healthy gut microbiota, compelling evidence for the symbiotic co-evolution of commensal species and the human immune system.

Most bacterial infections of gut tissue are caused by commensals, but relatively few bacterial groups are involved. Many potential pathogens belong to the facultatively anaerobic, Gram-negative phylum Proteobacteria, which includes *Salmonella*, *Shigella*, *Helicobacter*, and *Escherichia*. Pathogenic variants of these normally harmless bacteria arise as new genetic variants acquire properties called virulence factors that enable them to leave the gut lumen, breach the gut epithelium, and invade the underlying lamina propria.

A common childhood viral infection of the epithelial lining of the small intestine is caused by rotavirus, a double-stranded RNA virus. The infection causes an acute diarrhea, during which large numbers of stable and infectious virus particles are shed in the feces. Worldwide, 500,000 children die each year from rotavirus infection. In addition to bacteria and viruses, a spectrum of parasitic diseases are caused by helminth worms, as well as protozoans and other microorganisms that inhabit the gastrointestinal tract.

10-4 The gastrointestinal tract is invested with distinctive secondary lymphoid tissues

To provide prompt defense against infection, secondary lymphoid tissues and immune-system cells are spread throughout the gut and other mucosal tissues. The **gut-associated lymphoid tissues** (**GALT**) comprise two functionally distinct compartments. The lymphoid tissue directly beneath the mucosal epithelium is called the **inductive compartment**, because this is where interactions between antigen, dendritic cells, and lymphocytes induce adaptive immune responses. The underlying connective tissue, called the **lamina propria**, comprises the **effector compartment**, because this is where effector cells, including plasma cells, effector T cells, macrophages, mast cells, and eosinophils reside. Although not technically a part of the gut-associated lymphoid tissue, the **mesenteric lymph nodes**, the largest lymph nodes in the body, are dedicated to defending the gut. They form a chain within the mesentery, the membrane of connective tissue that holds the gut in place. Although the gut-associated lymphoid tissues come in a variety of sizes and forms, the microanatomy and organization of their inductive compartments into B-cell and T-cell zones are generally similar to those of other secondary lymphoid tissues. The secondary lymphoid tissues within the gut mucosa continuously sample and monitor the contents of the gut lumen, allowing adaptive immune responses to be quickly made against the gut microbiota and implemented locally before any prospective pathogen can invade the gut tissue. In contrast, a mesenteric lymph node can respond to infection only after the pathogen has invaded gut tissue and is then brought to the node in the draining lymph. This latter mechanism is like that used to respond to infections in the rest of the body, where adaptive immune responses are made in secondary lymphoid organs that are often distant from the site of infection.

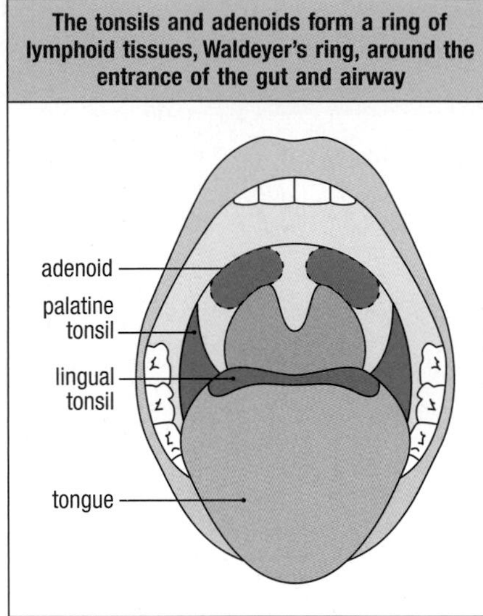

The tonsils and adenoids form a ring of lymphoid tissues, Waldeyer's ring, around the entrance of the gut and airway
adenoid
palatine tonsil
lingual tonsil
tongue

Figure 10.6 A ring of lymphoid organs guards the entrance to the gastrointestinal and respiratory tracts. Lymphoid tissues are shown in blue. The adenoids lie at either side of the base of the nose, and the palatine tonsils lie at either side of the palate at the back of the oral cavity. The lingual tonsils are on the base of the tongue.

At the back of the mouth and guarding the entrance to the gut and the airways are the **palatine tonsils**, **adenoids**, and **lingual tonsils**. These large aggregates of secondary lymphoid tissue are covered by a layer of squamous epithelium and form a ring known as Waldeyer's ring (Figure 10.6). In early childhood, when pathogens are being experienced for the first time and the mouth provides a conduit for all manner of extraneous material that is not food, the tonsils and adenoids can become painfully swollen because of recurrent infection. In the not-so-distant past, this condition was routinely treated by surgically removing the lymphoid organs, a procedure causing loss of immune capacity as reflected in the poorer secretory IgA response of such children, including the author of this book, to oral polio vaccination.

The small intestine is the major site of nutrient absorption, and its surface is deeply folded into finger-like projections called **villi** (singular **villus**), which greatly increase the surface area available for absorption. It is the part of the gut most heavily invested with lymphoid tissue. Characteristic secondary lymphoid organs of the small intestine are the Peyer's patches, which integrate into the intestinal wall and have a distinctive appearance, forming dome-like aggregates of lymphocytes that bulge into the intestinal lumen (Figure 10.7). The patches vary in size and contain between 5 and 200 B-cell follicles with germinal centers, interspersed with T-cell areas that also include dendritic cells. The small intestine also contains numerous **isolated lymphoid follicles**, each composed of a single follicle and consisting mostly of B cells. Isolated lymphoid follicles, but not Peyer's patches, are also a feature of the large intestine. A distinctive secondary lymphoid organ of the large intestine is the appendix (see Figure 10.2). It consists of a blind-ended tube about 10 cm in length and 0.5 cm in diameter that is attached to the cecum. It is packed with lymphoid follicles, and appendicitis results when it is overrun by infection. The only treatment for appendicitis is surgical removal of the appendix, to prevent it from bursting and causing life-threatening peritonitis—infection of the peritoneum, the membrane lining the abdominal cavity.

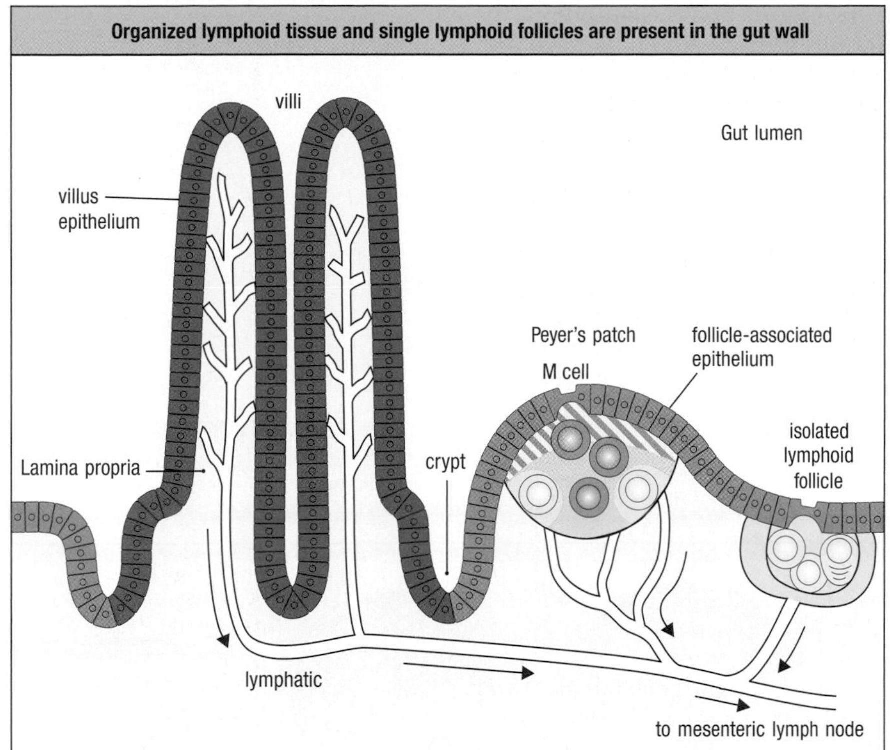

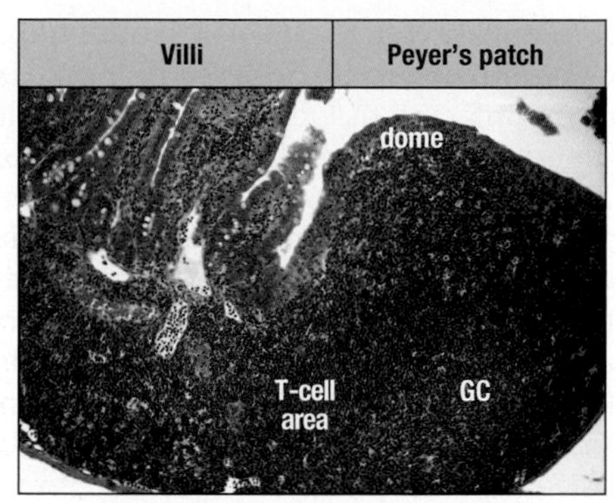

Figure 10.7 Gut-associated lymphoid tissues and lymphocytes. The diagram shows the structure of the mucosa of the small intestine. It consists of finger-like processes (villi) covered by a layer of thin epithelial cells (red) that are specialized for the uptake and further breakdown of already partly degraded food coming from the stomach. The tissue layer under the epithelium is the lamina propria, colored pale yellow in this and other figures in this chapter. Lymphatics arising in the lamina propria drain to the mesenteric lymph nodes, which are not shown on this diagram (the direction of lymph flow is indicated by arrows). Peyer's patches are secondary lymphoid organs that underlie the gut epithelium and consist of a T-cell area (blue), B-cell follicles (yellow), and a 'dome' area (striped blue and yellow) immediately under the epithelium that is populated by B cells, T cells, and dendritic cells. Antigen enters a Peyer's patch from the gut via the M cells. Peyer's patches have no afferent lymphatics, but they are a source of efferent lymphatics that connect with the lymphatics carrying lymph to the mesenteric lymph node. Also found in the gut wall are isolated lymphoid follicles consisting mainly of B cells. The light micrograph is of a section of gut epithelium and shows villi and a Peyer's patch. The T-cell area and a germinal center (GC) are indicated.

Anatomical changes
Enlarged cecum
Longer small intestine
Underdeveloped mesenteric lymph nodes
Underdeveloped Peyer's patches
Fewer isolated lymphoid follicles
Smaller spleen
Immunological effects
Reduction in secretory IgA and serum immunoglobulin
Reduction in systemic T-cell numbers and in their activation
Reduced cytotoxicity of CD8 T cells
Impaired lymphocyte homing to inflammatory sites
Reduced numbers of lymphocytes in mucosal tissues
Impaired responses of T_H17 CD4 T cells
Reduced ability of neutrophils to kill bacteria

Figure 10.8 In the absence of a microbiota, the immune system develops abnormally. Listed here are the differences distinguishing mice born and raised under sterile conditions from those raised under nonsterile conditions. The former have no microbiota, the latter have normal gut microbiota.

During early childhood, the human body and its immune system grow and mature in the context of the body's microbiota and the common pathogens in the environment. Like most other parts of the body, if the immune system is not used regularly it becomes impaired. This is well illustrated by laboratory mice that are born and raised under 'germ-free' (gnotobiotic) conditions. In comparison with control mice that have a normal gut microbiota, the gnotobiotic mice have stunted immune systems—with smaller secondary lymphoid tissues, lower levels of serum immunoglobulin, and a generally reduced capacity to make immune responses (Figure 10.8).

10-5 Inflammation of mucosal tissues is associated with causation not cure of disease

The systemic immune response to infection in non-mucosal tissues involves the activation of tissue macrophages, which by secreting inflammatory cytokines create a state of inflammation in the infected tissue. Neutrophils, NK cells, and other effector cells of innate immunity are recruited from the blood to the infected tissue, and dendritic cells migrate out of the infected tissue to the draining secondary lymphoid tissue to initiate adaptive immunity. Emerging from the adaptive immune response are effector T cells and pathogen-specific antibodies that travel to the infected tissue, where they work in conjunction with innate immunity to eliminate the pathogen and terminate the infection. Afterward, in the recovery phase, inflammation and immunity are suppressed, the damaged tissue is repaired, and both pathogens and effector cells of the immune system become excluded from the now healthy tissue. In effect, short violent episodes of localized and intense inflammation are the price paid to quash the sporadic infections of non-mucosal tissues (Figure 10.9, upper panels).

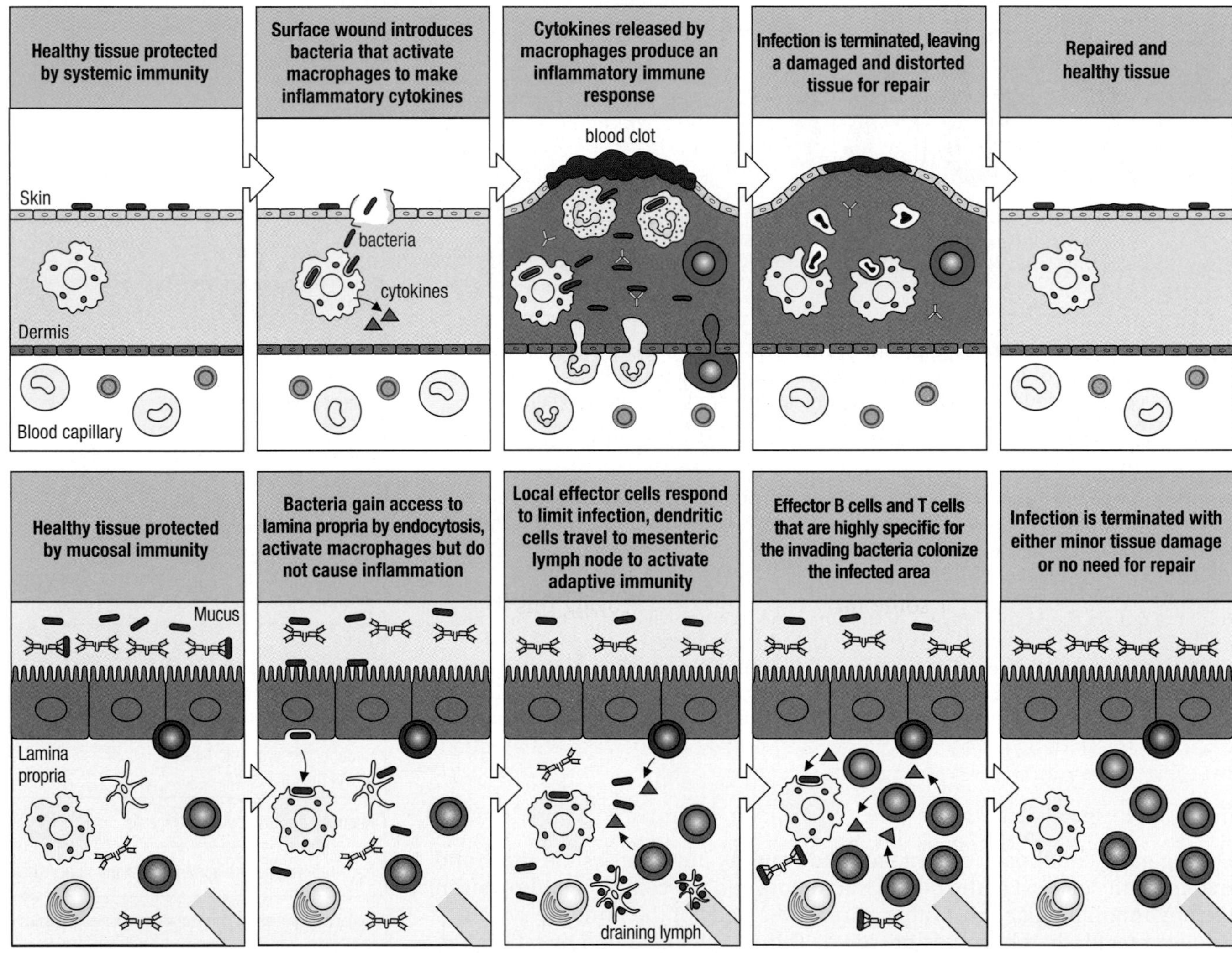

Figure 10.9 The systemic and mucosal immune systems use different strategies for coping with infections. Compared here are the immune responses made to infecting bacteria by the systemic immune system (upper panels) and the mucosal immune system (lower panels). As the systemic immune system cannot anticipate infection, it is necessary for macrophages to be activated by the invading bacteria and then to secrete cytokines that recruit effector cells to the infected tissue. This creates a state of inflammation in which the bacteria are killed, but at a cost to the structural integrity of the tissue. Infection is followed by an extensive period for repair and recovery of the damaged tissue (upper panels). In contrast, the mucosal immune system anticipates potential infections by continually making adaptive immune responses against the gut microbiota, which places secretory IgA in the gut lumen and the lamina propria, and effector cells in the lamina propria and the epithelium. When bacteria invade the gut tissue, effector molecules and cells are ready and waiting to contain the infection. In the absence of inflammation, a further adaptive immune response to the invading organism is made in the draining mesenteric lymphoid which augments that in the local lymphoid tissue. Little damage is done to the tissue, and repair occurs as part of the normal process by which gut epithelial cells are frequently turned over and replaced (lower panels).

In contrast to non-mucosal tissues, which interact only occasionally with the microbial world, the mucosal tissues have close and continuous contact with numerous and diverse commensal microorganisms, all of which are a potential source of pathogens. For the gut, any significant breach of the epithelial layer could lead to a massive influx of bacteria and infection of the type that occurs in peritonitis (see Section 10-4). To avoid this, the mucosal immune system adopts two complementary strategies. First, rather than being reactive like systemic immunity, the mucosal immune response is proactive and is constantly making adaptive immune responses against the microorganisms populating the gut. The result is that healthy gut tissue is populated with effector T cells and B cells that stand guard and are poised to respond to any invader from the gut lumen (Figure 10.9, lower panels). The advantage of a proactive strategy is that infections can be stopped earlier and with greater force than is possible in non-mucosal tissues.

The second strategy of the mucosal immune system is to be sparing in the activation of inflammation, because the molecular and cellular weapons of the inflammatory response inevitably cause damage to the tissues where they work, which for mucosal tissues, and particularly the gut, is more likely to exacerbate the infection than clear it up. Inflammation in the gut is the cause of a variety of chronic human diseases.

Of several strategies used to prevent inflammation in mucosal tissues, one is the use of regulatory T cells (CD4 T_{reg}) to turn off inflammatory T cells. IL-10 is a cytokine secreted by T_{reg} that suppresses inflammation by turning off the synthesis of inflammatory cytokines. Rare immunodeficient patients who lack a functional receptor for IL-10 suffer from a chronic inflammatory disease of the gut mucosa that resembles the more prevalent Crohn's disease and is mediated by inflammatory T_H1 and T_H17 subsets of CD4 T cells. Another inflammatory condition, **celiac disease**, is caused by an immune response in the gut lymphoid tissue that damages the intestinal epithelium and reduces the capacity of those affected to absorb nutrients from their food. This condition can arrest the growth and development of children, and in adults causes unpleasant symptoms including diarrhea and stomach pains and general ill health. Celiac disease is caused by an adaptive immune response to the proteins of gluten, a major component of grains such as wheat, barley, and rye, which are dietary staples for some human populations. Proving this cause-and-effect relationship, the symptoms of celiac disease disappear when patients adopt a strict gluten-free diet, but quickly come back if they consume gluten again. In healthy gut tissue a compromise is made between the competing demands of nutrition and defense. In celiac patients the truce is broken when a staple food is mistakenly perceived as a dangerous pathogen, which 'infects' the gut with every square meal.

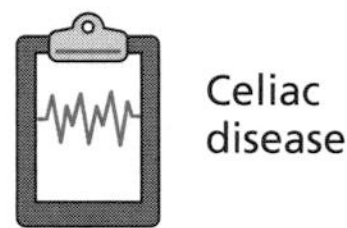
Celiac disease

The qualitatively different responses of the mucosal and systemic immune systems to microorganisms correlates with their developmental origin. During fetal development, the mesenteric lymph nodes and Peyer's patches differentiate independently of the spleen and the lymph nodes that supply systemic immunity. The distinctive development of the secondary lymphoid tissues of mucosal and systemic immunity occurs under the guidance of different sets of chemokines and receptors for cytokines in the tumor necrosis factor (TNF) family. The differences between the gut-associated lymphoid tissues and the systemic lymphoid organs are thus imprinted early on in life.

10-6 Intestinal epithelial cells contribute to innate immune responses in the gut

Intestinal epithelial cells are very active in the uptake of nutrients and other materials from the gut lumen. They also have Toll-like receptors on their apical and basolateral surfaces, for example TLR5, which recognizes flagellin, the

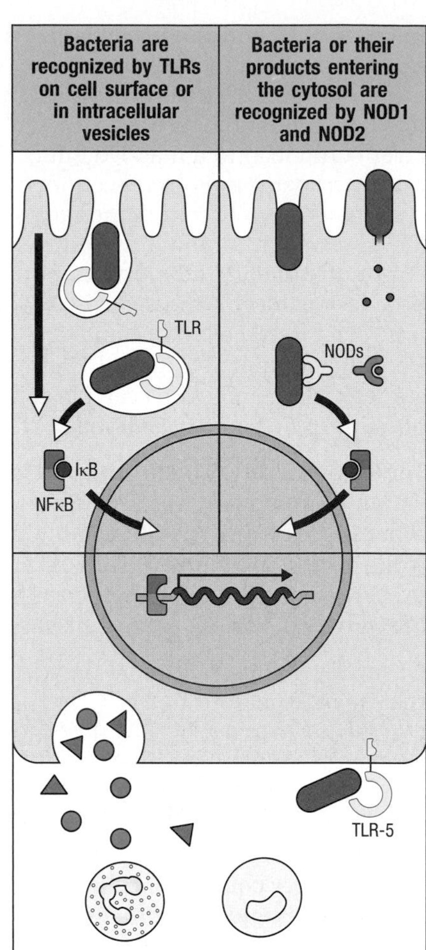

Bacteria are recognized by TLRs on cell surface or in intracellular vesicles
Bacteria or their products entering the cytosol are recognized by NOD1 and NOD2
TLR
NODs
IκB
NFκB
TLR-5
TLRs, NOD1, and NOD2 activate NFκB, inducing the epithelial cell to express inflammatory cytokines, chemokines, and other mediators. These recruit and activate neutrophils and monocytes

Figure 10.10 Epithelial cells contribute to the defense of mucosal tissue. As well as providing a barrier between the gut tissue and the contents of the gut lumen, the epithelial cells are also first responders to invading microorganisms. Epithelial cell receptors detect the invader and initiate the innate immune response by secreting cytokines and chemokines that recruit neutrophils and monocytes from the blood.

protein from which bacterial flagella are constructed. Toll-like receptors on the apical surface allow the cells to sense bacteria that overcome the defenses of the mucus and reach the epithelium; those on the basolateral surface sense invading bacteria that penetrate the epithelium. The cytoplasm of epithelial cells contains NOD1 and NOD2 receptors, which detect components of bacterial cell walls (see Section 3-5). Signals generated from NOD and Toll-like receptors lead to activation of NFκB and formation of the inflammasome by NOD-like receptor P3 (NLRP3). These events lead to the production and secretion of antimicrobial peptides, chemokines, and cytokines such as IL-1 and IL-6 by the epithelial cells (Figure 10.10). The defensins kill the bacteria, whereas the chemokines attract neutrophils (via the chemokine CXCL8), monocytes (via CCL3), eosinophils (via CCL4), T cells (via CCL5), and immature dendritic cells (via CCL20) from the blood.

In this way, epithelial cells respond to incipient infection with a quick and localized inflammatory response that is usually sufficient to eliminate the infection without causing lasting damage. If not, then an adaptive immune response is initiated in the draining mesenteric lymph node. Because gut epithelial cells turn over every 2 days, their inflammatory responses are tightly controlled and will only persist in the presence of infection.

10-7 Intestinal macrophages eliminate pathogens without creating a state of inflammation

In gut-associated lymphoid tissues the lamina propria is populated with intestinal macrophages that provide a first line of defense against microbial invasion. Although intestinal macrophages are proficient at phagocytosis and the elimination of microorganisms and apoptotic dying cells, they cannot perform other functions that characterize blood monocytes and macrophages present in non-mucosal tissues. These functions are those associated with the initiation and maintenance of a state of inflammation (Figure 10.11). Intestinal macrophages do not respond to infection by secreting inflammatory cytokines. Neither do they give a respiratory burst in response to inflammatory cytokines made by other cells. Although intestinal macrophages express MHC class II molecules, they lack B7 co-stimulators and also the capacity to make the cytokines needed to activate and expand naive T cells: IL-1, IL-10, IL-12, IL-21, IL-22, and IL-23. In short, the intestinal macrophage is not a professional antigen-presenting cell and cannot initiate adaptive immune responses. Neither are intestinal macrophages the instigators of inflammation like their counterparts in non-mucosal tissues, but they can fully perform their role of recognizing microorganisms and killing them in an environment free of inflammation. Because of these qualities, some immunologists describe the intestinal macrophages as 'inflammation-anergic' macrophages.

Intestinal macrophages live only for a few months, so their population is constantly being replenished through the recruitment of monocytes from the blood. These then differentiate into intestinal macrophages in the lamina propria. When the monocytes arrive at the intestines, they have all the inflammatory properties associated with macrophages in non-mucosal tissues. Under the influence of transforming growth factor (TGF)-β and other cytokines made

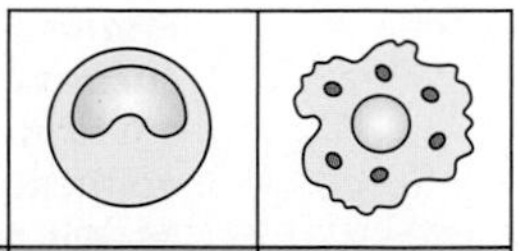

Figure 10.11 Blood monocytes reduce their inflammatory capacity when they develop into intestinal macrophages. Listed here are functions and cell-surface molecules that distinguish intestinal macrophages from the blood monocytes that give rise to them. TREM-1 is the triggering receptor expressed on myeloid cells 1. C5aR is the receptor for the anaphylotoxin C5a.

Function	Blood monocyte	Intestinal macrophage
Phagocytosis	+	+
Killing	+	+
Chemotaxis	+	–
Respiratory burst	+	–
Antigen presentation	+	?
Cytokine production	+	–
Co-stimulation	+	–
Phenotype		
CD4	+	low
CD11a, b, c	+	–
CD14	+	–
CD16, 32, 64	+ (subset)	–
CD18/integrin β_2	+	–
CD40	+	–
CD69	– (transient)	–
CD86/B7.2	+	–
CD88/C5aR	+	low
CD89/FcαR	+	–
CD123/IL-3Rα	+	–
CD354/TREM-1	+	–

by intestinal epithelium, stromal cells, and mast cells, the monocytes differentiate into intestinal macrophages by losing their inflammatory potential.

One way in which the inflammatory response of intestinal macrophages becomes attenuated is by preventing the expression of a subset of the cell-surface receptors and adhesion molecules that are used by macrophages in systemic immunity to generate inflammation. These include Fc receptors for IgA (CD89) and IgG (CD16, CD32, and CD64), the bacterial LPS receptor (CD14), complement receptors CR3 and CR4, the IL-2 and IL-3 receptors, and LFA-1. Another method of preventing inflammatory responses is modification of the signals sent by the cell-surface receptors of intestinal macrophages, for example TLR1 and TLR3–TLR9. This is achieved in various ways that all reach the same endpoint, the failure to activate NFκB, the master regulator of the inflammatory response (see Section 3-3). As a result of the selective disarming of the inflammatory response of monocytes when they become intestinal macrophages, the homeostatic environment in the healthy gut is one that is resistant to inflammation and the tissue disruption it inevitably causes. There is logic to this strategy, because damaged tissue provides the opportunity for invasion by the horde of microbes living just the other side of the gut epithelium.

10-8 M cells constantly transport microbes and antigens from the gut lumen to gut-associated lymphoid tissue

Whereas healthy skin is impermeable to microorganisms, healthy gut epithelium actively monitors the contents of the gut lumen. Absorption of nutrients by the small intestine is the function of the enterocytes in the epithelium of the villi. To aid in this task, the luminal face of an enterocyte (the surface facing into the gut lumen) is folded into numerous projections called microvilli—also called a 'brush border' from its appearance in the microscope. Interspersed between the enterocytes are goblet cells, which secrete mucus, and in the crypts between the villi are Paneth cells, which secrete defensins, lysozyme, and other antimicrobial factors. The villous epithelium is thus well defended against microbial invasion. By contrast, the **follicle-associated epithelium** that overlies lymphoid tissues in the small intestine is poorly defended. Goblet and Paneth cells are absent, and the enterocytes have a different phenotype, characterized by a reduced secretion of antimicrobial digestive enzymes such as alkaline phosphatase, and the possession of a thick glycocalyx on the brush border that shields the luminal cell surface from microorganisms and particles. These properties preserve approaching microorganisms intact and funnel them toward uniquely specialized cells of the follicle-associated epithelium called **microfold cells** (**M cells**). Their name comes from the widely spaced folds on the M cell's luminal surface, which lacks the brush border of an enterocyte (Figure 10.12). Strategically positioned over Peyer's patches and lymphoid follicles, M cells provide portals through which microorganisms and their antigens are transported from the gut lumen to the secondary lymphoid tissue by passage through the M cell in membrane vesicles.

The luminal (apical) surface of the M cell, with its characteristic folds, has adhesive properties that facilitate the endocytosis of microorganisms and

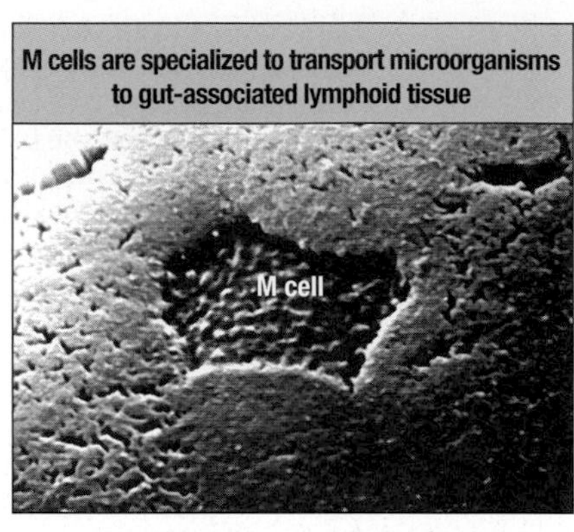

Figure 10.12 Microfold cells have characteristic membrane ruffles. This scanning electron micrograph of intestinal epithelium has a microfold or M cell in the center. It appears as a sunken area of the epithelium that has characteristic microfolds or ruffles on the surface. M cells capture microorganisms from the gut lumen and deliver them to Peyer's patches and the lymphoid follicles that underlie the M cells on the basolateral side of the epithelium. Magnification × 23,000.

particles. The surface also carries a variety of cell-surface receptors and adhesion molecules that recognize microbial antigens. On binding to cell-surface receptors, microorganisms and their antigens are internalized in endocytic vesicles that cross the M cell to fuse with the plasma membrane on the basolateral side. This process, called **transcytosis**, operates through several different mechanisms, which are used according to the size and physicochemical properties of the cargo. The distance traveled is short (1–2 μm), and the journey takes as little as 15 minutes because of the extensive invagination of the basolateral plasma membrane of the M cell to form the characteristic **intraepithelial pocket**. The pocket provides a local environment in the mucosal lymphoid tissue where the transported antigens and microorganisms can encounter dendritic cells, T cells, and B cells (Figure 10.13). Subsequent events in the secondary lymphoid tissue parallel those occurring in the systemic immune response.

10-9 Gut dendritic cells respond differently to food, commensal microorganisms, and pathogens

In the Peyer's patch, the dendritic cells that acquire antigens from M cells are present in the region of the subepithelial dome. These dendritic cells express CCR6, the receptor for the chemokine CCL20 produced by the follicle-associated epithelial cells. When taking up and processing antigens, these dendritic cells secrete IL-10, which prevents any production of inflammatory cytokines by the T cells that the dendritic cells activate.

In general, when soluble proteins and other macromolecules enter the body orally via the mouth and the alimentary canal they do not stimulate an antibody response. Thus the normal situation is that we do not make antibodies against the numerous degradation products of food that leave the stomach and pass leisurely through the intestines. This is called **oral tolerance**. In the healthy gut, potential antigens from food are transported through M cells and are taken up by a subset of CD103-expressing dendritic cells in the lamina propria that travel to the mesenteric lymph nodes. There the dendritic cells present the antigens to antigen-specific T cells and drive their differentiation into FoxP3-expressing regulatory T cells. These cells actively suppress the immune response to food antigens. Food antigens that are present at high concentrations on the dendritic cell surface can also induce antigen-specific T cells to become anergic.

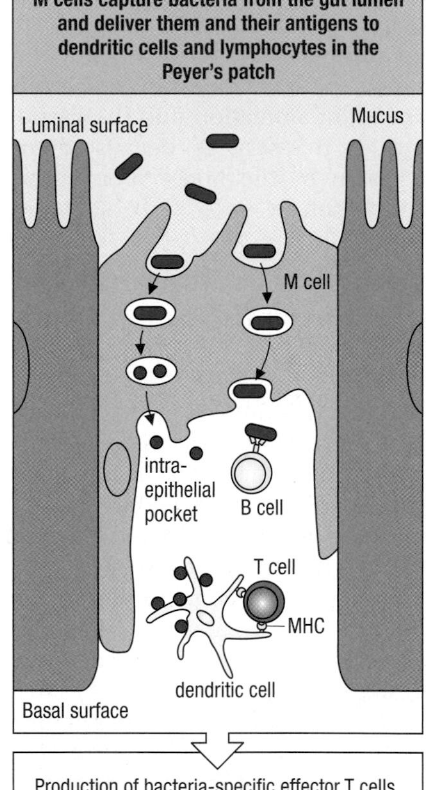

Figure 10.13 Uptake and transport of antigens by M cells. Adaptive immune responses in the gut are initiated and maintained by M cells that sample the gut's contents and deliver this material to the intra-epithelial pockets on the basolateral side of the M cell. Here, dendritic cells and B cells take up antigen and stimulate the proliferation and differentiation of antigen-specific T cells and B cells.

Commensal microorganisms are only beneficial to the human host if they live and multiply in the lumen of the gut. Any commensal organism that breaches the epithelial barrier is a potential pathogen, and is treated as such. To limit the size of the populations of commensal organisms in the gut lumen and to prevent them from infecting the tissues, specific IgA antibodies are made against the commensal species and these are constantly secreted into the gut lumen. In the healthy gut, small numbers of each commensal species enter the gut-associated lymphoid tissue. Dendritic cells take up the microbes and present their peptide antigens on MHC class II molecules to naive antigen-specific CD4 T cells. On activation and differentiation into helper CD4 T_{FH} cells, these helper T cells form conjugate pairs with antigen-specific B cells that have also taken up the microbes and are presenting their antigens on MHC class II molecules (see Figure 9.7, p. 237). This union drives the B cells to differentiate into plasma cells, which first secrete pentameric IgM and then switch the heavy-chain isotype to make secreted, dimeric IgA. By this mechanism the immune system is able to monitor the constitution of the gut microbiota and ensure that specific IgA is made against all its constituents. In episodes of change in the gut microbiota, such as occur after a course of antibiotic drugs, the synthesis of IgA will respond so that antibodies are made against new colonizing species but not against the species whose populations were exterminated by the drugs.

Although the delivery system of the M cells allows careful monitoring of the gut microbiota, it has the disadvantage of offering pathogenic agents, to which IgA has not been made, access to the tissues underlying the gut epithelium. The rapidity of M-cell transcytosis means that bacteria can survive the journey and establish an infection. For example, invasive species of *Shigella* exploit M cells to infect the colonic mucosa, causing widespread tissue damage. Poliovirus, which enters the human body by the oral route, binds to the CD155 molecule on M cells and is delivered to Peyer's patches, where it establishes local infections before spreading systemically.

The presence of infection within and around the gut-associated lymphoid tissue leads to dendritic cells carrying the pathogen and its antigens to the draining mesenteric lymph nodes, where an adaptive immune response is made. In the presence of infection, dendritic cells in the lamina propria and outside the organized lymphoid tissues become more mobile and capture pathogens independently of M cells. They move into the epithelial wall or send processes through it that capture microbes and antigens without disturbing the integrity of the epithelial barrier (Figure 10.14). Having obtained a cargo of antigens, the dendritic cells move into the T-cell area of the gut-associated lymphoid tissue, or travel in the draining lymph to the T-cell area of a mesenteric lymph node, to stimulate antigen-specific T cells.

Through this perpetual sampling of the gut lumen's contents, T cells specific for pathogenic microorganisms, commensal microorganisms, and food antigens are stimulated to become effector cells. Activated helper T cells then activate B cells to become plasma cells, as described in Chapter 9. These plasma cells secrete dimeric IgA specific for pathogens, commensals, and food antigens.

10-10 Activation of B cells and T cells in one mucosal tissue commits them to defending all mucosal tissues

On completing their development in the primary lymphoid organs, naive B cells and T cells enter the bloodstream to recirculate between blood, secondary lymphoid tissue, and lymph. Before encountering a specific antigen, these

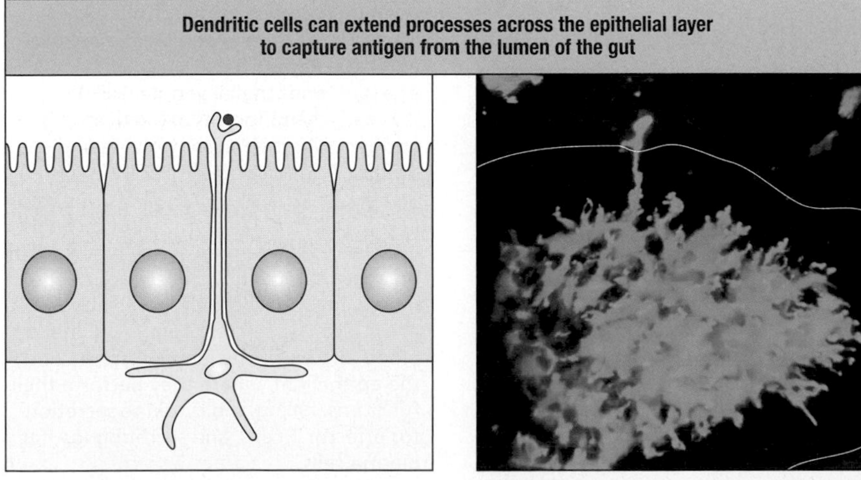

Dendritic cells can extend processes across the epithelial layer to capture antigen from the lumen of the gut

Figure 10.14 Capture of antigens from the intestine by dendritic cells. Dendritic cells in the lamina propria can sample the contents of the intestine by extending a process between the enterocytes without disturbing the barrier function of the epithelium (left panel). Such an event is captured in the micrograph (right panel), in which the mucosal surface is shown by the white line and the dendritic cell (stained green) in the lamina propria extends a single process over the white line and into the lumen of the gut. Magnification × 200. Micrograph courtesy of J.H. Niess.

naive lymphocytes can enter the secondary lymphoid tissues of both the systemic and the mucosal compartments of the immune system. Like the spleen and other lymph nodes, the Peyer's patches and mesenteric lymph nodes release chemokines CCL21 and CCL19, which bind to the chemokine receptor CCR7 expressed by naive B cells and T cells. This induces the naive lymphocytes to leave the blood at high endothelial venules and enter the secondary lymphoid tissue.

If specific antigen is not encountered in the Peyer's patch or mesenteric lymph node, the naive cells leave in the efferent lymph to continue recirculation. Lymphocytes that find their specific antigen are retained in the lymphoid tissue. Here, dendritic cells that are presenting specific antigens activate the naive T cells, causing them to proliferate and differentiate into effector T cells. This activation requires retinoic acid, a derivative of vitamin A that is made by the dendritic cells of mucosal tissue. The effector cells include helper CD4 T_{FH} cells, which activate naive antigen-specific B cells to become effector B cells.

After their activation in a Peyer's patch, effector B and T cells leave in the lymph and travel via the mesenteric lymph nodes to the thoracic duct and the blood (Figure 10.15). Cells activated in a mesenteric lymph node leave in the efferent

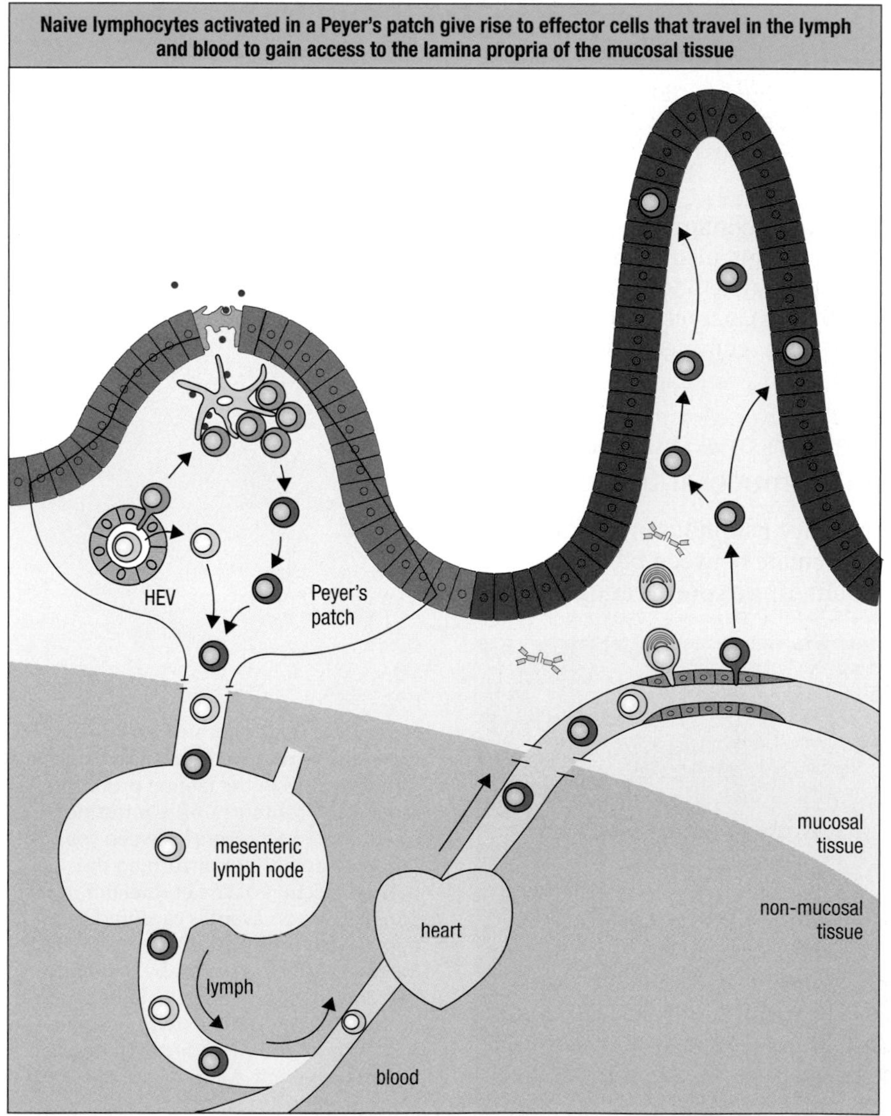
Naive lymphocytes activated in a Peyer's patch give rise to effector cells that travel in the lymph and blood to gain access to the lamina propria of the mucosal tissue
HEV
Peyer's patch
mesenteric lymph node
lymph
heart
blood
mucosal tissue
non-mucosal tissue

Figure 10.15 Lymphocytes activated in mucosal tissues return to those tissues as effector cells. Pathogens from the intestinal lumen enter a Peyer's patch through an M cell and are taken up and processed by dendritic cells. Naive T cells (green) and B cells (yellow) enter the Peyer's patch from the blood at a high endothelial venule (HEV). The naive lymphocytes are activated by antigen, whereupon they divide and differentiate into effector cells (blue). The effector cells leave the Peyer's patch in the lymph, and after passing through mesenteric lymph nodes they reach the blood, by which they travel back to the mucosal tissue where they were first activated. The effector cells leave the blood and enter the lamina propria and the epithelium, where they perform their functions: killing and cytokine secretion for effector T cells, and secreting IgA for plasma cells.

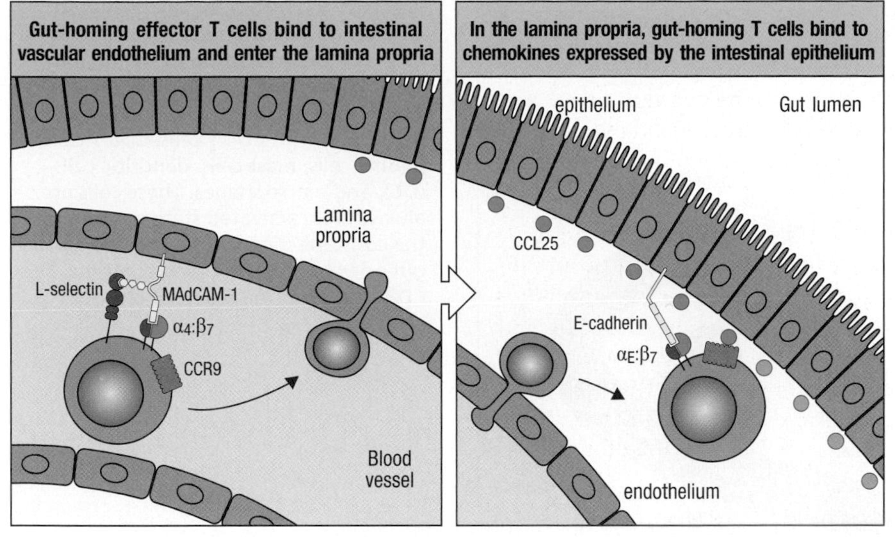
Gut-homing effector T cells bind to intestinal vascular endothelium and enter the lamina propria
In the lamina propria, gut-homing T cells bind to chemokines expressed by the intestinal epithelium
epithelium
Gut lumen
Lamina propria
CCL25
L-selectin
MAdCAM-1
α_4:β_7
CCR9
E-cadherin
α_E:β_7
Blood vessel
endothelium

Figure 10.16 Homing of effector T cells to the gut is controlled by adhesion molecules and chemokines. Antigen-activated T cells in mucosal lymphoid tissue become effector cells that leave in the lymph and then populate mucosal tissue from the blood. This homing is mediated by integrin α_4:β_7 on the effector T-cell binding to blood vessel MAdCAM-1 (left panel). In the lamina propria T cells are guided by chemokine CCL25, which is made by mucosal epithelium and binds to the CCR9 receptor on the T cells. Interaction with the gut epithelium is enhanced by T-cell integrin α_E:β_7 binding to epithelial cell E-cadherin.

lymph and similarly reach the blood. During differentiation, these lymphocytes lose expression of CCR7 and the cell-adhesion molecule L-selectin, and this prevents them from entering the secondary lymphoid tissues of the systemic immune system. The mucosa-derived effector cells express adhesion molecules and receptors that allow them to leave the blood at mucosa-associated lymphoid tissues (Figure 10.16). These molecules include integrin α_4:β_7, which binds specifically to the mucosal vascular addressin **MAdCAM-1** on endothelial cells of blood vessels in the gut wall, and CCR9, the receptor for chemokine CCL25 secreted by cells in the lamina propria (see Figure 10.16). The homing mechanisms are not specific to the particular mucosal tissue in which the effector cells were activated, but allow the effector B cells and T cells to enter and function in any mucosal tissue. For example, naive B and T cells activated in the gut-associated lymphoid tissue can thus enter and function in lymphoid tissue associated with the respiratory tract and vice versa. The benefit of this unifying arrangement is that the experience obtained by defeating infection in one mucosal tissue is used to improve the defenses of them all.

10-11 A variety of effector lymphocytes guard healthy mucosal tissue in the absence of infection

To avoid acute inflammatory responses of the type used to activate the systemic immune responses, the mucosal tissues are populated at all times with antigen-activated effector cells. This situation contrasts strongly with other tissues, which admit effector cells only when infected. Many of the effector cells in mucosal tissues were stimulated by antigens of commensal species, which likely account for most gut infections. Other effector lymphocytes arise from primary immune responses against pathogens, such as viruses, that are not normal inhabitants of the gut. The majority of the effector cells are T cells, there being more T cells in the gut-associated lymphoid tissue than in the rest of the body. The effector B cells are almost all plasma cells secreting either pentameric IgM or dimeric IgA, and they are mainly confined to the Peyer's patches. The T cells are heterogeneous and comprise both γ:δ T cells and α:β T cells, with CD8 T cells predominating in the epithelium and CD4 T cells in the lamina propria (Figure 10.17). In addition to CD4 T cells, the lamina propria also contains CD8 T cells and plasma cells—as well as dendritic cells and the occasional eosinophil or mast cell (see Figure 10.17). Neutrophils are rare in the healthy intestine, but they rapidly populate sites of inflammation and disease.

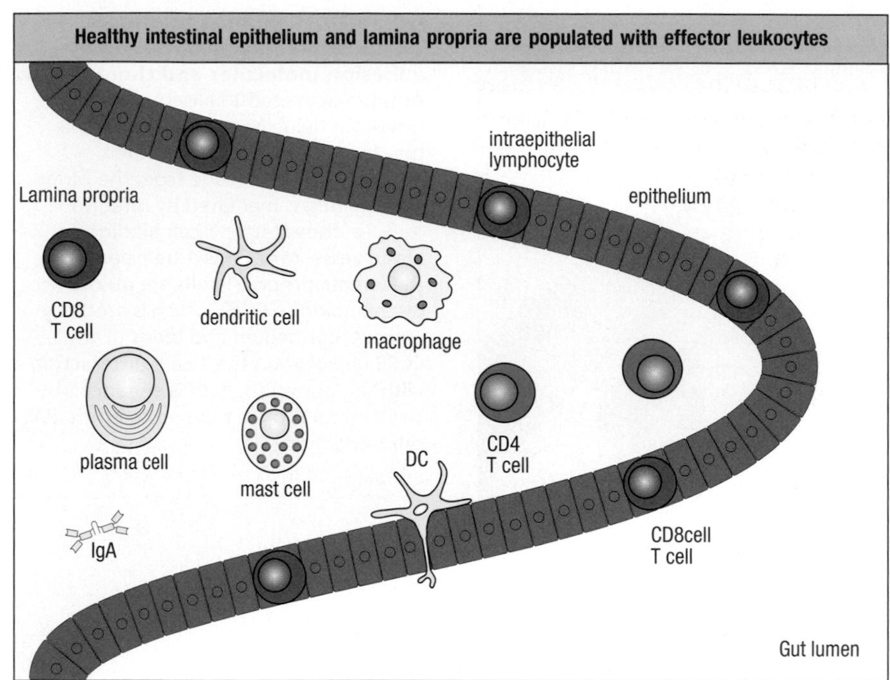

Figure 10.17 Most immune-system cells in mucosal tissues are activated effector cells. Outside the lymphoid tissues, the gut epithelium contains CD8 T cells, and the lamina propria is well populated with CD4 T cells, CD8 T cells, plasma cells, mast cells, dendritic cells (DC), and macrophages. These cells are always in an activated state because of the constant stimulation afforded by the gut's diverse and changing contents. The CD8 T cells include both $\alpha:\beta$ and $\gamma:\delta$ T cells.

Integrated into the epithelial layer of the small intestine is a distinctive type of CD8 T cell called the **intraepithelial lymphocyte**. On average, there is about one intraepithelial lymphocyte for every 7–10 epithelial cells (see Figure 10.17). Intraepithelial lymphocytes have already been activated by antigen and contain intracellular granules like those of CD8 cytotoxic T cells. The intraepithelial lymphocytes include both $\alpha:\beta$ CD8 T cells or $\gamma:\delta$ CD8 T cells. They express T-cell receptors with a limited range of antigen specificities, indicating that they were activated by a limited number of antigens, and they have a distinctive combination of chemokine receptors and adhesion molecules that enables them to occupy their unique position within the intestinal epithelium. Like other gut-homing T cells intraepithelial lymphocytes express the chemokine receptor CCR9, but instead of $\alpha_4:\beta_7$ integrin, they express the $\alpha_E:\beta_7$ integrin, which attaches the T cell to E-cadherin on the surface of epithelial cells (see Figure 10.16, right panel). This adhesive interaction enables intraepithelial lymphocytes to intercalate within the layer of intestinal epithelial cells while maintaining the epithelium's barrier function.

10-12 B cells activated in mucosal tissues give rise to plasma cells secreting IgM and IgA at mucosal surfaces

The mucosal surfaces of an adult human have a combined area of around 400 m^2. Defending these tissues is a coating of protective antibody that consists of secreted pentameric IgM and dimeric IgA and needs constant replenishment. Maintaining the antibody supply are 60 billion (6×10^{10}) mucosal plasma cells, comprising 80% of the body's plasma cells.

In the Peyer's patches and mesenteric lymph nodes defending the gut, activation of naive B cells by antigen and antigen-specific T_{FH} cells gives rise to an initial wave of effector B cells that leaves the lymphoid tissue in the efferent lymph en route to the bloodstream. In the blood the effector B cells travel to gut-associated lymphoid tissue, which they enter. This is achieved by the combined interactions of integrin $\alpha_4:\beta_7$ on the B cells with MAdCAM-1 on the intestinal vascular endothelium, and B-cell CCR9 binding to chemokine CCL25 emanating from the intestinal epithelial cells. Some B cells activated in gut-associated lymphoid tissue return to their tissue of origin, but most take up residence in other areas of the gut and in different mucosal tissues. This strategy enables all the mucosal tissues to benefit from the antibody produced in one of them. Effector B cells settle in the lamina propria, where they complete their differentiation into plasma cells that make pentameric IgM and secrete it into the subepithelial space. Here the J chain of the IgM molecule binds to the poly-Ig receptor expressed by immature epithelial cells, also called stem cells, located at the base of intestinal crypts (see Figure 2.18, p. 42). By transcytosis, the poly-Ig receptor carries the antibody from the basal side to the luminal side of the cell, where the IgM is released and bound by the mucus. This transport mechanism for secretory IgM is the same as that used for secretory IgA (see Figure 9.18, p. 248).

Only some of the antigen-activated B cells differentiate into plasma cells secreting IgM. The others remain in the B-cell area of the gut-associated lymphoid tissue, where they undergo affinity maturation and isotype switching. The switch is usually to the IgA isotype, the dominant class of immunoglobulin in mucosal secretions. Switching to IgA is orchestrated by TGF-β and uses the same genetic mechanisms as those described in Chapter 4 for isotype switching and somatic hypermutation in the spleen and lymph nodes (see Sections 4-14 and 4-15). Several other soluble factors enhance switching to the IgA isotype. These include inducible nitric oxide synthase (**iNOS**), which is produced by dendritic cells and induces increased expression of the B cells' TGF-β receptor, the vitamin A derivative retinoic acid, IL-4, IL-10, B-cell-activating factor (BAFF), and a proliferation-inducing ligand (APRIL). Both APRIL and BAFF are made by dendritic cells in gut lymphoid tissue and, in combination with IL-4, strongly bias isotype switching toward IgA.

Under the influence of these factors, effector B cells are programmed to make dimeric IgA that has higher affinities for antigen than the IgM antibodies made by the first wave of plasma cells. The isotype-switched cells constitute a second wave of effector B cells that, like the first, travel to the lamina propria of mucosal tissues throughout the body and differentiate into plasma cells. Plasma cells of the second wave make dimeric IgA that is secreted and transported across the mucosal epithelium by the poly-Ig receptor. Comparison of the sequences of IgA secreted by plasma cells of systemic and mucosal immunity shows a more extensive somatic hypermutation in the variable regions of mucosal IgA than in systemic IgA.

Dimeric IgA is the dominant immunoglobulin in tears, saliva, milk, and intestinal fluid. By contrast, IgG predominates in secretions of the nose, lower respiratory tract, and both the female and male urinogenital tracts. Monomeric IgG is actively transported into external secretions by the Fc receptor FcRn (Figure 10.18). IgE is transported across mucosal epithelial cells by the FcεIII receptor (CD23) and is present in small concentrations in saliva, the gut, and the respiratory tract. To some extent, all antibody isotypes are present in the secretions at mucosal surfaces, and the amounts increase at sites of inflammation and infection where the barrier function of the mucosal epithelium has been damaged.

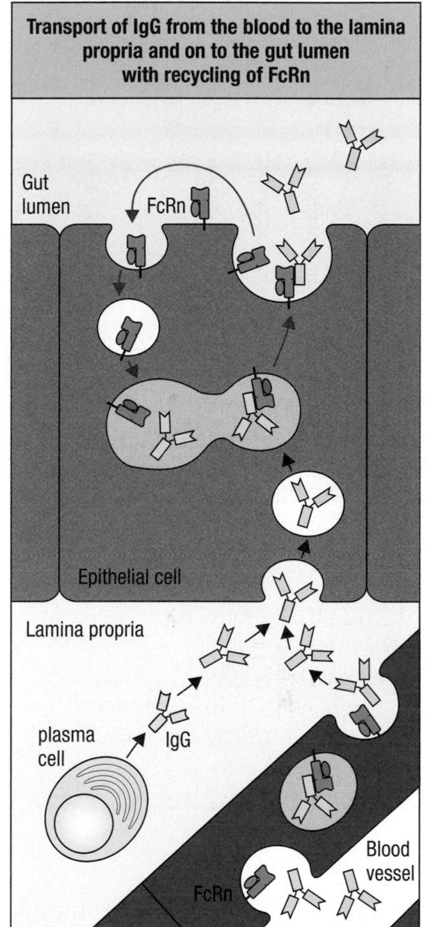

Figure 10.18 Transport of IgG from blood to mucosal secretions. FcRn is an Fc receptor for IgG and is used to shuttle IgG across cells. FcRn on endothelial cells lining blood vessels picks up IgG from the blood, transports it across the endothelial cell, and drops it in the lamina propria. FcRn on gut epithelial cells picks up IgG in the lamina propria, transports it across the epithelial cell, and drops it in the gut lumen. Although most plasma cells in the lamina propria make IgA, some make IgG. Unlike the poly-Ig receptor that transcytoses IgA, FcRn is not consumed in the process and can therefore be reused.

10-13 Secretory IgM and IgA protect mucosal surfaces from microbial invasion

M cells are continually sampling the gut contents, which activate B cells to make IgM and IgA antibodies specific for the gut microbiota. Transcytosis delivers the antibodies to the mucus layer on the luminal face of the gut epithelium. The antibodies are retained in the mucus by its inherent viscosity and by forming disulfide bonds with the mucin molecules (Figure 10.19). In keeping with the non-inflammatory environment in mucosal tissues, complement components are absent from mucosal secretions. Thus, unlike their systemic

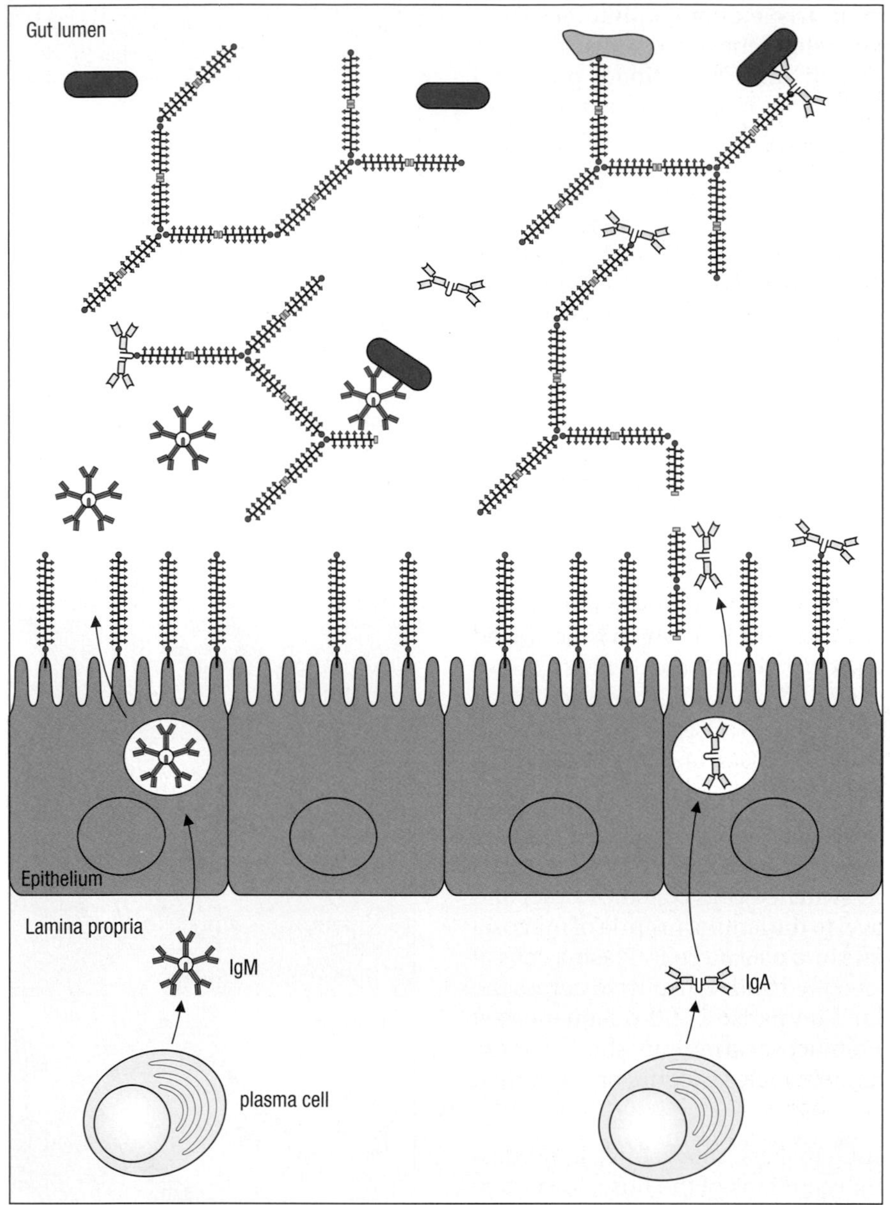

Figure 10.19 Secretory immunoglobulins become attached to the mucus, where they stand ready to bind commensal and pathogenic organisms. Secretory IgM and IgA are delivered to the luminal side of the gut epithelium by transcytosis. The α and μ heavy chains have cysteine residues in the C-terminal regions; these residues can form disulphide bonds with the free cysteine residues of mucin polypeptides. This tethers the antibodies to the mucus, where they can bind bacteria and prevent them from reaching the mucosal surface. Both secreted and membrane-associated mucin polypeptides are shown.

counterparts, the secretory immunoglobulins do not fix complement as a means of neutralizing pathogens, but instead they coat the microbial surface in ways that impede microbial invasion and proliferation. In approaching the gut epithelium, bacteria are slowed down by the mucus and exposed to the antibodies and antimicrobial peptides it contains. If the bacterium is of a species that has already been sampled by M cells and has stimulated an immune response, it is bound by antibody, prevented from reaching the gut epithelium, and killed by the antimicrobial peptides. Bacteria that reach the epithelium and gain access, via M cells or another route, to the lamina propria can be opsonized with antibody and targeted for phagocytosis by resident macrophages.

Some secretory antibodies are specific for the surface components that bacteria and viruses use to bind to epithelial cells and either infect them or exploit them to gain access to the underlying tissues. By binding to these surface molecules, the antibodies prevent the invasion and infection of gut tissue by such bacteria and viruses. Cholera, caused by the bacterium *Vibrio cholerae*, is a

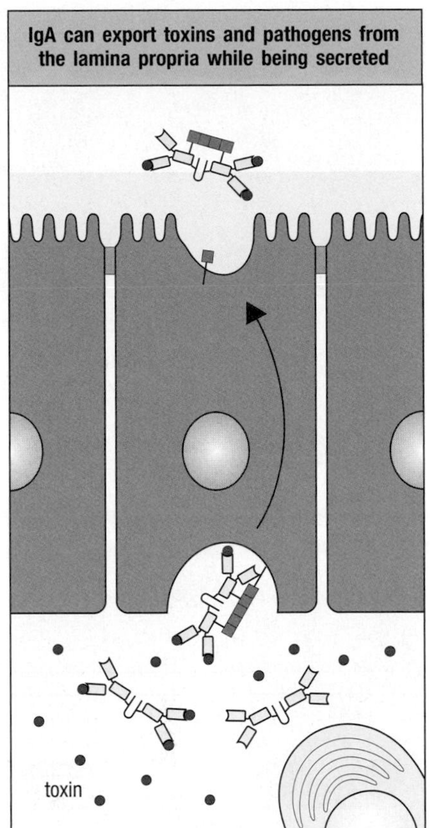

IgA can export toxins and pathogens from the lamina propria while being secreted
toxin

Figure 10.20 Secretory IgA can be used to remove pathogens and their products from the lamina propria. IgA that is secreted by plasma cells in the lamina propria can bind pathogens and antigens that are there and carry them to the intestinal lumen by transcytosis. In this way the toxins secreted by various species of bacteria, for example the cholera and diphtheria toxins, can be disposed of in the lumen and prevented from exerting their toxic effects.

life-threatening disease in which a toxin secreted by the bacterium perturbs the intestinal epithelium, causing chronic diarrhea and severe dehydration. To have these effects the cholera toxin must bind to the epithelial cells and be endocytosed. The toxin can be neutralized by specific, high affinity IgA that binds the toxin and covers up the site with which it binds to gut epithelial cells (Figure 10.20).

Secretory IgA has little capacity or opportunity to activate complement or act as an opsin, and it cannot induce a state of inflammation. Instead it has evolved to be a non-inflammatory immunoglobulin that limits the access of pathogens, commensals, and food products to mucosal surfaces in a manner that avoids unnecessary damage to these delicate and vital tissues. Antibodies specific for commensal bacteria are well represented in the IgA secreted into the gut. By restricting commensal organisms to the lumen of the gut, and limiting the size of their populations, these antibodies have a crucial role in maintaining the symbiotic relationship of the microbiota with its human host.

10-14 Two subclasses of IgA have complementary properties for controlling microbial populations

There are two subclasses of IgA—IgA1 and IgA2—that are both made as a systemic monomeric IgA and a secretory dimeric IgA. As we saw for the IgG subclasses (see Figure 4.33, p. 106), the two IgA subclasses differ mainly in the hinge region, which is twice as long in IgA1 (26 amino acids) than in IgA2 (13 amino acids) (Figure 10.21). The longer hinge in IgA1 makes it more flexible than IgA2 in binding to pathogens and thus more able to use multiple antigen-binding sites to bind to the same pathogen and deliver it to a phagocyte. The drawback to the longer IgA1 hinge is its greater susceptibility to proteolytic cleavage than the shorter IgA2 hinge. Major bacterial pathogens, including *Streptococcus pneumoniae*, *Neisseria meningitidis*, and *Haemophilus influenzae*, have evolved specific proteases that cleave the IgA1 hinge, thereby disconnecting the Fc and Fab regions. This prevents the antibody from targeting the bacteria to phagocyte-mediated destruction. Exactly the opposite effect can sometimes occur: bacteria coated with Fab fragments of IgA1 become more able to adhere to mucosal epithelium, penetrate the physical barrier, and gain access to the lamina propria to launch an infection.

In situations where IgA1 is ineffective because of the presence of specific proteases, the synthesis of IgA2 helps to control bacterial infection. Although the IgA2 hinge is less flexible, it is highly protected by covalently linked carbohydrate, and bacteria have so far failed to evolve a protease that can cleave IgA2. In the blood, lymphatics, and extracellular fluid of the connective tissue, where bacterial populations are small and the IgA1-specific proteases pose less of a threat, most of the IgA made (93%) is of the IgA1 isotype. In contrast, in the colon, where bacteria are present at the highest density and IgA1-specific proteases are ubiquitous, the majority of IgA made (60%) is of the IgA2 isotype (Figure 10.22).

The switch to IgA secretion normally goes from IgM to IgA1, but in the presence of the TNF-family cytokine APRIL, the isotype switches from IgM to IgA2. In the colon the epithelial cells make APRIL, which drives the switch to the IgA2 isotype in the resident B cells. In general, the synthesis of IgA2 is higher in

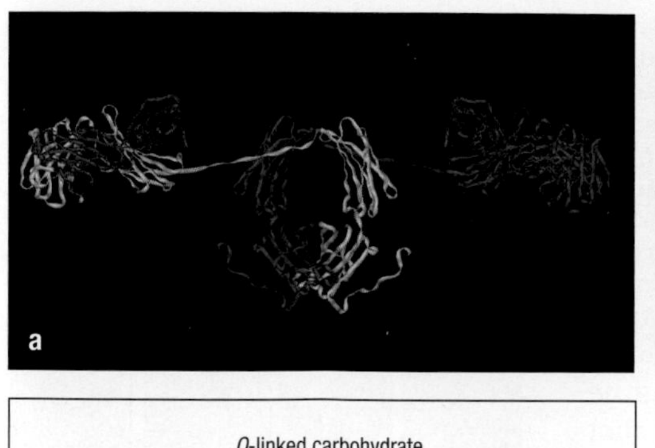

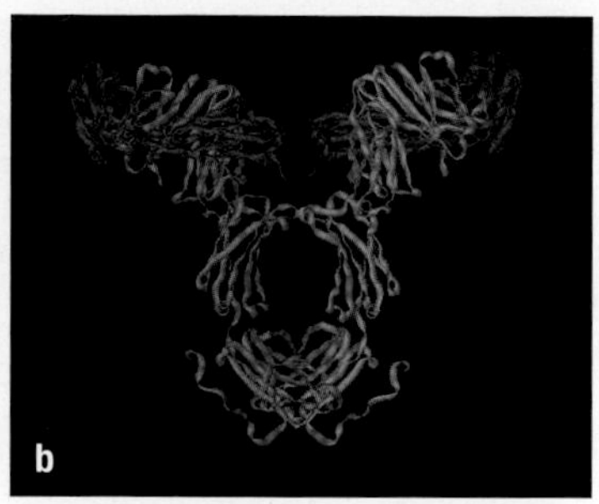

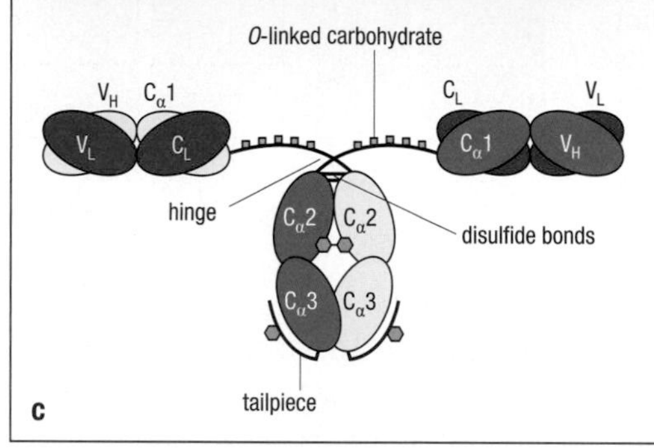

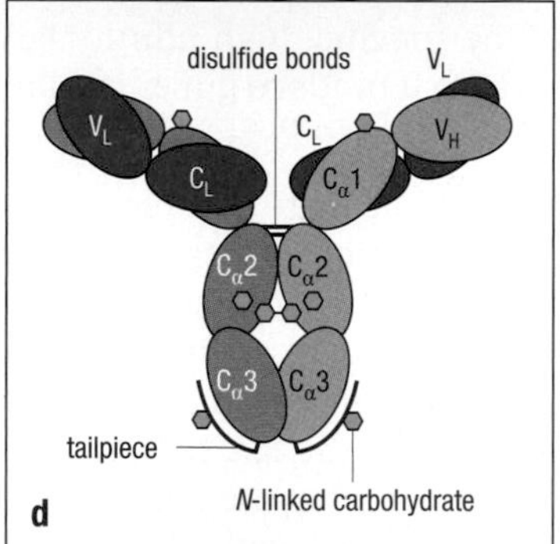

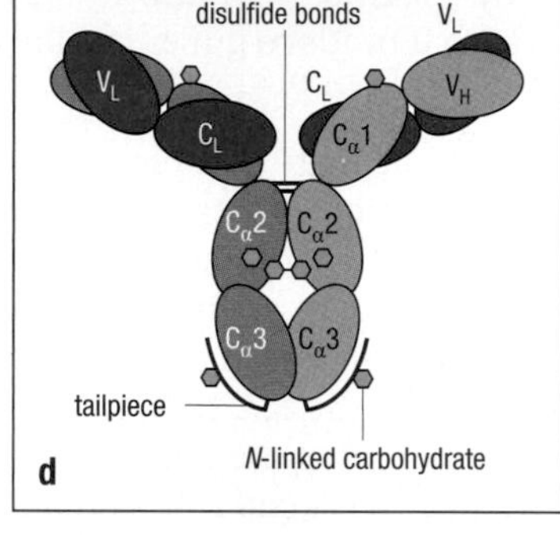

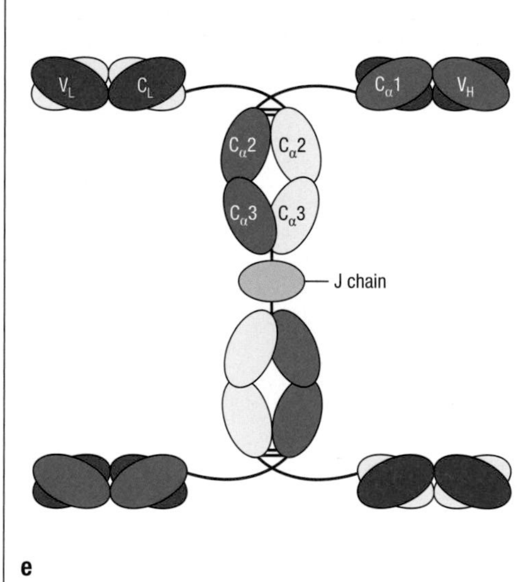

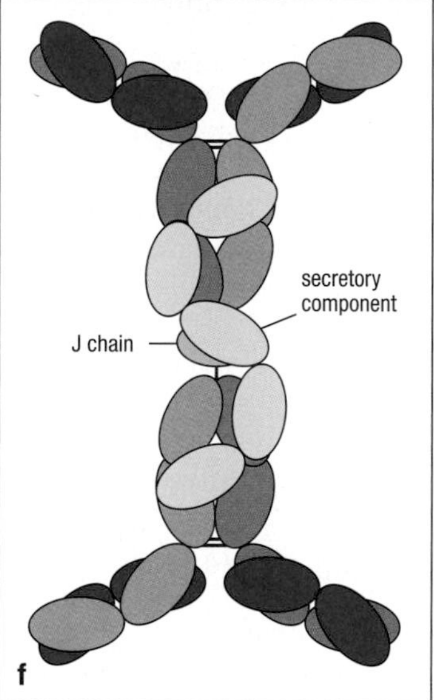

Figure 10.21 IgA1 and IgA2 have hinges of different lengths. Models of the three-dimensional structures of IgA1 (panel a) and IgA2 (panel b) are shown in the top panels, and diagrams corresponding to these models are shown in the middle panels: IgA1 (panel c) and IgA2 (panel d). Panel e shows the organization of a dimer of IgA1,with the attached J chain. Panel f shows a molecule of secreted IgA2, which contains both the J chain and the secretory piece of the poly-Ig receptor.

mucosal lymphoid tissues than in systemic lymphoid tissues, but the proportions of plasma cells making IgA1 and IgA2 also vary considerably between the different mucosal tissues (see Figure 10.22). The tissues most heavily populated with microorganisms—the large and small intestines, the mouth (supplied with IgA by the salivary glands), and the lactating breast (exposed to the heavily contaminated oral cavity of the suckling infant)—are those more focused on making IgA2. These differences show that the various mucosal tissues are not immunologically equivalent and that they face different challenges in balancing their burden of commensal and pathogenic microorganisms that make IgA1-specific proteases.

The two IgA subclasses are differentially produced in tissues			
Tissue	IgA1 %	IgA2 %	IgA1/A2 ratio
Spleen, peripheral lymph nodes, tonsils	93	7	13.3
Nasal mucosa	93	7	13.3
Bronchial mucosa	75	25	3.0
Lachrymal glands (tear ducts)	80	20	4.0
Salivary glands	64	36	1.8
Mammary glands	60	40	1.5
Gastric mucosa (stomach)	83	17	4.9
Duodenum–jejunum (upper small intestine)	71	29	2.4
Ileum (lower small intestine)	60	40	1.5
Colon (large intestine)	36	64	0.6

Figure 10.22 IgA1 and IgA2 are differentially expressed in mucosal tissues. Shown here are the relative proportions of plasma cells making IgA1 and IgA2 in each tissue. The number of plasma cells correlates directly with the amount of antibody made and secreted. Data courtesy of Per Brandtzaeg.

10-15 People lacking IgA are able to survive, reproduce, and generally remain healthy

Expert mucosal immunologists consider that IgA is probably the best-understood and most widely accepted mediator of mucosal immunity. Given the importance of mucosal immunity for human health, it is therefore surprising that many seemingly healthy people make little or no IgA. Because other immunoglobulin isotypes are not affected, this condition is called **selective IgA deficiency**. It occurs throughout the world but at frequencies that vary over two orders of magnitude (Figure 10.23). The cause of the IgA deficiency seems to be defective isotype switching from IgM to IgA. The condition has a genetic basis, as seen from the case of an infant with aplastic anemia and normal IgA who was treated with a bone marrow transplant from his HLA-identical, but IgA-deficient, sister. Upon reconstitution of his immune system by his sister's hematopoietic stem cells, the boy became IgA deficient but was otherwise healthy and no longer dependent upon medication. So why is IgA so dispensable?

Population	Incidence of selective IgA deficiency per million individuals
Saudi Arabian	6993
Spanish	6135
Nigerian	3968
US Caucasian	1667
English	1143
Brazilian	1036
Chinese	253
African-American	83
Japanese	60

Figure 10.23 Distribution of selective IgA deficiency among human populations.

As infants, IgA-deficient individuals need not make IgA because they can get it from their mothers, provided that the mother is not IgA-deficient. In nursing mothers, plasma cells derived from B cells activated in the gut, lungs, and other mucosae home to the lactating mammary gland to contribute their secretory IgA to the breast milk. The milk therefore contains all the different IgA antibodies the mother has recently made in responding to commensal microorganisms, infecting pathogens, and food antigens. On suckling, the infant's gut receives a portfolio of maternal IgA that provides protection against the gut microbiota and locally endemic pathogens. Until very recently in human history, mothers would have breastfed their children for 3–7 years after birth. In this most vulnerable period of life, most IgA-deficient children would have been protected by maternal IgA, which could explain how deleterious gene variants contributing to IgA deficiency have survived. The trend in present-day populations has been to shorten the period of breastfeeding. Although this is expected to increase the vulnerability of infants to infectious disease, it is mitigated by modern improvements in hygiene, nutrition, and vaccination that reduce the risks of infection.

Because they cannot switch isotype from IgM to IgA, plasma cells making other isotypes are more abundant in IgA-deficient people (Figure 10.24). For mucosal immunity, IgM is particularly important, because it has the J chain that interacts with the poly-Ig receptor and so can be secreted at mucosal surfaces, like IgA. Moreover, IgM always precedes IgA as the first secretory antibody in the adaptive responses of mucosal immunity. Increased secretion of pentameric IgM probably compensates for the absence of secretory IgA, at least in the relatively parasite-free environment of developed countries. Increased transport of IgG from lamina propria to the gut mucosa by FcRn could further augment defenses. IgA-deficient individuals are susceptible to bacterial infections of the lungs, and to intestinal infection by *Giardia lamblia*, a protozoan parasite. Thus the health and vigor of people with IgA deficiency today might in part reflect reduced pressure on the mucosal immune systems of human populations in modern industrialized societies. These people generally eat cooked, highly processed food, and are not infested with helminth worms and the other intestinal parasites that were prevalent in the past and

	Percentage of B cells making antibody of four different isotypes							
	Normal individuals				IgA-deficient individuals			
	IgA	IgM	IgG	IgD	IgA	IgM	IgG	IgD
Nasal glands	69	6	17	8	0	20	46	34
Lachrymal and parotid glands	82	6	5	7	0	21	22	57
Gastric mucosa	76	11	13	0	0	64	35	1
Small intestine	79	18	3	0	0	75	24	1

Figure 10.24 Selective IgA deficiency. The table compares individuals who have normal production of IgA with individuals who have selective IgA deficiency. The percentages of B cells making IgA, IgM, IgG, and IgD in four different mucosal tissues are shown. The amount of antibody made is proportional to the number of B cells. Data courtesy of Per Brandtzaeg.

still affect one-third of the world's population. By contrast, chronic lung disease is more frequent in people with IgA deficiency in industrialized countries. This suggests that the trend toward poorer air quality in the cities, in which most people live, makes the actions of IgA in the respiratory tract of increasing importance.

IgA deficiency is a clinically heterogeneous condition, and its epidemiology and segregation in families are complicated and remain unpredictable. The data cannot be explained by defects in a single gene, and indicate that IgA deficiency is caused by combinations of variants (alleles) of genes on different chromosomes, and that these combinations differ between human populations. As well as TGF-β and its receptor, retinoic acid, IL-4, IL-10, BAFF, and APRIL are all implicated in switching immunoglobulin isotype from IgM to IgA (see Section 10-12), and mutations in their genes are candidates for contributing to selective IgA deficiency.

10-16 T_H2-mediated immunity protects against helminth infections

Helminths are parasitic worms that live and reproduce in the intestines. They comprise three groups—nematodes, trematodes, and cestodes—that can all cause chronic and debilitating disease (Figure 10.25) by competing with the host for nutrients and causing local damage to the intestinal epithelium and blood vessels. With the exception of people in developed countries, virtually all humans are burdened with helminth infections. Because helminths are never commensal organisms but always pathogens, there are worldwide medical programs aimed at de-worming the entire human population. For similar reasons, the immune system has evolved a variety of mechanisms for the containment and elimination of helminth infections. The most effective immune response to a helminth depends on the particular parasite's life cycle. Some attach to the luminal side of the intestinal epithelium, others enter and colonize the epithelial cells, and yet others invade beyond the intestine and spend part of their life cycle in another tissue, such as the liver, lungs, or muscle.

In order to survive and flourish in the intestines, helminths must at all times avoid being cast into the flowing fluid of the gut lumen by the constant turnover and renewal of the enterocytes. Conversely, in countering helminth infections, the purpose of the immune system is to drive the worms into the gut lumen, from which they can be expelled in the feces. This can only be achieved by mounting an adaptive immune response that is dominated by T_H2 CD4 T cells and involves the production of the T_H2-associated cytokines IL-4, IL-9, IL-13, IL-25, and IL-33. Most counterproductive is an inflammatory response

	Some diseases caused by helminths		
Common name	Roundworms	Flukes	Tapeworms
Scientific name	Nematodes	Trematodes	Cestodes
Diseases caused	Ascariasis*, Dracunculiasis (guinea worm disease), Elephantiasis (lymphatic filariasis), Enterobiasis* (pinworm), Hookworm* Onchocerciasis (river blindness), Trichinosis*	Schistosomiasis, Fasciolopsiasis*	Tapeworm infection*

Figure 10.25 Helminths are major human pathogens that parasitize the intestines. The helminths comprise four major groups, three of which include human pathogens. *Caused by parasites that live in the gut lumen.

dominated by T_H1 cells and the production of interferon-γ (IFN-γ). This not only fails to eliminate the parasite but also exacerbates the infection and the likelihood of severe, persistent, and crippling disease. One deleterious effect of IFN-γ, for example, is to decrease the turnover rate of the epithelial cells, thereby making the intestinal environment a more stable one for the parasite.

Orchestrating the innate immune response to the invading helminth are the intestinal epithelial cells in the affected area of tissue, which detect the pathogen with their NOD and Toll-like receptors that then activate NFκB. When initiating a T_H2 response, the endothelial cells secrete IL-33, described as a T_H2 accelerator, and thymic stromal lymphopoietin (TSLP). These cytokines influence the local dendritic cells, which have taken up helminth antigens, to travel to the draining mesenteric lymph node and stimulate antigen-specific T cells to differentiate into CD4 T_H2 cells. This also produces CD4 T_{FH} cells that engage antigen-specific B cells and make them switch to the IgE isotype. A strong antigen-specific IgE response is one of several features that characterize an effective anti-parasite response (Figure 10.26).

Features of a protective T_H2-mediated immune response against infection by a helminth
Strong parasite-specific IgE response
Mucosal mast-cell hyperplasia
Eosinophilia in blood and intestine
Changes in intestinal muscle contractability
Goblet-cell hyperplasia
Crypt hyperplasia
Villous atrophy
Weight loss

Figure 10.26 Features that characterize a protective immune response to a helminth infection.

An abundance of mast cells in the helminth-infected tissue is another feature of a protective T_H2 response. IL-3 and IL-9 secreted by CD4 T_H2 cells recruit mast-cell precursors from the blood into the infected tissue, where they become fully differentiated mucosal mast cells. The high-affinity FcεRI receptor on mast cells binds IgE tightly in the absence of antigen. If helminth antigens then bind to the IgE and cross-link two FcεRI molecules, the mast cells become activated to release the contents of their preformed granules, which are full of highly active inflammatory mediators such as histamine. In the gut, these mediators induce the muscle spasms and watery feces that characterize diarrhea, a condition that thoroughly disrupts the environment in which the parasites live, and can evict them first into the gut lumen and then force them rapidly out of the body (see Section 9-13). CD4 T_H2 cells also secrete IL-5, which is the major cytokine controlling eosinophil development and function. During a helminth infection, the IL-5 increases the numbers of eosinophils in the blood and in the infected gut tissue. Like mast cells, eosinophils express FcεRI, which can bind parasite-specific IgE. This can then be cross-linked by the antigens on the worm's surface to activate the eosinophil. The antibody acts to tether the parasite to the surface of the eosinophil so that degranulation of the activated eosinophil will release the granules' toxic molecules, such as major basic protein (MBP), directly onto the worm's surface, where they can injure or kill the parasite. Under attack in these different ways, the parasite is less likely to survive for long in the gut epithelium.

IL-13 is a cytokine secreted by CD4 T_H2 cells that influences the dynamics of the intestinal epithelium. Hyperplasia in the stem cells of the crypt increases the production of goblet cells, which in turn increases the production of mucus. This makes it more likely that the worms will become enmeshed in mucus and more easily shed from the epithelium and flushed from the body. Increasing the production of enterocytes has the effect of increasing enterocyte turnover rate but not their abundance. The resultant halving of the enterocyte life-span perturbs the pathogens' environment, increasing the likelihood that the worms will be dislodged and shed into the gut lumen. Atrophy of the villi, reduced absorption of nutrients, and loss of weight by the host accompanies the T_H2-mediated immune response. This could represent a temporary channeling of resources into the immune response and away from other physiological functions and from the parasite.

Although the adaptive B-cell and T-cell responses are specific to the helminth causing the infection, there is little selectivity in the effector functions used. The immune response to helminth infections does not adapt to the differences distinguishing the life cycles of different species. The key difference in determining the fate of an infecting helminth and its human host is whether the

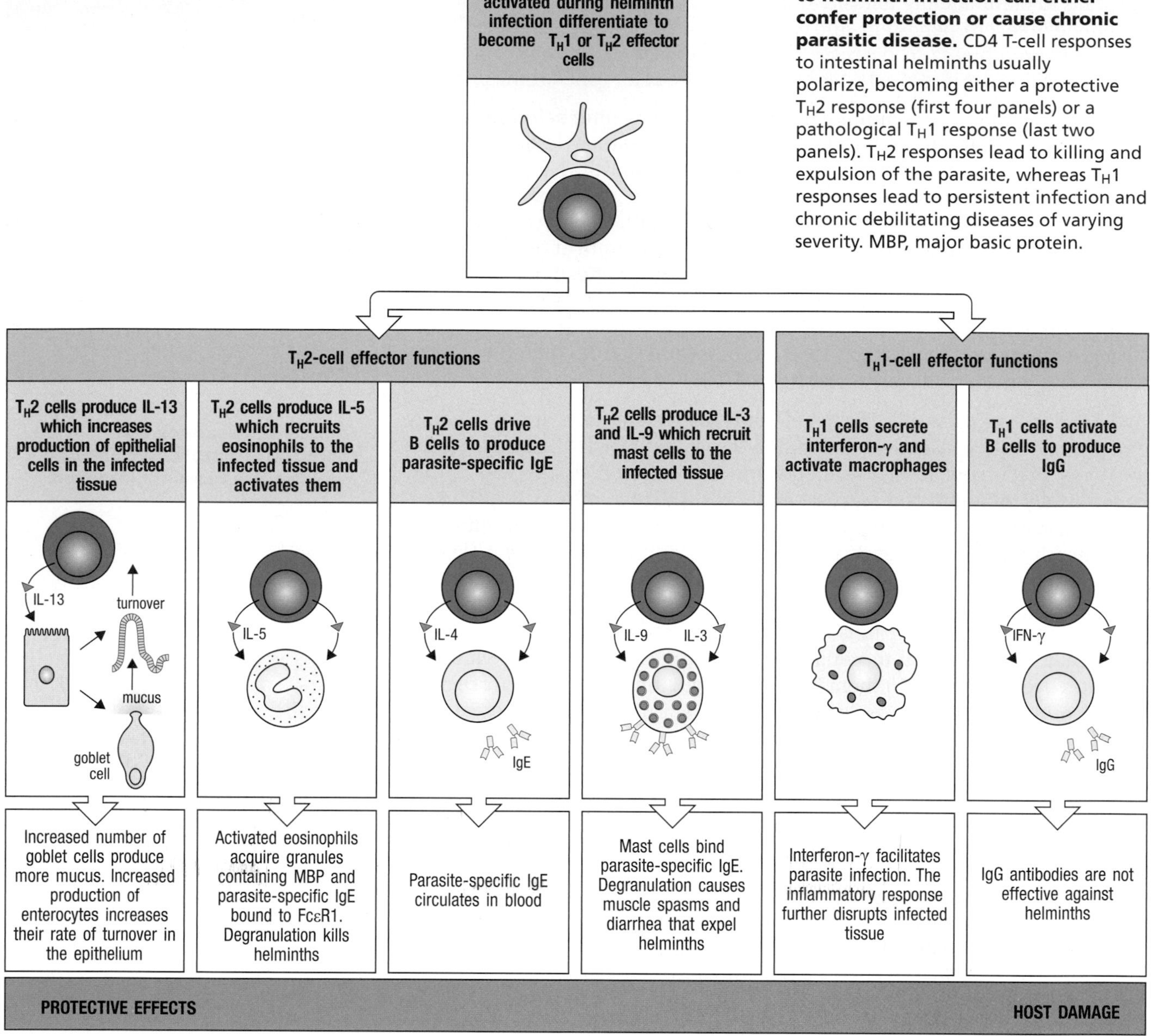

Figure 10.27 Human responses to helminth infection can either confer protection or cause chronic parasitic disease. CD4 T-cell responses to intestinal helminths usually polarize, becoming either a protective T_H2 response (first four panels) or a pathological T_H1 response (last two panels). T_H2 responses lead to killing and expulsion of the parasite, whereas T_H1 responses lead to persistent infection and chronic debilitating diseases of varying severity. MBP, major basic protein.

adaptive immune response becomes mainly T_H2 or mainly T_H1 in nature. A T_H2 response kills and eliminates the parasite to the benefit of the host, whereas a T_H1 response benefits the parasite at the expense of the health of the host (Figure 10.27).

Summary to Chapter 10

The mucosal surfaces of the body cover vital organs that communicate material and information between the human body and its internal environment. Because of these functions, the mucosal surfaces form a more fragile barrier than the skin and are more vulnerable to infection. The possibility of infection is increased even more by the greater area of the mucosal surfaces compared with the skin, and by the large, diverse populations of commensal microorganisms that inhabit mucosal surfaces, particularly those of the gut. Consequently,

Distinctive features of the mucosal immune system	
Anatomical features	Intimate interactions between mucosal epithelia and lymphoid tissues
	Discrete compartments of diffuse lymphoid tissue and more organized structures such as Peyer's patches, isolated lymphoid follicles, and tonsils
	Specialized antigen-uptake mechanisms provided by M cells in Peyer's patches, adenoids, and tonsils
Effector mechanisms	Activated effector T cells predominate even in the absence of infection
	Plasma cells are in the tissues where antibodies are needed
Immunoregulatory environment	Dominant and active downregulation of inflammatory immune responses to food and other innocuous environmental antigens
	Inflammation-anergic macrophages and tolerance-inducing dendritic cells

Figure 10.28 Distinctive features of adaptive immunity in mucosal tissues.

some 75% of the immune system's resources are dedicated to defending the mucosae. The mechanisms and character of adaptive immunity in mucosal tissue, as exemplified by the gut, differ in several important respects from adaptive immunity in other tissues (Figure 10.28).

Secondary lymphoid tissues, which are directly incorporated into the gut wall, continuously sample the gut's luminal contents and stimulate adaptive immune responses against pathogens, commensal organisms, and food. The effector T cells that are generated populate the epithelium and lamina propria of the gut, and the plasma cells produce dimeric IgA that is transcytosed to the lumen, where it coats the mucosal surface. In the healthy gut there is a chronic adaptive immune response that is not inflammatory in nature. This response, in combination with the mechanisms of innate immunity, ensures that microorganisms are confined to the lumen of the gut and prevented from breaching the mucosal barrier. Helminth worms are pathogens that inhabit the intestines and are controlled by adaptive immune responses made in the mesenteric lymph nodes and orchestrated by CD4 T_H2 cells. This response uses parasite-specific IgE to facilitate eosinophil-mediated killing of the worms and mast-cell mediated ejection of them from the body of the human host.

In conclusion, the strategy of the mucosal immune system is to avoid inflammation by being proactive and constantly making adaptive immune responses against potential pathogens before they cause infections. This approach contrasts with that of the systemic immune system, which avoids making an adaptive immune response unless it is absolutely necessary and then relies on inflammation to orchestrate that response.

Questions

10–1 Mucosae exist in all of the following anatomical locations except _____.

a. lactating breasts
b. urogenital tract
c. the limbs
d. gastrointestinal tract
e. salivary glands
f. lachrymal glands
g. the respiratory tract
h. the pancreas.

10–2 All of the following are characteristics of some or all mucosal surfaces except _____. (Select all that apply.)

a. the secretion of viscous fluid called mucus
b. reproductive activities
c. absorption of nutrients
d. participation in gas exchange
e. participation in sensory activities
f. collectively constitute approximately 25% of the body's immune activities
g. use of tight junctions to join epithelial layers
h. tissue regenerates about every 20–30 days.

10–3 Match the term in column A with its description in column B.

Column A	Column B
a. mucosae	1. constitute the gut microbiota
b. cecum	2. epithelial surfaces distributed throughout the body
c. systemic immune system	3. protective epithelial glycoproteins
d. mucins	4. located between the small and large intestines
e. commensal microorganisms	5. defends against pathogens that breach the skin

10–4 The main function of the mucus in the stomach is to _____.

a. trap and kill ingested microorganisms
b. enzymatically degrade complex nutrients
c. protect epithelial cells from the acidified environment
d. protect the microbiota from corrosive gastric juices
e. delay the digestive process to maximize absorption.

10–5 _____ arises from an adaptive immune response to the proteins of gluten.

a. Cholera
b. Celiac disease
c. Selective IgA deficiency
d. Crohn's disease.

10–6 The "M" used to name M cells of the gastrointestinal tract derives from _____.

a. mesenteric
b. microfold
c. monocytes
d. mucosa
e. mast cells.

10–7 Match the term in column A with its description in column B.

Column A	Column B
a. lamina propria	1. situated at the entrance of the gut and airway
b. Peyer's patch	2. a chain of lymph nodes in connective tissue of the gastrointestinal tract
c. Waldeyer's ring	3. dome-like bulging aggregates of lymphocytes that extend into the lumen of the gut
d. M cells	4. transport antigen to pockets on the basolateral side of the gut epithelium
e. mesenteric lymph nodes	5. CD8 T cells with limited range of antigen specificities
f. intraepithelial lymphocytes	6. connective tissue beneath the gut epithelium

10–8 Identify two ways in which the immune responses in gut mucosal tissues contrast with those initiated in systemic non-mucosal tissues.

10–9 Which of the following statements regarding intraepithelial lymphocytes is false? (Select all that apply.)

a. They comprise approximately 10% of the cells in the mucosal epithelium.
b. They are composed of both CD4 and CD8 T cells.
c. They are separated from the lamina propria by a basement membrane.
d. They are activated effector T cells with a narrow range of antigen specificities.
e. They do not include NK cells.
f. They express the $\alpha_4{:}\beta_7$ integrin that binds to E-cadherin on epithelial surfaces.

10–10 An important distinction between macrophages that populate the lamina propria of the gut and the macrophages that populate the skin is that the former _____.

a. cannot phagocytose and kill bacterial pathogens
b. do not present antigens to T cells
c. do not possess signaling receptors needed for production of inflammatory cytokines
d. express much higher levels of TLRs
e. are very rare in the gut mucosa.

10–11 Whereas _____ is the predominant immunoglobulin in intestinal fluid, _____ is the dominant immunoglobulin in the urogenital tract.

a. dimeric IgA; IgE
b. IgE; dimeric IgA
c. dimeric IgA; pentameric IgM
d. dimeric IgA; IgG
e. pentameric IgM; monomeric IgA.

10–12 Which of the following pairs is mismatched?

a. NOD1: a cytoplasmic receptor of intestinal epithelium
b. NLRP3: assists in the formation of an inflammasome
c. intestinal macrophages: professional antigen-presenting cells

d. TLR-5: detects flagellin on apical and basolateral epithelial surfaces
e. neutrophils: attracted by CXCL8.

10–13 A T lymphocyte activated in the GALT will subsequently home to all of the following except _____. (Select all that apply.)
a. mucosal lymphoid tissues of lactating mammary glands
b. the spleen
c. mucosal lymphoid tissues of the respiratory tract
d. systemic lymph nodes
e. mucosal lymphoid tissues of the gastrointestinal tract
f. lymphoid tissues of the skin.

10–14 Identify which of the following immune responses is pivotal to the killing and elimination of helminths.
a. killing by cytotoxic T cells in the lamina propria
b. TH1-induced inflammation
c. TH2-associated cytokines
d. phagocytosis by intestinal macrophages
e. systemic immune responses
f. B-cell secretion of IgG.

10–15 Richard Brennan began penicillamine therapy after he was diagnosed with Wilson's disease (manifested by copper accumulation in the tissues) at age 10 years. Ten months after beginning this treatment he began to experience multiple sinus infections, and one episode of pneumonia. Recently he came to the emergency room with acute diarrhea, vomiting, fever, and foul-smelling intestinal gas. Stool samples revealed the presence of trophozoites of Giardia. Blood tests showed normal levels of B and T cells and normal IgM and IgG concentrations, but markedly decreased IgA at 6 mg/dl (normal range 40–400 mg/dl). Richard was treated for his giardiasis with metronidazole. His selective IgA deficiency was associated with penicillamine, shown previously to be a complication in some patients with Wilson's disease. His IgA levels returned to normal when penicillamine was discontinued. This is an example of a drug-induced transient form of IgA deficiency. Which of the following antibodies that uses the same transport receptor as dimeric IgA would have been present in the lumen of the gastrointestinal tract and mucosal secretions of Richard while he was taking penicillamine?
a. IgD
b. IgM
c. IgG
d. IgE
e. none of the above.

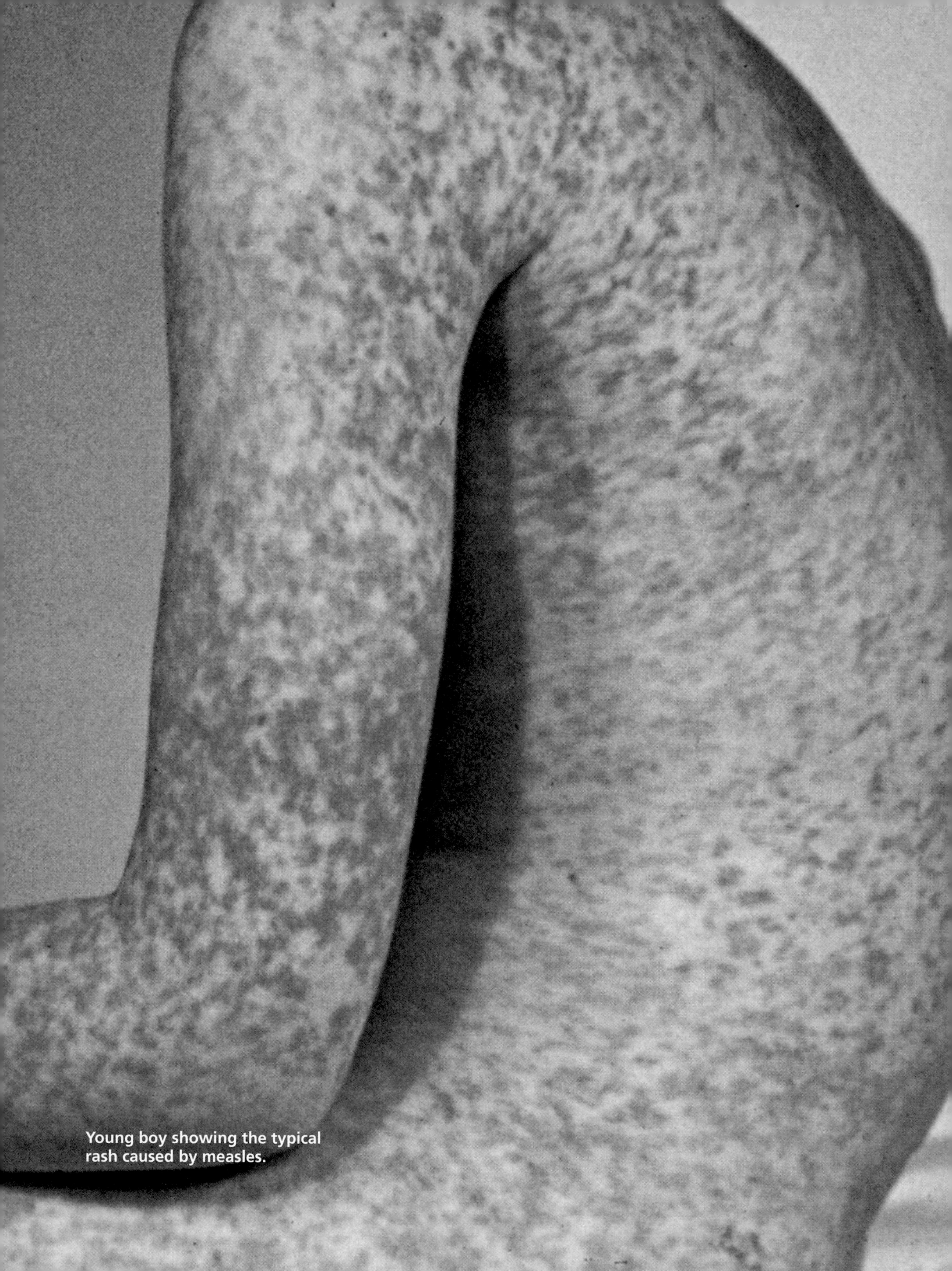

Young boy showing the typical rash caused by measles.

Chapter 11

Immunological Memory and Vaccination

This chapter examines the related topics of immunological memory and vaccination. During a successful primary immune response against a pathogen, two goals are achieved. The first is developing a powerful force of effector cells and molecules that ends the infection as rapidly as possible. The second is building up an immunological memory, a reserve of long-lived B cells and T cells called **memory cells**. This army will confront any future invasion by the pathogen with a **secondary immune response** of such speed and force that the infection will be cleared before it can harm the human host.

The first part of this chapter considers how immunological memory is developed during the primary response and becomes manifested in the secondary response. It is often said that the science of immunology began in antiquity with the Greek historian Thucydides observing the power of the secondary immune response. He wrote that survivors of the 'great plague of Athens,' in the 5th century BC, were spared when the plague returned years later. That may be the first recorded and surviving observation, but the connection had probably been made by thoughtful people for millennia before, and not only in Greece. The experiment occurs in every family. When children suffer infectious disease, they can be looked after by their parents and other adult relatives because the adults are immune, having survived the same disease in their childhood. And in the case of smallpox, for example, facial scars were a reliable way of identifying those who had had the disease.

In the second part of the chapter we examine how knowledge of immunological memory has been used in medical practice to improve human health and survival through the practice of **vaccination**. The aim of vaccination is to induce a primary immune response and immunological memory by immunizing people with a form of the pathogen, or a part of the pathogen, that stimulates a protective adaptive immune response but does not cause disease. If the vaccinated people subsequently encounter the pathogen, they make a secondary immune response that eliminates the pathogen before it takes hold. In poor countries where infectious diseases are endemic and vaccines expensive, there are campaigns to make vaccines more accessible. In rich countries, where vaccination programs have successfully eradicated much infectious disease, there are campaigns to stop vaccination because of the side effects, some of which are real and others imagined.

Immunological memory and the secondary immune response

In previous chapters we saw how an adaptive immune response is made against a pathogen that outruns the forces of innate immunity and successfully invades a person's body for the first time. In this circumstance, the infection causes disease and disability before the primary adaptive immune response has cleared the infection. Because the pathogen has invaded successfully on one occasion, it is more than likely to do it again, and with some regularity. The adaptive immune system, however, retains a memory of its battles with pathogens, which allows it to capitalize on past experience when confronting a pathogen that reinvades. This immunological memory allows a person to react to a pathogen's second infection with a secondary immune response that is quicker and stronger than the primary immune response. In most instances, the secondary response is so effective that the infection is cleared before it causes any significant symptoms of disease. In this part of the chapter we examine how immunological memory is formed during the latter stages of the primary immune response to a pathogen and how it is used to develop a secondary immune response to subsequent infections by the same pathogen.

11-1 Antibodies made in a primary immune response persist for several months and provide protection

With the termination of an infection by the primary immune response, raised levels of high-affinity pathogen-specific antibodies are present throughout the blood, lymph, and tissues, or at every mucosal surface. The antibodies are secreted by plasma cells residing in the bone marrow or in the tissue beneath a mucosal surface, and high levels are sustained for several months after the infection has been cleared (Figure 11.1). During this time, these antibodies provide **protective immunity**, ensuring that a subsequent invasion by the pathogen does not cause disease.

Many infectious diseases are seasonal. For example, during the winter you can be exposed repeatedly to the same cold virus over a period of weeks or months, as it is passed among family, friends, colleagues, and the community at large. During this period, antibodies raised against a cold caught early in the season prevent reinfection with the same virus later in the season. On invading again, the virus will immediately be coated with specific IgA or IgG. The virus will be neutralized by antibody and will fail to infect cells and replicate.

Figure 11.1 History of infection with a pathogen. Consider a student's history of infection with a pathogen. The student's first infection with the pathogen was not stopped by innate immunity, so a primary adaptive immune response developed. Production of effector T cells and antibodies terminated the infection. The effector T cells were soon inactivated, but antibody persisted, providing protective immunity that prevented reinfection despite frequent exposure to infected classmates. A year afterward, antibody levels had dropped and the pathogen would now be more likely to establish an infection. When a second infection did occur, a much faster and stronger secondary immune response was made; this eliminated the pathogen before it had a chance to disrupt tissue or cause signicant disease. This strong response was mediated by long-lived, pathogen-specific B cells and T cells that had been stockpiled during the primary immune response. The student's immune system had retained a 'memory' of that first infection.

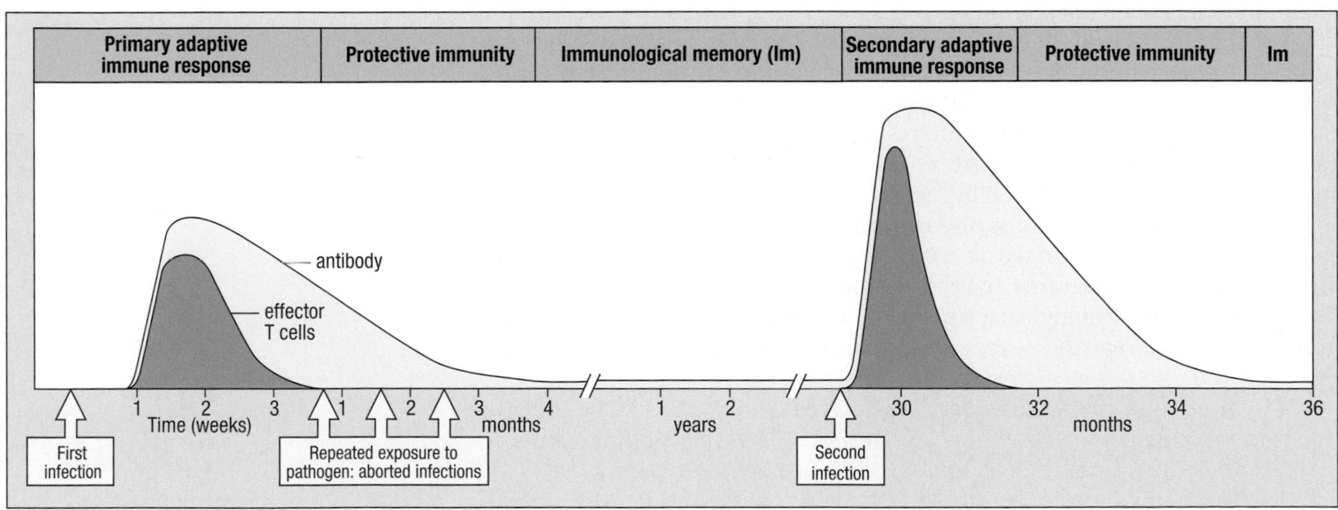

In a bacterial infection, bacteria opsonized by IgG or IgA is delivered to the Fc receptors and complement receptors of phagocytes. Parasites are killed or ejected by mast cells and eosinophils activated by parasite-specific IgE. In these circumstances,where specific antibody cooperates with all the effector functions of innate immunity, a pathogen gets little opportunity to grow and replicate. This means that the pathogen load does not reach the point at which a new adaptive immune response is activated (see Figure 11.1).

11-2 Low levels of pathogen-specific antibodies are maintained by long-lived plasma cells

Most plasma cells made in the primary response are short-lived. As a result, the amount of circulating pathogen-specific antibody gradually decreases over a period of a year and then reaches a low, steady-state level (see Figure 11.1) that is maintained for life by a small population of long-lived plasma cells in the bone marrow. Survival of these plasma cells is sustained by interactions with bone-marrow stromal cells and with the IL-6 secreted by the stromal cells. Short-lived plasma cells die as a result of several inhibitory mechanisms. In one, complexes of antigen and antibody bind to FcγRIIB1 and induce the plasma cell to die by apoptosis. A second cause of plasma-cell death is loss of contact with stromal cells and of the survival signals they give. This loss could result from competition with plasma cells activated by a more recent infection arriving at the bone marrow to seek succour from stromal cells.

The small, long-lived population of plasma cells is programmed to survive and to continue making pathogen-specific antibodies long after the pathogen and its antigens have been cleared from the body. These long-lived plasma cells and the antibodies they make form one component of the host's immunological memory of the pathogen. Any subsequent infection by the same pathogen will confront specific high-affinity antibody from the first moment. Here the major weapon of adaptive immunity is available to participate in the innate immune response by binding to the pathogen and efficiently delivering it to the effector cells of innate immunity. In some circumstances, the antibody in combination with innate immunity might be sufficient to end an infection. When that is not the case, the antibody speeds delivery of the pathogen and its antigens to the antigen-presenting cells that will initiate a secondary immune response.

11-3 Long-lived clones of memory B cells and T cells are produced in the primary immune response

The first goal of a primary adaptive immune response is to subdue the ongoing infection by a harmful pathogen that is outrunning innate immunity. This is accomplished by clonal expansion of pathogen-specific naive T cells and B cells to produce large populations of short-lived effector B cells and T cells that work together to eradicate the invading microorganisms. If this first goal is not achieved, the infected person either dies of the infection or suffers a chronic and often debilitating disease.

With attainment of the first goal, the second goal of the primary response is to ensure that future invasions by the pathogen will be met by an immune response of overwhelming force. Mediating such secondary immune responses are long-lived pathogen-specific **memory T cells** and **memory B cells**. These cells originate in the secondary lymphoid tissues during the primary response to the pathogen, and they form, with the long-lived plasma cells, the three components of immunological memory (Figure 11.2). The population of pathogen-specific memory cells mirrors that of the pathogen-specific effector cells; it can therefore include CD8 T cells, T_{FH}, T_H1, T_H2, and T_H17 CD4 T cells, and B cells programmed to become plasma cells secreting IgA,

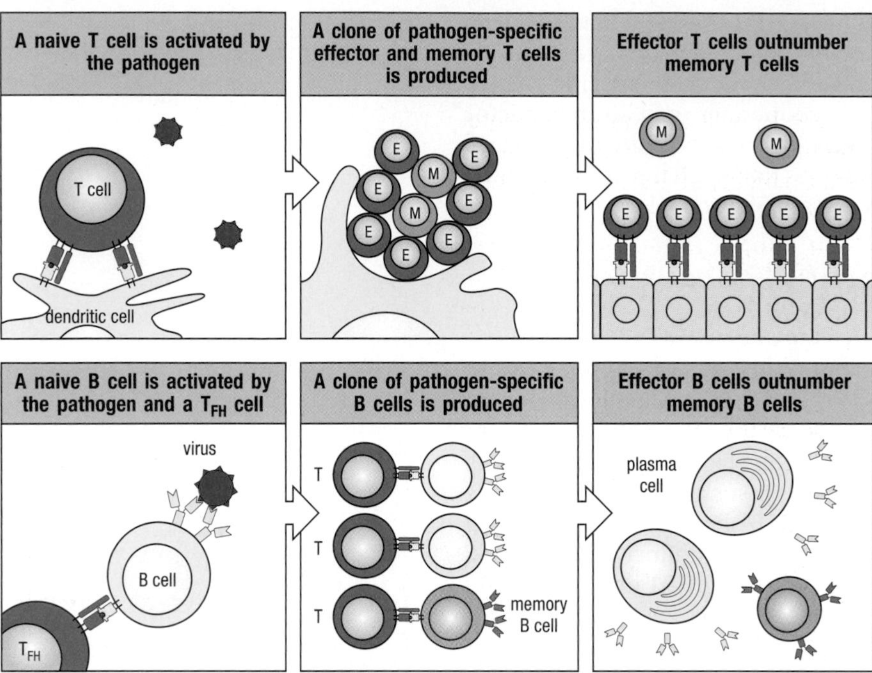

Figure 11.2 Both effector B and T cells and memory B and T cells are produced during a primary immune response.

IgG, and IgE antibodies. Both effector cells and memory cells are produced during the proliferation and differentiation of antigen-activated naive T and B cells in the secondary lymphoid tissue (see Chapters 8 and 9). At the beginning of a primary response, when infecting pathogens are most dangerous, effector lymphocytes are produced in much greater numbers than memory lymphocytes, but later on, when the pathogen is in defeat, the emphasis turns to making more memory cells.

A secondary immune response occurs when a pathogen successfully infects a person for the second time and is again not cleared by the combination of innate immunity and steady-state level of pathogen-specific antibody (see Figure 11.1). In responding to this second infection, memory cells have several advantages that enable them to respond more forcefully than was possible for naive lymphocytes during the primary response. First, pathogen-specific memory cells far outnumber their naive counterparts. Second, memory cells, like effector cells, are more readily activated than naive lymphocytes. Third, memory B cells have undergone isotype switching, somatic hypermutation, and affinity maturation (see Chapter 9). So, upon activation by the pathogen, the memory B cells make IgG, IgA, or IgE antibodies that are inherently better at binding the pathogen and delivering it for disposal than the antibodies made in the primary response, especially IgM. In the course of the secondary response, pathogen-activated memory B cells undergo further refinement through somatic hypermutation and affinity maturation of their immunoglobulins. Because of these improvements, the infecting pathogen is cleared more quickly by a secondary response, usually with few or no symptoms of disease (see Figure 11.1).

Affinity maturation in the secondary immune response produces a second generation of memory B cells that are superior to those that emerged from the primary response. Consequently, a third infection by the pathogen will be met by a tertiary antibody response that is even better than that made in the secondary response, and so on. In this way, successive infections with the same pathogen sharpen the defenses of adaptive immunity and immunological memory. Although immunologists sometimes use the terms tertiary immune response, quaternary response, and so on, it is more usual to refer to all memory responses as the secondary response. An older name, the 'anamnestic response,' means memory response.

11-4 Memory B cells and T cells provide protection against pathogens for decades and even for life

The phenomenon of immunological memory is well illustrated by Peter Panum's classic epidemiological study of the inhabitants of the Faroe Islands in the North Atlantic Ocean. The measles virus, a highly infectious and potentially life-threatening pathogen, was first introduced to the islands in 1781, when it caused a severe epidemic in which the entire human population was infected and suffered disease. More than 60 years later, in 1846, when the measles virus was again brought to the islands, almost all of the 5000 inhabitants who had been born since the first epidemic came down with the disease. But all the 98 survivors of the 1781 epidemic proved resistant: they had retained sufficient immunological memory to prevent their second exposure to the measles virus from becoming an established disease-causing infection.

Until the latter part of the 20th century, smallpox was, like measles, a much-feared killer of humankind: from 1850 to 1979 about 1 billion people died from smallpox infection. During this same period, worldwide vaccination programs progressively reduced the spread of smallpox virus to the point at which mass vaccination was discontinued in the United States in 1972, and by 1979 the smallpox virus was judged to have been eradicated worldwide. Just two immunizations with vaccinia virus, a close but benign relative of smallpox virus, induces a secondary immune response with immunological memory that also works for smallpox. So any subsequent encounter with the smallpox virus meets with a tertiary immune response that scotches the virus before it causes disease. At present, about half the population of the United States has been vaccinated against smallpox and half has not. Because neither group has ever been exposed to the smallpox virus, comparison of the two groups reveals much about the persistence of immunological memory in the absence of further stimulation by antigen.

After vaccination, the amount of vaccinia-specific antibody in the blood rapidly increases to a maximum level and then, over the next 12 months, decreases to about 1% of the maximum. This steady-state level is maintained for up to 75 years and possibly for life (Figure 11.3, top panel). Because an antibody molecule only survives for about 6 weeks in the blood, this antibody level must be maintained by memory plasma cells making vaccinia-specific antibody throughout a person's lifetime. After vaccination, the number of virus-specific B cells in the blood also increases rapidly to a maximum and then declines over a 10-year period to reach a stable level that is about 10% of the maximum. This pool of memory B cells is maintained for life in a state that can respond to infecting smallpox virus or to a further immunization with vaccinia. Vaccination also produces populations of memory CD4 T cells (Figure 11.3, middle panel) and CD8 T cells (Figure 11.3, bottom panel) that can also persist for up to 75 years. It is these pools of memory T cells, along with the memory B cells, that would respond to a smallpox virus infection or a further vaccination.

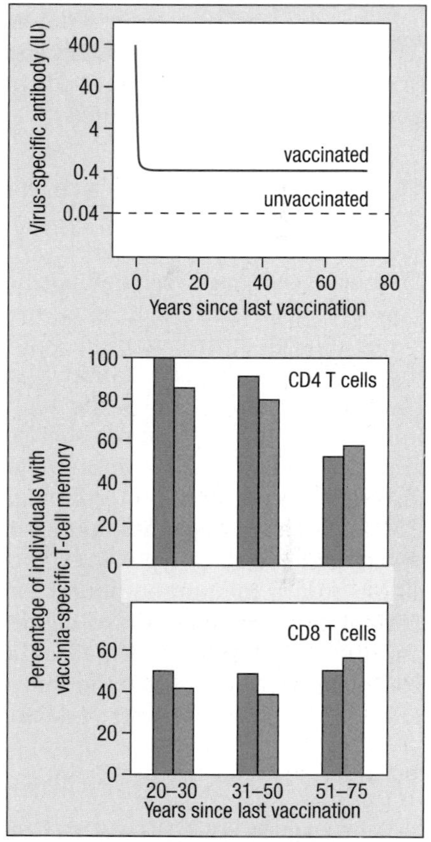

Figure 11.3 Retention of vaccinia-specific antibodies and T cells after vaccination against smallpox virus. Specific anti-vaccinia antibodies continue to be made for as long as 75 years after the last exposure to vaccinia virus, the smallpox surrogate that is used for vaccination (top panel). The numbers represent international units (IU) of antibody, a standardized way of measuring an antibody response. Many vaccinated individuals retain populations of vaccinia-specific CD4 T cells and CD8 T cells (bottom panel). Only small differences are observed for individuals who received one (blue bars) or two (pink bars) vaccinations. Courtesy of Mark Slifka.

Not all forms of protective immunity are as persistent as those induced by the smallpox vaccine or measles infection. After vaccination against diphtheria, a bacterial pathogen, the level of protective anti-diphtheria antibodies in the blood continues to decrease and is halved after 19 years. By contrast, the half-life of anti-measles protection is estimated to be 200 years.

11-5 Maintaining populations of memory cells does not depend upon the persistence of antigen

Lymphocytes have a general requirement for regular stimulation if they are to survive. When such signals are not received, lymphocytes die by apoptosis. During their development and recirculation, naive lymphocytes must receive survival signals through their antigen receptors (see Section 6-14). Memory lymphocytes are not bound by this restriction, as is evident from the persistence of vaccinia-specific lymphocytes in people having had no outside contact with vaccinia antigens for decades. Although one cannot rule out the existence of an internal depot that retains antigens from the time of vaccination, it is most unlikely (see Section 11-4).

Immunological memory is thus sustained by populations of long-lived lymphocytes that were induced on exposure to antigen but then persist in its absence. Although a memory population survives, individual memory cells have a limited lifespan. At any given time, most of the memory cells are in a quiescent state, but a small fraction are dividing and replenishing the population to make up for cells that have died. This antigen-independent activation and proliferation is driven by signals delivered by cytokines via their receptors on memory cells. The survival and proliferation of memory CD4 and CD8 T cells depends on signals from the IL-7 and IL-15 receptors. The renewal and replenishment of memory B cells and their cognate memory T cells is believed to occur in the bone marrow and to be driven by interactions with stromal cells and the cytokines they produce.

11-6 Changes to the antigen receptor distinguish naive, effector, and memory B cells

Memory B cells have been defined more precisely than memory T cells because they are clearly distinguishable from naive B cells: their immunoglobulin genes and cell-surface immunoglobulin have been altered by isotype switching and somatic hypermutation. Memory B cells are also clearly distinguishable from effector B cells—the plasma cells. Memory B cells have surface immunoglobulin and do not secrete antibody, whereas plasma cells secrete antibody and lack surface immunoglobulin. Memory B cells and plasma cells also have very different morphologies. In addition, memory B cells express CD27, which distinguishes them from naive and effector B cells. T-cell receptors do not switch isotype, undergo somatic hypermutation, or undergo transition from a membrane-bound form to a soluble form, and so it is more difficult for immunologists to define and distinguish between naive, effector, and memory T cells. This makes the investigation of T-cell memory a more complicated business than the study of B-cell memory. For this reason we will examine B-cell memory first, and then turn to T-cell memory.

11-7 In the secondary immune response, memory B cells are activated whereas naive B cells are inhibited

In the primary response, low-affinity IgM antibodies are made first, but then somatic hypermutation, affinity maturation, and isotype switching give rise to high-affinity IgG, IgA, and IgE (see Sections 4-14 and 4-15; see also Sections 9-8 and 9-9). Memory B cells are derived from the clones of B cells making antibodies with the highest affinity for antigen (see Section 9-10). Some weeks to months after an infection has cleared, memory B cells reach their maximum number, and this is sustained for life. At this point, the number of pathogen-specific memory B cells exceeds by 10–100-fold the number of naive pathogen-specific B cells that were activated in the primary response (Figure 11.4). To ensure that low-affinity antibodies and IgM are not made in the secondary response, the activation of naive pathogen-specific B cells is suppressed. This suppression is mediated by immune complexes composed of the pathogen or its antigens bound to antibodies made by the B cells activated in the primary response. These complexes bind to the B-cell receptor of pathogen-specific naive B cells and also to the inhibitory Fc receptor, FcγRIIB1, which is expressed by naive B cells but not by memory B cells. This

	Source of B cells	
	Unimmunized donor Primary response	Immunized donor Secondary response
Frequency of antigen-specific B cells	1 in 10^4 – 1 in 10^5	1 in 10^2 – 1 in 10^3
Isotype of antibody produced	IgM, IgG, IgA, IgE	IgG, IgA, IgE
Affinity of antibody	Low	High
Somatic hypermutation	Low	High

Figure 11.4 Comparison of the B-cell populations that participate in the primary and secondary adaptive immune responses. Key features that make the secondary response stronger than the primary response are the greater numbers of antigen-specific B cells present at the start of the secondary response and the preferential use of isotype-switched clones of B cells that express higher-affinity immunoglobulins as a result of somatic hypermutation and affinity maturation.

cross-linking of the B-cell receptor and the Fc receptor delivers a negative signal that inhibits activation of the pathogen-specific naive B cell and induces its death by apoptosis (Figure 11.5).

11-8 Activation of the primary and secondary immune responses have common features

A secondary immune response is made only if a reinfecting pathogen overcomes the combined defenses of innate immunity and the steady-state level of pathogen-specific antibody. In such circumstances, the pathogen expands its numbers at the site of infection and some get carried to the secondary lymphoid tissues by dendritic cells. Memory T cells differ from naive T cells in two ways that increase the speed of the secondary response. First, some recirculate to peripheral tissues rather than through secondary lymphoid organs, and so memory CD8 T cells and CD4 T_H1, T_H2, and T_H17 cells can be activated directly at a site of infection by dendritic cells and macrophages presenting their specific antigens. Second, their activation requirements are less demanding than those of naive T cells because they, like effector T cells, do not require

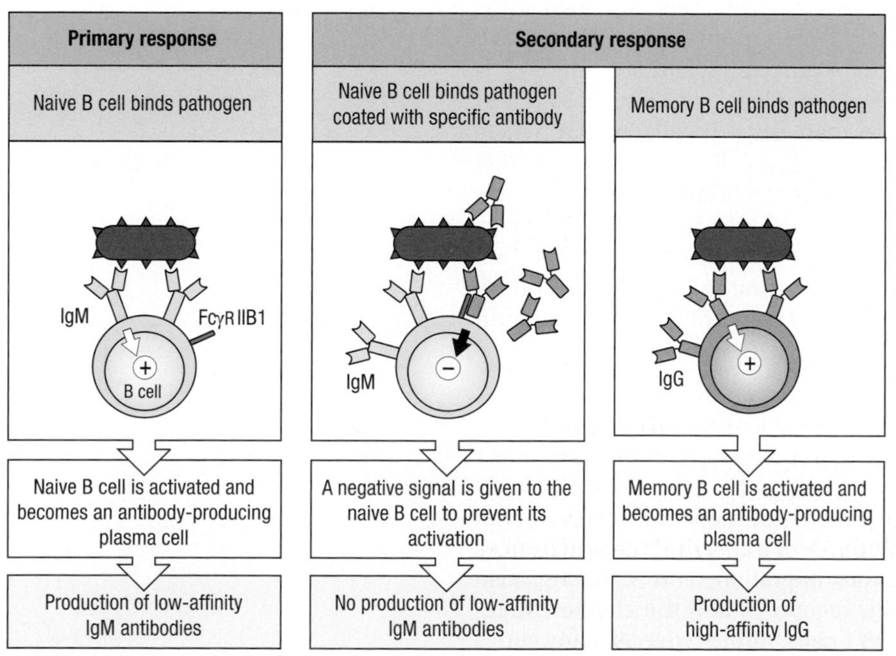

Figure 11.5 IgG antibody suppresses the activation of naive B cells by cross-linking the B-cell receptor and FcγRIIB1 on the B-cell surface. In a primary immune response, a pathogen binding to the antigen receptor of a naive B cell delivers a signal that activates the cell to become an antibody-producing plasma cell (left panel). In a secondary response, in which the antigen receptor and the inhibitory Fc receptor FcγRIIB1 on a naive B cell can be cross-linked by a pathogen coated with IgG, this delivers a negative signal that prevents the activation of the cell (center panel). Memory B cells do not express FcγRIIB1 and are activated by the pathogen binding to the IgG B-cell receptor. Most memory B cells make IgG1 (right panel).

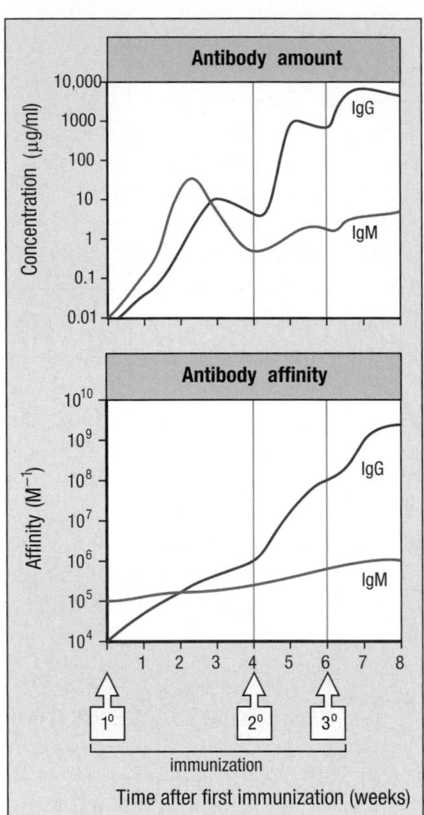

Figure 11.6 The amount and affinity of antibody increase after successive immunizations with the same antigen. This figure shows the results of an experiment on mice that mimics the development of specific antibodies when a person is given a course of three immunizations (1°, 2°, and 3°) with the same vaccine. The upper panel shows how the amounts of IgM (green) and IgG (blue) present in blood serum change over time. The lower panel shows the changes in average antibody affinity that occur. Note that the vertical axis of each graph has a logarithmic scale because the observed changes in antibody concentration and affinity are so large.

co-stimulation through CD28. Other memory T cells, including memory CD4 T_{FH} cells, are activated in the secondary lymphoid tissues by dendritic cells presenting the pathogen's antigens. Memory B cells recirculate between the blood and the lymph like naive B cells. As in the primary response, the secondary B-cell response begins in a secondary lymphoid tissue at the interface between the B-cell zone and the T-cell zone. There, the activation and proliferation of pathogen-specific memory B cells is driven by cognate interactions with the pathogen-specific effector CD4 T_{FH} cells.

Memory B cells that have bound antigen and internalized it by receptor-mediated endocytosis present peptide:MHC class II complexes to their cognate CD4 T_{FH} cells, which surround and infiltrate the germinal centers. Contact between the antigen-presenting B cells and CD4 T_{FH} cells leads to an exchange of activating signals and the proliferation of both the memory B cells and the T_{FH} cells. Competition for binding to antigen drives the selective activation of those B cells having B-cell receptors with the highest affinities for antigen. As in the primary response, some of these cells develop immediately into plasma cells, whereas others move to the follicles and participate in a germinal center reaction (see Sections 9-7 and 9-8). They enter a second round of proliferation, during which they undergo somatic hypermutation and further isotype switching, followed by affinity maturation. As a consequence, the average affinity of the antibodies made in the secondary response rises well above that of those made in the primary response (Figure 11.6).

Memory B cells are more sensitive than naive B cells to the presence of specific antigen, and their response is quicker than that of naive B cells. The high affinity of their antigen receptors makes memory B cells more efficient than naive B cells in binding and internalizing antigen for processing and presentation to CD4 T_{FH} cells. Memory B cells also express higher levels of MHC class II and co-stimulatory molecules on their surface than naive B cells, which makes their cognate interactions with antigen-specific T_{FH} cells more efficient. This has two effects. First, a smaller pathogen population is sufficient to trigger a B-cell response, which therefore occurs at an earlier stage in infection than in the primary response. Second, once activated, memory B cells take less time than activated naive B cells to differentiate into plasma cells. In the secondary response, new antibody is detectable in the blood after only 4 days, compared with 8 days in the primary response.

11-9 Combinations of cell-surface markers distinguish memory T cells from naive and effector T cells

Naive and effector T cells exhibit similar patterns of transcription for 95% of the genes they express. Prominent among the 5% of transcriptional differences are genes involved in activation, adhesion, migration, and signaling, also genes for cytokines, chemokines and their receptors, and the effector molecules that distinguish CD4 T cells from CD8 T cells. About twice as many genes distinguish CD8 memory cells from naive CD8 cells than distinguish CD4 memory T cells from their naive counterparts. As a consequence of the differences in gene expression, various cell-surface proteins are differentially expressed on naive T cells, effector T cells, and memory T cells, but the differences between effector and memory T cells are considerably less than the differences distinguishing them from naive T cells (Figure 11.7). The combination of CD45RA, CD45RO, L-selectin (CD62L), and CCR7 is commonly used to distinguish memory T cells from naive and effector T cells. The IL-7 receptor, which is essential for the renewal and survival of memory cells, also distinguishes memory cells from effector cells.

CD45 is a tyrosine phosphatase involved in antigen-activated signaling from the T-cell and B-cell receptors. Naive and memory T cells make different isoforms of the CD45 protein by alternative splicing of CD45 mRNA. Naive T cells express predominantly the CD45RA isoform, which functions poorly with the T-cell receptor complex and transduces weak signals when the T-cell receptor recognizes specific antigen. Memory T cells express CD45RO, which has a smaller extracellular domain than CD45RA, owing to three exons being spliced

Protein	Naive	Effector	Memory	Comments
CD44	+	+++	+++	Cell-adhesion molecule
CD45RO	+	+++	+++	Tyrosine phosphatase that modulates T-cell receptor signaling
CD45RA	+++	+	–	Tyrosine phosphatase that modulates T-cell receptor signaling
CD62L	+++	–	Some +++	Receptor for homing to lymph node
CCR7	+++	+/–	Some +++	Chemokine receptor for homing to lymph node
CD69	–	+++	–	Early activation antigen
Bcl-2	++	+/–	+++	Promotes cell survival
Interferon-γ	–	+++	+++	Effector cytokine; mRNA present and protein made on activation
Granzyme B	–	+++	+/–	Effector molecule in cell killing
FasL	–	+++	+	Effector molecule in cell killing
CD122	+/–	++	++	Part of receptor for IL-15 and IL-2
CD25	–	++	–	Part of receptor for IL-2
CD127	++	–	+++	Part of receptor for IL-7
Ly6C	+	+++	+++	GPI-linked protein
CXCR4	+	+	++	Receptor for chemokine CXCL12; controls tissue migration
CCR5	+/–	++	Some +++	Receptor for chemokines CCL3 and CCL4; tissue migration

Figure 11.7 **Proteins that are differentially expressed by naive T cells, effector T cells, and memory T cells.** GPI, glycosylphosphatidylinositol.

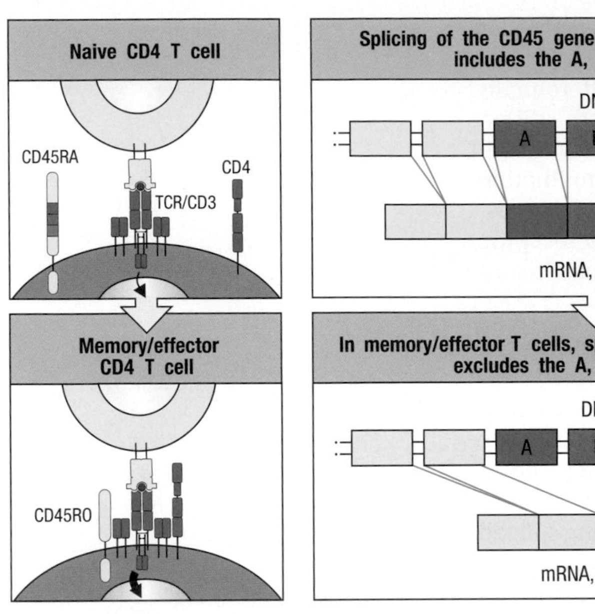

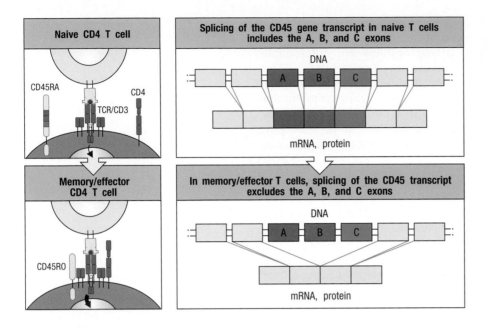

Figure 11.8 Memory CD4 T cells express an altered CD45 isoform that works more effectively with the T-cell receptor and co-receptors. CD45 is a transmembrane tyrosine phosphatase involved in T-cell activation and, by differential mRNA splicing, can be made in two isoforms, CD45RA and CD45RO, the former having a larger extracellular domain than the latter. Naive CD4 T cells express only CD45RA, effector T cells express predominantly CD45RO, and memory T cells express CD45RO (see Figure 11.7). The absence of the sequences encoded by exons A, B, and C in CD45RO enables it to associate with both the T-cell receptor and the CD4 co-receptor and improve the efficiency of signal transduction.

out of CD45RO mRNA. CD45RO interacts well with the T-cell receptor complex and transduces strong signals when the T-cell receptor recognizes specific antigen (Figure 11.8).

Healthy adult humans have around 10^{12} peripheral α:β T cells, comprising roughly equal numbers of naive T cells and memory T cells. Sequence analysis of T-cell receptors estimates that the naive T cells have 2.5×10^7 antigen specificities, whereas the memory T cells have only 1.5×10^5 antigen specificities. The acquisition of T-cell memory for a pathogen means that, on average, the number of memory T cells activated in the secondary response to a pathogen is a hundredfold greater than the number of naive T cells activated in the primary response. Unlike naive B cells, naive T cells can be activated in the secondary response. Their contribution, however, is a minor one.

11-10 Central and effector memory T cells recognize pathogens in different tissues of the body

Two subsets of memory T cells have been defined: **central memory T cells** (**T_{CM}**) and **effector memory T cells** (**T_{EM}**). The two subsets are distinguished by the tissues in which they reside and the tissues in which they respond to antigen. Central memory T cells express L-selectin (CD62L) and the chemokine receptor CCR7, which allow them, like naive T cells, to enter secondary lymphoid organs and be activated by antigens presented by dendritic cells. Before activation, central memory T cells exhibit only limited effector function, but they have a low threshold for activation combined with high potential for IL-2 production, cellular proliferation, and differentiation into effector cells.

Effector memory T cells do not circulate through the secondary lymphoid tissues, because they lack L-selectin and CCR7. Instead they express other chemokine receptors, such as CCR6, CCR4, CXCR3, and CCR5, that gain them entry to non-lymphoid tissues, including mucosal tissues, and inflamed tissues. Effector memory T cells are heterogeneous and represent the CD8 and T_H1, T_H2, and T_H17 subsets of primary effector cells. By patrolling the peripheral tissues, effector memory T cells can respond immediately to an infection at its site of origin. This talent complements the activation of central memory T cells in the draining lymphoid tissue, which is a slower process of activation but one that generates more effector T cells (Figure 11.9).

Central memory cells (T_{CM})	Effector memory cells (T_{EM})
L-selectin-positive	L-selectin-negative
CCR7-positive	CCR7-negative
Circulate in lymphoid organs	Circulate in non-lymphoid tissues
Stem-cell-like; can be activated by antigen and cytokines	Already differentiated; have high levels of effector molecules

Figure 11.9 The differences between central and effector memory T cells.

11-11 In viral infections, numerous effector CD8 T cells give rise to relatively few memory T cells

During the primary immune response to a viral infection, a vast army of virus-specific effector CD8 T cells is mobilized. Each antigen-activated naive T cell can give rise to as many as 50,000 cytotoxic T cells, which then work to kill off all the virus-infected cells. After clearance of the virus, some 95% of the CD8 T cells die by apoptosis, leaving 5% to constitute the memory CD8 T-cell population. These cells are not chosen at random but are those that express the IL-7 receptor. In number they exceed the naive CD8 T cells that contributed to the primary response by 100–1000-fold, ensuring that any future infection with the virus will be met with overwhelming force (Figure 11.10).

11-12 Immune-complex-mediated inhibition of naive B cells is used to prevent hemolytic anemia of the newborn

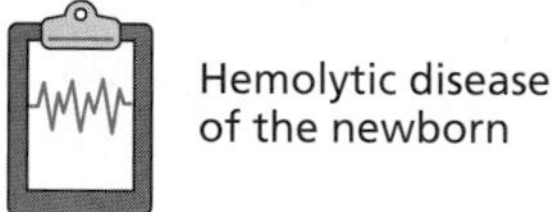

The inhibition of naive B cells by immune complexes is put to practical use in preventing **hemolytic anemia of the newborn**, also called **hemolytic disease of the newborn**. This syndrome affects families in which the father is positive for the polymorphic erythrocyte antigen called **Rhesus D** (**RhD**), and the mother is negative. During a first pregnancy with an RhD^+ baby, fetal erythrocytes cross the placenta and stimulate the mother's immune system to make anti-RhD antibodies. The antibodies made in this primary response cause little harm to the fetus, because they are mainly low-affinity IgM that cannot cross the placenta (Figure 11.11, left panel). During a second pregnancy with an RhD^+ baby, however, fetal red cells again cross the placenta and induce a secondary response to RhD. This produces more abundant antibody, which is now high-affinity IgG that is transported across the placenta by FcRn (see Section 9-14). These antibodies coat the fetal erythrocytes and cause the opsonized cells to be cleared from the circulation by macrophages in the spleen. When born, such babies have severe anemia (Figure 11.11, center panel), which can lead to other complications. Hemolytic anemia of the newborn is most common in Caucasians, where 16% of mothers are RhD^- and 84% of babies are RhD^+, than in Africans or Asians, where less than 1% of the population is RhD^-.

To prevent hemolytic anemia of the newborn, pregnant RhD^- women who have yet to make anti-RhD antibodies are infused with purified human anti-RhD IgG antibody, also called RhoGAM, during the 28th week of pregnancy. The amount of antibody infused is sufficient to coat all the fetal red cells that cross the placenta to enter the maternal circulation. Because all RhD antigen in the maternal circulation is in the form of complexes with human IgG, activation of the mother's RhD-specific naive B cells is prevented (Figure 11.11, right panel). In effect, the mother's immune system is tricked into responding to this primary exposure to RhD antigen as though it were a secondary exposure. Within 3 days after the baby's birth, the mother is given a second infusion of anti-RhD IgG antibody, because during the trauma of birth she will have

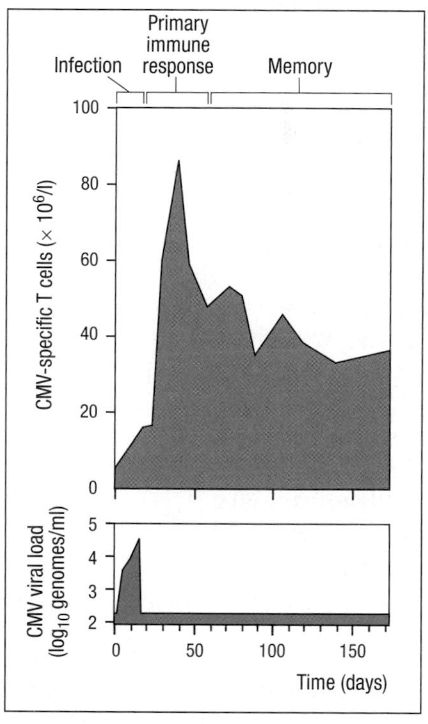

Figure 11.10 Generation of memory T cells during the response to a virus infection. Cytomegalovirus (CMV) is a latent herpesvirus that usually is quiescent but has episodes of activation that are quelled by the immune response. Such an episode is illustrated here for a CMV-carrying patient who underwent immunosuppressive treatment for cancer followed by a hematopoietic stem-cell transplant. The increase in viral load that occurs when the virus is reactivated (lower panel) triggers a rapid increase in the numbers of virus-specific effector CD8 T cells present in the blood (upper panel). This falls back once the virus has been brought under control, leaving a sustained lower level of long-lived, virus-specific memory T cells. Data courtesy of G. Aubert.

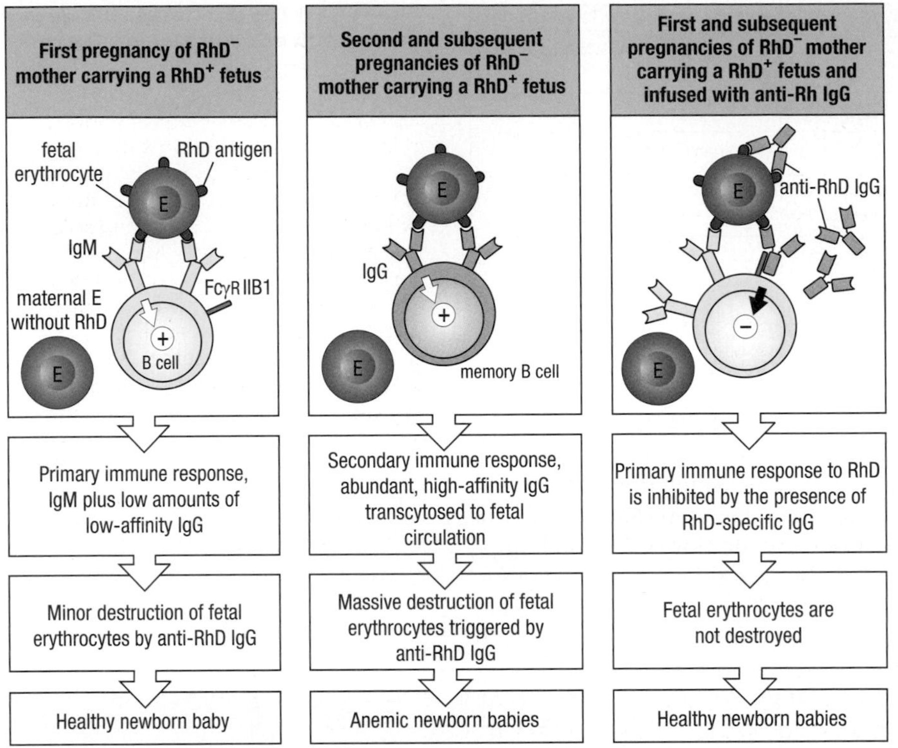

First pregnancy of RhD⁻ mother carrying a RhD⁺ fetus
fetal erythrocyte
RhD antigen
E
IgM
FcγRIIB1
maternal E without RhD
B cell
Primary immune response, IgM plus low amounts of low-affinity IgG
Minor destruction of fetal erythrocytes by anti-RhD IgG
Healthy newborn baby
Second and subsequent pregnancies of RhD⁻ mother carrying a RhD⁺ fetus
IgG
memory B cell
Secondary immune response, abundant, high-affinity IgG transcytosed to fetal circulation
Massive destruction of fetal erythrocytes triggered by anti-RhD IgG
Anemic newborn babies
First and subsequent pregnancies of RhD⁻ mother carrying a RhD⁺ fetus and infused with anti-Rh IgG
anti-RhD IgG
Primary immune response to RhD is inhibited by the presence of RhD-specific IgG
Fetal erythrocytes are not destroyed
Healthy newborn babies

Figure 11.11 Passive immunization with anti-Rhesus D antigen IgG prevents hemolytic anemia of the newborn. In human populations up to 16% of individuals lack the erythrocyte antigen RhD. Rh^- mothers carrying Rh^+ fetuses are exposed to fetal erythrocytes and make Rh-specific antibodies that pass to the fetal circulation and cause fetal red cells to be destroyed. In a first pregnancy of this type, the antibodies produced in the primary response cause only minor damage to the fetal red cells, and a healthy baby is born (left panel). In a second pregnancy, a secondary immune response ensues that produces antibodies causing massive destruction of fetal red cells, and at birth the baby is anemic (center panel). This disease can be prevented if in the first and subsequent pregnancies the mother is passively infused with purified human anti-Rh antibodies before she has made her own response. The immune complexes of fetal erythrocytes coated with IgG prevent a primary B-cell response from being made to the Rh antigen (right panel).

been further exposed to the baby's blood cells. This protects a future pregnancy from leading to hemolytic anemia of the newborn.

Although the amount of infused anti-RhD antibody (300 μg) is in excess of that needed to coat the fetal red cells in the maternal circulation, almost none of it will be transported across the placenta and into the fetal circulation, where it could damage fetal erythrocytes. This is because the anti-RhD is heavily diluted by the roughly 60 g of IgG in the mother's circulation that is not specific for the RhD antigen.

11-13 In the response to influenza virus, immunological memory is gradually eroded

The suppression of naive B-cell activation that occurs during the secondary response to a pathogen is a good strategy for dealing with conserved pathogens, such as the measles virus, which do not change their antigens, but has drawbacks when confronting highly mutable pathogens such as influenza virus. Every year, new influenza strains emerge that escape the protective immunity of some part of the human population. In these variant strains, one or more of the epitopes targeted by the preexisting antibodies has been lost. Having successfully terminated a first infection with influenza, you will have high-affinity antibodies against multiple epitopes of the viral capsid proteins. Together these antibodies neutralize the virus. During second and subsequent infections, the memory response limits the antibody response to the epitopes shared by the infecting strain and the original strain. With each passing year, you will be exposed to influenza viruses that have fewer and fewer epitopes to which your immunological memory can respond. This allows the virus to gradually escape your protective immunity and cause increasingly severe disease. At the same time your immune system is prevented from activating the many naive B cells that are capable of responding to the changes in the virus. The imprint made by the original strain is broken only on infection with a strain of influenza that lacks all the B-cell epitopes of the original strain

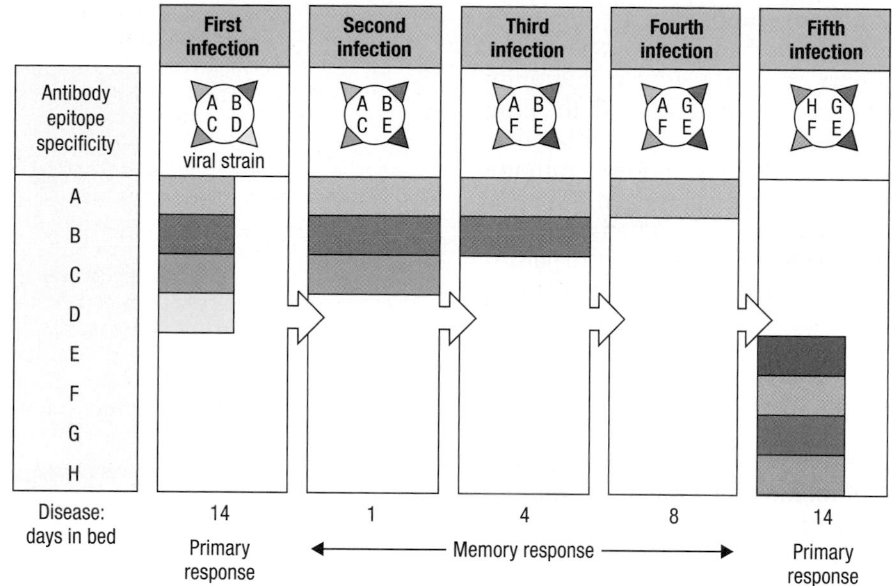

Figure 11.12 Highly mutable viruses such as influenza gradually escape from immunological memory without stimulating a compensatory immune response. A person's history of infection with influenza is shown here. The first infection is with a strain of influenza virus that elicits a primary antibody response to viral epitopes A, B, C, and D. The next four infections are with viruses that successively lose the epitopes of the first virus and gain in turn the epitopes E, F, G, and H. With each infection the strength of the person's memory response declines but cannot be compensated for by a new primary response until all the epitopes of the original strain are lost at the fifth infection. Then, no memory response is elicited, disease ensues, and a primary immune response against all the new epitopes is made.

(Figure 11.12). Such a strain will produce a full-blown case of influenza and stimulate a primary B-cell response targeted against the full complement of new influenza epitopes. This phenomenon, in which the first influenza strain to infect a person constrains the future response to other strains is described as **original antigenic sin**.

Summary

The primary immune response to a pathogen comprises a first phase of innate immunity and a second phase of adaptive immunity. When successful, the primary immune response serves three purposes: it clears the infection, temporarily strengthens defenses to prevent reinfection, and establishes a long-lasting immunological memory of the pathogen. This memory ensures that subsequent infections with the same pathogen will provoke a secondary immune response that is faster, stronger, and more effective than the primary response. Numerous factors synergize with each other to achieve these improvements (Figure 11.13). Immunological memory is mediated by long-lived populations of memory plasma cells that continue throughout life to make small amounts of high-affinity pathogen-specific antibodies, and by long-lived populations of pathogen-specific memory B cells and T cells. In the secondary response, the

Differences between the primary and secondary immune responses	
Primary response	Secondary response
Small number of pathogen-specific cells respond at the start	Large number of pathogen-specific cells respond immediately
Delay before pathogen-specific antibodies are produced	Pathogen-specific antibodies already present
Non-isotype-switched antibody having a mixture of affinities for the pathogen is produced at the start	Antibodies are isotype-switched and have high affinity for the pathogen
High threshold of activation	Lower threshold of activation
Delay before effector T cells are generated and are able to enter infected tissues	Effector T cells are present and can enter infected tissue immediately
Innate immunity works alone until an adaptive response is generated	Close cooperation between innate and adaptive immunity from the start

Figure 11.13 Differences between the primary and secondary immune responses.

lymphocytes and effector mechanisms of adaptive immunity are deployed from the very start of infection and in greater force than was possible in the primary response. Throughout the secondary response, innate immunity works in partnership with adaptive immunity, which decreases both the time taken to clear the infection and the collateral damage to tissues by inflammation. Because of the efficiency and effectiveness of the secondary immune response, the disease caused by a second infection with a pathogen is shorter and less severe than that arising from a first infection, and is often undetectable. In the course of the secondary response, further improvements to pathogen-specific immunity and memory are made, which decrease the probability of future infection by the pathogen.

Vaccination to prevent infectious disease

This part of the chapter considers **vaccination**, a routine medical procedure that provides people with an immunological memory of a pathogen without them having to be infected by the pathogen and suffer the disease it causes. This 'miracle' is achieved by immunizing people with a non-infectious material, called the **vaccine**, which contains the pathogen's antigens. Two or more immunizations are usually given. The first immunization mimics a primary infection with the pathogen and stimulates the primary immune response. Given at a later time, the second immunization stimulates the secondary immune response to the vaccine, and so on for additional immunizations. Because of the immunological memory induced by vaccination, any future infection with the pathogen is eliminated swiftly before it can cause disease. To protect both individuals and populations from infectious disease, it is best to vaccinate young children. Vaccination has protected billions of people against infectious diseases and is by far the most successful and frequently used manipulation of the immune response. It is even claimed that vaccination has saved more human lives than any other medical procedure.

The modern era of vaccination began in 1796 with Edward Jenner's published report on how material from animals infected with cowpox, a disease of cattle, could be used as a vaccine to protect against smallpox. Although Jenner's procedure was widely embraced, vaccines for other diseases did not emerge until well into the 19th century. That was when microorganisms were first isolated, grown in culture, and shown to cause disease. Largely by trial and error, methods were developed to inactivate pathogens, or impair their capacity to cause disease, and these crippled microbes were used as vaccines. Eventually, vaccines were developed against most of the epidemic diseases that then plagued human populations in western Europe and North America. With the discovery of antibiotic drugs in the mid-20th century it began to seem that the combination of vaccines and antibiotics would solve the problems of infectious disease for ever. Such optimism was soon tempered by the difficulties encountered in developing effective vaccines for certain diseases, and by the emergence of new pathogens and of antibiotic-resistant strains of old ones. The application of recombinant DNA technology and the sequencing of pathogen genomes have, however, opened up new and exciting prospects for making vaccines.

11-14 Protection against smallpox is achieved by immunization with the less dangerous cowpox virus

The first medically prescribed vaccine was against smallpox, a viral disease characterized by a rash of spots that develop into virus-loaded pustules that leave permanent scars on survivors. The rash frequently involved the face, and the disease was mainly spread by face-to-face contact (see Figure 1.1). The earliest vaccines actually contained pathogenic smallpox virus; they were made from dried pustules taken by physicians from patients experiencing a mild

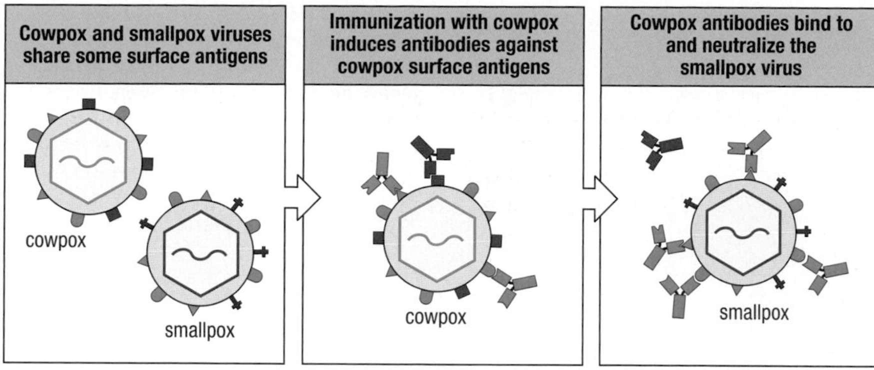
Cowpox and smallpox viruses share some surface antigens
cowpox
smallpox
Immunization with cowpox induces antibodies against cowpox surface antigens
cowpox
Cowpox antibodies bind to and neutralize the smallpox virus
smallpox

Figure 11.14 Vaccination with cowpox virus elicits neutralizing antibodies that react with antigenic determinants shared with smallpox virus. Shared antigenic determinants of cowpox also elicit protective T-cell immunity against smallpox (not shown here).

form of the disease. Small amounts of this material were given to healthy people, including children, either intranasally or intradermally through a scratch on the arm—a procedure known as **variolation**, from the word **variola**, the Latin name given both to the pustule and to the disease itself. Although successful in many cases, the drawback of variolation was the frequency with which it produced full-blown smallpox, resulting in the death of around 1 in 100 of those vaccinated. Despite the risk, variolation was widely used in the 18th century because the threat from smallpox was so much greater than that from variolation. At the time, smallpox killed one in four of those infected, and epidemics occurred on a regular basis. In London, for example, more than one-tenth of all deaths were due to smallpox.

Toward the end of the 18th century, Jenner's innovation was to use the related cowpox virus as a vaccine for smallpox. The cowpox or **vaccinia** virus causes very mild infections in people, but the immunity produced gives effective protection against both smallpox and cowpox because the two viruses have antigens in common (Figure 11.14). In the 19th century, Jenner's vaccine replaced variolation and was eventually responsible for the eradication of smallpox in the 20th century. The words vaccinia and vaccination derive from *vaccus*, the Latin word for cow. Although the term vaccination was once used specifically in the context of smallpox, it now refers to any deliberate immunization that induces protective immunity against a disease.

11-15 Smallpox is the only infectious disease of humans that has been eradicated worldwide by vaccination

Immunization with vaccinia induces long-lasting immunity (see Figure 11.3) and proved extremely effective against smallpox, leading to the eradication of the disease in 1979. After that, the immunization of children against smallpox began to be discontinued, and a smallpox vaccination is no longer part of the recommended childhood vaccinations in the United States (Figure 11.15). Smallpox is the only epidemic infectious disease so far to have been globally eradicated by vaccination, which was achieved through a coordinated international program. Several biological properties of smallpox also contributed to this success, of which we shall consider three. First, smallpox evolves slowly and its antigenic epitopes are conserved, so that immunity acquired as an infant is effective against smallpox if it infects later on in life. Second, the vaccine is a live virus that establishes an infection at the injection site in the skin, a tissue affected by natural smallpox infection. Thus the stimulation of the innate and adaptive immune responses by the vaccine mimics that caused by smallpox infection and produces memory cells that will provide effective defense against smallpox. The third helpful property of smallpox is that it infects only humans, meaning that there are no reservoirs of the virus in other animals. Once the chain of transmission between humans has been broken, the virus cannot survive.

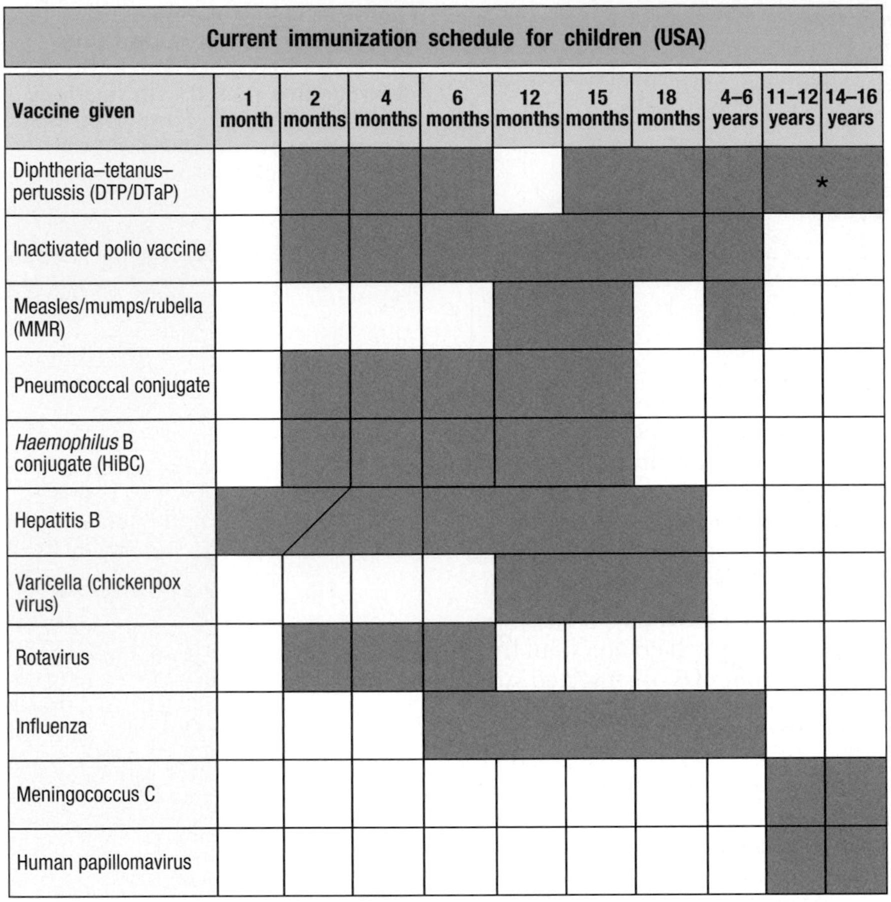

Current immunization schedule for children (USA)										
Vaccine given	1 month	2 months	4 months	6 months	12 months	15 months	18 months	4–6 years	11–12 years	14–16 years
Diphtheria–tetanus–pertussis (DTP/DTaP)									*	
Inactivated polio vaccine										
Measles/mumps/rubella (MMR)										
Pneumococcal conjugate										
Haemophilus B conjugate (HiBC)										
Hepatitis B										
Varicella (chickenpox virus)										
Rotavirus										
Influenza										
Meningococcus C										
Human papillomavirus										

Figure 11.15 Childhood vaccination schedules in the United States. Each red box denotes a time at which a vaccine dose should be given. Boxes spanning multiple months indicate a range of times during which the vaccine may be given. DTaP, diphtheria, tetanus, and an acellular pertussis vaccine. *Tetanus and diphtheria toxoids only.

An added benefit of smallpox vaccination was the protection it gave against other poxviruses, such as cowpox and monkeypox, which do infect humans and cause mild disease. In the absence of routine smallpox vaccination, the human population's immunity to the other poxviruses began to wane steadily, with the result that the frequency of human infections with cowpox and monkeypox began steadily to increase. At present, these viruses pose no obvious problem for human health, but they are being monitored to see if they become more virulent. Of concern is that such a progression could lead to the 'renaissance' of a smallpox-like virus.

11-16 Most viral vaccines are made from killed or inactivated viruses

Unfortunately, the Jennerian strategy for smallpox vaccination cannot be applied to most pathogenic viruses because they lack a natural 'safe' relative. Most viral vaccines contain the disease-causing virus, but in a form that does not cause severe disease. One type of vaccine consists of virus particles that cannot replicate because they have been treated chemically with formalin or physically with heat or irradiation. These are called **killed** or **inactivated virus vaccines**. The vaccines for influenza and rabies are of this type. This strategy only works if the viral nucleic acid can be completely and reliably inactivated, and it has the drawback that large amounts of pathogenic virus are produced in the manufacture of the vaccine.

A second type of vaccine consists of a mutant form of live virus that grows poorly in human cells and is no longer pathogenic to humans. Such vaccines, called **live-attenuated virus vaccines**, usually elicit a better protective

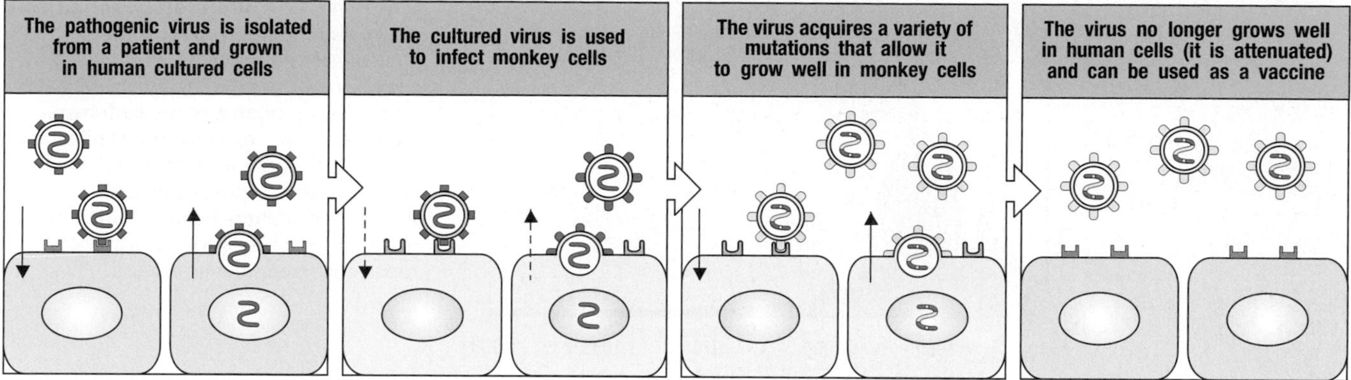

The pathogenic virus is isolated from a patient and grown in human cultured cells
The cultured virus is used to infect monkey cells
The virus acquires a variety of mutations that allow it to grow well in monkey cells
The virus no longer grows well in human cells (it is attenuated) and can be used as a vaccine

Figure 11.16 Attenuated viruses are selected by growing human viruses in non-human cells. To produce an attenuated virus, the virus is first isolated by growing it in cultured human cells. This in itself can cause some attenuation; the rubella vaccine, for example, was made in this way. In general, however, the virus is then grown in cells of a different species, such as a monkey, until it becomes fully adapted to those cells and grows only poorly in human cells. The adaptation is a result of mutation, usually a combination of several point mutations. An attenuated virus grows poorly in the human host: it produces immunity but not disease.

immunity than killed virus vaccines. This is because the attenuated virus can to some extent infect cells and replicate, and thus mimics a real infection. Most of the viral vaccines currently used to protect humans are live-attenuated vaccines. To attenuate a pathogenic human virus, it is grown in cells from another animal species. Such conditions select for variant viruses that grow faster in the non-human cells, becoming in the process less fit for growth in human cells (Figure 11.16). The measles, mumps, and yellow fever vaccines contain live-attenuated viruses. Attenuated viral strains can also arise naturally. As a viral infection passes though a human population, the virus can diversify by mutation, occasionally producing a strain with reduced pathogenicity. Such natural live-attenuated viruses are suitable candidates for vaccines, and one such strain of poliovirus (strain 2) is a part of the oral polio vaccine.

11-17 Both inactivated and live-attenuated vaccines protect against poliovirus

Although fewer than 1% of infections with poliovirus cause disease, they produce a variable paralysis called poliomyelitis or infantile paralysis that can kill or cripple the affected individuals, who are mostly children. Polio was a major epidemic disease in the 20th century and particularly common in the United States, where the worst recorded epidemic was in 1952. Whereas smallpox is transmitted by face-to-face contact, poliovirus is transmitted by the fecal–oral route, leading to various outbreaks of disease being associated with communal bathing in public swimming pools. The virus infects mucosal surfaces, indicating that an oral vaccine would be superior to a vaccine injected intradermally. However, one of the challenges in delivering an oral vaccine is to prevent its degradation in the gut before it can infect the intestinal epithelium. The first polio vaccine (the Salk vaccine) was approved in 1955; it contained killed virus of the three main strains and was injected into the skin. Because it contained three different polioviruses, this vaccine was said to be trivalent. By 1963, a similarly trivalent oral live-attenuated virus vaccine (the Sabin vaccine) began to be used. Because this trivalent oral polio vaccine (TVOP) gave better protection and was easier to administer, it started to replace the killed virus vaccine. Together, the two vaccines achieved an impressive reduction in the incidence of polio in the United States, where the last recorded case was in 1979 (Figure 11.17). The Global Polio Eradication Initiative began in 1988, when polio was endemic in 125 countries and caused the paralysis of 350,000

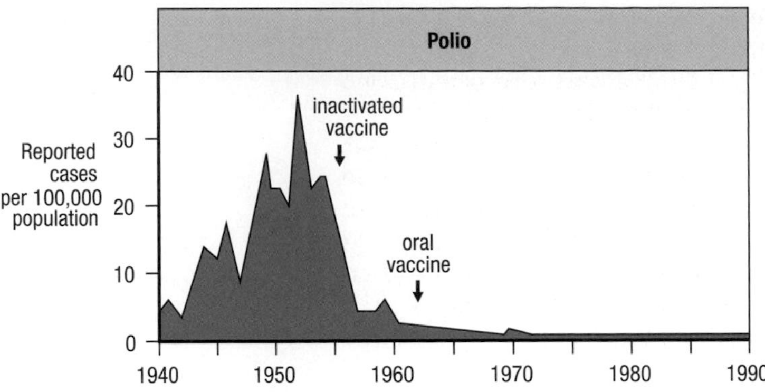

Figure 11.17 A successful vaccination campaign. Poliomyelitis has been virtually eliminated from the USA, as shown in the graph. The arrow indicates when the vaccination campaign began. Because the polio virus has not been eradicated worldwide and the volume of international travel is so high, immunization must be maintained in much of the population to prevent a recurrence of epidemic disease.

children each year. By the year 2000, the number of cases of polio was reduced by 99%, but since then there has been little further decrease, although endemic poliovirus has been eliminated from all countries of the world except Nigeria, Pakistan, and Afghanistan. Because poliovirus still persists in these countries and is highly contagious, children in other countries are still being vaccinated against polio (see Figure 11.15). Like smallpox, poliovirus infects only humans, so that if it is completely eliminated from the human species, the virus will then be extinct in the 'wild' and restricted to research laboratories.

11-18 Vaccination can inadvertently cause disease

Soon after the Salk polio vaccine was introduced in 1955, there was an outbreak of polio among 94 people who had received the vaccine and among 166 of their families and friends. This disaster immediately caused polio vaccination in the United States to be suspended for several months while an investigation to its cause was made. The infections were all traced back to one manufacturer and to two batches of vaccine containing pathogenic virus that was not properly inactivated. Thus vaccination had infected 94 people with pathogenic poliovirus, and they had passed the disease on to 70 unvaccinated others.

After its introduction and extensive use, the Sabin vaccine was found to induce polio and paralysis in three people per million vaccinated. These infections are all caused by strain 3, one of three live-attenuated polio strains that make up TVOP. All three of the strains are necessary to provide adequate protection against the natural range of poliovirus variants. Considering just the numbers, the inclusion of strain 3 in the vaccine prevents many more cases of disease than it causes. Although the attenuated strain 3 virus differs from a natural pathogenic strain of polio by 10 nucleotide substitutions, a back-mutation at just one of these positions is sufficient for strain 3 to revert and become pathogenic. Such a mutation in the viral genome can occur during manufacture of the vaccine, when it should be caught during quality control and eliminated, or after vaccination, when the virus replicates in vaccinated people.

In countries where poliovirus has been eradicated and the only cases of disease are caused by vaccination, there is often greater fear of the vaccination than of the disease. For this reason there has been considerable social pressure to improve the polio vaccine, and some parents and communities refuse to have their children vaccinated. One way to reduce the exposure to strain 3 is to use an inactivated polio vaccine (IPV) for the first immunization, and to follow this with TVOP for the second and subsequent immunizations. In this way the primary immunization cannot lead to infection and strain 3-specific antibodies made in the primary response will help prevent infection by any revertant of strain 3 in the subsequent immunizations. Because of the reversion rate, TVOP is no longer recommended for routine vaccination in the United States, and IPV is now the vaccine of choice (see Figure 11.15).

In 2007, the people of northern Nigeria experienced one of the largest recorded outbreaks of polio caused by a vaccine strain. Over a 2-year period, 69 children were paralyzed as a result of vaccination. In the three countries where polio remains endemic, episodes such as this have exacerbated the public's fear and distrust of vaccination and also of the people administering the vaccine. Exploiting the situation are political extremists, who shot dead some 30 polio workers in Pakistan and Nigeria during 2013. Although Syria had been free of polio since 1999, social disruption caused by civil war led to reduced vaccination rates and an outbreak of 10 cases of polio in 2013 among infants less than 2 years old. In such volatile environments the virus flourishes, reducing the likelihood that polio will be globally eradicated any time soon.

11-19 Subunit vaccines are made from the most antigenic components of a pathogen

In the immune response to hepatitis B virus (HBV) the protective, neutralizing antibodies are made against the surface protein of the virus, also called the HBV surface antigen. The hepatocytes of patients infected with HBV secrete the surface protein into the blood as minute particles. The first version of the anti-hepatitis B vaccine contained surface protein that was purified from the plasma of HBV-infected people. Because only one component or subunit of the virus is used for vaccination, this type of vaccine is called a **subunit vaccine**. In purifying the antigenic particles, the greatest concern was that infectious particles of HBV should be completely removed. If they were not, vaccination could infect people with the virus and give them disease, rather than serving the intended purpose of protecting them from the disease.

This concern, and the proven benefit of the vaccine in preventing life-threatening liver disease, led to the HBV subunit vaccine being one of the first vaccines to be made using recombinant DNA technology. Making the vaccine in this way did not involve viral particles, and it avoided any contact with them. The gene encoding the HBV surface antigen was inserted into the genome of baker's yeast. The recombinant yeast was then grown in mass cultures, from which the surface protein was purified in large quantity. The vaccine was introduced in 1986 and provides about 85% of people vaccinated with protective immunity against infection by HBV.

11-20 Invention of rotavirus vaccines took at least 30 years of research and development

The wheel-shaped rotavirus was discovered in 1973 and shown to be a major cause of severe childhood diarrhea. Almost all children are exposed to rotavirus, and the illness it causes is responsible for more than 2 million visits to hospital and 610,000 deaths per year worldwide (Figure 11.18). Efforts to make a vaccine began soon after the discovery of the virus, but it took more than 30 years to develop vaccines that were both effective and acceptable. There are now two of them. Along the way were promising vaccines that proved too weak, or too strong, or too likely to produce unacceptable side-effects.

The rotavirus genome consists of 11 molecules of double-stranded RNA. Viral coat proteins called VP4 and VP7 are the major targets for neutralizing antibodies. Rotaviruses undergo frequent mutation, with VP4 and VP7 being variable proteins that define different viral serotypes. These serotypes are further diversified by reassortment of the 11 genomic segments. Although there are 42 natural variants of rotavirus, just five of them account for 90% of the disease, and both vaccines incorporate antigens from these common strains. The Rotarix vaccine comprises an attenuated human rotavirus having common VP4 and VP7 variants. In the Jennerian tradition, the RotaTeq vaccine contains a mixture of five cattle rotaviruses that are nonpathogenic for humans, each of

Annual childhood mortality from rotavirus infection	
Country	Annual deaths (2008)
India	99,000
Nigeria	41,000
Pakistan	39,000
Democratic Republic of the Congo	33,000
Ethiopia	28,000
Afghanistan	25,000
Uganda	11,000
Bangladesh	10,000
Indonesia	10,000
Angola	9000

Figure 11.18 Childhood mortality from rotavirus infection in the 10 most affected countries. Source: Estimated rotavirus deaths for children under 5 years of age: 2008 (World Health Organization, 2013).

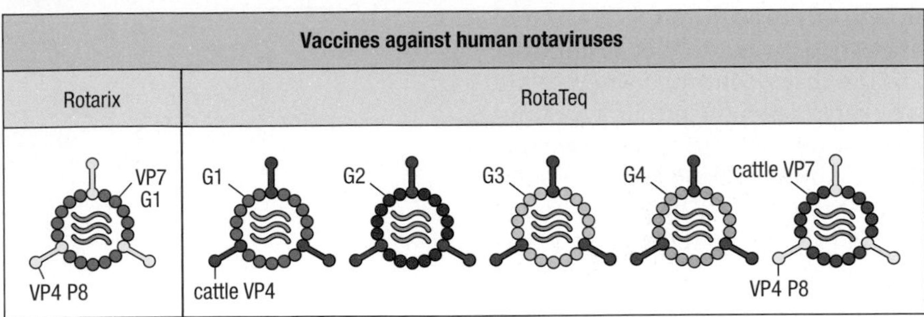

Figure 11.19 Two effective rotavirus vaccines. The Rotarix vaccine (left panel) is an attenuated human rotavirus that has common P8 and G1 variants of the VP4 and VP7 glycoproteins, respectively. The RotaTeq vaccine is based on a cattle rotavirus that is nonpathogenic to humans. The vaccine consists of five viral strains, four expressing different human VP7 variants (G1, G2, G3, and G4) and cattle VP4, and one expressing cattle VP7 and the human P8 variant of VP4. Cattle components are in purple.

which has been engineered to express a different human VP4 or VP7 glycoprotein (Figure 11.19). These vaccines give 85–98% protection against severe rotavirus diarrhea. Like the oral poliovirus vaccine, they can be made without sophisticated biotechnology by simple growth in tissue culture, and they are readily transferable to the poorer countries, where most mortality from rotavirus infection occurs (see Figure 11.18).

11-21 Bacterial vaccines are made from whole bacteria, secreted toxins, or capsular polysaccharides

The development of vaccines against bacterial pathogens used similar strategies to those outlined in the previous sections for viruses. Although much sought after, live-attenuated bacterial vaccines are few in number. The first one made, and the one most commonly used, is a tuberculosis vaccine. The Bacille Calmette–Guérin (BCG) strain used in the vaccine is not pathogenic to healthy humans and was derived from a bovine strain of *Mycobacterium tuberculosis*. The vaccine's efficacy varies with the human population, and although routinely given to children in some European countries, BCG has never been used in the United States. More recently, live-attenuated vaccines against species of *Salmonella* have been introduced for both medical and veterinary use. An attenuated strain of *Salmonella typhi*, the bacterium that causes typhoid fever, was made by mutagenesis and selection for the loss of a lipopolysaccharide necessary for pathogenesis. In this vaccine strain, an enzyme necessary for lipopolysaccharide synthesis is defective.

Some bacterial diseases are caused entirely by the effects of toxic proteins secreted by the bacteria. The two most important of these are diphtheria, caused by the **diphtheria toxin** of *Corynebacterium diphtheriae*, and tetanus, caused by the **tetanus toxin** of *Clostridium tetani*. To avoid contracting these diseases, infected individuals need to have supplies of high-affinity neutralizing antibodies that bind irreversibly to the toxin and inhibit its toxic activity. In the presence of such antibodies, the toxin molecules are neutralized immediately upon secretion, before they can influence the behavior of human cells and cause disease (see Section 9-16). Such vaccines are made by purifying the toxin and treating it with formalin to destroy its toxic activity. The inactivated proteins, called **toxoids**, retain sufficient antigenic activity to provide protection against disease. Diphtheria and tetanus vaccines are thus comparable to the viral subunit vaccines.

Many pathogenic bacteria have an outer capsule composed of polysaccharides that define species-specific and strain-specific antigens. For these encapsulated bacteria, which include the pneumococcus (*Streptococcus pneumoniae*), salmonellae, the meningococcus (*Neisseria meningitidis*), *Haemophilus influenzae, Escherichia coli, Klebsiella pneumoniae*, and *Bacteroides fragilis*, the capsule determines both the pathogenicity and the antigenicity of the organism. In particular, the capsule prevents the fixation of complement by the alternative pathway. Only when antibodies have bound to the capsule does complement fixation lead to bacterial clearance. Consequently, the aim of vaccination against such bacteria is to produce complement-fixing antibodies that bind to the capsule.

11-22 Conjugate vaccines enable high-affinity antibodies to be made against carbohydrate antigens

During the 1990s, there was an epidemic of meningitis in Europe and North America that was caused by a hypervirulent form of serogroup C *N. meningitidis*. This bacterium colonizes most people with no harmful effect, but in some individuals it invades the blood and the brain, causing disease that can be fatal or leave permanent damage to organs and tissues. The disease mainly affects young children of around 2 years of age and adolescents of 15–19 years. Protection against the infection is provided by high-affinity IgG antibodies that bind to the polysaccharides on the outer surface of the capsule and trigger complement-mediated killing of the bacteria. At the time of the epidemic, the existing vaccine was made from purified bacterial polysaccharide. However, it was ineffective in protecting infants, because it stimulated only a feeble T-cell-independent B-cell response that gave rise to low-affinity IgM antibodies and no memory B cells (see Section 9-3).

The fundamental problem with the polysaccharide vaccine was its inability to stimulate a CD4 T_{FH} cell response, because it contained no source of peptide epitopes for presentation by MHC class II molecules to naive CD4 T cells. Appreciation of the immunological principle that making high-affinity IgG antibodies requires the immunizing antigen be recognized by both B cells and T cells (see Section 8-19) led to invention of **conjugate vaccines**, in which different epitopes recognized by B cells and T cells are synthetically linked together (Figure 11.20). Effective *N. meningitidis* vaccines were made by conjugating the bacterial polysaccharide to either tetanus or diphtheria toxoid. These inactive forms of toxic bacterial proteins were already being used in the tetanus and diphtheria vaccines and were known to stimulate potent neutralizing IgG responses. On immunization, dendritic cells process the vaccine and present toxoid-derived peptides to naive CD4 T cells, which activate and differentiate into T_{FH} cells. Naive B cells specific for epitopes of the bacterial polysaccharide endocytose the intact conjugate, and after processing the toxoid they present its peptide antigens to the activated toxoid-specific T_{FH} cells. This interaction drives B-cell activation, leading to a germinal center reaction and the production of high-affinity, neutralizing IgG antibodies against the bacterial polysaccharide and the development of immunological memory (see Figure 11.20). In the year (1998/1999) in which vaccination with the conjugate was begun in the UK, there were 411 diagnosed infections with hypervirulent serogroup C *N. meningitidis* and 32 deaths for children aged 10 years or younger. In the ninth year of the vaccination program (2007/2008), there were

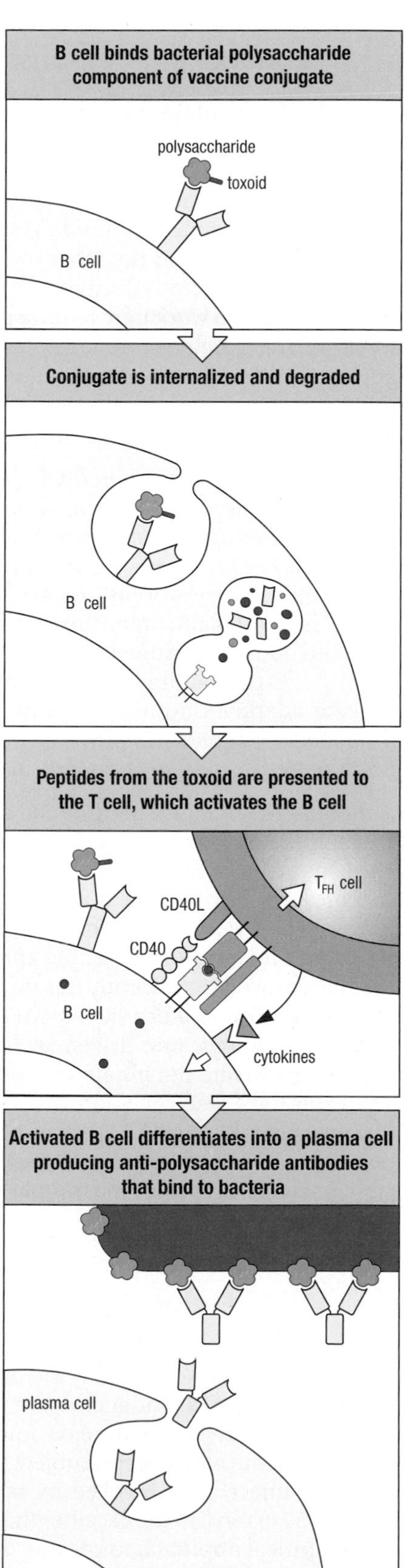

Figure 11.20 Molecular complexes recognized by both B and T cells make effective vaccines. The first panel shows a naive B cell's surface immunoglobulin binding a carbohydrate epitope on a vaccine composed of a *Haemophilus* polysaccharide (blue) conjugated to tetanus toxoid (red), a protein. This results in receptor-mediated endocytosis of the conjugate and its degradation in the endosomes and lyosomes, as shown in the second panel. Peptides derived from degradation of the tetanus toxoid part of the conjugate are bound by MHC class II molecules and presented on the B cell's surface. In the third panel, the receptor of a T_{FH} cell recognizes the peptide:MHC complex. This induces the T cell to secrete cytokines that activate the B cell to differentiate into plasma cells, which produce protective antibody against the *Haemophilus* polysaccharide (fourth panel).

only four infections and no deaths. Similar success was obtained with conjugate vaccines against two other pathogenic species of encapsulated bacteria: *Haemophilus influenzae* and *Streptococcus pneumoniae.*

11-23 Adjuvants are added to vaccines to activate and enhance the response to antigen

Vaccines that contain inactivated or attenuated microorganisms are able to stimulate the innate and adaptive immune responses to a pathogen, both of which are necessary for generating an adequate reserve of memory cells. Subunit and conjugate vaccines consisting of one or a few purified proteins do not activate the innate immune response, because they are not detected by Toll-like receptors and the other receptors of innate immunity that recognize the distinctive macromolecules of microbial pathogens (see Section 3-2). For these vaccines to work they require another component called an **adjuvant**, a word meaning helper, which serves to trigger the innate immune response and establish a state of inflammation at the site of vaccination. The inflammation is necessary for initiating an adaptive immune response against the antigens in the vaccine.

DTP is a widely administered vaccine that provides protection against three bacterial diseases: diphtheria, tetanus, and the whooping cough caused by *Bordetella pertussis* (see Figure 11.15). The components of the vaccine are diphtheria toxoid, tetanus toxoid, and inactivated *B. pertussis* bacteria. DTP is an example of a **combination vaccine** that provides protection against more than one pathogenic organism and disease. The bacteria in the vaccine have two functions. One function is to be the adjuvant that triggers the innate immune response; the other is to provide the antigens that will stimulate a specific adaptive immune response against *B. pertussis*. The inflammation induced by the bacteria provides the environment in which strong adaptive immune responses are also made against the diphtheria and tetanus toxoids.

Although research on experimental animals identified a variety of potent adjuvants, none of them is used in human vaccines, because of their toxicity. In 1924, alum—a form of aluminum hydroxide—was approved for use in human vaccines; for 73 years it was the only option. Although alum is safe, it is not a powerful adjuvant, as seen from the performance of vaccines containing diphtheria and tetanus toxoids as the antigen and alum as the adjuvant. These DT vaccines proved significantly less protective than the DTP vaccine. Approval in 1997 of the oil-in-water adjuvant MF59 initiated a new era in adjuvant research and development, one driven by knowledge of the mechanisms by which adjuvants activate the innate immune response. Since 1997, three additional adjuvants have received approval for use in human vaccines, and others are in development (Figure 11.21). The underlying strategy is to use microbial components recognized by Toll-like receptors as the active ingredient of the adjuvant, or small molecules that mimic their action.

11-24 Genome sequences of human pathogens have opened up new avenues for making vaccines

The original scientific strategy for vaccine development consisted of three steps: first, isolate and identify the pathogenic microorganism; second, inactivate the pathogen's virulence while preserving its immunogenicity; and third, inject the inactivated pathogen into experimental animals as a prelude to doing the same with human subjects. These three principles of "isolate, inactivate and inject," first invoked by Louis Pasteur, led to the discovery during 1885–1950 of most of the vaccines that are in common use today. Subsequently, this empirical approach to vaccine development proved less fruitful. Over the past 40 years, new strategies for vaccine discovery have emerged. These

Adjuvants				
Year licensed	**Name**	**Class**	**Contents**	**In vaccines against**
1924	Alum	Mineral salt	Aluminum phosphate or hydroxide	Many infectious diseases
1997	MF59	Oil-in-water emulsion	Squalene, polysorbate 80, sorbitan trioleate	Influenza
2000	Virosomes	Liposomes	Lipids, hemagglutinin	Influenza, hepatitis A
2005	AS04	Alum-absorbed TLR4 agonist	Aluminum hydroxide, monophosphoryl lipid A	Hepatitis B, human papilloma
2009	AS03	Oil-in-water emulsion	Squalene, polysorbate 80, α-tocopherol	Influenza
In development	CpG 7909	TLR9 agonist	CpG nucleotides	
	Imidazoquinolines	TLR7 and TLR8 agonist	Small molecules	
	PolyIC	TLR3 agonist	Double-stranded RNA analogs	
	Pam3Cys	TLR2 agonist	Lipopeptide	
	Flagellin	TLR5 agonist	Bacterial protein linked to antigen	

Figure 11.21 Some adjuvants licensed for use in human vaccines.

approaches involve rational design and have been made possible by the technical advances of recombinant DNA, genome sequencing, and macromolecular structure determination, combined with increasing knowledge of how the immune system works to respond to pathogens and remember them.

Sequencing the genomes of human pathogens has uncovered many unanticipated genes that were never found in previous studies of the pathogens' physiology and virulence. The sequences of the encoded proteins give clues to their location in the cell, their function, and the likelihood of their being a target for protective antibodies. A good example is *N. meningitidis* of serogroup B. In 2000, 29 previously unknown genes were identified in the sequenced genome of this bacterium that might encode antigenic targets for a protective antibody response. Analysis of these identified three superior antigens that were combined to make Bexsero, a vaccine against meningitis B that was approved for use in Europe in 2013. All three antigenic proteins—neisserial heparin-binding protein, *N. meningitidis* adhesion A (Nad A), and factor H-binding protein (fHbp)—function to increase the virulence of the bacterium. As an example, we shall consider fHbp, a lipoprotein expressed on the surface of the bacterium. As its name indicates, fHbp binds to factor H, an inhibitor of the alternative pathway of complement activation (see Section 2-5). By coating its outer surface with this inhibitor, the bacterium prevents complement activation at its own surface and thus avoids being lysed by the terminal components of complement (Figure 11.22). By coating its surface with a human protein, the bacterium also prevents an antibody response from being made against its own capsular antigens. Striking evidence for the coevolution of humans and *N. meningitidis* is that human fHbp does not bind to rat or mouse factor H. The invention and development of the Bexsero vaccine led to new knowledge of the physiology of the pathogen and its exploitation of the human immune system. By contrast, the traditional approaches to vaccine development were informed and driven by preexisting knowledge of the pathogen's physiology. Because of this difference, the modern approach has been called **reverse vaccinology**.

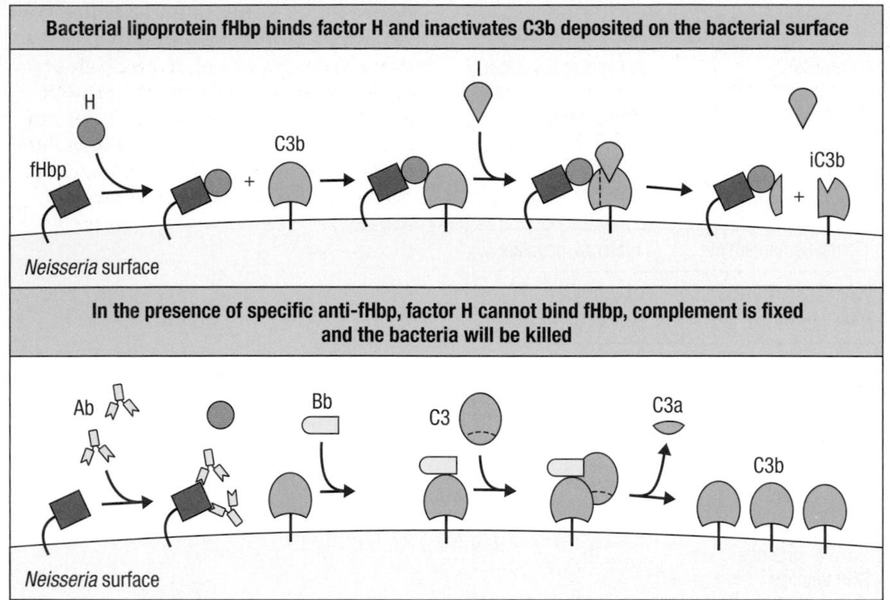

Figure 11.22 Vaccination with *Neisseria* factor H-binding protein (fHbp) prevents infection from taking hold. The upper panel shows how *Neisseria* uses fHbp to prevent complement fixation on its surface. fHbp is a surface lipoprotein that binds factor H and uses it to inactivate any C3b deposited on the bacterial surface. Inactivation involves factor I, which cleaves C3b to give iC3b. This prevents formation of the membrane-attack complex and the lysis of *Neisseria*. The lower panel shows how immunization with the Bexsero vaccine, which contains fHbp, counters this evasion. Vaccination generates high-affinity IgG antibodies against fHbp that cover up its binding site for factor H on fHbp and prevent its interference with complement activation by the alternative pathway. Coating the bacteria with anti-fHbp IgG also activates the classical pathway (not shown).

11-25 The ever-changing influenza virus requires a new vaccine every year

Because influenza is an RNA virus, the replication of its genome is more error-prone than the replication of DNA viruses such as smallpox and measles. The errors generate a diversity of mutant viruses, enabling influenza to evade human immunity; as a result, new strains of influenza spread through the human population every year. Consequently, immunological memory of an influenza infection decays quickly, having an effective lifetime of only a few years (see Section 11-13). The main targets for neutralizing antibodies are the hemagglutinin (H) and neuraminidase (N) surface glycoproteins that allow the virus to bind to cells of the respiratory epithelium and infect them. Good neutralizing antibodies block virus infection, prevent viral replication, and target viral particles for phagocytosis. Each winter, the new emerging strains of the virus exhibit sequence differences from the previous years' strains in the hemagglutinin, the neuraminidase, or both. Because of these differences, several hundred million cases of influenza occur each year, resulting in around 250,000 deaths, with children and the elderly being most severely affected. To counter the evolving changes in influenza, the World Health Organization (WHO) coordinates a program to identify each year's epidemic strains and send them to vaccine manufacturers. At some universities, including Stanford, where large segments of the population are either young or old, the university willingly gives an annual 'flu shot' to any student, teacher, or employee who wants one. By this method the university's collective immunological memory of influenza is refreshed each year.

The effectiveness of the WHO program depends on timing, the goal being to vaccinate populations before the influenza arrives to cause disease and death. This requires the early detection of the epidemic strains and the efficient manufacture and delivery of vaccine. In March 2009, cases of an unusually severe influenza-like illness started to appear in Mexico; by April the cause had been detected as a novel H1N1 strain of influenza, and the international authorities were notified. The first US case was confirmed on 15 April. The hemagglutinin and neuraminidase of this virus were found to be particularly divergent, causing concern, verging on alarm, that a particularly severe influenza pandemic would occur. On 7 June, manufacture of a vaccine began and at the same time the influenza pandemic began to gain momentum. Unfortunately, the pan-

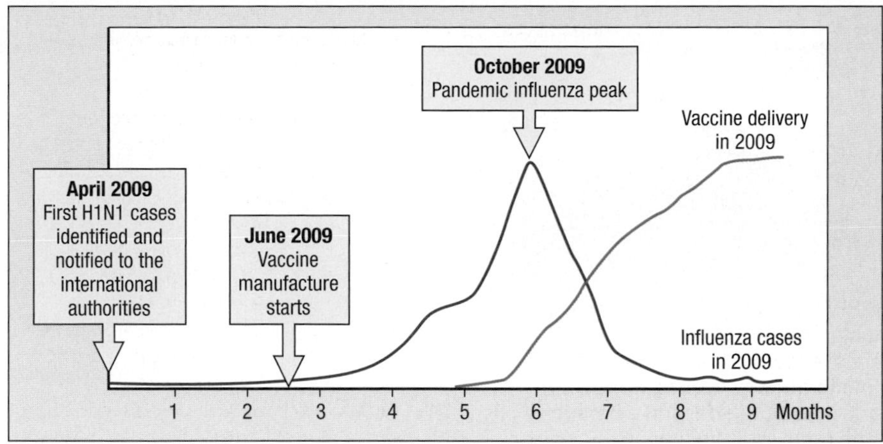

Figure 11.23 Time course of the H1N1 influenza pandemic of 2009 and the development of a vaccine against it. The red line represents the growth and decline in the number of influenza cases in the United States. The blue line represents the number of vaccine doses available against that specific strain. Data courtesy of D. Jordan/Centers for Disease Control.

demic peaked in October 2009 (in the United States) before the vaccine was generally available (Figure 11.23). Luckily, as it turned out, H1N1 did not cause severe illness or death in most of those affected, although mortality worldwide is estimated to have been in the hundreds of thousands. These events exposed the limitations of the current system for making influenza vaccines and stimulated the search for alternatives.

Four years later, the feasibility of a bold, innovative and provocative solution to the problem was reported. It was proposed that once a potentially epidemic strain of virus had been identified and its genome sequenced, the sequence would be sent immediately and electronically to manufacturers, who would synthesize the virus in less than a day using recombinant DNA technology. The synthetic virus would then be grown in large cultures of virus-infected cells, purified from the culture fluid, inactivated, and quality controlled. This would eliminate having to grow the virus before sending it to the manufacturer. It would also avoid the traditional method of growing the virus in chicken eggs, which involves an economically important avian species that is highly susceptible to influenza. The supply of vaccine would be effectively increased by adding an adjuvant, so that each dose needs less inactivated virus. With these changes, the protagonists proposed that a vaccine could be made available within weeks, rather than months, after detection of a new virus and well before any epidemic gained momentum. Although the proposal solves technological problems, the more formidable challenge seems to be persuading the regulatory agencies to change their methods of procedure and quality control, which were put in place in the 1950s and accounted for most of the delay in making the 2009 influenza vaccine.

11-26 The need for a vaccine and the demands placed upon it change with the prevalence of disease

In 18th-century Europe, the high probability of death or permanent facial scarring from smallpox made the risk of variolation acceptable to those who could afford it. Later, the side-effects of the cowpox vaccine were tolerated during a time when smallpox still posed a threat and terrified the population. This fear persisted into living memory, as is recounted in the eyewitness account on the next page by someone who had a brush with smallpox as a child. In the late 20th century, smallpox was eradicated and vaccination was discontinued. Smallpox vaccination was so successful a preventive medicine that it put itself out of business.

Concern for a vaccine's safety can sometimes lead to resurgence of a disease, as seen for whooping cough in the 1970s. At the beginning of the 20th century, 1 in 20 children in the USA died from whooping cough. The DTP vaccine

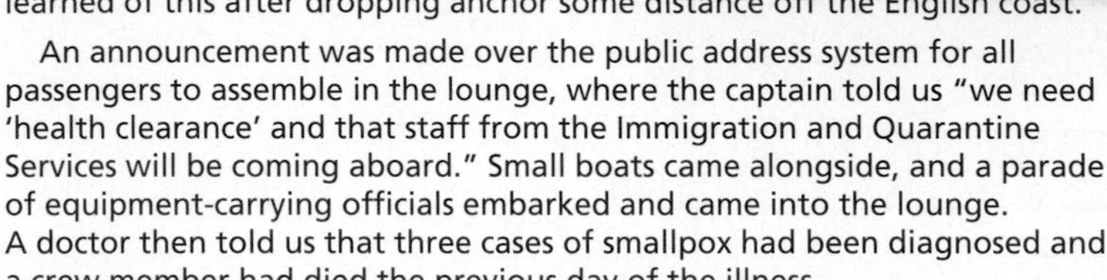

An eventful voyage from Sydney, Australia, to London, UK, on the SS Mooltan in 1949.

In 1949 my mother was taking myself, then 6 years old, and my younger sister 'home' from Australia on the SS Mooltan to see our English relatives. During our voyage from Sydney to London there was an outbreak of smallpox on board the ship, although we only learned of this after dropping anchor some distance off the English coast.

An announcement was made over the public address system for all passengers to assemble in the lounge, where the captain told us "we need 'health clearance' and that staff from the Immigration and Quarantine Services will be coming aboard." Small boats came alongside, and a parade of equipment-carrying officials embarked and came into the lounge. A doctor then told us that three cases of smallpox had been diagnosed and a crew member had died the previous day of the illness.

Tables were set up in the lounge in a semicircle with doctors and nurses at each. We lined up to be vaccinated. My memory is of a 'cork' with a needle set in its center being heated on a Bunsen burner, cooled and then a rough circle on our upper arms was scratched and criss-crossed just piercing the skin. At the next table a small piece of gauze impregnated with a white powder was placed on the scratches and rubbed in. At the third table a small dressing was applied. In time I had an incredibly painful upper arm at the vaccination site. A huge lobulated blister developed, surrounded by an extensive erythema as far as my elbow and shoulder. My whole upper arm was so grossly swollen that I had to wear a scarf as a sling. Most other passengers had the same reaction. As time went on many had burst blisters with nasty suppurating infections and high temperatures. Moving about was difficult. If someone bumped against the painful arm it could be agony and even getting dressed was difficult. Parents with babies had an especially difficult time.

Over the next few days the story of what had happened became clearer and spread throughout the ship. A Singalese crew member had become sick, but he had been hidden away for several days by his shipmates and his work shifts covered. He died of his illness, but by the time his death was reported several others were also seriously ill with high fever, delirium and the terrible rash. A woman in a cabin along the corridor from us became ill and was taken to the sick bay. She later died. As the days passed more people succumbed. We stayed anchored far from the shore until no new cases had occurred for several days. My mother told me of a terrible fear throughout the ship and of people confined to their tiny cabins except at meal times. I am not sure how long we stayed off-shore, but before coming into port at Tilbury, a huge amount of the ship's stock was dumped into the sea: china, glassware, cutlery, bedding, lounges and curtains were thrown overboard. It seemed no risks were being taken. Altogether eleven people had died and were buried at sea.

On going ashore we were taken straight to my grandparents' house. A quarantine official was assigned to visit us there every day and we were not to leave the house and garden for 10 days. The man who came was a tiny, nervous fellow, who was not the least reassuring to our family. In fact he seemed quite terrified about visiting us. He would ring the front door bell and then stand three or four metres up the path away from the door to ask if we were all well. My grandparents said they found out who their friends were at this time. Some people were kind and brought food and groceries, leaving them at the front gate and asking if we needed anything else. Others would walk along the street and cross to the other side of the road until well past the house. Luckily no symptoms appeared, and after two weeks we emerged to start leading a normal life.

About two decades later as a newly graduated physiotherapist, I was a witness to the last poliomyelitis epidemic in Australia and was involved in a rehabilitation programme for the many, mostly young people who had this dreadful disease. When I hear of opponents of vaccination, I do my best to tell them of my experiences and to urge them to do more research before refusing to vaccinate their children.

Iris Loudon

containing whole killed *B. pertussis* bacteria was introduced in the 1940s and was routinely given to infants at 3 months of age. The vaccination program produced a hundredfold decline in the annual incidence of whooping cough, from 2000 cases per million to 20 cases per million. But as people's fear of the disease abated, their concern for the vaccine's side-effects went up. Children vaccinated with pertussis vaccine inevitably develop inflammation at the injection site, and some develop a fever that induces persistent crying. Very rarely, the vaccinated child suffers fits and either a short-lived sleepiness or a transient floppy unresponsive state, all of which cause parental anxiety. In the 1970s, awareness of these established neurological side-effects was heightened by anecdotal reports of vaccination causing encephalitis and permanent brain damage, a connection that was never substantiated. Nonetheless, distrust of pertussis vaccination grew, most notably in Japan.

In Japan, DTP vaccination was introduced in 1947. By 1974, the incidence of pertussis had been reduced by more than 99% and in that year no deaths were ascribed to the disease. In the following year, two children died soon after vaccination, raising alarm that the deaths were caused by the vaccine. During the next five years, the number of Japanese children being vaccinated fell from 85% to 15% and, as a consequence, the incidence of whooping cough increased about twentyfold, as did the number of deaths from the disease. Japanese companies then developed vaccines containing antigenic components of pertussis instead of whole bacteria. In Japan, these acellular pertussis vaccines replaced the whole-cell vaccines in 1981. By 1989, the incidence of pertussis was again at the very low levels of 1974. Acellular vaccines are now being used in other countries because the side-effects of inflammation, pain, and fever are less than with DTP.

Measles is a highly infectious and potentially dangerous disease, which caused around 100 deaths per year in the UK before mass vaccination against measles began in 1968. The combined measles, mumps, and rubella (MMR) vaccine was introduced in 1988, and by the early 1990s more than 90% of children were being vaccinated against measles. Ten years later it was claimed, on the basis of 12 children diagnosed as autistic soon after vaccination, that there was a causal link between autism and the MMR vaccine. Although this claim was eventually shown to be false and its protagonist was discredited, distrust of MMR caused the proportion of children being vaccinated against measles to decrease steadily, with the result that outbreaks of measles in Britain increased in size and frequency (Figure 11.24). The effects of this unfounded distrust are still being felt. In the spring of 2013 an outbreak of more than 1000 cases of measles occurred in Swansea, a city where more than 6000 children had never received the MMR vaccine.

In 1998, the vast majority of the UK population had protective immunity against measles virus as a result of previous vaccination or infection. In this situation, a population has what is called **herd immunity**, which also indirectly protects the minority of people who have not been vaccinated. The

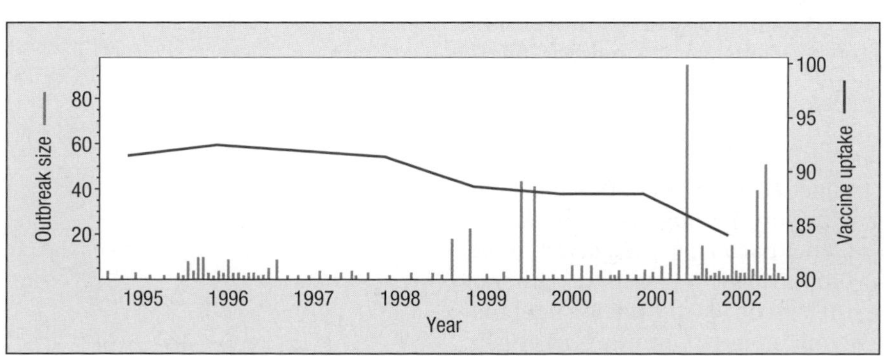

Figure 11.24 As fewer children were vaccinated against measles virus, the incidence of measles infections increased. Data are from the UK population. Since 1998, when the MMR vaccine was introduced, the number of children vaccinated (red line) decreased and outbreaks of measles (blue bars) increased in both frequency and size. The vaccine uptake is the percentage of children completing a primary course of the MMR vaccine at their second birthday. Data courtesy of V.A.A. Jansen and M.E. Ramsey.

Available vaccines for infectious diseases in humans			
Bacterial diseases	**Types of vaccine**	**Viral diseases**	**Types of vaccine**
Diphtheria (*Corynebacterium diphtheriae*)	Toxoid	Yellow fever	Attenuated virus
Tetanus (*Clostridium tetani*)	Toxoid	Measles	Attenuated virus
Pertussis (*Bordetella pertussis*)	Killed bacteria. Subunit vaccine composed of pertussis toxoid and other bacterial antigens	Mumps	Attenuated virus
Paratyphoid fever (*Salmonella paratyphi*)	Killed bacteria	Rubella	Attenuated virus
Typhus fever (*Rickettsia prowazekii*)	Killed bacteria	Polio	Attenuated virus (Sabin) or killed virus (Salk)
Cholera (*Vibrio cholerae*)	Killed bacteria or cell extract	Varicella (chickenpox)	Attenuated virus
Plague (*Yersinia pestis*)	Killed bacteria or cell extract	Influenza	Inactivated virus
Tuberculosis	Attenuated strain of bovine *Mycobacterium tuberculosis* (BCG)	Rabies	Inactivated virus (human). Attenuated virus (dogs and other animals). Recombinant live vaccinia-rabies (animals)
Typhoid fever (*Salmonella typhi*)	Vi polysaccharide subunit vaccines. Live-attenuated oral vaccine	Hepatitis A	Subunit vaccine (recombinant hepatitis antigen)
Meningitis (*Neisseria meningitidis*)	Purified capsular polysaccharide	Hepatitis B	Subunit vaccine (recombinant hepatitis antigen)
Bacterial pneumonia (*Streptococcus pneumoniae*)	Purified capsular polysaccharide Polysaccharide conjugated to protein	Human papillomavirus	Subunit vaccine (virus coat proteins)
Meningitis (*Haemophilus influenzae*)	*H. influenzae* polysaccharide conjugated to protein	Rotavirus	Attenuated virus Recombinant live virus

Figure 11.25 Diseases for which vaccines are available. Not all of these vaccines are equally effective, and not all are in routine use.

pathogen cannot create an epidemic, because of the low probability of finding susceptible individuals and creating a chain of infection. If the proportion of children being immunized against measles continues to decrease, however, the proportion of the UK population with no immunity to measles could become sufficiently large for herd immunity to be lost. An outbreak of measles could then more easily become an epidemic.

11-27 Vaccines have yet to be made against pathogens that establish chronic infections

The diseases for which we have effective vaccines (Figure 11.25) are ones in which the infection is acute and resolves in a matter of weeks, either by elimination of the pathogen or by the death of the host. Pathogens do not survive if they eliminate too many of their human hosts, so a majority of people do survive infections with common, familiar pathogens. At its worst, smallpox killed only one-third of the people it infected. Vaccines that protect against these diseases work because they stimulate immune responses that mimic the protective response provoked by natural infection with the pathogen.

By contrast, devising vaccines that work against chronic infectious diseases has proved difficult (Figure 11.26). After infection, pathogens such as the malarial parasite *Plasmodium falciparum*, the bacterium *Mycobacterium tuberculosis*, and the human immunodeficiency virus (HIV) interfere with the human immune system and make it work in their favor. From studying patients with these conditions, it is not always clear which type of immune response

Some diseases for which effective vaccines are not yet available		
Disease	**Estimated annual mortality**	**Estimated annual incidence* or total number of people infected†**
Malaria	600,000	220 million*
Schistosomiasis	23,000	200 million†
Tuberculosis‡	1.3 million	8.6 million*
Diarrheal disease	1.8 million	1.7 billion*
HIV/AIDS	1.6 million	2.3 million* 35.3 million†
Measles‡	160,000	20 million*
Hepatitis C	350,000§	3–4 million* 150 million†

Figure 11.26 Diseases for which better vaccines are needed. ‡The measles vaccines currently used are effective, but they are heat sensitive and require carefully controlled refrigeration, as well as reconstitution, before use. In some tropical countries this reduces their usefulness. The BCG vaccine against tuberculosis has proved of limited effectiveness and is not included in routine vaccination schedules. §Estimated mortality from the effects of chronic infection. Source of estimated mortality, incidence and prevalence data: World Health Organization (2011–2012 estimates).

would be effective if triggered by a vaccine. In this context, hepatitis C virus (HCV) is informative because it causes both acute and chronic infections. Depending on the population, 15–30% of infected people suffer mild symptoms of disease and make a primary immune response that clears the infection and bestows immunological memory of the virus (Figure 11.27). Nor does the acute infection ever kill the human host. The majority of people infected with HCV, some 130–170 million worldwide, make a suboptimal immune response that results in a chronic infection and persisting disease in which the infected liver goes through cycles of cellular destruction followed by tissue regeneration. This unsatisfactory situation can lead to cancer, liver failure, and the need for a liver transplant, for which patients with chronic HCV infection are the most numerous applicants. Despite much effort to make an HCV vaccine that goal has yet to be achieved but several new antiviral drugs are coming to the market.

Resolution of acute HCV infection correlates with an early-acting and robust innate immune response, followed by a delayed but forceful adaptive immune response. The latter involves strong effector CD4 and CD8 T-cell responses, and neutralizing antibodies that prevent the virus from infecting hepatocytes. In the innate immune response, the production of type I interferon seems critical, particularly IFN-λ, for which one genetic variant doubles the probability of resolution over that of a second variant. In the adaptive immune response, several HLA class I and II allotypes correlate either with resolved or chronic infection. In designing vaccines against HCV, the vaccinologists now have a better idea of what they should be aiming for.

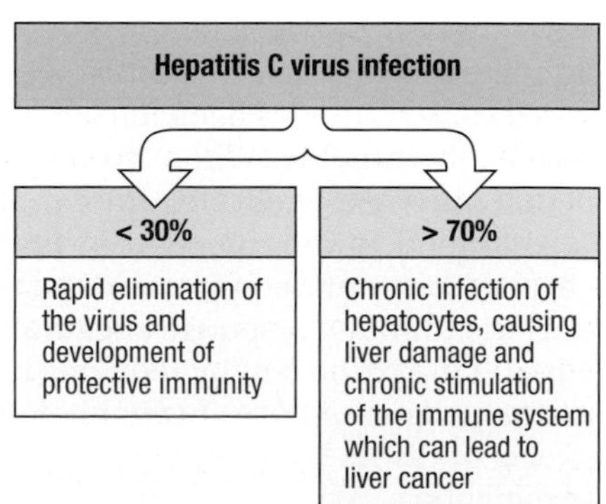

Figure 11.27 A minority of people exposed to hepatitis C virus resist the virus, whereas the majority develop chronic infection.

11-28 Vaccine development faces greater public scrutiny than drug development

Two big differences make the development and success of a vaccine inherently more difficult than for a drug. The first difference is that drugs are given to people who are ill and grateful for any alleviation of their condition, whereas vaccines are administered to healthy individuals, often the very young, for whom any adverse affects of the procedure can be of greater concern to their parents than the benefits. The second difference arises because drugs are prescribed, on a case-by-case basis, by a physician for a patient, whereas vaccination is usually part of a large program mandated by an authority and applied to the population at large. Exemplifying the latter are the worldwide vaccination programs aimed at eradicating smallpox and polio. If you have had a painful strep throat for two weeks and take a course of antibiotics, relief will come within days and you will give thanks for the benefit of such drugs. In contrast, if you have been vaccinated against poliovirus and never get the paralysis that made it so feared, you will never know whether protective immunity has spared you from disease or whether you have never been exposed to the virus. The only situation in which a vaccine's benefit can truly be felt is when exposure is obvious, because unvaccinated members of your community are suffering and succumbing to disease.

Drugs and vaccines all have unwanted side-effects, for which the risk must be weighed against the benefit. Because a vaccine's side-effects cause disease in healthy children, the standards set for vaccines are always higher than those for drugs. This sensitivity has also led to periodic episodes when disease or other syndromes are erroneously and irresponsibly attributed to be the consequence of vaccination, as autism and the MMR vaccine were so linked in 1998 (see Section 11-26). Because of the nature of herd immunity, individuals can gain the benefit of mass vaccination programs without sharing the risk. This is yet another difference from drugs, for which the cost and the benefit are inseparable. However, if enough individuals take this selfish strategy, then herd immunity is lost, and in the ensuing epidemic they could lose out, too.

Although vaccines are arguably the most successful of medical interventions, their acceptance and use have been greatly influenced by factors having nothing to do with science or medicine. For example, an early version of a rotavirus vaccine was abandoned in 1999 because of its side-effect, a bowel obstruction that, if not treated with an enema or surgery, could be fatal, which reportedly occurred in 1 out of 7000 children vaccinated. Although the decision to abandon the virus could be justified in the United States, where only 1 in 100,000 deaths is due to rotavirus infection, the vaccine would have greatly benefited the populations of other countries, such as India, where as many as 1 in 200 children were dying of rotavirus infection. Thus the decision to discontinue rotavirus vaccination in some countries led to many more deaths from infection than lives saved by halting vaccination

Summary

Vaccination is preventive medicine. It involves the deliberate immunization of healthy people with some form of a pathogen or its component antigens, and induces a protective immunity that prevents any future infection with the pathogen from causing disease. Vaccines consist of killed whole pathogens, live-attenuated strains, nonpathogenic species related to the pathogen, or of secreted or surface macromolecules from the pathogen. Vaccination has saved millions of lives and reduced the incidence of many common infectious diseases, particularly in the rich industrialized countries. The prevention of infectious disease by vaccination illustrates superbly how manipulation of the immune response can benefit public health. The pathogens for which effective vaccines have already been found are those that cause acute infections and are not highly mutable. When these pathogens cause epidemic disease, many people survive and develop immunity, showing that human immune systems can respond to the infection in productive ways. Death, when it occurs, is because the immune system was just too slow. What vaccination achieves is to start the immune response ahead of infection, giving the immune system an edge on the pathogen. By reducing the incidence of disease, successful vaccination programs have inevitably led to decreased public awareness of the effects of disease, and increased concern with the side-effects and safety of vaccination.

Until the approval of the hepatitis B virus subunit vaccine in 1981, vaccine development was largely a process of trial and error, one in which a knowledge of immunological mechanisms played little part and the guiding principle was for the vaccine to resemble the natural pathogen as closely as possible.

Although this approach worked for pathogens causing acute infections, it failed to discover vaccines against pathogens establishing chronic infections and causing chronic disease. The trouble is that we know little about what constitutes a successful immune response in the context of these diseases. Because of the difficulties of vaccine development, the costs, and the political and legal complications associated with their side-effects, both real and imaginary, commercial vaccine manufacture has been a declining industry for more than 40 years. For example, since 1967 the number of vaccine manufacturers in the United States declined from 26 to 5. When the previous edition of this book was published in 2009, the future of vaccinology looked gloomy. Since then, there has been a brightening, largely because of recombinant DNA technology. This revolution began to take hold in the 1980s, and was then expanded and refined to provide the high-throughput analyses required by the human genome project and its successors. This in turn has led to reverse vaccinology, by which a pathogen's genome is mined to identify possible antigens that are then studied to determine their biology and how they interact with the immune system. This approach increases our basic understanding of host–pathogen relationships, and at the same time engages in its application: the discovery of new vaccines. A better understanding of the innate immune response is leading to a new generation of adjuvants, the natural ligands of Toll-like receptors and other sensors of innate immunity. On the horizon is the possibility of speeding up the current methods for making vaccines against emerging epidemic and pandemic strains of influenza, so that populations will be vaccinated in good time, before the virus arrives.

Summary to Chapter 11

In earlier chapters of the book we saw how a primary adaptive immune response is developed against an invading pathogen that either evades or outruns the innate immune response. In those chapters we concentrated on the development of the effector mechanisms of immunity and the mechanisms by which they fight and defeat the pathogen. In the first part of this chapter we examined the primary immune response from a different perspective. At the same time that antibodies and effector T cells are being pressed into action, memory B cells and T cells are also being produced in the same lymphoid tissues as the effector cells. The memory cells do not contribute functionally to the primary response, but they will function in the future should the pathogen invade again.

After a successful primary response, the accumulated antibodies and effector cells prevent reinfection for up to a year, but then these defenses fade. If the pathogen subsequently reinvades, a secondary adaptive immune response is made and catalyzed by residual pathogen-specific antibody. Because memory cells are in greater numbers and are more quickly activated than naive cells, the secondary response is faster and more forceful than the primary response. Thus it is more likely to eliminate the pathogen at an early stage of infection, before symptoms of disease are felt. Each such encounter with the pathogen will refine and improve the capability of the adaptive immune defenses. The second part of the chapter shows how the biology of immunological memory has been translated into the widely used and highly beneficial procedure of vaccination. Vaccines can be either whole pathogens that have been manipulated to render them harmless, or non-infectious pathogen components. What they have in common is a lack of pathogenicity and the presence of pathogen antigens that stimulate a protective adaptive immune response. Young children given two or more immunizations with a vaccine develop a good immunological memory of the pathogen. If they are infected in the future with that pathogen, they will make a powerful secondary immune response that stops the infection in its tracks. In effect, the vaccine allows the immune system to practice on a surrogate pathogen before it sees the real thing. The first time one tries to do a new task it is never done well. With practice, however, one improves, and the more practice one has, the better one gets. Such tasks involve neurological memory, but immunological memory behaves in a remarkably similar fashion.

Questions

11–1 Indicate whether the following questions are true (T) or false (F).

a. Secondary immune responses take the same amount of time as primary immune responses to become effective.
b. On secondary exposure to an infectious agent there is reduced mortality.
c. Only immune responses made in mucosal secondary lymphoid tissues can provide protective immunity.
d. If an individual acquires a second cold in the same season it will most probably be caused by a different type of cold virus.
e. Plasma cells generated in a secondary immune response have longer life-spans than those made during a primary immune response.
f. During a primary immune response, only memory B cells are generated.

11–2 Which of the following pairs is mismatched?

a. activation of memory cells: secondary immune response
b. central memory T cells: restricted to non-lymphoid tissues
c. protective immunity: antibodies persist after primary immune response
d. attenuated vaccine: not pathogenic
e. vaccination: primary immune response.

11–3 All of the following statements regarding immunological memory are true except _____. (Select all that apply.)

a. During a primary immune response, effector B cells outnumber memory cells.
b. A small population of plasma cells secrete pathogen-specific antibody long after pathogen clearance from the body.
c. Memory T cells are produced during secondary immune responses, but not during primary immune responses.
d. Memory T cells and memory B cells originate in secondary lymphoid tissue through clonal expansion.
e. Memory B cells generated in secondary immune responses are more effective than memory B cells made in primary immune responses because of affinity maturation.
f. Protective immunity persists for the same period regardless of the type of pathogen.

11–4 Memory T cells _____. (Select all that apply.)

a. must all be activated in secondary lymphoid organs
b. do not require co-stimulation through CD28
c. are composed only of CD4 T cells
d. do not undergo somatic hypermutation
e. are usually short-lived
f. do not undergo isotype switching.

11–5 All of the following contribute to the establishment and maintenance of long-lived memory B cells except _____.

a. replenishment of memory population by cell division
b. isotype switching
c. somatic hypermutation
d. long-term persistence of antigen after primary immune response
e. interaction with stromal cells in the bone marrow.

11–6 Name two ways in which memory T cells differ from naive T cells that enable them to mount a more rapid immune response.

11–7 _____ describes the situation when the immune response generated against an influenza strain limits future antibody responses to variant epitopes of a different, but similar, strain.

a. Attenuation
b. Herd immunity
c. Variolation
d. Neutralization
e. Original antigenic sin.

11–8 Match the cell type in column A with its description in column B.

Column A	Column B
a. naive T cell	1. expresses CCR7 and has a low threshold for activation
b. T_{FH} cell	2. potentially cytotoxic and bears IL-7 receptor
c. central memory T cell (TCM)	3. expresses CD45RA and has a high threshold for activation
d. memory CD8 T cell	4. lacks L-selectin and CCR7 and recirculates to non-lymphoid tissues
e. effector memory T cell (TEM)	5. participates in efficient cognate interactions with memory B cells

11–9 All of the following are descriptive of inactivated virus vaccines except _____.

a. formalin-treated
b. grown in cells from another animal
c. heat-treated
d. irradiated
e. pathogenic virus required.

11–10

A. What is an adjuvant and why are they incorporated into human vaccines?
B. Provide several examples.

11–11 Match the virus in column A with its description in column B. Answers may be used more than once.

Column A	Column B
a. smallpox	1. vaccine uses genetically engineered yeast cells
b. hepatitis B virus	2. fecal–oral route of transmission
c. rotavirus	3. contains 11 genomic segments
d. poliovirus	4. face-to-face contact for transmission
e. influenza virus	5. short-lived immunological memory

11–12 All of the following statements about polioviruses are correct except _____.

a. Poliovirus vaccines are produced in both inactivated and live-attenuated forms.
b. The oral poliovirus vaccine is composed of three live-attenuated viral strains.
c. Poliovirus infects both humans and cattle.
d. If genetic reversion of strain 3 occurs when the virus is replicating in vaccinated people, pathogenesis may occur.
e. In the United States, the recommended vaccine for poliovirus is the inactivated poliovirus vaccine (IPV).

11–13 Indicate whether each of the following statements is true (T) or false (F).

a. An unsubstantiated association between the MMR vaccine and autism has led some parents to choose not to vaccinate their children.
b. The hepatitis B vaccine is associated with cancer of the liver.
c. Acellular vaccines have fewer side-effects than cellular vaccines.
d. Recombinant DNA technology can be used to reduce the virulence of pathogenic viruses.
e. The Rotarix vaccine is a subunit vaccine composed of purified VP4 and VP7 glycoproteins of the human rotavirus.
f. Rotavirus is a double-stranded DNA virus that lends itself well to recombinant DNA manipulation.

11–14 Madison Tavistock, a healthy 2-year-old living in Cincinnati, had attended the Wee Folks Daycare Facility for a year. Her parents joined an anti-vaccination group when she was 9 months old, 3 months before the recommended immunization schedule for the MMR vaccine, but after Madison had already been vaccinated with DTP. They strongly believe that the unsubstantiated risk of autism reported for MMR vaccination outweighs the benefits, and consequently have opted not to have their daughter immunized. The best explanation for why Madison has not contracted measles even though she has regular contact with other children in a large city is that _____.

a. Madison is tolerant to measles antigens.
b. The measles virus may have infected Madison but is dormant.
c. The DTP vaccine provides cross-protective immunity against measles.
d. The other children in the daycare center have been vaccinated and she has herd immunity.
e. The attenuated measles virus in the MMR vaccine received by the other children in her daycare facility was transmitted to Madison and she developed asymptomatic natural immunity.

11–15 Fatima Ahmed, a 25-year-old recent immigrant from Sudan, is brought to an obstetrician by her husband Samir for a first-time visit at about 38 weeks of pregnancy. This is Fatima's first pregnancy, and she and the baby are in excellent health. She has had no previous antenatal care, fearing that if her condition had been revealed she would not have been permitted entry into the United States. She is seen weekly by her physician and delivered a healthy baby girl without complication 18 days later. The justification for the obstetrician administering RhoGAM postnatally is that:

a. Fatima is RhD^+ and the baby is RhD^-.
b. Fatima is RhD^- and the baby is RhD^+.
c. Fatima is RhD^+ and Samir is RhD^-.
d. Samir is RhD^- and the baby is RhD^-.
e. Fatima is RhD^- and the baby is RhD^-.

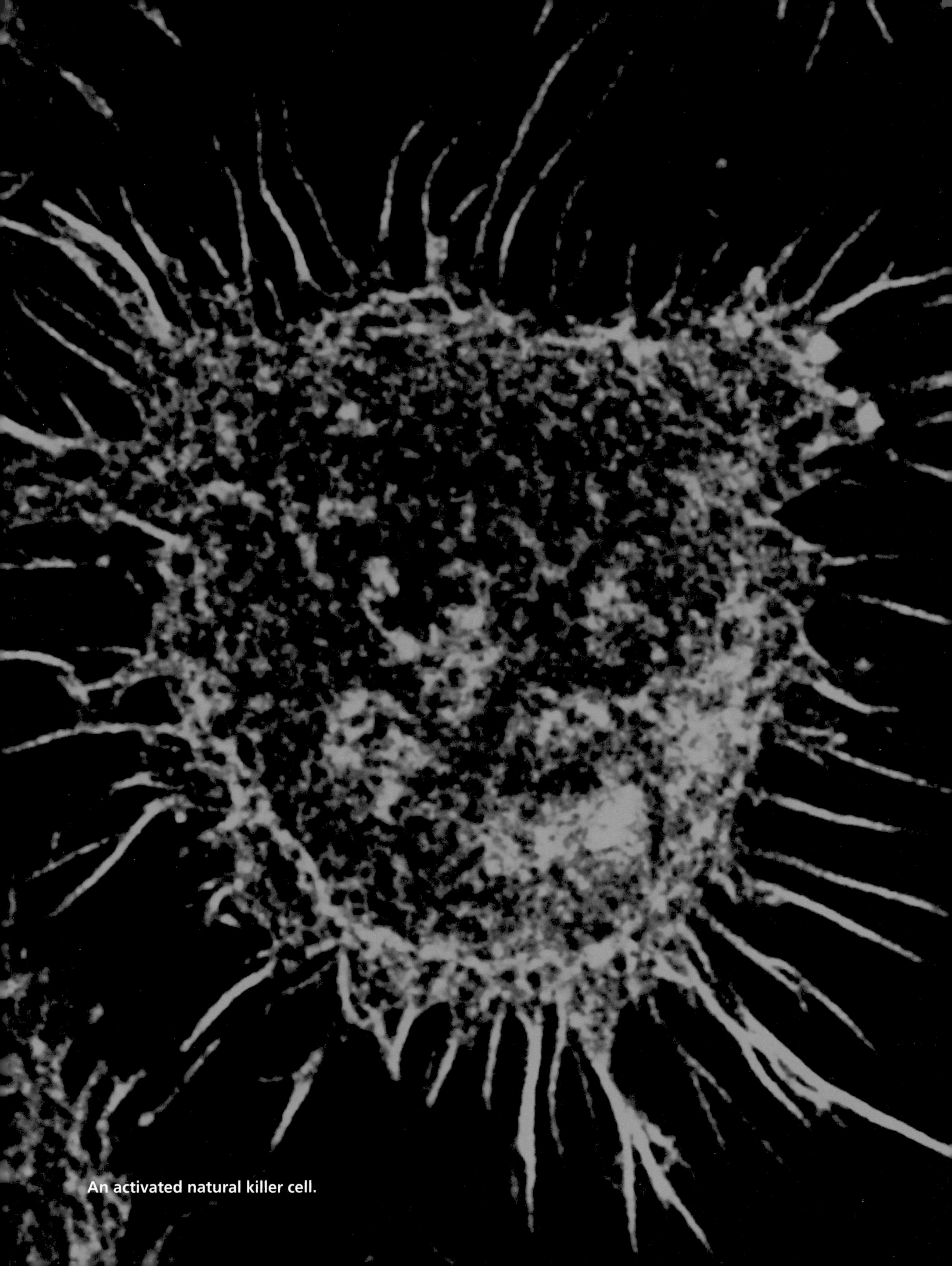

An activated natural killer cell.

Chapter 12

Coevolution of Innate and Adaptive Immunity

For much of the twentieth century, the community of immunologists was split into a minority who studied innate immunity, usually called inflammation, and a majority who studied adaptive immunity. This division originated with differences in concepts and methods, but then continued on the basis of precedent despite increasing evidence for connections between innate and adaptive immunity. The most important of these connections was the universal observation that an inflammatory response is a necessary prelude to making good antibodies. Although implemented in practice, this well-known fact was conveniently ignored in the thinking of most immunologists. That is, until the 1990s when the conceptual wall came down and the essential cooperation between innate and adaptive immunity was generally acknowledged.

Innate immunity is present in both vertebrates and invertebrates, whereas adaptive immunity is found only in vertebrates. Innate immunity therefore evolved before adaptive immunity, a progression reflected during the immune response, when the innate immune response precedes the adaptive immune response. The origin of innate immunity was 600 million years ago. Adaptive immunity began 200 million years later (400 million years ago), with the emergence of RAG proteins and receptors encoded by rearranging genes. For that innovation to have been useful, and thus subject to natural selection, it had to occur in the context of the existing cells and molecules of innate immunity. An example of this phenomenon is the classical pathway of complement activation, in which antibody was the innovation of adaptive immunity that hooked up with the preexisting complement components of innate immunity.

The overriding question we will address in this chapter is how innate immunity evolved once adaptive immunity had appeared. To help frame the question, three different models are considered in Figure 12.1. The first model is that adaptive immunity built upon innate immunity, as it existed 400 million years ago, and the constraints it imposed inhibited further innovation of innate immunity (Figure 12.1, left panel). In this scenario, innate immunity is like a living fossil. The second model proposes that, after the initial emergence of adaptive immunity, both innate and adaptive immunity continued to innovate, but independently (Figure 12.1, center panel). The third model also proposes that innate and adaptive immunity continued to evolve, but that it involved mutual exploitation of each other's cells and molecules, as well as independent innovation (Figure 12.1, right panel).

To seek answers to this question, this chapter considers three types of lymphocyte whose mix of innate and adaptive features throw light on the interdependence and coevolution of the innate and adaptive immune systems. Their discussion has been left until now because their place within the immune

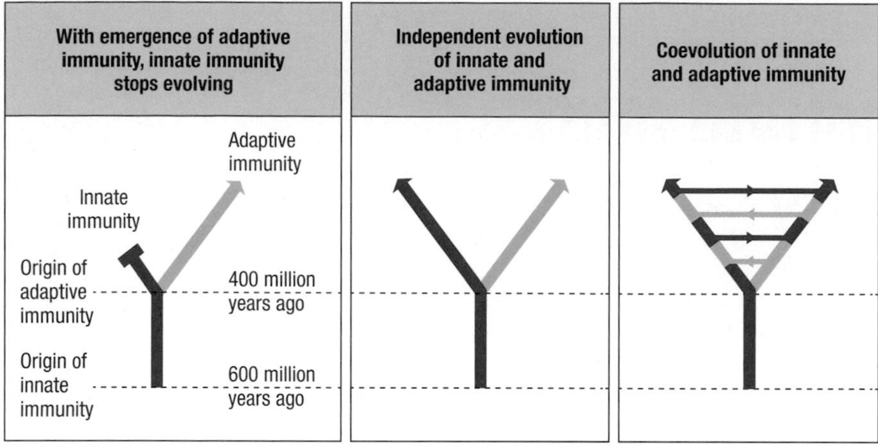

Figure 12.1 Three possible models for the evolution of innate and adaptive immunity. Each panel presents a scenario for the course of evolution of innate and adaptive immunity. The dashed horizontal lines indicate the origin of adaptive immunity some 400 million years ago and the origin of innate immunity at around 600 million years ago. The colored lines denote the components of innate immunity (red) and adaptive immunity (green). Left panel: the evolution of innate immunity slows down and stops after adaptive immunity has emerged and become established. Center panel, innate immunity and adaptive immunity evolve independently and at comparable rates once adaptive immunity has emerged and become established. Right panel: this model is like that in the center panel, except that the evolution of innate and the evolution of adaptive immunity are not independent but interactive. As indicated by the horizontal red and green lines, components of innate immunity continue to be incorporated into adaptive immunity, and components of adaptive immunity become part of innate immunity.

system is better understood once the basic principles of innate and adaptive immunity are appreciated. The first part of the chapter examines the natural killer (NK) cells. The cytotoxic and cytokine-secreting functions of these lymphocytes of innate immunity were introduced in Chapter 3; here we look at the receptors that NK cells use to distinguish healthy cells from infected and tumor cells. NK cells also make an important contribution to reproduction. The second part of the chapter discusses γ:δ T cells. Although the antigen receptors of γ:δ T cells are products of rearranging genes, the properties of these receptors and the antigens they recognize are different from those of α:β T-cell receptors. In the last part of the chapter we examine large and unusual subsets of α:β T cells that do not recognize peptide antigens and are not restricted by conventional MHC molecules.

Regulation of NK-cell function by MHC class I and related molecules

NK cells were introduced in Chapter 3 as cytotoxic lymphocytes of the innate immune response (see Sections 3-17 to 3-21). NK cells are recruited to sites of infection by the IL-12 secreted by resident macrophages. Once within an infected tissue, NK cells secrete cytokines, principally interferon (IFN)-γ, that increase macrophage activation; they also deploy their cytotoxic machinery to kill cells infected by intracellular pathogens, notably viruses. This chapter focuses on receptors that NK cells use to detect and respond to infection. We shall see how MHC class I molecules, previously discussed for their crucial role in cytotoxic T-cell development and function, are equally central to the development and function of NK cells. They lie at the very heart of the mechanisms by which NK cells combine tolerance toward healthy cells with the ability to kill cells compromised by infection, malignancy, and other forms of cellular stress (see Section 3-19). We shall see how NK cells participate in the adaptive immune response, and how they and their receptors for MHC class I molecules have been recruited to serve a necessary function in human reproduction.

12-1 NK cells express a range of activating and inhibitory receptors

NK cells do not rearrange immunoglobulin and T-cell receptor genes, a property that distinguishes them from all other lymphocyte populations. This suggests that NK cells are the modern representatives of an ancient line of lymphocytes, one that existed more than 400 million years ago, before the emergence of adaptive immunity in the vertebrates (see Section 5-3). Apart from the T-cell receptor and its associated CD3 complex, most proteins present on T-cell surfaces are also expressed by NK cells. Similarly, most of the NK-cell receptors are also expressed by some T cells. These similarities exist because NK cells and cytotoxic T cells have comparable effector functions—secreting cytokines and killing unhealthy cells—and are consistent with T cells having evolved from an NK-cell-like ancestor.

A major difference between T cells and NK cells is the way they detect antigens and respond to infection. Because T cells are activated only in the presence of their specific antigens, most naive T cells never become activated and used in an immune response. To reduce the body's investment in such unemployed cells, all T cells circulate in a dormant state from which they require several days of activation to become effector cells. T cells are also dependent on a single highly specific T-cell receptor to detect infection, whereas NK cells express a variety of activating receptors through which they can respond to infection (Figure 12.2).

Figure 12.2 NK cells use many activating and inhibitory receptors to distinguish unhealthy cells from healthy cells. The table gives the name and CD number of the receptor, the type and number of protein domains that form the binding site, the ligand or ligands for the receptor, and the type of protein domain that forms the ligand. The six UL-binding proteins (ULBPs) are so called because ULBP1, 2, and 6 bind to the cytomegalovirus UL16 protein. They have two extracellular domains resembling the MHC class I α_1 and α_2 domains and are encoded by a gene family on chromosome 6, apart from the MHC.

Activating NK-cell receptors			
Receptor	**Receptor structure**	**Ligand**	**Ligand structure**
NKp80	Lectin-like	NKp65	Lectin-like
CD94:NKG2C (CD159c)	Lectin-like	HLA-E with bound peptides derived from leader sequences of HLA-A, -B, and -C	MHC class I
KIR2DS1 (CD158h)	Ig-like (2 domains)	C2 epitope of HLA-C	MHC class I
KIR2DL4 (CD158d)	Ig-like (2)	HLA-G	MHC class I
NKG2D (CD314)	Lectin-like	MIC-A, MIC-B, ULBP1–6	MHC class I-like
CD16 (FcγRIIIA)	Ig-like (2)	IgG	Immunoglobulin
2B4 (CD244)	Ig-like (2)	CD48	Ig-like (2 domains)
DNAM-1 (CD226)	Ig-like (2)	Nectin-2 (CD112), Poliovirus receptor (PVR) (CD155)	Ig-like (3)
LFA-1 (CD11a)	Integrin	ICAM-1	Ig-like (5)
NKp30 (CD337)	Ig-like (1)	B7-H6 expressed by some tumors HLA-B-associated transcript 3 (BAT3)	Ig-like (2) Nuclear protein released by tumors
NKp46 (CD335)	Ig-like (2)	Viral hemagglutinins and neuraminidases Membrane protein of malarial parasite	Pathogen-derived antigens of varied structure
†NKp44 (CD336)	Ig-like (2)	Cancer-induced changes to self proteins; mixed-lineage leukemia protein 5 (MLL5)	MLL5 is an intracellular protein that becomes surface-associated in tumors
Inhibitory NK-cell receptors			
Receptor	**Receptor structure**	**Ligand**	**Ligand structure**
CD94:NKG2A (CD159a)	Lectin-like	HLA-E with bound peptides derived from HLA-A, -B, and -C leader sequences	MHC class I
KIR2DL1 (CD158a)	Ig-like (2 domains)	C2 epitope of HLA-C having methionine 80	MHC class I
KIR2DL2/3 (CD158b)	Ig-like (2)	C1 epitope of HLA-C having lysine 80, also HLA-B*46 and HLA-B*73	MHC class I
KIR3DL1 (CD158d)	Ig-like (3)	Bw4 epitope of HLA-A and HLA-B having RI/TALR motif at position 79–83	MHC class I
KIR3DL2 (CD158k)	Ig-like (3)	HLA-A*03 and HLA-A*11	MHC class I
LILRB1 (CD85j)	Ig-like (4)	Broad reactivity with HLA class I, strongest binding to HLA-G	MHC class I
†NKp44	Ig-like (2)	Proliferating cell nuclear antigen (PCNA), cancer associated	PCNA is recruited into the NK-cell synapse with tumor target cell and inhibits cell killing

The number in parentheses after Ig-like receptors and ligands is the number of Ig-like domains.
†NKp44 has both ITIM and ITAM motifs (see Section 12-3) and can mediate both activation and inhibition.

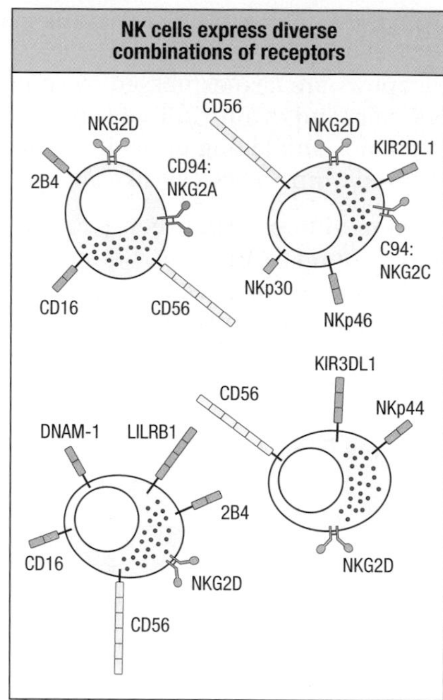

Figure 12.3 NK cells express diverse combinations of receptors. No single NK cell has all of the receptors listed in Figure 12.2 on its surface. Instead, individual NK cells express a subset of the receptors, which produces, as shown here, a diversity of NK-cell phenotypes. All the NK cells are shown expressing CD56, because the expression of this protein in the absence of a T-cell receptor is used to define NK cells (see Section 3-17). All NK cells also express an inhibitory receptor for self-MHC class I. CD94:NKG2A, KIR2DL1, KIR3DL1, and LILRB1 are examples of such receptors.

Some of these receptors, for example the Toll-like receptors TLR3, 7, and 8 (see Section 3-3), recognize categories of pathogens and their products, whereas others recognize alterations to human cell surfaces that are induced by infection and cellular stress. Because of the variety and versatility of NK-cell receptors, most NK cells that enter the circulation, if not all of them, will participate in fighting infection. Unlike T cells, NK cells circulate in a partly activated state that enables them to respond quickly to infection and be effective lymphocytes of innate immunity. To prevent them from attacking healthy cells, NK cells express inhibitory receptors that recognize ligands expressed on the surface of healthy tissue cells.

There are more than 30 different types of activating and inhibitory NK-cell receptors (see Figure 12.2). Individual NK cells express only a selection of these receptors, but the selection varies from cell to cell (Figure 12.3). The result is that NK cells with many thousands of different receptor phenotypes circulate in a person's blood. Every NK cell expresses both activating and inhibitory receptors, which together develop a balance of activating and inhibitory signals that ensures tolerance of healthy cells and a rapid response to unhealthy cells.

12-2 The strongest receptor that activates NK cells is an Fc receptor

During the innate immune response to a primary infection, activation of an NK cell requires intracellular signals to come from two or more of the activating receptors. This strategy, in which two receptors confirm the occurrence of infection, decreases the likelihood that NK cells will become activated inadvertently in the absence of infection.

During the adaptive immune response to an infection, when pathogen-specific IgG becomes available, NK cells can be activated by a single receptor (Figure 12.4). This receptor, **FcγRIIIA (CD16a)**, is an activating, low-affinity receptor for IgG (see Figure 9.35, p. 260), which is ubiquitously expressed by the CD56^{dim} subpopulation of NK cells (see Section 3-18). On binding to antigen-coated pathogens or immune complexes, the signals generated by this receptor activate NK cells without the assistance of a second activating receptor. Confirmation from a second receptor is unnecessary because the pathogen-specific antibody provides assurance that infection is truly present. The effect of this mechanism is to give NK cells a highly specific antigen receptor, which enables them to participate in the adaptive immune response and also provide accelerated innate immunity in the secondary immune response.

This adaptation can be interpreted in two ways: one is that the NK cell of innate immunity uses the Fc receptor to exploit the antibody molecule of adaptive immunity; the other is that the antibody molecule of adaptive immunity uses the Fc receptor to exploit the NK cell of innate immunity. Neither interpretation is correct, because this mechanism arose through a mutually beneficial coevolution, resulting in improvements to both innate and adaptive immunity. At some critical point in this history, a population of individuals whose NK cells had acquired FcγRIIIA were able to overtake and replace their

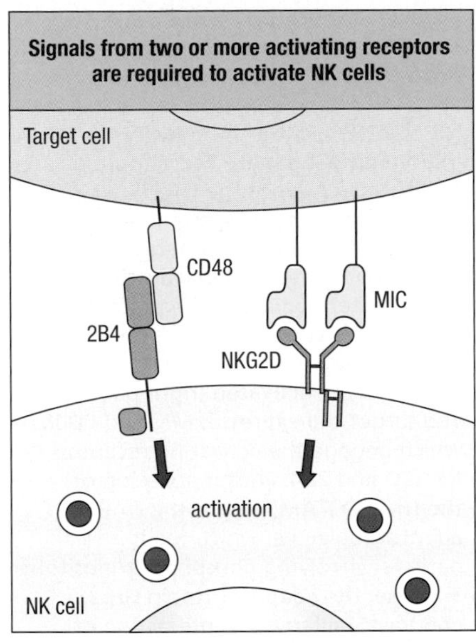

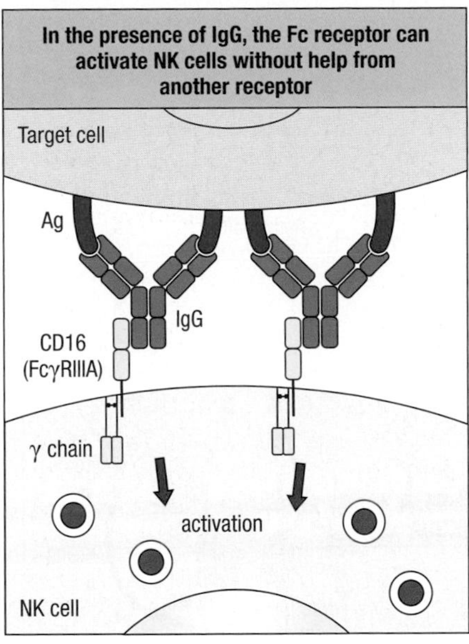

Figure 12.4 Activation of NK cells by receptors of innate and adaptive immunity. Left panel: activation of an NK cell by the 2B4 and NKG2D receptors of innate immunity. Both of these receptors must bind to their respective ligands on the target cell to produce signaling that will activate the NK cell. Right panel: an NK cell being activated by FcγRIIIA that has bound complexes of cell-surface antigens and antigen-specific IgG molecules. Signals from the Fc receptor alone are sufficient to activate the NK cell.

relatives whose NK cells had not acquired the receptor. The Fc receptors are a good example of the integration of innate and adaptive immunity because they enable almost all the cell types of innate immunity to contribute to adaptive immunity.

12-3 Many NK-cell receptors recognize MHC class I and related molecules

Because immunologists had previously associated NK cells with innate immunity and MHC molecules with adaptive immunity, it came as a surprise when the ligands for several NK-cell receptors were found to be MHC class I molecules or related proteins. Of these, the inhibitory NK-cell receptors tend to recognize conventional MHC class I molecules (see Figure 5.29, p. 136), whereas the activating receptors tend to recognize other proteins that are structurally and evolutionarily related to MHC class I; however, there are exceptions. Here we examine representative and functionally important examples of an inhibitory NK-cell receptor (CD94:NKG2A) and an activating NK-cell receptor (NKG2D).

CD94:NKG2A is an inhibitory receptor expressed by many NK cells and comprises a disulfide-linked heterodimer of the CD94 and NKG2A (CD159a) polypeptides. Each polypeptide has a C-type lectin domain, of the type present in many receptors of innate immunity (see Section 3-2). A characteristic of the C-type lectin domains in NK-cell receptors is that they bind to protein ligands and not to carbohydrate like many other receptors of innate immunity. HLA-E (see Section 5-18) is the ligand for CD94:NKG2A. This conserved HLA class I molecule has a ubiquitous tissue distribution like HLA-A, -B, and -C but a distinctive specificity for binding to peptides derived from the leader sequences of HLA-A, -B, and -C heavy chains. In binding to HLA-E, CD94:NKG2A contacts the peptide and the two α helices. Because an HLA-E molecule needs a bound peptide for proper assembly and delivery to the cell surface, the amount of HLA-E on a cell's surface provides a measure of the total amount of HLA-A, -B, and -C being made by the cell (Figure 12.5). This is also a measure of the cell's overall health, because infection and malignancy frequently reduce the cell-surface expression of these MHC molecules. In the extreme situation where a cell makes no HLA-A, -B, or -C, the HLA-E heavy chain and β_2-microglobulin are deprived of suitable binding peptides and remain stuck in the endoplasmic reticulum.

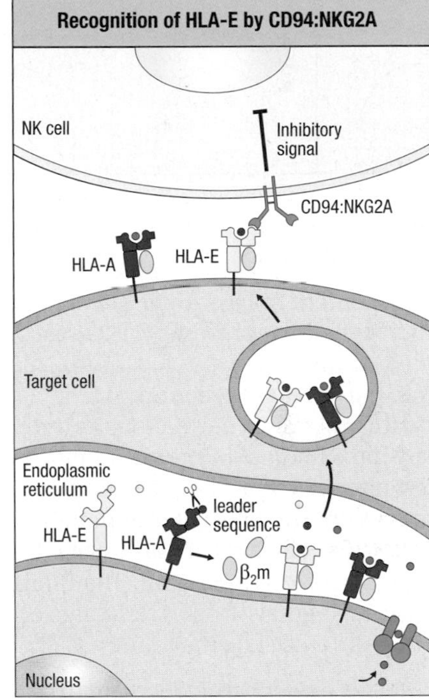

Figure 12.5 CD94:NKG2A recognizes complexes of HLA-E and peptides derived from the leader sequences of HLA-A, -B, and -C. During normal biosynthesis, the leader sequences of HLA-A, -B, and -C are cleaved off in the endoplasmic reticulum. These are further cleaved to give smaller peptides, for which the HLA-E peptide-binding site is highly specific. After binding one of these peptides, HLA-E travels to the cell surface and is recognized by the inhibitory NK-cell receptor CD94:NKG2A. For healthy target cells expressing normal levels of HLA-E, the inhibitory signals transduced by CD94:NKG2A tell the NK cell not to respond to the target cell. β_2m, β_2-microglobulin.

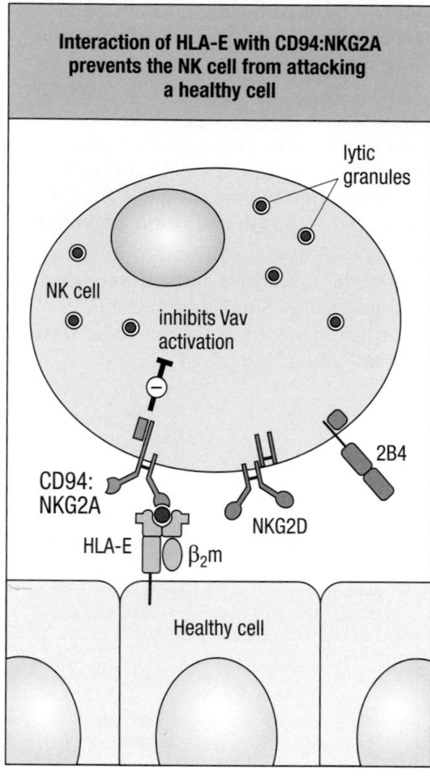

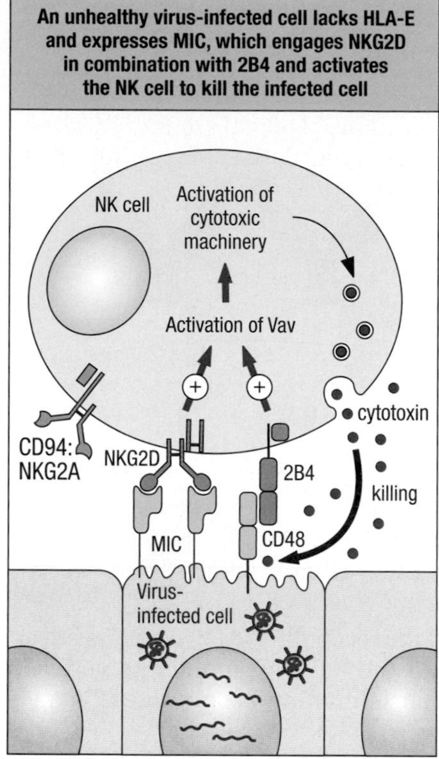

Figure 12.6 Signals from inhibitory receptors block the signaling pathways of NK-cell activation. When HLA-E on a healthy cell engages NK cell CD94:NKG2A, the NK cell is inhibited (first panel). The cytoplasmic tail of NKG2A binds the DAP12 adaptor molecule, which engages and activates the SHP-1 tyrosine phosphatase. SHP-1 dephosphorylates Vav inactivating this central intermediate in the pathways of NK-cell activation. When the same NK cell responds to a virus-infected cell, these pathways are activated (right panel). The target cell expresses MIC and CD48, which engage the activating receptors NKG2D and 2B4. Phosphorylation of the NKG2D ITAM recruits the DAP10 adaptor molecule, which feeds into a pathway involving phosphatidylinositol-3-kinase, the adaptor protein Grb2, and Vav. Similarly, the interaction of 2B4 with CD48, both members of the signaling lymphocyte activation molecule (SLAM) family, initiates signals leading to Vav phosphorylation. Phosphorylated Vav initiates further signaling and mobilization of NK-cell effector function.

When a CD94:NKG2A-expressing NK cell encounters a healthy tissue cell, the engagement of HLA-E on the tissue cell by CD94:NKG2A produces an inhibitory signal that prevents the NK cell from attacking the healthy cell. To do this, an **immunoreceptor tyrosine-based inhibitory motif** (**ITIM**) in the cytoplasmic tail of NKG2A recruits SHP-1, a tyrosine phosphatase. SHP-1 dephosphorylates the guanine-nucleotide-exchange factor Vav1, which inactivates this key intermediate in the signaling pathways for NK-cell activation (Figure 12.6, left panel). When this occurs, the NK cell detaches from the healthy cell and moves on to examine other potential targets. If the NK cell finds an unhealthy cell, with deficient expression of HLA-E, the CD94:NKG2A receptor is engaged either poorly or not at all. The inhibitory signal is then weak or nonexistent, allowing signals from activating receptors, such as NKG2D, to mobilize the NK-cell's cytotoxic machinery and kill the unhealthy cell.

NKG2D, an activating receptor expressed by all NK cells, is a homodimer of the CD314 polypeptide and similar to CD94:NKG2A in structure and organization. NKG2D has several types of ligand (see Figure 12.2). The **MIC glycoproteins** are one type, and are so called because they resemble MHC class I heavy chains and are encoded by genes in the class I region of the HLA complex. Unlike MHC class I molecules, the MIC proteins (**MIC-A** and **MIC-B**) are not present on healthy cells, do not associate with β_2-microglobulin, and do not bind peptides. MIC proteins belong to a group of **stress proteins** that are made only when cells are in trouble. They are expressed when cells become stressed, whether by infection, malignant transformation, elevated temperature, or other means. When NKG2D on an NK cell engages MIC on an infected cell, signals are induced that lead to the phosphorylation and activation of Vav1 and thus to the activation of NK-cell cytotoxicity. This results in the death of the infected cell and the recruitment of other effector cells to the diseased tissue (Figure 12.6, right panel).

From examining CD94:NKG2A and NKG2D, a balance is seen between the expression of either MHC class I or MIC molecules and the engagement of either inhibitory or activating NK-cell receptors. Healthy cells are associated with MHC class I expression and NK-cell inhibition, whereas diseased cells are associated with MIC expression and NK-cell activation. Evidence for a long-term partnership of MIC and MHC class I molecules is present in the genome. The HLA class I region consists of multiple genomic blocks of around 100 kilobases in length that arose through successive duplications from a primordial block containing one MIC gene and one MHC class I gene.

12-4 Immunoglobulin-like NK-cell receptors recognize polymorphic epitopes of HLA-A, HLA-B, and HLA-C

NK cells use the CD94:NKG2A receptor to assess the quantity of HLA-A, -B, and -C molecules in a potential target cell, as read out by the amount of HLA-E at the cell surface. In the human population, the CD94, NKG2A, and HLA-E polypeptides exhibit almost no genetic variation, so this strategy works equally well for all individuals, independently of their diverse HLA types. Its disadvantage is insensitivity to selective loss of expression of a particular HLA class I locus or allotype, which are common features of virus-infected cells and tumors. Because HLA-C is expressed at one-tenth of the level of HLA-A and -B, the selective loss of HLA-C expression reduces the total amount of HLA class I by only 9%; and for HLA-A and -B, the selective loss of one allotype reduces the total expression by 23%. For these reasons, there exist inhibitory NK receptors that recognize polymorphic determinants of HLA-A, -B, and -C molecules, and which are used to complement CD94:NKG2A. These inhibitory receptors are part of the family of **killer-cell immunoglobulin-like receptors** (**KIRs**).

The ligand-binding sites of KIRs comprise two or three immunoglobulin-like domains, and they are completely unrelated to the lectin-like binding sites of CD94:NKG2A and NKG2D. Like T-cell receptors and CD94:NKG2A, the inhibitory KIRs bind to the upturned face of the MHC class I molecule formed by the two α helices and the bound peptide (see Section 5-17). The difference is that KIRs (and CD94:NKG2A) cover less of the face than T-cell receptors, just that part involving the amino-terminal part of the α_2 helix and the carboxy-terminal parts of the bound peptide and the α_1 helix (Figure 12.7).

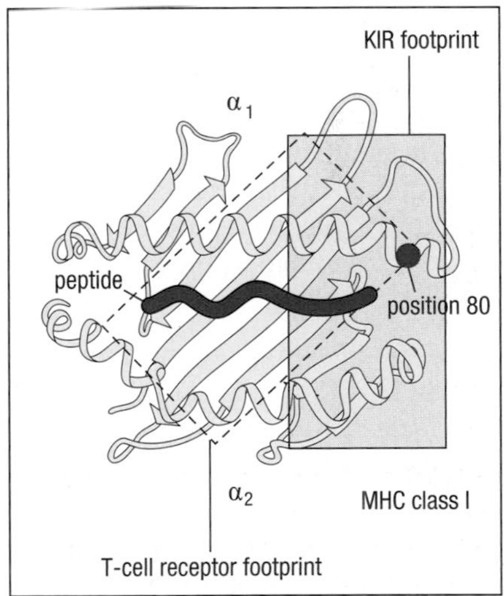

Figure 12.7 Killer-cell immunoglobulin-like receptors (KIRs) bind to the same face of the MHC class I molecule as the T-cell receptor. The ribbon diagram shows the structure of the α_1 and α_2 domains of an MHC class I molecule. The tops of the two α helices and of the bound peptide in the groove between them form the face that interacts with T-cell receptors and KIRs. The rectangle outlined with a dashed line shows the part of the MHC class I face that interacts with T-cell receptors; the rectangle outlined with a solid line shows the part that interacts with KIR.

Sequence variation in the carboxy-terminal part of the α_1 helix, notably at position 80 (see Figure 12.7), results in four alternative structures that are recognized by different inhibitory KIRs. These structures are also described as **KIR ligands** or **epitopes** of HLA class I. The primary source of KIR ligands is HLA-C, with HLA-A and HLA-B taking secondary roles. Reflecting this hierarchy, all HLA-C allotypes function as ligands for KIR, whereas only a minority of HLA-A and -B allotypes have this function. Every HLA-C allotype carries either the C1 or C2 epitope, the Bw4 epitope is carried by some HLA-A and -B allotypes, and the A3/11 epitope is carried by a different subset of HLA-A allotypes from those carrying the Bw4 epitope (Figure 12.8, top three panels).

Six of the KIRs are receptors for HLA class I molecules (Figure 12.8, bottom panel). These comprise four inhibitory receptors and two activating receptors. The KIRs are named according to the number of their Ig-like domains, either 2D or 3D, and the length of their cytoplasmic tail, either L for long or S for short. KIR3DL1 and KIR3DL2 are inhibitory receptors that recognize the Bw4 and A3/11 epitopes, respectively. The C1 epitope is recognized by inhibitory KIR2DL2/3, whereas the C2 epitope is recognized by inhibitory KIR2DL1 and activating KIR2DS1. The second activating receptor, KIR2DL4, is specific for HLA-G, which is expressed only by a subset of fetal trophoblast cells during pregnancy.

Because each of the four epitopes is carried only by a subset of HLA-A, -B, or -C allotypes, individuals can have between one and four of the epitopes. This is one way in which interactions between HLA class I and inhibitory KIRs diversify NK-cell reactivity and responses. For example, individuals who are homozygous for C1 and for the inhibitory C1-specific KIR are more likely to terminate an acute hepatitis C virus infection than other individuals. Like HLA-A, -B, and -C, the four inhibitory KIRs are highly polymorphic. The polymorphism affects several features of the KIR proteins: their level of expression at the cell surface, the strength of binding to HLA class I, and the strength of the inhibitory signal. Combination of KIR polymorphism with HLA class I polymorphism is a second way of diversifying human NK-cell reactivity and response. For example, among people infected with HIV the combination of Bw4 and the inhibitory Bw4-specific KIR slows disease progression, but to a variable extent depending upon the *KIR3DL1* allele and the HLA-A or -B allotype carrying the Bw4 epitope.

Although CD94:NKG2A and inhibitory KIR are structurally very different, the mechanisms by which NK cells use inhibitory KIRs to assess the integrity of potential target cells are the same as for CD94:NKG2A. The key difference is that CD94:NKG2A binding to HLA-E monitors the combined effect of all six HLA-A, -B, and -C allotypes expressed on the target cell's surface, whereas KIRs are monitoring between one and four of the six allotypes, depending on a person's HLA type. CD94:NKG2A and KIRs are inhibitory NK-cell receptors with complementary specificities for MHC class I. They act together to make NK cells tolerant of normal expression of MHC class I but responsive to pathological changes in that expression.

12-5 NK cells are educated to detect pathological change in MHC class I expression

NK cells acquire inhibitory receptors for MHC class I molecules at late stages of their development in the bone marrow. The CD94:NKG2A receptor is expressed first, followed by the KIRs. A key feature of the close-packed KIR gene family is the presence of competing bi-directional promoters, in which forward transcription leads to gene expression, and reverse transcription leads to gene silencing mediated by DNA methylation. This is a random process that generates NK cells expressing different numbers of KIR genes in different combinations. This **variegated expression** of the KIR genes produces individual NK cells that have stable KIR phenotypes and a population of NK cells having diverse KIR phenotypes.

To be effective in detecting changes in the expression of self-MHC class I molecules, an NK cell must express one inhibitory receptor that recognizes a self-MHC class I molecule. This receptor can be either CD94:NKG2A or an inhibitory KIR. Once developing NK cells acquire their inhibitory receptors, they test them out against the MHC class I molecules of other bone marrow cells. If an inhibitory receptor finds a cognate MHC class I molecule then it signals, initiating a process called **NK-cell education** that makes the NK cell sensitive to loss of self-MHC class I. The inhibitory signaling molecules produced during these initial interactions act on the signaling pathways from the NK cell's activating receptors to create a level balance between the activating and inhibitory signals (Figure 12.9). In this situation, where the NK cell is interacting with a healthy cell expressing normal levels of self-MHC class I, the cell

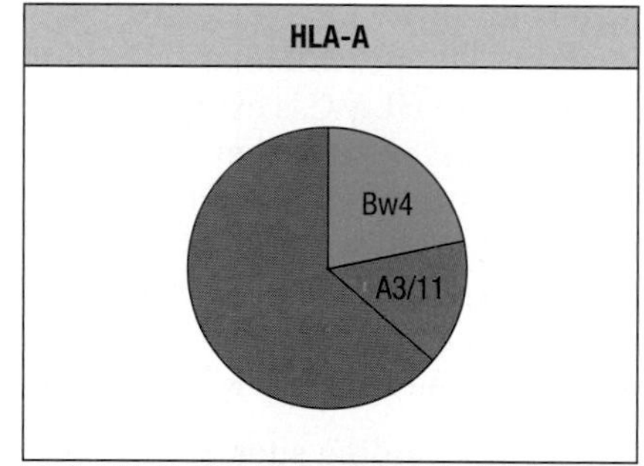

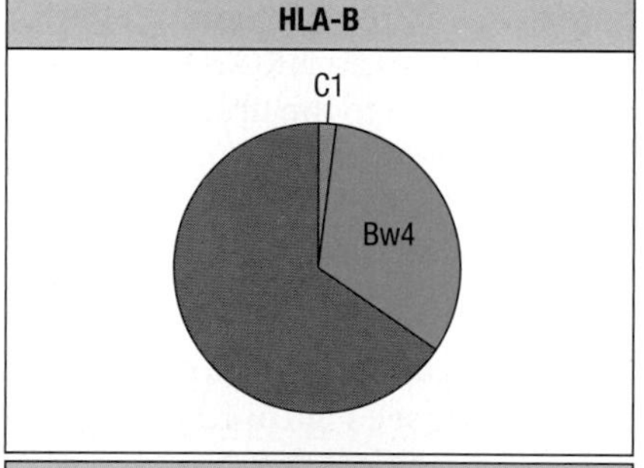

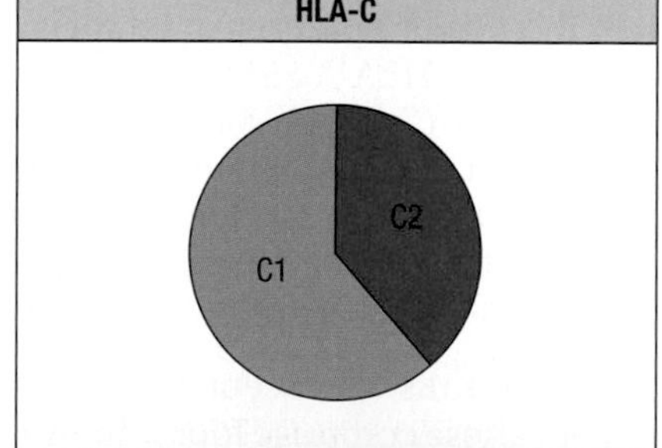

Cognate KIRs		
KIR	Epitope specificity	Signal
3DL1	Bw4	Inhibitory
3DL2	A3/11	Inhibitory
2DL1	C2	Inhibitory
2DL2/3	C1	Inhibitory
2DS1	C2	Activating
2DL4	HLA-G	Activating

Figure 12.8 Distribution among the HLA-A, -B, and -C allotypes of the four epitopes recognized by KIRs. The pie charts in the top three panels show the proportion of HLA-A, -B, and -C allotypes that carry the A3/11, Bw4, C1, and C2 epitopes. Two divergent HLA-B allotypes, HLA-B*46 and HLA-B*73 carry the C1 epitope. The bottom panel shows the KIR that recognizes each of the four epitopes and also HLA-G, and whether they have activating or inhibitory signaling function.

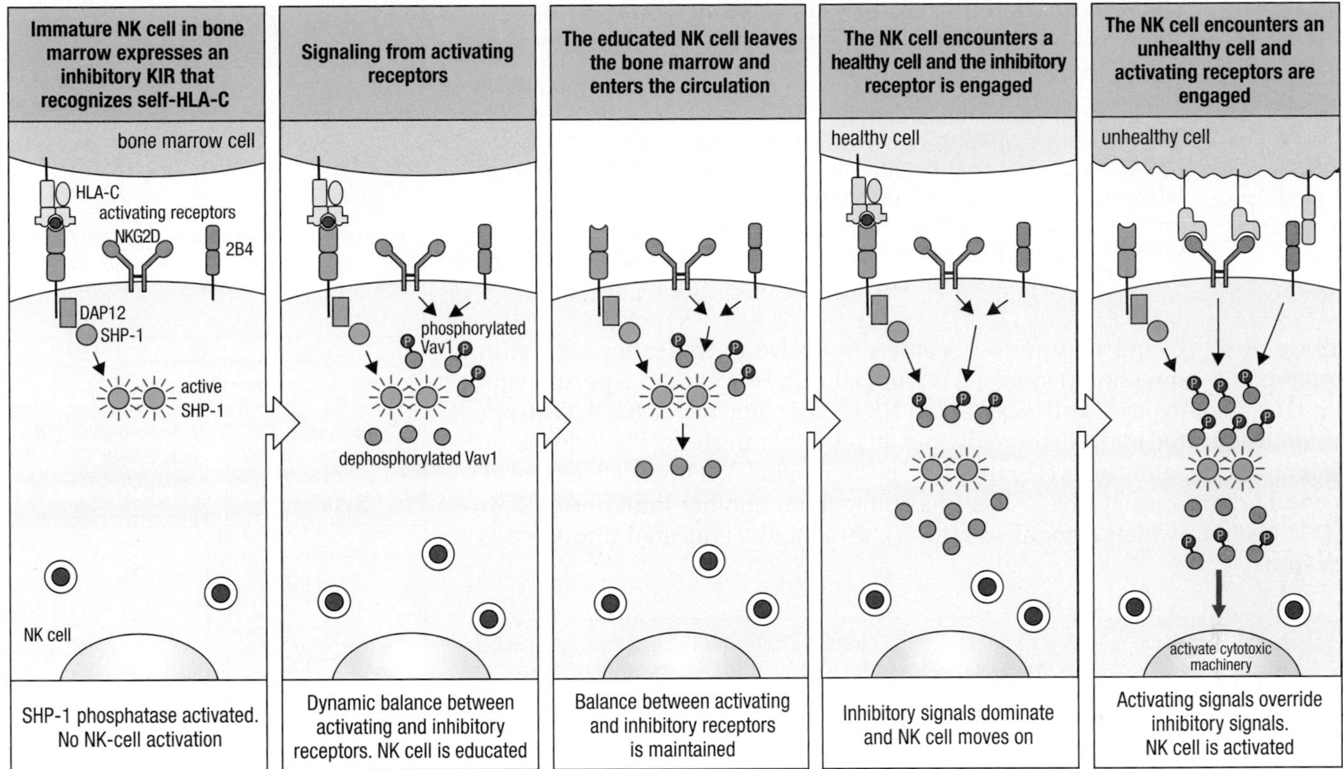

Figure 12.9 NK-cell education involves balancing the signaling pathways for inhibition and activation. The signals coming from the inhibitory receptors activate SHP-1 (first panel), which blocks the activation pathways by dephosphorylating Vav (second panel). This balancing of activation and inhibition is the education that matures the NK cell, which then leaves the bone marrow to enter the circulation and tissues (third panel). This balance is maintained when the NK cell interacts with healthy cells (fourth panel), but when an unhealthy cell is encountered (fifth panel), the inhibitory signals decrease and the activating signals increase, because ligands for the activating receptors are expressed on the unhealthy cell. The NK cell becomes fully activated and the unhealthy cell will soon be killed.

does not respond because signaling from the activating receptors is canceled out by signaling from the inhibitory receptors.

Once educated, the NK cell enters the circulation. If it then encounters an unhealthy cell with reduced MHC class I, the signals coming from the inhibitory receptor will be weaker than those generated during its education. In contrast, the activating signals will not be reduced and will probably be increased because of the presence on the unhealthy cell of stress proteins and pathogen antigens that provide ligands for activating receptors. The activating signals thus overcome the inhibitory signals and the NK cell is instructed to attack the unhealthy target cell.

An important consequence of NK-cell education is that the strength of an NK cell's inhibitory receptor directly correlates with the strength of its activation in response to a loss of MHC class I. In other words, stronger inhibitory receptors make for stronger NK cells. In this way, an NK cell's inhibitory receptor for self-MHC class I controls both its tolerance to self and its response to infection. CD94:NKG2A was the first of the inhibitory MHC class I receptors to evolve and is the first expressed during development. Cells that initially express CD94:NKG2A but are subsequently educated by KIRs can then lose expression of CD94:NKG2A (Figure 12.10).

To examine a specific example of education, consider a person who has HLA-C allotypes bearing the C1 epitope but not the C2 epitope. In the bone marrow, immature NK cells that carry the inhibitory C1-specific receptor KIR2DL3 are

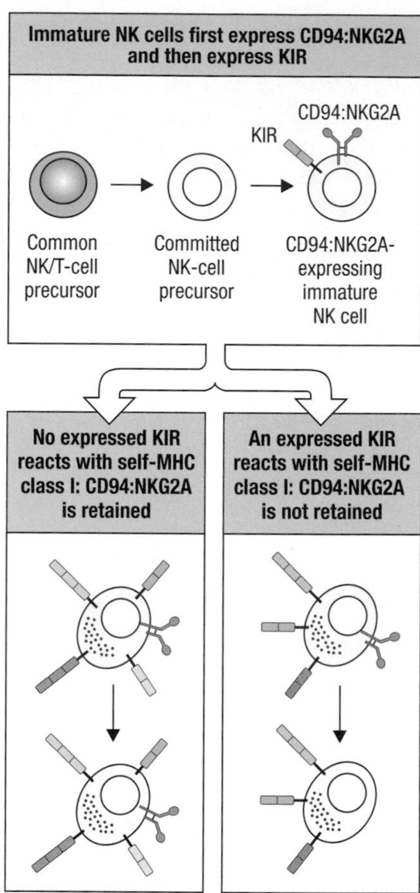

Figure 12.10 NK-cell acquisition of inhibitory receptors for self-MHC class I. In NK-cell development, the inhibitory receptor CD94:NKG2A, with broad MHC class I specificity is first expressed. Subsequently, the KIR genes are expressed. NK cells express different numbers and combinations of KIRs. If none of the expressed KIRs recognizes self-MHC class I, the NK cell retains expression of CD94:NKG2A as its only inhibitory receptor for self-MHC class I (bottom left panel). If one of the KIRs expressed by the NK cell recognizes self-MHC class I, then the cell usually, but not always, turns off CD94:NKG2A expression and uses KIR as the inhibitory receptor for self-MHC class I.

educated by C1 and mature to become responsive to unhealthy cells with impaired C1 expression (Figure 12.11, top panels). However, in a person who has HLA-C allotype C2 and lacks C1, an NK cell bearing only KIR2DL3 will not become educated and will therefore not attack either the body's healthy tissue cells or its unhealthy ones (Figure 12.11, bottom panels). If such cells express a second inhibitory self-MHC class I receptor, either another inhibitory KIR or CD94:NKG2A (which recognizes HLA-E), they can be educated through this receptor.

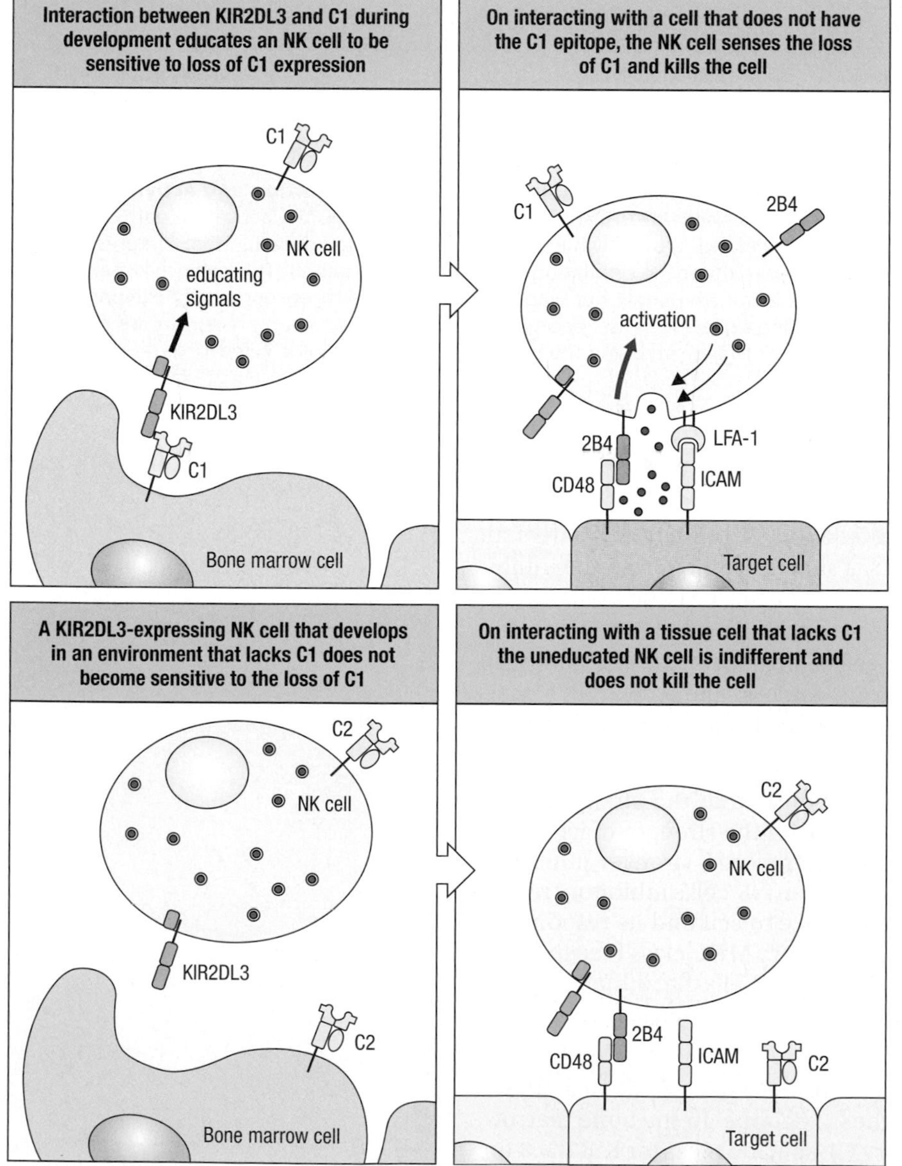

Figure 12.11 NK cells are educated by inhibitory receptors that recognize self-MHC class I. The upper panels show the education by C1 of an NK cell that expresses the C1-specific inhibitory receptor KIR2DL3. The interaction of C1 with KIR2DL3 develops an inhibitory signal that calibrates the NK cell to know what degree of inhibition is to be expected from healthy cells expressing the C1 epitope (left panel). The NK cell can then respond to tissue cells that lack C1. In the laboratory experiment shown in the right panel, the NK cell is killing cells isolated from a family member who does not have C1. Adhesive interactions between LFA-1 on the NK cell and ICAM on the target cell are crucial for the NK cell's response. The lower panels show the lack of education of a developing NK cell that has KIR2DL3 as the only inhibitory receptor in a person who does not have the epitope C1, only C2. As KIR2DL3 does not find C1, the NK cell receives educating inhibitory signal (left panel). If the NK cell subsequently interacts with self tissue cells that lack C1 it does not respond.

NK cells expressing CD94:NKG2A rarely express an inhibitory KIR that recognizes self-MHC class I, which indicates that CD94:NKG2A is often used as the back-up receptor when there is no inhibitory KIR that recognizes self-MHC class I. NK cells that are not educated by an inhibitory MHC class I receptor can leave the bone marrow and are found in the circulation. Their functions remain unknown. Research in progress suggests that education is not an irrevocable process that occurs only once in the life of an NK cell. Instead, the primary education received in the bone marrow can be supplemented or modified by secondary education when NK cells take up residence in other tissues.

Education of CD94:NKG2A by HLA-E is similar in all people and produces NK cells with intermediate strength to respond to a loss of self-MHC class I. In contrast, the education of KIR by HLA-A, -B, and -C allotypes is highly variable and produces NK cells with responses ranging both higher and lower than that achieved by CD94:NKG2A. This shows how the polymorphic variation of HLA class I and KIR combines to diversify the functional capacity of NK cells within individuals and the human population. The relative proportions of NK cells educated by CD94:NKG2A and KIRs varies with the number of KIR ligands brought by the particular HLA-A, -B, and -C tissue type. In people with one KIR ligand, usually C1, NK-cell education is dominated by CD94:NKG2A, whereas people with all four KIR ligands have a more even representation of NK cells educated by KIRs and CD94:NKG2A.

12-6 Different genomic complexes encode lectin-like and immunoglobulin-like NK-cell receptors

Lectin-like CD94:NKG2A and immunoglobulin-like KIRs both function to recognize MHC class I molecules and deliver inhibitory signals to NK cells. They are an example of **convergent evolution** in which natural selection acted independently, but similarly, on two different protein modules and produced the same function. This parallel evolution occurred at two different parts of the genome (Figure 12.12). The CD94 and NKG2A genes are on chromosome 12 in the **natural killer complex** (**NKC**), a region of gene families encoding lectin-like receptors of NK cells and other leukocytes, including NKG2D. In contrast, the **leukocyte receptor complex** (**LRC**) on chromosome 19 is packed with genes encoding immunoglobulin-like receptors expressed by NK cells and other leukocytes. This is the location of the KIR gene family, flanked on one side by the gene encoding the Fc receptor for IgA (FcαR) and on the other by the gene family encoding the leukocyte immunoglobulin-like receptors (LILRs). The LILRs include LILRB1, an inhibitory NK-cell receptor with broad specificity for HLA class I molecules.

Because the KIR and HLA class I genes are on different chromosomes, these polymorphic receptors and ligands segregate independently in families and populations. Consequently, parents frequently have ligand–receptor interactions that their children lack, and vice versa. Such segregation increases the

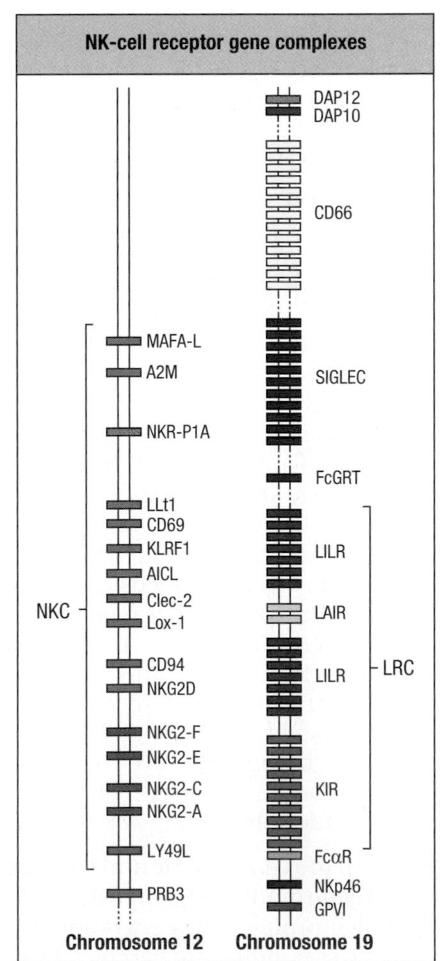

Figure 12.12 Many NK-cell receptors are encoded by genes in one of two genetic complexes. The natural killer complex (NKC) on human chromosome 12 encodes lectin-like NK-cell receptors, whereas the leukocyte receptor complex (LRC) on chromosome 19 encodes NK-cell receptors that are members of the immunoglobulin superfamily. The LRC comprises the killer-cell immunoglobulin-like receptor (KIR), the leukocyte immunoglobulin-like receptor (LILR), and the leukocyte-associated immunoglobulin-like receptor (LAIR) gene families. Flanking the LRC are genes encoding other immunoglobulin-like molecules of the immune system, including the NKp46 receptor of NK cells, and the DAP10 and DAP12 signaling adaptor molecules that transduce signals from activating NK-cell receptors. Figure based on data courtesy of John Trowsdale.

Similarities between human KIR and mouse Ly49 receptors
Recognize polymorphic determinants of MHC class I molecules
Educate NK cells to respond to missing self-MHC class I
Comprise polymorphic inhibitory receptors and non-polymorphic activating receptors
Use the same activating and inhibiting pathways of signal transduction
Exhibit variegated expression by which NK cells express different numbers and combinations of receptors
Organization of the gene families into two regions of gene-content variation bounded by conserved framework genes

Figure 12.13 Similarities between human KIRs and mouse Ly49 receptors. Although human KIRs have ligand-binding sites made from immunoglobulin-like domains and mouse Ly49 receptors are made from lectin-like domains, their biological properties and function are remarkably similar. This independent convergence on similar function is called convergent evolution.

population diversity of the KIR and HLA class I interactions but impedes improvement and optimization of the interaction between particular pairs of KIR and HLA class I molecules. As a consequence, this type of ligand–receptor system is inherently unstable, as is evident from a comparison of humans with other species.

Equivalents to the human KIR family are found only in higher primates, the apes and monkeys. In lower primates, the LRC contains a single non-functional KIR gene. The mouse LRC has no KIR genes, but in the NKC there is a gene family encoding lectin-like Ly49 receptors that recognize polymorphic MHC class I determinants and have other properties that almost precisely parallel those of human KIR (Figure 12.13). This again is the result of convergent evolution. In contrast, the human NKC has a single non-functional Ly49 gene. This pattern, in which a system of MHC class I receptors flourishes in one species but is dead in another, shows how these systems have the propensity to collapse, which is what would happen if the ligands and receptors became segregated away from each other. This situation could arise by chance or as the consequence of selection on other functions of MHC class I molecules, such as the generation of CD8 T-cell immunity. The overall effect is that these systems of variable NK-cell receptors for MHC class I are more rapidly evolving than other immune-system gene families, including the immunoglobulin and T-cell receptor gene families of adaptive immunity.

12-7 Human KIR haplotypes uniquely come in two distinctive forms

The MHC is the most diverse region of the human genome and the region most strongly associated with disease. The KIR gene family is also unusually diverse and strongly associated with human disease, often in combination with HLA class I genes. One component to the variation is that KIR genes exhibit allelic polymorphism; a second component is that KIR haplotypes differ in the number of genes, and the types of gene, they contain. This is called **gene-content variation**. Three genes common to all the haplotypes are called the framework genes because they define the basic organization of the KIR haplotype. Placed at the two ends of the locus and at the center, the framework genes flank two regions—the centromeric region and the telomeric region—that have variable gene content (Figure 12.14). The centromeric region contains genes encoding inhibitory HLA-C receptors; the telomeric region contains genes encoding inhibitory HLA-A and -B receptors. The genes are packed closely together and are separated by short homologous sequences that have been sites for asymmetric recombination events. Such events have created the gene-content diversity, because they can duplicate and delete KIR genes, and by swapping cytoplasmic tails they convert inhibitory receptors into activating receptors.

For both the centromeric and telomeric regions there are two distinctive types of KIR gene motif, designated A and B. The A motifs are conserved and enriched for polymorphic genes encoding inhibitory receptors for HLA class I molecules. The B motifs are variable and are enriched in non-polymorphic genes

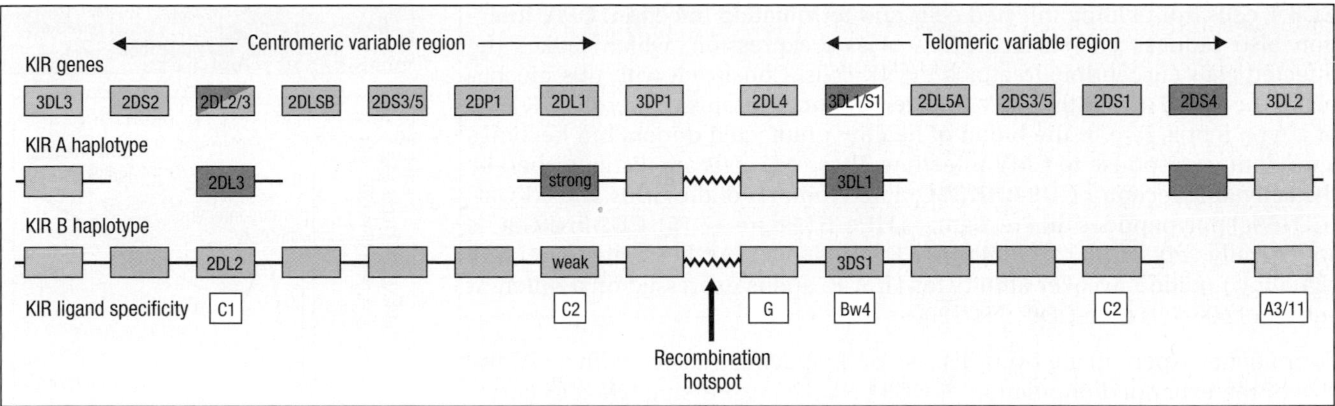

encoding activating receptors that recognize HLA class I molecules. The A motifs correlate with stronger education and NK cells that mediate stronger inflammatory responses, whereas the B motifs correlate with weaker education and NK cells that mediate weaker inflammatory responses. For most of the diseases associated with KIR, the correlations are with differences that distinguish between the A and B motifs. Such findings, and the fact that the KIR A and B motifs are present in all human populations, argue for a balancing selection that maintains the two motifs. The human KIR A motifs are similar to the KIR motifs found in chimpanzees and other apes, whereas the KIR B motifs are unique to the human species. The likely benefit of the KIR B haplotypes was the improvement they made to human reproduction, as described later, in Section 12-9.

Figure 12.14 Organization of the KIR genes and the two groups of KIR haplotypes. Each box corresponds to a different KIR gene. The top line shows the 15 KIR genes and their order in the locus. The next two lines of boxes depict typical KIR A (second line) and KIR B (third line) haplotypes. Framework genes are shaded gray; genes or alleles that are typical of KIR A and B haplotypes are colored red and blue, respectively. The white boxes in the bottom line show the HLA class I epitope recognized by that particular KIR. The site of recombination between the centromeric and telomeric regions of variable gene content is indicated by the jagged line. The 2DL1 alleles of the A haplotype encode strong inhibitory C2 receptors, whereas the 2DL1 alleles of the B haplotypes encode weak inhibitory C2 receptors.

12-8 Cytomegalovirus infection induces proliferation of NK cells expressing the activating HLA-E receptor

A fundamental characteristic of adaptive immunity is that an infecting pathogen permanently changes the composition of the B-cell and T-cell populations in the human host. In the course of the immune response, clones of antigen-specific cells proliferate and give rise to populations of long-lived memory cells. Given the many similarities between NK cells and T cells, it has been proposed that NK-cell populations also adapt to infections and become changed in the process.

So far, the one good example of this phenomenon comes from the analysis of NK cells in cytomegalovirus (CMV) infections. Like other human herpesviruses, CMV has undergone extensive coevolution with its human host and it has acquired mechanisms for evading human immune systems and establishing persistent infections that last for the life of the host. During this time, extended periods of latency are punctuated by episodes of active infection and synthesis of viral particles. Almost all CMV infections are asymptomatic, or cause only minor symptoms like those of a mild influenza infection. When severe disease occurs, it involves people with some degree of immunodeficiency: newborn infants, HIV-infected patients, patients taking immunosuppressive drugs after organ or tissue transplantation, and patients with genetic immunodeficiency. CMV is the most thoroughly studied human herpesvirus, because pregnant mothers can transmit the virus to the fetus and, in the developed world, CMV causes more problems with pregnancy and birth defects than any other virus. Infants born with congenital CMV infection can suffer gastrointestinal disease, low birth weight, hearing loss, and developmental disabilities.

The CMV genome encodes several proteins that interfere with the presentation of viral peptides by MHC class I molecules, and so prevent virus-specific

CD8 T cells from killing infected cells and terminating infection. CMV infection also reduces cell-surface MHC class I expression, which makes the infected cells susceptible to attack by NK cells. Consistent with this mechanism, the blood of healthy CMV-infected donors contains numerous NK cells of a type that is rare in the blood of healthy uninfected donors but becomes abundant in response to CMV infection. These NK cells are distinguished by the activating receptor **CD94:NKG2C**, a heterodimer of the CD94 and NKG2C (CD159c) polypeptides that recognizes HLA-E (Figure 12.15). CD94:NKG2C is structurally very similar to inhibitory CD94:NKG2A but has complementary signaling function, weaker affinity for HLA-E, and is expressed on a different subset of NK cells than CD94:NKG2A.

For patients experiencing overt disease on first being infected with CMV, the size of the expanded population of CD94:NKG2C-expressing NK cells correlates with the severity of the symptoms and can become 50% of the blood NK cells. Once infection is under control and health is recovered, the blood maintains a steady level of CD94:NKG2C-expressing NK cells. These NK cells have high cytotoxic potential but are unresponsive to cytokines and rarely proliferate. They also carry KIRs educated by self-HLA-C as well as CD57, a marker of terminal differentiation for cytotoxic CD8 T cells. These properties are characteristic of terminally differentiated effector NK cells, consistent with the purpose of the CD94:NKG2C-expressing NK cells being to control episodes of CMV reactivation by killing virus-infected and virus-producing cells. Expansion of this subset of NK cells represents an evolutionary adaptation of the human host to chronic CMV infection. Virus-specific NK-cell expansions are not seen for infections with other human herpesviruses, such as varicella-zoster virus (VZV), Epstein–Barr virus (EBV) and herpes simplex virus (HSV), even though severe infections with these viruses are the hallmark of patients who lack NK cells.

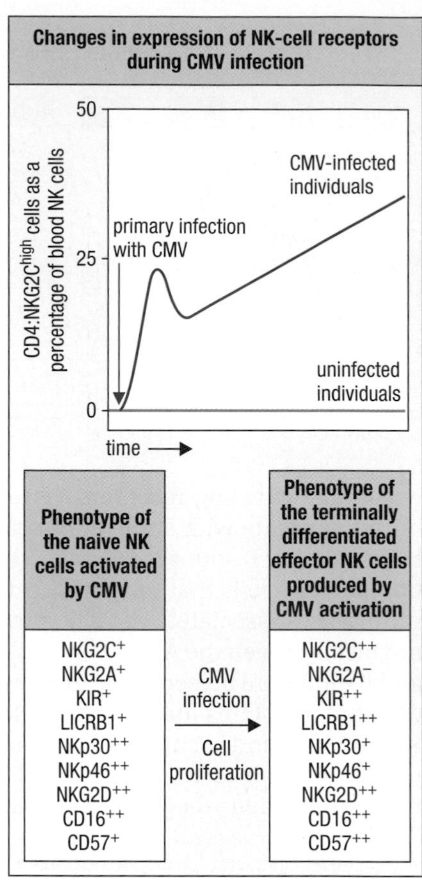

Figure 12.15 CMV infection causes the proliferation of NK cells expressing CD94:NKG2C, an activating receptor that recognizes HLA-E. NK cells expressing the activating HLA-E receptor CD94:NKG2C are rare in human blood and they have the phenotype shown in the lower left panel. The superscripts ++, +, and – denote cells with high, low, or no expression of the given cell-surface marker. When these cells are activated to proliferate and differentiate by CMV infection, they acquire the phenotype shown in the lower right panel. The graph in the upper panel shows the expansion in number of CD4:NKG2C^{++} cells over time after the start of CMV infection.

12-9 Interactions of uterine NK cells with fetal MHC class I molecules affect reproductive success

Comparison of mammalian genomes shows that genes of the reproductive and immune systems are the most variable genes: both within species and between species. This distinction stems in part from immune defense and reproduction being vital physiological functions that are essential for the survival of individuals, populations, and species. That is true for other physiological systems, but their genetics are not so variable. The immune system is distinguished by its perpetual interaction with diverse, rapidly evolving microorganisms. This inevitably puts it under strong, unpredictable selection pressures that are a threat to life and require constant change and adaptation by the many families of immune-system genes.

Whereas innate and adaptive immunity have coexisted for 400 million years, placental reproduction in mammals is a recent innovation, dating back around 125 million years. As is usual with biological invention, the evolution of placental reproduction exploited molecules, cells, and tissues that already existed. The NK cell is just one of various immune-system components that were co-opted by reproduction. This is evident in the regions of chromosome 19 that flank the LRC. They contain many genes contributing to immunity and/or to reproduction.

Placental reproduction can be viewed as a strategy to protect the very young when they are most vulnerable to infection and other environmental dangers. The developing fetus is sequestered from the outside world in a sterile, nurturing environment within the mother's body. The first placental mammals were small mouse-like creatures, but after the extinction of the dinosaurs, mammals were able to diversify and occupy numerous ecological niches. This involved intense competition between mammalian species, in which their

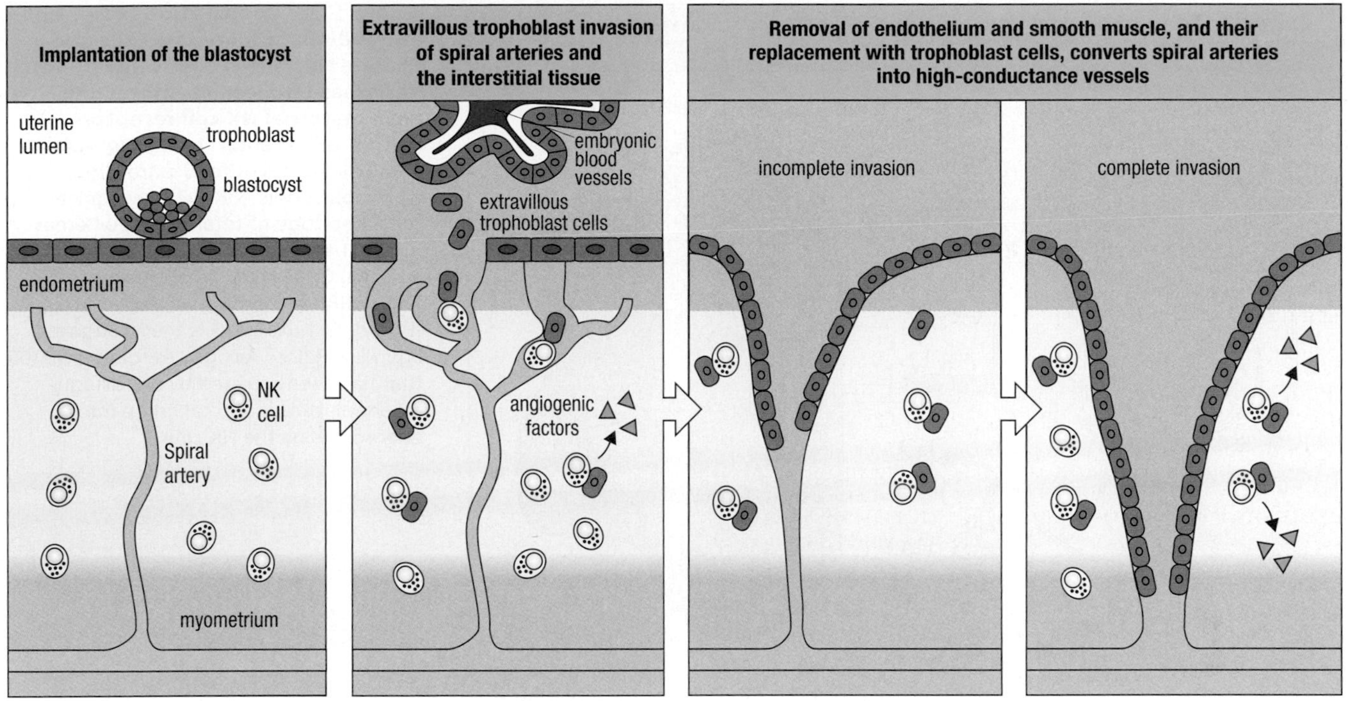

Figure 12.16 Uterine NK cells cooperate with fetal extravillous trophoblast cells to provide the placenta with an adequate supply of maternal blood. The first panel shows the state of the uterine wall before the blastocyst implants. The second panel shows a later stage at which the placenta is developing and the embryo has developed its own blood supply. Extravillous trophoblast cells have left the main layer of trophoblast and invaded the uterine tissue, where they interact with and are controlled by uterine NK cells. The third and fourth panels show the widening of arteries in the uterine wall by the actions of the extravillous trophoblast cells under the control of NK cells.

size, morphology, physiology, and behavior diversified. Keeping up with this progress required continual adjustment to the anatomy, physiology, and genes involved in reproduction, as reflected in the diversity of modern mammalian genomes.

As introduced in Chapter 3, uterine NK cells cooperate with fetal trophoblast cells during placenta formation at the beginning of pregnancy (see Section 3-18). The outer layer of the placenta, called the trophoblast, contains the only fetal cells that make direct contact with maternal blood and tissues. One type of trophoblast cell, the extravillous trophoblast cell, invades the uterine tissue and converts narrow maternal blood vessels, called spiral arteries, into the wider conduits needed to supply the placenta with maternal blood. This, in turn, allows the placenta to provide oxygen and nutrients to the growing fetus throughout pregnancy. Some extravillous trophoblasts invade the lumen of the spiral arteries, strip off the endothelium and replace it with trophoblast. Others similarly strip off and replace the smooth muscle from the outside of the arteries. The job of the uterine NK cells is to control the extravillous trophoblast cells, which are aggressive cells having properties in common with tumor cells. The depth to which the arteries are invaded is crucial, because complications of pregnancy are associated with invasion that is either too deep or too shallow (Figure 12.16).

Cooperation between the two cell types involves intimate contact in which NK-cell receptors bind MHC class I molecules on the trophoblast cell. Extravillous trophoblasts are the only trophoblast cells that express MHC class I molecules. The combination they express—HLA-C, -E, -F, and -G—is unique among the cells of the human body (Figure 12.17). Expressed only by extravillous trophoblast cells and dedicated to reproduction, HLA-G is expressed in a cell-surface form and in a secreted form that lacks the

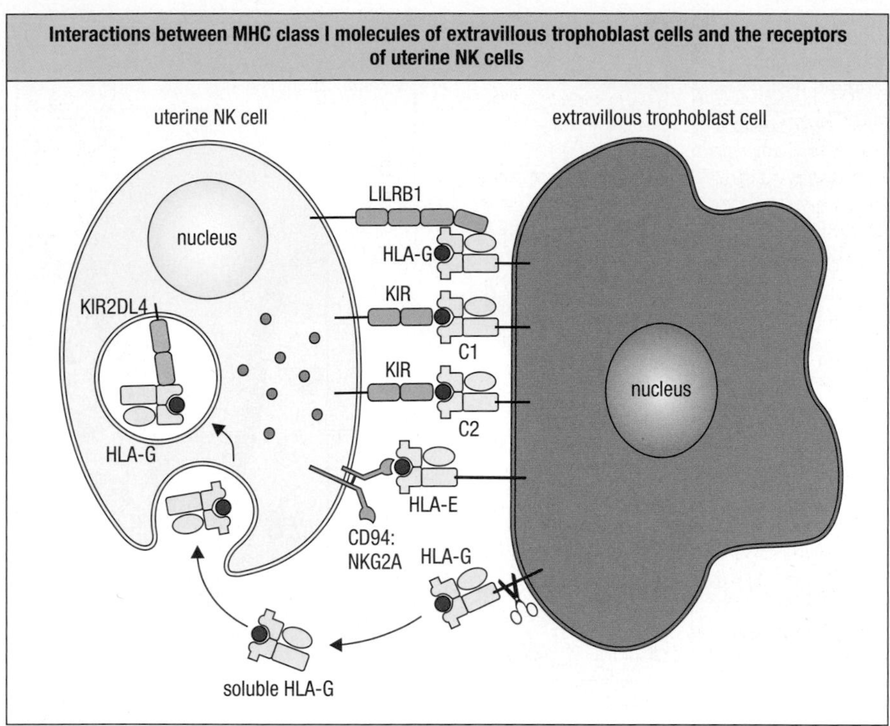

Figure 12.17 Cooperation between extravillous trophoblast cells and uterine NK cells is controlled by fetal HLA class I molecules interacting with maternal NK-cell receptors. Five NK-cell receptors recognize HLA class I molecules of fetal extravillous trophoblast cells. KIR2D2/3 recognizes the C1 epitope of fetal HLA-C, whereas KIRD2S1 and KIR2DL1 recognize the C2 epitope. CD94:NKG2A recognizes HLA-E, and LILRB1 recognizes cell-surface HLA-C, -E, and -G, but HLA-G is the strongest ligand. KIR2DL4 recognizes soluble HLA-G that has been secreted by extravillous trophoblast cells and taken up into endosomes by the NK cells.

transmembrane region. Soluble HLA-G gets taken up into the endosomes of an interacting NK cell, where it is bound by KIR2DL4, an activating receptor. Signals from KIR2DL4 and other activating receptors induce the NK cell to secrete cytokines and growth factors that promote the formation of new blood vessels; together these factors stimulate the demolition and remodeling of the maternal arteries. To balance the signals coming from KIR2DL4 and other activating receptors, uterine NK cells express inhibitory receptors that engage MHC class I molecules on extravillous trophoblast cells. LILRB1 recognizes cell-surface HLA-G, CD94:NKG2A recognizes HLA-E, and HLA-C is recognized by KIR2DL1, 2DL2, and 2DL3.

Children inherit one HLA haplotype from their mother and one from their father. In most pregnancies, the fetus expresses HLA class I and II allotypes that are not present in the mother and which would be recognized as foreign antigens by the mother's T cells if they were exposed to them. During the evolution of placental reproduction, such contact between fetal tissues and the cells of the maternal immune system was avoided. In addition, no trophoblast cells express HLA class II, and only extravillous trophoblast cells express HLA class I. Furthermore, although extravillous trophoblast cells make contacts with maternal blood cells, they lack HLA-A and HLA-B, the strong stimulators of T-cell alloreactivity. By comparison, HLA-C molecules stimulate T-cell alloreactions only weakly, because of their low cell-surface expression, their reduced sequence diversity, and their specialization as KIR ligands. In turn, uterine NK cells preferentially express KIRs that recognize the C1 and C2 epitopes of HLA-C, thus facilitating their interactions with the extravillous trophoblast cells (see Figure 12.17).

Shallow trophoblast invasion of the spiral arteries during implantation can lead later in pregnancy to pre-eclampsia, a condition in which the baby is undernourished and the mother responds by increasing the blood flow to the placenta. Without medical intervention the elevated blood pressure can lead to a hemorrhage (eclampsia) that threatens the life of both mother and child. Epidemiological studies show that the risk of pre-eclampsia is increased in pregnancies in which the mother is homozygous for C1 epitopes and KIR A haplotypes (see Section 12-7) and the fetus has a C2 epitope inherited from the

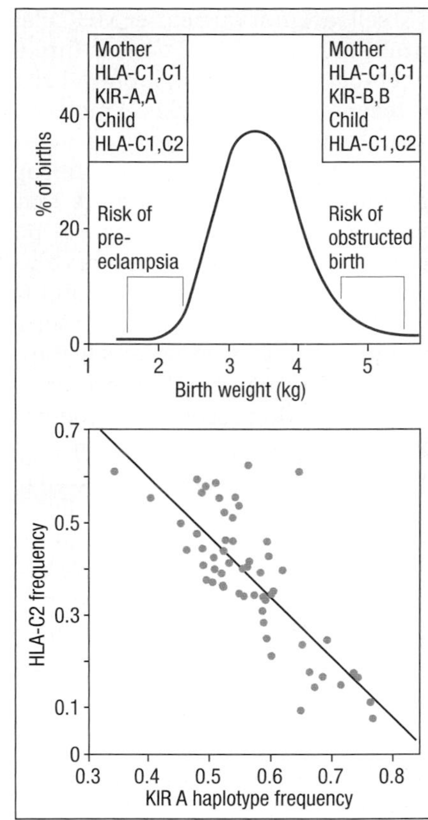

Figure 12.18 Certain combinations of fetal HLA-C and maternal KIR haplotypes are associated with complications of pregnancy and birth. Top panel: human birth weights form a normal distribution with a mean value of 3.4 kg. Two complications that are correlated with birth weight and with KIR and HLA-C haplotype are pre-eclampsia and obstructed birth. Pre-eclampsia is a condition of the last trimester of pregnancy and is associated with an underweight fetus and high maternal blood pressure. The genetic combination associated with pre-eclampsia is the mother being homozygous for KIR A and HLA-C bearing C1 and the fetus inheriting C2-bearing HLA-C from the father. Obstructed birth is the condition when the baby is too large and gets stuck in the birth canal while the mother is giving birth. The genetic combination associated with obstructed birth is when the mother is homozygous for KIR B and HLA-C bearing C1 and the fetus is heterozygous for HLA-C bearing C1 and C2. Both pre-eclampsia and obstructed birth can be avoided by using Cesarian section to deliver the baby. Pre-eclampsia can also be prevented from leading to eclampsia by inducing premature birth. Bottom panel: how the frequency of the C2 epitope in human populations is inversely correlated with the frequency of KIR A haplotypes. Each data point corresponds to a different human population. Included are 5 African, 23 Asian, 23 European, 1 Oceanian, 2 Amerindian, and 5 Hispanic populations. This correlation reveals the strong effect of natural selection due to pre-eclampsia and eclampsia. As a result of that selection, the frequency of pregnancies that have the genetic combination associated with pre-eclampsia is much reduced. There seems to have been a less strong selection exerted by obstructed births.

father (Figure 12.18). This association implies that interaction between trophoblast cell C2 and its inhibitory KIR receptor on uterine NK cells is the cause of inadequate invasion of the spiral arteries. The incidence of pre-eclampsia is less for pregnancies in which the fetus expresses C2 and the mother is C1 homozygous and has a KIR B haplotype. This is because the B haplotype, unlike the A, encodes an activating C2-specific receptor, which counterbalances the effect of the inhibitory receptor. Pre-eclampsia has had a major impact on human populations, as indicated by the inverse correlation between the frequencies of the KIR A haplotype and the C2 epitope throughout the world (Figure 12.18, lower panel). This outcome is the one expected if pregnancies involving KIR A/A mothers and C2 are more prone to pre-eclampsia, eclampsia, and mortality than pregnancies involving other combinations of maternal KIR haplotypes and fetal HLA-C epitopes. Currently, each year 76,000 pregnant women and 500,000 babies die from pre-eclampsia, almost all of these deaths being in the developing countries.

Another disorder of pregnancy is obstructed labor, which can also lead to the death of the mother and child in the absence of medical intervention. This occurs for babies who are too big because of deep placental invasion and overnourishment during pregnancy. The genetic combination associated with obstructed labor is a C1 homozygous mother with a KIR B haplotype and a fetus with paternally inherited C2 (see Figure 12.18). These associations with pregnancy disorders show how a balance of genetic factors helps maintain human birth weight within a certain range.

Summary

NK cells are lymphocytes of the innate immune response that have a wide range of activating and inhibitory receptors. These receptors are expressed in different combinations to produce a diverse repertoire of NK-cell phenotypes. The genes for a substantial number of NK-cell receptors are located in one of two genetic complexes. The natural killer complex (NKC) contains genes encoding lectin-like receptors, whereas the leukocyte receptor complex (LRC) encodes immunoglobulin-like receptors. Many of the ligands recognized by

NK-cell receptors are either MHC class I molecules or molecules that are structurally related to MHC class I. Through convergent evolution, both lectin-like receptors and immunoglobulin-like receptors have acquired similar functions as receptors for MHC class I molecules. Because NK cells circulate in a state of partial activation, there is an active mechanism to keep them tolerant of healthy human cells (Figure 12.19, upper panel). This tolerance is mediated by inhibitory lectin-like receptors that recognize HLA-E, and immunoglobulin-like receptors that recognize polymorphic determinants of HLA-A, -B, and -C. The inhibitory receptors also function in NK-cell development, when they educate NK cells to respond to unhealthy cells having a deficiency of MHC class I molecules on their surfaces. NK-cell attack is mediated by a variety of activating receptors and in the primary innate immune response requires signals to be generated by two or more of them (Figure 12.19, lower panel). Some activating receptors recognize pathogens and pathogen components, others recognize human proteins that are selectively expressed at the surface of stressed or unhealthy cells. A majority of circulating NK cells bear an activating Fc receptor that specifically binds to IgG. This receptor enables NK cells to participate in the adaptive immune response when signals from the Fc receptor are sufficient to activate NK cells. Herpesviruses are adept at exploiting human hosts and maintaining persistent infections, which are kept in check by NK cells. This is evident in people infected with cytomegalovirus, who maintain a large army of a type of NK cell that is barely detectable in

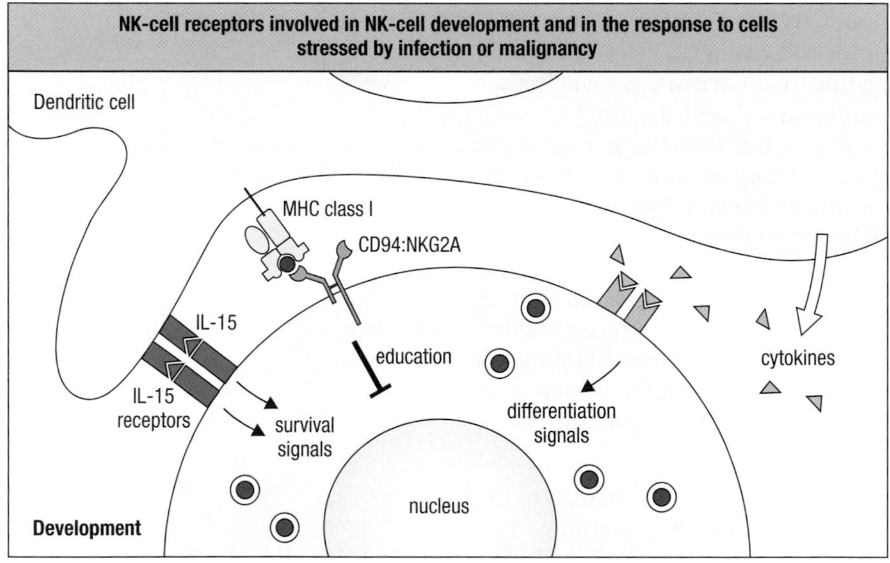

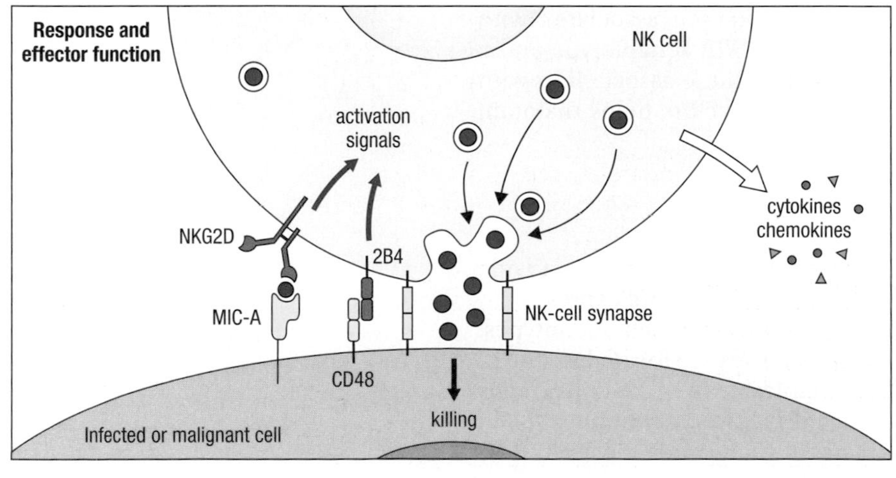

Figure 12.19 Various NK-cell receptors control NK-cell development, response, and effector function. The upper panel shows receptor–ligand interactions involved in NK-cell development. This involves a nurturing interaction with a dendritic cell and NK-cell education, mediated here by CD94:NKG2A. The lower panel shows an NK cell responding to an unhealthy cell. It becomes activated, mobilizes its cytotoxic granules, and kills the target cell. See Chapter 3 for more details of the interactions of NK cells with target cells (Section 3-19), macrophages (Section 3-20) and dendritic cells (Section 3-21).

uninfected people. NK cells in the uterus contribute to reproduction. The NK cells cooperate with fetal trophoblast cells in refashioning maternal blood vessels that will supply the fetus, via the placenta, with sustenance during pregnancy. This cooperation is mediated and modulated by NK-cell receptors that recognize conserved and polymorphic MHC class I molecules.

Maintenance of tissue integrity by γ:δ T cells

Knowledge that T-cell recognition of protein antigens is restricted to polymorphic MHC class I and II molecules stimulated the search for the α:β T-cell receptor and the subsequent discovery of its genes. In contrast, the genes encoding the γ:δ T-cell receptor were discovered serendipitously in the absence of any knowledge of γ:δ T cells and their receptors. From the beginning, the study of γ:δ T cells has always been focused on comparison and contrast with α:β T cells. The α:β T-cell receptors are highly diverse and they all recognize peptide antigens presented by polymorphic MHC molecules. The γ:δ T-cell receptors are far less diverse. They do not recognize peptide antigens but a wider range of chemical structures that distinguish stressed cells from healthy cells. On activation by antigen, the γ:δ T cells kill the distressed cells and secrete cytokines and growth factors that promote the repair of the damaged tissue.

12-10 γ:δ T cells are not governed by the same rules as α:β T cells

α:β T cells and γ:δ T cells derive from the same precursor cells and acquire their T-cell receptors in the thymus through similar processes of gene rearrangement (see Section 7-4). Despite their common developmental origin, the two T-cell types have important differences (Figure 12.20). The α:β T cells are

Comparison of the similarities and differences between α:β and γ:δ T cells		
	α:β T cells	**γ:δ T cells**
Site of development	Thymus	Thymus
Rearranging receptor genes	α and β genes	γ and δ genes
Germline repertoire of V gene segments	Large	Small
T-cell receptor diversity	Large	Small to medium
Positive selection	Yes	Unknown: probably no
Negative selection	Yes	Unknown: probably yes
Co-receptor	CD4 65% CD8 35%	$CD4^-$ $CD8^-$ ~70% CD8αα ~30% of gut IEL
Target antigens	Peptides presented by MHC class I or class II molecules	Self-proteins resembling MHC class I molecules Non-peptide small molecules presented by MHC class I-like and other cell-surface proteins
Abundance in blood	70% of blood lymphocytes	5% of blood lymphocytes
Abundance in tissues	Relatively rare and transient	Plentiful and resident
Activation	Circulate in inactive form that requires several days of activation	Present in tissues in a form that is quick to respond to infection and other forms of stress
Overall function	Adaptive immunity	Tissue homeostasis: surveillance, protection, and repair

Figure 12.20 Similarities and differences between α:β and γ:δ T cells.

subject to stringent positive and negative selection in the thymus, which ensures that each cell surviving selection has an MHC-restricted antigen receptor that recognizes complexes of a non-self peptide bound to a polymorphic self-MHC class I or II molecule (see Sections 7-8 and 7-11). γ:δ T-cell receptors are not restricted to peptides bound to polymorphic MHC molecules, although many of them do bind to proteins that are related in both structure and evolution to MHC class I. Thus γ:δ T cells are not subject to the same processes of thymic selection as α:β T cells. They probably do undergo some form of negative selection, but whether they are subject to their own form of positive selection is not yet known.

Antigen recognition by an α:β T cell is solely the responsibility of the T-cell receptor, but the cell's response depends on the co-receptor, either CD4 or CD8, and other co-stimulators for augmentation and amplification of the activation signal. γ:δ T cells do not depend on the CD4 and CD8 co-receptors. Most γ:δ T cells lack CD4 and CD8, and although some express homodimers of the CD8α chain, this is not a functional form of the co-receptor. Neither do γ:δ T cells express the CD28 co-stimulatory receptor that is crucial for the activation of α:β T cells (see Section 8-5).

Because the job of γ:δ T cells is to identify and respond to unhealthy cells, they are not limited to using the γ:δ T-cell receptor as the only receptor for antigen. Other types of receptor that recognize symptoms of cellular stress are expressed by γ:δ T cells and they can work in conjunction with the γ:δ T-cell receptor or independently of it. Thus the γ:δ T cell is less dependent on its T-cell receptor than the α:β T cell. Many of the other receptors expressed by γ:δ T cells are ones that are also expressed by NK cells (see Figure 12.2).

The lengths and sequences of γ:δ T-cell receptor CDR loops are more like those of immunoglobulins than those of α:β T-cell receptors, consistent with both B cells and γ:δ T cells not being subject to MHC restriction. The germline diversity of the gene segments encoding the γ and δ chains is much less than that of the α and β chains, and the massive potential for junctional diversity (see Section 5-5) is used only to a limited extent. All of these properties are consistent with the γ:δ T-cell receptors recognizing a different set of antigens or ligands from the peptides recognized by the α:β T cells.

People homozygous for a particular mutant allele of the *RAG1* gene have a reduced capacity for recombining the V, D, and J segments of the immunoglobulin and T-cell receptor genes, and suffer from immunodeficiency. These patients have almost no CD4 T cells, and much reduced numbers of CD8 T cells and B cells. They have normal numbers of γ:δ T cells, as well as NK cells, and normal blood concentrations of all the immunoglobulin isotypes. Although immunodeficient, these patients can make specific antibodies against some vaccines and infecting pathogens. The nature of the cellular deficiencies in these patients implies first, that γ:δ T cells are less dependent upon RAG-mediated recombination than are α:β T cells and B cells, and second, that the breadth of γ:δ T-cell functions can be maintained with a much smaller repertoire of receptors than is needed for α:β T cells and B cells. The study of these patients provides further evidence of similarity between γ:δ T cells and NK cells that distinguishes them from α:β T cells and B cells.

12-11 γ:δ T cells in blood and tissues express different γ:δ receptors

During early development of the human fetus, the presence of γ:δ T cells is detectable at 8 weeks of gestation, before the emergence of α:β T cells. The diversity of the γ:δ T-cell receptor repertoire is greatest at birth. During the first year of infancy, some T cells expressing particular combinations of γ and δ chains are stimulated to proliferate (clonal expansion), whereas others are not.

For example, at birth, γ:δ T cells using the $V_{\delta}1$ gene segment predominate in the blood but are subsequently replaced by cells expressing the $V_{\delta}2$ gene segment. The cause of these selective expansions is not known, but their timing coincides with the establishment of the microbiota, the commensal microorganisms that inhabit the body (see Chapter 10). Upon clonal expansion, the γ:δ T cells acquire a memory phenotype, develop cytotoxic granules and the capacity to kill cells, and secrete cytokines such as IFN-γ. Like memory α:β T cells (see Section 11-10), the memory γ:δ T cells are subdivided into effector memory and central memory cells by the expression of CD27 and CD45RA (Figure 12.21). In the blood, γ:δ T cells are a minority population comprising only 2–5% of the total number of T cells. In striking contrast, γ:δ T cells are the majority of the resident T cells in tissues. There, the ready-primed γ:δ T cells act as sentinels that can respond quickly to local infections as soon as they start, in the manner of an innate immune response.

Type of γ:δ T cell	Phenotype	
	CD27	CD45RA
Naive	+	+
Effector memory	–	–
Central memory	+	–
Terminally differentiated	–	+

Figure 12.21 The CD27 and CD45RA cell-surface markers are used to differentiate four developmental stages of γ:δ T cells.

The populations of γ:δ T cells that circulate in the blood and are resident in tissues are distinguished by their γ:δ T-cell receptors (Figure 12.22). In the tissues, most γ:δ T cells use the $V_{\delta}1$ gene segment, whereas almost all blood γ:δ T cells use the $V_{\delta}2$ gene segment in combination with $V_{\gamma}9$. Within the population of $V_{\gamma}9$:$V_{\delta}2$ T cells, the receptors are diversified through the use of different J and D segments and by junctional diversity (see Section 5-5). More than 80% of the circulating $V_{\gamma}9$:$V_{\delta}2$ T cells lack the chemokine receptor CCR7 and so cannot enter secondary lymphoid tissues as naive α:β T cells do. Instead, these γ:δ T cells express CCR5 and receptors for inflammatory cytokines, which allow them to enter infected tissues and respond to a range of pathogens, including influenza virus, *Mycobacterium tuberculosis*, and *Plasmodium falciparum*, the parasite that causes malaria.

A striking feature of γ:δ T cells is the extent to which they differ between animal species. For example, no γ:δ T-cell subpopulations are common to humans and mice. The total number of blood γ:δ T cells also varies between species and correlates inversely with the germline diversity of the genes encoding immunoglobulins and α:β T-cell receptors. Thus, species with highly diverse immunoglobulin and α:β T-cell receptor genes, such as humans and mice, have small numbers of γ:δ T cells (2–5% of blood lymphocytes). In contrast, chickens, rabbits, sheep, and cattle have larger numbers of γ:δ T cells (around 20–30% of blood lymphocytes) accompanied by low immunoglobulin and α:β T-cell receptor gene diversity (Figure 12.23). This correlation reflects the

Tissue	Predominant V gene segment	V(D)J diversity
Thymus	$V_{\delta}1$	High
Blood	$V_{\gamma}9$:$V_{\delta}2$	Intermediate
Spleen	$V_{\delta}1$	High
Liver	$V_{\gamma}3$ and $V_{\delta}1$	High
Gut epithelium	$V_{\gamma}3$ and $V_{\delta}1$	High
Dermis	$V_{\delta}1$	High
Uterus	$V_{\delta}1$	High

Figure 12.22 The γ:δ T cells in different tissues are distinguished by their γ:δ T-cell receptors.

Inverse correlation between the abundance of γ:δ T cells in the blood and the diversity of α:β T-cell receptor and immunoglobulin V-region genes in different species			
	γ:δ T cells: % of blood lymphocytes	Diversity of α:β T-cell receptor V-region genes	Diversity of Ig V-region genes
Human	5%	High	High
Mouse	5%		
Chicken	20%	Low	Low
Rabbit	20%		
Sheep	30%		
Cattle	30%		

Figure 12.23 Mammalian species with higher numbers of blood γ:δ T cells have reduced immunoglobulin and α:β T-cell receptor diversity.

versatility of γ:δ and α:β T cells and the overlap in their functional capability. The species differences also point to the rapid evolution of γ:δ T cells and their divergence in mammalian species with different ways of life and different selective pressures from pathogens.

12-12 $V_\gamma 9$:$V_\delta 2$ T cells recognize phosphoantigens presented on cell surfaces

The antigens recognized by the $V_\gamma 9$:$V_\delta 2$ T cells are not peptides or proteins but a variety of small molecules containing pyrophosphate that collectively are called **phosphoantigens**. They comprise various phosphorylated intermediates in the metabolic pathways of isoprenoid biosynthesis. One of these is hydroxymethyl-but-2-enyl-pyrophosphate (HMBPP), which is made by bacteria and some parasites but not by human cells. HMBPP is thus a foreign phosphoantigen and a feature of the bacterial and parasitic infections to which $V_\gamma 9$:$V_\delta 2$ T cells respond (Figure 12.24). An example of a self phosphoantigen is isopentyl pyrophosphate (IPP); it is present in humans and almost all other living organisms. $V_\gamma 9$:$V_\delta 2$ T cells are sensitive to the levels of IPP and will not respond to those of healthy cells. But the greater amounts of IPP made by infecting pathogens or by human cells stressed by infection or malignancy can activate the $V_\gamma 9$:$V_\delta 2$ T cells. As a phosphoantigen, HMBPP is 1000-fold stronger than IPP at stimulating γ:δ T cells. Another potent phosphoantigen, the synthetic compound bromohydrin pyrophosphate (BrHPP), enhances the capacity of γ:δ cells to perform antibody-dependent cell-mediated cytotoxicity, and in experimental systems it has been used to augment the capacity of therapeutic monoclonal antibodies such as rituximab to kill tumor cells (see Section 9-23).

A major function of γ:δ T cells is to repair damaged tissues through the secretion of cytokines. The use of aminobisphosphonate drugs, such as zoledronate, to treat osteoporosis, a degenerative disease caused by loss of bone mass, illustrates this function. The drugs cause the $V_\gamma 9$:$V_\delta 2$ T cells to accumulate IPP, which induces their activation and simulation of bone growth.

The $V_\gamma 9$:$V_\delta 2$ T-cell receptor does not bind to free phosphoantigens but to phosphoantigens presented by an isoform of CD277 called butyrophylin-3A1 (BTN3A1). BTN3A1 is structurally related to the B7 co-stimulator of dendritic cells (see Section 8-5) and is widely expressed by almost all types of human cell, including tumors. It consists of two immunoglobulin-like domains, one resembling a variable domain and the other a constant domain. The variable domain binds the phosphoantigen, and this complex of BTN3A1 and the phosphoantigen is then engaged by the $V_\gamma 9$:$V_\delta 2$ T-cell receptor. Such engagement produces signals that cause activation of the T cell's effector functions (Figure 12.25).

The γ:δ T cells of the blood represent an oligoclonal expansion of cells that all recognize pyrophosphate. The diversity of their $V_\gamma 9$:$V_\delta 2$ T-cell receptors (see Section 12-11) enables this population of T cells to bind a broad range of

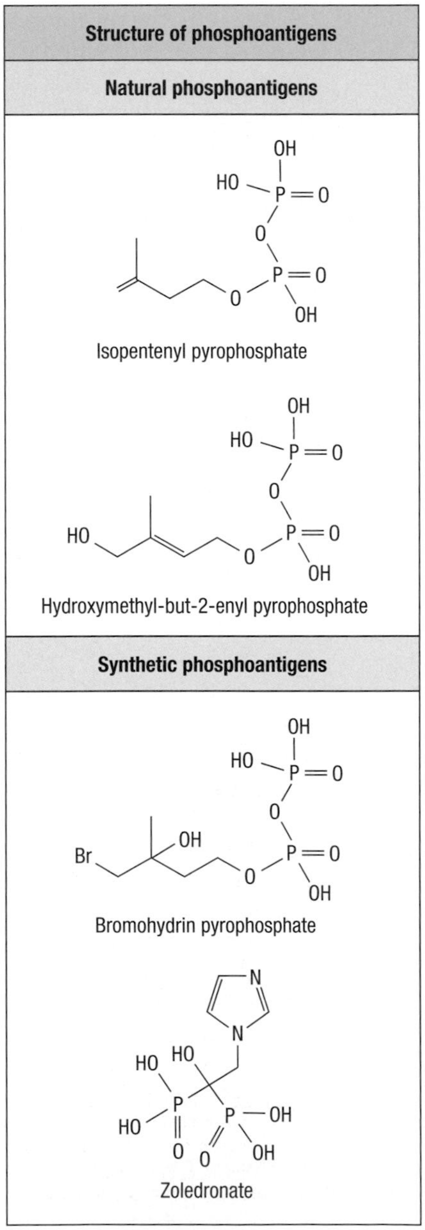

Structure of phosphoantigens

Natural phosphoantigens

Isopentenyl pyrophosphate

Hydroxymethyl-but-2-enyl pyrophosphate

Synthetic phosphoantigens

Bromohydrin pyrophosphate

Zoledronate

Figure 12.24 $V_\gamma 9$:$V_\delta 2$ receptors of blood γ:δ T cells recognize phosphoantigens. All phosphoantigens contain a pyrophosphate group, which is made up of two phosphates. The chemical structures of two natural and two synthetic phosphoantigens are shown here. Isopentenyl pyrophosphate is made by all living organisms. Hydroxymethyl-butenyl pyrophosphate is made by bacteria and some parasites but not by humans. Bromohydrin is a synthetic and potent phosphoantigen used in laboratory experiments to activate γ:δ T cells. Zoledronate is a synthetic phosphoantigen and a licensed drug; when given to osteoporosis patients, it stimulates their $V_\gamma 9$:$V_\delta 2$ γ:δ cells to repair eroded bone tissue.

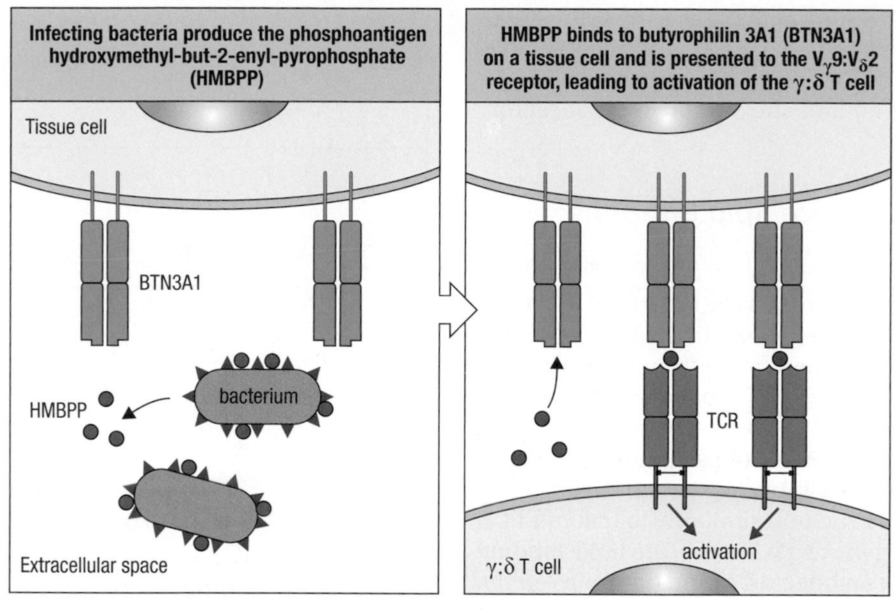

Figure 12.25 Butyrophilin 3A1 on tissue cells binds phosphoantigens and presents them to γ:δ T cells. Bacteria infecting a tissue release the phosphoantigen HMBPP (left panel). HMBPP is bound by the butyrophilin 3A1 molecule on the surface of cells in the infected tissue and is presented to the $V_\gamma9{:}V_\delta2$ receptors of γ:δ T cells. Signals coming from the T-cell receptor activate the γ:δ cells (right panel).

phosphoantigens. In this respect, the $V_\gamma9{:}V_\delta2$ T-cell receptors as a group resemble a receptor of innate immunity that recognizes a determinant shared by groups of pathogens. In addition to detecting the metabolic changes induced by pathogens, the $V_\gamma9{:}V_\delta2$ T-cell receptors also detect the metabolic changes in human cells and tissues caused by cancer, trauma, and other types of stress.

12-13 $V_\gamma4{:}V_\delta5$ T cells detect both virus-infected cells and tumor cells

A clinical situation in which cytomegalovirus (CMV) can cause disease (see Section 12-7) is when a CMV-negative patient is transplanted with a tissue graft from a CMV-positive donor. The virus in the graft infects the transplant recipient, who then makes an immune response against the virus. Characterizing this response is the expansion of T cells having $V_\gamma4{:}V_\delta5$ receptors, which can represent as many as 25% of the total T cells in the blood. The $V_\gamma4{:}V_\delta5$ T cells respond to CMV-infected cells by secreting inflammatory cytokines and killing the infected cells. In laboratory experiments, these T cells respond similarly to a variety of tumor cell lines, even though the T cells were never previously exposed to the tumor cells. The ligand for the $V_\gamma4{:}V_\delta5$ T-cell receptor is a protein called endothelial protein C receptor (EPCR) that is present on the surface of all the tumor and virus-infected target cells. (Long before this immunological function was discovered, EPCR was defined by its binding to protein C and its participation in the inhibitory regulation of blood clotting.) EPCR is structurally similar to the α_1 and α_2 domains of MHC class I molecules, but it binds phospholipids, not peptides, within the binding groove (Figure 12.26). The site on EPCR bound by the γ:δ T-cell receptor is away from the antigen-binding site, and distinct from the site on conventional MHC class I molecules that interacts with α:β T-cell receptors (Figure 12.26, right panel). When cDNAs encoding the $V_\gamma4$ and $V_\delta5$ chains were expressed in a

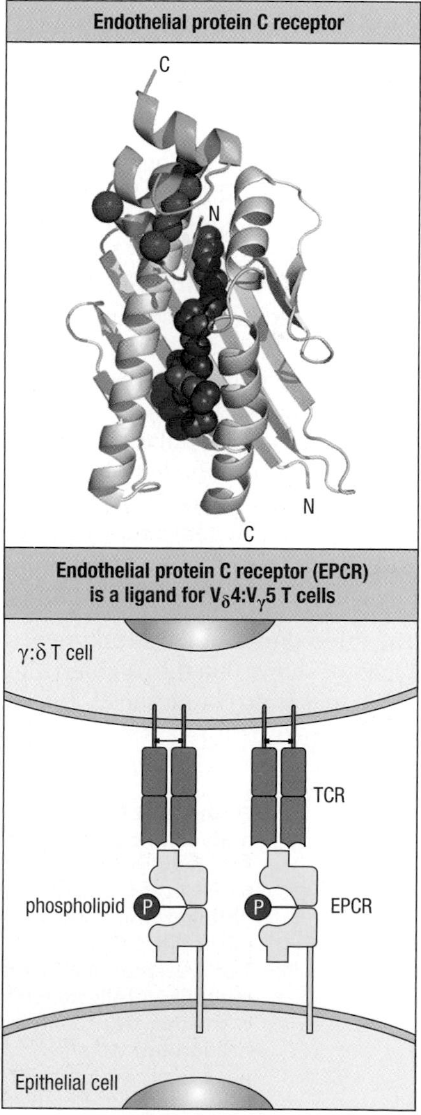

Figure 12.26 The endothelial protein C receptor is a ligand for $V_\gamma4{:}V_\delta5$ T cells. Upper panel: the endothelial protein C receptor (EPCR) is structurally like the α_1 and α_2 domains of MHC class I. This diagram shows how EPCR (yellow) has a groove that binds a phospholipid antigen (red and blue), not a peptide antigen, and a different site that binds protein C (purple). Lower panel: endothelial protein C is a ligand for $V_\gamma4{:}V_\delta5$ T cells. Structure courtesy of V. Oganesyan.

T-cell line that does not express its own T-cell receptor, they were sufficient to confer recognition of the tumor cell targets but not of the CMV-infected cells. This difference argues for recognition of CMV-infected cells being mediated by a combination of the $V_\gamma4{:}V_\delta5$ T-cell receptor and one or more other receptors on the surface of $V_\gamma4{:}V_\delta5$ T cells.

12-14 $V_\gamma{:}V_\delta1$ T-cell receptors recognize lipid antigens presented by CD1d

CD1 comprises five MHC class I-like heavy chains (CD1a, CD1b, CD1c, CD1d, and CD1e) that form heterodimers with β_2-microglobulin, like conventional MHC class I. The CD1 genes are on chromosome 1, away from the HLA complex on chromosome 6. CD1 molecules bind glycolipids of human and microbial origin and present them to T-cell receptors. On the basis of sequence similarity, the CD1 molecules are assigned to three groups: group 1 comprises CD1a, CD1b, and CD1c; group 2 comprises CD1d, and group 3 comprises CD1e. Only CD1d has a binding site with structural similarity to that of EPCR and, like EPCR, CD1d presents lipid antigens to $\gamma{:}\delta$ T cells. The lipid-binding sites of CD1 molecules have very different architectures from the groove of the peptide-presenting MHC class I molecules. The α_1 and α_2 domains form two or more hydrophobic channels running under the top surface of the protein. The long hydrophobic alkyl chains of glycolipid antigens (Figure 12.27) fit into these channels, leaving the lipid's hydrophilic headgroup to protrude and make contact with the T-cell receptor. Because $\gamma{:}\delta$ T-cell receptors are not subject to positive and negative selection in the thymus, $\gamma{:}\delta$ T cells can respond to lipids of human origin—self lipids—as well as to lipids of microbial origin.

Like MHC class I molecules, CD1d molecules assemble in the endoplasmic reticulum and require a bound lipid to give a stable structure. This first lipid bound is usually a self lipid. Once assembled, the CD1d molecule is transported to the cell surface. After reaching the surface, the CD1d molecules are internalized into endosomal vesicles, where they have the opportunity to exchange the bound self lipid for a lipid of microbial origin or another self lipid. After being recycled to the cell surface, the CD1d can then present the microbial lipid to $\gamma{:}\delta$ T cells. The epithelial cells in many tissues express CD1d. In particular, it is highly expressed by the enterocytes of intestinal epithelium, a tissue densely populated with $\gamma{:}\delta$ T cells. A majority of the intraepithelial lymphocytes and the lymphocytes in the lamina propria (see Section 10-11) are $\gamma{:}\delta$ T cells that have $V_\gamma{:}V_\delta1$ T-cell receptors. Sulfatide, a sulfated self lipid, is also at high concentration in intestinal epithelium. CD1d binds sulfatide within epithelial cells and brings the sulfatide to the cell surface, where it can engage with $V_\gamma{:}V_\delta1$ T-cell receptors (Figure 12.28).

The three-dimensional structure of a complex of sulfatide, CD1d, and a $V_\gamma{:}V_\delta1$ receptor shows that the $\delta1$ chain binds both CD1d and sulfatide, whereas the γ chain makes no contact at all. This explains how $V_\gamma{:}V_\delta1$ T cells with a variety of

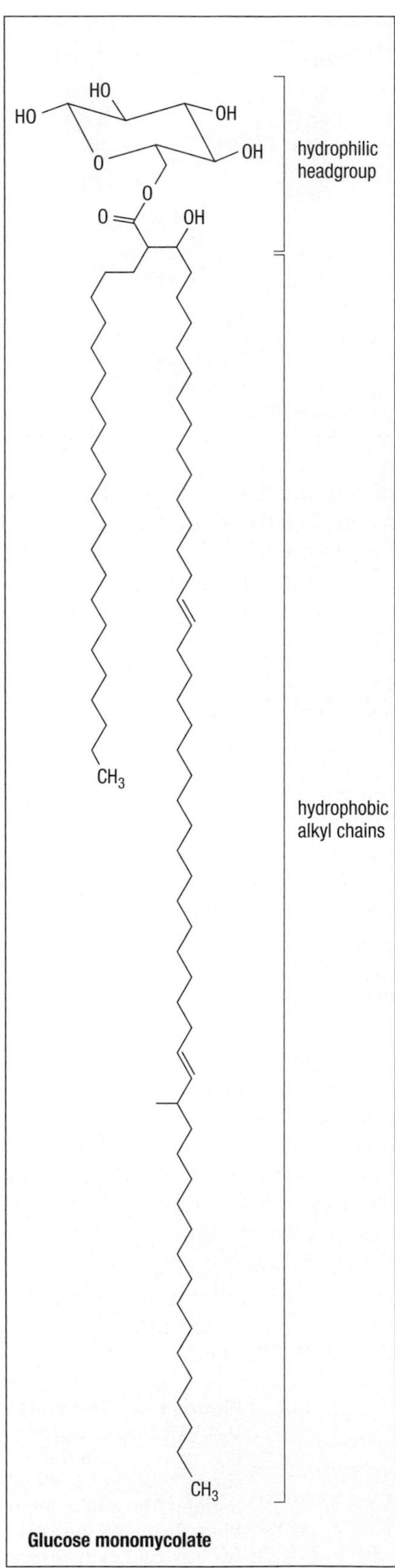

Figure 12.27 The chemical structure of glucose monomycolate, a glycolipid antigen from the bacterium *Mycobacterium phlei*. This glycolipid antigen is part of the cell wall of *M. phlei*. It is presented to $\alpha{:}\beta$ T cells by CD1b. Its physical and chemical properties are very different from those of the peptide antigens presented by MHC class I molecules. Because much of the glycolipid antigen is not seen by the T-cell receptor but is hidden in hydrophobic channels in the MHC class I molecule, T cells stimulated by one glycolipid antigen will cross-react with other structurally related lipids that have similar hydrophilic headgroups. *M. phlei* only causes disease for immunocompromised people, showing that most human immune systems are effective at controlling infection by these mycobacteria.

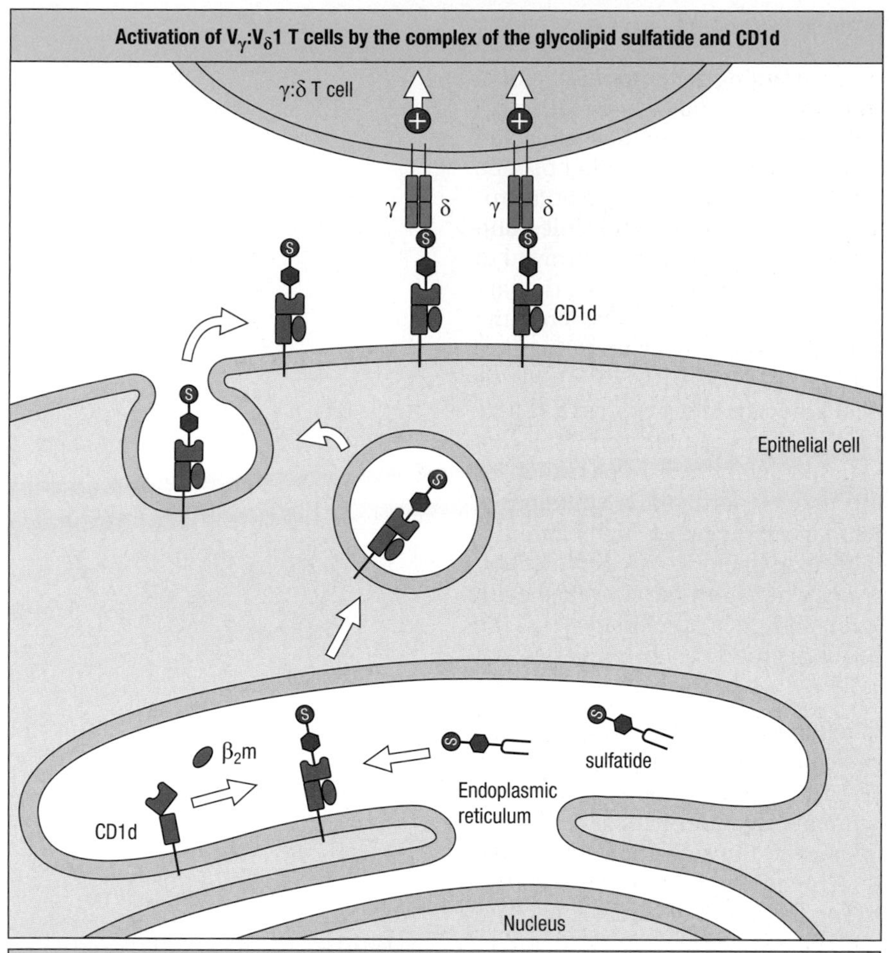

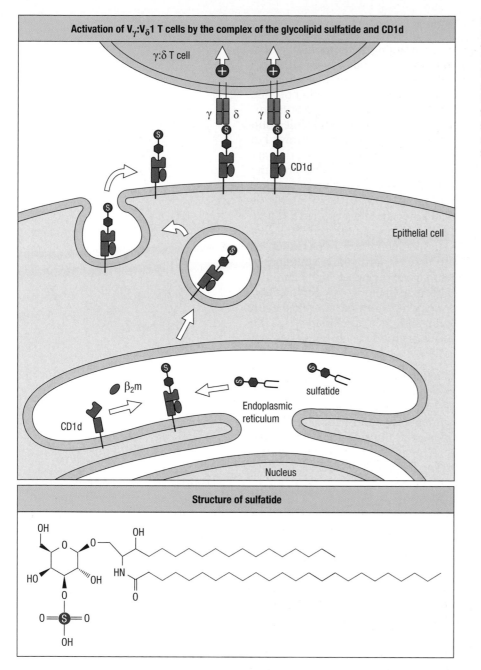

Figure 12.28 CD1d presents glycolipid antigens to the $V_\gamma:V_\delta 1$ receptors of γ:δ T cells. Sulfatide is a glycolipid made by human cells and is abundant in the intestinal epithelium. The upper panel shows sulfatide being bound by CD1d in the endoplasmic reticulum and then transported to the cell surface. There the sulfatide is recognized by the antigen receptor of a $V_\gamma:V_\delta$ T cell. The lower panel shows the chemical structure of sulfatide.

different γ chains have similar functions. This type of interaction with a single chain of the γ:δ T-cell receptor is very different from that of α:β T-cell receptors, in which both chains of the receptor make contact with both the antigen and its presenting molecule. All three CDR loops of the δ chain are involved in binding CD1d, and the CDR3 loop also makes ionic interactions with the sulfatide headgroup. Although the γ chain is not involved, there is a strong interaction between the receptor and the presented self antigen. The fact that the germline $V_\delta 1$ segment has an intrinsically strong interaction with CD1d indicates that they have coevolved to develop this interaction. The end result is that many of the $V_\gamma:V_\delta 1$ T cells are specific for lipid antigens presented by CD1d, consistent with observation that some 15% of the intraepithelial lymphocytes with $V_\gamma:V_\delta 1$ receptors are activated by sulfatide presented by CD1d.

Summary

Although γ:δ T cells and α:β T cells both use rearranging genes to make clonal antigen receptors, they differ markedly in other ways. γ and δ gene rearrangements are not used to produce a vast array of receptors that are then subjected to stringent positive and negative selection. Instead, a limited number of gene rearrangements are used to produce useful receptors that recognize non-proteinaceous antigens presented by various MHC class I-like molecules and other cell-surface molecules. Thus the germline γ and δ gene have evolved to produce useful receptors and to reduce the production of receptors that are not useful. Although rearranging receptor genes have been considered the hallmark of adaptive immunity, the γ:δ cells are in effect using them to make an assortment of receptors for innate immunity. Neither are γ:δ cells as dependent on their T-cell receptor as α:β T cells, because they express a variety of other activating receptors by which the cell can be activated. γ:δ T cells are markedly absent from secondary lymphoid tissues, where α:β T-cell responses are organized, but are present in blood and tissues, particularly epithelia. In these tissues, the γ:δ cells represent oligoclonal expansions, which can differ from one tissue to another. These cells are activated in the absence of infection and become memory cells that reside in tissues where they monitor the health of cells, through receptors that detect infection and signs of cellular stress. The γ:δ cells eliminate host cells and secrete fibroblast growth factor cytokines that facilitate tissue repair.

The properties and functions of γ:δ T cells parallel those of the B-1 subset of B cells (see Section 6-10), and both cell types contribute to the innate immune response. They both have a distinctive cell-surface phenotype and express a restricted repertoire of antigen receptors, so that large numbers of cells carry receptors for the same antigen. Such numbers, and the activation state of the γ:δ cells, means that γ:δ T cells and B-1 B cells can respond rapidly to infections and stress without first undergoing a lengthy period of clonal expansion and differentiation.

Restriction of α:β T cells by non-polymorphic MHC class I-like molecules

The classic α:β T cell recognizes a peptide antigen presented by an MHC molecule. In this section we examine three examples of α:β T cells that do not fit this paradigm. One comprises the α:β T cells that respond to lipid antigens presented by the MHC class I-like molecules CD1a, CD1b, and CD1c. These T-cell responses are a major part of human defenses against the mycobacteria that cause leprosy and tuberculosis. The other two types of non-conventional α:β T cell—NKT cells and MAIT cells—have similarity to γ:δ T cells because their T-cell receptors have invariant germline-encoded α chains, which each evolved to recognize a different MHC class I-like molecule. The receptor of NKT cells recognizes lipid antigens presented by CD1d. The receptor of MAIT cells recognizes antigens that are small organic molecules presented by the MHC class I-like molecule MR1. These antigens, which are products generated in vitamin B synthesis, enable MAIT cells to detect the pathogens that make the vitamin B on which humans depend.

12-15 CD1-restricted α:β T cells recognize lipid antigens of mycobacterial pathogens

Mycobacteria, such as those causing tuberculosis and leprosy, make many unusual lipids and glycolipids not made by human cells (Figure 12.29). In the response to mycobacterial infection, these unusual lipids serve as target antigens for effector CD4 T cells and CD8 T cells that express a diverse range of α:β

Mycobacterial lipid antigens are presented by CD1 molecules	
Antigen	**Antigen-presenting molecule**
Diacylsulfoglycolipid	CD1b
Glucose monomycolate	CD1b
Mycolic acid	CD1b
Lipoarabinomannan	CD1b
Lipomannan	CD1b
Phosphatidylinositol mannoside	CD1b
Mannosyl-β-1-phosphomycoketide	CD1c
Didehydroxymycobactin	CD1a

Figure 12.29 Presentation of mycobacterial antigens by CD1 molecules. The CD1 molecules are particularly important in the control and elimination of mycobacterial infections. Mycobacteria are distinguished by a large number of unusual and antigenic glycolipids that are bound by CD1. Examples of such glycolipids and the CD1 molecules that present them are shown here. The structure of glucose monomycolate is shown in Figure 12.27.

	Tissue distribution				Recognition by T-cell receptors	
CD1 genes	Developing thymocytes	Professional antigen-presenting cells	Other hematopoietic cells	Epithelium	α:β T-cell receptor	γ:δ T-cell receptor
CD1D	+	+	+	+	+	+
CD1A	+	+	–	–	+	+
CD1C	+	+	–	–	+	+
CD1B	+	+	–	–	+	+
CD1E	–	+	–	–	–	–

Figure 12.30 **The CD1 family of proteins that bind lipid antigens.** The order of the CD1 genes within the gene family on chromosome 1 is shown on the left. The group 1 CD1 genes, *CD1A*, *CD1B*, and *CD1C*, are together in the center and are flanked on one side by *CD1D*, the group 2 CD1 gene, and on the other side by *CD1E*, the group 3 CD1 gene. The tissue distribution of the five CD1 protein isoforms (CD1a, CD1b, CD1c, CD1d, and CD1e) is given, as is their capacity to present lipid antigens to the receptors of α:β and γ:δ T cells. CD1e is an intracellular lipid-transport protein that facilitates the loading of CD1b and CD1c with lipid antigens but does not itself present antigens to T cells.

receptors. What these receptors have in common is the recognition of lipid antigens presented by the MHC class I-like molecules CD1a, Cd1b, or Cd1c.

In comparison with CD1d (see Section 12-13), CD1a, CD1b, and CD1c have a limited tissue distribution (Figure 12.30). In the thymus they are expressed by double-positive thymocytes and mediate the selection of CD1-restricted α:β T cells. In the peripheral tissues, the group 1 CD1 molecules are expressed only by professional antigen-presenting cells, which are also the cells that mycobacteria target for infection. As ingested mycobacteria take up residence in intracellular vesicles, mycobacterial lipids are typically present in endosomes and other intracellular vesicles. As has been described for CD1d, a characteristic property of CD1 molecules is that they are continually cycling between the plasma membrane and endosomal vesicles. The endosomal vesicles contain lipid-transfer proteins, such as the saposins, that allow self lipids to be removed from CD1 molecules and replaced by mycobacterial lipid antigens. The CD1-lipid complexes are then taken to the cell surface and presented to CD1-restricted T cells. CD1a, CD1b, and CD1c form a team of complementary players. They present distinctive types of lipid antigen acquired by cycling through endosomal vesicles in different parts of the cell. Such differential traffic through the cell is mediated by adaptor proteins that load the CD1 molecules into particular transport vesicles. To perform these functions, the CD1 isoforms evolved antigen-binding sites of different size (Figure 12.31).

Functional properties of group 1 CD1 molecules			
	CD1a	CD1c	CD1b
Size of binding site (Å^3)	1350	1780	2200
Adaptor protein	None	AP2	AP2 and AP3
Depth of penetration of cell on cycling	Shallow	Intermediate	Deep
Site of lipid transfer	Cell surface Early endosomes	Intermediate endosomes	Late endosomes Lysosomes
State of binding site in absence of lipid	Intact and robust	Collapsed	Collapsed
Chaperoned by CD1e	No	Yes	Yes
Scaffolding lipids	No	No	Yes
Example of antigen	Dideoxymycobactin	Mycoketides	Diacylated sulfoglycolipids Mycolyl lipids

Figure 12.31 **Group 1 CD1 molecules have acquired distinctive properties that allow them to bind lipid antigens within different sets of intracellular vesicles.** These properties enable the CD1 molecules to scour all types of intracellular vesicles for mycobacterial antigens and to prevent mycobacteria from hiding from the immune defenses in any one type of vesicle.

CD1a has the smallest binding site and cycles in a shallow manner between the plasma membrane and the sorting endosomes. It alone can exchange lipids at the cell surface and is most promiscuous in the lipids it binds. The binding site of CD1c is of intermediate size, and in recycling it penetrates the cell to intermediate depth. This allows CD1c to broadly sample the contents of the intracellular vesicles and to present a wide range of antigens. CD1b has the largest binding site and penetrates the cell to greatest depth, binding lipid antigens in the harsh environment of the late endosomes and the lysosomes. This is achieved with the help of **lipid-transfer proteins**, including CD1e, which bind lipid antigens and then hands them over to CD1b. The binding site of CD1b is so large and flexible that it can accommodate multiple lipids, some of which are called scaffold lipids because they fill up the depths of the binding site and allow the antigenic lipid to lie on top and be accessible to the T-cell receptor. Patients with tuberculosis or leprosy make mycobacteria-specific α:β T-cell responses against lipid antigens presented by CD1a, CD1b, and CD1c.

12-16 NKT cells are innate lymphocytes that detect lipid antigens by using α:β T-cell receptors

The presentation of lipid antigens to γ:δ T cells by CD1d was described in Section 12-13. CD1d also presents lipid antigens to α:β T cells. These are not conventional α:β T cells but a subset of cells called **NKT cells** that express various receptors, such as NKG2D, that are considered NK-cell receptors (see Section 12-3). More importantly, what distinguishes NKT cells from other lymphocytes are their distinctive α:β T-cell receptors that exclusively recognize lipid and glycolipid antigens presented by CD1d. These T-cell receptors consist of a conserved α chain, $V_\alpha 24$–$J_\alpha 18$, associated with β chains made from the $V_\beta 11$ gene segment. Like NK cells and γ:δ T cells, NKT cells are ready for action, and after detecting infection they can deliver effector function within 4 hours. In the homogeneity of their antigen receptors and the speed of their response, NKT cells seem like and behave like cells of the innate immune response.

NKT cells develop in the thymus from the same precursor cell as the conventional α:β T cells that recognize peptide antigens and are restricted by HLA-A, -B, and -C molecules. NKT-cell development is defined by acquisition of the characteristic T-cell receptor and positive selection by self lipids presented by CD1d on double-positive thymocytes. In contrast, thymic epithelial cells mediate the positive selection of peptide-specific α:β T cells (see Figure 1.20, p. 21). The transcription factor promyelocytic leukaemia zinc finger protein (PLZF) is considered the master regulator of NKT-cell development, which is completed in the thymus and produces mature NKT cells that are fully equipped to respond to infection. NKT cells are widely distributed in tissues; they represent less than 1% of the T cells in the blood and do not recirculate between blood and lymph like naive conventional α:β T cells (see Section 1-10). NKT cells respond to antigenic differences that distinguish microbial glycolipids and phospholipids from self glycolipids and phospholipids.

The interaction of NKT-cell receptors with complexes of CD1d and lipid differs from that between CD8 T-cell receptors and peptide bound to MHC class I molecules. The α and β chains of the NKT-cell receptor lie parallel to the antigen-binding site, so that the conserved α chain covers the hydrophilic headgroup of the lipid antigen. Different lipid antigens are incorporated into these complexes by inducing conformational changes in the lipids to make them fit into the fixed binding site. This mechanism of induced fit contrasts with peptide presentation by MHC class I molecules, in which both the T-cell receptor and MHC class I molecule are chosen because they already fit the specific peptide.

Activation of NKT cells requires two signals: one coming from the T-cell receptor and one coming from a cytokine receptor. NKT-cell receptors can

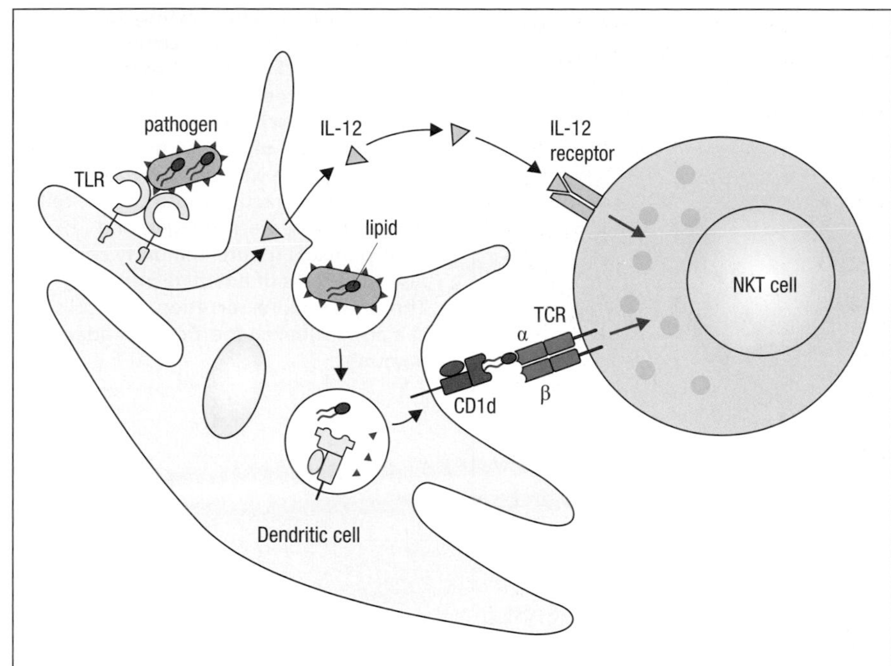

Figure 12.32 Activation of NKT cells requires two signals, one from the T-cell receptor and the other from a cytokine receptor. Uptake of pathogen and its degradation in the endosomes and lysosomes of a dendritic cell in infected tissue enable lipid antigens to be bound by CD1d and taken to the cell surface. Here they are engaged by the NKT cell's α:β T-cell receptor (TCR). This provides the first intracellular signal for NKT-cell activation. On sensing the presence of the pathogen, a Toll-like receptor (TLR) signals the dendritic cell to secrete the cytokine IL-12, which acts on the IL-12 receptor of the NKT cell to give the second intracellular signal for activation.

recognize lipids presented by CD1d on dendritic cells, macrophages, neutrophils, and B cells—cognate interactions that lead to mutual activation of the NKT cells and the antigen-presenting cell (Figure 12.32). NKT cells express high levels of the IL-12 receptor, and IL-12 is a major cytokine that activates NKT cells, as is also the case for NK cells (see Section 3-20). NKT cells can also be activated by IL-18, IL-23, and IL-25. Depending upon the antigen-presenting cell and the activating cytokines, NKT cells can be activated to produce a range of different effector responses that parallel those of the T cells of adaptive immunity (see Chapter 9). These can involve the production of T_H1 cytokines, T_H2 cytokines, T_H17 cytokines, and cellular cytotoxicity (Figure 12.33). Contributing to these functionally different responses is phenotypic heterogeneity in the NKT-cell population, such as in the expression of CD4 or CD8. NKT cells can enter sites of infection, where their interactions with pathogen-loaded dendritic cells lead to their mutual activation. Increased secretion of IFN-γ and upregulation of CD40 ligand by the NKT cells induce increased expression of IL-12 by the dendritic cells, which further activates the NKT cells. This helps to initiate the inflammatory response and the recruitment of neutrophils, NK cells, and monocytes to the site of infection.

CD1d is conserved and is present in both humans and mice. In contrast, mice have no equivalent to CD1a, CD1b, CD1c, and CD1e. The NKT cells that recognize CD1d are also very similar in humans and mice. These are therefore unusually conserved functions, which may well have been maintained throughout the evolution of placental mammals. Through germline evolution the $V_\alpha24$–$J_\alpha18$ segments have become a well-adapted CD1d receptor. These are properties that are typical of the receptors of innate immunity.

12-17 Mucosa-associated invariant T cells detect bacteria and fungi that make riboflavin

Mucosa-associated invariant T cells (MAIT cells) are a unique population of mature, effector CD8 α:β T cells that populate mucosal tissues, particularly the lungs, and are also present in the liver and to a smaller extent the blood. Like NKT cells (see Section 12-16), MAIT cells are defined by a closely related set of

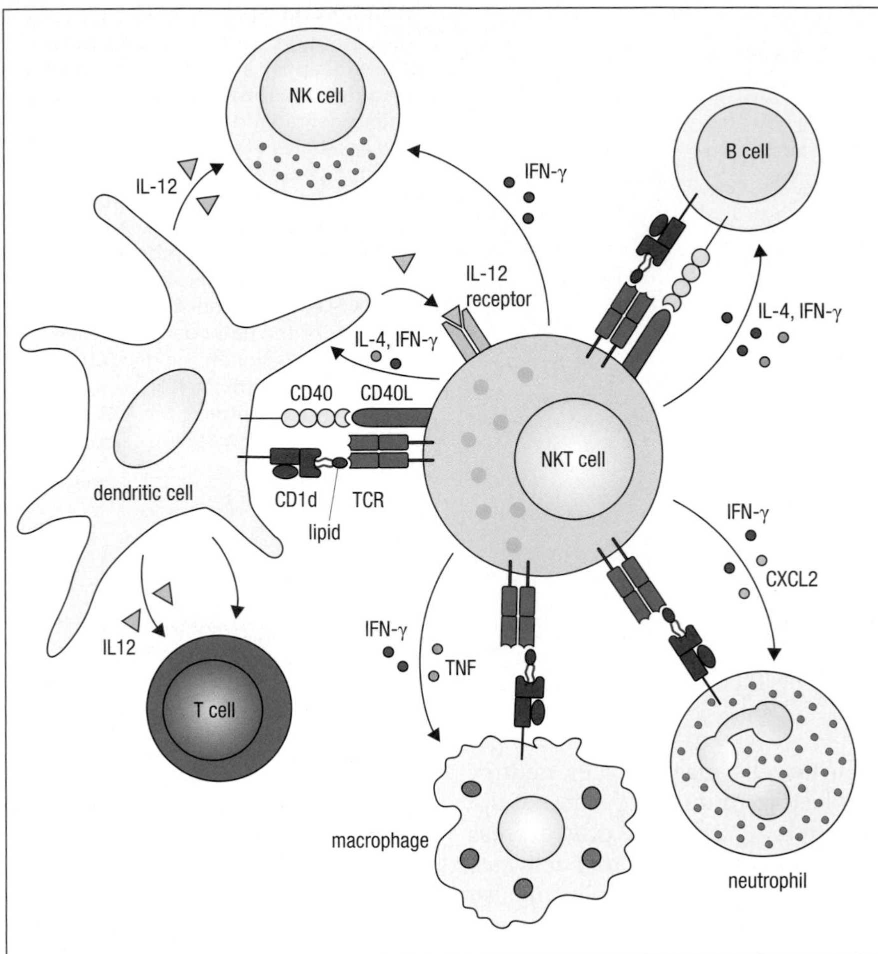

Figure 12.33 In responding to changes in self and microbial lipid antigens, NKT cells orchestrate an immune response through interactions with several different types of cells of innate and adaptive immunity. The NKT cell can have conjugate interactions with the NK cells, dendritic cells, macrophages, and neutrophils of innate immunity and also the B cells of adaptive immunity. Through cytokine secretion, NKT cells can also influence the T cells of adaptive immunity.

T-cell receptors: these comprise α chains with a germline-encoded $V_{\alpha}7.2$–$J_{\alpha}33$ rearrangement, and β chains with $V_{\beta}2$, $V_{\beta}13$, or $V_{\beta}22$ V gene segments. These T-cell receptors recognize antigens presented by MR1, an MHC class I-like molecule composed of β_2-microglobulin associated with a heavy chain encoded, like CD1, by a gene on chromosome 1. The antigen-binding site of MR1 is highly specialized for binding small heterocyclic organic molecules formed by bacteria and yeast during riboflavin biosynthesis. Riboflavin, commonly known as vitamin B_2, is an essential cofactor for many human enzymes, but it cannot be made by human cells and is acquired from both the diet and the microbiota. The antigen-binding cavity of the MR1 molecule is much smaller than those of peptide-binding MHC class I molecules, because of its lining of large basic and aromatic amino acid residues. The aromatic architecture of this cavity is complementary to the small, ringed metabolites that are MR1 ligands (Figure 12.34). In the complex of MR1, metabolite, and T-cell receptor, the metabolite is bound by the CDR3 loops of the α and β chains. The residues of the T-cell receptor α chain that bind to MR1 are all germline-encoded, and the somatic recombination process forms this α chain at high frequency. This is consistent with coevolution between MR1 and the germline segments encoding the T-cell receptor α chain.

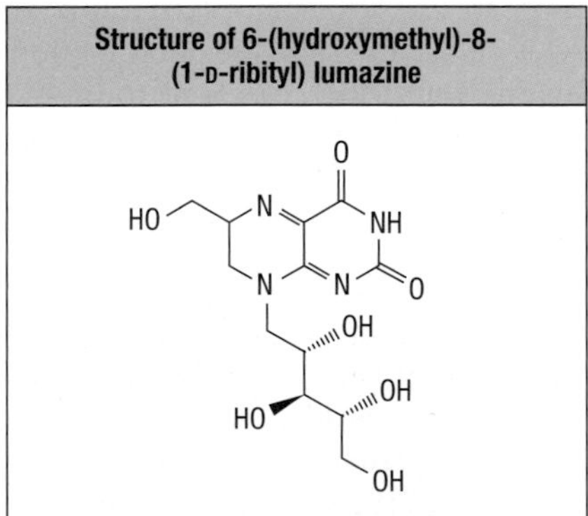

Figure 12.34 Mucosa-associated invariant T (MAIT) cells recognize by-products of the synthesis of riboflavin called pterins. Such substances were first isolated from butterfly wings. Hence the name pterin, which is derived from the Greek word for wing. The chemical structure of the pterin reduced 6-(hydroxymethyl)-8-(1-D-ribityl) lumazine is shown here.

MAIT cells detect and respond to infection by all species of bacteria and fungi that make riboflavin. They do not respond to species, such as the bacterium *Enterococcus faecalis*, that do not make riboflavin, and neither do they respond to viral infections.

Like NKT cells, MAIT cells develop in the thymus, where they are subject to positive selection by a small population of double-positive thymocytes that

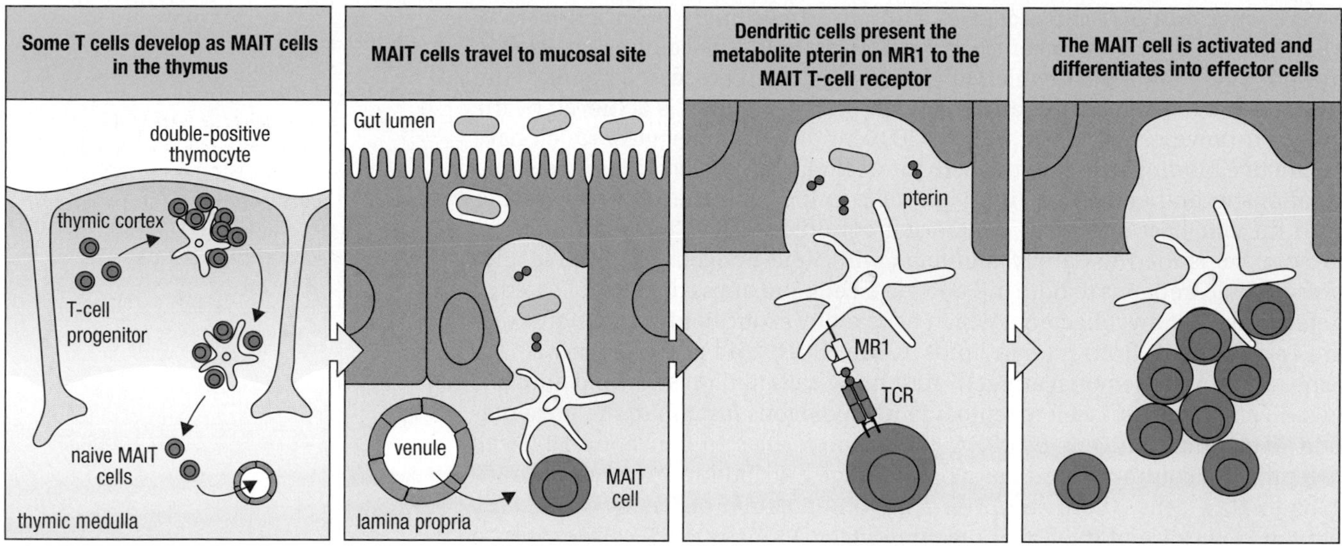

Some T cells develop as MAIT cells in the thymus
double-positive thymocyte
thymic cortex
T-cell progenitor
naive MAIT cells
thymic medulla
MAIT cells travel to mucosal site
Gut lumen
venule
MAIT cell
lamina propria
Dendritic cells present the metabolite pterin on MR1 to the MAIT T-cell receptor
pterin
MR1
TCR
The MAIT cell is activated and differentiates into effector cells

Figure 12.35 MAIT cell numbers and maturation are dependent on the microbiota. Like all other T cells, MAIT cells develop in the thymus, where they undergo positive selection on a population of double-positive thymocytes that express MR1. The naive MAIT cells leave the thymus in the blood and travel to mucosal tissue (first panel). In the mucosa, the MAIT cells interact with the resident dendritic cells (second panel). MR1 expressed on the dendritic cells has bound the pterins released by bacteria and yeast at the mucosal surface in the process of making riboflavin. The pterins are presented to the T-cell receptor of the MAIT cells (third panel). This stimulates the MAIT cells to divide and differentiate into effector T cells (fourth panel).

have unusually high levels of MR1 on the cell surface (Figure 12.35). Only small amounts of cell-surface MR1 can be detected on most cells, because most of the protein is within intracellular vesicles that shuttle between the cell's surface and the interior. When MAIT cells emerge from the thymus they are naive cells that then travel in the blood to take up residence in mucosal tissues and the liver. At these sites the naive MAIT cells are activated by exposure to the microbiota, which induces their proliferation and differentiation. At birth, just a few naive MAIT cells can be detected in the cord blood, but by 2 years of age they have become a large, oligoclonal population of effector cells with memory phenotype. MAIT cells comprise 1–10% of the T cells in blood and 20–40% of the lymphocytes in the liver. MAIT cells have antibacterial capacity, but the mechanisms by which this is achieved have yet to be worked out.

In contrast to the highly polymorphic MHC class I molecules and many of the MHC class I-like molecules that are ligands for NK-cell receptors, MR1 is exceptionally conserved throughout mammalian phylogeny, including in marsupials. For example, human and mouse MR1 have 90% sequence identity compared with around 70% for peptide-presenting MHC class I proteins. Because the structures of riboflavin and its metabolic intermediates have not changed since the human and mouse lineages separated 65 million years ago, any new variants of MR1 with altered antigen-binding sites have had no selective advantage. In contrast, the structures of the viral peptides that are presented by peptide-binding MHC class I molecules are changing all the time and providing continual but ever-changing selection pressures for new MHC class I variants. The conservation of MR1 has also enabled the germline-encoded $V_{\alpha}7.2$ and $J_{\alpha}33$ segments to coevolve with MR1.

Summary

Two whole-genome duplications occurred before the emergence of jawed vertebrates and their adaptive immune system based upon immunoglobulin domains and rearranging genes. Because MHC class I-like molecules pre-date the two duplications, MHC class I-like genes underwent the duplications and ended up at four different genomic locations. The CD1 gene family and the MHC class I genes are the present-day counterparts of two of those ancient sets of MHC class-like genes. In their properties and functions, the modern CD1 and MHC class I molecules are both contrasting and complementary. The MHC class I molecules became specialized in presenting processed peptides of defined size to α:β T-cell receptors, and they increased this capacity through a vast repertoire of T-cell receptors and an extraordinary polymorphism of MHC class I molecules. In contrast, the CD1 molecules became specialized in presenting lipid antigens, which are less diverse than processed peptides but more heterogeneous in size and shape. Lipid presentation is achieved by the four non-polymorphic CD1a, CD1b, CD1c, and CD1d molecules, which have distinctive binding sites and travel to intracellular vesicles in different cellular locations to find antigens to bind and take to the cell surface. In the vesicles with the harshest and most denaturing conditions, the CD1e lipid-transfer protein helps load its family members with lipid antigens. CD1 molecules present lipid antigens to both α:β and γ:δ T cells and are strongly implicated in defense against mycobacteria, which have many exotic lipid components that are very different from human lipids. CD1a, Cd1b, and CD1c are principally expressed on professional antigen-presenting cells and present lipid antigens to a diverse array of T-cell receptors in an analogous fashion to the presentation of peptide antigens by MHC class I molecules to conventional T-cell receptors. In contrast, CD1d has a broad tissue distribution and presents antigens to NKT cells. These comprise a large population of T cells with a very limited repertoire of α:β T-cell receptors. Like NK cells, NKT cells express a range of innate immunity receptors and they contribute early in the response to infection. MAIT cells are another large population of T cells with a limited repertoire of α:β T-cell receptors. They recognize small organic molecules made by bacteria and yeast but not by human cells, which are presented by MR1, the most conserved of the MHC and MHC class I-like molecules. MAIT cells patrol the mucosal surfaces, and their development and presence are dependent upon the microbiota.

Summary to Chapter 12

NK cells are the only lymphocytes that do not rearrange receptor genes. They are of innate immunity and ancient in origin. NK cells express a variety of receptors that are expressed differentially to give the NK-cell population a diverse repertoire of cell-surface phenotypes. The activating receptors either recognize pathogens and their products or human stress proteins that are expressed only by unhealthy cells. Enabling NK cells to contribute to adaptive immunity is their expression of an activating Fc receptor for IgG. Characteristic features of NK cells are inhibitory receptors that recognize MHC class I and educate NK cells to recognize cells that have a pathological loss of MHC class I expression. This function has arisen independently in two extended families of structurally different NK-cell receptors. These comprise lectin-like receptors encoded in the natural killer complex, and immunoglobulin-like receptors encoded in the leukocyte receptor complex. The inhibitory KIRs that recognize polymorphic determinants of HLA are encoded by a gene family that emerged only 60 million years ago and is thus a recent innovation of innate immunity. Even more recent is the division of KIR haplotypes into two functionally distinctive groups, a unique feature of human evolution that occurred only 5 million years ago. Infection with cytomegalovirus induces a massive proliferation of NK cells expressing an activating receptor that recognizes HLA-E. These cells, which are rare in uninfected people, are maintained throughout the persistent infection by this herpesvirus.

Because γ:δ T-cell receptors are made from rearranging genes, γ:δ T cells were first considered to be cells of adaptive immunity. However, the many differences between γ:δ and α:β T cells, and the similarities of γ:δ T cells to NK cells, question this assignment. γ:δ cells populate the blood and the tissues, where they are activated and develop effector function in the absence of infection. They monitor and maintain tissue integrity using a battery of receptors that detect infection and cellular distress. In performing these functions the γ:δ T-cell receptor does not play a unique role and is not essential. The diversity of γ:δ T-cell receptors is much less than that of α:β T-cell receptors, and each receptor recognizes a chemically related group of antigens that are neither peptide nor protein. These properties resemble those of innate immune receptors. On the other hand, several γ:δ T-cell receptors recognize antigens presented by molecules related to MHC class I, which is analogous to peptide presentation by MHC class I to CD8 T cells. The γ:δ T cells of different mammals differ in their abundance, tissue distribution, and receptors, showing how innovation in their functions evolved during divergence of these species. Overall, it looks as though the γ:δ T cell is a cell of innate immunity that uses its rearranging genes to make a number of different innate immune receptors.

Although the NKT cells and MAIT cells express α:β T-cell receptors, their properties are those of cells of innate immunity. The α-chain gene has evolved to enable the production of cells expressing a particular combination of V and J segments that produce α chains that bind complexes of antigen presented by conserved MHC class I-like molecules. NKT cells respond to lipid antigens; MAIT cells respond to species of bacteria and yeast that synthesize riboflavin. The NKT cells and MAIT cells provide good examples of innate immunity capturing and exploiting the cells of adaptive immunity. From this examination of NK cells, γ:δ cells, and unusual α:β T cells, we have found evidence for the third model of the evolution of innate and adaptive immunity (see Figure 12.1, third panel). Since its emergence 400 million years ago, there has been a co-evolution of adaptive and innate immunity that has integrated their components and their functions to where it is not always easy to draw a line between the two. What is readily seen from this chapter is the major role played by MHC class I and class I-like molecules as ligands for receptors of both innate and adaptive immunity. That is not the case for MHC class II molecules, which are molecules exclusively used in adaptive immunity.

Questions

12–1 NK cells express _____.

a. receptors to recognize alterations to human cell surfaces induced by infection
b. receptors to detect cellular stress
c. receptors for the IgG Fc region
d. Toll-like receptors
e. All of the above.

12–2 Explain the mechanism by which NK cells are able to engage a highly specific antigen receptor during their activation.

12–3 Unlike cytotoxic T lymphocytes, NK cells do not _____. (Select all that apply.)

a. secrete cytokines
b. participate in the adaptive immune response
c. rearrange T-cell receptor genes
d. use MHC class I molecules for their development and function
e. express CD3 components.

12–4 NK cells expressing high levels of _____ become abundant during CMV infection.

a. CD56
b. HLA-E
c. Ly49
d. CD94:NKG2A
e. CD94:NKG2C.

12–5 Match the term in column A with its description in column B.

Column A	Column B
a. CD94:NKG2A	1. on NK cells during CMV infection
b. ITIM	2. presents leader sequence peptides of HLA-A, -B, and -C
c. NKG2D	3. dephosphorylates Vav1
d. MIC-A and MIC-B	4. inhibitory receptor of many NK cells
e. SHP-1	5. activating receptor of all NK cells
f. HLA-E	6. stress proteins
g. CD94:NKG2C	7. located on cytoplasmic tail of NKG2A

12–6 What is the basic difference between the ligands recognized by CD94:NKG2A compared with inhibitory killer-cell immunoglobulin-like receptors (KIRs)?

12–7 An example of a lipid-transfer protein found in late endosomes and lysosomes is _____.

a. CD1a
b. CD1b
c. CD1c
d. CD1d
e. CD1e.

12–8 Match the term in column A with its description in column B.

Column A	Column B
a. NK-cell education	1. random expression patterns of KIR genes
b. gene-content variation	2. two different protein modules with the same function
c. framework genes	3. common to all KIR haplotypes
d. variegated expression	4. haplotypes differ in number and types of KIR genes
e. convergent evolution	5. engagement of inhibitory receptor with a cognate MHC class I molecule

12–9 The binding groove of endothelial protein C receptor (EPCR) binds to _____ and presents them to _____.

a. phosphoantigens; $V_{\gamma}9$:$V_{\delta}2$ T cells
b. MIC-A or MIC-B proteins; NK cells
c. sulfatides; V_{γ}:$V_{\delta}1$ T cells
d. phospholipids; $V_{\gamma}4$:$V_{\delta}5$ T cells
e. lipid antigens; $V_{\alpha}24$–$J_{\alpha}18$:$V_{\beta}11$ NKT cells.

12–10 Which of the following pairs is mismatched?

a. effector memory γ:δ T cells; CD45RA
b. splenic γ:δ T cells; $V_{\delta}1$
c. naive γ:δ T cells; CD27
d. isopentyl pyrophosphate; self phosphoantigen
e. hydroxyl-methyl-but-2-enyl-pyrophosphate; BTN3A1
f. glucose monomycolate; *Mycobacterium phlei.*

12–11 Which of the following statements regarding CD1 is correct?

a. CD1 binding grooves are very similar to those of MHC class I molecules.
b. Lipid antigens of CD1 bind with high affinity and are rarely exchanged.
c. CD1 comprises three MHC class I-like heavy chains known as CD1c, CD1d, and CD1e, each belonging to a different functional group.
d. CD1 molecules form heterodimers with β_2-microglobulin.
e. When a CD1d molecule presents sulfatide to a γ:δ T cell, all three CDR loops of both the γ and δ chains must be engaged to activate the T cell.
f. CD1 genes are located in the HLA complex on chromosome 6.

12–12 CD1-restricted α:β T cells respond to lipid antigens of mycobacterial pathogens using all except which of the following?

a. CD1a
b. CD1b
c. CD1c
d. CD1d.

12–13 All of the following regarding CD1a, b, and c are true except _____.

a. They have a more limited tissue distribution than CD1d.
b. CD1b associates with CD1e for the purpose of lipid transfer.
c. They cycle through endosomal vesicles in different locations of the cell with the assistance of adaptor proteins.
d. They possess different sizes of antigen-binding sites.
e. They all present lipid antigens.
f. They cycle continually between the cell surface and endosomal compartments, allowing exchange of antigens.
g. CD1b can accommodate more than one antigen simultaneously.
h. They are expressed by double-positive thymocytes and facilitate positive selection of CD1-restricted T cells.
i. They present antigens to NKT cells.

12–14 Identify the incorrect statement regarding NKT cells. (Select all that apply.)

a. They express α:β T-cell receptors.
b. They are positively selected in the thymus by self lipids presented by CD1a, b, c, and d on double-positive thymocytes.
c. They express NKG2D.
d. They do not recognize peptide antigens.
e. They are activated by lipid and glycolipid antigens.
f. They require two signals for activation; one from the T-cell receptor and one from CD28.
g. They make cognate interactions with a variety of leukocytes, including macrophages, dendritic cells, B cells, and neutrophils.
h. They have a limited effector response that parallels CD8 T cells.
i. They express CD4 or CD8, but not both simultaneously.
j. They are highly responsive to IL-12.
k. They express T-cell receptors consisting of a conserved $V_{\alpha}24$–$J_{\alpha}18$:$V_{\beta}11$.

12–15

A. Describe the ligand for the NK-cell receptor CD94:NKG2A.
B. Why is the concentration of this ligand on the target cell an effective measure of the presence or absence of classical MHC class I molecules?
C. Why is this ligand considered a broad mechanism for the NK-cell detection of unhealthy cells that is relatively insensitive to MHC class I polymorphisms?

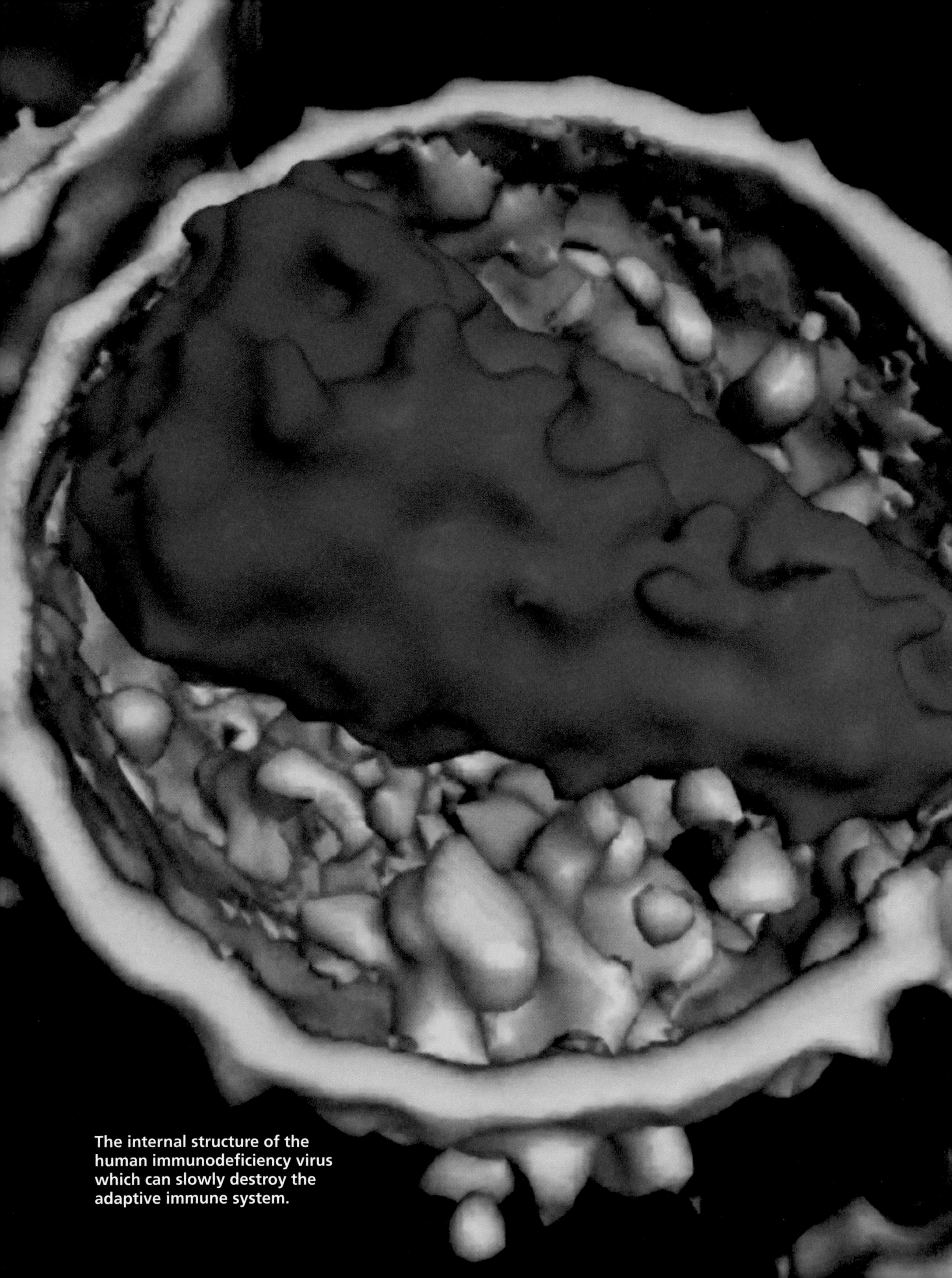

The internal structure of the human immunodeficiency virus which can slowly destroy the adaptive immune system.

Chapter 13

Failures of the Body's Defenses

Most pathogens that threaten the human body are prevented from establishing infection, and those infections that do occur are usually terminated by the actions of innate and adaptive immunity. In this situation there is strong pressure on pathogens to evolve ways of escaping or subverting the immune response. Microorganisms with such advantages compete successfully against other potential pathogens to exploit the resources of the human body. The first part of this chapter describes examples of the different types of mechanism they use.

The body's defenses against infection can also fail because of inherited deficiencies of the immune system. Some of these are described in the second part of the chapter. Within the human population there are mutant alleles for many of the genes encoding components of the immune system. These mutant genes cause **immunodeficiency diseases**, which vary in severity depending on which gene is defective and how the immune system is affected. Correlation of the molecular defects in immunodeficiency diseases with the types of infection to which patients become vulnerable reveals the effectiveness of the various arms of the immune response against different kinds of pathogen.

In the third part of the chapter we explore one particular host–pathogen relationship that combines themes from the first two parts of the chapter. This concerns the human immunodeficiency virus (HIV), which is extraordinarily effective at both escaping and subverting the immune response. During the course of an infection, which can last for decades, HIV gradually but inexorably wears down the immune system to the point at which it no longer works. The long-term consequence of HIV infection is that patients become severely immunodeficient and develop the fatal disease known as acquired immune deficiency syndrome (AIDS).

Evasion and subversion of the immune system by pathogens

The immune response to any pathogen involves complex molecular and cellular interactions between the pathogen and its host, and any stage in this interaction can be targeted by a pathogen and used for its own benefit. The systematic study of pathogen genomes reveals that most, if not all, pathogens have means of escaping or subverting immune defenses, and that some of them have many genes devoted to this purpose.

13-1 Genetic variation within some species of pathogens prevents effective long-term immunity

Antibodies directed against macromolecules on the surface of pathogens are the most important source of long-term protective immunity to many infectious diseases. Some species of pathogen evade such protection by existing as numerous different strains, which differ in the antigenic macromolecules on their outer surface. One such pathogen is the bacterium *Streptococcus pneumoniae*, which causes pneumonia. Genetic strains of *S. pneumoniae* differ in the structure of the capsular polysaccharides and compete with each other to infect humans. These strains, of which at least 90 are known, are called **serotypes** because antibody-based serological assays are used to define the differences between them. After resolution of infection with a particular serotype of *S. pneumoniae*, a person will have made antibodies that prevent reinfection with that type but will not prevent primary infection with another type (Figure 13.1). *S. pneumoniae* is a common cause of bacterial pneumonia because its genetic variation prevents individuals from developing effective immunological memory against all strains. Genetic variation in *S. pneumoniae* has evolved as a result of selection by the immune response of its human hosts.

13-2 Mutation and recombination allow influenza virus to escape from immunity

Some viruses also display genetic variation, influenza virus being a well-studied example. This virus infects the epithelia of the respiratory tract and passes easily from one person to another in the aerosols generated by coughs and sneezes. Protective immunity to influenza is provided principally by antibodies that bind to the hemagglutinin and neuraminidase glycoproteins of the viral envelope. These antibodies are made during the primary immune response to the virus. The course of a primary infection is short (1–2 weeks)

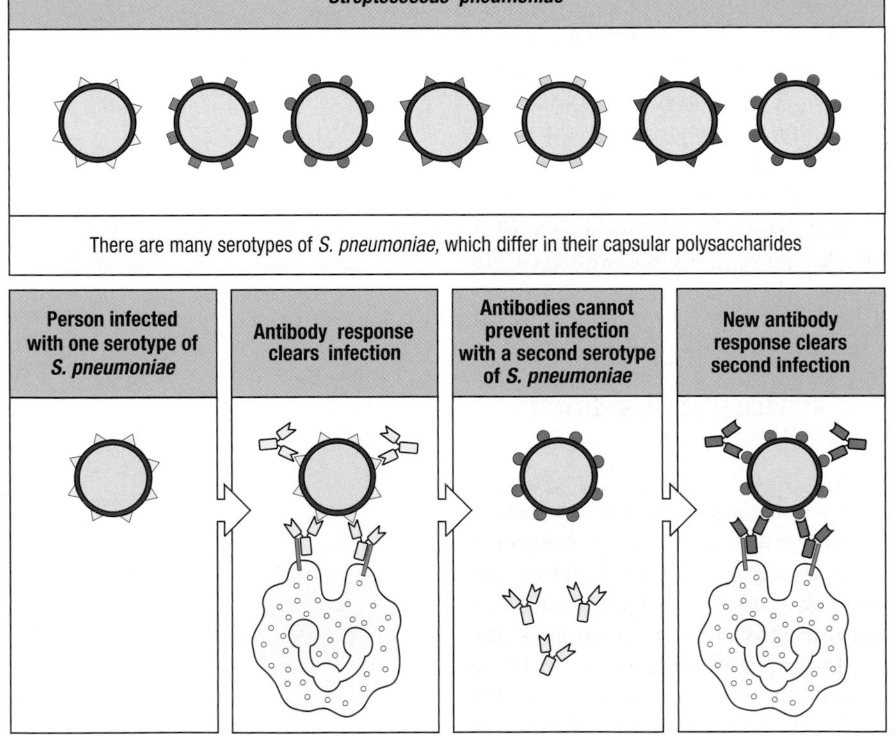

Figure 13.1 Protective immunity against *Streptococcus pneumoniae* is serotype-specific. Strains, or serotypes, of *S. pneumoniae* have antigenically different capsular polysaccharides, as shown in the upper panel. Antibodies against capsular polysaccharides opsonize the pathogen and enable it to be phagocytosed. A person infected with one serotype of *S. pneumoniae* clears the infection with type-specific antibody, as shown in the lower panels. These antibodies, however, have no protective effect when the same person is infected with a different serotype of *S. pneumoniae*. The second infection can be cleared only by making a new primary immune response that is directed specifically at the second serotype.

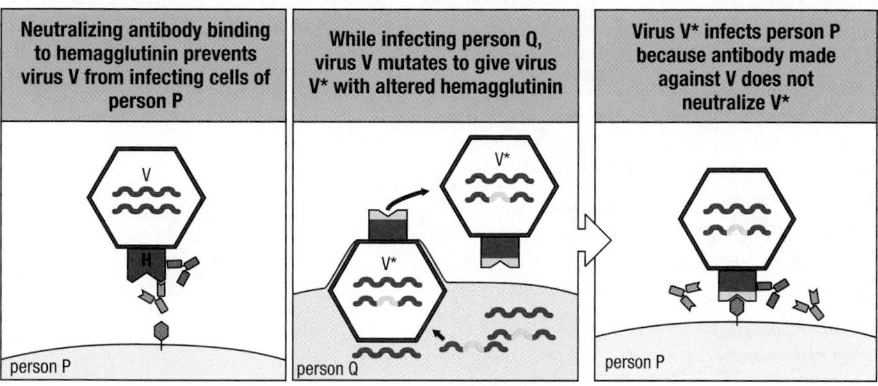

Neutralizing antibody binding to hemagglutinin prevents virus V from infecting cells of person P
While infecting person Q, virus V mutates to give virus V* with altered hemagglutinin
Virus V* infects person P because antibody made against V does not neutralize V*
V
H
V*
V*
person P
person Q
person P

Figure 13.2 Evolution of new influenza variants by antigenic drift. Upon infection with influenza strain V, person P produced antibodies against various epitopes of the viral hemagglutinin. Some antibodies are neutralizing (green); others are not (blue). When person P is further exposed to strain V, the neutralizing antibodies prevent the virus from infecting cells (left panel). In the course of infecting person Q, viral strain V mutates to give strain V*, which differs from V by one amino acid substitution (yellow) in the hemagglutinin (center panel). This amino-acid difference eliminates the epitope recognized by neutralizing antibodies made against strain V. Consequently, strain V* influenza virus can infect cells of person P without interference from the antibodies made against strain V (right panel). To clear this second influenza infection, person P must mount a primary immune response that makes neutralizing antibodies against strain V*. The viral neuraminidase (not shown) undergoes antigenic drift in a similar manner.

and the virus is cleared from the system by a combination of cell-mediated immunity and antibodies. The pattern of infection of influenza virus characteristically causes **epidemics**, in which the virus spreads rapidly through a local population and then abruptly subsides. Long-term survival of the influenza virus is ensured by the generation of new viral strains that evade the protective immunity acquired by human hosts during past epidemics.

Influenza is an RNA virus with a genome consisting of eight RNA molecules. RNA replication is relatively error-prone and generates many point mutations on which selection can act. New viral strains that lack the hemagglutinin or neuraminidase epitopes that induced protective immunity in the previous epidemic emerge regularly and cause an influenza epidemic every other winter or so. An individual's protective immunity to influenza is determined by the strain of virus to which they were first exposed—the phenomenon of 'original antigenic sin' (see Section 11-13). The history of exposure to particular strains of the virus differs within the population, largely according to age, and so there are subpopulations of people with differing degrees of immunity to the current strain of influenza. The people who suffer most at any particular time will be the very young, who have no previous exposure to influenza virus, and those whose protective immunity has been lost because of the new mutations present in the current strain. This type of evolution of influenza, which causes relatively mild and limited disease epidemics, is called **antigenic drift** (Figure 13.2).

In contrast, every 10–50 years an influenza virus emerges that is structurally quite different from its predecessors and is able to infect almost everyone. Besides spreading more widely to cause a **pandemic** (a worldwide epidemic), such viruses inflict more severe disease and a greater mortality than the viruses emerging from antigenic drift. The influenza strains that cause pandemics are recombinant viruses that derive some of their RNA genome from an avian influenza virus and the remainder from a human influenza virus. In these recombinant strains, the hemagglutinin and/or the neuraminidase are encoded by RNA molecules of avian origin and are antigenically very different

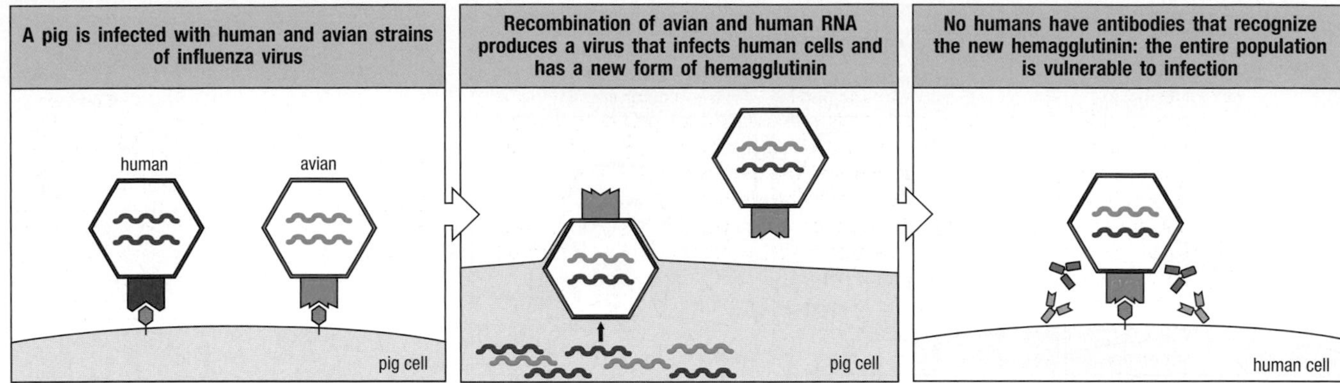

Figure 13.3 Evolution of new influenza virus variants by antigenic shift. Human (red) and avian (blue) influenza viruses can simultaneously infect pigs, which in this context are called secondary hosts. In a pig cell that is infected by both viruses (left panel), their RNA segments become reassorted to produce a variety of recombinant viruses. One type of recombinant virus has a hemagglutinin of avian origin (center panel). The avian hemagglutinin is antigenically very different from the hemagglutinins of the influenza viruses currently infecting the human population, including the one that infected the pig cell. Individual humans are highly susceptible to the recombinant virus because they do not have antibodies that bind to the hemagglutinin and prevent the virus from infecting cells (right panel). Because the entire human population is vulnerable to the recombinant virus, the latter has the potential to produce a pandemic influenza.

from the strains to which people have protective immunity. New pandemic strains often arise in parts of south-east Asia where farmers live in close proximity to their livestock such as pigs, chickens, and ducks. One theory is that the recombinant viruses arise in pigs that were simultaneously infected with both avian and human viruses. If such a recombinant virus jumps back into humans it has a tremendous competitive advantage, and in sweeping through the human population it will rapidly replace other influenza strains. Recombinant influenza viruses can similarly cause epidemics in bird populations and are greatly feared by poultry farmers. This mode of evolution is called **antigenic shift** (Figure 13.3).

13-3 Trypanosomes use gene conversion to change their surface antigens

Mutation and recombination are not the only methods by which pathogens can change the face they present to the immune system. Certain protozoans regularly change their surface antigens by a process of gene rearrangement. A good example is *Trypanosoma brucei*, the African trypanosome that is the major cause of sleeping sickness. The life cycle of the trypanosome involves both mammalian and insect hosts. Insect bites transmit trypanosomes to humans, in whom the parasites replicate in extracellular spaces. The trypanosome's surface is formed of a glycoprotein, of which there are numerous variants, each encoded by a different gene. The trypanosome genome contains more than 1000 genes encoding these **variable surface glycoproteins** (**VSGs**).

At any time, an individual trypanosome produces only one form of VSG. This is because rearrangement of a VSG gene into a unique site in the genome—the expression site—is required for gene expression. Rearrangement occurs by a process of **gene conversion** in which the gene in the expression site is excised and replaced by a copy of a different but homologous gene (Figure 13.4). The vast majority of the rapidly replicating trypanosomes that emerge after initial infection express the same dominant form of VSG. A small minority, however, has changed the expressed VSG gene and now expresses other forms. The host makes an antibody response to the dominant form of VSG, but not to the

Figure 13.4 Antigenic variation by African trypanosomes allows them to escape from adaptive immunity. The top three panels show the organization of genes in the *VSG* locus. In the top panel, the VSG^a gene (red box) is in the expression site, and the VSG^b (yellow box) and VSG^c gene (blue box) are inactive. In the second panel, gene conversion has replaced VSG^a with VSG^b at the expression site; in the third panel, VSG^c has replaced VSG^b at the expression site. The bottom panel shows how the patient's parasite burden (red, yellow, blue, and green lines) and antibody response (dotted black lines) co-evolve throughout the infection. In the first week, trypanosomes expressing VSG^a proliferate (red line), but as anti-VSG^a antibody is produced their numbers decline. The selection pressure imposed by anti-VSG^a antibody enables parasites expressing VSG^b to proliferate during the second week of infection (yellow line), and they then decline when anti-VSG^b antibody is made. In the third week, parasites expressing the VSG^c gene dominate, in the fourth week it is the turn of parasites expressing VSG^d to dominate, and so on, until the patient dies from all the damage caused by the succession of primary immune responses.

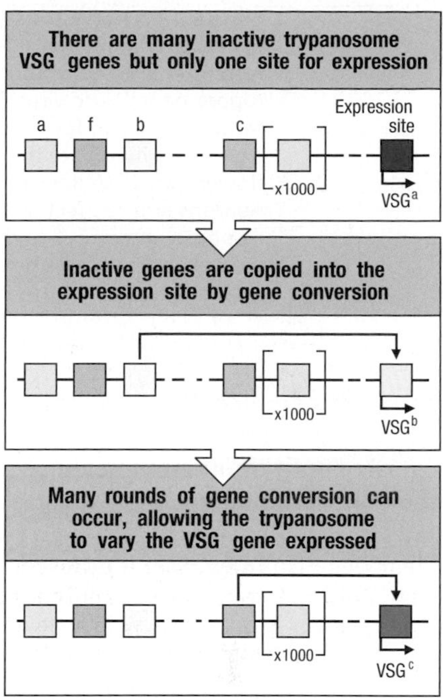

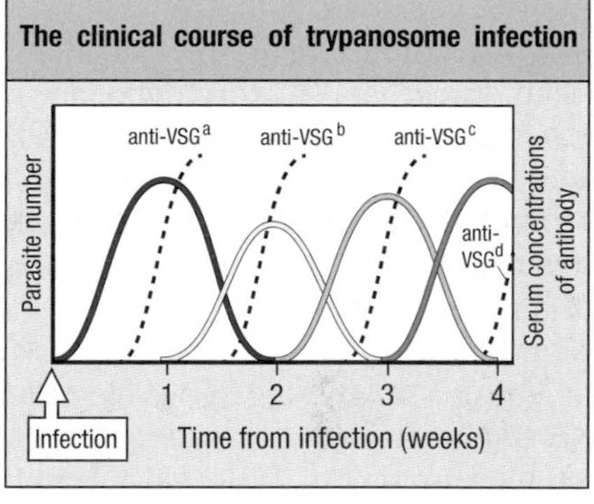

minority forms. Antibody-mediated clearance of trypanosomes expressing the dominant VSG facilitates the growth of those expressing the minority forms, one of which will come to dominate the trypanosome population. In time, the number of trypanosomes expressing the new dominant form is sufficient to stimulate the production of antibodies, which clear the new dominant form. This allows a further form to dominate, and so the cycle continues.

This mechanism of immune evasion causes trypanosome infections to produce a dramatic cycling in the number of parasites within an infected person (see Figure 13.4, bottom panel). The chronic cycle of antibody production and antigen clearance leads to a heavy deposition of immune complexes and inflammation. Neurological damage occurs and eventually leads to coma, the so-called sleeping sickness. Trypanosome infections are a major health problem for humans and cattle in large parts of Africa. Indeed, it is largely because of trypanosomes that wild populations of big game animals still survive in Africa and have not been completely replaced by cattle and other domesticated animals, as has occurred in many parts of the world. Malaria, another disease caused by a protozoan parasite that escapes immunity by varying its surface antigens, is also a major cause of human mortality in equatorial Africa.

Gene conversion enables similar strategies of antigenic variation to be used by several species of bacteria to escape the human immune response. This makes them successful pathogens and major problems for public-health. *Salmonella typhimurium*, a common cause of food poisoning, alternates the expression of two antigenically distinct flagellins, proteins of the bacterial flagella. This is achieved by reversible inversion of part of the promoter of one of the flagellin genes, which inactivates that gene and allows for expression of the second gene. *Neisseria gonorrhoeae*, the cause of the widespread sexually transmitted disease gonorrhea, has several variable antigens, the most impressive being the pilin protein, a component of the adhesive pili on the bacterial surface. Like the VSGs of African trypanosomes, pilin is encoded by a family of variant genes, only one of which is expressed at a time. Different versions of the pilin gene introduced into the expression site provide a minority population of variant bacteria. When the host's immune response places pressure on the dominant type, another is ready to take its place.

13-4 Herpesviruses persist in human hosts by hiding from the immune response

To terminate an established viral infection, infected cells must be killed by cytotoxic CD8 T cells. For this to occur, some of the peptides presented by

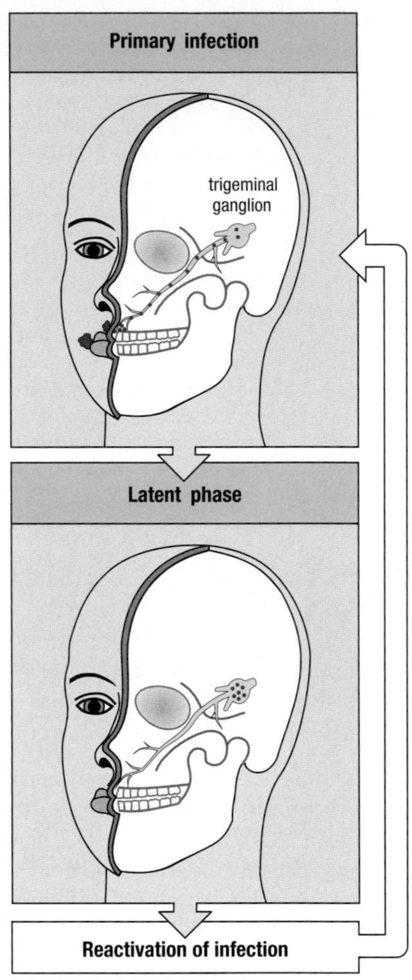

Figure 13.5 Persistence and reactivation of herpes simplex virus infection. Initial infection around the lips is cleared by the immune response, and the resulting tissue damage is manifested as cold sores (upper panel). The virus (small red dots) has meanwhile entered sensory neurons, for example those in the trigeminal ganglion with axons that innervate the lips, where the virus persists in a latent state (lower panel). Various forms of stress can cause the virus to leave the neurons and reinfect the epithelium. This reactivates the immune response and causes cold sores. People infected with herpes simplex viruses get cold sores periodically as a result of this process. During its active phase, the virus can pass from one person to another.

MHC class I molecules at the surface of infected cells must be of viral origin, a condition easily fulfilled by rapidly replicating viruses such as influenza. Consequently, influenza infections are efficiently cleared by the immune system by a combination of cytotoxic T cells and antibodies, the latter neutralizing extracellular virus particles. In contrast, some other viruses are difficult to clear because they enter a quiescent state within human cells, one in which they neither replicate nor generate enough virus-derived peptides to signal their presence to cytotoxic T cells. Development of this dormant state, called **latency**, which does not cause disease, is a favored strategy of the herpesviruses. Later on, when the initial immune response has subsided, the virus reactivates, causing an episode of disease.

Herpes simplex virus, the cause of cold sores, first infects epithelial cells and then spreads to sensory neurons serving the area of infection. The immune response clears virus from the epithelium, but the virus persists in a latent state in the sensory neurons. Various types of stress can reactivate the virus, including sunlight, bacterial infection, or hormonal change. After reactivation, the virus travels down the axons of the sensory neurons and reinfects the epithelial tissue (Figure 13.5). Viral replication in epithelial cells and the production of viral peptides restimulates CD8 T cells, which kill the infected cells, creating a new sore. This cycle can be repeated many times throughout life. Neurons are a favored site for latent viruses to lurk because they express very small numbers of MHC class I molecules, further reducing the potential for presentation of viral peptides to CD8 T cells.

The herpesvirus varicella-zoster (also called herpes zoster) remains latent in one or a few ganglia, chiefly dorsal root ganglia, after the acute infection of epithelium—chickenpox—is over. Stress or immunosuppression can reactivate the virus, which then moves down the nerve and infects the skin. Reinfection causes the reappearance of the classic varicella rash of blisters, which cover the area of skin served by the infected ganglia. The disease caused by reactivation of varicella-zoster is commonly known as shingles. In contrast to herpes simplex virus, reactivation of varicella-zoster usually occurs only once in a lifetime.

A third herpesvirus that causes persistent infection is the Epstein–Barr virus (EBV), to which most humans are exposed. First exposure in childhood produces a mild cold-like disease, whereas adolescents or adults encountering EBV for the first time develop infectious mononucleosis (also known as glandular fever), an acute infection of B lymphocytes. EBV infects B cells by binding to the CR2 component of the B-cell co-receptor complex (see Figure 9.3, p. 234). Most of the infected B cells proliferate and produce virus, leading in turn to the stimulation and proliferation of EBV-specific T cells. The result is an unusually large number of mononuclear white blood cells (lymphocytes, mostly T cells), which gives the disease its name. After some time, the acute infection is brought under control by CD8 cytotoxic T cells, which kill the virus-infected B cells. The virus persists in the body, however, because a minority of B cells become latently infected. This involves shutting off the synthesis of most viral proteins except EBNA-1, which maintains the viral genome in these cells. Latently infected cells do not present a target for attack by CD8 cytotoxic cells because the proteasome is unable to degrade EBNA-1 into peptides that are bound and presented by MHC class I molecules.

After recovery from the initial exposure to EBV it is unusual for reactivation of the virus to lead to disease. It seems likely that CD8 T cells quickly control episodes of viral reactivation. In immunosuppressed patients, however, reactivation of the virus can cause a disseminated EBV infection, and infected B cells can also undergo malignant transformation, causing B-cell lymphoproliferative disease.

13-5 Some pathogens sabotage or subvert immune defense mechanisms

Pathogens also exploit the immune-system cells that are ranged against them. *Mycobacterium tuberculosis* commandeers the macrophage's pathway of phagocytosis for its own purposes. On being phagocytosed, *M. tuberculosis* prevents the fusion of phagosome and lysosome, thus protecting itself from the bactericidal actions of the lysosomal contents. It then survives and flourishes within the cell's vesicular system. *Listeria monocytogenes*, in contrast, escapes from the phagosome into the macrophage's cytosol, where it grows and replicates. However, the intracytosolic way of life elicits cytotoxic CD8 T-cell responses against *L. monocytogenes*, which eventually terminate the infection.

The parasite *Toxoplasma gondii*, the cause of toxoplasmosis, creates its own specialized environment within the cells that it infects. This protozoan encloses itself in an impenetrable membrane-enclosed vesicle that does not fuse with other vesicles or membranes of the cell. Such isolation prevents *T. gondii*-derived peptides from binding to MHC molecules and stimulating a T-cell response to the parasite. The spirochete *Treponema pallidum*, the cause of syphilis, evades specific antibody by coating itself with human proteins. This is also a strategy pursued by the schistosome, *Schistosoma mansoni*, a parasitic helminth worm.

Of the four groups of pathogens (see Figure 1.4, pp. 6–7), viruses have evolved the greatest variety of mechanisms for subverting or escaping immune defenses. This is because their replication and life cycle depend completely on the metabolic and biosynthetic processes of human cells. Viral self-defense strategies include the capture of cellular genes encoding cytokines or cytokine receptors, which when expressed by the virus can divert the immune response. Also, the synthesis of proteins that inhibit complement fixation, and the synthesis of proteins that inhibit antigen processing and presentation by MHC class I molecules. Examples of defensive mechanisms used by herpesviruses and poxviruses are shown in Figure 13.6.

Major players in the immune response to viral infections are NK cells and CD8 T cells, killer lymphocytes that are dependent upon MHC class I molecules for their development and function. For this reason many viruses have evolved subversive mechanisms for interfering with the synthesis and expression of MHC class I. The herpesvirus human cytomegalovirus (CMV) is particularly rich in such mechanisms: it has 10 proteins that interfere in diverse ways to diminish the capacity of MHC class I molecules to stimulate NK-cell and CD8 T-cell responses against CMV-infected cells (Figure 13.7). One group of these saboteurs affect MHC class I molecules in various ways: by causing their degradation, by interfering with the proteasome, by interfering with the TAP or tapasin proteins, or by retaining MHC class I molecules in the endoplasmic reticulum—all mechanisms that prevent the presentation of viral antigens to CD8 T cells. Such mechanisms should favor an NK-cell response against the

Viral strategy	Specific mechanism	Result	Viral examples
Inhibition of humoral immunity	Virally encoded Fc receptor	Blocks effector functions of antibodies bound to infected cells	Herpes simplex Cytomegalovirus
	Virally encoded complement receptor	Blocks complement-mediated effector pathways	Herpes simplex
	Virally encoded complement control protein	Inhibits complement activation of infected cell	Vaccinia
Inhibition of inflammatory response	Virally encoded chemokine receptor homolog	Sensitizes infected cells to effects of some chemokines; advantage to virus unknown	Cytomegalovirus
	Virally encoded soluble cytokine receptor, e.g. IL-1 receptor homolog, TNF receptor homolog, IFN-γ receptor homolog	Blocks effects of cytokines by inhibiting their interaction with host receptors	Vaccinia
	Viral inhibition of adhesion molecule expression, e.g. LFA-3, ICAM-1	Blocks adhesion of lymphocytes to infected cells	Epstein–Barr virus
	Protection from NFκB activation by short sequences that mimic TLRs	Blocks inflammatory responses elicited by IL-1 or bacterial pathogens	Vaccinia
Blocking of antigen processing and presentation	Inhibition of MHC class I upregulation by IFN-γ	Impairs recognition of antigen-presenting cells by CD4 T cells	Herpes simplex Cytomegalovirus
	Inhibition of peptide transport by TAP	Blocks peptide association with MHC class I	Herpes simplex
Immunosuppression of host	Virally encoded cytokine homolog of IL-10	Inhibits T_H1 lymphocytes Reduces IFN-γ production	Epstein–Barr virus

Figure 13.6 Mechanisms by which herpesviruses and poxviruses subvert the immune response. Herpes simplex, cytomegalovirus, and Epstein–Barr viruses are herpesviruses; vaccinia is a poxvirus.

infected cells that are now lacking self-MHC class I (see Chapter 12). However, a second group of saboteurs interfere with the inhibitory NK-cell receptors CD94:NKG2A and LILRB1, which sense missing self-MHC class I, and with the activating NKG2D receptor, which recognizes the ligands MIC and ULBP (see Figure 12.2).

Human CMV is an extremely well-adapted human pathogen that infects more than half the population of the United States. Most of these 158 million people are unaware of their infection, because the virus causes few symptoms on initial infection and thereafter exists in a latent state, during which it is comfortably controlled by the combined activities of NK cells and CD8 T cells. In the healthy CMV-infected person there is a well-tuned balance in which the virus survives and multiplies with little expense to the host. In contrast, CMV causes life-threatening disease in immunocompromised individuals: the young, the elderly, the transplant recipient on immunosuppressive drugs, and people infected with HIV. CMV is the commonest infection affecting patients who have had a bone marrow transplant or hematopoietic stem-cell transplant, and if not treated with antiviral drugs it is fatal. CMV infects a wide range of human cell types and is spread via physical contact of bodily fluids (see also Section 12-8).

HCMV protein	Subversive effect on the immune response
US2	Transports HLA class I to cytoplasm for degradation by proteasome
US3	Retains HLA class I in the endoplasmic reticulum by blocking tapasin function
US6	Inhibits TAP ATPase activity and function
US10	Binds HLA class I and slows its transport from the endoplasmic reticulum to the cell surface
US11	Targets newly synthesized HLA class I heavy chains for degradation in the cytoplasm
UL16	Inhibits NK-cell recognition of infected cells by binding to the ULBP ligands of NKG2D
UL18	MHC class I heavy-chain homolog that binds to the inhibitory NK-cell receptor LILRB1
UL40	The UL40 leader peptide binds HLA-E and subverts CD94:NKG2A monitoring of HLA-A, -B, -C expression
UL83	Blocks access to the proteasome and the generation of MHC class I binding peptides
UL142	Downregulates the expression of the MIC-A and MIC-B ligands of NKG2D

Figure 13.7 Human cytomegalovirus interferes with the expression of MHC class I molecules in many different ways.

13-6 Bacterial superantigens stimulate a massive but ineffective CD4 T-cell response

Some species of Gram-positive bacteria, notably *Staphylococcus aureus* and *Streptococcus pyogenes*, secrete potent toxins, which at minuscule concentrations induce a violent disruption of an infected person's immune system. Because these small bacterial protein toxins activate so many different T-cell clones they are called **superantigens**. The mayhem is caused by the nonspecific activation of numerous clones of CD4 T cells, which involves 2–20% of the body's CD4 T cells and excessive production of IL-2, interferon (IFN)-γ, and TNF-α. Between them, the staphylococci and streptococci make more than 30 different superantigens. What they all have in common are binding sites that allow the superantigen to cross-link an MHC class II molecule on an antigen-presenting cell with the T-cell receptor on a CD4 T cell.

The soluble superantigen first forms a stable interaction with an MHC class II molecule on the antigen-presenting cell. Subsequently, when a circulating CD4 T cell investigates the antigen-presenting cell to search for its specific antigen, a second site on the superantigen binds to the variable domain of the T-cell receptor V_β chain. Then, at a third site, the superantigen binds to the T cell's CD28 co-stimulatory receptor and does so in a way that does not interfere with the interaction of CD28 with its B7 ligand (Figure 13.8). By this mechanism, superantigens can substitute for the specific antigenic peptide and assemble a complex of T-cell receptor, co-receptor, and co-stimulatory receptor that will fully activate the CD4 T cell (see Section 8-5). The various superantigens activate different subsets of CD4 T cells because they have different binding specificities for human V_β chains. Thus, *S. aureus* **toxic shock syndrome toxin-1 (TSST-1)** binds to β chains containing a $V_\beta2$ gene segment and can activate all CD4 T cells bearing T-cell receptors with this type of β chain. In contrast, ***S. aureus*** **enterotoxin B (SEB)** activates all CD4 T cells expressing $V_\beta1.1$, 3.2, 6.4, and 15.1 chains.

Eating food contaminated with *S. aureus* is a common cause of food poisoning. Several hours after enjoying the meal, the diner experiences different emotions while suffering episodes of violent vomiting and diarrhea as the mucosal immune system responds to the activation of CD4 T cells by bacterial superantigens. This response has the effect of flushing the bacteria and its toxins from the gastrointestinal tract, which benefits the host and raises the question of why bacteria have repeatedly evolved superantigens that target CD4 T cells, lymphocytes that are not involved in the immediate response to infection. One possibility is that a massive, and non-physiological, adaptive immune response at the onset of infection confuses the innate immune response. In this

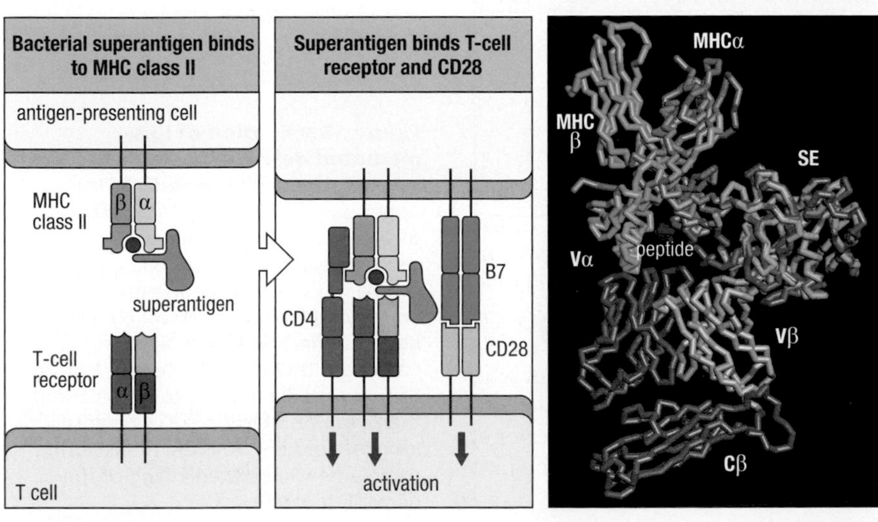

Figure 13.8 Bacterial superantigens activate CD4 T cells by cross-linking MHC class II molecules with α:β T-cell receptors and CD28 co-stimulatory molecules in the absence of antigenic peptides. The superantigen first binds MHC class II (left panel) and then engages the V_β chain and CD28 (center panel). Signals from the T-cell receptor, CD4 co-receptor, and CD28 combine to activate the T cell. In the molecular model (right panel), staphylococcal enterotoxin (SE, blue) interacts with an MHC class II molecule (yellow and green chains) and an α:β T-cell receptor (orange, gray, and pink chains). A peptide bound by the MHC class II molecule (red) is not recognized by the T-cell receptor. Molecular model courtesy of H.M. Li, B.A. Fields, and R.A. Mariuzza.

scenario, the cytokines poured out by the CD4 T cells interfere with the capacity of neutrophils and macrophages to phagocytose the bacteria that have crossed the mucosa and are about to establish infection in the tissues.

13-7 Subversion of IgA action by bacterial IgA-binding proteins

Staphylococcus aureus is an adept and opportunistic pathogen that possesses numerous virulence factors which interfere with human defenses and promote the bacterial colonization of invaded tissue. In addition to the superantigens, *S. aureus* secretes a family of structurally related **staphylococcal superantigen-like proteins** (**SSLPs**) that subvert and compromise human immunity in a variety of ways. The purpose of staphylococcal superantigen-like protein 7 (**SSLP7**) is to prevent monomeric IgA from delivering the bacteria to phagocytes. In the absence of SSLP7, IgA binds to a bacterium with its Fab arms and to FcαRI on neutrophils and macrophages with its Fc region. This activates the phagocyte to engulf and destroy the bacterium bound to the Fc receptor. SSLP7 has binding sites for the Fc region of IgA and for the C5 complement protein. These interactions create a large constrained complex in which IgA binding to FcαRI and complement-mediated killing of the bacterium are both prevented (Figure 13.9). Over evolutionary time, there has been

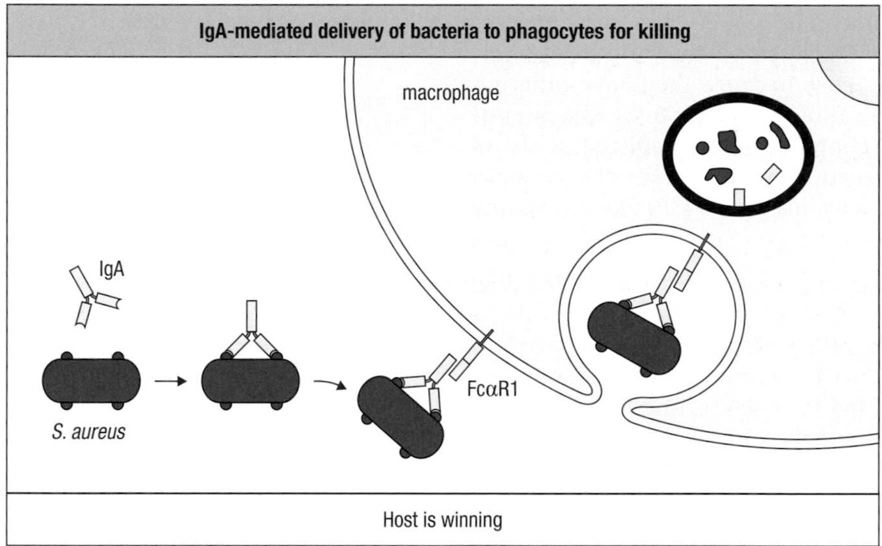

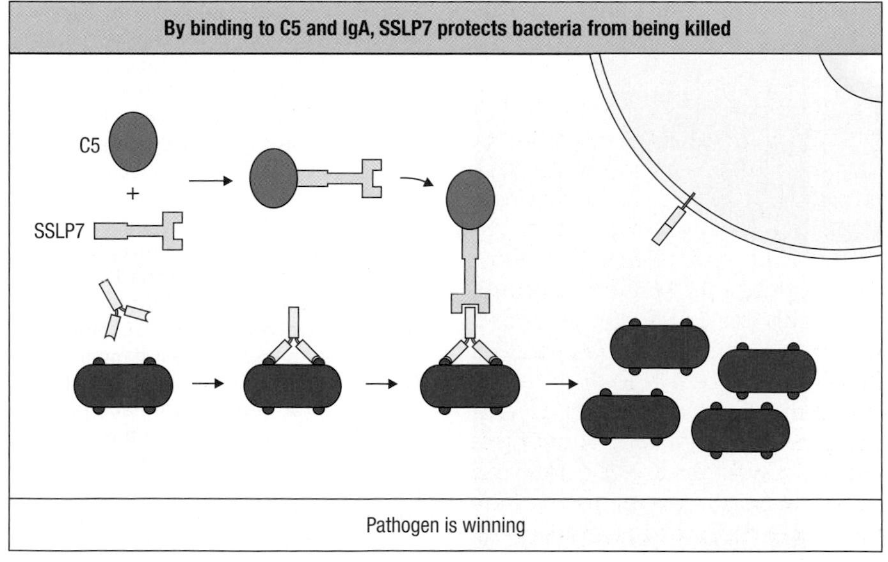

Figure 13.9 Evasion of IgA-mediated defense by staphylococcal superantigen-like protein 7. The upper panel shows how the combination of specific IgA and FcαRI on phagocytes causes the elimination of bacteria that cross a mucosal barrier and infect the underlying tissue. The lower panel shows how this mechanism for disposing of bacteria is thwarted by the SSLP7 protein of *S. aureus*. By binding to both C5 and the Fc region of IgA, SSLP7 prevents IgA from binding to FcαRI or activating complement-mediated killing of the bacteria (not shown).

a perpetual arms race between bacterial pathogens and their mammalian hosts. The host selects for new IgA variants that do not bind the bacterial SSLP, which in turn leads to selection for a new SSLP that binds to the latest version of IgA.

Summary

From the human perspective, the ideal immune system would be one that terminates infection before the pathogen damages tissues or saps the body's resources. In contrast, an ideal situation for a pathogen is one in which the immune system does not interfere with growth and replication, while other parts of the body provide food and shelter. To further their cause, pathogens have evolved ways of reducing the effectiveness of the human immune response. Antigenic variation in the pathogen prevents the maturation of the adaptive response and the development of useful immunological memory. Latency, a means of avoiding the immune response, allows viruses to lie low within cells until immunity has waned. More active strategies are for pathogens to interfere with key elements of the immune response, either to inhibit normal immune function or to recruit the response to the pathogen's advantage.

Inherited immunodeficiency diseases

Inherited defects in genes for components of the immune system cause **primary immunodeficiency diseases**, which reveal themselves by enhanced susceptibility to infection or autoimmunity. Primary immunodeficiency diseases are distinguished from **secondary immunodeficiency diseases**, which are due not to defective genes but to environmental factors, such as immunosuppressive drugs, that adversely impact the immune system. Before the advent of antibiotic therapy during the 1940s, most individuals with inherited immune defects died from infection during infancy or early childhood. Because so many normal infants also succumbed to infection in that earlier era, death from immunodeficiency did not stand out until the 1950s, when the first such disease was described. Since then, numerous inherited immunodeficiency diseases have been identified and correlated with susceptibility to particular classes of pathogen. Each disease is due to a defect in a particular protein or glycoprotein, and the precise symptoms depend on the role of that component in the immune response.

13-8 Rare primary immunodeficiency diseases reveal how the human immune system works

In dissecting the immune system of the laboratory mouse, immunologists 'knock out' a selected gene and examine the immunodeficiency syndrome that this causes. Equivalent human gene knockouts are provided by more than 200 primary immunodeficiency syndromes. Around 50 of these were described in the past 5 years, as the rate of discovery has increased with the application of complete genome sequencing. Treating and studying these patients have made invaluable contributions to knowledge of the human immune system. It is no coincidence that in almost all the previous chapters of this book, one or more primary immunodeficiency syndromes have been used to illustrate the functions of particular proteins and the effects of their absence (Figure 13.10). For some genes their deletion from human and mouse genomes leads to similar immunodeficiencies, whereas for other genes the knockout phenotypes are unpredictably different. Most of the human conditions are very rare and are caused by mutant genes that have no selective benefit for the people who carry them. They usually occur in small populations that are geographically or culturally isolated and in which there are traditions of marriage within the group. Recent advances in genetics and genomics have made the precise identification of the genes responsible for immunodeficiency syndromes much

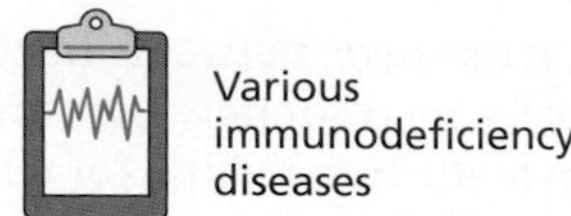

Inherited immunodeficiencies reveal the workings of the human immune system				
Name of deficiency	**Affected gene**	**Immune defect**	**Susceptibility**	**Reference (section)**
Asplenia	Not known	Absence of the spleen	Encapsulated extracellular bacteria	1-13, p. 23
C3 deficiency (see Figure 13.13 for other complement immunodeficiencies)	*C3*	Lack of C3	Recurrent infection with Gram-negative bacteria	2-3, p. 31
Factor I deficiency	*CFI* (complement factor I)	Absence of factor I and depletion of C3	Encapsulated bacteria	2-5, p. 34
Deficiencies of C5, C6, C7, C8, or C9	*C5, C6, C7, C8, C9*	Lack of complement-mediated lysis	Infections due to *Neisseria* species	2-7, p. 38
Paroxysmal nocturnal hemoglobinuria	Somatic and germline mutations in genes involved in phosphatidylinositol biosynthesis	Lack of complement-regulatory proteins DAF, HRF, and CD59	Lysis of erythrocytes by complement	2-7, p. 39
NEMO deficiency (X-linked hypohidrotic ectodermal dysplasia and immunodeficiency)	*IKBKG* (NEMO, inhibitor of κB kinase γ)	Impaired activation of NFκB	Chronic bacterial and viral infections; developmental defects	3-3, p. 53
Chronic granulomatous disease	*NOX1* (NADPH oxidase)	Impaired neutrophil function	Chronic bacterial and fungal infections	3-9, p. 61
MBL deficiency	*MBL* (mannose-binding lectin)	Lack of mannose-binding lectin	Meningitis due to *Neisseria meningitidis*	3-11, p. 65
NK-cell deficiency	*GATA2*	Absence of NK cells	Herpesvirus infections	3-17, p. 72
Hyper-IgM deficiency	Activation-induced cytidine deaminase (*AICDA*), or CD40 ligand (*CD40LG*), or *CD40*, or *IKBKG*	No isotype switching or somatic hypermutation in B cells	Extracellular bacterial and fungal infections	4-15, p. 102 9-10, p. 244
IgG2 deficiency	Not known	Lack of IgG2	Encapsulated bacteria	4-17, p. 106
SCID (see Figure 13.16 for the different causes of SCID)	Various	Absence of B-cell and T-cell function	All types of infection	5-2, p. 115
Omenn syndrome	*RAG1* or *RAG2* or *Artemis*	Impaired V(D)J recombination	All types of infection	5-2, p. 116
MHC class I deficiency (bare lymphocyte syndrome type I)	*TAP1* or *TAP2*	Low MHC class I expression	Respiratory virus infections	5-11, p. 127
Pre-B-cell receptor deficiency	*IGLL1* (λ5)	Lack of B cells and antibodies	Persistent bacterial infections	6-4, p. 154
X-linked agammaglobulinemia	*BTK* (Bruton's tyrosine kinase)	B cells blocked at pro-B-cell stage	Recurrent bacterial infections	6-8, p. 160
Complete DiGeorge syndrome	Not known	Absence of the thymus and T cells	All types of infection	7-1, p. 178
APECED (autoimmune poly-endocrinopathy–candidiasis–ectodermal dystrophy)	*AIRE* (autoimmune regulator)	Reduced T-cell tolerance to self antigens	Autoimmune diseases	7-12, p. 193
IPEX (immunodysregulation, poly-endocrinopathy, enteropathy, X-linked syndrome)	*FOXP3*	Lack of regulatory T cells and peripheral tolerance	Autoimmune diseases	16-2, p. 478
ZAP-70 deficiency	*ZAP70*	T cells that cannot signal through their antigen receptors	All types of infection	8-6, p. 208
C4A or C4B	*C4A, C4B*	Impaired clearance of immune complexes	Autoimmune disease and infections	9-18, p. 256
Selective IgA deficiency	Not known	Lack of IgA	No major susceptibility	10-15, p. 286

Figure 13.10 Mechanisms of human immunity are revealed by the study of inherited immunodeficiency syndromes. This figure shows the immunodeficiency syndromes mentioned previously in this book (with references to the relevant sections), their gene defects, and the effects they have on the immune system.

easier; now the more challenging task is for doctors in the field to recognize a novel form of immunodeficiency syndrome when they see it. International collaborations help in the identification and treatment of these patients.

Whereas most of the primary immunodeficiencies listed in Figure 13.10 were discovered in patients with severe disease and are caused by exceedingly rare mutant alleles, defects in other immune-system genes are more frequent and have less dramatic effects. Examples of the latter are defective MHC class I alleles, or the lack of an A or a B isotype of complement component C4 (see Section 9-18), which confers either an increased susceptibility to the autoimmune disease systemic lupus erythematosus (lack of A) or a somewhat heightened vulnerability to infection (lack of B). An extreme example is the gene-content variation of haplotypes encoding the KIR gene family of NK-cell receptors (see Section 12-7), which means that most people lack at least one KIR and its functions. Usually, those immune-system genes that can be lost with no drastic effect on immune function are members of multigene families in which another family member can compensate, to some extent, for the defective gene. In some cases this type of variability may represent a compromise, in that there are costs and benefits associated with having or lacking a particular gene.

13-9 Inherited immunodeficiency diseases are caused by dominant, recessive, or X-linked gene defects

Primary immunodeficiency diseases are of three types: dominant, recessive, or X-linked. Syndromes due to a **dominant** defective allele show up in children who inherit a normal, functional allele from one parent and the defective allele from the other parent. Disease occurs because the abnormal properties of the defective allele interfere with and dominate over the functions provided by the normal allele. In contrast, a disease caused by a **recessive** allele is only manifest in patients who inherit the defective allele from both parents. Individuals who have one defective allele and one normal allele are healthy and are called **carriers** of the disease trait. In recessive diseases, the defective allele does not interfere with the function of the normal allele. A key difference between dominant and recessive disease traits lies in the fate of heterozygous individuals: for a dominant trait they get the disease, for a recessive trait they do not.

X-linked diseases are caused by recessive defects in genes on the X chromosome. Because males have only one X chromosome and females have two, the disease occurs in all males who inherit an X chromosome with a defective allele, but it will not show up in their sisters even if they inherit the same X chromosome. Disease occurs in females only when they inherit a defective X chromosome from both parents. X-linked diseases, of which we have already met three (see Figure 13.9), are therefore far more frequent in boys than girls. For these traits only women serve as healthy carriers. Any disease caused by a dominant allele on an X chromosome would occur at equal frequency in boys and girls.

Dominance is most commonly seen when the defective gene encodes a protein that functions in a dimer or a larger protein complex. In such cases, the incorporation of one defective subunit can reduce or destroy the capacity of the complex to function. Before the 1950s, any dominant trait causing a severe immunodeficiency would probably have been eliminated from the population with the death of the child in whom the mutation first occurred. Thus, most of the severe inherited immunodeficiency syndromes that have been identified are due to recessive mutations in single genes. The known immunodeficiencies that are due to dominant mutations tend to be less severe and are caused by reduction in a function rather than its loss.

13-10 Recessive and dominant mutations in the IFN-γ receptor cause diseases of differing severity

IFN-γ, the major cytokine that activates macrophages, is made by NK cells during the innate immune response and by T_H1 CD4 T cells and CD8 cytotoxic T cells during the adaptive immune response. When IFN-γ binds to IFN-γ receptors on a macrophage surface, the cell is induced to make changes in gene expression and become better at engulfing and killing bacteria (see Section 8-18). The IFN-γ receptor is a dimer of two polypeptides, IFNγR1 and IFNγR2, which both associate with tyrosine kinases—Jak1 and Jak2, respectively. Functional IFN-γ is also a dimer, and binding of the dimeric cytokine to sites on the IFNγR1 polypeptide cross-links two molecules of the receptor to initiate the signaling cascade (Figure 13.11, first panel).

The response of macrophages to IFN-γ is crucial for defense against intravesicular bacteria, such as mycobacteria, and both dominant and recessive mutations in IFNγR1 have been identified in patients suffering from persistent mycobacterial infections (Figure 13.11, second and third panels). Both types of mutation cause the condition of **IFN-γ receptor deficiency**. The recessive alleles contain mutations that prevent any expression of IFNγR1 at the cell surface. The macrophages and monocytes of patients with two recessive alleles carry only IFNγR2 at their surfaces and are unresponsive to IFN-γ (see Figure 13.11, second and fourth panels). For this group of patients the disease is usually more severe and appears at an earlier age. Heterozygotes are healthy because the protein made from the defective allele does not interfere with that made from the normal allele, which assembles with IFNγR2 and moves to the cell surface as functional IFN-γ receptor (see Figure 13.11, first panel).

In the dominant mutants, IFNγR1 is truncated such that much of the cytoplasmic tail, which binds Jak1 and initiates signaling, is missing. The truncated IFNγR1 associates with IFNγR2 protein and is taken to the surface as a receptor that binds IFN-γ but cannot transduce a signal. At the cell surface, these

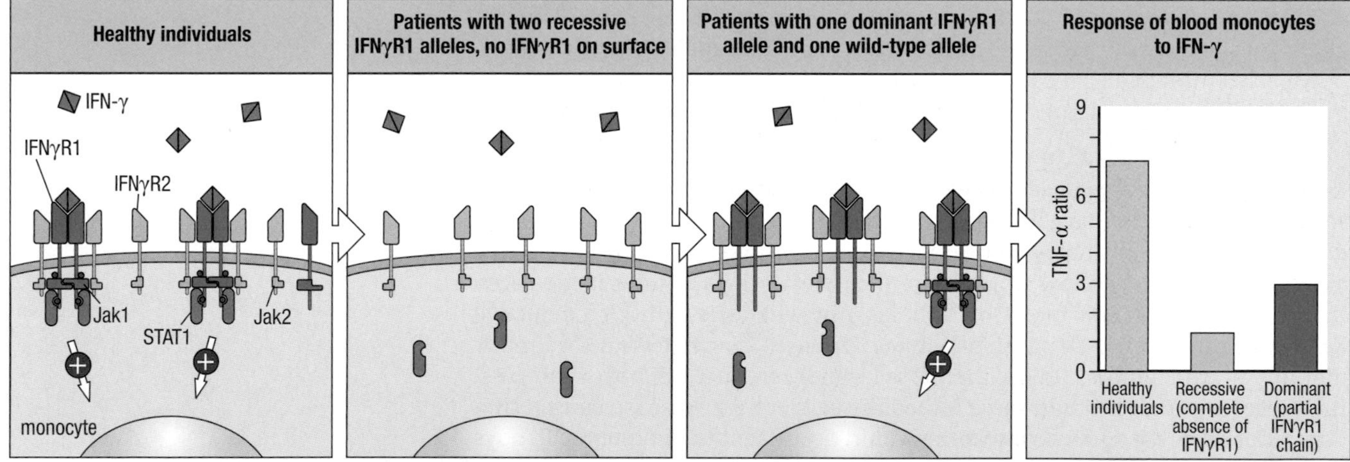

Figure 13.11 The impact of recessive and dominant mutations in the IFN-γ receptor on monocyte activation. Receptors for IFN-γ are composed of a dimer of IFNγR1 and IFNγR2. Two such dimers must be cross-linked by IFN-γ binding to the IFNγR1 chain for signaling to be triggered (first panel). Recessive mutant alleles of IFNγR1 produce a mutant chain that does not reach the surface. Thus, cells from patients homozygous for a recessive mutation have only IFNγR2 at the surface, lack IFNγR1 function, and cannot respond to IFN-γ (second panel). Heterozygotes for such a mutation produce sufficient numbers of wild-type chains to assemble enough functional receptors for a normal response to IFN-γ, as in the first panel. Dominant mutant alleles of IFNγR1 produce a mutant chain lacking a signaling domain. This chain can assemble into a dimer and bind IFN-γ but cannot signal (third panel). Heterozygotes for a dominant mutation make a small number of functional receptors composed entirely of wild-type chains, but most receptors are nonfunctional. Thus their response to IFN-γ is defective (third panel). The fourth panel compares the results of IFN-γ stimulation of blood monocytes from normal, homozygous recessive, and heterozygous dominant patients.

defective receptors compete for IFN-γ with the normal receptors that incorporate IFNγR1 made from the normal allele (see Figure 13.11, third panel). This competition is further weighted against the functional receptors because the absence of the cytoplasmic domain from IFNγR1 prevents the mutant receptor from being recycled by endocytosis. It therefore accumulates at the cell surface, at levels fivefold higher than the normal receptor. Because of the interference by the mutant receptors, the response of patients' macrophages and monocytes to IFN-γ is much reduced compared with healthy people, but it is greater than in patients carrying two recessive alleles (see Figure 13.11, fourth panel). Because of this difference, dominant mutants cause a less severe immunodeficiency, which tends to be detected at a later age.

13-11 Antibody deficiency leads to poor clearing of extracellular bacteria

The major threat to patients lacking antibodies is infection by pyogenic (pus forming) bacteria. These encapsulated bacteria, which include *Haemophilus influenzae*, *Streptococcus pneumoniae*, *Streptococcus pyogenes*, and *Staphylococcus aureus*, are not recognized by the phagocytic receptors of macrophages and neutrophils, so they frequently escape immediate elimination by the innate immune response. Such infections are normally cleared when the bacteria are opsonized by specific antibody and complement and then taken up and killed by phagocytes. For patients lacking antibodies, infections with pyogenic bacteria tend to persist unless treated with antibiotics.

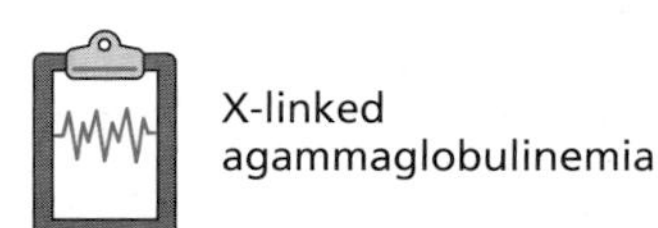

The first immunodeficiency disease to be described was characterized by antibody deficiency and X-linked inheritance and is named **X-linked agammaglobulinemia** (**XLA**). The defect in XLA is in a protein tyrosine kinase called Bruton's tyrosine kinase (BTK) to honor the discoverer of the syndrome. BTK contributes to intracellular signaling from the B-cell receptor and is necessary for the development and differentiation of pre-B cells (see Figure 6.12, p. 159). Males inheriting an X chromosome with a mutant allele of the *BTK* gene are unable to produce functional B cells. Although BTK is also expressed in monocytes and T cells, these cells are not obviously compromised by its absence in patients with XLA.

Women with one functional and one nonfunctional copy of the *BTK* gene are healthy, but they pass XLA on to half of their male children. In all females, one X chromosome is randomly inactivated in every cell. For all women, each X chromosome is inactivated in 50% of the pre-B cells. In women who are not XLA carriers both pools of pre-B cells develop to give a population of mature B cells in which each X chromosome is inactivated in 50% of cells. In contrast, for carriers of XLA, the only pre-B cells that can develop are those that express the functional *BTK* allele and inactivate the mutant allele. Consequently, all the mature B cells in XLA carriers have inactivated the same X chromosome. By using genetic markers that distinguish between the two X chromosomes, the women in families with a history of XLA, or of any other X-linked syndrome, can be typed as carriers or non-carriers of the genetic disease (Figure 13.12).

Patients who have immunodeficiencies confined to B-cell functions are able to resist many pathogens successfully, and those to which they are susceptible can be treated with antibiotics. Although pyogenic infections can be cured in this way, the successive rounds of infection and treatment sometimes lead to permanent tissue damage caused by the excessive release of proteases from both the infecting bacteria and the defending phagocytes. These effects are particularly pronounced in the airways, where the bronchi lose their elasticity and become sites of chronic inflammation. This condition, called **bronchiectasis**, can lead to chronic lung disease and eventual death. To prevent such developments, XLA patients are given monthly injections of intravenous

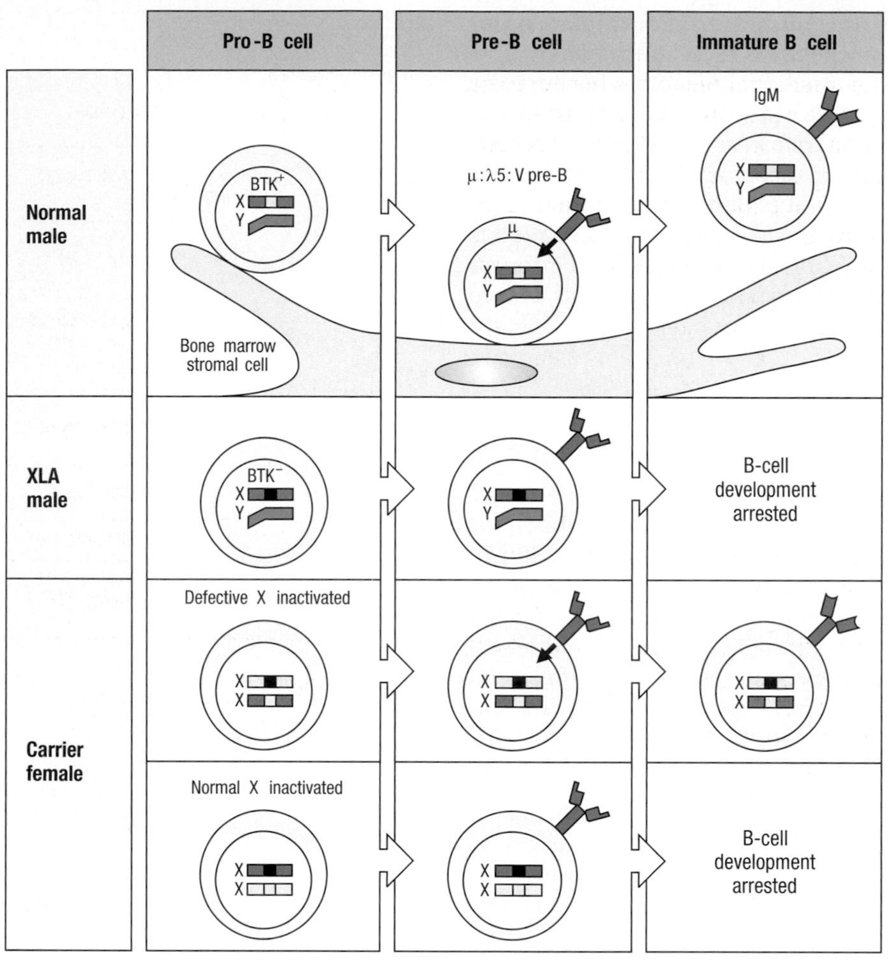

Figure 13.12 In patients with X-linked agammaglobulinemia (XLA), B cells do not develop beyond the pre-B cell stage. In XLA, Bruton's tyrosine kinase is defective. In patients with this disease, B cells become arrested at the pre-B-cell stage because intracellular signals cannot be generated by the pre-B-cell receptor. Most patients with XLA are males, because males have only one copy of the X chromosome. Heterozygous females are carriers of the disease trait although healthy themselves. During development, female cells randomly inactivate one of their X chromosomes. Consequently, half of the developing B cells in a female carrier become arrested at the pre-B-cell stage because they have inactivated the X chromosome that carries the good copy of the *BTK* gene; the other half develop to become functional B cells because they have inactivated the X chromosome that carries the bad copy of *BTK* and thus use the X chromosome having the good copy.

immunoglobulin (see page 90). Intravenous immunoglobulin contains antibodies against common pathogens and provides what is termed **passive immunity** against those pathogens.

13-12 Diminished production of antibodies also results from inherited defects in T-cell help

Diminished production of antibodies is also a symptom of defective genes encoding the membrane-associated cytokine CD40 ligand. As discussed in Chapter 9, interaction of CD40 ligand on activated T cells with B-cell CD40 is a crucial part of the T-cell help given to B cells. This stimulates B-cell activation, the development of germinal centers, and isotype switching. CD40 ligand is encoded on the X chromosome, so most patients with a hereditary deficiency in CD40 ligand are males. In the absence of CD40 ligand, virtually no specific antibody is made against T-cell dependent antigens and there are no germinal centers in the secondary lymphoid tissues (see Figure 9.11, p. 240). The blood of these patients has extremely low amounts of IgG, IgA, and IgE combined with abnormally high amounts of IgM. This latter characteristic led to the condition being named **X-linked hyper-IgM syndrome**. Patients with this immunodeficiency are inherently susceptible to infection with pyogenic bacteria. As for XLA patients (see Section 13-11), regular infusions of intravenous immunoglobulin help to prevent infections, and antibiotics are used to treat their infections.

Macrophage activation by effector CD4 T cells also depends on the interaction of CD40 on the macrophage with CD40 ligand on the T cell (see Section 8-18).

In patients with X-linked hyper-IgM syndrome this interaction does not occur, which impairs the macrophage production of granulocyte–macrophage colony-stimulating factor (GM-CSF), a cytokine that stimulates the development of neutrophils in the bone marrow and their release into the circulation. In immunocompetent individuals, the immune response to infection increases the number of white blood cells in the blood, a state called **leukocytosis**, but this does not occur in patients lacking CD40 ligand. On the contrary, their blood can become profoundly deficient in neutrophils, a state called **neutropenia** that often leads to severe sores and blisters in the mouth and throat. Being always infested with bacteria, the integrity of these mucosal tissues depends on perpetual surveillance of their microbiota by neutrophils. The sores and blisters experienced by patients with X-linked hyper-IgM syndrome can be cured by intravenous administration of GM-CSF.

Complement protein	Effects of deficiency
C1, C2, C4	Immune-complex disease
C3	Susceptibility to encapsulated bacteria
C5–C9	Susceptibility to *Neisseria*
Factor D, properdin (factor P)	Susceptibility to encapsulated bacteria and *Neisseria* but no immune-complex disease
Factor I	Similar effects to deficiency of C3
DAF, CD59	Autoimmune-like conditions including paroxysmal nocturnal hemoglobinuria
C1INH	Hereditary angioedema (HAE)

Figure 13.13 Diseases caused by deficiencies in the pathways of complement activation.

13-13 Complement defects impair antibody-mediated immunity and cause immune-complex disease

All the effector functions that antibodies recruit to clear pathogens and their antigens are facilitated by complement activation. Consequently, the spectra of infections associated with deficiencies in complement and antibody production overlap substantially. Defects in the activation of C3, and in C3 itself, result in susceptibility to a wide range of pyogenic infections, emphasizing the importance of C3 as an opsonin that promotes the efficient elimination of bacteria by phagocytes. In contrast, defects in C5–C9, the terminal complement components of the membrane-attack complex, have few effects. Of these susceptibility to *Neisseria* is the best example. The most effective defense against *Neisseria* is complement-mediated lysis of extracellular bacteria, and this requires all the components of the complement pathway. Figure 13.13 lists the effects of deficiency in complement components and complement inhibitory proteins.

The early components of the classical pathway are necessary for the elimination of immune complexes. As discussed in Section 9-20, the attachment of complement components to soluble immune complexes allows them to be transported, or ingested and degraded, by cells bearing complement receptors. Immune complexes are mainly transported by erythrocytes, which capture the complexes with their CR1 complement receptor that binds to C4b and C3b. Deficiencies in complement components C1–C4 impair the formation of C4b and C3b and lead to the accumulation of immune complexes in the blood, lymph, and extracellular fluid and to their deposition within tissues. In addition to directly damaging the tissues in which they deposit, immune complexes activate phagocytes, causing inflammation and further tissue damage. This condition is called **immune-complex** disease.

Deficiencies in the proteins that control complement activation can also have major effects. People deficient for factor I in effect lack C3. Because factor I is absent, the conversion of C3 to C3b runs unchecked, and supplies of C3 are rapidly depleted (see Section 2-4). Patients who lack properdin (factor P), a plasma protein that enhances the activity of the alternative pathway by stabilizing the C3 convertase, have a heightened susceptibility to *Neisseria*, because reduced C3 deposition impedes formation of the membrane-attack complex, the machinery used to kill the bacteria. In contrast, a deficiency in decay-accelerating factor (DAF) or CD59 causes an autoimmune-like condition. Lacking the protection conferred by DAF or CD59, the cells of these patients activate the alternative pathway of complement. The resultant complement-mediated lysis of erythrocytes causes the disease paroxysmal nocturnal hemoglobinuria (see Section 2-7).

Hereditary angioedema

Hereditary angioedema (**HAE**) is an autosomal dominant disease caused by a deficiency of the complement regulator **C1 inhibitor** (**C1INH**). The disease is

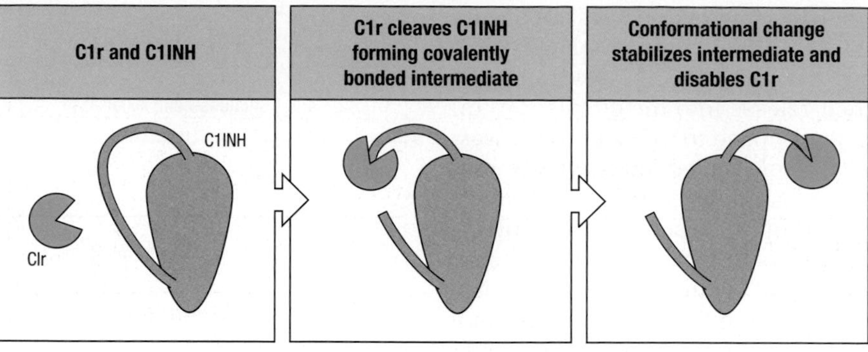
C1r and C1INH
C1r cleaves C1INH forming covalently bonded intermediate
Conformational change stabilizes intermediate and disables C1r
C1INH
Clr

Figure 13.14 C1INH permanently inhibits C1r and C1s. The inactivation of C1r is shown here. C1s is similarly inhibited. C1INH is a member of the serpin family of protease inhibitors. They act as pseudosubstrates. When they are cleaved by the protease, they form a permanent covalent linkage to the protease, which prevents it from releasing the pseudosubstrate and cleaving again. C1INH deficiency causes the syndrome hereditary angioedema.

characterized by bouts of subepithelial swelling of the face, larynx, and abdomen. The swelling around the larynx can lead to death by suffocation. C1 inhibitor affects serine proteases such as C1r and C1s by binding to the active site and forming a covalent bond that irreversibly inactivates the enzyme. In patients deficient in C1 inhibitor the classical pathway is overactive, causing abnormally low levels of C2 and C4 in the blood and unusually high production of the vasoactive C2a fragment. As well as participating in the regulation of complement activation, C1INH controls serine proteases involved in blood clotting. In patients with HAE, the blood-clotting pathways are also overactive, giving abnormally high levels of the vasoactive peptide bradykinin. The combined actions of C2a and bradykinin cause fluid to leak out of the blood into the tissues, causing the edema that characterizes HAE.

C1 inhibitor is a member of a large family of serine and cysteine protease inhibitors called **serpins**. By acting as a pseudosubstrate, each serpin molecule poisons the active site of a protease molecule (Figure 13.14). The genetic dominance of HAE does not arise from the C1 inhibitor being part of a multisubunit complex, as is the case for IFN-γ receptor deficiency (see Section 13-10). It occurs because one good copy of the *C1INH* gene cannot make sufficient inhibitor to control the complement and clotting cascades. HAE is treated with infusions of recombinant C1INH protein, some of which is purified from the milk of transgenic rabbits expressing the human *C1INH* gene.

13-14 Defects in phagocytes result in enhanced susceptibility to bacterial infection

Phagocytosis mediated by macrophages and neutrophils is the principal method by which the immune system gets rid of infecting bacteria and other microbes. Any deficiency that compromises phagocyte activity has profound effects on the immune system's capacity to clear infection (Figure 13.15). One such syndrome is due to mutation of the *CD18* gene and is known as **leukocyte adhesion deficiency**. CD18 is the β_2 subunit of the leukocyte integrins, which comprise the CR3, CR4, and LFA-1 adhesion molecules that are necessary for neutrophils and monocytes to leave the blood and enter sites of infection (see Section 3-8). In children with leukocyte adhesion deficiency, these cells are unable to enter infected tissues where they are needed to dispose of the infecting pathogen. Because CR3 and CR4 also function as complement receptors, a second defect is that the phagocytes are unable to engulf bacteria opsonized with complement (see Section 3-9). Children with leukocyte adhesion deficiency have recurrent and persistent infections with pyogenic bacteria. These respond poorly to antibiotics and are not cleared by seemingly normal B-cell and T-cell responses. Children with leukocyte adhesion deficiency suffer successive infections and usually succumb during the first 2 years of life.

Leukocyte adhesion deficiency

Defects in phagocyte function			
Syndrome	**Defective gene/protein**	**Functional effect**	**Clinical effect**
Leukocyte adhesion deficiency (LAD)	CD18 subunit of CR3, CR4, and LFA-1 adhesion molecules	Defective migration of monocytes and neutrophils to infected tissues Defective uptake of opsonized pathogens	Widespread infection with encapsulated bacteria
Chronic granulomatous disease (CGD)	Nicotinamide adenine dinucleotide phosphate (NADPH) oxidase	Defective respiratory burst Phagocytes unable to kill pathogens	Chronic bacterial and fungal infections, Granulomas
Glucose-6-phosphate dehydrogenase (G6PD) deficiency	Glucose-6-phosphate dehydrogenase		Chronic bacterial and fungal infections Some infections induce anemia
Myeloperoxidase deficiency (MPOD)	Myeloperoxidase		Chronic bacterial and fungal infections
Chédiak–Higashi syndrome (CHS)	Lysosomal trafficking regulator protein	Defective fusion of endosomes and lysosomes Defective phagocytosis	Recurrent and persistent bacterial infections, Granulomas Damaging effects to many organs

Figure 13.15 Genetic defects that affect phagocytes cause persistent bacterial infections.

The capacity of phagocytes to kill ingested bacteria can also be blunted by a single defective gene. In **chronic granulomatous disease (CGD)**, the antibacterial activity of phagocytes is compromised by their inability to produce the superoxide radical O_2^- (see Section 3-9). Mutations affecting any of the four proteins of the NADPH oxidase system can produce the phenotype. Patients with this disease suffer from chronic bacterial infections, often leading to granuloma formation. Deficiencies in the enzymes glucose-6-phosphate dehydrogenase and myeloperoxidase also impair intracellular bacterial killing, leading to a similar but less severe phenotype. A different phenotype characterizes **Chédiak–Higashi syndrome**, in which phagocytosed materials are not delivered to lysosomes because of a defect in the vesicle fusion mechanism. This lack of phagocyte function has effects in many different organs as well as leading to persistent and recurrent bacterial infections. The mutations causing this disease are in the *CHS1* gene on chromosome 1, which encodes the lysosomal trafficking protein that is critical for lysosome function.

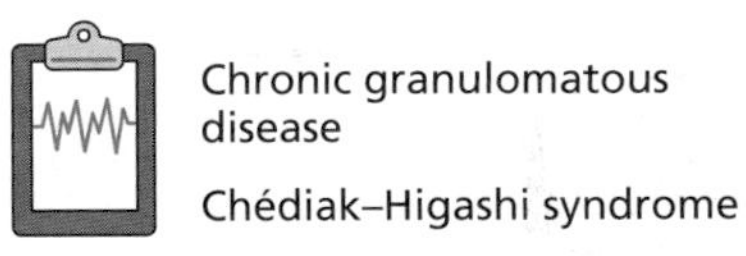

13-15 Defects in T-cell function result in severe combined immune deficiencies

Whereas B cells contribute only to the antibody response, T cells function in all aspects of adaptive immunity. This means that inherited defects in the mechanisms of T-cell development and T-cell function have a general depressive effect on the immune system's capacity to respond to infection. Patients with T-cell deficiencies tend to be susceptible to persistent or recurrent infections with a broader range of pathogens than patients with B-cell deficiencies (Figure 13.16). Those patients who make neither T-cell-dependent antibody responses nor cell-mediated immune responses are said to have **severe combined immune deficiency (SCID)**.

X-linked severe combined immunodeficiency

Adenosine deaminase deficiency

T-cell development and function depend on the action of many proteins, so the SCID phenotype can arise from defects in any one of a number of genes. Deficiency of **adenosine deaminase (ADA)** or **purine nucleoside phosphorylase (PNP)**, which are enzymes involved in purine degradation, account for 15% of SCID patients. The absence of these enzymes causes an accumulation of nucleotide metabolites in all types of human cell, but the effects are particularly toxic to developing T cells and, to much lesser extent, to developing B cells. Infants with these immunodeficiencies have an underdeveloped

Severe combined immunodeficiency syndromes and related conditions			
Syndrome	**Defective gene/protein**	**Functional effect**	**Clinical effect**
Adenosine deaminase deficiency	ADA/adenosine deaminase	Build-up of toxic metabolites, resulting in death of T cells and B cells	Extreme susceptibility to infections of all types, including opportunistic infections (SCID), which is lethal in infancy if not treated
Radiation-sensitive SCID (see Section 5-2)	RAG1, RAG2, and other proteins involved in V(D)J recombination	Failure of V(D)J recombination Non-production of T cells and B cells	
X-linked SCID	IL2RG/common γ chain	Defects in cytokine signaling, resulting in the non-development of T cells and NK cells	
Janus 3 kinase (Jak3) deficiency	JAK3		
Omenn syndrome (see Section 5-2)	RAG1, RAG2 (80% loss of function), and other proteins involved in V(D)J recombination	Deficient V(D)J recombination Non-production of T cells and B cells	SCID, chronic inflammation
Complete DiGeorge syndrome (see Section 7-1)	Associated with microdeletions of 22q11.2 region on chromosome 22	Defective thymus development and T-cell production	Cardiac defects, SCID
Wiskott–Aldrich syndrome (WAS)	WASP/WAS protein	Defective cytoskeletal function impairs lymphocyte function and interaction	Antibody deficiency, increased susceptibility to infections; thrombocytopenia with small platelets
MHC class II deficiency	Transcriptional activators acting at elements common to all HLA class II genes	Non-production of all HLA class II molecules, resulting in defective CD4 T-cell development in the thymus	Increased susceptibility to pyogenic and opportunistic infections
MHC class I deficiency	TAP1, TAP2	Impaired production of HLA class I molecules, resulting in failure of CD8 T cells to develop in the thymus	Repeated respiratory viral infections

Figure 13.16 SCID and other severe immunodeficiencies caused by absence of T-cell function.

thymus that contains few lymphocytes. ADA and PNP deficiencies are autosomally inherited (Figure 13.17).

Because of the unique inheritance pattern of the X chromosome, X-linked diseases are more easily discovered, and at least two forms of SCID are of this type. The one we consider here is due to mutation in the X-linked gene that encodes the **common gamma chain** (γ_c), a shared signaling component of the cell-surface receptors for IL-2, IL-4, IL-7, IL-9, and IL-15. (This γ chain of cytokine receptors is not the same as the γ chain associated with Fc receptors). When one of these cytokines binds to its receptor, γ_c interacts with the protein kinase Jak3 to produce intracellular signals (see Figure 8.22, p. 221). In the absence of a functional γ chain, none of the five cytokines can induce receptor signaling and so, predictably, the result is SCID. A very similar, but autosomally inherited, immunodeficiency occurs in patients who lack Jak3. The phenotype of SCID is so severe that affected infants survive only if their immune system is replaced by hematopoietic cell transplantation.

Another X-linked deficiency of T-cell function is **Wiskott–Aldrich syndrome** (**WAS**), a syndrome involving the impairment of platelets as well as lymphocytes. It shows up in childhood as a history of recurrent infections but is less immunologically severe than SCID. The patients have normal levels of T and B cells but they cannot make good antibody responses and are therefore kept on a course of intravenous immunoglobulin. The relevant gene on the X chromosome encodes the Wiskott–Aldrich syndrome protein (WASP). This protein is involved in the cytoskeletal reorganization that is needed before T cells can deliver cytokines and signals to the B cells, macrophages, and other target cells with which they form cognate interactions during their development and participation in the immune response (see Figure 8.26, p. 224).

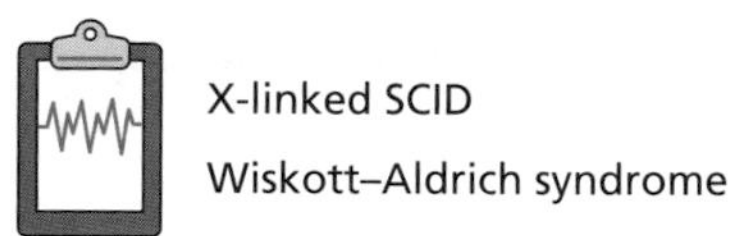

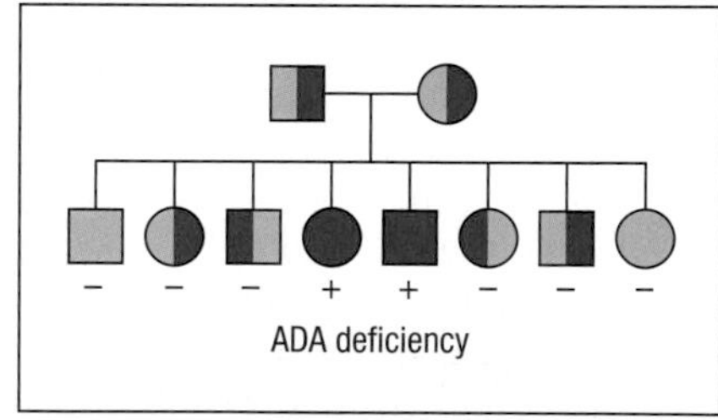

Figure 13.17 Inheritance of adenosine deaminase (ADA) deficiency in a family. Both parents are healthy carriers having one functional (red) and one defective (green) copy of the *ADA* gene. Two of the eight children inherited a defective copy of the *ADA* gene from each parent and have ADA deficiency (+). Males are indicated by squares and females by circles.

Lack of HLA class II molecules also causes a serious immunodeficiency. The deficiency was originally named 'bare lymphocyte syndrome' because the defect was first discovered on B lymphocytes, the major population of peripheral blood cells that expresses HLA class II, but is now more often called **MHC class II deficiency**. In these patients, CD4 T cells fail to develop (see Section 7-10), which compromises most aspects of adaptive immunity. MHC class II deficiency arises from defects in transcriptional regulators that are essential for the expression of all the HLA class II genes. A homozygous defect in any one of four proteins produces the condition. One protein is the class II transactivator (CIITA), the other three are components of RFX, a transcriptional complex that binds to a conserved sequence in the promoter of HLA class II genes called the X box.

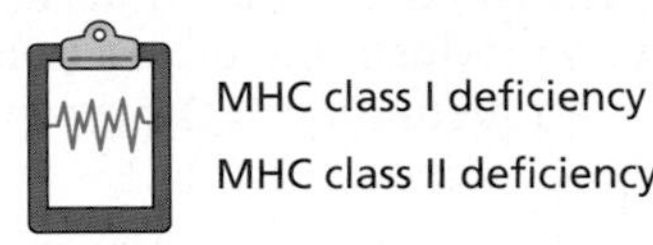

MHC class I deficiency

MHC class II deficiency

A defect in either of the two genes encoding the TAP peptide transporter (see Section 5-11) impedes the binding of peptides by HLA class I molecules, leading to an unusually low abundance of HLA class I molecules on cell surfaces. This form of immunodeficiency, called **MHC class I deficiency**, is less severe than the immunodeficiency caused by the absence of HLA class II, its principal effect being the selective loss of CD8 T cells (see Section 7-10) and of cytotoxic T-cell responses to intracellular infections.

Defects in various proteins and enzymes that contribute to the rearrangement of immunoglobulin and T-cell receptor genes cause autosomally inherited forms of SCID or a related immunodeficiency called Omenn syndrome, depending on the particular defect (see Section 5-2). These include the RAG-1 and RAG-2 proteins, the nuclease Artemis, and the DNA-dependent protein kinase (DNA-PK).

13-16 Some inherited immunodeficiencies lead to specific disease susceptibilities

Patients who lack the IFN-γ receptor suffer persistent and sometimes fatal infections of common intracellular bacteria, such as the ubiquitous nontuberculous strain of mycobacterium, *Mycobacterium avium* (see Section 13-10). **IL-12 receptor deficiency**, in which the receptor for the cytokine IL-12 is nonfunctional, produces a similar susceptibility to intracellular bacterial infections. In the innate immune response there is a mutual activation of NK cells and macrophages (Figure 13.18, left panel). This involves the IL-12 secreted by

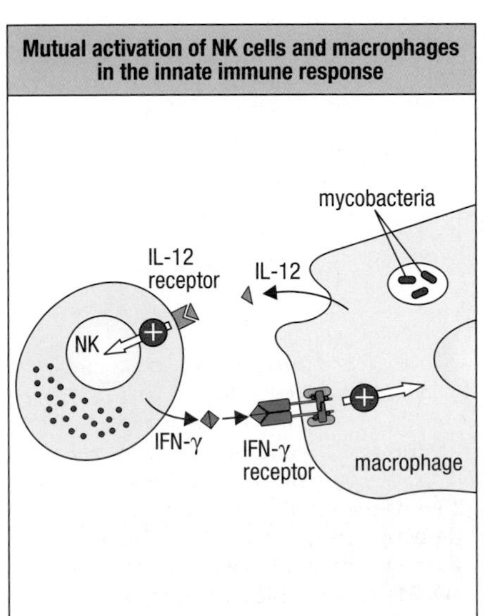

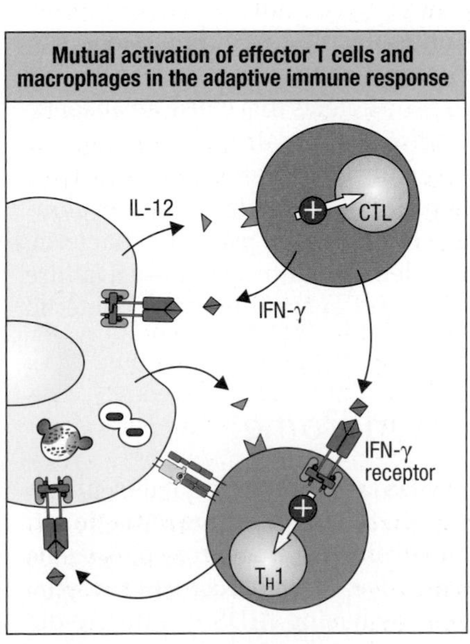

Figure 13.18 Mutual activation of macrophages and effector lymphocytes in the immune response to intracellular bacterial infections. In the innate immune response (left panel), macrophages are activated by the IFN-γ made by NK cells and in turn produce the cytokine IL-12. This binds to IL-12 receptors on the NK cells, inducing further secretion of IFN-γ and maintenance of macrophage activation. In the adaptive immune response (right panel), IL-12 secreted by macrophages acts on T_H1 cells, inducing their differentiation into IFN-γ-secreting T_H1 cells. CD8 cytotoxic T cells (CTLs) also respond to the IL-12 made by the macrophage and they too produce IFN-γ. In immunodeficient patients lacking the IL-12 receptor or the IFN-γ receptor, this cycle of mutual activation cannot proceed, so infection persists.

macrophages binding to the IL-12 receptor of NK cells and stimulating them to secrete IFN-γ. The IFN-γ then binds to the IFN-γ receptor on macrophages and activates phagocytosis and the secretion of proinflammatory cytokines. In the absence of a functioning IL-12 receptor, this cycle of mutual reinforcement cannot begin.

In the adaptive immune response, IL-12 secreted by macrophages binds to the IL-12 receptors of T cells, helping to induce the differentiation of T_H1 cells from activated antigen-specific naive CD4 T cells (see Section 8-10). On interacting with antigen on the macrophage surface, T_H1 cells secrete IFN-γ, which acts on the macrophage to strengthen its activation, and thus leads to the destruction of the intracellular bacteria (Figure 13.18, right panel). IL-12 also acts on cytotoxic T cells, inducing them to also produce IFN-γ and maintain both macrophage activation (see Section 8-16) and an environment favoring the differentiation of T_H1 cells. So, again, in the absence of a functioning IL-12 receptor this mutual activation of macrophage and effector T cells cannot get started. Unable to make an effective innate or adaptive immune response to intracellular bacteria, people lacking the IL-12 receptor suffer persistent infections with common strains of mycobacteria. When undiagnosed immunodeficient individuals, lacking either the IL-12 receptor or IFN-γ receptor, were vaccinated against tuberculosis with the Calmette–Guérin vaccine strain of live *Mycobacterium bovis*, the vaccine caused disseminated infection and disease. Even this weak and normally nonpathogenic strain of mycobacterium could not be controlled without functional IL-12 and IFN-γ receptors.

As we saw in Section 13-4, many healthy people maintain a persistent EBV infection of B cells, which is held in check by NK cells and EBV-specific T cells. For patients with a defect in the X-linked gene called *SH2D1A*, this balance is never achieved and childhood EBV infections can be severe and even progress to lymphoma. This immunodeficiency, which affects mostly boys, is called **X-linked lymphoproliferative syndrome** because it involves an ineffective proliferation of NK cells and T cells. Although the SH2D1A protein is believed to be a regulator of lymphocyte-activating signals, its precise functions and contribution to the control of EBV infection are not yet clear.

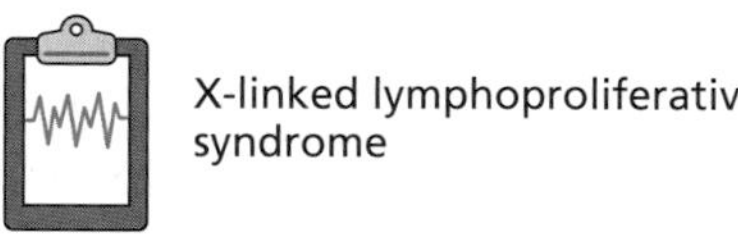
X-linked lymphoproliferative syndrome

Summary

The best-characterized gene defects affecting the immune system are those that show up in early childhood and confer exceptional vulnerability to common infections. The characterization of immunodeficiency diseases and the gene defects that cause them is almost the only way in humans of determining the relative importance of different cells and molecules in immune defenses, and of testing current models of how the human immune system works. The most severe immunodeficiencies are due to gene defects that cause an absence of all T-cell function and thus, directly or indirectly, impair B-cell function as well. Such deficiencies are known as severe combined immune deficiencies (SCID). The absence of antibodies due to genetic defects in B-cell development or function leads to particular susceptibility to pyogenic bacteria. Deficiencies in the early components of complement pathways cause a failure to opsonize pathogens. This results in increased susceptibility to bacterial infection, as do defects in phagocytes.

Acquired immune deficiency syndrome

Acquired immune deficiency syndrome (**AIDS**) was first described by physicians early in the 1980s. The disease is characterized by a massive reduction in the number of CD4 T cells, accompanied by severe infections from pathogens that rarely trouble healthy people, or by aggressive forms of Kaposi's sarcoma or B-cell lymphoma. All patients diagnosed as having AIDS eventually die

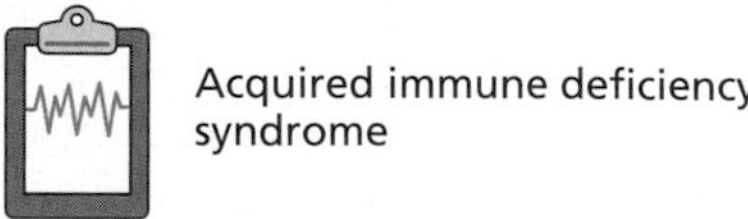
Acquired immune deficiency syndrome

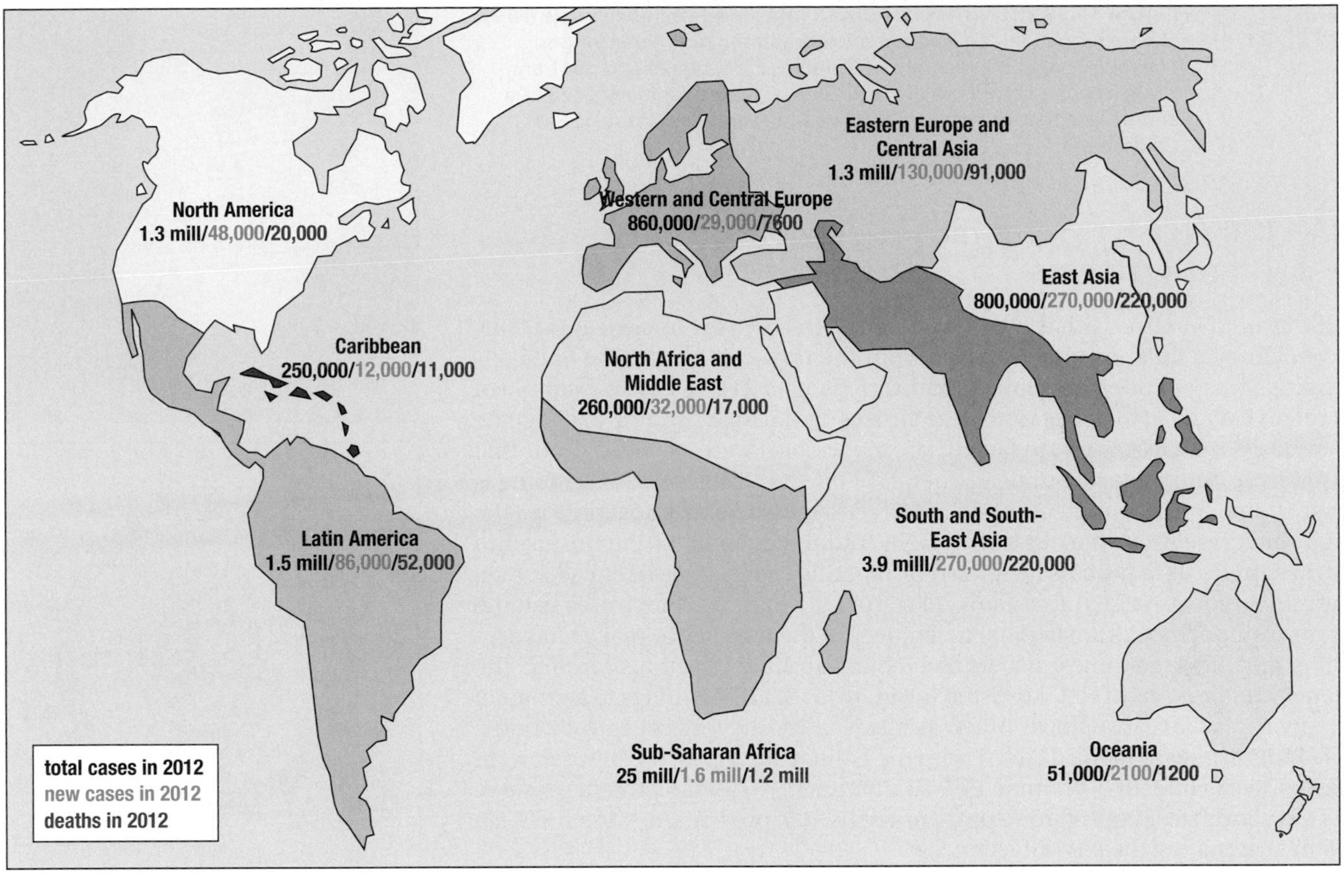

Eastern Europe and
Central Asia
1.3 mill/130,000/91,000
North America
1.3 mill/48,000/20,000
Western and Central Europe
860,000/29,000/7600
East Asia
800,000/270,000/220,000
Caribbean
250,000/12,000/11,000
North Africa and
Middle East
260,000/32,000/17,000
South and South-
East Asia
3.9 milll/270,000/220,000
Latin America
1.5 mill/86,000/52,000
Sub-Saharan Africa
25 mill/1.6 mill/1.2 mill
Oceania
51,000/2100/1200
total cases in 2012
new cases in 2012
deaths in 2012

Figure 13.19 HIV infection is widespread on all the inhabited continents. In 2012 there were about 35.3 million adults and children living with HIV/AIDS (black numbers) worldwide, including about 2.3 million new cases of HIV infection (blue numbers). About 1.6 million people died from AIDS in that year (red numbers). Data from *Global Report: UNAIDS Report on the Global AIDS Epidemic 2013* (UNAIDS; 2013).

from the effects of the disease. The virus now known to cause AIDS, the **human immunodeficiency virus (HIV)**, was first isolated in 1983. Two types of HIV are now distinguished—HIV-1 and HIV-2. In most countries, HIV-1 is the principal cause of AIDS. HIV-2 is less virulent, and causes a slower progression to AIDS. It is endemic to West Africa and has spread widely through Asia.

AIDS is a relatively new disease to the medical profession and also to the human species. The earliest evidence for HIV comes from samples from African patients obtained in the late 1950s. It is believed that the viruses first infected humans in Africa by jumping from other primate species—HIV-1 coming from the chimpanzee, HIV-2 from the sooty mangabey, a type of monkey. In neither of these species, and 39 other species of African monkey, does the endogenous HIV-related virus cause disease.

As commonly occurs when a naive host population is hit with a new infectious agent, the effects of HIV on the human population have been immense, and AIDS is now a disease of pandemic proportions (Figure 13.19). The World Health Organization currently estimates that some 35 million people are infected with HIV (Figure 13.20). Although advances continue to be made in understanding the nature of the disease and its origins, the number of people infected with HIV continues to grow—2.3 million new infections in 2012—and tens of millions of people will die from AIDS in the years to come.

People living with HIV
(millions, worldwide)
40
30
20
10
0
1991
1995
1999
2003
2007
2012
Year

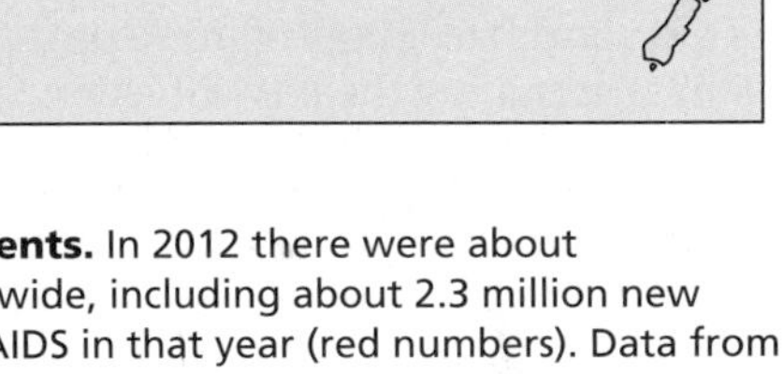

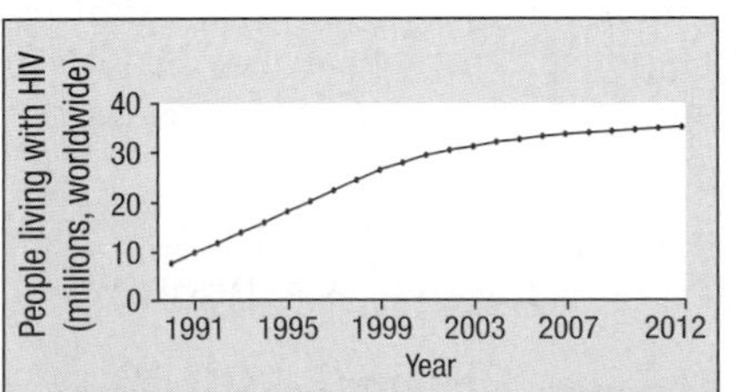

Figure 13.20 The number of people living with HIV infection worldwide is still increasing, but seems to be reaching a maximum.

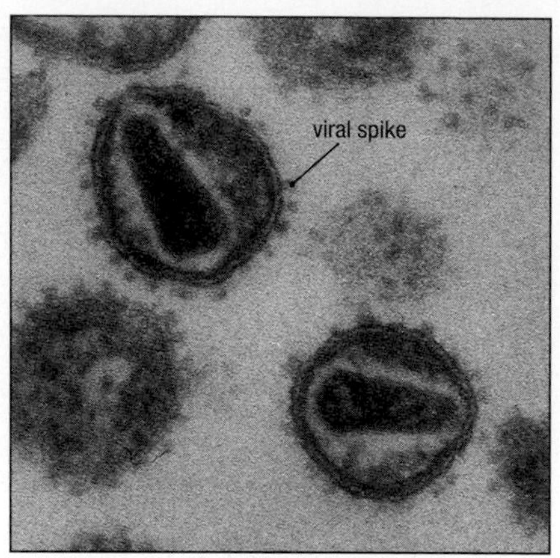

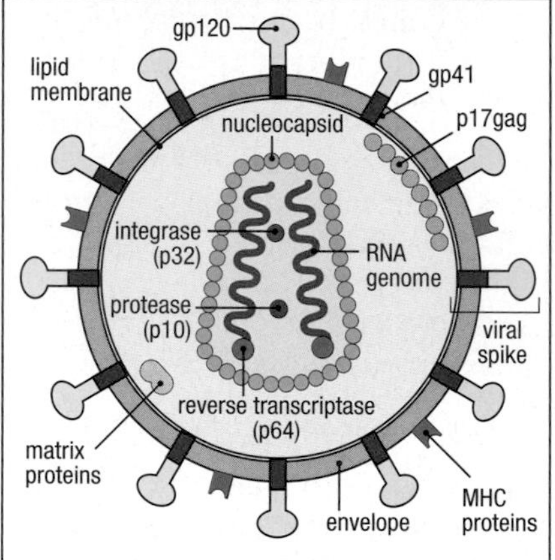

Figure 13.21 The virion of human immunodeficiency virus (HIV). The upper panel is an electron micrograph showing three virions. The lower panel is a diagram of a single virion. gp120 and gp41 are virally encoded envelope glycoproteins of molecular masses 120 kDa and 41 kDa that form the viral spike. Photograph courtesy of Hans Gelderblom.

13-17 HIV is a retrovirus that causes a slowly progressing chronic disease

HIV is an RNA virus with an RNA nucleoprotein core (the nucleocapsid) surrounded by a lipid envelope derived from the host-cell membrane and containing virus-encoded envelope proteins (Figure 13.21). HIV is an example of a **retrovirus**, so named because these viruses use an RNA genome to direct the synthesis of a DNA intermediate, a procedure backward or 'retro' from that used by most biological entities. When HIV infects a cell, the RNA genome is first copied into a complementary DNA (cDNA) by reverse transcriptase. The viral integrase then integrates the cDNA into the genome of the host cell to form a **provirus**, a process facilitated by repetitive DNA sequences called long terminal repeats (LTRs) that flank all retroviral genomes. Proviruses use the transcriptional and translational machinery of the host cell to make viral proteins and RNA genomes, which assemble into new infectious virions. The genes and proteins of HIV are listed in Figure 13.22. HIV belongs to a group of retroviruses that cause slowly progressing diseases. They are collectively called the **lentiviruses**, a name derived from the Latin word *lentus*, meaning slow. In the course of infecting a human cell, HIV recruits 273 human proteins to serve its own purpose. Many of these proteins are used to prevent the human system from terminating the HIV infection.

Around 8% of the human genome is made up of retrovirus-like sequences. These sequences are called **endogenous retroviruses** because they have become permanently integrated into the human genome and because transcripts from endogenous retroviruses are found in all human tissues. In contrast, HIV is an **exogenous retrovirus**. Like the herpesviruses (see Section 13-4), the retroviruses have a long history of exploiting humans and other primate species, and during that time they have evolved numerous and ingenious mechanisms for evading human immune systems. It is thus possible that the endogenous retroviruses, which are now part of self, actually contribute to the difficulties in halting infection by the exogenous and pathogenic retrovirus HIV.

In almost all people, HIV produces an infection that cannot be successfully terminated by the immune system and which continues throughout life. Although the initial acute infection is controlled to the point at which disease is not apparent, the virus persists and replicates in a manner that gradually exhausts the immune system, leading to immunodeficiency and death. From 1983 until 1997 there was no effective treatment for HIV, and during that period much was learnt of the natural course of HIV infection and the immune response that it provokes. At the present time, HIV is the most extensively studied human infectious disease.

13-18 HIV infects CD4 T cells, macrophages, and dendritic cells

Infection with HIV usually occurs after the transfer of bodily fluids from an infected person to an uninfected recipient. Provirus can be carried in infected CD4 T cells, dendritic cells, and macrophages, whereas virions can be transmitted via blood, semen, vaginal fluid, or mother's milk. Infection is commonly spread by sexual intercourse, intravenous administration of drugs with contaminated needles, breast-feeding, or transfusion of human blood or blood components from HIV-infected donors. Infections usually take place across a mucosal surface.

Macrophages, dendritic cells, and CD4 T cells are vulnerable to HIV infection because they express CD4, which the virus exploits as its cellular receptor. Chimpanzees, our closest living relatives, are resistant to HIV infection because of a small difference in the structure of their CD4 glycoprotein compared to ours. CD4 attaches to the spikes on the outer surface of the virus. These spikes are trimers of the envelope glycoprotein, which is a heterodimer of the gp41 and gp120 transmembrane glycoproteins. The two proteins are made as a single polypeptide, which is then cleaved by a host protease to give gp41 and gp120. The gp120 component of the spike binds tightly to human CD4, enabling virions to attach to CD4-expressing cells. Before the virus can enter the cell, gp120 must also bind a co-receptor in the host-cell membrane. Once the co-receptor is bound, the gp41 component of the envelope glycoprotein fuses the viral membrane with the host-cell membrane, allowing the viral genome and associated proteins to enter the cytoplasm of the now infected cell.

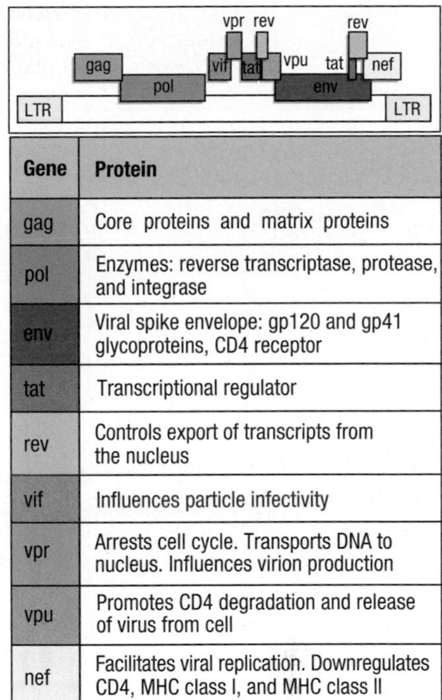

Gene	Protein
gag	Core proteins and matrix proteins
pol	Enzymes: reverse transcriptase, protease, and integrase
env	Viral spike envelope: gp120 and gp41 glycoproteins, CD4 receptor
tat	Transcriptional regulator
rev	Controls export of transcripts from the nucleus
vif	Influences particle infectivity
vpr	Arrests cell cycle. Transports DNA to nucleus. Influences virion production
vpu	Promotes CD4 degradation and release of virus from cell
nef	Facilitates viral replication. Downregulates CD4, MHC class I, and MHC class II

Figure 13.22 The genes and proteins of HIV-1. HIV-1 has an RNA genome consisting of nine genes flanked by long terminal repeats (LTRs). The products of the nine genes and their known functions are tabulated. Several of the viral genes are overlapping and are read in different frames. Others encode large polyproteins that after translation are cleaved to produce several proteins having different activities. The *gag*, *pol*, and *env* genes are common to all retroviruses, and their protein products are all present in the virion.

The viral co-receptors are normal human chemokine receptors that HIV subverts to further its own propagation. There are different variants of HIV, and the cell types that they infect depend largely upon which co-receptor they bind. The HIV variants that spread infection from one person to another bind to the co-receptor CCR5 present on macrophages, dendritic cells, and CD4 T cells. Although they infect several types of human cell, these HIV variants are called 'macrophage-tropic' for want of a better term. The HIV variants that infect activated CD4 T cells bind to the co-receptor CXCR4 and are called 'lymphocyte-tropic.' Whereas infection by macrophage-tropic HIV variants requires only modest levels of cell-surface CD4, infection by the lymphocyte-tropic viruses requires the higher levels present on activated CD4 T cells. Macrophages and dendritic cells at the site of virus entry are the first cells to be infected. Subsequently, the virus produced by the macrophages starts to infect the CD4 T-cell population. In about 50% of cases, the viral phenotype switches to the lymphocyte-tropic type late in infection. This is followed by a rapid decline in CD4 T-cell count and progression to AIDS. In their mutual interaction with the human population, the two types of HIV variant have complementary roles: the lymphocyte-tropic viruses cause the disease, while the macrophage-tropic variants spread the disease and make it a pandemic.

The production of infectious virions from HIV provirus requires the infected CD4 T cell to be activated. Activation induces the synthesis of the transcription factor NFκB, which binds to promoters in the provirus. This directs the RNA polymerase of the infected cell to transcribe viral RNAs. At least two of the proteins encoded by the virus serve to promote replication of the viral genome. Among other activities, the **Tat** protein binds to a sequence in the LTR of the viral mRNA, known as the transcriptional activation region (TAR), where it prevents transcription from shutting off and thus increases the transcription of viral RNA. The **Rev** protein controls the supply of viral RNA to the cytoplasm and the extent to which that RNA is spliced. At early times in infection, Rev delivers RNA that encodes the proteins necessary for making virions. Later, complete viral genomes are supplied, which assemble with the viral proteins to form complexes that bud through the plasma membrane to give infectious virions (Figure 13.23).

13-19 In the twentieth century, most HIV-infected people progressed in time to get AIDS

Immediately after becoming infected with HIV, a person can either be asymptomatic or experience a transient 'flu-like' illness. In either case, the virus becomes abundant in the blood, while the number of circulating CD4 T cells

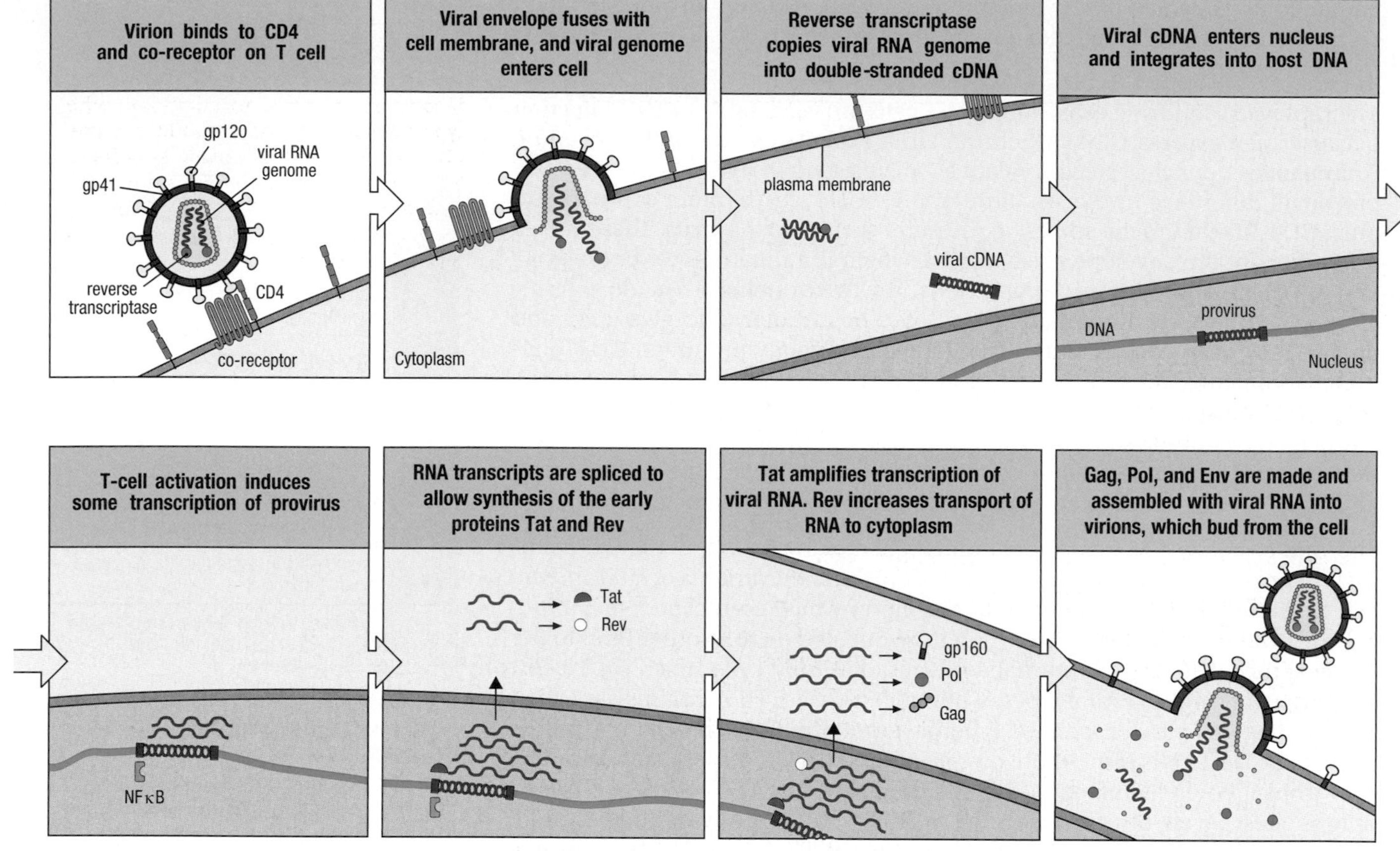

Figure 13.23 The life cycle of HIV in human cells. The gp120 envelope protein of the virus binds to CD4, enabling gp120 to also bind the chemokine co-receptor. This binding releases gp41, causing fusion of the viral envelope with the plasma membrane and release of the viral core into the cytoplasm. The RNA genome is released and reverse transcribed into double-stranded cDNA. This DNA migrates to the nucleus in association with the viral integrase, and becomes integrated into the cell genome, as a provirus (top panels). Activation of the T cell causes low-level transcription of the provirus that directs the synthesis of the early proteins Tat and Rev. These then expand and change the pattern of provirus transcription to produce mRNA encoding the protein constituents of the virion and RNA molecules corresponding to the HIV genome. Envelope proteins travel to the plasma membrane, whereas other viral proteins and viral genomic RNA assemble into nucleocapsids. New virus particles bud from the cell, acquiring their lipid envelope and envelope glycoproteins in the process (bottom panels).

declines markedly (Figure 13.24). This acute viremia is almost always accompanied by activation of an HIV-specific immune response, in which anti-HIV antibodies are produced and cytotoxic T cells become activated to kill virus-infected cells. This response reduces the load of virus carried by the infected person and causes a corresponding increase in the number of circulating CD4 T cells. When an infected person first exhibits detectable levels of anti-HIV antibodies in their blood serum, they are said to have undergone **seroconversion**. The amount of virus persisting in the blood after the symptoms of acute viremia have passed is directly correlated with the subsequent course of disease.

The initial phase of infection is followed by an asymptomatic period, also called 'clinical latency.' During this phase, which can last for 2–15 years, there is persistent infection and replication of HIV in CD4 T cells, causing a gradual decrease in T-cell numbers. Eventually, the number of CD4 T cells drops below that required to mount effective immune responses against other infectious agents. That transition marks the end of clinical latency, the beginning of the period of increasing immunodeficiency, and the onset of AIDS. Patients with

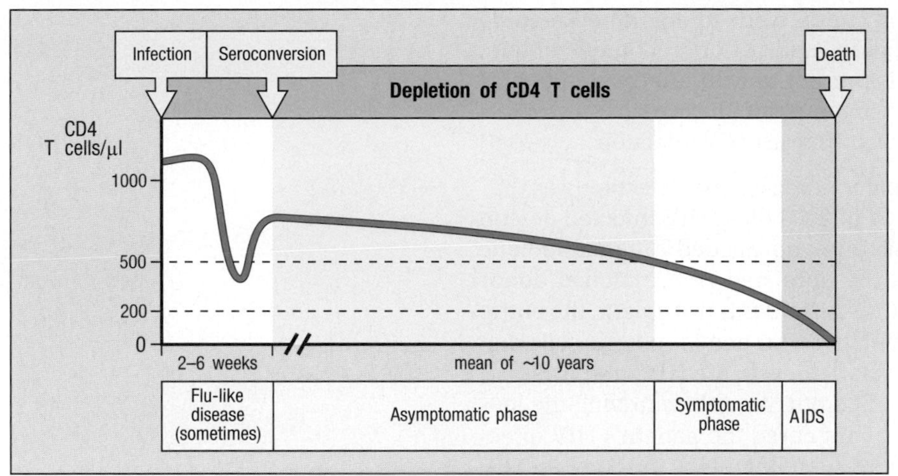

Figure 13.24 After infection with HIV there is a gradual extinction of CD4 T cells. The number of CD4 T cells (green line) refers to those present in peripheral blood. Opportunistic infections and other symptoms become more frequent as the CD4 T-cell count falls, starting at around 500 cells/μl. The disease then enters the symptomatic phase. When CD4 T-cell counts fall below 200 cells/μl, the patient is said to have AIDS.

AIDS become susceptible to a range of opportunistic infections and some cancers, and it is from the effects of these that they die.

At the beginning of the AIDS epidemic in North America and Europe, viral transmission through infected blood products caused hemophiliacs and other patients dependent on blood products to become infected with HIV. Because hemophiliacs are so dependent on the medical profession, whether they be HIV-infected or not, it was possible to study the progress of their HIV infections in a systematic and rigorous manner. The results, which were obtained during the time before effective treatments were available, show that most HIV-infected individuals are destined to progress to AIDS in the absence of effective medical intervention (Figure 13.25). Today, HIV infection through contaminated blood products has largely been eliminated in the richer countries by routine screening of individual units of blood for the presence of HIV.

13-20 Genetic deficiency of the CCR5 co-receptor for HIV confers resistance to infection

Some people who are heavily exposed to HIV never become infected. Similarly, their isolated macrophages and lymphocytes cannot be deliberately infected with macrophage-tropic variants of HIV, the viral strains responsible for the spread of infection. These people are resistant to HIV infection because none of their cells has CCR5, the co-receptor for macrophage-tropic HIV variants. The genetic basis for this defect is a mutated allele of the CCR5 gene in which

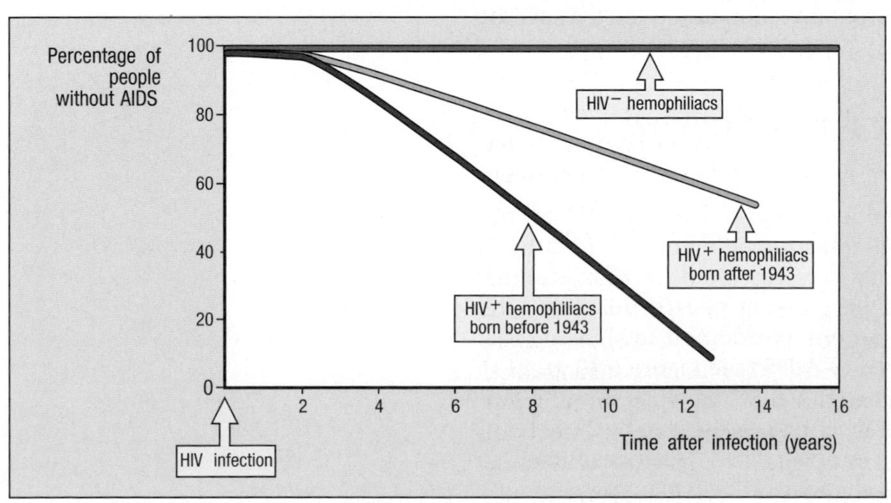

Figure 13.25 Once an HIV infection is established, it usually leads to AIDS. Hemophilia is an inherited disease in which the blood clots poorly. Hemophilia is treated by intravenous infusions of clotting factors purified from the blood of healthy donors. In the early 1980s, before the cause of AIDS was known, some seemingly healthy blood donors were infected with HIV. The virus from their donations contaminated some batches of clotting factor, and many hemophiliacs became infected with HIV. The graph shows the progression to AIDS of HIV-infected and uninfected hemophiliacs after infection. The rate of progression to AIDS increases with the patients' age because the capacity to generate new CD4 T cells decreases with age, due to the involution of the thymus (see Figure 7.4, p. 179).

a 32-nucleotide deletion from the coding region leads to an altered reading frame, premature termination of translation, and a nonfunctional protein. This deletion variant, called *CCR5*-Δ32, is present only in Caucasian populations, in which 10% of people are heterozygous for the variant and 1% is homozygous. Only the *CCR5*-Δ32 homozygotes resist HIV infection.

The importance of functional CCR5 for maintaining an HIV infection was dramatically demonstrated by a case reported in 2009 of an HIV-infected person who developed acute myelogenous leukemia and needed a hematopoietic stem-cell transplant. For this transplant the unrelated HLA-matched donor was also chosen to be homozygous for *CCR5*-Δ32. In this transplant, the entire immune system of the patient was repopulated with hematopoietic cells that lacked the CCR5 co-receptor. When this was achieved, the HIV in this patient ran out of cells to infect and so the virus died out. As well as driving the leukemia into remission, the stem-cell transplant cured the patient's HIV infection. Up to now, this is the only example of someone being cured of an HIV infection.

CCR5 is a receptor for the chemokines CCL3 (MIP-1α), CCL4 (MIP-1β), and CCL5 (RANTES). That almost everyone has a functional CCR5 gene argues for this receptor having made useful contributions to human immunity and survival. However, for individuals exposed to HIV, the advantage of not having CCR5 clearly outweighs that of having it. In the absence of the HIV pandemic, homozygosity for the CCR5 deletion would be considered a mild form of immunodeficiency, like some of those described in the previous part of this chapter, but in today's world it becomes an important asset of disease resistance. This type of evolutionary process, in which a component of the immune system that was useful in fighting past wars with pathogens becomes detrimental during a current conflict, has happened throughout human history. A pathogen subverts an immune-system component and consequently selects for the survival of those humans who have genetic variants that resist that subversion. What we experience is a never-ending arms race, forever selecting for polymorphism and change in the human immune system.

That the CCR5 deletion variant was at high frequency in Caucasians before the HIV epidemic argues for this mutation having enhanced human survival in Europe during previous epidemics of infectious disease. Both plague and smallpox have been advanced as the agent that first drove the CCR5 deletion to high frequency. In the current age, HIV will probably have its strongest selective effect in sub-Saharan Africa, where some 25 million people are infected with HIV and where HIV-infected individuals constitute more than 30% of the population of several countries. Because effective treatment is rarely available for African patients, mortality rates remain high and survival will be greatly enhanced by any genetic variants that prevent infection or reduce the severity of disease.

13-21 HLA and KIR polymorphisms influence the progression to AIDS

A feature of the early years of the AIDS epidemic was that a majority of those infected and diagnosed were young, well-educated people living in affluent countries. Thousands of these individuals were recruited into longitudinal studies that have correlated the clinical progression of HIV infection with genetic polymorphisms of the immune system. In addition to showing that HLA homozygosity speeds the progression to AIDS (see Figure 5.40, p. 143), these studies showed that the HLA-B*27 and HLA-B*57 allotypes slow the progression. Contributing to this effect, the HIV peptides presented by these HLA-B*27 and HLA-B*57 allotypes stimulate stronger CD8 T-cell responses to HIV-infected cells than peptides presented by other HLA-B allotypes.

Another property that the HLA-B*27 and HLA-B*57 allotypes have in common is the Bw4 epitope, the ligand for the NK-cell receptor KIR3DL1/S1. Combinations of Bw4$^+$ HLA-B allotypes and KIR3DL1/S1 allotypes are associated with different rates of progression to AIDS. The most favorable combinations are HLA-B*57 with high-expressing allotypes of KIR3DL1, and HLA-B*27 with low-expressing KIR3DL1 allotypes (Figure 13.26). These correlations suggest that the type of NK-cell response at the start of an HIV infection affects the relative success of the subsequent adaptive response.

About 1 in 300 of HIV-infected people are able to control the infection to the point at which HIV RNA is undetectable in their blood using the standard clinical assay. These people, who suppress the infection and maintain their health for decades, are called **elite controllers**. In addition, around 7% of people infected with HIV maintain a low viremia with 2000 copies or fewer of viral RNA per milliliter of blood. These people also maintain health and are called **viremic controllers**. Whole-genome comparison of the controllers with the progressors, who progress to AIDS, showed that the most significant factor is the HLA-B type. Of the controllers, 67% had HLA-B*13, -B*27, -B*57, or -B*58, compared with 37% of the progressors.

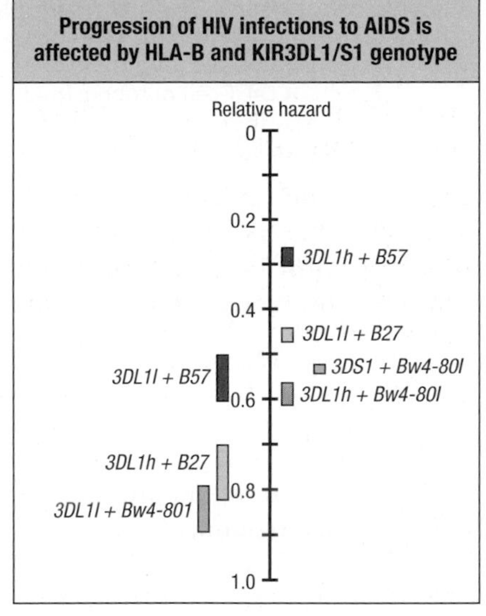

Figure 13.26 Combinations of KIR3DL1/S1 and HLA-B allotypes are associated with variable progression to AIDS for patients infected with HIV. KIR3DL1 and KIR3DS1 are alternative forms of the Bw4 receptor, with the former being inhibitory and the latter activating. Relative hazard is a measure of the speed of progression to AIDS, with 1.0 indicating a faster rate than 0. Bw4, the ligand for KIR3DL1 and KIR3DS1, is an epitope carried by about one-third of HLA-B allotypes and is determined by polymorphisms at positions 77–83 of the HLA-B heavy chain (see Section 12-4). HLA-B27 and HLA-B57 both have a Bw4 epitope that has isoleucine at position 80. Bw4-80I denotes all HLA-B allotypes that have this Bw4 epitope. The inhibitory KIR3DL1 allotypes are divided into two groups—3DL1l and 3DL1h—that have low and high cell-surface expression, respectively. Patients who lack a Bw4 epitope have a relative hazard of 1.0. Data courtesy of Mary Carrington.

13-22 HIV escapes the immune response and develops resistance to antiviral drugs by rapid mutation

People infected with HIV make adaptive immune responses that can prevent the overt symptoms of disease for many years. Included in these responses are T_H1 and T_H2 cells, B cells that make neutralizing antibodies, and CD8 cytotoxic T cells that kill virus-infected cells (Figure 13.27). However, the virus is rarely eliminated. One reason that keeps the virus ahead of the human immune response is its high rate of mutation throughout the course of an infection.

HIV and other retroviruses have high mutation rates because their reverse transcriptases lack proofreading mechanisms like those of the DNA polymerases that replicate the DNA of human cells. Consequently, reverse transcriptases are prone to making errors, and these nucleotide substitutions soon accumulate to give new variant viral genomes. Even though the infection of a person might start from a single viral species, mutation throughout the infection produces many viral variants, called quasi-species, which coexist within the infected person.

The presence of variant viruses increases the difficulty of terminating the infection by immune mechanisms. A successful neutralizing antibody selects for the survival of viral variants that lack the epitope recognized by the antibody. Similarly, pressure from virus-specific cytotoxic T cells selects for viruses in which the peptide epitope recognized by the cytotoxic T cell has changed. In some instances the homologous peptide derived from the variant virus

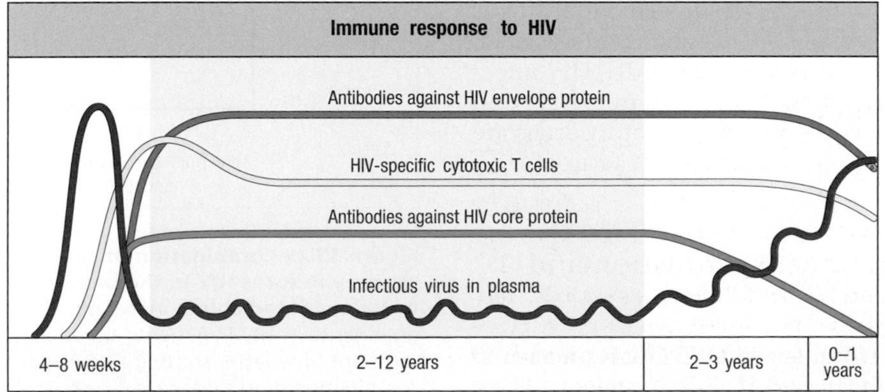

Figure 13.27 The natural course of an HIV infection and the immune response against it. In the early phase of HIV infection, while the adaptive immune response is being activated, the virus reaches high levels (red line). With production of HIV-specific antibodies (blue lines) and cytotoxic T cells (yellow line). The virus is controlled but is not eliminated. When the destruction of CD4 T cells outstrips their renewal, adaptive immunity declines and levels of virus increase again.

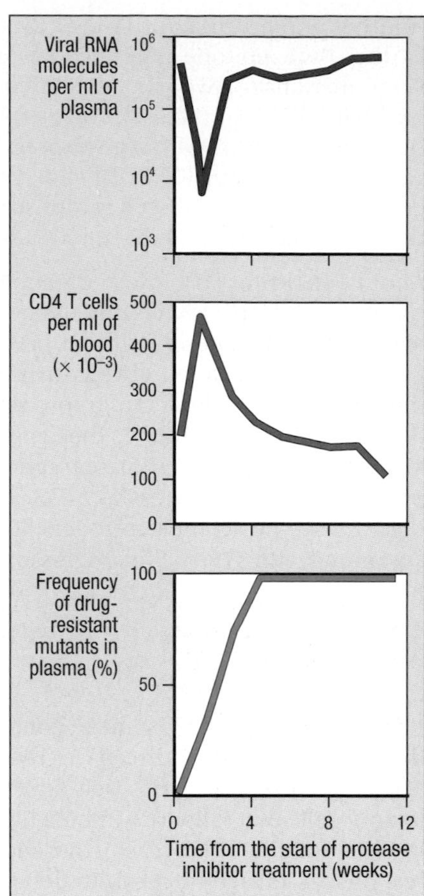

Figure 13.28 HIV can rapidly acquire resistance to protease inhibitor drugs. When a patient is treated with a single protease inhibitor drug the decrease in viral load (top panel) and increase in CD4 T-cell numbers (middle panel) is only transient. The drug's benefit is short-lived because it selects for existing minority populations of drug-resistant HIV variants that rapidly expand and then continue the progression to AIDS (bottom panel).

interferes with the presentation of the antigenic peptide from the original virus, thereby allowing both viral species to escape the cytotoxic T cell.

The high mutation rate of HIV greatly complicates the task of developing an effective vaccine. It also limits the effectiveness of antiviral drugs. Good targets for drugs are the reverse transcriptase, which is essential for the synthesis of provirus, and the viral protease that cleaves large viral polyproteins to give the individual enzymes and proteins. Inhibitors of reverse transcriptase and protease have been found, and these drugs prevent the further infection of healthy cells. Unfortunately, mutation inevitably produces HIV variants with proteins resistant to the action of the drugs (Figure 13.28). One situation in which a relatively short course of drug treatment has long-term benefit is pregnancy. Treating HIV-infected pregnant women with zidovudine, an inhibitor of the viral reverse transcriptase, can prevent mothers from transmitting HIV to their babies, both *in utero* and at birth.

Because HIV so easily escapes from the effects of any single drug, several antiviral drugs are now used together in what is called **combination therapy**. The ideal is to destroy the entire population of viruses before any one of them has accumulated enough mutations to resist all the drugs. Such combination therapy, also called **highly active anti-retroviral therapy** (**HAART**), was introduced in 1997 and has proved effective at reducing the abundance of virus (the viral load) (Figure 13.29) and retarding disease progression. The drugs do not stop virus production by cells that have already been infected, but they prevent new infections from forming a provirus and becoming productive. Two weeks after starting combination therapy, the amount of virus in the blood has decreased to about 5% of the level before treatment. The rapidity of this decline is because both activated CD4 T cells and free virions have short lifetimes (Figure 13.30). At this point no naive or effector CD4 T cells are making virus, and the level now present in the blood is due to the production of virus by longer-lived HIV-infected cells: macrophages, dendritic cells, and memory CD4 T cells. Continuing treatment further decreases the abundance of virus, but at a slower rate than before. Although the virus eventually becomes undetectable, this does not reflect its eradication, because the virus quickly re-emerges in patients who stop taking their medicine.

HAART has been a life-changing therapy for HIV patients in the industrialized countries where it is available. At present, a 20-year-old who becomes infected with HIV and is given HAART will not die from AIDS and can expect to live for 50 more years. However, HAART does not restore HIV-infected individuals completely to health because they have increased risk of bone, cardiovascular, kidney, liver, and neurological diseases. In the context of HAART, HIV infection becomes a chronic inflammatory disease that originates in the gut and spreads to other tissues and organs. A variety of anti-inflammatory drugs are under test for their capacity to prevent and alleviate this new form of HIV-mediated disease.

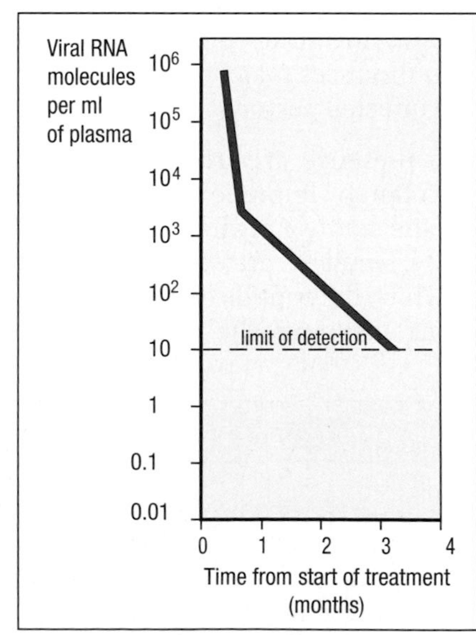

Figure 13.29 Combination drug therapy reduces HIV in the blood to below detectable levels. The abundance of HIV in patients' blood at different times after starting a course of combination drug therapy is shown.

13-23 Clinical latency is a period of active infection and renewal of CD4 T cells

The administration of antiviral drugs to HIV-infected individuals uncovered the active nature of the infection during the period of clinical latency. Within

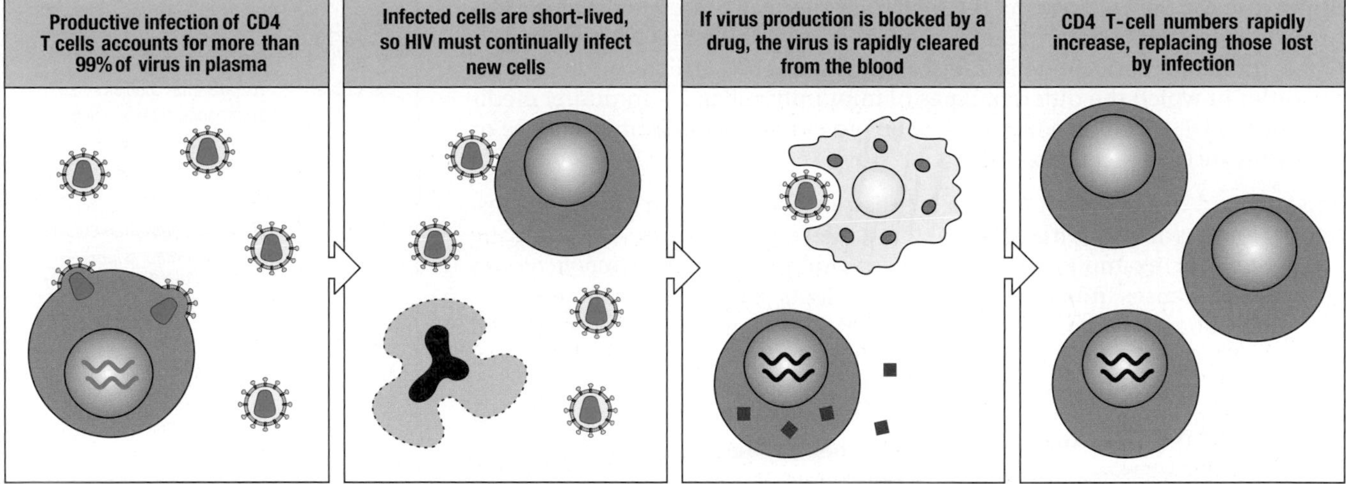

Productive infection of CD4 T cells accounts for more than 99% of virus in plasma
Infected cells are short-lived, so HIV must continually infect new cells
If virus production is blocked by a drug, the virus is rapidly cleared from the blood
CD4 T-cell numbers rapidly increase, replacing those lost by infection

Figure 13.30 Antiviral drugs rapidly clear virus from the blood and increase the number of circulating CD4 T cells. The first and second panels show that maintenance of HIV levels in the blood depends on the continual infection of newly produced CD4 T cells. This is because cells live for only a few days once infected. The third and fourth panels show the effects of administering a drug (red squares) that blocks the viral life cycle. The existing virions in the blood are rapidly cleared by the actions of neutralizing antibody, complement, and phagocytes. Newly produced CD4 T cells are not infected, whereupon they live longer and accumulate in the circulation.

2 days of starting a course of drugs, the amount of virus in the blood decreases markedly. At the same time, the number of CD4 T cells increases substantially (see Figure 13.28). This shows two things: first, that virus is being produced and cleared continuously within infected individuals, and second, that in the face of HIV infection the body continues to produce new CD4 T cells, which quickly become infected with HIV. Thus the period of clinical latency, when the overall numbers of CD4 T cells in the blood are gradually declining, is actually a time of immense immune activity. Vast numbers of T cells are produced and die, and virions are neutralized, the latter most probably through antibody-mediated opsonization and phagocytosis.

Although measurement of lymphocyte and virion numbers in the blood is used clinically to monitor the progress of HIV infection, the secondary lymphoid tissues are the sites where most lymphocytes are found and where CD4 T cells are activated and produce virus. In HIV-infected people, these tissues are loaded with virions, many of which are trapped on the surface of follicular dendritic cells.

11-24 HIV infection leads to immunodeficiency and death from opportunistic infections

Within a few days of being infected with HIV, a CD4 T cell dies. Three kinds of mechanism are thought to contribute to the death toll. One is direct killing as a result of the viral infection or virions binding to cell-surface receptors, the second is increased susceptibility of infected cells to apoptosis, and the third is killing by cytotoxic CD8 T cells specific for viral peptides presented by HLA class I molecules on the infected CD4 T cell.

Throughout clinical latency, the daily loss of CD4 T cells is for the most part compensated for by the supply of new CD4 T cells. With time, however, a steady decline in the number of CD4 T cells is evident, showing how the virus gradually wins this war of attrition. Eventually, CD4 T cell numbers become so low that immune responses to all foreign antigens are compromised and HIV-infected people become exceedingly susceptible to other infections—at this stage they have progressed to AIDS. Because CD4 T cells are central to every aspect of adaptive immunity, the vulnerability of AIDS patients to infection resembles that of children with inherited SCID.

The infections that most frequently affect patients with AIDS are caused by commensal microorganisms that live either in or on the body and are actively controlled by healthy people. When they infect, such agents are called

opportunistic pathogens, and the infections they cause are known as **opportunistic infections** (Figure 13.31). There is a rough hierarchy in the times at which particular opportunistic infections occur in AIDS, which correlates with the order in which the different types of immunity collapse. Immunity mediated by CD4 T_H1 cells is generally lost before either the antibody response or the cytotoxic CD8 T-cell response.

The oral and respiratory tracts are soft tissues loaded with microorganisms, and in many patients with AIDS they are the sites of the first opportunistic infections. For example, *Candida* causes oral thrush and *Mycobacterium tuberculosis* causes tuberculosis. Later, patients can suffer from diseases caused by the reactivation of latent herpesviruses that are no longer controlled by CD8 T cells. Such diseases include shingles caused by varicella-zoster, B-cell lymphoma caused by EBV, and an endothelial tumor called Kaposi's sarcoma, caused by the HHV8 herpesvirus. *Pneumocystis jirovecii*, a common environmental fungus that rarely troubles healthy people, is frequently the cause of pneumonia and death for AIDS patients. In the later stages of AIDS, reactivation of CMV can cause B-cell lymphoproliferative disease. Infection with the opportunistic pathogen *Mycobacterium avium* also becomes prominent. The opportunistic infections suffered by individual AIDS patients vary greatly. With the collapse of the immune system, only drugs and other interventions can be used to treat the opportunistic infections. These provide only temporary respite and also have their own deleterious effects. Eventually, the tissue damage that results from the combined effect of HIV infection, opportunistic infections, and medical intervention causes death.

Infections	
Parasites	*Toxoplasma* species *Cryptosporidium* species *Leishmania* species *Microsporidium* species
Bacteria	*Mycobacterium tuberculosis* *Mycobacterium avium intracellulare* *Salmonella* species
Fungi	*Pneumocystis jirovecii* *Cryptococcus neoformans* *Candida* species *Histoplasma capsulatum* *Coccidioides immitis*
Viruses	Herpes simplex Cytomegalovirus Varicella-zoster

Malignancies
Kaposi's sarcoma (associated with herpesvirus HHV8)
Non-Hodgkin's lymphoma, including EBV-positive Burkitt's lymphoma
Primary lymphoma of the brain

Figure 13.31 **Some opportunistic infections that kill patients with AIDS.** Listed are the most common opportunistic infections that kill AIDS patients in developed countries. The malignancies are listed separately but they are also the result of impaired responses to infectious agents.

13-25 A minority of HIV-infected individuals make antibodies that neutralize many strains of HIV

Although enormous resources have been invested in seeking a vaccine for HIV, the results have been generally disappointing. One encouraging result from a clinical trial reported in 2009 was that antibody recognizing the gp120 component of the envelope protein could provide protection against HIV infection. A second encouraging observation is that 1 in 500 HIV-infected individuals make small amounts of antibodies that bind and neutralize a broad range of HIV-1 strains. These individuals are called **elite neutralizers**, and the antibodies they make are termed **broadly neutralizing antibodies**. Such antibodies are not made during the primary response to HIV infection, but emerge only when a person has been infected for 2 years or longer, after their immune system has been stimulated by several HIV strains that successively emerge from the selection exerted by strain-specific antibodies. In following up these observations, techniques were developed for isolating B cells that make broadly neutralizing antibodies from the blood of elite neutralizers. Then by cloning and expressing cDNA encoding the heavy and light chains of these antibodies it became possible to make large quantities of many broadly neutralizing antibodies and define their specificity, structure, and common properties.

Most broadly neutralizing antibodies recognize one of four epitopes of the envelope glycoprotein of HIV-1 (Figure 13.32). These are all highly conserved regions that have biological importance for the virus. For example, one of the epitopes is the site on gp120 that binds CD4. A distinguishing property of broadly neutralizing antibodies is that their variable regions have many more somatic mutations than other antibodies. The heavy-chain variable regions of high-affinity antibodies typically have 10–20 mutations, whereas 40–80 mutations are present in the heavy-chain variable regions of broadly neutralizing antibodies. Such numbers indicate that the B cells making broadly neutralizing antibodies have over time been subject to several separate rounds of antigen-mediated somatic mutation (see Section 9-7). Unlike other antibodies, the broadly neutralizing antibodies acquire mutations in framework regions as well as in the CDR loops, and these substitutions contribute to the neutralizing function. A further difference is that variable regions of the broadly neutralizing antibodies acquire insertions and deletions, leading to some of the antibody heavy chains having unusually long CDR loops. These loops enable the antibody to penetrate the glycans that shield the viral spikes and gain access to bind to gp120. A feature of some broadly neutralizing antibodies is that they react with a variety of structurally unrelated antigenic molecules, and are said to be **polyreactive**. It is suggested that polyreactivity increases the probability that the antibodies will bind with both their antigen-binding sites—one site binding to gp120, the other to another component of the virus or the infected human cell. This is a potential problem because there are only 15 spikes and 45 envelope proteins on the surface of the virus. The small number of spikes and the disposition of the three copies of gp120 and gp41 within a spike decrease the probability that an IgG molecule can reach to bind two gp120 or gp41 epitopes with its Fab arms. One example of a polyreactive antibody binds to both CD4 and gp120 and can thus cross-link these two molecules by using one Fab arm to bind CD4 and the other to bind gp120.

Current research on broadly neutralizing antibodies is exploring two complementary clinical applications. The first is to design vaccines that will favor the production of these antibodies; the second is to infuse infected people, or people at risk of exposure to HIV, with the antibodies in a passive immunization. At the present time there is cautious optimism that successful application of broadly neutralizing antibodies will eventually bring the HIV epidemic under control.

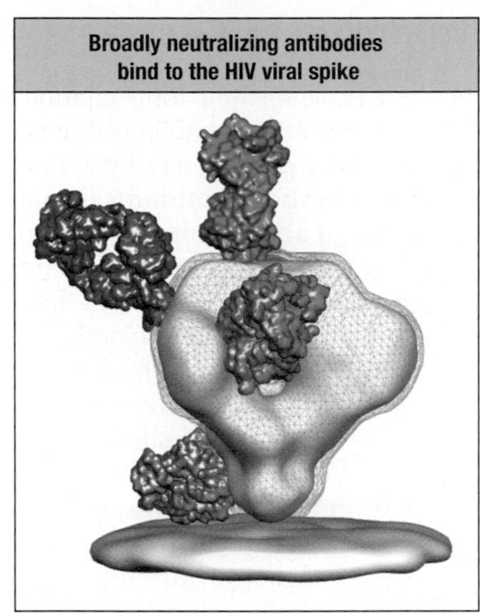

Broadly neutralizing antibodies bind to the HIV viral spike

Figure 13.32 Broadly neutralizing antibodies recognize four epitopes of the HIV envelope glycoprotein that trimerizes to form the viral spike. Broadly neutralizing antibodies protect against many different strains of HIV. These antibodies interact with conserved parts of the envelope glycoprotein, which are important for its functional interactions. The orange-colored antibody binds to the CD4-binding site; the green-colored antibody binds to conserved elements of the V1 and V2 hypervariable loop of gp120; the purple-colored antibody binds to the conserved elements of the glycan shield and the V3 hypervariable loop of gp120; and the grey-colored antibody binds gp41. Image courtesy of Louise Scharf.

Summary

The past 30 years has seen a worldwide epidemic of infectious disease caused by the human immunodeficiency virus (HIV), in which more than 25 million people have died. HIV infects CD4 T cells, macrophages, and dendritic cells, and uses the CD4 glycoprotein as its receptor and the CCR5 chemokine receptor as its co-receptor. The effects of HIV are due chiefly to a gradual and sustained destruction of CD4 T cells, which leads eventually to a profound T-cell immunodeficiency known as acquired immune deficiency syndrome (AIDS). CD4 T cells proliferate as part of their normal function and are continually being renewed; these properties allow HIV to maintain a long-lasting infection in which individuals remain relatively healthy for years. This property facilitated the spread of infection, which is mainly by sexual transmission. Patients with AIDS succumb to a range of opportunistic infections and to a much lesser extent from virus-associated cancers.

Some people are resistant to infection by HIV because they lack the CCR5 co-receptor. Of the people infected with HIV, a small fraction control the virus and can maintain their health. Particular HLA-B and KIR alleles correlate with this capacity to control HIV. Until 1997, to be infected with HIV was a death sentence. In that year, highly active anti-retroviral therapy (HAART) was introduced, which involves a combination of drugs, each inhibiting an enzyme essential for viral replication. This highly successful therapy prevents progression to AIDS and allows HIV-infected individuals to live a normal lifespan.

No person who becomes infected with HIV ever gets rid of the virus. HIV subverts the human genome to serve its own ends and rapidly mutates its own genome so that the immune response made by the human host can be easily evaded. These viral tactics have thwarted a 30-year quest for a vaccine against HIV. After several years of infection with HIV, some people make small amounts of antibodies that have the capacity to neutralize a broad range of HIV strains. Such antibodies give hope for new strategies for vaccination strategies and passive immunization.

Summary to Chapter 13

In the course of their long relationship with humans, successful pathogens have developed mechanisms that allow them to exploit the human body to the full. Indeed, a pathogen is, by definition, an organism that is habitually able to overcome the body's immune defenses to such an extent that it causes disease. One class of adaptations is those in which the pathogen changes itself or its behavior to evade the ongoing immune response. This prevents the immune system from adapting to the pathogen and improving the response. In a second type of adaptation, the pathogen is able to impair or prevent the immune response. Pathogens can have more than one such adaptation, and for some pathogens a considerable fraction of their genome is devoted to foiling the immune system. Highly successful pathogens are not necessarily the most virulent. Nonlethal ubiquitous host-pathogen relationships, such as that of humans and the cytomegalovirus virus, have generally evolved over a long period of association. However, even these pathogens can cause life-threatening disease when the immune system is compromised.

Inherited immunodeficiencies are caused by a defect in one of the genes necessary for the development or function of the immune system. Depending on the gene involved, immunodeficiencies range from manageable susceptibilities to particular pathogens to a general vulnerability created by the complete absence of adaptive immunity. The most severe inherited immunodeficiencies are very rare, which is evidence of the importance of the immune system to human survival.

Immunodeficiency can also be acquired as the result of infection. The human immunodeficiency virus (HIV) infects macrophages, dendritic cells, and CD4 T cells and eventually reduces the number of CD4 T cells to a level at which severe immunodeficiency results—the acquired immune deficiency syndrome (AIDS). This retroviral pathogen has been exploiting the human species for less than a century, but it already had an arsenal of effective adaptations that were acquired during its previous history in other primate hosts. Most people infected with HIV have no overt symptoms of disease for years, which facilitates the spread of HIV by sexual transmission, and HIV infection is now at pandemic proportions within the human population. Only a minority of people are resistant to HIV, either because their cells cannot be infected or because they have genetic variants of HLA, KIR, and other immune-system genes that allow them to control the infection and prevent the progress to AIDS. Combination therapy with anti-retroviral drugs prevents the progression to AIDS and allows HIV-infected people to lead long lives. No effective vaccine for HIV has been made, but new strategies for active and passive immunization, involving broadly neutralizing antibodies, are currently under intensive investigation.

Questions

13–1

A. Explain what is meant by serotype-specific immunity.

B. How does *Streptococcus pneumoniae* exploit serotype-specific immunity to evade detection?

13–2 The evolution of influenza virus that causes relatively mild and limited disease epidemics is referred to as _____.

a. latency
b. antigenic drift
c. antigenic shift
d. passive immunity
e. immune-complex disease.

13–3 Explain how cytomegalovirus (CMV) subverts the ability of (A) MHC class I molecules and (B) NK-cell receptors to mediate immune responses.

13–4

A. Explain why an inherited deficiency in complement components C3 or C4 can result in less efficient clearance of immune complexes from the blood and lymph than normal.

B. What clinical symptoms can this lead to?

C. What is the only known effect of deficiencies in complement components C5–C9? Explain this effect.

13–5 Match the pathogen in column A with its mode of evasion or subversion of the immune system.

Column A	Column B
a. *Trypanosoma brucei*	1. undergoes latency
b. Influenza viruses	2. carries out recombination of eight RNA molecules of the genome
c. *Toxoplasma gondii*	3. participates in genetic variation of capsular polysaccharides
d. Herpesviruses	4. encloses itself in an impenetrable membrane-bounded vesicle
e. *Streptococcus pneumoniae*	5. employs gene conversion
f. *Mycobacterium tuberculosis*	6. prevents fusion of phagosome with lysosome

13–6 Which of the following holds true if a woman carries one defective Bruton's tyrosine kinase (*BTK*) allele? (Select all that apply.)

a. She has no functional B cells and is therefore unable to produce antibodies.
b. She will transmit the defective allele to 50% of her children.
c. If her husband has a normal form of BTK, then 50% of her daughters will not be able to produce antibodies.
d. Her B cells will exhibit non-random X-chromosome inactivation.
e. She requires monthly injections of gamma globulin to prevent bronchiectasis.

13–7 Deficiencies in complement components C1–C4 are associated with _____.

a. neutropenia
b. chronic granulomatous disease
c. immune-complex disease
d. opportunistic infections
e. leukocytosis.

13–8 Explain the mechanism by which (A) macrophages and NK cells, and (B) macrophages and effector T cells become mutually activated in the course of an innate and adaptive immune response, respectively. (C) Which gene defects would impair these mutual activation processes?

13–9 Loss-of-function mutations in the common γ chain (γ_c) gene are particularly damaging to the immune system because _____.

a. homozygotes and heterozygotes fail to make sufficient γ_c protein for lymphocyte activation
b. T cells fail to develop in the thymus
c. T-cell dependent and T-cell independent B-cell responses are inhibited
d. multiple cytokine receptors cannot signal
e. MHC class II genes are unable to be expressed.

13–10 Which of the following is the most likely explanation for an individual who lacks CCR5 as a result of a homozygous defect in the CCR5 gene becoming infected with HIV?

a. The mutated CCR5 genes reverted to the normal form, rendering macrophages susceptible to macrophage-tropic HIV variants.
b. The macrophage-tropic HIV variant entered host cells using CD4 alone.
c. The viral nucleic acid alone was taken up by cells, as in cell transformation by bacterial DNA.
d. The individual had received a transplant of HIV-infected cells expressing normal CCR5.
e. The primary infection involved a lymphocyte-tropic strain of HIV that used CXCR4 as its co-receptor.

13–11 Identify the mismatched pair.

a. CXCR4: lymphocyte tropic
b. exogenous retrovirus: originates outside host's body
c. infectious virion production: activated CD4 T cells
d. tat: controls export of transcripts from the nucleus
e. lentivirus: long incubation period
f. CCR5: macrophage tropic.

13–12 Match the term in column A with its description in column B.

Column A	Column B
a. primary immunodeficiency disease	1. causes disease only in homozygotes
b. secondary immunodeficiency disease	2. caused by genetically inherited gene mutations that adversely affect the immune system
c. dominant defective allele	3. causes disease in heterozygotes
d. recessive defective allele	4. causes disease in males more commonly than in females
e. X-linked recessive disease	5. caused by non-genetic factors

13–13 The high degree of mutation in HIV, the accumulation of variant viruses, and the development of resistance to drug regimes are attributed to _____.

a. antigenic shift
b. gene conversion
c. absence of proofreading capability of reverse transcriptase
d. antigenic variation of variable surface glycoproteins (VSGs)
e. mutation of HIV RNA genome by host-cell RNA polymerase.

13–14 12-year-old Morgan Boyd had a 7-month history of several episodes of sudden onset of swelling of her tongue, lips, and eyes. These recurrences lasted 1–3 days before subsiding without medical intervention. Her older brother, Stephen, had had a similar problem when he was 16 years old. The edema was not itchy and there was no urticaria. Morgan did not experience airway obstruction or abdominal discomfort during any of her episodes. She was not taking any medications and did not have any drug allergies. A diagnosis of hereditary angioedema (HAE) was made on the basis of (1) blood measurements of key complement components after remission and again during a subsequent attack of angioedema, and (2) C1INH quantitative and qualitative tests. Which of the following complement proteins would be well below normal range in Morgan in both of her blood tests?

a. C1q
b. C3
c. C4
d. C5
e. C9

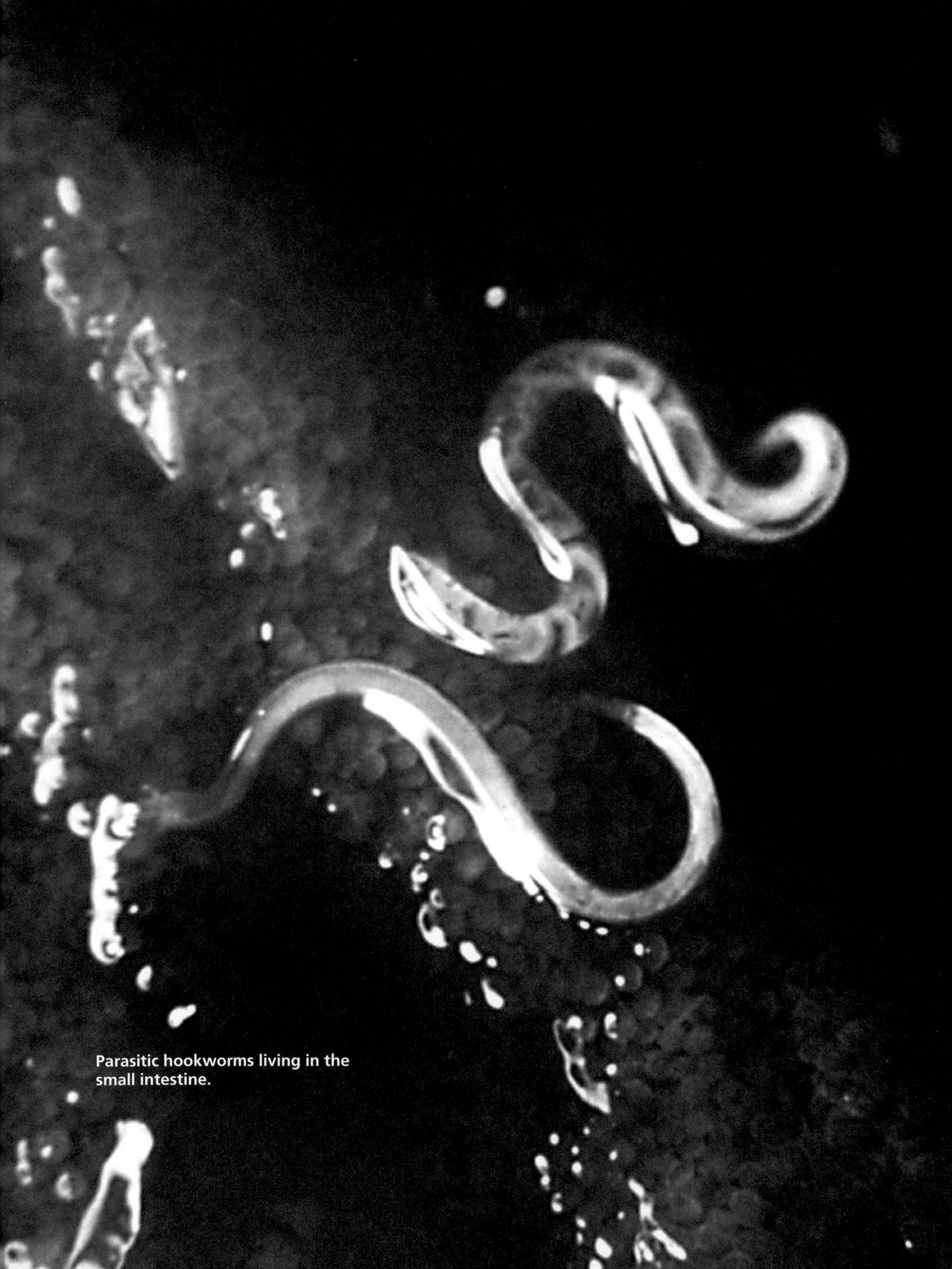
Parasitic hookworms living in the small intestine.

Chapter 14

IgE-Mediated Immunity and Allergy

The protective functions of immune defense are built upon mechanisms for distinguishing the constituents of infectious agents from the constituents of the human body. The immune system is thus able to be tolerant of self but recognize that pathogens and their antigens are foreign. Apart from infectious agents, humans come into daily contact with numerous macromolecules that are equally foreign but do not in any way threaten their health. Many of these potential antigens derive from the plants and animals that provide our food or are present in the environments where we live, work, and play. For most people for most of the time, contact with these molecules stimulates neither inflammation nor adaptive immunity. In these circumstances the immune system is seen to distinguish what is dangerous from what is not.

In other circumstances, however, certain kinds of harmless molecule can stimulate a primary adaptive immune response that generates immunological memory. On subsequent exposures to that antigen, the secondary immune response instigates inflammation and tissue damage, which is at best an irritation and at worst a threat to life. The person feels ill, as though fighting off an infection, when in reality no infection exists. These unnecessary overreactions of the immune system to innocuous environmental antigens are called **hypersensitivity reactions** or **allergic reactions**. The environmental antigens that cause such reactions are termed **allergens** (Figure 14.1). Allergens induce a state of hypersensitivity or **allergy**, the latter being of Greek derivation and meaning 'altered reactivity.' **Atopy** is another word derived from Greek that allergists use, meaning 'out of place' or 'peculiar.' Atopic individuals have a genetic predisposition to allergy, and 40% of the Caucasian populations of Europe and North America are atopic. During the past 30 years, the incidence of allergy has increased twofold to threefold, reaching a point where 10–40% of the inhabitants of the developed countries are now allergic to one or more environmental antigens. With this impressive trend, which is increasingly described as an **epidemic of allergy** by the medical profession as well as the media, each generation of children develops more allergies than their parents.

14-1 Different effector mechanisms cause four distinctive types of hypersensitivity reaction

All hypersensitivity reactions are the consequence of an adaptive immune response. They can be grouped into four types, according to the effector mechanism that mediates the hypersensitivity reaction (Figure 14.2). **Type I hypersensitivity reactions** are triggered by the interaction of an allergen with

Figure 14.1 Common causes of hypersensitivity reactions. Inhaled materials include plant pollens, the dander of domesticated animals, mold spores, and proteins in the feces of house dust mites. Injected materials include insect venoms, vaccines, drugs, and therapeutic proteins such as monoclonal antibodies. Ingested materials include some foods and orally administered drugs. Contacted materials include plant oils, industrial products of plant or synthetic origin, and metal ions, notably nickel.

Inhaled materials

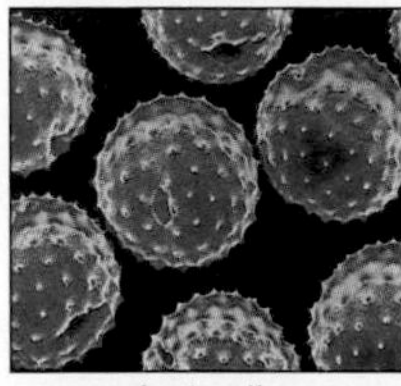
plant pollen

dust mite feces

Injected materials

insect venom

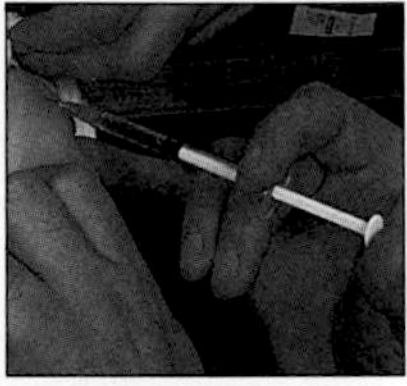
drugs

Ingested materials

peanuts

shellfish

Contacted materials

plant oil

metal

allergen-specific IgE bound to the FcεRI receptor of mast cells, basophils, and eosinophils. This interaction causes these cells to degranulate and release a potent mixture of inflammatory mediators (see Figure 14.2, first panels). A common type I hypersensitivity reaction is caused by inhaling particulate antigens such as plant pollens. Type I reactions have effects of varying severity, ranging from the irritation of a runny nose to stressful breathlessness and even death by asphyxiation. Type I hypersensitivity is also called **immediate hypersensitivity** because the reaction occurs immediately upon exposure to the allergen.

Type II hypersensitivity reactions are caused by an IgG response to chemically reactive small molecules that become covalently bound to the outside surface of cells. The chemical reaction modifies the structures of human cell-surface components, which are now perceived as foreign antigens by the immune system. B cells are stimulated to make IgG antibodies against the new epitopes. On binding to their specific cell-surface antigens, the antibodies cause the modified human cells to become subject to complement activation and phagocytosis. The ensuing result is an inflamed and damaged tissue. Penicillin is an example of a small reactive molecule that induces type II hypersensitivity reactions in some fraction of the patients who are prescribed this antibiotic drug (see Figure 14.2, second panels).

Type III hypersensitivity reactions are caused by small soluble immune complexes of antigen and specific IgG that form deposits in the walls of small blood vessels or the alveoli of the lungs. At these sites, the immune complexes activate complement and an inflammatory response that damages the tissue and impairs its function. When antibodies or other proteins derived from non-human animal species are given therapeutically to patients, type III hypersensitivity reactions are always a potential side effect (see Figure 14.2, third panels). For example, some diabetics who were regularly injected with bovine or porcine insulin developed type III hypersensitivity to the insulin. For this reason, diabetics are now mostly treated with synthetic human insulin.

Hypersensitivity reactions of types I, II, and III are all caused by antibody. In contrast, **type IV hypersensitivity reactions** are mediated by antigen-specific effector T cells, and in most instances by CD4 T_H1 cells (see Figure 14.2, fourth panels). For example, the red, itchy, and blistering skin rash that characterizes nickel allergy is caused by an infiltration of effector CD4 T_H1 cells. When the skin comes into contact with the nickel in objects such as coins, jewelry, and belt buckles, small amounts of the metal permeate the skin, and bivalent nickel ions are chelated by histidine residues in human proteins. Peptides containing the nickel-chelated residues are presented by HLA class II molecules and recognized as non-self by CD4 T cells that respond by attacking any area of the skin that comes into contact with nickel. Some type IV hypersensitivity reactions are due to CD8 T cells, an example being the allergic reaction to poison ivy. When the leaves of the plant are touched, an oily substance called pentadecacatechol permeates the skin and reacts chemically with human proteins. Degradation of these proteins yields abnormal peptides that are presented by HLA class I molecules and stimulate a cytotoxic CD8 T-cell response that ferociously attacks those areas of the skin that contacted the plant.

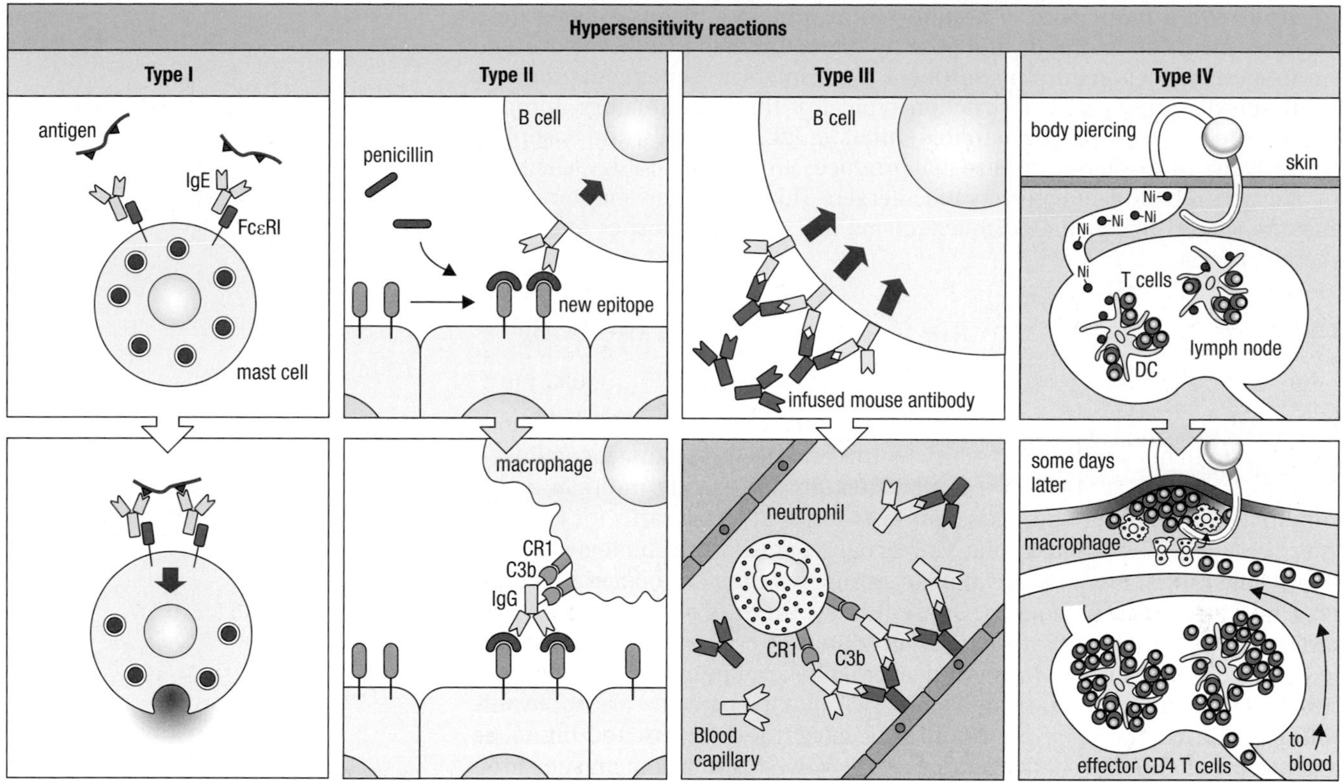

Figure 14.2 Hypersensitivity reactions fall into four classes based on the effector mechanism that causes them. Type I hypersensitivity reactions are caused by the interaction of soluble allergens with specific IgE that causes the degranulation of mast cells (first panels). Type II hypersensitivity reactions are mediated by IgG antibodies made against new epitopes of cell-surface proteins caused by chemical modification. Shown is the antibiotic drug penicillin forming a new epitope by chemically modifying a human cell-surface protein (second panels). Type III hypersensitivity reactions are caused by IgG antibodies made against a soluble foreign protein, such as a therapeutic mouse monoclonal antibody, which form immune complexes that are deposited in tissues and become subject to complement fixation and phagocyte attack (third panels). Most type IV hypersensitivity reactions are caused by CD4 T cells responding either to the epitopes of foreign proteins or to peptides derived from chemically modified human proteins, such as the nickel-modified peptides shown here, presented by MHC class II molecules (fourth panels).

Type IV hypersensitivity reactions are also called **delayed-type hypersensitivity (DTH)** because they only become apparent 1–3 days after exposure to antigen. In contrast, the symptoms of type II and III hypersensitivity reactions are usually felt within hours of exposure to the allergen. Type I hypersensitivities are unique to allergic disease, whereas reactions equivalent to type II, III, and IV hypersensitivities occur in inflammatory and autoimmune diseases as well as in transplantation of tissues and organs. For this reason, the rest of this chapter concentrates on IgE-mediated allergy; the other hypersensitivities will be examined in the next two chapters.

Shared mechanisms of immunity and allergy

The physiological role of IgE is to work with mast cells, basophils, and eosinophils in controlling infestation of the human body by parasitic multicellular organisms, of which helminth worms are major culprits (see Section 10-16). Parasite-specific IgE made in the primary response becomes bound to Fc receptors on mast cells. On subsequent exposure to the parasite, antigen-mediated cross-linking of IgE stimulates mast-cell degranulation and activates inflammatory and physical reactions that lessen the parasite load (see Figure 9.19, p. 249). In developed countries, where the common human parasites have been eradicated, the same mechanisms of recognition and effector function have a propensity to be directed at protein antigens derived from common plants and animals that pose no direct threat to health. Exposure to the antigen induces an immune response that confers no benefit but produces the tissue disruption and destruction typical of the inflammatory immune response to an acute infection. In this situation, IgE, mast cells, and basophils are the cause of an allergic disease that produces immunopathology whenever an allergic individual encounters the allergen. This part of the chapter examines the shared IgE-mediated mechanisms that control parasites and cause allergic diseases.

14-2 IgE-mediated immune responses defend the body against multicellular parasites

IgE is central to the T_H2 arm of adaptive immunity that is dedicated to defending the human body against helminth worms and the other parasitic multicellular animals (Figure 14.3). These parasites do not usually multiply in the human body like other pathogens but instead use it temporarily for a part of their life cycle. The multicellular (metazoan) parasites are biologically and chemically much, much, more similar to humans than are bacterial, fungal, viral, and protozoan pathogens. Consequently, they are inherently less antigenic and the antigens that provoke an immune response are the result of sequence differences in parasite proteins that have counterparts in other species, including humans. A further challenge is that the parasites are not readily killed by macrophages or neutrophils, because they are just too big to be phagocytosed. The alternative strategy that evolved is to eject parasites from the body through physical force. This is achieved by the familiar mechanisms of coughing, sneezing, nose blowing, vomiting, diarrhea, itch-induced scratching, and increasing the flow of mucus. The immune response to parasites that produces these effects occurs principally in mucosal tissues and involves T_H2 cells helping B cells to switch to the synthesis of IgE. The parasite-specific IgE is used to arm basophils, eosinophils, and mast cells, providing these inflammatory cells with specific antigen receptors that allow them to respond to specific parasite antigens. By working together, these cells produce the explosive inflammatory reactions that occur when antigen binding to IgE induces a concerted mast-cell degranulation. The violence dislodges embedded parasites from the tissues and ejects them from the human host. These mechanisms rarely eradicate the parasite from the body but they reduce both its population size and the severity of the disease they cause.

In tropical countries, parasitic infections are endemic; more than 1.5 billion people worldwide are heavily and persistently infected with helminths (Figure 14.4). A universal feature of helminth infection is the stimulation of CD4 T_H2-like responses that produce raised levels of IgE and increased numbers of eosinophils in blood and mast cells in tissues. The IgE is heterogeneous, and only a small fraction of the IgE molecules can be shown to be parasite-specific. This observation led to the idea that parasite infections induce a polyclonal activation of IgE-producing B cells that is not driven in an antigen-specific manner. Alternatively, it is possible that the response is driven by antigen but that the affinity of most of the IgE antibodies is below the limit of detection of the methods used. This debate has yet to be resolved. Despite having a relative abundance and variety of IgE in their systems, people with helminth infections are rarely afflicted by allergic disease.

Components of a typical adaptive immune response against helminth worms	
T cells	CD4 T_H2 cells
Cytokines	IL-3, IL-4, IL-5, IL-9, IL-10, IL-13
Antibodies	IgE, IgG1, IgG4
Effector cells	Expanded populations of eosinophils, basophils, and mast cells

Figure 14.3 Key cellular and molecular components that contribute to the T_H2 arm of adaptive immunity.

14-3 IgE antibodies emerge at early and late times in the primary immune response

In the primary response to an antigen, IgM is first switched either to IgG or IgE. The majority of responding B cells switch to IgG3 and the minority switch to

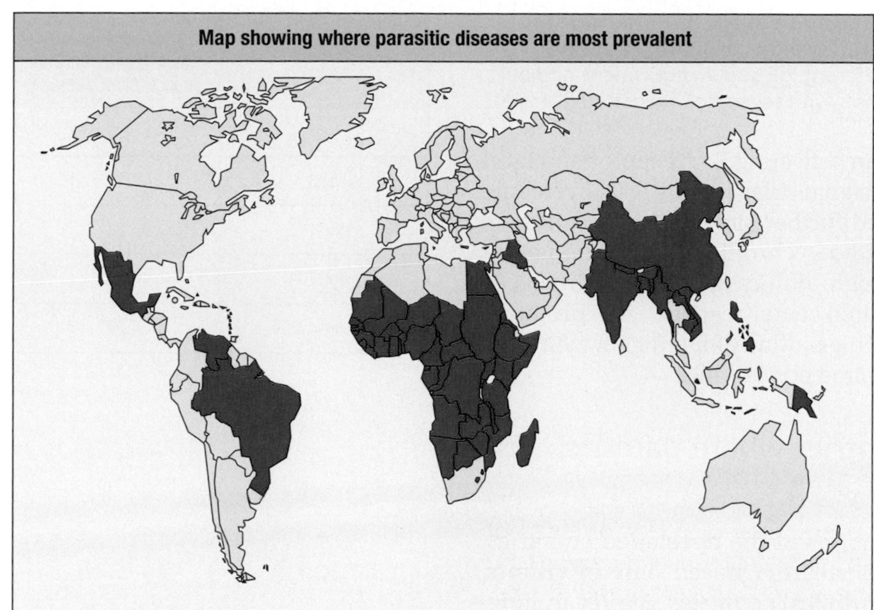

Map showing where parasitic diseases are most prevalent

Figure 14.4 Distribution of parasitic diseases worldwide. Red shading indicates countries in which more than five parasitic diseases are endemic.

IgE. The IgE-switched B cells quickly leave the germinal center to become plasma cells, and so the IgE they make at this early stage in the immune response has few or no somatic mutations and low or undetectable affinity for antigen. Switching to the four IgG isotypes occurs in the same order that the C-region genes are located in the heavy-chain locus (Figure 14.5). IgG3 is first, followed by IgG1, IgG2, and lastly IgG4. This order correlates with increasing duration of the germinal center reaction, increasing accumulation of somatic mutations, and increasing affinity of the antibody. In Figure 14.6 the differences in hypermutation between IgE and the four subtypes of IgG are compared for 10 hyperallergic individuals. In order of increasing mutation, the different isotypes are: IgE, IgG3 < IgG1 ~ IgG2 < IgG4. The early phase of the adaptive immune response involves IgM, IgE, and IgG3, which are short-lived (see Figure 4.29, p.103), and between them they favor the inflammatory effector functions of complement fixation, mast-cell degranulation, and phagocytosis (Figure 9.11, p. 249). In the later adaptive immune response, these isotypes give way to IgG1, IgG2, and IgG4, which are longer-lived, increasingly less inflammatory, and favor the formation of immune complexes. IgG4 is the last IgG isotype to be made; its synthesis is favored by the persistence of antigen, as occurs in parasite infections. IgG4 cannot fix complement and is preferentially bound by the inhibitory receptor FcγRIIB. When IgG4 molecules acquire Fab arms of different specificities (see Figure 4.35, p. 107), they can cross-link different antigens into immune complexes. Because IgG4 antibodies have the highest affinity of all the IgG isotypes, they can bind tightly to antigen with just one of their Fab arms, which drives cross-linking. When an

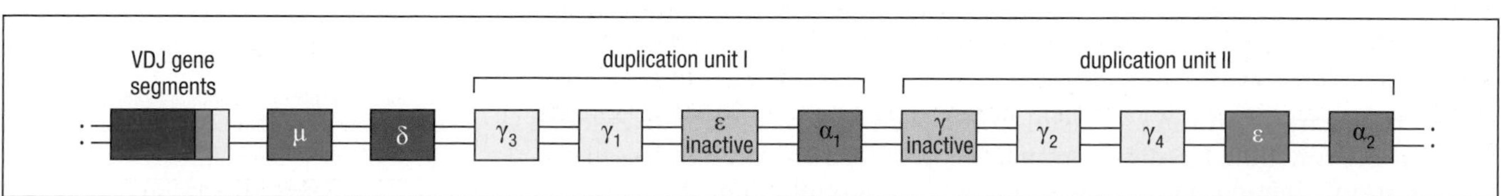

Figure 14.5 Organization of the immunoglobulin heavy-chain C-region genes on human chromosome 14. Downstream of the V, D, and J segments are nine C-region genes and two inactive pseudogenes. Following the μ and the δ genes are two duplicated units, each of which contains γ, ε, and α genes or pseudogenes. The genes in duplication unit 1 encode immunoglobulin isotypes that are more inflammatory in nature, whereas the isotypes encoded in duplication unit 2 are more associated with non-inflammatory T_H2 immunity.

Antibody isotype	Mean number of somatic mutations in heavy-chain V region
IgE	12
IgG3	18
IgG1	21
IgG2	22
IgG4	27

Figure 14.6 Levels of somatic mutation in IgE and the four IgG isotypes. Summary of the data from Wang et al. (2014) *PLoS ONE* 9:e89730, who analyzed ~6000 different IgE and an equivalent number of IgG sequences from 10 individuals with severe IgE-mediated allergies.

immune complex contains IgG1 bound to an activating Fcγ receptor and IgG4 bound to inhibitory FcγRIIB, the inhibitory signals triggered by IgG4 overcome the activating signals triggered by IgG1 and further slow down the immune response (Figure 14.7). Serum IgG4 is elevated in chronic helminth and other parasitic infections. Some B cells making high-affinity IgG, particularly IgG4, can switch to IgE at a late stage in the germinal center reaction. This produces plasma cells secreting IgE antibodies with a high affinity for antigen, which are used to arm mast cells, basophils, and activated eosinophils.

14-4 Allergy is prevalent in countries where parasite infections have been eliminated

The epidemic of allergy affects the populations of the developed countries, where parasitic infections are rare and usually diagnosed only in visitors, immigrants, or returning tourists from the tropical countries. Allergy is, however, a recent phenomenon in the developed countries, and its occurrence over the last 150 years correlates with the sanitization of food and water supplies, increasing personal hygiene, medical advances, and the elimination of creatures and habitats that spawn parasitic disease. Without the presence of helminth worms and other parasites in the human body, IgE-mediated immunity has lost its natural substrate and focus of attention. As a consequence of this ecological disruption, the T_H2 arm of immunity is more prone to develop in an unguided fashion, in which its powerful forces are no longer ranged against parasitic infection but are increasingly aimed at harmless environmental antigens. In human populations there is thus an inverse correlation between the incidence of parasitic infection and the incidence of allergic disease.

That this relationship was one of cause and effect was first put forward in 1989 in the **hygiene hypothesis**. This proposed that the epidemic of allergy was caused by improved hygiene, widespread vaccination to prevent infection, and increased use of antibiotic drugs to terminate infection (Figure 14.8). As children were exposed to fewer and less heavy infections than their parents had experienced, the children's immune systems were poorly educated, insufficiently exercised and less carefully controlled. In other words, a lack of practice in dealing with real-world infections led to the increased propensity to perceive danger in places where it does not exist. The epidemiological data that have accrued in the 25 years since the hygiene hypothesis was first proposed are all consistent with the allergy epidemic being the unanticipated and unwanted consequence of the successful elimination of infectious parasitic disease. With this realization, controlled and deliberate helminth infections are now being explored as potential treatments for severe allergic disease

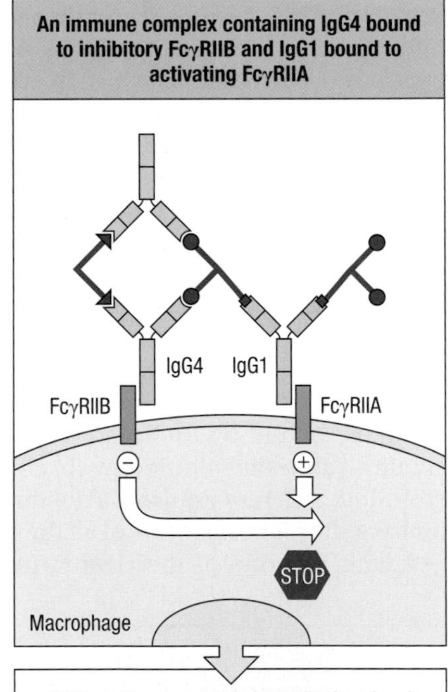
An immune complex containing IgG4 bound to inhibitory FcγRIIB and IgG1 bound to activating FcγRIIA
IgG4
IgG1
FcγRIIB
FcγRIIA
STOP
Macrophage
Inhibitory signals overwhelm activating signals and the macrophage does not secrete inflammatory cytokines

Figure 14.7 Inhibition of activating FcγRIIA by inhibitory FcγRIIB. IgG1 binds identical epitopes with its Fab arms, whereas IgG4 can have two different Fab arms that bind different epitopes.

14-5 IgE has distinctive properties that contrast with those of IgG

IgE and IgG share a common ancestor from which they have both diverged in structure and function. IgE differs from IgG and other antibody isotypes in several important characteristics. One is that IgE is not retained by the blood but instead concentrates in the tissues, where it binds to **FcεRI**, the high-affinity IgE receptor on mast cells (see Section 9-13). This receptor is a tetramer of a unit comprising one α chain that binds the IgE, and three signaling chains (one β and two γ chains) that activate the mast cell when antigen cross-links IgE (Figure 14.9).

A second important characteristic of IgE is that its three-dimensional structure differs greatly from that of IgG. Although both isotypes consist of two heavy chains and two light chains, IgG has a symmetrical, extended Y-shaped structure, whereas IgE has a bent, asymmetrical 'shrimp-like' structure in which the amino and carboxy terminals of the heavy chain are half as far apart as expected for an extended Y-shaped structure (see Figure 14.9). The source of this difference is the $C_\varepsilon2$ domain, which has no equivalent in IgG and replaces the flexible hinge of IgG with a more rigid structure. The Fab arms of IgE have less freedom of movement to capture antigens than their IgG counterparts. The binding of IgE to FcεRI involves one of the Cε3 domains of IgE and the two extracellular Ig-like domains of the receptor's α chain. The interaction surface is predominantly hydrophobic, which gives the interaction an affinity that is far greater ($K_d \approx 10^{10}$–10^{11} M^{-1}) than that of any other Fc receptor and is essentially irreversible. Unlike all other immunoglobulins, IgE binds to its high-affinity receptor in the absence of antigen, and in doing so the IgE molecule becomes even more bent. This conformation allows the antigen-binding sites to project out from the surface of the mast cell and be in advantageous positions for capturing antigens (see Figure 14.9).

The form of FcεRI expressed by mast cells is also expressed constitutively by basophils. These circulating cells, which enter inflamed tissues in order to function, have many properties in common with the sedentary mast cells. A second form of FcεRI, a trimer of units comprising one α and two γ chains, is expressed by a number of cell types after their activation by cytokines. Among these cells are eosinophils, monocytes, and smooth muscle cells, as well as platelets.

Tenets of the hygiene hypothesis
Excessive hygiene reduces childhood exposure to commensal and pathogenic microorganisms as well as to other humans and to animals.
Vaccination reduces the developing immune system's experience in facing natural infections, fighting them to a successful conclusion and terminating them.
Overreliance on antibiotics to terminate infections reduces the use of the immune system and its education in the discrimination of self from foreign pathogenic microorganisms. Antibiotics also select for resistant superbacteria, which also perturb the immune system.
↓
In childhood the development of the immune system is held back. It becomes poorly educated, exercised, and inexperienced in fighting real world infections. It turns to fighting imaginary infections and in the process causes a range of debilitating allergic diseases.

Figure 14.8 Tenets of the hygiene hypothesis.

14-6 IgE and FcεRI supply each mast cell with a diversity of antigen-specific receptors

After the primary IgE response has subsided and the antigen has been cleared by the usual means, all antigen-specific IgE molecules that did not encounter antigen become bound by their Fc regions to FcεRI on tissue mast cells. Because mast cells are long-lived, whereas basophils are short-lived, only the mast cells maintain a depot of antigen-specific IgE. This depot represents a memory of the primary immune response. The strength of the interaction between IgE and FcεRI is such that the FcεRI molecules of tissue mast cells are saturated with IgE of various specificities for antigen. This coating of IgE provides the mast cells with a diversity of antigen receptors. Here is another situation in which the tactics of innate and adaptive immunity have been combined (see Chapter 12). Unlike B cells, which commit to making antibody of just one specificity, mast cells passively acquire a diversity of IgE with different specificities (Figure 14.10). This is reminiscent of the multiple and diverse receptors for pathogens used by the macrophages, dendritic cells, and NK cells of innate immunity. This makes the mast cell more versatile than the B cell because it can respond to a variety of antigens, which can be diverse in their structure and biological origin. The mast cell is also a quicker and more efficient responder to antigen than the B cell. In the absence of antigen, the mast cell is loaded with granules containing preformed inflammatory mediators. When an antigen is encountered for the second time it will bind to the receptors and activate the cells to secrete their inflammatory mediators (Figure 14.11). This initial degranulation, which occurs immediately after IgE binds antigen, is followed by the induced synthesis of a wider range of inflammatory mediators.

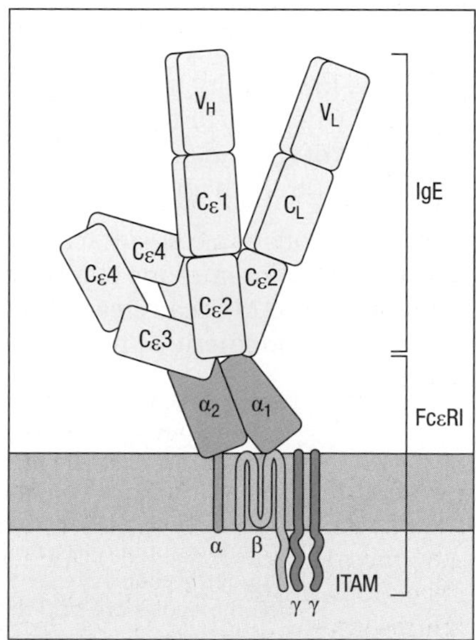

Figure 14.9 Structure of IgE bound to FcεRI.

14-7 FcεRII is a low-affinity receptor for IgE Fc regions that regulates the production of IgE by B cells

Whereas FcεRI is involved in mediating the effector functions of IgE, a second activating Fcε receptor has several roles in the production of IgE and its

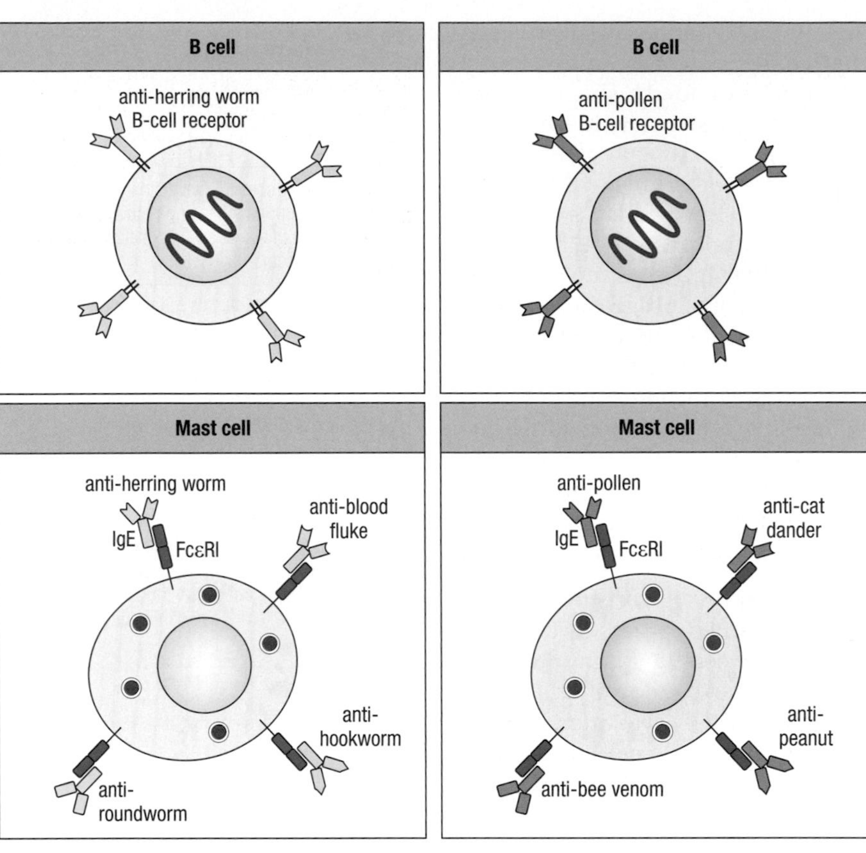

Figure 14.10 Comparison of the diversity of parasite-specific or allergen-specific antigen receptors on a B cell and a mast cell. Isotype-switched B cells are programmed to make one type of IgE molecule that is specific for a single antigen or allergen (upper panels). In contrast, a mast cell passively acquires IgE molecules with different specificities and made by different B cells, which bind to the cell-surface Fc receptor for IgE, FcεRI (lower panels). A mast cell displays around half a million FcεRI molecules on its surface and so can be armed with a variety of different IgEs at high density. By comparison, a B cell has only 50,000–100,000 B-cell receptors. The panels on the left depict anti-helminth IgEs in Africans infested with helminth worms. The panels on the right depict anti-allergen IgEs in Europeans suffering from allergies.

maintenance in the circulation. This is **FcεRII** (CD23), which has a quite different structure and functional properties from those of FcεRI (Figure 14.12). The two receptors bind to topographically separated sites on IgE Fc regions but they cannot bind simultaneously to the same IgE molecule because the conformation of the IgE Fc region that permits binding to FcεRII has no affinity for FcεRI, and vice versa. FcεRII is a homotrimer of the CD23 polypeptide, which has a C-type lectin-like domain and a long extended stalk (see Figure 14.12). The trimer resembles a small bouquet of flowers in which each binding-site domain is a separate flower, the three stalks oligomerize, and the cytoplasmic domains engage signaling molecules. Each of the three CD23 polypeptides binds with low affinity to a different IgE molecule. For this reason, FcεRII is called the low-affinity IgE receptor, to contrast it with the high-affinity FcεRI. However, when two or three binding sites on FcεRII are occupied by immune complexes of IgE and antigen, the avidity is much higher than the affinity, and the binding strength approaches that of FcεRI.

FcεRII functions as a cell-surface receptor and a soluble receptor for IgE and is also a receptor for immune complexes of antigen with IgE. A cell-surface protease called ADAM10 can cleave the CD23 polypeptide at different sites to produce either monomeric or trimeric forms of soluble CD23. ADAM10 is one of

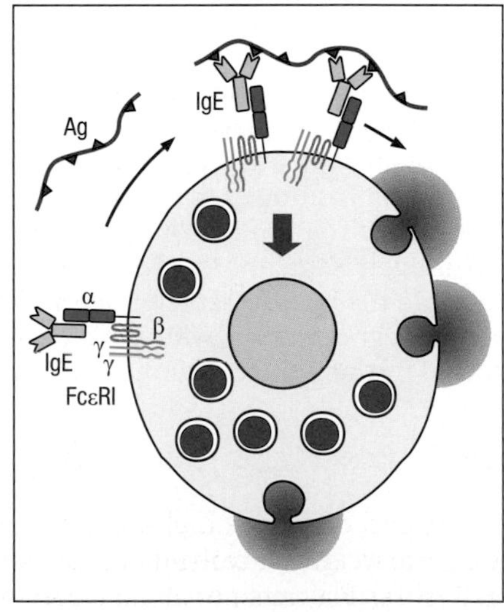

Figure 14.11 Cross-linking of FcεRI on the surface of mast cells by antigen and IgE causes mast-cell activation and degranulation. The high-affinity IgE receptor (FcεRI) is made up of one α chain, one β chain, and two γ chains. The α chain contains the IgE-binding site, which is formed by the two extracellular immunoglobulin-like domains. The β chain and the two disulfide-bonded γ chains are largely intracellular and contribute to signaling. IgE molecules are bound to mast-cell FcεRI in the absence of antigen. When antigen binds to the IgE molecules, they and their associated receptor molecules become cross-linked. This event transmits a signal to the mast cell that leads quickly to the release of preformed granules (purple) containing histamine and other inflammatory mediators.

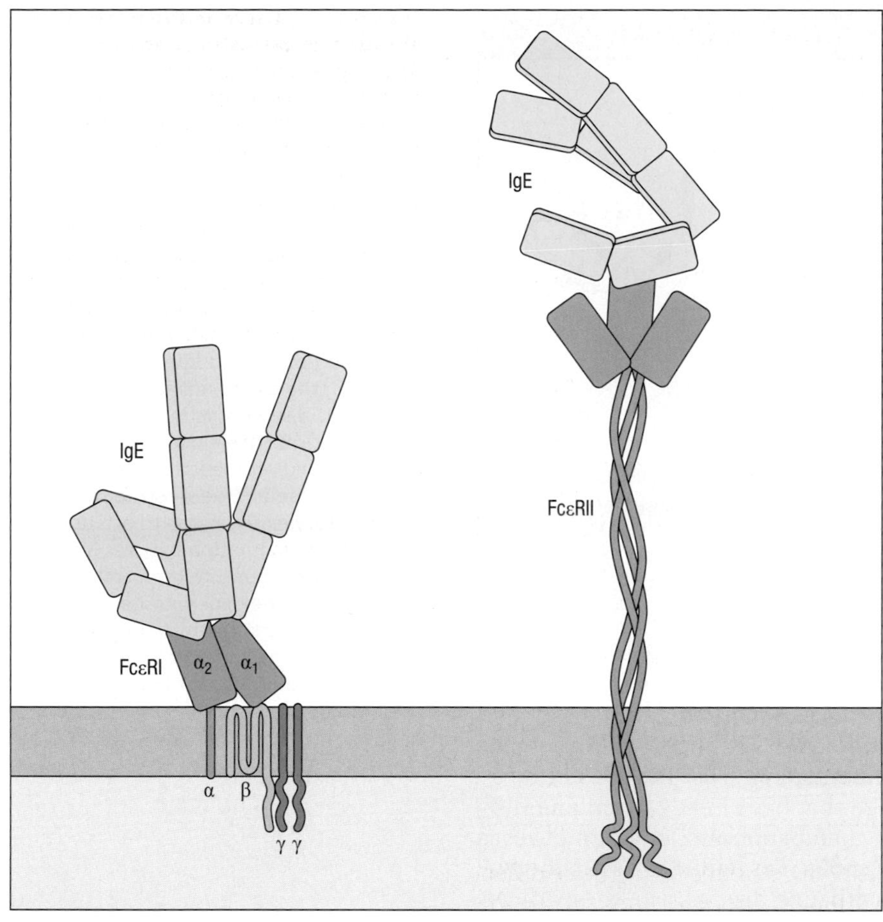

Figure 14.12 Comparison of the structure of FcεRI and FcεRII. FcεRI is composed of a single IgE-binding polypeptide (the α chain) that is associated with a complexity of signaling polypeptide chains (the β and γ chains). Fcε RII is a trimer of one polypeptide that contains both an extracellular IgE-binding domain and an intracellular signal-transduction domain.

a family of such proteins that are collectively called **sheddases** because they enable cells to shed soluble forms of their membrane proteins. The soluble forms of FcεRII have cytokine-like activities that act in an autocrine or paracrine fashion. Their potential effects are wide-ranging because FcεRII is expressed by B cells and by cells with which they interact, including T cells, monocytes, follicular dendritic cells, and bone marrow stromal cells. In addition to its binding site for IgE, FcεRII has sites that bind to the B-cell co-receptor (CR2), MHC class II molecules, and integrins.

When a B cell in a germinal center reaction commits to the synthesis of IgE, it begins to produce and then shed FcεRII. In the absence of antigen, a complex is formed at the B-cell surface in which soluble FcεRII interacts with both the B-cell receptor (surface IgE) and the B-cell co-receptor. The trimeric form of soluble FcεRII generates synergistic signaling that promotes the differentiation of the B cell into an IgE-secreting plasma cell (Figure 14.13), whereas the monomeric form of soluble FcεRII inhibits such differentiation.

Because FcεRII functions in a soluble trimeric form, it is absolutely crucial that FcεRI and FcεRII cannot bind simultaneously to the same IgE molecule. If that were possible, soluble trimeric FcεRII could activate mast cells and basophils in the absence of antigen by cross-linking the IgE molecules bound to FcεRI.

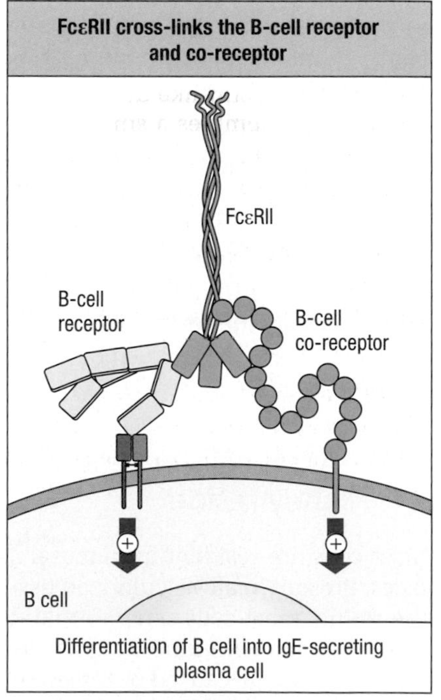

Figure 14.13 FcεRII cross-links the B cell receptor and the B cell co-receptor to stimulate the production of IgE.

14-8 Treatment of allergic disease with an IgE-specific monoclonal antibody

In the past decade, a high-affinity humanized monoclonal IgG that is specific for human IgE (omalizumab) has been shown to alleviate the symptoms of asthma and a range of allergic conditions. The key feature of this antibody is

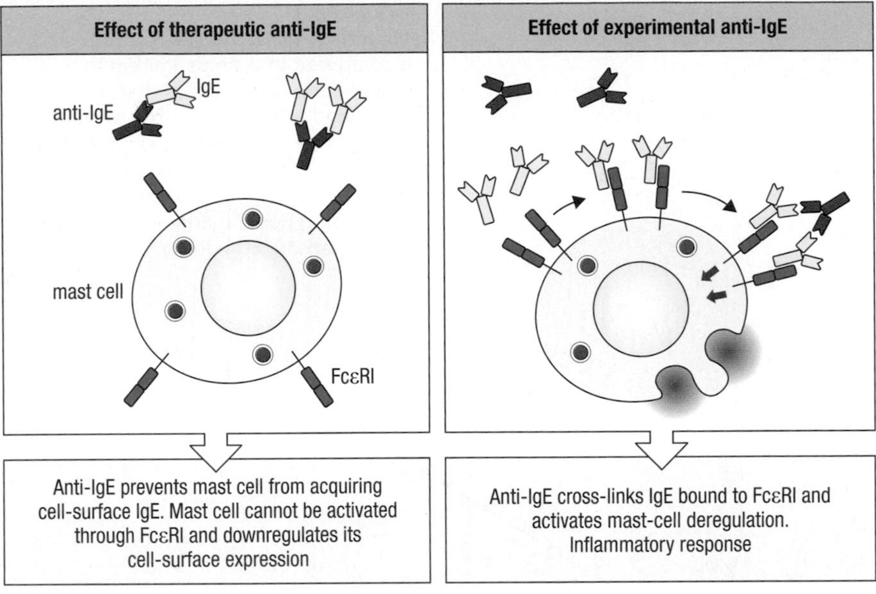

Figure 14.14 Therapeutic application of monoclonal anti-IgE antibody. Omalizumab is a monoclonal antibody used in clinical practice to reduce the symptoms of IgE-mediated allergic diseases such as asthma. On binding to IgG, omalizumab covers up the site on IgE that binds to FcεRI on mast cells, basophils, and eosinophils (left panel). The antibody thus prevents IgE from arming these effector cells and enabling them to respond to antigens and allergens that have stimulated the production of specific IgE. When first proposed, the use of anti-IgE as a therapy for allergic disease was met with concern and skepticism. These negative reactions were founded on the knowledge that anti-IgE antibodies were routinely used in laboratory experiments to activate mast-cell degranulation by cross-linking IgE molecules bound to mast cell FcεRI (right panel). The experimental anti-IgE antibodies that activate mast cells bind to epitopes on IgE that are away from the binding site for FcεRI, whereas the therapeutic antibody covers up the binding site for FcεRI.

that its target epitope corresponds to the site on IgE bound by FcεRI. This gives omalizumab two important properties. The first is that it binds soluble IgE and prevents it from binding to FcεRI on the surface of mast cells, basophils, and activated eosinophils. This prevents the interaction of antigens and allergens with specific IgE from activating these inflammatory cells. The second property is that omalizumab cannot bind to IgE that has already been bound by FcεRI on a cell surface. This means that the therapeutic antibody cannot act as a surrogate allergen and cross-link two IgE molecules bound to FcεRI, thereby activating the inflammatory cells and exacerbating disease (Figure 14.14). An additional benefit is that IgE bound to omalizumab is unable to bind to FcεRII, which further suppresses the functionality of IgE. The reduced binding of IgE to the two IgE receptors inhibits IgE-mediated allergic responses and also leads to reduced expression of FcεRI on the surface of mast cells and basophils.

The study of patients receiving anti-IgE therapy has led to biological and clinical insight into the functions of IgE. After 10 years of anti-IgE therapy, no susceptibility to infection or other disease has been observed, which demonstrates that people living in the developed countries do not depend on IgE for immune protection. This is additional supporting evidence for the importance and specialization of IgE-mediated immunity in the defense against helminth worms and other parasites. In the clinic, the success or not of anti-IgE therapy is helping clinicians to distinguish between asthma due to IgE-mediated allergy and that due to other causes.

14-9 Mast cells defend and maintain the tissues in which they reside

Mast cells are resident in mucosal and epithelial tissues lining the body surfaces. Present in all vascularized tissues except the central nervous system and the retina, mast cells serve to maintain the integrity of the tissue where they reside, alert the immune system to local trauma and infection, and facilitate the repair of damage caused by infection or wounds. Morphologically, the defining characteristic of the mast cell is the cytoplasm, which is filled with 50–200 large granules that contain inflammatory mediators. The name **mast cell** derives from the German word *Mastzellen*, meaning 'fattened cells' or 'well-fed cells,' which is how they appear in the microscope (Figure 14.15). Mast-cell granules contain histamine, heparin, tumor necrosis factor-α (TNF-α), chondroitin sulfate, neutral proteases, and other degradative enzymes and inflammatory mediators. Heparin, an acidic proteoglycan, is largely responsible for the characteristic staining of the mast-cell granules with basic dyes.

Although mast cells are best known for their destructive effects in IgE-mediated allergic reactions, they also respond constructively to a wider range of stimuli. As well as expressing FcεRI, mast cells express Toll-like receptors and Fc receptors for IgA and IgG. Mast cells can thus contribute to both the innate and the adaptive immune responses to infection. Whereas the IgE-mediated activation of mast cells via FcεRI is limited to degranulation and to the synthesis of small inflammatory mediators known as eicosanoids, signaling through other receptors induces the manufacture and secretion of cytokines that recruit neutrophils, eosinophils, and effector T cells to the infected tissue, and also induces the secretion of growth factors that promote the repair of tissue damage. Mast cells can thus tailor their cytokine response to the type of infecting pathogen (Figure 14.16).

Two kinds of human mast cell are distinguished by their location and the type of protease they produce: a **mucosal mast cell**, which produces the protease tryptase, and a **connective tissue mast cell**, which produces chymotryptase. In patients with T-cell immunodeficiencies, only connective tissue mast cells are present, indicating that the development of mucosal mast cells depends on the effector T cells that populate the mucosal tissues (see Section 10-4). Such tissue-specific differentiation of the mucosal mast cells in the gut probably specializes the mast cells for stimulating non-inflammatory responses against the gut microbiota and IgE-mediated responses against intestinal parasites.

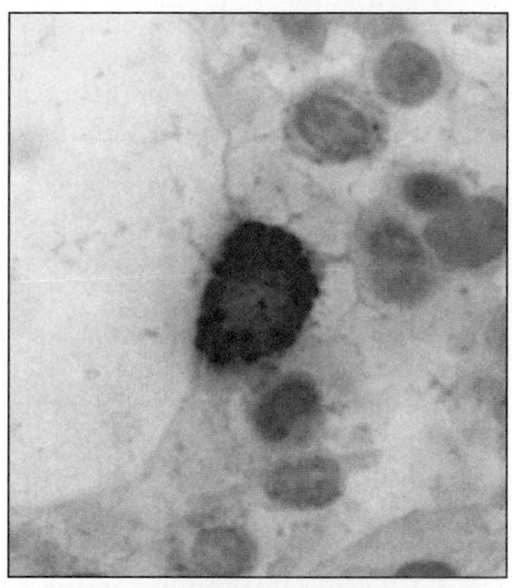

Figure 14.15 Light micrograph of a mast cell stained for the granule protease chymase to show the numerous cytoplasmic granules. The mast cell is in the center of the picture: its nucleus is stained pink and the cytoplasmic granules are stained red. Magnification ×1000. Photograph courtesy of D. Friend.

14-10 Tissue mast cells orchestrate IgE-mediated reactions through the release of inflammatory mediators

Mast-cell activation occurs in the presence of any antigen that can cross-link the IgE molecules bound to FcεRI at the cell surface. This can be accomplished

Class of product	Product	Biological effect
Enzyme	Tryptase, chymase, cathepsin G, carboxypeptidase	Remodel connective tissue matrix
Toxic mediator	Histamine, heparin	Poison parasites Increase vascular permeability Cause smooth muscle contraction
Cytokine	TNF-α (some stored preformed in granules)	Promotes inflammation, stimulates cytokine production by many cell types, activates endothelium
	IL-4, IL-13	Stimulate and amplify T_H2-cell response
	IL-3, IL-5, GM-CSF	Promote eosinophil production and activation
Chemokine	CCL3	Attracts monocytes, macrophages, and neutrophils
Lipid mediator	Leukotrienes C_4, D_4, and E_4	Cause smooth muscle contraction Increase vascular permeability Cause mucus secretion
	Platelet-activating factor	Attracts leukocytes Amplifies production of lipid mediators Activates neutrophils, eosinophils, and platelets

Figure 14.16 Effector molecules released by mast-cell degranulation. The boxes shaded in red show the inflammatory mediators (proteases, histamine, heparin, and TNF-α) that are prepackaged in granules and released immediately after mast cells have been stimulated by antigen binding to FcεRI-associated IgE. TNF-α is also synthesized after mast-cell activation. The inflammatory mediators listed in the white boxes are synthesized and released only as a consequence of mast-cell activation.

by antigens with repetitive epitopes that cross-link IgE molecules of the same specificity, or by antigens possessing two or more different epitopes that cross-link IgE molecules of different specificities. Once a mast cell's receptors have been cross-linked, degranulation occurs within a few seconds, releasing the stored mediators into the immediate extracellular environment (see Figure 14.11).

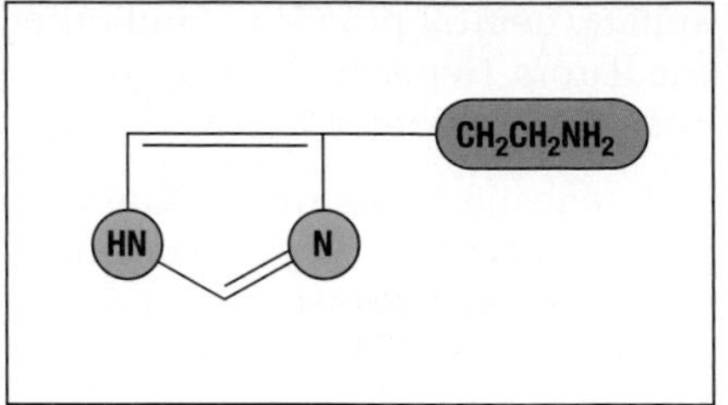

Figure 14.17 Chemical structure of histamine.

Prominent among the mediators is histamine, an amine derivative of the amino acid histidine (Figure 14.17). Histamine exerts a variety of physiological effects through three kinds of histamine receptor—H1, H2, and H3—which have been defined on different cell types. Acute allergic reactions involve the binding of histamine to the H1 receptor on nearby smooth muscle cells and on endothelial cells of blood vessels. Stimulation of the H1 receptor on endothelial cells induces vessel permeability and the entry of other cells and molecules into the allergen-containing tissue, causing inflammation. Smooth muscle cells are induced to contract on binding histamine, which constricts the airways, for example, and histamine also acts on mucosal epithelia to increase the secretion of mucus. All these actions produce different effects depending on the tissue exposed to allergen. Sneezing, coughing, wheezing, vomiting, and diarrhea can all be symptoms induced in the course of an allergic reaction.

Other molecules released from mast-cell granules include mast-cell chymotryptase, tryptase, and other neutral proteases that activate metalloproteases in the extracellular matrix. Collectively, these enzymes break down extracellular matrix proteins. The action of histamine is complemented by that of TNF-α, another cytokine released from mast-cell granules. TNF-α activates endothelial cells, causing an increased expression of adhesion molecules, thus promoting leukocyte traffic from the blood into the increasingly inflamed tissue (see Figure 3.9, p. 54). Mast cells are unique in their capacity to store TNF-α and release it on demand. At the start of an inflammatory response, mast cells are usually the main source of this inflammatory cytokine.

Besides the preformed inflammatory mediators in granules, mast cells synthesize and secrete other mediators in response to activation (see Figure 14.16). These include chemokines, cytokines—IL-4 and more TNF-α—and eicosanoids. The latter, which include the prostaglandins and the leukotrienes, are lipids synthesized from fatty acids (Figure 14.18, top panel). All these mediators act locally within the area of tissue surrounding the activated mast cells. The leukotrienes have activities similar to those of histamine, but are more than 100 times more potent on a molecule-for-molecule basis. The two kinds of mediator are therefore complementary. Histamine provides a rapid response while the more potent leukotrienes are being made. In the later stages of allergic reactions, leukotrienes are principally responsible for inflammation, smooth muscle contraction, the constriction of airways, and the secretion of mucus from mucosal epithelium. Mast cells also secrete prostaglandin D_2 (PGD_2), which promotes the dilation and increased permeability of blood vessels and also acts as a chemoattractant for neutrophils. Aspirin reduces inflammation by inactivating prostaglandin synthase, the first enzyme in the cyclooxygenase pathway. The inactivation is irreversible because aspirin binds covalently to the active site of the enzyme (Figure 14.18, bottom panel).

The combined effect of the chemical mediators released by mast cells is to attract circulating leukocytes to the site of mast-cell activation, where they amplify the reaction initiated by antigen and IgE on the mast-cell surface. These effector leukocytes include eosinophils, basophils, neutrophils, and T_H2 lymphocytes. In parasitic infections these cells work together to promote explosive reactions that either kill the parasite or expel it from the body (see Section 9-13). In this situation, the benefit of eliminating the pathogen is worth suffering the damage caused by the immune response. In allergic reactions aimed at harmless antigens, the immune response provides no benefit at all,

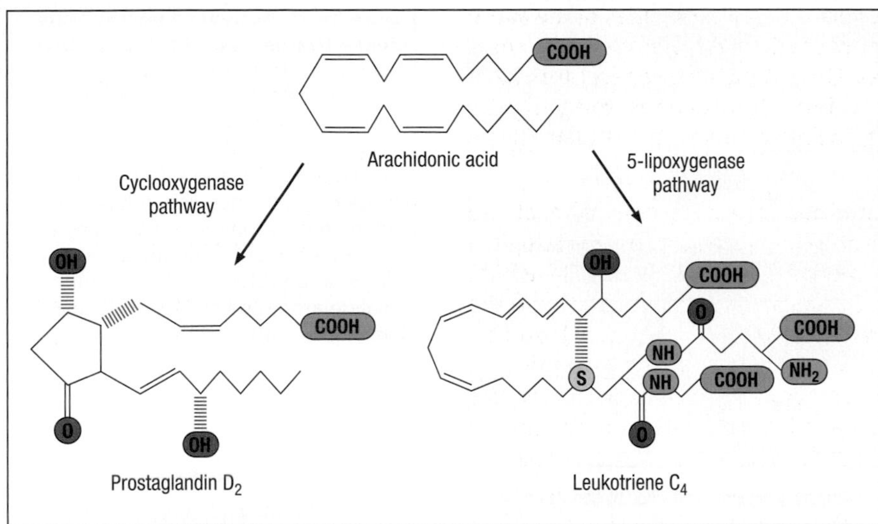

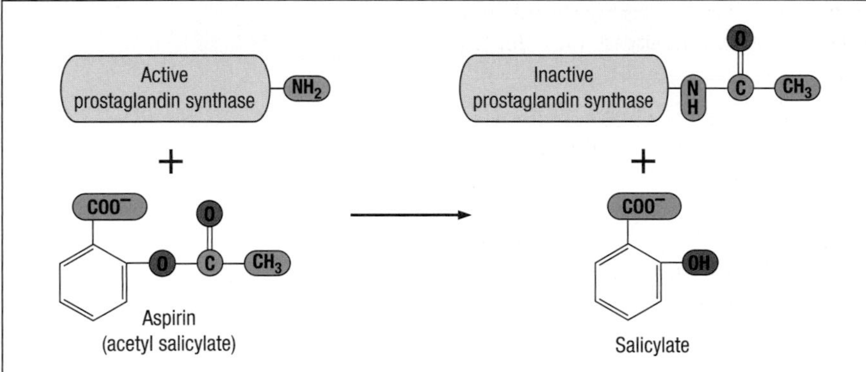

Figure 14.18 Mast cells synthesize prostaglandins and leukotrienes from arachidonic acid by different enzyme pathways. Arachidonic acid, an unsaturated fatty acid, is used as the substrate for the cyclooxygenase pathway to make prostaglandins, and as the substrate for the 5-lipoxygenase pathway to make leukotrienes, as shown in the top panel. Arachidonate is itself made from linolenic acid, one of the essential unsaturated fatty acids that cannot be made by human cells and must therefore be provided in the diet. Aspirin (acetyl salicylate) prevents the cyclooxygenase pathway from synthesizing prostaglandins by irreversibly inhibiting the enzyme prostaglandin synthase, as shown in the bottom panel.

only detrimental side-effects—damage to the body's tissues and impairment of their function. For allergens that pervade the environment and provide a continual stimulation, the allergic response will gather in strength as increasing numbers of mast cells and eosinophils are recruited to the affected tissue and join in the frustrated attempts to eliminate the 'pathogen.'

14-11 Eosinophils are specialized granulocytes that release toxic mediators in IgE-mediated responses

Eosinophils are granulocytes whose granules contain arginine-rich basic proteins. In histological sections these proteins stain heavily with eosin, hence the name eosinophil, meaning 'liking eosin' (Figure 14.19). Only a minority of eosinophils are present in the blood. Most are resident in tissues, especially in the connective tissue immediately underlying the epithelia of the respiratory,

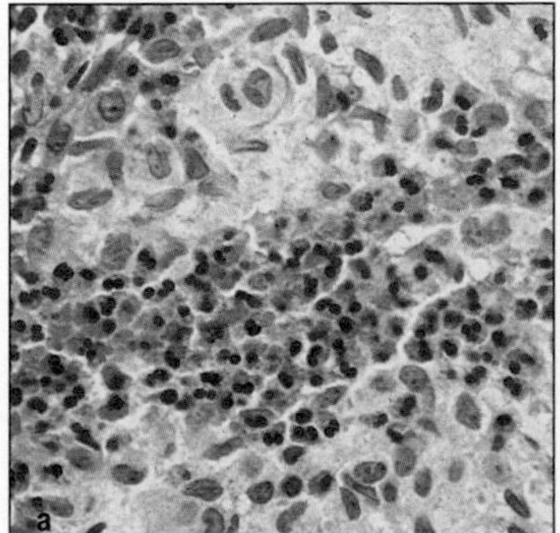

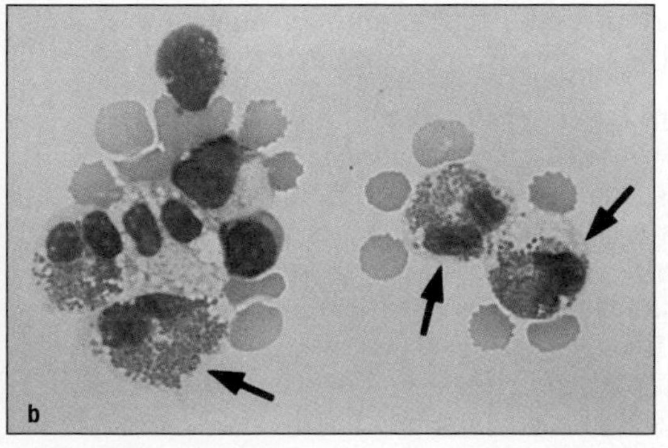

Figure 14.19 Eosinophils have a characteristic pattern of staining in histological sections. The light micrograph in panel a shows a section through skin affected by Langerhans cell histiocytosis—a condition in which Langerhans cells and eosinophils mutually stimulate each others' proliferation. The eosinophils have bilobed nuclei and stain pink with the stain eosin. The higher-power light micrograph in panel b is of a blood smear in which partly degranulated eosinophils (arrows) are surrounded by erythrocytes. Left panel courtesy of T. Krausz; right panel courtesy of F. Rosen and R. Geha.

Class of product	Product	Biological effect
Enzyme	Eosinophil peroxidase	Poisons parasites and mammalian cells by catalyzing halogenation Triggers histamine release from mast cells
	Eosinophil collagenase	Remodels connective tissue matrix
Toxic protein	Major basic protein	Poisons parasites and mammalian cells Triggers histamine release from mast cells
	Eosinophil cationic protein	Poisons parasites Neurotoxin
	Eosinophil-derived neurotoxin	Degradation of RNA Antiviral effects
Cytokine	IL-3, IL-5, GM-CSF	Amplify eosinophil production by bone marrow Cause eosinophil activation
Chemokine	CXCL8	Promotes influx of leukocytes
Lipid mediator	Leukotrienes C_4, D_4, and E_4	Cause smooth muscle contraction Increase vascular permeability Cause mucus secretion
	Platelet-activating factor	Attracts leukocytes Amplifies production of lipid mediators Activates neutrophils, eosinophils, and platelets

Figure 14.20 Activated eosinophils release toxins, cytokines and other inflammatory mediators. The red boxes denote the enzymes and toxic proteins that are preformed, packaged in the granules, and released immediately on eosinophil activation. The white boxes denote the cytokines, chemokines, and lipid mediators that are synthesized as a consequence of eosinophil activation. Halogenation refers to the addition of halide atoms, such as chlorine and bromine, to molecules.

gastrointestinal, and urogenital tracts. As with mast cells, the activation of eosinophils by external stimuli leads to the staged release of toxic molecules and inflammatory mediators. Preformed chemical mediators and proteins present in the granules are released first (Figure 14.20, top panel). The normal function of these highly toxic molecules is to kill invading microorganisms and parasites directly. This is followed by a slower induced synthesis and secretion of prostaglandins, leukotrienes, and cytokines, which amplify the inflammatory response by the activation of epithelial cells and leukocytes, including more eosinophils (Figure 14.20, bottom panel).

The eosinophil response is highly toxic and is potentially damaging to the host as well as to parasites. Various control mechanisms limit the damage. When the body is healthy, the number of eosinophils is kept low by restricting their production in the bone marrow. When infection or antigenic stimulation activates T_H2 cells, the IL-5 and other cytokines released by the T_H2 cells stimulate the bone marrow to increase both the production of eosinophils and their release into the circulation. The migration of eosinophils into tissues is controlled by a group of chemokines (CCL5, CCL7, CCL11, and CCL13) that bind to the receptor CCR3 expressed by eosinophils. Of these chemokines, CCL11, also called **eotaxin**, is particularly important in the migration of eosinophils and is produced by activated endothelial cells, T cells, and monocytes.

The activity of eosinophils is further controlled by modulation of their sensitivity to external stimuli. This is achieved through the regulated expression of FcεRI. In the resting state, eosinophils do not express FcεRI, do not bind IgE, and therefore cannot be induced to degranulate by antigen. Once an inflammatory response has been set in motion, cytokines and chemokines in the inflammatory site induce eosinophils to express FcεRI. The expression of Fcγ receptors and complement receptors on the eosinophil surface also increases, facilitating the binding of eosinophils to pathogen surfaces coated with IgG and complement.

The capacity of eosinophils to cause tissue damage is clearly seen by their effects on patients who have abnormally high numbers of these cells, a condition called **eosinophilia**. Certain T-cell lymphomas, for example, constitutively secrete IL-5, which increases production of eosinophils and creates a state of eosinophilia. Patients with eosinophilia suffer damage to the endocardium of the heart and to nerves, which can lead to heart failure and neuropathy. Both of these effects are caused by the cytotoxic and neurotoxic proteins released from eosinophil granules.

In localized allergic reactions, mast-cell degranulation and the activation of T_H2 cells cause activated eosinophils at the site. Their presence is characteristic of chronic allergic inflammation to accumulate and eosinophils are the principal cause of the airway damage that characterizes chronic asthma.

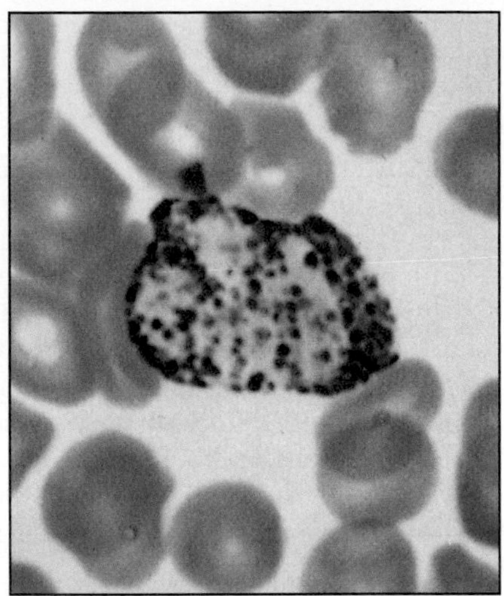

Figure 14.21 Light micrograph of a basophil stained with Wright's Giemsa. The granular basophil in the center is surrounded by erythrocytes in this blood smear. Magnification ×1000. Photograph courtesy of D. Friend.

14-12 Basophils are rare granulocytes that initiate T_H2 responses and the production of IgE

Basophils are granulocytes having granules that stain with basic dyes such as hematoxylin (Figure 14.21). These granules contain a similar, but not identical, set of mediators to those of mast cells. Although basophils resemble mast cells in several ways, developmentally they are more closely related to eosinophils. They have a common stem-cell precursor and require similar growth factors, including IL-3, IL-5, and GM-CSF. The production of eosinophils and basophils is reciprocally regulated, so that a combination of TGF-β and IL-3 promotes the maturation of basophils while suppressing that of eosinophils. Basophils normally constitute less than 1% of the white blood cells, a scarcity that has made them much more difficult to study than other leukocytes.

Knowledge of basophil biology is now advancing and points to the basophil as the key cell that initiates T_H2 responses. This is achieved by the basophil's unique capacity to secrete the T_H2 cytokines IL-4 and IL-13 at the beginning of an immune response. During the innate immune response, basophils are recruited to infected tissue, where they are activated through their Toll-like receptors and other receptors of innate immunity. In the infected tissue, basophils contribute effector functions similar to those of eosinophils. At the beginning of a T_H2 adaptive immune response, basophils are recruited to the secondary lymphoid tissues, where their secretion of IL-4 and IL-13 biases the response toward T_H2 (see Section 8-10).

Basophils express CD40 ligand, which binds to the CD40 of antigen-stimulated B cells and, in combination with secreted IL-4 and IL-13, drives isotype switching to IgE and IgG4. Once specific IgE has been made, it binds to FcεRI on basophils and provides the main means for activating basophils in the secondary response to the pathogen.

Mast cells, eosinophils, and basophils often act in concert. Mast-cell degranulation initiates the inflammatory response, which then recruits eosinophils and basophils. Eosinophil degranulation releases **major basic protein** (see Figure 14.20), which in turn causes the degranulation of mast cells and basophils. This latter effect is augmented by one or more of the cytokines—IL-3, IL-5, and GM-CSF—that affect the growth, differentiation, and activation of eosinophils and basophils.

Summary

The purpose of the T_H2 arm of adaptive immunity is to control the populations of metazoan parasites that infest human bodies. Prominent among these parasites are the helminth worms that live and persist in the gut. Antibodies of the IgE isotype are specialized in providing immunity against these parasites. IgE is bound by the FcεRI receptor of basophils, eosinophils, and mast cells, which provide the violent effector functions that are needed to dislodge parasites from mucosal tissues, kill them, and eject them from the body. In a primary immune response, an early wave of low-affinity IgE antibodies can be followed later by a wave of high-affinity IgE. Orchestrating the IgE response of B cells is the FcεRII receptor (CD23). In the absence of antigen, the IgE made in the primary response is used to arm tissue mast cells, giving them parasite-specific receptors that ensure a strong secondary response to subsequent incursions by the parasite. In very recent human history, the populations of developed countries have effectively eliminated parasite infections. Correlating with this change is a marked increase in the incidence of allergic diseases, which are caused by inappropriate T_H2 responses targeted to protein antigens derived from plants and animals that are present in the environment but pose no direct threat to human health. In the developing countries, where parasite infections are endemic, there has been no such increase in allergic disease. But it is beginning to happen as these countries adopt a Western style of life.

IgE-mediated allergic disease

In this part of the chapter we examine what types of protein antigen can act as allergens, the nature of diseases they cause, and how these allergic diseases can be treated, cured, and prevented. Allergic diseases vary in their symptoms and their severity, depending on the route of allergen entry and the tissues and organs affected.

14-13 Allergens are protein antigens, some of which resemble parasite antigens

Symptoms of allergic disease are felt only after a person has made IgE antibodies that react with the allergen and those antibodies have armed mast cells in the tissue exposed to the allergen. In this situation, the person has become sensitized to the allergen. The allergens that provoke type I hypersensitivity responses are always proteins. Some are foreign proteins from plants or animals, whereas others are human proteins that are chemically modified, for example by a drug such as penicillin. Because allergens represent only a minute proportion of all the proteins that enter human bodies, a central focus of research on allergy is to define the individual protein allergens and identify what characteristics distinguish them as a group from non-allergenic proteins. Definitive answers have yet to be found, but certain properties characterize the major inhaled allergens (Figure 14.22).

Features of inhaled allergens that promote priming of the T_H2 cells that drive the IgE response	
Molecular type	Proteins, because only they induce T-cell responses
Function	Many allergens are proteases
Low dose	Favors activation of IL-4-producing CD4 T cells
Low molecular mass	Allergen diffuses out of particle into mucus
High solubility	Allergen is readily eluted from particle
High stability	Allergen survives in desiccated particles
Contains peptides that bind host MHC class II	Necessary for T-cell priming

Figure 14.22 Properties of inhaled allergens.

Many allergens are small, soluble proteins that are present in dried-up particles of material derived from plants and animals. Examples are pollen grains, the mixture of dried cat skin and saliva that forms dander, and the dried feces of the house dust mite *Dermatophagoides pteronyssimus*. The light dry particles become airborne and are inhaled by humans. Once inhaled, the particles become caught in the mucus bathing the epithelia of the airways and lungs. They then rehydrate, releasing the antigenic proteins. These antigens are carried to professional antigen-presenting cells within the mucosa. The antigens are then processed and presented by these cells to CD4 T cells, stimulating a T_H2 response that leads to IgE being made and bound by mast cells (Figure 14.23). Small soluble protein antigens are more efficiently leached out of particles and penetrate the mucosa. A substantial proportion of allergens are proteases, and it is likely that their enzymatic activities facilitate the breakdown of the particle, the release of allergen, and the generation of peptides that bind to MHC class II molecules and stimulate T_H2 cells.

The major allergen responsible for more than 20% of the allergies in the human population of North America is a cysteine protease derived from *D. ptero-*

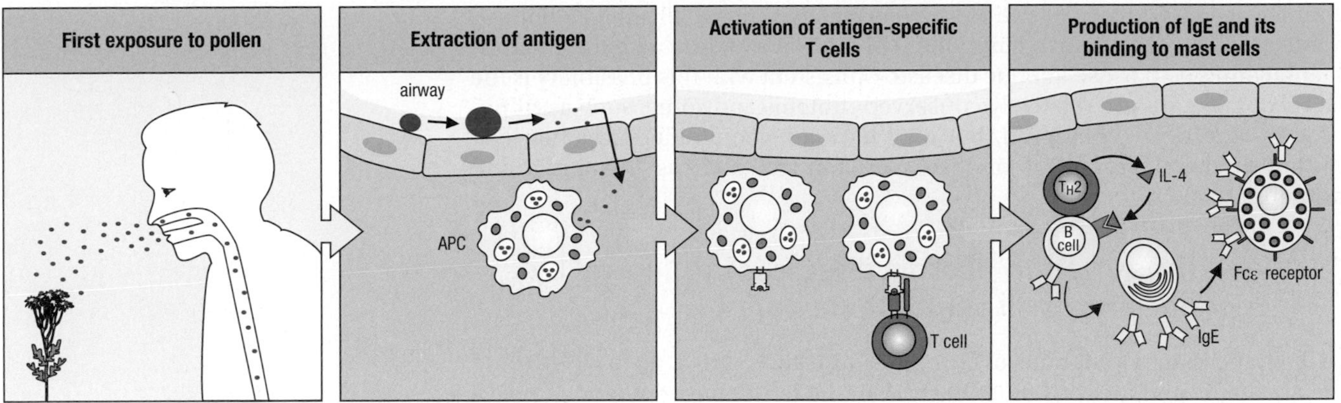

Figure 14.23 Sensitization to an inhaled allergen. Inhaled pollen particles become trapped at the mucosal surface of the airways (first panel). Mucosal antigen-presenting cells (APC), either dendritic cells or macrophages, take up the antigens and allergens that leach out of the pollen (second panel). The APCs activate antigen-specific naive T cells, driving them to become IL4-secreting T_H2 helper cells (third panel). The T_H2 cells activate pollen-specific naive B cells. IL-4 secreted by the T_H2 cell binds to the B cell's IL-4 receptor, driving the B cell to switch its immunoglobulin isotype to IgE and to become an IgE-secreting plasma cell. Pollen-specific IgE binds to FcεRI on mast cells (fourth panel). During this first exposure to the pollen, no ill effects are felt. By becoming armed, or sensitized, with allergen-specific IgE, the mast cells become acutely sensitive to subsequent inhalation of pollen. Extensive mast cell degranulation will occur as soon as pollen antigens bind to mast-cell associated IgE and cause disease (not shown).

nyssimus. Advances in the heating and cooling of homes, offices, and other buildings are believed to be responsible for the prevalence of this allergy because they provide an environment that encourages both the growth of *D. pteronyssimus* and the desiccation of its feces. The air currents created by forced-air heating, air conditioners, and vacuum cleaners all help to move the particles into the air, where they will be breathed in by the human inhabitants of the buildings. The cysteine protease of *D. pteronyssimus* is related to the cysteine protease papain, which comes from the papaya fruit and is used in cooking as a meat tenderizer. Workers involved in the commercial production of papain become allergic to the enzyme, an example of an occupational immunological disease. Similarly, the protease subtilisin, the 'biological' component of some laundry detergents, causes allergy in laundry workers. Chymopapain, which is used in medicine to degrade the intervertebral disks of patients with sciatica, can also stimulate an allergic response.

Systematic studies are under way to identify the helminth proteins recognized by IgE in the serum of helminth-infected individuals and to compare them with allergens recognized by IgE in the serum of allergic individuals. In several cases the protein that serves as an allergen is structurally related to an antigenic helminth protein (Figure 14.24). For example, papain, a plant cysteine protease, is structurally related to the cathepsin-β-like cysteine protease of the blood fluke *Schistosoma mansoni*. These connections suggest that allergies

Helminth species	Helminth antigen	Common related allergen
Herring worm (*Anisakis simplex*)	Cystatin	Minor cat allergen
	Cytochrome *c* oxidase subunit 3	Bermuda grass pollen antigen
	Troponin C	German cockroach allergen
Blood fluke (*Schistosoma mansoni*)	Cathepsin B-like cysteine protease	Papain
Roundworm (*Ascaris lumbricoides*)	Glutathione *S*-transferase	Dust mite allergen
Malayan filaria (*Brugia malayi*)	Tropomyosin	Tropomyosin

Figure 14.24 Structural similarities between helminth antigens and allergens. Extensive analysis of the helminth antigens recognized by IgE in helminth-infected people and of the allergens recognized by IgE in allergic individuals shows that they have many similarities. Allergens represent a relatively small number of the total plant and animal proteins to which human systems are exposed. Six examples are shown.

could be a consequence of antibody cross-reactivity, in which an IgE antibody made against a parasite or a component of food cross-reacts with the environmental antigen to cause allergic disease. Consistent with this possibility is the fact that allergens are relatively conserved proteins and some, such as tropomyosin, are highly conserved, but they have antigenic differences between distantly related groups of multicellular animals such as helminths and humans.

14-14 Predisposition to allergic disease is influenced by genetic and environmental factors

In the Caucasian populations of Europe and North America, up to 40% of people are atopic and are thus more likely than the rest of the population to have an allergy. Atopic people also have, on average, higher blood levels of IgE and eosinophils than non-atopic people. Family studies and genomic analyses show there is a genetic basis for atopy. For example, half the risk of getting asthma is due to genetics and half to environmental factors such as smoking.

Unlike the simplicity of the inherited immunodeficiencies, in which disease is due to defects in one gene, the genetics of susceptibility to asthma is complicated and involves polymorphisms of many different genes that contribute to adaptive immunity. Of the nine genes shown in Figure 14.25, six are involved directly in the production and function of antigen-specific IgE. These genes encode MHC class II molecules, the T-cell receptor α chain, IL-4, the IL-4 receptor, the TIM proteins (implicated in balancing T_H1 and T_H2 responses), and FcεRI. The other three genes encode an enzyme involved in repairing damaged bronchial tissue (ADAM33), the β_2-adrenergic receptor that influences bronchial activity and for which epinephrine (adrenaline) and norepinephrine are ligands, and 5-lipoxygenase, an enzyme involved in synthesis of the leukotrienes released by mast cells, eosinophils, and basophils.

A cluster of genes on chromosome 5 includes the genes for IL-4, IL-3, IL-5, IL-9, IL-12, IL-13, and GM-CSF. These cytokines are all directly involved in

Gene	Nature of polymorphism	Possible mechanism of association
MHC class II genes	Structural variants	Enhanced presentation of particular allergen-derived peptides
T-cell receptor α locus	Non-coding variant	Enhanced T-cell recognition of certain allergen-derived peptides
TIM gene family	Promoter and structural variants	Regulation of the T_H1/T_H2 balance
IL4 (IL-4)	Promoter variant	Variation in expression of IL-4
IL4R (IL-4 receptor α chain)	Structural variant	Increased signaling in response to IL-4
High-affinity IgE receptor β chain	Structural variant	Variation in consequences of IgE ligation by antigen
ALOX5 (5-lipoxygenase)	Promoter variant	Variation in leukotriene production
β_2-Adrenergic receptor	Structural variants	Increased bronchial hyper-reactivity
ADAM33	Structural variants	Variation in airway remodeling

Figure 14.25 Genes associated with susceptibility to asthma.

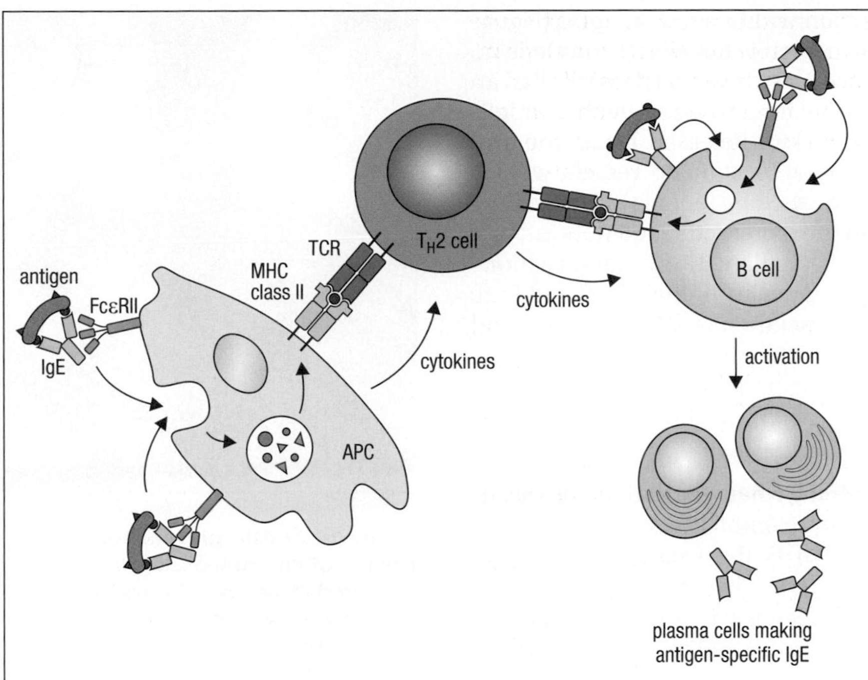

Figure 14.26 Transfer of IgE and antigen from mother to child. During pregnancy, IgE and IgE bound to antigen can be transferred from the mother's circulation to that of the fetus. Antigens can also bind to this maternal IgE in the fetus. Complexes of antigen and IgE are bound by the FcεRII on fetal antigen-presenting cells (APC) and are then processed and presented to naive antigen-specific T cells. This gives rise to T_H2 cells that help antigen-specific B cells that have taken up antigen to become plasma cells secreting antigen-specific IgE. By this mechanism, the mother could provide the fetus with protective immunity against helminths before the baby is born and first exposed to these parasites. Conversely an allergic mother can pass on the allergy to her children.

isotype switching, eosinophil survival, and mast-cell proliferation. Different subsets of these genes correlate with asthma and atopic dermatitis, another allergic disease. Examination of the same disease in different ethnic groups also gives correlations with different subsets of these genes. This complexity reflects the fact that allergic disease is caused by slight, but significant, perturbations of the balance between the T_H1 and T_H2 responses, which can be caused by many different combinations of gene polymorphisms as well as by environmental factors.

A child is more at risk of allergic disease—infantile dermatitis, for example—if the mother is atopic than if the father is atopic. This bias may have a genetic contribution, but it is also due to the maternal environment in which the child's early development takes place. During pregnancy, maternal IgE cannot cross the placenta but it is present in amniotic fluid, which the fetus ingests in quantity, up to 700 ml per day by term, as a source of nourishment. Because the epithelium of the fetal gut is leaky, the maternal IgE gains access to the fetal mucosa and associated lymphoid tissues. Suggested, but not proven, is that complexes of IgE with antigen or allergen are transferred in this way from mother to child and that they then bind to macrophage and B-cell FcεRII for processing and presentation to CD4 T_H2 cells (Figure 14.26). Maternal IgE could thus stimulate the fetus to make IgE of the same antigenic specificities made by the mother. In a family carrying helminth parasites, this mechanism of education by the mother could provide the fetus with antibody-mediated defense against the worms before being born and exposed to them. Similarly, in a family without parasites and an allergic mother, the fetus could be born already making anti-IgE antibodies against common allergens before being exposed to them. Breast milk also contains maternal IgE, and so such processes of maternal education can continue after birth.

14-15 IgE-mediated allergic reactions consist of an immediate response followed by a late-phase response

One way in which clinical allergists assess a person's sensitivity to allergens is to inject small quantities of common allergens into the skin. Substances to which the person is sensitive produce a characteristic inflammatory reaction called a **wheal and flare** at the site of injection within a few minutes (Figure 14.27, arm on left). Substances to which the person is not allergic produce no such reaction. Because of its rapid appearance, the wheal and flare is called an **immediate reaction**. Such reactions are the direct consequence of IgE-mediated degranulation of mast cells in the skin. Released histamine and other mediators cause increased permeability of local blood vessels, whereupon fluid leaves the blood and produces local swelling (edema). The swelling produces the wheal at the injection site, and the increased blood flow into the surrounding area produces the redness that is the flare. Immediate reactions can last for up to 30 minutes, and the relative intensity of the wheal and flare varies. Some 6–8 hours after the immediate reaction has subsided, a second reaction—the **late-phase reaction**—occurs at the site of injection (Figure 14.27, arm on right). This consists of a more widespread swelling and is due to the leukotrienes, chemokines, and cytokines synthesized by mast cells after IgE-mediated activation.

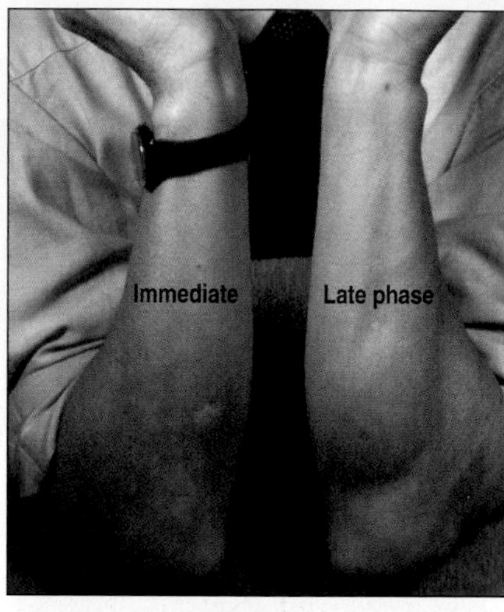

Figure 14.27 Allergic reactions consist of an immediate reaction followed after some hours by a late-phase reaction. The photograph shows a volunteer who has received an intradermal injection of grass-pollen extract in each arm. One injection (left) was made 15 minutes before the photograph was taken and shows the characteristic wheal and flare of the immediate reaction. The wheal is the raised area of skin with the injection site at the center; the flare is the redness (erythema) spreading out from the wheal. The other injection (right) was made 6 hours before the photograph was taken and shows the late-phase reaction. The swelling has spread from the site of injection to involve surrounding tissue. Photograph courtesy of S.R. Durham.

Although skin tests are a good way of determining whether a patient has developed an allergen-specific IgE response, they cannot assess, for example, whether the response to a particular allergen is the cause of a patient's asthma—an allergic reaction in which the airways become inflamed, constricted, and blocked with mucus. To do this, allergists make direct measurements of a person's breathing capacity in the absence and presence of inhaled allergen (Figure 14.28). On inhalation of an allergen to which a patient is sensitized, mucosal mast cells in the respiratory tract that are armed with allergen-specific IgE will degranulate. The released mediators cause immediate constriction of the bronchial smooth muscle, which results in the expulsion of material from the lungs by coughing, and difficulty in breathing. As in the skin test, the immediate response in the lungs finishes within an hour, to be followed some 6 hours later by the late-phase response, which is due to leukotrienes and other mediators. In allergies to inhaled antigens, such as chronic asthma, the late-phase reaction is most damaging. It induces the recruitment of leukocytes, particularly eosinophils and T_H2 lymphocytes, into the site and, if antigen persists, the late-phase response can easily develop into a chronic inflammatory response in which allergen-specific T_H2 cells promote IgE production and a state of eosinophilia.

14-16 The effects of IgE-mediated allergic reactions vary with the site of mast-cell activation

When sensitized people are re-exposed to an allergen, the effect of the IgE-mediated reaction varies depending on the allergen and tissue where it

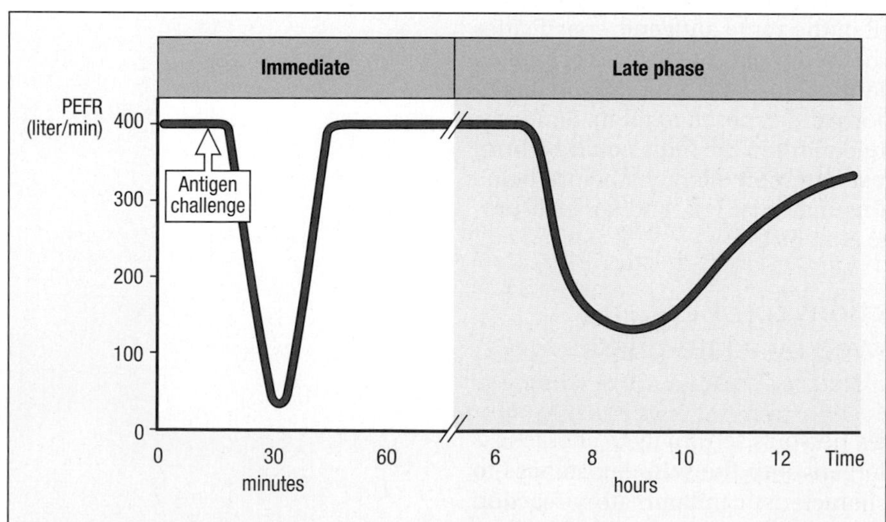

Figure 14.28 The course of an asthmatic response to inhaled allergen. The response is measured in terms of breathing capacity: the peak expiratory flow rate (PEFR). This is a measurement of the maximum speed with which a person can blow air out of their lungs in one burst. The immediate asthmatic response in the lungs finishes within an hour, to be followed some 6 hours later by the late-phase response.

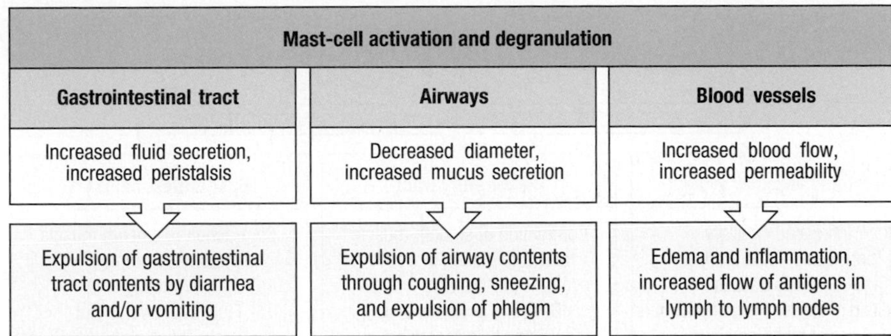

Mast-cell activation and degranulation
Gastrointestinal tract
Airways
Blood vessels
Increased fluid secretion, increased peristalsis
Decreased diameter, increased mucus secretion
Increased blood flow, increased permeability
Expulsion of gastrointestinal tract contents by diarrhea and/or vomiting
Expulsion of airway contents through coughing, sneezing, and expulsion of phlegm
Edema and inflammation, increased flow of antigens in lymph to lymph nodes

Figure 14.29 The physical effects of IgE-mediated mast-cell degranulation vary with the tissue exposed to allergen.

activates mast cells armed with IgE. Only the mast cells at the site of exposure degranulate, and once released, the preformed mediators are short-lived. Their effects on blood vessels and smooth muscles are therefore confined to the immediate vicinity of the activated mast cells. The more sustained effects of the late-phase response are also restricted to the site of allergen exposure, because the leukotrienes and other induced mediators are also short-lived. The anatomy of the site of contact also determines how quickly the inflammatory reaction subsides. The tissues most commonly exposed to allergens are the mucosae of the respiratory and gastrointestinal tracts, the blood, and connective tissues. Airborne allergens irritate the respiratory tract, and food-borne antigens irritate the gastrointestinal tract; the blood and connective tissues receive allergens through insect bites and other wounds and also from absorption via the gut and respiratory mucosa (Figure 14.29).

For gut parasites, interactions between antigen, IgE, and mast cells trigger violent muscular contractions that can expel worms from the gastrointestinal tract and increase fluid flow to wash them out. In the lungs, the muscular spasms and increased secretion of mucus that result from mast-cell activation can be seen as a way to expel organisms attached to respiratory epithelium, such as lung flukes. In an allergic reaction, this crude defense is mistakenly ranged against non-threatening particles and proteins that are often smaller than any microorganism. The violence and power of the IgE-mediated response means that the clinical manifestations of an allergic reaction can vary markedly depending on the amount of IgE that a sensitized person has made, the amount of allergen triggering the reaction, and the route by which it entered the body. In the next four sections we examine how allergic reactions vary when they occur in different tissues.

14-17 Systemic anaphylaxis is caused by allergens in the blood

When an allergen enters the bloodstream it can cause widespread activation of the connective tissue mast cells associated with blood vessels. This causes a dangerous hypersensitivity reaction called **systemic anaphylaxis**. During systemic anaphylaxis, disseminated mast-cell activation causes both an increase in vascular permeability and a widespread constriction of smooth muscle. Fluid leaving the blood causes the blood pressure to drop drastically, a condition called **anaphylactic shock**, and the connective tissues to swell. Damage is sustained by many organ systems and their function is impaired. Death is usually caused by asphyxiation due to constriction of the airways and swelling of the epiglottis (Figure 14.30). In the United States, more than 180 deaths a year are the result of anaphylaxis, which is the most extreme allergic reaction of the body's defenses. Because this form of immunity is fatal rather than protective, it was called '*ana*phylaxis,' meaning anti-protection, to contrast with the '*pro*phylaxis' afforded by protective immunity.

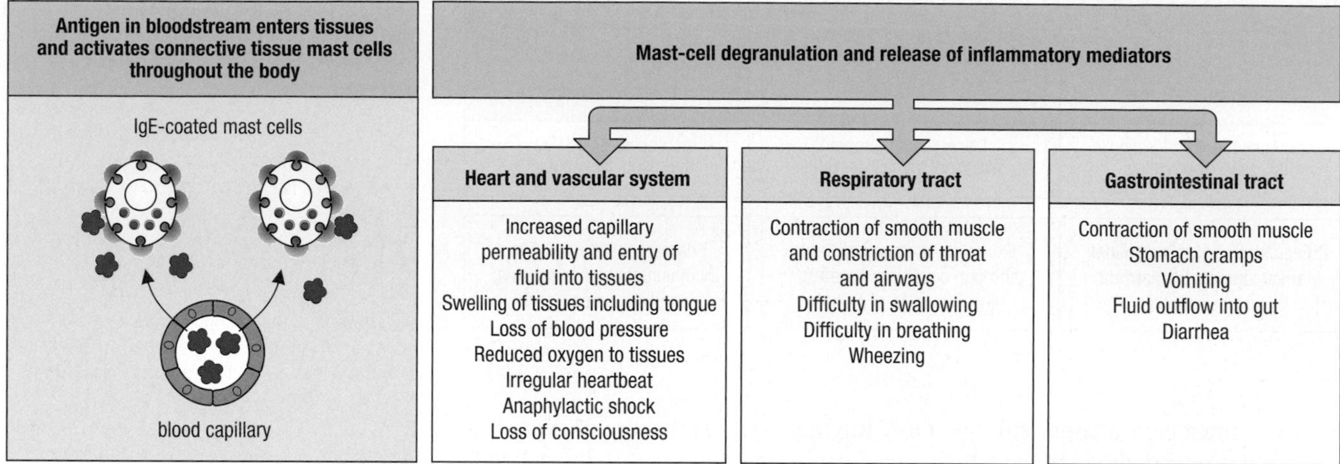

Figure 14.30 Systemic anaphylaxis is caused by allergens that reach the bloodstream and activate mast cells throughout the body.

Potential allergens are introduced directly into the blood by stings from wasps, bees, and other venomous insects; these account for about a quarter of the fatalities from anaphylaxis in the United States. Drug injections can also lead to anaphylaxis. Systemic anaphylaxis can also be provoked by food or drugs taken orally, if the allergens they contain are absorbed rapidly from the gut into the blood. Foods that can cause anaphylaxis include peanuts and brazil nuts. Anaphylactic reactions can be fatal, but treatment with an injection of epinephrine will usually bring them under control. Epinephrine stimulates the reformation of tight junctions between endothelial cells. This reduces their permeability and prevents fluid loss from the blood, diminishing tissue swelling and raising blood pressure. Epinephrine also relaxes constricted bronchial smooth muscle and stimulates the heart (Figure 14.31). The danger of anaphylaxis is such that patients with known anaphylactic sensitivity to insect venoms or food are advised to carry an auto-injector containing epinephrine at all times.

The most common cause of systemic anaphylaxis is allergy to penicillin and related antibiotics, which accounts for about 100 fatalities a year in the United States. Penicillin is a small organic molecule with a reactive β-lactam ring. On ingestion or injection of the drug, the β-lactam ring can be opened up to produce covalent conjugates with proteins of the body, creating new 'foreign' epitopes (Figure 14.32). In some individuals, the response is dominated by T_H2 cells that help B cells produce IgE specific for the new epitopes. If penicillin is given to a person who has been sensitized in this manner, it causes anaphylaxis and even death (Figure 14.33). For this reason, clinicians take care to avoid prescribing any drug to patients who have a history of allergy to that drug. Unfortunately, penicillin cannot be modified to remove its allergic potential because the reactive β-lactam ring is essential for its antibiotic activity.

Reactions resembling anaphylaxis can occur in the absence of specific interaction between an allergen and IgE. Such **anaphylactoid reactions** are caused by other stimuli that induce mast-cell degranulation and can be occasioned by exercise or by certain drugs and chemicals. Anaphylactoid reactions are also treated with epinephrine.

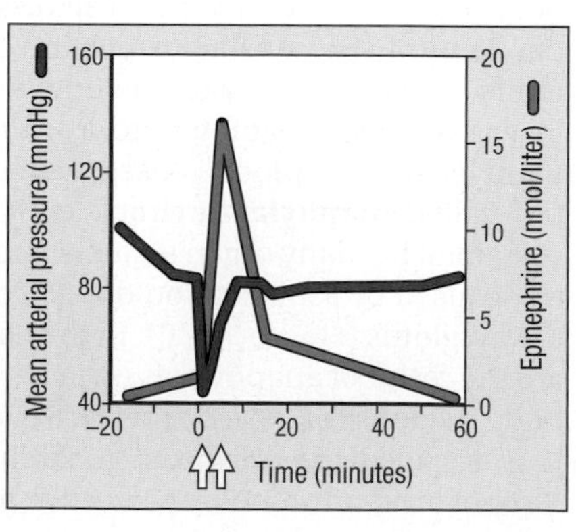

Figure 14.31 The change in blood pressure during systemic anaphylaxis and its treatment with epinephrine. Time 0 indicates the time at which the anaphylactic reaction was reported by the patient. The arrows indicate the times at which injections of epinephrine were given.

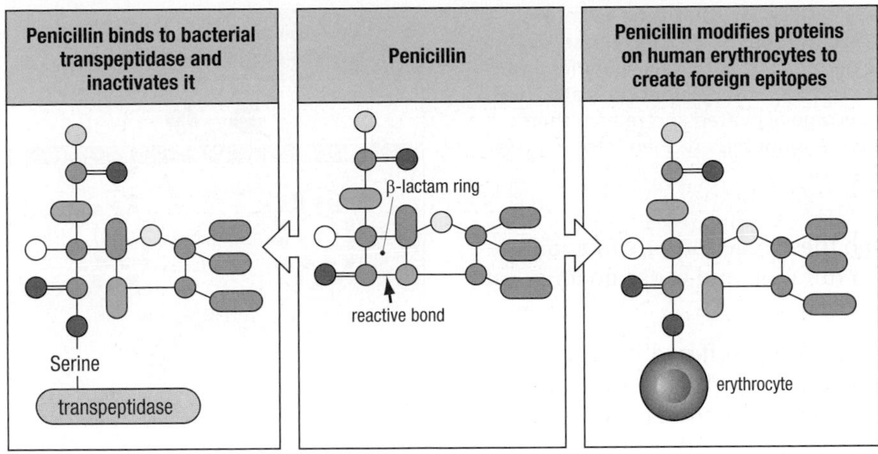

Figure 14.32 Penicillin and other drugs cause human cells to display foreign antigens. Penicillin exerts its antibacterial action by mimicking the substrate for bacterial transpeptidase that is essential for bacterial cell wall synthesis. On binding transpeptidase, a reactive bond in the β-lactam ring of penicillin opens and binds covalently to the active site of the transpeptidase. This kills the enzyme (lower left panel). The same mechanism causes penicillin molecules to be covalently bonded to surface proteins of human cells (lower right panel). This creates new epitopes that are seen as foreign antigens. Erythrocytes (red blood cells) are the cells most commonly modified in this way.

14-18 Rhinitis and asthma are caused by inhaled allergens

Allergens most commonly enter the body by inhalation. Mild allergies to inhaled antigens are common, being manifested as violent bursts of sneezing and a runny nose, a condition called **allergic rhinitis** or hay fever. It is caused by allergens that diffuse across the mucous membrane of the nasal passages and activate mucosal mast cells beneath the nasal epithelium (Figure 14.34). Allergic rhinitis is characterized by local edema, leading to obstruction of the nasal airways and a nasal discharge of mucus that is rich in eosinophils. There is also a generalized irritation of the nose due to histamine release. The reaction can extend to the ear and throat, and the accumulation of fluid in the blocked sinuses and eustachian tubes is conducive to bacterial infection. The same exposure to allergen that produces rhinitis can affect the conjunctiva of the eyes, where the reaction is called **allergic conjunctivitis**. It produces

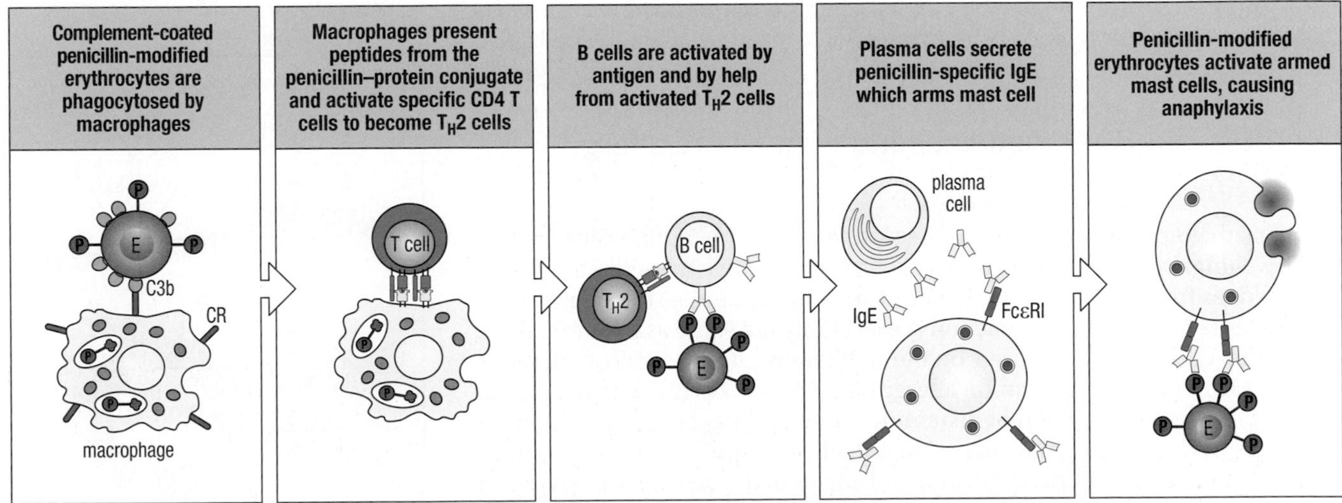

Figure 14.33 In some individuals, penicillin–protein conjugates stimulate the production of anti-penicillin antibodies. Erythrocytes (E) that have been covalently bonded to penicillin (P) are phagocytosed by macrophages, which process the penicillin-modified proteins and present peptide antigens to specific CD4 T cells (first and second panels). These are activated to become effector T_H2 cells, which stimulate antigen-specific B cells to produce IgE antibodies against the penicillin-modified epitope, which then arm tissue mast cells (third and fourth panels). When an individual sensitized in this way is next prescribed penicillin, the penicillin-conjugated red cells will activate the mast cells (fifth panel), causing an anaphylactic reaction. If not treated this can be fatal.

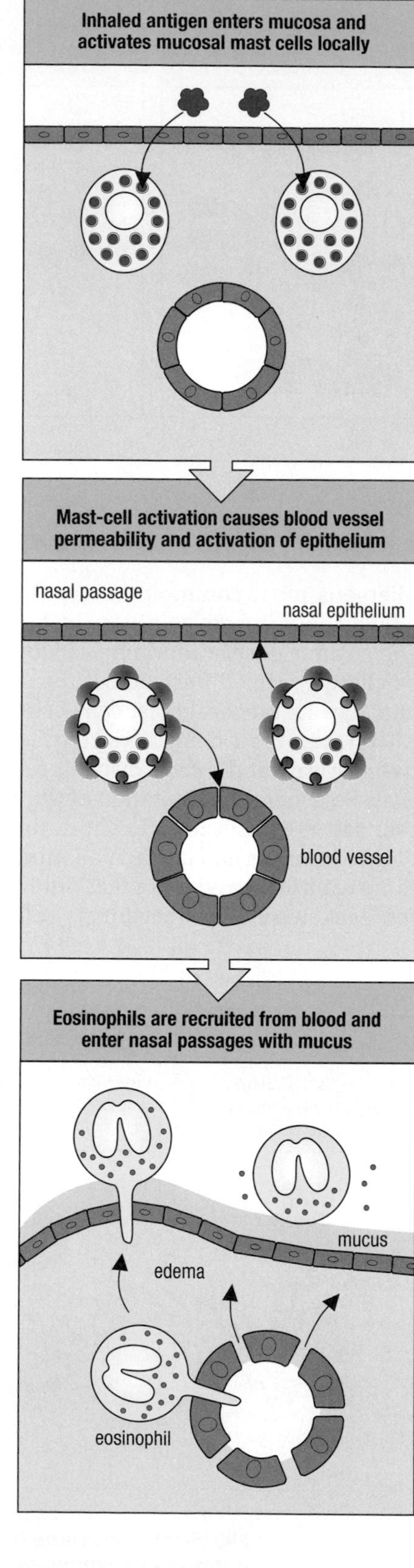

Figure 14.34 Allergic rhinitis is caused by allergens entering the respiratory tract. Histamine and other mediators released by activated mast cells increase the permeability of local capillaries and activate the nasal epithelium to produce mucus. Eosinophils attracted into the tissues from the blood become activated and release their inflammatory mediators. Activated eosinophils are shed into the nasal passages.

itchiness, tears, and inflammation. Although these reactions are uncomfortable and distressing, they are usually of short duration and cause no long-lasting tissue damage.

Much more serious is **allergic asthma**, a condition suffered by 300 million people worldwide, in which allergic reactions cause chronic difficulties in breathing, such as shortness of breath and wheezing. Asthma is triggered by allergens activating submucosal mast cells in the lower airways of the respiratory tract. Within seconds of mast-cell degranulation there is an increase in the fluid and mucus being secreted into the respiratory tract, and bronchial constriction caused by contraction of the smooth muscle surrounding the airway. Chronic inflammation of the airways is a characteristic feature of asthma, involving a persistent infiltration of leukocytes, including T_H2 lymphocytes, eosinophils, and neutrophils (Figure 14.35). The overall effect of the asthmatic attack is to trap air in the lungs, making breathing more difficult. Patients with allergic asthma often need treatment, and asthmatic attacks can prove fatal, with 250,000 deaths worldwide each year attributed to the disease.

Although allergic asthma is initially driven by a response to a specific allergen, the chronic inflammation that subsequently develops seems to be perpetuated in the absence of further exposure to the allergen. In **chronic asthma** the airways can become almost totally occluded by plugs of mucus (Figure 14.36). A generalized hyper-responsiveness or hyper-reactivity in the airways also develops, and environmental factors other than re-exposure to specific allergen can trigger asthmatic attacks. Typically, the airways of chronic asthmatics are hyper-reactive to chemical irritants commonly present in air, such as cigarette smoke and sulfur dioxide. Disease can be exacerbated by immune responses to bacterial or viral infections of the respiratory tract, especially when responses are dominated by T_H2 cells. For this reason, chronic asthma is classified as a type IV hypersensitivity reaction caused by effector T cells (see Figure 14.2).

14-19 Urticaria, angioedema, and eczema are allergic reactions in the skin

Allergens that activate mast cells in the skin to release histamine cause raised itchy swellings called **urticaria** or **hives**. Urticaria means 'nettle-rash' and the word derives from *Urtica*, the Latin name for stinging nettles; the origin of the word hives is unknown. This reaction is essentially the same as the immediate wheal-and-flare reaction caused by the deliberate introduction of allergens into the skin in tests to determine allergy (see Section 14-15), which can also be produced by the injection of histamine alone (Figure 14.37). Activation of mast cells in deeper subcutaneous tissue leads to a similar but more diffuse swelling called **angioedema**. Urticaria and angioedema can arise as a result of any allergy to a food or drug if the allergen gets carried to the skin by the bloodstream, and they are among the many reactions that occur during systemic anaphylaxis. Insect bites are a common cause of urticaria, and such local reactions usually occur without inducing a more general anaphylaxis.

A more prolonged allergic response in the skin is observed in some atopic children. This condition is called **atopic dermatitis** or **eczema**. The word eczema is of Greek derivation and means to 'break out' or 'boil over.' The condition is

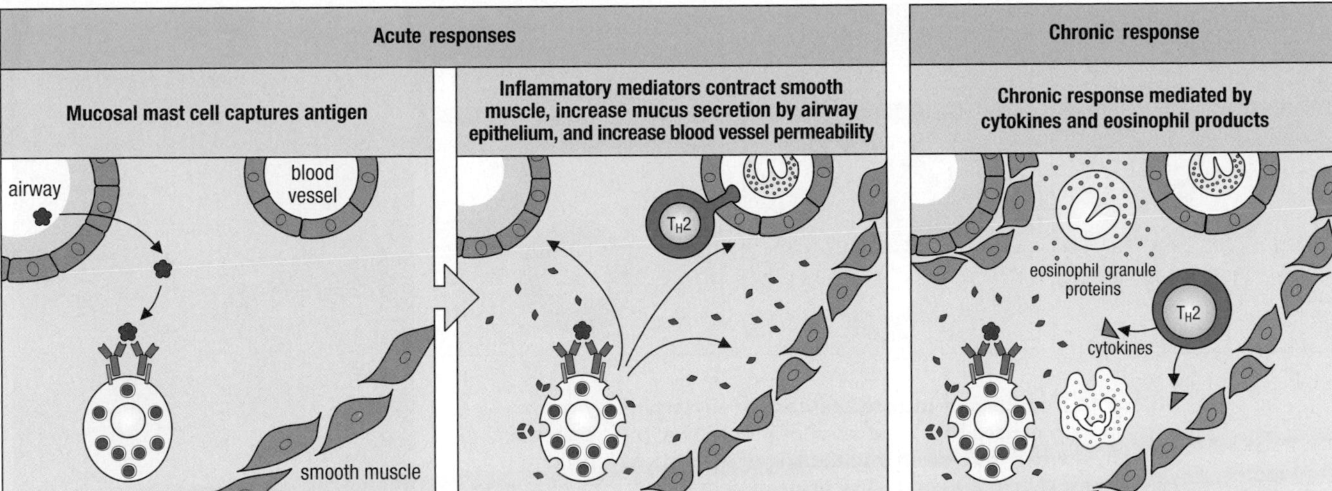

Figure 14.35 The acute response in allergic asthma leads to T_H2-mediated chronic inflammation of the airways. In sensitized individuals, mast cells carrying IgE specific for the allergen are present in the mucosa of the airways (first panel). Cross-linking of specific IgE on the surface of mast cells by inhaled allergen triggers them to secrete inflammatory mediators, causing bronchial smooth muscle contraction and an increased secretion of mucus from mucosal epithelium, which together lead to airway obstruction (second panel). Increased blood vessel permeability, also caused by inflammatory mediators, leads to edema and an influx of inflammatory cells, including eosinophils and T_H2 lymphocytes. Activated mast cells and T_H2 cells secrete cytokines that also augment eosinophil activation and degranulation, which causes further tissue injury and influx of inflammatory cells (third panel). The end result is chronic inflammation, which can then cause irreversible damage to the airways.

characterized by an inflammatory response that causes a chronic and itching skin rash with associated skin eruptions and fluid discharge. This response has similarities to that occurring in the bronchial walls of asthmatics. Eczema frequently presents in families with a history of asthma and allergic rhinitis, and is often associated with high IgE levels. However, the severity of the dermatitis is not readily correlated with exposure to particular allergens or to the levels of allergen-specific IgE. Thus the etiology of eczema remains poorly understood. Neither is it understood why eczema usually clears in adolescence, whereas rhinitis and asthma more often persist throughout life.

Eczema is believed to be due to loss of the protective barrier function of the skin, allowing allergens to enter the skin and stimulate a T_H2-cell mediated immune response. Consistent with this idea, individuals who inherit a weakness in skin barrier function have an increased likelihood of developing

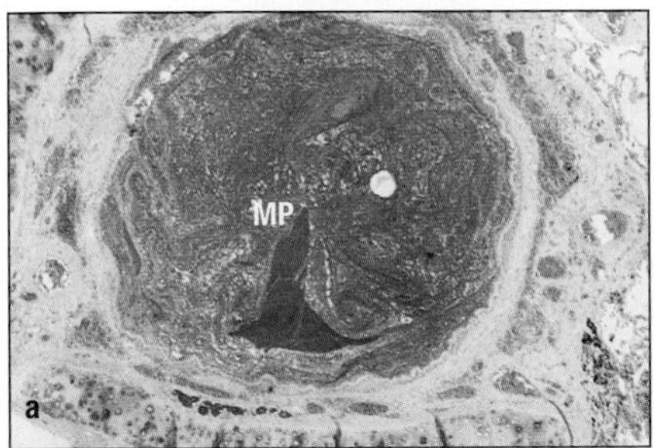

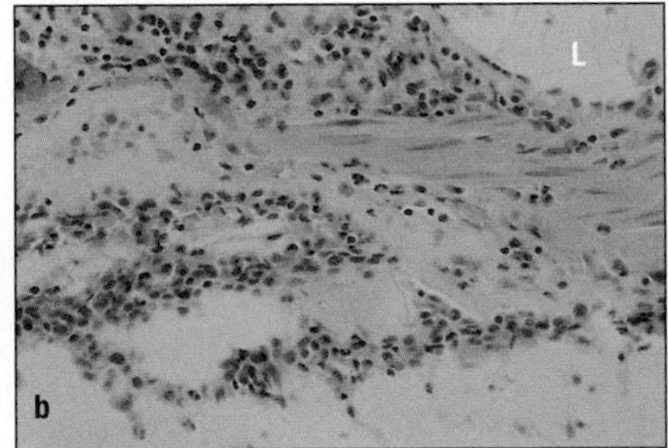

Figure 14.36 Inflammation of the airways in chronic asthma restricts breathing. Panel a shows a light micrograph of a section through the bronchus of a patient who died of asthma; there is almost total occlusion of the airway by a mucous plug (MP). The small white circle is all that is left of the lumen of the bronchus. In panel b, a light micrograph at higher magnification gives a closer view of the bronchial wall. It shows injury to the epithelium lining the bronchus, accompanied by a dense inflammatory infiltrate that includes eosinophils, neutrophils, and lymphocytes. L, lumen of the bronchus. Photographs courtesy of T. Krausz.

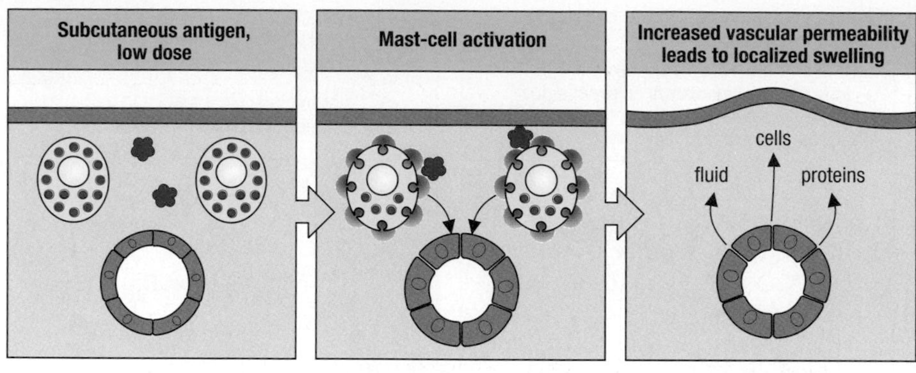

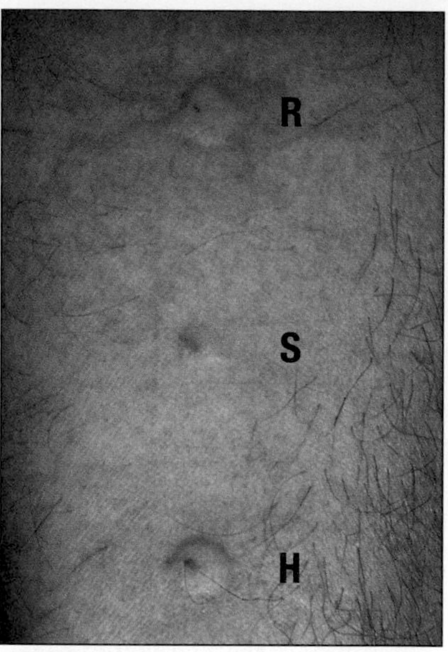

Figure 14.37 Allergen-induced release of histamine by mast cells in skin causes localized swelling. As shown in the panels on the left, allergen introduced into the skin of a sensitized individual causes mast cells in the connective tissue to degranulate. The histamine released dilates local blood vessels, causing rapid swelling due to the leakage of fluid and proteins into tissues. The photograph shows the raised swellings (wheals) that appear 20 minutes after intradermal injections of ragweed pollen antigen (R) or histamine (H) into a person who is allergic to ragweed. The small wheal at the site of saline injection (S) is due to the volume of fluid injected into the dermis. Photograph courtesy of R. Geha.

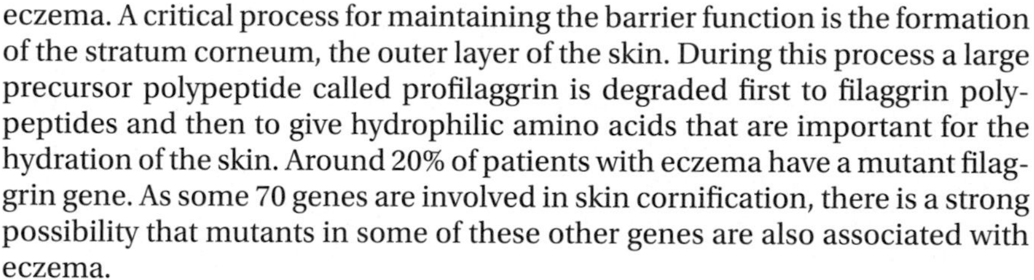

eczema. A critical process for maintaining the barrier function is the formation of the stratum corneum, the outer layer of the skin. During this process a large precursor polypeptide called profilaggrin is degraded first to filaggrin polypeptides and then to give hydrophilic amino acids that are important for the hydration of the skin. Around 20% of patients with eczema have a mutant filaggrin gene. As some 70 genes are involved in skin cornification, there is a strong possibility that mutants in some of these other genes are also associated with eczema.

14-20 Food allergies cause systemic effects as well as gut reactions

Humans probably eat a greater variety of foodstuffs than any other animal. All human food is derived from plants and animals and contains a million or more different proteins, many of which are potentially immunogenic. As food passes down the gastrointestinal tract, the proteins are degraded by proteases into peptides of ever-decreasing size. These are a potential source of peptides for presentation to T_H2 cells. Despite the quantity and variety of food that humans eat, IgE is made against an extremely small proportion of the proteins ingested. However, people sensitized to a particular protein will be allergic to any food containing that protein. Foods that commonly cause allergies include grains, nuts, fruits, legumes, fish, shellfish, eggs, and milk.

Once sensitized to a food allergen, any subsequent intake causes a marked and immediate reaction that can drive the person from the dining room. The allergen passes across the epithelial wall of the gut and binds to IgE on the mucosal mast cells associated with the gastrointestinal tract. The mast cells degranulate, releasing their mediators, principally histamine. The local blood vessels become permeable, and fluid leaves the blood and passes across the gut epithelium into the lumen of the gut. Meanwhile, contraction of smooth muscle of the stomach wall produces cramps and vomiting; the same reaction in the intestine produces diarrhea (Figure 14.38). These reactions evolved to expel gut parasites; in the allergic reaction they serve to expel the allergen-containing food from the human body. This goal is indeed accomplished, but at the expense of dehydration, weakness, and waste of food.

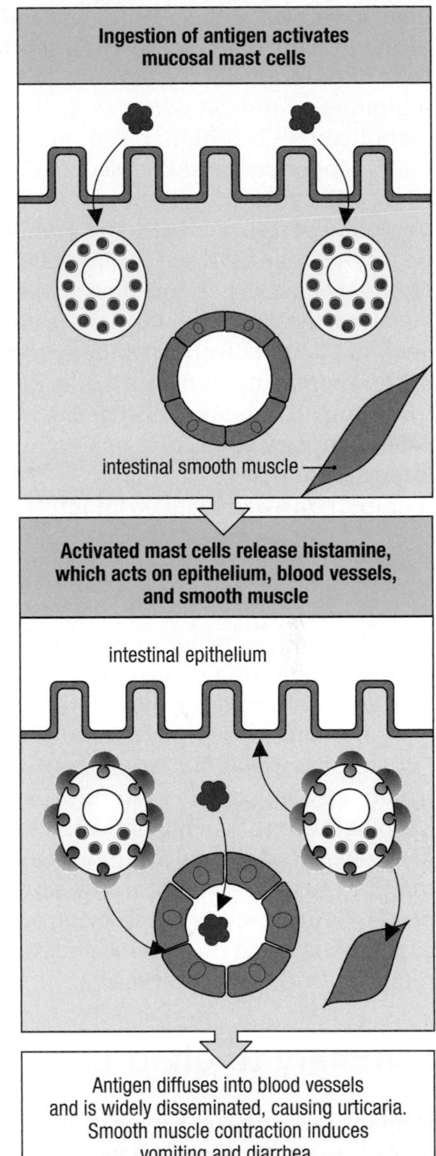

Figure 14.38 Ingested allergen can cause vomiting, diarrhea, and urticaria. Localized reactions are caused by the action of histamine on intestinal epithelium, blood vessels, and smooth muscles. Urticaria is caused by antigen that enters blood vessels and is carried to the skin.

In addition to allergic reactions localized to the gut, food allergens also cause reactions in other tissues, notably the skin. Depending on the timing and course of the gut reaction and of the uptake of allergen from the gut, allergen can enter the circulation and be transported elsewhere in the body. Mast cells in the connective tissue in the deeper layers of the skin tend to be activated by such blood-borne allergens, and their degranulation produces urticaria and angioedema. In this context, orally administered drugs behave similarly to food: they, too, can produce intestinal reactions, urticaria, and angioedema in sensitized individuals.

14-21 Allergic reactions are prevented and treated by three complementary approaches

Three distinct strategies are used to reduce the effects of allergic disease. The first is one of prevention—to modify a patient's behavior and environment so that contact with the allergen is avoided. Allergen-containing foods are avoided, houses are refurnished in ways that discourage mites, pets are kept outside, and desert vacations or sea cruises are taken during the pollen season.

The second strategy is pharmacological—to use drugs that reduce the impact of any contact with allergen. In this category is the use of monoclonal anti-IgE that prevents the binding of IgE to FcεRI and FcεRII (see Section 14-18). Other drugs block the effector pathways of the allergic response and limit the inflammation after IgE-induced activation of mast cells, eosinophils, and basophils. Antihistamines reduce rhinitis and urticaria by preventing histamine from binding to H_1 histamine receptors on vascular endothelium and thus increasing vascular permeability. Inhaled drugs used for treating asthma include low-dose steroids as an immunosuppressant, and also fast-acting agonists of the β-adrenergic receptor, such as albuterol, that dilate the bronchioles and facilitate breathing. Epinephrine is used to treat anaphylactic reactions.

The third strategy in the treatment of allergy is immunological—to prevent the production of allergen-specific IgE. One way of achieving this is to modulate the antibody response so that it shifts away from one dominated by IgE to one dominated by IgG. A procedure called **desensitization**, which was first described in 1911 and is used in a similar way today, can achieve this for some patients and some allergies. Patients are given a series of allergen injections in which the dose is initially very small and is gradually increased. Successful treatments are associated with the production of allergen-specific antibodies of the IgG4 isotype—which, like IgE, is a product of a T_H2 response—and increased levels of IL-10. IgG4 is functionally monovalent and forms complexes with antigen that do not recruit effector cells (see Section 4-17). An occasional consequence of the injections used for desensitization is anaphylaxis, because the patient is being exposed to the allergen to which they are sensitized. For this reason 'allergy shots' are given under carefully controlled conditions in which patients are monitored for the early symptoms of systemic anaphylaxis and, if need be, given epinephrine. Another and exploratory form of desensitization is to introduce helminth worms into the guts of allergy and asthma patients. This aims in effect to reworm some people in the developed countries of the world.

Because allergies are such a prevalent and increasing problem in the developed countries, there is much interest within the biotechnology and pharmaceutical industries in developing new approaches to the relief or cure of allergy. One potential set of targets for drugs are the signaling pathways that cells of the immune system use to enhance the IgE response. For example, inhibitors of the cytokines IL-4, IL-5, or IL-13 could block such pathways. Mepolizumab is a humanized monoclonal antibody with specificity for IL-5 that has been successfully used to treat patients with eosinophilia (see Section 14-11). Because the development of eosinophils is dependent on IL-5, this treatment severely reduces the number of blood and tissue eosinophils, but it does not eliminate them completely. Patients treated with anti-IL-5 for as long as 6 years have experienced no adverse effects due to their drug-induced deficiency of eosinophils. Like the asthma patients treated with anti-IgE (see Section 14-8), such observations support the view that the T_H2 arm of adaptive immunity is singularly aimed at helminths and other metazoan parasites. Consequently, people in societies in which these parasites have been eradicated can survive and be healthy without some cells or molecules that are important for the T_H2 response. Asthma can also be alleviated by anti-IL-5 therapy, consistent with eosinophils being a major cause of this disease.

Summary

Allergies induced by IgE antibodies that bind to harmless environmental antigens exhibit a wide range of disease symptoms and severity. Influencing factors are the nature of the allergen, its abundance and distribution in the environment, its site of entry to the human body, and human genetics. In Caucasian populations, around 40% of people have a genetic susceptibility to becoming allergic, and in some societies the incidence of allergy is approaching that number. The high incidence, severity, and chronicity of asthma in the developed world has driven the search for new drugs that will be effective in treating these diseases. Monoclonal antibodies that interfere with the T_H2 arm of adaptive immunity are showing promise because they alleviate allergic disease while having no adverse effects on immunity against infection by bacterial, viral, and fungal pathogens.

Summary to Chapter 14

Antibodies of the IgE isotype evolved specifically to cooperate with mast cells, basophils, and eosinophils in controlling the populations of helminth worms and other multicellular parasites that naturally inhabit vertebrate bodies and have coevolved with human immune systems for more than one million years. Because the helminths are far bigger than bacterial, fungal, viral, and protozoan pathogens, different effector mechanisms are used to control the size of the parasite population and the sites in the body where the parasites reside. Because parasite genomes are more similar to human genomes than microbial genomes, there is less scope for making antibodies that are highly parasite-specific. In human populations in the developing world, where helminths remain endemic, children become infested with helminths soon after birth and receive parasite-specific IgE from their mothers. The child's T_H2 arm of immunity then develops in the context of the helminths, which are likely to play a critical role in guiding that development.

During the past 150 years, the human population in what is now the developed world embarked unknowingly on the experiment of ridding their bodies of helminths. A result of that experiment has been the evolution of a variety of highly inflammatory allergic diseases, which are caused by IgE antibodies that bind to antigens from plants and animals that are not human parasites or pathogens. These antigens are often related genetically and structurally to the helminth antigens that provoke an IgE response in helminth-infected individuals. The absence of helminths means that in these populations the T_H2 arm of

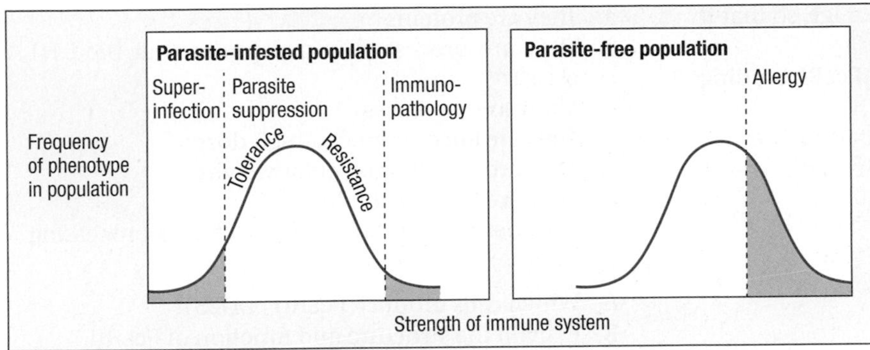

Figure 14.39 Evolution of the epidemic of allergy. In a human population endemically infected with parasites, there is a balance between extremes (left panel). At one end of the spectrum are people too tolerant of parasites who become superinfected and suffer parasitic disease. At the other end of the spectrum are people who make an overly strong anti-parasite response that damages mucosal tissues and causes immunopathology. The majority of people have a balance between tolerance of and resistance to parasites that makes them relatively healthy. In a population that has eradicated the parasites (right panel), the immune system does not develop in the presence of parasites and parasite-specific IgE and this balance is lost. There is an increased tendency for the generation of inflammatory IgE-mediated immune responses directed against environmental antigens that are structurally similar to parasite antigens. These responses cause allergy.

human immunity is likely to be developing in an abnormal way that has never previously been subject to natural selection, and which may not function well in helping to balance inflammation caused by the T_H1 and T_H17 arms of immunity. In addition, the absence of helminths will have eliminated natural selection for helminth immunity in these populations, which could have increased the selection for polymorphisms in immune-system genes that favor inflammatory T_H1-based and T_H17-based immunity (Figure 14.39). The current 'epidemic of allergy' is likely to have arisen from a combination of these crucial developmental and genetic factors.

Questions

14–1 Identify the mismatched pair. (Select all that apply.)

a. type I hypersensitivity: mast cells, basophils, and eosinophils
b. type II hypersensitivity: penicillin allergy
c. type III hypersensitivity: modified cell-surface components
d. type IV hypersensitivity: CD4 T_H1 or CD8 T cells
e. type III hypersensitivity: nonhuman therapeutic proteins
f. type IV hypersensitivity: modified intracellular human proteins
g. type II hypersensitivity: immediate hypersensitivity
h. type I hypersensitivity; IgE cross-linking.

14–2 If B cells making high-affinity IgG4 are unable to switch to IgE at a late stage in the germinal center reaction during a helminth infection, then _____. (Select all that apply.)

a. a chronic infection will ensue
b. IgG4 will activate macrophages to secrete inflammatory cytokines
c. a type III hypersensitivity reaction will result
d. mast cells, basophils, and eosinophils will not be activated
e. a delayed-type hypersensitivity reaction is required to eradicate the infection.

14–3 All of the following are tenets of the hygiene hypothesis except _____.

a. widespread use of antibiotics interferes with the use and education of the immune system
b. childhood exposure to microorganisms is reduced
c. a decrease in allergic diseases in developed countries compared to developing countries
d. antibiotic-driven selection of superbacteria perturbs the immune system
e. widespread immunization practices compromise the development of a fully functional immune response.

14–4 In the context of IgE, explain why mast cells are more versatile than B cells in reference to their response to antigen.

a. Because mast cells possess multiple and diverse receptors for pathogens.
b. Because B cells express FcεRII which is an inhibitory receptor that suppresses B cells making IgE antibody.
c. Because mast cells can not only be induced to release inflammatory mediators upon cross-linking of IgE on their cell surface, they can also engulf antigen bound to IgE through receptor-mediated endocytosis.
d. Because mast cells bind to a different IgE molecules with various specificities for antigens that can be diverse in their structure and biological origin.

14–5 The reason why therapeutic anti-IgE is able to suppress allergic responses is that _____.

a. anti-IgE stimulates endocytosis of IgE from the surface of mast cells, removing its availability to antigen

b. anti-IgE changes the conformation of IgE so that it can bind only to FcεRII
c. when IgE is bound to anti-IgE, its FcεRI binding domain is masked
d. anti-IgE induces cleavage of FcεR1 from the cell surface but only if FcεRI is occupied by IgE.

14–6 _____ has/have biological activities in common with histamine but are more potent.
a. prostaglandins
b. arachidonic acid
c. carboxypeptidases
d. eotaxin
e. leukotrienes.

14–7 Eotaxin (CCL11) is a chemokine of key importance to the emigration of eosinophils into tissues. Which of the following cells produces this chemokine? (Select all that apply.)
a. activated endothelial cells
b. mucosal epithelium
c. T cells
d. smooth muscle cells
e. monocytes
f. bone marrow stroma cells
g. mast cells.

14–8 IL-4 and IL-13 are important cytokines that assist in the polarization of antigen-stimulated T cells toward a T_H2 response. Which of the following cell types secrete these cytokines during IgE-mediated immune responses? (Select all that apply.)
a. eosinophils
b. B cells
c. mast cells
d. basophils
e. neutrophils.

14–9 Match the term in column A with its description in column B.

Column A	Column B
a. allergic asthma	1. eye exposure to allergen
b. allergic rhinitis	2. mild allergy to inhaled antigen
c. urticaria	3. hyperreactivity to chemical irritants
d. chronic asthma	4. diffuse swelling involving deep subcutaneous tissue
e. allergic conjunctivitis	5. skin eruptions and fluid discharge
f. angioedema	6. raised itchy swellings of the skin
g. eczema	7. allergies involving submucosal mast cells of lower airways

14–10 Inhaled allergens possess which of the following features that promote the priming of T_H2 cells that drive IgE responses? (Select all that apply.)
a. They are proteins
b. They are processed into peptides that bind HLA class I
c. Many are proteases
d. They are encountered at high dose
e. They are of high molecular weight
f. They are highly soluble
g. They have low solubility, which favors processing.

14–11
A. Which cells produce FcεRII (CD23)?
B. Explain the structure and function of FcεRII.
C. Explain the role of FcεRII in promoting IgE production by B cells.

14–12 The _____ occurs 6–8 hours after an initial type I allergic response in which mast cells produce leukotrienes, chemokines, and cytokines.
a. Arthus reaction
b. late-phase reaction
c. delayed-type hypersensitivity reaction
d. condition known as anaphylactic shock
e. skin rash known as chronic urticaria.

14–13 Systemic anaphylaxis is caused by the presence of allergen in _____.
a. the gastrointestinal tract
b. the respiratory tract
c. the circulation
d. the skin
e. mismatched blood transfusions.

14–14 It has been suggested that complexes of IgE with allergen can be transferred from mother to child and render the child more predisposed to allergic disease.
A. Identify two ways in which this type of sensitization might occur.
B. Which molecule, found on the surface of the antigen-presenting cell and the B cell, has a key role in mediating the production of allergen-specific IgE by fetal B cells?

14–15 Identify three pharmacological agents that are used to treat allergic reactions and explain how they mediate their effects.

14–16 Vienna Coombs, 12 years old, regularly experiences respiratory problems associated with allergic asthma. Since 2 years of age she has had bouts of wheezing and coughing when she comes into contact with her uncle's cat and also each spring and late summer. Oral anti-histamines and albuterol usually control these episodes, but yesterday she had a severe attack and was admitted to the hospital. She recovered after treatment with inhaled bronchodilator and intravenous steroids. She returned home with instructions to complete a 1-week course of the oral corticosteroid prednisone and to inhale albuterol as needed for wheezing or chest tightness after that. She was explicitly instructed not take the nonsteroidal anti-inflammatory drugs (NSAIDs) aspirin or ibuprofen. Which of the following best explains why patients with allergic asthma should not take NSAIDs?

a. Because NSAIDs will interfere with the wheal-and-flare reaction if the patient needs to undergo skin testing.
b. Because NSAIDs will inhibit the production of leukotriene receptors.
c. Because NSAIDs inhibit the cyclooxygenase pathway and consequently direct arachidonic acid to the lipoxygenase pathway, causing more leukotrienes to be made.
d. Because NSAIDs increase the rate at which mast cells regranulate.
e. Because NSAIDs favor IgE isotype switching.

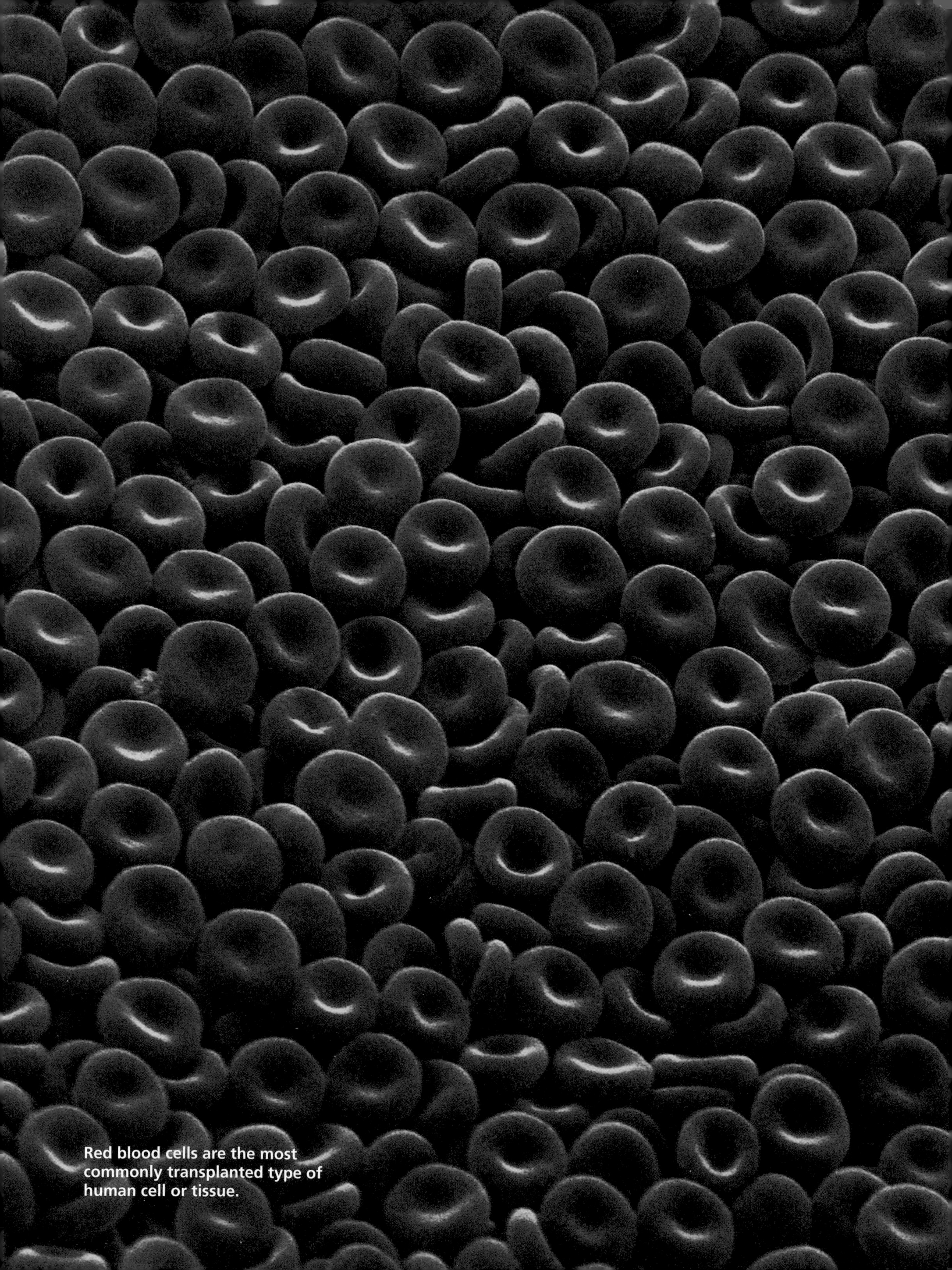

Red blood cells are the most commonly transplanted type of human cell or tissue.

Chapter 15

Transplantation of Tissues and Organs

The replacement of diseased, damaged, or worn-out tissue was for centuries a dream of the medical profession. Achieving the reality required overcoming three fundamental problems. First, transplants must be introduced in ways that allow them to perform their normal physiological functions. Second, the health of both recipient and transplant must be maintained during surgery and other procedures used during transplantation. Third, the recipient's immune system must be prevented from responding to the transplant and causing its rejection.

During the past 60 years, solutions to these problems have been found, and transplantation has progressed from being an experimental procedure to the treatment of choice for a variety of conditions. In clinical practice, selective suppression of the response to the transplanted tissue has yet to be achieved, and so nonspecific suppression is accomplished by using a variety of drugs and antibodies. In contrast to vaccination, which selectively stimulates immunity to a particular pathogen, transplantation involves manipulations that cause widespread inactivation of the immune response.

The first part of this chapter looks at blood transfusion, the most common form of tissue transplantation, and also shows how the detrimental responses that the immune system makes against transplanted tissue are examples of the type II, III, and IV hypersensitivity reactions. Transplants can either be liquid, like a blood transfusion, or solid, like a kidney transplant, and they differ in how they are performed and how they stimulate damaging immune responses. The second part of the chapter examines solid organ transplantation, with emphasis on the kidney, the most frequently transplanted organ. The last part of the chapter is concerned with liquid transplants that contain hematopoietic stem cells obtained from bone marrow, peripheral blood, or umbilical cord blood. As a group these procedures are called hematopoietic cell transplants or hematopoietic stem-cell transplants.

Allogeneic transplantation can trigger hypersensitivity reactions

In most types of transplantation there is considerable benefit if the donor and recipient have compatible tissue types that can coexist without provoking too strong an immune response. In this situation, the donor and recipient are described as being **histocompatible**. The antigens important in blood transfusion are the ABO and Rhesus D blood group antigens of the erythrocyte surface. In the transplantation of organs and hematopoietic cells it is important to match for HLA class I and II antigens. When donor and recipient are not histocompatible, a hypersensitivity reaction is likely to occur.

15-1 Blood is the most common transplanted tissue

In 1812, a blood transfusion was the first clinical transplant to save a human life, and since then the use and success of the procedure has increased dramatically. Blood is by far the most commonly transplanted tissue in clinical medicine, being given to one in four people at some time during their life. It is used when trauma, surgery, childbirth, or disease causes severe blood loss and patients need the immediate replacement of fluid, proteins, and red blood cells. Today, donated blood is commonly separated into erythrocytes, plasma, and platelets, so transfusions are made only of those components a patient needs. Plasma replaces fluid and prevents bleeding, erythrocytes improve respiration and metabolism, and platelets facilitate clotting to prevent bleeding.

Four properties of blood have aided its extensive use as a transplanted tissue. First, blood is readily donated by healthy individuals, and at regular intervals, without compromising their health. Second, the procedure of blood transfusion is simple and inexpensive compared with solid organ transplantation, involving only the intravenous infusion of a liquid allograft. Third, blood components are usually required to function for only a limited time, because within a few weeks the patient's bone marrow will make up the loss. This demand is less stringent than that placed on transplanted organs, such as hearts and kidneys, which need to function for years. The fourth property is that erythrocytes, the beneficial cells in a blood transfusion, do not express polymorphic major histocompatibility complex (MHC) class I or class II molecules, the major genetic barriers to the transplantation of other tissues and organs.

15-2 Before blood transfusion, donors and recipients are matched for ABO and the Rhesus D antigens

For successful blood transfusion, the donor and recipient must be compatible for the **ABO system** of polymorphic antigens present on the erythrocyte surface. The molecular basis for these antigens is structural polymorphism in the carbohydrate component of glycolipids and of a glycoprotein called band 3 of the red cell's membrane (Figure 15.1). Of the three antigens, A and B can cause problems when present on transfused erythrocytes, whereas the O antigen cannot.

Blood group antigens A and B, but not the O antigen, have structural similarity to cell-surface carbohydrates of common commensal bacteria to which everybody is exposed. People who lack the A or the B antigen, or both, make antibodies against the corresponding carbohydrates of commensal bacteria. Thus, the serum of people with blood group O invariably contains IgG antibodies against the A and B antigens. If such a person is mistakenly transfused with group A or group B blood, the antibodies bind to the transfused erythrocytes, causing complement fixation and rapid clearance of the red cells from the circulation. Four blood types—O, A, B, and AB—account for the vast majority of people. Consequently there are 16 possible combinations of donor and recipient, of which 9 are compatible in blood transfusion and 7 are not (Figure 15.2). Also detrimental to the outcome of blood transfusion is incompatibility for the Rhesus (Rh) system of blood group antigens, particularly the presence and absence of the Rhesus D (RhD) antigen. Transfusion of recipients who lack RhD with blood from a RhD positive donor induces an antibody response against RhD that makes any further transfusion with RhD positive blood dangerous. For RhD negative women it makes future pregnancies potentially susceptible to hemolytic disease of the newborn (see Section 11-12).

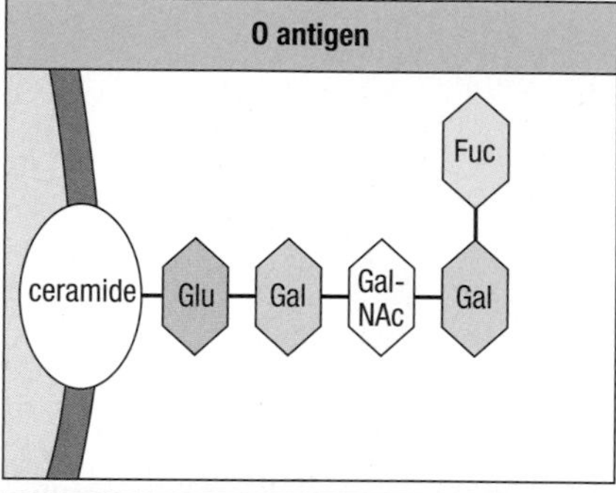

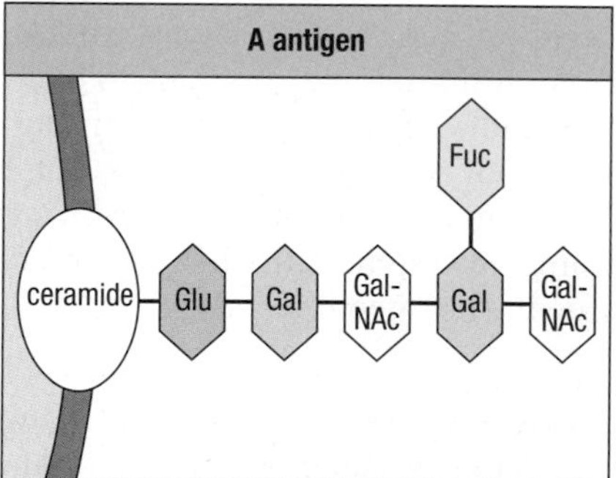

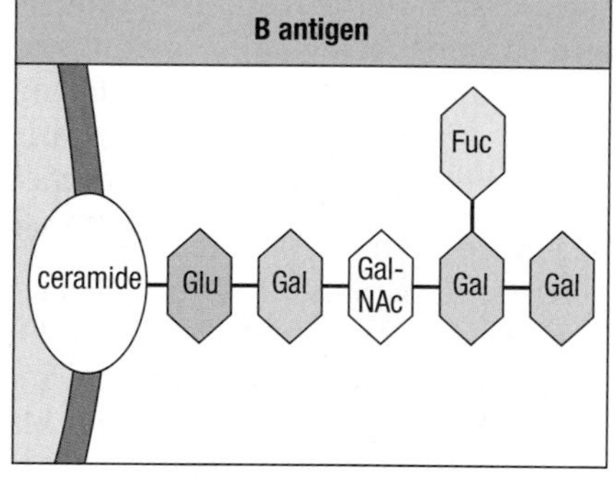

Figure 15.1 Structures of the ABO blood group antigens. The ABO antigens arise from a family of glycolipids at the erythrocyte surface. Their core structure consists of the lipid ceramide, attached to an oligosaccharide consisting of glucose (Glu), galactose (Gal), *N*-acetyl galactosamine (GalNAc), galactose, and fucose (Fuc). In people of blood group O this is the only glycolipid made. People of blood group A have an enzyme that adds an additional *N*-acetyl galactosamine, forming the A antigen. People of blood group B have an enzyme that adds an additional galactose to the core structure, forming the B antigen. The erythrocytes of people of blood groups A and B also express the core structure alone, which is why alloantibodies against O are not made.

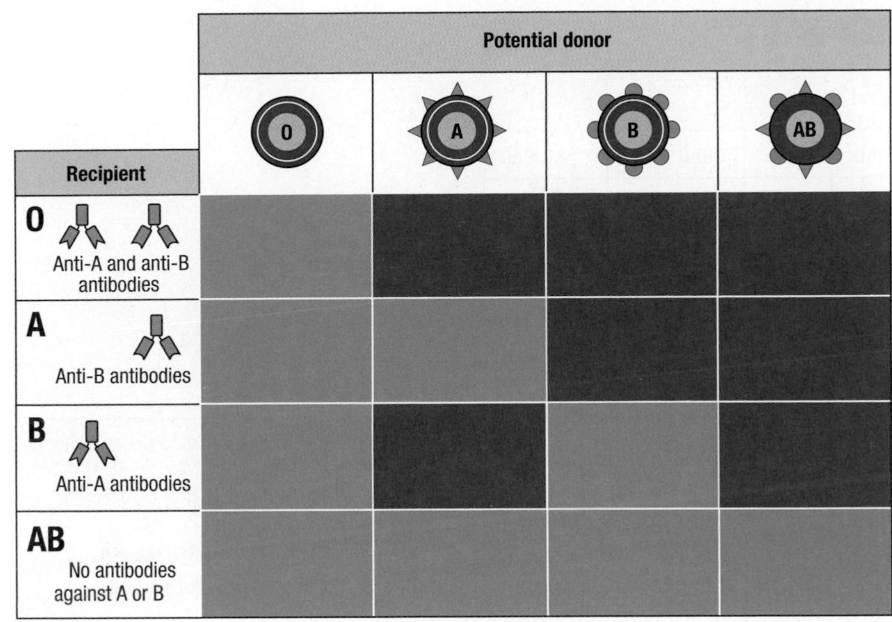

Figure 15.2 Donors and recipients for blood transfusion must be matched for the ABO system of blood group antigens. Common gut bacteria bear antigens that resemble the blood group A and B antigens. For individuals who lack the A, B, or both A and B blood group antigens, the bacterial antigens stimulate an antibody response against the corresponding blood group antigen (left column); thus, type O individuals, who lack A and B, have both anti-A and anti-B antibodies, whereas type AB individuals have neither. The combinations of donor and recipient blood groups that allow blood transfusion are indicated by green boxes; the combinations that would result in an immune reaction, and must be avoided, are indicated by red boxes.

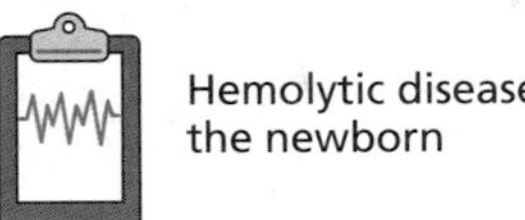

15-3 Incompatibility of blood group antigens causes type II hypersensitivity reactions

The underlying immunological mechanism provoked by ABO incompatibility is identical to that of a type II hypersensitivity reaction (see Section 14-1). In type II hypersensitivity reactions to penicillin, the foreign antigen on the erythrocytes is the result of chemical modification induced by the drug, whereas the foreign antigen on a transfused red cell is an additional sugar that is present on the cell-surface glycolipids. As well as thwarting the purpose of the blood transfusion, these hemolytic reactions can cause fever, chills, shock, renal failure, and even death. To prevent these dangerous type II hypersensitivity reactions, all patients needing a blood transfusion and all donated blood are serologically typed for the ABO and RhD antigens. They are then only transfused with blood of compatible ABO and RhD type (Figure 15.3). Of the 64 combinations of donor and recipient, 27 are compatible for blood transfusion. Matching blood group antigens for blood transfusion is much simpler than matching HLA antigens for transplantation of other tissues.

Some 30 other systems of polymorphic blood group antigens are known, but compared with ABO and Rhesus their differences are less likely to cause incompatibility for blood transfusion. Instead of typing donors and recipients for these other blood groups, a different approach is taken to prevent transfusions being performed when antibodies against these antigens could cause complications. To identify such antibodies, a patient's serum is tested, before transfusion, for reactivity with the red cells from different donors that have been selected for transfusion on the basis of their ABO and RhD types (see Figure 15.3). Only red cells that do not bind antibodies in the patient's blood are used for transfusion. This direct assessment of compatibility is called a **cross-match test**. Cross-match tests are not performed between the patient's red cells and serum from the donor's blood, because when donor antibodies from transfused blood bind to the recipient's erythrocytes they have no detrimental effect. The quantity of donor antibody transfused is so small compared with the number of the recipient's erythrocytes that the amount of antibody is insufficient to trigger hemolysis. Antibodies against the non-ABO, non-Rhesus blood group antigens are made in patients who have had multiple blood

Histocompatibility in blood transfusion									
		Blood donor type							
		O RhD$^-$	O RhD$^+$	A RhD$^-$	A RhD$^+$	B RhD$^-$	B RhD$^+$	AB RhD$^-$	AB RhD$^+$
Blood recipient type	O RhD$^-$	■							
	O RhD$^+$	■	■						
	A RhD$^-$	■		■					
	A RhD$^+$	■	■	■	■				
	B RhD$^-$	■				■			
	B RhD$^+$	■	■			■	■		
	AB RhD$^-$	■		■		■		■	
	AB RhD$^+$	■	■	■	■	■	■	■	■
Frequency (%)		6.6	37.4	6.3	35.7	1.5	8.5	0.6	3.4

Figure 15.3 Matching the ABO and Rhesus erythrocyte antigens in blood transfusion. Although there are many different red-cell antigens, the important ones in clinical practice are A, B, O, and the Rhesus D antigen (RhD). O does not represent a specific antigen but the absence of both the A and B antigens. The green squares show the combinations of blood donor and recipient that permit successful transfusion. O RhD$^-$ people lack all three antigens (A, B, and Rhesus D) and are 'universal donors' who can provide a transfusion to any other person but can receive blood only from other O RhD$^-$ donors. In contrast, AB RhD$^+$ people have all three antigens and are 'universal recipients' who can receive blood from any donor but can donate only to other AB RhD$^+$ donors. The bottom line gives the frequency of the eight blood types in the US population.

transfusions. Some of these patients have made so many different alloantibodies that it becomes difficult to find any blood donor to whom they are not reactive.

15-4 Hyperacute rejection of transplanted organs is a type II hypersensitivity reaction

The ABO antigens are present on the endothelial cells of blood vessels as well as on erythrocytes. This is a crucial factor in the transplantation of solid organs, particularly for the highly vascularized kidney. If a type O recipient receives a kidney graft from a type A donor, then anti-A antibodies in the recipient's circulation quickly and extensively bind to the blood vessels of the graft. By fixing complement throughout the graft's vasculature, IgG antibodies cause rapid rejection of the graft (Figure 15.4). This type of rejection, called **hyperacute rejection**, can occur even before a transplanted patient has left the operating

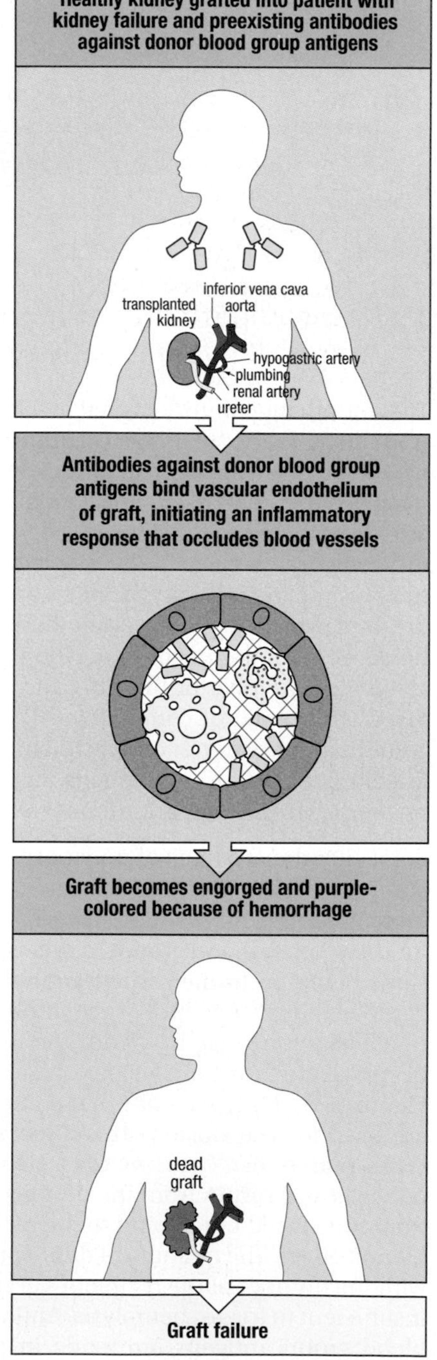

Healthy kidney grafted into patient with kidney failure and preexisting antibodies against donor blood group antigens
inferior vena cava
aorta
transplanted kidney
hypogastric artery
plumbing
renal artery
ureter
Antibodies against donor blood group antigens bind vascular endothelium of graft, initiating an inflammatory response that occludes blood vessels
Graft becomes engorged and purple-colored because of hemorrhage
dead graft
Graft failure

Figure 15.4 Hyperacute rejection is a type II hypersensity reaction caused by preexisting antibodies binding to the graft. Before transplantation, some recipients have made antibodies that react with donor ABO or HLA class I antigens. When the donor organ is grafted into such a recipient, the antibodies immediately bind to vascular endothelium, initiating the complement and clotting cascades. Blood vessels in the graft become obstructed by clots and leak, causing hemorrhage of blood into the graft. The graft becomes engorged, turns purple from the presence of deoxygenated blood, and dies. In hematopoietic cell transplantation ABO compatibility is not required.

room. The only possible treatment is to remove the kidney and return the patient to dialysis. Hyperacute rejection is the most devastating type of rejection for organ grafts and is an extreme form of type II hypersensitivity reaction (see Section 14-1) caused by the vasculature becoming thoroughly coated with antibodies. To avoid hyperacute rejection, transplant donors and recipients are typed and cross-matched for the ABO blood group antigens.

Because HLA class I molecules are expressed constitutively on vascular endothelium, preexisting antibodies against HLA class I variants can also cause hyperacute rejection. It is therefore essential that transplant recipients do not have antibodies that bind to the HLA class I allotypes of the transplanted organ. Antibodies against HLA class II molecules also contribute to hyperacute rejection, but to lesser extent. HLA class II molecules are not normally present on vascular endothelium, but their expression can be induced by infection, inflammation, and trauma, all of which occur during clinical transplantation.

Because no reliable method for reversing hyperacute rejection has been found, this type of rejection is avoided by choosing compatible transplant donors and recipients. Compatibility is assessed with a cross-match test, in which blood serum from the prospective recipient is assessed for the presence of antibodies that bind to the lymphocytes of prospective donors. The traditional cross-match test detects antibodies in the patient's serum that trigger complement-mediated lysis of the donor's lymphocytes. The assay is usually performed on separated B cells and T cells so that reactivity with HLA class I or II molecules can be distinguished: anti-HLA class I antibodies react with both B cells and T cells, whereas antibodies against HLA class II react only with B cells. A more sensitive cross-match assay uses flow cytometry to examine the binding of the patient's antibodies to the prospective donor's lymphocytes (see Figure 4.13, p. 89). This allows all antibody isotypes to be detected, not just those that fix complement.

15-5 Anti-HLA antibodies can arise from pregnancy, blood transfusion, or previous transplants

In some cases, prospective transplant patients have already made anti-HLA class I and II antibodies. The natural situation in which this happens is pregnancy. In almost all pregnancies the fetus expresses paternal HLA allotypes that are not part of the mother's HLA type and have the potential to stimulate an alloreactive immune response. During gestation, the anatomy of the placenta segregates the fetal and maternal circulations and prevents the mother's B cells and T cells from being stimulated by fetal alloantigens (Figure 15.5). During the trauma associated with birth, when the mother, child, and placenta are forcibly separated, cells and material of fetal origin enter the maternal

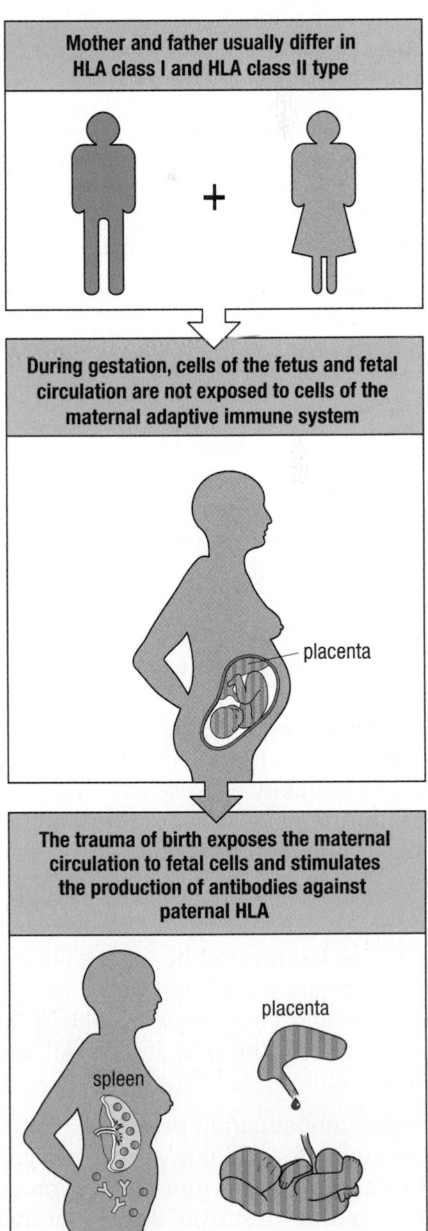

Figure 15.5 Pregnancy is the one natural situation that leads to the production of anti-HLA antibodies. With very few exceptions, the mother and father in human families have different HLA types (top panel). When the mother becomes pregnant, she carries for 9 months a fetus that expresses one HLA haplotype of maternal origin (pink) and one HLA haplotype of paternal origin (blue) (center panel). Although the paternal HLA class I and II molecules expressed by the fetus are alloantigens against which the mother's immune system has the potential to respond, the fetus does not provoke such a response during pregnancy and is protected from preexisting alloreactive antibodies or T cells. The trauma associated with childbirth, which physically separates the mother, the child, and the placenta, allows fetal cells and antigens to enter the maternal circulation and stimulate an adaptive immune response to the paternally inherited HLA molecules (bottom panel).

circulation and stimulate an immune response. Antibodies can be made against any paternal HLA allotype expressed by the baby and not carried by the mother. With successive pregnancies, increasing levels of anti-HLA antibodies can develop, making multiparous women the main source of the anti-HLA sera used in serological HLA typing. The presence of anti-HLA antibodies in the maternal circulation has no detrimental effect on subsequent pregnancies, but will complicate any future search for a compatible organ transplant should one be needed.

Blood transfusions can also lead to the production of anti-HLA antibodies. In routine blood transfusion, HLA type is not assessed, the matching being restricted to just the ABO and Rhesus D types. Infusion of HLA-incompatible leukocytes and platelets in a blood transfusion can therefore generate antibodies specific for the donor HLA allotypes. Patients who received multiple blood transfusions have been stimulated by numerous HLA allotypes and can develop antibodies that react with the cells of most other people in the population. To determine the extent to which a patient needing a transplant has been sensitized against potential donors, the patient's serum is tested for reactivity with white blood cells obtained from a panel of individuals representing the population. The number of positive reactions is expressed as a percentage **panel reactive antibody** (**PRA**). The higher the value of a patient's PRA, the more difficult it is to find a compatible transplant donor.

A third way in which patients develop anti-HLA antibodies is from a previous organ transplant. Now that transplantation has been routine practice for more than 40 years, many patients have had more than one transplant. As with blood transfusions, the more transplants a person has had, the higher their percentage PRA tends to be.

15-6 Transplant rejection and graft-versus-host disease are type IV hypersensitivity reactions

Immune responses against transplanted tissues or organs are caused by genetic differences between the donor and the recipient, the most important being the antigenic differences between highly polymorphic HLA class I and class II molecules. That is why these antigens were called both the **transplantation antigens** and the major histocompatibility antigens. The complex of genes that encodes the transplantation antigens has the general name of the major histocompatibility complex (MHC), although it goes under different names in different species (for example, the HLA complex in humans and the H-2 complex in mice) and its central role in T-cell functions is described in Chapter 5. Self antigens, such as the MHC molecules, which vary between members of the same species, are called **alloantigens**, and the immune responses they provoke are called alloreactions (see Section 5-23). A subfield of immunology, called **immunogenetics**, is devoted to the genetics of alloantigens and their impact on the immune system.

In clinical transplantation, two distinct types of alloreaction can occur, depending on the type of tissue transplanted. Transplantation of solid organs, such as kidney and heart, is a surgical procedure in which the patient's diseased organ is replaced with a healthy organ from the recipient. In this situation the alloreactions developed by the recipient's immune system are directed at the cells of the graft and can kill them, an outcome called **transplant rejection** (Figure 15.6, left panel).

A different situation pertains in hematopoietic stem-cell transplantation, the treatment for patients with severe genetic immunodeficiency or hematopoietic cancers. In this non-surgical procedure, the recipient's hematopoietic system is first destroyed by chemotherapy and irradiation. The patient then receives, by intravenous infusion, a liquid graft containing healthy hemato-

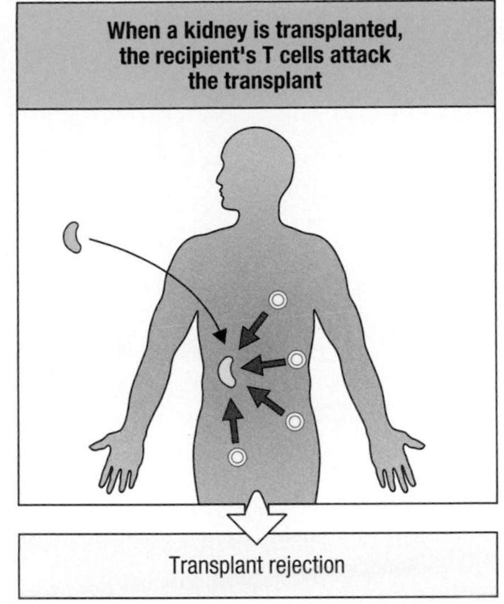

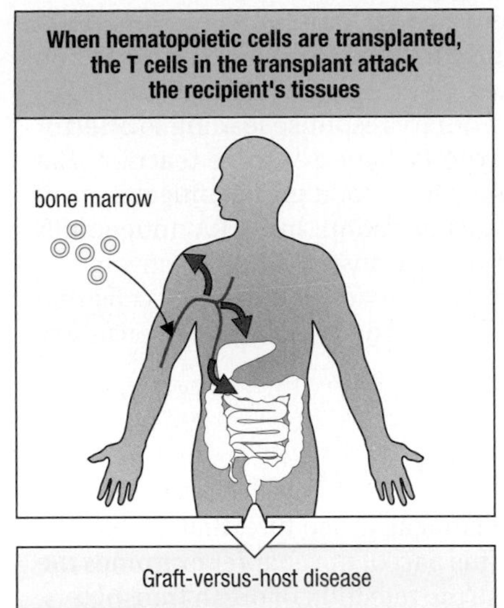

Figure 15.6 Alloreactions in transplant rejection and graft-versus-host reaction. As shown in the left panel, rejection of a transplanted organ occurs when the recipient makes an immune response against it. Graft-versus-host disease occurs when T cells in transplanted bone marrow attack the tissues of the recipient or host, principally the skin, liver, and intestines, as shown in the right panel.

poietic stem cells from the donor. Over the following year, the stem cells from the graft gradually reconstitute a new hematopoietic system in the patient.

The major alloreactions that occur after hematopoietic cell transplantation are due to donor T cells in the graft that respond to and attack the recipient's healthy tissues. This type of alloreactive response by donor lymphocytes is called a **graft-versus-host reaction** (**GVHR**). It causes **graft-versus-host disease** (**GVHD**), which, to a varying degree, affects almost all patients who receive a hematopoietic stem-cell transplant (Figure 15.6, right panel). For many years the only source of a hematopoietic stem-cell transplant was an aspirate, taken surgically, of a donor's bone marrow. The procedure was then called **bone marrow transplantation**. Subsequently, techniques have been developed for obtaining hematopoietic stem cells from a donor's peripheral blood or from umbilical cord blood after giving birth. Because of these various sources for hematopoietic stem cells, the procedures are collectively called **hematopoietic stem-cell transplantation** or **hematopoietic cell transplantation**.

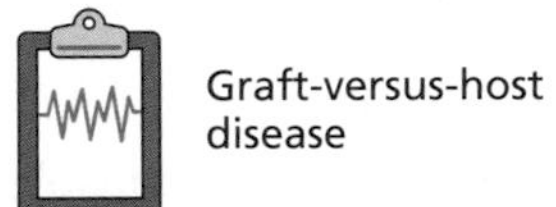

Because people do not normally make an immune response to their own tissues, tissues transplanted from one site to another on the same person are not rejected. This type of transplant, called an **autograft**, is used to treat patients who have suffered burns. Skin from unaffected parts of the body is grafted into the burnt areas, where it facilitates wound healing. Immunogenetic differences are also avoided when tissue is transplanted between identical twins. The first successful kidney transplant, in 1954, involved the donation of a kidney by a healthy twin to his brother, who was suffering from kidney failure. A transplant between genetically identical individuals is called a **syngeneic** transplant or an **isograft**. A transplant made between two genetically different individuals is called an **allograft** or an **allogeneic** transplant.

Summary

Type II hypersensitivity reactions occur after a person with IgG antibodies against A, B, or Rhesus D blood group antigen is transfused with red cells carrying the antigen. A massive hemolytic reaction with systemic effects is induced. Type II hypersensitivity reactions also occur when a person with IgG antibodies against the A or B blood group antigen is transplanted with a kidney expressing the target antigen on vascular endothelium. On coating the blood vessels, the IgG initiates a massive inflammatory reaction that instantly destroys the transplanted tissue and threatens the recipient's life. Such hyperacute rejection also occurs if the recipient has IgG antibodies against HLA class I antigens expressed by the kidney endothelium. Recipients transplanted with an HLA-mismatched kidney who do not have preexisting antibodies against the foreign HLA antigens make a primary response leading to effector T cells attacking the grafted kidney in a type IV hypersensitive reaction that can lead to its acute rejection. Antibodies against blood group antigens are of no consequence in hematopoietic cell transplantation, but HLA antigen differences can lead to donor-derived T cells in the transplant being activated by the recipient's HLA antigens. This produces alloreactive effector T cells that attack many tissues of the body, causing a systemic type IV hypersensitivity reaction called graft-versus-host disease.

Transplantation of solid organs

Organ transplantation is now a routine clinical procedure that saves and extends many thousands of human lives. This part of the chapter examines the immunological mechanisms that can cause the rejection of organ transplants, and the genetic strategies, immunosuppressive drugs, and other therapies that are used to avoid, prevent, and treat rejection. Emphasis will be on the kidney because it is frequently transplanted and is relatively vulnerable to rejection.

15-7 Organ transplantation involves procedures that inflame the donated organ and the transplant recipient

Patients receiving an organ transplant usually have a history of disease in which the organ to be replaced has gradually degenerated. Often there is an immune component to this decay, such as the deposition of immune complexes that leads to kidney damage and failure. Before transplantation, patients with renal failure are maintained by dialysis, a procedure that induces inflammation through the interaction of dialysis membranes and serum proteins. Consequently, the transplant patient is already inflamed before transplantation, a state that is exacerbated by the damage and disruption caused by the surgery involved in transplantation. Thus, on receiving an organ transplant, the recipient's body is both prepared and ready to direct innate and adaptive immunity toward the transplanted tissue.

Donated organs are also inflamed. In the case of cadaveric organs—that is, those taken from a dead donor—the donor will usually have died in a violent or stressful manner, and the procedures used to collect organs and transport them to the transplant center add to the stress (Figure 15.7). During this time, the organs are deprived of blood, a state called **ischemia**. Ischemia causes

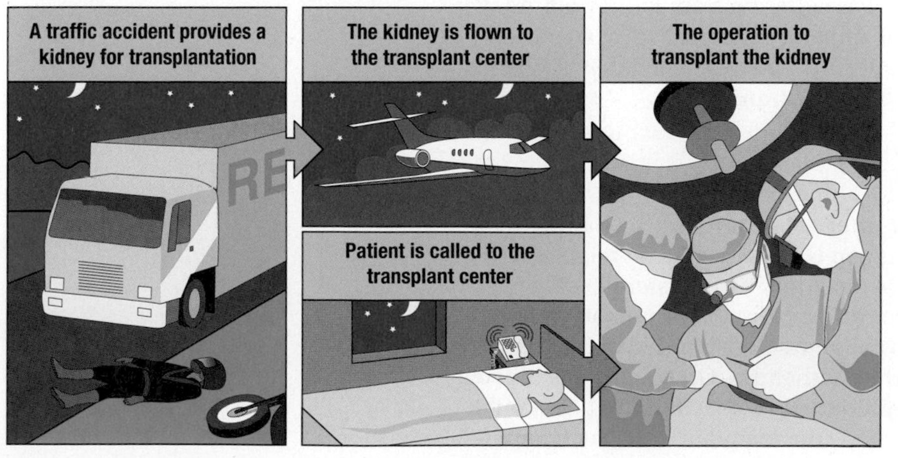

A traffic accident provides a kidney for transplantation
The kidney is flown to the transplant center
The operation to transplant the kidney
Patient is called to the transplant center

Figure 15.7 Clinical transplantation involves both a donated organ and a transplant recipient that are stressed and inflamed. This cartoon depicts a typical sequence of events before transplantation of a cadaveric organ transplant.

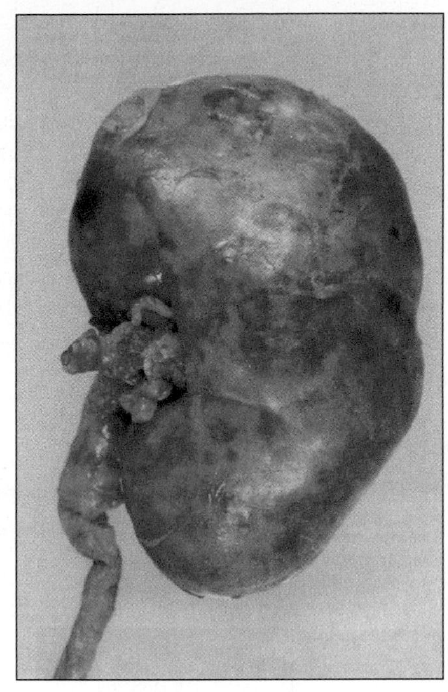

Figure 15.8 Gross appearance of an acutely rejected kidney. The rejected graft is swollen and has deep-red areas of hemorrhage and gray areas of necrotic tissue. Courtesy of B.D. Kahan.

damage to the blood vessels and tissue of the organs through activation of the endothelium and of the complement system, infiltration with inflammatory leukocytes, and the production of cytokines. The success of transplantation depends heavily on limiting the damage caused by ischemia. For kidney or liver transplantation it is possible to use a living healthy donor, and this confers great advantage. Donation and transplantation are performed at the same time and in the same place, with minimal time of ischemia. This approach has mainly been used for related donors and recipients, but it is now extended to unrelated family members, for example husbands and wives. Because of the reduced inflammation and tissue damage associated with organs from living healthy donors, the success of the transplantation is less sensitive to HLA mismatch than transplantation from cadaveric donors.

15-8 Acute rejection is a type IV hypersensitivity caused by effector T cells responding to HLA differences between donor and recipient

Most organ transplants are performed across some HLA class I and/or II difference. In this situation the transplant recipient's naive T-cell population includes clones of alloreactive T cells that recognize HLA allotypes of the transplanted tissue that are not shared with the recipient. CD8 T cells respond to the HLA class I differences, and CD4 T cells respond to the HLA class II differences. The alloreactive T-cell response produces effector CD4 and CD8 T cells, both of which can attack the organ graft and destroy it (Figure 15.8). This is called **acute rejection** and is a form of type IV hypersensitivity reaction. Unlike hyperacute rejection, it takes several days to develop, during which the naive T cells are activated to divide and differentiate into effector cells. This gives the transplant physician time during which acute rejection can be reduced or prevented by intervention. To prevent acute rejection, all transplant patients are conditioned with immunosuppressive drugs before transplantation and maintained on them after transplantation. Patients are carefully monitored for early signs of acute rejection and are treated with additional immunosuppressive drugs or anti-T-cell antibodies when they occur.

The inflamed state of the transplanted organ activates the organ's dendritic cells. These donor-derived dendritic cells migrate to the draining secondary lymphoid tissue, where they settle into the T-cell zone and present complexes of donor MHC and donor self peptides to the recipient's circulating T cells. Because different HLA allotypes bind different sets of self peptides, each allotype selects a different repertoire of T-cell receptors during thymic selection. Consequently, the T-cell repertoire selected by the recipient's HLA type contains numerous T-cell clones that can respond to the self-peptide:HLA complexes presented by donor cells of a different HLA type. For this reason, alloreactive T-cell responses stimulated by HLA differences are stronger than T-cell responses to a vaccine or to a pathogen. Many clones of alloreactive T cells have a memory phenotype, revealing that they were originally stimulated and expanded in response to pathogens and are cross-reactive with allogeneic HLA. This type of alloreactive response, in which recipient T cells are stimulated by direct interaction of their receptors with the allogeneic HLA molecules expressed by donor dendritic cells, is called the **direct pathway of allorecognition** (Figure 15.9). It produces effector T cells that migrate to the transplanted tissue, where T_H1 cells activate the resident macrophages to inflame the tissue further and CD8 T cells systematically kill the cells of the transplanted tissue.

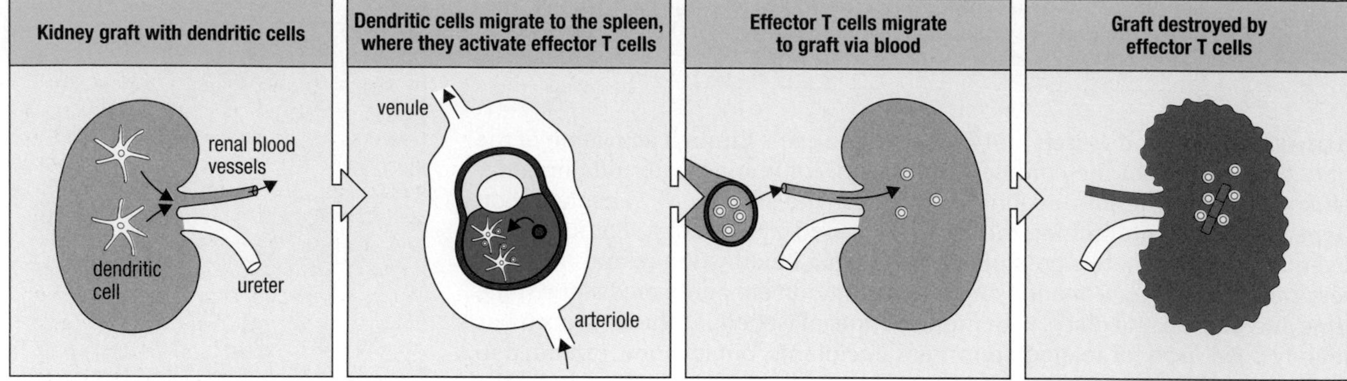
Kidney graft with dendritic cells
renal blood vessels
dendritic cell
ureter
Dendritic cells migrate to the spleen, where they activate effector T cells
venule
arteriole
Effector T cells migrate to graft via blood
Graft destroyed by effector T cells

Figure 15.9 Acute rejection of a kidney graft through the direct pathway of allorecognition. Donor dendritic cells in the graft carry complexes of donor HLA molecules and donor peptides on their surfaces. The dendritic cells travel to the spleen, where they move to the T-cell areas. Here, they activate the recipient's alloreactive T lymphocytes. After activation, the effector T cells travel in the blood to the grafted organ, where they attack cells expressing the complexes of peptide and either HLA class I or HLA class II recognized by their T-cell receptors.

15-9 HLA differences between transplant donor and recipient activate numerous alloreactive T cells

We have seen how a serological cross-match test is used prospectively to examine how a patient's antibodies could react with transfused blood or transplanted tissue. The **mixed lymphocyte reaction** is an analogous cellular test that has been used to assess the extent to which a patient's T cells could respond to a transplanted organ from a live donor. **Peripheral blood mononuclear cells** (**PBMC**), comprising mainly lymphocytes and monocytes, from the patient needing a transplant are cultured together for five days with cells similarly isolated from the prospective donor. In addition the donor's cells are lethally irradiated, so that they do not respond to the patient's cells but are still capable of stimulating the patient's T cells (Figure 15.10, top panel). During the five day culture, a primary immune response is made in which the naive T cells having receptors that recognize allogeneic HLA class I and II are stimulated to divide and differentiate into effector CD8 and CD4 T cells. In the course of the culture the proliferation of T cells was measured (Figure 15.10, bottom left panel). At the end of the culture the population of effector T cells was assayed for its capacity to kill cells from the donor. (Figure 15.10, bottom right panel). Thus the mixed-lymphocyte reaction is an *in vitro* model of acute graft rejection. Because of the time taken to perform the test, the mixed lymphocyte reaction is now rarely used in matching transplant donors and recipients. In the research laboratory, however, the mixed lymphocyte reaction has been very informative and was instrumental in the discovery of MHC class II, which involved distinguishing it from MHC class I.

In a mixed lymphocyte reaction between HLA disparate individuals, around 5% of the T cells that are restricted by a particular HLA class I or class II allotype can be activated by an allogeneic HLA class I allotype of the same locus.

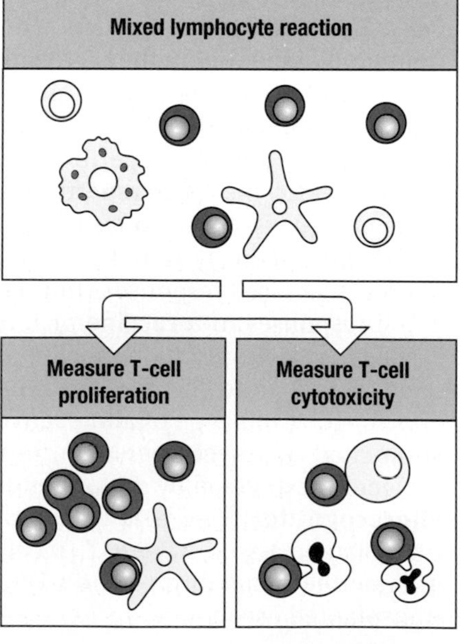
Mixed lymphocyte reaction
Measure T-cell proliferation
Measure T-cell cytotoxicity

Figure 15.10 The mixed lymphocyte reaction is a cellular test of HLA difference between transplant donor and recipient and its potential for graft rejection. Blood lymphocytes, monocytes, and dendritic cells are isolated from a patient seeking a kidney graft (blue) and from a possible kidney donor (yellow). The donor's cells are irradiated so they act only as stimulators and not as responders (upper panel). The patient's cells and the donor's cells are cultured together for five days. During this time alloreactive T cells of the patient are activated by the allogeneic HLA class I and II molecules of the donor. After 3 or 4 days of culture the proliferation of the differentiating T cells is measured (bottom panel, left). After 5 days of culture the capacity of effector CD8 T cells to kill donor cells is assessed (bottom panel, right). Proliferation measures the magnitude of the alloreactive response, killing of donor cells measures its capacity for graft rejection.

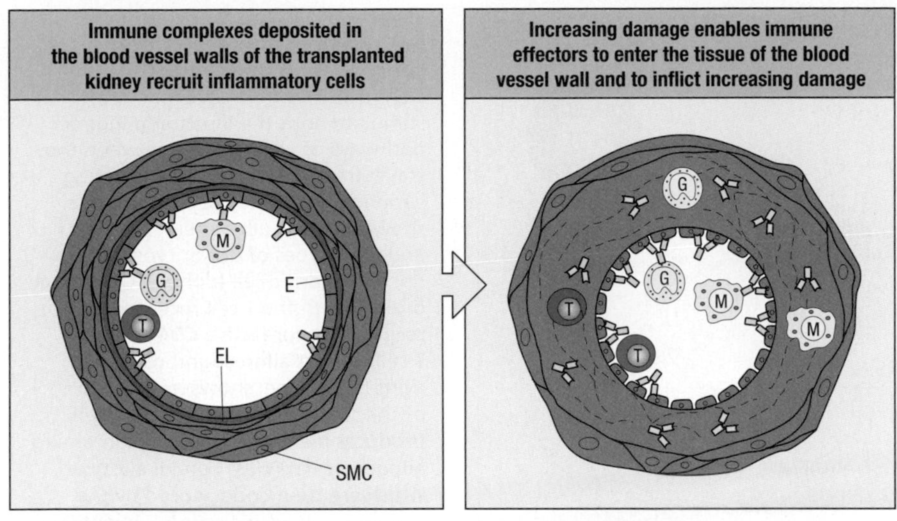

Immune complexes deposited in the blood vessel walls of the transplanted kidney recruit inflammatory cells
Increasing damage enables immune effectors to enter the tissue of the blood vessel wall and to inflict increasing damage
M
G
E
T
EL
SMC
G
T
M
T
M

Figure 15.11 Chronic rejection of a kidney transplant. Left panel: chronic rejection is caused by complexes of HLA molecules and anti-HLA antibodies that deposit in the blood vessels of the organ transplant. Immune complexes bound to endothelial cells (E) recruit Fc receptor-bearing monocytes and neutrophils. EL: internal elastic lamina; SMC: smooth muscle cells. Right panel: accumulating damage leads to EL thickening and infiltration of the underlying intima with SMCs, macrophages (M), granulocytes (G), alloreactive T cells (T), and antibodies. This narrows the blood vessel lumen and creates chronic inflammation that intensifies tissue repair. In time, the vessel becomes obstructed, ischemic, and fibrotic.

This involves T cells responding to numerous complexes of the same allogeneic HLA molecule with many different peptides. The strength of the response is comparable to that induced by a bacterial superantigen (see Section 13-6) and emphasizes the benefit of HLA matching and the necessity for potent immunosuppressive drugs.

15-10 Chronic rejection of organ transplants is caused by a type III hypersensitivity reaction

As well as acute and hyperacute rejection, transplanted human organs can be subject to **chronic rejection**. This phenomenon, which only becomes manifest some months or years after transplantation, is characterized by reactions in the vasculature of the graft that cause thickening of the vessel walls and a narrowing of their lumens (Figure 15.11). The blood supply to the graft gradually becomes inadequate, causing ischemia, loss of function and eventually death of the graft. Chronic rejection causes the failure of more than half of all kidney and heart grafts within 10 years after transplantation. Chronic rejection is a type III hypersensitivity reaction caused by IgG antibodies made against the allogeneic HLA class I molecules of the graft, forming immune complexes that deposit in the blood vessels of the transplanted kidney. Consistent with antibodies causing chronic rejection, grafts undergoing chronic rejection are infiltrated with CD40-expressing B cells and T helper cells expressing CD40 ligand. The anti-B-cell antibody rituximab is used to treat chronic rejection.

The CD4 helper T cells initiating the response that leads to chronic rejection of organ transplants do not recognize their antigens by the direct pathway of allorecognition, but by a different route, the **indirect pathway of allorecognition**. The differences in mechanism between the two pathways of allorecognition are shown in Figure 15.12. In the indirect pathway, some of the donor-derived dendritic cells that migrate to the draining lymphoid tissue die there by apoptosis. Membrane fragments containing HLA molecules from these apoptotic cells are taken up by dendritic cells of the transplant recipient and processed so that peptides derived from donor HLA allotypes are presented by the recipient's HLA allotypes. Because of the endocytic mode of uptake, most of these peptides, which can be derived from HLA class I or II molecules, are presented by the HLA class II allotypes of the recipient. If the peptides differ in amino acid sequence from those formed by degradation of the recipient's HLA allotypes, then these complexes will stimulate a CD4 T-cell alloreaction. The responding CD4 T cells are specific for the complex of a peptide derived from a donor HLA allotype bound to a recipient HLA class II allotype. This way of stimulating alloreactive T cells is called the indirect pathway

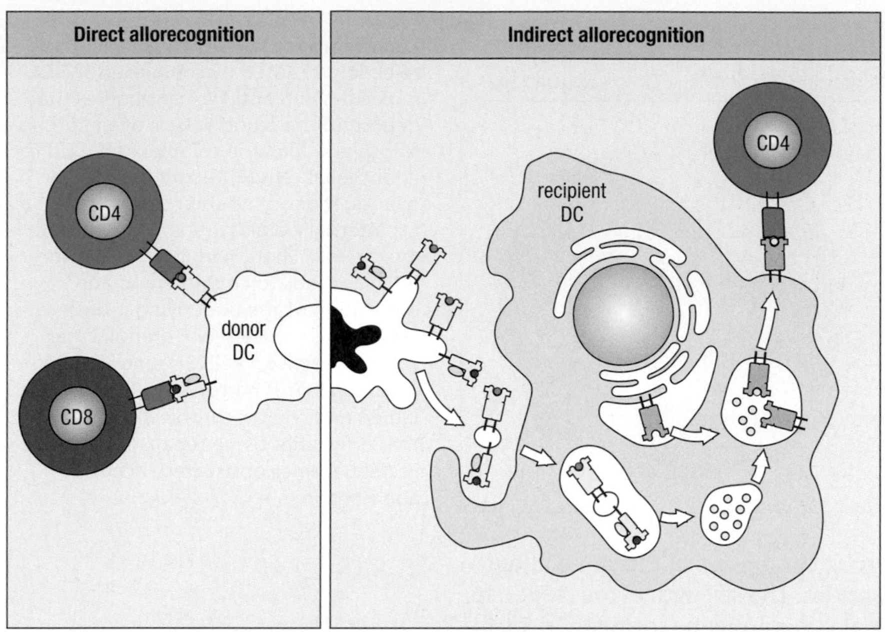

Figure 15.12 Direct and indirect pathways of allorecognition contribute to graft rejection. Dendritic cells from an organ graft stimulate both the direct and indirect pathways of allorecognition when they travel from the graft to the draining lymphoid tissue. The left-hand panel shows how the allogeneic HLA class I and II allotypes of donor type on a donor dendritic cell (donor DC) interact directly with the T-cell receptors of the recipient's alloreactive CD4 and CD8 T cells (direct allorecognition). The right-hand panel shows how death of the same antigen-presenting cell produces membrane vesicles containing allogeneic HLA class I and II allotypes, which are then endocytosed by the recipient's dendritic cells (recipient DC). Peptides derived from the donor's HLA molecules (yellow) are then presented by the recipient's HLA molecules (orange) to peptide-specific T cells (indirect allorecognition). Presentation by HLA class II molecules to CD4 T cells is shown here. Peptides derived from donor HLA are also presented by recipient HLA class I molecules to CD8 T cells (not shown).

because the alloreactive T cells do not directly recognize the transplanted cells but instead recognize subcellular material that has been processed and presented by the recipient's own cells (see Figure 15.12).

The indirect pathway of allorecognition is a special case of the normal mechanism by which T cells recognize the protein antigens of pathogens; in transplantation, the foreign peptide antigens come from the proteins of another human body. T cells stimulated by the indirect pathway of allorecognition can also contribute to the acute rejection of transplanted organs, although they are usually less numerous than those stimulated by the direct pathway.

The alloreactive T-cell response stimulated by the direct pathway wanes with time after transplantation. This correlates with the elimination of dendritic cells of donor origin and the repopulation of the transplanted organ with immature dendritic cells of recipient origin. These latter cells can, however, still increase the stimulation of alloreactive T cells through the indirect pathway of allorecognition. After cross-match tests, transplant recipients are selected for their lack of serum antibody that could react with the transplanted organ. This indicates they also lack memory B cells that could respond to the transplant by making antibodies directed at the allogeneic HLA of the graft. It does not, however, preclude the possibility that the transplant recipients have naive B cells capable of making antibodies that react with the allogeneic HLA antigens. After transplantation, activation of helper CD4 T cells through the indirect pathway could thus initiate an antibody response against the allogeneic HLA (Figure 15.13). These T cells then help the naive B cells to become plasma cells making antibodies specific for the foreign HLA class I allotypes of the graft. The HLA-specific CD4 T cells could also provide help to B cells specific for other alloantigens present in the HLA-containing subcellular fragments. By these means the strength and breadth of the antibody response could increase, leading to increasingly impaired function of the transplanted kidney.

The indirect pathway of allorecognition can also give rise to regulatory CD4 T cells that suppress alloreactive CD4 and CD8 effector T cells and improve the clinical outcome after kidney transplantation. Such regulatory T cells seem more active in patients who previously received blood transfusions that, fortuitously, shared an HLA-DR allotype with the transplanted kidney. This phenomenon of previous blood transfusions improving the outcome of organ transplantation is known as the **transfusion effect**.

Figure 15.13 The indirect pathway of allorecognition is responsible for stimulating the production of the anti-HLA antibodies that cause chronic graft rejection. The processing and presentation of allogeneic HLA class I by a dendritic cell (DC) of the recipient is shown. The dendritic cell activates helper CD4 T cells, which in turn activate B cells that have bound and internalized allogeneic donor HLA molecules. Shown here is a cognate interaction that leads to the production of an anti-HLA class I antibody. Anti-HLA class II antibodies can be produced similarly. Because activated endothelium expresses both HLA class I and II molecules, antibodies against both classes of HLA molecule can contribute to chronic rejection.

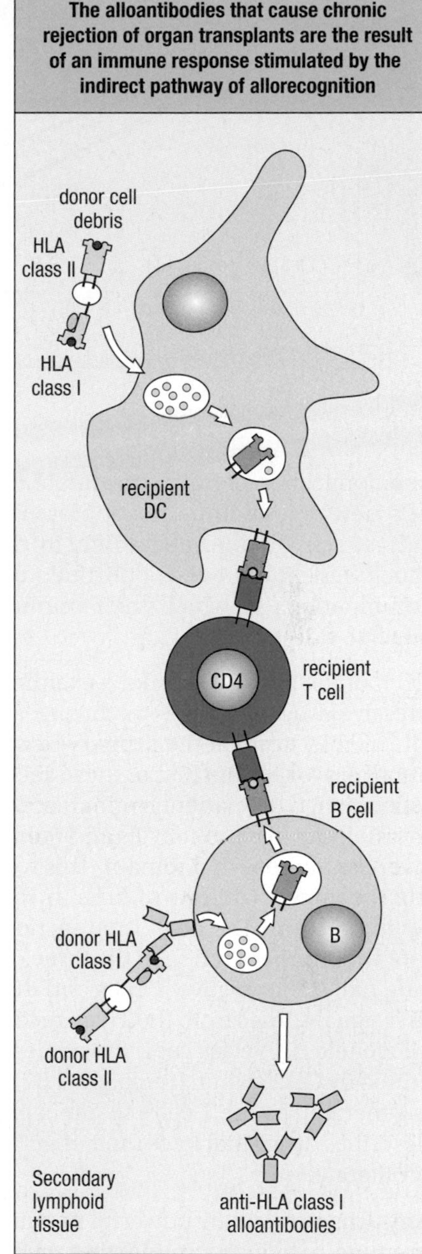

15-11 Matching donor and recipient HLA class I and II allotypes improves the success of transplantation

The first successful organ transplant involved a kidney donor and recipient who were identical twins. In this situation there was no risk of alloreactivity and graft rejection. Because very few patients with kidney disease have an identical twin, other approaches were needed to make transplantation a more generally available therapy. A combination of two complementary tactics achieved this goal. The first is selection of a transplant donor who has an HLA type as similar as possible to that of the recipient. This reduces the number of alloreactive T cells that become activated by the transplant. The second tactic is administration of immunosuppressive drugs that interfere with the activation of alloreactive T cells.

Clinical transplantation was pioneered with the kidney graft because of two reasons of practicality. First, patients with failed kidneys could be sustained by the established procedure of dialysis. This meant that neither graft failure nor graft rejection need lead to the death of the recipient. The second reason is the simple fact that everyone has two kidneys but can manage with one, so healthy and altruistic relatives could give a kidney to their needy kin. Immunogenetic differences within a family are far smaller than in the population at large, so the probability of finding an HLA-matched person within the family is higher. Analysis of the fate of kidney transplants performed between HLA-identical and non-identical family members was instrumental in showing that the better the HLA class I and II match, the better the clinical outcome. HLA-A, -B, -C, and -DR allotypes are the most important ones to match. DNA typing has superseded serological methods for matching HLA antigens, but serology is still essential for the cross-match test.

After transplantation success was demonstrated between living relatives, methods were developed to transplant kidneys from unrelated donors who had been killed in accidents (cadaveric donors). More than 100,000 such kidney transplants have been performed worldwide, and a statistical analysis of these data shows that both graft performance and long-term health of the recipient increase with the degree of HLA match (Figure 15.14).

15-12 Immunosuppressive drugs make allogeneic transplantation possible as routine therapy

Because the number of donated organs is small, the number of patients on waiting lists is large, and HLA matching is not always a high priority, the majority of patients transplanted with cadaveric kidneys are mismatched at one or more HLA genes. In all of these transplants there is the potential for stimulating alloreactive T cells that could cause acute rejection of the kidney graft. **Immunosuppressive drugs** prevent this from occurring. They are of various chemical types and interfere with T-cell activation and differentiation in different ways. During the past 50 years almost all major advances in the success

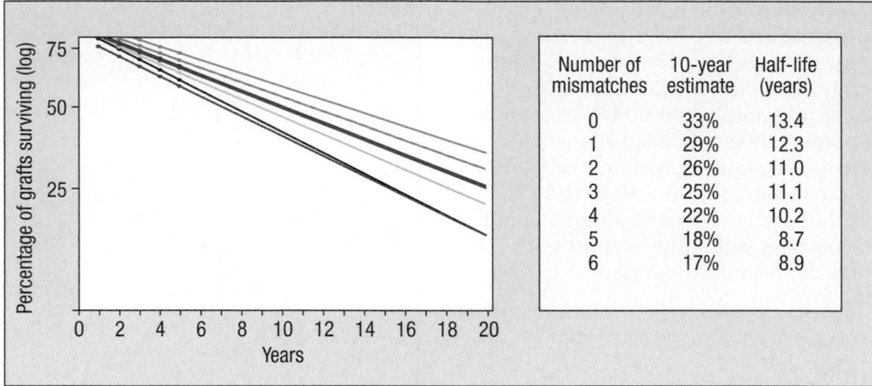

Number of mismatches	10-year estimate	Half-life (years)
0	33%	13.4
1	29%	12.3
2	26%	11.0
3	25%	11.1
4	22%	10.2
5	18%	8.7
6	17%	8.9

Figure 15.14 HLA matching improves the survival of transplanted kidneys. The colored lines in the left panel represent the actual (to 5 years) and projected survival rates of kidney grafts in patients with no (blue), 1 (orange), 2 (red), 3 (dark blue), 4 (green), 5 (black), and 6 (brown) HLA mismatches, plotted on a semi-logarithmic scale. Data courtesy of G. Opelz.

and application of organ transplantation have been heralded by the discovery of a new type of immunosuppressive drug. Each of these drugs has adverse side-effects, but the effects differ from one drug to another. This means there is much benefit in having a portfolio of different drugs, so they can be used in combinations in which their immunosuppressive effects add up but their adverse side-effects do not.

Each of the next five sections examines a different aspect of the immunosuppressive drugs used before, during, and after kidney transplantation, and the effects they have on the primary immune response that the foreign graft inevitably provokes. The first of these sections describes the day before transplantation, when the patient begins a course of drugs that nonspecifically depletes the majority of lymphocytes and monocytes and generally inhibits the responsiveness of those that remain. This weakens the patient's immune system, giving the graft a window of time in which to recover from surgery and regain renal function. The next three sections examine drugs that specifically interfere with the generation of the three different signals that are necessary to activate naive alloreactive T cells and drive them to become effector T cells. The first signal comes from the T-cell receptor on recognizing antigen presented by allogeneic HLA class I or II molecules. Second is the co-stimulatory signal generated by CD28 when it engages a B7 co-stimulatory molecule. The third signal is generated by the T cell's IL-2 receptor when it binds to IL-2. The fifth section describes drugs that have their effect after activated naive T cells have begun to proliferate.

Any drug sufficiently powerful to inhibit alloreactions also inhibits the normal immune responses to infecting pathogens. Consequently, administration of these drugs, which is greatest during the period immediately before and after transplantation, renders transplant patients highly susceptible to infection. Patients are initially cared for in conditions under which their exposure to pathogens is reduced. As their immune systems accommodate to the graft, the dose of immunosuppressive drugs is gradually reduced to 'maintenance levels' that prevent rejection while sustaining active defenses against infection. Patients are carefully monitored for early symptoms of acute rejection which are treated with additional immunosuppression. The longer the time that a kidney is successfully protected from acute rejection, the more likely is onset of chronic rejection.

Certain side-effects emerge only after patients have taken immunosuppressive drugs for long periods. These include a higher incidence of certain malignant diseases, particularly carcinomas of the skin and the genital tract, lymphoma, and Kaposi's sarcoma. On average, the incidence of cancer in transplant recipients is three times that of similarly aged people who have not received a transplant.

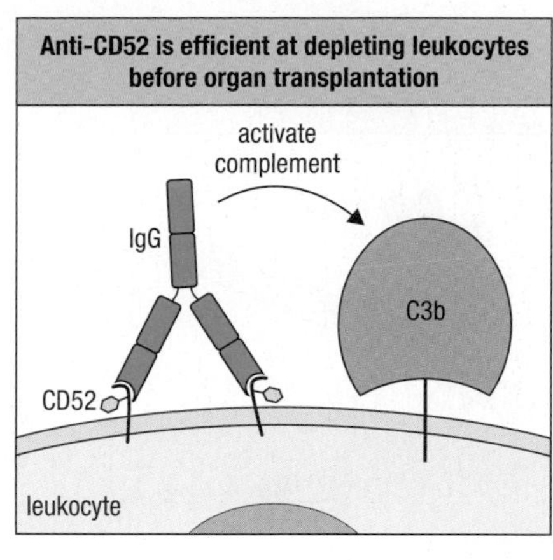

Figure 15.15 Anti-CD52 is used to deplete leukocytes from patients undergoing organ transplantation. Anti-CD52 is efficient at fixing complement on leukocyte surfaces and consigning them to phagocytosis. This property is attributed to the small size of CD52, which has a molecular weight of 8 kDa compared to 160 kDa for IgG, and 180 kDa for C3b. On binding CD52, anti-CD52 is brought close to the cell surface, increasing the likelihood that C3b produced by complement activation will bind covalently to the leukocyte surface.

15-13 Some treatments induce immunosuppression before transplantation

Antibodies that broadly react with white blood cells are given to patients before and after transplantation to deplete these cells and generally weaken the immune system. One of the antibody preparations is purified from the serum of rabbits immunized with human thymocytes. This **rabbit antithymocyte globulin** (**rATG**) is a polyclonal mixture of high-affinity antibodies that binds to T cells, B cells, NK cells, and dendritic cells, as well as to endothelial cells. The bound rabbit IgG fixes human complement well and delivers the leukocytes to be killed by phagocytes. Another reagent used for this purpose is a humanized rat monoclonal IgG that is specific for CD52 and called **alemtuzumab** or **anti-CD52**. CD52 is expressed on almost all lymphocytes, monocytes, and macrophages, and anti-CD52 antibodies induce a profound, long-lasting lymphopenia. The cell-surface complex of CD52 and anti-CD52 is unusually efficient in fixing complement, which is attributed to the distinctive structure of CD52. It is a tiny protein of only 49 amino acids that is tethered to cell membranes by a glycophosphatidylinositol tail (Figure 15.15). Thus antibodies bound to CD52 are very close to the membrane, increasing the probability that the exposed thioester bonds of activated C3 will react with membrane components rather than water molecules (see Section 2-3). The physiological function of CD52 is not known.

Prednisone, the immunosuppressive drug most extensively used in organ transplantation, is a synthetic derivative of hydrocortisone, also called cortisol, the principal steroid made by the adrenal cortex. For more than 55 years, hydrocortisone has been used clinically to reduce inflammation. Four times more potent than hydrocortisone, prednisone is an example of a **pro-drug**, which means that the drug given to patients is inactive, and only in the human body does it become converted to the active form. Prednisone has no immunosuppressive activity until it is enzymatically converted *in vivo* to **prednisolone** (Figure 15.16). By itself, prednisone is insufficiently immunosuppressive to prevent graft rejection, but it works effectively in combination with other drugs.

Corticosteroids have wide-ranging physiological effects and affect all white blood cells, as well as other cells. Steroid hormones do not have cell-surface receptors. They diffuse across the plasma membrane to bind specific receptors in the cytoplasm. In the absence of steroid, the receptors associate with another cytoplasmic polypeptide called Hsp90 (**h**eat-**s**hock **p**rotein of 90 kDa molecular weight). Binding steroid induces conformational change in the receptor, causing dissociation of Hsp90. This allows the receptor:steroid complex to turn on the transcription of selected genes (Figure 15.17). The transcription of about 1% of genes is influenced by corticosteroids. Because of the

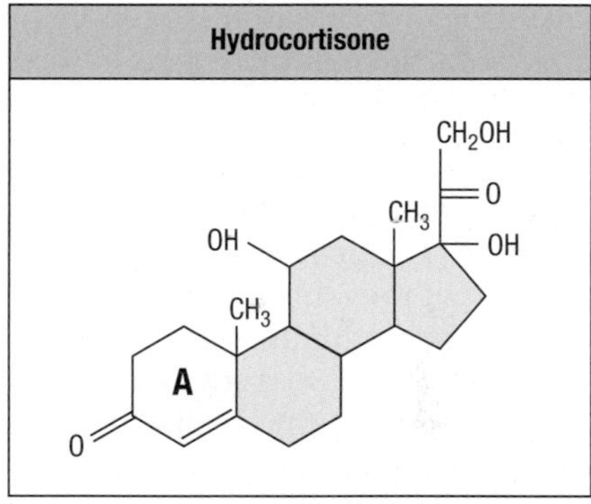

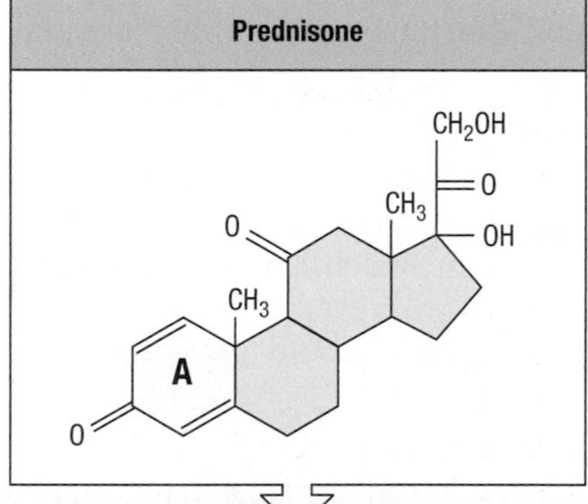

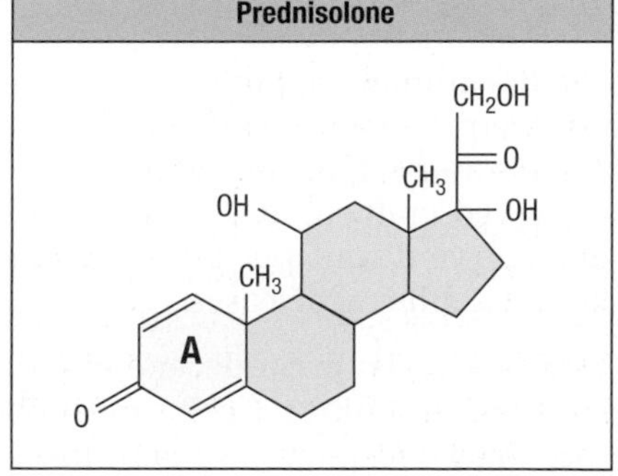

Figure 15.16 Chemical structures of hydrocortisone, prednisone, and prednisolone. Prednisone is a synthetic analog of the natural adrenocorticosteroid hydrocortisone, or cortisol. It is converted *in vivo* to the biologically active form, prednisolone. Introduction of the 1,2 double bond into the A ring of prednisone increases anti-inflammatory potency approximately fourfold compared to hydrocortisone.

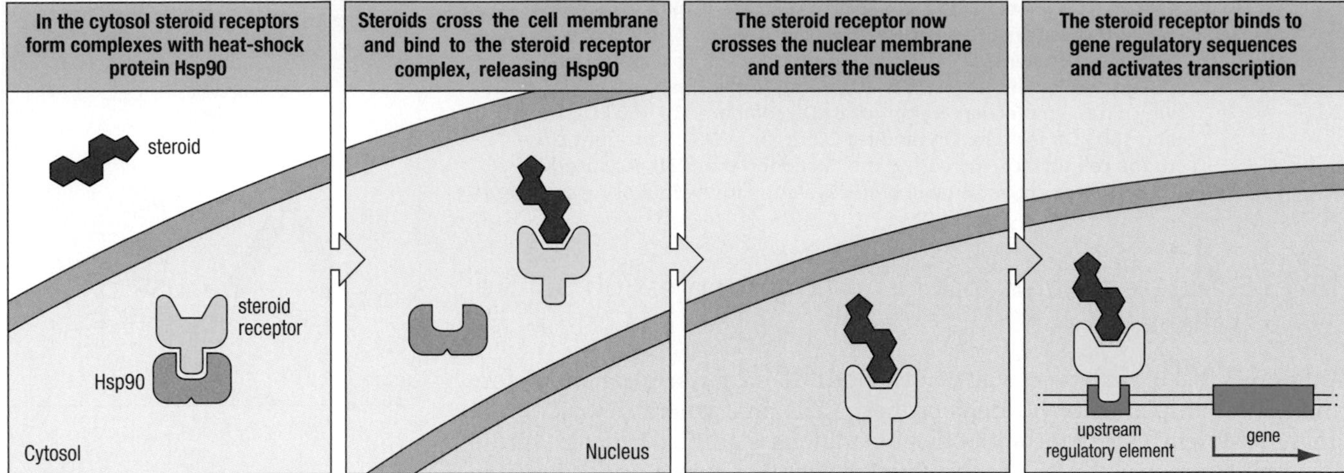

In the cytosol steroid receptors form complexes with heat-shock protein Hsp90
steroid
steroid receptor
Hsp90
Cytosol
Steroids cross the cell membrane and bind to the steroid receptor complex, releasing Hsp90
Nucleus
The steroid receptor now crosses the nuclear membrane and enters the nucleus
The steroid receptor binds to gene regulatory sequences and activates transcription
upstream regulatory element
gene

Figure 15.17 Steroids change patterns of gene transcription. Corticosteroids are lipid-soluble compounds that diffuse across the plasma membrane and bind to their receptors in the cytosol. Binding of corticosteroid to the receptor displaces a heat-shock protein named Hsp90, exposing the DNA-binding region of the receptor, which then enters the nucleus and binds to specific DNA sequences in the promoter regions of steroid-responsive genes. Corticosteroids modulate the transcription of a wide variety of genes.

way in which they work, corticosteroids are most effective as immunosuppressive drugs when administered before the transplant is performed. With this approach, the pattern of gene expression in the recipient is already changed by the time of transplantation and alloantigenic challenge.

A key immunosuppressive effect of hydrocortisone and prednisolone is to prevent the action of NFκB, the central transcription factor of the inflammatory response. Corticosteroids increase the production of IκBα, the inhibitory regulator that prevents NFκB from gaining access to the nucleus and turning on the cytokine and other genes that create a state of inflammation (see Section 3-3) (Figure 15.18). Another critical effect is that prednisolone alters lymphocyte homing, so that lymphocytes are barred from entering secondary lymphoid tissues and sites of inflammation. Instead, they congregate in the bone marrow. In this way, naive lymphocyte cannot be activated by alloantigens, and effector T cells cannot enter and attack the graft.

Because of their multifarious effects on gene expression and cellular metabolism, corticosteroid drugs have many adverse side-effects, including fluid retention, weight gain, diabetes, loss of bone mineral, and thinning of the skin. They are also used as an acute immunosuppressive agent during episodes of rejection, which are often caused by infection, but their prolonged use is avoided wherever possible. The severity of the side-effects has been a major driving force behind the search for and discovery of other immunosuppressive drugs.

Corticosteroid therapy	
Activity	**Effect**
↓ IL-1, TNF-α, GM-CSF IL-3, IL-4, IL-5, CXCL8	↓ Inflammation caused by cytokines
↓ NOS	↓ NO
↓ Phospholipase A_2 ↓ Cyclooxygenase type 2 ↑ Lipocortin-1	↓ Prostaglandins ↓ Leukotrienes
↓ Adhesion molecules	Reduced emigration of leukocytes from vessels
Induction of endonucleases	Induction of apoptosis in lymphocytes and eosinophils

Figure 15.18 Effects of corticosteroids on the immune system. Corticosteroids alter the expression of many genes, to achieve anti-inflammatory effects. First, they reduce the production of inflammatory mediators, including some cytokines, prostaglandins, and nitric oxide (NO). By their effects on other cytokines, corticosteroids also decrease the synthesis of IL-2 by activated lymphocytes. Second, they prevent inflammatory cell migration to sites of inflammation by inhibiting the expression of adhesion molecules. Third, corticosteroids promote apoptotic death of leukocytes and lymphocytes. NOS, nitric oxide synthase.

15-14 T-cell activation can be targeted by immunosuppressive drugs

The first immunosuppressive drugs known to selectively inhibit T-cell activation were the microbial products cyclosporin and tacrolimus. These drugs had a major impact on clinical transplantation in the 1980s and 1990s, leading to improved graft survival, a broader range of tissues and organs being transplanted, and transplantation becoming the recommended treatment for a larger number of diseases.

Cyclosporin is a cyclic decapeptide derived from the soil fungus *Tolypocladium inflatum*. It inhibits T-cell activation by disrupting signal transduction from the T-cell receptor. Normally these signals cause membrane lipids to be hydrolyzed, producing inositol trisphosphate and leading to the release of Ca^{2+} from intracellular stores. Elevated cytosolic Ca^{2+} activates the cytoplasmic serine/threonine phosphatase **calcineurin**, which in turn activates the transcription factor NFAT (see Section 8-6). In resting T cells, this transcription factor is in the cytoplasm and phosphorylated. By removing the phosphate, calcineurin permits NFAT to enter the nucleus, where it binds the transcription factor AP-1 to form a complex that initiates transcription of the gene encoding IL-2. Cyclosporin interferes with calcineurin activity. It diffuses across the plasma membrane to the cytosol, where it binds peptidyl-prolyl isomerase enzymes called **cyclophilins**. The complex of cyclosporin A and cyclophilin binds calcineurin, inhibiting its phosphatase activity and preventing it from activating NFAT. In the presence of cyclosporin, IL-2 is never made, and T-cell activation is shut down at an early stage (Figure 15.19).

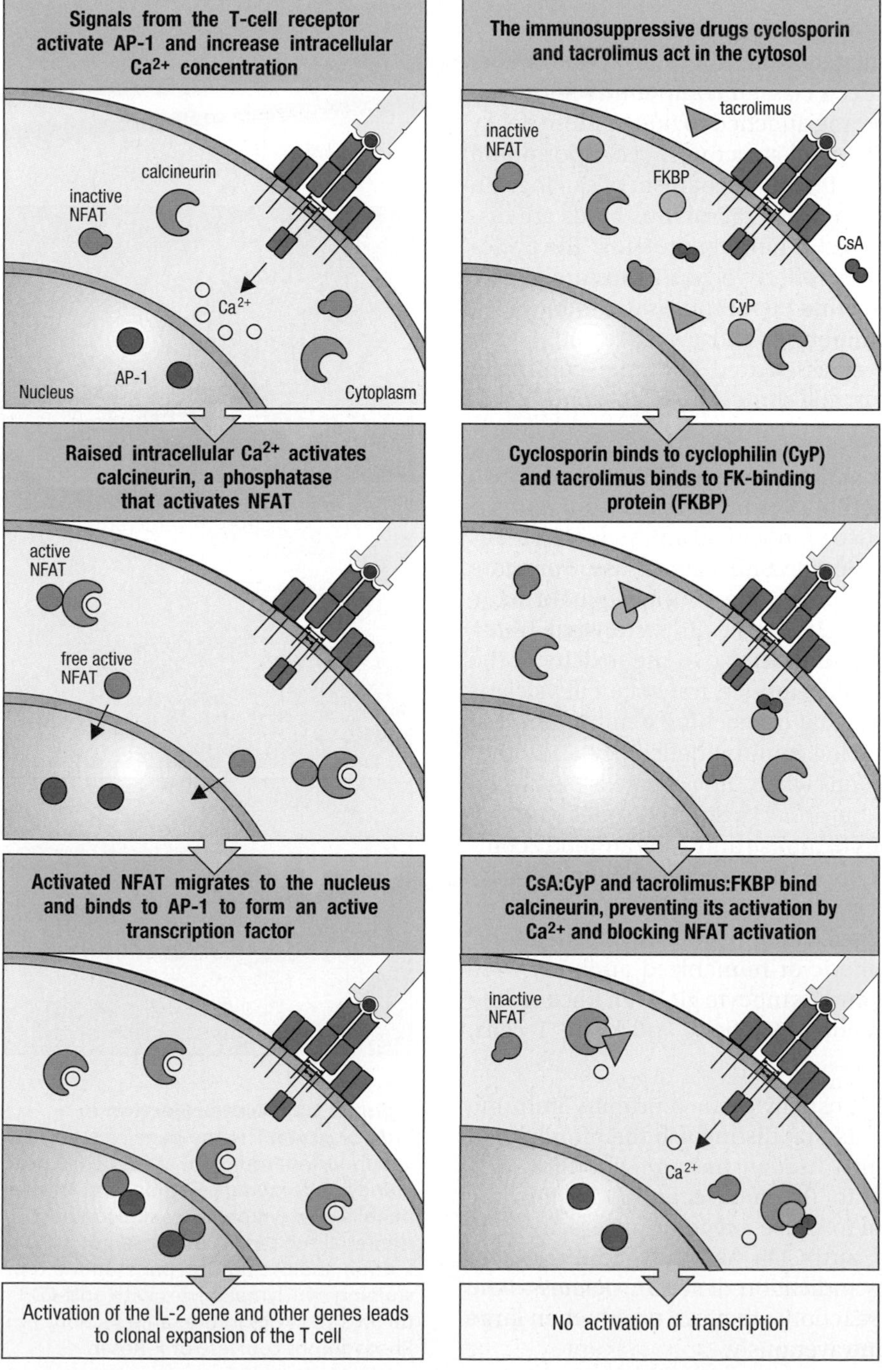

Figure 15.19 Cyclosporin and tacrolimus inhibit T-cell activation by interfering with the serine/threonine phosphatase calcineurin. Signaling via T-cell receptor-associated tyrosine kinases leads to the activation of the transcription factor AP-1 (left panels). The Ca^{2+} binds to calcineurin, thereby activating it to dephosphorylate the cytoplasmic form of the nuclear factor of activated T cells (NFAT). Once dephosphorylated, the active NFAT migrates to the nucleus to form a complex with AP-1; the NFAT:AP-1 complex then induces the transcription of genes required for T-cell activation, including the gene encoding IL-2. Cyclosporin and tacrolimus interfere with the activation of AP-1 (left panels). Cyclosporin binds to cyclophilin and tacrolimus binds to FK-binding protein (FKBP). The complex of cyclophilin with cyclosporin binds to calcineurin, blocking its ability to activate NFAT. The complex of tacrolimus with FKBP binds to calcineurin at the same site, also blocking its activity.

Cell type	Effects of cyclosporin and tacrolimus
T lymphocyte	Reduced expression of IL-2, IL-3, IL-4, GM-CSF, TNF-α Reduced cell division because of decreased IL-2 Reduced Ca^{2+}-dependent exocytosis of cytotoxic granules Inhibition of antigen-driven apoptosis
B lymphocyte	Inhibition of cell division because T-cell cytokines are absent Inhibition of antigen-driven cell division Induction of apoptosis after B-cell activation
Granulocyte	Reduced Ca^{2+}-dependent exocytosis of granules

Figure 15.20 Immunological effects of cyclosporin and tacrolimus.

Tacrolimus, also called FK506, was isolated from the soil actinomycete *Streptomyces tsukabaensis*. It is a macrolide, a class of compound with structures based on a many-membered lactone ring attached to one or more deoxy sugars. Although having very different chemical structures, cyclosporin and tacrolimus suppress T-cell activation by the similar mechanism of calcineurin inhibition. The peptidyl-prolyl isomerases to which tacrolimus binds are distinct from the cyclophilins and are known as FK-binding proteins. The cyclophilins and FK-binding proteins are known collectively as **immunophilins**. Although the principal effect of cyclosporin and tacrolimus is to inhibit T-cell activation, the activation of B cells and granulocytes is also suppressed (Figure 15.20). Nephrocytotoxicity is a side-effect associated with the continued administration of cyclosporin or tacrolimus, and some patients become sensitized to these drugs.

Antibodies that bind to the T-cell receptor or the associated CD3 proteins also have the potential to inhibit the signaling that normally results from antigen recognition. In 1986, the CD3-specific mouse monoclonal antibody OKT3 was the very first monoclonal antibody to be licensed for clinical use. For more than 20 years it was used as a treatment to counter emerging episodes of acute transplant rejection. At the time of surgery and for some time afterward, transplant patients are heavily immunosuppressed. Because of the toxicity of the drugs and the vulnerability to infection that they induce, transplant physicians gradually lower the dose of drugs with the goal of reaching a minimum that preserve tolerance of the graft while reducing immunodeficiency and other adverse effects. Inevitably, there are occasions when the balance is perturbed and early symptoms of rejection appear (Figure 15.21). Such episodes could be halted with a course of 5–15 daily injections of mouse anti-CD3 antibody combined with prednisone. Anti-CD3 worked not by fixing complement and having the T cells removed by phagocytes, but by causing the T-cell receptors to be internalized and unable to recognize antigen. Attempts to convert the therapeutic mouse anti-CD3 into either a chimeric or humanized antibody were unsuccessful. Once anti-CD52 and rabbit antithymocyte globulin became the preferred reagents, because they had broader specificity and killed T cells, OKT3 was taken off the market in 2008.

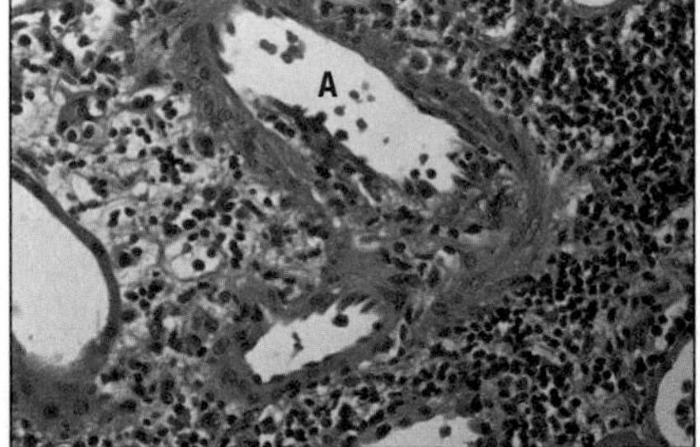

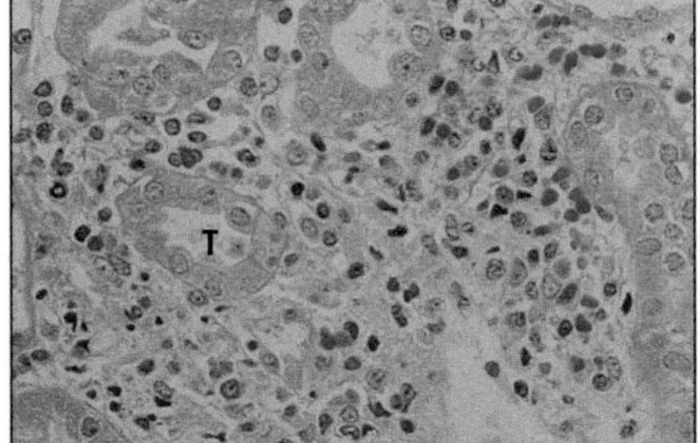

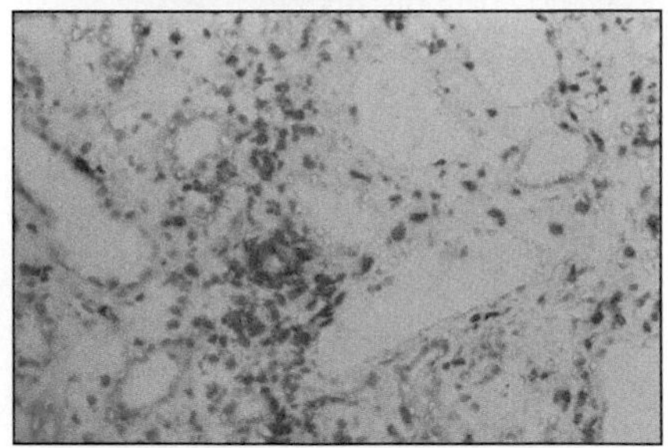

Figure 15.21 Acute rejection in a kidney graft. The top panel shows lymphocytes around an arteriole (A) in a kidney undergoing rejection. The middle panel shows lymphocytes surrounding the renal tubules (T) of the same kidney, and the bottom panel shows the staining of T lymphocytes with anti-CD3 (brown staining) in the same section. Photographs courtesy of F. Rosen.

During a course of OKT3 therapy a patient usually made a primary immune response against the antigenic determinants that distinguish the monoclonal mouse immunoglobulin from the patient's human immunoglobulins. The antibodies made by the patient bound to the mouse antibody, forming immune complexes that had the potential to cause a type III hypersensitivity reaction. Because of the quantities of anti-CD3 antibody administered, patients were carefully monitored for any indication of serum sickness—the extreme form of type III hypersensitivity reaction—that can occur when large amounts of foreign proteins are injected intravenously.

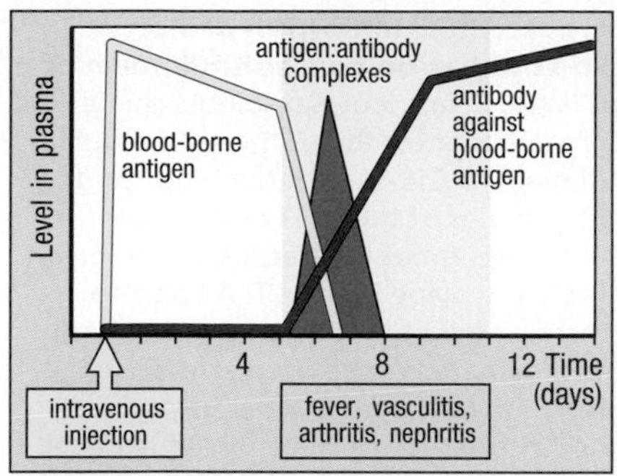

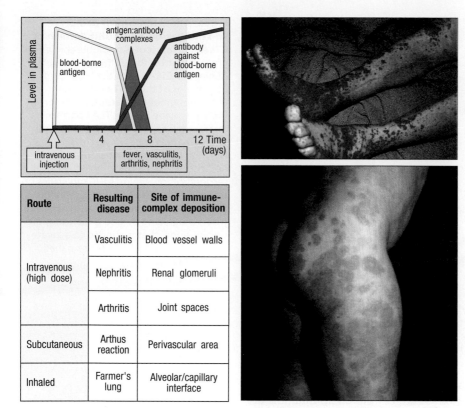

Route	Resulting disease	Site of immune-complex deposition
Intravenous (high dose)	Vasculitis	Blood vessel walls
	Nephritis	Renal glomeruli
	Arthritis	Joint spaces
Subcutaneous	Arthus reaction	Perivascular area
Inhaled	Farmer's lung	Alveolar/capillary interface

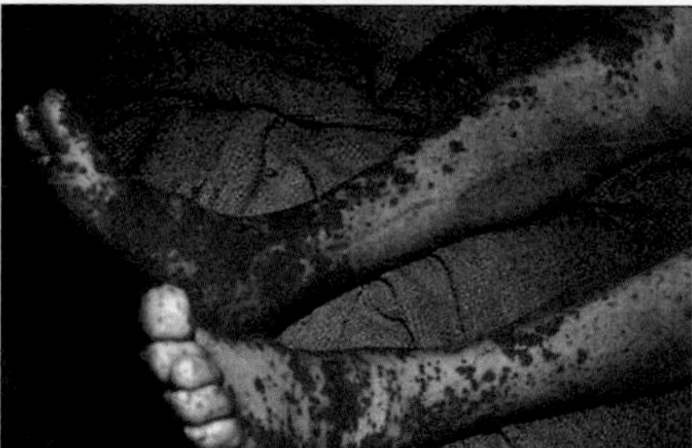

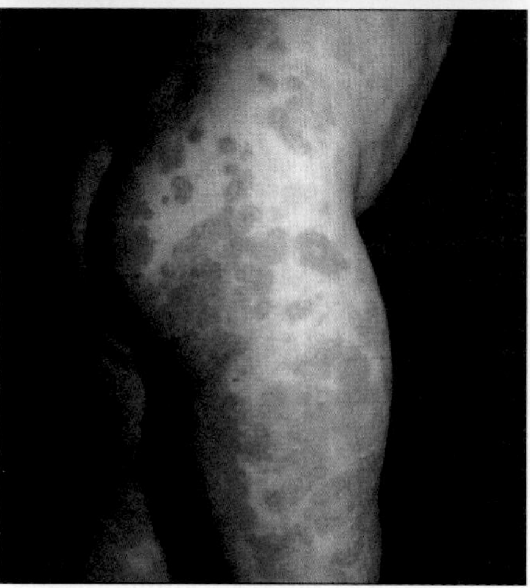

Figure 15.22 Serum sickness is a classic example of a type III hypersensitivity reaction. Upper left panel: an injection of a large amount of an antigenic protein, for example a mouse monoclonal antibody (yellow curve) into the circulation leads to a primary immune response with antibody production (red curve). The antibodies form immune complexes with the antigen (blue shaded area). The complexes are deposited in small blood vessels and activate complement and phagocytes, inducing fever and the symptoms of vasculitis, nephritis, and arthritis (lower left panel). These effects are transient and resolve once the antigen has been cleared. The photographs show hemorrhage in the skin (upper panel) and urticarial rash (lower panel) resulting from serum sickness. Photographs courtesy of R. Geha.

Serum sickness occurs some 7–10 days after administration of the therapeutic antibody and is characterized by chills, fevers, rash, arthritis, vasculitis, and sometimes glomerulonephritis (Figure 15.22).

The timing of serum sickness coincides with the synthesis of high-affinity human IgG antibodies, which form immune complexes with the mouse monoclonal antibody. Because the circulation is loaded with the mouse antibody, large quantities of small immune complexes are formed and then disperse throughout the body, where they become deposited in the walls of blood capillaries. The complexes fix complement and activate leukocytes bearing Fc receptors or complement receptors. These activated cells create numerous localized areas of inflammation that in aggregate cause widespread damage.

Serum sickness is of limited duration because the formation of immune complexes induces clearance of the antigenic protein by normal phagocytic pathways. After symptoms of serum sickness have abated, an additional injection of the therapeutic antibody would induce a secondary response, producing exacerbated disease symptoms within a day or two of injection. For these reasons transplant patients were always restricted to a single course of OKT3 antibody. With the use of humanized and human monoclonal antibodies, the risk of serum sickness is much reduced.

15-15 Alloreactive T-cell co-stimulation can be blocked with a soluble form of CTLA4

Activation of a naive alloreactive T cell by a dendritic cell requires one set of activating signals coming from the T-cell receptor and CD3 complex and a second set of activating signals coming from the CD28 co-stimulatory receptor. The latter signals are generated when CD28 interacts with B7 co-stimulatory molecules expressed on the dendritic cell. CTLA4 is the inhibitory receptor that recognizes B7 and is used as a negative regulator of T-cell activation (see

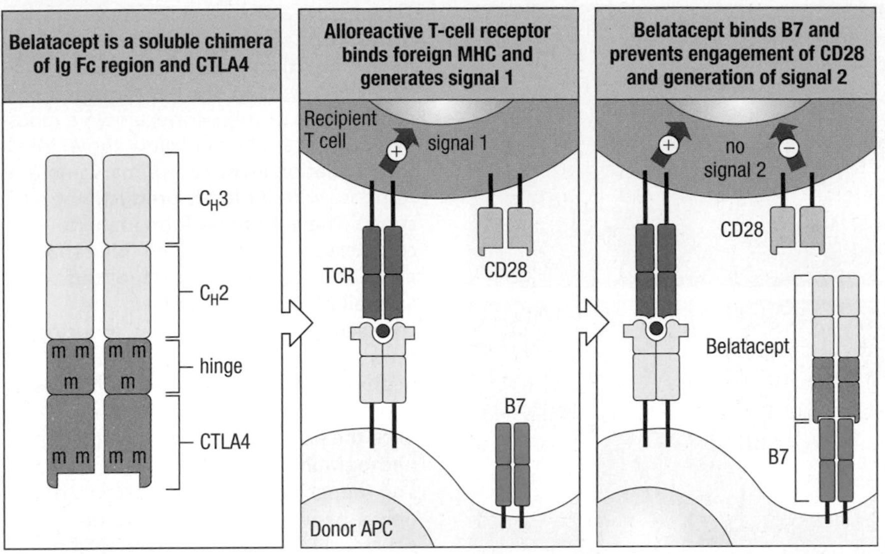
Belatacept is a soluble chimera of Ig Fc region and CTLA4
C_H3
C_H2
hinge
CTLA4
m
Alloreactive T-cell receptor binds foreign MHC and generates signal 1
Recipient T cell
signal 1
TCR
CD28
B7
Donor APC
Belatacept binds B7 and prevents engagement of CD28 and generation of signal 2
no signal 2
CD28
Belatacept
B7

Figure 15.23 Inhibition of T-cell co-stimulation by a soluble form of CTLA4. Belatacept is a soluble chimeric protein that combines the extracellular domain of CTLA4 with the hinge and Fc domains of the IgG1 heavy chain. In the design of belatacept, mutations (m) were made in the CTLA4 part to increase avidity for B7 and in the hinge to eliminate Fc-mediated effector function (left panel). Belatacept does not interfere with the alloreactive T-cell recognition of foreign MHC molecules on transplanted tissues (center panel), but on binding tightly to B7 it prevents CD28 from recognizing B7 and generating the second signal that is needed to activate naive alloreactive T cells (right panel).

Section 8-5). To facilitate this function, CTLA4 binds B7 with 20 times the strength of CD28, so that when CTLA4 is expressed it outcompetes CD28 in binding B7.

Belatacept is the generic name given to an immunosuppressive drug that was approved for use in transplantation in 2011. It targets the co-stimulation of T cells and is a synthetic fusion protein that consists of the extracellular B7-binding domains of CTLA4 with the Fc fragment of IgG1. This soluble protein uses its CTLA4 component to bind tightly to the B7 molecules expressed by activated dendritic cells, which are presenting alloantigens. This interaction prevents the CD28 receptors of alloreactive T cells from binding to B7 and generating the co-stimulatory signal necessary for their activation (Figure 15.23). As measured by graft survival, belatacept works as well as the best of the older, established drugs and is better than them in preserving kidney function. On the downside, belatacept is associated with increased incidence of episodes of acute rejection, illustrating how every immunosuppressive drug that gains approval has both strengths and weaknesses.

15-16 Blocking cytokine signaling can prevent alloreactive T-cell activation

Once a T cell has received signals from the antigen receptor and the co-stimulatory receptor, it needs one further set of signals from a cytokine receptor if activation is to proceed. The dominant cytokine of T-cell activation is IL-2, which is made by T cells and acts in both autocrine and paracrine fashion by binding to the IL-2 receptor expressed by T cells (see Section 8-7). Naive alloreactive T cells express a low-affinity receptor for IL-2 that consists of the β and γ chains. On recognition of alloantigen, signals from the T-cell receptor direct the synthesis of the γ chain (CD25), which associates with β and γ chains to form the high-affinity IL-2 receptor.

Chimeric basiliximab and humanized daclizumab are monoclonal IgG1 antibodies that are specific for CD25 and have been used as immunosuppressive drugs in transplantation for more than 15 years. These **anti-CD25 antibodies** bind tightly to the high-affinity IL-2 receptors on T cells undergoing activation and prevent the receptor from binding IL-2 and generating the intracellular signals that are necessary to continue the activation (Figure 15.24). The antibodies are first given to transplant recipients just before the transplant is performed so that antibodies are systemically present to bind to CD25 when it first

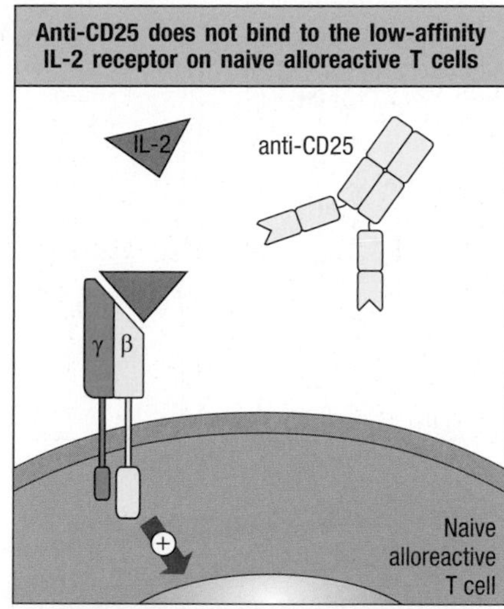

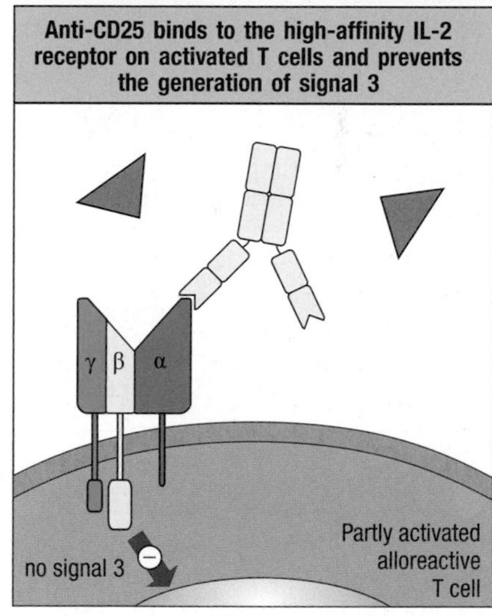

Figure 15.24 Inhibition of alloreactive T-cell activation by monoclonal CD25-specific antibody. CD25 is not part of the low-affinity IL-2 receptor of naive alloreactive T cells (top panel) but is the α chain of the high-affinity IL-2 receptor of alloreactive T cells that have been activated by signals 1 and 2. On binding to the high-affinity IL-2 receptor, anti-CD25 prevents binding of IL-2 to the receptor (bottom panel). This prevents the generation of signal 3 and interferes with the further activation, proliferation, and differentiation of the T cells. Anti-CD25 antibodies used clinically are chimeric basiliximab and humanized daclizumab.

gets to the T-cell surface. Subsequent infusions of anti-CD25 are given during the first 2 months after transplantation, the period when acute rejection is most likely. A single dose of the antibody will saturate all the CD25 molecules in the body within 24 hours, the half-life of the antibody is 2 weeks, and its suppressive effect can last for more than a month.

The benefit in using anti-CD25 as an immunosuppressive drug is that it is highly specific for T cells embarking on activation, and does not affect the majority of circulating T cells or any other type of cell. The beauty is that a knowledge of T-cell biology has enabled the drugs to be designed for their purpose. Thus basiliximab and daclizumab do not have the wide-ranging effects of a drug such as prednisone. Of greatest importance, they do not induce significant immunodeficiency in the transplant patient, who is then less likely to suffer from infections during convalescence and recovery.

Rapamycin (also called **sirolimus**) is an immunosuppressive macrolide that was isolated from *Streptomyces hygroscopicus*, a soil bacterium found on Easter Island. The island's Polynesian name, 'Rapa ui,' was used to name the drug. Although rapamycin binds to FK-binding proteins, it does not interfere with calcineurin but blocks T-cell activation at a later stage by preventing signal transduction from the IL-2 receptor. Rapamycin is more toxic than either cyclosporin A or tacrolimus but is a useful component of combination therapy.

15-17 Cytotoxic drugs target the replication and proliferation of alloantigen-activated T cells

One class of immunosuppressive drugs, the **cytotoxic drugs**, kill alloantigen-activated T cells by interfering with their replication and proliferation. A cytotoxic drug much used in kidney transplantation is **azathioprine**, a prodrug that is first converted *in vivo* to 6-mercaptopurine and then to 6-thioinosinic acid (Figure 15.25). The latter inhibits the production of inosinic acid, an intermediate in the biosynthesis of adenine and guanine nucleotides, essential components of DNA. Azathioprine has no effect until cells attempt to replicate their DNA, which is impossible and so the cells die. Like all cytotoxic drugs, azathioprine has no selectivity and its worst side-effect is the collateral damage to those tissues that are always engaged in cell division. Principally affected are bone marrow, intestinal epithelium, and hair follicles, leading to anemia, leukopenia, thrombocytopenia, intestinal damage, and loss of hair. When pregnant women have to take cytotoxic drugs, fetal development can be adversely affected. A major reason that cyclosporin and tacrolimus had such an impact when first introduced was that they did not have these severe adverse affects. Mycophenolate mofetil, a product of the fungus *Penicillium stoloniferum*, is a more recently developed drug with similar effects to those of azathioprine. It is metabolized in the liver to **mycophenolic acid** (see Figure 15.25), which prevents cell division by inhibiting inosine monophosphate dehydrogenase, an enzyme necessary for guanine synthesis. Because cytotoxic drugs do not act until after alloreactive T cells have recognized graft alloantigens, they are usually administered only after transplantation.

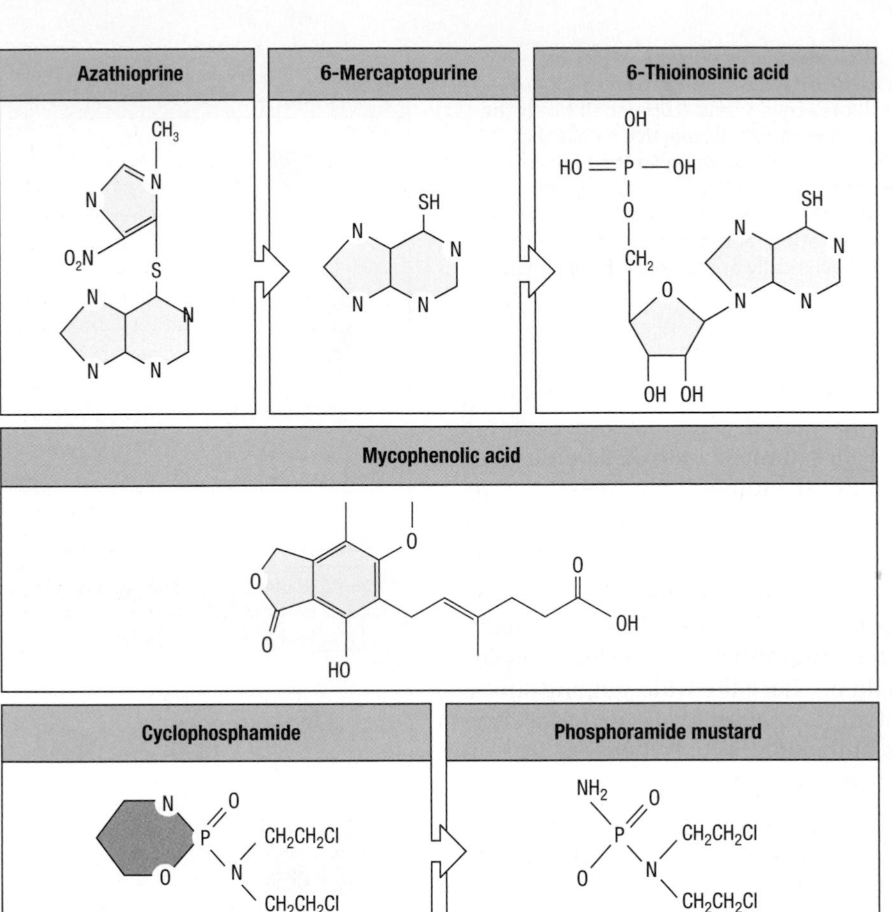

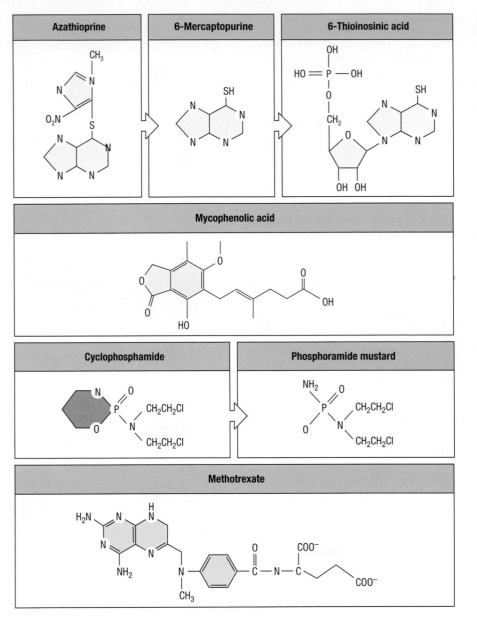

Figure 15.25 The chemical structures and metabolism of cytotoxic drugs. Azathioprine was developed as a modification of the anticancer drug 6-mercaptopurine; by blocking the reactive thiol group, the metabolism of this drug is slowed down. It is slowly converted *in vivo* to 6-mercaptopurine, which is then metabolized to 6-thioinosinic acid; this blocks the pathway of purine biosynthesis. Mycophenolate is a newer drug that also blocks purine biosynthesis after being metabolized to mycophenolic acid. Cyclophosphamide was developed as a stable pro-drug, which is activated enzymatically in the body to phosphoramide mustard, a powerful and unstable DNA-alkylating agent. Methotrexate blocks DNA synthesis by interfering with thymidine synthesis.

Cyclophosphamide was originally developed as a chemical weapon and saw heavy use during World War I. It is a pro-drug that is metabolized to phosphoramide mustard, a compound that alkylates and cross-links DNA molecules (see Figure 15.25). Such disruption renders cells incapable of DNA replication and also affects transcription. Cyclophosphamide has toxic effects that limit its clinical application. As well as the damage to tissues with actively dividing cells, cyclophosphamide specifically damages the bladder, sometimes causing cancer or a condition called hemorrhagic cystitis. Unlike azathioprine, cyclophosphamide is not particularly toxic to the liver, and it is a useful alternative for patients who have sustained liver damage or become otherwise sensitized to azathioprine. Cyclophosphamide is most effective when used in short courses of treatment.

Methotrexate was one of the first cytotoxic drugs shown to be effective in treating cancer cells. It prevents DNA replication by inhibiting dihydrofolate reductase, an enzyme essential for the cellular synthesis of thymidine. Methotrexate is the drug of choice for inhibiting GVHD in bone marrow transplant recipients (see Figure 15.25).

Figure 15.26 The need for tissue transplants outruns the supply of donated organs. Patients who are eligible for a transplant have to wait 2–3 years on average in the United States or the UK before they receive the transplant. The numbers of waiting patients in the United States for three time points in the period 1999–2008 are shown. In 2007, a total of 28,353 transplants were performed. Each year, more than 6000 patents on the waiting list die before receiving a transplant. Data courtesy of the United Network for Organ Sharing.

	Patients on waiting list		
Organ	1999 August	2003 November	2014 May
Kidney	42,875	83,284	100,431
Liver	13,698	17,237	15,735
Heart	4287	3556	3958
Lung	3343	3907	1660
Pancreas	502	1404	1195

15-18 Patients needing a transplant outnumber the available organs

Kidney transplantation has advanced and it is now possible to transplant cadaveric kidneys across greater HLA mismatches. This progress helped the development of heart transplantation, for which only cadaveric donors could be considered. Heart transplantation is inherently difficult, because when a grafted heart fails the patient dies, whereas when kidney grafts fail the patient goes back on dialysis. Cyclosporin and tacrolimus increased the success of heart transplantation, mainly by preventing death from acute rejection or infection during the first few months after transplantation. As a consequence, the number of heart transplants increased after 1979. In the United States, some 2000 patients each year receive a heart transplant, and more than half of them are projected to be alive 10 years after the operation.

Liver transplantation similarly progressed with the introduction of cyclosporine and tacrolimus, from being a risky procedure to one offering considerable benefit. In 1979 only 30–40% of patients survived a liver transplant for more than a year; today 70–90% of patients are surviving after 1 year and 60% are alive after 5 years. Similar improvement has been seen for lung transplantation.

The very success of solid organ transplantation creates its own problem, namely that the number of patients who could benefit from a kidney, heart, or liver transplant far exceeds the organs available from live and deceased donors (Figure 15.26). Patients are therefore placed on waiting lists and chosen for transplantation on the basis of various criteria, including the disease severity and the HLA match with available organs. To increase the organ supply, some countries have instituted a policy whereby cadaveric organs from accident victims become automatically available for clinical transplantation unless the person has deliberately opted out. Other countries retain the policy that organs can be used for transplantation only if an accident victim has deliberately opted in by previously signing a consent form. Even then, relatives can overrule the victim's wishes. In providing organs for transplantation, the 'opt out' system works much better than 'opt in' (Figure 15.27).

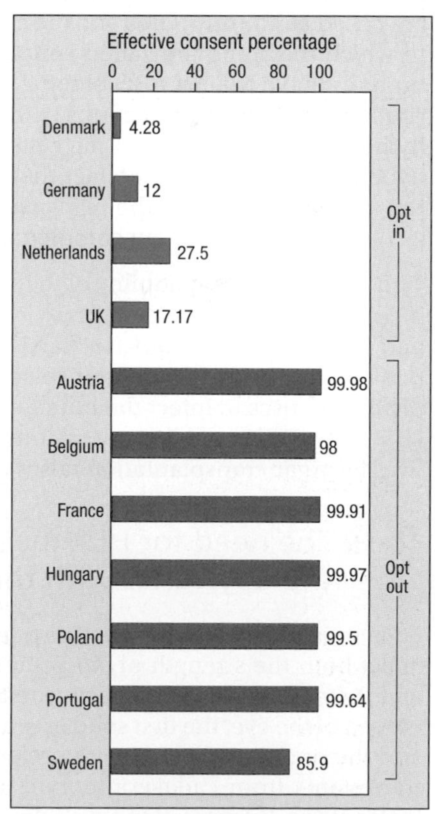

Figure 15.27 The supply of cadaveric donor organs is higher in countries where opting out of donation requires effort than in countries where opting in requires effort. Data courtesy of E.J. Johnson and D. Goldstein.

The unsatisfied demand for kidneys has inevitably led to a flourishing and unregulated international trade in human kidneys, generally involving donors from the poorest countries selling one of their kidneys for transplantation to patients from the richest countries (Figure 15.28). These operations can involve so-called 'transplant tourism' to medical centers in countries of intermediate wealth, where the donor, the recipient, and the surgeon are brought together by a broker. Although many national and international organizations prohibit such trade, it is now estimated to account for 10% of the 80,000 kidneys transplanted each year.

The limiting supply of organs could, in principle, be overcome by using organs from animals. This type of transplantation, in which donor and recipient are of different species, is called **xenotransplantation**, and the grafted tissue a **xenograft**. Pigs are considered the most suitable donor species for humans: first because pig and human organs are of similar sizes, and second because they

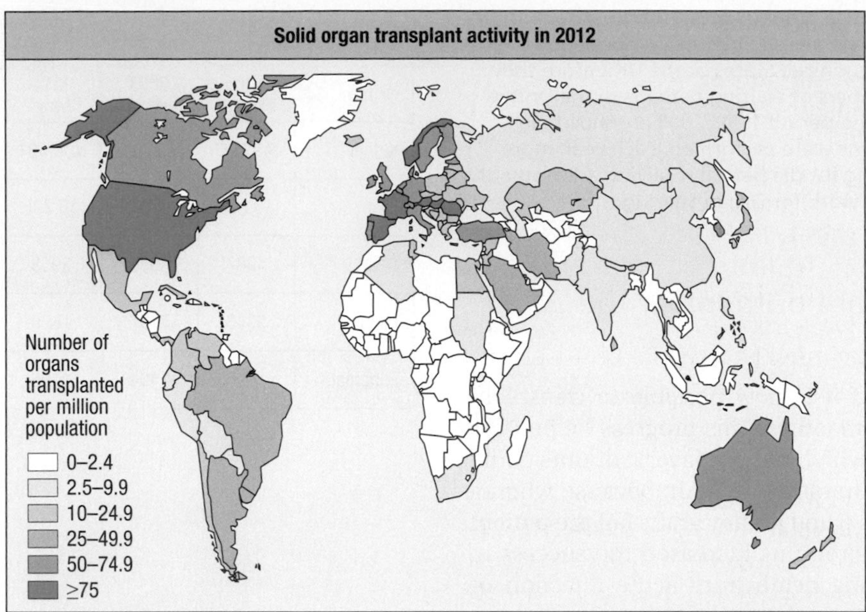

Figure 15.28 Frequency of solid organ transplantation throughout the world. The frequency of solid organ transplants per million inhabitants is shown for each country. Approximately 80% of the transplantations are of kidneys.

are already farmed, slaughtered, and consumed by humans in large numbers. The immunological barriers facing xenotransplantation are formidable. As a start, most humans have circulating antibodies that bind to pig endothelial cells and would cause hyperacute rejection. In this context these antibodies are called **xenoantibodies** and the carbohydrate antigens on pig endothelium to which they bind are called **xenoantigens**. As with the alloantibodies that humans make against ABO blood group alloantigens, the xenoantibodies are probably induced by infections with common bacteria whose surface carbohydrates resemble those of pig cells. Exacerbating the potential problems of hyperacute rejection is the fact that the complement regulatory proteins on the surface of pig cells (the pig versions of CD59, DAF, and MCP; see Section 2-5) do not inhibit human complement.

To improve the compatibility of their organs with the human immune system, pigs would need to be genetically modified with several human genes. Over and above the immunological barrier is the concern that immunosuppressed patients receiving pig xenotransplants could provide a route for endogenous pig retroviruses to infect the human population and have an effect like that of HIV. As on many previous occasions, the development of procedures to save lives by organ transplantation raises complicated ethical questions.

15-19 The need for HLA matching and immunosuppressive therapy varies with the organ transplanted

Each organ has a unique anatomy, function, and vasculature, properties that affect both the strength of antigenic stimulation induced by transplantation and the nature of the alloreactive response. At one end of the spectrum is the cornea of the eye, the first solid organ to be transplanted successfully (in 1905) and the one transplanted with greatest frequency: more than 30,000 corneal transplants from cadaveric donors are performed each year in the United States alone. Because of the eye's distinctive immunological environment, corneal transplants are transplanted with 90% success in the absence of HLA matching or immunosuppressive therapy (Figure 15.29).

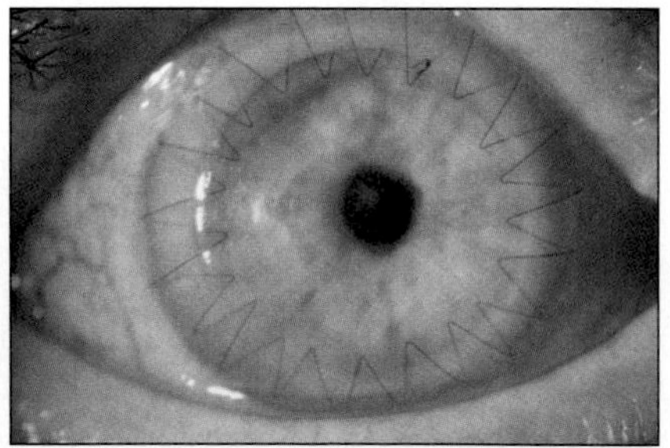

Figure 15.29 A successful corneal allograft. Successful transplantation of corneas from cadaveric donors requires no assessment of HLA type and no administration of immunosuppressive drugs. The lack of any rejection response is because of the naturally immunosuppressive environment in the anterior chamber of the eye and the lack of blood vessels in the cornea. Courtesy of Jerry Y. Niederkorn and James P. McCulley.

For the eye to function properly, the cornea and the anterior chamber of the eye must remain clear and allow a precise passage of light onto the retina. Because any inflammation can impair vision, the eye has evolved an immunological environment that suppresses inflammation while maintaining sufficient protection against pathogens. The cornea is transparent and lacks vasculature, while the aqueous humor of the anterior chamber contains immunomodulatory factors that inhibit the activation of T cells, macrophages, neutrophils, and complement. In particular, the cytokine TGF-β causes the resident dendritic cells to downregulate T-cell co-stimulatory factors such as CD40 and prevents the secretion of IL-12. When antigens are introduced into the eyes of experimental animals, they are carried to the spleen by these dendritic cells, where they generate a T-cell response that is skewed toward regulatory T cells and the production of IL-4 and TGF-β. This capacity to generate an active and systemic state of tolerance to foreign antigens in the eye is called **anterior chamber-associated immune deviation** (**ACAID**). It allows the eye to tolerate corneal grafts of all HLA types and provides a natural example of the specific tolerance that transplant physicians aspire to induce against other transplanted tissues.

Liver is also transplanted successfully across major HLA class I and II differences. It has even been claimed that liver transplant outcome is inversely correlated with the degree of HLA match. As a consequence, HLA type or cross-match are not assessed before liver transplantation; ABO type is the only genetic factor affecting donor selection. Clinical experience indicates that the liver is relatively refractory to either acute or hyperacute rejection, yet the use of cyclosporin and tacrolimus has markedly improved the success of liver transplants. The liver has a specialized architecture and vasculature, and hepatocytes express very low levels of HLA class I proteins and no HLA class II. These properties, and the daily exposure of liver cells to the digestion products of a myriad of foreign proteins from the intestines, with their characteristic anti-inflammatory environment (see Chapter 10), could all contribute to the distinct immunobiology of the transplanted allogeneic liver.

At the other end of the spectrum from eye and liver is the bone marrow, for which transplantation is most sensitive to HLA disparity. The reasons for this are explained in the next part of this chapter.

Summary

Of the problems faced by transplant physicians, the most challenging are the recipient's immune responses against ABO and HLA alloantigens. Hyperacute rejection triggered by preformed antibodies is avoided by selecting donor organs that do not react with the recipient's alloantibodies. Acute rejection arises through the direct pathway of allorecognition, whereby the recipient's T cells respond to the foreign HLA class I and II allotypes of the transplanted organ. The strength of the alloreactive T-cell response is reduced by selecting a donor who is as closely as possible HLA-matched to the recipient. Complementing this approach is a portfolio of immunosuppressive drugs and monoclonal antibodies that are used to prevent and treat acute rejection (Figure 15.30). During the year after transplantation there is a gradual accommodation between the organ graft and the immune system of the recipient that allows the dose of immunosuppressive drugs to be reduced to level that protects the graft while not making the recipient immunodeficient and unable to fend off infection. A small number of patients become fully tolerant of the kidney graft and no longer require any immunosuppression. In the longer term, chronic rejection mediated by antibodies that bind to the allogeneic HLA molecules of the graft are the major threat to health. The antibody response to HLA alloantigens occurs by the indirect pathway of allorecognition, in which the recipient's dendritic cells process allogeneic HLA molecules of the graft and present them to T cells. The potential for rejection varies with the type of organ transplanted, as does the need to match donor and recipient for HLA type. Organ transplantation has become such a successful treatment

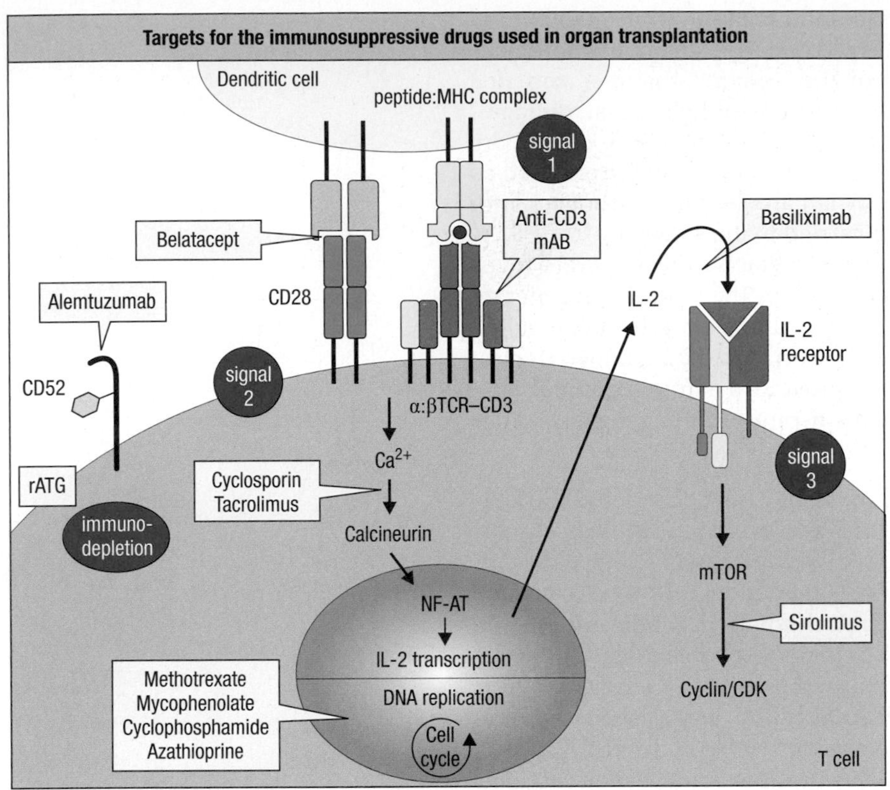

Figure 15.30 Immunosuppressive drugs act at different stages in the activation of alloreactive T cells. Rabbit antithymocyte globulin (rATG) and anti-CD52 monoclonal antibody (alemtuzumab) are used to deplete T cells and other leukocytes before transplantation. Anti-CD3 monoclonal antibody prevents the generation of signal 1 from the T-cell receptor complex, whereas cyclosporin and tacrolimus interfere with the delivery of signal 1, inhibiting the action of calcineurin. The CTLA4:Fc fusion protein belatacept binds B7 and prevent the generation of signal 2 from CD28, the co-stimulatory receptor. The anti-CD25 antibody binds to the high-affinity IL-2 receptor on partly activated T cells and prevents the generation of signal 3. Sirolimus interferes with the delivery of signal 3. Azathioprine, mycophenolate, methotrexate, and cyclophosphamide sabotage the replication and proliferation of activated T cells.

for disease that the majority of patients who could benefit from an organ transplant are not able to receive this treatment, because of the limiting supply of donated organs.

Hematopoietic cell transplantation

Whereas solid organ transplantation involves a partnership between transplant surgeons who carry out the procedures and transplant physicians who care for the recipients afterward, hematopoietic cell transplantation involves collaboration between hematologists, oncologists, and radiologists. Hematopoietic cells are given in a liquid transplant that is infused intravenously into patients whose own hematopoietic system has been weakened or destroyed by chemotherapy and irradiation (Figure 15.31). The key component in a hematopoietic cell transplant is the hematopoietic stem cell, which in time repopulates the entire hematopoietic system of the recipient. Hematopoietic cell transplantation is the therapy of choice for an increasing variety of genetic and malignant diseases and, in contrast to solid organ

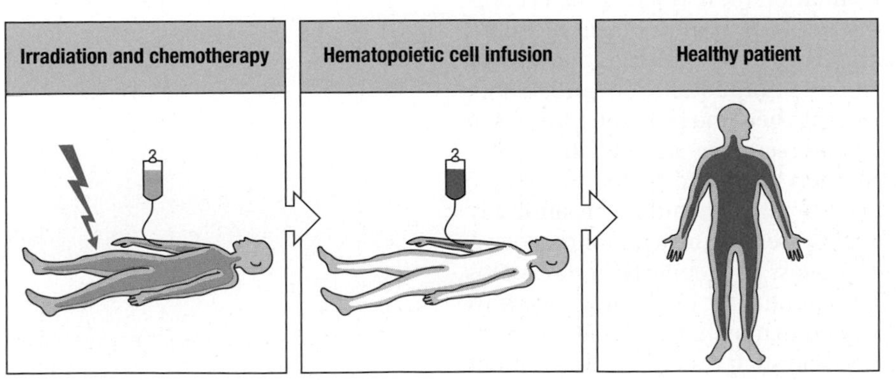

Figure 15.31 Bone marrow transplantation is a therapy for genetic and malignant diseases of hematopoietic cells. The patient's diseased hematopoietic system is destroyed by chemotherapy and irradiation. An infusion of hematopoietic stem cells obtained from a healthy HLA-matched donor is then given. Over a period of months the hematopoietic stem cells in the graft reconstitute the patient with a healthy hematopoietic system.

transplantation, the number of willing hematopoietic donors far exceeds the number of patients needing a transplant. But the greater sensitivity of hematopoietic cell transplantation to HLA difference means that many patients are unable to find a donor with an optimal HLA match.

15-20 Hematopoietic cell transplantation is a treatment for genetic diseases of blood cells

Hematopoietic cell transplantation was first developed as a treatment for life-threatening genetic diseases in which the function of one or more types of hematopoietic cell is impaired. In Chapter 13 we noted how this therapy is used to reconstitute healthy immune systems in children with inherited immunodeficiencies; it is also the therapy of choice for red-cell deficiencies, such as thalassemia major, sickle-cell anemia, and Fanconi's anemia (Figure 15.32). After hematopoietic cell transplantation, the hematopoietic stem cells reconstitute the patient's immune system and also their red cells, platelets, and bone marrow. In 2–3 weeks after a successful transplant, new circulating blood cells begin to be produced from the transplanted marrow. This is a sign that the pluripotent stem cells have colonized the bones, the process known as **engraftment**. With time, the transplant replaces the defective hematopoietic system with one that is normal. Complete reconstitution can take a year or more.

The logistics of liquid hematopoietic cell transplantation are different from those of solid organ transplantation. The donors are alive and healthy and, like a blood transfusion, the transplanted tissue is given by intravenous infusion that involves no surgery. The immunology of hematopoietic cell transplantation is also different from that of organ transplantation. A patient receiving a kidney transplant retains his or her own immune system, which becomes subject to control by immunosuppressive drugs, in which the goal is to prevent graft rejection while maintaining the capacity to fight infection. A patient

Genetic diseases treatable by bone marrow transplantation	
Disease	Deficiency
Wiskott–Aldrich syndrome	Defective leukocytes and platelets
Fanconi anemia	Failure of bone marrow to make blood cells
Kostmann syndrome	Low neutrophil count (neutropenia)
Osteopetrosis	Defective bone modeling and remodeling by osteoclasts
Ataxia telangiectasia	Neurological impairment and immunodeficiency
Diamond–Blackfan syndrome	Low erythrocyte count (anemia)
Mucocutaneous candidiasis	Ineffective T-cell response to fungal infections
Cartilage–hair hypoplasia	Short limbs, fine sparse hair and immunodeficiency
Mucopolysaccharidosis	Various deficiencies of lysosomal enzymes
Gaucher's syndrome	Deficiency of the lysosomal enzyme glucocerebrosidase
Thalassemia major	Defective hemoglobin, impaired erythrocyte function
Sickle-cell anemia	Defective hemoglobin, impaired erythrocyte function

Figure 15.32 Genetic diseases for which bone marrow transplantation is a therapy. As well as the diseases listed here, bone marrow transplantation is a therapy for many genetically determined immunodeficiencies, such as SCID (see Chapter 11).

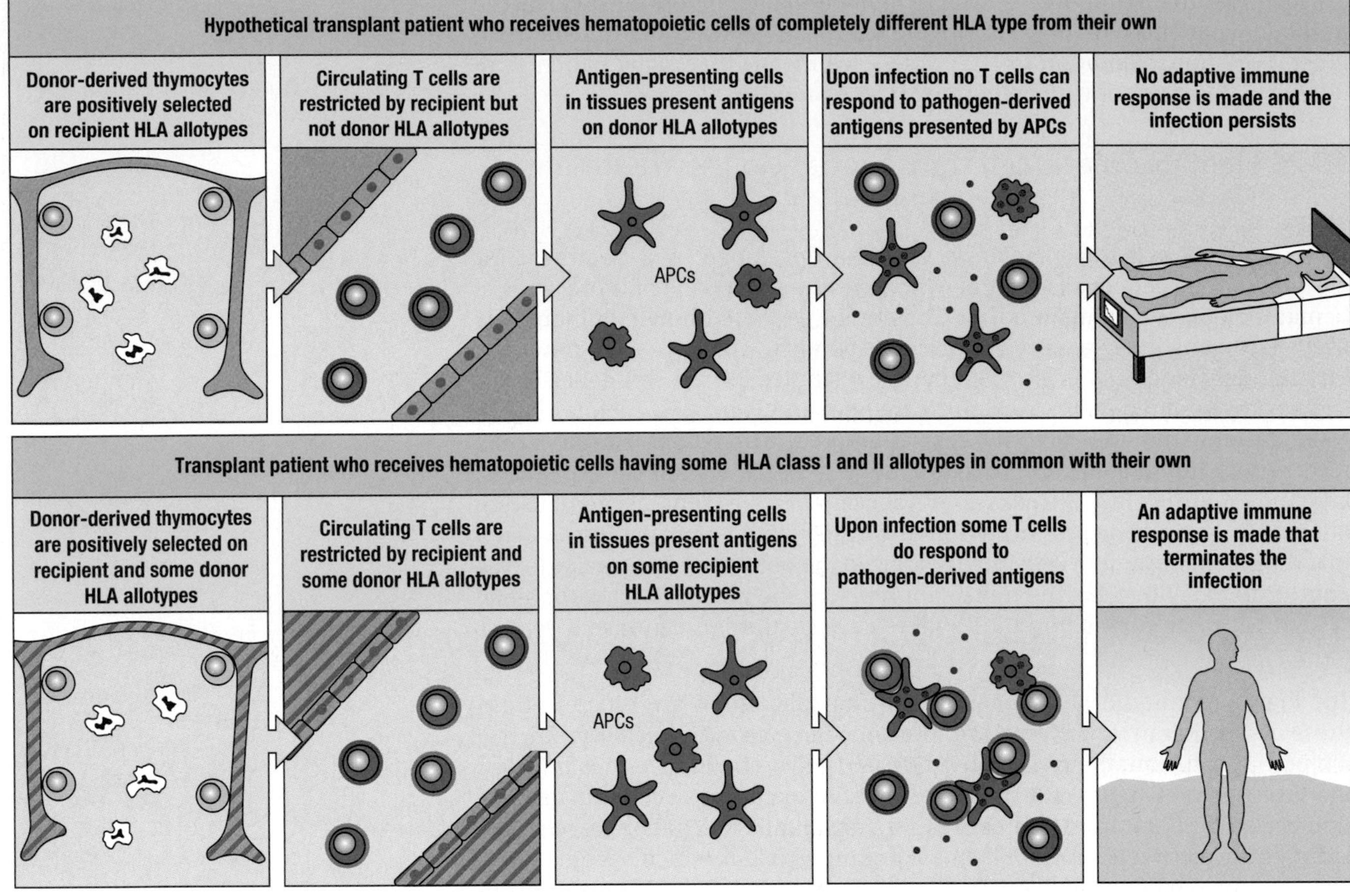

Figure 15.33 The donor and recipient of a bone marrow transplant must share HLA class I and II molecules if they are to reconstitute T-cell function in the recipient. After bone marrow transplantation, donor-derived thymocytes are positively selected on the HLA molecules carried by the recipient's thymic epithelium. The top panels show the hypothetical situation in which none of the recipient's HLA allotypes (red) are the same as the donor's HLA allotypes (blue). In this situation, the recipient could not reconstitute a working T-cell system and would suffer from severe combined immunodeficiency. The lower panels show the situation when the recipient and donor share the HLA allotypes indicated by blue. In clinical practice, bone marrow transplant donors and recipients are chosen to share as many HLA class I and II allotypes as possible. APC, antigen-presenting cell.

receiving a hematopoietic cell transplant is first subjected to a combination of cytotoxic drugs and irradiation that annihilates the immune system. This conditioning regimen, called **myeloablative therapy** because it destroys the bone marrow, serves two purposes. First, the crippling of the recipient's immune system rules out any possibility that the grafted stem cells will be subject to immunological rejection; second, killing all the hematopoietic cells in the recipient's bone marrow provides necessary space for the transplanted stem cells to interact with stromal cells, flourish, and create a new immune system.

Within a few weeks after transplantation, the hematopoietic system begins to reconstitute. The cells of innate immunity, for example granulocytes and NK cells, recover more quickly than the B cells and T cells of adaptive immunity. When the immune system has been fully reconstituted, the patient is a chimera in which the hematopoietic cells are of donor genotype and the rest of the cells are of recipient genotype. The patient's new immune system becomes tolerant of both donor and recipient HLA allotypes. A critical feature for the 'new' T cells is that they are positively selected by recipient HLA allotypes on thymic epithelium, but in responding to infection they will be presented with peptide antigens by donor HLA allotypes expressed by dendritic cells and other antigen-presenting cells of donor origin (Figure 15.33). For this to work, the recipient and donor must have HLA class I and II allotypes in common; the more HLA allotypes they share, the more robust will be the patient's T-cell response to any pathogen.

15-21 Allogeneic hematopoietic cell transplantation is the preferred treatment for many cancers

As well as being used to replace defective hematopoietic systems with functional ones, hematopoietic cell transplantation is an important treatment for many cancers, particularly malignancies of immune-system cells. In treating patients' cancers with chemotherapy and irradiation, the advantages gained by killing off the malignant cells have to be balanced against the damage caused to normally proliferating vital tissues. Of these, the bone marrow is the most susceptible. Hematopoietic cell transplantation offers the opportunity to increase anticancer treatment beyond the point at which it becomes lethal for both cancer and patient, after which the patient is rescued with an allogeneic hematopoietic cell transplant from a healthy HLA-matched donor. Since the first bone marrow transplants were performed more than 35 years ago, the procedure has been used against an increasing range of cancers (Figure 15.34).

Malignant diseases treatable by hematopoietic cell transplantation	
Allogeneic transplant	Autologous transplant
Aplastic anemia Leukemia Acute myelogenous leukemia (AML) Acute lymphocytic leukemia (ALL) Chronic myelogenous leukemia (CML) Myelodysplasia Multiple myeloma Non-Hodgkin's lymphoma Hodgkin's disease	Leukemia Acute myelogenous leukemia (AML) Acute lymphocytic leukemia (ALL) Multiple myeloma Non-Hodgkin's lymphoma Hodgkin's disease Solid tumors Ovarian Testicular Neuroblastoma

Figure 15.34 Cancers for which bone marrow transplantation is a therapy.

15-22 After hematopoietic cell transplantation, the patient is attacked by alloreactive T cells in the graft

Because of the severe conditioning regimen, rejection mediated by the recipient's T cells responding to HLA allotypes in a hematopoietic cell graft is not a problem. Instead, the main cause of damaging alloreactions is mature donor-derived CD4 and CD8 T cells in the transplant that respond to the recipient's HLA allotypes in a graft-versus-host reaction (GVHR) (Figure 15.35). This GVHR is the major cause of morbidity and mortality after hematopoietic cell transplantation. The condition it causes, acute graft-versus-host disease (GVHD), can attack almost every tissue of the body but principally involves the skin, the intestines, and the liver. GVHD is essentially a systemic type IV hypersensitivity reaction, and one that can prove fatal. The severity of acute GVHD varies, four grades being defined in clinical diagnosis (Figure 15.36). The characteristic skin rash of GVHD tends to develop with the kinetics of a primary immune response during the 10–28 days after transplantation. The fine, diffuse erythematous rash begins on the palms of the hands, the soles of the feet, and on the head, and then spreads to the trunk. The intestinal reaction causes cramps and diarrhea, and inflammation of the bile ducts in the liver causes hyperbilirubinemia and an increase in the levels of liver enzymes in the blood. Methotrexate in combination with cyclosporin A is used to reduce the incidence and severity of GVHD.

Allogeneic hematopoietic cell transplant contains mature and memory T cells

T cells circulate in blood to secondary lymphoid tissues. Alloreactive cells interact with dendritic cells and proliferate

Effector CD4 and CD8 T cells enter tissues inflamed by the conditioning regimen and cause further tissue damage

efferent lymph

Lymph node

Skin

Blood vessel

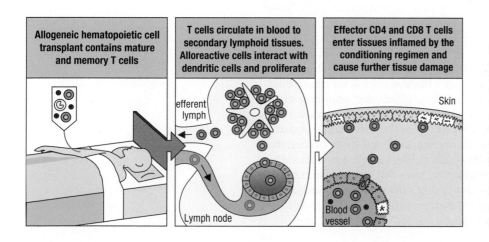

Figure 15.35 Graft-versus-host disease is due to donor T cells in the graft that attack the recipient's tissues. After bone marrow transplantation, any mature donor CD4 and CD8 T cells present in the graft that are specific for the recipient's HLA allotypes become activated in secondary lymphoid tissues. Effector CD4 and CD8 T cells move into the circulation and preferentially enter and attack tissues that have been most damaged by the conditioning regimen of chemotherapy and irradiation: skin, intestines, and liver.

Tissue reactions in the four grades of graft-versus-host disease			
Grade	Skin	Liver	Gastrointestinal tract
I	Maculopapular rash on <25% of body surface	Serum bilirubin 2–3 mg/dl	>500 ml diarrhea/day
II	Maculopapular rash on <25–50% of body surface	Serum bilirubin 3–6 mg/dl	>1000 ml diarrhea/day
III	Generalized erythroderma	Serum bilirubin 6–15 mg/dl	>1500 ml diarrhea/day
IV	Generalized erythroderma with blistering and desquamation	Serum bilirubin 15 mg/dl	Severe abdominal pain with or without intestinal obstruction

Figure 15.36 Characteristics of the four grades of graft-versus-host disease.

Mature T cells from the transplant will circulate in the recipient's blood and enter secondary lymphoid tissues, where they interact with the recipient's dendritic cells. Alloreactive T cells will be stimulated to divide and differentiate into effector cells, which will travel via the lymph and the blood to inflamed tissues. The conditioning regimen, which seeks to destroy the rapidly dividing bone marrow cells, also damages other tissues in which there is normally much cellular proliferation. Prominent among these are the skin, the intestinal epithelium, and the hepatocytes of the liver: the targets for GVHD. The conditioning regimen preferentially creates inflammation in these tissues, which has been described as a 'cytokine storm.' This environment causes dendritic cells to activate and migrate from the inflamed tissues to draining secondary lymphoid tissue, where they stimulate alloreactive T cells. More importantly, it makes these tissues more accessible to alloreactive effector T cells.

Because the number of mature T cells present in the transplant is limited, the duration of acute GVHD is usually restricted to the first few months after transplantation. The chronic rejection reactions seen in kidney transplant patients have their analog in the chronic GVHD experienced by 25–45% of bone marrow transplant patients who survive longer than 6 months after the transplant. The cumulative course of this chronic and gradually worsening condition is to produce severe immunodeficiency, leading to recurrent life-threatening infections.

15-23 HLA matching of donor and recipient is most important for hematopoietic cell transplantation

To some extent, almost all patients receiving a hematopoietic cell transplant suffer from GVHD. The severity of the disease correlates strongly with the extent of HLA mismatch (Figure 15.37). Because of the potentially fatal consequences of GVHD, the clinical outcome and success of hematopoietic cell transplantation are far more sensitive to HLA mismatching than is the case for solid organ transplantation. The best donor for a hematopoietic cell transplant is an identical twin; the second best donor is an HLA-identical sibling. In families, the likelihood of two siblings being HLA-identical is 25%, which means in practice that numerous hematopoietic cell transplants are performed with donors and recipients who are HLA-identical siblings. For patients who have no HLA-identical sibling donor, there are national and international registries of HLA-typed people who are willing to be donors, and which can be searched to identify those who provide the best HLA match. Worldwide, some 20 million potential donors have been HLA typed for this purpose. There is an unusual degree of international cooperation in identifying suitable donors and providing the samples of hematopoietic cells to perform transplantations with the best possible HLA match. Whereas donors are in short supply for organ

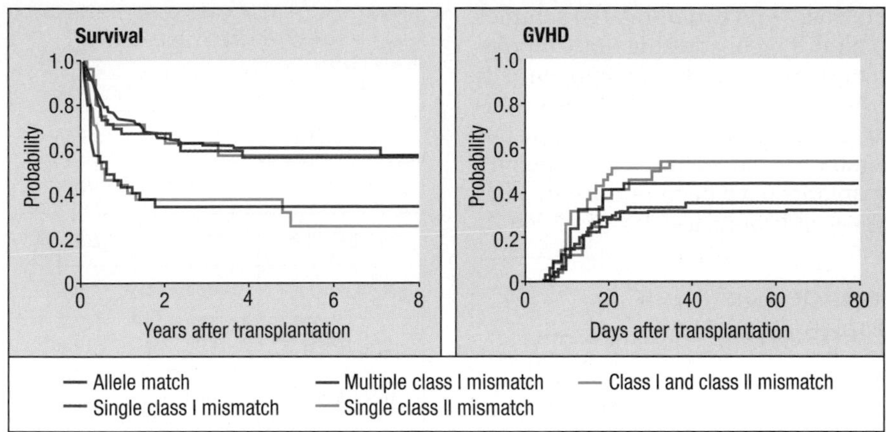

Figure 15.37 The success of hematopoietic cell transplantation correlates with the extent of the HLA match. Two parameters of clinical outcome are shown: survival of the patient (left panel) and the incidence of severe GVHD (right panel). Severe GVHD is defined as grades III and IV (see Figure 15.36). An 'allele match' means the donor and recipient have identical HLA-A, -B, -C, and -DR alleles. A single mismatch means that the donor and recipient differ by one HLA class I or II allele. Transplants with class I and class II mismatch have one mismatched allele for HLA class I and one for class II. Overall, the fewer the mismatches, the better the survival and health of the transplanted patient. Data courtesy of E. Petersdorf.

transplantation, numerous potential hematopoietic cell transplant donors are available, although only a small minority are ever called upon to make a donation.

Despite the large number of registered donors, about 30% of the cancer patients who are clinically eligible for a hematopoietic cell transplant do not find a suitable HLA-matched donor. To help these patients, a procedure called **autologous hematopoietic cell transplantation** can be used. A sample of the patient's bone marrow is taken before the remainder is ablated in the course of chemotherapy and irradiation given to treat the cancer. The hematopoietic stem cells in the bone marrow sample are purified away from any tumor cells and are then given back to the patient when the chemotherapy and irradiation treatments are over. Because autologous transplants are perfectly histocompatible, the patients do not experience GVHD and they are not given immunosuppressive drugs. The limitation of the autologous hematopoietic cell transplant is the frequency with which the cancer returns, a phenomenon called **relapse**. The rate of relapse is significantly higher for autologous transplants than for allogeneic transplants.

In pioneering hematopoietic cell transplantation, bone marrow was the source of hematopoietic stem cells. This required a surgical procedure and anasthesia in which the donor's bone marrow was aspirated from the iliac crests of the pelvis. Hematopoietic stem cells for autologous and allogeneic transplantation are now obtainable by less invasive procedures than bone marrow aspiration. One approach is to recruit hematopoietic stem cells from the donor's bone marrow into the peripheral blood and then isolate the stem cells from the blood. The stem cells are mobilized from the marrow by treating the donor with the growth factors granulocyte colony-stimulating factor (G-CSF) and granulocyte–macrophage colony-stimulating factor (GM-CSF). Leukocytes are then selectively removed from the blood by a process called leukapheresis that uses an apheresis machine to which the donor's circulation is connected for several hours. Antibody specific for CD34, the best marker for human hematopoietic stem cells, is then used to extract a cell fraction enriched for CD34-expressing stem cells from the mixed leukocyte population. Between a quarter and half a billion CD34-positive cells are needed to ensure prompt engraftment after transplantation.

A second approach is to use umbilical cord blood, obtained from the placenta after birth as the source of hematopoietic stem cells. Umbilical cord blood has the combined benefits of being rich in stem cells and having fewer alloreactive T cells than are present in the bone marrow and peripheral blood preparations used for transplantation. Although engraftment is slower with a cord blood transplant, there is less GVHD, and a greater HLA disparity can be tolerated than in transplantation with bone marrow or blood-derived stem cells. Because of the limited volume of cord blood samples, a common practice is to transplant patients with the combined stem cells from two unrelated samples of cord blood. After transplantation, the patient's reconstituting hematopoietic system becomes a chimera of cells from the two donors, but in time the cells from one donor overwhelm the cells of the second. There is benefit to this competition, because transplants using two cord blood samples fare better than those with one. Since its first use in 1988 to treat a child with Fanconi's anemia, more than 25,000 cord blood transplants have been performed, and cord blood banks have been established in several countries.

15-24 Minor histocompatibility antigens trigger alloreactive T cells in recipients of HLA-identical transplants

When seeking an allogeneic hematopoietic cell donor, the 'gold standard' is a healthy HLA-identical sibling. Thousands of such transplants have been performed and the clinical data extensively analyzed. Despite HLA compatibility with their donor, many of these patients were affected by GVHD, particularly brothers receiving a transplant from an HLA-identical sister. In this circumstance, GVHD is caused by T cells specific for self-peptide antigens presented by an HLA class I or II molecule on the brother's cells, but not by that same HLA molecule on his sister's cells. In the sister, such T cells are not deleted during thymic selection and will be transferred to the brother in the transplant. The peptide antigens involved are known as H-Y antigens because they derive from proteins encoded on the Y chromosome, which is carried only by males, and differ in sequence from the homologous protein encoded on the X chromosome. Several H-Y antigens are defined and include peptides presented by HLA-A, HLA-B, HLA-DQ, and HLA-DR (Figure 15.38). Alloantigens such as the H-Y antigens, in which the allogeneic difference is due to the bound peptide and not to the MHC molecule, are called **minor histocompatibility antigens**, and the genes encoding them **minor histocompatibility loci**.

Minor histocompatibility antigens arise not only because of the differences between the sexes but also from the wide variety of polymorphic human proteins encoded by autosomal genes for which the allotypic differences affect the way in which peptides derived from the self protein are processed and presented by HLA class I or II (Figure 15.39). The majority of minor histocompatibility antigens are presented by HLA class I, which is consistent with these

Minor histocompatibility H-Y antigens		
Peptide antigen	HLA restriction	Gene
IVD $\frac{C^+}{S}$ LTEMY	A*01	*USP9Y*
FIDSYIC$^+$QV	A*02	*SMCY*
EVLLRPGLHFR	A*33	*TMSB4Y*
SP $\frac{S}{A}$ VDKA $\frac{R}{Q}$ AEL	B*07	*SMCY*
LP $\frac{H}{R}$ NHT $\frac{D}{N}$ L	B*08	*UTY*
$\frac{R}{G}$ ESEE $\frac{E}{A}$ S $\frac{V}{P}$ SL	B*4001	*UTY*
TIRYPDP $\frac{V}{L}$ I	B*52	*RPS4Y1*
HIE $\frac{N}{S}$ FSD $\frac{FD}{VE}$ MGE	DQB*05	*DOX3Y*
SKGRYIPPHLR	DRB1*1501	*DOX3Y*
VIKVNDTVQI	DRB3*0301	*RPS4Y1*

Figure 15.38 The minor histocompatibility H-Y antigens are diverse and expressed only by males. For each peptide antigen listed here, its amino acid sequence, the HLA class I or class II molecule that presents it (HLA restriction), and the gene encoding the protein from which it is derived are shown. Where known, the amino acid substitutions that distinguish the H-Y antigen (upper letters) from the homologous sequence encoded by the X chromosome (lower letters) are indicated. The H-Y antigens presented by A*01 and A*02 have an additional cysteine that transforms disulfide bond with the cysteine indicated by C⁺ in the peptide sequence.

Autosomal minor histocompatibility antigens				
Name	Peptide antigen	HLA restriction	Chromosome	Gene
HA-3	V $\frac{T}{M}$ EPGTAQY	A*01	15	LBC oncogene
HA-2	YIGEVLVS $\frac{V}{M}$	A*02	7	Myosin 1G
HA-8	$\frac{R}{P}$ TLDKVLEV	A*02	9	*KIAA020*
HA-1	VL $\frac{H}{R}$ DDLLEA	A*02	19	*KIAA0223*
ACC-1	DYLQ $\frac{Y}{C}$ VLQI	A*24	15	*BCL2A1*
UGT2B17	AELCNPFLY	A*29	13	*UGT2B17*
LRH-1	TPNQRQNVC	B*07	17	*P2X5*
ACC-2	KEFED $\frac{D}{G}$ IINW	B*44	15	*BCL2A1*
HB-1	EEKRGSL $\frac{H}{Y}$ VW	B*44	5	Unknown

Figure 15.39 Minor histocompatibility antigens encoded by autosomal genes. For each antigen listed here, the amino acid sequence, the HLA class I or class II molecule that presents it (HLA restriction), the chromosomal location of the gene that encodes it, and the name of the gene are given. Where known, the amino acid substitutions that distinguish the antigenic variant of the protein (upper letters) from the non-antigenic variant (lower letters) are indicated.

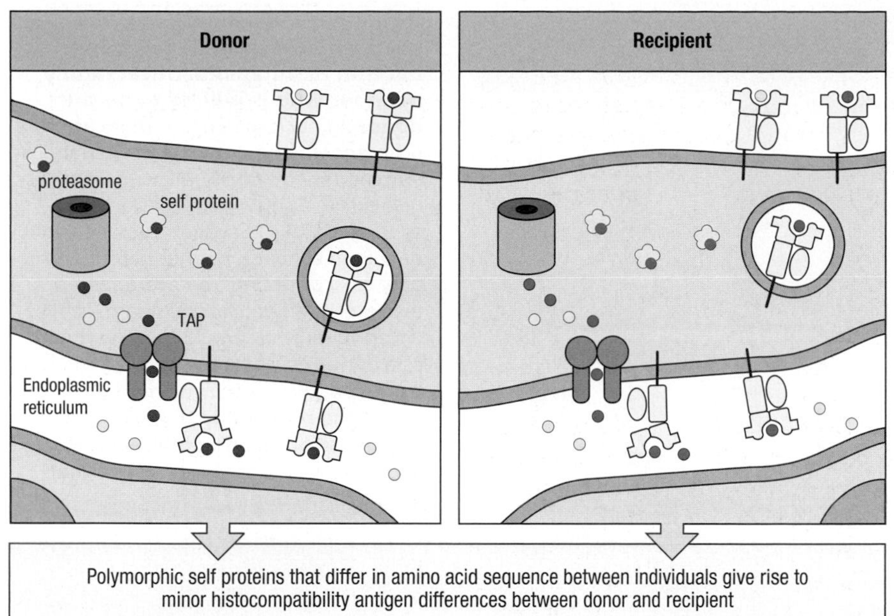
Donor
Recipient
proteasome
self protein
TAP
Endoplasmic reticulum
Polymorphic self proteins that differ in amino acid sequence between individuals give rise to minor histocompatibility antigen differences between donor and recipient

Figure 15.40 Minor histocompatibility antigens are peptides derived from polymorphic proteins other than HLA class I and class II molecules. Self proteins are routinely digested by proteasomes within the cell's cytosol, and peptides derived from them are delivered to the endoplasmic reticulum, where they bind to MHC class I molecules and are delivered to the cell surface. If a polymorphic protein differs between the graft donor (shown in red on the left) and the recipient (shown in blue on the right), it can give rise to an antigenic peptide (red on the donor cell) that can be recognized by the recipient's T cells as non-self and elicit an immune response. Such antigens are called minor histocompatibility antigens.

peptides being breakdown products of intracellular proteins (Figure 15.40). For some, the allelic difference determines whether or not the peptide binds to the HLA molecule, but for others it alters the stability of the peptide, its post-translational modification, and whether a protein is actually made.

15-25 Some GVHD helps engraftment and prevents relapse of malignant disease

A way to reduce the severity of GVHD is to use a specific antibody or lectin to remove mature T cells from the hematopoietic stem cell preparation before it is given to the patient. Although markedly reducing GVHD, this procedure leads to a higher incidence of graft failure and, for cancer patients, a higher incidence of disease relapse. T-cell alloreactions are thus seen to make two contributions to the success of transplantion. T-cell alloreactions help engraftment by subduing residual activity of the recipient's immune system; they also eliminate malignant cells that survived the intensive conditioning regimen. Two other observations support this concept of helpful graft-versus-host reactions. First, graft failure and disease relapse are more likely for transplants in which there is no alloreactivity—transplants with an autologous or identical twin donor. Second, acute GVHD due to minor histocompatibility antigens improves the long-term clinical outcome for transplants between HLA-identical siblings.

When alloreactive T cells in the graft help rid the patient of residual leukemia cells it is called a **graft-versus-leukemia (GVL) effect** or a **graft-versus-tumor (GVT) effect**. A current movement in hematopoietic stem-cell transplantation is protocols that promote the GVL reaction and place less emphasis on tumor elimination by chemotherapy and irradiation. As less severe conditioning regimens are used, the patient's hematopoietic system is damaged but not destroyed, so that after treatment the patient is less immunodeficient and recovers more quickly. Patients treated with the older protocols would often need to be isolated for several weeks in intensive care, whereas patients receiving these so-called 'mini-transplants' need not be isolated. They spend less time in hospital, and some can even be treated as outpatients. The mini-transplant has the potential to provide treatment to many cancer patients who for reasons of age or previous treatment history are unlikely to survive any therapy that requires ablation of the bone marrow.

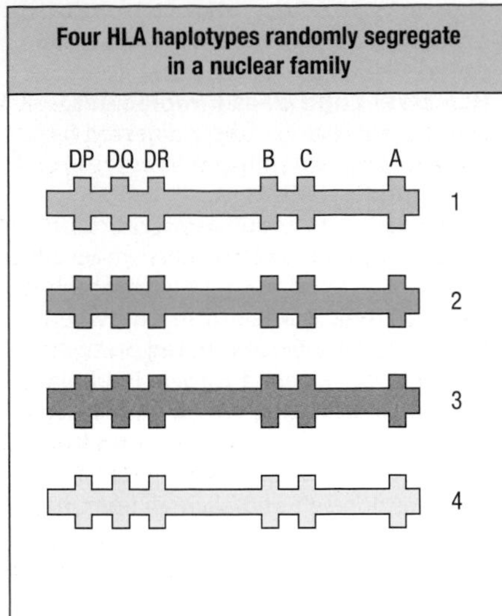

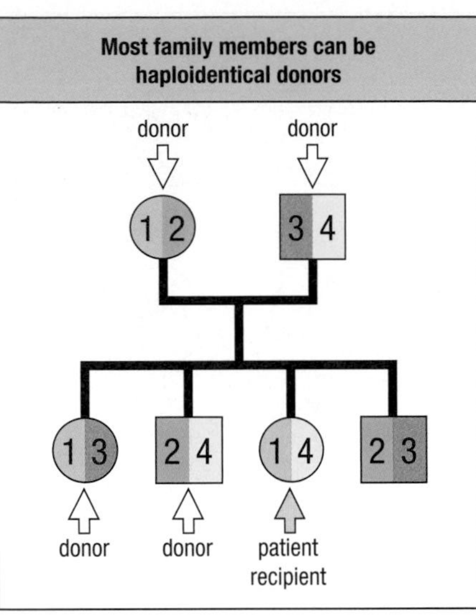

Figure 15.41 Most patients who need a hematopoietic cell transplant have an HLA-haploidentical family member who is willing to be the donor. In nuclear families, there are four different HLA haplotypes (left panel). For any child in the family, either parent can give an HLA-haploidentical transplant, as can 50% of siblings, on average (right panel). In contrast, neither parent and only 25% of siblings, on average, can give an HLA-identical transplant. In determining the suitability of siblings as a donor for a patient needing a transplant, it is not uncommon to detect the presence of HLA haplotypes contributed by neither official parent. Such discoveries are treated with the utmost discretion.

One approach to producing a GVL effect is to give transplant patients transfusions of donor lymphocytes or T cells after they have received the hematopoietic stem-cell graft. Such donor lymphocyte transfusions are given at a time when the inflammation caused by the conditioning regimen has subsided and the likelihood of severe GVHD is diminished.

15-26 NK cells also mediate graft-versus-leukemia effects

Although the trend in HLA-matched transplants is to use gentler protocols with nondestructive conditioning regimens, the opposite strategy has helped some of the 30% of patients with leukemia who cannot find an HLA-matched donor. Most of these patients have a willing donor in their family who shares one HLA haplotype with the patient but differs in the second. Potential donors are mothers, fathers, and 50% of siblings. This type of transplant is called a **haploidentical transplant** (Figure 15.41). Incompatibility for a complete HLA haplotype has the potential to generate strong, fatal alloreactions. To prevent GVHD, the graft is stringently depleted of T cells and the recipient is infused with anti-T-cell antibodies. Graft rejection is prevented through a combination of an intensive conditioning regimen and a larger than normal dose of hematopoietic stem cells.

Patients given a haploidentical transplant get little or no GVHD and require no further immunosuppressive treatment after transplantation. In general, reconstitution of NK cells occurs more rapidly than that of T cells (Figure 15.42), and as the patients' immune systems reconstitute, alloreactive NK cells can emerge. These provide a GVL effect that reduces the incidence of leukemic relapse. The occurrence and specificity of the NK-cell-mediated alloreactions are determined by the interactions of inhibitory KIR with HLA-B and HLA-C ligands and are predictable from the HLA types of the donor and recipient (see Sections 12-4 and 12-5) (Figure 15.43). As a rule, NK-cell alloreactions occur when the recipient's HLA class I allotypes provide ligands for fewer types of inhibitory KIR than the donor's HLA class I allotypes. Patients with acute myelogenous leukemia benefit from such NK-cell alloreactions, whereas patients with acute lymphocytic leukemia (ALL) do not. The alloreactive NK-cell response wanes and is undetectable 4 months after transplantation. With full reconstitution of the immune system, the NK-cell population becomes tolerant of both recipient and donor cells.

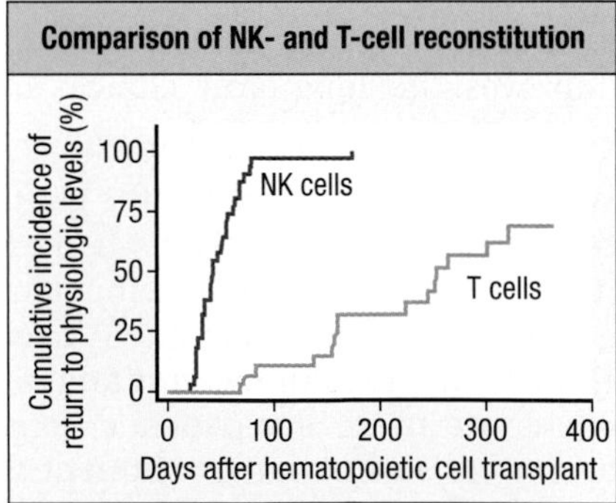

Figure 15.42 After bone marrow irradiation, populations of NK cells and T cells are reconstituted at different rates. Data from B.M. Triplett, E.M. Horwitz, R. Iyengar et al., *Leukemia* 23:1278–1287, 2009.

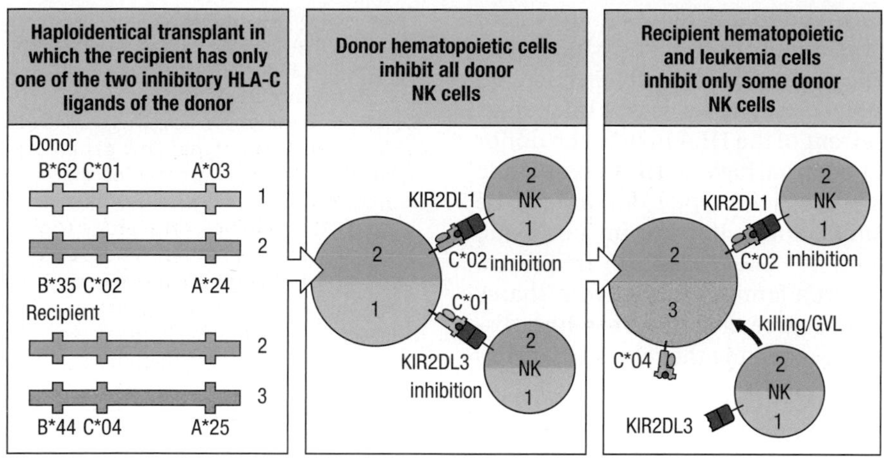

Figure 15.43 Alloreactive NK cells can provide a graft-versus-leukemia effect in patients receiving a haploidentical hematopoietic cell transplant. The HLA class I genotypes of donor and recipient, a patient with acute myelogenous leukemia, are shown in the left panel. The key difference is at HLA-C. The donor has C*01 (with asparagine at position 80) and C*02 (with lysine at position 80); the recipient has C*02 and C*04, both of which have Lys80. The NK cells of donor genotype that emerge after transplantation divide into two groups according to whether they are inhibited by C*02 interacting with KIR2DL1 or by C*01 interacting with KIR2DL3 (center panel). Recipient hematopoietic cells, including residual leukemia cells, inhibit only the first group of NK cells and not the second group. The latter kill the residual leukemia cells (right panel).

15-27 Hematopoietic cell transplantation can induce tolerance of a solid organ transplant

For about 8% of pairs of non-identical twins, the blood circulations become joined during gestation. After birth, the hematopoietic system of each twin is a chimera that contains cells of both genotypes. Their immune systems become tolerant of the other twin's cells, as seen by the acceptance of an HLA-disparate skin graft from the other twin and the lack of any stimulation in mixed lymphocyte cultures (Figure 15.44).

An analogous situation to that of the twins arises for some patients who have had an allogeneic hematopoietic cell transplant and then at a later date have a kidney transplant donated by the same family member that provided the hematopoietic cell transplant. As a result of the hematopoietic cell transplant, the patient is tolerant of the kidney graft and does not require chronic maintenance on immunosuppressive drugs, unlike other recipients of organ grafts. This precedent suggested that combining solid organ transplantation with some mild form of hematopoietic cell transplant from the same donor could induce a more robust and stable tolerance of the solid organ and eliminate the necessity for long-term immunosuppression. Promising results have been obtained in an exploratory trial of combined hematopoietic cell and kidney transplants involving family members differing by one HLA haplotype. Before kidney transplantation, the patients were given a non-myeloablative conditioning regimen consisting of cyclophosphamide, cyclosporin, anti-CD2 antibody, and thymic irradiation. After kidney transplantation, the patients were infused with donor bone marrow. This protocol produced a transient hematopoietic chimerism in the patients, induced long-term tolerance of the kidney grafts for up to 10 years, and allowed immunosuppressive therapy to be discontinued after 9–14 months.

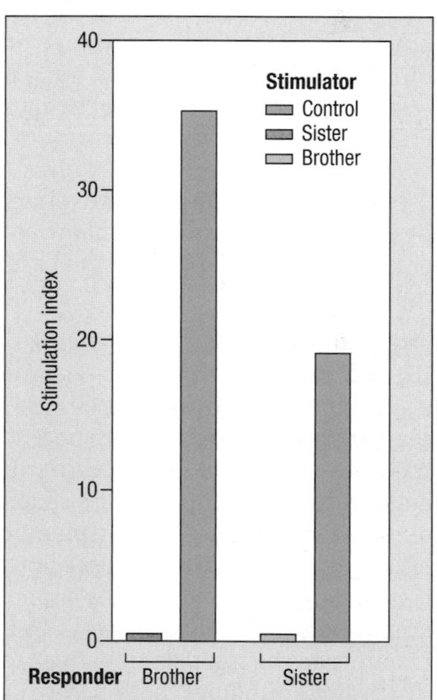

Figure 15.44 Dizygotic twins who have had a common blood circulation during gestation are tolerant of each other's tissues. The stimulation index is derived from the results of mixed lymphocyte reactions (MLRs) in which the lymphocytes of each twin are stimulated by the other twin and an unrelated control. When the sister's lymphocytes are used as stimulators in the MLR against the brother's, and vice versa, a negligible response is obtained in comparison with the strong response that both make to lymphocytes from an unrelated control. Data courtesy of Frans Claas.

Summary

Hematopoietic stem-cell transplantation is used to correct genetic immunodeficiencies and eliminate hematopoietic cell tumors, including leukemias and lymphomas. Sources of hematopoietic stem cells can be bone marrow or peripheral blood donated by healthy donors or the umbilical cord blood collected after the birth of a baby. Before transplantation, the patient undergoes a conditioning regimen that destroys the existing bone marrow, as well as the tumor cells in cancer patients. The patient is also given immunosuppressive drugs and antibodies that prevent rejection of the graft and allow the hematopoietic stem cells to seed the bone marrow, where they divide and differentiate. The immunosuppression also reduces the graft-versus-host disease caused by mature T cells in the graft that respond to differences in the major and minor histocompatibility antigens of the patient and donor. Minor antigens are peptides derived from polymorphic parts of human proteins that are presented to T cells by MHC class I or II molecules. The success of bone marrow transplantation is correlated with the extent of the HLA match: the donor of choice is an HLA-identical sibling. For patients that have no HLA-compatible sibling donor, large registries listing millions of HLA-typed donors can be searched to find an unrelated HLA-compatible donor. If that is unsuccessful, an autologous hematopoietic transplant is one possible treatment. Alternatively, a haploidentical transplant from a family member who shares one HLA haplotype can be performed. Some degree of graft-versus-host disease is helpful in establishing the graft and in preventing the relapse of malignant disease.

Summary to Chapter 15

Many human diseases involve the malfunction of a single organ or tissue that causes incapacitation and can often lead to death. With transplantation, a diseased tissue is replaced by a healthy one, leading to improved health and longer life. Although blood transfusions are short-lived, they demonstrated the clinical value of transplantation long before it was possible for other tissues and organs to be transplanted. Similarly, the observation of life-threatening transfusion reactions and the serological analysis of polymorphic red-cell antigens revealed the importance of immunogenetics. For transplantation of other tissues, the proof of principle came from successful transplantations between genetically identical twins, but for such procedures to have any general application it was necessary to deal with the highly polymorphic histocompatibility antigens—HLA class I and class II. By presenting peptides to T cells, they provoke adaptive alloreactive immune responses that reject and destroy a solid organ transplant or can kill the recipient by causing severe graft-versus-host disease in the case of hematopoietic stem-cell transplantation.

The alloreactions that occur after tissue transplantation are examples of type II, III, and IV hypersensitivity reactions. A general principle in the practice of transplantation is to avoid preformed antibodies that would bind to the transplanted tissue. Antibodies are potent weapons and can neutralize a transplant with the same effect with which they neutralize a pathogen. Avoidance is achieved empirically by cross-match tests. The potential for stimulating an alloreactive response in transplantation is reduced by finding a donor whose HLA type matches that of the recipient. For patients with common HLA types this is easily achieved, but for patients with one of the many rare HLA types a compromise is often necessary. To prevent the stimulation of an alloreactive response to the transplant, patients are given a variety of immunosuppressive drugs and antibodies, both before and after transplantation, that prevent the activation and proliferation of T cells. Patients are carefully monitored and any evidence of graft rejection or graft-versus-host disease is treated with additional immunosuppression. The drug doses are gradually lowered to a maintenance level.

The impressive improvement in transplantation success over the past 30 years has come principally from more effective drugs and antibodies that nonspecifically suppress T-cell activity. Throughout this period, the goal of transplant immunologists has been to selectively suppress the immune response to the allogeneic transplant and so release the transplant patient from a lifetime of immunosuppressive drugs. Various lines of evidence suggest that this might be achieved by establishing a chimeric hematopoietic system in the transplant recipient that is populated by hematopoietic stem cells of both donor and recipient origin.

Questions

15–1 _____ are universal recipients who can receive blood from any donor, but can donate only to individuals with their blood type.

a. AB RhD +
b. AB RhD –
c. O RhD +
d. O RhD –
e. A RhD +.

15–2 In the context of blood type and blood transfusions, match the term in Column A with its description in Column B.

Column A	Column B
a. blood-group antigen O	1. individuals express *N*-acetyl galactosamine
b. HLA class I antigen	2. associated with hemolytic disease of the newborn
c. Rhesus D antigen	3. bear(s) structural similarity to cell-surface carbohydrates of common commensal bacteria
d. blood-group antigen A	4. individuals possess anti-A and anti-B antibodies
e. blood-group antigens A and B	5. stimulate(s) antibody production during pregnancy but not found on erythrocytes

15–3 Patients who have had multiple blood transfusions _____. (Select all that apply.)

a. have developed immunological tolerance to a broad range of blood antigens
b. are considered universal donors because they do not contain alloantibodies in their serum
c. have an increased likelihood that they have antibodies against the non-ABO, non-Rhesus blood groups antigens
d. pose problems when finding suitable donors because they are reactive to mltiple blood types
e. should never donate blood because of the increased risk of hyperacute rejection
f. will have a higher percentage panel reactive antibody (PRA) value than patients who have had no blood transfusions.

15–4 Identify the mismatched pair. (Select all that apply.)

a. cross-match test: blood transfusion
b. multiparous women: anti-HLA sera
c. type II hypersensitivity reaction: organ transplant
d. MHC molecules: alloantigens
e. antibody-mediated organ rejection: immune complex deposition
f. transplant rejection: bone marrow attacks recipient's tissues
g. isograft: syngeneic transplant
h. xenograft: donor and recipient are of different species.

15–5 Identify the mismatched pair. (Select all that apply.)

a. direct pathway of allorecognition: donor antigen-presenting cells
b. prednisone: pro-drug
c. autograft: skin transplant from own tissues
d. hyperacute rejection: pre-existing antibodies
e. mixed lymphocyte reaction: irradiated donor cells
f. xenograft: transplant between two genetically different individuals of the same species
g. matching HLA antigens: DNA typing and serology
h. anterior chamber-associated immune deviation (ACAID): increased propensity to reject HLA-mismatched corneal grafts.

15–6 Before its withdrawal from the market in 2008, OKT3 was used as an immunosuppressive drug for transplant patients. Which of the following best describes the major complication encountered with this drug?

a. uncontrolled upregulation of gene subsets that exacerbated the inflammatory state in transplant patients
b. profound, long-lasting lymphopenia that rendered the patient susceptible to opportunistic infection during convalescence and recovery
c. adverse side-effects including fluid retention, bone loss, and weight gain
d. induction of type III hypersensitivity reaction caused by immune complexes of human IgG antibodies with mouse monoclonal antibodies
e. collateral damage to tissues engaged in cell division including bone marrow, intestinal epithelium, and hair follicles.

15–7 All of the following are associated with a type III hypersensitivity reaction except _____.

a. immune complex deposition in blood vessels
b. generation of foreign epitopes on the cell surface of grafted tissue by drug-induced modification
c. complement fixation
d. arthritis, vasculitis and glomerulonephritis
e. inflammation
f. mouse monoclonal antibodies.

15–8 Identify which of the following is not a characteristic of belatacept.

a. contains part of CTLA4 that binds with high affinity to B7
b. prevents delivery of second signal to naive alloreactive T cells
c. contains the hinge and Fc domains of IgG1
d. interferes with alloreactive T-cell recognition of foreign MHC molecules on transplanted tissues
e. facilitates complement fixation on the dendritic-cell surface
f. exists as a soluble therapeutic agent.

15–9

A. What is myeloablative therapy?

B. Explain two reasons for carrying out myeloablative therapy in hematopoietic stem cell recipients.

15–10 All of the following statements are true regarding GVHD except _____.

a. The main tissues involved in GVHD include the skin, the intestines, and the liver.
b. The severity of GVHD correlates with the degree of HLA mismatch.
c. GVHD involves alloreactions of donor-derived CD4 and CD8 T cells toward recipient tissues.
d. Major histocompatibility antigens, but not minor histocompatibility antigens, stimulate GVHD.
e. GVHD is a systemic type IV hypersensitivity reaction.
f. GVHD is not experienced in patients receiving an autologous hematopoietic cell transplant.

15–11 What is the predicted outcome if a patient receives an allogeneic hematopoietic cell transplant from a family member, and then receives a kidney transplant several months later from the same donor?

a. Acute kidney rejection because the patient's hematopoietic system has been fully reconstituted with patient-derived alloreactive T cells and NK cells.
b. Graft-versus-host disease mediated by alloreactive T cells in the transplanted kidney.
c. Secondary immune response by patient-derived memory T lymphocytes that were initially activated by the hematopoietic cell transplant.
d. The patient is tolerant to alloantigens on the kidney graft and does not require long-term immunosuppressive drugs.

15–12 Other than bone marrow aspiration, identify two other sources of stem cells for autologous and allogeneic hematopoietic cell transplantation.

15–13 Barriers to using pig organs in human transplantation are _____. (Select all that apply.)

a. the transfer of endogenous pig retroviruses to humans
b. that pig organs secrete high levels of cytokines that induce inflammation in humans
c. that humans contain preexisting alloantibodies that bind to pig carbohydrate antigens on pig endothelium
d. that pig MHC molecules stimulate strong xenoreactive T-cell responses
e. that pig complement regulatory proteins do not inhibit human complement.

15–14 Identify the different (i) physiological effects and (ii) side-effects mediated by corticosteroids when they are administered to transplant patients.

15–15 Vanad Patel arrived unconscious by ambulance at the emergency room following a motorcycle accident. He had severe injuries including a compound fracture of his tibia and needed immediate surgery and a blood transfusion. Routine blood-group typing categorized him as O$^+$ because his red blood cells showed no reaction with anti-A or anti-B antibodies. But cross-matching with the different bags of O blood in the blood bank showed cross-reactivity or incompatibility with them all. The blood-bank staff reported that the patient had a rare blood type called the Bombay blood group. His twin brother Nadeesh, who arrived at the emergency room within hours of the family being notified, donated the blood his brother needed. It was recommended that he and his brother wear medical alert bracelets in case they need emergency blood transfusions in a situation where thorough cross-matching is not possible. Which of the following is consistent with the Bombay phenotype?

a. The Bombay phenotype results from the inability to synthesize the O antigen and therefore these individuals are not tolerant of the O antigen.
b. The Bombay phenotype is caused by a mutation in the RhD antigen, and therefore these individuals are not tolerant of the RhD antigen.
c. Sera of individuals with the Bombay blood type can only be given to individuals who are Rhesus-negative.
d. Sera from a type O individual would cause agglutination of red blood cells from an individual with the Bombay blood type.
e. The enzyme that adds either *N*-acetyl galactosamine (type A) or galactose (type B) to the core glycolipid found on O antigens, is defective in individuals with the Bombay phenotype.

15–16 Carter Petersen received an apparently successful bone marrow transplant from his HLA-identical sister. Twenty-five days later he developed watery diarrhea, a patchy rash on his face and neck that spread to his trunk, and jaundice. He showed improvement after treatment with cyclosporin and methotrexate. What is the most likely explanation for his symptoms?

a. acute rejection of the graft
b. graft-versus-host disease
c. cytomegalovirus infection
d. type II hypersensitivity reaction
e. host-versus-graft disease.

A tick burrowing into human skin.

Chapter 16

Disruption of Healthy Tissue by the Adaptive Immune Response

We have seen how hypersensitivity to harmless environmental antigens leads to acute or chronic disease depending on the nature of the antigen and the frequency with which it is encountered (see Chapter 14). In this chapter we consider a related set of chronic diseases, ones caused by adaptive immune responses that become misdirected at healthy cells and tissues. Such diseases are known as **autoimmune diseases**, with more than 100 different types being clinically described. Like IgE-mediated allergies, autoimmune diseases are more common in the affluent, industrialized countries. Their frequencies, now 5–10%, have increased like the allergies and they, too, are attributed to recent and ongoing changes in human habits and lives, as invoked in the hygiene hypothesis (see Figure 14.8, p. 407). Autoimmune diseases vary widely in the tissues they attack and the symptoms they cause. Some focus on a particular organ or cell type, others act systemically. For most autoimmune diseases the incidence differs between females and males, with females being more commonly affected.

Autoimmune diseases are the result of antibodies and effector T cells that attack healthy cells and tissue as though they were infected with a pathogen and produce reactions resembling the type II, III, and IV hypersensitivity reactions. Once an **autoimmune response** has started, it usually continues throughout the life of the patient and tends to broaden in scope with increasing severity of disease. With cellular death and disruption of tissue, physiological function becomes compromised and in some conditions is the direct cause of death. Before insulin injections were introduced as therapy for type 1 diabetes, teenagers and young adults routinely died from this prevalent autoimmune disease. Because the disease-causing antibodies and effector cells recognize self antigens, an autoimmune response represents a collective failure of the mechanisms that maintain **self-tolerance**. Although much is known of the effects of autoimmune disease, little is known of the singular events that break self-tolerance and initiate the autoimmune response. This ignorance persists because symptoms of an autoimmune disease are usually perceived by patients and diagnosed by doctors only years after the start of the autoimmune response, during which time it has expanded and evolved into a chaotic civil war that belies its own origin. Onset of the symptoms of type 1 diabetes, for example, occurs only when the insulin-secreting cells of the pancreas have almost all been killed off. Although a satisfactory description for the molecular and cellular events that lead to autoimmune diseases has yet to be achieved, there is general consensus that such a description will involve genetic, developmental, and environmental factors.

16-1 Every autoimmune disease resembles a type II, III, or IV hypersensitivity reaction

Like the hypersensitivity reactions described in Chapters 14 and 15, autoimmune diseases can be classified according to the effector mechanism causing the disease (Figure 16.1). Three kinds of autoimmune disease correspond to the type II, type III, and type IV hypersensitivities (see Section 14-1); no

Autoimmune disease	Autoantigen	Consequence
Antibody against cell-surface or matrix antigens (type II)		
Autoimmune hemolytic anemia	Rh blood group antigens, I antigen	Destruction of red blood cells by complement and phagocytes, anemia
Autoimmune thrombocytopenia purpura	Platelet integrin gpIIb:IIIa	Abnormal bleeding
Goodpasture's syndrome	Non-collagenous domain of basement membrane collagen type IV	Glomerulonephritis, pulmonary hemorrhage
Pemphigus vulgaris	Epidermal cadherin	Blistering of skin
Pemphigus foliaceus	Desmoglein	Mild blistering of skin
Acute rheumatic fever	Streptococcal cell wall antigens. Antibodies cross-react with cardiac muscle	Arthritis, myocarditis, late scarring of heart valves
Graves' disease	Thyroid-stimulating hormone receptor	Hyperthyroidism
Myasthenia gravis	Acetylcholine receptor	Progressive weakness
Type 2 diabetes (insulin-resistant diabetes)	Insulin receptor (antagonist)	Hyperglycemia, ketoacidosis
Hypoglycemia	Insulin receptor (agonist)	Hypoglycemia
Immune-complex disease (type III)		
Subacute bacterial endocarditis	Bacterial antigen	Glomerulonephritis
Mixed essential cryoglobulinemia	Rheumatoid factor IgG complexes (with or without hepatitis C antigens)	Systemic vasculitis
Systemic lupus erythematosus	DNA, histones, ribosomes, snRNP, scRNP	Glomerulonephritis, vasculitis, arthritis
T cell-mediated disease (type IV)		
Type 1 diabetes (insulin-dependent diabetes mellitus)	Pancreatic β-cell antigen	β-cell destruction
Rheumatoid arthritis	Unknown synovial joint antigen	Joint inflammation and destruction
Multiple sclerosis	Myelin basic protein, proteolipid protein	Brain degeneration. Paralysis

Figure 16.1 A selection of autoimmune diseases, the symptoms they cause, and the autoantigens associated with the immune response. The autoimmune diseases are classified as types II, III, and IV because their tissue-damaging effects are like those of hypersensitivity reactions types II, III, and IV, respectively (see Chapter 14). snRNP, small nuclear ribonucleoprotein; scRNP, small cytoplasmic ribonucleoprotein.

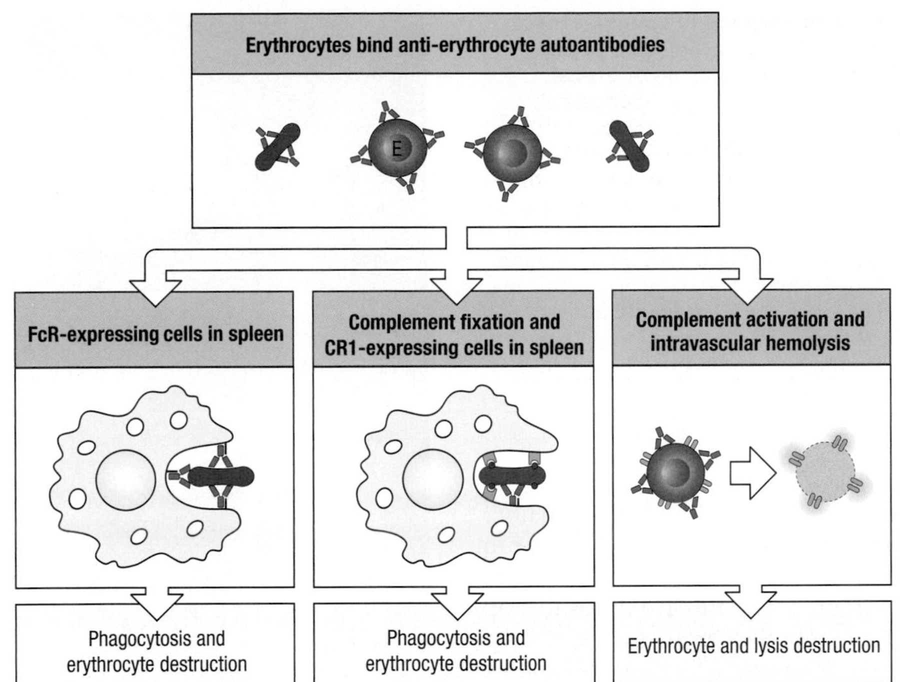

Figure 16.2 Three mechanisms destroy erythrocytes in autoimmune hemolytic anemia. Erythrocytes opsonized with IgG can be bound and engulfed by phagocytes in the spleen that bear an Fcγ receptor (lower left panel) a complement receptor (lower middle panel) or both types of receptor (not shown). Complement fixation on the erythrocyte surface can also lead to complement-mediated lysis of the opsonized erythrocyte.

autoimmune disease is mediated by IgE, the cause of type I hypersensitivity reactions. The autoimmune diseases corresponding to type II hypersensitivity are mediated by antibodies directed against components of cell surfaces or the extracellular matrix; those corresponding to type III hypersensitivity are mediated by soluble immune complexes deposited in tissues; those corresponding to type IV hypersensitivity are mediated by effector T cells.

Autoimmunity corresponding to the type II hypersensitivity reaction frequently targets blood cells. In **autoimmune hemolytic anemia**, IgG and IgM antibodies bind to components of the erythrocyte surface, where they activate complement by the classical pathway. This leads to assembly of the membrane-attack complex and hemolysis—the lysis of red cells. Alternatively, erythrocytes coated with antibody and C3b are cleared from the circulation, principally by the Fc and complement receptors of phagocytes in the spleen (Figure 16.2). These mechanisms induce a deficiency of red blood cells. The condition is called **anemia**, a term derived from Greek words meaning 'without blood.'

Autoimmune hemolytic anemia

White blood cells are also targets for autoantibodies and complement activation. Because nucleated leukocytes are less susceptible to complement-mediated lysis than erythrocytes, the main effect of complement fixation on leukocyte surfaces is opsonization. As the opsonized leukocytes circulate through the spleen they are removed and degraded by the resident macrophages. For example, patients who develop autoantibodies against neutrophil surface antigens suffer a deficiency of circulating neutrophils—a state called **neutropenia**. Because leukocytes opsonized with antibody and complement are still functional, a treatment for patients with persistent autoimmunity to white blood cells is splenectomy. After removal of the spleen, opsonized leukocytes survive longer in the circulation.

Antibody responses to components of the extracellular matrix are uncommon, but are damaging when they occur. In **Goodpasture's syndrome**, the autoantibody is specific for the α_3 chain of type IV collagen, the collagen of basement membranes throughout the body. The IgG is deposited along the basement membranes of renal glomeruli and renal tubules, where they elicit an

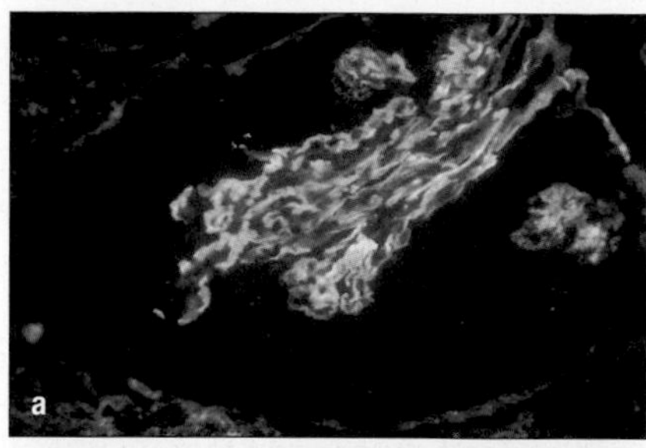

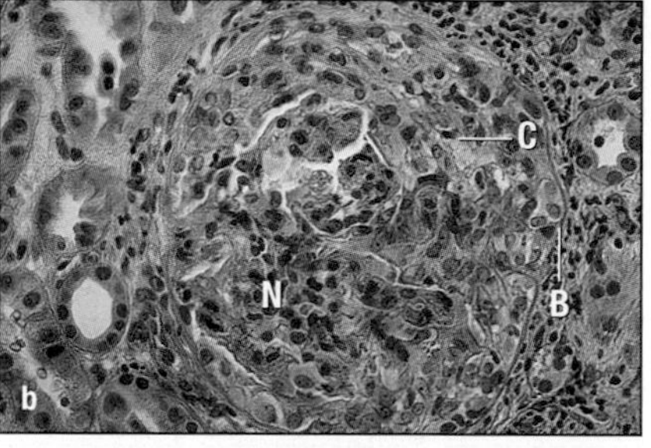

Figure 16.3 Autoantibodies specific for type IV collagen react with the basement membranes of kidney glomeruli, causing Goodpasture's syndrome. The panels show sections of a renal corpuscle in serial biopsies taken from a patient with Goodpasture's syndrome. Panel a, glomerulus stained for IgG deposition by immunofluorescence. Antibody against glomerular basement membrane is deposited in a linear fashion (green staining) along the glomerular basement membrane. The autoantibody causes the local activation of cells bearing Fc receptors, the activation of complement, and the influx of neutrophils. Panel b, hematoxylin and eosin staining of a section through a renal corpuscle shows that the glomerulus is compressed by the formation of a crescent (C) of proliferating mononuclear cells within the Bowman's capsule (B) and that there is an influx of neutrophils (N) into the glomerular tuft. Photographs courtesy of M. Thompson and D. Evans.

inflammatory response (Figure 16.3). These basement membranes are essential for the blood-filtering mechanism of the kidney. As IgG and inflammatory cells accumulate, kidney function becomes progressively impaired, leading to kidney failure and death if not treated. Treatment involves plasma exchange to remove existing antibodies, together with immunosuppressive drugs to stop new ones from being made. Although autoantibodies are deposited in the basement membranes of other organs, the high-pressure filtering of blood by renal glomeruli is the vital function most sensitive to their effects. Indeed, kidney damage by the immune system is responsible for one-quarter of all cases of end-stage renal failure.

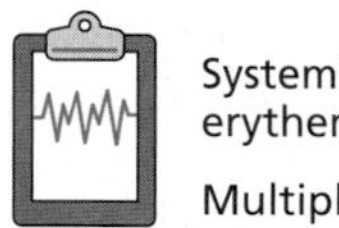

Systemic lupus erythematosus

Multiple sclerosis

Systemic lupus erythematosus (**SLE**) is a disease in which IgG is made against a wide range of cell-surface and intracellular self antigens that are common to many cell types. The immune complexes formed by these antigens and antibodies are deposited in various tissues, where they cause inflammatory reactions resembling type III hypersensitivity reactions. The deposits can cause glomerulonephritis in the kidneys, arthritis in the joints, and a butterfly-shaped skin rash on the face, which gave the disease its name (Figure 16.4). Because the rash, or erythema, gives the face an appearance of a wolf's head (*lupus* is Latin for wolf) the disease was first described clinically as 'lupus erythematosus.' Later, with appreciation of its systemic nature, the name was expanded to 'systemic lupus erythematosus.' The disease presents in diverse ways, with only a proportion of patients getting the facial rash. SLE can be a very severe disease, in which the unwanted reactions of autoimmunity stimulate further autoimmunity that sends the immune system careering down a path of ever-increasing and uncontrolled destruction. SLE is particularly common in women of African or Asian origin, 1 in 500 of whom has the disease.

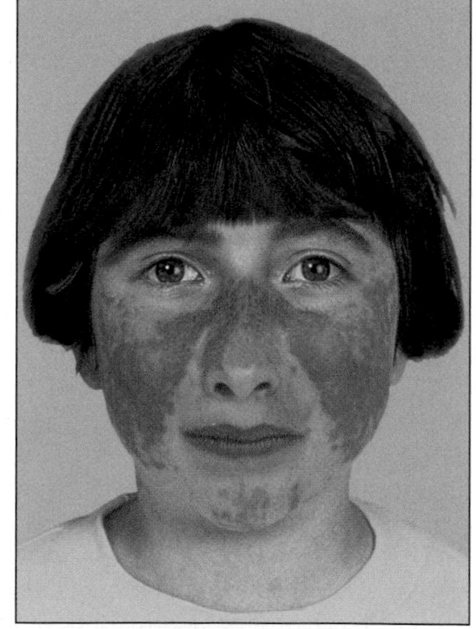

Figure 16.4 The characteristic facial rash of systemic lupus erythematosus. Historically, this butterfly-shaped rash was first used to define and diagnose the disease. Now that the disease is defined immunologically, it is recognized that a proportion of patients who have the disease do not get the rash. Photograph courtesy of M. Walport.

In **multiple sclerosis**, autoimmune effector cells attack the myelin sheath of nerve cells to produce sclerotic plaques of demyelinated tissue in the white matter of the central nervous system. Disease symptoms include motor weakness, impaired vision, lack of coordination, and spasticity (excessive contraction of muscles). The effects of activated T_H1 CD4 cells and the interferon-γ (IFN-γ) they secrete are the cause of multiple sclerosis, which resembles a T cell-mediated type IV hypersensitivity reaction. T_H1 CD4 cells are enriched in the blood and cerebrospinal fluid. They activate macrophages that release proteases and cytokines, which cause the demyelination and sclerotic plaque formation. The importance of IFN-γ in mediating multiple sclerosis was convincingly demonstrated by a clinical trial of this cytokine as a possible treatment for the disease; unfortunately, the disease only worsened in patients treated with IFN-γ.

Like SLE, multiple sclerosis is a highly variable disease. It can take a slow progressive course or it can alternate between acute attacks of exacerbating disease and periods of gradual recovery. The incidence of multiple sclerosis reaches 1 in 1000 in some populations, with onset typically occurring between the ages of 25 and 35. In its extreme form, severe disability or death occurs within a few years, whereas some patients with mild disease experience little neurological impediment. In 90% of multiple sclerosis patients, the sclerotic plaques contain plasma cells that secrete oligoclonal IgG into the cerebrospinal fluid. The autoantigens recognized by these antibodies are the structural proteins of myelin. These include myelin basic protein, proteolipid protein, and myelin oligodendrocyte glycoprotein. Regular subcutaneous injection of IFN-β_1 reduces the incidence of disease attacks and the appearance of plaques. Disease attacks are treated with high doses of immunosuppressive drugs.

16-2 Autoimmune diseases arise when tolerance to self antigens is lost

In previous chapters of this book we encountered various mechanisms that prevent healthy cells and tissues of the body from stimulating an autoimmune response (Figure 16.5). That most people never suffer autoimmune disease shows that the combination of these mechanisms is usually effective in maintaining a state of self-tolerance. Autoimmune diseases occur on rare occasions when self-tolerance to one or a few self antigens is lost and adaptive immunity becomes inadvertently directed toward normal components of the healthy human body. In effect, the autoimmune response strives to eliminate the target autoantigens from the body, and until that is achieved or the patient dies, a chronic state of inflammation and lymphocyte infiltration exists in the tissues where the autoantigens are found. This severely interferes with tissue function, and also provides an environment in which there can be further loss of tolerance and an autoimmune response of increasing breadth and strength. Although the regulatory mechanisms of the immune system respond to the situation and can provide temporary relief between episodes of disease, autoimmune diseases are rarely resolved or cured. They are chronic diseases in which the immune response remains suspended in its destructive phase and never reaches the phase at which inflammation is replaced by tissue repair and reconstruction.

The extraordinary array of autoimmune disease symptoms presented by patients with immunodeficiency who lack the transcription factor AIRE is compelling evidence for the connection between immunological tolerance and autoimmunity (Figure 16.6). The function of AIRE is to induce the deletion of thymocytes that recognize peptides cleaved from tissue-specific proteins expressed by one or a small number of cells or tissues. AIRE ensures that these proteins are expressed in the thymus, where their peptide antigens contribute to negative selection of the T-cell repertoire (see Section 7-12). Although defective alleles of the *AIRE* gene are infrequent worldwide, they are sufficiently common in some populations—Finns, Sardinians, and Iranian Jews—for some children to inherit two defective alleles. For these infants, the normal array of tissue-specific proteins is not expressed in the thymus and so negative selection of the T-cell repertoire is incomplete. In these children's circulation are clones of naive T cells specific for peptides derived from tissue-specific proteins and presented by self-MHC molecules. Starting in infancy, these self-reactive CD4 and CD8 T cells respond to tissue-specific self antigens. The effector CD8 T cells kill cells expressing the tissue-specific proteins, and the effector CD4 T cells help B cells make high-affinity antibodies against them. The B-cell and T-cell responses are directed against various tissues, including

Mechanisms that contribute to immunological self-tolerance
Negative selection of B cells in the bone marrow
Expression of tissue-specific proteins in the thymus so that they participate in negative selection of T cells
Negative selection of T cells in the thymus
Exclusion of lymphocytes from certain peripheral tissues: brain, eye, testis
Induction of anergy in autoreactive B and T cells that reach the peripheral circulation
Suppression of autoimmune responses by regulatory T cells

Figure 16.5 Mechanisms that contribute to immunological self-tolerance.

APECED patients suffer a variety of autoimmune diseases and candidiasis	
Symptom	Frequency in Finnish patients (%)
Endocrine glands	
Hypoparathyroidism	85
Adrenal failure	72
Ovarian failure	60
Insulin-dependent diabetes mellitus	18
Testicular atrophy	14
Parietal cell atrophy	13
Hypothyroidism	6
Other tissues	
Candidiasis	100
Dental enamel hypoplasia	77
Nail dystrophy	52
Tympanic membrane calcification	33
Alopecia	27
Keratopathy	22
Vitiligo	13
Hepatitis	13
Intestinal malabsorption	10

Figure 16.6 Patients with deficiency of the autoimmune regulator protein AIRE suffer symptoms that are characteristic of a wide range of autoimmune diseases. This condition is called autoimmune polyendocrinopathy–candidiasis–ectodermal dystrophy (APECED) or inherited autoimmune polyglandular disease (APD).

most endocrine glands. These immunodeficient children exhibit a diversity of symptoms, each typical of one or more autoimmune diseases. This syndrome of inherited **autoimmune polyglandular disease** (**APD**), also called **autoimmune polyendocrinopathy–candidiasis–ectodermal dystrophy** (**APECED**), strongly suggests that all autoimmune diseases are due to a loss of T-cell tolerance. The ectodermal dystrophy manifests itself as abnormalities of teeth, hair, and fingernails (Figure 16.7).

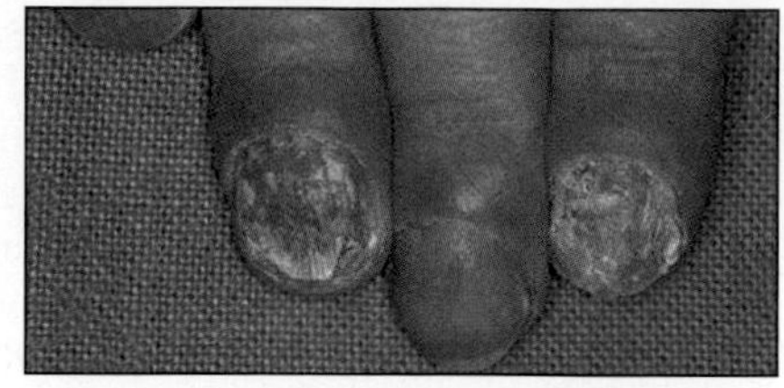

Figure 16.7 Dystrophic fingernails in a patient with APECED. Such visible symptoms of the disease can help in the diagnosis of children with APECED. Photograph courtesy of Mark S. Anderson.

Despite the range and severity of their symptoms, patients with APECED have substantial life-spans, the most life-threatening complications of the condition being squamous-cell carcinoma and fulminant autoimmune hepatitis. Thus, the absence of self-tolerance to major components of several organs does not cause an acute and catastrophic disease, but instead one that combines features of common autoimmune diseases, such as diabetes and SLE. Like those diseases, the course of APECED is highly heterogeneous, indicating the involvement of other genetic and environmental factors. In Finnish, but not Iranian, patients, for example, the autoimmune symptoms are foreshadowed by chronic infection with the fungus *Candida albicans*, despite making a strong antibody response to *Candida* antigens. Persistent *Candida* infection is the only symptom of disease that is common to all Finnish APECED patients (see Figure 16.6).

Autoimmune polyendocrinopathy-candidiasis-ectodermal dystrophy

Immune dysregulation, polyendocrinopathy, enteropathy X-linked disease

A second rare immunodeficiency disease reveals the importance of regulatory T cells (T_{reg}) in preventing autoimmune disease. Uniquely defining T_{reg} is their use of the transcriptional repressor protein FoxP3; all T_{reg}, but no other cells, express FoxP3 (see Section 8-10). Mutation of the *FOXP3* gene on the X chromosome causes an immunodeficiency that principally affects boys and is called **immune dysregulation, polyendocrinopathy, enteropathy, and X-linked syndrome** (**IPEX**). Although rare, 136 patients and 63 different mutations in the *FOXP3* gene have been studied since diagnosis of the first patients in the 1980s. Although showing no abnormalities at birth, children with IPEX rapidly develop enteritis with intractable diarrhea, type 1 diabetes, and eczema within the first months of life. Over time, other organs become subject to autoimmunity—the thyroid, for example. Because of their inflamed and disordered intestines, the children do not thrive. They also suffer recurrent infections that exacerbate the autoimmunities. If infants with IPEX are not transplanted with hematopoietic stem cells from an HLA-identical sibling, they die within the first year of life.

Phenotypically the T-cell populations in IPEX patients appear normal, including the $CD4^+CD25^+FoxP3^+$ T cells that in healthy individuals are the T_{reg}. Most of the mutations in the *FOXP3* gene do not interfere with transcription and translation, and so the mRNA and the FoxP3 protein are detectable, although both the protein and the cells are functionally impaired. In the absence of functional T_{reg}, there are increases in inflammatory T_H17 cells and eosinophils associated with a high level of IgE. In contrast, the levels of IgA, IgG, and IgM are diminished by the diarrhea. What this experiment of nature tells us is that T_{reg} are vital components of the immune system that prevent inflammatory T cells from breaking tolerance and causing autoimmune disease.

16-3 HLA is the dominant genetic factor affecting susceptibility to autoimmune disease

Susceptibility to autoimmune diseases runs in families and varies between populations and ethnic groups. Such clinical observations first indicated that genetic factors confer differential susceptibility to autoimmune diseases. The most important genetic factors influencing susceptibility are genes in the HLA complex, and almost every autoimmune disease has an HLA association. HLA genes account for 50% of the genetic predisposition to autoimmune conditions, and numerous other genes make small contributions to the other 50%.

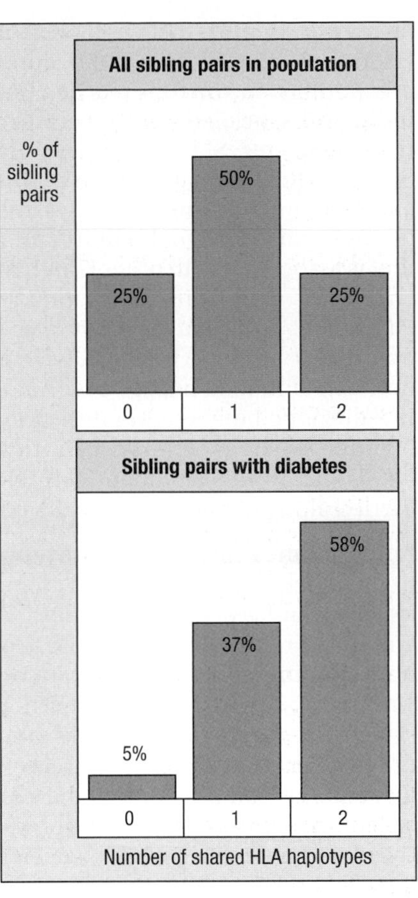

Figure 16.8 Family studies reveal that HLA type correlates with susceptibility to type 1 diabetes. The top panel shows the frequency with which two siblings share HLA haplotypes in the population as a whole. The percentages are those expected from a simple Mendelian segregation of the two maternal and two paternal HLA haplotypes. In the bottom panel the analysis has been confined to pairs of siblings who both have type 1 diabetes. In these sibling pairs, the frequency distribution of HLA haplotypes differs greatly from that expected from simple Mendelian segregation. Pairs of siblings with the disease are much more likely to have the same HLA type than are pairs of siblings who are healthy.

A simple but powerful way to assess the influence of HLA genes on autoimmune disease is to study families. Only four HLA haplotypes segregate in nuclear families, two maternal haplotypes and two paternal haplotypes. On average, 50% of siblings in healthy families share one HLA haplotype, 25% share two HLA haplotypes, and 25% are HLA disparate (Figure 16.8, upper panel). The distribution changes when pairs of siblings with type 1 diabetes are similarly compared: the frequency of diabetic sibling pairs who share two HLA haplotypes more than doubles, whereas the frequency of diabetic pairs having one or no HLA haplotype in common is depressed (Figure 16.8, lower panel). Such analysis demonstrates that HLA type influences susceptibility to type 1 diabetes, and other autoimmune diseases give similar results.

Although family analysis shows clearly that HLA genes are risk factors for autoimmune disease, it is not good at determining which HLA genes are functionally involved in the disease. This is better done by using large panels of unrelated patients and controls in which numerous HLA haplotypes having different combinations of HLA polymorphisms are analyzed. Such studies show that the strong associations are with the polymorphic HLA class I and class II genes (Figure 16.9), with weaker effects due to other genes of the HLA complex, for example the gene encoding TNF-α. More diseases are associated with HLA class II than with HLA class I, although the two strongest associations are with class I molecules—HLA-B27 with ankylosing spondylitis, a form of arthritis, and HLA-A29 with birdshot retinopathy, an autoimmune disease of the eye.

The assignment of HLA genes to disease susceptibilities is complicated by the fact that particular alleles of the different polymorphic genes are combined in HLA haplotypes at frequencies higher than expected by chance. This phenomenon of preferential allele associations is called **linkage disequilibrium**. An

HLA-associated risk factors for autoimmune disease				
Disease	HLA allotype	Frequency (%)		Relative risk
		Patients	Control	
Ankylosing spondylitis	B27	> 95	9	> 150
Birdshot chorioretinopathy	A29	> 95	4	> 50
Narcolepsy	DQ6	> 95	33	> 40
Celiac disease	DQ2 and DQ8	95	28	30
Type 1 diabetes	DQ2 and DQ8	81	23	14
Subacute thyroiditis	B35	70	14	14
Multiple sclerosis	DQ6	86	33	12
Rheumatoid arthritis	DR4	81	33	9
Juvenile rheumatoid arthritis	DR8	38	7	8
Psoriasis vulgaris	Cw6	87	33	7
Addison's disease	DR3	69	27	5
Graves' disease	DR3	65	27	4
Myasthenia gravis	DR3	50	27	2
Type 1 diabetes	DQ6	< 0.1	33	0.02

Figure 16.9 Associations of HLA allotypes with autoimmune disease. The data are derived from the Norwegian population. In the column 'Relative risk,' numbers greater than 1 indicate that the HLA allotype confers increased susceptibility relative to the general population; numbers less than 1 indicate increased protection. Data courtesy of Erik Thorsby.

extreme example is the A1–B8–DR3–DQ2 haplotype, which includes alleles for *HLA-A, -B, -C, -DR,* and *-DQ* (Figure 16.10) and is characteristic of Caucasian populations, in which it reaches frequencies of up to 11%. This haplotype is uniquely associated with several common autoimmune diseases, including type 1 diabetes, SLE, myasthenia gravis, autoimmune hepatitis, and primary biliary cirrhosis. In determining which polymorphic HLA genes are functionally important for which diseases, the A1–B8–DR3–DQ2 haplotype is not helpful. Achieving that goal requires an analysis of the frequency of each disease that is associated with other HLA haplotypes that have some, but not all, of the HLA class I and II alleles present on the A1–B8–DR3–DQ2 haplotype. Particularly informative are studies of non-Caucasian populations, in which the alleles of the A1–B8–DR3–DQ2 haplotype are present but distributed between different haplotypes. Analysis of Caucasians failed to distinguish whether type 1 diabetes was functionally associated with HLA-DQ2, HLA-DR3, or both, whereas analysis of Africans and Asians clearly showed it to be HLA-DQ2. In this situation HLA-DR3 is said to be hitchhiking, because it is carried along in the analysis by its linkage disequilibrium with HLA-DQ2.

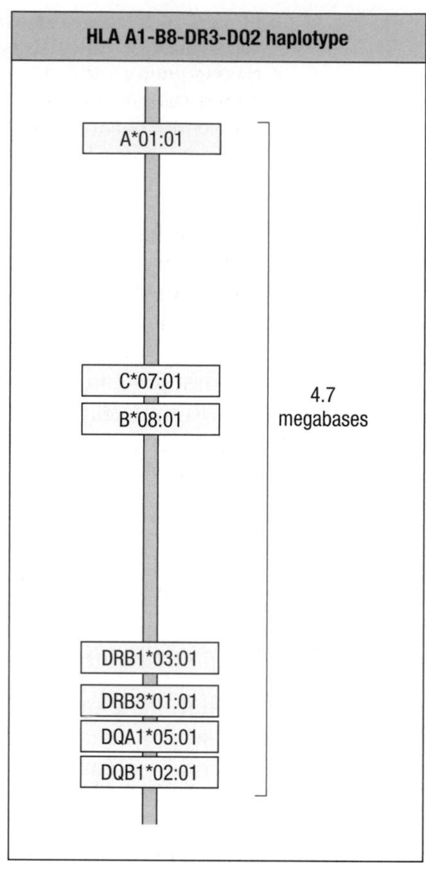

Figure 16.10 Key genes of the A1–B8–DR3–DQ2 HLA haplotype.

Autoimmune diseases are more prevalent in Caucasians than in other populations, which is in part the effect of the HLA-DQ2, HLA-DR3 haplotype. Despite its disadvantages, this haplotype is carried by more than 10 million Europeans and is well represented in the white populations of other continents. This illustrates an important general point, which is that the large majority of people who have a disease-associated HLA type will never suffer from an autoimmune disease associated with that type. Conversely, not all patients who develop an HLA-associated disease have an HLA type that is correlated with disease susceptibility. Thus, the associations of common HLA alleles with autoimmune diseases do not derive from 'good' versus 'bad' alleles, as is the case for genetic diseases such as APECED and IPEX (see Section 16-2). Even for the strongest HLA association with disease—HLA-B27 and ankylosing spondylitis—only 2% of people with HLA-B27 develop the disease, and the number increases only to 20% in families with a history of ankylosing spondylitis.

Independently of HLA type, susceptibility to autoimmune disease varies significantly between men and women (Figure 16.11). Ankylosing spondylitis is more prevalent in men, whereas most autoimmune diseases are more common in women. Genetic differences between women and men must therefore contribute to these differential disease susceptibilities. In studies on laboratory mice, females are also more susceptible to autoimmunity than males, but this difference goes away when the mice are raised in a germ-free environment. These observations suggest that there are mutual influences between host genetics and the microbiota that affect susceptibility to autoimmunities, and they do so differently in males and females.

16-4 HLA associations reflect the importance of T-cell tolerance in preventing autoimmunity

The functions of polymorphic HLA class I and II molecules are to present peptide antigens to CD8 and CD4 T cells, respectively. Although some autoimmune diseases are mediated by antibodies and others by effector T cells, they are all associated with either HLA class I or class II and these associations are orders of magnitude stronger than with any other genes. The number, nature and strength of the disease associations argues strongly that loss of T-cell tolerance to a self antigen is the necessary first step in the development of an autoimmune disease. Moreover, the fact that single HLA allotypes are associated with disease suggest that this first step involves one peptide bound to one HLA allotype being recognized by one clone of T cells. The disease associations with HLA are also consistent with T-cell tolerance being more important for preventing autoimmunity than B-cell tolerance.

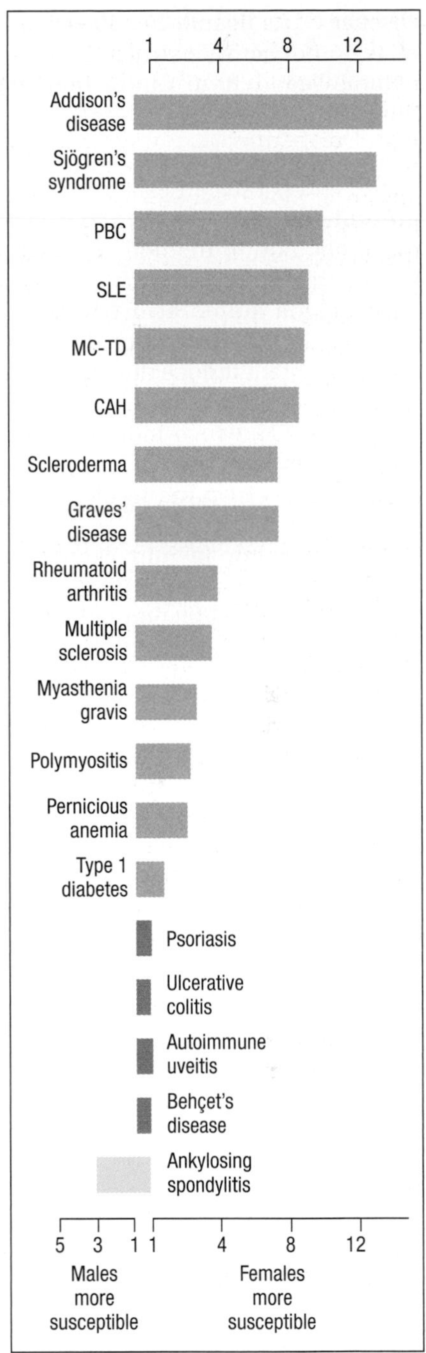

Figure 16.11 Relative incidences of autoimmune disease in females and males. The numbers on the two scales give the ratio of men to women with a disease (on the left) and the ratio of women to men with a disease (on the right). PBC, primary billiary cirrhosis; CAH, chronic active hepatitis; MC-TD, mixed connective tissue disease.

The antibodies that contribute to autoimmune diseases are products of B cells that have switched isotype and been subject to affinity maturation in a germinal center reaction, which means that they received help from antigen-specific CD4 T cells. Thus, B-cell autoimmunity is contingent on a loss of T-cell tolerance. Mechanisms for eliminating autoreactive cells from the naive B-cell population are less efficient than thymic selection of T cells because B cells are only selected against the self antigens expressed in bone marrow and the circulation, and not against the self antigens in the tissues. Thus, there are numerous circulating autoreactive B cells that can be brought into play once T-cell tolerance has been breached.

That many more autoimmune diseases are associated with HLA class II than with HLA class I indicates that CD4 T cells are inherently more likely to lose tolerance to a self antigen than are CD8 T cells. This propensity fits with naive CD4 T cells having a lower threshold than naive CD8 T cells for activation by foreign antigens (see Section 8-12). Type IV hypersensitivity reactions exhibit the same bias, the majority being mediated by CD4 T_H1 cells and a minority by CD8 T cells (see Section 14-1).

When tissues become inflamed, cells are induced to increase the expression of MHC class I and class II molecules. These changes, which increase the number of different peptide antigens presented by a cell and their density on the cell surface, can lead to interactions with clones of T cells that were not sensitive to the lower levels of antigen presentation. The greatest changes occur when the inflammatory cytokine IFN-γ induces MHC class II expression on cells that normally do not express these MHC molecules. Such cells, which include thyroid cells, pancreatic β cells, astrocytes, and microglia, are conspicuously present in the tissues targeted by autoimmunity.

16-5 Binding of antibodies to cell-surface receptors causes several autoimmune diseases

Several autoimmune diseases are caused by autoantibodies that bind to a cell-surface receptor and perturb its normal function and regulation (Figure 16.12). Autoantibodies that are **receptor agonists** mimic the natural ligand of the receptor and cause the receptor to transduce activating signals in the

Diseases mediated by antibodies against cell-surface receptors				
Syndrome	**Antigen**	**Antibody**	**Consequence**	**Target cell**
Graves' disease	Thyroid-stimulating hormone receptor	Agonist	Hyperthyroidism	Thyroid epithelial cell
Myasthenia gravis	Acetylcholine receptor	Antagonist	Progressive muscle weakness	Muscle
Insulin-resistant diabetes	Insulin receptor	Antagonist	Hyperglycemia, ketoacidosis	All cells
Hypoglycemia	Insulin receptor	Agonist	Hypoglycemia	All cells

Figure 16.12 Diseases mediated by antibodies against cell-surface receptors. Antibodies act as agonists when they stimulate a receptor on binding it, and as antagonists when they block a receptor's function on binding it.

absence of its ligand. In contrast, autoantibodies that are **receptor antagonists** do not activate signaling on binding to the receptor and they block the natural ligand from binding to the receptor and activating its signaling function.

Graves' disease is caused by an autoimmune response that affects the thyroid gland, and is thus an example of an **organ-specific** or **tissue-specific autoimmune disease**. The thyroid is an endocrine gland that regulates the basal metabolic rate of the body through the secretion of two related hormones, tri-iodothyronine and tetra-iodothyronine (thyroxine), small iodinated derivatives of the amino acid tyrosine. When increased cellular metabolism is required, for example when the ambient temperature drops, signals from the nervous system induce the pituitary, another endocrine gland, to secrete thyroid-stimulating hormone (TSH). Thyroid epithelial cells express receptors that bind TSH, which induces the production and secretion of thyroid hormones. The hormones induce cellular metabolism that raises the body temperature. This in turns feeds back to the pituitary and shuts down further release of TSH. Graves' disease is caused by agonist autoantibodies specific for the TSH receptor. By mimicking the natural ligand, the antibodies bound to the TSH receptor cause chronic overproduction of thyroid hormones that is independent of regulation by TSH and insensitive to the metabolic needs of the body (Figure 16.13). This **hyperthyroid** condition causes heat intolerance, nervousness, irritability, warm moist skin, weight loss, and enlargement of the thyroid.

Other aspects of Graves' disease are outwardly bulging eyes and a characteristic stare. This condition, called Graves' ophthalmopathy, is due to autoantibodies that bind to the eye muscles. These antibodies were made against a thyroid protein and they cross-react with an eye-muscle protein. Short-term treatment for Graves' disease is provided by methimazole and propylthiouracil. These drugs inhibit the production of thyroid hormones by reducing the uptake of iodine by the thyroid. In the long term, the disease is treated by

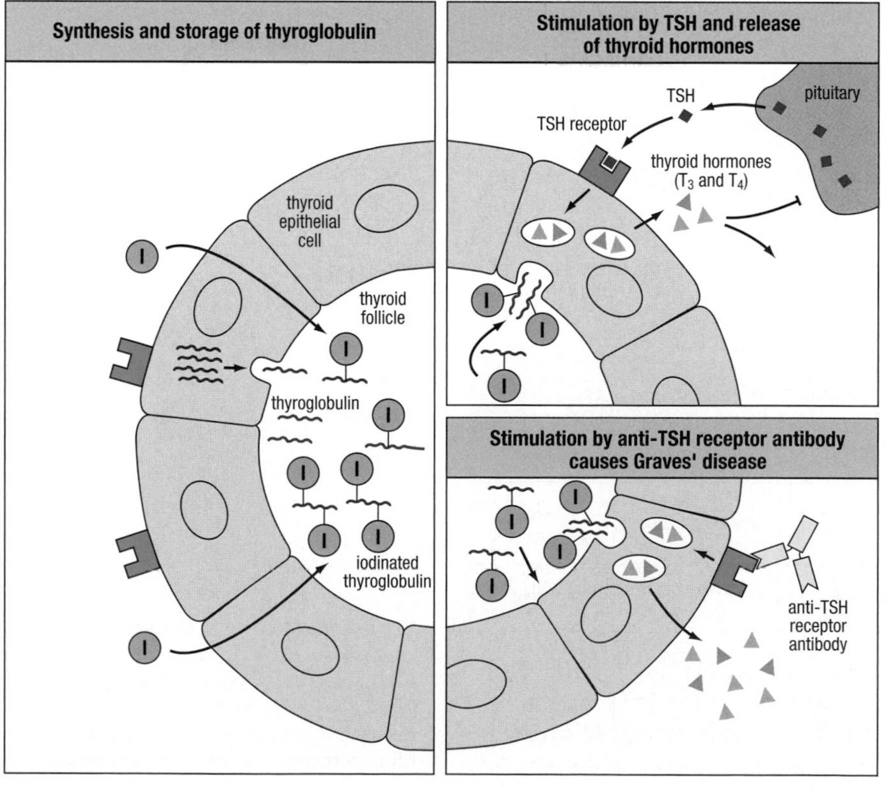

Figure 16.13 Autoantibodies against the TSH receptor cause overproduction of thyroid hormones and Graves' disease. Thyroid epithelial cells make thyroglobulin, a glycoprotein that is stored in follicles formed by the spherical arrangement of thyroid cells. Iodide (green circles) is taken up and used to iodinate and cross-link tyrosine residues of thyroglobulin (left half of the figure). When thyroid hormones are needed. Thyroid-stimulating hormone (TSH) from the pituitary gland binds to the TSH receptor on thyroid cells, inducing the endocytosis and breakdown of iodinated thyroglobulin, with release of the thyroid hormones tri-iodothyronine (T_3) and thyroxine (T_4). As well as regulating metabolism, T_3 and T_4 signal the pituitary to stop releasing TSH (upper right panel). In Graves' disease, autoantibodies bind to the TSH receptor of thyroid cells, mimicking TSH and inducing the continuous synthesis and release of thyroid hormones. In patients with Graves' disease, the production of thyroid hormones becomes independent of the presence of TSH and of the body's requirements for thyroid hormones (lower right panel).

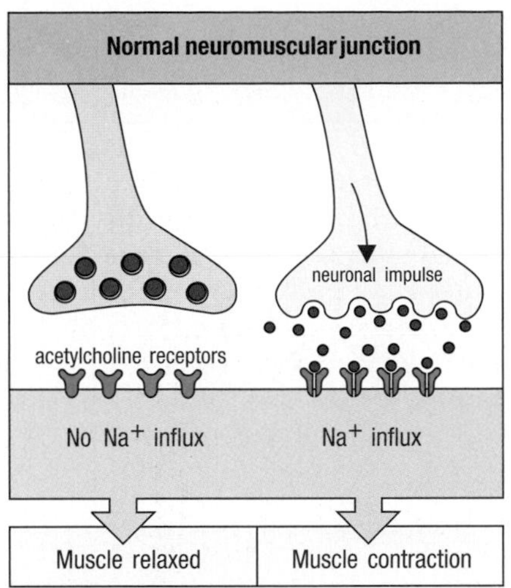

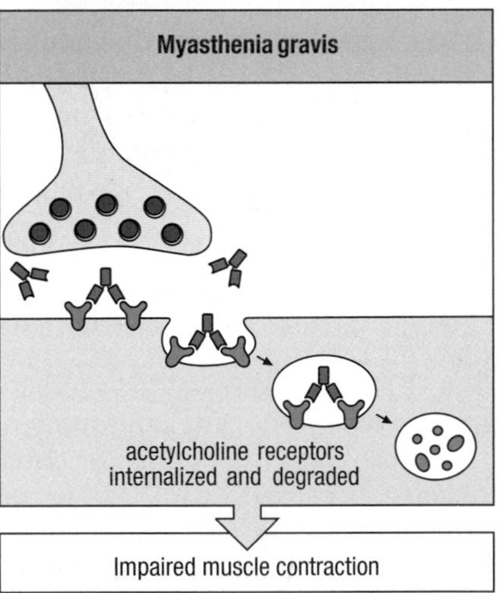

Figure 16.14 Autoantibodies against the acetylcholine receptor cause myasthenia gravis. In a healthy neuromuscular junction, signals generated in nerves cause the release of acetylcholine, which binds to the acetylcholine receptors of the muscle cells, causing an inflow of sodium ions that indirectly causes muscle contraction (upper panel). In patients with myasthenia gravis, autoantibodies specific for the acetylcholine receptor reduce the number of receptors on the muscle-cell surface by binding to the receptors and causing their endocytosis and degradation (lower panel). Consequently, the efficiency of the neuromuscular junction is reduced, which is manifested as muscle weakening.

completely stopping thyroid function, either by surgical removal of the gland or by its irradiation on administration of the radioactive iodine isotope ^{131}I. Thyroid function is then replaced by daily doses of synthetic thyroid hormones. The formation of autoantibodies in patients with Graves' disease is driven by a CD4 T_H2 response. Graves' disease is associated with HLA-DR3 (see Figure 16.10), which points to this autoimmune response being initiated by a T_H2 cell that recognizes a TSH-derived peptide presented by HLA-DR3. Such a T cell could then provide help to B cells specific for epitopes of TSH.

Myasthenia gravis is an autoimmune disease in which signaling from nerve to muscle across the neuromuscular junction is impaired (Figure 16.14). Antagonistic autoantibodies bind to the acetylcholine receptors on muscle cells, inducing their endocytosis and intracellular degradation in lysosomes. The loss of cell-surface acetylcholine receptors makes the muscle less sensitive to neuronal stimulation. Consequently, patients with myasthenia gravis suffer progressive muscle weakening as levels of autoantibody rise; the name of the disease means severe ('*gravis*') muscle ('*myo*') weakness ('*asthenia*'). Early symptoms of the disease are droopy eyelids and double vision. With time, other facial muscles weaken and similar effects on chest muscles impair breathing. This makes patients susceptible to respiratory infections and can even cause death. Treatment for myasthenia gravis is the drug pyridostigmine, an inhibitor of the enzyme cholinesterase, which degrades acetylcholine. By preventing acetylcholine degradation, pyridostigmine increases the capacity of acetylcholine to compete with the autoantibodies for the receptors. During crises of severe muscle weakening, immunosuppressive drugs, principally azathioprine but also others, are used to inhibit production of the autoantibodies. Like Graves' disease, myasthenia gravis is associated with HLA-DR3 (see Figure 16.10). This suggests that myasthenia gravis is initiated by CD4 T cells that recognize an acetylcholine-derived peptide presented by HLA-DR3 and then give help to B cells specific for the acetylcholine receptor.

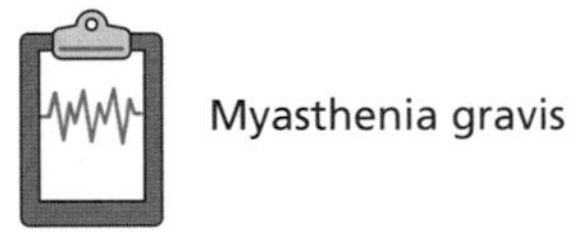
Myasthenia gravis

Agonistic and antagonistic autoantibodies can be made against the insulin receptor, and they lead to different symptoms. Because insulin receptors are present on all cells, the diseases caused by antibodies against insulin receptors affect the entire body and are examples of **systemic autoimmune diseases**. The cells of patients with antagonistic autoantibodies are unable to take up glucose, which accumulates in the blood, causing hyperglycemia and a form of diabetes mellitus that is resistant to treatment with insulin. In contrast, in patients with agonistic antibodies, the antibodies mimic insulin and cause the cells to continuously remove glucose from the blood. Blood glucose decreases to an abnormally low level, inducing a state of hypoglycemia and light-headedness caused by insufficient glucose reaching the brain. These conditions can be treated with immunosuppressive drugs and with anti-CD20 antibody to eliminate the B cells making the autoantibodies.

If a pregnant woman has an IgG-mediated autoimmune disease, for example Graves' disease, her autoantibodies are transported by FcRn from the maternal blood across the placenta to the fetal blood (see Section 9-14). At birth the baby will exhibit symptoms of disease, but with time the maternal IgG in the

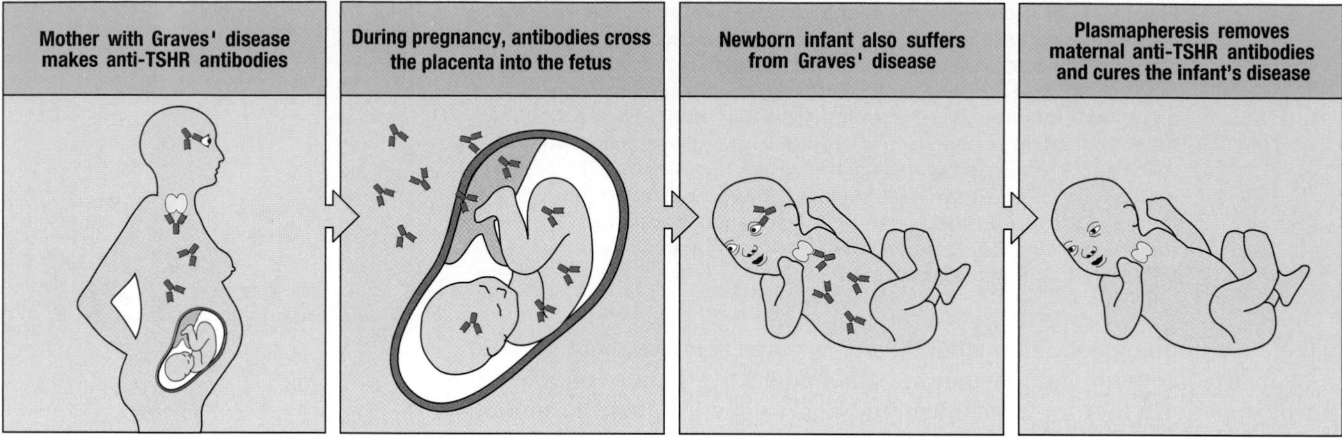

Mother with Graves' disease makes anti-TSHR antibodies
During pregnancy, antibodies cross the placenta into the fetus
Newborn infant also suffers from Graves' disease
Plasmapheresis removes maternal anti-TSHR antibodies and cures the infant's disease

Figure 16.15 Temporary symptoms of antibody-mediated autoimmune diseases can be passed from affected mothers to their newborn babies. The mother has Graves' disease and Graves' ophthalmopathy, which causes her eyes to bulge. IgG autoantibodies against the thyroid-stimulating hormone receptor (TSHR) pass from the mother to the fetus *in utero* and passively give the baby a temporary Graves' disease that disappears with the degradation of maternal IgG in the infant's circulation.

infant's circulation degrades and the disease symptoms gradually go away (Figure 16.15). Because B cells making the autoantibodies cannot cross the placenta, such babies suffer only a transient autoimmune disease. For diseases that could affect the infant's growth, the recommended treatment is to remove the autoantibody by total exchange of blood plasma.

16-6 Organized lymphoid tissue sometimes forms at sites inflamed by autoimmune disease

The CD4 T_H2 response that mediates Graves' disease produces little inflammation or lymphocytic infiltration of the thyroid tissue, which retains its normal morphology. By contrast, **chronic thyroiditis**, also called **Hashimoto's disease**, is caused by a CD4 T_H1 response, which produces both antibodies and effector CD4 T cells that are specific for thyroid antigens. Lymphocytes infiltrate the thyroid, causing a progressive destruction of the normal thyroid tissue and a corresponding loss of the capacity to make thyroid hormones. Patients with Hashimoto's disease become **hypothyroid** and eventually are unable to make thyroid hormone. Treatment for Hashimoto's disease is replacement therapy with synthetic thyroid hormones taken orally on a daily basis.

A characteristic feature of Hashimoto's disease is that the lymphocytes and other cells infiltrating the thyroid gland become organized into structures resembling the typical microanatomy of secondary lymphoid organs (Figure 16.16). These structures, called **ectopic lymphoid tissues** or tertiary lymphoid organs, contain T-cell and B-cell areas, dendritic cells, follicular dendritic cells, and macrophages. The process by which they form, termed lymphoid neogenesis, resembles the formation of secondary lymphoid tissues and is similarly driven by lymphotoxin (see Section 6-14). Unlike a lymph node, the ectopic lymphoid tissue is not encapsulated, lacks lymphatics, and is exposed to the inflammatory environment of the autoimmune response. Ectopic lymphoid tissue also functions like a secondary lymphoid tissue. Within the organized structure, B cells and T cells are stimulated by antigen to give effector cells, and in germinal center reactions B cells undergo isotype switching and somatic hypermutation to produce plasma cells making high-affinity autoantibodies. Targets for the autoantibodies are the proteins thyroglobulin, thyroid peroxidase, the TSH receptor, and the thyroid iodide transporter, all of which are uniquely expressed in thyroid cells.

Ectopic lymphoid tissue is formed in other autoimmune diseases, including rheumatoid arthritis, Graves' disease, and multiple sclerosis, but with less regularity than for Hashimoto's disease. Ectopic lymphoid tissue is also formed in

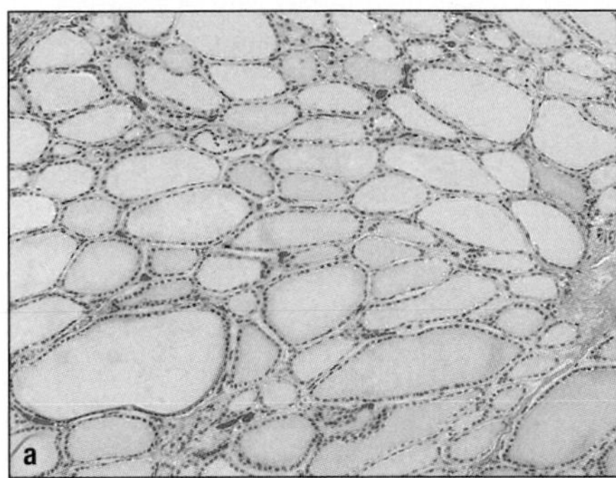

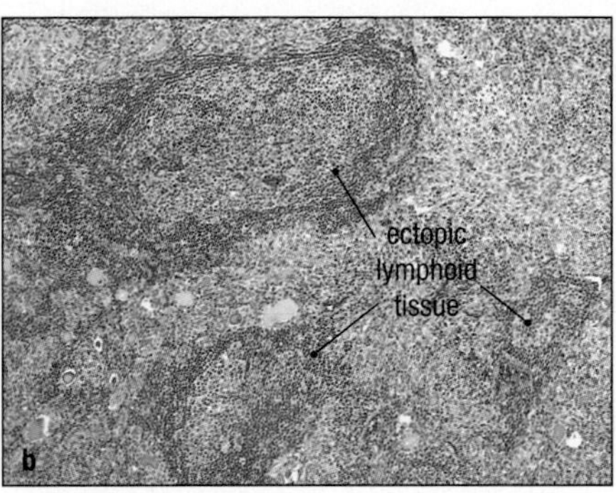

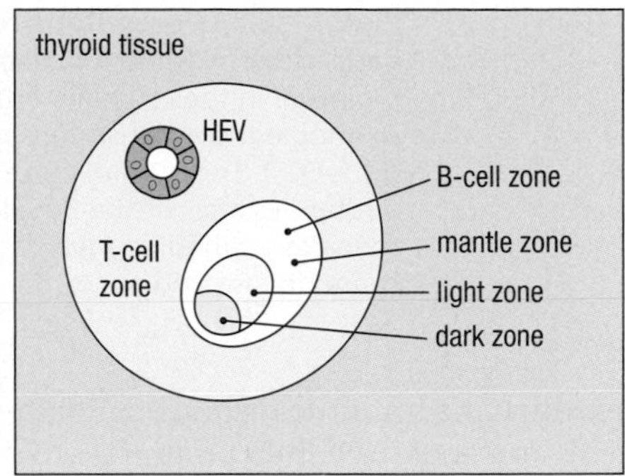

Figure 16.16 Hashimoto's thyroiditis. In a healthy thyroid gland, the epithelial cells form spherical follicles containing thyroglobulin (panel a). In patients with Hashimoto's thyroiditis the thyroid gland becomes infiltrated with lymphocytes, which destroy the normal architecture of the thyroid gland and can become organized into structures resembling secondary lymphoid tissue (panel b), as shown in the schematic diagram at the right. Micrographs courtesy of Yasodha Natkunam.

tissues that are chronically infected with a pathogen, for example in the liver during chronic hepatitis C infection. In Chapter 10 we saw how the secondary lymphoid tissues in the gut are placed close to the sites of heavy infestation with microbes; the formation of ectopic lymphoid tissue may be a similar strategy that permits amplification of the response to a persistent infection. For a tissue sustaining an autoimmune response, this strategy can only exacerbate the disease.

Characteristic properties of endocrine glands make them particularly prone to autoimmune disease. First, the function of an endocrine gland—the synthesis and secretion of a particular hormone—involves tissue-specific proteins not expressed in other cells or tissues but made in large quantities. Second, because they secrete their hormones into the blood, endocrine tissues are well vascularized, which facilitates interactions with cells and molecules of the immune system. Loss of endocrine function has drastic systemic effects, causing pronounced disease and in some cases death. Each autoimmune disease of endocrine tissue is due to an impaired function of a single type of epithelial cell within an endocrine gland (Figure 16.17).

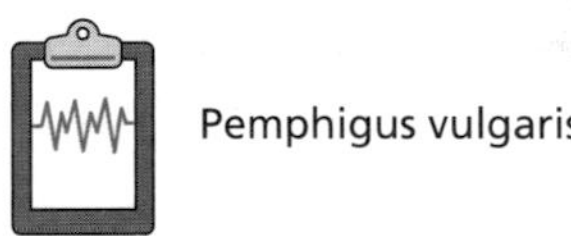

16-7 The antibody response to an autoantigen can broaden and strengthen by epitope spreading

Pemphigus vulgaris and the milder variant **pemphigus foliaceus** are autoimmune conditions characterized by blistering of the skin. They are caused by IgG antibodies made against desmogleins, adhesion molecules of the cell junctions called desmosomes that bind skin keratinocytes tightly to each other. The antibodies impair this binding and with it the integrity of the skin. Genetic and environmental factors make pemphigus foliaceus endemic to some rural Brazilian communities, up to 1 in 30 people being affected by the disease. Long-term study of these people shows that the antibody response evolves in a specific manner that correlates with clinical progression. The extracellular (EC) part of desmoglein comprises four structurally similar domains, EC1–EC4, and a structurally divergent fifth domain, EC5. People first make IgG against epitopes of EC5, the domain closest to the cell membrane, and these antibodies are not associated with disease. Neither do they bind to cell-surface desmoglein, nor do they induce disease when injected into mice. The onset of pemphigus coincides with the presence of antibodies against the EC1 and EC2 domains (Figure 16.18). These IgGs bind to cell-surface desmoglein, and cause skin blistering disease when injected into mice. The process

Autoimmune diseases of endocrine glands	
Thyroid gland	Hashimoto's thyroiditis Graves' disease Subacute thyroiditis Idiopathic hypothyroidism
Islets of Langerhans (pancreas)	Type 1 diabetes (insulin-dependent diabetes, juvenile-onset diabetes) Type 2 diabetes (insulin-resistant diabetes, adult-onset diabetes)
Adrenal gland	Addison's disease

Figure 16.17 Autoimmune diseases of endocrine glands.

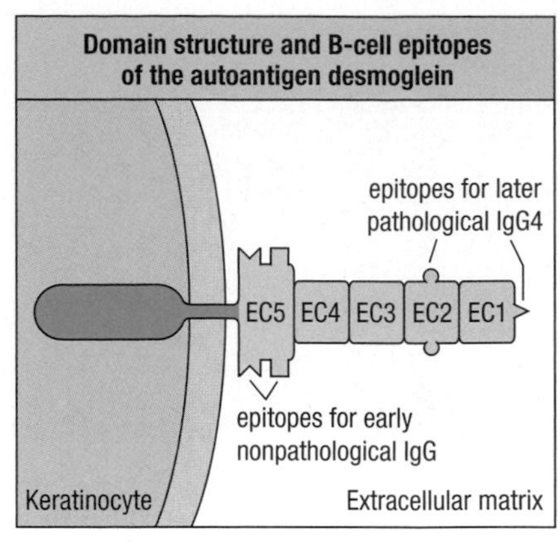

Figure 16.18 Pemphigus foliaceus is a skin blistering disease caused by autoantibodies specific for desmoglein. An adhesion molecule in the cell junctions that hold keratinocytes together, desmoglein is a cell-surface protein with five extracellular domains (EC1–EC5). The autoimmune response starts by making harmless antibodies against the EC5 domain; over time, the response can spread to make antibodies against the EC1 and EC2 domains. These antibodies cause disease and are of the IgG4 isotype.

by which the immune response initially targets epitopes in one part of an antigenic molecule and then progresses to different epitopes of the same molecule is called **intramolecular epitope spreading.** The anti-desmoglein response in pemphigus foliaceus involves the spreading of B-cell epitopes and it is only with epitope spreading that disease-causing antibodies are made. One way in which this could happen is for the initiating antigen to be soluble forms of desmoglein, in which EC5 epitopes are dominant, and to give a first wave of anti-EC5 IgG. Binding of these antibodies to soluble desmoglein then forms immune complexes in which the EC5 epitopes are covered up and the EC1 and EC2 epitopes are exposed. The immune complexes can bind to the antigen receptors of B cells specific for EC1 and EC2, thereby giving rise to a second wave of IgG that binds EC1 and EC2 and causes pemphigus foliaceus (Figure 16.19).

Intramolecular epitope spreading also occurs for the T-cell response in autoimmunity. The T-cell epitopes to which the response spreads are frequently ones to which the immune system is not tolerant because those self peptides are not normally presented by MHC molecules at sufficient levels. These epitopes are called **cryptic epitopes** because they are normally hidden from the immune system and are revealed only under inflammatory conditions.

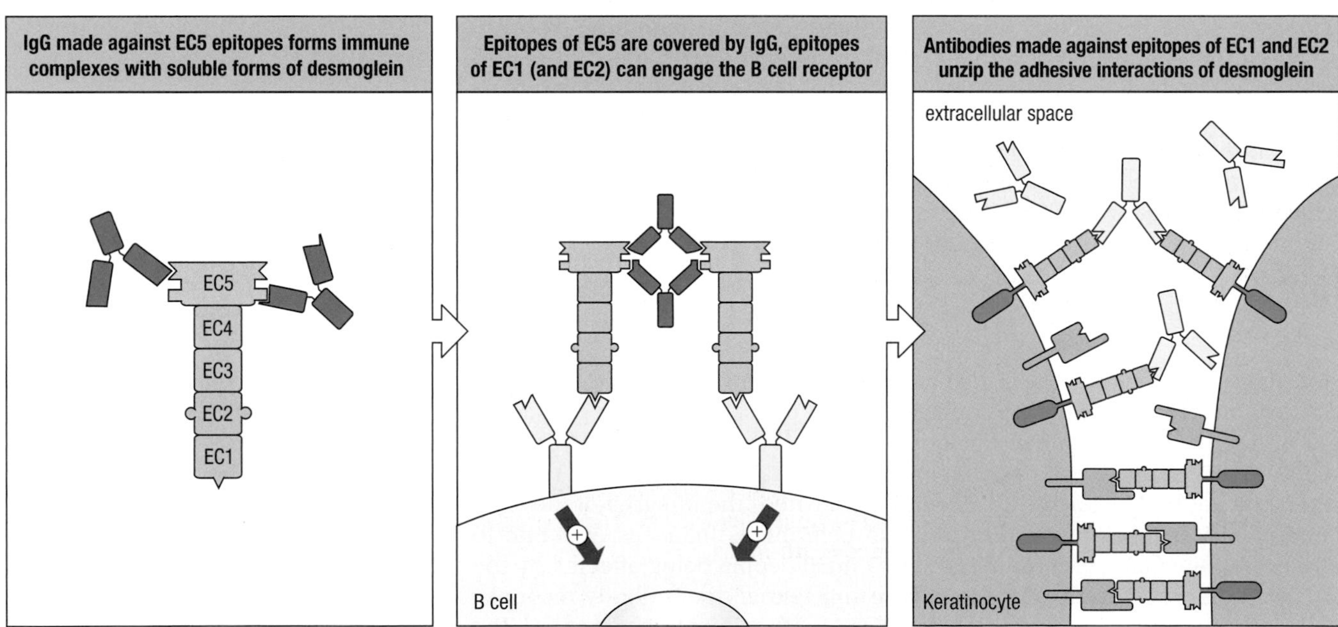

Figure 16.19 How antibodies against desmoglein cause skin blistering. In the early phase of the autoimmune response to desmoglein, antibodies are made against epitopes of the EC5 domain. These epitopes are not accessible to antibody in functional membrane-associated desmoglein, but the antibodies can bind to soluble degradation products of desmoglein (left panel). Soluble immune complexes of antibody desmoglein are bound and processed by B cells specific for epitopes of the EC1 and EC2 domains (center panel). This causes epitope spreading in the later phase of the autoimmune response and the synthesis of high-affinity IgG4 antibodies specific for the EC1 and EC2 epitopes. These epitopes of membrane-associated desmoglein are accessible to antibody, which interferes with the physiological adhesive interactions of desmoglein that are necessary for maintaining skin integrity. Consequently, the antibodies cause the outer layers of the skin to separate, giving blisters (right panel).

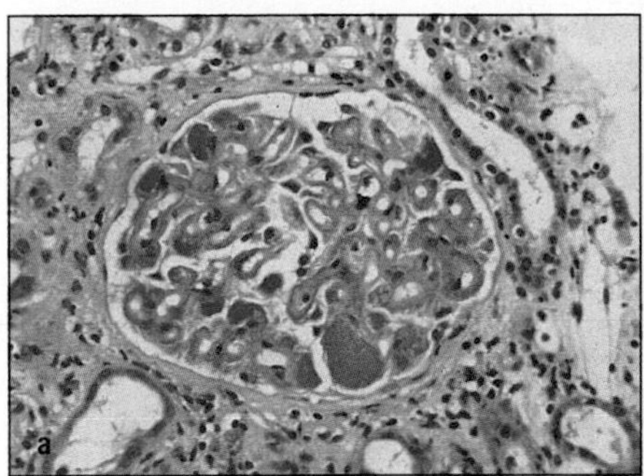

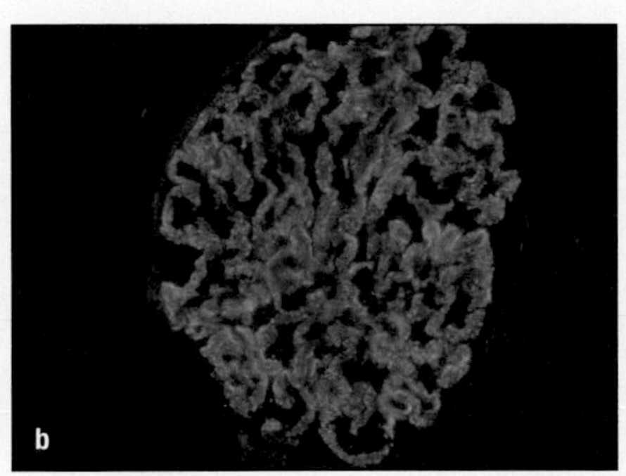

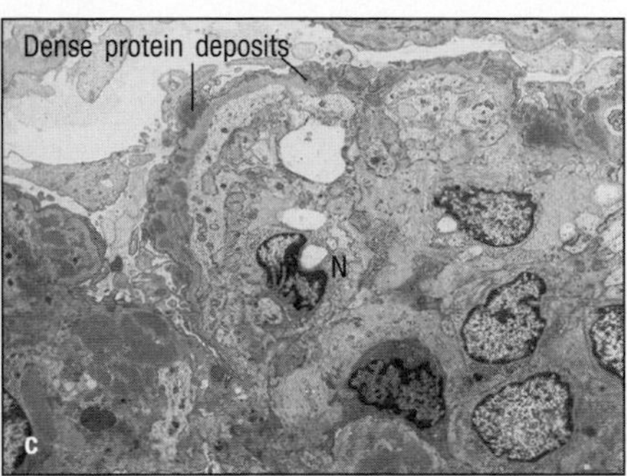

Figure 16.20 Deposition of immune complexes in the kidney glomeruli in systemic lupus erythematosus (SLE). Panel a shows a section through a glomerulus of a patient with SLE. Deposition of immune complexes causes thickening of the basement membrane. In panel b a similar kidney section is stained with fluorescent anti-immunoglobulin antibodies, revealing the presence of immunoglobulin in the basement membrane deposits. Panel c is an electron micrograph of part of a glomerulus. Dense protein deposits are seen between the glomerular basement membrane and the renal epithelial cells. Neutrophils (N) are also present, attracted by the deposited immune complexes. Photographs courtesy of H.T. Cook and M. Kashgarian.

16-8 Intermolecular epitope spreading occurs in systemic autoimmune disease

Systemic lupus erythematosus (SLE) is an autoimmune disease characterized by IgG antibodies specific for a wide range of self antigens present in numerous cells of the body (see Section 16-1). These include constituents of cell surfaces, cytoplasm, and nucleus, including nucleic acids and nucleoprotein particles. This broad and destructive antibody response develops gradually by a general process of **epitope spreading** that involves intramolecular epitope spreading but also an expansion of the number of cellular constituents that become autoantigens. The latter process is called **intermolecular epitope spreading**. At the beginning of the autoimmune response, the binding of autoantibodies to cell-surface components initiates inflammatory reactions that cause cell killing and initiate tissue disruption. Damaged and dying cells release soluble cellular antigens that form soluble immune complexes. On being deposited in blood vessels, kidneys, joints, and other tissues, such complexes initiate further inflammatory reactions (Figure 16.20). Damaged and dying cells also release particles containing a variety of macromolecules that are potential targets for autoantigens. Once an antibody has been made against one component of a particle, that antibody can deliver the particle to cells and facilitate the development of antibodies against the other components. This process of escalation, by which the autoimmune response is expanded in scope and strength, can affect every tissue of the body. The disease commonly follows a course in which outbreaks of intense inflammation alternate with periods of relative calm. For individual patients, the course of the disease is highly variable, both in its severity and in the organs and tissues involved. Many patients with SLE eventually die of the disease because of failure of vital organs such as the brain or the kidneys (see Figure 16.20).

Characteristic of SLE are antibodies against nucleic acids and protein components of nucleoprotein particles. The antibody specificities depend on the HLA class II type. HLA-DR3 confers the greatest susceptibility to SLE, and HLA-DR3 patients make antibodies against proteins of a small cytoplasmic ribonucleoprotein complex. In contrast, HLA-DR2 patients make antibodies against double-stranded DNA, and HLA-DR5 patients make antibodies against the spliceosome, a nuclear ribonucleoprotein complex. In each of these responses, the initiating event is a loss of T-cell tolerance and the activation of a clone of autoreactive T cells that is specific for a peptide derived from the nucleoprotein complex and presented by the HLA-DR molecule. Cells from

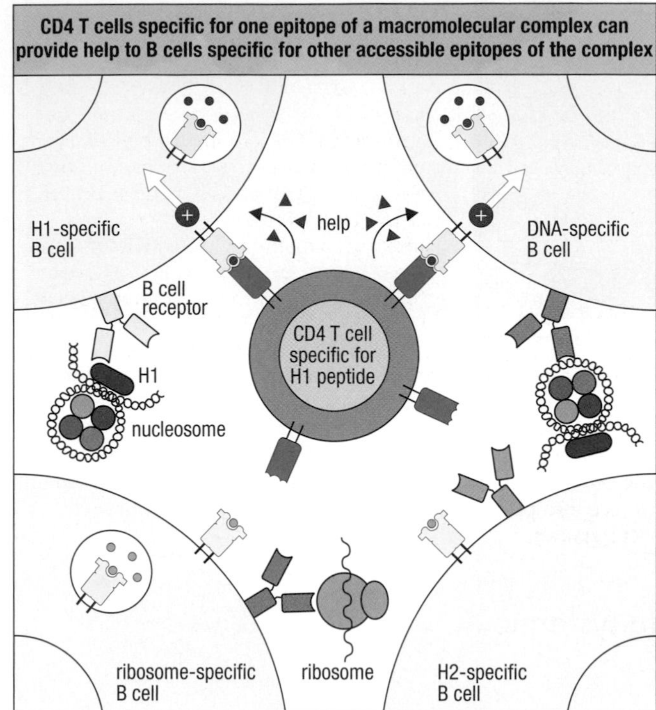

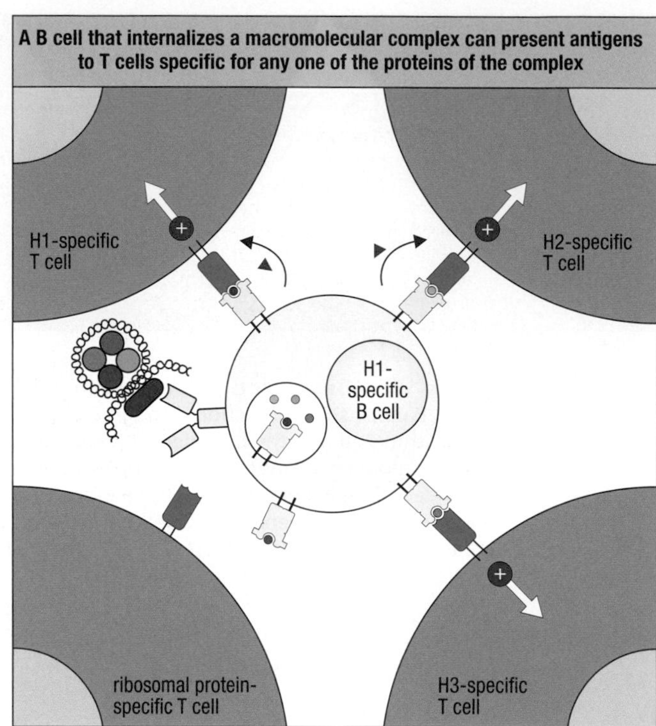

Figure 16.21 In systemic lupus erythematosus (SLE) the immune response is broadened in a antigen-specific manner. In patients with SLE, an ever-broadening immune response is made against nucleoprotein antigens such as nucleosomes, which consist of histones and DNA and are released from dying and disintegrating cells. The left panel shows how the emergence of a single clone of autoreactive CD4 T cells can lead to a diverse B-cell response to nucleosome components. The T cell in the center is specific for a specific peptide (red circles) from the linker histone H1, which is present on the surface of the nucleosome. The B cells at the top are specific for epitopes on the surface of a nucleosome, on H1 and DNA, respectively, and thus bind and endocytose intact nucleosomes, process the constituents, and present the H1 peptide to the helper T cell. Such B cells will be activated to make antibodies, which in the case of the DNA-specific B cell will be anti-DNA antibodies. The B cell at the bottom right is specific for an epitope on histone H2, which is hidden inside the intact nucleosome and is thus inaccessible to the B-cell receptor. This B cell does not bind the nucleosome and does not become activated by the H1-specific helper T cell. A B cell specific for another type of nucleoprotein particle, the ribosome (which is composed of RNA and specific ribosomal proteins), will not bind nucleosomes (bottom left) and will not be activated by the T cell. In reality, a T cell interacts with one B cell at a time, but different members of the same T-cell clone will interact with the B cells of different specificity. The right panel shows the broadening of the T-cell response to the nucleosome. The H1-specific B cell in the center has processed an intact nucleosome and is presenting a variety of nucleosome-derived peptide antigens on its MHC class II molecules. This B cell can activate a T cell specific for any of these peptide antigens, which will include those from the internal histones H2, H3, and H4 as well as those from H1. This H1-specific B cell will not activate T cells specific for peptide antigens of ribosomes, because ribosomes do not contain histones.

this one clone of T cells are able to activate B cells specific for many different surface epitopes of the complex, the only requirement being that the B-cell receptor binds and internalizes the complex and presents the peptide recognized by the T cell (Figure 16.21, left panel). This mechanism allows peptide-specific T cells to help B cells make high-affinity antibodies against nucleic acids, macromolecules not recognized by T cells.

Once activated, B cells will broaden the T-cell response. By presenting peptides derived from all the proteins in the nucleoprotein complex, the B cells activate T-cell clones specific for those peptides. This activation is achieved either through cognate interactions between B cells and T cells (Figure 16.21, right panel) or by antibody binding to the nucleoprotein complex and facilitating its uptake, processing, and presentation by a macrophage or a dendritic cell.

Uses of intravenous immunoglobulin in autoimmune disease		
Benefit	**Disease**	**Symptoms**
Definitely beneficial	Graves' ophthalmopathy	Bulging eyes
	Immune thrombocytopenia	Loss of platelets, bleeding, poor blood clotting
Probably beneficial	Dermatomyositis and polymyositis	Muscle weakness, skin rash
	Autoimmune uveitis	Inflamed eye, blurred vision
Possibly beneficial	Severe rheumatoid arthritis	Joint erosion, pain, loss of mobility
	Type 1 diabetes	Loss of insulin production, severe metabolic disorder
	Systemic lupus erythematosus	Joint pain and swelling, butterfly rash, fatigue
	Post-transfusion purpura	Loss of platelets after a blood transfusion
	Autoimmune neutropenia	Loss of neutrophils, increased susceptibility to infection
	Autoimmune hemolytic anemia	Loss of red blood cells, fatigue
	Autoimmune hemophilia	Bleeding into tissues and joints

Figure 16.22 Intravenous immunoglobulin as a treatment for autoimmune disease. Although there is an extensive literature describing the use of intravenous immunoglobulin as a treatment for autoimmune disease, definitive clinical trials are few and thus the treatment is officially approved for only a few conditions. It is, however, often prescribed 'off label.' Shown here is a summary of autoimmune diseases for which intravenous immunoglobulin treatment is definitely beneficial, probably beneficial, and possibly beneficial.

16-9 Intravenous immunoglobulin is a therapy for autoimmune diseases

In previous chapters we saw how **intravenous immunoglobulin (IVIG)** is given to patients with inherited immunodeficiencies that prevent them from making their own antibodies (see Section 13-11). Until 1980, the antibody-replacement function of IVIG was its only application. But around that time, treatment of a child with Wiskott–Aldrich syndrome by IVIG was seen to cause a significant increase in the number of blood platelets. This observation led to the use of IVIG as a treatment for **immune thrombocytopenia**, a rare bleeding disorder in which autoantibodies and immune complexes cause platelet destruction and perturb platelet production. If the number of platelets falls below 10^9 per liter of blood, severe, spontaneous bleeding ensues. The success of this treatment stimulated the application of IVIG to other autoimmune diseases, with varying results (Figure 16.22).

The beneficial effects of IVIG on autoimmune disease require doses of 1–3 g per kilogram body, weight in contrast with the dose of 0.5 g/kg used for antibody replacement. This high dose is required because IVIG alleviates autoimmune disease by complete saturation of the binding sites for IgG on all the Fcγ receptors. This prevents the autoantibodies and their complexes with antigen from engaging effector functions. In effect, phagocytosis mediated by Fcγ receptors is shut down. The function of FcRn is similarly overwhelmed by IVIG, which has the effect of reducing the half-life of IgG and increasing the rate at which self-reactive IgG antibodies are cleared from the circulation. By similar saturation of the binding sites for IgG on C1q, complement activation is also inhibited. IVIG also induces increased expression of the inhibitory receptor FcγRIIB, which can contribute to the inhibition of phagocytosis. These functions of IVIG as a treatment for autoimmunity are due to the Fc parts of the immunoglobulin molecules, and the effects can be replicated when equivalent quantities of Fc fragments alone are injected intravenously (Figure 16.23).

The abnormally high level of circulating IgG that is achieved with IVIG has a generally suppressive effect on immunoglobulin synthesis, so that the production of autoantibodies is reduced. Antigen presentation by B cells is also suppressed, meaning that epitope spreading and expansion of the autoimmune

Functions of intravenous immunoglobulin
Saturates Fc receptors and inhibits Fc receptor-mediated phagocytosis
Saturates FcRn, inhibits recycling of IgG, increases clearance of IgG and reduces its half-life in the blood
Upregulates expression of inhibitory FcγRIIB and further inhibits phagocytosis
Contains anti-idiotypic antibodies that neutralize autoantibodies made by the patient
Suppresses immunoglobulin production including production of autoantibodies
Contains helpful autoantibodies, for example anti-BAFF, that prevent B-cell survival
Downregulates antigen presentation
Attenuates complement activation

Figure 16.23 The immunomodulatory effects of intravenous immunoglobulin.

response are inhibited. IVIG, which is a mind-boggling mixture of antibodies from as many as 60,000 blood donors, contains some antibodies with specificities that contribute to suppressing the autoimmunity. One example is antibody that binds to the antigen-binding site of an autoantibody and prevents it from binding to the autoantigen. Antibodies that bind to the antigen-binding site of another immunoglobulin are called **anti-idiotypic antibodies**. A second example is an antibody that binds to BAFF, the cytokine that promotes B-cell survival (see Section 6-14). The antibody inhibits BAFF activity, causing premature death of the B cells making autoantibodies, as well as B cells making other antibodies.

IVIG alleviates the symptoms of autoimmune disease in several ways. As described above, it attenuates the function of existing autoantibody by decreasing its half-life in the circulation and by preventing it from recruiting effector functions. It prevents B cells and plasma cells from making fresh supplies of existing autoantibodies, and it suppresses the activation of naive autoreactive B cells. IVIG is an exceptionally complicated therapeutic—the opposite of a monoclonal antibody—and one for which applications and immunomodulatory effects are being intensively explored. Since the discovery of the effect of IVIG on autoimmune disease, the amount of IVIG used worldwide has steadily increased from 300 kg in 1980 to 100,000 kg in 2010.

Rheumatic diseases caused by autoimmunity
Systemic lupus erythematosus (SLE) Rheumatoid arthritis Juvenile arthritis Sjögren's syndrome Scleroderma (progressive systemic sclerosis) Polymyositis–dermatomyositis Behçet's disease Ankylosing spondylitis Reiter's syndrome Psoriatic arthritis

Figure 16.24 Rheumatic diseases are autoimmune in nature.

16-10 Monoclonal antibodies that target TNF-α and B cells are used to treat rheumatoid arthritis

Common sites of disease in SLE are the joints, where the deposition of immune complexes causes inflammation and arthritis. More than 90% of patients with SLE suffer from arthritis, and it is often the first symptom to be noticed. SLE is one of several rheumatic diseases, almost all of which are autoimmune in nature (Figure 16.24).

Rheumatoid arthritis is the most common rheumatic disease, affecting 1–3% of the US population, with women outnumbering men by three to one (see Figure 16.10). The disease involves chronic and episodic inflammation of the joints (Figure 16.25), usually starting between 20 and 40 years of age. Some 80% of patients with rheumatoid arthritis make IgM, IgG, and IgA antibodies specific for the Fc region of human IgG. **Rheumatoid factor** is the name given to these anti-immunoglobulin autoantibodies.

The synovium of an arthritic joint is infiltrated with leukocytes. These include neutrophils, macrophages, CD4 and CD8 T cells, B cells, lymphoblasts, and plasma cells making rheumatoid factor. Prostaglandins and leukotrienes are major mediators of the inflammation. Neutrophils also release lysosomal enzymes into the synovial space, causing tissue damage and inducing proliferation of the synovium. Dendritic cells activate autoimmune CD4 T cells, and they in turn activate macrophages. The activated macrophages accumulate in the inflamed synovium, where they secrete inflammatory cytokines that recruit additional effector cells into the joints, all of which adds to the tissue erosion. Proteinases and collagenases produced by inflammatory cells in a joint can extend the damage to cartilage, to supporting structures such as ligaments and tendons, and eventually to the bones.

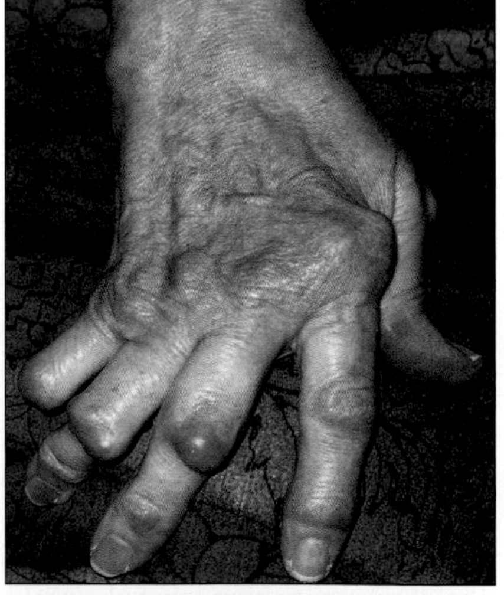

Figure 16.25 Inflamed joints in the hand of a patient with rheumatoid arthritis. Photograph courtesy of J. Cush.

Rheumatoid arthritis is a chronic, painful, and debilitating disease, which patients can suffer for many decades of their lives. Treatment usually combines physiotherapy with anti-inflammatory and immunosuppressive drugs. Infusion with monoclonal antibodies specific for TNF-α alleviates the symptoms of arthritis by eliminating this inflammatory cytokine, which reduces joint swelling and pain (Figure 16.26). For a few patients, anti-TNF-α therapy has terminated their arthritis disease. Widely used anti-TNF-α antibodies are infliximab, a chimeric antibody, and adalimumab, a human antibody. The

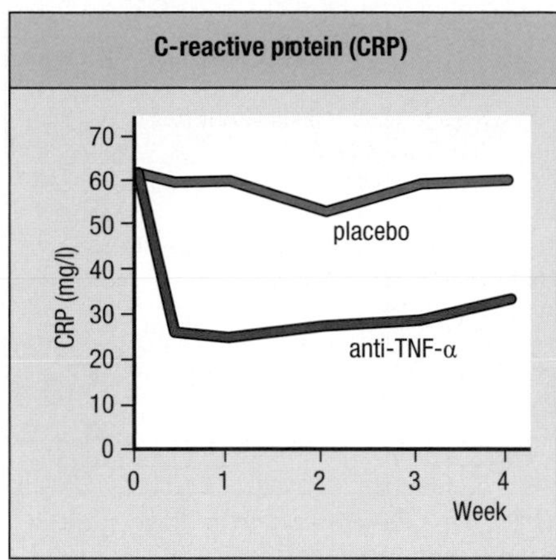

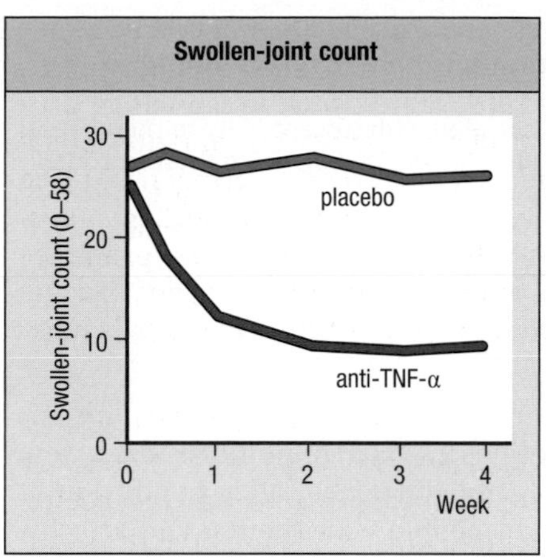

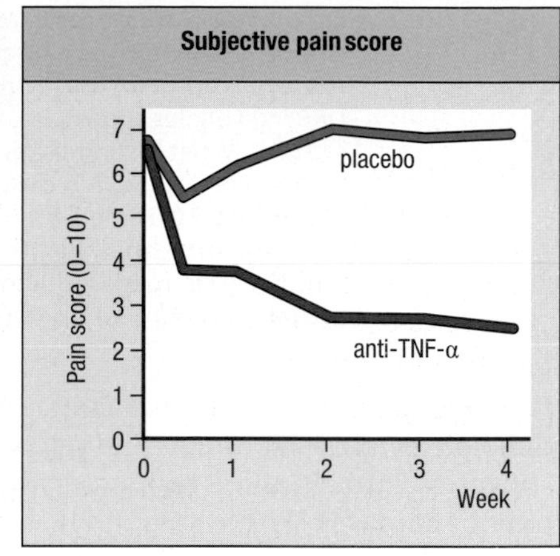

Figure 16.26 The effects of treatment of rheumatoid arthritis with anti-TNF-α. For each parameter measured (level of C-reactive protein, swollen joints, and pain), the values for patients given a placebo treatment are depicted by the blue curve and the values for patients treated with anti-TNF-α antibody are depicted by the red curve.

anti-CD20 monoclonal antibody rituximab is also used as a treatment for rheumatoid arthritis. This antibody, which binds to B cells and makes them targets for NK-cell killing (see Section 9-23), reduces the population of circulating B cells by 98% and gives major benefit to 80% of patients and some benefit to another 30%. These clinical results show that antibodies make a significant contribution to rheumatoid arthritis. As anti-TNF-α and anti-CD20 therapies are increasingly prescribed, their side-effects, such as reduced resistance to infection, are being observed and studied.

16-11 Rheumatoid arthritis is influenced by genetic and environmental factors

The discovery of rheumatoid factor first led to the idea that rheumatoid arthritis was an antibody-mediated disease. Challenging this theory were subsequent observations showing that rheumatoid factor is absent from 20% of patients with rheumatoid arthritis and present in patients with other diseases—30% of SLE patients, for example. Later correlation of the risk of rheumatoid arthritis with the DR4 allotype of the HLA class II molecule HLA-DR promoted the hypothesis that effector CD4 T cells mediate the disease. However, therapies targeted at T cells had no effect on rheumatoid arthritis, whereas therapy targeted at B cells has been successful. These observations are consistent with the activation of an autoreactive T cell being the critical first step of the autoimmune response, whereas antibodies are a major cause of the disease that emerges many years later.

The HLA-DR4 allotype confers susceptibility to rheumatoid arthritis (see Figure 16.9). The α chain of HLA-DR is invariant and the β chain is polymorphic. The DR4 allotype encompasses a set of subtypes that differ by a few amino acid substitutions in the β chain (Figure 16.27). Although small in number, these substitutions influence which peptides are bound by each DR4 subtype and which T-cell receptors can recognize them. Consequently, these substitutions vary the extent to which each subtype correlates with disease susceptibility and resistance. Figure 16.27 shows the differences between DR4 subtypes HLA-DRB1*04:01 and HLA-DRB1*04:04, which confer susceptibility to rheumatoid arthritis, and HLA-DRB1*04:02, which does not and may even be protective. Distinguishing DRB1*04:02 from DRB1*04:01 and *04:04 are negatively charged residues at positions 70 and 71 that confer different peptide-presenting characteristics. It is therefore likely that HLA-DRB1*04:02 cannot present the self antigen that is presented by both HLA-DRB1*04:01 and *04:04 and is recognized by the autoreactive T cell that breaks tolerance and initiates the autoimmune response leading to rheumatoid arthritis.

Figure 16.27 Basic residues in the peptide-binding groove of the DRβ*04 chain are necessary to confer susceptibility to rheumatoid arthritis. With the exception of *DRB1*04:02*, all the *DRB1*04* alleles shown are associated with susceptibility to rheumatoid arthritis. What distinguishes the DRβ*04:02 chain from those encoded by the other alleles is a cluster of amino acid substitutions at positions 67, 70, and 71. These substitutions change the localized charge environment within the peptide-binding groove by removing a basic (positively charged) residue (shown in blue) and inserting two acidic (negatively charged) residues (shown in red).

*DRB1*04 allele*	Amino acid position in DRβ chain		
	67	70	71
**04:01*	L	Q	K
**04:02*	I	D	E
**04:04*	L	Q	R
**04:05*	L	Q	K
**04:08*	L	Q	R

In some people with rheumatoid arthritis, the autoimmune response is directed toward self proteins and self peptides in which arginine residues were converted into citrulline residues by inducible enzymes called peptidyl arginine deiminases (PADs) (Figure 16.28). In people with rheumatoid arthritis who make antibodies against citrullinated epitopes, the association with HLA-DR4 is strong, but in those who lack such antibodies there is no association. This means that two different immunological mechanisms are causing the disease diagnosed as rheumatoid arthritis. Such heterogeneity could explain why only 50% of rheumatoid arthritis patients get major benefit from anti-TNF-α or anti-CD20 therapy.

Citrullinated proteins provide a source of peptide antigens to which the T-cell repertoire is not tolerant. Presentation of citrullinated self peptides by MHC class II molecules could therefore activate specific CD4 T cells. In turn, the CD4 T cells could give help to B cells specific for the citrullinated self protein from which the T-cell epitope was cleaved. The activated B cells need not be specific for citrullinated epitopes, because helper T cells specific for one epitope of a protein can activate B cells specific for all other epitopes of the protein (the phenomenon of linked recognition; see Section 8-19). In the context of this mechanism, HLA-DR4 is expected to be particularly good at presenting citrullinated peptides to T cells. Citrullination makes proteins more susceptible to proteolysis, which also enhances their capacity to stimulate an autoimmune response.

Smoking is the major environmental factor associated with rheumatoid arthritis (Figure 16.29). This effect, however, is only seen for the subset of patients who have antibodies against citrullinated self proteins. Thus smoking, HLA-DR4, and an immune response to citrullinated proteins are all tied together in the same disease-causing mechanism. A working model is that the damage caused by smoking induces the expression of PAD in the respiratory tract, and that this stimulates an autoimmune response to citrullinated self proteins. This immune response is not immediately destined to attack the joints, consistent with observations that antibodies against citrullinated self proteins appear years before the symptoms of arthritis. The joints are attacked later, when some independent trauma such as a wound or an infection induces a state of inflammation in the joint and the activation of PAD. In these circumstances, effector and memory lymphocytes specific for citrullinated self proteins enter the inflamed joint tissue and respond to their specific antigens. The actions of effector T cells and the deposition of immune complexes exacerbate the inflammation and lead to the symptoms of rheumatoid arthritis.

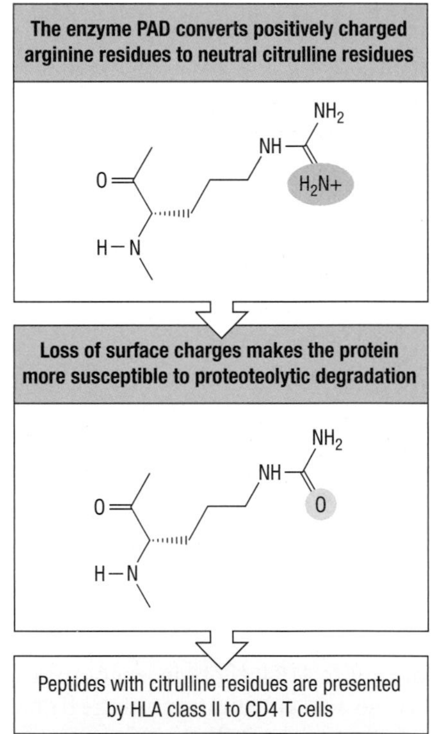
The enzyme PAD converts positively charged arginine residues to neutral citrulline residues
NH_2
NH
O
H_2N+
H–N
Loss of surface charges makes the protein more susceptible to proteoteolytic degradation
NH_2
NH
O
O
H–N
Peptides with citrulline residues are presented by HLA class II to CD4 T cells

Figure 16.28 The enzyme peptidyl arginine deiminase converts the arginine residues of tissue proteins to citrulline. In tissues stressed by wounds or infection, peptidyl arginine deaminase (PAD) activity is induced. By converting arginine residues to citrulline, PAD destabilizes proteins and makes them more susceptible to degradation. It also introduces novel B-cell and T-cell epitopes into tissue proteins that can stimulate an autoimmune response.

16-12 Autoimmune disease can be an adverse side-effect of an immune response to infection

In the previous section we saw how tissue damage caused by habitual smoking sets the stage for an autoimmune response. In analogous fashion, pathogens are also environmental factors that cause inflammation and tissue disruption whenever they invade and trigger an immune response. For almost every autoimmune disease there are clinical observations suggesting that the

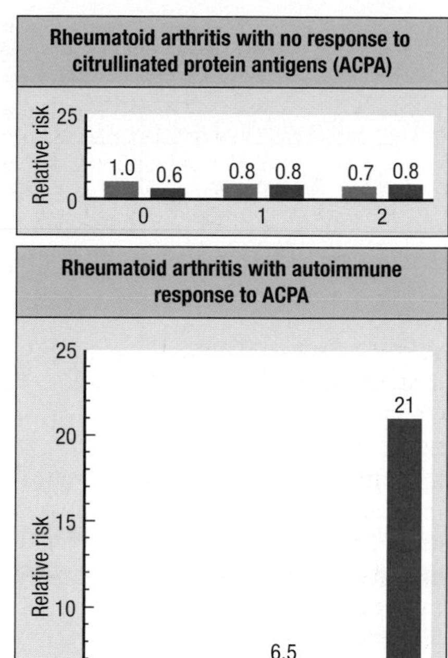

Figure 16.29 Patients with rheumatoid arthritis form two distinct groups. Upper panel: the relative risk of developing rheumatoid arthritis in which no autoimmune response to citrullinated protein antigens (ACPA) is made does not correlate with the presence of *HLA-DRB1*04* 'susceptibility' alleles or with smoking. The red columns indicate smokers, the blue columns nonsmokers. The number above each column is the relative risk of disease for that group. Lower panel: in contrast, the relative risk of developing rheumatoid arthritis in which an autoimmune response to ACPA has been made is increased by the presence of *HLA-DRB1*04* susceptibility alleles and by smoking. At highest risk are smokers who have any two *HLA-DRB1*04* susceptibility alleles (see Figure 16.27). These data are from a cohort of Swedish patients with rheumatoid arthritis. Data courtesy of Lars Klareskog.

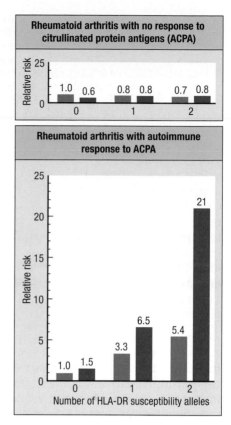

autoimmune response began in the context of the protective immune response made against an infecting pathogen. For some diseases the evidence is compelling; for others it can be correlative, circumstantial, or anecdotal.

Experimental evidence that infection is necessary for inducing autoimmunity comes from the experience of immunologists in inducing autoimmune disease in animals by the injection of tissues, tissue extracts, or purified autoantigens. Injection of these preparations alone does not produce an autoimmune response. However, when the self components are mixed with microbial products that induce inflammation at the site of injection, autoimmunity is reliably produced. In models of spontaneous autoimmune disease, in which no manipulation of the animals is required for the induction of disease, the incidence and severity of disease vary with the conditions under which the animals are housed, and the microorganisms to which they are exposed; there seem to be similar effects for people. The incidence of autoimmune disease in human populations increases with financial wealth, industrial development, and the changing ways of life that follow them.

A simple and well-defined example of an autoimmune disease that is a by-product of the specific response to infection is **rheumatic fever**. This involves inflammation of the heart, joints, and kidneys, which can follow 2–3 weeks after a throat infection with certain strains of *Streptococcus pyogenes*. In defeating the bacterial infection, antibodies specific for cell-wall components of *S. pyogenes* are made. Some of these antibodies happen to react with epitopes present on human heart, joint, and kidney tissue. On binding to the heart they activate complement and generate an acute and widespread inflammation—rheumatic fever—which sometimes causes heart failure (Figure 16.30). Such a chance resemblance of pathogen and host antigens is called **molecular mimicry**. This example shows how health-improving

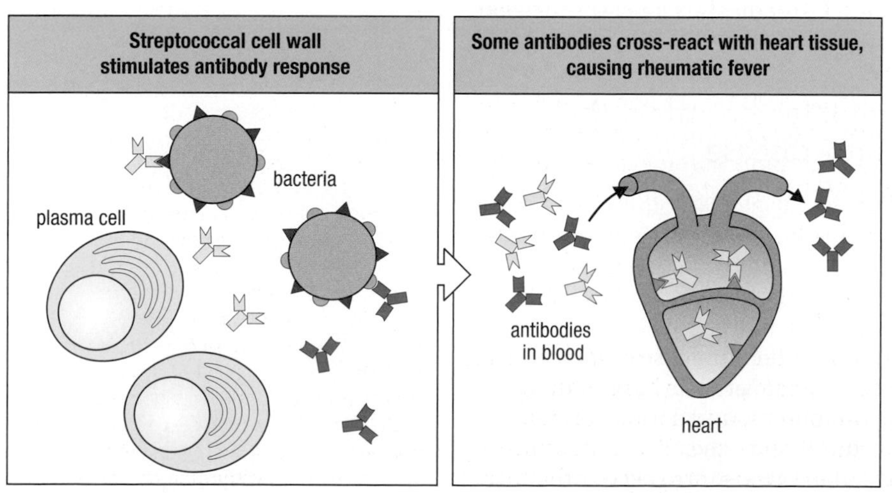

Figure 16.30 Antibodies against streptococcal cell-wall antigens cross-react with antigens on heart tissue. The immune response to the bacteria produces antibodies against various epitopes of the bacterial cell surface. Some of these antibodies (yellow) cross-react with the heart, whereas others (blue) do not. An epitope in the heart (orange) is structurally similar, but not identical, to a bacterial epitope (red).

Associations of infection with autoimmunity		
Infection	**HLA association**	**Consequence**
Group A *Streptococcus*	Not known	Rheumatic fever (carditis, polyarthritis)
Chlamydia trachomatis	HLA-B27	Reiter's syndrome (arthritis)
Shigella flexneri, *Salmonella typhimurium*, *Salmonella enteritidis*, *Yersinia enterocolitica*, *Campylobacter jejuni*	HLA-B27	Reactive arthritis
Borrelia burgdorferi	HLA-DR2, DR4	Chronic arthritis in Lyme disease
Coxsackie A virus, Coxsackie B virus, echoviruses, rubella	HLA-DQ2, HLA-DQ8 DR4	Type 1 diabetes

Figure 16.31 Infections associated with the start of autoimmunity. Lyme disease, a form of arthritis, is caused by *Borrelia* bacteria that are transmitted from rodents to humans by tick bites of the type shown on the opening page of this chapter.

immune responses that successfully control infections can inadvertently become life-threatening autoimmunities. Rheumatic fever is a transient autoimmune disease; lack of T-cell help means that the autoantigens cannot continue to stimulate antibody synthesis. This situation arises because the CD4 T cells that helped in the antibacterial response are not stimulated by the autoantigens. The incidence of rheumatic fever has greatly diminished since streptococcal infections began to be treated with antibiotics.

The transient nature of rheumatic fever emphasizes the necessity of T cells for the chronic autoimmunity that characterizes most autoimmune diseases. T-cell activation occurs only in the presence of inflammation, which is itself instigated by infection. It is therefore plausible that all destructive autoimmune reactions have their origins in infection. Although less well established than for rheumatic fever, bacterial infections have been implicated in Reiter's syndrome and reactive arthritis, two of the arthritic diseases associated with HLA-B27 (Figure 16.31). When cohorts of people are inadvertently infected with the same strain of food-poisoning bacteria, the frequency of autoimmune disease that later develops among those carrying HLA-B27 is higher for the infected group than for uninfected people carrying HLA-B27 in the population as a whole. Coxsackie B virus, which infects the β cells of the pancreas, has been implicated in triggering type 1 diabetes. Candidiasis is a possible trigger of autoimmunity in APECED (see Figure 16.6).

16-13 Noninfectious environmental factors affect the development of autoimmune disease

The fact that genetically predisposed individuals develop autoimmune disease with a maximum frequency of 20% points to the importance of environmental factors in determining who actually gets ill. The habit of smoking tobacco is a nongenetic factor that damages the mucosae of the airways and exacerbates many diseases. All patients with Goodpasture's syndrome (see Section 16-1) develop glomerulonephritis, but only those who habitually smoke cigarettes develop pulmonary hemorrhage as well. In nonsmokers, the basement membranes of lung alveoli are inaccessible to antibodies, and so there is neither deposition of antibody nor disruption of the tissue. Alveoli in the lungs of smokers are chronically damaged from their daily exposure to cigarette smoke.

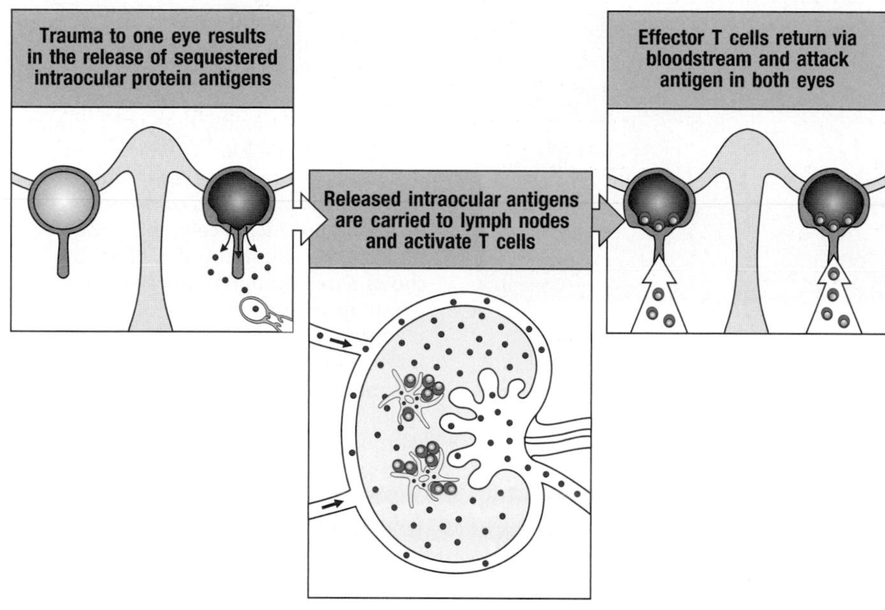

Figure 16.32 Physical trauma to one eye initiates autoimmunity that can destroy vision in both eyes.

This lack of integrity gives circulating antibodies access to the basement membranes, where deposition of immune complexes activates complement, causing blood vessels to burst and hemorrhage.

Physical trauma can expose sites in the body, such as the anterior chamber of the eye, that are normally not accessed by cells of the immune system and contain autoantigens to which they are not normally exposed. If a punch to the face ruptures an eye, proteins unique to the eye drain from the anterior chamber to the local lymph node and there induce an autoimmune response. On occasion, this response causes the damaged eye to become blind. Unfortunately, the undamaged eye also becomes accessible to the autoimmune effector cells and antibodies and will cause blindness there unless treatment is given. This condition is called **sympathetic ophthalmia** (Figure 16.32). That both eyes are attacked shows the immunological privilege of eyes is achieved by mechanisms that prevent the induction of an immune response, rather than by a rigid barrier that prevents all access of immune cells and molecules. To preserve the patient's undamaged eye it is necessary to remove the damaged eye, to prevent further flow of antigens to the lymph node, and to give immunosuppressive drugs to interfere with the ongoing immune response.

16-14 Type 1 diabetes is caused by the selective destruction of insulin-producing cells in the pancreas

Insulin is secreted by the pancreas in response to the increased blood glucose after a meal. On binding to surface receptors, insulin stimulates the body's cells to take up and incorporate glucose into carbohydrates and fats. **Type 1 diabetes**, also called insulin-dependent diabetes mellitus (IDDM) or juvenile-onset diabetes, is caused by the selective autoimmune destruction of the insulin-producing cells of the pancreas. As a major regulator of cellular metabolism, insulin is essential for the normal growth and development of children. Disease symptoms usually manifest themselves in childhood or adolescence, and they rapidly progress to coma and death in the absence of treatment. Type 1 diabetes principally affects populations of European origin, in which one person in 300 is affected. This distribution, and the impact of the disease on young children, has made type 1 diabetes a major target and driver of

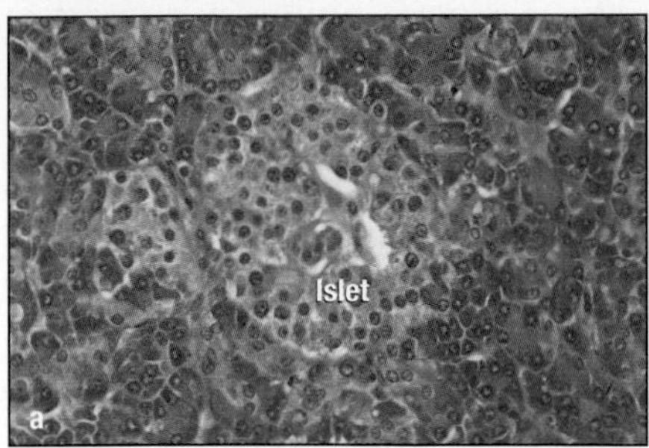

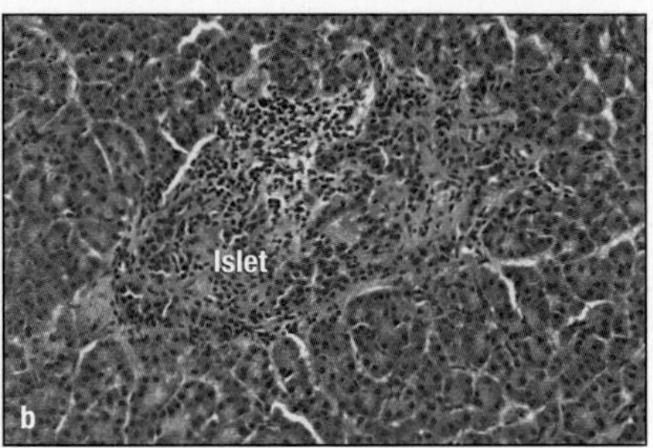

Figure 16.33 Comparison of histological sections of a pancreas from a healthy person and a patient with type 1 diabetes. Panel a is a micrograph of healthy human pancreas, showing a single islet. The islet is the discrete light-staining area in the center of the photograph. It is composed of hormone-producing cells, including the β cells that produce insulin. Panel b shows a micrograph of an islet from a patient with acute onset of type 1 diabetes. The islet shows insulitis, an infiltration of lymphocytes from the islet periphery toward the center. The lymphocytes are the clusters of cells with darkly staining nuclei. Both tissue sections are stained with hematoxylin and eosin; magnification ×250. Photographs courtesy of G. Klöppel.

research in the countries of Western Europe, North America, and Australia. It is no coincidence that insulin was the first protein to have its complete amino acid sequence determined.

Scattered within the exocrine tissue of the pancreas are the **islets of Langerhans**, small clumps of endocrine cells that make the hormones insulin, glucagon, and somatostatin. The pancreas contains about half a million islets, each consisting of a few hundred cells. Each islet cell is programmed to make a single hormone: α cells make glucagon, β cells make insulin, and δ cells make somatostatin. In patients with type 1 diabetes, antibody and T-cell responses are made against insulin, glutamic acid decarboxylase, and other specialized proteins of the pancreatic β cell. Which responses cause the disease remains uncertain. Antigen-specific CD8 T cells are believed to mediate β-cell destruction, gradually reducing the number of insulin-secreting cells. Individual islets become successively infiltrated with lymphocytes, a process called **insulitis**. The β cells comprise about two-thirds of the islet cells; as they die, the architecture of the islet degenerates. A healthy person has about 10^8 β cells, providing insulin-making capacity much greater than the body needs. Because of this excess, and the low rate of β-cell destruction, disease symptoms do not manifest until years after the autoimmune response begins. Disease commences when there are insufficient β cells to provide the insulin necessary to control the level of blood glucose (Figure 16.33). For patients with type 1 diabetes, the usual treatment is daily injection with synthetic human insulin.

16-15 Combinations of HLA class II allotypes confer susceptibility and resistance to type 1 diabetes

Polymorphisms of HLA-DQ and HLA-DR are associated with susceptibility and resistance to type 1 diabetes. The effects of DQ allotypes are stronger and are modified by DR allotypes. Several DQ allotypes are associated with type 1 diabetes, whereas the influence of HLA-DR is limited to the DR4 allotype. We learnt in Chapter 5 that the HLA-DQ α and β chains are both polymorphic and encoded by neighboring genes. In a heterozygous individual, the HLA-DQ molecules are formed both by the α and β chains encoded on the same HLA haplotype and by the α and β chains encoded by different HLA haplotypes. There is, however, preferential association of α and β chains derived from the same haplotype, so that these comprise most, but not all, of the HLA-DQ molecules. Because of this preference, the names of HLA-DQ allotypes—HLA-DQ2, for example—refers specifically to the protein formed by the α and β chains encoded on the same haplotype.

Common Caucasian HLA haplotypes that encode either the DQ2 or the DQ8 allotype confer susceptibility to type 1 diabetes. Heterozygous individuals who have both the DQ2 and DQ8 haplotypes are more susceptible to diabetes than individuals who have only one of them. This augmented susceptibility is due to a novel HLA-DQ heterodimer, assembled only in heterozygous individuals and consisting of the DQ8 α chain DQA1*03 and the DQ2 β chain DQB1*02:01

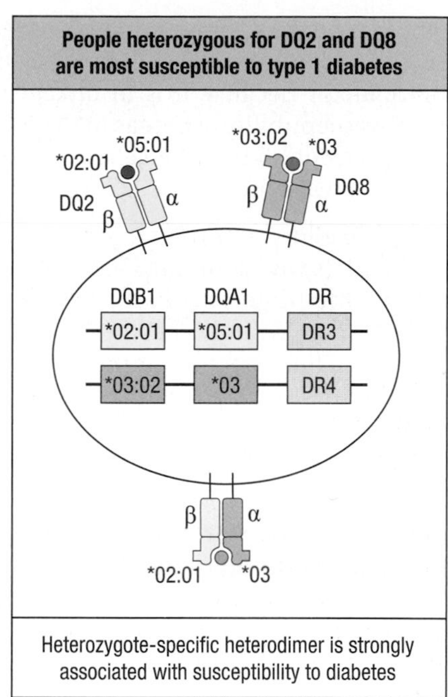

Figure 16.34 Certain HLA heterozygous individuals are more susceptible to diabetes than homozygous individuals. The person shown here has two HLA haplotypes that are independently associated with susceptibility to type 1 diabetes. The *DR3* haplotype contains *DQ* genes that encode the DQα*05:01 chain and the DQβ*02:01 chain; the *DR4* haplotype contains genes that encode the DQα*03 chain and the DQβ*03:02 chain. The two α chains and two β chains made in this person's cells can associate in different combinations to form four different DQ isoforms, of which three (those shown in the figure) are associated with susceptibility to diabetes. The DQ heterodimer associated with the highest susceptibility is that comprising the DQα*03 chain made from the *DR4* haplotype and the DQβ*02:01 chain made from the *DR3* haplotype. This heterodimer can be made only in *DR3*/*DR4* heterozygous individuals, whereas the two heterodimers with weaker disease association are also made in homozygous individuals: heterodimers of the DQα*05:01 and DQβ*02:01 chains (called the DQ2 molecule) in *DR3* haplotype homozygotes and heterodimers of the DQα*03 and DQβ*03:02 chains (called the DQ8 molecule) in *DR4* haplotype homozygotes. The heterozygote is therefore more susceptible to disease than either homozygote. As a general rule, heterozygosity gives increased fitness over homozygosity, but in this situation the reverse is true.

(Figure 16.34). In northern Europeans, this combination of α and β chains is never encoded by genes on the same HLA haplotype, so it is only made by heterozygotes. In contrast, some African HLA haplotypes have the *DQA1*03* allele paired with the *DQB1*02:01* allele on the same haplotype, and these haplotypes confer a similar susceptibility to type 1 diabetes as the combination of DQ2 and DQ8 haplotypes in Caucasians. The DQ4 and DQ9 allotypes also confer susceptibility to type 1 diabetes, but to lesser extent than either DQ2 or DQ8.

The disease susceptibility conferred by the DQA1*03:DQB1*02:01 heterodimer is strongly influenced by the presence of the DR4 allotype, which is in linkage disequilibrium with DQ8. Only the HLA-DR β chain is polymorphic, simplifying its disease associations compared with HLA-DQ. As we saw in the setting of rheumatoid arthritis, DR4 consists of multiple subtypes that differ from each other by a few amino acid substitutions in the β chain (see Section 16-11). As seen in Figure 16.35, single amino acid substitutions can determine

Risk of type 1 diabetes	HLA locus			Amino acid position in DRβ chain			
	DQB1	*DQA1*	*DRB1*	67	71	74	86
Protective	*03:02 *02:01	*03 *05:01	*04:03 *03	L	R	E	V
Moderate	*03:02 *02:01	*03 *05:01	*04:04 *03	L	R	**A**	V
High	*03:02 *02:01	*03 *05:01	*04:05 *03	L	R	**A**	**G**
High	*03:02 *02:01	*03 *05:01	*04:01 *03	L	**K**	**A**	**G**

Figure 16.35 HLA-DR4 subtypes modify the susceptibility to type 1 diabetes conferred by the DQα*03:DQβ*0201 heterodimer. The leftmost panels show the risk of type 1 diabetes for individuals having the haplotype combinations shown in the adjacent column. All four individuals are heterozygotes for *DR3* and *DR4* haplotypes and make the DQα*03:DQβ*02:01 heterodimer. These individuals have different *DRB1*04* alleles, as indicated by the colored boxes. The amino acid substitutions that distinguish the four DRβ*04 chains are shown on the right. The common amino acid residue at each position is shown in gray; the rarer residue is highlighted in black. The associated HLA-DR4 subtype modifies the risk of type 1 diabetes conferred by DQα*03:DQβ*02:01 in qualitative and quantitative ways.

whether a DR4 subtype confers susceptibility or resistance to type 1 diabetes. In Caucasians, the DQ8 allotype (*DQB1*03:02:DQA1*03*) is a strong susceptibility factor because it is in linkage disequilibrium with DR4 subtypes that favor susceptibility, whereas in the Chinese, DQ8 is poorly associated with disease susceptibility because it is in linkage disequilibrium with DR4 subtypes favoring resistance.

The DQ6 allotype confers strong resistance to type 1 diabetes. In heterozygous individuals who combine DQ6 with a second DQ allotype that confers susceptibility, the protection that DQ6 confers is dominant. Consequently, people who have DQ6 rarely get type 1 diabetes. This type of dominance has not been described for other autoimmune diseases. It may be unique to type 1 diabetes, or it could simply reflect the greater breadth and depth to which the immunogenetics of type 1 diabetes has been investigated. The DQ7 allotype is associated with a weak resistance, which in heterozygotes is not dominant over susceptibility allotypes.

The combination of maternal and paternal haplotypes at the HLA-DQ and HLA-DR genes produces a diversity of human genotypes that range from resistance to susceptibility in their contribution to genetic susceptibility and resistance to type 1 diabetes.

16-16 Celiac disease is a hypersensitivity to food that has much in common with autoimmune disease

Celiac disease is an inflammatory disease of the gut mucosa for which the interplay between genetic and environmental factors is understood in some detail. 'Celiac' means pertaining to the abdominal cavity and derives from the Greek word for 'hollow.' The disease affects up to 20% of Caucasians, and is caused by an immune response to the gluten proteins of wheat flour, or the related proteins of barley and rye, all of which are major components of Western diets. CD4 T cells that respond to gluten-derived peptides in the gut-associated lymphoid tissues activate tissue macrophages, which secrete pro-inflammatory cytokines that produce inflammation and tissue damage in the small intestine. With persistent intake of gluten, the inflammation becomes chronic and eventually causes atrophy of the intestinal villi, malabsorption of nutrients, and diarrhea (Figure 16.36). Children with the disease fail to thrive; adults can become anemic, depressed, and prone to other diseases, including intestinal cancer. Once diagnosed, the disease can be halted by strict adherence to a gluten-free diet, with complete recovery of the normal morphology and function of the gut mucosa. However, the disease returns whenever gluten-containing food is eaten. It is quite clear that the major environmental factor that determines the onset of celiac disease is dietary gluten. For this reason the condition is also called gluten-sensitive enteropathy. Characterization of the immune response that causes celiac disease shows that it has much in common with the autoimmune responses in the diseases examined in this chapter (Figure 16.37). The main difference is that the environmental entity that triggers celiac disease—gluten—is known, whereas that is not the case for the autoimmune diseases. In this context of ignorance, celiac disease provides an informative guide, if not a precedent.

16-17 Celiac disease is caused by the selective destruction of intestinal epithelial cells

Genetic predisposition to celiac disease is strong, as shown by the 75% concordance between monozygotic twins. HLA-DQ accounts for 50% of the genetic predisposition, and 39 non-MHC genes account for a further 14%. Many of these genes are also associated with autoimmune diseases. A majority (about 80%) of people with celiac disease have the DQ2 allotype, and most of

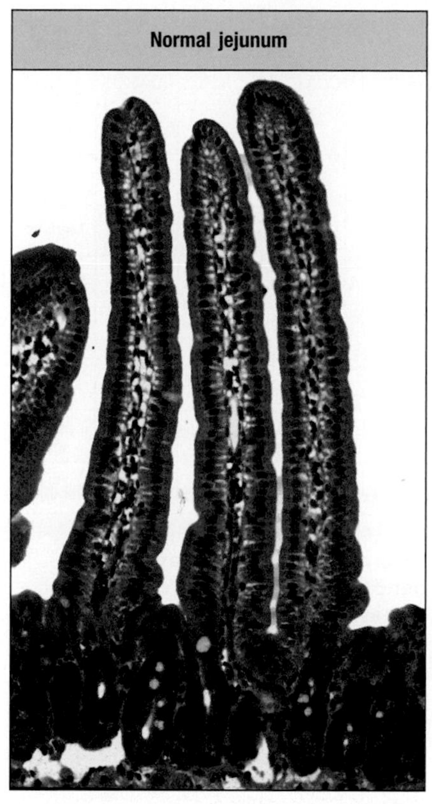

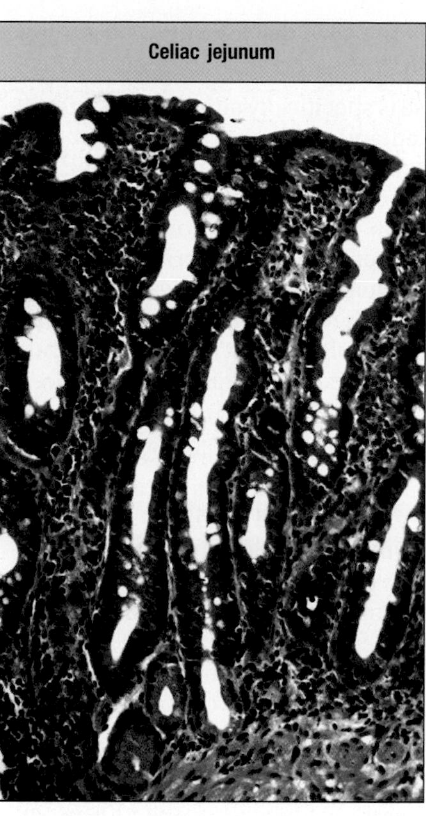

Figure 16.36 Comparison of healthy and celiac intestinal mucosa. Left: the surface of the normal small intestine is folded into finger-like villi, which provide an extensive surface for nutrient absorption. Right: in celiac disease the inflammation and immune response damage the villi. There is lengthening and increased cell division in the underlying crypts to produce new epithelial cells. There are greater numbers of lymphocytes in the epithelial layer and an increase in effector CD4 T cells, plasma cells, and macrophages in the lamina propria. The damage to the villi reduces the person's ability to utilize food and can cause life-threatening malabsorption and diarrhea. Right photograph courtesy of Allan Mowat.

the rest have DQ8. These are the same DQ allotypes that predispose to type 1 diabetes (see Section 16-15). Some 90% of celiac disease patients have the DQ2 molecule comprising the DQB1*02 β chain and the DQA1*05 α chain, and most of the rest have the common form of DQ8, comprising DQB1*03 and

	Immune mechanism								
Type of disease	MHC class II molecules	Auto-antibodies	T_H1 type immunity	Post-translational modifications	Type I IFN	IL-15	IL-21	NK-cell receptors	Conserved MHC class I molecules
Celiac disease	Evidence	Evidence	Evidence	Evidence	Evidence	Evidence	Evidence	Evidence	Evidence
Rheumatoid arthritis	Evidence	Evidence	Evidence	Evidence	Evidence	Evidence	Evidence	Evidence	Evidence
Type 1 diabetes	Evidence	Evidence	Evidence	Evidence	Evidence	Evidence	Evidence	Unclear	Unclear
Multiple sclerosis	Evidence	Evidence	Evidence	Evidence	No evidence	Evidence	Unclear	Unclear	Unclear
Autoimmune thyroiditis	Evidence	Evidence	Evidence	Unclear	Evidence	Evidence	Unclear	Unclear	Unclear
Systemic lupus erythematosus	Evidence	Evidence	Evidence	Unclear	Evidence	Unclear	Evidence	Unclear	Unclear
Primary biliary cirrhosis	Evidence	Evidence	Evidence	Unclear	No evidence	No evidence	No evidence	Unclear	No evidence
Psoriasis or psoriatic arthritis	No evidence	Unclear	Unclear	Unclear	Unclear	Evidence	Evidence	Evidence	Unclear
Inflammatory bowel disease	Evidence	Evidence	Evidence	Unclear	No evidence	Evidence	Evidence	Evidence	Evidence

Evidence for involvement; Unclear evidence for involvement; No evidence for involvement

Figure 16.37 Shared characteristics of the immune mechanisms causing celiac disease and autoimmune diseases.

DQA1*03. The relative risk of susceptibility to celiac disease varies according to the precise HLA genotype (Figure 16.38).

In celiac disease, the autoimmune response is specifically aimed at the intestinal epithelial cells. These cells are attacked and destroyed, but basal membrane is left intact and there are no ulcerations of the tissue. Infiltrating the lamina propria are CD4 T_H1 cells that respond to gluten-derived peptides presented by DQ2 or DQ8 allotypes and secrete cytokines that orchestrate the inflammatory response. There are cells presenting gluten-derived peptides and plasma cells secreting anti-transglutaminase IgA antibodies. Unusually large numbers of cytotoxic CD8 cells infiltrate the epithelium. These cell populations are undetectable in the gut tissue of people without the condition, nor are they found in the gut tissue of patients who have embraced a gluten-free diet and no longer have symptoms of disease.

DQ2		DQ8		Relative risk of celiac disease
DQB1*02	DQA1*05	DQB1*03	DQA1*03	
+	+	+	+	High
+ +	+			High
		+	+	Medium
+	+			Medium
+ +				Medium
+				Low
	+			Very low
				Very low

Figure 16.38 Comparison of genotypes and risk for celiac disease. Shown here are how various combinations of two DQα chains and two DQβ chains give genotypes with a range of relative risk for celiac disease.

Gluten gives bread dough its elastic, gluey properties and comprises two families of proteins rich in glutamine and proline—the glutenins and the gliadins. Present in celiacs are T cells that recognize glutenin- and gliadin-derived peptides presented by HLA-DQ. Distinguishing these peptide epitopes are chemical modifications made in the gut. Particular glutamine residues in the wheat proteins are converted to glutamate by the human enzyme tissue transglutaminase. These changes, which give the peptides additional negative charges, allow the peptides to bind to the positively charged binding pockets of the DQ2 and DQ8 allotypes. The modified gluten peptides in celiac disease are analogous to the citrullinated peptides in rheumatoid arthritis (see Section 16-11). Important instigators of disease in some patients are T cells directed against epitopes contained in an exceptionally proline-rich 33-residue gliadin fragment. This fragment resists proteolysis, both by the extracellular digestive enzymes and the intracellular proteases of antigen processing, and binds strongly to tissue transglutaminase. After transglutaminase has converted particular glutamine residues to glutamate, the 33-residue fragment binds to HLA-DQ and activates inflammatory effector T cells (Figure 16.39). By binding to HLA-DQ in different registers, the fragment presents distinctive epitopes to different clones of T cells; these have a synergistic effect in initiating the disease-causing immune response. Once an inflammatory response has begun, it becomes exacerbated by increased production of transglutaminase and of the peptide antigens recognized by disease-causing T cells.

All patients with celiac disease make IgG or IgA autoantibodies specific for tissue transglutaminase. The presence of such antibodies is diagnostic for the disease, which is confirmed by biopsy of the small intestine. HLA-DQ typing

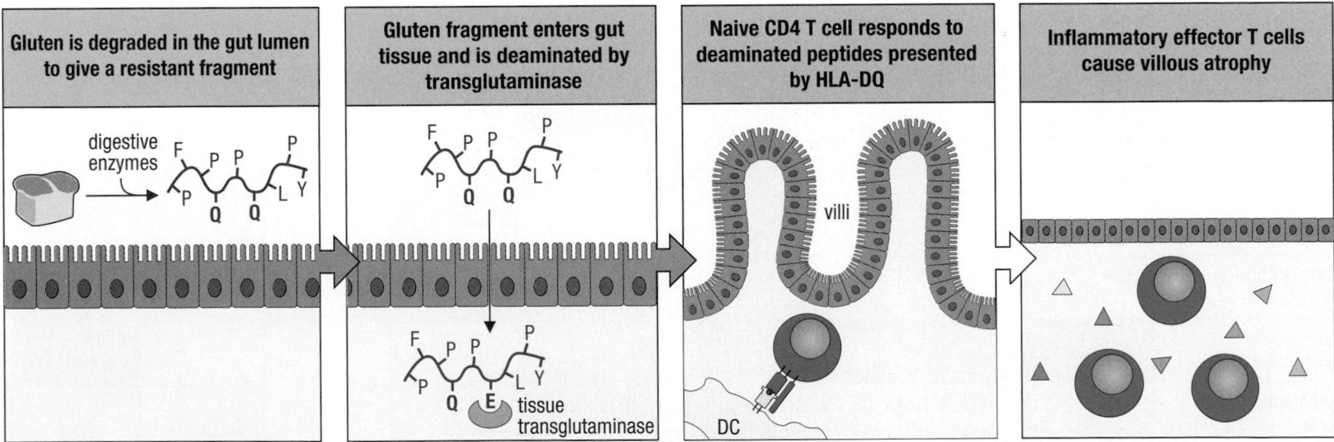

Figure 16.39 The mechanism of celiac disease. In celiac disease, inflammation of the small intestine is caused by a CD4 T-cell response. The T cells are specific for gluten-derived peptides that have been deaminated by tissue transaminase and presented by HLA-DQ8 or HLA-DQ2 molecules. Only part of the peptide epitope is shown. DC, dendritic cell.

can also be informative in diagnosis, because patients who lack DQ2 or DQ8 do not have celiac disease but have some other intestinal disorder. Most celiacs also make anti-gliadin antibodies. These antibodies are likely to be the products of B cells expressing surface immunoglobulins that bind and internalize complexes of transglutaminase and a gliadin fragment and then present modified gliadin fragments on HLA-DQ molecules to T cells.

Although gluten-specific T cells are found specifically in the gut of celiac patients but not in healthy controls, the peripheral blood T cells of both patients and controls contain cells that respond to gluten in culture. Unlike the disease-associated T cells of the gut, the gluten-specific T cells of the peripheral blood recognize a wider range of glutenin and gliadin peptides. These peptides are also presented by a wider range of HLA class II isoforms and are not necessarily modified by transglutaminase. This comparison emphasizes the importance of studying disease-causing mechanisms in the tissues affected by the disease.

As celiac-like diseases do not commonly arise from the consumption of rice or other dietary staples of non-Caucasian populations, the disease is practically specific to Caucasians. But although bread is a staple for almost all Caucasian peoples and about 30% have the DQ2 allotype, only 0.5–1% of them get celiac disease. Other factors apart from gluten must therefore contribute to the disease. Although the concordance rate for celiac disease in monozygotic twins is impressively high at around 75%, the 25% discordance also points to additional environmental factors that set the stage for gluten to trigger an inflammatory T-cell response.

Repeated infections with rotavirus have been linked to early childhood onset of celiac disease, and adults receiving IFN-γ therapy for hepatitis appear more likely to develop celiac disease. In these circumstances, antibody or memory T cells that emerge from an infection could cross-react with gluten epitopes and start a disease-causing immune response. Between 1985 and 1996 there was an epidemic of celiac disease among Swedish children in which the incidence increased fourfold. This was ascribed to changes in infant feeding, with gluten being introduced into the diet at a later time, when the infants were no longer protected by the maternal IgA they had previously obtained in breast milk. The implication is that infants exposed to dietary gluten early and in the presence of maternal IgA become tolerant of gluten, whereas infants exposed later and in the absence of maternal IgA are more likely to respond to gluten as though it were an infecting pathogen.

Our species has existed for some 180,000 years and for most of that time gluten was not a significant part of any human's diet. Only with the adoption of farming around 16,000 years ago were cereals available in quantity to be baked into bread. Thus to human mucosal immune systems gluten would have been perceived as a completely foreign protein to which they were not attuned. The existence, prevalance, and increase of celiac disease 16,000 years later, indicates that human immune systems have still not got completely used to gluten.

16-18 Senescence of the thymus and the T-cell population contributes to autoimmunity

Maintaining T-cell tolerance to self is essential for preventing autoimmune disease. An unanswered question is how this self-tolerance is affected by the gradual decay of the thymus, the source of all T cells. Involution of the human thymus begins soon after birth. Although remaining the same size, there is an inexorable reduction in the part of the thymus that produces new naive T cells, which is first replaced by tissue containing mature T cells and then by fat. At

50 years of age the capacity to produce new T cells has declined to 20% of that at birth, and by age 70 it is practically gone (see Figure 7.4, p. 179). By their nature, T-cell populations are dynamic: T cells must divide periodically to survive, with around 1% of the body's T cells being replaced each day. Once the thymus can no longer fulfill the demand for naive T cells, the immune system compensates by expanding the size of existing T-cell clones and by altering the properties of some T cells to make them more resistant to apoptosis. This latter type of T cell lacks CD28 and expresses KIR and other NK-cell receptors.

Rheumatoid arthritis is a disease associated with people over 50 years of age. With age there is an inverse correlation between the incidence of rheumatoid arthritis and the T-cell-making capacity of the thymus (Figure 16.40). The T cells of people with rheumatoid arthritis have antigen receptors that are one-tenth as diverse as the T cells from healthy people of similar age. In the blood and affected joints of the patients can be found large clones of expanded auto-reactive CD4 T cells that lack CD28 and express NK-cell receptors, notably the activating receptor KIR2DS2. Although these cells are insensitive to co-stimulation, they are not anergic and can be activated through KIR2DS2 to produce large amounts of IFN-γ. These highly inflammatory CD4 T cells are implicated in sustaining the disease. In people with rheumatoid arthritis, the T-cell population seems prematurely aged.

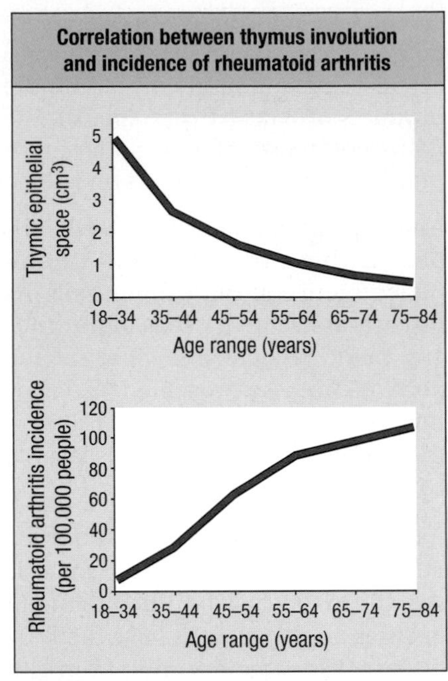

Figure 16.40 Correlation between thymus involution and rheumatoid arthritis. With age there is an inverse correlation between the decreasing capacity of the thymus to make new T cells and the increasing incidence of rheumatoid arthritis. Data courtesy of C.M. Weyand and J.J. Goronzy.

16-19 Autoinflammatory diseases of innate immunity

So far in this chapter we have considered tissue-disrupting diseases caused by the antibodies or effector T cells of the adaptive immune response. With increasing knowledge of innate immunity and advances in genetic analysis, another class of tissue-disrupting diseases has begun to be studied in depth. These diseases, which do not involve antibodies or effector T cells, are characterized by chronic and recurrent bouts of systemic inflammation mediated by cells of innate immunity. To distinguish these diseases from the autoimmune diseases they are called **autoinflammatory diseases**. Like the autoimmune diseases, the autoinflammatory diseases tend to run in families and be more prevalent in some populations and geographical regions than in others. These observations indicate a genetic component to their etiology.

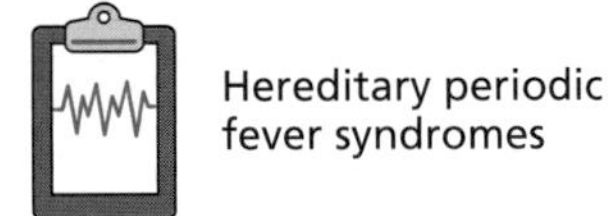

Genetic analyses have identified 12 genes associated with various autoinflammatory diseases (Figure 16.41). The exemplar for this work has been the study of Familial Mediterranean Fever (FMF), a common disease of Middle Eastern populations. Up to one third of a population can be carriers of the autosomal recessive variants of the *MEFV* gene that cause the disease. FMF usually starts in childhood, even in the neonatal period, and is characterized by episodes of fever that last for 1–3 days. Also diagnostic of FMF is a reddish and raised skin rash on the foot, ankle and lower leg, called erysipeloid erythema. The rash is caused by massive influx of neutrophils into the skin. The large quantities of serum amyloid A made in the acute phase response can lead to the deposition of amyloid fragments in vital organs, which accumulate over over time and can compromise their function. Transitory arthritis also occurs during the course of an episode of fever. Initiating attacks can be stress, immunizations or trauma.

The *MEFV* gene encodes a protein called pyrin. This is a component of an inflammasome that is different from the one containing NLRP3 (also called cryopyrin) (see Figure 3.11, p. 56) but has similar function in amplifying production of IL-1β. More than 200 variants of the *MEFV* gene have been found and they involve many different positions in the protein sequence. These variants can change and expand pyrin function, as well as cause loss of function. Although such studies are just beginning, the evidence points to *MEFV* being a highly polymorphic gene that diversifies pyrin function in human populations, as well as producing some combinations of substitutions that predispose to autoinflammatory disease.

Disease	Protein	Dominant (D)/ Recessive (R)	Skin rash	Arthritis	Other
Familial Mediterranean fever	Pyrin	R	+	+	SAA amyloidosis
TNF receptor-associated periodic syndrome	TNF-receptor 1, (CD120a)	D	+	+	SAA amyloidosis
Hyper-IgD syndrome	Mevalonate kinase	R	+	+	Lymph node involvement
Familial cold autoinflammatory syndrome 1	NLRP3 (cryopyrin)	D	+	+	Occular inflammation
Blau syndrome	NOD2	D	+	+	Occular inflammation
Majeed syndrome	Lipin-2	R	+	+	Osteomyelitis
Familial cold autoinflammatory syndrome 2	NLRP12	D	+	+	Early childhood onset
Deficiency of IL-1-receptor agonist	IL-1 receptor agonist	R	+	+	Lymph node involvement
Pyogenic arthritis, pyodema gangrenosum and acne syndrome	CD2 binding protein 1	D	+	+	Early childhood onset
Early-onset enterocolitis (inflammatory bowel syndrome)	IL-10 receptor α IL-10 receptor β	R	+	–	Early infancy onset
Joint contractures, muscle atrophy, macrocytic anemia and panniculitis-induced lipodystrophy syndrome	Proteasome subunit β5i	R	+	+	Chronic

Figure 16.41 Autoinflammatory diseases that have been associated with mutation and variation of genes contributing to innate immunity and inflammation.

Several genes associated with autoinflammatory diseases encode proteins involved in the production, amplification and regulation of IL-1β (see Figure 16.41). These include the genes encoding NLRP3 and CD2 binding protein 1, both inflammasome components, as well as the IL-1 receptor antagonist. Given the central role of IL-1β in innate immunity (see Section 3-6) and the benefit of diversity in the immune response it would be surprising if this aspect of innate immunity is not functionally diverse within the human population.

FMF is an example of a periodic fever. Other periodic fevers are distinguished by the duration of the episodes of fever, as well as by the types of skin rash and arthritis that they cause. Intensive study of inflammatory autoimmune diseases is just beginning and in the next ten years they will likely be seen to be as numerous and as diverse, as the autoimmune diseases.

Summary to Chapter 16

This chapter and the previous two chapters cover three different situations in which the immune system acts in unhelpful ways that cause disease and make human bodies unhealthy. This is achieved by aiming the cells and molecules of adaptive immunity at antigens that pose no threat to human health. The diseases examined in Chapter 14 are the allergies to innocuous plant and animal proteins in the environment and the food we eat, as well as to chemicals such as drugs that we take to improve our health. The responses examined in Chapter 15 are those to the blood transfusions and tissue transplants that are given as therapy to people who would otherwise die, or suffer terminal disease, from loss of blood loss, organ failure or cancer. This chapter examines the chronic autoimmune diseases caused by the activation of T and B cells that recognize normal components of the healthy human body. Although there is considerable clinical diversity in autoimmune disease, allergic disease, and

Hypersensitivity reaction	Allergy	Transplantation	Autoimmunity
Type I	Peanut	–	–
Type II	Chronic urticaria	Hyperacute rejection	Autoimmune hemolytic anemia
Type III	Serum sickness	Chronic rejection	Systemic lupus erythematosus
Type IV	Poison ivy	Acute rejection	Type 1 diabetes

Figure 16.42 Comparison of allergy, transplantation, and autoimmunity. Allergic diseases, the diseases arising from transplantation, and autoimmune diseases all involve effector mechanisms that correspond to the type II, III, and IV hypersensitivity reactions. Unique to allergic disease is the type I hypersensitivity reaction mediated by IgE.

tissues transplanted, the underlying mechanisms are limited to four basic types that correspond to the four classical hypersensitivity reactions originally defined by allergists.

Type I hypersensitivity reactions are IgE-mediated allergies that have no equivalent transplant reaction or autoimmune disease. Chronic urticaria, hyperacute transplant rejection, and autoimmune hemolytic anemia are examples of type II hypersensitivity reactions, which are caused by IgG binding to antigens on plasma membranes, connective tissue and other surfaces. Serum sickness, chronic transplant rejection and systemic lupus erythematosus are examples of type III hypersensitivity reactions and are caused by the deposition of soluble antigen:antibody complexes in blood vessel walls. Poison ivy allergy, acute transplant rejection and type 1 diabetes are all examples of type IV hypersensitivity reactions and are mediated by effector T cells (Figure 16.42). The striking diversity in the clinical manifestation of autoimmune diseases is not a function of the autoimmune response but of the cells and tissues that are attacked, the effect of the attack on their physiological function, and how that impacts the body as a whole. In the affluent, industrialized countries, the incidence of both autoimmune disease and allergy have been increasing in recent decades, and similar trends are being observed in developing countries as they become more affluent and industrialized and adopt a more Western lifestyle.

Some autoimmune diseases are caused by a response that targets one particular organ, tissue, or cell type in a very specific manner. Others, in contrast, are systemic diseases that involve broad immune response directed against components common to all types of human cell. Autoimmune diseases are failures of adaptive immunity and of the central and peripheral mechanisms that make B-cell and T-cell populations tolerant of self. The fact that a large majority of people never suffer from autoimmune disease demonstrates the stringency of the selective and suppressive mechanisms of tolerance. Confirming this assessment is the breadth and severity of the autoimmune diseases affecting immunodeficiency patients who lack regulatory CD4 T cells or have impaired negative selection of autoreactive T cells in the thymus. T-cell tolerance is the gatekeeper of autoimmunity because it is enforced more strongly than B-cell tolerance. Thus the necessary first step in developing an autoimmune response is for one clone of activated T cells to recognize a self antigen. How this happens is uncertain, but a likely possibility is that it occurs inadvertently as a by-product of the immune response to an infection. In the inflammatory and cytokine-rich environment that accompanies infection the presentation of self antigens by MHC molecules is increased and the requirements for T-cell activation are relaxed. In this situation some self-peptide that is normally presented at a low level that goes undetected by T cells might inadvertantly reach a level of presentation that stimulates a T-cell clone. The genetic associations of HLA class I and II polymorphism with autoimmune disease reflect the crucial role that breaking T-cell tolerance has in starting an autoimmune response. The fact that HLA class II associations outnumber HLA class I associations indicates that most autoimmune diseases are started by a self-reactive CD4 T cell.

Autoimmune diseases are complicated multifactorial diseases that arise through a combination of genetic, developmental, and environmental factors. About half the genetic predisposition is determined by HLA type, and the other half consists of much smaller contributions from numerous other immune-system genes. Even for HLA genes, the correlations with disease susceptibility are weak in absolute terms, so that most people with a predisposing HLA type do not develop disease, and not all patients with the disease have the associated HLA type. The same is true for the environmental factors that affect autoimmune disease. The environmental factors are ones that damage the integrity of human tissues and stimulate the adaptive immune system. These comprise trauma, chemicals, irradiation, smoking, and certain foods, all of which may also be accompanied by some form of infection. Identifying the infections that trigger autoimmune diseases is inherently difficult, first because the infections are successfully resolved and rarely recorded, and second because the onset of disease symptoms occurs years to decades after the event that triggered the autoimmune response.

The fact that autoimmune diseases take so long to emerge probably reflects the modest origin of the response with one or a few T clones and the resilience of the human body to the damage they cause. Evidence for the latter is that many autoimmune diseases are characterized by episodes of acute disease that are interspersed with periods of recovery. Because B-cell tolerance is inherently leaky, the autoreactive T cells that break tolerance will immediately be able to activate existing naive autoreactive B cells to become plasma cells secreting autoantibodies. Synergy between the B and T cells allows the breadth of the response to increase by epitope spreading. As the autoimmune response gains strength and begins to damage tissue, a state of chronic inflammation develops. In these circumstances there is increased loss of tolerance, recruitment of additional autoreactive B and T cells, and expansion of the autoimmune response to an increasing range of autoantigens, which can include DNA, RNA, and nucleoproteins.

A major challenge in treating autoimmune diseases is that patients usually come to medical attention only when the destructive effects of autoimmunity are at an advanced stage and when the initiating autoimmune response has become expanded and diversified. With better knowledge of the mechanisms operating in human autoimmune disease it should be possible to identify patients at earlier stages of disease, when tissue damage is limited, the autoimmune response is more narrowly defined, and it should be easier to control the response without resorting to nonspecific immunosuppression.

A newly defined category of diseases are the autoinflammatory dieases that are caused by dysregulation of innate immunity and the inflammatory response. Underlying these conditions are mutations or polymorphism of genes, particularly those involved in the production of IL-1.

Questions

16–1 Identify the mismatched pair.
a. multiple sclerosis: demyelination in central nervous system
b. Graves' disease: receptor antagonist antibodies
c. myasthenia gravis: pyridostigmine treatment
d. Hashimoto's disease: ectopic lymphoid tissues
e. ectodermal dystrophy: hair, teeth, and fingernail abnormalities
f. immune dysregulation polyendocrinopathy enteropathy and X-linked syndrome: requires hematopoietic stem cell transplant
g. almost all autoimmune diseases: HLA complex associations.

16–2 All of the following are associated with ectopic lymphoid tissues except _____.
a. they form by a process known as lymphoid neogenesis
b. they are a characteristic feature of Hashimoto's disease
c. they are encapsulated and contain lymphatics
d. they resemble and function like secondary lymphoid tissues
e. they are also called tertiary lymphoid organs.

16–3 All of the following describe mechanisms that influence tolerance to self-antigens except _____.
a. induction of anergy in peripheral circulation of T and B cells
b. induction of tissue-specific protein expression in the thymus
c. exclusion of T and B cells from tissues protected by immunological privilege
d. negative selection of B cells in the bone marrow
e. negative selection of T cells in the peripheral circulation
f. immunosuppression by T_{reg}.

16–4 When particular alleles of different polymorphic genes are found in HLA haplotypes at higher than expected frequencies, this is referred to as _____.
a. epitope spreading
b. linkage disequilibrium
c. molecular mimicry
d. self-tolerance
e. antigenic shift.

16–5
A. What is the rationale for categorizing the autoimmune diseases as types II, III, and IV? Be specific.
B. In this categorization, why is there no type I autoimmune disease?

16–6 From the list below, identify which disease is not mediated by antibodies that bind to and alter the normal function of cell-surface receptors.
a. insulin-resistant diabetes
b. Goodpasture's syndrome
c. Graves' disease
d. hypoglycemia
e. myasthenia gravis.

16–7 Match the autoimmune disease in column A with its cause in column B.

Column A	Column B
a. APECED	1. loss of acetylcholine receptors
b. Graves' disease	2. loss of FoxP3 function
c. IPEX	3. loss of AIRE function
d. pemphigus vulgaris	4. loss of skin integrity
e. myasthenia gravis	5. overproduction of thyroid hormones.

16–8 _____ is the process involving the generation of antibodies to one part of an antigenic molecule, but with time the antibodies produced bind to different parts of the same molecule.
a. linkage disequilibrium
b. systemic autoimmunity
c. intramolecular epitope spreading
d. intermolecular epitope spreading
e. lymphoid neogenesis.

16–9 Identify the mismatched pair.
a. Reiter's syndrome: coxsackie B virus
b. rheumatoid arthritis: citrullinated proteins
c. sympathetic ophthalmia: physical trauma
d. rheumatic fever: molecular mimicry
e. Goodpasture's syndrome: pulmonary hemorrhage.

16–10 All of the following are correct statements about type 1 diabetes except _____. (Select all that apply.)
a. insulin in produced by the β cells in the islets of Langerhans, while glucagon is made by the δ cells
b. in patients with type 1 diabetes, there are B cells and T cells that bear receptors for specialized proteins of the pancreas
c. DZA1*03:DQB1*02:01 heterodimers confer increased susceptibility to type 1 diabetes and this in strongly influenced by the DR4 allotype
d. DQ6 confers strong resistance to type 1 diabetes, even in heterozygotes who possess a second DQ allotype that confers susceptibility
e. because of differences in linkage equilibrium involving DR4 and DQ8 allotypes, Caucasians are less susceptible to type 1 diabetes than Chinese people.

16–11 All of the following statements regarding celiac disease are true except _____.
a. celiac disease is caused by an immune response to

the glutenins and gliadins found in wheat, rice, barley, and rye
b. there is a very high concordance rate for celiac disease in monozygotic twins, at around 75%
c. all patients with celiac disease make autoantibodies specific for tissue transglutaminase
d. celiac disease does not develop in individuals who lack the DQ2 and DQ8 allotypes
e. early onset of celiac disease in children has been linked to repeated rotavirus infection.

16–12 All of the following are associated with an increased likelihood of developing autoimmune disease except _____.
a. modification of peptides by deamination or citrullination
b. age-related thymic involution
c. smoking
d. possession of particular HLA allotypes
e. isotype switching to IgE influenced by autoreactive T cells
f. having an identical twin who has an autoimmune disease.

16–13 The primary function of the transcription factor AIRE is to _____.
a. facilitate the apoptosis of self-reactive B lymphocytes in the bone marrow
b. activate the expression of tissue-specific proteins in the thymus
c. activate regulatory T cells
d. induce a state of nonresponsiveness in self-reactive T lymphocytes
e. participate in gene expression events required for positive selection of the developing T-cell repertoire.

16–14
A. In general, which genes have been found to correlate best with susceptibility or resistance to autoimmune diseases? (Do not specify individual genes or diseases.)
B. What general hypothesis has been proposed to account for this association?

16–15 Michelle Gilmartin, 10 years old, was seen by a physician 12 months ago because of a petechial rash and a history of nose bleeds (epistaxis). Purpura and petechiae were present in her oral mucosa, extremities, face, and trunk. Blood tests revealed a platelet count of only 10×10^9/l (normal range 150–350 $\times 10^9$/l); all other lab findings were near normal. A diagnosis of acute immune thrombocytopenic purpura (ITP) was supported by raised numbers of megakaryocytes in a bone marrow aspirate and anti-platelet antibodies in the serum. After initially successful treatment with prednisone, Michelle came back to the physician last week with petechial rash over her body, epistaxis that required nasal packing, gross hematuria, and multiple bruises on her anterior shins. Her platelet count was 9×10^9/l. IVIG was administered at 1 g/kg as a single infusion in combination with prednisone. Her platelet numbers increased to 28×10^9/l after 24 hours, 64×10^9/l after 48 hours and 130×10^9/l after 72 hours. All of the following are functions associated with IVIG except _____.
a. blockage of Fc receptors on macrophages in the reticuloendothelial system, especially the spleen
b. neutralization of anti-idiotypic antibodies
c. attenuation of complement activation
d. upregulation of FcRn with the effect of increasing the half-life of IgG in the IVIG treatment
e. provision of anti-BAFF antibodies that leads to decreased survival and premature death of B cells.

16–16 Courtney Povlosky, 19 years old, came to her general practitioner with a 2-month history of increased sweating, insomnia, weight loss, palpitations, and an enlarged thyroid gland. She had hand tremors, and a resting pulse of 140/minute. Her physician ordered thyroid function tests including free thyroxine (T_4), triiodothyronine (T_3), thyroid-stimulating hormone (TSH), and anti-TSH receptor antibodies. Courtney's T_3 and free T_4 levels were very significantly elevated above normal range, whereas her TSH levels were below normal range. The antibody test confirmed the presence of anti-TSH receptor antibodies in her serum. A diagnosis of Graves' disease was made. She was treated with an antithyroid drug (ATD), carbimazole, to control her autoimmune condition. Surgery and radioiodine ablation were not considered because of her favorable response to the ATD. Which of the following explains why patients with Graves' disease have low levels of TSH despite the array of symptoms that are consistent with hyperthyroidism?
a. Antibodies directed against the TSH receptor on the pituitary gland directly suppress the production of TSH, but thyroid hormones are unaffected.
b. In Graves' disease, the overproduction and secretion of non-iodinated thyroglobulin mediates inhibition of pituitary gland function.
c. Excess T_3 and free T_4 stimulate atrophy of the thyroid follicular tissue, which in turn disrupts the negative feedback loop for TSH production.
d. In Graves' disease the thyroid gland is not stimulated by TSH, but instead it is stimulated by agonist antibodies directed against the TSH receptor.
e. Antibodies directed against the TSH receptor inhibit thyroid gland function, which results in reduced secretion of TSH from the thyroid gland.

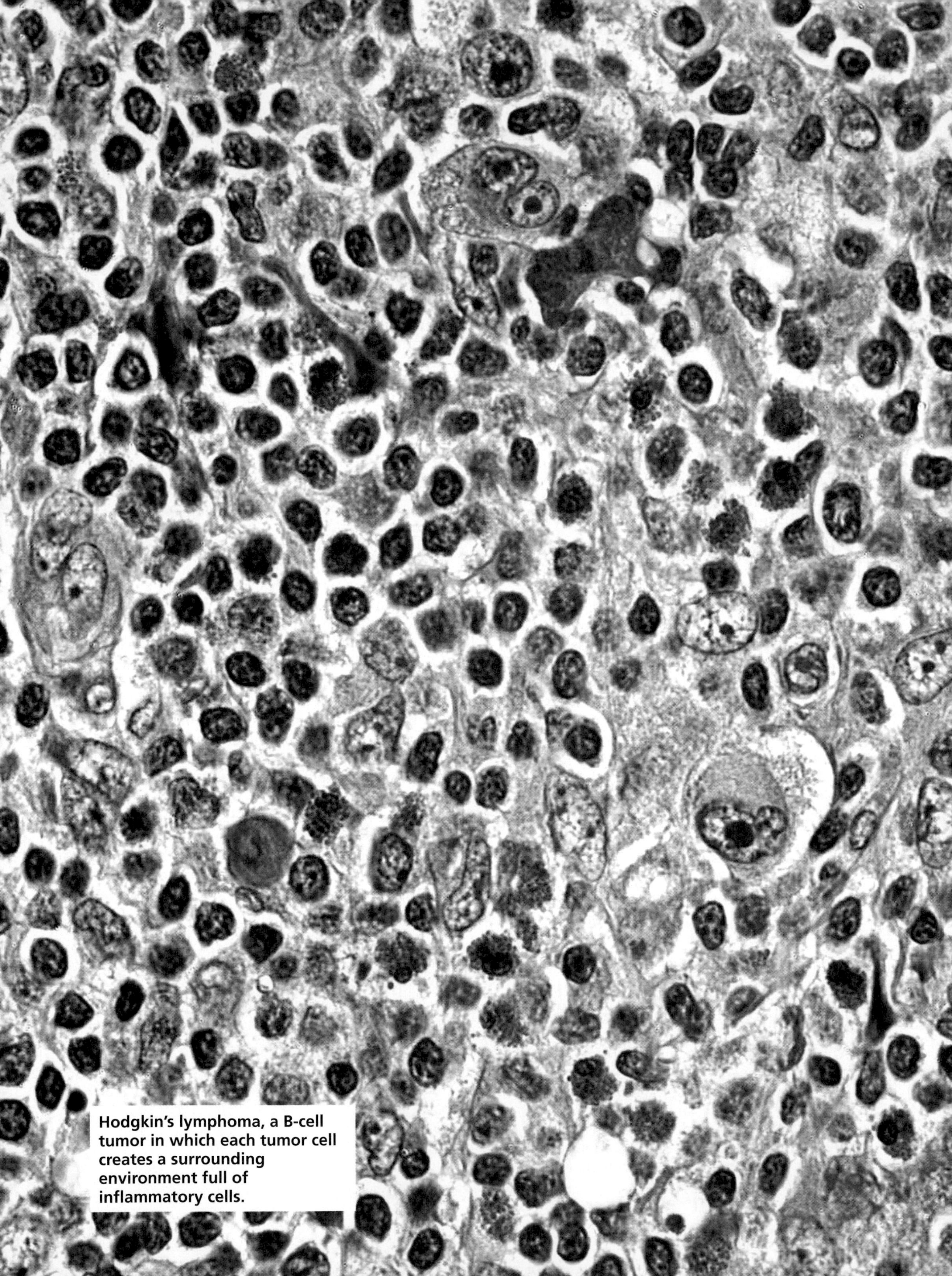
Hodgkin's lymphoma, a B-cell tumor in which each tumor cell creates a surrounding environment full of inflammatory cells.

Chapter 17

Cancer and Its Interactions With the Immune System

Cancer is a diverse collection of life-threatening diseases caused by abnormal and invasive cell proliferation. It accounts for about 20% of deaths in the industrialized countries; worldwide, there are some 14 million new cases of cancer in adults each year, and half of these people will die from the disease. Cancer is largely a disease of older people—the median age for cancer is 70 years—and in the industrialized societies, where both life expectancy and the average age of the population are increasing, so is fear of cancer.

For a normal cell to become a disease-causing cancer involves a lengthy process. It entails the accumulation of several independent mutations, which individually have all escaped the cell's intrinsic safety mechanisms that repair mutated DNA and cause abnormally behaving cells to die by apoptosis. Cancer cells resemble virus-infected cells in that their internal affairs are perturbed and reordered from those of normal cells, and indeed some cancers are the consequences of viral infection. The immune system can identify, control, and eliminate cancer cells by using the same cells and molecules that are employed against virus-infected cells. Most emerging cancers are likely to be eliminated by the immune system before they are detectable or cause any symptoms. It is a minority of cancers that defeat the immune system and progress to cause the diseases that are so feared.

In treating cancer, physicians resort to surgery, radiation, and cytotoxic drugs, sometimes referred to 'slash, burn, and poison.' Although these treatments give remission or cure to some patients, more often they are limited by the incomplete elimination of cancer cells and the deleterious side effects of the treatment. Today, fewer than 10% of patients with metastatic colon cancer can expect to survive for more than 5 years after diagnosis, and for pancreatic cancer the percentage is 5%. By contrast, 85% of patients diagnosed and treated for stage 1 breast cancer survive for at least 10 years. For more than a century, cancer immunologists have sought to harness a patient's immune system to augment conventional therapies. Although ideas for boosting cancer immunity have regularly been shown to work to some extent in animal models, the development of useful routine immunotherapies for human cancer is still a work in progress. Increased understanding of the mechanisms by which an immune response to a cancer is started and sustained has led to several promising new initiatives. In this chapter we first consider the processes by which cancers emerge in the body. Then we examine the immune response to cancer. Finally, we turn to the ways in which the immune system is being manipulated and stimulated to make stronger and more specific responses to cancer cells.

17-1 Cancer results from mutations that cause uncontrolled cell growth

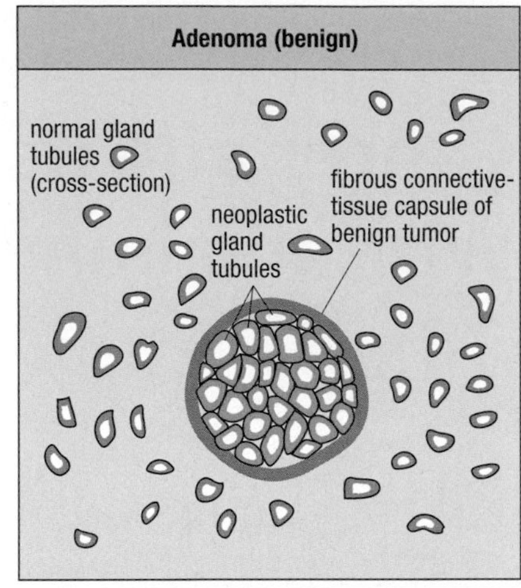

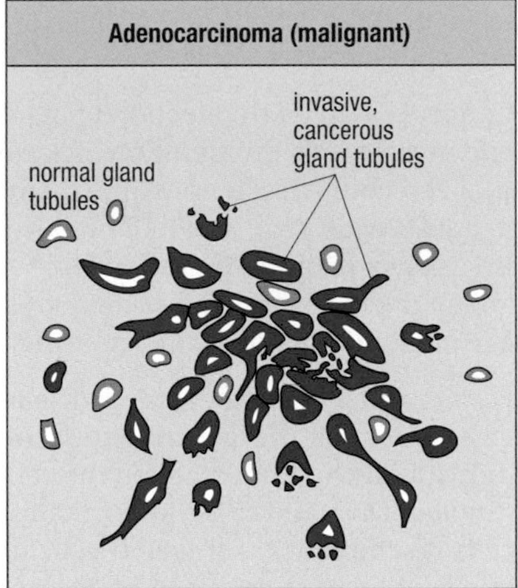

Figure 17.1 Benign and malignant tumors can arise from the same type of tissue. The diagram illustrates tumors found in the breast. Adenoma is the general name given to a benign tumor of glandular tissue; malignant tumors of glandular tissues are called adenocarcinomas.

Maintaining the human body involves continuing cell division to replace worn-out cells, to repair damaged tissue, and to mount immune responses against invading pathogens. On the order of 10^{16} cell divisions are estimated to occur in a human body during a lifetime. An essential preparation for cell division is DNA replication, which makes two identical copies of the diploid genome. Although the enzymes that replicate DNA are extremely accurate and use proofreading mechanisms to correct mistakes, rare errors are still made. Additional changes in DNA arise from chemical damage that escapes the DNA repair machinery.

Changes in DNA are called **mutations**. They include the substitution, insertion, and deletion of nucleotides, recombination between different members of a gene family, and chromosomal rearrangements. Mutations in the germline—the eggs and sperm—provide the variation that allows the human species to diversify and evolve. On average, a child's germline carries 60 mutations not present in either parent's genome. In contrast, mutations in somatic cells affect only the individual in which they arise. Most somatic mutations are never noticed, because their effects, if any, are manifest in a single cell. There are, however, certain mutations that abolish the normal controls on cell division and cell survival. When that happens, the mutant cell proliferates to form an expanding population of mutant cells that eventually disrupts the body's physiology and organ function, causing the diseases that are collectively called cancer.

Tumor, which means swelling, and **neoplasm**, which means new growth, are words used synonymously to describe tissue in which cells are multiplying abnormally. The branch of medicine that deals with tumors is called **oncology**, the prefix *onco* deriving from the Greek word for swelling, *ogkos*. Not all tumors are malignant: **benign tumors**, such as warts, are encapsulated, localized, and limited in size; the **malignant tumors**, in contrast, can continually increase their size by breaking through basal laminae and invading adjacent tissues (Figure 17.1).

The word **cancer**, meaning an evil spreading in the manner of a crab (Latin, *cancer*), is used to describe the diseases caused by malignant tumors. In addition to spreading out locally from their site of origin, cancer cells can be carried by lymph or blood to distant sites, where they initiate new foci of cancerous growth. This mode of spreading is called **metastasis**, the site of origin being called the primary tumor and the sites of spreading being called secondary tumors. Cancers arise most commonly in tissues that are actively undergoing cell division and are thus most likely to accumulate mutations as a result of errors in DNA replication. These tissues include the epithelial linings of the gastrointestinal tract, urogenital tract, and mammary glands (Figure 17.2). Cancers of epithelial cells are known as **carcinomas**; cancers of other cell types as **sarcomas**. Cancers of immune system cells are known as **leukemias** when they involve circulating cells, **lymphomas** when they involve solid lymphoid tumors, and **myelomas** when they involve bone marrow (see Section 6-16).

17-2 A cancer arises from a single cell that has accumulated multiple mutations

The best defenses against cancer lie within each cell of the human body and not with the specialized cells of the immune system. The integrity of the body is so dependent on well-controlled cell division that many mechanisms have evolved to ensure this. These include the repair of certain types of DNA damage as well as mechanisms that prevent the survival and division of cells with

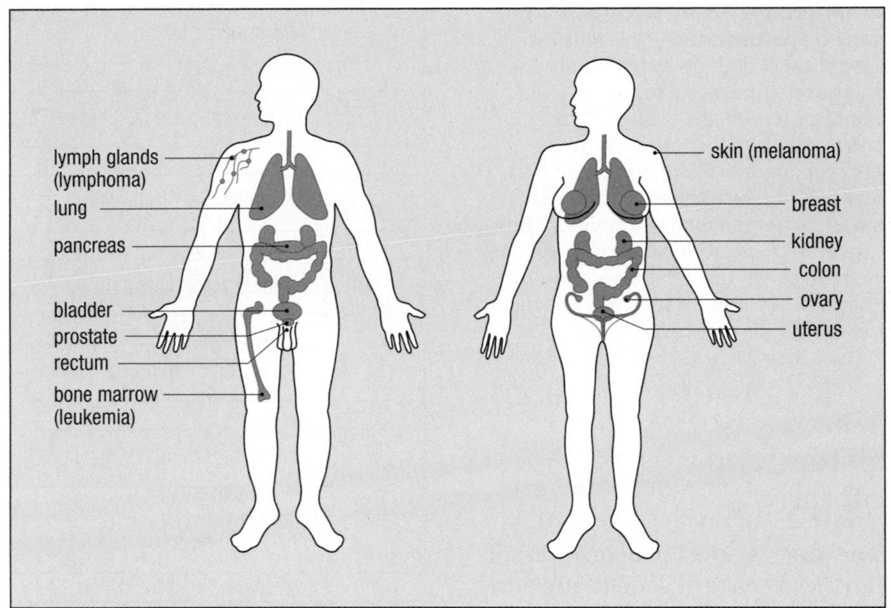

Figure 17.2 Tissues giving rise to the most common cancers in the US population. For simplicity, tissues common to males and females are not all shown on both figures.

badly damaged DNA. Consequently, the control of cell division never depends on the function of just one protein, and a cell cannot become cancerous by a mutation in just one gene. For a cell to give rise to a cancer, it must first accumulate multiple mutations, and these must occur in genes concerned with the control of cell multiplication and cell survival. When a cell becomes able to form a cancer it is said to have undergone **malignant transformation**.

There are two main classes of gene that, if mutated or misexpressed, contribute to malignant transformation. **Proto-oncogenes** are genes that contribute to the normal processes of cell division that are carried out every day in the human body. There are more than 100 human proto-oncogenes; they encode growth factors and their receptors, and also proteins involved in signal transduction and gene transcription. The mutant forms of proto-oncogenes that contribute to malignant transformation are called **oncogenes**.

The second class of genes involved in cellular transformation are called the **tumor suppressor genes** because in their normal functional state they encode proteins that prevent the unwanted proliferation of mutant cells. Exemplary is the tumor suppressor gene that encodes the p53 protein. This protein is expressed in response to DNA damage and causes the damaged cell to die by apoptosis. Loss of the *p53* gene or mutations that interfere with its protective function are the most frequent mutations present in human cancers. More than 50% of cases of human cancer have a mutation in *p53*, revealing its importance in protecting the body from cancer. Indeed, *p53* might well have evolved as an internal cellular defense against cancer.

It has been estimated that a cell must accumulate at least five or six independent mutations before it can become cancerous. The actual number depends on the cell type and the particular genes in which the mutations occur (Figure 17.3). Because of the random manner in which mutations occur, the combination of genetic lesions in each cancer is unique.

The low frequency of mutation, together with the requirement for multiple mutations, means that each cancer inevitably arises from a single cell that has undergone malignant transformation. This explains why in paired organs, such as the lungs, cancer initially affects only one of them. In the course of a lifetime, a person's cells accumulate mutations, and the probability that a cell somewhere in the body will have the right combination of mutations to cause a cancer increases in a nonlinear fashion. For this reason, the incidence of cancer increases with age, and cancers are largely diseases of older people.

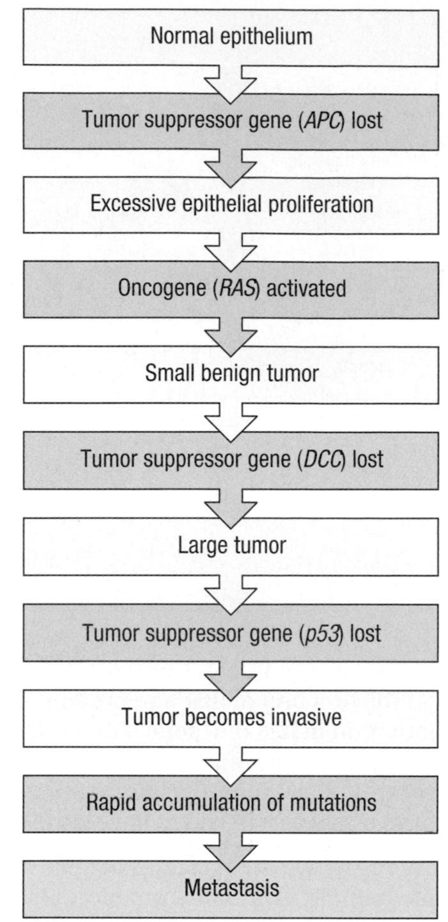

Figure 17.3 A typical series of mutations acquired by a cell while progressing to malignant transformation. This illustration shows the development of colorectal cancer by successive mutations in different genes. The morphological changes accompanying each change are indicated. *RAS* is an oncogene; *APC*, *DCC*, and *p53* are tumor suppressor genes. Both copies of a tumor suppressor gene must be mutated to contribute to malignant transformation, so that seven mutations occur here. These would typically be accumulated over a period of 10–20 years or more. After malignant transformation, the tumor cells rapidly accumulate more mutations. Some of these contribute to making the cells more invasive, but others are thought to be simply a consequence, rather than a cause, of the cell's becoming cancerous. With further mutation the tumor cells become genetically heterogeneous. Natural selection will then favor those cells that divide faster and are more invasive.

17-3 Exposure to chemicals, radiation, and viruses facilitates progression to cancer

Cancer is not, however, an inevitable outcome of aging. Most people never suffer from cancer, even in old age. This reveals the existence of genetic and environmental factors that influence the risk of developing a cancer. Genetic factors include the possession of a germline mutation in one copy of a tumor suppressor gene. Such a mutation in the *p53* gene is the underlying cause of Li–Fraumeni syndrome, a condition in which there is an unusually strong predisposition to multiple cancers that arise at a relatively early age. This is because only one new mutation is needed, in the 'good' copy of *p53*, to abolish all p53 protein function. In other people, at least two new mutations, one in each copy of *p53*, would be needed.

Of greater importance to the population at large are environmental insults that increase the number of mutations suffered by the body. Chemical and physical agents that damage DNA in such a way as to cause an increased rate of mutation are called **mutagens**. Many known mutagens can increase the risk of cancer and are known as **carcinogens**. People who have had heavy or prolonged exposure to carcinogenic agents, which include certain chemicals, ultraviolet light, and other forms of radiation, are more at risk of developing cancer than those who have not been exposed. Most marked is the increased frequency of lung cancer among people who smoke cigarettes.

Chemical carcinogens tend to cause mutations that are due to single nucleotide substitutions in DNA. Radiation, in contrast, tends to produce grosser forms of damage such as DNA breaks, cross-linked nucleotides, abnormal recombination, and chromosome translocations. The radiation released by the atomic bombs that were exploded at Hiroshima and Nagasaki in 1945 caused an increased incidence of leukemia in those who survived the acute effects of the blasts. Less marked, but more pervasive, is the increasing incidence of skin cancer caused by overexposure to ultraviolet radiation from the sun in the pursuit of work, fun, and fashion. However, despite the fear of cancer in developed societies and the well-known risk factors, people in those societies still deliberately engage in behavior that they know increases their chances of developing cancer in later years.

Certain viruses have the potential to transform cells, and viruses are associated with some 15% of human cancers (Figure 17.4). Such viruses are called **oncogenic viruses.** The known oncogenic viruses that affect humans are DNA viruses, except for the RNA retrovirus HTLV-1, which is associated with adult T-cell leukemia.

Human oncogenic viruses typically set up chronic infections in a cell, producing novel virus-encoded proteins that override or interfere with the cell's normal mechanisms for regulating cell division. Infected cells therefore start to

Viruses associated with human cancers		
Virus	**Associated tumors**	**Areas of high incidence**
DNA viruses		
Papillomavirus (many distinct strains)	Genital warts (benign) Carcinoma of uterine cervix	Worldwide Worldwide
Hepatitis B virus	Liver cancer (hepatocellular carcinoma)	Southeast Asia Tropical Africa
Epstein–Barr virus	Burkitt's lymphoma (cancer of B lymphocytes) Nasopharyngeal carcinoma B-cell lymphoproliferative disease	West Africa Papua New Guinea Southern China Greenland (Inuit) Immunosuppressed or immunodeficient patients
RNA viruses		
Human T-cell leukemia virus type 1 (HTLV-1)	Adult T-cell leukemia/ lymphoma	Japan (Kyushu) West Indies
Human immunodeficiency virus (HIV-1) and human herpesvirus 8 (HHV8)	Kaposi's sarcoma	Central Africa

Figure 17.4 Viruses associated with human cancers. Only a minority of the people infected with an oncogenic virus progress to develop cancer. Thus the viruses must act in conjunction with other factors. Some of the viruses probably contribute to cancer only indirectly. For example, Kaposi's sarcoma is restricted to people who are immunosuppressed for other reasons, for example an HIV-1 infection or an organ transplant.

proliferate. For example, the Epstein–Barr virus induces infected B cells to divide repeatedly (see Section 13-4), and if the infected cells are not cleared rapidly, a proportion of them go on to become transformed. Certain strains of papillomavirus predispose to cancer of the cervix. They encode proteins that prevent the normal tumor suppressor mechanisms from acting within the infected cells. Viral proteins bind to p53 protein and to another tumor suppressor protein called retinoblastoma protein (Rb), blocking their functions and enabling the virus-infected epithelial cells to proliferate. Other chronic infections can lead to cancer because the tissue damage they cause necessitates a high rate of cell renewal and, as a consequence, mutations accumulate more rapidly. Of this type are liver cancers arising from chronic infections with hepatitis B or C virus, and the stomach cancers associated with ulcers caused by infection with the bacterium *Helicobacter pylori.*

17-4 Certain common features distinguish cancer cells from normal cells

Cancer comprises a highly heterogeneous set of diseases distinguished by their tissue of origin and their state of differentiation. In Section 6-16, for example, we saw that there are cancers representing all stages of B-cell differentiation. Despite their heterogeneity, all successful and life-threatening cancers share a core set of seven characteristics. Six of these properties have already been mentioned: cancer cells stimulate their own growth; they ignore growth-inhibiting signals from other cells; they avoid death by apoptosis; they become well connected to the blood; they leave their site of origin to invade other tissues; and they inexorably expand their population. The seventh requirement of a successful cancer is for it to outwit the immune system (Figure 17.5).

Necessary characteristics of cancer cells
They stimulate their own growth
They ignore growth-inhibiting signals
They avoid death by apoptosis
They develop a blood supply: angiogenesis
They leave their site of origin to invade other tissues: metastasis
They replicate constantly to expand their numbers
They evade and outrun the immune response

Figure 17.5 All cancer-causing cells have seven key features in common.

An increased incidence of cancer among transplant patients maintained on immunosuppressive drugs shows that the immune system can control or eliminate cancer cells. A study of around 6000 kidney transplant patients in

Scandinavia revealed higher incidences of colon, lung, bladder, kidney, ureteric, and endocrine cancer than in the general population. In these immunosuppressed patients, latent Epstein–Barr virus, which is no threat to normal individuals, can reactivate and become malignant. Similarly, people with immunodeficiencies and lacking key components of the immune system are more likely to get cancer than immunocompetent individuals. On the positive side, the survival of cancer patients correlates directly with the extent to which there is an ongoing adaptive immune response within the tumor mass. Together, these observations point to the immune system's being able to recognize cancer cells and eliminate them before they threaten to cause disease. This function of the immune system—to survey the body for cancer—is called **immunosurveillance** or **cancer immunosurveillance**. The contribution of immunosurveillance to preventing cancer was once disputed, but after decades of skepticism, evidence from the study of humans and mice has now turned opinion in its favor.

17-5 Immune responses to cancer have similarities with those to virus-infected cells

Like viral infections, malignant transformation alters protein expression in cancer cells in ways that make the cells appear foreign to the immune system. Notably, there are changes in the expression of MHC class I molecules that can be detected by the NK cells of innate immunity and the cytotoxic T cells of adaptive immunity. Thus the mechanisms for cancer immunosurveillance are the same as those used for detecting and responding to viral infection.

An important difference is in the tempo of the immune response to cancer compared with that to infection. At mucosal surfaces there is a chronic response to the local microbial populations that will readily activate both innate and adaptive immunity when potential pathogens breach a mucosal barrier. In other tissues, the innate and adaptive responses to an infection are started by inflammation produced at the initial site of infection. These mechanisms make innate immune mechanisms available within hours after the start of infection, and adaptive immune responses are fully operational within 2 weeks. In comparison, the malignant transformation of a single cell that marks the start of a cancer might go unnoticed by the body. For years the growth of a cancer might cause no damage of the sort that triggers inflammation and an immune response (Figure 17.6). In time, the growth of the cancer will begin to damage healthy tissue and raise the alarm of inflammation. In some cases, the immune response will be able to control or even eliminate the cancer, but in others the tumor load will be so great that the immune system will be overwhelmed.

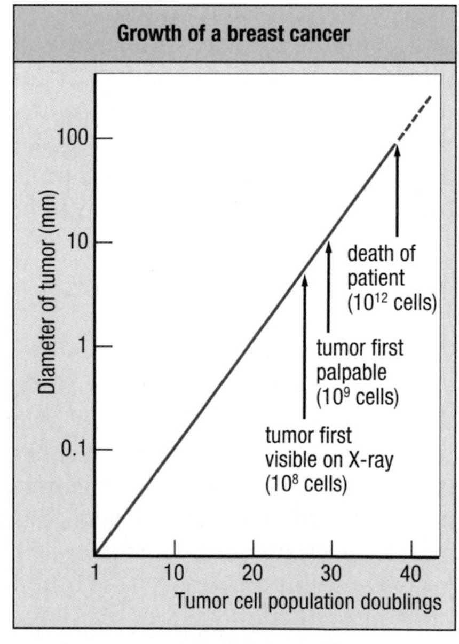

Figure 17.6 The growth and lifespan of a common human tumor. The diameter of the breast cancer is plotted on a logarithmic scale. Many years can elapse before the tumor becomes noticeable to the patient or her immune system.

Early in the twentieth century it was noted that some cancer patients contracting bacterial infections experienced regression—reduction in mass—of their tumors. It was surmised that the inflammation induced by infection had stimulated an immune response to the tumor as well as to the pathogen. A more recent application of this phenomenon is the established treatment for superficial bladder cancer in which chronic inflammation of the bladder is achieved by introduction of the BCG vaccine (see Section 11-21) through a procedure known as intravesicular BCG therapy, in which viable *Mycobacterium bovis* cells are introduced into the patient by bladder instillation. The component of the vaccine that gives it this antitumor effect is the unmethylated CpG-containing DNA of the mycobacteria, the ligand for the Toll-like receptor TLR9 (see Section 3-13). The cells implicated in this antitumor effect are γ:δ T cells secreting IL-17.

As with infections, it is likely that many of the potentially cancerous cells that arise in the human body are detected by immunosurveillance and eliminated at an early stage before they have had any perturbing effect on the anatomy and function of the human body. Only a small minority of such cells are able to grow, multiply, and give rise to clinical cancer.

17-6 Allogeneic differences in MHC class I molecules enable cytotoxic T cells to eliminate tumor cells

Tumors of laboratory mice grow when transplanted to mice of identical MHC type but fail to take hold when transplanted to mice of different MHC type (Figure 17.7). In these animals, the tumor cells are killed by alloreactive CD8 T cells that recognize the allogeneic MHC class I molecules. Similarly, human tumor cells are easily killed by alloreactive CD8 T cells but will also survive passage to an HLA-compatible person—as has occurred in some transplant patients. In one such case, the two kidneys from a deceased woman were given to different, unrelated HLA-matched recipients. Within two years, both the transplant recipients had developed metastatic melanoma. This was nos coincidence. It turned out that 16 years before transplantation the donor had been successfully treated for primary melanoma and had been considered free of tumor cells. On the contrary, however, she must have retained some tumor cells that were controlled by her immune system. When these tumor cells were transferred in the donated kidneys to the immunosuppressed transplant patients they were released from immunological restraint and free to grow uncontrollably. Some cancers are believed to be propagated by minority subpopulations of **cancer stem cells** that are self-renewing and more resistant to the toxins and radiation that are used to treat cancer. Thus treatment for melanoma may have eliminated all the donor's tumor cells except for some cancer stem cells, which remained quiescent and in low numbers until after transplantation.

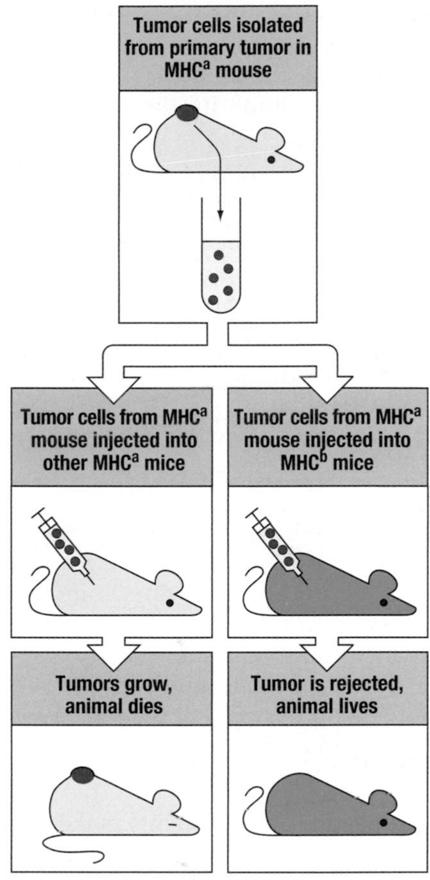

Figure 17.7 When tumors are transplanted between MHC-incompatible mice they are rejected by the alloreactive immune response to MHC differences. The left panels depict transplantation of a tumor between two mice of the same MHC type. The tumor grows in the recipient. In the right panels, the tumor is transplanted to a mouse of a different MHC type and is rejected. Experiments of precisely this type led to the discovery of the MHC, although their intent was to discover tumor-specific antigens.

In general a tumor dies with the death of the individual in which it arose. Tumors that can be passed from one individual to another are uncommon, but they do exist. Passage of venereal sarcoma between domestic dogs occurs in the course of copulation. In the wild, a fatal facial tumor of Tasmanian devils is contagious and usually passed when these aggressive marsupials fight and bite each other's faces (Figure 17.8). The cells of all these tumors are characterized by distinctive chromosomal aberrations showing they are all part of the same clone that arose originally from a transformed Schwann cell in one individual. First observed in 1996, the facial tumor can kill an infected Tasmanian devil in three months, has reduced several populations by up to 95% and threatens the species with an imminent extinction in the wild. Because this tumor is contagious, and of epidemic scale, the disease it causes is now called Devil Facial Tumor Disease (DTFD). Two factors have been identified that likely contribute to the tumor's success. First is that the Tasmanian Devil has a limited MHC polymorphism, which means that the tumor can spread among individuals of identical or similar MHC type. This would have been particularly important during the early development of the tumor's infectivity. The second factor is that the tumor cells express almost no MHC class I at the cell surface. This is not a consequence of structural mutations, but subversion of the regulatory mechanisms of MHC class I gene expression and also of proteins like TAP and β_2-microglobulin that are essential for MHC class I to be

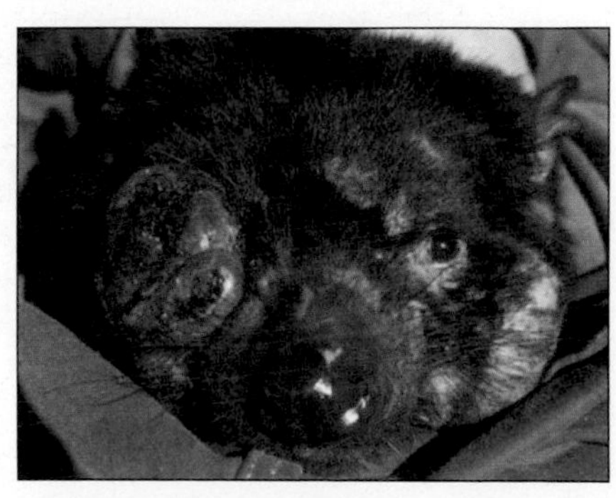

Figure 17.8 The 'infectious' facial tumor of the Tasmanian devil. A native of Tasmania, the smaller of the two main Australian islands, the Tasmanian devil is an aggressive carnivorous marsupial, hence its name. The tumor that disfigures the right eye of this individual is caused by a clone of tumor cells, which passes from one Tasmanian devil to another when they fight and bite each others faces. The transferred tumor cells propagate because of the limited MHC diversity in the population of Tasmanian devils. The tumor cells are MHC-compatible with many individuals and are recognized as self by their immune systems. Photograph courtesy of Menna Jones.

expressed. Because these genes retain their functional potential, treating the tumor cells with autologous interferon-γ induces the expression of Tasmanian Devil MHC class I. Such knowledge is opening up ways for treating infected devils and countering the epidemic of DTFD.

Although human individuals and populations are prone to aggressive behavior that is the equal of the Tasmanian Devil, the high polymorphism of human HLA class I and II genes provides a formidable barrier that prevents the initial growth and adaptation of any tumor cells that are passed from one person to another. Situations in which such transfer could potentially occur are the same as those that serve to spread HIV: intimate contact in love and war, blood transfusion, and sharing of syringes and needles in the non-medical use of drugs.

17-7 Mutations acquired by somatic cells during oncogenesis can give rise to tumor-specific antigens

Somatic cells that undergo malignant transformation have genomic differences that distinguish them from every other cell in the body. Cells transformed by oncogenic viruses are genetically the most distinctive, while at the other end of the spectrum are tumors arising from a transformed cell with point mutations in some half-dozen oncogenes or tumor suppressor genes. As a tumor grows, further mutations occur, introducing genetic heterogeneity into the tumor-cell population. Selection for variant cells with faster growth or increased metastatic potential causes the genomes of the tumor and host to diverge even further.

Almost all cancer patients make an adaptive immune response against their tumor. The antigens to which they respond are called **tumor antigens**. By studying the tumor-reactive antibodies and CD8 T cells made by patients with different types of cancer, more than 1000 different tumor antigens have been defined. Antigens present on tumor cells but not on normal cells are called **tumor-specific antigens**. Antigens expressed on tumor cells but also found on certain normal cells, often in smaller amounts, are called **tumor-associated antigens** (Figure 17.9). Tumor-specific antigens have structures that are not

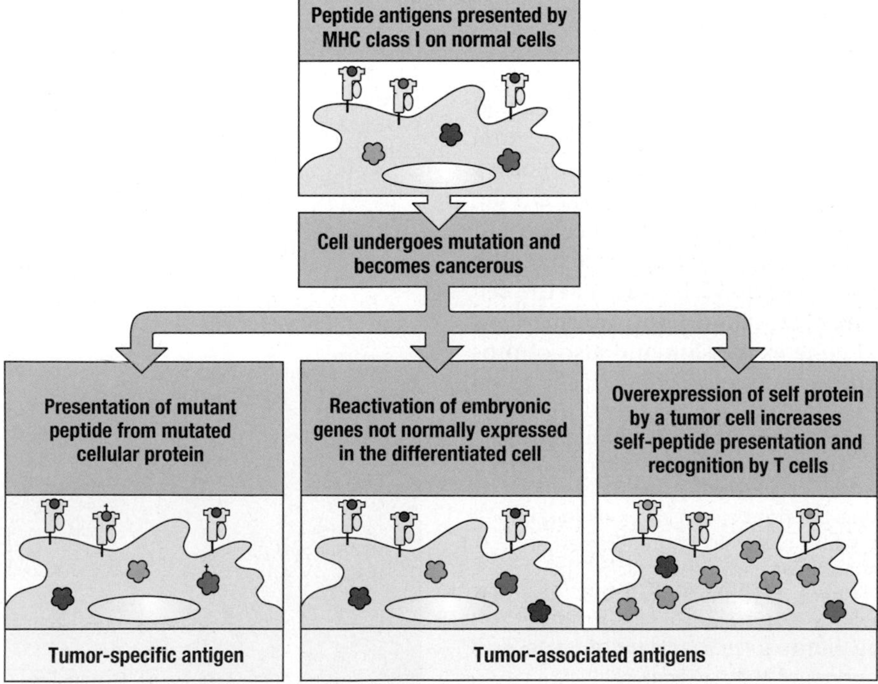

Figure 17.9 Tumor-specific and tumor-associated antigens.

Mutated tumor antigens			
Source of peptide antigen	**Disease**	**HLA restriction**	**Peptide antigen**
MART2	Melanoma	A1	FLEGNEVGKTY
ME1	Non-small cell lung carcinoma	A2	FLDEFMEGV
p53	Head and neck squamous-cell carcinoma	A2	VVCEPPEV
KIAA0205	Bladder tumor	B44	AEPINIQTW
Triosephosphate isomerase	Melanoma	DR1	GELIGILNAKVPAD
BCR–ABL fusion protein	Chronic myelogenous leukemia	DR4 B8 A2	ATGFKQSS KALQRPVAS GFKQSS KAL SS KALQRPV ▲ point of fusion

Figure 17.10 Tumor-specific antigens contain tumor-specific mutations. Shown here are tumor antigens discovered in the course of studying the T-cell response to different types of cancer: five are peptide antigens containing a point mutation (in magenta), and three are peptides derived from the novel protein produced by the fusion of the *BCR* and *ABL* genes that characterizes chronic myelogenous leukemia. *ABL* is an oncogene and *BCR* encodes a kinase (not the B-cell receptor). The HLA molecule that presents each peptide antigen, and thus 'restricts' the immune response, is shown. These data are taken from the Cancer Immunity Peptide Database website [http://cancerimmunity.org/peptide/], which contains many more examples of mutant tumor antigens.

present in any normal cell. They can derive from viral proteins, from the mutated parts of mutant cellular proteins, or from amino acid sequences spanning tumor-specific recombination sites between genes. Some examples of this class of mutant tumor antigens are shown in Figure 17.10. Peptides containing the mutations in proteins such as p53 are bound by HLA class I and II molecules and presented to CD8 and CD4 T cells. Associated with 95% of the cases of chronic myelogenous leukemia is a chromosomal rearrangement that involves a fusion of the kinase-encoding *BCR* gene on chromosome 22 with the *ABL* proto-oncogene on chromosome 9. The protein encoded by the fused genes is instrumental in the progression to cancer, but it also provides novel peptide antigens that span the point of fusion and can be presented by both HLA class I and II molecules. Tumor antigens can also derive from abnormal patterns of protein modification, including glycosylation and phosphorylation, or by translation of unusually spliced mRNAs.

Another way in which 'foreign' peptide antigens can be produced from self proteins is by **peptide splicing**. Several peptide antigens recognized by tumor-specific T cells do not correspond to a continuous linear sequence of amino acid residues in the protein from which the peptide is derived. One clone of CD8 T cells obtained from a melanoma patient recognizes a nonamer peptide derived from a melanocyte glycoprotein that is presented by HLA-A*32:01. This peptide corresponds to residues 40–42 and 47–52 of the glycoprotein. In the course of the glycoprotein's degradation by the proteasome, cleavage after residues 42 and 46 removes residues 43–46, which is followed by the formation of a peptide bond between residues 42 and 47. This peptide does not correspond to a sequence in the native protein (Figure 17.11). In other tumor-specific antigens, the peptides are stitched together in an order that is different from that in the protein providing the epitope.

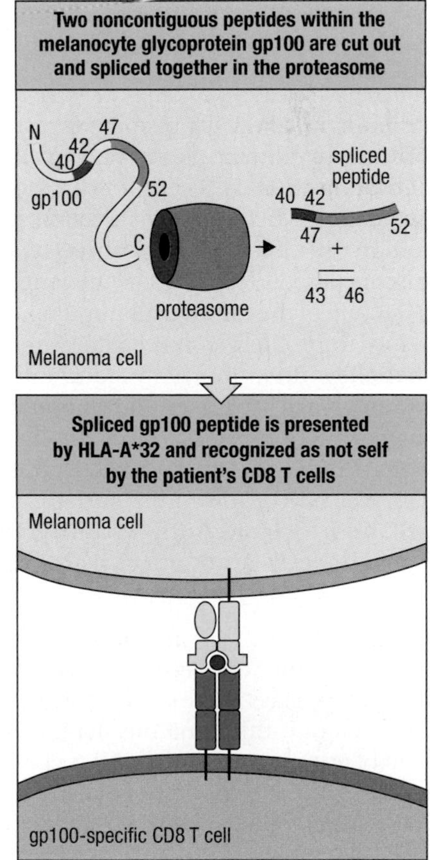

Figure 17.11 Production of tumor-specific antigens by 'cutting and pasting' peptides from self proteins.

17-8 Cancer/testis antigens are a prominent type of tumor-associated antigen

Tumor-associated antigens derive from normal cellular proteins to which the immune system is not tolerant and which become immunogenic when expressed by the tumor. They can be proteins that are normally made in immunologically privileged sites, or proteins that are normally not produced in an amount sufficient to be detected by T cells. Many tumor-associated antigens correspond to human proteins normally expressed only by immature sperm in the testis or by the trophoblast (the cells of the very early embryo that contribute to the placenta; see Section 12-9). The immune system is not tolerant to either of these cell types because they do not express HLA-A, HLA-B, or HLA class II molecules. The genes encoding these tumor-associated antigens are not expressed by any somatic cells but become turned on in cancer cells. This subset of tumor-associated antigens is called the **cancer/testis antigens** or **CT antigens**. Some 90 different genes, including several gene families, encode CT antigens, and about half of them are on the X chromosome (Figure 17.12). This bias, which is unlikely to be fortuitous, indicates that genes with functions that are normally restricted to reproduction and fetal development can be hijacked in somatic cells to produce cancers. During the course of pregnancy, the mother's energy and resources become increasingly devoted to promoting the growth and development of the fetus. Similarly, as a cancer grows and differentiates, it, too, consumes an increasing part of the body's energy and nourishment—but unlike pregnancy, the natural endpoint is not birth but death.

Cancer/testis antigens			
Antigen	Alternative name	Chromosome	Number of genes
CT1	MAGEA	X	11
CT2	BAGE	13	5
CT3	MAGEB	X	4
CT4	GAGE1	X	8
CT5	SSX	X	4
CT6	NY-ESO-1 LAGE	X	2
CT7	MAGEC	X	2
CT8	SYPC1	1	1
CT9	BRDT	1	1
CT10	MAGEC2	X	1

Figure 17.12 Genes encoding cancer/testis antigens are concentrated on the X chromosome. Thirty-eight different types of cancer/testis (CT) antigen have been defined and numbered in a CT series analogous to the CD series of differentiation antigens. Shown here are the first 10, and best-characterized, CT antigens, the number of genes that encode them, and the chromosomal location of the genes. Genes encoding 17 of the 38 CT antigens are on the X chromosome; the others are distributed between 11 autosomes, none of which has genes for more than three CT antigens. This information and more can be found on the Cancer Immunity CT Gene Database website [http://cancerimmunity.org/resources/ct-gene-database/].

17-9 Successful tumors evade and manipulate the immune response

When an immune response is made against a tumor, it imposes selection on the population of tumor cells. Variant cells that have low expression of tumor antigens, or of mutant epitopes, and are no longer recognized by effector T cells or antibodies may be able to evade the immune response. The longer a cancer grows, expands its cell population, and colonizes different sites and environments within the human body, the more genetic variation it acquires and the less likely it becomes that the immune response can delay or terminate the disease. Thus the immune system has its best chance of eliminating cancer in the early stages, when the population of cancer cells is small and has yet to adapt to the immune response.

When epithelial cells become transformed, they increase the amount of the MIC stress proteins on their surface. The MIC glycoproteins are ligands for the activating NKG2D receptor, which is expressed on NK cells, γ:δ T cells, and cytotoxic CD8 T cells (see Section 12-3). Expression of MIC by the nascent tumor cells enables these three types of lymphocyte to attack and kill the tumor cells, and if it is done efficiently the tumor is driven to extinction. If only a fraction of the tumor cells population is killed, the remainder have the opportunity to proliferate and acquire mutations and changes in gene expression that allow them to evade the immune response. Some successful epithelial tumors make proteases that cleave MIC from their own cell surfaces, producing a soluble form that binds to the NKG2D receptor of infiltrating lymphocytes. Binding of MIC to NKG2D induces receptor-mediated endocytosis that removes NKG2D from the surface and speeds its degradation. By removing MIC from their own surfaces and removing NKG2D from lymphocyte surfaces, the tumor cells evade attack by NK cells, γ:δ T cells, and cytotoxic CD8 T cells (Figure 17.13).

Cytotoxic CD8 T cells are the best effector cells for killing tumor cells. One way in which tumor cells escape a cytotoxic T cell is to stop expressing the autologous HLA class I molecule that presents the tumor antigen to the T cell. Between one-third and one-half of all human tumors have defective expression of one or more of their HLA class I allotypes (Figure 17.14). What this tells us is that many patients have made a CD8 T-cell response against their cancer, but variant cells lacking particular HLA class I allotypes have escaped that immune response and have proliferated to keep the cancer going.

Loss of HLA class I expression can in some circumstances make a tumor susceptible to attack by NK cells. The killing of acute myelogenous leukemia cells by alloreactive NK cells in an HLA-haploidentical bone marrow transplant

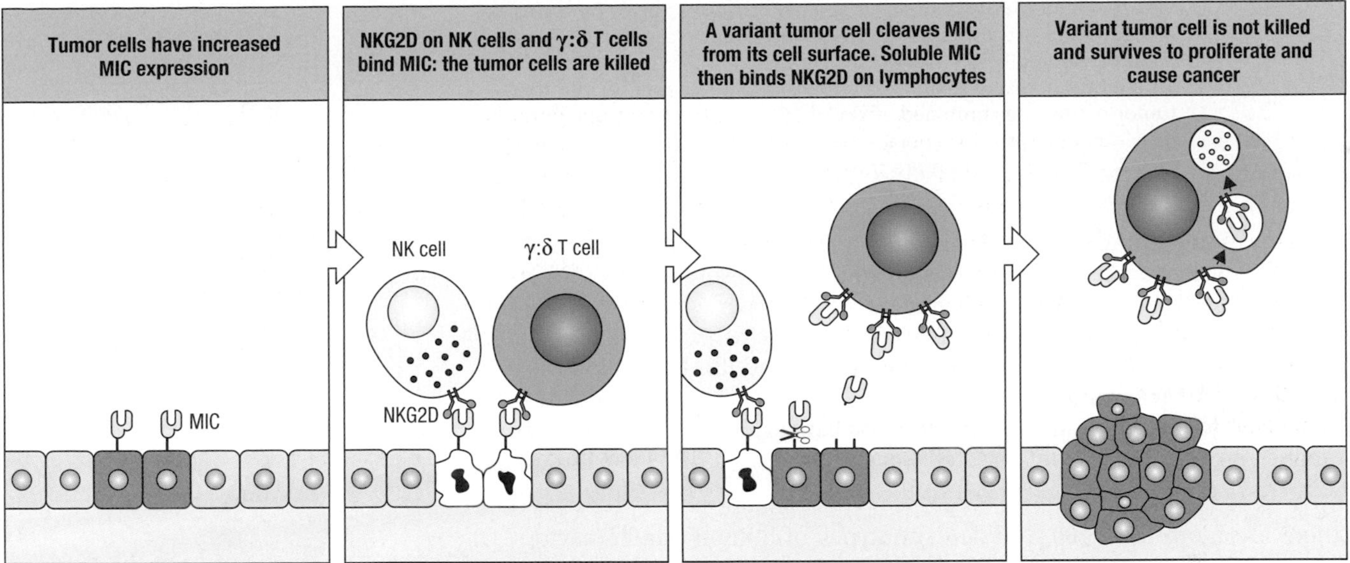

Tumor cells have increased MIC expression
NKG2D on NK cells and γ:δ T cells bind MIC: the tumor cells are killed
A variant tumor cell cleaves MIC from its cell surface. Soluble MIC then binds NKG2D on lymphocytes
Variant tumor cell is not killed and survives to proliferate and cause cancer
MIC
NK cell
γ:δ T cell
NKG2D

Figure 17.13 Human epithelial tumors inhibit the response of lymphocytes expressing NKG2D. Malignant transformation of epithelial cells induces the expression of MIC proteins. These are recognized by the NKG2D receptors of NK cells, γ:δ cells, and CD8 T cells, which enables these cells to kill the tumor cells. Variant tumor cells can evade this response by making a protease that cleaves MIC from the cell surface. This has two effects to the tumor's advantage: first, the variant tumor now lacks the ligand for NKG2D; and second, the binding of soluble MIC to NKG2D on the lymphocyte surface causes endocytosis and degradation of the receptor:MIC complex.

(see Section 15-26) occurs because the recipient's tumor cells lack an HLA class I allotype that the donor-derived NK cells see as self. This is an example of NK cells attacking a cell that has a reduced expression of MHC class I (see Figure 12.6, p. 334).

As well as evading the immune response, tumors can manipulate the immune response in their favor. In the absence of inflammation, tumor-cell antigens can be processed and presented to T cells by dendritic cells that lack B7 co-stimulators. As a result, the tumor-specific T cells will become anergic (see Section 8-8). Some tumors secrete cytokines such as TGF-β that create an immunosuppressive environment in the tumor, which can be reinforced by the recruitment of regulatory T cells (Figure 17.15). As tumors grow, they become heavily infiltrated with leukocytes. By studying these cells in biopsies and surgical specimens, correlations have been made with clinical outcome. The more numerous the regulatory T cells are, the poorer is the clinical prognosis. Conversely, increasing numbers of cytotoxic T cells, T_{FH} cells, memory T cells, and B cells are associated with a better prognosis.

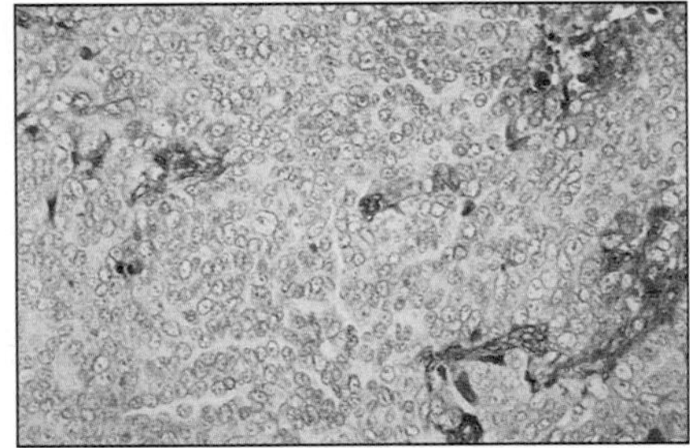

Figure 17.14 Loss of HLA class I expression in a cancer of the prostate gland. The section is of a human prostate cancer that has been stained with a monoclonal antibody specific for HLA class I molecules. The antibody is conjugated to horseradish peroxidase, which produces a brown stain wherever the antibody binds. The stain and HLA class I molecules are not seen on the tumor mass but are restricted to lymphocytes infiltrating the tumor and tissue stromal cells. Photograph courtesy of G. Stamp.

17-10 Vaccination against human papillomaviruses can prevent cervical and other genital cancers

Persistent infections with oncogenic human papillomaviruses (HPVs) are the most common sexually transmitted diseases. These viruses can cause anal, mouth, and throat cancers, and genital warts in both women and men, cancers of the cervix, vulva and vagina in women, and cancer of the penis in men. Approximately 50% of sexually active individuals experience an HPV infection. Worldwide, some 250,000 women die of cervical cancer each year, making it one of the most frequent cancers for women. The spread of HPV and the incidence of these cancers can be reduced by immunizing young women and men with an HPV vaccine before they become sexually active.

HPV has a small 8-kilobase DNA genome encoding eight proteins. These comprise two structural proteins that form the viral capsid, plus six non-structural

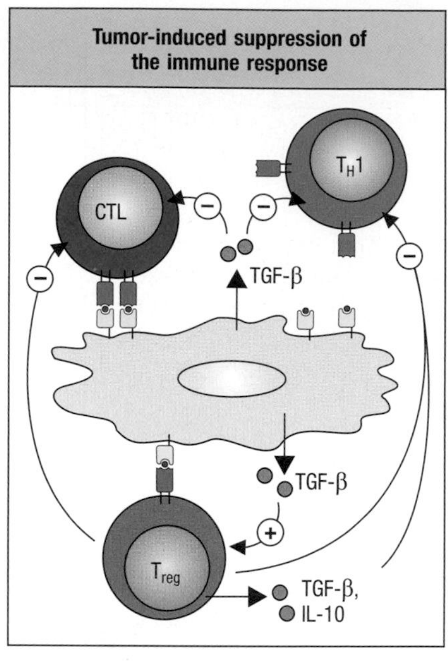

Figure 17.15 Manipulation of the immune response by a tumor. Tumors can protect themselves from the immune response by secreting immunomodulatory cytokines, such as TGF-β, which suppress the inflammatory response and recruit regulatory T cells (T_{reg}) into the tumor tissue. The combined effect of TGF-β and IL-10 made by the regulatory T cells is to suppress the actions of effector CD8 T cells and CD4 T_H1 helper cells that are specific for tumor antigens.

proteins used for viral replication and subverting host defenses (Figure 17.16). HPV establishes persistent infections, of which 95% resolve after 1–2 years. In the infections that persist longer, some infected cells integrate HPV DNA into their genomes and express the viral E6 and E7 proteins, which bind and inactivate the human tumor suppressor proteins p53 and Rb. This allows the cells to acquire and accumulate mutations, which over years or decades can eventually lead to malignant transformation and cancer. The cancer cell's genome retains the HPV DNA, but the cell usually expresses only the E6 and E7 proteins.

There are more than 100 different genotypes of human papillomavirus. Of these, 30 infect genitalia, with different strains giving rise to genital warts and cancers. The immune response to a natural HPV infection is directed toward the variable regions of L1, the major component of the viral capsid. The antibody responses are weak, with only 50% of infected individuals making detectable antibodies. The viral component of the HPV vaccine is recombinant L1 protein, which assembles spontaneously into noninfectious particles that in structure and appearance resemble the virus. Routine vaccination against HPV began in 2006 and has used two different vaccines: a bivalent vaccine containing the HPV16 and HPV18 genotypes that are responsible for 70% of cervical cancers, and a tetravalent vaccine that also includes the HPV6 and HPV11 genotypes responsible for 90% of genital warts (Figure 17.17). Both vaccines include alum as the adjuvant, and the bivalent vaccine also includes a derivative of endotoxin (bacterial cell-wall lipopolysaccharide), which binds TLR4 and further activates innate immunity. The bivalent vaccine is given only to women, whereas the tetravalent vaccine is given to both women and men. Vaccination elicits HPV-specific antibodies in amounts 80–100-fold greater than those stimulated by natural infections. As in natural infections, the antibodies are predominantly genotype-specific. The vaccines are successful at preventing infection by HPV, but they have no effect on an established infection. Consequently, it is important for teenagers to be vaccinated before they become sexually active. A current recommendation is for three immunizations over a 6-month period for females aged 11–26 years and for males aged 13–21 years.

Papillomavirus proteins and their functions	
Protein	**Function**
L1	Major capsid protein
L2	Minor capsid protein
E1	DNA helicase
E2	Transcription factor
E4	Structural protein expressed by infected keratinocytes
E5	Cofactor for epidermal growth factor, regulates MHC class I expression
E6	p53 degradation, telomerase activation
E7	Binding to Rb, E2F release

Figure 17.16 Papillomaviruses have only eight proteins. The viral capsid is composed of 72 pentamers of the L1 protein and has icosahedral symmetry. The L2 protein is incorporated into this structure, but at a proportion of less than one molecule per L1 pentamer. The gray shading indicates the proteins that contribute to the process of malignant transformation.

17-11 Vaccination with tumor antigens can cause cancer to regress but it is unpredictable

Knowing the immune system has the potential to kill tumor cells but is poorly triggered to make such responses, tumor immunologists are exploring ways to stimulate and enhance human immune responses to cancer. They hope to devise cancer vaccines that can be used for both treatment and prophylaxis. Several studies and clinical trials have focused on patients with melanoma, an aggressive skin cancer for which there is no reliable treatment. The first CT antigens were discovered as targets for cytotoxic T cells obtained from melanoma patients, and were called MAGEA1 and MAGEA3 (*m*elanoma *a*ntigen *e*ncoding). Figure 17.18 shows the effects of vaccination with an epitope of the CT1 antigen MAGEA3, which is presented to CD8 T cells by HLA-A1. Before vaccination, the patient had had surgery to remove a cutaneous melanoma, and 2 months later developed metastases that expressed the MAGEA3 antigen. Over a period of 2 years the patient was vaccinated 11 times with either a recombinant virus encoding the epitope or a synthetic peptide of the same sequence. As a result of the vaccination, the frequency of MAGEA3-specific cytotoxic CD8 T cells was increased 30-fold, commensurate with a steady regression of the tumor to produce a complete remission that lasted for more than 2 years (see Figure 17.18). For reasons that are not understood, tumor regression was seen in only 20% of the vaccinated patients, and remission was achieved only in 10%. Several other clinical trials focused on vaccination with the CT6 antigen NY-ESO-1, which is well expressed by various types of cancer and stimulates strong T-cell and B-cell responses. Although initial results were promising, they were not borne out when larger numbers of patients were studied. This lack of beneficial effect was correlated with vaccine-specific regulatory T cells that infiltrated the tumors.

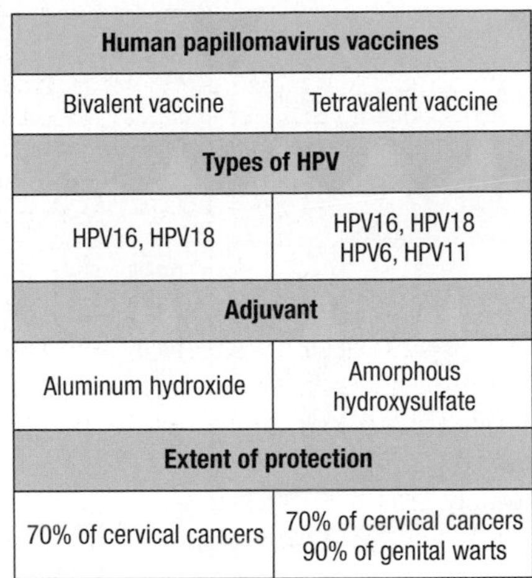

Human papillomavirus vaccines	
Bivalent vaccine	Tetravalent vaccine
Types of HPV	
HPV16, HPV18	HPV16, HPV18 HPV6, HPV11
Adjuvant	
Aluminum hydroxide	Amorphous hydroxysulfate
Extent of protection	
70% of cervical cancers	70% of cervical cancers 90% of genital warts

Figure 17.17 Components of the papilloma vaccines.

17-12 Monoclonal antibodies that interfere with negative regulators of the immune response can be used to treat cancer

Melanoma accounts for 75% of the deaths due to skin cancer. There were 9500 melanoma-related deaths in the United States in 2013, and in Australia, which has the highest incidence of melanoma in the world, some 15,000 people will be diagnosed with the cancer in 2014. Of patients diagnosed with advanced melanoma, around 75% survive for less than 1 year. Although most cancer patients make an adaptive immune response against their tumor cells, that response is usually too weak to control or eliminate the tumor. One recent and promising approach for improving the antitumor response is to increase the strength and persistence of T-cell co-stimulation by treating the patients with a human anti-CTLA4 monoclonal antibody. One such antibody, ipilimumab, was made by immunizing a mouse engineered to have human heavy-chain and light-chain genes with human CTLA4. In 2011 ipilimumab was approved

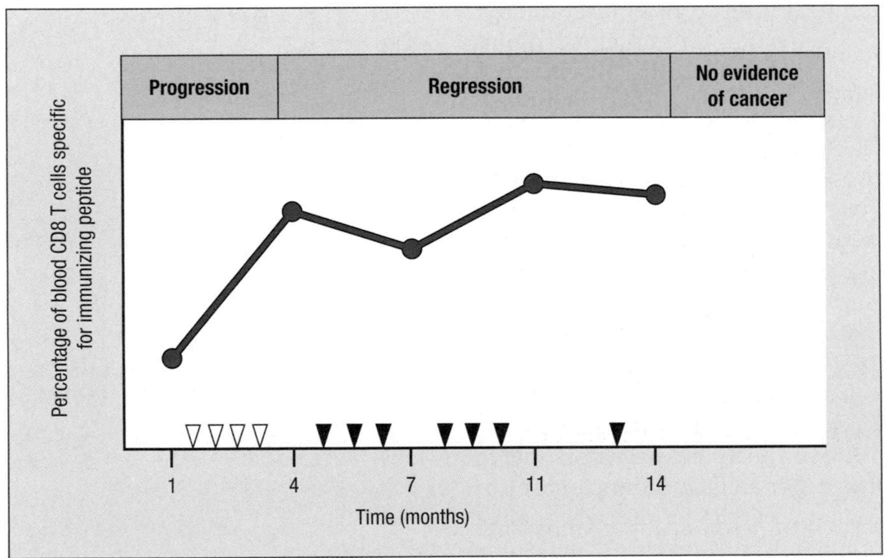

Figure 17.18 The results of vaccination of a cancer patient with a tumor antigen. A patient with metastatic melanoma was vaccinated with an antigenic peptide from the MAGEA3 protein that is presented by the patient's HLA-A1 allotype. The first four immunizations used a recombinant viral vaccine (open arrowheads); the last seven immunizations (black arrowheads) were with synthetic peptides. The red line shows the percentage of total CD8 T cells in the peripheral blood that were specific for the peptide. The status of the cancer's growth is indicated in the blue box above the graph. Data courtesy of Pierre Coulie.

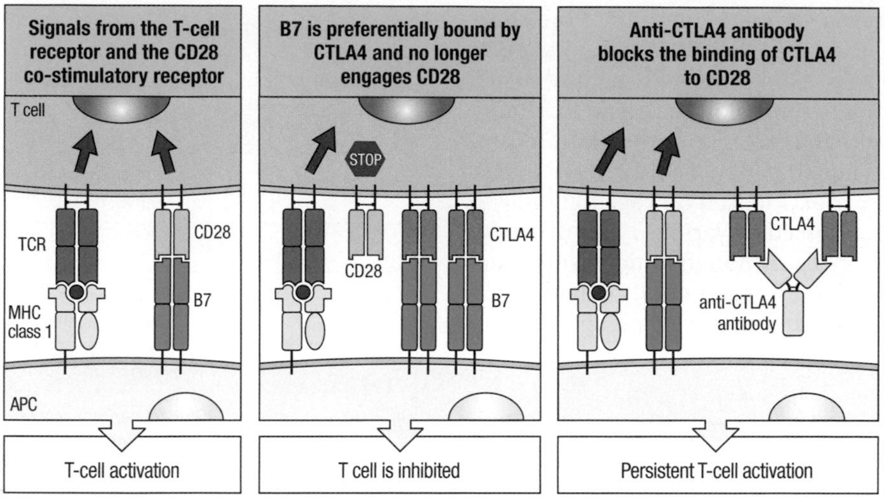

Figure 17.19 Blocking the inhibitory effects of CTLA4 with a human monoclonal antibody. Activation of T cells by antigen is dependent on signals coming from the co-stimulatory receptor CD28 on engaging the B7 co-stimulatory molecule on an antigen-presenting cell (APC) (first panel). CTLA4 negatively regulates T-cell activation by competing with CD28 for binding to B7 (second panel). A therapeutic anti-CTLA4 monoclonal antibody, such as ipilimumab, prevents CTLA4 from competing with B7 and thus promotes T cell activation (third panel). This antibody is prescribed to melanoma patients as a means of boosting their T-cell response to the tumor.

in North America and Europe as a therapy for advanced cases of melanoma. By binding to CTLA4, the negative regulator of T-cell co-stimulation, and preventing its binding to CD28 (Figure 17.19), ipilimumab maintains T-cell effector functions at a higher level. This promotes the killing and elimination of melanoma cells. More than 17,000 patients have been treated with the antibody, which has improved the 2-year survival rate to 24%.

BRAF is an enzyme in an intracellular signaling pathway that stimulates cell proliferation. Tumor cells acquire mutations in BRAF that cause the pathway to be always 'on,' leading to unregulated cell proliferation and escape from apoptosis. Vemurafenib is a small organic molecule that inhibits BRAF and was licensed in 2011 as treatment for metastatic melanoma with mutant BRAF. The effect of therapy that combines ipilimumab and vemurafenib is being explored.

Predictably, the adverse side effects of inhibiting CTLA4 are to increase inflammation and to lower the threshold for T-cell activation. Common symptoms are diarrhea and colitis in the gut, and itching skin rashes. Also observed are autoimmune reactions in the endocrine glands and the liver. These complications are treated with corticosteroid drugs.

The protein programmed death 1 (PD-1, CD279) is a member of the same family of co-stimulatory molecules as B7. The expression of PD-1 increases when NK cells, B cells, T cells, dendritic cells, and monocytes are activated. The two ligands for PD-1, called PD-L1 (CD274) and PD-L2 (CD273), are present on many types of cancer cell and have immunosuppressive effects on engaging the PD-1 on activated leukocytes. These effects include inducing apoptosis of cytotoxic CD8 T cells, inducing effector CD4 T cells to differentiate into regulatory T cells (T_{reg}), and impairment of NK-cell activation, conjugate formation, cytotoxicity, and cytokine production. Monoclonal antibodies specific for PD-1, PD-L1, or PD-L2 prevent these suppressive effects and have potential to be useful anti-cancer therapies. Nivolumab is a human IgG4 monoclonal antibody specific for PD-1 that is undergoing clinical trials for efficacy against renal-cell carcinoma, lung cancer, melanoma, and advanced or metastatic solid tumors.

17-13 T-cell responses to tumor cells can be improved with chimeric antigen receptors

Although most cancer patients make effector CD4 T-cell and CD8 T-cell responses against their tumors, these responses are usually too weak to eliminate the tumor and cure the patient. One characteristic of this weakness is that the T-cell receptors have affinities for their target peptide:MHC complexes that are only 1–10% as strong as the affinities typically seen for T cells that protect against viral infection. A strategy that seeks to address this problem is to isolate T cells from a blood sample, engineer them to express a high-affinity receptor that recognizes the tumor, and then put them back into the patient.

One such technique starts with a natural low-affinity T-cell receptor from a tumor-specific T cell and uses mutagenesis to increase the receptor's affinity for the complex of peptide and HLA class I molecule it recognizes. The improved T-cell receptor is then expressed in T cells isolated from the patient. The T cells, now expressing two types of T-cell receptor, are given back to the patient. There are two limitations to this approach: first, the tumor cells can still escape from the T cells by downmodulating the antigen-presenting HLA class I allotype; and second, each improved T-cell receptor can only be given to patients having a particular HLA class I allotype.

With a different approach, these two limitations are circumvented. This method involves making a **chimeric antigen receptor** (**CAR**) that has an antigen-binding site made from the heavy-chain and light-chain variable domains (V domains) of an antibody of defined specificity. This construct is known as **F_V (variable fragment)**. The two V domains are engineered to be part of the same polypeptide, which is connected by a transmembrane region to a cytoplasmic tail that contains three types of signaling domain. These derive from the ζ chain of the T-cell receptor complex (see Figure 5.6, p. 118), the CD28 co-stimulatory receptor, and another activating receptor such as CD137, a member of the TNF-α receptor family (Figure 17.20). This robust signaling capacity ensures that the T cells are not rendered anergic when they interact with an antigen through the chimeric receptor. The chimeric antigen receptor therefore has a monomeric antigen-binding site like that of an IgG Fab fragment (see Figure 4.3, p. 83) and is designed to have high affinity for its target antigen, and a signaling capacity comparable to, if not greater than, that of the T-cell receptor and co-stimulatory receptor. The gene for the chimeric antigen receptor is introduced into T cells that have been isolated from the patient, and the engineered T cells, now expressing the receptor, are given back to the patient. The process of isolating T cells from a patient, manipulating them and then reintroducing them into the patient is called **adoptive T-cell transfer**.

The application of chimeric antigen receptors is being explored in the context of treating B-cell tumors with T cells expressing chimeric antigen receptors that recognize the CD19 component of the B-cell co-receptor (Figure 17.21). CD19 is expressed only by cells of the B-cell lineage and by follicular dendritic cells, so the therapy has limited side effects and no direct effects on cells not involved in B-cell function. This approach requires no consideration of HLA restriction, and a single DNA construct encoding the chimeric antigen receptor can be used to transduce T cells from any patients with B-cell malignancy. Twenty-seven clinical trials testing the efficacy of CD19-specific chimeric antigen receptors are in progress.

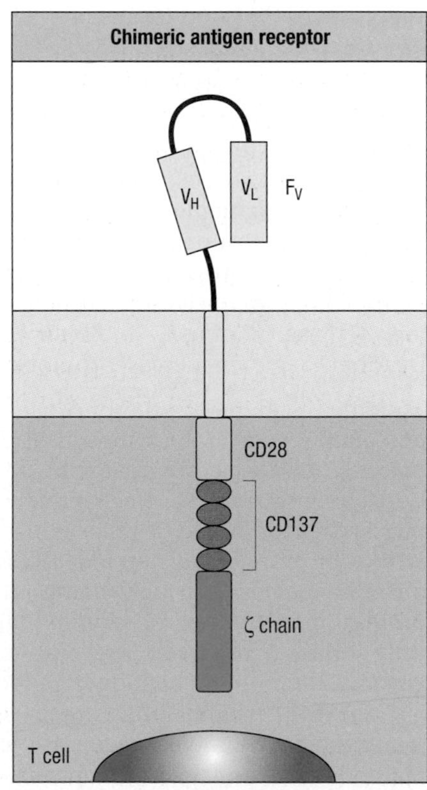

Figure 17.20 Structure of a chimeric antigen receptor. These artificial and engineered T-cell receptors are designed to have the binding strength and specificity of immunoglobulins. The chimeric antigen receptor is a single polypeptide that contains domains derived from several different immune-system proteins. The extracellular binding site consists of one immunoglobulin light-chain variable region (V_L) and one immunoglobulin heavy-chain variable region (V_H). This is connected via a transmembrane region to three signaling domains. One is derived from CD28, a second is derived from CD137, and the third is derived from the ζ chain of the CD3 complex. Thus the chimeric antigen receptor is capable of transducing all the signals needed to activate the T cell.

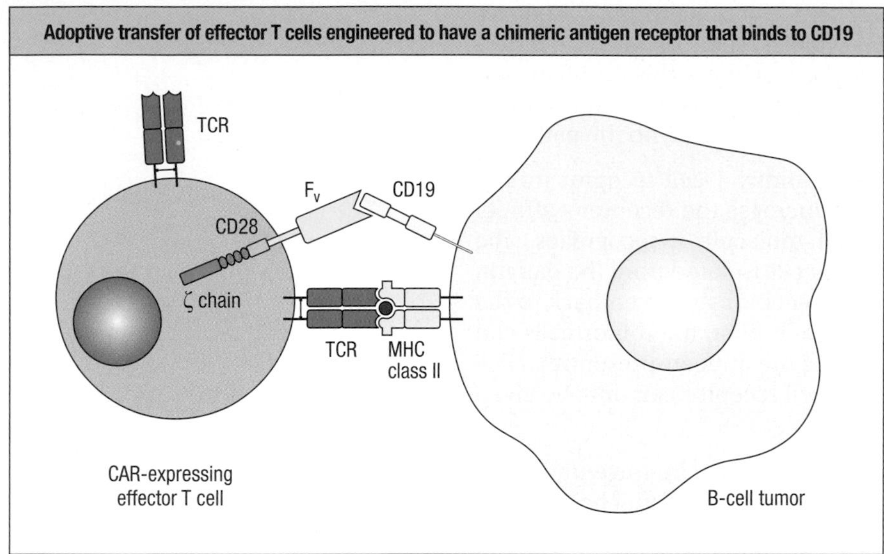

Figure 17.21 Mechanism of action of an anti-CD19 chimeric antigen receptor. In exploring the potential of chimeric antigen receptors (CARs) in cancer immunotherapy, B-cell tumors have been the major target. Many of the CARs are specific for the CD19 component of the B-cell co-receptor. T cells isolated from a patient's blood are engineered to express the CAR and are then infused back into the patient. The CAR on the T cells binds tightly to CD19 on the tumor cell and provides strong signals that activate the T cell's effector functions. Shown here is a T cell that has a tumor-specific T-cell receptor, which will be augmented by the activity of the CAR. T cells that are not specific for the tumor will also be targeted at the tumor by the CAR (not shown).

17-14 The antitumor response of γ:δ T cells and NK cells can be augmented

Two properties of γ:δ T cells make them attractive cells to manipulate to enhance a patient's immune response to cancer. One is that γ:δ T cells express diverse combinations of receptors that enable them to recognize and respond to differences between malignant and healthy cells (see Section 12-10). The other is that they can be activated either *in vivo* or *in vitro* by the simple administration of a phosphoantigen such as zoledronate (see Figure 12.24, p. 350). Although several studies have been able to increase the numbers and activation of γ:δ T cells in patients with various cancers, significant clinical benefits have yet to be attained. Clinical application is limited by our incomplete knowledge and understanding of the antigens that γ:δ T-cells recognize on tumor cells and how they recognize them.

The mechanism by which therapeutic monoclonal antibodies kill tumor cells is NK cell-mediated **antibody-dependent cellular cytotoxicity** (**ADCC**). By binding to the tumor-cell antigen with the two Fab arms and to the FcγRIII on the NK cell with Fc, the monoclonal antibody creates a strong adhesion between the two cells (Figure 17.22). The activating signals generated by FcγRIII are augmented by signals coming from NKG2D and other NK-cell receptors to activate the cytotoxic machinery of the NK cell. An unwanted feature of ADCC, however, is that activation of the NK cell is followed by FcγRIII being shed from the NK-cell surface through the action of the ADAM33 metalloproteinase. Adding an inhibitor of the metalloproteinase to the patient's treatment regimen increases tumor-cell killing by ADCC.

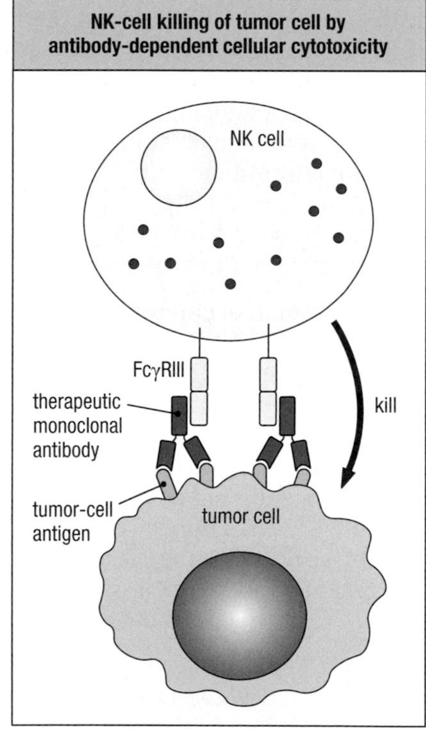

Figure 17.22 Many therapeutic anti-cancer monoclonal antibodies work by delivering the tumor cells to NK cell-mediated ADCC. Such antibodies bind to a cell-surface antigen of the tumor cells, for example CD20. The Fc regions of the antibodies engage FcγRIII on an NK cell, which then becomes activated to kill the tumor cell.

Multiple myeloma is a tumor of plasma cells that disrupts the bone marrow and often results in the synthesis of large amounts of a monoclonal antibody. There is no satisfactory treatment, and the median survival time is 3–7 years. Consequently, there is a need for new approaches to treatment. Current therapies are lenalidomide (a small-molecule drug related to thalidomide), which blocks the formation of new blood vessels (angiogenesis) and has varied effects on lymphocytes, and dexamethasone, an immunosuppressive steroid. The proteasome inhibitor bortezomib is also used to treat multiple myeloma. This drug inhibits tumor-cell proliferation, induces tumor-cell apoptosis, and by limiting the supply of peptides it reduces the surface expression of MHC class I and makes the tumor cells more susceptible to killing by NK cells. Another treatment for multiple myeloma is the humanized anti-CD319 monoclonal antibody elotuzumab, which opsonizes the tumor cells and makes them good targets for NK cells. Augmenting this therapeutic function, anti-CD319 inhibits the nurturing interactions between tumor cells and bone marrow cells that promote the proliferation and survival of the tumor cells. Combining bortezomib with elutuzumab improves the clinical benefit over either drug alone.

A potential strategy for increasing the response of NK cells to tumor cells is to use a monoclonal antibody that binds to inhibitory KIRs (see Section 12-4) on the NK cell and prevents their engagement of MHC class I ligands. A humanized monoclonal antibody (lirilumab) that binds to all HLA-C specific KIRs and blocks ligand binding and inhibitory signaling is currently being tested in clinical trials as a treatment for multiple myeloma.

17-15 T-cell responses to tumors can be improved by adoptive transfer of antigen-activated dendritic cells

One possible way of improving the immune response to tumor antigens is to isolate a patient's dendritic cells, culture them *in vitro* with a tumor antigen and then infuse the antigen-primed dendritic cells back into the patient. This form of immunotherapy is being used as a treatment for patients with late-stage metastatic prostate cancer. Monocytes isolated from the patient's blood are cultured with a recombinant fusion protein called sipuleucel-T that consists of prostatic acid phosphatase, a major antigen of prostate tumor cells, attached to the cytokine granulocyte–macrophage colony-stimulating factor (GM-CSF). When GM-CSF binds to its receptor on the monocytes, it activates the cells and drives them along a pathway of differentiation to become more like dendritic cells. When the receptor and the bound GM-CSF are internalized and taken to the endosomes and lysosomes for degradation, the prostatic acid phosphatase is carried with them. This facilitates the processing of the antigen into peptides and their presentation on the monocyte surface by MHC class II molecules.

At this point the antigen-primed monocytes are infused intravenously into the patient, where they home to the spleen and can present the peptide antigens to naive antigen-specific CD4 T cells (Figure 17.23). It is also possible that cross-presentation of peptides derived from the sipuleucel-T antigen also occurs in the *in vitro* culture, and in that event a CD8 T-cell response would also be made. The effect of the therapy is to extend the lives of the patients by a median of 4 months. Although seemingly a modest effect, the patients being treated with sipuleucel-T have such short life expectancies that 4 months is a significant time. In 2010, sipuleucel-T was approved for use in the United States and is the first approved example of what has been called a therapeutic **cancer vaccine**.

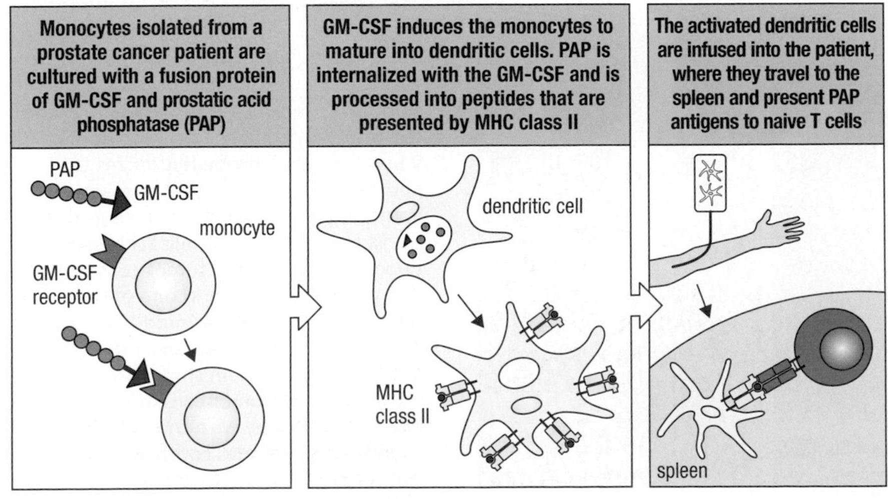

Monocytes isolated from a prostate cancer patient are cultured with a fusion protein of GM-CSF and prostatic acid phosphatase (PAP)
PAP
GM-CSF
monocyte
GM-CSF receptor
GM-CSF induces the monocytes to mature into dendritic cells. PAP is internalized with the GM-CSF and is processed into peptides that are presented by MHC class II
dendritic cell
MHC class II
The activated dendritic cells are infused into the patient, where they travel to the spleen and present PAP antigens to naive T cells
spleen

Figure 17.23 Adoptive transfer of dendritic cells loaded *in vitro* with tumor antigens. Monocytes are isolated from a blood sample obtained from the patient with prostate cancer and are cultured with sipuleucel-T, a fusion protein of a prostate cancer antigen (prostatic acid phosphatase) and a growth factor for monocytes, GM-CSF (left panel). The monocytes differentiate into dendritic cells and present peptides derived from the cancer antigen on MHC class II molecules (center panel). The dendritic cells are harvested and infused intravenously back into the patient (right panel).

17-16 Monoclonal antibodies are valuable tools for the diagnosis of cancer

The use of monocolonal antibodies and other forms of molecular analysis have enormously improved the precision with which physicians can identify a tumor and distinguish it from other closely related tumors. This is of great importance because a therapy effective against one type of tumor is not always the best therapy for a related tumor. An example of a family of related tumors is given by the B-cell tumors that correspond to all stages of B-cell differentiation and are treated in different ways (see Figure 6.23, p. 171).

Burkitt's lymphoma is derived from germinal center B cells and, like the centroblast, this tumor is fast growing and thus aggressive, but potentially curable with a course of chemotherapy. To be effective, such therapy benefits from early and accurate diagnosis. Figure 17.24 shows how tissue from a tumor biopsy can be stained in three complementary ways to diagnose, or confirm a diagnosis of, Burkitt's lymphoma. Classical histological staining with hematoxylin and eosin dyes gives a diagnostic 'starry-sky' appearance arising from two types of cell. The majority of cells are small fast-dividing tumor cells; the minority cells are large macrophages that phagocytose the tumor cells (Figure 17.24, left panel).

A diagnostic feature of Burkitt's lymphoma is the causative translocation between the *MYC* proto-oncogene on chromosome 8 and one of the immunoglobulin loci; either the heavy-chain locus on chromosome 14, the κ light-chain locus on chromosome 2 or the λ light-chain locus on chromosome 22 (see Figure 6.14, p. 161). These translocations arise from aberrant somatic hypermutation. The presence of a translocation in the tumor cells is assessed by using DNA probes that hybridize to sequences on either side of the breakpoint in the *MYC* proto-oncogene. A normal chromosome 8 reacts with both of the probes whereas a translocated chromosome 8 reacts with only one of them (Figure 17.24, center panel). In a normal cell the probes react with only the two copies of chromosome 8, whereas in a malignant cell the probes react with an additional chromosome, which has either an immunoglobulin light-chain or heavy-chain locus. Thus three chromosomes become fluorescent in tumor cells compared to two chromosomes in normal cells. The chimeric gene that is created by the translocation produces a fusion protein called a MYC oncoprotein. Thus a third way to assess potential tumor cells is to stain them with a monoclonal antibody specific for the MYC oncoprotein. Tumor cells react strongly with this antibody, independently showing that they are indeed tumor cells. By contrast, the macrophages are devoid of MYC and do not stain with the antibody (Figure 17.24, right panel).

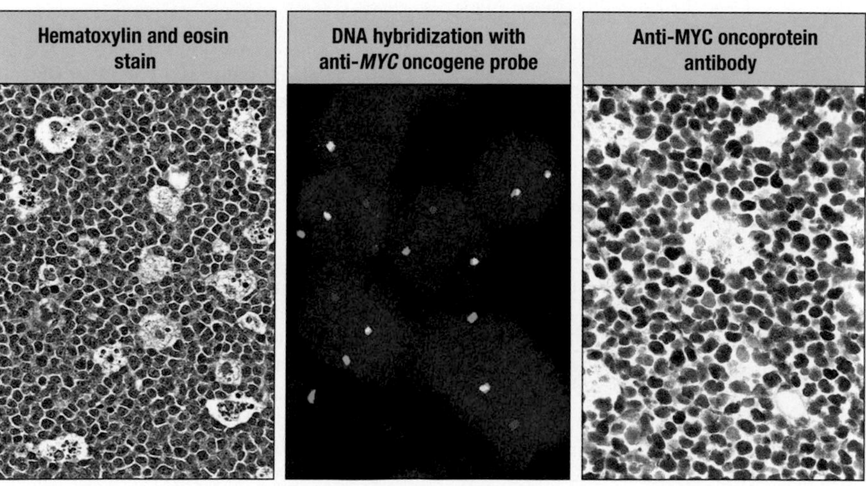

Figure 17.24 Characterization of Burkitt's lymphoma by analysis of cells, genes and proteins. Classical histology of a tumor section with hematoxylin and eosin staining gives a 'starry-sky' in which the mass of lymphoma cells form the sky and the tumor-eating macrophages are the stars (left panel). Probing the genome of the tumor cells with red and green fluorescent probes that recognize sequences near the ends of the *MYC* proto-oncogene on chromosome 8 distinguishes cells that have the characteristic translocations of Burkitt's lymphoma. In a normal chromosome 8 both the red and green probes hybridize with the intact *MYC* gene giving a yellow color. In a translocated chromome 8, the red probe hybridizes to chromosome 8 but the green probe hybridizes to another chromosome that contains the immunoglobulin locus involved in the translocation. Normal cells will have two yellow spots and tumor cells have one yellow, one red and one green spot (center panel). Staining a section of tumor with antibody specific for the MYC oncoprotein shows intensive staining of the tumor cells, but not the macrophages, with the antibody (right panel). Images courtesy Yasodha Natkunam.

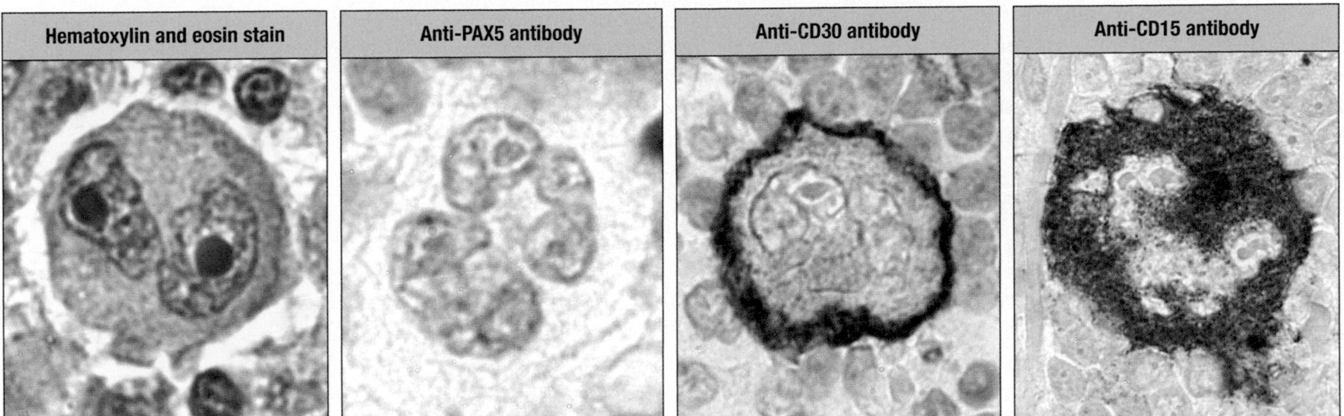

Figure 17.25 Hodgkin's lymphoma is caused by B cells that have lost B-cell characteristics. In the tumors of this disease the malignant B cells are a small minority of the cells. They have a characteristic morphology with 'owl eyes' and are separated from each other and surrounded by inflammatory cells (first panel). Downregulation of PAX5 is characterictic of Hodgkins' lymphoma (second panel). Lack of this transcription factor abrogates signaling from the B-cell receptor. Cell-surface expression of CD30 (third panel) and CD15 (fourth panel) provide targets for immunotherapy. Images courtesy Yasodha Natkunam.

Although Hodgkins's disease was one of the first tumors to be successfully treated, by radiotherapy the tumor's origin as a germinal center B cell was discovered only recently. This is because only 1% of the mass of the Hodgkin's lymphoma is contributed by the tumor cells and they are morphologically and functionally very different from B cells. The tumor cells, called Reed-Sternberg cells, have a characteristic appearance with eyes like an owl (Figure 17.25, first panel). These cells tend to be separated from each other and surrounded by a local microenvironment of inflammatory cells that they create (see Chapter Opener). A characteristic property of Hodgkin's lymphoma is the absence of PAX5 (Figure 17.25, second panel), the defining transcription factor of the B-cell lineage (see page 159). As a consequence there is an absence of B-cell signaling pathways including signaling from the B-cell receptor. The tumor cells express CD30 (Figure 17.25, third panel) and CD15 (Figure 17.25, fourth panel), markers associated with antigen-presenting cells, which helps distinguish Hodgkin's lymphoma from other large B-cell lymphomas.

Anaplastic large cell lymphoma (ALCL) is a T-cell lymphoma distinguished by large horseshoe-shaped cells (Figure 17.26, left panel). The key mutation that defines this tumor is a translocation between the anaplastic lymphoma kinase (ALK) gene on chromosome 2 and the nucleophosmin gene on chromosome 5.

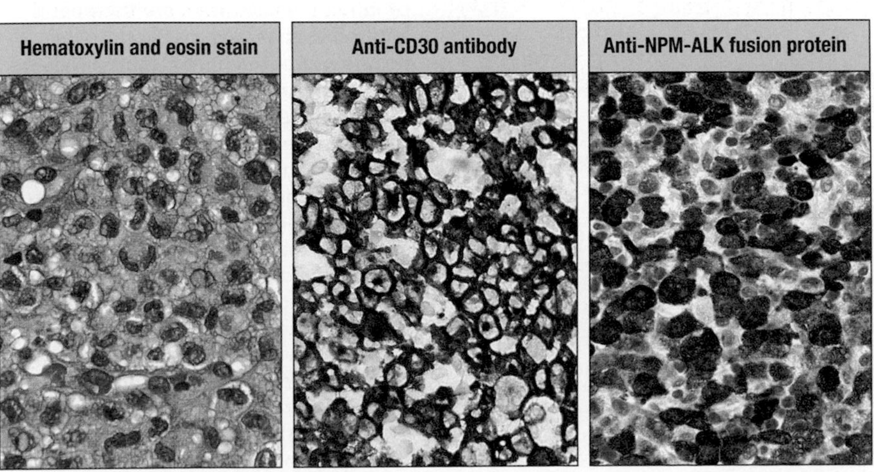

Figure 17.26 Anaplastic large-cell lymphoma is a T-cell tumor with distinctive morphology and chromosomal translocation. The cells of this tumor are large and characteristically horseshoe shaped (left panel). They have a translocation between the anaplastic lymphoma kinase gene and the nucleophosmin gene. The fusion protein that is made from the chimeric gene is oncogenic and accumulates in nucleus and cytoplasmin (center panel). The tumor expresses CD30 at the cell surface, which is used as a target for therapy that involves anti-CD30 antibodies coupled to a toxin. Images courtesy Yasodha Natkunam.

This chimeric gene encodes an oncogenic fusion protein that accumulates within the nucleus and cytoplasm of the tumor cells, giving them a characteristic and diagnostic staining pattern (Figure 17.26, center panel). The tumor cells express CD30 at the cell-surface, which is being used as a target for immunotherapeutic monoclonal antibodies.

17-17 Monoclonal antibodies against cell-surface antigens are increasingly used in cancer therapy

Humanized monoclonal antibodies that recognize cell-surface components of tumor cells are seeing increased use as cancer treatments. They can be given alone as 'naked antibodies' or as conjugates with a toxin or a radioactive isotope. Naked antibodies can reduce tumor growth by inhibiting the functions of signaling receptors and by opsonizing the tumor cells, which enables complement fixation and phagocytosis (see Section 17-14), or killing by NK cells. The anti-CD52 antibody can also induce apoptosis. All except one of the antibodies listed in Figure 17.27 is directed at a cell-surface component. The exception, an antibody specific for vascular endothelial growth factor (VEGF) has a different effect: it neutralizes this cytokine and prevents the angiogenesis that is necessary for tumors to grow.

A second use of monoclonal antibodies in cancer therapy does not rely on the natural effector functions of the antibodies. Instead, the antibodies are used to deliver a toxic agent that kills the cells to which the antibody attaches. The antibody brentuximab is specific for CD30, a marker of activated B cells. As a conjugate, brentuximab–vedotin, in which brentuximab is joined to the cytotoxin auristatin by a cathepsin-cleavable linker, has been approved as a last-resort treatment for Hodgkin's lymphoma and systemic anaplastic large-cell lymphoma. On binding to CD30 on the surface of the lymphoma cells, the conjugate is endocytosed. In the lysosomes the drug is cleaved from the antibody by cellular cathepsin. Thus released, the drug enters the nucleus and prevents

Human monoclonal antibodies used for treating cancer		
Naked antibodies		
Trastuzumab IgG1	Breast cancer	Inhibition of ERBB2 signaling
Bevacizumab IgG1	Colon cancer	Inhibition of VEGF signaling
Cetuximab IgG1	Squamous cell carcinoma	Inhibition of EGFR signaling, ADCC
Panitumab IgG2	Colorectal carcinoma	Inhibition of EGFR signaling
Ipilimumab IgG1	Melanoma	Inhibition of CTLA4 signaling
Rituximab IgG1	$CD20^+$ non-Hodgkin's B-cell lymphoma, chronic lymphocytic leukemia	ADCC via binding to CD20
Alemtuzumab IgG1	Single treatment for B-cell chronic lymphocytic leukemia	Binding to CD52, apoptosis, and complement-dependent cytotoxicity (CDC)
Ofatumumab IgG1	B-cell chronic lymphocytic leukemia	Binding to CD20, ADCC, CDC
Conjugated antibodies		
Brentuximab–vedotin IgG1	Hodgkin's lymphoma, anaplastic large-cell lymphoma	Binding to CD30, delivery of auristatin toxin
^{90}Y-labeled ibritumomab IgG1	Follicular non-Hodgkin's lymphoma	CD20 binding and delivery of radioactive ^{90}Y isotope

Figure 17.27 Humanized monoclonal antibodies used in the treatment of cancer. Shown here are therapeutic antibodies that have been approved for use as treatments for cancer. Naked antibodies are not modified and exert their therapeutic effect by binding to the tumor cells and delivering them to Fc receptor-mediated killing by NK cells or phagocytes, or to complement-mediated elimination. Conjugated antibodies have been chemically modified so that they are linked to an inactive form of a toxin, such as vedotin, or to a radioactive isotope, such yttrium-90. Conjugated antibodies deliver the toxin or radioactive isotope specifically to the surface of cancer cells, where they poison or irradiate the cells and bring about their death.

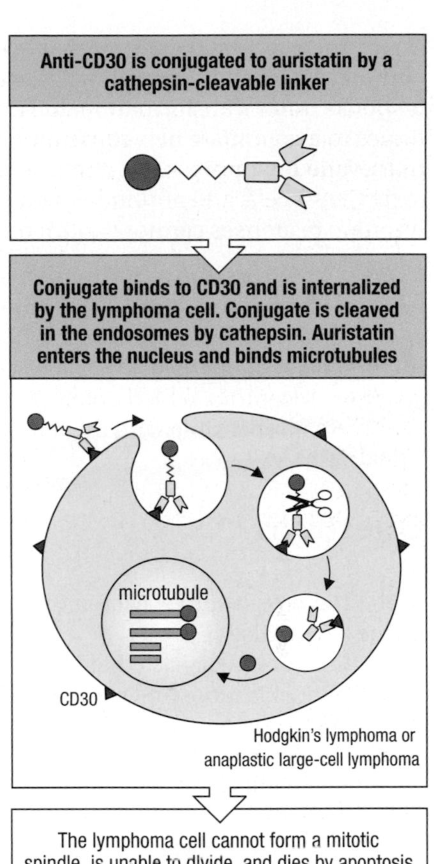

Figure 17.28 Antibodies are used to target toxins to tumor cells. In this example, the anti-CD30 antibody brentuximab is coupled to the cytotoxic drug auristatin by a cathepsin-cleavable linker. Together, the linker and the auristatin are named vedotin. The brentuximab–vedotin conjugate is used as a last-resort treatment for Hodgkin's lymphoma and anaplastic large-cell lymphoma. After the antibody has bound the conjugate to CD30 on the tumor-cell surface, it is internalized into endosomes, where cathepsin cleaves the linker and releases the auristatin. This potent anti-mitotic drug passes to the nucleus, where it binds to microtubules and prevents them from forming a mitotic spindle. This prevents cell division and induces the tumor cell to die by apoptosis.

cell division by blocking the polymerization of tubulin (Figure 17.28). Conjugates of a biological toxin and an antibody are called **immunotoxins**. The potential advantage of immunotoxins over conventional chemotherapy is that the destructive power of the toxin is more specifically targeted toward the tumor cells and away from proliferating normal tissues. Auristatin is so toxic that it can only be administered when conjugated to an antibody.

Ibritumomab is an anti-CD20 antibody that is conjugated to a radioactive isotope and is used to treat non-Hodgkin's lymphoma. When conjugated to indium-111, it is used to target and image the tumor by detecting the indium-111 γ-radiation. For therapy, the antibody is conjugated with yttrium-90, which kills the tumor cells with its β-radiation. The potential advantage of these radioactive antibody conjugates over conventional radiation therapy is their greater precision. The radiation does not have to pass through and damage other tissues to reach the tumor but is generated directly at the surface of the tumor cells (Figure 17.29).

Summary to Chapter 17

Cancer, a disease that can affect any tissue in the body, results from the uncontrolled growth of human cells. Each case of cancer derives from a single cell that has undergone malignant transformation, a process requiring mutations in several genes that are important for cell survival and cell division. Cancer cells become divorced from the mechanisms that maintain tissue integrity, and they gradually compete with normal cells for food and space within the human body. In time, the effect of the competition becomes manifested as disease and ultimately death. Every cancer is unique and dies with the person in which it arose.

The principal defenses against cancer are intrinsic to every cell of the body. These are the normal cell processes that monitor DNA integrity, and control DNA replication and cell division; they kill cells growing abnormally and ensure that multiple mutations are necessary for malignant transformation to occur. Environmental factors such as chemicals, radiation, and viruses that mutate, rearrange, and change the expression of human genes all increase the probability of cancer.

The second line of defense against cancer is the immune system, which can detect the tumor antigens by the abnormalities in the transformed cells with the same set of mechanisms that are used to respond to virus-infected cells. The immune system surveys the body for nascent cancer and probably eliminates most transformed cells that escape the intrinsic defenses at an early

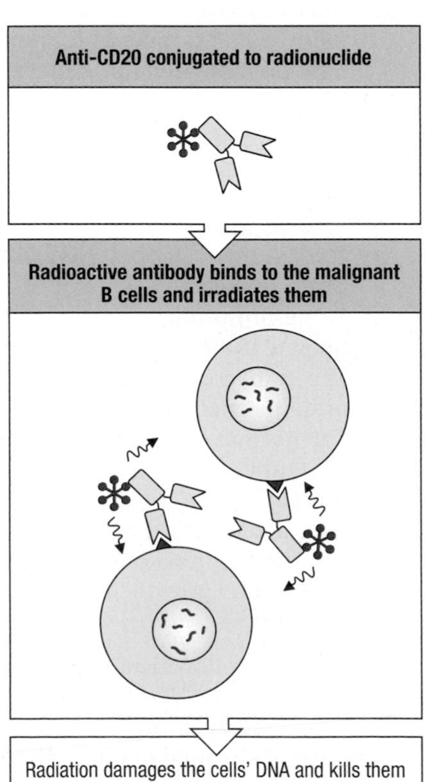

Figure 17.29 Antibodies can target radioactive isotopes to the tumor-cell surface. Conjugates of anti-CD20 antibodies with radioactive isotopes are used to irradiate and kill non-Hodgkin's B-cell lymphoma.

stage. The successful cancers that eventually manifest themselves as disease and are diagnosed by oncologists are those that have defeated the immune response. After transformation, the expanding population of cancer cells continues to accumulate new mutations, which can yield new tumor cell variants that evade or suppress the immune response. Although tumor-specific cytotoxic CD8 T cells and antibodies can be detected in patients with cancer, these immune responses cannot control or eliminate the disease.

In the past 8 years, the use of human monoclonal antibodies in cancer therapy has gone from being an emerging and experimental field to becoming a successful and established practice. What is now emerging are experimental and exciting ways of engineering T cells and NK cells to give them additional and synthetic receptors, which enable them to bind to tumors with much greater avidity than is possible with the natural lymphocyte receptors.

Questions

17–1 Indicate whether each of the following statements is true (T) or false (F).

a. Cancer is a disease that affects mainly the elderly.
b. Cancer is more prevalent in industrialized societies that have longer life expectancies.
c. Cancer cells form when a single mutation arises that abolishes the normal controls of cell proliferation and apoptosis.
d. Cancer cells are usually eliminated by immune mechanisms resembling antiviral defenses.
e. Most cancer treatments lead to complete remission.
f. The majority of cancers are caused by oncogenic viruses.
g. Individuals with immunodeficiencies and transplant patients taking immunosuppressive drugs have an increased risk of developing cancer.

17–2 What are the seven critical features of cancer cells that contribute to their capacity to establish disease?

17–3 All of the following regarding proto-oncogenes are correct except _____.

a. The human genome encodes more than 100 proto-oncogenes.
b. Some proto-oncogenes encode proteins that participate in signal transduction.
c. The transcription of some genes is regulated by proteins encoded by proto-oncogenes.
d. The normal function of proto-oncogenes is to prevent unwanted replication of mutant cells.
e. Mutated proto-oncogenes that give rise to cancer are called oncogenes.
f. A mutated proto-oncogene may alter the manner in which growth factors or their receptors function.

17–4 All of the following are associated with CT antigens except _____.

a. expressed by immature sperm in the testis
b. half are encoded on the X chromosome
c. generated by peptide splicing
d. expressed by trophoblast cells during fetal development
e. classified as tumor-associated antigens.

17–5 All of the following are examples of how tumors or the microenvironment in which they develop suppress immune responses except _____.

a. induction of T-cell anergy
b. peptide splicing of self proteins
c. cleavage of MIC glycoproteins from tumor-cell surfaces
d. TGF-β-induced recruitment of regulatory T cells
e. release of TGF-β and IL-10 by regulatory T cells.

17–6 When primary tumor cells are transplanted from a donor to an MHC-incompatible recipient, which of the following is likely to occur?

a. The recipient's somatic cells become malignantly transformed.
b. A secondary localized tumor develops in the recipient at the site of infection.
c. The recipient's alloreactive T cells kill the tumor cells.
d. Cancer stem cells in the transplant remain quiescent with the potential to reactivate if the recipient become immunocompromised.
e. Serum proteases in the recipient cleave MIC from the surface of the tumor cells, enabling these cells to evade immune detection.

17–7 Which of the following statements is not true with reference to human papillomaviruses (HPV)?

a. HPV causes genital, anal, mouth, and throat cancers and genital warts.
b. There are more than 100 different genotypes of HPV but vaccines are made against only two (bivalent) or four (quadrivalent) genotypes.
c. HPV has a small DNA genome.
d. HPV encodes E6 and E7 proteins, which incapacitate p53 and Rb; over time this may cause malignant transformation.
e. About 95% of HPV-infected individuals make

antibodies against the variable region of the L1 viral capsid.

f. HPV vaccines are ineffective in individuals who have an established infection.

17–8 Match the term in column A with its description in column B.

Column A	Column B
a. tumor-associated antigens	1. more resistant to chemotherapy and radiation than tumor cells
b. peptide splicing	2. novel peptide antigens arising from viruses, genomic and protein alterations, aberrant mRNA splicing, and peptide splicing
c. cancer stem cells	3. formation of peptide bonds in the proteasome producing a non-contiguous amino acid sequence not found in the native protein
d. soluble MIC glycoproteins	4. arise from reactivation of normally silenced genes or overexpression of genes
e. tumor-specific antigens	5. enhance removal of NKG2D from lymphocyte surfaces

17–9 ADAM33 metalloproteinase inhibitors are sometimes co-administered with anti-tumor monoclonal antibodies in order to _____.

a. block angiogenesis
b. inhibit proteasome function in tumor cells
c. enhance NK-cell killing by ADCC
d. promote differentiation of dendritic cells
e. reduce the surface expression of MHC class I on tumor cells.

17–10

A. Explain the mechanism by which ipilumumab exerts its inhibitory function in the context of cancer treatment.
B. (i) Which type of cancer has been treated with ipilumumab? (ii) With what success?

17–11

A. Describe two different strategies for introducing tumor-specific high-affinity receptors into T cells for adoptive T-cell transfer.
B. What feature is associated with one of these strategies that restricts its application to a small cohort of patients?
C. Describe a specific clinical application associated with the other strategy.

17–12 All of the following are characteristics of sipuleucel-T except _____.

a. it consists of GM-CSF
b. it inhibits nurturing interactions between tumor cells and bone marrow cells
c. it drives monocytes to differentiate into dendritic cells
d. it was the first approved therapeutic cancer vaccine
e. it consists of prostatic acid phosphatase
f. it gives rise to peptides that are presented by MHC class II molecules.

17–13 Describe the difference between (A) 'naked antibodies' and (B) conjugated antibodies in the context of using humanized monoclonal antibodies for tumor killing.

17–14 Monoclonal antibodies are used in cancer treatment because of their ability to _____. (Select all that apply.)

a. target tumor cells for immune responses such as ADCC or opsonization
b. suppress regulatory cells
c. deliver radioactive molecules to track the status of metastasis
d. enhance the expression of tumor-specific antigens
e. be conjugated to cytotoxic drugs to kill tumor cells.

17–15 Daichi Yamada, 65 years old, immigrated to California to live with his son after the Fukushima tsunami and nuclear power plant disaster destroyed his community in 2011. Recently he began complaining of fatigue, left-side pain, abdominal fullness, and night sweats. His son scheduled a check-up for him, and a routine blood screening test showed elevated basophils and eosinophils. Bone marrow tests showed the Philadelphia chromosome by fluorescence *in situ* hybridization (FISH), confirming rearrangement between chromosomes 9 and 22, which occurs in 95% of cases of chronic myelogenous leukemia (CML). He also had an enlarged spleen and liver. A diagnosis of CML was made and treatment with the tyrosine kinase inhibitor imatinib mesylate (Gleevec) began. Over the next 4–8 weeks Daichi's bone marrow showed regrowth of normal stem-cell populations, and stable white blood cell proportions were restored. Which of the following genes are associated with the chromosomal translocation in CML?

a. *p53* and *Rb*
b. *MIC-A* and *NKG2D*
c. *MAGEA1* and *MAGEA3*
d. *PD-L1* and *PD-L2*
e. *BCR* and *ABL*.

17–16 At 63 years old, Lauren Brooks was successfully treated with chemotherapy and radiation for cancer of the urinary bladder epithelium. At a later follow-up appointment with her oncologist, however, tests revealed that the cancer had returned. Her physician opted to try a different strategy aimed at inducing an *in situ* state of chronic inflammation intended to stimulate an antitumor immune response. Which of the following is the most likely course of action?

a. intramuscular vaccination with the BCG vaccine
b. intradermal vaccination with tumor antigens

derived from the patient's tumor cells from a bladder biopsy

c. intravenous infusion of a monoclonal antibody specific for IL-10, an anti-inflammatory cytokine
d. intravenous infusion of the patient's tumor cells transfected *ex vivo* with a gene encoding IL-13, an anti-inflammatory cytokine
e. intravesical adjuvant therapy with BCG vaccine via bladder instillation.

Answers

Chapter 1

1-1

A. The primary (or central) lymphoid tissues are the bone marrow (and liver in the fetus) and the thymus. The main secondary (or peripheral) lymphoid tissues are the lymph nodes, spleen, and mucosa-associated lymphoid tissues (MALT). The last of these include gut-associated lymphoid tissue (GALT), such as the tonsils, adenoids, appendix, and Peyer's patches, and bronchial-associated lymphoid tissues (BALT).

B. Primary (or central) lymphoid tissues are the anatomical locations where lymphocytes complete their development and reach the state of maturation required for the recognition of, and response to, a potential pathogen. B cells mature in the bone marrow and fetal liver, and T cells mature in the thymus. Both lymphocyte lineages are derived from a common hematopoietic stem cell. Secondary (or peripheral) lymphoid tissues provide the anatomical sites where lymphocytes encounter antigen and immune responses are induced. Antigen is delivered to the secondary lymphoid tissues through an afferent lymphatic vessel, and is retained in the lymphoid tissue for encounter with lymphocytes bearing antigen-specific receptors.

1-2 d

1-3 a—2; b—4; c—1; d—3; e—5

1-4 a

1-5 a

1-6 a

1-7 e

1-8 d

1-9 a

1-10 b

1-11 In most cases, a prior innate response to infection is necessary for lymphocytes to be activated to produce an adaptive immune response. In the innate response, macrophages activated by pathogen-associated molecules release cytokines that promote inflammation, slow the spread of infection, and help activate an adaptive immune response in peripheral lymphoid organs.

1-12 **Rationale**: The correct answer is c. Congenital asplenia runs in families and results from insufficient production of the ribosomal protein SA, which is required for the spleen to develop. Janice inherited the defective gene from one of her natural parents. Infections with encapsulated bacteria such as *Streptococcus pneumoniae*, which are surrounded by a thick polysaccharide capsule and are resistant to phagocytosis, are particularly problematic in children with congenital asplenia, because the spleen is the lymphoid organ that filters the blood and prevents such bacteria from entering the bloodstream and causing sepsis. Defensin is one of many antimicrobial substances that defend epithelial surfaces, so its absence is unlikely to overwhelm host defenses, as seen in this case. A deficiency in immunoglobulin or T-cell receptor production, or the inability to activate complement, would cause a increased risk for a number of bacterial infections, not just encapsulated bacteria in the bloodstream, and numerous infections would have been observed in Janice since her birth if any of these proteins had not been functional.

Chapter 2

2-1 b

2-2 c

2-3 d

2-4 The innate immune response must be of the appropriate form to access the particular compartment of an ongoing infection. If a pathogen exists in extracellular spaces or surfaces, then soluble molecules such as complement and antimicrobial peptides are effective in eradicating the pathogen. Some pathogens, however, reside in intracellular compartments, such as the nucleus, cytosol, or vesicles. If the pathogens are enclosed within a vesicle, the innate immune response instructs the infected cell to direct antimicrobial mechanisms to the interior of the vesicle to eradicate viable microbes. Macrophages carry out this type of antimicrobial activity. If the pathogens are replicating free in the nucleus or cytosol, then innate mechanisms are more aggressive and are aimed at killing the infected cell.

2–5 b

2–6 d

2–7 c

2–8 d

2–9 a—1; b—2; c—3; d—4; e—5

2–10 e

2–11 c

2–12 a—3; b—1; c—4; d—2; e—5

2–13 a

2–14 **Rationale**: The correct answer is d. In the absence of the complement control protein CD59 (protectin), which normally prevents the recruitment of C9 by the complex of C5b, C6, C7, and C8 on human cells, the formation of the membrane-attack complex by all three complement pathways is not fully regulated, and the host cell is more susceptible to lysis. Because C5 must first be converted to C5b to initiate the formation of this complex, if C5 is blocked by eculizumab, then C5b will not be available, and the red blood cell surface will be spared from lysis. Option a is incorrect because it is C5a, not C5, that is an anaphylatoxin, and it is the formation of a transmembrane pore, not inflammation, that causes red cell lysis. Option b is incorrect because C5 convertase production does not require C5; $C3b_2Bb$ is the C5 convertase of the alternative pathway, and C4b2a3b is the C5 convertase of the lectin and classical pathways. Finally, option c is incorrect because GPI anchor formation is not inhibited by C5; normal individuals with functional C5 are able to form GPI anchors provided that the metabolic pathway for doing so is not defective, as is the unfortunate case for John Binstead.

2–15 **Rationale**: The correct answer is b. In the absence of factor I, the alternative C3 convertase C3bBb is forming continuously, even in the absence of infection. This results in depletion of the C3 pool in the blood, lymph, and other extracellular fluids, lowering its concentration to levels below the threshold needed to mount effective innate immune responses that rely on the alternative pathway of complement activation. Encapsulated bacteria pose a particular threat because their eradication is dependent on opsonization, in which the opsonin C3b has a key role.

Chapter 3

3–1 b, d, g

3–2 b

3–3 a—2; b—5; c—1; d—1, 5, 6; e—3; f—4, 5; g—7

3–4 a

3–5 c

3–6 Step 1: dilation of blood capillaries, combined with the binding of neutrophil sialyl-Lewis^x carbohydrate to selectins, a type of adhesion molecule, expressed on activated endothelium, slows down neutrophils as they roll along the endothelium, binding reversibly to the endothelial selectins, a process called rolling adhesion. Step 2: in response to the chemokine CXCL8 through binding to CXCR1 and CXCR2 chemokine receptors, the neutrophil integrins LFA-1 and CR3 increase their affinity for the endothelial adhesion molecules ICAM-1 and ICAM-2 through conformational changes. Neutrophils are now engaged in tight binding, are immobilized on the endothelium, and the rolling stops. Step 3: the neutrophil squeezes between the endothelial cells, a process known as extravasation. When the basement membrane is encountered, proteases (such as elastase) are secreted by the neutrophil causing degradation of the laminins and collagens of the basement membrane, and the process of extravasation is complete. Step 4: migration to the focus of infection is mediated through a CXCL8 gradient established by activated macrophages in the infected tissue. Neutrophils bearing CXCL8 receptors will travel toward the highest concentration of CXCL8 until they arrive at the focus of infection, and will also secrete their own CXCL8, which will assist with the migration of even more neutrophils to the infected location.

3–7

A. a—1, 5; b—1, 2, 5; c—1; d—5; e—1; f—2; g—6; h—4, 6; i—3; j—3; k—5; l—5

B. (i) The cytokines IL-1, IL-6, CXCL8, IL-12, and TNF-α are produced by macrophages. (ii) Type I interferons can be produced by many different types of cell when infected with a virus. Specialized cells called plasmacytoid dendritic cells produce large amounts of type I interferons.

3–8 d

3–9 f

3–10 b, g

3–11 Cytokines are relatively short-lived, which means that they must mediate their effects soon after their release as soluble proteins if they are to be functional. This property usually limits the actions of cytokines to within the immediate vicinity of their production. Sometimes the effects of cytokines are localized even more by direct contact between the cytokine-secreting cell and the responding cell bearing the specific cytokine receptor. These properties minimize systemic effects and confine immune responses to restricted locations, which affects a limited number of responding cells.

3–12 First, IL-1β is synthesized as an inactive precursor called proIL-1β, which is retained in the cytoplasm. It is activated only after cleavage by caspase 1. Second, caspase 1 is also made as an inactive precursor called procaspase 1. Upon macrophage activation and assembly of the inflammasome, procaspase 1 accumulates there as oligomers favoring the autoproteolysis of procaspase 1 to form caspase 1. Caspase 1 then facilitates the maturation of IL-1β in the cytoplasm and also in specialized secretory granules, and IL-1β is then subsequently secreted in large quantities by the macrophage.

3–13 e

3–14 **Rationale**: The correct answer is b. Mutations in NEMO (a subunit of IKK) compromise the activity of IKK, which is required for the phosphorylation of IκB and its subsequent release from NFκB in the cytosol. NFκB is thereby inhibited from translocating to the nucleus, where as a transcription factor it normally directs the synthesis of TNF-α and other inflammatory cytokines, as well as other proteins involved in development and immunity. Answer a is the consequence of a defect in NADPH oxidase causing chronic granulomatous disease. Answer c is associated with a defect in forming DAF, HRF, and CD59 as a result of the inability to form their GPI linkages. Paneth cells in the intestinal tract, not macrophages, make cryptdins, and their synthesis is independent of NFκB, so answer d is not correct. Finally, answer e is not correct because the expression of IFN-α is independent of NFκB and is regulated instead by interferon-response factor 7 (IRF7).

Chapter 4

4–1 c

4–2 a

4–3 a—6; b—5; c—4; d—2; e—3; f—1

4–4 c

4–5 d, b, e, a, c

4–6 c

4–7 c, d

4–8 e

4–9 b

4–10

A. IgG3 is most efficient at complement activation.
B. The hinge region of IgG3 is the longest of all the IgG subclasses. Its length and flexibility make the IgG3 Fc region more accessible for binding C1 compared with the other IgG subclasses.
C. IgG4 fails to activate complement because its Fc region binds C1 poorly.

4–11 a—1, 2, 5, 9; b—7; c—6, 7, 10; d—1, 2, 3, 4, 6, 8; e—2

4–12 b

4–13 IgG4 would be considered the most desirable in principle because it lacks the ability to activate complement, and so will not trigger complement-mediated inflammation, which could cause damage to the patient. However, the monoclonal IgG4 will swap immunoglobulin chains with the patient's IgG4 at high frequency, thus becoming functionally monovalent, compromising its ability to bind as strongly to antigen as the other, stable, IgG subclasses. One solution would be to alter the CH3 region of IgG4 genetically in such a way that it cannot swap heavy:light chain units with other IgG4 antibodies but retains the inability to activate complement.

4–14 d

4–15 **Rationale**: The correct answer is d. In the absence of functional RAG-1 or RAG-2, which are required for V(D)J recombination, the rearrangement of both immunoglobulin and T-cell receptor genes is impaired. This interferes with the normal development of B cells and T cells, accounting for the markedly lower number of lymphocytes in the total white blood cell count. The diagnosis is supported by the absence of a thymic shadow on the chest X-ray, which would be visible if T-cell development were proceeding normally. AID deficiency would compromise immunoglobulin class switching and a hyper IgM syndrome would be expected. MHC class I deficiency would only affect the development of CD8 T cells because its expression on thymic epithelium is necessary for intrathymic maturation of CD8 T cells. Toll-like receptors and defensins have important roles in innate immune responses, but a deficiency of either would not explain the failure of B and T lymphocyte development.

4–16 **Rationale**: The correct answer is f. Aliya has IgM antibodies, so this is clearly not a case of X-linked agammaglobulinemia, in which a complete absence of immunoglobulin is expected. The father has normal immunoglobulin levels and the patient is a girl, so X-linked hyper IgM syndrome is highly unlikely. IgA deficiency can be ruled out because she does not also make IgG. Patients with acute lymphoblastic leukemia do not have an isotype-switching defect, and children with severe combined immunodeficiency die in their first year without treatment. A defect in both copies of the gene encoding activation-induced cytidine deaminase (AID) is most likely; AID is required for both isotype switching and somatic hypermutation, which accounts for Aliya's ability to make IgM but her inability to make any other antibody isotype. Both of her parents would be heterozygous carriers of the AID defect but would produce sufficient amounts of AID to escape immunodeficiency.

Chapter 5

5–1 c, e

5–2 a—2; b—3; c—5; d—1; e—4

5–3 b

5–4 e

5–5 c

5–6 a—4; b—1; c—2; d—5; e—3

5–7 a

5–8 a—5; b—4; c—3; d—2; e—1; f—6

5–9 a, c, e, f, h

5–10 b

5–11 c

5–12 c

5-13 d

5-14 **Rationale**: The correct answer is b. CIITA is the major transcriptional control factor for MHC class II gene expression. In the absence of functional CIITA, the expression of all MHC class II genes and the invariant-chain gene is compromised. This in turn negatively affects the intrathymic development and functionality of CD4 T cells, accounting not only for the severely reduced levels of CD4 T cells but also for the hypogammaglobulinemia observed in Christina. A mutation in HLA-DQ would not explain the failure to detect other MHC class II isotypes. The existence of both B and T cells rules out a RAG-1 defect, which would result in the inability to rearrange both immunoglobulin and T-cell receptor genes. TAP-1 is not associated with MHC class II expression but instead influences MHC class I expression. Finally, a defect in CD3ε would result in impairment of all T-cell responses, but this was not observed in the *in vitro* functionality tests involving allogeneic B cells.

5-15 **Rationale**: The correct answer is d. A deficiency in either TAP-1 or TAP-2, which function to transport peptide fragments to the lumen of the endoplasmic reticulum, would inhibit the expression of MHC class I on the cell surface because peptide must be bound to an MHC class I molecule to permit its transport from the endoplasmic reticulum to the cell membrane. Low levels of MHC class I account for the low levels of CD8 T cells; MHC class I-restricted CD8 T cells would fail to undergo positive selection in the thymus if MHC class I levels were abnormally low. Deficiencies of HLA-DM, invariant chain (which gives rise to CLIP), or CIITA would affect the MHC class II pathway of antigen presentation, but would have no effect on the MHC class I pathway.

Chapter 6

6-1 a—4; b—1; c—5; d—2; e—6; f—3

6-2 b

6-3 c

6-4 a

6-5 d

6-6 a

6-7 c

6-8 e

6-9 e

6-10 a

6-11 a—5; b—1; c—2; d—3; e—4

6-12 a

6-13

A. A B-cell tumor comprises cells derived from a single cell that has undergone transformation, resulting in uncontrolled growth. No further maturation of the B cell occurs after transformation. If the B cell has rearranged the heavy-chain and light-chain genes before transformation, then immunoglobulin will be expressed at the cell surface. Because all of the cells in the tumor belong to the same clone, the immunoglobulin on all of them will be made of the same heavy and light chains.

B. Pre-B-cell leukemia is characterized by transformation before the rearrangement of light-chain genes. If transformation occurs at the large pre-B-cell stage, the immunoglobulin on the cell surface will be composed of μ:VpreBλ5. If transformation occurs at the small pre-B-cell stage, there would be little or no immunoglobulin on the cell surface because surrogate light chain expression is turned off at this stage and the μ heavy chains are retained in the endoplasmic reticulum. Normal immature B cells that have not undergone transformation express IgM containing μ plus a κ or λ light chain.

6-14 **Rationale**: The correct answer is b. Multiple myeloma results from the outgrowth of a single plasma cell that has undergone malignant transformation in the bone marrow. Tumor masses expand to the point at which the relatively limited amount of space available in the bone becomes filled with tumor cells, constraining the development of erythrocytes and neutrophils and causing anemia and neutropenia, respectively. Serum IgG would be predominantly monoclonal, because the tumor derives from a single plasma cell. Similarly, the Bence-Jones protein would be of the λ type, not κ, because this is an IgG λ multiple myeloma. Patients with multiple myeloma would be susceptible to pyogenic infections because the diversity of their immunoglobulins is limited, making these patients immunocompromised.

6-15 **Rationale**: The correct answer is b. XLA is an X-linked recessive condition affecting males and is characterized by the absence of B cells and serum immunoglobulins owing to a defect in B-cell development. Specifically, Bruton's tyrosine kinase (Btk) is defective, which compromises normal signaling properties needed for progression through the stages of B-cell development. The absence of serum IgG and IgM makes these patients particularly susceptible to infections with bacteria that are encapsulated, because such patients are unable to phagocytose these pyogenic bacteria by means of Fc receptors, having to rely only on complement activation and opsonization for clearance. The absence of IgM rules out hyper-IgM syndrome. CGD, although an X-linked disease, is not associated with hypogammaglobulinemia; persistent antigenic stimulation in CGD causes hypergammaglobulinemia. The presence of CD4 and CD8 T cells rules out both DiGeorge syndrome and bare lymphocyte syndrome type I (MHC class I deficiency), in which either a complete absence of T cells or an absence of CD8 T cells is expected, respectively.

Chapter 7

7-1 c

7-2 b

7-3 a, c

7-4 e

7-5 The first checkpoint occurs after the rearrangement of the β-chain locus, which tests for the ability of the β-chain to associate with pTα, the surrogate light chain, and form the pre-T-cell receptor on the cell surface. The second checkpoint occurs after the rearrangement of the α-chain locus, which tests for the ability of the α-chain to associate with the β-chain and form the T-cell receptor on the cell surface.

7-6 a—3; b—4; c—1; d—5; e—3

7-7 a

7-8 a, c, d, e

7-9 b

7-10 c

7-11 a—3; b—4; c—1; d—2; e—5

7-12 e

7-13

A. Cells that express MHC class II are the professional antigen-presenting cells (B cells, macrophages, and dendritic cells), thymic epithelial cells, neural microglia, and activated T cells (in humans).

B. Macrophages, dendritic cells, and thymic epithelial cells either populate the thymus or circulate through it. Cortical thymic epithelial cells participate in positive selection by presenting MHC class I and class II molecules with self peptides to double-positive (CD4 CD8) thymocytes. These are developing T cells that have successfully rearranged the TCRα and TCRβ genes. Only T cells that have T-cell receptors that can interact with self MHC are positively selected, thus shaping a T-cell repertoire that is specific for self-MHC molecules. If the affinity for self-peptide:-self-MHC is too weak, the T cells die by neglect, through apoptosis. Cortical and medullary thymic epithelium may also participate in negative selection by inducing the apoptosis of thymocytes that bear a T-cell receptor with high affinity for self MHC, self peptide or a combination of the two. Thymic epithelium, circulating macrophages, and dendritic cells found primarily at the cortico-medullary junction participate in negative selection and aid in the elimination of potentially self-reactive T cells bearing high-affinity T-cell receptors for complexes of self peptides:self MHC. Because these cells are circulating between tissues, organs, and the thymus, a heterogeneous array of self peptides will be transported to the thymus and displayed. This is important because some self peptides are expressed in locations other than the thymus and may be encountered in secondary lymphoid organs.

C. Although circulating dendritic cells and macrophages are effective at endocytosing cellular debris in extrathymic locations and then presenting self peptides derived from this debris to thymocytes, some self proteins are excluded from this housekeeping function. Some of these self proteins are located in immunologically privileged sites where leukocytes do not usually circulate. These sites are normally MHC class II negative, which is important because it avoids the presentation by MHC class II of any self peptides that were not presented in the thymus during negative selection. It should be pointed out, however, that some of these tissues can be induced to express MHC class II molecules under the influence of certain cytokines, for example interferon-γ, during an inflammatory response. This is believed to be one mechanism by which tolerance can be broken, resulting in autoimmunity.

7-14 b, c, d

7-15 b

7-16 **Rationale**: The correct answer is b. APECED is a rare autosomal condition affecting both males and females and more common in clusters in certain homogeneous ethnic populations (such as Finns) as a result of consanguineous marriages or clusters of descendants of a common family. The underlying cause is a loss-of-function mutation in AIRE, resulting in the inability to express tissue-specific antigens in the thymus during T-cell development. This leads to a failure in negative selection of autoreactive T cells, which subsequently become activated in the periphery when presented with self-derived antigens. Therefore it is the lack of AIRE, and not the abundance of this transcriptional regulator, that is responsible. Blood tests ruled out a deficiency in B-cell numbers, and the only opportunistic infection that has been a medical problem for Raija is candidiasis. A CD40 ligand deficiency would lead to hyper IgM, but blood tests showed normal levels of isotype-switched immunoglobulins (IgG and IgA). The presence of T cells demonstrates that she has a functional thymus, and this rules out DiGeorge syndrome as an option.

Chapter 8

8-1 a

8-2 b, e

8-3 e

8-4 d

8-5 a

8-6 e

8-7 d

8-8 b

8-9 d

8-10 In an immune response, the binding of IL-2 to a high-affinity IL-2 receptor composed of α, β, and γ chains drives the proliferation and differentiation of T cells that occurs after they have encountered their specific antigen. The activated T cell will divide two to three times daily for about a week, producing a clone of thousands of identical antigen-specific effector T cells. The α chain is required to form a complex with preexisting γ and β chains to make the high-affinity receptor. The receptor formed by the γ and β chains only has a low affinity for IL-2. In the absence of IL-2 and its high-affinity receptor, T cells will not be fully activated, will not differentiate, and will not undergo clonal expansion. Thus, by preventing the production of IL-2 and its receptor, cyclosporin A prevents the clonal expansion of T cells specific for the foreign antigens on the graft and their differentiation into effector T cells, and thus suppresses the immune response directed against the graft.

8–11 c

8–12 b

8–13 a—3; b—5; c—4; d—1; e—2

8–14 b

8–15 **Rationale**: The correct answer is b. Leprosy, caused by *Mycobacterium leprae*, can come in two forms. Lepromatous leprosy is characterized by a bias toward T_H2 cells, which produce the cytokines IL-4, IL-5, and IL-10. In contrast, tuberculoid leprosy is associated with a bias toward T_H1 cells, which characteristically produce IL-2, IFN-γ, and lymphotoxin (LT). The location of the lesions on Vijay's extremities is also consistent with an infection with *M. leprae* rather than with *M. tuberculosis* (which causes tuberculosis), because *M. leprae* grows optimally at 30°C whereas *M. tuberculosis* grows best at 37°C and primarily causes lung disease. Vijay most probably contracted *M. leprae* in an area of South India where leprosy is endemic before his emigration to the United States.

Chapter 9

9–1 The B-cell co-receptor, made up of CD21 (complement receptor 2), CD19, and CD81, cooperates with the B-cell receptor in B-cell activation and increases the sensitivity of the B cell to antigen 1000–10,000-fold. CD21 binds to iC3b and C3d, factor I cleavage products of C3b, deposited on the surface of a pathogen or on soluble antigens. CD19 provides the long cytoplasmic tail involved in signaling. CD81 associates with CD19, facilitates the expression of CD19 on the surface of the B cells, and aids in the cooperation between the B-cell receptor and co-receptor within microdomains in the B-cell plasma membrane. When the B-cell receptor and co-receptor are ligated by antigen and C3d (or iC3b), respectively, the protein kinase Lyn and the cytoplasmic tail of CD19 come close together. Lyn is bound to the immunoreceptor tyrosine-based activation motifs (ITAMs) of Igα, which phosphorylate CD19 when it is nearby. The phosphorylated CD19 tail will initiate activation signals that complement those generated by the B-cell receptor complex.

9–2 a

9–3

A. After B-cell receptor and co-receptor cross-linking by antigen and complement, the B cell undergoes changes in the chemokine receptors and adhesion molecules present on its surface. These molecules help to guide the B cell into the T-cell area of the lymph node, where it is temporarily trapped by adhesive interactions with T cells.

B. Being trapped in the T-cell area facilitates interaction with newly differentiated helper T_{FH} cells that are specific for the same antigen. If the B cell presents the appropriate peptide antigen on its surface MHC class II molecules, a stable B cell–T_{FH} cell conjugate pair forms. Mutual signaling leads to expression of cytokines by the helper T cell; these drive the B cell to undergo proliferation and differentiation into antibody-producing cells.

9–4 e

9–5 e

9–6 a, c

9–7 c

9–8 a—4; b—2; c—3; d—1; e—5; f—6

9–9 e

9–10 e

9–11 a, d

9–12 b, c, e, f

9–13 c

9–14

A. IgG and FcRn on the apical (luminal) side of an endothelial cell lining a capillary wall are first taken up by fluid-phase pinocytosis. As the endosome acidifies, IgG associates with two FcRn molecules, thereby sparing it from degradation in lysosomes. The complex is transcytosed to the basolateral face of the endothelial cells, where on encountering a more basic pH, IgG is released into the extracellular space.

B. IgG is transported into the extracellular fluids of connective tissue, and across the placenta into the fetal circulation during pregnancy.

9–15 **Rationale**: The correct answer is a. Anthony had normal levels of T cells and B cells, so a defect in either RAG1 or CD3 can be ruled out because a RAG1 deficiency would interrupt somatic recombination in both B cells and T cells, and a CD3 defect would inhibit T-cell development. Elevated IgM in the serum indicates the ability of B cells to become activated, so the B-cell co-receptor components CD19 and CD81 are likely not to be affected. Isotype switching is induced by helper T cells and requires the ligation of CD40 on the surface of B cells by CD40 ligand on the surface of T cells. In the absence of this interaction, B cells would not be able to switch from IgM to the other immunoglobulin isotypes, so abnormally high levels of IgM would be produced. Furthermore, CD40 ligand has an important role during granulopoiesis and the development of neutrophils, accounting for Anthony's neutropenia.

Chapter 10

10–1 c

10–2 f, h

10–3 a—2; b—4; c—5; d—3; e—1

10–4 c

10–5 b

10-6 b

10-7 a—6; b—3; c—1; d—4; e—2; f—5

10-8
(i) Mucosal immunity is proactive rather than reactive. Adaptive responses are continually made against the diverse array of commensal microorganisms that reside in the gut. This is facilitated through effector T cells and B cells that populate the lamina propria and are ready to respond if the mucosal epithelium is breached.
(ii) Mucosal immunity is more prudent in the activation of inflammation, to avoid tissue damage to the gut mucosae. One way in which this is accomplished is through the use of CD4 T_{reg} that secrete IL-10, which suppresses inflammation and the production of inflammatory cytokines.

10-9 b, f

10-10 c

10-11 d

10-12 c

10-13 b, d, f

10-14 c

10-15 **Rationale**: The correct answer is b. Like dimeric IgA, pentameric IgM also contains a J chain, which permits its translocation across mucosal surfaces using the poly-Ig receptor. The lack of IgA will be compensated for by increases in other antibody isotypes, including IgM. Richard had only a temporary form of selective IgA deficiency, but many patients with genetic deficiencies causing permanent IgA deficiency are asymptomatic, because they are protected by the mucosal secretion of pentameric IgM.

Chapter 11

11-1 a—F; b—T; c—F; d—T; e—F; f—F

11-2 b

11-3 c, f

11-4 b, d, f

11-5 d

11-6
(i) Some memory T cells recirculate to peripheral tissues instead of secondary lymphoid organs, allowing immediate activation by antigen-presenting cells at infected sites.
(ii) The activation requirements of memory T cells are less stringent and are not dependent on co-stimulatory signals.

11-7 e

11-8 a—3; b—5; c—1; d—2; e—4

11-9 b

11-10
A. Adjuvants are components included in human vaccines that are recognized by Toll-like receptors and induce nonspecific antigen-independent inflammation by activating innate immune responses. This enhances the adaptive immune response to specific antigens in the vaccine and stimulates the production of memory cells.
B. Examples of adjuvants used in human vaccines include alum, a form of aluminum phosphate or aluminum hydroxide; MF59 and AS03, squalene-oil-water emulsions; virosomes, liposomes containing hemagglutinin; AS04, aluminum hydroxide plus monophosphoryl lipid A; and bacterial components included as part of some vaccines, for example whole *Bordetella pertussis* as part of the DTP (diphtheria, tetanus, pertussis) vaccine.

11-11 a—4; b—1; c—2, 3; d—2; e—4, 5

11-12 c

11-13 a—T; b—F; c—T; d—T; e—F; f—F

11-14 **Rationale**: The correct answer is d. This is an example of herd immunity, a type of immunity that is conferred in which immunization of a large portion of the population (or herd) provides protection to the minority who are unvaccinated. The number of children in her daycare facility who have been immunized with MMR (attenuated measles, mumps, and rubella viruses) is likely to be at the level needed to achieve herd immunity, thereby minimizing the probability of contracting the disease. Tolerance to measles antigen does not explain her lack of infection because tolerance is a state of immunological unresponsiveness, which would prove lethal if Madison became infected with measles. 'Dormancy' can occur with the measles virus, leading to a condition known as subacute sclerosing panencephalitis, but this occurs 2–10 years after the original measles illness. The DTP vaccine does not contain any determinants that provide cross-protection against measles. Finally, measles, mumps, and rubella vaccine viruses are not transmitted from vaccinated persons, and in fact pose no risk even during pregnancy, so it is not possible that Madison was infected through contact with children recently vaccinated.

11-15 **Rationale**: The correct answer is b. This case involves hemolytic disease of the newborn, which is only a problem in families where the mother is negative for the Rhesus D (RhD) antigen and the father is positive. If Fatima were RhD^+ there would be no risk of hemolytic disease of the newborn and no need to administer RhoGAM. If Fatima's baby were RhD^- there would also be no concern, because even if fetal blood did enter the maternal circulation there would be no alloimmunization against the Rhesus antigen. If Samir were RhD^- and the baby RhD^+, then, assuming marital fidelity, Fatima must be RhD^+ and thus tolerant to Rhesus antigen. If Fatima is RhD^- and the baby RhD^+, however, alloimmunization could occur, putting a second pregnancy with a RhD^+ baby at greater risk of hemolytic disease. Maternal anti-Rh IgG alloantibodies would cross the placenta during pregnancy, enter the fetal circulation and cause hemolysis of fetal RhD^+ erythrocytes, with resulting severe anemia in the newborn baby.

Chapter 12

12–1 e

12–2 NK cells do not rearrange either T-cell receptor or immunoglobulin genes and therefore do not express highly specific antigen receptors on their cell surface. They do, however, express FcγRIIIA (CD16a), which will bind to the Fc portion of IgG when IgG is bound to its specific receptor on the surface of a target cell. In this way, CD16a is able to participate in the adaptive immune response in a highly specific manner.

12–3 c, e

12–4 e

12–5 a—4; b—7; c—5; d—6; e—3; f—2; g—1

12–6 They both recognize HLA-A, -B, and -C, but through different mechanisms. CD94:NKG2A recognizes the peptides derived from cleavage of the leader sequences of HLA-A, -B, and -C that are subsequently presented by HLA-E on the cell surface. KIRs interact with particular allotypes of intact HLA-A, -B, or C molecules on the cell surface by binding to a region encompassing the tops of their two α helices and bound peptides called the KIR footprint.

12–7 e

12–8 a—5; b—4; c—3; d—1; e—2

12–9 d

12–10 a

12–11 d

12–12 d

12–13 i

12–14 b, f, h

12–15

A. The ligand for CD94:NKG2A is the non-classical MHC class I molecule HLA-E bound to conserved peptides derived from the leader sequence of the heavy chains of classical MHC class I molecules HLA-A, -B, and -C.

B. This ligand will be produced only if there is a steady supply of HLA-A, -B, and -C heavy chains in the cell. If the supply of these proteins is interrupted, for example during a viral infection in which the cell's ribosomes are used primarily for viral protein synthesis, then the leader peptides will not be provided in the lumen of the endoplasmic reticulum (ER) for binding to HLA-E. HLA-E will be retained in the ER and its levels on the cell surface will decrease.

C. The mechanism works effectively despite the high degree of classical MHC class I polymorphism because HLA-E itself is essentially monomorphic and because only peptides from the leader sequences of the classical MHC molecules are needed, and these are relatively well conserved between the different isoforms.

Chapter 13

13–1

A. Some pathogens, such as *Streptococcus pneumoniae*, exist in several antigenically different strains known as serotypes. Antibodies and memory B cells generated during an infection with one serotype will protect the host from reinfection with the same serotype but will not protect against a first infection with a different serotype of the same pathogen, because different epitopes are expressed. Thus, the immunity generated is serotype-specific.

B. *Streptococcus pneumoniae* has evolved at least 90 different serotypes that differ in their capsular polysaccharide antigens. A new infection with a different serotype provokes a new primary response rather than the more effective secondary immune response. This is advantageous to the pathogen because it prolongs the period of survival in the host and hence increases the likelihood of transmission to a new host.

13–2 b

13–3

A. CMV produces proteins that interfere with MHC class I molecule function by (i) causing the degradation of MHC class I by transporting them to the proteasome, (ii) disrupting peptide production by interfering with proteasome accessibility, and (iii) disrupting transport and loading of peptides onto MHC class I molecules by interfering with the activity of TAP or tapasin proteins, respectively, which results in the retention of MHC class I molecules in the endoplasmic reticulum.

B. CMV produces proteins that interfere with NK-cell killing by (i) producing a protein that is a homolog of MHC class I heavy chains that binds to NK-cell inhibitory receptor LILRB1, (ii) enabling the monitoring of HLA-A, -B, and -C expression by the NK-cell inhibitory receptor CD94:NKG2A by producing a CMV-derived leader peptide that binds to HLA-E, and (iii) producing a protein that reduces the expression of MIC-A and MIC-B ligands of the NK-cell activating receptor NKG2D. These effects subvert the NK cell's ability to sense missing self-MHC class I or to engage in killing mechanisms.

13–4

A. A lack of C3 or C4 function means that immune complexes of antibody and antigen do not bind C3 or C4 and thus cannot bind to the complement receptors on phagocytes that facilitate their clearance from the circulation.

B. Immune complexes accumulate in the circulation and are deposited in tissues. They damage tissues directly and also activate phagocytes (via the binding of antibody Fc regions to Fc receptors), causing inflammation and further tissue damage.

C. Increased susceptibility to bacteria of the genus *Neisseria*. C5–C9 are required to form the membrane-attack complex on the bacterial membrane that leads to bacterial lysis. An absence of any of these components means that the complex cannot be formed. Complement-mediated lysis is an important means of protection against *Neisseria*.

13-5 a—5; b—2; c—4; d—1; e—3; f—6

13-6 b, d

13-7 c

13-8

A. During an innate immune response, NK cells produce IFN-γ, which activates macrophages, which then produce IL-12. NK cells are activated by IL-12 and in turn release additional IFN-γ that maintains macrophages in an activated state. The macrophages engage phagocytic effector mechanisms and secrete pro-inflammatory cytokines.
B. During an adaptive immune response, T cells differentiate into T_H1 cells under the influence of IL-12 secreted by macrophages. T_H1 cells interacting with macrophages, via engagement of the T-cell receptor with peptides presented by MHC class II, secrete IFN-γ, which activates macrophages and thus causes the destruction of intravesicular pathogens. IL-12 also induces CD8 T cells to produce IFN-γ, which further strengthens macrophage maintenance.
C. Defects in either the IFN-γ receptor or the IL-12 receptor would disrupt these mutual activation processes.

13-9 d

13-10 e

13-11 d

13-12 a—2; b—5; c—3; d—1; e—4

13-13 b

13-14 **Rationale**: The correct answer is c. HAE is an autosomal dominant immune deficiency caused by insufficient production of C1INH. Left unregulated, the serine proteases of C1r and C1s lead to hyperactive cleavage and vascular consumption of C2 and C4, and the production of the vasoactive C2a fragment. C1INH also regulates serine proteases needed for blood clotting; in HAE, bradykinin—another vasoactive substance—is overproduced. The combined actions of C2a and bradykinin mediate increased vascular permeability and consequential edema. A decline in serum complement C4, a marker of complement activation, is a hallmark finding in HAE. C4 levels will be low not only during attacks but also during remission. C1q and C3, although early components in the classical pathway like C2 and C4, are found at normal levels in HAE. C5 and C9, terminal complement components of the membrane-attack complex, are also normal in HAE

Chapter 14

14-1 c, g

14-2 d

14-3 c

14-4 d

14-5 c

14-6 e

14-7 a, c, e

14-8 c, d

14-9 a—7; b—2; c—6; d—3; e—1; f—4; g—5

14-10 a, c, f

14-11

A. FcεRII is synthesized by B cells, T cells, monocytes, follicular dendritic cells, and bone-marrow stromal cells.
B. It is produced as a membrane trimer and can be cleaved by ADAM10 to produce soluble monomers and trimmers that possess cytokine-like activities that activate (trimer) or inhibit (monomer) B-cell differentiation.
C. When activated B cells start to produce produce IgE, they shed the trimeric form of FcεRII from their surface; it can then form a complex with the B-cell receptor (surface IgE) and the B-cell co-receptor on the B-cell surface. This in turn sends a signal that helps to promote the differentiation into IgE-secreting plasma cells.

14-12 b

14-13 c

14-14

A. (i) Ingestion of amniotic fluid during pregnancy. (ii) Ingestion of breast milk after birth.
B. FcεRII.

14-15 (i) Antihistamines: interfere with the ability of histamine to bind to the H_1 histamine receptors on vascular endothelium. (ii) Corticosteroids: generalized suppression of leukocyte function. (iii) Epinephrine: mediates vasoconstriction and smooth muscle relaxation of airways.

14-16 **Rationale**: The correct answer is c. NSAIDs bind irreversibly to the enzyme prostaglandin synthase and thus interrupt the cyclooxygenase pathway that leads to the production of prostaglandins from arachidonic acid. If prostaglandin is not made, additional arachidonic acid precursor is available for the 5-lipoxygenase pathway, with resulting increased production of leukotrienes. Leukotrienes have activities similar to histamine, but are 100 times more potent on a molecule-for-molecule basis. Although histamine is the inflammatory mediator made immediately during an allergic response, leukotrienes are synthesized in the later stages of allergic reactions and cause inflammation, smooth muscle contraction, constriction of airways, and the secretion of mucus from mucosal epithelium. These physiological responses could be dangerously exacerbated during an allergic episode if an individual was taking NSAIDs.

Chapter 15

15-1 a

15-2 a—4; b—5; c—2; d—1; e—3

15–3 c, d, f

15–4 c, f

15–5 f, h

15–6 d

15–7 b

15–8 d

15–9

A. Myeloablative therapy is the annihilation of the immune system of a patient before receiving a hematopoietic cell transplant. It involves subjecting the patient to both cytotoxic drugs and irradiation that destroy the patient's bone marrow.

B. This conditioning regimen is conducted before intravenous infusion of the transplant, so as (i) to incapacitate the patient's immune system to avoid graft rejection, and (ii) to provide the space needed for the donor stem cells to interact with the stromal cells of the bone marrow and begin the process of reconstituting the hematopoietic system.

15–10 d

15–11 d

15–12 (i) Hematopoietic stem cells from the peripheral blood of a donor or of the patient themself are an alternative to bone marrow. Before collection, the person is treated with G-CSF and GM-CSF to mobilize stem cells from the bone marrow into the blood. These samples are enriched for CD34-expressing stem cells before infusion. (ii) Umbilical cord blood from the placenta, often involving combined stem cells from two unrelated cord blood samples because of limited volume, is another source of hematopoietic stem cells.

15–13 a, c, e

15–14 (i) Corticosteroids, such as prednisone, induce the transcription of a subset of about 1% of genes. One of these genes is IκBα, an inhibitory regulator of NFκB. Consequently, NFκB is retained in the cytoplasm and is unable to translocate to the nucleus, where it would otherwise function as a key transcription factor of the inflammatory response in activated cells. Corticosteroids also inhibit lymphocyte homing to secondary lymphoid tissues and sites of inflammation, thereby preventing naive T-cell activation and the entry of effector T cells into the grafted organ. In addition, corticosteroid therapy activates endonucleases in target cells, promoting apoptosis and death in lymphocytes and eosinophils. (ii) Corticosteroids cause side-effects, including fluid retention, weight gain, diabetes, bone demineralization, and thinning of the skin.

15–15 **Rationale**: The correct answer is a. The O antigen (also called the H antigen) present in people of blood type O is related to the A and B antigens, and in people of genetic type A, B, or AB, is acted upon by enzymes to produce either the A or the B antigen or both, respectively. Individuals with the Bombay phenotype (hh) do not produce O antigen and therefore cannot synthesize either A or B antigen, even if they actually carry functional genes for the enzymes that would otherwise give rise to these antigens. Therefore, hh individuals make antibodies not only against A and B antigens but also against O antigens, because they are not tolerant of the O antigen. Their sera are therefore incompatible with all red cells (A, B, and O) except those of the same rare blood group. Vanad initially tested as O +, indicating that he does produce the RhD antigen on the surface of his red blood cells, so the RhD status of a blood donor would be irrelevant because Vanad is tolerant of this antigen.

15–16 **Rationale**: The correct answer is b. This is a case of graft-versus-host disease (GVHD) involving differences in minor histocompatibility antigens between Carter and his sister. The presence of a maculopapular rash and jaundice are characteristic of tissue reactions involved in GVHD, not of type II hypersensitivity reactions or cytomegalovirus infection. The myeloablative conditioning regimen would have destroyed Carter's immunocompetent cells, abolishing the possibility of graft rejection. However, mature T cells in the transplanted bone marrow may react to differences in major and minor histocompatibility antigens. In this case, because the HLA was identical, and the donor is a female, the donor would have mature T cells with specificity toward male-specific or polymorphic protein antigens to which she is not tolerant, for example, H-Y antigens presented by HLA class I or class II on the brother's cells. Because this is a bone marrow transplant, and not an organ transplant, acute rejection of the graft and host-versus-graft reactions are not considerations.

Chapter 16

16–1 b

16–2 c

16–3 e

16–4 b

16–5

A. Autoimmune diseases are distinguished by the type of immune reaction that is responsible for the disease. Three types of autoimmune response parallel the effector mechanisms described in Chapters 14 and 15 for hypersensitivity reactions of types II, III, and IV. The same system used for differentiating hypersensitivity reactions is also used for autoimmune diseases.

Specifically, type II autoimmunity is caused by antibodies directed against cell-surface or extracellular matrix self antigens. Type III autoimmunity is the result of the deposition of small, soluble immune complexes in tissues. The third type of autoimmune disease that corresponds to a hypersensitivity reaction is type IV, which is mediated by effector T cells.

B. Autoimmune diseases are chronic conditions in which effector T cells or antibodies that either fix complement or recruit phagocytes slowly erode one or more tissues of the body. IgE, which recruits the effector functions of mast cells and activated eosinophils, causes rapid violent reactions that have the capacity to quickly kill a person, as in anaphylaxis. Unlike IgE-mediated allergic reactions, for which allergic individuals can survive by avoiding exposure to the allergen, any IgE response to a self antigen, which cannot be avoided, would probably cause death before either the unfortunate person or their physician had any sense of what was going on.

16–6 b

16–7 a—3; b—5; c—2; d—4; e—1

16–8 c

16–9 a

16–10 a, e

16–11 a

16–12 e

16–13 b

16–14

A. Genes of the HLA complex, specifically the HLA class I and class II genes. Associations with HLA class II genes are most common. Various alleles of these genes are associated with a higher or lower susceptibility to particular autoimmune diseases compared with the incidence of the diseases in the population as a whole.

B. The polymorphic HLA genes encode the proteins that present peptide antigens to T cells. It has been proposed that particular alleles of these genes are associated with particular autoimmune diseases because of their ability to present the required peptide epitope(s) to autoreactive T cells. Associations with HLA class II genes are more common than those with HLA class I because CD4 T cells rather than CD8 T cells are most commonly involved in autoimmunity.

16–15 **Rationale**: The correct answer is d. FcRn normally protects IgG from degradation, thereby extending its half-life in the serum. IVIG at the doses used will overwhelm and incapacitate the normal function of FcRn. This will have the effect of reducing the half-life of circulating IgG and increasing the rate of clearance of autoantibodies from the circulation.

16–16 **Rationale**: The correct answer is d. TSH is secreted by the pituitary gland (not the thyroid gland) when there are low levels of T_3 and T_4 in the circulation (which would be the case if thyroid follicular tissue was compromised). TSH is not produced when there are high levels of T_3 and T_4, because of an inhibitory feedback process that shuts off TSH production by the pituitary gland. The receptor for TSH is expressed on thyroid epithelium (not pituitary tissue), and when bound to TSH the thyroid gland is induced to synthesize more T_3 and T_4 from an iodinated thyroglobulin precursor. Persistent autoantibodies against the TSH receptor mimic the effect of TSH stimulation, thereby causing continuous secretion of T_3 and T_4; these in turn feed back on the pituitary gland and shut off the production of TSH. In this manner, patients with Graves' disease have low levels of TSH, yet they experience hyperthyroidism by uninterrupted signaling through the TSH receptor.

Chapter 17

17–1 a —T; b—T; c—F; d—T; e—F; f—F; g—T

17–2 Cancer cells

1. activate their own proliferation
2. do not yield to growth-inhibiting signals
3. escape apoptosis-inducing signals
4. become vascularized by the formation of new blood vessels
5. spread to other tissues at distant sites
6. undergo clonal expansion through repeated cell division
7. evade eradication by immune cells.

17–3 d

17–4 c

17–5 b

17–6 c

17–7 e

17–8 a—4; b—3; c—1; d—5; e—2

17–9 c

17–10

A. Ipilumumab is a human anti-CTLA-4 monoclonal antibody. When blocked by ipilumumab, CTLA-4 (a negative regulator of T cell co-stimulation), is unable to bind to its ligand, CD28. Consequently, the T cell is not inhibited and is able to exert its effector function for longer periods in helping to strengthen the immune response against tumor cells.

B. (i) More than 17,000 melanoma patients have been treated with ipilumumab. (ii) Around 75% of untreated patients survive less than 1 year, but treatment has extended this to a 2-year survival rate in 24% of patients.

17–11

A. (i) One strategy is to engineer a tumor-specific T-cell receptor to improve its affinity for the complex of tumor peptide and HLA class I molecule it recognizes. The patient's T cells are then treated *ex vivo* to achieve expression of the engineered T-cell receptor, before they are returned to the patient. (ii) The other strategy uses a chimeric antigen receptor (CAR) that is encoded by a single gene that encodes: (1) a tumor-antigen-binding site made from antibody variable fragments (the variable domains of the heavy and light chains that are part of the same polypeptide and held together by a linker peptide); (2) a transmembrane anchor; and (3) a cytoplasmic tail containing multiple signaling domains from CD28, CD137, and the ζ chain of the T-cell receptor complex. After *ex vivo* manipulation of the patient's T cells, those cells expressing the CAR are reintroduced to the patient.

B. The T-cell receptor strategy can only be administered to patients who have an HLA-class I allotype that is compatible with the MHC restriction profile of the T cell from which the original T-cell receptor genes were isolated.

C. The CAR strategy has been used in 27 clinical trials to treat B-cell tumors in which the variable fragments have specificity for CD19.

17–12 b

17–13

A. 'Naked antibodies' function in the same way as antibodies do during response to virally infected cells. They are not conjugated to toxins or radioactive isotopes. Upon binding to their target, they mediate a variety of effects including modulation of signaling by cell surface receptors, opsonization of tumor cells (leading to complement fixation and phagocytosis), or NK-cell killing by antibody-dependent cellular cytotoxicity (ADCC). Some 'naked antibodies', such as the anti-CD52 antibody, also induce apoptosis.

B. Conjugated antibodies are chemically linked to either a toxin or a radioactive isotope. Toxins, such as auristatin, are internalized by tumor cells by endocytosis, where they selectively mediate their cytotoxic effect. Radioactive isotopes, such as yttrium-90, are similarly delivered with precision to the target cell, where DNA damage and cell death occur via radioactive decay (β-radiation).

17–14 a, c, e

17–15 **Rationale**: The answer is e. The Philadelphia chromosome is the result of a fusion of two different chromosomes involving the *BCR* and *ABL* genes. *BCR* on chromosome 22 encodes a tyrosine kinase, which becomes fused to the *ABL* proto-oncogene on chromosome 9. The protein encoded by the fused genes promotes cancer progression, but when inhibited by long-term treatment with Gleevec, remission occurs.

17–16 **Rationale**: The answer is e. The *Mycobacterium bovis* BCG vaccine is an attenuated producer of unmethylated CpG-containing DNA, the bacterial-specific ligand for intracellular Toll-like receptor 9 (TLR9). TLR9 is expressed by macrophages and dendritic cells and transduces signals for inflammatory cytokines; this is integral to the *in situ* inflammatory response sought for stimulation of an antitumor immune response as well as an antimycobacterial response. So it must be applied directly to the site of the tumor to have this effect.

Glossary

α:β T cell the majority type of T cell, which expresses an antigen receptor composed of an α chain and a β chain. These T cells include all T cells that recognize peptide antigen presented by MHC class I and II molecules.

α:β T-cell receptor one of two classes of **T-cell receptor**, which is the antigen receptor carried by the majority of T cells in the circulation. The other is the γ:δ T-cell receptor.

α_2-macroglobulin a protease inhibitor in plasma that is a component of innate immunity. It inhibits proteases that are produced or acquired by bacteria to aid their invasion.

α-defensin a subset of the secreted antimicrobial peptides called **defensins**.

β_2-microglobulin invariant polypeptide that is common to all MHC class I molecules. Also called the light chain of MHC class I molecules.

β-defensin a subset of the secreted antimicrobial peptides called **defensins**.

γ:δ T cell a minority type of circulating T cell that expresses an antigen receptor composed of a γ chain and a δ chain. These cells recognize different types of antigen from those recognized by α:β T cells, including phosphoantigens and proteins, and are not dependent on the presentation of antigen by MHC molecules. γ:δ T cells are more abundant in the gut and other tissues.

λ5 the surrogate light chain of the pre-B-cell receptor produced during B-cell development.

12/23 rule the fact that V(D)J recombination can only occur between gene segments with particular lengths of spacer in the recombination signal sequences, which means that a V_H region cannot be joined directly to a J_H region without the involvement of D_H. Two types of spacer are 12 and 23 nucleotides in length.

ABO system set of blood group antigens on red blood cells that have to be appropriately matched between donor and recipient for successful blood transfusion or organ transplantation.

ACAID *see* **anterior chamber-associated immune deviation**.

acquired immune deficiency syndrome (**AIDS**) disease caused by infection with the **human immunodeficiency virus** (HIV). It involves a gradual destruction of the CD4 T-cell population and increasing susceptibility to infection.

acquired immunity alternative term for **adaptive immunity**; pathogen-specific immunity acquired as a consequence of infection or vaccination.

activated dendritic cell a dendritic cell that takes up pathogens at a site of infection and is stimulated to travel to secondary lymphoid tissue and present antigens to T cells.

activation-induced cytidine deaminase (**AID**) enzyme that deaminates DNA at cytosine resides, converting them to uracil. The activity of this enzyme and the consequent repair of the damaged DNA are the basis of somatic hypermutation and isotype switching in activated B cells.

acute rejection rejection of transplanted cells, tissues, or organs from a genetically dissimilar donor that is due to a T-cell response stimulated by the 'foreign' HLA antigens on the transplant.

acute-phase protein any of a number of diverse plasma proteins that are rapidly made in large amounts by the liver in response to infection. They include mannose-binding lectin (MBL), C-reactive protein (CRP), and fibrinogen. Their activities contribute to the innate immune response in a variety of ways, including complement activation.

acute-phase response innate immune response that occurs soon after the start of an infection and involves the synthesis of acute-phase proteins by the liver and their secretion into the blood.

adaptive immune response the response of antigen-specific B and T lymphocytes to antigen, including the development of immunological memory.

adaptive immunity the state of resistance to infection that is produced by the adaptive immune response.

adaptor protein protein in an intracellular signaling pathway that has no enzymatic or other activity in itself but brings together two other components of the pathway so that they can interact.

ADCC acronym for **antibody-dependent cell-mediated cytotoxicity**. A function of NK cells.

adenoids mucosa-associated secondary lymphoid tissues located in the nasal cavity.

adenosine deaminase (**ADA**) enzyme involved in purine breakdown. Its absence leads to the accumulation of toxic purine nucleosides and nucleotides, resulting in the death of most developing lymphocytes within the thymus. The disease that results from a genetic lack of adenosine deaminase is called **adenosine deaminase deficiency**. Its symptoms are **severe combined immunodeficiency**.

adenosine deaminase deficiency *see* **adenosine deaminase**.

adhesin any of various molecules on the surfaces of bacteria that bind to epithelial cells, enabling the bacteria to colonize epithelial surfaces.

adhesion molecule any of a number of cell-surface proteins that enable human cells to bind to each other. *See also* **selectins**; **integrins**; **vascular addressins**; **immunoglobulin superfamily**.

adjuvant substances used in experimental immunology and in vaccines to enhance the adaptive immune response to an antigen. The adjuvant activates innate immunity and must be injected along with the antigen to be effective. Without an adjuvant most protein antigens are not immunogenic.

adoptive T-cell transfer any procedure in which T cells are transfused into a patient for therapeutic purposes. In cancer immunotherapy, they are usually the patient's own T cells that have been isolated, manipulated in culture (for example, to make them more responsive to the patients' tumor), and then reinfused.

afferent lymphatic vessel vessel that brings lymph into a lymph node.

affinity measure of the strength with which one molecule binds to another at a single binding site. *See also* **avidity**.

affinity maturation the increase in affinity of the antigen-binding sites of antibodies for antigen that occurs during the course of an adaptive immune response. It is the result of somatic hypermutation of the rearranged immunoglobulin V-gene region and the consequent selection of mutated B cells that make antigen receptors of higher affinity for their antigen.

agammaglobulinemia an inability to make antibodies, reflected by abnormally low amounts or absence of antibodies in the blood. *See also* **X-linked agammaglobulinemia**.

agglutinate to clump together small particles to form larger particles. Usually refers to antibody or some other multivalent molecule that crosslinks antigens on more than one particle. Such particles are said to be **agglutinated**. When the particles are red blood cells, the phenomenon is called hemagglutination. *See also* **influenza hemagglutinin**.

AID *see* **activation-induced cytidine deaminase**.

AIDS *see* **acquired immune deficiency syndrome**.

AIRE *see* **autoimmune regulator**.

alemtuzumab *see* **anti-CD52**.

allelic exclusion in reference to B-cell development, the fact that each mature B cell expresses only one of the two immunoglobulin heavy-chain or light-chain alleles. For each locus, half of the cells in the B-cell population express the maternal allele, and the remaining cells express the paternal allele.

allergen any antigen that elicits hypersensitivity or allergic reactions. Allergens are usually innocuous proteins that do not in themselves threaten the integrity of the body.

allergic asthma disease caused by an IgE-mediated hypersensitivity reaction to inhaled antigen, in which the bronchi constrict and the patient has difficulty in breathing.

allergic conjunctivitis IgE-mediated hypersensitivity reactions in the conjunctiva of the eye, causing inflammation and tears. It is usually caused by airborne antigens such as grass and tree pollens.

allergic reaction the result of a secondary immune response to an otherwise innocuous environmental antigen, or **allergen**, causing a wide variety of unpleasant and sometimes life-threatening symptoms. The description 'allergic' is often reserved for reactions involving IgE (type I hypersensitivity reactions); other types of immunological hypersensitivity reaction involve other types of antibody (such as IgG) or effector T cells.

allergic rhinitis IgE-mediated hypersensitivity reaction manifested in the nasal mucosa that causes runny nose, sneezing, and tears, often caused by airborne antigens such as grass and tree pollens. Often seen in conjunction with **allergic conjunctivitis**. It is also known as hay fever.

allergy state of immunological hypersensitivity to a normally innocuous environmental antigen. The term allergy is often reserved for type I hypersensitivity reactions, that is, those mediated by IgE, but may refer in a general way to all four types of immunological hypersensitivity reaction (*see* **hypersensitivity reaction**). It results from the interaction between the antigen and antibodies or T cells produced by previous exposure to the same antigen and causes a variety of unpleasant and sometimes life-threatening symptoms, depending on the route of entry of the antigen and the type of effector mechanisms involved. *See also* **allergic asthma**; **allergic conjunctivitis**; **allergic rhinitis**.

alloantibody antibody that is made by immunization of one member of a vertebrate species with antigen derived from another member of the same species. Alloantibodies recognize antigens that are the result of allelic variation at polymorphic genes. Common types of alloantibody are those recognizing blood group antigens, and MHC class I and class II molecules.

alloantigen an antigen that differs between members of the same vertebrate species, such as HLA molecules and blood group antigens. Alloantigens are determined by the different alleles of polymorphic genes.

allogeneic describes two members of the same species who are genetically different.

allograft tissue graft made between genetically non-identical members of the same species.

alloreaction adaptive immune response made by one member of a species to an allogeneic antigen from another member of the same species.

alloreactive T cell T cell in one member of a species that responds to an allogeneic antigen from another member of the same species.

allotype a naturally occurring variant of a protein. Allotypes are encoded by different alleles of the same gene. *See also* **isotype**.

alternative C3 convertase the C3 convertase of the alternative pathway of complement activation. It is composed of C3 or C3(H_2O) bound to proteolytically active Bb (C3Bb or C3(H_2O)Bb) and cleaves C3 into C3a and C3b.

alternative C5 convertase the C5 convertase of the alternative pathway of complement activation. It is composed of two molecules of C3b bound to Bb ($C3b_2Bb$) and cleaves C5 into C5a and C5b.

alternative pathway of complement activation one of three pathways of complement activation. It is triggered by the presence of infection but does not involve antibody. The early stages leading to cleavage of C3 involve iC3b, factor B, and factor D. *See also* **classical pathway of complement activation; lectin pathway of complement activation**.

anaphylactic shock the result of an IgE-mediated hypersensitivity reaction to an antigen that enters the bloodstream (for example insect venom entering the bloodstream via a sting). Rapid systemic collapse of the circulation (shock) ensues, and suffocation due to rapid swelling of the trachea can cause death. *See also* **systemic anaphylaxis**.

anaphylactoid reactions reactions that are clinically similar to anaphylactic shock but do not involve IgE.

anaphylatoxin general name for complement fragments C3a and C5a, which are produced during complement activation. They induce inflammation, recruiting fluid and inflammatory cells to sites of antigen deposition. In some circumstances, C3a and C5a can induce anaphylactic shock.

anaphylaxis *see* **systemic anaphylaxis; anaphylactic shock**.

anchor residue amino acid residues in MHC-binding peptides that interact with pockets in the peptide-binding groove of the MHC molecule. Peptides that bind to a given MHC allotype have the same or similar anchor residues.

anemia a deficiency of red blood cells. It can have many causes, including the destruction of red blood cells that occurs in some hypersensitivity and autoimmune diseases, and genetic deficiencies of red-cell production (for example Fanconi's anemia).

anergy state of non-responsiveness to an antigen. People are said to be anergic when they cannot mount delayed-type hypersensitivity reactions on challenge with an antigen. T and B cells are said to be anergic when they cannot respond to their specific antigen.

angioedema swelling of the skin as a result of IgE-mediated allergic reactions, which cause increased permeability of subcutaneous blood vessels with consequent leakage of fluid into the skin.

anterior chamber-associated immune deviation (ACAID) active and systemic state of tolerance to foreign antigens that exists in the eye and which allows it to tolerate corneal grafts of all HLA types. It is due to immunomodulatory factors in the aqueous humor of the anterior chamber of the eye.

antibody the secreted form of the immunoglobulin expressed by a B cell. It is produced by the plasma cells that differentiate from B cells.

antibody repertoire the total number of different specific antibodies that can be made by an individual, estimated at around 10^9.

antibody-dependent cell-mediated cytotoxicity (ADCC) the killing of antibody-coated target cells by NK cells having the receptor FcγRIII (CD16), which recognizes the Fc region of the bound antibody.

anti-CD25 antibody monoclonal antibody against the γ chain (CD25) of the IL-2 receptor, which blocks activation of the receptor and thus blocks the action of IL-2 in stimulating the proliferation and differentiation of antigen-activated lymphocytes. It is used as an immunosuppressant pretreatment in transplant patients to prevent rejection responses.

anti-CD52 monoclonal antibody against the CD52 antigen, which is present on leukocytes. It is used therapeutically to deplete leukocytes from patients undergoing organ transplantation.

antigen any molecule or molecular fragment that either is recognized by an antibody or B-cell receptor or can be bound by an MHC molecule and presented to a T-cell receptor. *See also* **epitope**.

antigen presentation the display of antigen as peptide fragments bound to MHC molecules on the surface of cells. This is the form in which antigen is recognized by α:β T cells.

antigen processing the intracellular degradation of proteins into peptides that bind to MHC molecules for presentation to α:β T cells.

antigen receptor the highly variable cell-surface receptor on lymphocytes that recognizes antigen. For a B cell, the receptor is a cell-surface immunoglobulin; for a T cell, the receptor is a rather similar molecule called the T-cell receptor. All the antigen receptors on an individual lymphocyte are identical and recognize the same antigen epitope.

antigen-binding site the site on an immunoglobulin or T-cell receptor molecule that binds specific antigen.

antigenic determinant *see* **epitope**.

antigenic drift process by which point mutations in influenza virus genes cause alterations in the structure of viral surface antigens. This causes year-to-year antigenic differences in strains of influenza virus.

antigenic shift process by which influenza viruses reassort their segmented genomes and change their surface antigens radically. New viruses arising by antigenic shift are the usual cause of influenza pandemics.

antigen-presenting cell any cell displaying peptide antigen on its surface in combination with MHC molecules so that it can be recognized by an antigen-specific T cell. *See also* **professional antigen-presenting cell**.

anti-idiotypic antibody antibody that binds to an epitope in the antigen-binding site of another immunoglobulin.

antiserum (plural **antisera**) fluid component of clotted blood from an immune individual that contains antibodies against a given antigen. An antiserum contains a heterogeneous collection of antibodies that bind the antigen.

AP-1 family of transcription factors, some of which participate in lymphocyte activation.

APECED acronym for **autoimmune polyendocrinopathy–candidiasis–ectodermal dystrophy**.

apoptosis mechanism of cell death in which the cells to be killed are induced to degrade themselves from within, in a tidy manner. Also called programmed cell death.

appendix gut-associated secondary lymphoid tissue located at the beginning of the colon.

atopic dermatitis an IgE-mediated allergic disease that affects the skin and runs in families. Also called **eczema**.

atopy the genetically determined tendency of some people to produce IgE-mediated hypersensitivity reactions (allergic reactions) against innocuous substances.

autocrine describes a cytokine or other secreted molecule that acts on the same type of cell as the one that secreted it.

autograft graft of tissue taken from one anatomical site and transplanted to another in the same individual.

autoimmune disease disease in which the pathology is caused by an adaptive immune response to normal components of healthy tissue.

autoimmune hemolytic anemia disease characterized by reduced numbers of red blood cells (anemia). The reduction is caused by autoantibodies that bind to surface antigens of red blood cells, resulting in destruction of the cells.

autoimmune polyendocrinopathy–candidiasis–ectodermal dystrophy (**APECED**) autoimmune disease caused by a lack of a protein called **autoimmune regulator** (**AIRE**), which results in the production of T cells reactive to a number of tissues in the body. Also known as autoimmune polyglandular disease (APD).

autoimmune polyglandular disease (**APD**) *see* **autoimmune polyendocrinopathy-candidiasis-ectodermal dystrophy**.

autoimmune regulator (**AIRE**) transcription factor that causes several hundred tissue-specific genes to be transcribed by a subpopulation of epithelial cells in the thymus, and which thus enables the developing T-cell population to become tolerant of antigens that normally occur only outside the thymus.

autoimmune response adaptive immune response directed at an antigenic component of the responder's own body.

autoinflammatory disease type of disease characterized by chronic and recurrent bouts of systemic inflammation mediated by cells of innate immunity, and which does not involve antibodies or effector T cells. Some autoinflammatory diseases have a genetic component, such as familial Mediterranean fever and the hereditary periodic fevers.

autologous describes cells, HLA molecules, and other components, that derive from the individual or person in question.

autologous hematopoietic cell transplantation transplantation of bone marrow or other source of hematopoietic stem cells in which the donor and recipient are the same person. In such cases, bone marrow is removed from the patient, treated in some way to remove diseased or harmful cells, and then reinfused.

autoreactive another word for **self-reactive**.

avidity the overall strength of binding of an antibody with multiple binding sites to an antigen (also with multiple sites), in contrast to the affinity, which is the strength of binding at a single site.

azathioprine immunosuppressive drug that kills dividing cells. It is used in transplantation to help suppress rejection reactions.

azurophilic granule *see* **primary granule**.

B cell one of the two main types of lymphocyte responsible for adaptive immunity (the other is the T cell). The B-cell arm of the immune system is dedicated to making immunoglobulins in the form of cell-surface antigen receptors (the B-cell receptors) and secreted antibodies. Also known as **B lymphocyte**.

B lymphocyte alternative name for **B cell**.

B-1 cell a minority type of B cell, which does not require T-cell help, does not undergo affinity maturation, and participates in early defense, producing antibody of broad specificity and relatively low affinity. It expresses the CD5 glycoprotein and is also known as a CD5 B cell.

B-2 cell the majority type of B cell. It undergoes somatic hypermutation and affinity maturation, does not express the CD5 glycoprotein, and makes antibody of narrow specificity.

B7 molecule either of the B7.1 and B7.2 proteins, which are co-stimulatory molecules present on the surface of professional antigen-presenting cells such as dendritic cells. With other related proteins they form the B7 family of co-stimulatory molecules.

bacteria (singular bacterium) diverse prokaryotic microorganisms that are responsible for many infectious diseases of humans and other animals. Some bacterial pathogens live only extracellularly, colonizing tissue surfaces and intercellular spaces; others can invade cells and live intracellularly.

BAFF B-cell activating factor in the TNF family, a cytokine that promotes B-cell survival.

balancing selection type of evolutionary selection that acts to maintain a variety of phenotypes (such as different variants of MHC molecules) in a population.

bare lymphocyte syndrome type 1 rare genetically determined immunodeficiency disease in which MHC class I molecules are not present on cell surfaces, as a result of non-functional TAP proteins. The result is a deficiency of CD8 T-cell function. Also known as **MHC class I deficiency**.

basophil white blood cell present in small numbers in the blood; it is one of the three types of granulocyte. Basophils contain granules that stain with basic dyes; hence their name.

B-cell co-receptor complex of the polypeptides CD19, TAPA-1, and complement receptor 2 (CR2) that binds to complement on the antigen bearing target cell and so augments the B-cell receptor's response to specific antigen.

B-cell receptor the antigen receptor on B cells, which is a membrane-bound immunoglobulin molecule. Each B cell is programmed to make a single type of immunoglobulin. The cell-surface form of this immunoglobulin serves as the B-cell receptor for specific antigen. Associated in the membrane with the immunoglobulin are the signal transduction molecules Igα and Igβ.

belatacept a chimeric protein of CTLA4 and the Fc portion of an IgG molecule. It acts as an immunosuppressant drug by the CTLA4 binding to B7 molecules on T cells and preventing their co-stimulation by CD28 on antigen-presenting cells, and by the Fc portion activating complement on the surface of dendritic cells and eliminating them.

benign tumor cellular growth, such as a wart, which is caused by an abnormal proliferation of cells but is localized and contained by epithelial barriers.

bone marrow tissue in the center of certain bones that is the major site of generation of all the cellular elements of the blood (hematopoiesis).

bone marrow transplantation the replacement of a person's diseased or functionally deficient blood and immune systems by the infusion of healthy bone marrow, which contains hematopoietic stem cells, from a donor.

broadly neutralizing antibody in reference to HIV-1, a protective antibody response made by a small proportion of infected people that can neutralize a wide range of HIV-1 strains.

bronchial-associated lymphoid tissue (**BALT**) the lymphoid cells and organized lymphoid tissues of the respiratory tract.

bronchiectasis chronic inflammation of the bronchioles of the lung.

C domain *see* **constant domain**.

C region *see* **constant region**.

C1 complement protein C1, the first complement protein activated in the classical pathway of complement fixation. It is a complex of proteins C1q, C1s, and C1r and binds to C-reactive protein or antibody coating an antigen surface. Activated C1 has serine protease activity and cleaves complement component **C4** to produce C4b, and complement component **C2** to produce C2a, which together form the classical C3 convertase C4b2a.

C1 inhibitor (**C1INH**) regulatory protein in plasma that inhibits the enzyme activity of activated complement component C1. C1INH deficiency causes the disease **hereditary angioedema**, in which spontaneous complement activation causes episodes of epiglottal swelling and other symptoms.

C2 complement protein C2; for details *see* C1.

C3 *see* **complement component C3**.

C3 convertase any of the proteolytic enzymes that are formed during complement activation and cleave complement component C3 to C3b and C3a, thereby enabling C3b to bond covalently to antigens. *See also* **alternative C3 convertase**; **classical C3 convertase**.

C4 complement protein C4; for details *see* C1.

calcineurin a cytosolic serine/threonine phosphatase that contributes to T-cell activation. The immunosuppressive drugs cyclosporin A and tacrolimus act by inhibiting calcineurin.

calnexin membrane protein of the endoplasmic reticulum that facilitates the folding of newly synthesized MHC molecules and other glycoproteins.

calreticulin soluble chaperone protein in the endoplasmic reticulum that is part of the peptide-loading complex that loads peptides onto MHC class I molecules in the endoplasmic reticulum.

cancer an umbrella term for a wide variety of diseases caused by abnormal and invasive cell proliferation.

cancer immunosurveillance *see* **immunosurveillance**.

cancer stem cell one of a minority population of cells in a tumor that are self-renewing and more resistant to the cytotoxic drugs and radiation that are commonly used to treat cancer.

cancer vaccine preparation of the patient's own white blood cells that have been engineered to enhance the immune response against cancer antigens.

cancer/testis antigen (**CT antigen**) a subset of tumor-associated antigens that normally are expressed only in immature sperm in the testis.

CAR *see* **chimeric antigen receptor**.

carcinogen any chemical or physical agent that increases the risk of cancer for people who are exposed to them.

carcinoma cancer of epithelial cells.

carrier person who carries one copy of a recessive allele for a hereditary disease and does not show symptoms. Such a person can pass the allele on to successive generations.

caspase type of protease that is involved in generating some cytokines from their inactive pro-proteins as part of a protein complex called an inflammasome. Caspases also have many nonimmune functions.

caspase-recruitment domain a domain in NOD-like receptors that recruits other signaling proteins (in this case not caspases) to the NOD receptor signaling pathway.

catalytic antibody antibody that binds an antigen, chemically changes it, and then releases it.

CCL19 chemokine that is secreted by stromal cells and dendritic cells in secondary lymphoid tissues and attracts leukocytes and dendritic cells to the tissue.

CCL21 chemokine that is secreted by stromal cells and dendritic cells in secondary lymphoid tissues and attracts leukocytes and dendritic cells to the tissue.

CCP module *see* **complement control protein module**.

CCR7 receptor on leukocytes for chemokines CCL19 and CCL21.

CD19 component of the B-cell co-receptor.

CD2 adhesion molecule on T cells that binds to the LFA-3 adhesion molecule on antigen-presenting cells.

CD21 another name for complement receptor 2 (CR2). It is a component of the B-cell receptor.

CD28 the low-affinity receptor on T cells that interacts with B7 co-stimulatory molecules to promote T-cell activation.

CD3 complex complex of signaling proteins that associates with T-cell receptors. It consists of CD3γ, δ, and ε chains, and ζ chains.

CD34 vascular addressin that is expressed on high endothelial venules in lymph nodes and is involved in the extravasation of white blood cells.

CD4 T cell T cell that expresses the CD4 co-receptor and recognizes peptide antigens presented by MHC class II molecules. One of the two main subclasses of α:β T cell, the other being the CD8 T cell.

CD40 cell-surface glycoprotein on B cells that interacts with CD40 ligand on T cells triggering B-cell proliferation, differentiation, and isotype switching.

CD40 ligand a transmembrane protein on T cells that is the ligand for CD40 on B cells and triggers their proliferation and differentiation, and their ability to switch isotype. A defective CD40 ligand gene is one cause of hyper-Ig syndrome, in which only IgM antibodies are made.

CD5 B cells an alternative name for human **B-1 cells**.

CD59 alternative name for **protectin**, a complement control protein.

CD8 T cell T cell that expresses the CD8 co-receptor and recognizes peptide antigens presented by MHC class I molecules. One of the two main subclasses of α:β T cell, the other being the CD4 T cell.

CD81 cell-surface protein that is a component of the B-cell receptor and also a cellular receptor for hepatitis C virus. Also called TAPA-1.

CD94:NKG2A an inhibitory receptor expressed on a subset of NK cells. Its ligand is HLA-E.

CD94:NKG2C an activating receptor expressed on a subset of NK cells. Its ligand is HLA-E.

CDR *see* **complementarity-determining region**.

celiac disease inflammatory hypersensitivity disease of the gut mucosa caused by an immune response to the gluten proteins present in some cereals such as wheat and barley but not rice.

cell-mediated immunity any adaptive immune response in which antigen-specific effector T cells dominate. It is defined operationally as all adaptive immunity that cannot be transferred to a naive recipient with serum antibody.

central lymphoid tissue *see* **primary lymphoid tissue**.

central memory T cell ($\mathbf{T_{CM}}$) one of two subsets of memory T cells (the other being effector memory T cells) that are distinguished by different activation requirements. Central memory T cells have a preference for the T-cell zones of secondary lymphoid tissues and take longer than effector memory T cells to mature into functioning effector T cells after encounter with their specific antigen.

central MHC the class III region of the major histocompatibility complex, located between the class I and II regions.

central tolerance tolerance to self antigens that is generated in B-cell and T-cell populations during their development in the primary lymphoid organs (bone marrow and thymus).

centroblast large dividing B cell present in germinal centers. Somatic hypermutation occurs in these cells, and antibody-secreting and memory B cells derive from them.

centrocyte nondividing B cell in germinal centers. Centrocytes have undergone isotype switching and somatic hypermutation.

CGD acronym for **chronic granulomatous disease**.

checkpoint any one of a number of stages in lymphocyte development at which the production of potentially functional immunoglobulin chains and T-cell receptor chains from rearranged genes is tested by the cell.

Chédiak–Higashi syndrome genetic disease in which phagocytes malfunction. Their lysosomes fail to fuse properly with phagosomes, and killing of ingested bacteria is impaired.

chemokine any member of a large group of small proteins involved in guiding white blood cells to sites where their functions are needed. They have a central role in inflammatory responses.

chimeric antigen receptor (**CAR**) chimeric cell-surface protein composed of a single antigen-binding site spliced to a series of intracellular signaling domains. It has potential applications in cancer immunotherapy.

chimeric monoclonal antibody monoclonal antibody that combines mouse variable regions with human constant regions.

chronic asthma disease characterized by chronic inflammation of the airways and difficulty in breathing. Although probably initiated by exposure to an allergen (**allergic asthma**), chronic asthma can be perpetuated in its absence and is exacerbated by smoking.

chronic granulomatous disease (**CGD**) immunodeficiency disease in which multiple granulomas form as a result of defective elimination of bacteria by phagocytic cells. It is caused by a defect in the NADPH oxidase system of enzymes, which generates the superoxide radical involved in bacterial killing.

chronic rejection rejection of organ grafts that occurs years after transplantation and is characterized by degeneration and occlusion of the blood vessels in the graft. It is caused by an antibody response to the HLA class I alloantigens of the graft.

chronic thyroiditis autoimmune disease that causes progressive destruction of the thyroid gland. Also called Hashimoto's thyroiditis or **Hashimoto's disease**.

class an alternative term for **isotype**, particularly as it pertains to immunoglobulins. The immunoglobulin classes are IgA, IgD, IgE, IgG, IgM.

class I region the part of the major histocompatibility complex that contains the MHC class I heavy-chain genes.

class II region the part of the major histocompatibility complex that contains the MHC class II α- and β-chain genes.

class II-associated invariant-chain peptide (CLIP) peptide of variable length cleaved from the invariant chain protein by proteases in the endosomal pathway. CLIP remains unstably bound in the peptide-binding cleft of an MHC class II molecule until it is removed by the **HLA-DM** protein and replaced by an antigen peptide.

class III region the region of the major histocompatibility complex between the class I and II regions. Also known as the central MHC, it contains no genes for class I or II MHC molecules.

class switching *see* **isotype switching**.

classical C3 convertase surface-associated serine protease of the classical pathway of complement activation. It is composed of the complement components C4b2a and cleaves C3 into C3a and C3b.

classical pathway of complement activation one of three pathways of complement activation. It is activated by antibody bound to antigen, and involves complement components C1, C4, and C2 in the generation of the classical C3 and C5 convertases. *See also* **alternative pathway of complement activation; lectin pathway of complement activation**.

clonal deletion the elimination during lymphocyte development of immature lymphocytes that bind to self antigens. It is the main mechanism that produces self-tolerance.

clonal expansion the multiplication of lymphocytes after their activation by antigen, so that large clones of rare antigen-specific lymphocytes are generated to fight the infecting pathogen.

clonal selection the central principle of adaptive immunity. It is the mechanism by which adaptive immune responses derive only from individual antigen-specific lymphocytes, which are stimulated by the antigen to proliferate and differentiate into antigen-specific effector cells.

clusterin complement-regulatory protein that prevents the soluble complex of C5b with C6 and C7 from associating with cell membranes.

coagulation system collection of enzymes and other proteins in blood that function to form blood clots. The coagulation system is activated by damage to blood vessels.

coding joint the joint between the ends of two rearranged immunoglobulin or T-cell receptor gene segments.

cognate pair a CD4 effector T cell bound via its antigen receptor to its target cell (either a macrophage or a B cell). Also used for other pairs of interacting cells.

collectin any member of a family of calcium-dependent sugar-binding proteins (lectins) containing collagen-like sequences. An example is mannose-binding lectin.

combination therapy antiviral therapy (for HIV, for example) in which several antiviral drugs are used together to try to avoid the rapid generation of mutant viruses resistant to one of the drugs alone.

combination vaccine vaccine that contains antigens derived from more than one pathogen and is designed to provide protection against more than one disease.

commensal a microorganism that habitually lives on or in the human body; one that normally causes no disease or harm and can be beneficial.

commensal microorganism *see* **commensal**.

common gamma chain (γ_c) a protein chain that is the signaling component of several different cytokine receptors, including those for IL-2, IL-4, IL-7, IL-9, and IL-15.

complement collection of plasma proteins that act in a cascade of reactions to attack extracellular forms of pathogens in extracellular spaces and the blood. Pathogens become coated with complement proteins, which can either kill the pathogen directly or facilitate its engulfment and destruction by phagocytes. It is involved in both innate and adaptive immunity and is activated either directly or indirectly by the presence of infection.

complement activation the initiation of a series of reactions involving the complement proteins present in plasma and extracellular fluid, and leading to the death and elimination of the pathogen. It can be triggered either directly or indirectly by the presence of a pathogen. *See also* **alternative pathway of complement activation; classical pathway of complement activation; complement; lectin pathway of complement activation**.

complement component 3 (C3) the central and most important component of the complement system, which is cleaved into C3a and C3b in the complement reactions. C3b becomes bound covalently to pathogen surfaces and acts as an opsonin, facilitating the phagocytosis of the pathogen. C3a is an anaphylatoxin.

complement control protein any of a diverse group of proteins that inhibit complement activation at various stages and by different mechanisms. *See also* **C1 inhibitor; C4-binding protein; decay-accelerating factor; factor I; membrane cofactor protein; protectin**.

complement control protein module family of structurally similar protein modules found in many of the proteins that regulate complement activity.

complement fixation the covalent attachment of C3b or C4b to pathogen surfaces, which is a central feature of the action of complement because it facilitates phagocytosis of the pathogen.

complement receptor 1 (**CR1**) receptor present on macrophages and other cells that binds the C3b fragment of complement deposited on a pathogen surface, thus aiding phagocytosis of the pathogen.

complement receptor 3 (**CR3**) receptor present on macrophages and other cells that binds the iC3b fragment of complement if present on a pathogen surface, thus aiding phagocytosis of the pathogen.

complement receptor 4 (**CR4**) receptor present on macrophages and other cells that binds the iC3b fragment of complement if present on a pathogen surface, thus aiding phagocytosis of the pathogen.

complement system a system of some 30 soluble and cell-surface proteins, comprising the complement proteins and the complement control proteins. It provides a major mechanism in innate and adaptive immunity for identifying and eliminating pathogens and their products. *See also* **complement**.

complementarity-determining region (**CDR**) short region of high diversity in amino acid sequence within the variable region of immunoglobulin and T-cell receptor chains. There are three CDRs (CDR1, CDR2, and CDR3) in each variable region, which collectively contribute to the antigen-binding site and determine the antigenic specificity. The CDRs are the most variable parts of the variable domains and are also called hypervariable regions.

conformational epitope an antigenic structure on a protein that is formed from several separate regions in the primary sequence of a protein brought together by protein folding. Antibodies that bind conformational epitopes bind only to native folded proteins. Also called a **discontinuous epitope**.

conjugate pair *see* **cognate pair**.

conjugate vaccine vaccine made from capsular polysaccharides bound to an immunogenic protein such as tetanus toxoid. The protein provides peptide epitopes that stimulate CD4 T cells to help B cells specific for epitopes of the polysaccharide.

connective tissue mast cell one of two types of **mast cell** in humans (the other is the **mucosal mast cell**). It is found in connective tissues throughout the body.

constant domain (**C domain**) protein domain that makes up the constant regions of the immunoglobulin and T-cell receptor polypeptide chains. Constant regions of T-cell receptor chains and immunoglobulin light chains contain one constant domain; immunoglobulin heavy chains contain two or three, depending on the isotype.

constant region (**C region**) that part of an immunoglobulin or T-cell receptor (or of its constituent polypeptide chains) that is of identical amino acid sequence in molecules of the same isotype but different antigen-binding specificities.

constitutive proteasome the version of the proteasome present in normal uninfected cells.

convergent evolution the independent evolution of two proteins toward the same function.

co-stimulator molecule, **co-stimulatory molecule** cell-surface protein on an antigen-presenting cell that delivers signals to an interacting naive lymphocyte that are required in addition to the antigen-binding signal for the lymphocyte to respond. Co-stimulator molecules include the B7.1 and B7.2 proteins on professional antigen-presenting cells, which engage molecules CD28 and CTLA-4 on T cells. CD40 ligand on T cells serves a co-stimulatory role when it binds to CD40 on B cells.

co-stimulatory receptor the CD28 cell-surface protein on naive T cells that binds to the B7 co-stimulatory molecules on dendritic cells, and delivers an additional activating signal to the T cell.

co-stimulatory signal any signal that is required for the activation of a naive lymphocyte in addition to the signal delivered via the antigen receptor. *See also* **co-stimulator molecule**.

CR1, CR2, CR3, CR4 *see* **complement receptor 1; complement receptor 2; complement receptor 3; complement receptor 4**

CR3 *see* **complement receptor 3**.

CR4 *see* **complement receptor 4**.

C-reactive protein (**CRP**) soluble acute-phase protein that binds to phosphorylcholine, a surface constituent of various bacteria. CRP binds to bacteria, opsonizing them for uptake by phagocytes. It can also activate the classical pathway of complement fixation and bind to Fc receptors.

cross-match test test used in blood typing and histocompatibility testing to determine whether donor and recipient have antibodies against each other's cells that might interfere with successful transfusion or transplantation.

cross-presentation process that occurs within dendritic cells whereby antigen of extracellular origin is presented by MHC class I molecules.

cross-priming initiation of an adaptive immune response by cross-presentation of antigen.

CRP *see* **C-reactive protein**.

cryptdin α-defensin HD5 or HD6, which are antibacterial proteins secreted by the Paneth cells of the small intestine.

cryptic epitope antigenic determinant on a molecule that is normally hidden from the immune system but becomes revealed under conditions of infection or inflammation.

CT antigen *see* **cancer/testis antigen**.

CTLA4 high-affinity inhibitory cell-surface receptor on T cells that interacts with B7 co-stimulatory molecules.

C-type lectin carbohydrate-binding protein in which the binding depends on the presence of calcium. The binding domain is called a **C-type lectin domain**.

CXCL13 chemokine secreted by follicular dendritic cells that attracts B cells into lymphoid follicles.

CXCL8 chemokine produced by activated macrophages and involved in the migration of neutrophils from the blood into infected tissue.

cyclophilin any member of a family of cytoplasmic proteins that bind to the immunosuppressive drugs cyclosporin A and tacrolimus. The complex of drug and cyclophilin binds calcineurin, and this prevents T-cell activation.

cyclophosphamide alkylating agent used as an immunosuppressive drug. It acts by killing rapidly dividing cells, including lymphocytes proliferating in response to antigen.

cyclosporin A immunosuppressive drug that specifically prevents T-cell activation and effector function. Also called cyclosporine.

cytokine any of a large number of proteins secreted by cells that act locally to change the behavior of neighboring cells. Cytokines act by binding to specific receptors on their target cells. Cytokines made by lymphocytes are often called lymphokines or interleukins (abbreviated IL).

cytotoxic drug one of a class of drugs that specifically kill dividing and proliferating cells. Cytotoxic drugs are used as immunosuppressants in transplantation and in autoimmune diseases, and as an anti-cancer treatment.

cytotoxic T cell type of effector T cell that kills its target cells. It expresses the CD8 co-receptor and recognizes peptide antigens presented by MHC class I molecules. It is important in host defense against viruses and other cytosolic pathogens because it can recognize and kill the infected cells.

cytotoxin a protein made by cytotoxic T cells that participate in the destruction of target cells. Perforins, granzymes, and granulysin are examples of cytotoxins.

D gene segment short DNA sequence present in multiple versions in immunoglobulin heavy-chain loci and in T-cell receptor β- and δ-chain loci. In the rearranged functional genes at these loci, a D gene segment connects the V and J gene segments. The D stands for 'diversity,' because the D gene segments provide additional diversity in these receptor chains.

DAF *see* **decay-accelerating factor**.

dark zone the part of a germinal center in secondary lymphoid tissue that contains dividing centroblasts.

DC-SIGN adhesion molecule unique to activated dendritic cells that binds ICAM-3 on T-cell surfaces. Also called CD209.

decay-accelerating factor (**DAF**) cell-surface protein that prevents complement activation on human cells. DAF binds to C3 convertases of both the alternative and classical pathways of complement activation and, by displacing Bb and C2a, respectively, prevents their action.

dectin-1 phagocytic receptor on macrophages that recognizes carbohydrate ligands on pathogens.

defensin any member of a large family of small antimicrobial peptides 35–40 amino acids long that can penetrate microbial membranes and disrupt their integrity. They are present at epithelial surfaces and in neutrophil granules.

delayed-type hypersensitivity (**DTH**) form of cell-mediated immunity elicited by antigen in the skin and mediated by CD4 T_H1 cells. It is called delayed-type hypersensitivity because the reaction appears hours to days after antigen is injected.

dendritic cell professional antigen-presenting cell with a branched, dendrite-like morphology that are present in tissues. It is derived from the bone marrow and is distinct from the follicular dendritic cell that presents antigen to B cells. **Immature dendritic cells** take up and process antigens but cannot yet stimulate T cells. **Mature** or **activated dendritic cells** are present in secondary lymphoid tissues and are able to stimulate T cells.

desensitization therapeutic procedure in which an allergic individual is exposed to increasing doses of allergen with the goal of inhibiting their allergic reactions. It probably works by shifting the balance between CD4 T_H1 and T_H2 cells, thus changing the antibody produced from IgE to IgG.

DiGeorge syndrome recessive genetic immunodeficiency disease in which thymic epithelium fails to develop.

diphtheria toxin cytotoxic protein secreted by the bacterium *Corynebacterium diphtheriae*, the cause of diphtheria, and which causes the disease symptoms. The diphtheria vaccine consists of an inactive form of the toxin called diphtheria toxoid.

direct pathway of allorecognition type of alloreactive response in which T cells of the recipient of a transplant are stimulated by direct interaction of their receptors with the allogeneic HLA molecules expressed by dendritic cells from the donor, present in the transplant.

directional selection type of natural selection that replaces older alleles with newer variants (for example, in the MHC). Its characteristic outcome is change.

discontinuous epitope *see* **conformational epitope**.

diversity gene segment *see* **D gene segment**.

DN thymocyte *see* **double-negative thymocyte**.

dominant describes an allele that influences the phenotype in homozygous and heterozygous individuals. Usually refers to disease-causing alleles that encode proteins with functional defects.

double-negative thymocyte (**DN thymocyte**) immature T cell in the thymus that expresses neither CD4 nor CD8.

double-positive thymocyte (**DP thymocyte**) T cell at an intermediate stage of development in the thymus. It expresses both CD4 and CD8.

DP thymocyte *see* **double-positive thymocyte**.

draining lymph node the lymph node nearest to a site of infection, to which extracellular fluid containing antigen and cells from the site is transported.

early pro-B-cell an early stage of differentiation in the development of B cells, after the B-cell precursor but preceding the pre-B-cell stage.

ectopic lymphoid tissue tissue resembling secondary lymphoid organs that forms in diseased and inflamed organs that do not normally contain lymphoid tissues, for example in the thyroid in Hashimoto's disease.

eczema common allergic skin disease of children, appearing as scaly, reddened and itchy patches on the skin. Also called atopic dermatitis.

edema abnormal accumulation of fluid in connective tissue, leading to swelling.

effector cell a terminally differentiated activated lymphocyte that can kill pathogens or remove them from the body without the need for further differentiation.

effector compartment in the mucosal immune system, a tissue such as the gut lamina propria in which the majority of effector cells reside.

effector mechanism any of the physiological and cellular processes used by the immune system to destroy pathogens and remove them from the body.

effector memory T cell (T_{EM}) one of two subsets of memory T cells (the other being central memory T cells) that are distinguished by different activation requirements. Effector memory T cells have a preference for inflamed tissues and are activated more quickly than central memory cells to mature into functioning effector T cells after encounter with their specific antigen.

efferent lymphatic vessel vessel by which lymph and lymphocytes leave a lymph node en route to the blood.

elite controller the 1 in 300 of HIV-infected people who are able to reduce the virus in their blood to undetectable levels, and maintain their health for decades.

elite neutralizer the 1 in 500 HIV-infected individuals that make small amounts of antibodies that bind and neutralize a broad range of HIV-1 strains.

encapsulated bacteria bacteria that possess thick carbohydrate coats that protect them from phagocytosis. Encapsulated bacteria cause extracellular infections and can be dealt with by phagocytes only if the bacteria are first coated with antibody and complement.

endogenous retrovirus retrovirus that has been integrated into the genome of a species.

endoplasmic reticulum aminopeptidase (**ERAP**) enzyme in the endoplasmic reticulum that removes amino acids from the amino-terminal end of peptides bound to MHC class I molecules to improve their fit, a process known as peptide editing.

endosome intracellular membrane-enclosed vesicle that is formed by the invagination and pinching-off of a portion of plasma membrane, bringing extracellular material into the cell.

endothelium epithelium lining the interior of blood vessels.

engraftment the time at which a bone marrow transplant is making new blood cells.

eosinophil white blood cell that is one of the three types of granulocyte. It contains granules that stain with eosin (hence their name) and whose contents are secreted when the cell is stimulated. Eosinophils contribute chiefly to defense against parasitic infections.

eosinophilia a greater than normal number of eosinophils in the blood. It is present in a variety of diseases and, if activated, the eosinophils can cause tissue damage by release of their toxic granules.

eotaxin chemokine that attracts eosinophils. Also called CXCL11.

epidemic outbreak of infectious disease that affects many individuals within a population.

epidemic of allergy the increase in the incidence of allergy over the past 30 years in developed countries.

epithelium (plural **epithelia**) general name for a type of supracellular structure composed of a single layer (for example the lining of the gut and respiratory tract) or multiple layers (for example the epidermis) of cells bound tightly to each other. They are generally found as surface layers that communicate between the rest of a tissue and its environment. Epithelia line the internal and external cavities of the body, and are also part of many internal organs.

epitope the portion of an antigenic molecule that is bound by an antibody or gives rise to the MHC-binding peptide that is recognized by a T-cell receptor. Also called an antigenic determinant.

epitope spreading the process by which the immune response initially targets epitopes in one part of an antigenic molecule and then progresses to different epitopes. *See also* **intermolecular epitope spreading**; **intramolecular epitope spreading**.

ERp57 thiol reductase protein that is part of the peptide-loading complex that loads peptides onto MHC class I molecules in the endoplasmic reticulum.

erythrocyte red blood cell.

exogenous retrovirus infective retrovirus that originates outside the species being infected.

extracellular infection invasion of the body by pathogens that live outside cells in extracellular spaces, on the surfaces of epithelia, or in the blood.

extravasation the movement of cells or fluid from within blood vessels to the surrounding tissues.

Fab fragment a proteolytic fragment of IgG that consists of the light chain and the amino-terminal half of the heavy chain held together by a disulfide bond between the chains. It is called Fab because it is the 'fragment with antigen binding' specificity. In the intact IgG molecule, the parts corresponding to the Fab fragment are often called Fab or Fab arms.

factor P *see* **properdin**.

factor B plasma protein of the alternative pathway of complement. It binds to C3(H_2O) or C3b and is cleaved to form part of the alternative C3 convertases (C3(H_2O)Bb and C3bBb).

factor D protease that cleaves factor B to Bb and Ba in the alternative pathway of complement activation.

factor H complement regulatory protein in plasma that inactivates the C3 convertase of the alternative pathway and C5 convertases by binding to C3b and rendering it susceptible to cleavage by factor I to produce inactive iC3b.

factor I protease that regulates complement action by cleaving C3b and C4b into inactive forms.

factor J soluble complement control protein that prevents the complex of C5b with C6 and C7 from associating with host-cell membranes and initiating an attack on them.

Fc fragment fragment of an antibody that consists of the carboxy-terminal halves of the two heavy chains disulfide-bonded to each other by the residual hinge region. It is produced by proteolytic cleavage of antibody. It is called Fc for 'fragment crystallizable.' In an intact antibody the part corresponding to the Fc fragment is called **Fc**, **Fc region**, or Fc piece.

Fc receptor cell-surface receptor for the Fc portion of immunoglobulin isotypes. There are different Fc receptors for the different isotypes and subtypes. Examples are the Fcγ receptor FcγRI (that binds IgG) and the Fcε receptor **FcεRI** that binds IgE.

Fc region an alternative term for the **Fc fragment** of an antibody.

FcεRI receptor present on the surface of mast cells, basophils, and activated eosinophils that binds free IgE with very high affinity. When antigen binds to IgE and cross-links FcεRI, it causes cellular activation and degranulation.

FcεRII low-affinity receptor for IgE that is present in both membrane-bound and soluble forms. It is expressed by B cells, T cells, and follicular dendritic cells and is involved in the maintenance of IgE production by B cells.

FcγRI receptor present on macrophages, neutrophils, and eosinophils that binds the Fc regions of IgG antibodies with high affinity, stimulating the internalization and destruction of antigen:antibody complexes. It binds most strongly to IgG3.

FcγRII class of receptor present on various myeloid cells that binds the Fc regions of IgG antibodies with relatively low affinity. The class comprises an activating receptor, **FcγRIIA**, which promotes the internalization of bound antigen:antibody complexes, and two inhibitory receptors (**FcγRIIB1** and **FcγRIIB2**), for which binding of IgG inhibits the activation of the receptor-bearing cell.

FcγRIIIA an activating receptor for IgG, with relatively low affinity. It is the only Fc receptor expressed by NK cells, in which it is responsible for **antibody-dependent cell-mediated cytotoxicity** (**ADCC**). Also expressed on macrophages, neutrophils, and eosinophils. Also called CD16a.

FcRn an Fc receptor that transports IgG across epithelia and has a structure resembling an MHC class I molecule.

FDC *see* **follicular dendritic cell**.

fever rise of body temperature above the normal range. It is caused by cytokines produced in response to infection.

flow cytometry technique in which individual cells can be counted or identified by their cell-surface molecules after fluorescent labeling.

follicle *see* **primary lymphoid follicle**; **secondary lymphoid follicle**.

follicle-associated epithelium the epithelium that overlies a mucosal lymphoid tissue or organ, and through which microorganisms are transported.

follicular center cell lymphoma a malignancy of mature B cells.

follicular dendritic cell (**FDC**) non-lymphoid cell characteristic of follicles in secondary lymphoid tissues. It has long branching processes that make intimate contact with B cells and have Fc and complement receptors that hold antigen:antibody:complement complexes on their surfaces for long periods. These cells are crucial in selecting antigen-binding B cells during antibody responses.

framework region the relatively invariant amino acid sequences within the variable domains of immunoglobulins and T-cell receptors that provide a structural scaffold for the antigen-binding complementarity-determining regions.

fungi a group of single-celled and multicellular eukaryotic organisms, including the yeasts and molds, that can cause a variety of diseases. Immunity to fungi involves both antibody-mediated and cell-mediated responses.

F_v (**variable fragment**) a single polypeptide engineered to contain a heavy-chain V domain and a light-chain V domain that combine to form an antigen-binding site of predefined specificity. It has therapeutic applications as an alternative to monoclonal antibodies.

GALT acronym for **gut-associated lymphoid tissue**, the most extensive secondary lymphoid tissues in the human body.

gamma globulin antibody-containing preparation made from the plasma of healthy blood donors. Such preparations contain antibodies against a range of common pathogens. *See also* **intravenous immunoglobulin**.

gene conversion process whereby one copy of a gene, or part of a gene, is replaced by a different version of that gene, or part of a gene.

gene family set of genes encoding proteins of similar structure, and often of similar function, such as the MHC class I genes.

gene segment short DNA sequence that occurs in multiple slightly variable copies in the immunoglobulin and T-cell receptor genes, and from which the V-region gene is assembled. There are three different types: **V gene segment**, **J gene segment**, and **D gene segment**.

gene-content variation type of genetic variation, seen for example in KIR haplotypes, in which different haplotypes differ in the numbers and types of KIR genes they contain.

genetic polymorphism variation in a population owing to the existence of two or more alleles of a given gene.

germinal center area in secondary lymphoid tissue that is a site of intense B-cell proliferation, selection, maturation, and cell death. Germinal centers form around follicular dendritic cell networks when activated B cells migrate into lymphoid follicles. The cellular and morphological events that form the germinal center and take place there are called the **germinal center reaction**.

germline configuration the unrearranged organization of the immunoglobulin and T-cell receptor genes in the DNA of germ cells and in somatic cells other than T cells and B cells.

germline form *see* **germline configuration**.

GlyCAM-1 vascular addressin present on the high endothelial venules of secondary lymphoid tissues. It is an important ligand for the L-selectin adhesion molecule on naive lymphocytes and directs these cells to leave the blood and enter the lymphoid tissues.

Goodpasture's syndrome autoimmune disease in which autoantibodies against type IV collagen of the basement membrane of blood vessel endothelium cause extensive vasculitis.

graft-versus-host disease (**GVHD**) pathological condition caused by the graft-versus-host reaction (GVHR), which is the response of mature donor-derived T cells in transplanted bone marrow or other type of hematopoietic cell transplant to the alloantigens of the recipient's tissues.

graft-versus-host reaction (**GVHR**) *see* **graft-versus-host disease**.

graft-versus-leukemia (**GVL**) **effect** in hematopoietic cell transplantation as therapy for leukemia, some degree of genetic incompatibility between donor and recipient is thought to help T cells or NK cells from the transplant to eliminate residual leukemia cells in the recipient. Also known as the graft-versus-tumor effect.

graft-versus-tumor (GVT) effect *see* **graft-versus-leukemia effect**.

granulocyte white blood cell with irregularly shaped, multilobed nuclei, and cytoplasmic granules. There are three types of granulocyte: neutrophil, eosinophil, and basophil. Granulocytes are also known as **polymorphonuclear leukocytes**.

granuloma site of chronic inflammation usually triggered by persistent infectious agents such as mycobacteria, or by a non-degradable foreign body. Granulomas have a central area of macrophages, often fused into multinucleate giant cells, surrounded by T lymphocytes.

granulysin membrane-perturbing protein present in the granules of cytotoxic T cells and NK cells. With perforin and serglycin it is thought to make pores in the target cell's membrane.

granzyme any of a group of serine esterases present in the granules of cytotoxic T cells and NK cells. On entering the cytosol of a target cell, granzymes induce apoptosis. Also called fragmentins.

Graves' disease autoimmune disease in which antibodies against the thyroid-stimulating hormone receptor cause the overproduction of thyroid hormone and the symptoms of hyperthyroidism.

gut-associated lymphoid tissue (**GALT**) all lymphoid tissue closely associated with the gastrointestinal tract, including the palatine tonsils, Peyer's patches in the intestine, and layers of intraepithelial lymphocytes.

HAART acronym for **highly active anti-retroviral therapy**.

HAE acronym for **hereditary angioedema**.

half-life in reference to cellular life span, the period of time during which a population of cells reduces to half its original size.

haploidentical transplant tissue or organ transplant from a donor who shares one HLA haplotype with the patient but differs in the second.

haplotype in reference to a linked cluster of polymorphic genes, the set of alleles carried on a single chromosome. Every person inherits two (usually different) haplotypes for such genes, one from each parent. The term was first used in connection with the genes of the **major histocompatibility complex**.

Hashimoto's disease autoimmune disease characterized by persistent high levels of antibodies against thyroid-specific antigens. These antibodies recruit NK cells to the thyroid, leading to damage and inflammation; secondary lymphoid tissue forms in the thyroid, replacing functional thyroid tissue. Also known as **Hashimoto's thyroiditis**.

heavy chain (**H chain**) the larger of the two types of polypeptide in an immunoglobulin molecule. It consists of one variable domain and a number of constant domains. Immunoglobulin heavy chains come in a variety of heavy-chain isotypes, or classes, each of which confers a distinctive effector function on the antibody molecule.

helper T cell general name for effector CD4 T cells that function to help other immune-system cells perform their roles, such as to help B cells produce antibody.

hematopoiesis the generation of the cellular elements of blood, including the red blood cells, white blood cells, and platelets. These cells all originate from pluripotent **hematopoietic stem cells** whose differentiated progeny divide under the influence of various **hematopoietic growth factors**.

hematopoietic cell any blood cell or its precursor cell types.

hematopoietic cell transplantation, hematopoietic stem-cell transplantation type of transplantation in which the role of the graft is to replace the hematopoietic system. Sources of hematopoietic stem cells include bone marrow, peripheral blood, and umbilical cord blood.

hemolytic anemia of the newborn a potentially fatal disease caused by maternal IgG antibodies directed toward paternal antigens expressed on fetal red blood cells. The usual target of this response is the Rh blood group antigen. Maternal anti-Rh IgG antibodies cross the placenta to attack the fetal red blood cells. Also called erythroblastosis fetalis.

hemolytic disease of the newborn *see* **hemolytic anemia of the newborn**.

herd immunity the phenomenon whereby those people in a population who have no protective immunity against a pathogen are largely protected from infection when the majority of the population has protective immunity and is resistant to the pathogen.

hereditary angioedema (**HAE**) genetic disease that results from deficiency of the C1 inhibitor of the complement system (C1INH). In the absence of C1 inhibitor, spontaneous activation of the complement system causes diffuse fluid leakage from blood vessels, the most serious consequence of which is epiglottal swelling leading to suffocation.

heterozygous describes individuals who have inherited different forms (alleles) of a given gene from their two parents.

highly active anti-retroviral therapy (**HAART**) combination therapy for HIV infection, in which several antiviral drugs are used together to try and avoid the rapid generation of drug-resistant mutant viruses that occurs when one of the drugs is used alone.

highly polymorphic describes genes that have many alleles and for which most individuals in a population are heterozygous.

histamine vasoactive amine stored in mast-cell granules. Histamine is released when antigen binds to IgE molecules on mast cells. It causes the dilation of local blood vessels and the contraction of smooth muscle, producing some of the symptoms of immediate hypersensitivity reactions. Antihistamines are drugs that counter histamine action.

histocompatible describes two individuals with the same or near-identical HLA types and so who will not mount strong immune rejection reactions against each other's tissues.

HIV acronym for **human immunodeficiency virus**.

hives itchy, raised, red swellings on the skin, often all over the body, which arise when an allergen to which an individual is sensitized gets carried to the skin by the bloodstream. Also called **urticaria** or nettle rash.

HLA class I molecule the name for the human version of an **MHC class I molecule**.

HLA class II molecule the name for the human version of an **MHC class II molecule**.

HLA complex *see* **human leukocyte antigen complex**.

HLA type the combination of HLA class I and class II allotypes that a person expresses.

HLA-A, **HLA-B**, **HLA-C** the highly polymorphic human MHC class I genes.

HLA-DM an invariant MHC class II molecule in humans that is involved in the intracellular loading of MHC class II molecules with peptides.

HLA-DO a relatively invariant human MHC class II molecule. The function of HLA-DO is to counteract that of HLA-DM.

HLA-DP, **HLA-DQ**, **HLA-DR** the highly polymorphic human MHC class II molecules. Each class II molecule is made from α and β chains encoded by *A* and *B* genes, respectively. For example, the HLA-DPα and HLA-DPβ chains are encoded by the *HLA-DPA* and *HLA-DPB* genes, respectively.

HLA-E, HLA-G relatively invariant human MHC class I molecules that form ligands for NK-cell receptors.

HLA-F a monomorphic human MHC class I molecule that is a chaperone for HLA class I molecules that lose their bound peptide at the cell surface.

Hodgkin's disease malignant disease caused by transformed germinal center B cells.

homing the movement of cells from the circulation to a tissue. For example, naive T cells home secondary lymphoid tissues and effector T cells home to a site of infection.

homologous restriction factor (**HRF**) cell-surface complement control protein that prevents the recruitment of C9 by the complex of C5b, C6, C7, and C8 and formation of the damaging **membrane-attack complex**.

homozygous describes an individual who has inherited the same form (allele) of a given gene from both parents.

human immunodeficiency virus (**HIV**) causative agent of the **acquired immune deficiency syndrome** (AIDS). HIV is a retrovirus of the lentivirus family that infects CD4 T cells, leading to their slow depletion, which eventually results in immunodeficiency and death from infection.

human leukocyte antigen complex (**HLA complex**) the name for the major histocompatibility complex in humans. It is located on the short arm of chromosome 6 and encodes the polymorphic HLA class I and class II molecules, as well as other proteins involved in immune-system function. Individual genes are designated by capital letters, such as HLA-A, and alleles are designated by numbers, such as HLA-A*0201.

humanize refers to the process by which a mouse monoclonal IgG having high affinity and specificity for an antigen of clinical interest is converted into an antibody that resembles a human IgG but retains the specificity and affinity of the mouse antibody. By means of genetic engineering the heavy and light chain CDR loops of a human IgG are replaced by the corresponding CDR loops from the mouse monoclonal antibody. This gives a humanized mouse monoclonal antibody. When administered to a patient the humanized antibody does not provoke a strong anti-IgG response, as is the case for a mouse antibody. Thus patients can be given multiple doses of a humanized antibody over extended periods of time, with much less risk of adverse side effects such as serum sickness.

humoral immunity immunity that is mediated by antibodies and can therefore be transferred to a non-immune recipient by serum.

humors old-fashioned term for the body fluids.

hybridoma hybrid cell lines that make monoclonal antibodies of defined specificity. They are formed by fusing a specific antibody-producing B lymphocyte with a myeloma cell that grows in tissue culture and does not make any immunoglobulin chains of its own.

hygiene hypothesis hypothesis advanced to explain the increasing incidence over the past 50 years of hypersensitivity and autoimmune diseases in developed countries. It is proposed that the increase has occurred because widespread hygiene, vaccination, and antibiotic therapy have prevented children's immune systems from becoming used to dealing appropriately with either natural infections or innocuous environmental antigens.

hyper IgM immunodeficiency genetically determined immunodeficiency disease in which B cells cannot switch their immunoglobulin heavy-chain isotype, and so make unusually high amounts of IgM and no other isotypes. It leads to abnormal susceptibility to infections with pyogenic bacteria, particularly in the sinuses, ears, and lungs. It can be due to several different underlying mutations. Also called hyper IgM syndrome.

hyperacute rejection rapid rejection of an allogeneic tissue graft as a result of preformed antibodies that react against ABO blood group antigens or HLA class I antigens on the graft. The antibodies bind to endothelium and trigger the blood-clotting cascade, leading to **ischemia** and death of the transplanted organ or tissue.

hypersensitivity reactions immune responses to innocuous antigens that lead to symptomatic reactions on re-exposure. These can cause **hypersensitivity diseases** if they occur repetitively. This state of heightened reactivity to an antigen is called **hypersensitivity**. Hypersensitivity reactions are classified by mechanism: type I hypersensitivity reactions involve the triggering of mast cells by IgE antibodies; type II hypersensitivity reactions involve IgG antibodies against cell-surface or matrix antigens; type III hypersensitivity reactions involve the formation of antigen:antibody complexes; and type IV hypersensitivity reactions are mediated by effector T cells.

hyperthyroid describes an abnormally high production of thyroid hormones by the thyroid gland.

hypervariable region *see* **complementarity-determining region**.

hypothyroid describes an abnormally low production of thyroid hormone by the thyroid gland.

iC3 or C3(H_2O) the product formed when the thioester bond of complement component C3 is hydrolyzed by water. Alternatively named C3(H_2O).

ICAM ICAM-1, ICAM-2, ICAM-3: adhesion molecules that are involved in the homing of leukocytes to infected tissues, the homing of lymphocytes to secondary lymphoid organs, and interactions between leukocytes.

ICOS acronym for **inducible T-cell co-stimulator**.

IFN-γ receptor deficiency general name for a genetically determined immunodeficiency caused by a lack, or low levels, of the IFN-γ receptor on macrophages and monocytes, and thus a deficit in their function. The disease is characterized by an inability to clear intracellular bacteria, especially mycobacteria.

Ig domain *see* **immunoglobulin domain**.

Igα, Igβ polypeptide chains associated with the B-cell receptor that transduce signals to the interior of the B cell when the B-cell receptor binds antigen.

IgA *see* **immunoglobulin A**.

IgD *see* **immunoglobulin D**.

IgE *see* **immunoglobulin E**.

IgG *see* **immunoglobulin G**.

IgM *see* **immunoglobulin M**.

IL general abbreviation for **interleukin**.

IL-12 receptor deficiency genetically determined immunodeficiency caused by a lack of the IL-12 receptor on macrophages and monocytes, and thus a deficit in their function. The disease is characterized by an inability to clear intracellular bacteria, especially mycobacteria.

IL-2 abbreviation for **interleukin-2**.

immature B cell specifically, a developing B cell that has rearranged a heavy-chain gene and a light-chain gene and expresses surface IgM but not IgD. It can also be used in a general way to describe any developing B cell before it has reached maturation.

immature dendritic cell dendritic cell present in tissues, which takes up antigen but does not express co-stimulatory molecules and cannot yet act as a professional antigen-presenting cell to naive T cells. After activation by antigen it develops into a **mature dendritic cell**.

immediate hypersensitivity an alternative term for a type I or IgE-mediated **hypersensitivity reaction**, as the reaction in a sensitized person occurs within minutes of exposure to antigen, as a result of the activation of mast cells by antigen cross-linking of bound preexisting IgE. Also called **immediate reaction**.

immediate reaction *see* **immediate hypersensitivity**.

immune resistant to infection.

immune complex protein complex formed by the binding of antibodies to soluble antigens. The size of immune complexes depends on the relative concentrations of antigen and antibody. Large immune complexes are cleared by phagocytes bearing Fc and complement receptors. Small soluble immune complexes tend to be deposited on the walls of small blood vessels, where they can activate complement and cause damage.

immune dysregulation, polyendocrinopathy, enteropathy, and X-linked syndrome (**IPEX**) an immunodeficiency disease caused by lack of FoxP3, a transcription factor necessary for the development of regulatory T cells.

immune system the tissues, cells, and molecules involved in the defense of the body against infectious agents.

immune thrombocytopenia rare bleeding disorder in which autoantibodies and immune complexes cause platelet destruction and perturb platelet production.

immunity the ability to resist a specific infection.

immunization the deliberate provocation of an adaptive immune response by introducing antigen into the body.

immunodeficiency disease any inherited or acquired disorder in which some part or parts of the immune system are either absent or defective, resulting in failure to mount an effective immune response to pathogens.

immunogenetics subfield of immunology that deals with the polymorphism and population genetics of genes and proteins that are important for immune system function.

immunoglobulin the general name for antibodies and B-cell antigen receptors.

immunoglobulin A (IgA) the class of immunoglobulin having α heavy chains. Dimeric IgA antibodies are the main antibodies present in mucosal secretions. Monomeric IgA is present in the blood.

immunoglobulin D (IgD) the class of immunoglobulin having δ heavy chains. It appears as surface immunoglobulin on mature naive B cells but its function is unknown. Transcription of IgD is coordinated with that of IgM.

immunoglobulin domain (Ig domain) protein structure module consisting of about 100 amino acids that fold into a sandwich of two β sheets held together by a disulfide bond. Immunoglobulin heavy and light chains are made up of a series of immunoglobulin domains. Similar domains are present in numerous other immune system proteins.

immunoglobulin E (IgE) the class of immunoglobulin having ε heavy chains. IgE is involved in reactions against internal parasites, particularly helminth worms, and in allergic reactions.

immunoglobulin G (IgG) the class of immunoglobulin having γ heavy chains. IgG is the most abundant class of immunoglobulin in plasma.

immunoglobulin M (IgM) the class of immunoglobulin having μ heavy chains. IgM is the first immunoglobulin to appear on the surface of B cells and the first antibody secreted during an immune response. It is secreted in pentameric form.

immunoglobulin superfamily the name given to all the proteins that contain one or more immunoglobulin or immunoglobulin-like domains.

immunoglobulin-like domain protein domain that resembles the immunoglobulin domain in structure but is present in a variety of other proteins, including MHC class I and II molecules, CD4, and CD8.

immunological memory the capacity of the immune system to make quicker and stronger adaptive immune responses to successive encounters with an antigen. It is specific for a particular antigen and is long-lived.

immunophilin intracellular protein with peptidyl-prolyl *cis-trans* isomerase activity that binds the immunosuppressive drugs cyclosporin A, tacrolimus, and sirolimus (rapamycin)

immunoreceptor tyrosine-based activation motif (ITAM) sequence in the cytoplasmic domains of membrane receptors that are sites of tyrosine phosphorylation and of association with tyrosine kinases and phosphotyrosine-binding proteins involved in signal transduction. Related motifs with opposing effects are **immunoreceptor tyrosine-based inhibitory motifs (ITIMs)**, which recruit phosphatases that remove the phosphate groups added by tyrosine kinases.

immunoreceptor tyrosine-based inhibitory motif (ITIM) sequences in the cytoplasmic domains of membrane receptors that inhibit signaling by recruiting phosphatases that remove activating phosphate groups.

immunosuppressive drug any drug that prevents an immune response, usually by interfering with T-cell activation, proliferation, and differentiation.

immunosurveillance the capacity of the immune system to recognize cancer cells at an early stage and eliminate them before they cause disease.

immunotoxin conjugate composed of a specific antibody chemically coupled to a toxic protein usually derived from a plant or microbe. The antibody is designed to bind specifically to target cells, such as cancer cells, and deliver the toxin to kill them.

inactivated virus vaccine *see* **killed virus vaccine**.

indirect pathway of allorecognition one means by which alloreactive T cells in the recipient of a transplant can be stimulated to react against the transplant. The alloreactive T cells do not directly recognize the transplanted cells but recognize subcellular material from these cells that has been processed and presented by the recipient's own dendritic cells.

induced regulatory T cell regulatory T cell that emerges during the course of an immune response by the differentiation of naive CD4 T cells. *See also* **natural regulatory T cell**.

inducible T-cell co-stimulator (ICOS) cell-surface protein on T_{FH} cells, structurally related to CD28 and CTLA4, that binds to the ICOS ligand on dendritic cells and transduces stimulating signals.

inductive compartment in the mucosal immune system, the lymphoid tissue directly beneath the mucosal epithelium, where interactions between antigen, dendritic cells, and lymphocytes induce adaptive immune responses.

inflammasome protein complex formed by the NOD-like receptor NLRP3 and caspase in response to pathogens. The caspase cleaves pro-IL-1β to give the active inflammatory cytokine IL-1β.

inflammation general term for the local accumulation of fluid, plasma proteins, and white blood cells that is initiated by physical injury, infection, or a local immune response. It is also known as an **inflammatory response**. The cells that invade tissues and help mediate inflammation are often called **inflammatory cells**. Those cytokines that promote inflammation are known as **inflammatory cytokines**.

inflammatory cell *see* **inflammation**.

inflammatory cytokine *see* **inflammation**.

inflammatory mediator any substance released by various cell types that contribute to the production of inflammation at a site of infection or trauma.

inflammatory response *see* **inflammation**.

influenza hemagglutinin glycoprotein in the coat of the influenza virus that binds to certain carbohydrates on human cells, the first step in infection of cells with the virus. Changes in the hemagglutinin are the major source of the antigenic shift that heralds a major new pandemic. The protein is called a hemagglutinin because it can agglutinate red blood cells.

innate immune response immune response that is initiated immediately on infection and does not depend on lymphocytes. It depends on host defenses such as complement, neutrophils, macrophages, and NK cells, which provide nonspecific defense against a wide range of pathogens. This response does not generate immunological memory.

innate immunity the host defense mechanisms that act from the start of an infection and do not adapt to a particular pathogen or generate immunological memory.

iNOS inducible nitric oxide synthase produced by dendritic cells; it induces increased expression of the B cells' TGF-β receptor, aiding switching to the IgA isotype.

insulitis infiltration of the pancreatic islets of Langerhans with lymphocytes and other leukocytes. This is a symptom of early type 1 diabetes, but is also seen in the non-autoimmune type 2 diabetes.

integrin any member of a class of cell-surface glycoproteins that mediate adhesive interactions between cells and the extracellular matrix. They comprise two subunits, α and β.

interallelic conversion mechanism of genetic recombination between two alleles of a gene in which a segment of one allele is replaced with the homologous segment from the other. This mechanism is used to generate new HLA class I and II alleles.

interferon general name for a family of cytokines that act specifically to induce cells to resist virus infection. The main interferons are **interferon-α (IFN-α)** and **interferon-β (IFN-β)**, which are known as type I interferons and are produced by leukocytes and fibroblasts, respectively, as well as by other cells. *See also* **interferon-γ**.

interferon response this refers to changes in the expression of a variety of genes in cells exposed to interferon that help the cells resist virus infection.

interferon-γ (IFN-γ) a cytokine structurally and functionally unrelated to the type I interferons. It is a product of CD4 T_H1 cells, CD8 T cells, and NK cells, and acts principally to activate macrophages in both innate and adaptive immunity.

interleukin generic term used for many of the cytokines produced by leukocytes. *See also* **IL-1**; **IL-2**; etc.

interleukin-1β (IL-1β) *see* **interferon**.

interleukin-12 (IL-12) cytokine released by macrophages that activates NK cells.

interleukin-2 (IL-2) cytokine produced by activated T cells that is essential for the proliferation of activated T cells and the development of an adaptive immune response.

interleukin-6 (IL-6) cytokine released by macrophages, which with IL-1 and TNF-α induces a wide range of inflammatory responses at early times in infection.

intermolecular epitope spreading process by which the immune response initially reacts against an epitope of one antigenic molecule and then progresses to epitopes on different proteins.

intracellular infection invasion of the body by a pathogen that can replicate inside human cells.

intraepithelial lymphocyte distinctive types of CD8 T cell and γ:δ T cell that are integrated into the epithelial layer of the small intestine.

intraepithelial pocket invagination of the basolateral plasma membrane of the M cell in the gut epitheliuam, which enables the transported antigens and microorganisms to encounter dendritic cells, T cells, and B cells in the underlying lymphoid tissue.

intramolecular epitope spreading process by which the immune response initially reacts against epitopes in one part of an antigenic molecule and then progresses to respond to other, non-cross-reactive, epitopes in the same molecule.

intravenous immunoglobulin (IVIG) preparation of serum gamma globulin containing many different antibodies that is used as a treatment to replace antibodies and increase platelets in immunodeficiency and autoimmune diseases.

invariant chain (Ii) polypeptide that associates with major histocompatibility complex (MHC) class II proteins in the endoplasmic reticulum and prevents them from binding peptides there. Ii is degraded in endosomes, enabling MHC class II molecules to bind endosomal peptides.

IPEX (immune dysregulation, polyendocrinopathy, enteropathy, and X-linked syndrome) an immunodeficiency disease caused by lack of FoxP3, a transcription factor necessary for the development of regulatory T cells.

ischemia deficient blood supply due to obstructed blood vessels.

islets of Langerhans the endocrine hormone-producing tissue of the pancreas, which includes the β cells that produce insulin.

isoform general term for a particular MHC protein.

isograft graft of a tissue or organ from one genetically identical individual to another.

isolated lymphoid follicle secondary lymphoid tissue in the gut wall that resembles a lymphoid follicle and is composed mainly of B cells.

isotype the class of an immunoglobulin—that is, IgM, IgG, IgD, IgA, or IgE—each of which has a distinct heavy-chain constant region encoded by a different constant-region gene. The heavy-chain constant region determines the effector properties of each antibody class.

isotype switching the process by which a B cell changes the class of immunoglobulin it makes while preserving the antigenic specificity of the immunoglobulin. Isotype switching involves a somatic recombination process that attaches a different heavy-chain constant-region gene to the existing variable-region exon. Also called class switching.

ITAM *see* **immunoreceptor tyrosine-based activation motif**.

ITIM *see* **immunoreceptor tyrosine-based inhibitory motif**.

IVIG *see* **intravenous immunoglobulin**.

J gene segment relatively short DNA sequence present in multiple different copies at all immunoglobulin and T-cell receptor loci. At gene rearrangement, a J and a V gene segment are joined directly in immunoglobulin light-chain genes and T-cell receptor α- and γ-chain genes, and via a D gene segment in the immunoglobulin heavy-chain gene and T-cell receptor β- and δ-chain genes. The J stands for 'joining.'

JAK (Janus kinase) family of tyrosine kinases that transduce activating signals from cytokine receptors.

joining gene segment *see* **J gene segment**.

junctional diversity diversity present in immunoglobulin and T-cell receptor polypeptides that is created during the process of gene rearrangement by the addition or removal of nucleotides at the junctions between gene segments.

kappa (κ) one of the two types of immunoglobulin light chain; the other is λ.

killed virus vaccine a vaccine which contains viral particles that have been deliberately killed by heat, chemicals, or radiation. Also called inactivated virus vaccine.

killer-cell immunoglobulin-like receptor (KIR) member of a family of receptors on NK cells that bind to MHC class I molecules on target cells. Some transmit activating signals and some transmit inhibitory signals to the NK cell.

kinin system enzymatic cascade of plasma proteins that is triggered by tissue damage and helps to facilitate wound healing.

KIR ligand epitope of an HLA class I molecule that binds to a killer-cell immunoglobulin-like receptor (KIR) on NK cells.

Kupffer cell phagocytic cell in the liver that lines the hepatic sinusoids.

lambda (λ) one of the two types of immunoglobulin light chain; the other is κ.

lamina propria connective tissue underlying the epithelium and lymphoid tissues in the gut.

large granular lymphocyte an alternative and earlier name for the natural killer cell (NK cell). Microscopy revealed two types of blood lymphocyte: the small lymphocytes (B cells and T cells) and the large granular lymphocytes (NK cells).

large pre-B cell the less mature of the two pre-B-cell stages. A large pre-B cell has successfully rearranged a heavy-chain gene and makes a μ heavy chain that assembles with the surrogate light chain to form the pre-B-cell receptor. It has stopped rearrangement of the heavy-chain genes but has yet to commence rearrangement of the light-chain genes. *See also* **small pre-B cell**.

late pro-B cell an early stage in the development of B cells, after the B-cell precursor but preceding the pre-B-cell stage. Heavy-chain gene rearrangement is under way. D_H to J_H joining has occurred in the early pro-B-cell stage, whereas V_H to DJ_H joining occurs at this stage.

latency state adopted by some viruses in which they have entered cells but do not replicate.

late-phase reaction that part of a type I hypersensitivity reaction that occurs 7–12 hours after contact with antigen and is resistant to treatment with antihistamine.

Lck intracellular protein tyrosine kinase associated with the CD4 and CD8 co-receptors of T cells.

lectin general name for proteins that recognize carbohydrates.

lectin pathway of complement activation one of the three pathways of complement activation. It is activated by the binding of a mannose-binding lectin present in blood plasma to mannose-containing peptidoglycans on bacterial surfaces. *See also* **alternative pathway of complement activation**; **classical pathway of complement activation**.

lentiviruses group of slow retroviruses that includes the **human immunodeficiency virus** (HIV). They have a long incubation period, and disease can take years to become apparent.

leucine-rich repeat region (LRR) motif repeated multiple times in the pathogen-recognition domains of Toll-like receptors and NOD-like receptors.

leukemia disease resulting from the unrestrained proliferation of a malignant white blood cell and characterized by abnormally high numbers of white cells in the blood. A leukemia can be lymphocytic, myelocytic, or monocytic, depending on the type of white blood cell that became malignant.

leukocyte general term for a white blood cell. Lymphocytes, granulocytes, and monocytes are all leukocytes.

leukocyte adhesion deficiency immunodeficiency disease in which the common β chain of the leukocyte **integrins** is not made. The resulting deficiency of integrins mainly affects the ability of leukocytes to enter sites infected with extracellular pathogens, and so such infections cannot be effectively eradicated.

leukocyte function-associated antigen-3 *see* **LFA-3**.

leukocyte receptor complex (LRC) complex of genes encoding immunoglobulin-like receptors on human NK cells, including the killer-cell immunoglobulin-like receptors (KIRs).

leukocytosis condition in which there are increased numbers of leukocytes in the blood. It is commonly seen in acute infection.

LFA-3 leukocyte function-associated antigen-3, cell adhesion molecule of the immunoglobulin superfamily that is expressed on antigen-presenting cells (and other types of cell) and mediates their adhesion to T cells.

light chain (L chain) the smaller of the two types of polypeptide chain in an immunoglobulin molecule. It consists of one variable and one constant domain, and is disulfide-bonded to a heavy chain in the immunoglobulin molecule. There are two classes of light chain, known as κ and λ.

light zone the part of a germinal center in secondary lymphoid tissue that contains non-dividing centrocytes interacting with follicular dendritic cells.

linear epitope antigenic structure in a protein that consists of a linear sequence of amino acids within the protein's primary structure.

lingual tonsil aggregates of secondary lymphoid tissue at the back of the tongue.

linkage disequilibrium the situation when particular alleles of two or more polymorphic genes (for example those comprising an HLA haplotype) are inherited together at frequencies higher than expected by chance.

linked recognition the mechanism whereby B cells and T cells specific for different epitopes of the same antigen can cooperate. The B cell internalizes and degrades antigen bound to its B-cell receptor, and presents antigenic peptides on its MHC molecules. Helper T cells recognizing these peptide antigens help activate the B cell, which then produces antibodies against its specific antigen.

lipid-transfer protein proteins in endosomal vesicles that are involved in the removal of self lipids from CD1 molecules and their replacement by mycobacterial lipids.

live-attenuated virus vaccine a vaccine composed of live viruses having an accumulation of mutations that impede their growth in human cells and their capacity to cause disease.

L-selectin an adhesion molecule of the selectin family found on lymphocytes. It binds to CD34 and GlyCAM-1 on high endothelial venules to initiate the migration of naive lymphocytes into secondary lymphoid tissue.

lymph mixture of extracellular fluid and cells that is carried by the lymphatic system.

lymph node a type of secondary lymphoid tissue found at many sites in the body where lymphatic vessels converge. Antigens are delivered by the lymph and presented to lymphocytes within the lymph node to initiate adaptive immune responses.

lymphatic vessel (**lymphatic**) thin-walled vessel that carries lymph. Lymphatic vessels are part of the lymphatic system that transports lymph from tissues to secondary lymphoid tissues (with the exception of the spleen) and from secondary lymphoid tissues to the thoracic duct.

lymphocyte general name for a class of white blood cells consisting of three main cell types. B lymphocytes (B cells) and T lymphocytes (T cells) are small cells that bear variable cell-surface receptors for antigen and are responsible for adaptive immune responses. Natural killer (NK) cells are large granular cells and are the lymphocytes of innate immunity.

lymphocyte recirculation the continual migration of naive lymphocytes from blood to secondary lymphoid tissues to lymph and back to the blood. An exception to this pattern is traffic to the spleen: lymphocytes both enter and leave the spleen in the blood.

lymphoid follicle an aggregation of mainly B cells in secondary lymphoid tissues. Naive B cells must pass through follicles for their survival; after B cells have been activated by antigen they enter the follicles, where they proliferate and undergo somatic hypermutation and isotype switching.

lymphoid lineage all types of lymphocyte, and the bone marrow cells that give rise to them.

lymphoid organ any of numerous organized tissues that contain large numbers of lymphocytes held in a non-lymphoid stroma. The primary lymphoid organs, where lymphocytes are generated, are the thymus and bone marrow. The main secondary lymphoid tissues, in which adaptive immune responses are initiated, are the lymph nodes, spleen, and mucosa-associated lymphoid tissues such as tonsils, Peyer's patches, and isolated lymphoid follicles.

lymphoid progenitor stem cell in bone marrow cell that gives rise to all lymphocytes.

lymphoid tissue tissue that contain large numbers of lymphocytes, including the permanent organized lymphoid organs and more diffuse and transient collections of lymphocytes that arise in mucosal tissues.

lymphokine any cytokine produced by lymphocytes.

lymphoma tumor of lymphocytes that grows in lymphoid and other tissues but in which the malignant lymphocytes do not enter the blood in large numbers.

lysosome digestive intracellular organelle that contains degradative enzymes and breaks down macromolecules.

lytic granule intracellular storage granule of cytotoxic T cells and NK cells that contain the cytotoxins perforin, granulysin, and granzymes. These proteins are released when the lymphocytes interact with their target cells and are responsible for killing the target cells.

Microfold cell (M cell) specialized cell type in intestinal epithelium through which antigens and pathogens enter gut-associated lymphoid tissue from the intestines.

macrophage large mononuclear phagocytic cell resident in most tissues that bears receptors for many components of pathogens. Macrophages are derived from blood monocytes and contribute to innate immunity and early nonadaptive phases of host defense, as well as being effector cells in adaptive immune responses. They function as professional antigen-presenting cells and as phagocytic scavenger cells, and produce cytokines that recruit a variety of immune-system cells to an ongoing immune response.

macrophage activation stimulation of macrophages, which increases their phagocytic, antigen-presenting, and bacterial killing functions. It occurs in the course of infection.

macrophage receptor with collagenous structure (**MARCO**) a type of scavenger receptor on the surface of macrophages that binds both Gram-negative and Gram-positive bacteria.

macropinocytosis the nonspecific uptake of large amounts of extracellular fluid by endocytosis, a characteristic of dendritic cells.

MAdCAM-1 mucosal cell adhesion molecule-1, a mucosal addressin that is recognized by the lymphocyte surface proteins L-selectin and VLA-4. This interaction mediates the specific homing of lymphocytes to mucosal tissues.

MAIT cell *see* **mucosal-associated invariant T cell.**

major basic protein constituent of eosinophil granules that is released on eosinophil activation. It acts on mast cells to cause their degranulation.

major histocompatibility complex (**MHC**) large cluster of mainly immune-system genes present in all vertebrates that encodes, among other proteins, the highly polymorphic MHC class I and class II molecules, which present peptide antigens to T cells and are the basis of an individual's tissue type.

malignant transformation the changes that occur in a cell to make it cancerous.

malignant tumor tumor that is capable of uncontrolled and invasive growth.

mannose receptor cell-surface receptor on dendritic cells, macrophages, and other leukocytes that binds to mannose residues on the surfaces of pathogens.

mannose-binding lectin (**MBL**) soluble acute-phase protein in the blood that binds to mannose residues on pathogen surfaces and, when bound, activates the complement system by the lectin pathway.

mast cell large bone marrow derived cell resident in connective tissues throughout the body. Mast cells contain large granules that store a variety of chemical mediators including **histamine**. Mast cells have high-affinity Fcε receptors (FcεRI) that bind free IgE. Antigen binding to IgE associated with mast cells triggers mast-cell activation and degranulation, producing a local or systemic immediate **hypersensitivity reaction**. Mast cells have a crucial role in **allergic reactions**.

master regulator transcription factor that controls the differentiation of a naive CD4 T cell into a particular type of effector T cell. These are T_H1 cells, T_H2 cells, T_H17 cells, T_{FH} cells, and T_{reg} cells.

mature B cell in reference to B-cell development, a B cell that has IgM and IgD on its surface and is able to respond to antigen.

mature dendritic cell dendritic cell in secondary lymphoid tissues that expresses co-stimulatory molecules and other cell-surface molecules that enable it to present antigen to naive T cells and activate them.

MBL *see* **mannose-binding lectin**.

MCP *see* **membrane cofactor protein**.

MDA-5 a **RIG-I-like receptor** in the cytoplasm of many cells that recognizes viral RNAs.

medullary sinus macrophage highly phagocytic macrophage resident in the sinuses of the lymph node medulla. These macrophages filter lymph before it leaves the node, removing any pathogens and their antigens.

megakaryocyte large cell of the erythroid lineage, produced in the bone marrow and resident there, from which blood platelets are produced by fragmentation.

membrane cofactor protein (**MCP**) membrane-associated complement regulatory protein on human cells that promotes the inactivation of C3b and C4b by factor I.

membrane-attack complex the complex of terminal complement components that forms a pore in the membrane of the target cell, damaging the membrane and leading to cell lysis.

memory B cell long-lived antigen-specific B cell that is produced from activated B cells during the primary immune response to an antigen. On subsequent exposure to their specific antigen, memory B cells are reactivated to differentiate into plasma cells as part of the secondary and subsequent immune responses to that antigen.

memory cell general term for a lymphocyte that is responsible for the phenomenon of immunological memory.

memory T cell long-lived antigen-specific T cell that is produced from activated T cells during the primary response to an antigen. On subsequent exposure to their specific antigen, memory T cells are activated to differentiate into effector T cells as part of the secondary and subsequent immune responses to that antigen.

mesenteric lymph node one of a chain of lymph nodes in the mesentery, the membrane that holds the gut in place. These lymph nodes connect by lymphatics to the lymphoid tissues of the gut, and pathogen-bearing dendritic cells are transported there to initiate additional adaptive immune responses against gut pathogens.

metastasis the spread of a tumor from its site of origin to other tissues. Some cells from the primary tumor invade other tissues, and grow and divide to become secondary tumors.

methotrexate cytotoxic drug used to inhibit **graft-versus-host reactions** in hematopoietic cell transplant recipients.

MHC class II deficiency a genetically determined immunodeficiency in which MHC class II molecules are not produced and which results in a lack of CD4 T-cell function and severe immunodeficiency.

MHC class II transactivator a transcriptional activator of MHC class II genes. If defective, it results in the absence of MHC class II molecules on all cells, causing a severe immunodeficiency (**MHC class II deficiency**) characterized by a lack of CD4 T-cell function.

MHC class I deficiency rare genetically determined immunodeficiency disease in which MHC class I molecules are not present on cell surfaces, as a result of nonfunctional TAP proteins. The result is a deficiency of CD8 T-cell function. Also known as **bare lymphocyte syndrome type 1**.

MHC restriction the property that a given T-cell receptor recognizes its peptide antigen only when the peptide is bound to a particular form of MHC molecule.

MIC glycoprotein MIC-A and MIC-B are stress-induced glycoproteins that appear on the surface of epithelial cells in response to infection, damage, or other stresses, and which are recognized by the NKG2D receptor of NK cells.

MIC-A and MIC-B *see* **MIC glycoprotein**.

microbiota the microorganisms that habitually live in or on the human body; they normally do not cause disease and in many cases provide positive benefit.

microfold cell (**M cell**) cell in the gut mucosa that facilitates the transport of pathogens and antigens from the gut lumen to secondary lymphoid tissues underlying the gut epithelium.

micropinocytosis the internalization of small volumes of extracellular fluid in membrane vesicles.

MIIC acronym for MHC class II compartment and pronounced 'em-two-see.' An endosomal compartment in professional antigen-presenting cells where MHC class II molecules load with peptides derived from pathogens and antigens that have been taken up from the extracellular environment.

minor histocompatibility antigen any antigen derived from a polymorphic cellular protein that can lead to the rejection of a tissue or organ transplant in a genetically non-identical individual.

minor histocompatibility locus a gene encoding a protein that can act as a minor histocompatibility antigen.

mixed lymphocyte reaction (**MLR**) cellular assay for detecting MHC differences between two individuals. The T cells from one individual proliferate in response to allogeneic MHC molecules on the cells of the other individual.

MLR *see* **mixed lymphocyte reaction**.

molecular mimicry antigenic similarity between a pathogen antigen and a cellular antigen, which results in the induction of antibodies or T cells that act against the pathogen but also cross-react with the self antigen.

monoclonal antibody antibody produced by a single clone of B lymphocytes so that all the antibody molecules are identical in structure and antigen specificity.

monocyte phagocytic white blood cell with a bean-shaped nucleus. It is the precursor of the tissue macrophage.

monomorphic having only one form. The term is used, for example, for genes that have only one allele.

mucin any of a family of very large glycoproteins that are secreted by mucosal epithelia and are the main components of the mucus that protects these surfaces and helps keep bacteria away from them.

mucosa (plural **mucosae**) mucus-secreting epithelium such as that lining the respiratory, intestinal, and urogenital tracts. The mammary glands and the conjunctiva of the eye are also considered in this category. Mucosal epithelium communicates with the external environment and is the route of entry of most pathogens.

mucosa-associated lymphoid tissue (**MALT**) aggregations of lymphoid cells in mucosal epithelia and in the lamina propria beneath. The main mucosa-associated lymphoid tissues are the **gut-associated lymphoid tissue** (**GALT**) and the **bronchial-associated lymphoid tissue** (**BALT**).

mucosal mast cell one of two types of **mast cell** in humans (the other is the **connective tissue mast cell**). It is found in mucosal tissues throughout the body.

mucosal surface *see* **mucosa**.

mucosal-associated invariant T cell (**MAIT cell**) population of effector CD8 α:β T cells that populate mucosal tissues, particularly the lungs, and are also present in the liver and to a smaller extent the blood. Their T-cell receptors recognize metabolites of riboflavin bound to the MR1 protein.

mucus slimy protective secretion composed of glycoproteins, proteoglycans, peptides, and enzymes that is produced by the goblet cells in many internal epithelia.

multiple sclerosis chronic progressive neurological disease characterized by patches of demyelination in the central nervous system and lymphocyte infiltration of the brain. It is believed to be an autoimmune disease.

multivalent having more than one binding site for the same or different ligands.

mutagen any agent, such as a chemical or radiation, that can cause a mutation.

mutation an alteration in the DNA sequence of a gene.

myasthenia gravis autoimmune disease in which autoantibodies against the acetylcholine receptor on skeletal muscle cells cause a block of signal transmission from nerve to muscle at neuromuscular junctions, leading to progressive muscle weakness and eventually to death.

mycophenolic acid the active derivative of the immunosuppressive and cytotoxic drug mycophenolate mofetil, which is metabolized to mycophenolic acid in the liver. It acts by inhibiting guanine synthesis.

myeloablative therapy conditioning regime used before some types of hematopoietic cell transplant. The recipient's immune system is destroyed by a combination of cytotoxic drugs and irradiation. Myeloablative means destroying the bone marrow.

myeloid lineage the granulocytes, monocytes, macrophages, mast cells and dendritic cells, and the bone marrow cells that give rise to them.

myeloid progenitor the stem cell in bone marrow that gives rise to the myeloid lineage (granulocytes, monocytes, macrophages, mast cells, and dendritic cells).

myeloma tumor of plasma cells resident in bone marrow.

N nucleotides non-templated nucleotides added at the junctions between gene segments of T-cell receptor and immunoglobulin heavy-chain variable-region sequences during somatic recombination. The N nucleotides contribute to the diversity of these molecules. They are not encoded in the gene segments but are inserted at random by the enzyme **terminal deoxynucleotidyltransferase** (**TdT**).

NADPH oxidase multisubunit enzyme that produces superoxide radicals and contributes to the killing of internalized pathogens in neutrophils. A deficiency in any of the NADPH subunits can be a cause of chronic granulomatous disease, an immunodeficiency disease.

naive B cell mature B cell that has left the bone marrow but has not yet encountered its specific antigen.

natural killer cell (**NK cell**) large, granular, cytotoxic lymphocyte that circulates in the blood and is important in innate immunity to intracellular pathogens such as viruses. NK cells do not have variable receptors for antigen but have numerous other receptors by which they can recognize and kill virus-infected cells and certain tumor cells. Also known as **large granular lymphocyte**.

natural killer complex (**NKC**) a genomic region of gene families encoding lectin-like receptors of NK cells and other leukocytes.

natural regulatory T cell regulatory CD4 T cells that are specified as such during their development in the thymus.

negative selection process in the thymus whereby developing T cells that recognize self antigens are induced to die by apoptosis.

NEMO deficiency alternative name for **X-linked hypohidrotic ectodermal dysplasia and immunodeficiency**.

neoplasm a tumor, which may be benign or malignant.

netosis mechanism by which the death of neutrophils produces neutrophil extracellular traps (NETs) that trap and kill pathogens. The neutrophil nucleus swells and bursts, the chromatin dissolves and becomes extruded from the cell in a network of decondensed DNA.

neutralization the mechanism by which antibodies binding to sites on pathogens prevent growth of the pathogen and/or its entry into cells. The toxicity of bacterial toxins can similarly be **neutralized** by bound antibody.

neutralizing antibody high-affinity IgA and IgG antibodies that bind to pathogens and prevent their growth or entry into cells.

neutropenia a condition of abnormally low numbers of neutrophils in the blood, which can be due to various causes such as genetic deficiencies in neutrophil production, or autoimmunity directed against surface antigens on white blood cells.

neutrophil phagocytic white blood cell that enters infected tissues in large numbers and engulfs and kills extracellular pathogens. It is a type of granulocyte and contains granules that stain with neutral dyes; hence its name. It is by far the most abundant white blood cell.

neutrophil extracellular trap (**NET**) *see* **netosis**.

NFAT (**nuclear factor of activated T cells**) transcription factor that is activated as a result of signaling from the T-cell receptor. It functions as a complex of the NFAT protein with the dimer of Fos and Jun proteins known as AP-1.

NK cell *see* **natural killer cell**.

NKC *see* **natural killer complex**.

NK-cell education a process that makes the NK cell sensitive to loss of expression of self-MHC class I.

NK-cell synapse a structure formed between an NK cell and its target in which interacting sets of receptors and ligands become concentrated at localized regions on the surfaces of the two cells, where they help hold the cells together and where signals and material are exchanged.

NKG2D an activating receptor expressed by all NK cells. It has several types of ligand, including the **MIC glycoproteins**.

NKT cells subset of α:β T cells expressing receptors that recognize lipid antigens presented by the MHC class I-like protein CD1d.

NOD1 *see* **NOD-like receptor**.

NOD2 *see* **NOD-like receptor**.

NOD-like receptor (**NLR**) any of a class of intracellular receptors (such as NOD1 and NOD2) that detect products derived from the intracellular degradation of phagocytosed pathogens, such as cell-wall components of bacteria.

nonproductive rearrangement a gene rearrangement at an immunoglobulin or T-cell receptor locus in developing lymphocytes that does not translate into a useful protein chain.

non-self peptide any peptide that derives from a protein not present in a person's own body. It can refer to a peptide derived from a pathogen protein or from a protein from a genetically unrelated person. Such peptides are recognized as foreign and provoke an immune response.

non-self protein any protein that is not present in a person's own body. It can refer to a pathogen protein or to a protein from a genetically unrelated person that is recognized as foreign. Such proteins are recognized as foreign and provoke an immune response.

nuclear factor κB (**NFκB**) a transcription factor activated by signaling from the Toll-like receptors and many other receptors, and that helps turn on the expression of many immune-system genes.

oligomorphic describes genes that have only a small number of different alleles in the population.

Omenn syndrome a genetically determined severe immunodeficiency caused by missense mutations that produce RAG proteins with impaired enzymatic activity, leading to a failure of lymphocyte development.

oncogene mutant or unregulated form of a gene that can cause cells to proliferate abnormally and form a tumor. Most oncogenes are derived from genes that control cell growth and division.

oncogenic virus a virus that is involved in causing cancer.

oncology the clinical discipline that deals with the study, diagnosis, and treatment of cancer.

opportunistic infection infection that can be caused by a normally commensal microorganism in people whose immune systems are compromised.

opportunistic pathogen a microorganism that causes disease only in individuals whose immune systems are in some way compromised.

opsonin general name for antibodies and complement proteins that coat pathogens, thereby facilitating their phagocytosis by neutrophils or macrophages carrying receptors for the opsonin.

opsonization the coating of the surface of a pathogen or other particle with any molecule that makes it more readily ingested by phagocytes. Antibody and complement opsonize extracellular bacteria for phagocytosis by neutrophils and macrophages because the phagocytic cells carry receptors for these molecules.

oral tolerance the characteristic tolerance of the immune system for 'foreign' antigens, such as food, ingested into the gastrointestinal tract.

organ-specific or tissue-specific autoimmune disease autoimmune disease targeted at a particular organ, such as the thyroid gland in Graves' disease.

original antigenic sin a bias seen in successive immune responses to structurally related antigens such as those on different strains of influenza virus. On infection for the second time with influenza, the antibody response is restricted to epitopes that the second strain shares with the first strain to which the person was exposed. Other highly immunogenic epitopes on the second and subsequent viruses are ignored.

P nucleotides 'palindromic' nucleotides added into the junctions between immunoglobulin and T-cell receptor gene segments during the somatic recombinations that generate a rearranged variable-region sequence. They are an inverse repeat (a palindrome) of the nucleotide sequence at the end of the adjacent gene segments.

palatine tonsils aggregates of secondary lymphoid tissue at the sides of the throat.

pandemic outbreak of an infectious disease that spreads worldwide. For example, the "Spanish flu" following World War I.

panel reactive antibody (**PRA**) a measure of the likelihood that a patient seeking a transplant is sensitized to potential donors. The patient's serum is tested against tissue from a representative panel of individuals for antibodies that would cause the immediate hyperacute rejection of a graft. The PRA is the percentage of individuals in the panel whose cells react with the patient's antibodies.

paracrine describes a cytokine that is released from one type of cell and acts on another type of cell.

parasite general name for the unicellular protozoa and multicellular worms that infect animals and humans and live within them, causing disease.

paroxysmal nocturnal hemoglobinuria a genetic disease in which the complement regulatory proteins CD59 and DAF are defective, so that complement activation leads to episodes of spontaneous hemolysis. The defect is in the attachment of CD59 and DAF to cell membranes by a glycolipid anchor.

passive immunity immunity to a given pathogen that has been acquired by the injection of preformed antibodies, antiserum, or T cells.

passive immunization the injection of specific antibodies to provide protection against a pathogen or toxin. The administered antibodies may derive from human blood donors, immunized animals, or hybridoma cell lines.

passive transfer of immunity transfer of immunity to a nonimmune individual by the injection of specific antibody, immune serum, or T cells.

pathogen an organism, most commonly a microorganism, that can cause disease.

PBMC see **peripheral blood mononuclear cells**.

pemphigus foliaceus an autoimmune disease that results in blistering of the skin. Caused by IgG antibodies made against desmogleins, adhesion molecules of the cell junctions called desmosomes that bind skin keratinocytes tightly to each other.

pemphigus vulgaris a severe autoimmine disease resulting in blistering of the skin. It is caused by IgG antibodies made against desmogleins, adhesion molecules of the cell junctions called desmosomes that bind skin keratinocytes tightly to each other.

pentraxin any member of a family of pentameric proteins that circulate in the blood and lymph and can bind to the surfaces of a variety of pathogens and target them for destruction. C-reactive protein is a pentraxin.

peptide editing the selection of peptides by the peptide-loading complex that bind strongly to MHC class I molecules, together with the trimming of bound peptides to the correct length by the **endoplasmic reticulum aminopeptidase** (**ERAP**).

peptide splicing the joining together in the cell of peptide fragments derived from protein degradation in such a way as to form a novel antigen not present in the native protein. Some tumor antigens appear to originate in this way.

peptide-binding motif the combination of anchor residues in an MHC protein that correspond to the amino acid sequences of peptides that can bind to that protein.

peptide-loading complex a protein complex in the endoplasmic reticulum membrane that loads peptides from the TAP transporter onto MHC class I molecules.

perforin one of the proteins released by cytotoxic T cells on contact with their target cells. It forms pores in the target cell membrane that contribute to cell killing.

peripheral blood mononuclear cells (**PBMC**) the cellular fraction of a blood sample that contains mainly lymphocytes and monocytes: that is, cells with a round unlobed nucleus. They are extracted from whole blood by density gradient fractionation in the presence of the hydrophilic polysaccharide Ficoll-Paque.

peripheral lymphoid tissue *see* **secondary lymphoid tissue**.

peripheral tolerance tolerance to self antigens that is acquired by the lymphocyte population outside the primary lymphoid organs (thymus and bone marrow).

Peyer's patch organized gut-associated lymphoid tissue present in the wall of the small intestine, especially the ileum.

phagocyte a cell specialized to perform phagocytosis. The principal phagocytic cells in mammals are neutrophils and macrophages.

phagocytic receptor any of numerous pathogen-recognition receptors on macrophages and neutrophils that stimulate the phagocytosis of their bound ligand.

phagolysosome intracellular vesicle formed by fusion of

a phagosome with a lysosome, in which the phagocytosed material is broken down by degradative lysosomal enzymes.

phagosome intracellular vesicle containing material taken up by phagocytosis.

phosphoantigen any of a number of phosphorylated small molecules produced in the bacterial isoprenoid biosynthesis pathway that are recognized by the T-cell receptors of γ:δ T cells.

plasma cell terminally differentiated B lymphocyte that secretes antibody.

plasmacytoid dendritic cell a type of dendritic cell that produces copious amounts of interferon in response to the detection of viral infection though its Toll-like receptors.

platelet a small non-nucleated cell fragment, derived from megakaryocytes, that is present in large numbers in the blood and is involved in blood clotting.

pluripotent hematopoietic stem cell the stem cell in bone marrow that gives rise to all the cellular elements of the blood.

polarized describes cells in which one side of the cell is recognizably and functionally different from the other. Gut epithelial cells are a classic example.

polymeric immunoglobulin receptor (**poly-Ig receptor**) a receptor present on the basolateral membrane of mucosal epithelial cells that binds polymeric immunoglobulins, especially dimeric IgA but also IgM, that have been produced in mucosal tissues and transports them across the epithelium by transcytosis.

polymorphism (*adj.* **polymorphic**) the existence of different variants of a gene or trait in the population. Genetic polymorphism is defined as the existence of two or more forms (alleles) of a given gene within the population, with the variant alleles each occurring at a frequency greater than 1%.

polymorphonuclear leukocyte an alternative name for a **granulocyte** (a class of white blood cells comprising neutrophils, eosinophils, and basophils), describing the varied morphology of the nuclei.

polyreactive describes antibodies that can bind to a range of structurally different antigens.

polyspecificity the ability to bind to a range of different antigens, a property shown by some antibodies. It is also known as polyreactivity.

positive selection the process in the thymus that selects for developing T cells with receptors that recognize peptide antigens presented by self-MHC molecules. Only cells that are positively selected are allowed to continue their maturation.

PRA *see* **panel reactive antibody**.

pre-B cell a developing B cell that has rearranged its heavy-chain genes but not its light-chain genes. It expresses an immunoglobulin-like receptor called the **pre-B-cell receptor**. Pre-B cells develop from pro-B cells; if they survive, they develop into immature B cells. *See also* **large pre-B cell**; **small pre-B cell**.

pre-B-cell receptor immunoglobulin-like receptor that is expressed on the surface of pre-B cells. It consists of μ heavy chains in association with surrogate light chains formed from **λ5** and **VpreB**. Its appearance signals the cessation of heavy-chain gene rearrangement.

prednisolone the biologically active compound to which the immunosuppressive drug prednisone is converted *in vivo*.

prednisone a synthetic steroid drug with powerful anti-inflammatory and immunosuppressive activity. It is used in organ transplantation and to treat autoimmune disease.

pre-T cell developing T cell that has productively rearranged a T-cell receptor β-chain gene and has produced a β chain that can form a complex with the surrogate α chain, **pTα**.

pre-T cell receptor receptor that is present on the surface of some immature thymocytes. It consists of a TCRβ chain associated with a surrogate α chain called **pTα**.

primary focus temporary aggregate of proliferating activated antigen-specific B cells and T cells that forms during a secondary lymphoid tissue at the beginning of an adaptive immune response.

primary granule one type of secretory granule in neutrophils. Such granules are similar to lysosomes in the types of hydrolytic and destructive enzymes that they contain; they are distinguished from the other types of granule by containing myeloperoxidase. Also called azurophilic granule.

primary immune response the adaptive immune response that follows a person's first exposure to an antigen.

primary immunodeficiency disease disease in which there is a failure of immunological function as a result of a defect in one or more genes encoding components of the immune system.

primary lymphoid follicle a B-cell area in a secondary lymphoid tissue in the absence of an immune response. It contains resting B lymphocytes. *See also* **secondary lymphoid follicle**.

primary lymphoid tissue anatomical site of lymphocyte development. In humans, these are the bone marrow (where B lymphocytes develop) and the thymus (where T lymphocytes develop from precursor cells that have migrated from the bone marrow). Also called **central lymphoid tissue**.

pro-B cell an early stage of B-cell development when the B cell first expresses B-cell marker proteins and rearranges its heavy-chain genes. It is followed by the pre-B-cell stage. *See also* **late pro-B cell**; **early pro-B cell**.

pro-drug biologically inactive compound that is metabolized in the body to produce an active drug.

productive rearrangement gene rearrangement at immunoglobulin or T-cell receptor loci in a developing lymphocyte which leads to a gene that directs the synthesis of a functional polypeptide chain.

professional antigen-presenting cell refers primarily to dendritic cells, but also to macrophages and B cells, all of which take up, process and present antigens from the environment, and once activated, are able to activate naive or partly activated T cells.

programmed cell death an alternative term for **apoptosis**.

pro-inflammatory cytokine any cytokine that acts to promote a state of inflammation in the infected tissue. Another name for inflammatory cytokine.

promiscuous binding specificity, **promiscuous specificity** the ability of a protein to bind a number of different ligands; an example of this is the MHC molecules, which can each bind numerous peptides of different sequences.

properdin a plasma protein that helps to activate the alternative pathway of complement activation. It binds to and stabilizes the alternative pathway C3 and C5 convertases on the surfaces of bacterial cells. Also called factor P.

protease inhibitor proteins such as α_2-macroglobulin that can bind to proteases and inhibit their enzymatic activity.

proteasome large multisubunit protease present in the cytoplasm of all cells that degrades cytoplasmic proteins. It generates the peptides presented by MHC class I molecules.

protectin a protein on the surface of human cells that prevents the assembly of the complement membrane-attack complex on the cell surface, thus protecting human cells against complement-mediated lysis. Also called CD59.

protective immunity the specific immunological resistance to a pathogen that is present in an individual during months after either vaccination or recovery from an infection with the pathogen, and which is due to pathogen-specific antibodies and effector T cells produced during the primary response.

proto-oncogene a gene that regulates the control of cell growth and division and that, when mutated or aberrantly expressed, can contribute to the malignant transformation of cells, leading to cancer.

provirus the DNA form of a retrovirus when it is integrated into the host cell genome. In this state it can remain transcriptionally inactive for a long time.

pTα the surrogate α chain that combines with the TCRβ chain to form the pre-T-cell receptor.

purine nucleoside phosphorylase (**PNP**) enzyme involved in purine metabolism. Its deficiency results in the accumulation of purine nucleosides, which are toxic for developing T cells. This leads to **severe combined immunodeficiency**.

pus thick yellowish-white fluid that is formed in infected wounds. It is composed of dead and dying white blood cells (principally neutrophils), tissue debris, and dead microorganisms.

pyogenic causing the formation of pus.

pyrogen any molecule that induces fever, such as some bacterial products and certain cytokines.

rabbit antithymocyte globulin (**rATG**) antibody preparation purified from the serum of rabbits immunized with human thymocytes, which is used to induce pre-transplantation immunosuppression in transplant patients. The antibodies bind to T cells, B cells, NK cells, and dendritic cells and fix complement, opsonizing the leukocytes to be killed by phagocytes.

RAG-1, RAG-2 the recombination proteins that form the complex that carries out V(D)J recombination.

rapamycin alternative and earlier name for the immunosuppressive drug sirolimus, which is used to prevent the rejection of organ transplants.

receptor agonist a ligand, such as an autoantibody, that binds to a receptor in such a way that it mimics the natural ligand and activates the receptor. Autoantibodies produced in the autoimmune disease Graves' disease act as receptor agonists and activate the receptor for thyroid-stimulating hormone.

receptor antagonist a ligand, such as an autoantibody, that binds to a receptor in such a way that it prevents receptor activation. Antagonistic autoantibodies against the acetylcholine receptor on muscle cells are produced in myasthenia gravis and cause internalization and degradation of the receptor.

receptor editing the replacement of a self-reactive antigen receptor in a developing B cell with a non-self-reactive receptor, by further rounds of light-chain gene rearrangement that can produce a new light chain to replace the self-reactive one.

receptor-mediated endocytosis internalization of extracellular material that is bound to a receptor at the cell surface.

recessive describes a variant allele that determines the phenotype only when present in two copies (that is, on both homologous chromosomes) or when no other copy of the allele is present (for example, for genes on the X chromosome in males). Usually refers to disease-causing alleles that encode proteins with functional defects.

recombination signal sequence (**RSS**) short stretch of DNA flanking each of the gene segments that are rearranged to generate V-region exons. They are the sites at which somatic recombination occurs.

recombination-activating genes (***RAG1, RAG2***) two genes that are essential for the rearrangement of immunoglobulin and T-cell receptor genes in B cells and T cells, respectively. They encode the proteins RAG-1 and RAG-2, which form a protein complex that catalyzes the recombination process.

regulator of complement activation any of a group of proteins that regulate the activity of complement and contain one or more copies of a particular structural motif of ~60 amino acid residues called the CCP (complement control protein) motif. The CCP motif is also called a sushi domain because of its similarity in shape to a slice of a sushi roll.

regulatory CD4 T cell *see* **regulatory T cell.**

regulatory T cell (**T$_{reg}$**) an antigen-specific CD4 T cell that functions to suppress or limit immune responses. Also called suppressor T cell.

relapse the return of a cancer or other disease after a period of remission in which it seemed to have been successfully treated.

respiratory burst metabolic change accompanied by a transient increase in oxygen consumption that occurs in neutrophils and macrophages after they have taken up opsonized particles. It leads to the generation of toxic oxygen metabolites and other antibacterial substances that attack the phagocytosed material.

retrovirus a type of RNA virus, which replicates via a DNA intermediate that inserts into the host cell genome. The **human immune deficiency virus** (HIV) is a retrovirus.

Rev the protein encoded by the **human immunodeficiency virus** (HIV) that controls the supply of viral RNA to the cytoplasm and the extent to which that RNA is spliced.

reverse vaccinology an innovative approach to vaccine development that discovers new knowledge of the physiology of the pathogen and how it exploits the human immune system.

Rhesus (Rh) a series of human blood group antigens. The Rh antigens were originally named Rhesus because they were incorrectly believed to be identical to antigens detected on the surface of erythrocytes of the rhesus monkey, a type of macaque.

Rhesus D antigen (RhD) the one rhesus antigen that must be matched for successful blood transfusion. RhD mismatch between fetus and mother is the cause of **hemolytic anemia of the newborn**.

rheumatic fever autoimmune disease involving inflammation of the heart, joints, and kidneys, which can follow 2–3 weeks after a throat infection with certain strains of *Streptococcus pyogenes*. It is caused by antibodies made against bacterial antigens that cross-react with components of heart tissue. The antibodies form immune complexes that become deposited in the affected tissues.

rheumatoid arthritis common inflammatory disease of joints that is due to an autoimmune response.

rheumatoid factor IgM antibody with specificity for human IgG that is produced in some people with rheumatoid arthritis.

RIG-I-like receptor (RLR) any of a class of intracellular receptor proteins present in most cells that recognize viral RNAs in the cytoplasm. Examples are RIG-1 and MDA-5.

RSS *see* **recombination signal sequence**.

S protein soluble regulatory protein of complement that prevents the soluble complex of C5b with C6 and C7 from associating with cell membranes.

***S. aureus* enterotoxin B (SEB)** a toxin secreted by strains of staphylococcus that acts on the gut to cause the symptoms of food poisoning. It is also a superantigen.

S1P *see* **sphingosine 1-phosphate**.

sarcoma tumor that arises from a cell of connective tissue.

scavenger receptor any of a group of disparate phagocytic cell-surface receptors on macrophages that binds to an assortment of negatively charged ligands, including sulfated polysaccharides, nucleic acids, and the phosphate-containing lipoteichoic acids in the cell walls of Gram-positive bacteria.

SCID acronym for **severe combined immune deficiency**.

secondary follicle B-cell area of a secondary lymphoid tissue that is responding to antigen. It contains proliferating and differentiating B cells and develops a germinal center in which B cells undergo somatic hypermutation and isotype switching. Also called **secondary lymphoid follicle**.

secondary granule one type of secretory granule in neutrophils; such granules are distinguished from the other types by containing lactoferrin. Also called specific granule.

secondary immune response the adaptive immune response provoked by a second exposure to an antigen. It differs from the primary response by starting sooner and building more quickly, and is due to the presence of long-lived memory B cells and T cells specific for the antigen.

secondary immunodeficiency disease disease in which there is a failure of immunological function as a result of infection or the use of immunosuppressive drugs, rather than as a result of defects in genes encoding components of the immune system.

secondary lymphoid follicle B-cell area of a secondary lymphoid tissue that is responding to antigen. It contains proliferating and differentiating B cells and develops a germinal center in which B cells undergo somatic hypermutation and isotype switching. *See also* **primary lymphoid follicle**.

secondary lymphoid tissue the lymph nodes, spleen, and mucosa-associated lymphoid tissues. These are the tissues in which immune responses are initiated. The highly organized tissues such as lymph nodes and spleen are also often known as **secondary lymphoid organs**. Also called **peripheral lymphoid tissue**.

secretory component, secretory piece the fragment of the poly-Ig receptor that is left attached to dimeric IgA that has been secreted across a mucosal epithelium.

secretory IgA (SIgA) dimeric IgA molecule that is produced by plasma cells in mucosal tissues and secreted across the mucosal surface.

segmental exchange *see* **interallelic conversion**.

selective IgA deficiency a common condition in which little or no IgA is made but the production of all other isotypes is normal. It usually causes no apparent symptoms of immunodeficiency.

self antigen any normal constituent of the body to which the immune system would respond were it not for the mechanisms of tolerance that destroy or inactivate self-reactive B and T cells.

self MHC a person's own MHC class I and class II molecules.

self peptide any peptide produced from the body's own proteins. In the absence of infection, these peptides occupy the peptide-binding sites of MHC molecules on cell surfaces and are usually not recognized as antigenic by that person's immune system.

self protein any protein of a person's own body. These proteins are usually not recognized as antigenic by that person's immune system.

self renewal the ability of a population of cells, such as stem cells, to maintain itself permanently by multiplication and limited differentiation.

self-reactive able to respond to self antigens.

self-tolerance the normal situation whereby a person's immune system does not respond to constituents of his or her body. The circulating lymphocyte population in an individual is thus said to be **self-tolerant**.

self-tolerant describes the normal situation in which a person's immune system, in particular the circulating lymphocyte population, does not make an immune response to the constituents of that person's body.

septic shock a shock syndrome that is frequently fatal, caused by the systemic release of the cytokine TNF-α after bacterial infection of the bloodstream, usually with Gram-negative bacteria.

serglycin proteoglycan that forms a complex with granulysin and perforin in the lytic granules of T cells and NK cells.

seroconversion phase of an infection when antibodies against the infecting agent are first detectable in the blood.

serotype antigenically different strain of a bacterium or other pathogen that can be distinguished by immunological means, for example by antibody-based detection tests. Also used to describe different types of human alloantigen such as HLA and blood group antigens.

serpin class of protease inhibitor proteins that inhibit serine and cysteine proteases. The C1 inhibitor is an example of a serpin.

serum amyloid A protein an acute-phase protein that associates with high-density lipoprotein protein particles and can interact with Toll-like receptors and the CD36 scavenger receptor to activate cells to produce inflammatory cytokines.

severe combined immunodeficiency disease (SCID) a severe immune deficiency disease in which neither antibody nor T-cell responses are made. It is usually the result of genetic defects that lead to T-cell deficiencies, and is fatal in childhood if not treated by hematopoietic cell transplantation.

sheddase an enzyme that cleaves a cell-surface protein to produce a free soluble form of the protein.

SIgA *see* **secretory IgA**.

signal joint the joint present in the circle of DNA that is formed as a byproduct of V(D)J recombination and discarded.

signal transducers and activators of transcription *see* **STAT**.

single-positive thymocyte a T cell at a late stage of development in the thymus, characterized by the cell-surface expression of either the CD4 or the CD8 co-receptor.

sirolimus the proprietary name for the immunosuppressive drug rapamycin. Both names are in common use.

SLE acronym for **systemic lupus erythematosus**.

small lymphocyte the general name for a recirculating resting T or B lymphocyte.

small pre-B cell pre-B cell in which light-chain gene rearrangements are being made. *See also* **large pre-B cell**.

somatic hypermutation mutation that occurs at high frequency in the rearranged variable-region DNA segments of immunoglobulin genes in activated B cells, resulting in the production of variant antibodies, some of which have a higher affinity for the antigen.

somatic recombination DNA recombination that occurs between gene segments in the immunoglobulin loci and T-cell receptor loci in developing B cells and T cells, respectively. It generates a complete exon composed of a V gene segment and a J gene segment (and a D gene segment at the immunoglobulin and T-cell receptor heavy-chain loci) that encodes the variable region of an immunoglobulin or T-cell receptor polypeptide chain.

specific describes the property of antibodies and other antigen-binding molecules and immune system receptors that interact selectively with only one type or a few types of molecule or cell.

specific granule *see* **secondary granule**.

sphingosine 1-phosphate (S1P) a chemotactic lipid that binds to a receptor on naive T cells. Naive T cells in secondary lymphoid tissues that have not encountered specific antigen are drawn back into the peripheral circulation by a gradient of sphingosine 1-phosphate.

spleen an organ situated adjacent to the cardiac end of the stomach. One function of the spleen is to remove old or damaged red blood cells from the circulation; the other is as a secondary lymphoid organ in which immune responses to blood-borne pathogens and antigens are initiated.

SR-A a scavenger receptor that recognizes the lipopolysaccharide of Gram-negative bacteria, teichoic acids of Gram-positive bacteria, and CpG-rich bacterial DNA.

SR-B a scavenger receptor that recognizes microbial lipids.

SSLP7 staphylococcal superantigen-like protein 7, which prevents monomeric IgA from delivering bound bacteria to phagocytes.

staphylococcal superantigen-like protein (SSLP) any of a number of superantigen-like proteins secreted by *Staphylococcus aureus* that subvert and compromise human immunity, either by acting as superantigens or in other ways. *See also* ***S. aureus*** **enterotoxin B; SSLP7**.

STATs acronym for signal transducers and activators of transcription. Proteins that are activated to become transcription factors through the Janus kinase signaling pathway from cytokine receptors.

stress protein any of various proteins, such as the **MIC glycoproteins**, that are made only when cells are stressed by infection, malignant transformation, elevated temperature, or other means.

stromal cell cell type that provides the supporting framework of a tissue or organ.

subcapsular sinus macrophage specialized macrophage resident in the subcapsular sinus of the lymph node. It captures complement-tagged antigens from the afferent lymph and holds them at the cell surface.

subunit vaccine a vaccine composed only of isolated antigenic components of a pathogen and not the pathogen itself, either alive or dead.

superantigen molecule that, by binding nonspecifically to MHC class II molecules and T-cell receptors, stimulates the polyclonal activation of T cells.

surrogate light chain protein that mimics an immunoglobulin light chain. It is made up of two subunits, **VpreB** and **λ5**, and is produced by pro-B cells. Together with the μ heavy chain it forms the pre-B-cell receptor in pre-B cells.

survival signal in immunology, generally refers to the signals that developing and naive mature B cells and T cells must receive if they are to survive.

switch region short DNA sequence preceding each heavy-chain constant-region gene, at which somatic recombination occurs when B cells switch from producing one immunoglobulin isotype to another.

switch sequence *see* **switch region**.

sympathetic ophthalmia autoimmune response that sometimes follows damage to an eye and affects both the damaged eye and the healthy eye.

syngeneic genetically identical. A syngeneic tissue graft is one made between two genetically identical individuals. It does not provoke an immune response or a rejection reaction.

systemic anaphylaxis rapid-onset and potentially fatal IgE-mediated allergic reaction, in which antigen (such as insect venom) in the bloodstream triggers the activation of mast cells throughout the body, causing circulatory collapse and suffocation due to tracheal swelling.

systemic autoimmune disease autoimmune disease that involves reactions to common components of the body and whose effects are not confined to one particular organ.

systemic lupus erythematosus (**SLE**) a systemic autoimmune disease in which autoantibodies made against DNA, RNA, and nucleoprotein particles form immune complexes that damage small blood vessels.

T cell one of the two main classes of lymphocyte responsible for adaptive immunity (the other is the B cell). T cells originate in the bone marrow, develop in the thymus, and are responsible for cell-mediated immunity. Their cell-surface antigen receptor is called the T-cell receptor. Effector T cells comprise various subtypes including the **cytotoxic T cell** (which is responsible for killing virus-infected cells), the **regulatory T cell**, and a number of subsets of helper T cells that cooperate with other immune-system cells to aid macrophage activation and antibody production. Also known as **T lymphocyte**.

T follicular helper cell (**T_{FH}**) effector CD4 T cell present in lymphoid follicles that cooperates with B cells to help antibody production.

T lymphocyte alternative name for **T cell**.

tacrolimus immunosuppressive polypeptide drug that inactivates T cells by inhibiting signal transduction from the T-cell receptor. It is widely used to suppress transplant rejection. It is also known as FK506.

TAP acronym for the **transporter associated with antigen processing**.

tapasin TAP-associated protein, a chaperone protein involved in the assembly of peptide:MHC class I molecule complexes in the endoplasmic reticulum.

target cell any cell that is acted on directly by effector T cells, effector cells, or molecules. For example, virus-infected cells are the targets of cytotoxic T cells, which kill them, and naive B cells are the targets of effector CD4 T cells, which help stimulate them to produce antibodies.

Tat protein encoded by the **human immunodeficiency virus** (HIV) that prevents transcription from shutting off and thus increases the transcription of viral RNA.

T-cell activation the stimulation of mature naive T cells by antigen presented to them by professional antigen-presenting cells. It leads to their proliferation and differentiation into effector T cells.

T-cell anergy state of nonresponsiveness that is induced in naive T cells if their antigen receptor is engaged in the absence of co-stimulation.

T-cell area that part of a secondary lymphoid tissue where the lymphocytes are predominantly T cells. Also called the T-cell zone.

T-cell co-receptor either of the cell-surface glycoproteins CD4 or CD8, which cooperate with the T-cell receptor to determine which target cells a given T cell will interact with.

T-cell precursor undifferentiated cell that gives rise to α:β T cells and γ:δ T cells in the thymus.

T-cell priming the activation of mature naive T cells by antigen presented to them by professional antigen-presenting cells.

T-cell receptor the highly variable antigen receptor of T lymphocytes. On most circulating T cells it is composed of a variable α chain and a variable β chain and is known as the α:β T-cell receptor. This receptor recognizes peptide antigens derived from the breakdown of proteins. A minority of circulating T cells carry a γ:δ T-cell receptor, in which the two chains are called γ and δ and are generally less diverse in their specificity and can recognize antigens other than peptides, such as phosphoantigens and glycolipids. Both types of receptor are present at the cell surface in association with a complex of invariant CD3 chains and ζ chains, which have a signaling function.

T-cell receptor α chain (**TCRα**) one of the two polypeptide chains that make up the most common form of T-cell receptor, the α:β T-cell receptor.

T-cell receptor β chain (**TCRβ**) one of the two polypeptide chains that make up the most common form of T-cell receptor, the α:β T-cell receptor.

T-cell receptor γ chain (**TCRγ**) one of the two polypeptide chains that make up the minority form of T-cell receptor, the γ:δ T-cell receptor.

T-cell receptor δ chain (**TCRδ**) one of the two polypeptide chains that make up the minority form of T-cell receptor, the γ:δ T-cell receptor.

T-cell receptor complex the combination of T-cell receptor and invariant CD3 and ζ chains that makes a functional antigen receptor on the T-cell surface that can signal to the interior when antigen is bound.

T-cell synapse the localized area of contact between a T cell and an interacting B cell or macrophage where receptors and their ligands congregate, signals are transduced, and cytokines are exchanged.

T-cell zone *see* **T-cell area**.

T_{CM} *see* **central memory T cell**.

T_{EM} *see* **effector memory T cell**.

terminal deoxynucleotidyl transferase (**TdT**) enzyme that inserts non-templated nucleotides (N nucleotides) into the junctions between gene segments during the rearrangement of the T-cell receptor and immunoglobulin genes.

tertiary granule one type of secretory granule in neutrophils; such granules are distinguished from the other types by containing gelatinase. Also called gelatinase granule.

tetanus toxin protein neurotoxin produced by the bacterium *Clostridium tetani*. The cause of the disease tetanus.

T_{FH} *see* **T follicular helper cell**.

T_H1 cell a type of effector CD4 T cell that is characterized by the production of inflammatory cytokines. They are involved mainly in activating macrophages. Also called inflammatory T cell.

T_H17 cell a type of effector CD4 T cell that is characterized by the production of IL-17 and the promotion of inflammatory responses. These cells are associated with autoimmune conditions.

T_H2 cell a type of effector CD4 T cell that is characterized by the production of non-inflammatory cytokines.

thymectomy the surgical removal of the thymus gland.

thymic anlage the tissue from which the thymic stroma develops during embryogenesis.

thymic stroma the reticular epithelial cells and connective tissue of the thymus, which form the essential microenvironment for T-cell development.

thymocyte the name given to a T cell when it is developing in the thymus.

thymus a primary lymphoid organ in the upper part of the middle of the chest, just behind the breastbone. It is the site of T-cell development.

thymus-dependent lymphocyte *see* **T cell**.

thymus-independent antigen (**TI antigen**) antigen that can elicit antibody production in the absence of T cells. There are two types of TI antigen: **TI-1 antigens**, which have intrinsic B-cell activating activity, and **TI-2 antigens**, which have multiple identical epitopes that cross-link B-cell receptors.

tingible body macrophage phagocytic cell that is seen in histological sections engulfing apoptotic B cells in germinal centers.

TIR an intracellular signaling domain found in Toll-like receptors and the receptor for the cytokine interleukin-1.

TLR *see* **Toll-like receptor**.

TNF-α *see* **tumor necrosis factor-α**.

Toll-like receptor (**TLR**) any of a family of receptors of innate immunity that are present in many types of leukocyte, especially macrophages, dendritic cells, and neutrophils. Each type of TLR is specific for a different type of common pathogen component, such as bacterial lipopolysaccharide, bacterial flagellin, and CpG-rich bacterial DNA. TLRs include both cell-surface receptors and receptors that are located in endosome membranes. The responses of macrophages and dendritic cells to stimulation via TLRs are also essential in enabling the initiation of an adaptive immune response.

tonsils large aggregates of lymphoid cells lying on each side of the pharynx.

toxic shock syndrome toxin-1 (**TSST-1**) superantigen secreted by *Staphylococcus aureus* that causes toxic shock, a systemic shock reaction resulting from the overproduction of cytokines by CD4 T cells activated by the superantigen.

toxoid a toxin that has been deliberately inactivated by heat or chemical treatment so that it is no longer toxic but can still provoke a protective immune response as a vaccine.

transcytosis the transport of molecules from one side of an epithelium to the other. This involves endocytosis into vesicles within the epithelial cells at one face of the epithelium and release of the vesicles at the other.

transfusion effect improved outcome of transplantation if the recipient has previously been given blood transfusions from people who share an **HLA-DR** allotype with the organ with which the patient is subsequently transplanted.

translocation chromosomal abnormality in which a piece of one chromosome has become joined to another chromosome. The rearranging genes of B and T cells are often the sites of translocations in B-cell and T-cell tumors.

transplant rejection immune reaction that is directed toward transplanted tissue from a genetically non-identical donor and leads to death of the graft.

transplantation antigen name historically given to the MHC molecules because they are the main antigens that provoke the rejection of transplanted organs.

transporter associated with antigen processing (**TAP**) ATP-binding protein in the endoplasmic reticulum membrane that transports peptides from the cytosol to the lumen of the endoplasmic reticulum. It is composed of two subunits, TAP-1 and TAP-2. It supplies MHC class I molecules with peptides.

T_{reg} *see* **regulatory T cell**.

tumor a growth arising from uncontrolled cell proliferation. It may be benign and self-limiting, or malignant and invasive.

tumor antigen cell-surface component of tumor cells that can elicit an immune response in the affected individual. *See also* **tumor-associated antigens**; **tumor-specific antigens**.

tumor necrosis factor-α (**TNF-α**) cytokine produced by macrophages and T cells that has several functions in the immune response and is the prototype of the TNF family of cytokines. These cytokines function as cell-associated or secreted proteins that interact with receptors of the tumor necrosis factor receptor (TNFR) family.

tumor suppressor gene gene encoding a cellular protein that normally functions to prevent cells from becoming cancerous.

tumor-associated antigen antigen characteristic of certain types of tumor cells, and which is also present on some types of normal cell.

tumor-specific antigen antigen that is characteristic of tumor cells and is not expressed by normal healthy cells.

type 1 diabetes an autoimmune disease in which insulin-secreting β cells of the pancreatic islets of Langerhans are gradually destroyed. Also called insulin-dependent diabetes mellitus (IDDM) or juvenile-onset diabetes (because it usually appears early in life, in childhood or adolescence).

type I hypersensitivity reaction an allergic reaction that occurs within minutes of exposure to antigen in a sensitized person, as a result of the activation of mast cells by antigen cross-linking specific IgE that was made in response to an initial encounter with the antigen, and which has become bound to Fc receptors on the mast cell. Symptoms can range from the unpleasant but relatively minor, as in hay fever (**allergic rhinitis**), to the life-threatening reaction of **systemic anaphylaxis**.

type I interferon collective name for interferon-α and interferon-β, which are cytokines produced by virus-infected cells that interfere with viral replication by the infected cell, and also signal to neighboring uninfected cells to prepare for resisting infection.

type II hypersensitivity reaction tissue-damaging immune reaction caused by the secondary immune response to small chemically reactive molecules that modify cell-surface components and stimulate the production of specific IgG antibodies.

type II interferon interferon-γ, which is a cytokine with multiple pro-inflammatory functions in an immune response.

type III hypersensitivity reaction tissue-damaging immune reaction caused by immune complexes formed during the secondary immune response to soluble proteins of non-human origin.

type IV hypersensitivity reaction a delayed-type tissue-damaging immune reaction caused by the secondary response of T cells specific for peptides derived from human proteins that have been modified by small chemically reactive molecules from the environment.

urticaria the technical term for hives, which are itchy, raised, red swellings on the skin, often all over the body, and which arise when an allergen to which the individual is sensitized gets carried to the skin by the bloodstream.

uterine NK cells (**uNK**) specialized NK cells present in the uterus, where they regulate the activities of fetal trophoblast cells during formation of the placenta.

V domain *see* **variable domain**.

V gene segment DNA sequence in the immunoglobulin or T-cell receptor genes that encodes the first 95 or so amino acids of the V domain. There are multiple different V gene segments in the germline genome. To produce a complete exon encoding a V domain, one V gene segment must be rearranged to join up with a J or a rearranged DJ gene segment.

V region *see* **variable region**.

V(D)J recombinase the set of enzymes needed to recombine V, D, and J segments in the rearrangement of immunoglobulin and T-cell receptor genes. It includes the RAG-1 and RAG-2 proteins.

V(D)J recombination the mechanism of gene rearrangement by which V, D, and J segments of immunoglobulin or T-cell receptor genes are brought together, with elimination of intervening segments of DNA, to make functional variable-region genes. *See also* **somatic recombination**.

vaccination the deliberate induction of protective immunity to a pathogen by the administration of killed or non-pathogenic forms of the pathogen, or its antigens, to induce an immune response.

vaccine any preparation made from a pathogen that is used for vaccination and provides protective immunity against infection with the pathogen.

vaccinia the cowpox virus. It causes a limited infection in humans that leads to immunity to the human smallpox virus. Was used as the vaccine against smallpox that led to eradication of the disease.

variable domain (**V domain**) the amino-terminal domain of immunoglobulin and T-cell receptor polypeptide chains. Paired variable domains make up the antigen-binding site.

variable fragment *see* **F_V**.

variable gene segment *see* **V gene segment**.

variable region (**V region**) the part of an immunoglobulin or T-cell receptor, or of its constituent polypeptide chains, that varies in amino acid sequence between isoforms with different antigenic specificities. It is responsible for determining antigen specificity.

variable surface glycoprotein (**VSG**) any of a family of glycoproteins that form the surface coat of African trypanosomes. The trypanosome can repeatedly change its glycoprotein coat by expressing different glycoprotein genes by a process akin to gene conversion.

variegated expression random process that generates NK cells expressing different numbers of KIR genes in different combinations, thus producing a population of NK cells having diverse KIR phenotypes.

variola name given to both the smallpox virus and the disease it causes: smallpox.

variolation historical procedure for immunization against smallpox in which a small amount of live smallpox virus was introduced through scarification of the skin.

villus (plural **villi**) finger-like projection in the wall of the small intestine.

viremic controller HIV-infected individuals who maintain a low viremia with 2000 copies or fewer of viral RNA per milliliter of blood, and maintain health.

virus submicroscopic pathogen composed of a nucleic acid genome enclosed in a protein coat. Viruses replicate only inside living cells because they do not possess all the metabolic machinery required for independent life. A viral particle is called a virion.

VLA-4 integrin present on the surface of T lymphocytes that binds to mucosal cell adhesion molecules.

VpreB one of the proteins that forms the surrogate light chain of the **pre-B-cell receptor**. The other is **λ5**.

WAS acronym for **Wiskott–Aldrich syndrome**.

wheal and flare the reaction observed when small amounts of allergen are injected into the dermis of an individual who is allergic to the antigen. It consists of a raised area of skin containing fluid (the wheal) with a spreading, red, itchy reaction around it (the flare).

Wiskott–Aldrich syndrome (**WAS**) genetic immunodeficiency disease in which interactions between T cells and B cells are defective, owing to a defective protein (Wiskott–Aldrich syndrome protein, or WASP) that is normally required for cytoskeletal reorganization. Antibody responses in particular are deficient.

XLA *see* **X-linked agammaglobulinemia.**

xenoantibody antibody produced in one species of animal against an antigen of another animal species.

xenoantigen antigen on the tissues of one species of animal that provokes an immune response in members of another animal species.

xenograft an organ from one animal species transplanted into another animal species.

xenotransplantation the transplantation of organs from one animal species into another. It is being explored as a possible solution to the shortage of human organs for clinical transplantation.

X-linked agammaglobulinemia (**XLA**) genetic disorder in which B-cell development is arrested at the pre-B-cell stage and neither mature B cells nor antibodies are formed. The disease is due to a defect in the gene encoding the protein tyrosine kinase BTK, which is located on the X chromosome.

X-linked disease heritable diseases caused by recessive mutations in genes carried on the X chromosome. Because males have only one copy of the X chromosome, such diseases show up much more frequently in males than in females.

X-linked hyper-IgM syndrome a type of **hyper IgM immunodeficiency**.

X-linked hypohidrotic ectodermal dysplasia and immunodeficiency genetically determined condition characterized by developmental defects and immunodeficiency, caused by a defect in a component of the NFκB signaling pathway called NEMO (NFκB essential modulator). The NEMO gene is on the X chromosome and so the condition mostly affects boys. Also called NEMO deficiency.

X-linked lymphoproliferative syndrome immunodeficiency due to a defect in the *SH2D1A* gene on the X chromosome, resulting in an inability to control infections with Epstein–Barr virus (EBV).

ZAP-70 acronym for zeta-chain-associated protein kinase of 70 kDa molecular weight. It is a cytoplasmic tyrosine kinase in T cells that forms part of the signal transduction pathway from T-cell receptors.

zymogen the functionally inactive form of a protease, which requires cleavage by another protease to become active.

Figure Acknowledgments

Chapter 1

Opener image: S. Schuller, Wellcome Images.

Figure 1.3 panel a © Omikron/Science Photo Library; panel b, c, e, i, j, k, © Eye of Science/Science Photo Library; panel d © Juergen Berger/Science Photo Library; panel f © A. Ryter/Institut Pasteur; panel g © Philiple Gontier/Science Photo Library; panel h © A.B. Dowsett/Science Photo Library; panel l © Sinclair Stammers/Science Photo Library.

Figure 1.13 from F. Geissman, *Nat. Immunol.*, **8**:558–560. © 2007 With permission from Macmillan Publishers Ltd.

Chapter 2

Opener image: University of Edinburgh, Wellcome Images.

Figure 2.13 from S. Bhakdi et al., *Blut,* **60**(6):309–318. © 1990 With permission from Springer-Verlag.

Chapter 3

Opener image: David Phillips/Photo Researchers.

Figure 3.21 from J.D. Gitlin and H.R. Colten, *Lymphokine,* **14**:123–153. © 1980 With permission from Elsevier.

Figure 3.27 from S.Thiel and K.B.M. Reid, *FEBS Lett.,* **250**:78–84. © 1989 With permission from Federation of the European Biochemical Societies ,

Figure 3.31 from J-L Casanova, L. Abel and L. Quintana-Murei, *Ann. Rev. Immunol.,* **29**:447–491. © 2011 With permission from Annual Reviews.

Figure 3.35 from F.P. Siegal et al., *Science,* **284**(5421):1835–1837. © 1999 With permission from AAAS.

Chapter 4

Opener image: R. Dourmashkin, Wellcome Images.

Figure 4.6 panel a from L. Harris et al., *Nature,* **360**:369–372. © 1992 With permission from Macmillan Publishers Ltd.

Figure 4.27 from A.C. Davis et al., *Euro. J. Immunol.,* **18**:1001–1008. © 1988 With permission from Wiley-VCH.

Chapter 5

Opener image: Paul Duprex, Wellcome Images.

Figure 5.2 from K.C. Garcia et al., *Science,* **274**:209–219. © 1996 American Association for the Advancement of Science. With permission from AAAS.

Figure 5.4 panel a from I. Roitt et al, *Immunology* 5th Edition. © 1998 With permission from Elsevier.

Figure 5.26 from K. C. Garcia et al., *Science,* **274**:209–219. © 1996 With permission from AAAS.

Chapter 6

Opener image: Dennis Kunkel Microscopy, Inc.

Chapter 7

Opener image: Copyright 2008 Joseph R. Siebert, Ph.D.—Custom Medical Stock Photo, All Rights Reserved.

Figure 7.8 from C.D. Surh and J. Sprent, *Nature,* **372**:100–103. © 1994 With permission from Macmillan Publishers Ltd.

Chapter 8

Opener image: Yasodha Natkunam, Department of Pathology, Stanford University School of Medicine.

Figure 8.2 from P. Perre et al., *Nature,* **388**:787–792. © 1997 With permission from Macmillan Publishers Ltd.

Figure 8.16 from G. Kaplan and Z.A. Cohn, *Int. Rev. Exp. Pathol.,* **28**:45–78. © 1986 With permission from Elsevier.

Figure 8.23 panel c from P.A. Henkart and E. Martz (eds), *Second International Workshop on Cell Mediated Cytotoxicity.* © 1985 Kluwer/Plenum Publishers. With permission from Springer Science and Business Media.

Chapter 9

Opener image: Copyright 2008 SPL—Customer Medical Stock Photo, All Rights Reserved.

Figure 9.5 from A.K. Szakal et al., *J. Immunol.,* **134**:1349–1359. |© 1985 With permission from The American Association of Immunologists, Inc.

Figure 9.10 from D. Zucker-Franklin et al., *Atlas of Blood Cells: Function and Pathology* 2nd Edition. © 1988 With permission from Lippincott, Williams and Wilkins.

Chapter 10

Chapter opener: © Martin Oeggerli, supported by School of Life Sciences, FHNW.

Figure 10.3 from M.S. Ali and J.P. Pearson, *Laryngoscope,* **117**:932–938. © 2007 The Triological Society. With permission from John Wiley and Sons.

Figure 10.5 adapted from P.D. Smith, T.T. MacDonald, and R.S. Blumberg (eds), *Principles of Mucosal Immunology*, © 2013 Garland Science.

Figure 10.7 from A. Mowat and J. Viney, *Immunol. Rev.*, **156**:145–166, 1997. With permission from John Wiley and Sons.

Figures 10.8 and 10.10 adapted from P.D. Smith, T.T. MacDonald, and R.S. Blumberg (eds), *Principles of Mucosal Immunology*. © 2013 Garland Science.

Figure 10.11 from P.D. Smith et al., *Mucosal Immunol.*, **4**:31–42. © 2010 With permission from Macmillan Publishers Ltd.

Figure10.12 from A. Mowat and J. Viney, *Immunol. Rev.*, **156**: 145–166. © 1997 With permission from John Wiley and Sons.

Figure 10.14 from J.H. Niess, *Science*, **307**:254–257. © 2005 With permission from AAAS.

Figure 10.21 from P.D. Smith, T.T. MacDonald, and R.S. Blumberg (eds), *Principles of Mucosal Immunology*. © 2013 Garland Science.

Figures 10.22 and 10.24 from P. Brandtzaeg and F.E. Johansen, *Immunol.Rev.*, **206**:32–63. © 2005 With permission from John Wiley and Sons.

Chapter 11

Opener image: from L.K. Pickering and C.J. Baker (eds.) Section 3: summaries of infectious diseases; measles. Red Book® Online. © 2012 With permission of the American Academy of Pediatrics.

Figure 11.23 from R. Rappouli and P.R. Dormitzer, *Science*, **336**:1531–1533. © 2012.

Figure 11.24 from V.A.A. Jansen et al., *Science*, **301**:804. © 2003 With permission from AAAS.

Chapter 12:

Opener image: from E.M. Mace and J.S. Orange, *PNAS*, **11**:6708–6713. © 2014 With permission of the National Academy of Sciences of the United States of America.

Figures 12.21 and 12.24 adapted from S. Kalyan and D. Kabelitz, *Cell. Mol. Immunol.*, **10**:21–29. © 2013 With permission from Macmillan Publishers Ltd.

Figures 12.33 and 12.34 adapted from P. J. Brennan et al., *Nat. Rev. Immunol.*, **13**:101–117. © 2013 With permission from Macmillan Publishers Ltd.

Chapter 13

Opener image: Stephen Fuller, Wellcome Images.

Figure 13.26 adapted from M. P. Martin, et al., *Nat. Genet.*, **39**:733–740. © 2008 With permission from Macmillan Publishers Ltd.

Chapter 14

Opener image: from the Centers for Disease Control and Prevention.

Figure 14.1 pollen © E.Gueho/CNRI/Science Photo Library; house dust mite © K.H. Kjeldsen/Science Photo Library; wasp © Claude Nuridsany and Marie Perennon/Science Photo Library; vaccine © Chris Priest and Mark Clarke / Science Photo Library; peanuts © Adrienne Hart-Davis / Science Photo Library; shellfish © Sheila Terry / Science Photo Library; poison ivy © Ken Samuelson / Tony Stone Images Images ® Copyright 1999 Photodisc, Inc.

Figure 14.4 from K. Steven, *The Biochemist*, **131**:17. © 2009 With permission from The Biochemical Society.

Figures 14.9 and 14.13 from H.J. Gouldand and B.J. Sutton, *Nat. Rev. Immunol.*, **8**:205–217. © 2008.

Figure 14.39 adapted from J. E. Allen and R.M. Maizels, *Nat. Rev. Immunol.*, **11**:375–389. © 2011 With permission from Macmillan Publishers Ltd.

Chapter 15

Opener image: Annie Cavanagh, Wellcome Images.

Figure 15.8 from B.D. Kahan and C. Ponticelli, *Principles and Practice of Renal Transplantation* 1st edition. © 2000 With permission from Thomson Publishing.

Figure 15.27 from E. J. Johnson and D. Goldstein, *Science*, **302**:1338–1339. © 2003 With permission from the AAAS.

Figure 15.29 from J.P. McCulley and J.Y. Niederkorn, *Curr. Eye Res.*, **32**:1005–1016. © 2007 With permission from Taylor and Francis Group Ltd. (http://www.tandf.co.uk/journals)

Figure 15.38 from E.W. Petersdorf et al., *Blood*, **92**:3515–3520. © (1998) American Society of Hematology, and P. Parham and K. McQueen, *Nat. Rev. Immunol.* **3**(2):108–122. © 2003 With permission from Macmillan Publishers Ltd.

Figure 15.42 from B.M. Triplett et al., *Leukemia*, **23**:1278–1287. © 2009 With permission from Nature Publishing Group.

Figure 15.44 from F. Claas, *Hum. Immunol.*, **61**:190–192. © 2000 With permission from American Society for Histocompatibility and Immunogenetics.

Chapter 16

Opener image: Ashley Prytherch, Royal Surrey County Hospital NHS Foundation Trust, Wellcome Images.

Figure 16.11 adapted from U. Nussinovitch and Y. Shoenfeld, *Autoimmun. Rev.*, **11**:A337–A385. © 2012 With permission from Elsevier.

Figure 16.22 adapted from J.S. Orange et al., *J. Allergy Clin. Immunol.*, **117**:S525–S553. © 2006.

Figure 16.37 from L.M. Sollid and B. Jabri, *Nat. Rev. Immunol.*, **13**:294–302. © 2013 With permission from Macmillan Publishers Ltd.

Figure 16.38 adapted from F. Megiorni amd A. Pizzuti, *J. Biomed. Sci.*, **19**:88. © 2012.

Figure 16.40 from C. M. Weyand et al., *Exp. Gerontol.*, **38**:833–841. © 2003 With permission from Elsevier.

Chapter 17

Opener image: Yasodha Natkunam, Department of Pathology, Stanford University School of Medicine.

Figure 17.8 from H. McCallum and M. Jones, *PLOS Biol.*, **4**:e342. © 2006.

Figure 17.18 from M.J. Scanlan.et al., *Immunol. Rev.*, **188**:22–32. © 2002. With permission from John Wiley and Sons.

Index

2B4, NK-cell activation, **Fig. 12.2**, **Fig. 12.4**, **Fig. 12.6**
8D6, germinal center formation, 239
12/23 rule, 93, **Fig. 4.19**

A

ABO blood group antigens, 434, **Fig. 15.1**
  hyperacute graft rejection, 436–437, **Fig. 15.4**
  incompatible blood transfusion, 434, 435
  matching, blood transfusion, 435–436, **Fig. 15.2**, **Fig. 15.3**
Acetylcholine receptor autoantibodies, 483, **Fig. 16.14**
Acquired immune deficiency syndrome (AIDS), 386–397
  see also Human immunodeficiency virus
  CD4 T cells *see* CD4 T cells, HIV infection
  opportunistic infections, 395–396, **Fig. 13.31**
  progression to, 390–391, **Fig. 13.24**, **Fig. 13.25**
Acquired immunity *see* Protective immunity
Activation-induced cell death, T cells, 193
Activation-induced cytidine deaminase (AID), 100, 101, **Fig. 4.28**
  induction of synthesis, 239
  inherited deficiency, 102, **Fig. 13.10**
Acute lymphoblastic leukemia (ALL), 466, **Fig. 6.23**
Acute myelogenous leukemia (AML), 466, 518–519
Acute-phase proteins, 62–63, **Fig. 3.21**
Acute-phase response, 62–63, **Fig. 3.24**
Adalimumab, 91, 490–491
ADAM10, 408–409
ADAM33, 418, **Fig. 14.25**
Adaptive immune response, 2, 10–12, **Fig. 1.8**, **Fig. 3.1**
  disruption of healthy tissue, 473–505
  initiation, 20–23, **Fig. 1.23**
    NK cell–dendritic cell interactions, 77, **Fig. 3.41**
  mucosal *vs.* non-mucosal tissues, 290, **Fig. 10.28**
  primary *see* Primary immune response
  secondary *see* Secondary immune response
Adaptive immunity, 10
  coevolution with innate immunity, 329–361, **Fig. 12.1**
  importance, 12, **Fig. 1.10**
Adaptor proteins, 52
Addison's disease, **Fig. 16.9**
Addressins, vascular *see* Vascular addressins
Adenocarcinoma, **Fig. 17.1**
Adenoids, 25, 272, **Fig. 10.6**
Adenoma, **Fig. 17.1**
Adenosine deaminase (ADA) deficiency, 383–384, **Fig. 13.16**, **Fig. 13.17**
Adhesins, bacterial, 252–253
Adhesion molecules (cell-adhesion molecules; CAMs), 57, **Fig. 3.12**
  B-cell development, 152, **Fig. 6.5**
  effector T cells, 218–219, **Fig. 8.18**, **Fig. 8.19**
  mucosa-derived effector lymphocytes, 281, **Fig. 10.16**
  naive T cell homing, 205, **Fig. 8.6**
  neutrophil recruitment, 57–58, **Fig. 3.13**
Adjuvants, 211, 316, **Fig. 11.21**
Adoptive transfer
  antigen-activated dendritic cells, 525, **Fig. 17.23**
  T cells, 523, **Fig. 17.21**
β_2-Adrenergic receptor, 418, **Fig. 14.25**
Afferent lymphatic vessels, 22, **Fig. 1.22**
Affinity, antibody, 87, 103
Affinity maturation, 100–101, **Fig. 4.26**
  germinal centers, 169, 241–243, **Fig. 9.11**
  secondary immune response, 298, 302, **Fig. 11.6**
Agammaglobulinemia, 81
  X-linked *see* X-linked agammaglobulinemia
Agglutination, 251
AID *see* Activation-induced cytidine deaminase
AIDS *see* Acquired immune deficiency syndrome
AIRE *see* Autoimmune regulator
Alemtuzumab (anti-CD52)
  cancer, **Fig. 17.27**
  transplant recipients, 447, **Fig. 15.15**, **Fig. 15.30**
Alleles, 135–136
Allelic exclusion
  immunoglobulin genes, 96–97, 154–155, **Fig. 6.8**
  T-cell receptor β-chain locus, 185, 190–191
Allergens, 401, 416–418, **Fig. 14.1**
  avoidance, 427
  blood-borne, 421–422, **Fig. 14.30**
  desensitization, 427
  immediate response, 420, **Fig. 14.27**
  ingested, 426–427, **Fig. 14.38**
  inhaled, 416–417, **Fig. 14.22**
    course of asthmatic response, 420, **Fig. 14.28**
    diseases caused, 423–424, **Fig. 14.34**
    sensitization to, 416–417, **Fig. 14.23**
  late-phase response, 420, **Fig. 14.27**
  similarities to parasite antigens, 417–418, **Fig. 14.24**
Allergic asthma *see* Asthma
Allergic conjunctivitis, 423–424
Allergic disease, IgE-mediated, 416–428
  antigens triggering *see* Allergens
  genetic predisposition, 418–419, **Fig. 14.25**
  immediate and late-phase responses, 419–420, **Fig. 14.27**, **Fig. 14.28**
  localized mast-cell activation, 420–421, **Fig. 14.29**
  monoclonal antibody therapy, 90, 409–410, 428, **Fig. 14.14**
  prevention and treatment, 427–428
Allergic reactions *see* Hypersensitivity reactions
Allergic rhinitis, 423–424, **Fig. 14.34**
Allergy, 401, 403–416
  epidemic of, 401, 406, 429, **Fig. 14.39**
  hygiene hypothesis, 406, **Fig. 14.8**
  vs. transplantation and autoimmunity, 503–504, **Fig. 16.42**
Alloantibodies, 144
Alloantigens, 438
Allogeneic MHC molecules, 143
Allogeneic transplant, 439
Allograft, 439
Alloreactions, 143–144, 438–439, **Fig. 15.6**
  hematopoietic stem-cell transplantation, 461–462, **Fig. 15.35**
  organ transplantation *see* Transplant rejection
Alloreactive T cells, 143
  acute graft rejection, 441, **Fig. 15.9**
  chronic graft rejection, 443–444, **Fig. 15.13**
  graft-versus-host disease, 461–462, **Fig. 15.35**

HLA differences activating, 442–443, **Fig. 15.10**
immunosuppressive agents targeting, 448–454
minor histocompatibility antigens triggering, 464–465
Allorecognition
direct pathway, 441, **Fig. 15.9**, **Fig. 15.12**
indirect pathway, 443–444, **Fig. 15.12**, **Fig. 15.13**
Allotypes, 68, 136
α:β T-cell receptors *see under* T-cell receptor(s)
α:β T cells, 118, 119, **Fig. 5.7**
see also T cell(s)
development, 181–186, **Fig. 7.7**, **Fig. 7.9**, **Fig. 7.15**
restriction by non-polymorphic MHC class I-like molecules, 354–360
selection in thymus, 188–195
vs. γ:δ T cells, 347–348, **Fig. 12.20**
α heavy chains, immunoglobulin, 83, **Fig. 4.5**
Alternative RNA splicing
coexpression of IgM and IgD, 96, **Fig. 4.22**
membrane-bound and secreted immunoglobulins, 98–100, **Fig. 4.24**
Alum, 316, **Fig. 11.21**
Amyloid A protein, serum, 63, **Fig. 3.21**
Amyloid P component, serum (SAP), 43, **Fig. 2.20**
Anakinra, 56
Anamnestic response, 298
Anaphylactic shock, 40, 421
Anaphylactoid reactions, 422
Anaphylatoxins, 40, 65, **Fig. 2.15**
Anaphylaxis, systemic, 421–422, **Fig. 14.30**
induced by desensitization, 427
role of IgE, 250
treatment, 422, **Fig. 14.31**
Anaplastic large cell lymphoma (ALCL)
diagnosis, 527–528, **Fig. 17.26**
treatment, 528–529, **Fig. 17.27**, **Fig. 17.28**
Anchor residues, 139–140, **Fig. 5.34**
Anemia, 475
Anergy
B cell, 166, 169, **Fig. 6.19**
T cell, 193, 210–211, **Fig. 8.13**
tumor-specific T cells, 519
Angioedema, 424, 427
hereditary (HAE), 381–382
Anisakis simplex, **Fig. 14.24**
Ankylosing spondylitis, HLA-B27 association, 479, 480, **Fig. 16.9**
Anterior chamber-associated immune deviation (ACAID), 457
Anthrax, **Fig. 9.25**
Antibiotics, effect on gut microbiota, 3, **Fig. 1.2**
Antibodies, 16, 18–19, 81–109
see also Immunoglobulin(s)
affinities, 87, 103
affinity maturation *see* Affinity maturation
antigen-binding site, 83, 85–88, **Fig. 4.2**
hypervariable regions, 85–86, **Fig. 4.8**
noncovalent interactions with antigens, 87
production of diversity, 92–96
somatic hypermutation, 100, **Fig. 4.26**
variations in shape and size, 86–88, **Fig. 4.11**
autoreactive *see* Autoantibodies
avidity, 103
catalytic, 88
complement activation, 255–258, **Fig. 4.30**
diversity, 81–109
mechanisms of generation, 91–107
structural basis, 82–88
effector functions, 18–19, 231, 245–263, **Fig. 1.18**
inherited deficiencies, 379–381, **Fig. 13.12**
isotypes *see* IgA; IgD; IgE; IgG; IgM
isotype switching *see* Isotype switching
monoclonal *see* Monoclonal antibodies
neutralization by, 18, **Fig. 1.18**
neutralizing *see* Neutralizing antibodies
opsonization, 19, 103, 259–260, **Fig. 1.18**, **Fig. 4.30**
polyspecific, 161–162
primary response, 169, 296–297, **Fig. 11.1**
production, 98–100, 231–245, **Fig. 4.24**
changes over time, 297, **Fig. 11.1**
long-lived plasma cells, 297
newly differentiated plasma cells, 169
repertoire, 81
secondary response, 170, 298, **Fig. 11.1**
amount and affinity, 302, **Fig. 11.6**
vs. primary response, 300–301, **Fig. 11.4**
somatic hypermutation *see* Somatic hypermutation
specificity, 81, 96–97
structure, 82–88
vs. B-cell and T-cell receptors, 16–17, **Fig. 1.17**
Antibody-dependent cell-mediated cytotoxicity (ADCC), 261–262, **Fig. 9.36**
cancer therapy augmenting, 524–525, **Fig. 17.22**
Anti-CD3 monoclonal antibodies, 90, 450–451, **Fig. 15.21**, **Fig. 15.30**
Anti-CD19 chimeric antigen receptors, 523, **Fig. 17.21**
Anti-CD20 monoclonal antibodies
antibody-dependent cell-mediated cytotoxicity, 261–262, **Fig. 9.36**
non-Hodgkin's lymphoma, 90
radioactive isotope conjugates, 529, **Fig. 17.29**
rheumatoid arthritis, 491
Anti-CD25 monoclonal antibodies, 452–453, **Fig. 15.24**, **Fig. 15.30**
Anti-CD28 monoclonal antibodies, 207
Anti-CD30 antibodies, conjugated, 528–529, **Fig. 17.27**, **Fig. 17.28**
Anti-CD52 *see* Alemtuzumab
Anti-CD319 monoclonal antibodies, 524–525
Anti-CTLA4 monoclonal antibodies, 521–522, **Fig. 17.19**
Antigen(s), 17
see also Autoantigens; Self antigens
binding to antibodies, 87
capture
dendritic cells, 200–201, **Fig. 8.1**
follicular dendritic cells, 235–236, **Fig. 9.6**
gut dendritic cells, 279, **Fig. 10.14**
M cells, 277–278, **Fig. 10.13**
causing allergy *see* Allergens
encounter by B cells, 17, 98, 168–170, **Fig. 6.22**
encounter by T cells, 17, 193–194, 203–204, **Fig. 8.4**
linked recognition, 226
lipid *see* Lipid antigens
multivalent, 86, **Fig. 4.9**
recognition by T cells, 113–145
thymus-independent (TI), 235, **Fig. 9.4**
transplantation, 438
tumor *see* Tumor antigens
Antigen-binding sites
antibodies *see* Antibodies, antigen-binding site
T-cell receptors, 114–115, 132, **Fig. 5.2**
Antigenic determinants *see* Epitopes
Antigenic drift, 367, **Fig. 13.2**
Antigenic shift, 368, **Fig. 13.3**
Antigenic variation, pathogens, 368–369, **Fig. 13.4**
Antigen presentation, 120–135, **Fig. 5.10**
by B cells to T_{FH} cells, 225–226, **Fig. 8.28**
to CD4 T cells, 122–123, **Fig. 5.15**
to CD8 T cells, 122, **Fig. 5.15**
cross-presentation, 131–132, 203, **Fig. 8.3**
by dendritic cells *see under* Dendritic cells
HLA class II region genes, 138, **Fig. 5.32**
by memory B cells, 302
MHC molecules, 121–135
MHC polymorphism and, 138–140, **Fig. 5.33**
peptide:MHC:T-cell receptor complex, 132–133, **Fig. 5.26**
Antigen-presenting cells (APCs)
cross-presentation, 131–132
effector T-cell activation, 219
MHC class II molecules, 132
NKT-cell activation, 357, **Fig. 12.32**
professional, 132, 207
see also B cell(s); Dendritic cells; Macrophages
co-stimulatory molecules, 207
IgE-mediated allergy, 416, **Fig. 14.23**
Antigen processing, 120–135, **Fig. 5.10**
dendritic cells, 202–203, **Fig. 8.3**
HLA class II region genes, 138, **Fig. 5.32**
intracellular pathways, 125–126, 134–135, **Fig. 5.17**, **Fig. 5.27**
MHC class I-specific, 126–128, **Fig. 5.20**
MHC class II-specific, 129–130, **Fig. 5.22**
Antigen receptors, 17
see also B-cell receptors; T-cell receptor(s)
chimeric, 522–523, **Fig. 17.20**, **Fig. 17.21**
Antihistamines, 427
Anti-HLA antibodies
chronic graft rejection, 444, **Fig. 15.13**
hyperacute rejection, 437
sources of pre-existing, 437–438, **Fig. 15.5**
Anti-idiotypic antibodies, 490
Anti-IgE monoclonal antibodies, 90, 409–410, **Fig. 14.14**
Anti-IL-5 monoclonal antibodies, 428
Antimicrobial peptides/proteins, 6, 41–43
see also Defensins
neutrophils, 60
Anti-RhD IgG antibodies, 305–306, **Fig. 11.11**
Antisera, 88
Antithymocyte globulin, rabbit (rATG), 447, **Fig. 15.30**
Anti-TNF-α antibodies, 91, 490–491, **Fig. 16.26**
Antiviral drugs, HIV infection, 394, **Fig. 13.28**, **Fig. 13.29**, **Fig. 13.30**
AP-1
cyclosporin/tacrolimus and, 449, **Fig. 15.19**
T-cell signaling, 209, **Fig. 8.11**
virus-infected cells, 69, **Fig. 3.33**

APC gene, **Fig. 17.3**
APE1 endonuclease, 101–102
APECED *see* Autoimmune polyendocrinopathy-candidiasis-ectodermal dystrophy
Apoptosis, 223, **Fig. 8.25**
anergic B cells, 168
centrocytes, 241, **Fig. 9.12**
induced by cytotoxic T cells, 223–224, **Fig. 8.26**
neutrophils, 61, **Fig. 3.16**
NK cell-mediated, 74–75
pro-B cells, 153, 154
self-reactive immature B cells, 165, **Fig. 6.18**
thymocytes, 182–183, 190, **Fig. 7.8**
Appendicitis, 272
Appendix, 25, 272
APRIL, 283, 285–286
Artemis, 385, **Fig. 13.10**
Arthus reaction, **Fig. 15.22**
Ascaris lumbricoides, **Fig. 14.24**
Aspirin (acetyl salicylate), 412, **Fig. 14.18**
Asplenia, 23–24, **Fig. 1.25**, **Fig. 13.10**
Asthma (allergic), 424
chronic, 424, **Fig. 14.35**, **Fig. 14.36**
genetic predisposition, 418–419, **Fig. 14.25**
immediate and late-phase responses, 420, **Fig. 14.28**
monoclonal antibody therapy, 90, 409–410
role of IgE, 250, **Fig. 14.35**
treatment, 427, 428
Ataxia telangiectasia, **Fig. 15.32**
Atopic dermatitis (or eczema), 424–426
Atopy, 401, 418
Auristatin, 528–529, **Fig. 17.28**
Autoantibodies
causing autoimmune disease, 475–476, **Fig. 16.1**
cell-surface receptors, 481–484, **Fig. 16.12**
epitope spreading, 486–488
transfer to fetus, 483–484, **Fig. 16.15**
Autoantigens, 477, **Fig. 16.1**
Autocrine action, 220
IL-2, 209
interferon-β, 69, **Fig. 3.33**
Autograft, 439
Autoimmune diseases, 473–505
antibody binding to cell-surface receptors, 481–484, **Fig. 16.12**
ectopic lymphoid tissue, 484–485, **Fig. 16.16**
effector mechanisms, 474–477, **Fig. 16.1**
endocrine glands, 485, **Fig. 16.17**
environmental factors, 492–495
epitope spreading, 485–488
gender differences, 480, **Fig. 16.11**
genetic susceptibility, 478–481, **Fig. 16.8**, **Fig. 16.9**
infections triggering, 492–494, **Fig. 16.31**
intravenous immunoglobulin therapy, 489–490, **Fig. 16.22**, **Fig. 16.23**
loss of self-tolerance, 473, 477–478
organ-specific, 482
systemic, 483, 487–488
T-cell senescence, 501–502, **Fig. 16.40**
tissue-specific, 482
type II, 475, **Fig. 16.1**
type III, 475, 476, **Fig. 16.1**
type IV, 475, 476–477, **Fig. 16.1**
vs. allergy and transplantation, 503–504, **Fig. 16.42**
Autoimmune polyendocrinopathy-candidiasis-ectodermal dystrophy (APECED), 193, 477–478, **Fig. 13.10**, **Fig. 16.6**, **Fig. 16.7**
Autoimmune polyglandular syndrome (APD) *see* Autoimmune polyendocrinopathy-candidiasis-ectodermal dystrophy
Autoimmune regulator (AIRE), 192–193
inherited deficiency, 193, 477–478, **Fig. 13.10**, **Fig. 16.6**
Autoimmune response, 473
Autoimmunity, 473
Autoinflammatory diseases, 502–503, **Fig. 16.41**
Autologous hematopoietic stem-cell transplantation, 461
Autologous MHC molecules, 143
Autoreactive B cells *see* B cell(s), autoreactive/self-reactive
Autoreactive cells/receptors, 163
Autoreactive T cells *see* T cell(s), autoreactive/self-reactive
Avian influenza viruses, 367–368, **Fig. 13.3**
Avidity, antibody, 103
Azathioprine, 453, **Fig. 15.25**, **Fig. 15.30**

B

B-1 cells, 161–162, **Fig. 6.15**
B-2 cells, 161, **Fig. 6.15**
B7
absence
effector T cells, 219, **Fig. 8.20**
inducing T-cell anergy, 211, **Fig. 8.13**
activation of naive T cells, 206–207, **Fig. 8.8**
Ba, 33, 34
Bacille Calmette–Guérin vaccine (BCG vaccine), 314, **Fig. 11.25**
disseminated infections, 386
intravesicular, bladder cancer, 514
Bacteria, 4, **Fig. 1.4**
commensal *see* Commensal microorganisms
encapsulated *see* Encapsulated bacteria
evasion of immune response, 366, 369, **Fig. 13.1**
neutralizing antibodies, 251–253, **Fig. 9.24**
subversion of immune response, 371
superantigens, 373–374, **Fig. 13.8**
vaccines against, 314–316
Bacteroides fragilis, 314–315
BAFF, 168, 239, 283
Balancing selection, MHC haplotypes, 141, **Fig. 5.37**
Bare lymphocyte syndromes, 192
type 1 (MHC class I deficiency), 127, 385, **Fig. 13.10**, **Fig. 13.16**
type 2 (MHC class II deficiency), 192, 385, **Fig. 13.16**
Barriers against infection, physical, 4–8, 29, **Fig. 1.5**, **Fig. 2.1**
Basiliximab, 452–453, **Fig. 15.24**, **Fig. 15.30**
Basophils, 15, 415, **Fig. 1.11**
Fcε receptors, 248, 407
IgD receptor, 104
initiation of T_H2 response, 213, 415
microscopic appearance, 415, **Fig. 14.21**
numbers in blood, **Fig. 1.14**
Battlefield injuries, 4
Bb, 33, 37, **Fig. 2.6**, **Fig. 2.12**
B cell(s), 16–17
activated
see also Centroblasts; Centrocytes
differentiation, 168–170, 244, **Fig. 6.22**, **Fig. 9.15**
helper T_{FH} cell interactions, 236–238
somatic hypermutation/isotype switching, 239–241, **Fig. 9.11**
activation
memory B cells, 169–170, 300–301, 302
naive B cells *see* Naive B cells, activation
anergy, 166, 169, **Fig. 6.19**
antibody production, 231–245
antigen processing, 129
antigen receptors *see* B-cell receptors
autoreactive/self-reactive, 163, 164–166
inactivation, 166, 241, **Fig. 6.19**
receptor editing, 165–166, **Fig. 6.18**
retention in bone marrow, 164, **Fig. 6.17**
circulation through lymph nodes, 167, **Fig. 6.20**
clonal deletion, 165
clonal expansion, 238–239, **Fig. 9.9**
co-stimulatory molecules, 207
development, 149–173, **Fig. 6.24**
see also Bone marrow
B-cell tumors and, 170, **Fig. 6.23**
bone marrow, 150–163
maturation phase, 167–170
numbers of cells produced, 168
phases, 149–150, **Fig. 6.1**
purging self-reactive B cells, 164–166
secondary lymphoid tissues and circulation, 150, 164, 167–170, **Fig. 6.2**
sites, 20, **Fig. 1.19**
differentiation into plasma cells, 17, 168–170, **Fig. 6.22**
effector *see* Plasma cells
half-life, 166, 168
help from T cells, 122, 225–226, **Fig. 5.13**, **Fig. 8.28**
immature, 151, 163–168, **Fig. 6.4**, **Fig. 6.16**
bone marrow stromal cell interactions, 152, **Fig. 6.5**
development, 156–157
maturation, 167–168, **Fig. 6.21**
purging of self-reactive, 164–166
inherited deficiency, 88–90, 154
malignant transformation *see* B-cell tumors
mature, 164, 168
memory *see* Memory B cells
monospecificity, 96–97
mucosal lymphoid tissues, **Fig. 10.15**
naive *see* Naive B cells
precursor cells, 150–151, **Fig. 6.3**
recirculation, 167, 168, **Fig. 6.21**
self-tolerance, 164, 166
splenic white pulp, **Fig. 1.24**
subsets, 161–162, **Fig. 6.15**
B-cell-activating factor *see* BAFF
B-cell co-receptor, 232–234, **Fig. 9.2**, **Fig. 9.3**
B-cell receptors
see also Immunoglobulin(s)
cross-linking, B-cell activation, 232, **Fig. 9.1**
editing, self-reactive B cells, 165–166, **Fig. 6.18**
formation, 97–98, 151, 156, **Fig. 4.23**

secreted form *see* Antibodies
selection and development of repertoire, 163–172
signal transduction, 232, **Fig. 9.1**, **Fig. 9.3**
vs. antibodies and T-cell receptors, 16–17, **Fig. 1.17**
vs. pre-B-cell receptor, **Fig. 6.7**
B-cell tumors
adoptive T-cell transfer, 523, **Fig. 17.21**
chromosomal translocations, 160–161, **Fig. 6.14**
stages of B-cell development, 170, **Fig. 6.23**
BCG vaccine *see* Bacille Calmette–Guérin vaccine
Bcl-2 protein, 161, **Fig. 11.7**
BCL2 proto-oncogene, 161
Bcl6, T_{FH} cell development, 212, 214
Bcl-x_L, 241, **Fig. 9.12**
BCR-ABL fusion, 517, **Fig. 17.10**
Bee stings, 421
Belatacept, 452, **Fig. 15.23**, **Fig. 15.30**
Benign tumors, 510, **Fig. 17.1**
Bevacizumab, **Fig. 17.27**
Birdshot chorioretinopathy, 479, **Fig. 16.9**
Bladder cancer, 514, **Fig. 17.10**
Blau syndrome, **Fig. 16.41**
Blk, B-cell receptor signaling, 232, **Fig. 9.1**
Blood, effects of allergens entering, **Fig. 14.30**
Blood-borne infections
antibody functions, 246
splenic function, 23, **Fig. 1.24**
Blood transfusion, 434–436
anti-HLA antibodies after, 438
blood group matching, 434, **Fig. 15.2**, **Fig. 15.3**
cross-match tests, 435–436
incompatible, 435
organ transplantation after prior, 444
Blood vessels
see also Endothelium, vascular
changes in chronic graft rejection, 443, **Fig. 15.11**
IgE-mediated allergic reactions, 421, **Fig. 14.29**
B lymphoblasts, 238–239
B-lymphocyte-induced maturation protein 1 (BLIMP-1), 238–239
B lymphocytes *see* B cell(s)
Bone marrow
B-cell development, 150–163
checkpoints, 157, **Fig. 6.11**
dependence on stromal cells, 151–152, **Fig. 6.5**
proteins regulating, 157–160, **Fig. 6.12**, **Fig. 6.13**
stages, 150–151, **Fig. 6.3**, **Fig. 6.4**
development of central tolerance, 166
hematopoiesis, 12
migration of plasma cells to, 170
retention of self-reactive immature B cells, 164, **Fig. 6.17**
Bone marrow transplantation, 439, 461
see also Hematopoietic stem-cell transplantation
Bordetella pertussis, 50, 316
Bortezomib, 524
Botulinum toxin, **Fig. 9.25**
Botulism, **Fig. 9.25**
BP-1, **Fig. 6.12**
Bradykinin, 40, 382
BRAF inhibition, melanoma, 522
Brambell receptor *see* FcRn
Breast cancer, **Fig. 17.6**, **Fig. 17.27**
Breast milk, dimeric IgA, 250, 287, **Fig. 9.21**
Brentuximab-vedotin conjugate, 528–529, **Fig. 17.27**, **Fig. 17.28**
Bromohydrin pyrophosphate (BrHPP), 350, **Fig. 12.24**
Bronchial-associated lymphoid tissue (BALT), 25
Bronchiectasis, 379
Brugia malayi, **Fig. 14.24**
Bruton's tyrosine kinase *see* Btk
Btk
B-cell development, 160, **Fig. 6.12**
inherited deficiency, 160, 379, **Fig. 13.10**, **Fig. 13.12**
Burkitt's lymphoma, **Fig. 6.23**
chromosomal translocations, 161, **Fig. 6.14**
diagnosis, 526, **Fig. 17.24**
Butyrophylin-3A1 (BTN3A1), 350, **Fig. 12.25**

C

C1, 66
inherited deficiency, 381, **Fig. 13.13**
interaction with C-reactive protein, 66, **Fig. 3.28**
structure, 66, **Fig. 3.27**
C1 inhibitor (C1INH), 381–382, **Fig. 13.14**
inherited deficiency, 381–382, **Fig. 13.13**
C1q, 66, **Fig. 3.27**
IgG binding, 257, **Fig. 9.31**
IgM binding, 255, **Fig. 9.28**
C1r, 66, 255, **Fig. 3.27**
inhibition, 382, **Fig. 13.14**
C1s, 66, 255, **Fig. 3.27**
inhibition, 382, **Fig. 13.14**
C2, 64–65
cleavage, 65, 66, **Fig. 3.25**
inherited deficiency, 381, **Fig. 13.13**
C2a, production, 65, **Fig. 3.25**
C2b, production, 65
C3, 31–32
alternative complement activation pathway, 32–33
cleavage, 31, 32, 33, **Fig. 2.4**, **Fig. 2.6**
on pathogen surface, 34, **Fig. 2.8**
hydrolyzed (iC3 or C3(H_2O)), 32–33, **Fig. 2.6**
inherited deficiency, 31, 381, **Fig. 13.10**, **Fig. 13.13**
lectin/classical complement activation pathway, 65, **Fig. 3.25**
thioester bond, 31–32, **Fig. 2.4**
C3a, 31, **Fig. 2.3**
formation, 32, **Fig. 2.4**
induction of inflammation, 40, **Fig. 2.15**
receptors, 39
C3b, 31, **Fig. 2.3**
alternative C3 convertase *see* C3 convertase, alternative
alternative C5 convertase, 37, **Fig. 2.12**
amplification of C3 cleavage, 34, **Fig. 2.8**
attachment to pathogens, 32, 33, 255–256, **Fig. 2.4**, **Fig. 9.29**
B-cell activation, 233, 235–236, **Fig. 9.2**
formation, 32, 34, **Fig. 2.4**, **Fig. 2.6**
immune complexes, 258, **Fig. 9.32**
receptors on phagocytes, 36–37, **Fig. 2.10**
regulation of deposition, 34–36, **Fig. 2.9**
C3 convertases, 34
alternative (C3bBb), 34, **Fig. 2.7**, **Fig. 3.26**
amplified formation, 34, **Fig. 2.8**
IgM-mediated assembly, 256
regulation of activity, 34–36, **Fig. 2.9**
classical (C4bC2a), 65, **Fig. 3.26**
formation, 65, 66, **Fig. 3.25**
IgM-mediated assembly, 255–256
iC3bBb, 33–34, **Fig. 2.6**
C3d
B-cell activation, 233, 235–236, **Fig. 9.3**, **Fig. 9.6**
production, 233, **Fig. 9.2**
C4, 64–65
A and B forms, 256–257, **Fig. 9.30**
inherited deficiencies, 256–257, 377, **Fig. 13.10**
cleavage, 65, 66, **Fig. 3.25**, **Fig. 3.28**
inherited deficiencies, 381, **Fig. 13.13**
C4a, 65, **Fig. 3.28**
C4b
binding to pathogens, 255, **Fig. 9.29**
production, 65, **Fig. 3.28**
C5, 37–38, **Fig. 2.11**, **Fig. 2.12**
C5a
induction of inflammation, 40, **Fig. 2.15**
production, 37, 39, **Fig. 2.12**
receptors, 39
C5b, 37–38, **Fig. 2.12**, **Fig. 2.13**
C5–C9, inherited deficiencies, 38, 381, **Fig. 13.10**, **Fig. 13.13**
C5 convertase, alternative, 37, **Fig. 2.12**, **Fig. 2.13**
C6, 37–38, **Fig. 2.11**, **Fig. 2.13**
C7, 37–38, **Fig. 2.11**, **Fig. 2.13**
C8, 38, **Fig. 2.11**, **Fig. 2.13**
C9, 38, **Fig. 2.11**, **Fig. 2.13**
Calcineurin
inhibition by cyclosporin/tacrolimus, 449, **Fig. 15.19**
T-cell receptor signaling, **Fig. 8.11**
Calnexin, 127, **Fig. 5.20**
Calreticulin, 127–128, **Fig. 5.20**
Cancer, 509–530
see also Tumors
causation, 510–513, **Fig. 17.3**
cells, distinctive features, 513–514, **Fig. 17.5**
chimeric antigen receptors, 522–523, **Fig. 17.21**
common locations, 510, **Fig. 17.2**
defined, 510
diagnosis, 526–528
environmental factors, 512–513
genetic factors, 512
hematopoietic cell transplantation, 461, 463, 465–466, **Fig. 15.34**
immune response, 514–519
immunosurveillance, 514–515
immunotherapy, 520–525, 528–529
monoclonal antibody therapy, 528–529, **Fig. 17.27**
augmenting NK cell responses, 524–525, **Fig. 17.22**
radioactive isotope conjugates, 529, **Fig. 17.29**
targeting negative regulators, 521–522, **Fig. 17.19**
toxin conjugates, 528–529, **Fig. 17.28**
stem cells, 515
in transplant recipients, 446, 513–514
vaccination for prevention, 519–520

vaccination therapy, 520–521, **Fig. 17.18**
vaccine, 525
Cancer/testis (CT) antigens, 517–518, **Fig. 17.12**
vaccination with, 520–521, **Fig. 17.18**
Candida
chronic infections, 478
macrophage receptors, 50
opportunistic infections, 396
Candida albicans, morphology, **Fig. 1.3**
Candidiasis, mucocutaneous, **Fig. 15.32**
Capsular polysaccharides (CPs), bacterial
see also Encapsulated bacteria
macrophage receptors, 50, **Fig. 3.3**
vaccines targeting, 314–316
Carcinogens, 512
Carcinomas, 510
CARD (caspase-recruitment domain), 55, 69
Carriers, inherited immunodeficiency diseases, 377
Cartilage-hair hypoplasia, **Fig. 15.32**
Caspase 1, 55–56, **Fig. 3.11**
Caspase-recruitment domain (CARD), 55, 69
Caspases, 55
Catalase, 61, **Fig. 3.17**
Catalytic antibodies, 88
Cathepsin L, 192
Cathepsin S, 130
CCL11 (eotaxin), 414
CCL19
immature B cells, 167, **Fig. 6.21**
mucosal tissues, 280
naive T-cell homing, 204–205
CCL20, 278
CCL21
dendritic cell maturation, 203
immature B cells, 167, **Fig. 6.21**
mucosal tissues, 280
naive T-cell homing, 204–205
CCL25, 281, 282–283, **Fig. 10.16**
CCR5
deficiency, resistance to HIV, 391–392
γ:δ T cells, 349
as HIV co-receptor, 389
CCR6, 278
CCR7
activated naive B cell–T_{FH} cell interactions, 237
dendritic cells, 203
immature B cells, 167
memory T cells, 303, 304, **Fig. 11.7**
mucosa-derived lymphocytes, 280, 281
naive T cells, 205
CCR9, 281, 282–283, **Fig. 10.16**
CD1, 352
evolution, 359–360
mycobacterial antigen presentation, 354–356, **Fig. 12.29**
proteins and genes, 352, 355, **Fig. 12.30**, **Fig. 12.31**
CD1a, **Fig. 12.30**
functional properties, **Fig. 12.31**
lipid antigen presentation, 355–356, **Fig. 12.29**
CD1b, **Fig. 12.30**
functional properties, **Fig. 12.31**
lipid antigen presentation, 355–356, **Fig. 12.27**, **Fig. 12.29**
CD1c, **Fig. 12.30**
functional properties, **Fig. 12.31**
lipid antigen presentation, 355–356, **Fig. 12.29**
CD1d, **Fig. 12.30**
conserved function, 357
lipid presentation to γ:δ T cells, 352–353, **Fig. 12.28**
lipid presentation to NKT cells, 356–357, **Fig. 12.32**
CD1e, 356, **Fig. 12.30**
CD2
developing T cells, 188, **Fig. 7.14**
effector T cells, 219, **Fig. 8.18**
T-cell commitment and, 180, **Fig. 7.5**
T cell–dendritic cell adhesion, 205
CD3 complex, 118, **Fig. 5.6**
γ:δ T-cell receptors, 183
pre-T-cell receptors, 184, **Fig. 7.10**
CD3 proteins, 118, **Fig. 5.6**
expression on thymocytes, 188, **Fig. 7.14**
inherited defects, 118
initiation of TCR signaling, 208, **Fig. 8.10**
monoclonal antibodies, 90, 450–451, **Fig. 15.21**, **Fig. 15.30**
CD4, 122–123
expression by γ:δ T cells, 348
expression on thymocytes, 182, 185, 188, **Fig. 7.7**, **Fig. 7.14**
as HIV receptor, 122–123, 389, **Fig. 13.23**
initiation of TCR signaling, 208
MHC class II binding, 124, **Fig. 5.15**
negative thymocytes, 180, **Fig. 7.5**
structure, **Fig. 5.12**
CD4 CD8 double-positive T cells *see* Double-positive (DP) thymocytes
CD4 T cells, 122
see also Helper CD4 T cells
activation by superantigens, 373–374, **Fig. 13.8**
antigen presentation to, 122–123, **Fig. 5.15**
autoimmune disease, 481, 483, 484
CD8 T-cell activation, 216, **Fig. 8.17**
celiac disease, 500, **Fig. 16.39**
chronic graft rejection, 443–444
development, 191–192, **Fig. 7.17**
effector, 122, 194, 218–222
acute graft rejection, 441
cell-surface molecules, 218–219, **Fig. 8.18**
differentiation, 211–215, **Fig. 8.14**
functions, 122, 218–222, 224–227, **Fig. 5.13**
polarized response, 214–215
effector molecules, **Fig. 8.21**
HIV infection, 390
antiviral drug therapy and, 394, **Fig. 13.30**
circulating numbers, 389–390, 395, **Fig. 13.24**
clinical latency period, 390, 394–395
virus entry into cells, 122–123, 389
virus life cycle, 389, **Fig. 13.23**
intracellular signaling pathways, **Fig. 8.11**
memory, cell-surface molecules, 303–304, **Fig. 11.8**
regulatory (T_{reg}) *see* Regulatory T cells
self-reactive, 193, **Fig. 7.19**
type IV hypersensitivity reactions, 402, **Fig. 14.2**
CD5, T cells, 180, **Fig. 7.5**
CD5 B cells, 161–162
CD8, 122
expression by γ:δ T cells, 348
expression on thymocytes, 182, 185, 188, **Fig. 7.7**, **Fig. 7.14**
initiation of TCR signaling, 208
MHC class I binding, 124, **Fig. 5.15**
negative thymocytes, 180, **Fig. 7.5**
structure, **Fig. 5.12**
CD8 T cells, 122
see also Cytotoxic CD8 T cells
activation of naive, 215–216, **Fig. 8.17**
antigen cross-presentation to, 131–132
antigen presentation to, 122, **Fig. 5.15**
development, 191–192, **Fig. 7.17**
effector, 122, 194, 218–224
activation requirements, 219, **Fig. 8.20**
acute graft rejection, 441
cytokines and cytotoxins, 220–221, **Fig. 8.21**
functions, 122, 218–224, **Fig. 5.13**
generation of memory cells, 305, **Fig. 11.10**
MAIT cells, 357–359
evasion by viruses, 369–372
intracellular signaling pathways, **Fig. 8.11**
memory, 303, 305, **Fig. 11.10**
mucosal tissues, 281–282, **Fig. 10.17**
role in autoimmune disease, 481
type IV hypersensitivity, 402
CD11b/CD18 *see* CR3
CD14 (LPS receptor)
cells expressing, **Fig. 3.3**, **Fig. 3.15**
TLR4 co-receptor function, 51, **Fig. 3.6**, **Fig. 3.7**
CD15, in Hodgkin's disease, 527, **Fig. 17.25**
CD16 *see* FcγRIII
CD18, inherited deficiency, 382, **Fig. 13.15**
CD19
B-cell co-receptor, 233–234, **Fig. 9.2**, **Fig. 9.3**
chimeric antigen receptors against, 523, **Fig. 17.21**
developing B cells, 152, **Fig. 6.12**
CD20, monoclonal antibodies against *see* Anti-CD20 monoclonal antibodies
CD21 *see* CR2
CD23 *see* FcεRII
CD24, developing B cells, **Fig. 6.12**
CD25 (IL-2 receptor α chain)
developing B cells, **Fig. 6.12**
developing T cells, **Fig. 7.14**
monoclonal antibodies, 452–453, **Fig. 15.24**, **Fig. 15.30**
regulatory T cells, 193, 226
CD27
γ:δ T cells, 349, **Fig. 12.21**
memory B cells, 300
CD28
independence, effector T-cell activation, 219, **Fig. 8.20**
monoclonal antibody against, 207
naive T-cell activation, 206–207, **Fig. 8.8**
stabilization of IL-2, 210
superantigen interactions, 373, **Fig. 13.8**
CD30
conjugated antibodies targeting, 528–529, **Fig. 17.27**, **Fig. 17.28**
lymphomas expressing, 527, 528, **Fig. 17.25**, **Fig. 17.26**
CD31, 58, **Fig. 3.13**
CD32 *see* FcγRII
CD34

hematopoietic stem cells, 150, 463
naive T cell homing, 205, **Fig. 8.6**
thymocytes, 180, **Fig. 7.5**
CD40
B cell–helper T_{FH} cell interaction, 225, 238, **Fig. 8.28**, **Fig. 9.8**
inherited deficiency, **Fig. 13.10**
isotype switching, 243–244
macrophage activation, 225, **Fig. 8.27**
CD40 ligand
B cell–helper T_{FH} cell interaction, 225, 238, **Fig. 8.28**, **Fig. 9.8**
inherited deficiency, 243–244, 380–381, **Fig. 13.10**
isotype switching, 243–244
macrophage activation, 224–225, **Fig. 8.27**
CD43, developing B cells, **Fig. 6.12**
CD44
effector T cells, **Fig. 8.18**
memory T cells, **Fig. 11.7**
thymocytes, **Fig. 7.5**
CD45
developing B cells, **Fig. 6.12**
memory T cells, 303–304, **Fig. 11.7**
CD45RA
γ:δ T cells, 349, **Fig. 12.21**
naive T cells, 303–304, **Fig. 11.7**, **Fig. 11.8**
CD45RO
effector T cells, **Fig. 8.18**
memory T cells, 303–304, **Fig. 11.7**, **Fig. 11.8**
CD52, 447, **Fig. 15.15**
CD56, 71
CD57, 342
CD59
complement regulation, 38–39, **Fig. 2.14**
inherited deficiency, 39, 381, **Fig. 13.13**
CD62L *see* L-selectin
CD64 *see* FcγRI
CD69, 206, 237, **Fig. 11.7**
CD81, 233–234, **Fig. 9.2**
CD94:NKG2A, 333–334, **Fig. 12.2**
inhibition of NK-cell activation, 334, **Fig. 12.6**
ligand binding, 333, **Fig. 12.5**
NK-cell education, 337, 339, **Fig. 12.10**
placental formation, 344, **Fig. 12.17**
vs. inhibitory KIRs, 336
CD94:NKG2C, **Fig. 12.2**
cytomegalovirus infection, 342, **Fig. 12.15**
CD127 *see* Interleukin-7 (IL-7) receptor
C domains *see* Constant domains
CDRs *see* Complementarity-determining regions
Celiac disease, 274, 498–501
HLA association, 498–501, **Fig. 16.9**, **Fig. 16.38**
immune mechanisms, 498, 500–501, **Fig. 16.37**, **Fig. 16.39**
intestinal epithelial cell destruction, 498, 500, **Fig. 16.36**
Cell-adhesion molecules (CAMs) *see* Adhesion molecules
Cell-mediated immunity, 199–228
polarized T_H1 response, 214
Cell-surface receptors
autoantibodies, 481–484, **Fig. 16.12**
innate immunity, 47–49, **Fig. 3.2**
Central memory T cells (T_{CM}), 304, **Fig. 11.9**
Central-supramolecular activation complex (c-SMAC), 208, **Fig. 8.9**
Central tolerance, 166, 193
Centroblasts, 169, 239–240, 241, **Fig. 9.11**
Centrocytes, 169, 240–241, **Fig. 9.11**
antigen-mediated selection, 241–243, **Fig. 9.12**
apoptosis, 241, **Fig. 9.12**
differentiation into plasma/memory cells, 244, **Fig. 9.15**
Cervical cancer, 513, 519
Cetruximab, **Fig. 17.27**
C genes *see* Constant region genes
Chaperones, 127
Chédiak–Higashi syndrome, 383, **Fig. 13.15**
Chemicals, carcinogenic, 512
Chemokine receptors, 57–58, **Fig. 3.14**
as HIV co-receptors, 389
neutrophil recruitment, 57–58, **Fig. 3.13**
Chemokines
B-cell guidance, 167–168, **Fig. 6.21**
eosinophil guidance, 414
naive T-cell guidance, 205, **Fig. 8.6**
neutrophil recruitment, 54, 57–58, **Fig. 3.13**
Chickenpox, 370
Children
immunization schedule, **Fig. 11.15**
rotavirus infections, 313, **Fig. 11.18**
Chimeric antigen receptors (CAR), 522–523, **Fig. 17.20**, **Fig. 17.21**
Chimpanzees, resistance to HIV, 389
Cholera, 284–285, **Fig. 9.25**
vaccine, **Fig. 11.25**
Chromosomal translocations
anaplastic large cell lymphoma, 527–528, **Fig. 17.26**
B-cell tumors, 160–161, **Fig. 6.14**
detection in Burkitt's lymphoma, 526, **Fig. 17.24**
Chronic granulomatous disease (CGD), 61–62, 383, **Fig. 13.10**, **Fig. 13.15**
associated infections, 62, **Fig. 3.19**
Chronic lymphocytic leukemia (CLL), 162, **Fig. 6.23**, **Fig. 17.27**
Chronic myelogenous leukemia, 517, **Fig. 17.10**
Chymopapain, 417
Chymotryptase, mast cell, 411, 412
CIITA (class II transactivator), 138, 385
Citrullinated proteins
antibodies in rheumatoid arthritis, 492, **Fig. 16.29**
generation, 492, **Fig. 16.28**
Class II-associated invariant chain (CLIP), 130–131, **Fig. 5.23**
Class II transactivator (CIITA), 138, 385
Class switching *see* Isotype switching
Clonal deletion, B cells, 165
Clonal expansion, 11
B cells, 238–239, **Fig. 9.9**
γ:δ T cell populations, 348–349
large pre-B cells, 155, 157
Clonal selection, 11
see also Negative selection; Positive selection
Clostridium difficile, 3, 29, **Fig. 1.2**
Clostridium perfringens, **Fig. 9.25**
Clostridium tetani, 314
Clusterin, 38
Coagulation system, 40
Coding joint, 95, **Fig. 4.19**
Coevolution, innate and adaptive immunity, 329–361, **Fig. 12.1**
Cognate pairs (conjugate pairs)
see also T-cell synapse
B cell–effector T_{FH} cell, 225–226, 238, **Fig. 8.28**, **Fig. 9.7**, **Fig. 9.8**
dendritic cell–naive T cell, 206
effector T cell–target cell, 211, 218
macrophage–T_H1 cell, 225
Cold autoinflammatory syndromes, familial, **Fig. 16.41**
Cold sores, 370, **Fig. 13.5**
Cold viruses, 296
Colicins, 3
Collagenase, eosinophil, **Fig. 14.20**
Collagen type IV, autoantibodies, 475–476, **Fig. 16.3**
Collectins, 64
Colorectal cancer, **Fig. 17.3**, **Fig. 17.27**
Commensal microorganisms, 2–3, 29–30
gut *see under* Gastrointestinal tract
protection against infection, 3, 29
Common gamma chain (γ_c), 384
Complement, 31–39
activation, 31–32, **Fig. 2.3**
alternative pathway, 32–34, **Fig. 2.6**
antibody isotypes mediating, 255, **Fig. 9.27**
antibody-mediated, 255–258, **Fig. 4.30**
autoimmune diseases, 475
classical pathway, 32, 66, **Fig. 3.25**
IgG, 257–258, **Fig. 9.31**
IgM, 255–256, **Fig. 9.28**
immune complexes, 257–258, **Fig. 9.31**
lectin pathway, 32, 63–66, **Fig. 3.25**
pathways, 32–34, 66, **Fig. 2.5**
components
see also specific components
inherited deficiencies, 381–382, **Fig. 13.13**
evasion strategies, 36
fixation, 31, 256–257, **Fig. 2.4**, **Fig. 9.29**
functions, 8, **Fig. 1.6**
inflammatory response, 40, **Fig. 2.15**
terminal components, 37–39, **Fig. 2.11**
see also Membrane-attack complex
inherited deficiencies, 38, 381
regulation, 38–39, **Fig. 2.14**
Complementarity-determining regions (CDR1–CDR3)
see also Hypervariable regions
antibodies, 86, **Fig. 4.8**
generation of diversity, 92, 95–96, **Fig. 4.20**
gene segments encoding, 92
somatic hypermutation, 100, **Fig. 4.26**
humanized monoclonal antibodies, 90
T-cell receptors, 114–115, **Fig. 5.2**
peptide:MHC complex interactions, 133, **Fig. 5.26**
Complement control protein (CCP) modules, 36
Complement control proteins, 34–36, **Fig. 2.9**
Complement receptors
see also CR1; CR2; CR3; CR4
antigen capture in B-cell follicles, 235–236, **Fig. 9.6**
macrophages, 36–37, **Fig. 2.10**
Complement system, 31
Conjugate pairs *see* Cognate pairs
Conjunctivitis, allergic, 423–424
Connective tissue mast cells, 411

Constant domains (C domains)
antibodies, 84, **Fig. 4.6**, **Fig. 4.7**
T-cell receptors, 114, **Fig. 5.1**
Constant region genes (C genes), 92, **Fig. 4.15**, **Fig. 14.5**
recombination in isotype switching, 101–102, **Fig. 4.28**
segments, 92, **Fig. 4.17**
switch sequences/regions, 101, 243, **Fig. 4.28**
T-cell receptors, 115
Constant regions (C regions)
antibodies, 83, **Fig. 4.2**
T-cell receptors, 114, **Fig. 5.1**
Convergent evolution, NK-cell receptors, 339, 340
Corneal transplantation, 456–457, **Fig. 15.29**
Corticosteroids
chemical structure, **Fig. 15.16**
effects on immune system, 448, **Fig. 15.18**
mechanisms of action, 447–448, **Fig. 15.17**
transplant recipients, 447–448
Corynebacterium diphtheriae, 314
Co-stimulatory molecules, 206–207
Co-stimulatory receptors, 206–207
Co-stimulatory signals
absence, inducing T-cell tolerance, 211, **Fig. 8.13**
IL-2 production, 210, **Fig. 8.12**
independence of effector T-cell activation, 219, **Fig. 8.20**
induction by adjuvants, 211
naive T-cell activation, 206–207, 215–216, **Fig. 8.8**
Cowpox, 309, 310
Cowpox virus *see* Vaccinia virus
CpG-rich bacterial DNA, 50, **Fig. 3.3**
unmethylated, 67, **Fig. 3.29**
CR1 (complement receptor 1; CD21)
antigen capture in B-cell follicles, 235–236
B-cell activation, 233, **Fig. 9.2**
immune complex clearance, 258, **Fig. 9.32**
macrophages, 36, **Fig. 2.10**
CR2 (complement receptor 2; CD21)
antigen capture in B-cell follicles, 235, **Fig. 9.6**
B-cell co-receptor, 233, **Fig. 9.2**, **Fig. 9.3**
CR3 (complement receptor 3; Mac-1; CD11b/CD18)
inherited defects, 382
macrophages, 36–37, 50, **Fig. 3.3**
neutrophil recruitment, 57–58
NK cell cytotoxicity, 74
CR4 (complement receptor 4)
inherited defects, 382
macrophages, 36–37, 50, **Fig. 3.3**
neutrophils, **Fig. 3.15**
C-reactive protein (CRP), 62–63, **Fig. 3.21**
binding to Fc receptors, 263
C1 binding, 255
complement activation, 66, **Fig. 3.28**
rheumatoid arthritis, **Fig. 16.26**
structure, **Fig. 3.22**
C regions *see* Constant regions
Cross-match tests, 435–436, 437
Cross-presentation, antigen, 131–132, 203, **Fig. 5.24**, **Fig. 8.3**
Cross-priming, 131–132
CRP *see* C-reactive protein
Cryoglobulinemia, mixed essential, **Fig. 16.1**
Cryopyrin (NLRP3), 55, 56, **Fig. 3.11**
Cryptdins, 41, **Fig. 2.18**
CT antigens *see* Cancer/testis (CT) antigens
CTLA4, 207, 451–452
monoclonal antibodies against, 521–522, **Fig. 17.19**
synthetic fusion protein targeting, 452, **Fig. 15.23**, **Fig. 15.30**
C-type lectin domains (CTLD), 50
C-type lectins, 50
CXCL8, 53, 54, **Fig. 3.9**
neutrophil recruitment, 57–58, **Fig. 3.13**
NK-cell recruitment, **Fig. 3.40**
CXCL13
B-cell homing, 168, **Fig. 6.21**
T_{FH} cell guidance, 214
CXCR1, neutrophils, 57–58
CXCR2, neutrophils, 57–58
CXCR5, T_{FH} cells, 214
Cyclophilins, 449, **Fig. 15.19**
Cyclophosphamide, 454, **Fig. 15.25**, **Fig. 15.30**
Cyclosporin A, 210, 448–450, 455
immunological effects, 450, **Fig. 15.20**
mechanism of action, 448–449, **Fig. 15.19**, **Fig. 15.30**
Cytokine receptors
common gamma chain (γ_c), 384
signaling pathways, 221, **Fig. 8.22**
Cytokines, 53
see also specific cytokines
CD4 T-cell differentiation, 213–214, **Fig. 8.14**
centrocyte differentiation, 244, **Fig. 9.15**
chromosome 5 gene cluster, 418–419
control of isotype switching, 243–244, **Fig. 9.13**
corticosteroid actions, **Fig. 15.18**
effector T cells, 220, 221–222, **Fig. 8.21**
eosinophils, **Fig. 14.20**
inflammatory (pro-inflammatory)
induction of synthesis, 53, **Fig. 3.7**, **Fig. 3.9**
recruitment of neutrophils, 57–58
role in inflammation, 53–54, **Fig. 3.9**
systemic effects, 62–63, **Fig. 3.20**
viral infection, 71–72, **Fig. 3.32**
innate immune response, 9
macrophages, **Fig. 1.16**
mast cells, 412, **Fig. 14.16**
mechanisms of action, 221–222, **Fig. 8.22**
polarized T_H1/T_H2 responses, 214–215, **Fig. 8.15**
Cytomegalovirus (CMV)
γ:δ T-cell response, 351–352
memory T-cell response, **Fig. 11.10**
NK-cell response, 341–342, **Fig. 12.15**
reactivation in AIDS, 396
subversion of immune response, 371–372, **Fig. 13.6**, **Fig. 13.7**
Cytosol
antigen processing, 126–127, **Fig. 5.18**
compartment, 125, **Fig. 5.17**
Cytotoxic CD8 T cells, 17, 215, 218–224
see also CD8 T cells
activation requirements, 219, **Fig. 8.20**
antigen presentation to, 122, **Fig. 5.15**
cancer surveillance, 515
differentiation, 194, 215–216, **Fig. 1.23**, **Fig. 8.17**
effector functions, 122, 218–224, **Fig. 5.13**
effector molecules, 220–221, 223, **Fig. 8.21**
evasion by tumor cells, 518, **Fig. 17.13**
induction of apoptosis, 223–224, **Fig. 8.26**
killing of target cells, 222–224, **Fig. 8.23**, **Fig. 8.24**
lytic granules, 220, 222–223, **Fig. 8.23**, **Fig. 8.26**
response to infection, 218–219
Cytotoxic drugs, 453–454, **Fig. 15.25**
Cytotoxicity, NK cell, 72–74, **Fig. 3.38**
Cytotoxins, 220–221, **Fig. 8.21**
induction of apoptosis, 224, **Fig. 8.26**
targeted delivery, 222–223, **Fig. 8.23**

D

Daclizumab, 452–453, **Fig. 15.24**
DAF *see* Decay accelerating factor
Dander, animal, 416
DAP12, **Fig. 12.6**
DCC gene, **Fig. 17.3**
DC-SIGN, 205
Decay-accelerating factor (DAF), 35–36, **Fig. 2.9**
inherited deficiency, 39, 381, **Fig. 13.13**
modular structure, 36
Dectin-1, 50, **Fig. 3.3**
Defensins, 41–43
α-defensins, 41, 42, **Fig. 2.18**, **Fig. 2.19**
β-defensins, 41, 42, **Fig. 2.19**
binding by mucins, 269
intestinal wall, 41, **Fig. 2.18**
mechanisms of microbial destruction, 41, **Fig. 2.17**
Delayed-type hypersensitivity (DTH), 403
δ heavy chains, immunoglobulin, 83, **Fig. 4.5**
generation, 96, **Fig. 4.22**
Dendritic cells, 15, 200–204, **Fig. 1.11**
activation, 203
adoptive transfer, 525, **Fig. 17.23**
antigen presentation
to naive CD8 T cells, 215–216, **Fig. 8.17**
to naive T cells, 203–204, **Fig. 8.3**
antigen processing, 129–130, 202–203, **Fig. 8.3**
antigen uptake and transfer to lymph nodes, 200–201, **Fig. 8.1**
co-stimulatory molecules, 207, **Fig. 8.8**
donor organ-derived
direct allorecognition, 441, **Fig. 15.9**, **Fig. 15.12**
indirect allorecognition, 443, **Fig. 15.12**, **Fig. 15.13**
follicular *see* Follicular dendritic cells
gut, antigen uptake and processing, 278–279, **Fig. 10.14**
HIV infection, 389
immature, 200, **Fig. 8.2**
initiation of adaptive immunity, 22, **Fig. 1.23**
maturation, 200–201, 203, **Fig. 8.2**
mature (activated), 200–201, **Fig. 8.2**
naive T-cell adhesion, 205–206, **Fig. 8.7**
negative selection of T cells, 192, **Fig. 7.18**
NK cell interactions, 76–77, **Fig. 3.41**
plasmacytoid, 71, **Fig. 3.35**
splenic, 23
T_H1 cell differentiation, 213
thymus, 178, **Fig. 7.3**
transport to lymph nodes, 22
Dermatitis, atopic, 424–426
Dermatophagoides pteronyssimus, 416–417
Desensitization, 427

Desmoglein autoantibodies, 485–486, **Fig. 16.18**, **Fig. 16.19**
Devil Facial Tumor Disease (DTFD), 515–516, **Fig. 17.8**
D gene segments *see* Diversity (D) gene segments
Diabetes
  type 1 (insulin-dependent), 495–498, **Fig. 16.1**
    genetic susceptibility, 479, 496–498, **Fig. 16.8**
    HLA allotype associations, 496–498, **Fig. 16.9**, **Fig. 16.34**, **Fig. 16.35**
    infections triggering, 494, **Fig. 16.31**
    pancreatic β-cell destruction, 496, **Fig. 16.33**
  type 2 (insulin-resistant), 483, **Fig. 16.1**, **Fig. 16.12**
Diacylsulfoglycolipid, **Fig. 12.29**
Diamond–Blackfan syndrome, **Fig. 15.32**
Diapedesis, **Fig. 3.13**, **Fig. 8.6**
Diarrhea
  food allergies, 426, **Fig. 14.38**
  helminth infections, 289
Diarrheal disease, **Fig. 11.26**
Didehydroxymycobactin, **Fig. 12.29**
DiGeorge syndrome, 178–179, 234–235, **Fig. 13.10**, **Fig. 13.16**
Diphtheria
  toxin, 253, 314, **Fig. 9.25**
  vaccination, 299
  vaccine, 253, 314, **Fig. 11.25**
Diphtheria, tetanus and pertussis (DTP) vaccine, 316, 319–321, **Fig. 11.15**
Diphyllobothrium latum, 248
Directional selection, MHC haplotypes, 142, **Fig. 5.37**
Diversity (D) gene segments
  immunoglobulins, 92, **Fig. 4.15**, **Fig. 4.17**
  recombination signal sequences (RSSs), 93, **Fig. 4.18**
  somatic recombination, 92–95, **Fig. 4.16**
  T-cell receptors, 115, **Fig. 5.3**, **Fig. 5.8**
  T-cell *vs.* B-cell receptors, **Fig. 5.9**
DNA
  antibodies against, 166
  hairpins, 95, **Fig. 4.20**
  mutations *see* Mutations
  palindromic sequences, 95, **Fig. 4.20**
  viral, recognition, 69, **Fig. 3.29**
DNA-dependent protein kinase (DNA-PK), 385
DNAM-1, **Fig. 12.2**
Dominant diseases, 377
Double-negative (DN) thymocytes, 180, **Fig. 7.5**
  γ:δ or pre-T-cell receptor expression, 183–184, **Fig. 7.9**
  gene expression, **Fig. 7.14**
  lineage commitment, 181–182, **Fig. 7.7**
Double-positive (DP) thymocytes, 182, **Fig. 7.15**
  gene expression, **Fig. 7.14**
  positive selection, 188–191
  TCR gene rearrangements, 185–186
Drug allergy, 422, **Fig. 14.32**
DTP vaccine, 316, 319–321, **Fig. 11.15**

E

E2A, 152, 153, **Fig. 6.12**
EBF, 152, 153, **Fig. 6.12**
EBNA-1, 371
Ebola virus, 4
E-cadherin, 282, **Fig. 10.16**
Ectopic lymphoid tissues, 484–485
Eczema, atopic, 424–426
Edema, 9
Effector B cells *see* Plasma cells
Effector cells, 17
  adaptive immunity, 10–11, **Fig. 1.9**
  differentiation in lymph nodes, 22–23, **Fig. 1.23**
  Fcγ receptors, 258–261
  innate immunity, 8, **Fig. 1.6**
  mucosal tissues, 280–282, **Fig. 10.17**
Effector mechanisms
  adaptive immune response, 10
  innate immune response, 8, **Fig. 1.6**
Effector memory T cells (T_{EM}), 304, **Fig. 11.9**
Effector T cells, 17, 199, 218–227
  see also Cytotoxic CD8 T cells; Helper CD4 T cells; Regulatory T cells
  activation, 219, **Fig. 8.20**
  acute graft rejection, 441
  cell-surface molecules, 218–219, **Fig. 8.18**
  effector molecules, 220–222, **Fig. 8.21**
  exit from lymph nodes, 206, 218
  functions, 122, **Fig. 5.13**
  generation in lymph nodes, 22–23, **Fig. 1.23**
  induction of differentiation, 194, 209–210, **Fig. 8.11**
  memory (T_{EM}), 304
  mucosal tissues, 281–282, **Fig. 10.17**
    commitment to, 280–281, **Fig. 10.15**
    homing, 281, **Fig. 10.16**
  regulation of responses, 225–226
  vs. memory T cells, 302–303, **Fig. 11.7**
Efferent lymphatic vessels, 22, **Fig. 1.22**
Eicosanoids, 412, **Fig. 14.18**
Elotuzumab, 524–525
Encapsulated bacteria, 24, 35
  inherited susceptibility to infections, 89, 379–381
  vaccines against, 314–316
Endemic diseases, 4
Endocarditis, subacute bacterial, **Fig. 16.1**
Endocrine glands, autoimmune diseases, 485, **Fig. 16.17**
Endocytic vesicles, 125
  antigen uptake and processing, 129–130
  MHC class II delivery to, 129–130, **Fig. 5.22**
Endocytosis
  pathogenic proteins/particles, 129
  receptor-mediated *see* Receptor-mediated endocytosis
Endoplasmic reticulum, 125, **Fig. 5.17**
  MHC class II assembly, 130, **Fig. 5.23**
  peptide loading of MHC class I, 127–128, **Fig. 5.20**
  transport of peptides to, 127, **Fig. 5.19**
Endoplasmic reticulum aminopeptidase (ERAP), 128, **Fig. 5.21**
Endosomes, 50, 129
Endothelial protein C receptor (EPCR), 351–352, **Fig. 12.26**
Endothelium, vascular
  adhesion molecules, 57, **Fig. 3.12**
  changes in inflammation, 9
  neutrophil recruitment, 57–58, **Fig. 3.13**
Endotoxins, **Fig. 9.25**
  see also Lipopolysaccharide
Engraftment, 459
Enterocytes *see* Intestinal epithelial cells
Envelope proteins, HIV, 388, **Fig. 13.21**, **Fig. 13.22**
  see also gp41; gp120
  broadly neutralizing antibodies, 396–397, **Fig. 13.32**
Eosinophil cationic protein, **Fig. 14.20**
Eosinophilia, 415
Eosinophils, 15, 413–415, **Fig. 1.11**
  allergic reactions, 415
  basophil interactions, 415
  Fcε receptors, 248
  intestinal helminth infections, 289, **Fig. 10.27**
  microscopic appearance, 413, **Fig. 14.19**
  numbers in blood, **Fig. 1.14**
  parasitic infections, 250, **Fig. 9.20**
  substances secreted by, 414, **Fig. 14.20**
Eotaxin, 414
Epidemics, influenza, 367
Epidermophyton floccosum, **Fig. 1.3**
Epinephrine, anaphylactic reactions, 422, **Fig. 14.31**
Epithelia
  see also Mucosal surfaces
  defensins, 41, 42, **Fig. 2.18**
  mucin secretion, 269
  physical barriers to infection, 4–8, 29–30, **Fig. 2.1**
Epithelium, 4
Epitopes (antigenic determinants), 86–87, 113, **Fig. 4.9**
  cryptic, 486
  discontinuous, 87, **Fig. 4.11**
  linear, 87, **Fig. 4.11**
  modification by penicillin/drugs, **Fig. 14.32**
  recognized by naive B cells and T_{FH} cells, 225–226
  spreading, 485–488
    intermolecular, 487–488, **Fig. 16.21**
    intramolecular, 486, **Fig. 16.19**
  types, 86–87, **Fig. 4.10**
ε heavy chains, immunoglobulin, 83, **Fig. 4.5**
Epstein–Barr virus (EBV)
  evasion/subversion of immune response, 370–371, **Fig. 13.6**
  oncogenicity, 513, **Fig. 17.4**
  reactivation in immunosuppressed patients, 514
  X-linked lymphoproliferative syndrome, 386
ERAP (endoplasmic reticulum aminopeptidase), 128, **Fig. 5.21**
ERp57, 127–128, **Fig. 5.20**
Erythrocytes, 12–14, **Fig. 1.11**
  ABO antigens *see* ABO blood group antigens
  destruction, autoimmune hemolytic anemia, 475, **Fig. 16.2**
  immune complex clearance, 258, **Fig. 9.32**
Erythroid lineage cells, 12–14, **Fig. 1.13**
Erythroid progenitor cells, 12–14, **Fig. 1.13**
Escherichia coli, 3, 62, 314–315
Evolution
  allergic diseases, 428–429, **Fig. 14.39**
  HIV resistance, 392
  host–pathogen relationships, 4
  influenza viruses, 366–368, **Fig. 13.2**, **Fig. 13.3**
  innate and adaptive immunity, 329–361, **Fig. 12.1**

MHC class I-like molecules, 359–360
NK-cell receptors, 339, 340
placental reproduction, 342–345
RAG genes, 117, **Fig. 5.5**
Exotoxins, **Fig. 9.25**
Extracellular matrix, autoantibodies to, 475–476
Extracellular pathogens, 30–31, **Fig. 2.2**
antigen presentation, 122
cross-presentation, 131–132
antigen processing, 126, 129–130, 134–135, **Fig. 5.22**, **Fig. 5.27**
inherited susceptibility to, 379–381
innate immune response, 47–68
Extravasation, neutrophil, 58, **Fig. 3.13**
Eye
immunosuppressive environment, 457, **Fig. 15.29**
trauma-induced autoimmunity, 495, **Fig. 16.32**

F

Fab fragments, 83, **Fig. 4.3**
vs. T-cell receptors, 114, **Fig. 5.1**
Factor B, 33
alternative C3 convertase, 34, **Fig. 2.8**
iC3 actions, 33, **Fig. 2.6**
Factor D, 33–34, **Fig. 2.6**
alternative C3 convertase, **Fig. 2.8**
inherited deficiency, **Fig. 13.13**
Factor H, 34–36, **Fig. 2.9**
Factor H-binding protein (fHbp), 317, **Fig. 11.22**
Factor I, 34–35, **Fig. 2.9**
inherited deficiency, 35, 381, **Fig. 13.10**, **Fig. 13.13**
Factor J, 38
Factor P (properdin), 34, **Fig. 2.9**
inherited deficiency, 381, **Fig. 13.13**
FADD, **Fig. 3.32**
Familial Mediterranean fever (FMF), 502, **Fig. 16.41**
Fanconi anemia, 464, **Fig. 15.32**
Farmer's lung, **Fig. 15.22**
Fas ligand (FasL), memory T cells, **Fig. 11.7**
FcαRI, 262–263, **Fig. 9.37**
FcεRI, 248
β chain gene polymorphism, 418, **Fig. 14.25**
comparative features, **Fig. 9.37**
IgE binding, 406–407, **Fig. 14.9**
mast-cell activation, 248, 249–250, 407, **Fig. 9.19**, **Fig. 14.11**
mast cells, 407, **Fig. 14.10**
omalizumab actions, 410, **Fig. 14.14**
regulation of eosinophils, 414
vs. FcεRII, 408, **Fig. 14.12**
FcεRII (CD23), 407–409
mucosal surfaces, 283
stimulation of IgE production, 409, **Fig. 14.13**
structure, 408, **Fig. 14.12**
Fc fragments, 83, **Fig. 4.3**
FcγRI, 259–260, **Fig. 9.35**
binding to IgG subclasses, 259, **Fig. 9.33**
comparative features, 262, **Fig. 9.37**
C-reactive protein binding, 263
FcγRII, 260–261, 263
FcγRIIA, 261, **Fig. 9.35**, **Fig. 9.37**
immune complex formation, **Fig. 14.7**
FcγRIIB, immune complex formation, 405–406, **Fig. 14.7**
FcγRIIB1, 261, **Fig. 9.35**, **Fig. 9.37**
regulation of naive B cells, 300–301, **Fig. 11.5**
FcγRIIB2, 261, **Fig. 9.35**, **Fig. 9.37**
FcγRIII (CD16), 260, 261–262, **Fig. 9.37**
FcγRIIIA (CD16a), 261, **Fig. 9.35**
NK-cell activation, 332, **Fig. 12.4**
NK cells, 332–333, **Fig. 12.2**
FcγRIIIB, 261, **Fig. 9.35**
Fc receptors, 246, 258–263
common γ chain, 259, 262, **Fig. 9.33**, **Fig. 9.37**
comparison of features, 262–263, **Fig. 9.37**
C-reactive protein binding, 263
Fc region (Fc piece), 83
FcRn, 246, 250, **Fig. 9.17**
IgG transport to mucosal secretions, 283, **Fig. 10.18**
vs. Fc receptors, 258
Fetus
see also Newborn babies; Pregnancy
hematopoiesis, 12, **Fig. 1.12**
HLA haplotype, 344
IgE transfer to, 419, **Fig. 14.26**
IgG transfer to, 250, **Fig. 9.21**
maternal alloantibodies, 144
transfer of autoimmune disease to, 483–484, **Fig. 16.15**
Fever, 62
Fibrinogen, **Fig. 3.21**, **Fig. 3.24**
Fibronectin, bacterial adhesion, 252, **Fig. 9.24**
Filaggrin gene mutations, 426
FK506 *see* Tacrolimus
FK-binding protein (FKBP), 450, **Fig. 15.19**
Flagellin, 369, **Fig. 3.29**
Flow cytometry, 88–89, **Fig. 4.13**
FLT3, **Fig. 6.12**
Follicle-associated epithelium, small intestine, 277
Follicular center cell lymphoma, 170, **Fig. 6.23**
Follicular dendritic cells (FDCs), 235–236, **Fig. 9.5**
antigen capture, 235–236, **Fig. 9.6**
B-cell interactions, 167–168, **Fig. 6.21**
germinal centers, 240, 241, **Fig. 9.11**
induction of centroblast formation, 239
rescue of centrocytes from apoptosis, 241
Follicular helper T cells *see* T_{FH} cells
Food
allergies, 426–427, **Fig. 14.38**
anaphylactic reactions, 422
antigens, tolerance to, 278
poisoning, 369, 373, **Fig. 9.25**
Fos, 209, **Fig. 8.11**
FoxP3
inherited deficiency, 226–227, 478, **Fig. 13.10**
regulatory T cells, 193, 212, 214, **Fig. 8.14**
Framework regions, antibodies, 85–86, **Fig. 4.8**
Fungi, 4, **Fig. 1.4**
F_V (variable fragment), 523
Fyn, B-cell receptor signaling, 232, **Fig. 9.1**

G

GALT *see* Gut-associated lymphoid tissues
γ chain, common
cytokine receptors ($γ_c$), 384
Fc receptors, 259, 262, **Fig. 9.33**, **Fig. 9.37**
γ:δ T-cell receptors *see under* T-cell receptor(s)
γ:δ T cells, 118–119, 347–354, **Fig. 5.7**
antigens recognized, 119, 350–353
cancer immunotherapy, 524
development, 181–184, 188, **Fig. 7.7**, **Fig. 7.9**
developmental stages, 348–349, **Fig. 12.21**
evasion by tumor cells, 518, **Fig. 17.13**
species differences, 349–350, **Fig. 12.23**
subpopulations, 349, **Fig. 12.22**
$V_γ4$:$V_δ5$-expressing, 351–352, **Fig. 12.26**
$V_γ9$:$V_δ2$-expressing, 349, **Fig. 12.22**
antigens recognized, 350–351, **Fig. 12.24**, **Fig. 12.25**
$V_γ$:$V_δ1$-expressing, 352–353, **Fig. 12.22**, **Fig. 12.28**
vs. α:β T cells, 347–348, **Fig. 12.20**
γ heavy chains, immunoglobulin, 83, **Fig. 4.5**
protein domains, 84, **Fig. 4.6**
Gas gangrene, **Fig. 9.25**
Gastrointestinal tract, 269–270, **Fig. 10.4**
see also Small intestine
commensal microorganisms, 3, 30, 268, 269–271
benefits to humans, 271, **Fig. 10.5**
effects of antibiotics, 3, **Fig. 1.2**
immune responses, 278–279
pathogenic variants, 271
role in immune system development, 273, **Fig. 10.8**
specific IgA antibodies, 278–279, 285
IgE-mediated allergic reactions, 421, **Fig. 14.29**
infections, 271, 279
physical barriers to infection, 6, **Fig. 2.1**
secondary lymphoid tissues, 25, 272–273, **Fig. 10.7**
see also Gut-associated lymphoid tissues
GATA3
T-cell development, 188, **Fig. 7.14**
T_H2 development, 213, **Fig. 8.14**
Gaucher's syndrome, **Fig. 15.32**
Gelatinase, neutrophil, 59, 60
Gender differences, autoimmune diseases, 480, **Fig. 16.11**
Gene-content variation, KIR haplotypes, 340–341, **Fig. 12.14**
Gene conversion
new MHC allele generation, **Fig. 5.38**
pathogen antigens, 368–369, **Fig. 13.4**
Gene rearrangements
see also under Immunoglobulin gene(s); T-cell receptor(s)
evolution, 117, **Fig. 5.5**
trypanosome proteins, 368–369, **Fig. 13.4**
Gene segments
immunoglobulin, 91–92, **Fig. 4.15**
T-cell receptor, 115, **Fig. 5.3**, **Fig. 5.8**
Genetic diseases
hematopoietic stem-cell transplantation, 459, **Fig. 15.32**
immunodeficiency *see* Inherited immunodeficiency diseases
Genetic polymorphism, 68, 113–114
Genetic variation, pathogens, 366–369
Genome sequencing, 317
Germinal center reaction, 169, 240
Germinal centers, 22, 239–240, **Fig. 1.22**, **Fig. 9.9**

absence, hyper-IgM syndrome, 243–244, **Fig. 9.14**
activated B cells, 169
anatomy, 239–240, **Fig. 9.11**
antigen-mediated selection of centrocytes, 241–243, **Fig. 9.12**
Giardia lamblia, 287
Glandular fever, 370
Gliadin peptides, 500, 501
Glomerular basement membrane autoantibodies, 475–476, **Fig. 16.3**
Glucans, 50, **Fig. 3.3**
Glucose-6-phosphate dehydrogenase (G6PD) deficiency, 383, **Fig. 13.15**
Glucose monomycolate, **Fig. 12.27**, **Fig. 12.29**
γ-Glutamyl diaminopimelic acid, 54–55
Gluten, 500, 501
Gluten-sensitive enteropathy *see* Celiac disease
GlyCAM-1, naive T-cell homing, 205, **Fig. 8.6**
Glycolipid antigens, 352, **Fig. 12.27**
mycobacterial, 354–355, **Fig. 12.29**
Glycosylphosphatidylinositol, 39, **Fig. 3.29**
Gnotobiotic mice, 273, **Fig. 10.8**
Goblet cells, small intestine, 277
Golgi apparatus, **Fig. 5.17**
Gonorrhea, 369
Goodpasture's syndrome, 475–476, **Fig. 16.1**
effects of smoking, 494–495
type IV collagen autoantibodies, 475–476, **Fig. 16.3**
gp41, **Fig. 13.21**, **Fig. 13.22**
broadly neutralizing antibodies, 397, **Fig. 13.32**
HIV life cycle, 389, **Fig. 13.23**
gp120, **Fig. 13.21**, **Fig. 13.22**
broadly neutralizing antibodies, 396–397, **Fig. 13.32**
HIV life cycle, 389, **Fig. 13.23**
G-protein-coupled receptors, **Fig. 3.14**
Graft rejection *see* Transplant rejection
Graft-versus-host disease (GVHD), 461–462, **Fig. 15.35**
alloreactions, 439, **Fig. 15.6**
benefits in malignant disease, 465–466
grading of severity, 461, **Fig. 15.36**
HLA matching and, 462, **Fig. 15.37**
minor histocompatibility antigens causing, 464
Graft-versus-host reaction (GVHR), 439, 461, **Fig. 15.6**
Graft-versus-leukemia (GVL) effect, 465–466
NK cells mediating, 466, **Fig. 15.42**, **Fig. 15.43**
Graft-versus-tumor (GVT) effect, 465
Gram-negative bacteria
defensins, 42
recognition by macrophages, 50, 51–52
septic shock, 68
Gram-positive bacteria, 42, 50
Granulocyte colony-stimulating factor (G-CSF), 463
Granulocyte macrophage colony-stimulating factor (GM-CSF), 381, 463, 525
Granulocytes, 14–15
Granulomas, 61
Granulysin, 220–221, 224
Granzymes, 220, 224, **Fig. 11.7**
Graves' disease, 482–483, **Fig. 16.1**
autoantibodies, 482–483, **Fig. 16.12**, **Fig. 16.13**
HLA association, 483, **Fig. 16.9**
newborn babies, 483–484, **Fig. 16.15**
Growth factors, B-cell development, 152, **Fig. 6.5**
Gut *see* Gastrointestinal tract
Gut-associated lymphoid tissues (GALT), 25, 272–273, **Fig. 1.26**, **Fig. 10.7**
B- and T-cell commitment to, 280–281, **Fig. 10.15**
effector compartment, 272
inductive compartment, 272
migration of plasma cells to, 170
transport of microbes to, 277–278, **Fig. 10.13**, **Fig. 10.14**

H

Haemophilus influenzae
childhood vaccination, **Fig. 11.15**
IgA1 cleavage, 285
increased susceptibility to, 24, 160, 379
vaccine, 314–315, 316, **Fig. 11.20**, **Fig. 11.25**
Hairpins, DNA, 95, **Fig. 4.20**
Haploidentical hematopoietic stem cell transplantation, 466, **Fig. 15.41**, **Fig. 15.43**
Hashimoto's disease, 484, **Fig. 16.16**
HBD1, 42, **Fig. 2.19**
HBD2, 42, **Fig. 2.19**
HBD3, 42, **Fig. 2.19**
HBD4, **Fig. 2.19**
HD5, 41, **Fig. 2.18**, **Fig. 2.19**
HD6, 41, **Fig. 2.18**, **Fig. 2.19**
Heart transplantation, 455, **Fig. 15.26**
Heat-shock proteins, corticosteroid actions, 447, **Fig. 15.17**
Heavy chains (H chains), 82–83, **Fig. 4.2**
see also α heavy chains; δ heavy chains; ε heavy chains; γ heavy chains; μ heavy chains
alternative RNA splicing, 98–99, **Fig. 4.24**
constant domain (C_H), 84, **Fig. 4.6**
framework regions, 85
gene locus, 91–92, **Fig. 4.15**
allelic exclusion, 154–155, **Fig. 6.8**
gene rearrangements, 92–95, 96, **Fig. 4.16**, **Fig. 4.21**
B-cell precursors, 151, 154, **Fig. 6.4**
productive and nonproductive, 152–153, **Fig. 6.6**
gene segments, 91–92, **Fig. 4.15**, **Fig. 4.17**
gene transcription, 96, **Fig. 4.22**
regulation, 159, **Fig. 6.13**
hypervariable regions (HV), 85–86
isotypes, 83, **Fig. 4.5**
pre-B-cell receptor, 153–154, **Fig. 6.7**
variable domain (V_H), 84, **Fig. 4.6**
Helminth parasites, 288–290
antigens
similarity to allergens, 417–418, **Fig. 14.24**
transfer from mother to fetus, 419, **Fig. 14.26**
conferring resistance to allergies, 428–429, **Fig. 14.39**
diseases caused by, **Fig. 1.4**, **Fig. 10.25**
T_H2-mediated immunity, 288–290, **Fig. 10.26**, **Fig. 10.27**
Helper CD4 T cells, 17, 199
see also CD4 T cells; T_H1 cells; T_H2 cells; T_H17 cells
antigen presentation to, 122–123, **Fig. 5.15**
B-cell differentiation, 244, **Fig. 9.15**
development, 194, 211–213, **Fig. 1.23**
differentiation into subsets, 212–213, **Fig. 8.14**
effector functions, 122, 218–219, 224–226, **Fig. 5.13**
follicular *see* T_{FH} cells
Hemagglutinin, influenza, 318, 366–368
antigenic drift, 367, **Fig. 13.2**
antigenic shift, 368, **Fig. 13.3**
neutralizing antibodies, 251, **Fig. 9.23**
Hematopoiesis, 12–16, **Fig. 1.12**
Hematopoietic cells, 12, **Fig. 1.11**
Fc receptors, 258–263
Hematopoietic cell transplantation *see* Hematopoietic stem-cell transplantation
Hematopoietic stem cells, 12–14
CD34 marker, 150, 463
cells derived from, 12–16, **Fig. 1.13**
pluripotent, 12, 150, **Fig. 1.13**
self renewal, 12
sources for transplant, 439, 463–464
Hematopoietic stem-cell transplantation, 458–468, **Fig. 15.31**
allogeneic, 461–463, 464–467
alloreactions, 438–439, **Fig. 15.6**
autologous, 463
cancer, 461, 463, 465–466, **Fig. 15.34**
genetic diseases, 459, **Fig. 15.32**
graft-versus-host disease, 461–462, **Fig. 15.35**, **Fig. 15.36**
haploidentical, 466, **Fig. 15.41**, **Fig. 15.43**
HLA allotype sharing, 460–461, **Fig. 15.33**
HLA matching, 462–464, **Fig. 15.37**
inducing tolerance of solid organ transplants, 467, **Fig. 15.44**
mini-transplant protocols, 465
minor histocompatibility antigens, 464–465
procedure, 459–460
Hemoglobinuria, paroxysmal nocturnal, 39, **Fig. 13.10**
Hemolytic anemia
autoimmune, 475, **Fig. 16.1**, **Fig. 16.2**
of the newborn, 305–306, **Fig. 11.11**
Hemophiliacs, HIV infection, 391, **Fig. 13.25**
Hepatitis A vaccine, **Fig. 11.25**
Hepatitis B virus (HBV)
chronic infection, 226
oncogenicity, 513, **Fig. 17.4**
vaccine, 313, **Fig. 11.15**, **Fig. 11.25**
Hepatitis C virus
antigen presentation, 131
chronic infection, 323, **Fig. 11.27**
KIR polymorphism and, 336
vaccine, 323, **Fig. 11.26**
Herd immunity, 321–322, 324
Hereditary angioedema (HAE), 381–382
Herpes simplex virus, 370, **Fig. 13.5**, **Fig. 13.6**
Herpesviruses
evasion of immune response, 369–371
reactivation in AIDS, 396
subversion of immune response, 371–372, **Fig. 13.6**
Herpes zoster, 370
Heterozygous, 136
High endothelial venules (HEV)
B-cell homing, 167, **Fig. 6.20**, **Fig. 6.21**
naive T-cell homing, 203, 205, **Fig. 8.6**

Highly active anti-retroviral therapy (HAART), 394, **Fig. 13.29**
Histamine, 412, **Fig. 14.17**
allergic reactions, 412
reactions in skin, 420, 424, **Fig. 14.37**
release from mast cells, 249, **Fig. 9.19**
Histocompatibility, 433–434
Histoplasma, macrophage receptors, 50
HIV *see* Human immunodeficiency virus
Hives, 424
HLA, 135–138
see also MHC
alleles
generation of new, 142–143, **Fig. 5.38**
numbers per locus, 136, **Fig. 5.29**
allotypes, 136, **Fig. 5.29**
autoimmune disease, 478–481, **Fig. 16.8**, **Fig. 16.9**
celiac disease, 498–501, **Fig. 16.38**
sharing, hematopoietic cell transplantation, 460–461, **Fig. 15.33**
type 1 diabetes, 496–498, **Fig. 16.34**
antibodies *see* Anti-HLA antibodies
class I molecules *see* HLA class I molecules
class I region, 137, 138, **Fig. 5.30**
class II molecules *see* HLA class II molecules
class II region, 137, 138, **Fig. 5.30**, **Fig. 5.32**
class III region, 137, **Fig. 5.30**
complex, 135, 137, 438, **Fig. 5.30**
differences, transplant donor/recipient, 441–443
acute graft rejection, 441
chronic graft rejection, 443–444, **Fig. 15.12**
mixed lymphocyte reaction, 442–443, **Fig. 15.10**
gene organization, 137–138, **Fig. 5.30**
haplotypes, 137
fetus, 344
linkage dysequilibrium, 479–480, **Fig. 16.10**
matching
haploidentical hematopoietic cell transplants, 466, **Fig. 15.41**
hematopoietic cell transplants, 462–464, **Fig. 15.37**
organ transplants, 445, 457, **Fig. 15.14**
polymorphism, 135–136, **Fig. 5.28**, **Fig. 5.29**
see also MHC, polymorphism
HIV progression and, 142–143, 392–393, **Fig. 5.40**, **Fig. 13.26**
importance for human survival, 142–143, **Fig. 5.39**
natural selection generating, 140–143, **Fig. 5.37**
transplant rejection and, 143–144
restriction, 140, **Fig. 5.35**
minor histocompatibility antigens, 464–465, **Fig. 15.39**
type, 143
HLA-A, 136, **Fig. 5.28**
derived peptides, recognition by CD94:NKG2A, 333, **Fig. 12.5**
KIR ligands or epitopes, 335–336, **Fig. 12.8**
peptide-binding motif, **Fig. 5.34**
polymorphism, 136, **Fig. 5.29**
HLA-B, 136, **Fig. 5.28**
derived peptides, recognition by CD94:NKG2A, 333, **Fig. 12.5**
KIR ligands or epitopes, 335–336, **Fig. 12.8**
peptide-binding motif, **Fig. 5.34**
polymorphism, 136, **Fig. 5.29**
HLA-B27
ankylosing spondylitis, 479, 480
Reiter's syndrome/reactive arthritis, 494
HLA-C, 136, **Fig. 5.28**
derived peptides, recognition by CD94:NKG2A, 333, **Fig. 12.5**
KIR ligands or epitopes, 335–336, **Fig. 12.8**
NK-cell education, 337–339, **Fig. 12.11**
pregnancy complications and, 344–345, **Fig. 12.18**
polymorphism, 136, **Fig. 5.29**
HLA class I molecules, 135–136
see also MHC class I molecules; *specific molecules*
acute graft rejection, 441
gene organization, 137, 138, **Fig. 5.30**
hyperacute graft rejection, 437, **Fig. 15.4**
inherited deficiency, 385
isotypes, 136, **Fig. 5.28**
loss of expression by cancer cells, 518–519, **Fig. 17.14**
matching, organ transplantation, 445
minor histocompatibility antigens and, **Fig. 15.39**
trophoblast cells, 344
HLA class II molecules, 135–136
see also MHC class II molecules; *specific molecules*
acute graft rejection, 441
gene organization, 137, 138, **Fig. 5.30**, **Fig. 5.32**
hyperacute graft rejection, 437
inherited deficiency, 385
isotypes, 136, **Fig. 5.28**
matching, organ transplantation, 445
minor histocompatibility antigens and, **Fig. 15.39**
HLA-DM, 136, **Fig. 5.28**
function, 130–131, **Fig. 5.23**
genes, 137, **Fig. 5.30**
polymorphism, **Fig. 5.29**
HLA-DO, 136, **Fig. 5.28**
function, 131
genes, 137, **Fig. 5.30**
polymorphism, **Fig. 5.29**
HLA-DP, 136, **Fig. 5.28**
genes, 137, **Fig. 5.30**
polymorphism, 136, **Fig. 5.29**
HLA-DQ, 136, **Fig. 5.28**
allotypes
autoimmune disease and, 480
celiac disease, 498–501, **Fig. 16.38**
type 1 diabetes, 496–498, **Fig. 16.34**, **Fig. 16.35**
genes, 137, **Fig. 5.30**
peptide-binding motif, **Fig. 5.34**
polymorphism, 136, **Fig. 5.29**
HLA-DR, 136, **Fig. 5.28**
allotypes
autoimmune disease and, 480
Graves' disease, 483
myasthenia gravis, 483
rheumatoid arthritis, 491–492, **Fig. 16.27**
systemic lupus erythematosus, 487–488
type 1 diabetes, 496–498, **Fig. 16.34**, **Fig. 16.35**
genes, 137, **Fig. 5.30**, **Fig. 5.31**
peptide-binding motif, **Fig. 5.34**
polymorphism, 136, **Fig. 5.29**
HLA-E, 136, **Fig. 5.28**
CD94:NKG2A interactions, 333–334, **Fig. 12.5**, **Fig. 12.6**
NK-cell activation in cytomegalovirus infection, 342, **Fig. 12.15**
NK-cell education, 339
polymorphism, 136, **Fig. 5.29**
HLA-F, 136, **Fig. 5.28**, **Fig. 5.29**
HLA-G, 136, **Fig. 5.28**, **Fig. 5.29**
NK cell KIR, 335, **Fig. 12.8**
trophoblast-uterine NK cell interactions, 343–344, **Fig. 12.17**
HNP1, **Fig. 2.19**
HNP2, **Fig. 2.19**
HNP3, **Fig. 2.19**
HNP4, **Fig. 2.19**
Hodgkin's disease, 170, **Fig. 6.23**
diagnosis, 527, **Fig. 17.25**
treatment, 528–529, **Fig. 17.27**, **Fig. 17.28**
Homing
gut-specific effector lymphocytes, 281, **Fig. 10.16**
naive B cells, 167, **Fig. 6.21**
naive T cells, 204–206, **Fig. 8.6**
neutrophils, 53–54, 57–58, **Fig. 3.13**
Homologous restriction factor (HRF), 38–39, **Fig. 2.14**
Homozygous, 136
House dust mite, 416–417
Hsp90, corticosteroid actions, 447, **Fig. 15.17**
Human cytomegalovirus (HCMV) *see* Cytomegalovirus
Human herpes virus 8 (HHV8), **Fig. 17.4**
Human immunodeficiency virus (HIV), 365, 387–397
see also Acquired immune deficiency syndrome
antiviral drug resistance, 394, **Fig. 13.28**
broadly neutralizing antibodies, 396–397, **Fig. 13.32**
CD4 T cells *see* CD4 T cells, HIV infection
combination drug therapy, 394, **Fig. 13.29**
entry into cells, 122–123, 389, **Fig. 13.23**
escape from immune response, 393–394, **Fig. 13.27**
genes and proteins, 388, **Fig. 13.22**
infection, 388–397
clinical latency period, 390, 394–395
elite controllers, 393
elite neutralizers, 396
global spread, 387, **Fig. 13.19**, **Fig. 13.20**
HLA polymorphisms and, 142–143, 392–393, **Fig. 5.40**, **Fig. 13.26**
induction of immunodeficiency, 395–396
KIR polymorphisms and, 336, 393, **Fig. 13.26**
natural history, 389–391, **Fig. 13.24**
progression to AIDS, 390–391, **Fig. 13.24**, **Fig. 13.25**
resistance to, 391–392
seroconversion, 390, **Fig. 13.24**
viremic controllers, 393
lymphocyte-tropic variants, 389
macrophage-tropic variants, 389
morphology, 388, **Fig. 1.3**, **Fig. 13.21**
oncogenic effects, **Fig. 17.4**
provirus, 388, **Fig. 13.23**

transmission, 388–389
type 1 (HIV-1), 387, **Fig. 13.22**
type 2 (HIV-2), 387
vaccine, 322–323, 396, **Fig. 11.26**
virions, 389, **Fig. 13.21**
Human leukocyte antigens *see* HLA
Human papillomaviruses (HPV), 519–520
oncogenicity, 513, **Fig. 17.4**
proteins, 519–520, **Fig. 17.16**
vaccination, 518–519, **Fig. 11.15**
vaccines, 520, **Fig. 11.25**, **Fig. 17.17**
Human T-cell leukemia virus type 1 (HTLV-1), 512, **Fig. 17.4**
Humoral immunity, 18, 231–264
polarized T_H2 response, 214
Humors, 18
H-Y antigens, 464, **Fig. 15.38**
Hybridomas, 88, 154–155, **Fig. 4.12**
Hydrocortisone, 447, 448, **Fig. 15.16**
Hydrogen peroxide, 60, **Fig. 3.17**
Hydroxymethyl-but-2-enyl-pyrophosphate (HMBPP), 350, **Fig. 12.24**, **Fig. 12.25**
Hygiene hypothesis, 406, **Fig. 14.8**
Hyper-IgD syndrome, **Fig. 16.41**
Hyper-IgM syndrome, 102, **Fig. 13.10**
absence of germinal centers, 243–244, **Fig. 9.14**
X-linked, 380–381
Hypersensitivity reactions, 401–403, **Fig. 14.2**
allergy, transplant rejection and autoimmunity compared, 504, **Fig. 16.42**
allogeneic transplantation, 433–440
common causes, **Fig. 14.1**
type I, 401–402, 403, **Fig. 14.2**
see also Allergic disease, IgE-mediated
type II, 402, 403, **Fig. 14.2**
autoimmune diseases, 475, **Fig. 16.1**
hyperacute graft rejection, 436–437, **Fig. 15.4**
incompatible blood transfusions, 435
type III, 402, 403, **Fig. 14.2**
autoimmune diseases, 475, 476, **Fig. 16.1**
chronic graft rejection, 443–444
serum sickness, 450–451, **Fig. 15.22**
type IV, 402–403, **Fig. 14.2**
autoimmune diseases, 475, 476–477, **Fig. 16.1**
transplant recipients, 438–439, 441, 461
Hyperthyroidism, 482
Hypervariable regions (HVs)
see also Complementarity-determining regions
antibodies, 85–86, **Fig. 4.8**
T-cell receptors, 114–115, **Fig. 5.2**
Hypoglycemia, autoimmune, 483, **Fig. 16.1**, **Fig. 16.12**
Hypohydrotic ectodermal dysplasia and immunodeficiency, X-linked, 53, **Fig. 3.8**, **Fig. 13.10**
Hypothyroidism, 484

I

Ibritumomab, 529, **Fig. 17.27**
iC3, 32–33, **Fig. 2.6**
iC3b, 34, **Fig. 2.9**
B-cell activation, 233
macrophage interactions, 36
production, 233, **Fig. 9.2**
ICAM-1
B cell–helper T_{FH} cell interaction, 238, **Fig. 9.8**
effector T-cell interactions, 219
naive T-cell homing, 205, **Fig. 8.6**
neutrophil recruitment, 57–58, **Fig. 3.13**
peripheral supramolecular activation complex, 208, **Fig. 8.9**
ICAM-2, 57–58, 205
ICAM-3, 205
ICOS (inducible T-cell co-stimulator), 214
Ig *see* Immunoglobulin(s)
IgA, 83, 103–104
dimeric (secretory), 104, 268, **Fig. 4.31**
binding by mucins, 269
breast milk, 250, 287, **Fig. 9.21**
commensal bacteria-specific, 278–279, 285
effector functions, 246–247
mucosal secretions, 282, 283
neutralizing toxins and venoms, 253–255
poly-Ig receptor, 247, **Fig. 9.18**
preventing pathogen entry to cells, 251, 252–253, **Fig. 9.23**, **Fig. 9.24**
protection at mucosal surfaces, 283–285, **Fig. 10.19**, **Fig. 10.20**
transcytosis, 247, **Fig. 9.18**
effector functions, 103–104, **Fig. 4.30**
evasion by staphylococcal superantigen-like proteins, 374–375, **Fig. 13.9**
isotype switching to, 243, 283, 285, **Fig. 9.13**
monomeric, 103
effector function, 246
Fc receptor, 262–263, **Fig. 9.37**
secretory component (secretory piece), 247, **Fig. 9.18**
selective deficiency, 286–288, **Fig. 10.23**, **Fig. 10.24**, **Fig. 13.10**
structure, 84, **Fig. 4.5**
subclasses, 103, 285–286, **Fig. 4.29**
IgA1, 285–286
differential expression, 285–286, **Fig. 10.22**
hinge region, 285, **Fig. 10.21**
physical properties, **Fig. 4.29**
IgA2, 285–286
differential expression, 285–286, **Fig. 10.22**
hinge region, 285, **Fig. 10.21**
physical properties, **Fig. 4.29**
Igα, 97, 156, **Fig. 4.23**
pre-B-cell receptor, **Fig. 6.7**
regulation of expression, 152, 158, **Fig. 6.12**
signal transduction, 232, **Fig. 9.1**
Igβ, 97, 156, **Fig. 4.23**
pre-B-cell receptor, 153, **Fig. 6.7**
regulation of expression, 158, **Fig. 6.12**
signal transduction, 232, **Fig. 9.1**
IgD, 83
anergic B cells, 166, **Fig. 6.19**
biased use of λ light chains, 156
effector function, 104, **Fig. 4.30**
expression by maturing B cells, 164, 167
isotype switching from, 101–102, **Fig. 4.28**
mature naive B cells, 96, **Fig. 4.22**
membrane-bound form, 97–98
physical properties, **Fig. 4.29**
secreted antibodies, 98–99
structure, 84, **Fig. 4.5**
IgE, 83, 104, 406–410
effector functions, 104, 247–250, **Fig. 4.30**
Fc receptors, 248, 262, 407–409, **Fig. 9.37**
high affinity *see* FcεRI
low-affinity *see* FcεRII
important characteristics, 406–407
intestinal helminth infections, 289, **Fig. 10.27**
isotype switching to, 243, 404–406, **Fig. 9.13**, **Fig. 14.5**
mast-cell activation, 248, 249–250, 407, **Fig. 9.19**, **Fig. 14.11**
mediated allergic disease, 250, 402, 416–428
monoclonal antibody targeting, 90, 409–410, **Fig. 14.14**
parasite-specific responses, 248–250, 404, **Fig. 9.20**
physical properties, **Fig. 4.29**
somatic hypermutation levels, 405, **Fig. 14.6**
structure, 84, **Fig. 4.5**
transfer from mother to child, 419, **Fig. 14.26**
transport across mucosae, 283
IgG, 83, 104–105
complement activation, 257–258, **Fig. 9.31**
conformational flexibility, 104–105, **Fig. 4.32**
domains, 84, **Fig. 4.6**
effector functions, 104–107, 246, **Fig. 4.30**
Fc receptors, 258–262, **Fig. 9.35**, **Fig. 9.37**
FcRn receptor, 246, **Fig. 9.17**
hinge region, 83, **Fig. 4.2**, **Fig. 4.4**
infants, 250–251, **Fig. 9.22**
isotype switching to, 243, **Fig. 9.13**
negative regulation of naive B cells, **Fig. 11.5**
neutralizing function, 253–255, **Fig. 9.26**
proteolytic cleavage, 83, **Fig. 4.3**
structure, 82–84, **Fig. 4.2**, **Fig. 4.5**
subclasses, 103, **Fig. 4.29**
effector functions, 105–107, **Fig. 4.30**, **Fig. 4.34**
FcγRI binding, 259, **Fig. 9.33**
hinge region structures, 105–106, **Fig. 4.33**
IgE-mediated allergies, 405–406, **Fig. 14.6**
isotype switching to, 243, 405, **Fig. 9.13**, **Fig. 14.5**
transfer from mother to fetus, 250, **Fig. 9.21**
transport to mucosal secretions, 283, **Fig. 10.18**
type II hypersensitivity reactions, 402, **Fig. 14.2**
vs. IgE, 406–407
IgG1, 105–106, **Fig. 4.29**
effector functions, 105–106, **Fig. 4.30**, **Fig. 4.34**
FcγRI binding, 259, **Fig. 9.33**
immune complex formation, 405–406, **Fig. 14.7**
isotype switching to, 101, 405, **Fig. 4.28**
structure, **Fig. 4.33**
IgG2, 105–106, **Fig. 4.29**
deficiency, 106, **Fig. 13.10**
effector functions, 105–106, **Fig. 4.30**, **Fig. 4.34**
Fc receptor, 261
isotype switching to, 405
structure, **Fig. 4.33**
IgG3, 106, **Fig. 4.29**
deficiency, 106

effector functions, 106, **Fig. 4.30**, **Fig. 4.34**
FcγRI binding, 259, **Fig. 9.33**
isotype switching to, 404–405
structure, 106, **Fig. 4.33**
IgG4, 106–107, **Fig. 4.29**
allergen-specific, 427
effector functions, 106–107, **Fig. 4.30**, **Fig. 4.34**
functionally monovalent, 106–107, **Fig. 4.35**
immune complex formation, 405–406, **Fig. 14.7**
isotype switching to, 405
structure, **Fig. 4.33**
therapeutic use, 107
IgM, 83
anergic B cells, 166, **Fig. 6.19**
cell-surface (monomeric), 97–98, **Fig. 4.24**
cross-linking, B-cell activation, 232, **Fig. 9.1**
maturing B cells, 164, 165, 167
structure, 101, **Fig. 4.27**
synthesis, 108, 156, **Fig. 4.36**, **Fig. 6.10**
timing of expression, 151, **Fig. 6.4**
complement activation, 255–256, **Fig. 9.28**
developing B cells, 151, **Fig. 6.10**
DiGeorge syndrome, 234–235
domain structure, 84, **Fig. 4.5**
effector functions, 103, 246, **Fig. 4.30**
IgA-deficient individuals, 287, **Fig. 10.24**
isotype switching, 101–102, **Fig. 4.28**
naive B cells, 96, **Fig. 4.22**
physical properties, **Fig. 4.29**
secreted (pentameric), 282
poly-Ig receptor, 247
production, 98–100, 169, 238, **Fig. 4.24**, **Fig. 9.9**
protective function, 283–284, **Fig. 10.19**
structure, 101, **Fig. 4.27**
transport mechanism, 283
staple form, C1q binding, 255, **Fig. 9.28**
IκB, 52–53, 448, **Fig. 3.7**
Ikaros, 188, **Fig. 7.14**
IKK (inhibitor of κ kinase)
γ subunit (IKKγ), inherited deficiency, 53, **Fig. 3.8**
NOD receptor signaling, 55, **Fig. 3.10**
TLR4 signaling, 52–53, **Fig. 3.7**
Immediate hypersensitivity *see* Hypersensitivity reactions, type I
Immune-complex disease, 381
Immune complexes
autoimmune diseases due to, **Fig. 16.1**
chronic graft rejection, 443, **Fig. 15.11**
complement activation, 257–258, **Fig. 9.31**
deposition in kidney, 258, **Fig. 16.20**
preventing hemolytic disease of newborn, 305–306, **Fig. 11.11**
regulation of naive B cells, 300–301, **Fig. 11.5**
removal from circulation, 258, **Fig. 9.32**
serum sickness, 451, **Fig. 15.22**
systemic lupus erythematosus (SLE), 258, 476, 487, **Fig. 16.20**
type III hypersensitivity reactions, 403, **Fig. 14.2**
Immune dysregulation, polyendocrinopathy, enteropathy, X-linked syndrome (IPEX), 478, **Fig. 13.10**
Immune response
see also Adaptive immune response; Innate immune response
evasion/manipulation by tumors, 518–519, **Fig. 17.13**, **Fig. 17.14**, **Fig. 17.15**
evasion/subversion by pathogens, 365–375
phases, 47, **Fig. 3.1**
primary *see* Primary immune response
secondary *see* Secondary immune response
Immune system, 1
Immunity, 1
Immunization, 1–2
see also Vaccination
childhood schedule, **Fig. 11.15**
passive, 255, 380
Immunocompromised individuals
cancer risk, 513–514
cytomegalovirus infections, 372
Immunodeficiency diseases, 24, 365
see also Acquired immune deficiency syndrome
inherited *see* Inherited immunodeficiency diseases
primary, 375
secondary, 375
Immunogenetics, 438
Immunoglobulin(s), 16–17, 81–109
antigen-binding site *see* Antibodies, antigen-binding site
diversity, 81–109
mechanisms of generation, 91–107
structural basis, 82–88
vs. T-cell receptors, 120, **Fig. 5.9**
domains, 83–85, **Fig. 4.6**, **Fig. 4.7**
Fc receptors, 246, 258–263
genes *see* Immunoglobulin gene(s)
intravenous *see* Intravenous immunoglobulin
isotype exclusion, 156–157
isotypes (classes), 82, 83, 103
see also IgA; IgD; IgE; IgG; IgM
distribution in body, **Fig. 9.21**
effector functions, 103–107, 245–250, **Fig. 4.30**
physical properties, **Fig. 4.29**
structure, 83, **Fig. 4.5**
switching *see* Isotype switching
transfer from mother to baby, 250, **Fig. 9.21**
membrane-bound form, 97–98, 99, **Fig. 4.23**, **Fig. 4.24**
proteolytic fragments, 83, **Fig. 4.3**
secreted form *see* Antibodies
structure, 81–109, **Fig. 4.2**
vs. T-cell receptors, 113–114, **Fig. 5.1**
synthesis and membrane binding, 97–98, **Fig. 4.23**
Immunoglobulin gene(s), 91–92
allelic exclusion, 96–97, 154–155, **Fig. 6.8**
chromosome locations, 91–92, **Fig. 4.15**
control of transcription, 159, **Fig. 6.13**
germline configuration, 91, **Fig. 4.15**, **Fig. 6.4**
rearrangements, 92–96, 108, **Fig. 4.36**
B-1 cells, 161–162
developing B cells, 150–151, **Fig. 6.4**
D to J, 95–96, **Fig. 4.20**
nonproductive and productive, 152–153, **Fig. 6.6**
recombination enzymes, 93–95, **Fig. 4.19**
regulation, 96–97, 157, 159, **Fig. 6.13**
summary, 162–163, **Fig. 6.16**
segments, 91–92, **Fig. 4.15**
somatic recombination, 92–95, **Fig. 4.16**
summary of changes, 108–109, **Fig. 4.37**
translocations in B-cell tumors, 160–161, **Fig. 6.14**
Immunoglobulin-like domains, 85
Fc receptors, 260
MHC molecules, 123, 124, **Fig. 5.14**
NK-cell receptors, 335, **Fig. 12.2**
T-cell receptors, 114, **Fig. 5.1**
Immunoglobulin superfamily, 85, **Fig. 3.12**
Immunological memory *see* Memory, immunological
Immunological privilege, eye, 495
Immunological tolerance *see* Tolerance, immunological
Immunology, 1
Immunophilins, 450
Immunoproteasome, 127
Immunoreceptor tyrosine-based activation motifs (ITAMs)
B-cell receptor, 232, **Fig. 9.1**
Fc receptors, 259
T-cell receptor, 208, **Fig. 8.10**
Immunoreceptor tyrosine-based inhibitory motifs (ITIMs)
Fc receptors, 261, **Fig. 9.37**
NK-cell receptors, 334
Immunosuppressive drugs, 210
autoimmune disease, 483
mechanisms of action, **Fig. 15.30**
organ transplant recipients, 441, 445–454, 457–458
side-effects, 446
Immunosurveillance, cancer, 514–515
Immunotoxins, 529, **Fig. 17.28**
Indigenous American populations, HLA diversity, 142, **Fig. 5.39**
Inducible T-cell co-stimulator (ICOS), 214
Infants
see also Newborn babies
dietary gluten exposure, 501
protective antibodies, 250–251, 287, **Fig. 9.22**
Infections
causing cancer, 512–513, **Fig. 17.4**
effector T-cell responses, 218–219
establishment, 20
extracellular *see* Extracellular pathogens
initiation of T-cell activation, 200–201, **Fig. 8.1**
intracellular *see* Intracellular pathogens
mucosal *vs.* systemic immune responses, 273–275, **Fig. 10.9**
opportunistic, AIDS, 395–396, **Fig. 13.31**
persistent, 369–371, **Fig. 13.5**
recovery from, 23
triggering autoimmune disease, 492–494, **Fig. 16.31**
Infectious diseases, 1, **Fig. 1.4**
with available vaccines, **Fig. 11.25**
chronic, vaccines against, 322–323
organisms causing *see* Pathogens
protective immunity, 296–297, **Fig. 11.1**
selecting for MHC diversity, 140–143, **Fig. 5.37**
Infectious mononucleosis, 370
Inflammasome, 55–56, **Fig. 3.11**
Inflammation, 9, **Fig. 1.7**
airway, allergic asthma, 424, **Fig. 14.35**, **Fig. 14.36**

allergic reactions, 412
cancer-induced, 514
complement-mediated, 40, **Fig. 2.15**
IgE-mediated, 249
induction by macrophages, 53–54, **Fig. 3.9**
mucosal immune system, 273–275, 276–277, **Fig. 10.9**
rheumatoid arthritis, 490, **Fig. 16.25**
role of NFκB, 52–53, **Fig. 3.7**
suppressive effects of corticosteroids, 448, **Fig. 15.18**
systemic effects, 62–63, **Fig. 3.20**
transplanted organs and recipients, 440–441, **Fig. 15.7**
viral infections, 71–72, **Fig. 3.32**
Inflammatory cells, 9
Inflammatory cytokines *see* Cytokines, inflammatory
Inflammatory mediators
eosinophils, 414, **Fig. 14.20**
mast cells, 249, 411–413, **Fig. 14.16**
parasitic infections, 250
Infliximab, 490–491, **Fig. 16.26**
Influenza, 4, 11
deaths, 23
epidemics, 367
H1N1 pandemic, 318–319, **Fig. 11.23**
immune response, 11, 370
immunization schedule, **Fig. 11.15**
pandemics, 367–368
vaccines, 318–319, **Fig. 11.25**
Influenza virus
evasion of immune response, 366–368, **Fig. 13.2**, **Fig. 13.3**
extracellular and intracellular forms, 30–31
hemagglutinin *see* Hemagglutinin, influenza
immunological memory, 306–307, **Fig. 11.12**
morphology, **Fig. 1.3**
neutralizing antibodies, 251–252, **Fig. 9.23**
Inherited immunodeficiency diseases, 375–386, **Fig. 13.10**
antibody deficiencies, 379–381, **Fig. 13.12**
complement deficiencies, 381–382, **Fig. 13.13**
genetics, 377
IFN-γ receptor mutations, 378–379, **Fig. 13.11**
phagocyte defects, 382–383, **Fig. 13.15**
specific disease susceptibilities, 385–386
T-cell defects, 178–179, 380–381, 383–385, **Fig. 13.16**
Inhibitor of κB *see* IκB
Inhibitor of κB kinase *see* IKK
Innate immune response, 8–9
effector mechanisms, 8, **Fig. 1.6**
failure to contain infection, 10, 12, 77, **Fig. 3.41**
immediate, 29–44, **Fig. 3.1**
importance, 12, **Fig. 1.10**
induced, 47–78, **Fig. 3.1**
inflammation, 9, **Fig. 1.7**
pathogen recognition mechanisms, 8
vs. adaptive immunity, 10, **Fig. 1.8**
Innate immunity, 2, 8
autoinflammatory diseases, 502–503
cellular receptors, 47–49, **Fig. 3.2**
coevolution with adaptive immunity, 329–361, **Fig. 12.1**
immediate defenses, 29–44
importance, 12
induced defenses, 47–78
Insects, venomous, 422
Insulin, 495, 496
Insulin-dependent diabetes mellitus (IDDM) *see* Diabetes, type 1
Insulin receptor autoantibodies, 483, **Fig. 16.12**
Insulitis, 496, **Fig. 16.33**
Integrins, 50, **Fig. 3.12**
effector T cells, 219, **Fig. 8.19**
mucosa-specific lymphocyte homing, 281, 282, **Fig. 10.16**
Interallelic conversion, 142, **Fig. 5.38**
Intercellular adhesion molecules *see* ICAM-1; ICAM-2; ICAM-3
Interferon(s), 68–71
coordinate gene regulation, 138
response, 70, **Fig. 3.33**, **Fig. 3.34**
type I, 68–69
functions, 70–71, **Fig. 3.34**
induction of synthesis, 69, 75, **Fig. 3.32**, **Fig. 3.33**, **Fig. 3.39**
NK-cell activation, 71, 74, **Fig. 3.34**, **Fig. 3.38**
plasmacytoid dendritic cells, 71
therapeutic use, 71
type II, 76
Interferon-α (IFN-α)
functions, 70–71, **Fig. 3.34**
induction of synthesis, 75, **Fig. 3.34**, **Fig. 3.39**
NK-cell activation, 71, 74, **Fig. 3.38**
Interferon-β (IFN-β)
functions, 70–71, **Fig. 3.34**
induction of synthesis, 69, 75, **Fig. 3.33**, **Fig. 3.39**
NK-cell activation, 71, 74, **Fig. 3.38**
therapy in multiple sclerosis, 477
Interferon-γ (IFN-γ), 76
celiac disease and, 501
control of HLA-DM/HLA-DO balance, 131
control of isotype switching, 243, **Fig. 9.13**
coordinate gene regulation, 138
intestinal helminth infections, 289, **Fig. 10.27**
macrophage activation, 224–225, **Fig. 8.27**
multiple sclerosis, 476
production by NK cells, 76, **Fig. 3.40**
proteasome modification, 126–127
T cells expressing, 223, **Fig. 11.7**
T_H1 development, 213, **Fig. 8.14**
Interferon-γ (IFN-γ) receptors, 378, **Fig. 13.11**
deficiency, 378–379, 385
dominant and recessive mutations, 378–379, **Fig. 13.11**
intracellular infections, 386, **Fig. 13.18**
Interferon-response factor 3 (IRF3), 69, **Fig. 3.32**, **Fig. 3.33**
NK-cell activation, 75, **Fig. 3.39**
Interferon-response factor 7 (IRF7), 70–71, **Fig. 3.34**
NK-cell activation, 75, **Fig. 3.39**
Interleukin-1 (IL-1), 55
Interleukin-1β (IL-1β), 53, **Fig. 3.9**
amplification of production, 55–56, **Fig. 3.11**
autoinflammatory syndromes and, 503
induction of inflammation, 53–54
precursor protein (proIL-1β), 55–56, **Fig. 3.11**
systemic effects, 62–63, **Fig. 3.20**
Interleukin-1 receptor antagonist (IL-1RA), 56
deficiency, **Fig. 16.41**
Interleukin-2 (IL-2)
activated T cells, 209–210, **Fig. 8.12**
CD8 T-cell differentiation, 216, **Fig. 8.17**
Interleukin-2 (IL-2) receptors
activated T cells, 210, **Fig. 8.12**
α chain *see* CD25
CD8 T cells, 216, **Fig. 8.17**
monoclonal antibody inhibiting, 452–453, **Fig. 15.24**, **Fig. 15.30**
Interleukin-3 (IL-3), intestinal helminth infections, 289, **Fig. 10.27**
Interleukin-4 (IL-4)
control of isotype switching, 243, **Fig. 9.13**
gene polymorphism, 418, **Fig. 14.25**
inhibition of macrophage activation, 225
memory B-cell development, 244, **Fig. 9.15**
secretion by basophils, 415
sensitization to inhaled allergens, **Fig. 14.23**
T_H2 development, 213, **Fig. 8.14**
Interleukin-4 (IL-4) receptor, gene polymorphism, 418, **Fig. 14.25**
Interleukin-5 (IL-5)
B-cell differentiation, 238
control of isotype switching, **Fig. 9.13**
intestinal helminth infections, 289, **Fig. 10.27**
monoclonal antibodies, 428
regulation of eosinophils, 414, 415
Interleukin-6 (IL-6), 53, 54, **Fig. 3.9**
B-cell effects, 238, 239
systemic effects, 62–63, **Fig. 3.20**
T_{FH} cell development, 214
Interleukin-7 (IL-7)
B-cell development, 152, **Fig. 6.5**
T-cell development, 180
Interleukin-7 (IL-7) receptor
B cells, **Fig. 6.5**, **Fig. 6.12**
memory T cells, 300, 303, 305
thymocytes, 180, **Fig. 7.5**
Interleukin-9 (IL-9), intestinal helminth infections, 289, **Fig. 10.27**
Interleukin-10 (IL-10)
B-1 cell survival, 162
deficiency, 275
gut dendritic cells, 278
inhibition of macrophage activation, 225
plasma cell development, 244, **Fig. 9.15**
Interleukin-12 (IL-12)
NK cell activation, 54, 75–76, **Fig. 3.40**
NKT cell activation, 357, **Fig. 12.32**
production by macrophages, 53, **Fig. 3.9**
T_H1 development, 213, **Fig. 8.14**
Interleukin-12 (IL-12) receptors, 75–76
deficiency, 385–386
Interleukin-13 (IL-13)
inhibition of macrophage activation, 225
intestinal helminth infections, 289, **Fig. 10.27**
secretion by basophils, 415
Interleukin-15 (IL-15)
germinal center formation, 239
NK-cell activation, 76, **Fig. 3.40**
Interleukin-15 (IL-15) receptors, memory T cell survival, 300

Interleukin-16 (IL-16), 214
Interleukin-17 (IL-17), 213–214
Interleukin-21 (IL-21), 214
Interleukin-33 (IL-33), 289
Interleukins, 220
Intestinal epithelial cells (enterocytes), 277
  see also Intraepithelial lymphocytes; M cells
  destruction in celiac disease, 498, 500, **Fig. 16.36**
  innate immune responses, 276–277, **Fig. 10.10**
  response to helminth infections, 289
Intestine *see* Large intestine; Small intestine
Intracellular compartments, 125–126, **Fig. 5.17**
Intracellular pathogens, 30–31, **Fig. 2.2**
  antigen presentation, 122–123
  antigen processing, 126–128, 134, **Fig. 5.19**, **Fig. 5.27**
  cytotoxic CD8 T-cell responses, 222–224
  innate and adaptive immune response, 385–386, **Fig. 13.18**
  interferon response, 68–71
  susceptibility to infection, 385–386
Intraepithelial lymphocytes (IEL), 282, **Fig. 10.17**
Intraepithelial pockets, 278, **Fig. 10.13**
Intravenous immunoglobulin (IVIG)
  autoimmune disease, 489–490, **Fig. 16.22**, **Fig. 16.23**
  inherited immunodeficiencies, 90, 160, 379–380
Invariant chain, 130, **Fig. 5.23**
  gene locus, 138
IPEX, 478, **Fig. 13.10**
Ipilimumab, 521–522, **Fig. 17.19**, **Fig. 17.27**
IRAK4, 52, **Fig. 3.7**
Ischemia, donor organs, 440–441
Islets of Langerhans, 496, **Fig. 16.33**
Isograft, 439
Isopentyl pyrophosphate (IPP), 350, **Fig. 12.24**
Isotype switching, 101–102, **Fig. 4.28**
  B-cell zone of lymph node, 239–241, **Fig. 9.11**
  regulation by cytokines, 243–244, **Fig. 9.13**
ITAMs *see* Immunoreceptor tyrosine-based activation motifs
ITIM *see* Immunoreceptor tyrosine-based inhibitory motifs

J

Jak1, 70, 378, **Fig. 13.11**
Jak2, 378, **Fig. 13.11**
Jak3, 384
  deficiency, **Fig. 13.16**
Janus kinases (JAKs), 221, **Fig. 8.22**
J chain, **Fig. 4.27**, **Fig. 4.31**
Jenner, Edward, 1–2, 308, 309
J gene segments *see* Joining gene segments
Joining (J) gene segments
  immunoglobulins, 92, **Fig. 4.15**, **Fig. 4.17**
  recombination signal sequences (RSSs), 93, **Fig. 4.18**
  somatic recombination, 92–95, **Fig. 4.16**
  T-cell receptors, 115, **Fig. 5.3**, **Fig. 5.8**
  T-cell *vs.* B-cell receptors, **Fig. 5.9**
Jun, 209
Junctional diversity
  immunoglobulins, 95–96, **Fig. 4.20**
  T-cell receptors, 115, 119
  T-cell *vs.* B-cell receptors, **Fig. 5.9**
Juvenile rheumatoid arthritis, HLA association, **Fig. 16.9**

K

Kaposi's sarcoma, 396, **Fig. 17.4**
κ light chains, immunoglobulin, 83
  gene locus, 91–92, **Fig. 4.15**
  gene rearrangements, 151, 155–156, **Fig. 6.9**
  gene segments, 93, **Fig. 4.17**
  utilization by B cells, 156
Kidney
  glomerular basement membrane autoantibodies, 475–476, **Fig. 16.3**
  immune complex deposition, 258, **Fig. 16.20**
Kidney transplantation
  see also Organ transplantation
  acute rejection, **Fig. 15.8**, **Fig. 15.9**
  anti-T-cell antibody therapy, 90, 450, **Fig. 15.21**
  baseline inflammatory state, 440–441, **Fig. 15.7**
  cadaveric donors, 445
  cancer risk after, 513–514
  chronic rejection, 443
  combined with hematopoietic cell transplant, 467
  HLA matching, 445, **Fig. 15.14**
  hyperacute rejection, 436–437, **Fig. 15.4**
  immunosuppressive drugs, 445–446
  live organ donors, 441, 445
  mechanism of graft rejection, 143
  supply of donor organs, 454, **Fig. 15.26**
  syngeneic, 439
  transfer of tumors, 515
Killer-cell immunoglobulin-like receptors (KIR), 335–336
  see also specific KIRs
  cancer therapy targeting, 525
  gene organization, 339–340, **Fig. 12.12**
  haplotypes
    gene-content variation, 340–341, **Fig. 12.14**
    pregnancy complications and, 344–345, **Fig. 12.18**
  ligand-binding sites, 335, **Fig. 12.7**
  ligands or epitopes of HLA class I, 335–336, **Fig. 12.8**
  NK-cell education, 336, 337–339, **Fig. 12.9**, **Fig. 12.10**
  polymorphisms
    and HIV progression, 336, 393, **Fig. 13.26**
    and human disease, 336, 340, 377
  similarities to Ly49 receptors, 340, **Fig. 12.13**
  variegated gene expression, 336
Kinin system, 40
KIR *see* Killer-cell immunoglobulin-like receptors
KIR2DL1, 335, 336, **Fig. 12.2**
  placental formation, 344, **Fig. 12.17**
KIR2DL2/3, 335, 336, **Fig. 12.2**
  NK-cell education, 337–339, **Fig. 12.11**
  placental formation, 344, **Fig. 12.17**
KIR2DL4, 335, 336, **Fig. 12.2**
  placental formation, 344, **Fig. 12.17**
KIR2DS1, 335, 336, **Fig. 12.2**
KIR2DS2, senescent T cells, 502
KIR3DL1, 335, 336, **Fig. 12.2**
KIR3DL2, 335, 336, **Fig. 12.2**
Kit
  B-cell development, 152, **Fig. 6.5**, **Fig. 6.12**
  T-cell development, **Fig. 7.14**
Klebsiella pneumoniae, 314–315
Kostmann syndrome, **Fig. 15.32**
Kupffer cells, 36, 131–132

L

Lactoferrin, 60
λ5, 153–154, **Fig. 6.7**
  inherited defects, 154, **Fig. 13.10**
  regulation of expression, 158, **Fig. 6.12**
λ light chains, immunoglobulin, 83
  gene locus, 92, **Fig. 4.15**
  gene rearrangements, 151, 155–156, **Fig. 6.9**
  gene segments, 93, **Fig. 4.17**
  utilization by human B cells, 156
Lamina propria, 272, **Fig. 10.7**
  effector lymphocytes, 281, 283, **Fig. 10.16**
  IgA production, 247, **Fig. 9.18**
  macrophages, 276–277
Langerhans cell histiocytosis, **Fig. 14.19**
Langerhans cells, **Fig. 8.1**
Large granular lymphocytes, 16
Large intestine, lymphoid tissues, 272
Latent infections, 370–371, **Fig. 13.5**
Lck
  expression in developing T cells, 188, **Fig. 7.14**
  pre-T-cell receptor signaling, 184
  role in TCR signaling, 192, 208, **Fig. 8.10**
Leader peptide (L), genes, 92, **Fig. 4.15**
Lectin-like domains, NK-cell receptors, 333, **Fig. 12.2**
Lectins, 50
Leishmania, macrophage receptors, 50
Lenalidomide, 524
Lentiviruses, 388
Leprosy
  lepromatous/tuberculoid
    macrophage activation, 225
    polarized T_H1 response, 215, **Fig. 8.15**, **Fig. 8.16**
  T-cell responses to lipid antigens, 354–356
Leucine-rich repeat region (LRR), 51
Leukapheresis, 463
Leukemia, 510
  see also Graft-versus-leukemia (GVL) effect
  acute lymphoblastic (ALL), 466, **Fig. 6.23**
  acute myelogenous (AML), 466, 518–519
  chronic lymphocytic (CLL), 162, **Fig. 6.23**, **Fig. 17.27**
  chronic myelogenous, 517, **Fig. 17.10**
  hematopoietic stem-cell transplant, 465–466
  pre-B-cell, **Fig. 6.23**
  radiation-induced, 512
Leukocyte adhesion deficiency, 382, **Fig. 13.15**
Leukocyte-associated immunoglobulin-like receptors (LAIR), **Fig. 12.12**
Leukocyte immunoglobulin-like receptors (LILR), 339, **Fig. 12.12**
Leukocyte-receptor complex (LRC), 263, 339, 340, **Fig. 12.12**
Leukocytes (white blood cells), 12
  adhesion to vascular endothelium, 57, **Fig. 3.12**

autoantibodies against, 475
in inflammation, 9
numbers in blood, **Fig. 1.14**
origins and development, 12–16
Leukocytosis, 381
Leukotrienes, 412, **Fig. 14.18**
LFA-1
B cell–helper T_{FH} cell interaction, 238, **Fig. 9.8**
effector T cells, 219, **Fig. 8.18**, **Fig. 8.19**
inherited defects, 382
naive T-cell homing, 205–206, **Fig. 8.6**
neutrophil recruitment, 57–58, **Fig. 3.13**
NK cells, 74, **Fig. 12.2**
peripheral supramolecular activation complex, 208, **Fig. 8.9**
LFA-3, T cell–dendritic cell adhesion, 205
Li–Fraumeni syndrome, 512
Light chains (L chains), 82–83, **Fig. 4.2**
biased use by B cells, 156
constant domain, 84, **Fig. 4.6**, **Fig. 4.7**
framework regions, 85–86, **Fig. 4.8**
gene loci, 91–92
gene rearrangements, 92–93, **Fig. 4.16**
B-cell precursors, 151, 155–157, **Fig. 6.4**, **Fig. 6.9**
self-reactive immature B cells, 165–166, **Fig. 6.18**
gene segments, 91–92, **Fig. 4.15**, **Fig. 4.17**
hypervariable regions (HV), 85–86, **Fig. 4.8**
isotypes, 83
κ *see* κ light chains
λ *see* λ light chains
surrogate, 153–154, **Fig. 6.7**
synthesis, 151, **Fig. 6.4**
variable domain (V_L), 84, **Fig. 4.6**, **Fig. 4.7**
LILRB1, 339, **Fig. 12.2**
placental formation, 344, **Fig. 12.17**
Linkage disequilibrium, HLA allele associations, 479–480
Lipid antigens
chemical structure, **Fig. 12.27**
presentation to γ:δ T cells, 352–353, **Fig. 12.28**, **Fig. 12.32**
presentation to NKT cells, 356–357
recognized by CD1-restricted α:β T cells, 354–356, **Fig. 12.29**
Lipid mediators
eosinophils, **Fig. 14.20**
mast cells, 412, **Fig. 14.16**, **Fig. 14.18**
Lipid-transfer proteins, 355, 356
Lipoarabinomannan, **Fig. 12.29**
Lipomannan, **Fig. 12.29**
Lipopeptides, 50, **Fig. 3.3**, **Fig. 3.29**
Lipopolysaccharide (LPS)
downstream signaling, 52–53, **Fig. 3.7**
macrophage receptors recognizing, 50, **Fig. 3.3**
neutrophil receptors, **Fig. 3.15**
receptor *see* CD14
recognition by TLR4, 51–52, **Fig. 3.6**, **Fig. 3.29**
Lipopolysaccharide-binding protein (LBP), **Fig. 3.6**
Lipoteichoic acid (LTA), **Fig. 3.3**, **Fig. 3.29**
5-Lipoxygenase, **Fig. 14.25**
Lirilumab, 525
Listeria monocytogenes, 371, **Fig. 1.3**
Liver transplantation
factors affecting success, 457
organ donors, 441, 455, **Fig. 15.26**
LMP2, 138, **Fig. 5.32**
LMP7, 138, **Fig. 5.32**
LPS *see* Lipopolysaccharide
L-selectin (CD62L)
effector T cells, 219, **Fig. 8.18**
memory T cells, 303, 304, **Fig. 11.7**
mucosa-derived effector lymphocytes, 281
naive T-cell homing, 205, **Fig. 8.6**
Lung carcinoma, non-small cell, **Fig. 17.10**
Lung transplantation, **Fig. 15.26**
Lupus erythematosus, 476
see also Systemic lupus erythematosus
Ly49 receptors, 340, **Fig. 12.13**
Lyme disease, chronic arthritis, **Fig. 16.31**
Lymph, 20
naive T cells, 203, **Fig. 8.5**
transport of pathogens, 21–22, **Fig. 1.21**
Lymphatics, 20
Lymphatic vessels, 20
afferent, 22, **Fig. 1.22**
efferent, 22, **Fig. 1.22**
Lymph nodes, 20, 21–23
activation of adaptive immunity, 22–23, **Fig. 1.23**
anatomy, **Fig. 1.22**
B-cell activation, 168–169, **Fig. 6.22**, **Fig. 9.7**
B-cell circulation, 167, **Fig. 6.20**
B-cell maturation, 167–168, **Fig. 6.21**
B-cell zone, 239–241
draining, 21, **Fig. 1.21**, **Fig. 1.23**
exit of T cells from, 206
mantle zone, 239–240, **Fig. 9.11**
medullary cords, 238–239, **Fig. 9.9**
migration of plasma cells to, 170
naive T-cell recirculation, 206, **Fig. 8.5**
swelling during infections, 22, 240
T-cell activation, 200–201, **Fig. 8.1**
T-cell zone (T-cell area), 203
activation of naive B cells, 236–237, **Fig. 9.7**
transport of pathogens to, 20–21, **Fig. 1.21**
Lymphoblasts, 169
Lymphocytes, 10
see also B cell(s); T cell(s)
activation by pathogens, **Fig. 1.21**
antigen receptors, 17
cell-surface receptors, 10
circulating, populations, 71
effector *see* Effector cells
gut-associated, 272
intraepithelial, 282, **Fig. 10.17**
large granular, 16
lymph nodes, 21–22, **Fig. 1.22**
mucosal lymphoid tissues, 25, **Fig. 1.26**
numbers in blood, **Fig. 1.14**
origins, 16
recirculation, 20, **Fig. 1.20**
recognition of pathogens, 10–11
selection by pathogens, 11, **Fig. 1.9**
small, 16, 20, **Fig. 1.11**, **Fig. 1.20**
spleen, 23, **Fig. 1.24**
thymus-dependent, 177
tissue locations, 19–20, **Fig. 1.19**
Lymphoid follicles, 22, **Fig. 1.22**
B-cell maturation, 167–168, **Fig. 6.21**
isolated, intestinal, 272, **Fig. 10.7**
primary, 167, **Fig. 6.20**
antigen capture and display, 235–236, **Fig. 9.6**
entry of immature B cells, 167, **Fig. 6.21**
germinal center formation, 239, **Fig. 9.9**
movement of B cells to, 239–240
secondary, 169, 239–240
spleen, **Fig. 1.24**
Lymphoid lineage cells, 16, **Fig. 1.13**
Lymphoid neogenesis, 484
Lymphoid organs, 19
Lymphoid progenitor cells, 16, 150, **Fig. 1.13**, **Fig. 6.3**
bone marrow stromal-cell interactions, **Fig. 6.5**
Lymphoid tissues, 19–20
primary (central), 19–20, 178, **Fig. 1.19**
secondary (peripheral) *see* Secondary lymphoid tissues
tertiary (ectopic), 484–485, **Fig. 16.16**
Lymphokines, 220
Lymphoma, 510
see also Hodgkin's disease; *specific lymphomas*
diagnosis, 526–528
treatment, 528–529, **Fig. 17.27**
Lymphoproliferative syndrome, X-linked, 386
Lymphotoxin (LT), B cell–follicular dendritic cell interactions, 167–168
Lyn, B-cell receptor signaling, 232, 233, **Fig. 9.1**
Lysosomes
antigen processing, 126
fusion with phagosome, 50, 60–61, **Fig. 3.16**
Lysozyme, 6
neutrophils, 60
Paneth cells, 41, **Fig. 2.18**

M

Mac-1 *see* CR3
α_2-Macroglobulins, 40–41, **Fig. 2.16**
Macrophage receptor with collagenous structure (MARCO), 50, **Fig. 3.3**
Macrophages, 15, **Fig. 1.11**
activation, 224–225, **Fig. 8.27**
activity in lymph nodes, 22, **Fig. 1.23**
antigen processing, 129–130, 131, **Fig. 5.22**
cell-surface receptors, 49–51, **Fig. 3.3**
complement receptors, 36–37, **Fig. 2.10**
co-stimulatory molecules, 207
cytokines, **Fig. 1.16**
induction of synthesis, 51–53, **Fig. 3.7**
inflammatory effects, 53–54, **Fig. 3.9**
help from CD4 T cells, 122, **Fig. 5.13**
HIV infection, 389
induction of inflammation, 53–54, **Fig. 3.9**
inflammation-anergic, 276
inherited defects, 382–383, **Fig. 13.15**
intestinal, 276–277, **Fig. 10.11**
intracellular pathogens, 371
lipopolysaccharide recognition, 51–52, **Fig. 3.6**
lymph node, functions, 201–202
medullary sinus, 236
negative selection of T cells, 192, **Fig. 7.18**
NK-cell interactions, 75–76, **Fig. 3.40**
phagocytosis, 15, **Fig. 1.16**
as first line of cellular defense, 36–37
and pathogen degradation, 50, **Fig. 3.4**
receptors mediating, 49–50, **Fig. 3.3**
role of complement, 36–37, **Fig. 2.10**
responses to pathogens, 15, **Fig. 1.16**
signaling receptors, 50–51, **Fig. 3.3**
splenic, 23
subcapsular sinus, 236

thymic, 178, 182–183, **Fig. 7.3**, **Fig. 7.8**
tingible body, 241
TLR4 signaling pathways, 51–53, **Fig. 3.7**
Macropinocytosis, by dendritic cells, 202, **Fig. 8.3**
MAdCAM-1, 281, 282, **Fig. 10.16**
MAGEA1/MAGEA3 antigens, 520–521, **Fig. 17.18**
MAIT (mucosa-associated invariant T) cells, 354, 357–359, **Fig. 12.34**
Majeed syndrome, **Fig. 16.41**
Major basic protein, 415, **Fig. 14.20**
Major histocompatibility complex *see* MHC
Malaria, 322–323, 369, **Fig. 11.26**
Malignant transformation, 511, 516, **Fig. 17.3**
Malignant tumors, 510, **Fig. 17.1**
see also Cancer
MALT *see* Mucosa-associated lymphoid tissue
Mannose-binding lectin (MBL), 63–66, **Fig. 3.24**
acute-phase response, **Fig. 3.21**
complement activation, 63–66, **Fig. 3.25**
inherited deficiency, 65–66, **Fig. 13.10**
structure, 63, **Fig. 3.23**
Mannose-binding lectin associated serine protease 1 (MASP-1), 64, **Fig. 3.23**, **Fig. 3.25**
Mannose-binding lectin associated serine protease 2 (MASP-2), 64–65, **Fig. 3.23**, **Fig. 3.25**
Mannose receptors, 50, **Fig. 3.3**
Mannosylated lipoarabinomannan, **Fig. 3.3**
Mannosyl-β-1-phosphomycoketide, **Fig. 12.29**
Mannuronic acid, **Fig. 3.3**
Mantle cell lymphoma, **Fig. 6.23**
Mantle zone, 239–240, **Fig. 9.11**
MARCO (macrophage receptor with collagenous structure), 50, **Fig. 3.3**
MART2, **Fig. 17.10**
MASP *see* Mannose-binding lectin associated serine protease
Mast cells, 15–16, 410–413, **Fig. 1.11**
activation/degranulation, 411–412
IgE binding to FcεRI, 407, **Fig. 14.11**
IgE function, 248, 249–250, **Fig. 9.19**
localized allergic reactions, 420–421, **Fig. 14.29**
mediators released, 411–413, **Fig. 14.16**
recruitment of basophils, 415
connective tissue, 411
effector functions, 411, **Fig. 14.16**
Fcε receptors, 248, 249–250, **Fig. 9.19**
FcεRI, 407, **Fig. 14.10**
granules, 410–411, **Fig. 14.15**
intestinal helminth infections, 289, **Fig. 10.27**
mucosal, 411
Master regulators
CD4 T-cell differentiation, 212, 213, 214
NKT-cell development, 356
MAVS (mitochondrial antiviral signaling) protein, 69, **Fig. 3.32**
MBL *see* Mannose-binding lectin
M cells (microfold cells), 25, 277–278, **Fig. 1.26**
antigen uptake and transport, 277–278, **Fig. 10.13**
appearance, 277, **Fig. 10.12**
exploitation by pathogens, 279
MD2, TLR4 association, 51–52, **Fig. 3.6**, **Fig. 3.7**
MDA-5, 69
ME1, **Fig. 17.10**
Measles, 321–322
acquired immunity, 11
immunological memory, 299
incidence over time, 321, **Fig. 11.24**
vaccine, **Fig. 11.25**, **Fig. 11.26**
Measles, mumps and rubella (MMR) vaccine, 321, **Fig. 11.15**
Mediterranean fever, familial (FMF), 502, **Fig. 16.41**
Medullary sinus macrophages, 236
MEFV gene, 502
Megakaryocytes, 12, 14, **Fig. 1.11**
Melanoma
monoclonal antibody therapy, 521–522, **Fig. 17.19**, **Fig. 17.27**
transplant-mediated transfer, 515
tumor antigens, 517, **Fig. 17.10**, **Fig. 17.11**
vaccination, 520–521, **Fig. 17.18**
Membrane-attack complex, 37–38, **Fig. 2.11**
inherited deficiency, 38, 381
membrane pore formation, 38, **Fig. 2.13**
regulation of formation, 38–39, **Fig. 2.14**
Membrane co-factor protein (MCP), 35–36, **Fig. 2.9**
Memory, immunological, 11–12, 295, 296–308
influenza virus escape, 306–307, **Fig. 11.12**
vaccination inducing, 299, **Fig. 11.3**
Memory B cells, 297–300
activation, 300–301, 302
antigen-independent survival, 299–300
differentiation, 169–170, 244, **Fig. 9.15**
duration of protection, 299, **Fig. 11.3**
production, 297–298, **Fig. 11.2**
vs. naive and effector B cells, 300, 302
Memory cells, 11, 295
Memory T cells, 297–300, 302–305
activation, 301–302
antigen-independent survival, 299–300
antigen specificities, 304
cell-surface proteins, 302–304, **Fig. 11.7**
central (T_{CM}), 304, **Fig. 11.9**
duration of protection, 299, **Fig. 11.3**
effector (T_{EM}), 304, **Fig. 11.9**
γ:δ, 349, **Fig. 12.21**
generation in virus infections, 305, **Fig. 11.10**
production, 297–298, **Fig. 11.2**
Meningitis
Neisseria meningitidis, 66, 315–316
vaccine, 315–316, **Fig. 11.25**
Meningococcus *see Neisseria meningitidis*
Mepolizumab, 428
6-Mercaptopurine, 453, **Fig. 15.25**
Mesenteric lymph nodes, 272, **Fig. 10.7**
B- and T-cell activation, **Fig. 10.15**
transport of pathogens to, 279
Metastasis, 510
Methotrexate, 454, 461, **Fig. 15.25**, **Fig. 15.30**
MF59, 316, **Fig. 11.21**
MHC (major histocompatibility complex), 113–114, 135–144, 438
see also HLA
alleles, 135–136
generation of new, 142–143, **Fig. 5.38**
allotypes, 136, **Fig. 5.29**
antigen presentation, 121–135, **Fig. 5.27**
class I molecules *see* MHC class I molecules
class I region, 137, 138, **Fig. 5.30**
class II molecules *see* MHC class II molecules
class II region, 137, 138, **Fig. 5.30**
class III region (central), 137, **Fig. 5.30**
diversity, 135–136
generation, 140–143, **Fig. 5.35**
importance for human survival, 142–143, **Fig. 5.39**
gene families, 135
gene organization, 137–138, **Fig. 5.30**
heterozygote advantage, 141, **Fig. 5.36**
heterozygotes, 136, 137
highly polymorphic genes, 136, **Fig. 5.28**, **Fig. 5.29**
homozygotes, 136, 137
isoforms, 136
autologous/allogeneic, 143
isotypes, 135, **Fig. 5.28**
monomorphic genes, 136, **Fig. 5.28**
oligomorphic genes, 136, **Fig. 5.28**
polymorphism, 113–114, 135–136, **Fig. 5.28**, **Fig. 5.29**
see also HLA, polymorphism
antigen presentation effects, 138–140, **Fig. 5.33**
natural selection generating, 140–143, **Fig. 5.37**
transplant rejection, 143–144
restriction, 140, **Fig. 5.35**
self *see* Self-MHC molecules
MHC class I-like molecules, non-polymorphic
see also CD1; MR1
evolution, 359–360
restricting α:β T cells, 354–360
MHC class I molecules, 122, **Fig. 5.11**
see also HLA class I molecules
anchor residues, 139, **Fig. 5.34**
antigen cross-presentation, 131–132, **Fig. 5.24**
antigen processing for, 126–127, **Fig. 5.19**
CD8 binding site, 124, **Fig. 5.15**
cells expressing, 132, **Fig. 5.25**
cytomegalovirus (CMV) and, 341–342
deficiency, 127, 385, **Fig. 13.10**, **Fig. 13.16**
genes, 135–136, **Fig. 5.30**
isotypes, 135–136, **Fig. 5.28**
KIR ligands or epitopes, 335–336, **Fig. 12.8**
NK-cell education detecting changes in, 336–339, **Fig. 12.9**
NK-cell receptors recognizing, 333–339, **Fig. 12.2**
killer-cell immunoglobulin-like receptors, 335–336, **Fig. 12.7**
NK-cell inhibition, 333–335, **Fig. 12.6**
peptide binding, 124–125, 127–128
allotype variability and, 139, **Fig. 5.33**
intracellular compartments, 126
peptide-loading complex, 127–128, **Fig. 5.20**
promiscuity, 124
T-cell receptor complex, 132–133, **Fig. 5.26**
peptide-binding groove, 124, **Fig. 5.16**
peptide-binding motifs, 139, **Fig. 5.34**
recycling to endoplasmic reticulum, 128
structure, 123, **Fig. 5.14**
trophoblast–uterine NK cell interactions, 343–344, **Fig. 12.17**
tumor cells, 515–516
virus interference mechanisms, 371–372, **Fig. 13.7**

MHC class II compartment (MIIC), 130–131
MHC class II molecules, 122, **Fig. 5.11**
  see also HLA class II molecules
  allergic disease susceptibility and, 418, **Fig. 14.25**
  anchor residues, 139, **Fig. 5.34**
  antigen processing for, 126, 129–130, **Fig. 5.22**
  CD4 binding site, 124, **Fig. 5.15**
  cells expressing, 132, **Fig. 5.25**
  deficiency, 192, 385, **Fig. 13.16**
  effector T-cell activation, 219
  genes, 135–136, **Fig. 5.30**, **Fig. 5.32**
  invariant chain association, 130, **Fig. 5.23**
  isotypes, 136, **Fig. 5.28**
  naive B-cell activation, 237–238
  peptide binding, 124–125, 126
    allotype variability and, 139, **Fig. 5.33**
    promiscuity, 124
    regulation, 130–131, **Fig. 5.23**
    self peptides, 126
    T-cell receptor complex, 133, **Fig. 5.26**
    vesicular system, 129–130, **Fig. 5.22**
  peptide-binding groove, 124, **Fig. 5.16**
  peptide-binding motifs, 139, **Fig. 5.34**
  structure, 123–124, **Fig. 5.14**
  superantigen interactions, 373, **Fig. 13.8**
MHC class II transactivator (CIITA), 138, 385
MIC glycoproteins (MIC-A and MIC-B)
  activation of NK cells, 334–335, **Fig. 12.6**
  tumor cells, 518, **Fig. 17.13**
Microbiota, 3, 30
Microfold cells *see* M cells
β_2-Microglobulin, 123, **Fig. 5.14**
  gene locus, 137
  MR1, 358
  peptide-loading complex, 127, 128, **Fig. 5.20**
Microorganisms
  causing human disease, **Fig. 1.4**
  commensal *see* Commensal microorganisms
  pathogenic *see* Pathogens
  physical barriers against, 4–8, 29, **Fig. 1.5**
MIIC (MHC class II compartment), 130–131
Minor histocompatibility antigens, 464–465, **Fig. 15.38**
  derivation from self proteins, 465, **Fig. 15.40**
  HLA molecules presenting, 464, **Fig. 15.39**
Minor histocompatibility loci, 464
Mixed essential cryoglobulinemia, **Fig. 16.1**
Mixed lymphocyte reaction (MLR), 442–443, **Fig. 15.10**, **Fig. 15.44**
MLL5 (mixed-lineage leukemia protein 5), **Fig. 12.2**
Molecular mimicry, 493–494
Monkeypox, 310
Monoclonal antibodies, 88–91
  allergic disease, 90, 409–410, 428, **Fig. 14.14**
  cancer diagnosis, 526–528
  cancer therapy *see* Cancer, monoclonal antibody therapy
  cell-surface markers, 88, **Fig. 4.13**
  chimeric, 90, **Fig. 4.14**
  conjugated, 528–529, **Fig. 17.27**
  human, 90–91, **Fig. 4.14**
  humanized, 90, **Fig. 4.14**
  IgG4 subclass, 107
  production, 88, 154–155, **Fig. 4.12**
  rheumatoid arthritis, 490–491, **Fig. 16.26**
  therapeutic uses, 90–91, **Fig. 4.14**
  transplant recipients, 447, 450–451, 452–453
Monocytes, 15, **Fig. 1.11**
  development into intestinal macrophages, 276–277, **Fig. 10.11**
  numbers in blood, **Fig. 1.14**
  recruitment to infected tissue, 54
Monomorphic, 136
Montagu, Lady Mary Wortley, 1
MR1, 354, 358–359, **Fig. 12.35**
mRNA splicing, alternative *see* Alternative RNA splicing
μ heavy chains, immunoglobulin, 83, **Fig. 4.5**
  allelic exclusion, 154–155, **Fig. 6.8**
  alternative RNA splicing, 99–100, **Fig. 4.24**
  B-cell precursor cells, 150–151
  elimination of B cells unable to make, 153–154
  synthesis, 96, 150–151, **Fig. 4.22**, **Fig. 6.4**
Mucins, 269, **Fig. 10.2**, **Fig. 10.3**
Mucopolysaccharidosis, **Fig. 15.32**
Mucosa-associated invariant T (MAIT) cells, 354, 357–359, **Fig. 12.34**
Mucosa-associated lymphoid tissue (MALT), 25
  see also Gut-associated lymphoid tissues
  B- and T-cell commitment to, **Fig. 10.15**
  dimeric IgA synthesis, 247
Mucosae, 267
Mucosal immunity, 267–307
  vs. systemic immunity, 273–275, **Fig. 10.9**
Mucosal mast cells, 411
Mucosal surfaces
  see also Epithelia; Gastrointestinal tract
  commensal microorganisms, 29–30
  distinctive features of adaptive immunity, 290, **Fig. 10.28**
  distribution in body, 267–268, **Fig. 10.1**
  effector lymphocytes, 280–282, **Fig. 10.17**
  lymphoid tissues, 25
  mucins, 269, **Fig. 10.3**
  neutralizing antibodies, 252–253, **Fig. 9.24**
  physical barriers to infection, 6–8, 29, **Fig. 1.5**, **Fig. 2.1**
  secretory immunoglobulins *see* IgA, dimeric; IgM, secreted
  vulnerability to infection, 267–268
Mucus, 6, 267
  intestinal helminth infections, 289
  mucins and viscoelastic properties, 269, **Fig. 10.2**
  secretory immunoglobulins, 283–284, **Fig. 10.19**
Multiple sclerosis, 476–477, **Fig. 16.1**, **Fig. 16.9**
Mumps vaccine, **Fig. 11.25**
Muramyl dipeptide (MDP), NOD receptor, 55
Mutagens, 512
Mutations, 510
  cancer causation, 510–513, **Fig. 17.3**
  HIV, 393–394, **Fig. 13.28**
  influenza virus genes, 366–368, **Fig. 13.2**
  tumor-specific, 516–517, **Fig. 17.10**
Myasthenia gravis, 483, **Fig. 16.1**, **Fig. 16.12**, **Fig. 16.14**
  HLA association, 483, **Fig. 16.9**
Mycobacteria
  evasion mechanisms, 130
  lipid antigens presented by CD1, 354–356, **Fig. 12.29**
  susceptibility to infection, 378, 385–386
Mycobacterium avium, 385, 396
Mycobacterium bovis, 386, 514
Mycobacterium leprae, 215, **Fig. 8.16**
Mycobacterium phlei, glucose monomycolate, **Fig. 12.27**
Mycobacterium tuberculosis, 322–323, 371, **Fig. 1.3**
  see also Tuberculosis
Mycolic acid, **Fig. 12.29**
Mycophenolate mofetil, 453, **Fig. 15.30**
Mycophenolic acid, 453, **Fig. 15.25**
MYC proto-oncogene, 161, **Fig. 6.14**
  diagnostic monoclonal antibodies, 526, **Fig. 17.24**
MyD88
  independent TLR signaling, 75, **Fig. 3.39**
  NK cell activation, 74, **Fig. 3.39**
  TLR4 signaling, 52, **Fig. 3.6**, **Fig. 3.7**
Myeloablative therapy, 460
Myeloid lineage cells, 14–15, **Fig. 1.13**
Myeloid progenitor cells, 14, **Fig. 1.13**
Myeloma, multiple, 170, **Fig. 6.23**
  monoclonal antibody production, **Fig. 4.12**
  treatment, 524–525
Myeloperoxidase deficiency, 383, **Fig. 13.15**
Myxoma virus, rabbit, **Fig. 13.6**

N

NADPH oxidase
  inherited defects, 61–62, 383, **Fig. 13.10**, **Fig. 13.15**
  microbial killing, 60–61, **Fig. 3.16**, **Fig. 3.17**
Naive B cells, 168
  activation, 232–238
    antigens captured by follicular dendritic cells, 235, **Fig. 9.6**
    B-cell co-receptors, 232–234, **Fig. 9.2**, **Fig. 9.3**
    cross-linking of surface immunoglobulin, 232, **Fig. 9.1**
    in lymph nodes, 22
    morphological effects, 239, **Fig. 9.10**
    mucosal tissues, 280, **Fig. 10.15**
    plasma cell differentiation, 168–169, **Fig. 6.22**
    role of T_{FH} cells, 225–226, **Fig. 8.28**
    secondary response, 300–301, **Fig. 11.5**
    T-cell-independent, 235, **Fig. 9.4**
  commitment to mucosal tissues, 279–281, **Fig. 10.15**
  encounter with antigen, 98, 168–169, **Fig. 6.22**
  help from T_{FH} cells, 225–226, **Fig. 8.28**
  IgM and IgD production, 96, **Fig. 4.22**
  suppression in secondary response, 300–301, **Fig. 11.5**
  vs. memory B cells, 300, 302
  vs. plasma cells, 245, **Fig. 9.16**
Naive T cells, 203–210
  activation, 193–194, 199–218
    co-stimulatory signals, 206–207, **Fig. 8.8**
    dendritic cell adhesion, 205–206, **Fig. 8.7**
    inducing proliferation and differentiation, 209–210
    inducing self tolerance, 211, **Fig. 8.13**
    intracellular signaling pathways, 208–209, **Fig. 8.10**, **Fig. 8.11**
    mucosal tissues, 280, **Fig. 10.15**
  adhesion to dendritic cells, 205–206, **Fig. 8.7**

commitment to mucosal tissues, 279–281, **Fig. 10.15**
differentiation of activated, 211–216
encounter with antigen, 203–204, **Fig. 8.4**
homing to secondary lymphoid tissues, 204–206, **Fig. 8.6**
induction of anergy, 210–211, **Fig. 8.13**
recirculation through lymph nodes, 203–204, 206, **Fig. 8.5**
vs. memory T cells, 302–303, **Fig. 11.7**
Narcolepsy, **Fig. 16.9**
Natural killer (NK)-cell receptors, 330–347, **Fig. 12.2**
diversity of expression, 332, **Fig. 12.3**
Fc receptors, 261, 332–333, **Fig. 12.4**
genomic complexes encoding, 339–340, **Fig. 12.12**
immunoglobulin-like, 335–336, 339–340, **Fig. 12.2**
lectin-like, 333, 339–340, **Fig. 12.2**
MHC class I and related molecules, 333–339
NKT cells, 356
summary of regulatory function, 345–347, **Fig. 12.19**
Natural killer (NK) cells, 16, 71–77, **Fig. 1.11**
activation
by IL-12, 75–76, **Fig. 3.40**
inhibition by CD94:NKG2A, 334, **Fig. 12.6**
receptors of innate immunity, **Fig. 12.4**
by type I interferons, 71, 74, **Fig. 3.34**, **Fig. 3.38**
via Fc receptor, 332, **Fig. 12.4**
via NKG2D, 334, **Fig. 12.6**
via TLRs, 74, **Fig. 3.39**
antibody-dependent cytotoxicity, 261–262, **Fig. 9.36**
cancer immunotherapy targeting, 524–525, **Fig. 17.22**
$CD56^{bright}$, 72
$CD56^{dim}$, 72
cytomegalovirus infection, 341–342, **Fig. 12.15**
cytotoxicity, 72–74, **Fig. 3.38**
deficiency, 72, **Fig. 13.10**
dendritic cell interactions, 76–77, **Fig. 3.41**
differentiation from T cells, 72
education, 336–339, **Fig. 12.9**
evasion by tumor cells, 518–519, **Fig. 17.13**
evasion by viruses, 371–372
graft-versus-leukemia effect, 466, **Fig. 15.42**, **Fig. 15.43**
macrophage interactions, 75–76, **Fig. 3.40**
phenotypic diversity, 332, **Fig. 12.3**
recirculation, 71–72, **Fig. 3.37**
recruitment to infected tissue, 54
regulation of function, 330–347, **Fig. 12.19**
response to virus infections, 71–72, **Fig. 3.36**
similarity to γ:δ T cells, 348
subpopulations, 72–73
uterine (uNK), 71, 342–345
Natural killer (NK)-cell synapse, 74, **Fig. 3.40**
Natural killer complex (NKC), 339, 340, **Fig. 12.12**
Natural selection, MHC haplotypes, 140–143, **Fig. 5.37**
Negative selection
B cells, 149–150, 164, **Fig. 6.1**
T cells, 192–193, **Fig. 7.18**
Neisseria
complement activation, 256
increased susceptibility, 38, 381
Neisseria gonorrhoeae, antigenic variation, 369
Neisseria meningitidis (meningococcus), 315–316
factor H-binding protein (fHbp), 317, **Fig. 11.22**
IgA1 cleavage, 285
immunization schedule, **Fig. 11.15**
increased susceptibility, 66
protective role of IgG2, 261
vaccines, 315–316, 317, **Fig. 11.25**
NEMO deficiency, 53, **Fig. 3.8**, **Fig. 13.10**
Neoplasms, 510
Netosis, 61, **Fig. 3.18**
Neuraminidase, influenza, 318, 366–368
Neurotoxin, eosinophil-derived, **Fig. 14.20**
Neutralization, antibody-mediated, 18, **Fig. 1.18**
Neutralizing antibodies, 103, 231
HIV, 396–397, **Fig. 13.32**
isotypes acting as, **Fig. 4.30**
microbial toxins and animal venoms, 253–255, **Fig. 9.26**
preventing pathogen entry to cells, 251–253, **Fig. 9.23**, **Fig. 9.24**
Neutropenia, 381, 475
Neutrophil extracellular traps (NETs), 61, **Fig. 3.18**
Neutrophils, 14–15, 56–62
apoptosis and removal, 61, **Fig. 3.16**
azurophilic (primary) granules, 59, **Fig. 3.16**
defensins, 41–42, **Fig. 2.19**
extravasation, 58, **Fig. 3.13**
gelatinase (tertiary) granules, 59, 60
inherited defects, 382–383, **Fig. 13.15**
microbial killing, 59–61, **Fig. 3.15**, **Fig. 3.16**, **Fig. 3.17**
morphology, **Fig. 1.11**
netosis, 61, **Fig. 3.18**
numbers in blood, 57, **Fig. 1.14**
phagocytosis, 59, **Fig. 3.15**, **Fig. 3.16**
receptors for microbial products, 59, **Fig. 3.15**
recruitment to sites of infection, 53–54, 57–58, **Fig. 3.13**
specific (secondary) granules, 59, 60, **Fig. 3.16**
storage and mobilization, **Fig. 1.15**
Newborn babies
see also Fetus; Infants
colonization with commensals, 29–30
transient autoimmune disease, 483–484, **Fig. 16.15**
NFAT
activation in T cells, 209, **Fig. 8.11**
induction of IL-2, 210
inhibition by cyclosporin/tacrolimus, 449, **Fig. 15.19**
NFκB
activation of inflammation, 52–53, **Fig. 3.7**
corticosteroid actions, 448
defective, IKKγ deficiency, 53, **Fig. 3.8**
NOD receptor signaling, 55, **Fig. 3.10**
T-cell signaling, 209, **Fig. 8.11**
virus-infected cells, **Fig. 3.32**, **Fig. 3.33**
Nickel allergy, 402
Nitric oxide synthase, inducible (iNOS), 283
Nivolumab, 522
NK cells *see* Natural killer (NK) cells
NKG2A *see* CD94:NKG2A
NKG2D, 334–335, **Fig. 12.2**
evasion by tumor cells, 518, **Fig. 17.13**
NK-cell activation, 334, **Fig. 12.4**, **Fig. 12.6**
NKp30, **Fig. 12.2**
NKp44, **Fig. 12.2**
NKp46, **Fig. 12.2**
NKp80, **Fig. 12.2**
NKT cells, 354, 356–357
activation, 356–357, **Fig. 12.32**
effector functions, 357, **Fig. 12.33**
NLRP3, 55, 56, **Fig. 3.11**
NLRP3 gene mutations, 56
N nucleotides, 95–96, 115, **Fig. 4.20**, **Fig. 5.9**
NOD-like receptors (NLRs), 54–55, **Fig. 3.10**
NOD proteins (NOD1; NOD2), 54–55
intestinal epithelial cells, 276, **Fig. 10.10**
Non-Hodgkin's lymphoma
antibody-targeted radioisotope therapy, 529, **Fig. 17.27**, **Fig. 17.29**
monoclonal antibody therapy, 90, **Fig. 17.27**
Non-self, discrimination from self, 47–49, **Fig. 3.2**
Non-self peptides, 126
Non-self proteins, 126
Notch1, 180–181, **Fig. 7.6**, **Fig. 7.14**
NPM-ALK fusion protein, 527–528, **Fig. 17.26**
Nuclear factor κB *see* NFκB
Nuclear factor of activated T cells *see* NFAT

O

Obstructed labor, 345, **Fig. 12.18**
Ofatumumab, **Fig. 17.27**
OKT3 monoclonal antibody, 450–451
Oligoadenylate synthetase, 70
Oligomorphic, 136
Omalizumab, 90, 409–410, **Fig. 14.14**
Omenn syndrome, 116, 385, **Fig. 5.4**, **Fig. 13.10**, **Fig. 13.16**
Oncogenes, 161, 511
Oncogenic viruses, 512–513, **Fig. 17.4**
Oncology, 510
Opportunistic infections, AIDS, 395–396, **Fig. 13.31**
Opportunistic pathogens, 3, 396
Opsonins
acute-phase proteins, 63, 64, **Fig. 3.24**
antibodies, 103
Opsonization
antibody-mediated, 19, 103, 259–260, **Fig. 1.18**, **Fig. 4.30**
complement-mediated, 36
Oral tolerance, 278
Organ donors
cadaveric, 440–441, 455, **Fig. 15.7**, **Fig. 15.27**
living, 441, 455
shortages, 455–456, **Fig. 15.26**, **Fig. 15.27**
Organ transplantation, 440–458
baseline inflammatory state, 440–441, **Fig. 15.7**
global distribution, 455, **Fig. 15.28**
graft rejection *see* Transplant rejection
HLA matching, 445, **Fig. 15.14**
immunosuppressive drugs, 445–454
induced tolerance for, 467, **Fig. 15.44**
mixed lymphocyte reaction, 442, **Fig. 15.10**
organ-specific differences, 456–457

previous, anti-HLA antibodies, 438
transfusion effect, 444
Original antigenic sin, 307, 367
Osteopetrosis, **Fig. 15.32**

P

p53 gene mutations, 511, 512, **Fig. 17.3**
p53 protein, 511, 513, **Fig. 17.10**
Palindromic DNA sequences, 95, **Fig. 4.20**
Pancreas
β-cell destruction, 496, **Fig. 16.33**
transplantation, **Fig. 15.26**
Pandemics, influenza, 367–368
Panel reactive antibody (PRA), 438
Paneth cells, 41, 277, **Fig. 2.18**
Panitumumab, **Fig. 17.27**
Papain, 417
Paracrine action, 209–210, 220
interferon-β, 69, **Fig. 3.33**
Parasites, 4, **Fig. 1.4**
see also Helminth parasites; Protozoan parasites
adaptive immune response, 404, **Fig. 14.3**
antigens, similarity to allergens, 417–418, **Fig. 14.24**
conferring resistance to allergy, 428–429
global distribution, 404, **Fig. 14.4**
IgE-mediated responses, 248–250, 404, **Fig. 9.20**
T_H2-mediated responses, 213, 288–290
Paratyphoid fever vaccine, **Fig. 11.25**
Paroxysmal nocturnal hemoglobinuria, 39, **Fig. 13.10**
Passive immunization, 255, 380
Passive transfer of immunity, 250
Pathogens, 3–4
antibody-mediated inactivation or destruction, 18–19
complement fixation, 31, 34, **Fig. 2.8**
complement-mediated lysis, 37–38
destruction mechanisms, 8, **Fig. 1.6**
diversity, 4, **Fig. 1.3**, **Fig. 1.4**
evasion of immune response, 365–375
genetic variation, 366–369
hidden (latent) infection, 369–371, **Fig. 13.5**
sabotage or subversion, 371–372, **Fig. 13.6**
evolutionary relationship with host, 4
extracellular *see* Extracellular pathogens
initiation of adaptive immunity, 20–23, **Fig. 1.21**
intracellular *see* Intracellular pathogens
opportunistic, 3, 396
recognition mechanisms, 8, 10–11, **Fig. 1.6**, **Fig. 1.9**
routes of entry into host, 8
selection of lymphocytes, 11, **Fig. 1.9**
Pax-5, 153
expression in Hodgkin's disease, 527, **Fig. 17.25**
function, 159, **Fig. 6.13**
pattern of expression, 159, **Fig. 6.12**
PD-1 (programmed death 1), 522
PD-L1/PD-L2, 522
Pemphigus foliaceus, 485–486, **Fig. 16.1**, **Fig. 16.18**, **Fig. 16.19**
Pemphigus vulgaris, 485–486, **Fig. 16.1**
Penicillin allergy, 422, **Fig. 14.32**, **Fig. 14.33**
Pentraxins, 43, **Fig. 2.20**, **Fig. 3.22**
Peptide-binding motifs, 139–140, **Fig. 5.34**
Peptide editing, 128
Peptide-loading complex, 127–128, **Fig. 5.20**
Peptide:MHC complexes
see also Self-peptide:self-MHC complexes
naive CD8 T-cell activation, 215–216, **Fig. 8.17**
T-cell receptor binding, 132–133, **Fig. 5.26**
T-cell receptor signaling, 208–209, **Fig. 8.10**
transport to cell surface, 130, **Fig. 5.27**
Peptides (antigenic)
binding by MHC molecules, 124–125, **Fig. 5.16**
presentation *see* Antigen presentation
production *see* Antigen processing
recognition by T cells, 113, 120–135, **Fig. 5.10**
self and non-self, 126
transport to endoplasmic reticulum, 127, **Fig. 5.19**
tumor-specific, 517, **Fig. 17.10**, **Fig. 17.11**
Peptide splicing, 517, **Fig. 17.11**
Peptidoglycan, 54–55, **Fig. 3.3**
Peptidyl arginine deiminases (PADs), 492, **Fig. 16.28**
Perforin, 220, 224
Periarteriolar lymphoid sheath (PALS), **Fig. 1.24**
Periodic fevers, 503
Peripheral blood mononuclear cells (PBMC), 442
Peripheral-supramolecular activation complex (p-SMAC), 208, **Fig. 8.9**
Peripheral tolerance, 166, 193
Peroxidase, eosinophil, **Fig. 14.20**
Persistent infections, 369–371, **Fig. 13.5**
Pertussis toxin, **Fig. 9.25**
Pertussis vaccine, 316, 319–321, **Fig. 11.25**
see also DTP vaccine
Peyer's patches, 25, 272, **Fig. 1.26**, **Fig. 10.7**
B- and T-cell activation, 280, **Fig. 10.15**
dendritic cells, 278
Phagocytes, 14, 56
see also Macrophages; Neutrophils
antigen processing, 129–130
Fc receptors, 259–260, **Fig. 9.34**
inherited defects, 382–383, **Fig. 13.15**
innate immune response, **Fig. 1.6**
Phagocytic receptors
macrophages, 49–50, **Fig. 3.3**
neutrophils, 59, **Fig. 3.15**
Phagocytosis
macrophages *see* Macrophages, phagocytosis
neutrophils, 59, **Fig. 3.15**, **Fig. 3.16**
opsonized pathogens, 19, **Fig. 1.18**
pathogenic proteins/particles, 129
Phagolysosomes
antigen processing, 129–130
macrophages, 50, **Fig. 1.16**
neutrophils, 60–61, **Fig. 3.16**
Phagosomes, 129, **Fig. 1.6**
macrophages, 50, **Fig. 1.16**
neutrophils, 59, **Fig. 3.16**
Phosphatidylinositol mannoside, **Fig. 12.29**
Phosphoantigens, 350, **Fig. 12.24**
recognition by γ:δ T cells, 350–351, **Fig. 12.25**
Pigs, as organ donors, 455–456
Pilin, 369
Placenta
evolution, 342–343
formation, 343–344, **Fig. 12.16**
IgG transfer, 250
Plague vaccine, **Fig. 11.25**
Plasma cells, 16, 81, **Fig. 4.1**
differentiation, 168–170, **Fig. 6.22**
from activated naive B cells, 238–239, **Fig. 9.9**
from centrocytes, 241, 244, **Fig. 9.12**, **Fig. 9.15**
initiating adaptive immunity, 22–23, **Fig. 1.23**
regulation by cytokines, 244, **Fig. 9.15**
secondary immune response, 302
synthesis of secreted antibodies, 99–100, **Fig. 4.24**
long-lived population, 297
morphological features, 239, **Fig. 1.11**, **Fig. 9.10**
mucosal tissues, 281, 282–283
prospective, migration into lymphoid tissues, 170
vs. memory B cells, 300
vs. naive B cells, 245, **Fig. 9.16**
Plasmacytoid dendritic cells (PDCs), 71, **Fig. 3.35**
Plasma proteins
acute-phase response, 62–63
innate immune response, 39–41
Platelets, 14
functions, 40
origins, 12, 14
Pluripotent hematopoietic stem cells, 12, 150, **Fig. 1.13**
Pneumococcus *see Streptococcus pneumoniae*
Pneumocystis jirovecii (Pneumocystis carinii), 396, **Fig. 1.3**
Pneumonia vaccine, **Fig. 11.25**
P nucleotides, 95–96, 115, **Fig. 4.20**, **Fig. 5.9**
Poison ivy, 402
Poliomyelitis (polio), 311–313
progress towards elimination, 311–312, **Fig. 11.17**
vaccination-associated, 312–313
vaccines, 311–313, **Fig. 11.7**, **Fig. 11.15**
Poliovirus, 279
Pollen allergy, **Fig. 14.23**, **Fig. 14.37**
Polyadenylation, heavy chain, **Fig. 4.22**, **Fig. 4.24**
Poly-Ig receptor, 247, 283, **Fig. 9.18**
Polymorphism *see* Genetic polymorphism
Polymorphonuclear leukocytes, 14, 56
see also Neutrophils
Polyspecific antibodies, 161–162
Positive selection
B cells, 150, **Fig. 6.1**
MAIT cells, 358–359, **Fig. 12.35**
NKT cells, 356
T cells, 189–192, **Fig. 7.16**
Pre-B-cell leukemia, **Fig. 6.23**
Pre-B-cell receptor, 153–154, **Fig. 6.7**
allelic exclusion due to, 154–155, **Fig. 6.8**
inherited deficiency, 154, **Fig. 13.10**
pro-B cell selection, 157, **Fig. 6.11**
regulation of expression, 159
Pre-B cells, 151, **Fig. 6.16**
bone marrow stromal cell interactions, **Fig. 6.5**
large, 151, 153, **Fig. 6.4**

clonal expansion, 155, 157
small, 151, **Fig. 6.4**
checkpoint, 157, **Fig. 6.11**
IgM expression, 156, **Fig. 6.10**
light chain gene rearrangements, 155–157, **Fig. 6.9**
Prednisolone, 447, 448, **Fig. 15.16**
Prednisone, 447, **Fig. 15.16**
Pre-eclampsia, 344–345, **Fig. 12.18**
Pregnancy
see also Fetus
alloreactions, 144
anti-HLA antibodies, 437–438, **Fig. 15.5**
complications, 344–345, **Fig. 12.18**
hemolytic anemia of newborn, 305–306
HIV infection, 394
trophoblast-uterine NK cell interactions, 343–345
Pre-T-cell receptors
expression in thymocytes, 183–184, **Fig. 7.9**
structure, 184, **Fig. 7.10**
Pre-T cells, 184
proliferation, 185
TCR α-chain gene rearrangements, 185–186, **Fig. 7.12**
Primary focus of clonal expansion, 238–239, **Fig. 9.9**
Primary follicles *see* Lymphoid follicles, primary
Primary immune response, 11–12
antibody persistence after, 296–297, **Fig. 11.1**
B-cell activation, 169, 231–238
B-cell populations participating, 300–301, **Fig. 11.4**
common features with secondary response, 301–302
IgE production, 404–406, **Fig. 14.5**
memory T and B cell production, 297–298, **Fig. 11.2**
T-cell activation, 199–200, 203–204
vs. secondary response, 307–308, **Fig. 11.13**
Priming, T cell, 199
Pro-B cells, 150–151, **Fig. 6.3**, **Fig. 6.16**
apoptosis, 153, 154
bone marrow stromal cell interactions, **Fig. 6.5**
early, 150, **Fig. 6.4**
heavy-chain gene rearrangements, 152–153, **Fig. 6.6**
late, 150, **Fig. 6.4**
checkpoint, 157, **Fig. 6.11**
pre-B-cell receptor, 153–154, **Fig. 6.7**
program of protein expression, 158–159, **Fig. 6.12**
Procaspase 1, 55, **Fig. 3.11**
Pro-drugs, 447
Programmed cell death *see* Apoptosis
Programmed death 1 (PD-1) protein, 522
Prointerleukin-1β (proIL-1β), 55–56, **Fig. 3.11**
Proliferating cell nuclear antigen (PCNA), **Fig. 12.2**
Promyelocytic leukemia zinc finger protein (PLZF), 356
Properdin (factor P), 34, **Fig. 2.9**
inherited deficiency, 381, **Fig. 13.13**
Prostaglandins, 412, **Fig. 14.18**
Protease inhibitor drugs, HIV resistance, 394, **Fig. 13.28**
Protease inhibitors, 40–41, **Fig. 2.16**
serpin family, 382, **Fig. 13.14**
Proteases
allergic reactions, 416–417
mast cell, 412
microbial, 40
Proteasome, 126–127, **Fig. 5.18**
constitutive, 127
genes, 138, **Fig. 5.32**
Protectin *see* CD59
Protective immunity, 11
long-term, 299, **Fig. 11.3**
primary immune response, 296–297, **Fig. 11.1**
Protein F, 252, **Fig. 9.24**
Protein kinase C-θ, 209, **Fig. 8.11**
Protein kinase R (PKR), 70
Protein tyrosine kinases
B-cell activation, 232, **Fig. 9.1**
T-cell signaling, 208–209, **Fig. 8.10**
Proto-oncogenes, 161, 511
translocations involving, 161, **Fig. 6.14**
Protozoan parasites, **Fig. 1.4**
evasion/subversion of immune response, 368–369, 371
Provirus, 388
Pseudomembranous colitis, 3
Psoriasis, **Fig. 16.9**
pTα, 184, **Fig. 7.10**
pattern of expression, 187, **Fig. 7.14**
Pterins, **Fig. 12.34**
PTX3, 43, **Fig. 2.20**
Purine nucleoside phosphorylase (PNP) deficiency, 383–384
Pus, 14–15, **Fig. 1.16**
Pyogenic infections, inherited susceptibility, 379–381
Pyridostigmine, 483
Pyrin, 502
Pyrogens, 62

R

Rabbit antithymocyte globulin (rATG), 447, **Fig. 15.30**
Rabbit myxoma virus, **Fig. 13.6**
Rabies vaccine, **Fig. 11.25**
Radiation, carcinogenic effects, 513
Radioactive isotope-monoclonal antibody conjugates, 529, **Fig. 17.29**
RAG-1/RAG-2 *see* Recombination-activating genes
Rapamycin (sirolimus), 210, 453, **Fig. 15.30**
Ras, T-cell signaling, 209, **Fig. 8.11**
Rb protein, 513, 520
Reactive arthritis, 494, **Fig. 16.31**
Receptor editing
B cells, 165–166, **Fig. 6.18**
T cells, 191
Receptor-mediated endocytosis
B cells, 129, 225–226
dendritic cells, 202, **Fig. 8.3**
macrophages, 50
Recessive diseases, 377
Recombination
enzymes, 94–96, 115
influenza virus RNA, 367–368, **Fig. 13.3**
isotype switching, 101–102, **Fig. 4.28**
somatic *see* Somatic recombination
V(D)J *see* V(D)J recombination
Recombination-activating genes (RAG-1 and RAG-2), 94–95, 115
B-cell development
pattern of expression, 158, **Fig. 6.12**
regulation of expression, 152, 154, 155
evolution, 117, **Fig. 5.5**
inherited defects, 115–116, 348, 385, **Fig. 5.4**, **Fig. 13.10**, **Fig. 13.16**
initiation of recombination, 94–95, **Fig. 4.19**
T-cell development, 185, 187, 190, **Fig. 7.14**
Recombination signal sequences (RSSs), 93, **Fig. 4.18**
evolution, 117, **Fig. 5.5**
initiation of recombination, 94–95, **Fig. 4.19**
T-cell receptor genes, 115
Red blood cells *see* Erythrocytes
Reed–Sternberg cells, 170, 527, **Fig. 17.25**
Regulators of complement activation (RCA), 36
Regulatory T cells (T_{reg}), 17, 193
development, 212, **Fig. 8.14**
effector functions, 214, 226–227
effector molecules, 226, **Fig. 8.21**
induced, 214
induction of tolerance, 193, **Fig. 7.19**
mucosal tissues, 275
natural, 214
oral tolerance, 278
prevention of autoimmunity, 478
recruitment by tumors, 519, **Fig. 17.15**
transfusion effect in transplant recipients, 444
Reiter's syndrome, 494, **Fig. 16.31**
Respiratory burst, 61, **Fig. 3.17**
Respiratory tract
IgE-mediated allergy, 421, 423–424, **Fig. 14.29**
inhaled allergens, 416–417, **Fig. 14.23**
lymphoid tissues, 25
physical barriers to infection, 6, **Fig. 2.1**
Retinoblastoma (Rb) protein, 513, 520
Retroviruses, 388
endogenous, 388
Reverse transcriptase inhibitors, HIV resistance, 394
Reverse vaccinology, 317
Rev protein, 389, **Fig. 13.22**, **Fig. 13.23**
RFX deficiency, 385
Rhesus (Rh) antigens
hemolytic disease of newborn, 305–306, **Fig. 11.11**
incompatible blood transfusion, 434
matching, for blood transfusion, 435–436, **Fig. 15.3**
Rheumatic diseases, autoimmune nature, 490, **Fig. 16.24**
Rheumatic fever, 493–494, **Fig. 16.1**, **Fig. 16.30**, **Fig. 16.31**
Rheumatoid arthritis, 490–492, **Fig. 16.1**, **Fig. 16.25**
citrullinated protein antibodies, 492, **Fig. 16.29**
genetic and environmental factors, 491–492
HLA associations, 491–492, **Fig. 16.9**, **Fig. 16.27**
monoclonal antibody therapy, 91, 490–491, **Fig. 16.26**
thymic involution and, 502, **Fig. 16.40**
Rheumatoid factor, 490
Rhinitis, allergic, 423–424, **Fig. 14.34**
Riboflavin, 358, **Fig. 12.34**
Ricin, 50
RIG-I, 69

RIG-I-like receptors (RLRs), 69, **Fig. 3.32**
RIPK2, 55, **Fig. 3.10**
Rituximab
antibody-dependent cell-mediated cytotoxicity, 261–262, **Fig. 9.36**
malignant disease, 90, **Fig. 17.27**
rheumatoid arthritis, 491
RNA
alternative splicing *see* Alternative RNA splicing
viral, recognition, 69, 74, **Fig. 3.29**, **Fig. 3.32**, **Fig. 3.39**
RORγT, 212, 214
Rotavirus, 271
celiac disease and, 501
childhood immunization, **Fig. 11.15**
childhood mortality, **Fig. 11.18**
vaccines, 313–314, 324, **Fig. 11.19**, **Fig. 11.25**
R-type lectin, 50
Rubella vaccine, **Fig. 11.25**

S

Salmonella enteritidis, **Fig. 1.3**
Salmonella typhimurium, antigenic variation, 369
Salmonella typhi vaccine, 314, **Fig. 11.25**
Sarcomas, 510
Scarlet fever, **Fig. 9.25**
Scavenger receptors, 50, **Fig. 3.3**
Schistosoma mansoni
cathepsin-β-like cysteine protease, 417, **Fig. 14.24**
IgE-mediated response, 250, **Fig. 9.20**
microscopic appearance, **Fig. 1.3**
Schistosomiasis, **Fig. 11.26**
SCID *see* Severe combined immunodeficiency syndrome
Secondary follicles, 169, 239–240
Secondary immune response, 12, 295, 296–308, **Fig. 11.1**
antibody response, 170, 298
B-cell populations participating, 300–301, **Fig. 11.4**
common features with primary response, 301–302
hemolytic anemia of newborn, 305, **Fig. 11.11**
memory T and B cells, 298
suppression of naive B cells, 300–301, **Fig. 11.5**
vs. primary response, 307–308, **Fig. 11.13**
Secondary lymphoid tissues, 20–25, **Fig. 1.19**
see also Lymph nodes
antigen encounter by naive T cells, 203–204, **Fig. 8.4**
B-cell activation, 168–169, **Fig. 6.22**, **Fig. 9.7**
B-cell development, 150, 164, 167–170, **Fig. 6.2**, **Fig. 6.21**
B-cell homing, 167–168, **Fig. 6.21**
gut, 25, 272–273, **Fig. 10.7**
HIV infection, 395
initiation of adaptive immunity, 20–23
mucosal surfaces, 25
naive T-cell activation, 200–201, **Fig. 8.1**
naive T-cell homing, 204–206, **Fig. 8.6**
T-cell differentiation, 193–194
Secretory component (secretory piece), 247, **Fig. 9.18**
Secretory IgA *see* IgA, dimeric
Segmental exchange (interallelic conversion), 142, **Fig. 5.38**
Selectins, 57, **Fig. 3.12**, **Fig. 3.13**
Self, discrimination from non-self, 47–49, **Fig. 3.2**
Self antigens, 163
loss of T-cell tolerance, 480–481
monovalent, immature B cells specific for, 166
multivalent, immature B cells specific for, 164–166
Self-MHC molecules, 143
negative selection of T cells, 143, 192, **Fig. 7.18**
NK-cell education, 336–339, **Fig. 12.9**, **Fig. 12.10**, **Fig. 12.11**
positive selection of T cells, 189–190, **Fig. 7.16**
Self peptides, 126
Self-peptide:self-MHC complexes
determination of CD4 *vs.* CD8 T cells, 191–192, **Fig. 7.17**
positive selection of T cells, 189–190
Self proteins, 126
producing tumor-specific antigens from, 517, **Fig. 17.11**
Self-reactive B cells *see* B cell(s), autoreactive/self-reactive
Self-reactive cells/receptors, 163
Self-reactive T cells *see* T cell(s), autoreactive/self-reactive
Self renewal, 12
Self-tolerance
see also Tolerance, immunological
B-cell repertoire, 164, 166
failure of, 473, 477–478, 480–481
mechanisms, **Fig. 16.5**
Sensitization, to allergens, 416–417, **Fig. 14.23**
Septicemia, 246
Septic shock, 68
Serglycin, 220–221, 224
Serotypes, 366, **Fig. 13.1**
Serpins, 382, **Fig. 13.14**
Serum amyloid A protein, 63, **Fig. 3.21**
Serum amyloid P component (SAP), 43, **Fig. 2.20**
Serum sickness, 450–451, **Fig. 15.22**
Severe combined immunodeficiency syndrome (SCID), 383–385, **Fig. 13.10**, **Fig. 13.16**
RAG defects (radiation-sensitive), 115–116, **Fig. 5.4**, **Fig. 13.16**
X-linked forms, 384, **Fig. 13.16**
SH2D1A gene defects, 386
Sheddases, 409
Shigella, 279
Shingles, 370
SHP-1, 334, **Fig. 12.6**, **Fig. 12.9**
Sialic acid, 35–36, 269
Sialyl-Lewis^x, 57, **Fig. 3.13**
Sickle-cell anemia, **Fig. 15.32**
Signal joint, 95, **Fig. 4.19**
Sipuleucel-T, 525, **Fig. 17.23**
Sirolimus, 210, 453, **Fig. 15.30**
Skin
allergic reactions, 424–425, **Fig. 14.37**
blistering diseases, 484–485, **Fig. 16.18**, **Fig. 16.19**
cancer, 512
commensal microorganisms, 29–30
grafts, 439
infections, 200, **Fig. 8.1**
physical barrier to infection, 4–8, 29, **Fig. 1.5**, **Fig. 2.1**
Skin tests (intradermal), 419–420, **Fig. 14.27**, **Fig. 14.37**
SLE *see* Systemic lupus erythematosus
Sleeping sickness, 369
Small intestine
see also Gastrointestinal tract; Gut-associated lymphoid tissues
B- and T-cell activation, 280–281, **Fig. 10.15**
defensins, 41, **Fig. 2.18**
effector lymphocytes, 281–282, **Fig. 10.16**, **Fig. 10.17**
epithelial cells *see* Intestinal epithelial cells
follicle-associated epithelium, 277
secondary lymphoid tissues, 272, **Fig. 10.7**
Small lymphocytes, 16, 20, **Fig. 1.11**, **Fig. 1.20**
Smallpox, 299, 319
eradication, 309–310, **Fig. 1.1**
eyewitness account, 320
vaccination, 308–310, **Fig. 11.14**
changing demand for, 319
discovery, 1–2, 308–309
immunological memory, 299, **Fig. 11.3**
Smoking
Goodpasture's syndrome, 494–495
rheumatoid arthritis causation, 492, **Fig. 16.29**
Snake venom, 254–255
SOCS (suppressors of cytokine signaling) proteins, 221–222
Somatic hypermutation, 100–101, **Fig. 4.25**
B-cell zone of lymph node, 239–241
IgE *vs.* IgG isotypes, 405, **Fig. 14.6**
secondary immune response, 298, 302
Somatic recombination
immunoglobulin gene segments, 92–95, **Fig. 4.16**
T-cell receptor gene segments, 115
Sore throat, 252, **Fig. 9.24**
Specificity, antibody, 81, 96–97
Sphingosine 1-phosphate (S1P), 206, 237
Spleen, 20, 23–25, **Fig. 1.24**
Splenectomy, 25
S protein, 38
SR-A (scavenger receptor A), 50, **Fig. 3.3**
SR-B (scavenger receptor B), 50, **Fig. 3.3**
SSLP7, 374–375, **Fig. 13.9**
Staphylococcus aureus, 379
complement evasion strategy, 36
enterotoxin B (SEB), 373, **Fig. 13.8**
exotoxins, **Fig. 9.25**
increased susceptibility to, 160
morphology, **Fig. 1.3**
superantigen-like proteins (SSLPs), 374–375, **Fig. 13.9**
superantigens, 373
STATs, 221, **Fig. 8.22**
Stem-cell factor (SCF), 152, **Fig. 6.5**
Stem cells
cancer, 515
hematopoietic *see* Hematopoietic stem cells
Streptococcus pneumoniae (pneumococcus), 23–24, 379
C-reactive protein and, 63, 263
genetic variation/serotypes, 366, **Fig. 13.1**
IgA1 cleavage, 285
increased susceptibility to, 160

morphology, **Fig. 1.3**
vaccine, 314–315, 316, **Fig. 11.15**, **Fig. 11.25**
Streptococcus pyogenes, 379
complement evasion strategy, 36
exotoxins, **Fig. 9.25**
increased susceptibility to, 160
neutralizing antibodies, 252–253, **Fig. 9.24**
protease, 40
rheumatic fever, 493–494, **Fig. 16.30**
superantigens, 373
Stress proteins, 334
Subcapsular sinus macrophages, 236
Subtilisin, 417
Sulfatide, 352–353, **Fig. 12.28**
Superantigens, bacterial, 373–374, **Fig. 13.8**
Superoxide dismutase, 60, **Fig. 3.17**
Superoxide radicals, 60
Suppressors of cytokine signaling (SOCS) proteins, 221–222
Supramolecular activation complex
central (c-SMAC), 208, **Fig. 8.9**
peripheral (SMAC), 208, **Fig. 8.9**
Surfactant proteins A and D (SP-A and SP-D), 64
Surrogate α chain, pre-T-cell receptor, 184, **Fig. 7.10**
Surrogate light chain, 153, **Fig. 6.7**
Survival signals, 153
Switch sequences (regions), 101, 243, **Fig. 4.28**
"Swollen glands," 22, 240
Syk, B-cell receptor signaling, 232, **Fig. 9.1**
Symbiotic relationships, gut bacteria, 30
Sympathetic ophthalmia, 495, **Fig. 16.32**
Synapse
natural killer (NK) cell, 74, **Fig. 3.40**
T cell *see* T-cell synapse
Syngeneic transplant, 439
Syphilis, 371
Systemic lupus erythematosus (SLE), 476, **Fig. 16.1**
anti-DNA antibodies, 166
complement component deficiency, 257
facial rash, 476, **Fig. 16.4**
immune complex deposition, 258, 476, 487, **Fig. 16.20**
intermolecular epitope spreading, 487–488, **Fig. 16.21**
joint involvement, 490

T

Tacrolimus, 210, 450, 455
immunological effects, 450, **Fig. 15.20**
mechanism of action, **Fig. 15.19**, **Fig. 15.30**
TAK1, 55, **Fig. 3.10**
Talin, 208, **Fig. 8.9**, **Fig. 9.8**
TAP *see* Transporter associated with antigen processing
Tapasin, 127, **Fig. 5.20**
gene, 138, **Fig. 5.32**
Target cells, 211
cytotoxic CD8 T-cell killing, 222–224, **Fig. 8.23**, **Fig. 8.24**
effector T cells, 219, 220
Tasmanian devil, facial tumors, 515–516, **Fig. 17.8**
Tat protein, 389, **Fig. 13.22**, **Fig. 13.23**
T-bet, T_H1 development, 213, **Fig. 8.14**
T cell(s), 16–17
activation
effector T cells, 219, **Fig. 8.20**
immunosuppressive drugs targeting, 448–453
induced cell death, 193
naive T cells *see* Naive T cells, activation
adoptive transfer, 523, **Fig. 17.21**
alloreactive *see* Alloreactive T cells
α:β *see* α:β T cells
anergy, 193, 210–211, **Fig. 8.13**
antigen receptors *see* T-cell receptor(s)
antigen recognition, 113–145
autoimmune diseases mediated by, 476, **Fig. 16.1**
autoreactive/self-reactive, 193, 481, **Fig. 7.19**
cryptic epitopes, 486
induction of anergy, 210–211, **Fig. 8.13**
systemic lupus erythematosus, 487–488, **Fig. 16.21**
CD4 *see* CD4 T cells
CD8 *see* CD8 T cells
classes, 118, **Fig. 5.7**
development, 177–196
see also Thymocytes
checkpoints, 184, 186, **Fig. 7.15**
early phase, 177–188, **Fig. 7.1**, **Fig. 7.15**
patterns of gene expression, 186–188, **Fig. 7.14**
selection of T-cell repertoire, 188–193
sites, 20, **Fig. 1.19**
stages, 194–195, **Fig. 7.21**
differentiation after encounter with antigen, 17, 193–194
effector *see* Effector T cells
γ:δ *see* γ:δ T cells
help for B cells, 122, 225–226, **Fig. 5.13**, **Fig. 8.28**
IL-2-induced differentiation and proliferation, 209–210, **Fig. 8.12**
immunosuppressive drugs targeting, 448–454
inherited defects, 178–179, 380–381, 383–385, **Fig. 13.16**
lineages, 118, 177–178
memory *see* Memory T cells
MHC restriction, 140, **Fig. 5.35**
naive *see* Naive T cells
negative selection, 192–193, **Fig. 7.18**
positive selection, 189–192, **Fig. 7.16**
precursors, 178, **Fig. 7.1**
priming, 199
regulatory *see* Regulatory T cells
senescence, autoimmunity and, 501–502, **Fig. 16.40**
splenic white pulp, **Fig. 1.24**
tolerance, 193, 211, **Fig. 8.13**
failure, autoimmune disease, 480–481
T-cell area *see under* Lymph nodes
T-cell co-receptors, 122
see also CD4; CD8
binding to MHC molecules, 124, **Fig. 5.15**
clustering, T-cell activation, 208
γ:δ T cells, 348
T-cell receptor(s) (TCR), 16–17, 113, 114–120
α:β, 118, **Fig. 5.7**
expression on cell surface, 117–118, **Fig. 5.6**
MAIT cells, 357–358
synthesis, 116–117
antigen-binding site, 114–115, 132, **Fig. 5.2**
antigen recognition, 121–135, **Fig. 5.10**
MHC polymorphism effects, 139, **Fig. 5.33**
peptide:MHC complexes, 132–133, **Fig. 5.26**
chimeric, cancer immunotherapy, 522–523, **Fig. 17.21**
clustering on T-cell activation, 208, **Fig. 8.10**
complex, 118, **Fig. 5.6**
dendritic cell adhesion and, 205–206, **Fig. 8.7**
diversity, 114–120
mechanisms of generation, 115–117
vs. immunoglobulins, 120, **Fig. 5.9**
editing, 191
expression on cell surface, 117–118, **Fig. 5.6**
γ:δ, 118–119, **Fig. 5.7**
antigens recognized, 350–353
expression in thymocytes, 183–184, **Fig. 7.9**
γ:δ T-cell subpopulations, 349–350, **Fig. 12.22**
lack of diversity, 347, 348
gene organization, 115, 118–119, **Fig. 5.3**, **Fig. 5.8**
gene rearrangements, 115, 119, **Fig. 5.3**
initiation, 180–181, **Fig. 7.5**
lineage commitment and, 181–184, **Fig. 7.7**, **Fig. 7.9**
productive and nonproductive, 183
regulation, 186, **Fig. 7.14**
thymocytes, 183–186
NKT cells, 356
signaling pathways, 208–209, **Fig. 8.11**
structure, 114–115, **Fig. 5.1**
vs. antibodies and B-cell receptors, 16–17, 120, **Fig. 1.17**
T-cell receptor α chain (TCRα), 114, **Fig. 5.1**
gene polymorphism, allergic disease, 418, **Fig. 14.25**
gene rearrangements, 115, 185–186, **Fig. 5.3**
δ-chain gene deletion, 186, **Fig. 7.13**
during positive selection, 190–191
role in lineage commitment, 183–184, **Fig. 7.9**
successive, 185–186, **Fig. 7.12**
genes, 115, **Fig. 5.3**, **Fig. 5.8**
surrogate, 184, **Fig. 7.10**
synthesis, 116–117, 186
TCR assembly, 116–117, 186
T-cell receptor β chain (TCRβ), 114, **Fig. 5.1**
gene rearrangements, 115, **Fig. 5.3**
rescue of nonproductive, 184–185, **Fig. 7.11**
role in lineage commitment, 183–184, **Fig. 7.9**
genes, 115, **Fig. 5.3**
pre-T-cell receptor, 184, **Fig. 7.10**
synthesis, 116–117
TCR assembly, 116–117, 186
T-cell receptor δ chain (TCRδ), 118–119
gene deletion, 186, **Fig. 7.13**
gene loci, 118–119, **Fig. 5.8**
gene rearrangements, 119, 183–184, **Fig. 7.9**
T-cell receptor γ chain (TCRγ), 118
gene loci, 119, **Fig. 5.8**
gene rearrangements, 119, 183–184, **Fig. 7.9**
T-cell synapse
see also Cognate pairs
B cell–effector T_{FH} cell, 238, **Fig. 9.8**

cytotoxic CD8 T cells, 222–223, **Fig. 8.23**
effector T cells, 218, 220, 221
naive T-cell activation, 207–208, **Fig. 8.9**
T-cell zone *see under* Lymph nodes
TCR *see* T-cell receptor(s)
TdT *see* Terminal deoxynucleotidyl transferase
Teichoic acids, 50
Terminal deoxynucleotidyl transferase (TdT), 95, **Fig. 4.20**
B-cell development, 158, **Fig. 6.12**
T-cell development, 187, **Fig. 7.14**
Tertiary immune response, 298
Tetanus
toxin, 253, 314, **Fig. 9.25**
vaccine, 253, 314, **Fig. 11.25**
T_{FH} cells (T follicular helper cells), 232
cytokines regulating isotype switching, 243–244, **Fig. 9.13**
development, 212, 214, **Fig. 8.14**
effector functions, 218, 225–226, **Fig. 8.28**
effector molecules, **Fig. 8.21**
naive B-cell activation, 234, 236–238, **Fig. 9.7**, **Fig. 9.8**
rescue of centrocytes from apoptosis, 241, **Fig. 9.12**
secondary immune response, 302
TGF-β *see* Transforming growth factor-β
T_H1 cells
development, 212, 213, **Fig. 8.14**
effector functions, 213, 218–219, 224–225
effector molecules, 224–225, **Fig. 8.21**
helminth infections, 289, **Fig. 10.27**
macrophage activation, 224–225, **Fig. 8.27**
multiple sclerosis, 476
polarized response, 214–215, **Fig. 8.15**, **Fig. 8.16**
type IV hypersensitivity reactions, 402
T_H2 cells
allergic asthma, 424, **Fig. 14.35**
basophil-mediated induction, 213, 415
development, 212, 213, **Fig. 8.14**
effector functions, 213, 218–219
effector molecules, **Fig. 8.21**
eosinophil recruitment, 414, 415
IgE-mediated allergic disease, 416, **Fig. 14.22**, **Fig. 14.23**
inhibition of macrophage activation, 225
intestinal helminth infections, 288–290, **Fig. 10.26**, **Fig. 10.27**
parasitic infections, 404, **Fig. 14.3**
penicillin allergy, 422, **Fig. 14.33**
polarized response, 214–215, **Fig. 8.15**
T_H17 cells
development, 212, 213–214, **Fig. 8.14**
effector functions, 218–219
effector molecules, **Fig. 8.21**
Thalassemia major, **Fig. 15.32**
6-Thioinosinic acid, 453, **Fig. 15.25**
Thoracic duct, 20
Th-POK, 188, 192, **Fig. 7.14**
Thrombocytopenia, immune, 489
Thrombocytopenic purpura, autoimmune, **Fig. 16.1**
Thymectomy, 180
Thymic anlage, 178
Thymic epithelial cells, 178, **Fig. 7.2**
antigen processing and presentation, 192–193
positive selection of T cells, 189–190, **Fig. 7.16**
Thymic stroma, 178
Thymic stromal lymphopoietin (TSLP), 289
Thymocytes, 178, **Fig. 7.2**
apoptosis, 182–183, 190, **Fig. 7.8**
cell-surface markers, 180, **Fig. 7.5**
commitment to T-cell lineage, 180–181, **Fig. 7.5**
double-negative *see* Double-negative (DN) thymocytes
double-positive *see* Double-positive (DP) thymocytes
locations, 178, **Fig. 7.3**
negative selection, 192–193, **Fig. 7.18**
patterns of gene expression, 186–188, **Fig. 7.14**
positive selection, 189–192, **Fig. 7.16**
single-positive, 191–192
Thymus, 20, 178–180
absent, 178–179
cellular organization, 178, **Fig. 7.3**
early T-cell development, 177–188, **Fig. 7.1**, **Fig. 7.15**
expression of tissue-specific genes, 192–193
involution, 179–180, **Fig. 7.4**
autoimmune disease and, 501–502, **Fig. 16.40**
MAIT cell development, 358–359, **Fig. 12.35**
NKT cell development, 356
selection of T-cell repertoire, 188–193
Thymus-dependent lymphocytes, 177
Thymus-independent (TI) antigens, 235, **Fig. 9.4**
Thyroglobulin, **Fig. 16.13**
Thyroid gland
autoimmune diseases, 482, 484, **Fig. 16.17**
ectopic lymphoid tissue, 484, **Fig. 16.16**
Thyroiditis
chronic (Hashimoto's), 484, **Fig. 16.16**
subacute, **Fig. 16.9**
Thyroid-stimulating hormone (TSH), 482, **Fig. 16.13**
Thyroid-stimulating hormone (TSH) receptor autoantibodies, 482, **Fig. 16.13**
Thyroxine (T_4), 482, **Fig. 16.13**
TIM proteins, 418, **Fig. 14.25**
Tingible body macrophages, 241
TIR domain
IL-1 receptor, 55
Toll-like receptors, 51–52, **Fig. 3.5**
Tissue transglutaminase, 500–501, **Fig. 16.39**
TLR1:TLR2 heterodimer, **Fig. 3.29**, **Fig. 3.30**
TLR2:TLR6 heterodimer, **Fig. 3.29**
TLR3, 67, 74, **Fig. 3.29**
cellular location, **Fig. 3.30**
signaling pathway, 75, **Fig. 3.39**
TLR4, 51–53, **Fig. 3.29**
allotypes, septic shock risk, 68
cellular location, 66, **Fig. 3.30**
recognition of lipopolysaccharide, 51–52, **Fig. 3.6**
signaling pathways, 52–53, **Fig. 3.7**
structure, 51, **Fig. 3.5**
TLR5, **Fig. 3.29**
TLR7, 71, 74, **Fig. 3.29**, **Fig. 3.39**
TLR8, 74, **Fig. 3.29**
TLR9, 67, 71, **Fig. 3.29**
TLR10, **Fig. 3.29**
TLRs *see* Toll-like receptors
T lymphocytes *see* T cell(s)
TNF-α *see* Tumor necrosis factor-α
Tolerance, immunological
see also Self-tolerance
to allografts, induced, 467, **Fig. 15.44**
B cell, 164, 166
central, 166, 193
failure, autoimmune disease, 480–481
oral, 278
peripheral, 166, 193
T cell, 193, 211, **Fig. 8.13**
Toll-like receptors (TLRs), 66–68
see also specific receptors (TLR1–TLR10)
cellular locations, 66–67, **Fig. 3.30**
dendritic cells, 203
genetic variation, 67–68, **Fig. 3.31**
intestinal epithelial cells, 275–276, **Fig. 10.10**
macrophages, 51–52, **Fig. 3.3**
NK cells, 74
recognition of microbial products, 66–67, **Fig. 3.29**
structure, 51, **Fig. 3.5**
Toll receptor-associated activator of interferon (TRIF), 75, **Fig. 3.39**
Tonsillitis, 242
Tonsils, 25, 272, **Fig. 10.6**
Toxic-shock syndrome, **Fig. 9.25**
Toxic-shock syndrome toxin-1 (TSST-1), 373
Toxin-monoclonal antibody conjugates (immunotoxins), 529, **Fig. 17.28**
Toxins, microbial
as cause of disease, 253, **Fig. 9.25**
neutralization by antibodies, 253–254, **Fig. 9.26**
superantigens, 373–374, **Fig. 13.8**
vaccines against, 314
Toxoids, 253, 314
Toxoplasma gondii, 371
TRADD, **Fig. 3.32**
TRAF6, 52, **Fig. 3.7**, **Fig. 3.32**
Transcytosis, 278
dimeric IgA, 247, **Fig. 9.18**
exploitation by pathogens, 279
secretory IgM, 283
Transforming growth factor-β (TGF-β)
anterior chamber of eye, 457
control of isotype switching, 283, **Fig. 9.13**
inhibition of macrophage activation, 225
regulatory T cells, 226
T_H17-cell development, 214
tumors secreting, 519, **Fig. 17.15**
Transfusion *see* Blood transfusion
Transfusion effect, organ transplant recipients, 444
Transglutaminase, 500–501, **Fig. 16.39**
Translocations *see* Chromosomal translocations
Transpeptidase, penicillin binding, **Fig. 14.32**
Transplantation, 433–468
alloreactions, 438–439, **Fig. 15.6**
antigens, 438
cancer risk after, 446, 513–514
hypersensitivity reactions to allogeneic, 433–440
solid organ *see* Organ transplantation
tumors, 515, **Fig. 17.7**
Transplant rejection, 438, **Fig. 15.6**
acute, 441, **Fig. 15.8**, **Fig. 15.9**
monoclonal antibody therapy, 450–451, **Fig. 15.21**
chronic, 443–444, **Fig. 15.11**

direct and indirect allorecognition, 443–444, **Fig. 15.12**
hyperacute, 436–437, **Fig. 15.4**
MHC polymorphism and, 143–144
monoclonal antibodies for, 90
vs. allergy and autoimmunity, 503–504, **Fig. 16.42**
Transplant tourism, 455
Transporter associated with antigen processing (TAP), 127, **Fig. 5.19**
genes, 138, **Fig. 5.32**
inherited defects, 127, 385, **Fig. 13.10**, **Fig. 13.16**
peptide-loading complex, 127–128, **Fig. 5.20**
Transposase, 117, **Fig. 5.5**
Transposon, 117, **Fig. 5.5**
Trastuzumab, **Fig. 17.27**
Trauma, initiating autoimmunity, 495, **Fig. 16.32**
T_{reg} *see* Regulatory T cells
Treponema pallidum, 371
TRIF, 75, **Fig. 3.39**
Tri-iodothyronine (T_3), 482, **Fig. 16.13**
Trophoblast cells
HLA expression, 344
inadequate invasion, 344–345, **Fig. 12.18**
uterine NK-cell interactions, 343–344, **Fig. 12.16**, **Fig. 12.17**
Trypanosoma brucei, **Fig. 1.3**
Trypanosomes, antigenic variation, 368–369, **Fig. 13.4**
Tryptase, mast cell, 411, 412
TSH *see* Thyroid-stimulating hormone
Tuberculosis, **Fig. 11.26**
see also Mycobacterium tuberculosis
AIDS patients, 396
T-cell responses to lipid antigens, 354–356
vaccine *see* Bacille Calmette–Guérin vaccine
Tumor antigens, 516–518, **Fig. 17.9**, **Fig. 17.10**
γ:δ T-cell response, 351–352
primed dendritic cells, cancer therapy, 525, **Fig. 17.23**
vaccination, cancer therapy, 520–521, **Fig. 17.18**
Tumor-associated antigens, 516, 517–518, **Fig. 17.9**, **Fig. 17.12**
Tumor necrosis factor-α (TNF-α), 53–54
monoclonal antibodies, 91, 490–491, **Fig. 16.26**
neutrophil recruitment, 57
release by mast cells, 412, **Fig. 14.16**
role in inflammation, 53–54, **Fig. 3.9**
secretion by macrophages, 53
systemic effects, 62–63, 68, **Fig. 3.20**
Tumor necrosis factor (TNF) receptor-associated periodic syndrome, **Fig. 16.41**
Tumors, 510
see also Cancer
benign, 510, **Fig. 17.1**
contagious, 515–516, **Fig. 17.8**
evasion of immune response, 518–519, **Fig. 17.13**, **Fig. 17.14**
growth, 514, **Fig. 17.6**
malignant, 510, **Fig. 17.1**
manipulation of immune response, 519, **Fig. 17.15**
transplantation studies, 515, **Fig. 17.7**
Tumor-specific antigens, 516–517, **Fig. 17.9**, **Fig. 17.10**, **Fig. 17.11**
Tumor suppressor genes, 511
Twins
dizygotic, tolerance between, 467, **Fig. 15.44**
identical, transplants between, 439, 445
Tyk2, 70
Typhoid vaccine, 314, **Fig. 11.25**
Typhus fever vaccine, **Fig. 11.25**

U

UL-binding proteins (ULBPs), 372, **Fig. 12.2**
Ultraviolet radiation, overexposure to, 512
Umbilical cord blood, hematopoietic stem cells, 463–464
Uracil-DNA-glycosylase (UNG), 101
Urogenital tract, physical barriers to infection, 6
Urticaria, 424, 427, **Fig. 14.38**
Uterine natural killer cells (uNK), 71, 342–345
cooperation with trophoblast cells, 343, **Fig. 12.16**
fetal MHC class I interactions, 343–344, **Fig. 12.17**
pregnancy complications and, 344–345, **Fig. 12.18**

V

Vaccination, 1–2, 295, 308–325
asplenia/post-splenectomy, 24
cancer prevention, 519–520
cancer therapy, 520–521, **Fig. 17.18**
changing demands, 319–322
childhood schedule, **Fig. 11.15**
disease inadvertently caused by, 312–313
immunological memory, 12, 299, **Fig. 11.3**
Vaccines, 308
adjuvants, 211, 316, **Fig. 11.21**
bacterial, 314–316
cancer, 525
combination, 316
conjugate, 315–316, **Fig. 11.20**
currently available, 322, **Fig. 11.25**
development, 2, 316–317, 323–324
diseases needing better, 322–323, **Fig. 11.26**
killed/inactivated virus, 310
live-attenuated bacterial, 314
live-attenuated virus, 310–311, **Fig. 11.16**
neutralizing antibodies, 231
public acceptance, 319–322, 324
side-effects, 324
subunit, 313, 316
toxoid, 253, 314
viral, 310–314
Vaccinia virus (cowpox virus), 2, 308–309, **Fig. 11.14**
immunological memory, 299, **Fig. 11.3**
subversion of immune response, **Fig. 13.6**
Variable (V) domains
antibodies, 84, **Fig. 4.6**, **Fig. 4.7**
framework regions, 85–86
hypervariable regions (HV), 85–86, **Fig. 4.8**
T-cell receptors, 114–115, **Fig. 5.1**
engineered, 523, **Fig. 17.20**
Variable (V) gene segments
immunoglobulins, 92, **Fig. 4.15**, **Fig. 4.17**
recombination signal sequences (RSSs), 93, **Fig. 4.18**
somatic recombination, 92–95, **Fig. 4.16**
T-cell receptors, 115, **Fig. 5.3**, **Fig. 5.8**
T-cell *vs.* B-cell receptors, **Fig. 5.9**
Variable (V) regions
antibodies, 83, **Fig. 4.2**
construction from gene segments, 92, **Fig. 4.16**
genes, 91–92, **Fig. 4.15**
recombination of gene segments, 92–95, **Fig. 4.16**
somatic hypermutation *see* Somatic hypermutation
T-cell receptors, 114, **Fig. 5.1**
Variable surface glycoproteins (VSGs), trypanosome, 368–369, **Fig. 13.4**
Varicella vaccine, **Fig. 11.15**, **Fig. 11.25**
Varicella-zoster virus, 370
Variola, 309
Variolation, 309
Vascular addressins, **Fig. 3.12**
mucosal (MAdCAM-1), **Fig. 10.16**
naive T cell homing, 205
Vasodilation, in inflammation, 9
Vav1 protein, 334, **Fig. 12.6**, **Fig. 12.9**
VCAM-1, 219, **Fig. 6.5**, **Fig. 8.19**
V(D)J recombinase, 94–95
V(D)J recombination
immunoglobulins, 94–96, **Fig. 4.19**, **Fig. 4.20**
T-cell receptors, 115, 117, **Fig. 5.3**
Vemurafenib, 522
Venomous animals/insects, 254–255, 422
Vesicular system, intracellular, 125, **Fig. 5.17**
antigen processing, 126, 129–130, **Fig. 5.22**
CD1 protein recycling, 352, 355, **Fig. 12.28**
pathogens exploiting, 130
peptide:MHC class I complex transport, 128
Vibrio cholerae, 284–285
Villi, small intestinal, 272, 277, **Fig. 10.7**
Viruses, 4, **Fig. 1.4**
antigen cross-presentation, 131–132, **Fig. 5.24**
antigen processing, 126, 202–203, **Fig. 8.3**
cytotoxic T cell responses, 222, 223, 224
evasion of immune response, 366–369, 369–371, **Fig. 13.2**
γ:δ T-cell response, 351–352
innate immune response, 68–77
interferon response, 70–71, **Fig. 3.33**, **Fig. 3.34**
memory CD8 T-cell response, 305, **Fig. 11.10**
naive CD8 T-cell activation, 215–216, **Fig. 8.17**
neutralizing antibodies, 251–253, **Fig. 9.23**
NK-cell activation, 71–72, **Fig. 3.36**
NK cell cytotoxicity, 72–74, **Fig. 3.38**
NK cell–dendritic cell interactions, 77, **Fig. 3.41**
NK cell–macrophage interactions, 75–76, **Fig. 3.40**
oncogenic, 512–513, **Fig. 17.4**
plasmacytoid dendritic cell responses, 71
RIG-1-like receptors recognizing, 69, **Fig. 3.32**
subversion of immune response, 371–372, **Fig. 13.6**
TLRs recognizing, 66–67, **Fig. 3.29**
vaccines against, 310–314
Vitamin K, 271, **Fig. 10.5**

VLA-4
  developing B cells, **Fig. 6.5**
  effector T cells, 219, **Fig. 8.18**, **Fig. 8.19**
Vomiting, food allergies, 426, **Fig. 14.38**
VpreB
  pre-B-cell receptor, 153–154, **Fig. 6.7**
  regulation of expression, 158, **Fig. 6.12**

W

Waldenström's macroglobulinemia, **Fig. 6.23**
Waldeyer's ring, 272, **Fig. 10.6**
Warfare, 4
Wasp stings, 421
Wheal and flare, 420, **Fig. 14.27**
White blood cells *see* Leukocytes
Whooping cough, 316, 319–321, **Fig. 9.25**
  vaccine *see* Pertussis vaccine
Wiskott–Aldrich syndrome, 384, **Fig. 13.16**, **Fig. 15.32**
Wiskott–Aldrich syndrome protein (WASP), 384
Worms *see* Helminth parasites

X

Xenoantibodies, 456
Xenoantigens, 456
Xenograft, 455
Xenotransplantation, 455–456
X-linked agammaglobulinemia (XLA), 160, 379–380, **Fig. 13.10**
  passive immunization, 380
  typing of carriers, 379, **Fig. 13.12**
X-linked diseases, 377
X-linked hypohydrotic ectodermal dysplasia and immunodeficiency, 53, **Fig. 3.8**, **Fig. 13.10**
X-linked lymphoproliferative syndrome, 386

Y

Yellow fever vaccine, **Fig. 11.25**

Z

ZAP-70, 188, **Fig. 7.14**
  deficiency, 208, **Fig. 13.10**
  initiation of TCR signaling, 208–209, **Fig. 8.10**, **Fig. 8.11**
ζ chain, 118, **Fig. 5.6**
  ITAMs, 208, **Fig. 8.10**
  pre-T-cell receptors, 184
ζ chain-associated protein of 70 kDa *see* ZAP-70
Zidovudine (AZT), 394
Zoledronate, 350, **Fig. 12.24**
Zymogens, 31, 64
Zymosan, **Fig. 3.29**